PETROLEUM EXPLORATION HANDBOOK

McGRAW-HILL HANDBOOKS

ALJIAN · Purchasing Handbook
AMERICAN INSTITUTE OF PHYSICS · American Institute of Physics Handbook
AMERICAN SOCIETY OF MECHANICAL ENGINEERS · ASME Handbooks:
  Engineering Tables — Metals Engineering—Processes
  Metals Engineering—Design — Metals Properties
BAUMEISTER AND MARKS · Standard Handbook for Mechanical Engineers
BEEMAN · Industrial Power Systems Handbook
BERRY, BOLLAY, AND BEERS · Handbook of Meteorology
BLATZ · Radiation Hygiene Handbook
BRADY · Materials Handbook
BURINGTON · Handbook of Mathematical Tables and Formulas
BURINGTON AND MAY · Handbook of Probability and Statistics with Tables
CARROLL · Industrial Instrument Servicing Handbook
COCKRELL · Industrial Electronics Handbook
CONDON AND ODISHAW · Handbook of Physics
CONSIDINE · Process Instruments and Controls Handbook
CROCKER · Piping Handbook, 4th ed.
CROFT, CARR, AND WATT · American Electricians' Handbook, 8th ed.
DAVIS · Handbook of Applied Hydraulics, 2d ed.
ETHERINGTON · Nuclear Engineering Handbook
FACTORY MUTUAL ENGINEERING DIVISION · Handbook of Industrial Loss Prevention
FINK · Television Engineering Handbook
FINK AND CARROLL · Standard Handbook for Electrical Engineers
GAYLORD AND GAYLORD · Structural Engineers' Handbook
GUTHRIE · Petroleum Products Handbook
HARRIS · Handbook of Noise Control
HARRIS AND CREDE · Shock and Vibration Handbook
HAVERS AND STUBBS · Handbook of Heavy Construction
HENNEY · Radio Engineering Handbook
HUNTER · Handbook of Semiconductor Electronics
HUSKEY AND KORN · Computer Handbook
JASIK · Antenna Engineering Handbook

Juran · Quality-control Handbook
Kallen · Handbook of Instrumentation and Controls
King · Handbook of Hydraulics, 4th ed.
Koelle · Handbook of Astronautical Engineering
Korn and Korn · Mathematical Handbook for Scientists and Engineers
Kurtz · The Lineman's Handbook
La Londe and Janes · Concrete Engineering Handbook
Landee, Davis, and Albrecht · Electronic Designers' Handbook
Lange · Handbook of Chemistry
Le Grand · The New American Machinist's Handbook
Magill, Holden, and Ackley · Air Pollution Handbook
Manas · National Plumbing Code Handbook
Mantell · Engineering Materials Handbook
Markus · Handbook of Electronic Control Circuits
Markus and Zeluff · Handbook of Industrial Electronic Circuits
Maynard · Industrial Engineering Handbook
Merritt · Building Construction Handbook
Morrow · Maintenance Engineering Handbook
Perry · Chemical Engineers' Handbook
Shand · Glass Engineering Handbook
Society of Manufacturing Engineers:
Die Design Handbook
Handbook of Fixture Design
Manufacturing Planning and Estimating Handbook
Tool Engineers Handbook
Staniar · Plant Engineering Handbook
Streeter · Handbook of Fluid Dynamics
Terman · Radio Engineers' Handbook
Truxal · Control Engineers' Handbook
Urquhart · Civil Engineering Handbook
Watt · NFPA Handbook of the National Electrical Code
Woods · Highway Engineering Handbook
Yoder, Heneman, Turnbull, and Stone · Handbook of Personnel Management and Labor Relations

PETROLEUM EXPLORATION HANDBOOK

A PRACTICAL MANUAL SUMMARIZING THE APPLICATION OF EARTH SCIENCES TO PETROLEUM EXPLORATION

GRAHAM B. MOODY, Editor

Petroleum Consultant. Chairman of the Board, Buttes Gas & Oil Co. Formerly Chief Reserves Engineer, Standard Oil Company of California. A Past President of AAPG

A McGRAW-HILL CLASSIC HANDBOOK REISSUE

FIRST EDITION

McGRAW-HILL BOOK COMPANY

New York St. Louis San Francisco Düsseldorf Johannesburg Kuala Lumpur London Mexico Montreal New Delhi Panama Paris São Paulo Singapore Sydney Tokyo Toronto

1982 Reissue

PETROLEUM EXPLORATION HANDBOOK

Library of Congress Catalog Card Number: 60-14046

42867

567890 VBVB 8987654321

CONTRIBUTORS

JOHN A. S. ADAMS, Professor of Geology, Rice University, Houston, Tex. *(Chapter 12: Geochemical Surveys)*

EDWIN M. BEDFORD, Independent specialist in leasing. Formerly Manager, Land Department, Western Gulf Oil Company, Los Angeles, Calif. *(Chapter 16: Leases, Concessions, Options, Permits)*

B. W. BEEBE, President, Exploration, Inc. Petroleum Exploration Consultant. Formerly Vice-president of Production and Director of Keating Drilling Company. A Past Vice-president of AAPG. Boulder, Colo. *(Chapter 1: Philosophy of Exploration; Chapter 17: Drilling the Exploratory Well)*

GEORGE S. BUCHANAN, Senior Vice-president, Husky Oil Company. A Past President of AAPG. Cody, Wyo. *(Chapter 2: Purpose and Obligation of Exploration)*

HENRY L. BUTHMAN, Regional Photogeologist, California Oil Company (Western Division), Denver, Colo. *(Chapter 9: Photogeology)*

CHARLES V. CAMPBELL, Research Geologist, Jersey Production Research Company, Tulsa, Okla. *(Chapter 22: Laboratory Procedures in Exploration)*

JOHN L. P. CAMPBELL, Head Office Sales Engineer–Geologist, Lane-Wells Company, Houston, Tex. *(Chapter 20: Radioactivity Well Logging)*

FRANK B. CARTER, Consulting Petroleum Geologist. Formerly Senior Administrative Geologist, General Petroleum Corporation, Glendale, Calif. *(Chapter 14: Economic Analysis of Exploratory Project)*

C. G. CLAYPOOL, Group Head, Jersey Production Research Company, Tulsa, Okla. *(Chapter 22: Laboratory Procedures in Exploration)*

H. G. DOLL, Chairman of the Board, Schlumberger Limited. New York, N.Y. *(Chapter 19: Electrical Logging)*

KARL DYK, Chief Geophysicist, Pan American Petroleum Corporation. A Past Vice-president of the Society of Exploration Geophysicists. Tulsa, Okla. *(Chapter 11: Geophysical Surveys)*

WILLIAM R. EVITT, Research Associate, Jersey Production Research Company, Tulsa, Okla. *(Chapter 22: Laboratory Procedures in Exploration)*

ED J. HAMNER, Consulting Geologist. Houston, Tex. *(Chapter 8: Planning Exploration Project)*

WILLIAM S. HOFFMEISTER, Group Leader (Paleontology), Jersey Production Research Company, Tulsa, Okla. *(Chapter 22: Laboratory Procedures in Exploration)*

FREDERIC H. LAHEE, Consulting Geologist. Formerly Geological Counselor, Sun Oil Company. A Past President of AAPG. Dallas, Tex. (*Chapter 4: Probability of Success in Exploration*)

HARVEY W. LEE, Administrative Assistant, Foreign Operations, Union Oil Company of California, Los Angeles, Calif. (*Chapter 3: Exploration Personnel; Chapter 24: Exploration Scouting*)

JOHN L. LOFTIS, Jr., Regional Exploration Manager, Southwest Region, Humble Division, Humble Oil & Refining Company, Houston, Tex. (*Chapter 8: Planning Exploration Project*)

F. B. LOOMIS, Chief Geologist, Clark Oil & Refining Corporation, Denver, Colo. (*Chapter 13: Subsurface Geology*)

M. MARTIN, Executive Vice President, Géopétrole, Paris, France. (*Chapter 19: Electrical Logging*)

CLARK MILLISON, Consulting Geologist, Englewood, Colo. (*Chapter 18: Coring*)

KARL A. MYGDAL, Coordinator, Producing Property Acquisitions, The Pure Oil Company, Denver, Colo. (*Chapter 15: Writing the Report*)

A. J. PEARSON, Petroleum Engineer, Continental Illinois National Bank and Trust Company of Chicago, Chicago, Ill. (*Chapter 21: Miscellaneous Well Logs*)

R. DANA RUSSELL, Supervisor, Geology Department, Denver Research Center, The Ohio Oil Company, Littleton, Colo. (*Chapter 23: Exploration Research*)

H. J. SCHAETTI, Section Head, Jersey Production Research Company, Tulsa, Okla. (*Chapter 22: Laboratory Procedures in Exploration*)

F. SEGESMAN, Publication and Interpretation Department, Schlumberger Well Surveying Corporation, Ridgefield, Conn. (*Chapter 19: Electrical Logging*)

DEAN H. SHELDON, Petroleum Engineer, South Pasadena, Calif. (*Chapter 7: Petroleum Reservoirs*)

KENNETH G. SMITH, Continental Oil Company, Durango, Colo. (*Chapter 10: Surface Geology*)

NORMAN C. SMITH, Consulting Photogeologist, Dallas, Tex. (*Chapter 9: Photogeology*)

M. P. TIXIER, Manager of Field Interpretation, Schlumberger Well Surveying Corporation, Houston, Tex. (*Chapter 19: Electrical Logging*)

CHARLES E. WEBBER, Director, Natural Gas Administration, Sun Oil Company. President and Director of Natural Gasoline Association of America. Philadelphia, Pa. (*Chapter 25: Estimation of Petroleum Reserves Discovered*)

LEWIS G. WEEKS, Geological Consultant (worldwide). Formerly Chief Geologist, Standard Oil Company of New Jersey. A Past President of AAPG. Westport, Conn. (*Chapter 5: Origin, Migration, and Occurrence of Petroleum*)

SHERMAN A. WENGERD, Petroleum Consultant. Professor of Geology, University of New Mexico, Albuquerque, N. Mex. (*Chapter 10: Surface Geology*)

MILTON T. WHITAKER, Staff Reserves and Evaluation Engineer, Mobil Oil Company, Los Angeles Division, Los Angeles, Calif. (*Chapter 14: Economic Analysis of Exploratory Project*)

DAVID G. WILLIS, Manager, Research Computer Systems, Lockheed Missiles and Space Division. Research Associate in Geophysics, Stanford University. Atherton, Calif. (*Chapter 6: Entrapment of Petroleum*)

PREFACE

The purpose of this handbook is to summarize and to correlate all the activities involved in petroleum exploration. There are many excellent textbooks on particular techniques of exploration, such as those used in field geology and subsurface geology. This book is not intended to compete with the textbooks but rather to excite readers' interest to such an extent that they will become eager seekers for more information about exploration.

The opening chapters discuss fundamental concepts of exploration: the diagnostic traits of the explorationist; the organization required to prosecute the hunt for oil; the risks involved; how oil is formed; how and why oil is trapped in some places but not in others. The succeeding chapters describe the application of earth sciences and various techniques to petroleum exploration and the implementation of an exploratory project—from the original planning thereof to the drilling and logging of an exploratory well. The final chapters cover complementary activities, the extreme importance of which is often overlooked.

The person who reads this book carefully will gain a comprehensive understanding of the complex, highly specialized but mutually dependent divisions of exploration. It can be read with profit by those actively engaged in the technical phases of exploration, by those in management, by bankers—in fact, by anyone who wants to know about exploration. Each chapter gives a summary of current thinking about the subject treated. The material is presented, of course, in less detail than is given in textbooks covering the same topic. Those who wish to specialize in a particular phase of exploration should consult the books and papers cited in chapter references and in the bibliography.

Exploration is dynamic, never static. The application of basic methods is being improved continually by better interpretation of old information. Research in the earth sciences introduces improved techniques for attacking and solving the innumerable problems associated with the search for oil. Ideas and conclusions which are entirely logical, based on careful analysis of all data available at the time of formulation and writing, can become obsolete before they appear in print. It will be a long time before the "last word" can be written about the interpretation of scientific data applicable to exploration.

The reader will note that some authors differ in their opinions concerning the value of certain types of exploratory effort. This is to be expected and indicates a healthy condition in our profession. So long as explorationists argue about techniques and interpretations, they are progressing. When they agree that they have the *final* answer to a problem, not susceptible to revision in the future, they begin to lose value as hunters for oil.

The legal and tax aspects of the oil business are so intricate that it would require several volumes to give a comprehensive summary of their impact on exploration. The canny explorationist will leave such matters to the lawyers and tax experts.

The contributing authors represent, collectively, close to a thousand years of experience in exploration effort. They have expended several thousand man-hours of painstaking work on their manuscripts. Each one writes with authority about his subject. They merit the hearty thanks of our profession and of the oil industry at large for their contributions. The editor personally is most grateful to them for their hard work, patience, and cooperation that have made possible the completion of this book.

Graham B. Moody

Berkeley, Calif.

CONTENTS

Chapter 1

PHILOSOPHY OF EXPLORATION

By B. W. Beebe

President, Exploration, Inc. Petroleum Exploration Consultant. Formerly Vice-president of Production and Director of Keating Drilling Company. A Past Vice-president of AAPG

Boulder, Colo.

The petroleum geologist is the keystone of the arch of the petroleum industry. Industry, commerce, and our civilization depend on adequate supplies of cheap crude oil and gas. The petroleum geologist, with the cooperation of others, supplies this demand. Petroleum geology can actually be divided into several classifications, the most important of which is the exploration geologist—the prospector, the pioneer, the man in whose mind an oil or gas field must be visualized before a discovery can be made. No other professional individual has such a unique and peculiar position, the blending of precise scientific and engineering observations with judgment, experience, philosophy, and intuition. Thus we speak of the art of petroleum prospecting.

The petroleum industry furnishes most of the energy required for our rapidly growing and expanding, highly technical civilization, a civilization the foundations of which are based on the availability of cheap, efficient sources of energy. These sources are for the most part the fossil fuels, primarily oil and gas. A little more than 100 years ago, 90 per cent of our energy sources was furnished by human and animal muscle; today, 98 per cent of our energy comes from inanimate sources, and more than two-thirds of this total is supplied by oil and gas. Demand for these products, furthermore, is increasing at an unbelievably rapid rate. Few of us ever stop to realize how completely dependent we are in every phase of our everyday living on oil and gas and the multitude of products derived from them. The acceptance of geology by the petroleum industry and its rise to its current preeminent position were products of sheer necessity. The pressure of demand was such that the haphazard methods of exploring for oil and gas were no longer adequate.

The application of the earth sciences to the problem of exploration for hydrocarbons barely covers a half century, although many of the basic principles governing the search for oil have been recognized for well over 100 years. Two famous pioneer geologists, H. D. Rogers and W. B. Rogers, noted as early as 1840 that the greatest amounts of oil came from warm saline springs along the crests of anticlines. Shortly thereafter, Sir William Logan, director of the Canadian Geological Survey, in a report published in 1844, recognized the relationship of oil accumulation and anticlinal structure. The first clearly printed statement of the anticlinal theory was made in 1861 by T. Sterry Hunt, geologist and chemist of the Canadian Survey. Through these developments and the subsequent work of geologists E. W. Evans, E. B. Andrews, and Alexander Winchell, most of the basic principles of the occurrence and accumulation of hydrocarbons were recognized by 1870 (Howell, 1934, p. 10).* However, little attempt was made to apply these principles practically. It remained for Israel C. White, justly recognized as the father of petroleum geology, to win recognition of the application of geology by the oil industry. White resigned from the Pennsylvania Survey in 1883 and became a consulting geologist. In 1885, he published a short paper in *Science* which, together with his demonstrations by concrete example of the positive value of geology, finally won acceptance of its application to exploration problems.

His early successes, however, did not result in immediate response, and he struggled for years, not only with oil men, but with fellow members of the profession. He was ably supported by Edward Orton with his monumental studies of oil and gas fields in northwestern Ohio. By 1900, the U.S. Geological Survey was studying oil-producing areas in Ohio and Pennsylvania, and it was generally recognized that oil, under certain circumstances, could be found in synclines. Neither White nor his predecessors believed or stated that all anticlines could be productive or that in any particular area an anticline was the best place to look for deposits of petroleum. It is unfortunate that the term "anticlinal theory" was applied to the results of White's studies. He recognized that anticlines were only one of many factors involved in searching for commercial deposits of oil and gas. All this pioneer work, however, gave a firm foundation to the application of geology to problems of exploration.

E. T. Dumble was employed as a geologist for Southern Pacific Railroad in Texas in 1897, W. W. Orcutt established the geological department for Union Oil Company the following year, and the Santa Fe Railroad employed H. B. Goodrich as a geologist in the Ardmore Basin area of southern Oklahoma in 1903. Ralph Arnold, W. W. Orcutt, Joseph A. Taff, G. C. Gester, and others were recommending drilling of anticlines mapped on the surface in California early in the twentieth century. Moody (1954, p. 1905, Fig. 11) shows that the peak of oil discovery resulting from drilling exposed surface anticlinal structures was reached in California in 1910.

It remained, however, for the increased demand for petroleum brought about by the development of the automobile in the early twentieth century to place geology on a sound footing in industry. Perhaps modern petroleum geology dates from about 1913, when Gypsy Oil Company, now Gulf Oil Corporation, and Empire Oil and Gas Company, now Cities Service Oil Company, established formal geological departments. In 1915, oil was discovered at Eldorado, Kansas, on a large dome which had been mapped geologically, and there was no doubt that geology had won acceptance by the oil industry. By 1917, Empire had a large extensive department with the first subsurface section. The structural period of prospecting had commenced, and the anticlinal theory became so thoroughly ingrained in the thinking of geologists and the petroleum industry that our progress still suffers from such restrictive application. The next 25 years marked the golden age of structural prospecting, and numerous methods and techniques were developed to aid the search for favorable structure. These included surface mapping, core drilling, magnetic and gravimetric prospecting, subsurface mapping, the refraction and the reflection seismograph. But by the late 1930s, it became apparent to thinking geologists that there was a finite number of closed anticlinal structures, that each one found reduced those remaining, and that

* References appear at the end of the chapter.

each discovery made the next one more elusive. Obviously the larger, more obvious and prolific structures were found first, and ultimately the law of diminishing returns was reflected in the exploration effort. It became more and more expensive to find less oil and gas. There was a plethora of methods and a paucity of large, prolific structures.

By 1940 this situation had become quite apparent. Many geologists were discovering that oil could occur in traps other than structural, and that these stratigraphic traps could be found only by careful, detailed, analytical geologic studies. The same type of struggle faced by White and his supporters has again been raging in the industry and the profession for the past 20 years. Although the industry and the profession pay lip service to the stratigraphic trap concept, little real progress and understanding, with accompanying development of methods and techniques to discover such traps, have been made. Our ability to furnish our domestic economy with the required hydrocarbons is directly dependent on our ability to develop and apply principles, methods, and techniques which will result in the discovery of stratigraphic oil and gas reservoirs.

There is no magic in the term "petroleum exploration geologist." The petroleum exploration geologist must first be a well- and broadly trained geologist skilled in the application of many branches of the science to the problem of finding oil and gas. To borrow a term from medicine, a true exploration geologist is a general practitioner, applying the principles of structural geology, sedimentation, stratigraphy, paleontology, mineralogy, petrology, geomorphology, and historical geology to the problem of finding deposits of oil and gas. The finding of hydrocarbons is statistical in character and is based on the principle of continuous exposure to situations and circumstances under which hydrocarbons are known to occur. Oil and gas are not found by flashes of genius but are a product of rigorous observation and tenacious, dogged, often dreary intensive work and study. Like most other scientific enterprises, the background of a discovery consists of 1 per cent inspiration and 99 per cent perspiration. Levorsen (1954, p. 640) has stated so aptly: "The drilling of dry holes, holes that find only barren traps, or no traps at all, is a calculated risk of petroleum exploration."

The essential ingredients in any successful exploration program are capital, time, and technical skill. The exploration geologist furnishes the most important of these ingredients: the technical skill, blended with an optimistic philosophy, perseverance, judgment, intuition, and perhaps a bit of mysticism. His function is to employ properly the other two ingredients, capital and time. His greatest enemies are dogmatism and prejudice. It is most difficult with a vast quantity of miscellaneous data to separate actual facts from what appear to be facts, and there is a fundamental difference. It is often difficult to classify and analyze properly, then to arrive at a clear, logical conclusion. Unless the geologist is unusually objective and careful in his thought processes, constantly employing the scientific method of reasoning, he will succeed only in rearranging his prejudices. The history of exploration geology is replete with examples of large, potentially productive areas which failed to yield to conventional methods of exploration and were condemned by the industry, only to afford production when attacked later by an operator not bound by prejudice and misconception. One cannot help recognizing that, despite the conquest of the oil industry by geologists, many important oil fields have been found without technical advice in areas condemned by the profound manifestos of prominent members of the profession. No geologist is infallible in his judgment.

Oil fields and oil provinces are found by positive, constructive thinking. No oil or gas was ever found by condemning a prospect or an area. A geologist may be correct in his recommendation most of the time by condemning prospects, but the end result is the liquidation of his company's assets through the pipeline.

The fundamental thesis of the exploration geologist is that oil is abundant in favorable areas and that any area of thick marine sediments is a potential producing province. Failure to find oil in such areas through an exploration program is usually the fault of the program; oil is not necessarily absent. Timing is also an important ingredient. Favorable areas often experience several cycles of exploration, until finally enough information has been gathered to reveal the flaws in the earlier exploration programs. A discovery results, and a new productive province has been found. In many cases, reasons for failure are so elementary that it seems inconceivable that so much time and effort have been required. It is fundamental that the more precise and positive the data and the more varied the indications of a favorable prospect, the better the prospect. Discovery of a new field which opens a new province is the ultimate achievement in the career of a petroleum geologist. The reaching of this goal involves a coordinated research: gathering, classifying, analyzing, and applying reason and experience to a great diversity of data.

What then are the essential characteristics of this paragon of scientific virtue, the successful exploration geologist? First and foremost, he must be a dedicated geologist, an individual to whom geology is not a means of making a living but a way of life, a man to whom there is nothing more interesting and fascinating than geology. He constantly is amazed that anything could be so intriguing and stimulating and yet be financially rewarding. The exploration geologist is the elite member of the geological profession, the pioneer, bold, adventuresome, courageous, ever ready to explore new vistas.

The same attributes that characterized our pioneer ancestors must be his. He must be an optimist, resourceful, curious, inventive, tenacious, with an abundance of initiative. Above all, he must have vivid but controlled imagination, remembering that the oil field must first be visualized in the mind of the prospector. Coupled with these qualities, he must have a prodigious memory and the ability to recognize the importance of data which at first glance might seem trivial. Then he must have the ability to recognize and appraise clearly changing situations and the mental adaptability to alter an opinion, no matter how dear to his heart. The importance of salesmanship cannot be overlooked, because an exploration geologist is continuously selling himself and his ideas. This means that he must be able to express himself fluently, clearly, and concisely both orally and in writing. Finally, he must have stamina, the drive which results in the maximum output per unit of time and the ability to spend uncounted hours in intensive concentration. Finding hydrocarbons is not an 8-hour-per-day, 40-hour-per-week job. The effort has nothing to do with retirement plans, vacations, fringe benefits. Security for the exploration geologist lies in the opportunity to take calculated risks. No other profession is so fraught with hazard, with complete disaster lurking in every venture, but no other profession offers the opportunity for such tangible and intangible rewards for successful pioneering.

There are approximately 15,000 petroleum geologists employed by industry, yet only a small percentage of these are real exploration geologists. Some are in exploitation, administration, evaluation, and related fields. There are also the few truly scientific specialists in various types of research. But unfortunately far too many are the gatherers of information, the artisans, the "geological clerks." All these categories support the exploration geologist in his efforts. Exploration geology is a highly individualistic and creative enterprise, not adaptable to the electronic calculator or the assembly line.

The exploration geologist nears the end of an era of prospecting and, with ever-increasing demand for the product of his efforts, must advance boldly into a new era. Exploration, for the most part, since its acceptance by industry, has been based on the search for structure, primarily anticlinal. The watchword has been: "Find a closed structure!" Most of the instruments, methods, and techniques which have been developed and successfully used have been designed to find abnormal structure in the surface and subsurface. The search for structure has been the practice despite the recognition by many thoughtful geologists that there are many important oil fields in which structure plays little or no part in the accumulation, that there are many other factors equally or more important. There are, for instance, unconformities, facies changes, depositional traps; these are the really determining factors controlling oil and gas accumulation. The anticlinal theory on which the industry has based its exploratory operations for so many years does not state that a local closed anticlinal structure is the best place to find deposits of hydrocarbons, and yet this is the assumption which has prevailed. Anticlines are the easiest type of trap to find, and, more than this, are the easiest to depict and sell to management. Many years ago an essay was written by the late E. DeGolyer (1923) entitled "The Seductive Influence of the Closed Contour." To quote Wallace E. Pratt (1942, p. 25): "Geologists appear to be infatuated with anticlines. The lure of anticlines has blinded them to the real controls in the occurrence of oil pools." A. I. Levorsen (1936, p. 523) summed up the situation most succinctly as early as 1936: "We have all, therefore, drifted along from one structure-finding method to another until we are now apparently approaching the real danger of an oversupply of methods and an undersupply of anticlines."

Most of the obvious, easily found structures within the continental United States have been drilled. The current high cost of finding and capturing a barrel of oil and the small reserve per productive discovery are ample evidence that our entire philosophy of exploration must undergo a complete transformation. The day of the geologist searching for oil with a structural or isopachous map is ending. The exploration geologist of the future must pay far more attention to sedimentation, facies changes, geological history, tectonic control of sedimentation, and environmental conditions and must learn to depict stratigraphic traps as he has so superbly learned to illustrate anticlinal structure. There are many undrilled anticlinal structures in areas outside the United States, but there are also many prolific oil fields to be found within this country. Drilling for oil in this country has given more geological information about North America than is known of any other area. There are still many potentially productive, relatively unexplored areas beckoning the pioneer. It remains only for them to be attacked by geologists with vision, courage, perseverance, and initiative. Those who blend the finest in geology with the knowledge that the end result must be converted to profits in terms of dollars will assure exploration geology its preeminent position in the industry.

REFERENCES

Ball, Max W.: "This Fascinating Oil Business," The Bobbs-Merrill Company, Inc., Indianapolis, 1940.

Ball, Max W. (ed.): Possible Future Oil Provinces of North America, *Bull. AAPG*, vol. 35, pp. 141–485, February, 1951.

Beebe, B. W.: The Development of Petroleum Geology, in "A History of the Geological Sciences," unpublished symposium, California Institute of Technology, Division of Geological Sciences, 1940.

Beebe, B. W.: "A Philosophy of Exploration," unpublished address, 1956.

Beebe, B. W.: "Geology Now Unemployed, Future Unlimited," unpublished address, 1958.

Buckley, Stuart E. (ed.): "Petroleum Conservation," chap. 3, pp. 77–83, American Institute of Mining and Metallurgical Engineers, 1951.

Clapp, Frederick G.: Fundamental Criteria for Oil Occurrence, *Bull. AAPG*, vol. 11, pp. 683–703, July, 1927.

Clifford, Jr., O. C.: An Oil Finder Looks at His Profession and Scales It Down to Size, *Oil and Gas J.*, vol. 56, pp. 156–160, May 26, 1958.

Cram, Ira H.: Definitions of Geology and Subsurface Geology, *Bull. AAPG*, vol. 29, p. 470, April, 1945.

Cram, Ira H.: Resources and Resourcefulness, *Bull. AAPG*, vol. 29, pp. 857–871, July, 1945.

Cram, Ira H.: Geology Is Useful, *Bull. AAPG*, vol. 32, pp. 1–10, January, 1948.

DeGolyer, E.: The Seductive Influence of the Closed Contour, *Econ. Geol.*, vol. 23, pp. 681–682, 1923.

DeGolyer, E.: Future Position of Petroleum Geology in the Oil Industry, *Bull. AAPG*, vol. 24, pp. 1389–1399, August, 1940.

DeGolyer, E.: "The Development of the Art of Prospecting," Princeton University Press, Princeton, N.J., 1940.

Heald, K. C.: Essentials for Oil Pools, in "Elements of the Petroleum Industry," pp. 26–62, Seeley W. Mudd Series, American Institute of Mining and Metallurgical Engineers, 1940.

Heroy, W. B.: Petroleum Geology, in "Geology 1888–1938," pp. 511–548, Geological Society of America, 1941.

Howell, J. V.: Historical Development of Structural Theory of Accumulation of Oil and Gas, in "Problems of Petroleum Geology," pp. 1–23, *AAPG*, 1934.

Lahee, Frederick H.: Degrees of Success in Wildcat Drilling, *Bull. AAPG*, vol. 35, p. 138, January, 1951.

Levorsen, A. I.: Stratigraphic versus Structural Accumulation, *Bull. AAPG*, vol. 20, pp. 521–530, May, 1936.

Levorsen, A. I.: "Geology of Petroleum," chap. 1, pp. 3–12; chap. 14, pp. 606–635; chap. 15, pp. 636–651; W. H. Freeman and Co., San Francisco, 1954.

Moody, Graham B.: Oil in California, *Bull. AAPG*, vol. 38, pp. 1888–1910, September, 1954.

Noble, Earl B.: Geological Masks and Prejudices, *Bull. AAPG*, vol. 31, pp. 1109–1117, July, 1947.

Pratt, Wallace E.: "Oil in the Earth," University of Kansas Press, Lawrence, Kans., 1942.

Price, Paul H.: Evolution of Geologic Thought in Prospecting for Oil and Natural Gas, *Bull. AAPG*, vol. 31, pp. 673–697, April, 1947.

Sloan, Raymond D.: "The Future of the Exploration Geologist, Can He Meet the Challenge," unpublished address, 1957.

Weeks, L. G.: Factors of Sedimentary Basin Development That Control Oil Occurrence, *Bull. AAPG*, vol. 36, pp. 2071–2124, November, 1952.

Chapter 2

PURPOSE AND OBLIGATION OF EXPLORATION

By George S. Buchanan

Senior Vice-president, Husky Oil Company. Past President of AAPG
Cody, Wyo.

Purpose of Exploration

Our great dynamic oil and gas industry, which now furnishes, and will continue to furnish, a goodly portion of the world's energy needs, adds to the efficiency of our machines by proper lubrication, and furnishes the raw materials for scores of petrochemical products, can continue to thrive and benefit mankind only by a constant and vigilant search for new reserves. While incipient oil and gas fields may be forming today, it is the geologically ancient oil and gas accumulations which have or will have a bearing on our immediate and near-future economy.

Whether or not any particular oil company is fully integrated, the industry as a whole breaks down into three principal functions: production, refining or manufacturing, and marketing. The production end of the business may be separated chronologically into, first, land and lease acquisition; second, exploration; and finally exploitation operations. In this chapter we are primarily concerned with the exploration phase, since the finding of new reserves is essential and imperative to all other phases of the industry.

Approaches and techniques to the finding of oil have varied widely over the past decades and will continue to change in the future, but the urge to find new oil in order to help replace that which has already been produced or is now being produced has been a vital challenge and a form of pioneering to those engaged in the finding effort. Hence, since the turn of the century a small group of men classifying themselves as oil geologists has now grown to the startling size of nearly 25,000 explorationists in the oil and gas industry throughout the world. They include geologists, paleontologists and micropaleontologists, mineralogists, stratigraphers, geophysicists, geochemists, petroleum engineers, hydrologists, and many other specialists in phases of these earth science fields. Whatever the choice of the area in which these earth scientists are engaged, their primary function at all times is to aid in the finding of new reserves to replace those which are being consumed on an ever-expanding scale.

A great deal of attention has been applied in recent years to increase the recovery of known reserves. Through many years of production history it has been found that in some fields only a small part of the oil in place has been or could be recovered by so-called primary methods. In other reservoirs with an adequate and natural water drive or through proper control of the gas which may be associated with the oil and is often the prime mover, particularly in flowing wells, primary recoveries have reached as high as 50 or 60 per cent of the original oil in place. Adding only a few per cent of recoverable oil to some reservoirs can and does in many instances amount to millions

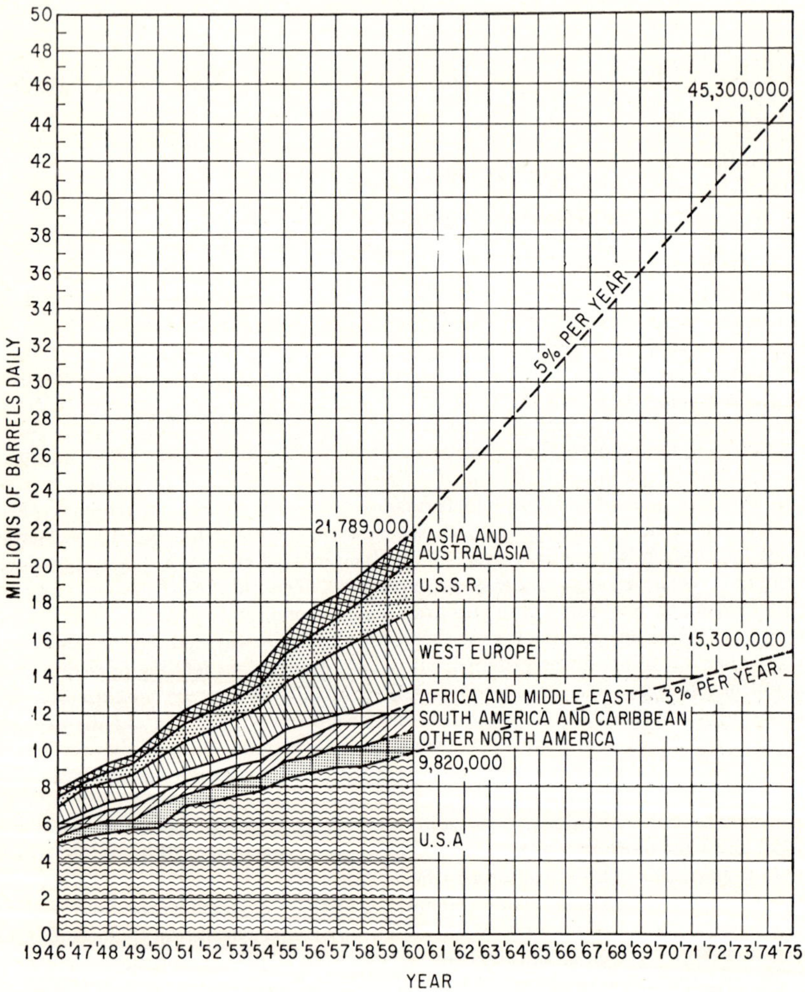

FIGURE 2-1. World demand: actual for past 14 years, estimated through 1975.

of barrels which might otherwise be lost to our use. The introduction of water, gas, LPG, or other solvents or a combination of any of these at the proper time is of great importance. It is known that some reservoirs that are being produced should commence so-called secondary methods of recovery simultaneously with the so-called primary methods. Not only is the timing of great importance in the element of recovery, but the methods employed are equally so. The basic background data for such study have fallen to earth scientists. Proper structural mapping of the producing reservoirs; their chemical and mineralogical make-up, including size, orientation, distribution, packing of mineral particles, the nature of and amount of cementation; the fluid content with associated pressure measurements and direction of fluid movement; permeability; porosity; and temperature are some of the vital factors to which attention is directed to find the proper method of increasing recovery through so-called secondary means. This phase of exploration, while not so romantic perhaps as the finding of a virgin reservoir, is nonetheless highly rewarding in terms of achievement.

To emphasize the need for finding new reserves of oil and gas, total United States and world demand graphs are shown for the past 14 years and the future estimated demand through 1975 (see Figure 2-1). These estimates of demand were presented at the Annual Meeting of the American Petroleum Institute in November, 1959, and were published in the January issue of *World Oil*.

The total demand for the United States in 1960 was estimated at 9.82 million barrels daily, and the total world demand for 1960 was estimated at 21,789,000 barrels, or an over-all increase of 6 per cent over 1959. It will be noted on the chart that the percentage increase, particularly in the last 5 years, has been higher throughout the rest of the world than the percentage increase in the United States, and everything indicates this will be the pattern in the years ahead. Assuming, however, an increase of 3 per cent per year for the United States, we should have a total demand in 1975 of 15.30 million barrels daily. If we assume a 5 per cent per year increase in demand throughout the world for the same period, we should have a total demand of 45.3 million barrels per day in 1975. The producing capacity of the United States at the present time is probably in the neighborhood of 10 million barrels daily, and it is questionable that this producing capacity will be increased or maintained in the years ahead. It also appears that in the years through 1975 there will be an adequate world-wide source of petroleum as such without resorting to other fossil fuels to any great extent. Coal hydrogenation and recovery of oil from tar sands and from oil shales will be a part of the picture only in areas where they may be competitive on a cost basis. Of course, the latter have very little or no finding costs, so that the basis of their cost is principally related to their developmental costs.

If the productive capacity of the United States remains for some time at around 10 million barrels daily, it will be readily seen that imports will play an increasing roll in our domestic oil economy. This means that the exploration branch of our industry will have to be more conscious of and take a greater part in foreign exploration than it has in the past.

Obligation of Exploration

Having expressed some ideas which should show clearly that the primary purpose of exploration is to find petroleum, it might be well to consider the obligation of the exploration branch to the industry and to the respective companies and individuals who furnish the risk capital that makes an exploratory program possible.

Whether or not any exploration program is limited to a portion of the United States or includes all of North America or is world-wide in its scope, the basic approach is the same, namely, to investigate large provinces which are classified as possible or probable petroliferous areas. This investigation would include at least superficial studies to determine whether possible or favorable source beds may occur in the province and whether suitable reservoir rocks are suspected or known to be present. It should also be determined whether the sedimentary section present is a thin mantle or of considerable thickness. Techniques that might be employed in this connection are discussed in considerable detail in following chapters.

Having determined that a particular, large province or area is generally attractive, it is also the obligation of the exploration branch to localize favorable prospects for exploratory drilling within the province and of the land department to secure sufficient leases to cover the prospect as fully as possible.

In order to convince management that an exploratory well should be drilled on any prospect it is also the responsibility of the exploration branch to give its best estimate of probable costs and an educated guess as to the possible yield in the event the exploratory well is successful. Some prospects may call for more than one exploratory well to test thoroughly or condemn the prospect in question. This would be particularly true of stratigraphic traps, hydrodynamic accumulations, or truncated structural features. These facts should be clearly spelled out at the beginning of exploratory drilling so that a second or third well may not come as a shock to those less qualified to judge the merits of any particular prospect.

The exploration branch and the production operations branch, working in close association and cooperation, should jointly submit a program for the drilling of the first exploratory well on each prospect and subsequent exploratory wells if deemed necessary. If an exploratory well is successful, joint planning with the same group in the development of the area should be undertaken. In the event the well is unsuccessful, all the information gained from drilling the well should be studied immediately after its completion while the information is still available and fresh in everyone's mind. Following this study a new recommendation should be forthcoming for further exploratory effort or abandonment of the project.

Conclusion

It is still possible with all the tools and techniques which today's exploration manager has at his call that he would not be fortunate enough to make a discovery. On the other hand, his group would have unraveled the geology much more completely and quickly, located and drilled the test well faster, with less effort, and with a great deal less chance of having passed up an oil field than his predecessors not many years ago.

And what of tomorrow's tools?

1. They will probably consist of those in use today but with considerable improvements.

2. It is doubtful if a short-cut or direct method of oil and gas finding is just around the corner. Electrical methods, soil analysis, and air-borne radioactive surveys have been heralded in some quarters as the panacea for direct oil finding, but scientific application has found their shortcomings.

3. The greatest tool of all will be the continuous imaginative thinking of those explorationists dedicated to their profession.

Chapter 3

EXPLORATION PERSONNEL

By HARVEY W. LEE

Administrative Assistant, Foreign Operations, Union Oil Company of California Los Angeles, Calif.

The search for petroleum is carried on by small independent wildcatters with limited capital on the one hand and by the large fully integrated oil companies on the other, the latter with large exploratory budgets. There may be a difference between them in the magnitude and scope of their operations, available resources, and number of technical and scientifically trained personnel, but the ultimate goal of both is to find oil or gas that can be produced at a profit.

INDEPENDENT OPERATOR

Let us first consider the case of the independent operator in relation to his wildcat operations. He may be a druggist, lawyer, doctor, geologist, truck driver, driller, or a member of other numerous professions or occupations. Provided he is familiar with industry procedure and technically trained along certain lines, he may be able to handle some of the details of this venture himself with very little outside assistance. In most cases, however, this is unlikely, and it probably will be necessary for him to obtain, at least initially, some legal advice and counsel regarding his rights and privileges, the proper lease form, and the drawing of subsequent contracts. It is probable, furthermore, that the services of a capable geologist and an engineer will be needed to handle the many details too numerous to mention, both before the project is started and after the well starts drilling. Some consultants are sufficiently experienced and trained in both the geological and engineering professions that a single person can serve as both a geologist and a petroleum engineer. Many of the problems confronting the independent operator can be resolved by informal discussions with one or more of the technical men. Eventually, the operator is faced with accounting and tax problems that are best solved by experts in their chosen profession.

The following details should be carefully considered by the independent operator. These are not necessarily listed in the order of importance and do not include many of the conditions that may be encountered in special cases.

1. Proper lease form, pooling clauses, offset requirements, royalty interest, drilling commitments, regulation of production by state agencies.

2. Geologic report of the property, evaluation of prospect, location of the well to be drilled, producing formations expected, and estimated depths.

3. Determine offering price for land or lease, actual negotiation and acquisition of property, obtaining rights of way.

4. Applications and permits to drill a well, erect a derrick; posting of bonds; adequate insurance coverage.

5. Proposed drilling program or "well prognosis," including hole sizes, casing sizes and depths, drilling mud type and control, coring intervals, electrical logging, well surveys, testing of indicative showings; estimate of drilling costs.

6. Request for bids from drilling contractors. Decision as to who will furnish certain materials, such as mud, water, fuel, drillpipe, bits, etc.; should the well be contracted on a footage basis or should bids ask for daywork with different prices quoted for drilling time, coring, testing, fishing, cementing, standby, etc. (These factors depend upon how much previous drilling has already been done in the vicinity. A drilling contractor is usually unwilling to make a footage bid on exploratory wells located in virgin areas because no drilling history or cost figures are available.)

7. With the spudding of the well it will be necessary to maintain continuous supervision from both a geological and an engineering standpoint. An agent of the operator should be present at all times to be sure that all drilling operations are conducted in an efficient and workmanlike manner and that the best interests of the operator are satisfied.

8. It will often be necessary to employ or contract for certain outside services, such as a well-logging truck, mud control, cementing, formation testing, electrical logging.

9. If successful, there is the selection and additional cost of flowing or pumping equipment, storage facilities, gas traps, etc.

10. Once production is established, the negotiation of contracts begins for the purchase of crude oil, the sale of gas, and treatment of the gas for its liquid content.

11. There is the accounting and the tax problem—both of prime importance.

12. Once on production there is the routine gauging of the wells together with necessary maintenance and well repair.

13. When the commitment for the second and subsequent wells becomes due, the majority of the above details will again have to be considered.

It becomes quite evident, therefore, that the consultant probably plays an important role in the venture of an independent operator.

Technical Fees

Many geologists have chosen to enter the consulting field after obtaining necessary experience elsewhere. Their fees are rather flexible, possibly a contractual flat sum for the project plus expenses, a per diem basis with field expenses, a contingent basis, reimbursement of expense with an overriding royalty interest, or any combination thereof. Similarly, the consulting petroleum engineer advises and supervises the drilling operations on a comparable basis. Quite frequently one person or a consulting firm can serve in both capacities. In many states all types of scientific and professional engineers are required to be licensed to assure the public of their competence and qualifications in their chosen profession. The independent operator should make certain that he obtains competent professional geologists and engineers to advise him.

SMALL COMPANY

A small company, typical of many in the exploration and producing phase of the oil business, may consist of only four people (see Figure 3-1). This small organization is an outgrowth of the independent operator and represents a diversification of individual effort and responsibility, as well as a sharing of the risk involved.

From an exploratory standpoint, an executive or an administrator and a technically trained geologist or engineer comprise the nucleus of a small company. In this case the executive would be cognizant of his company's policies and probably be familiar with finance, land, and legal problems, while the technically trained person would be responsible for all the geological problems and engineering details. Even this does not preclude the possibility of some outside technical assistance, as the drilling of any well is a round-the-clock effort and some of the necessary work to be done is other than at the well site. In addition, the importance of a competent file clerk cannot be underestimated. In small companies the officers and employees have multiple and often overlapping duties. Decisions can usually be made in small informal conferences on short notice and immediate action taken thereafter without the usual regimentation or protocol so evident in large companies. This gives the small company the opportunity to act at once; a big company needs more time to make decisions. It is quite unlikely that the independent wildcatter would have more than one exploratory project under way at any one time. However, a small company that is active in the exploratory field may have several wildcat plays or prospects in various stages of development simultaneously. A new prospect may be in readiness to drill upon

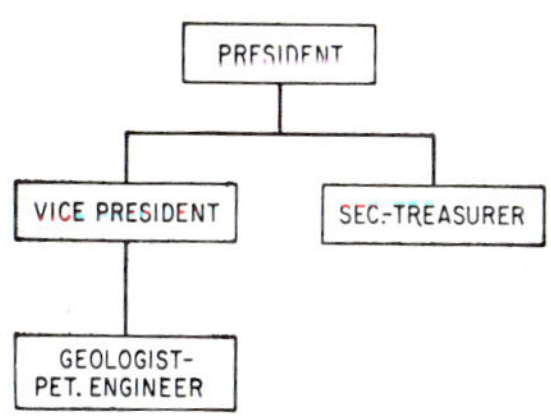

FIGURE 3-1. Organization chart of small company.

the completion or abandonment of the current drilling project. The geologist or engineer concerned with more than one active project at a time is likely to need additional assistance, and the possibility that these projects may be in widely separated areas would also add to the numerous duties.

The same exploratory details listed previously as confronting the independent operator must also be considered by the small company. However, the presence of additional personnel, instead of part-time consultants, permits a greater distribution of the work load.

LARGE COMPANY

A large oil company with millions of dollars budgeted for exploration has a considerable advantage over an independent operator or a small company. The many active projects under way at one time, varying from regional reconnaissance of potential interest areas to recommendations and negotiations leading up to the actual drilling stage, necessitate a large exploratory staff. With a larger number of people employed in exploration, the respective duties of each person become more narrowly defined along a specialized line.

Practically every branch of science that could be of possible service in finding oil is being exploited by most large companies. With large exploratory staffs there will be experts in almost every branch of scientific endeavor, and in many cases pure research is involved.

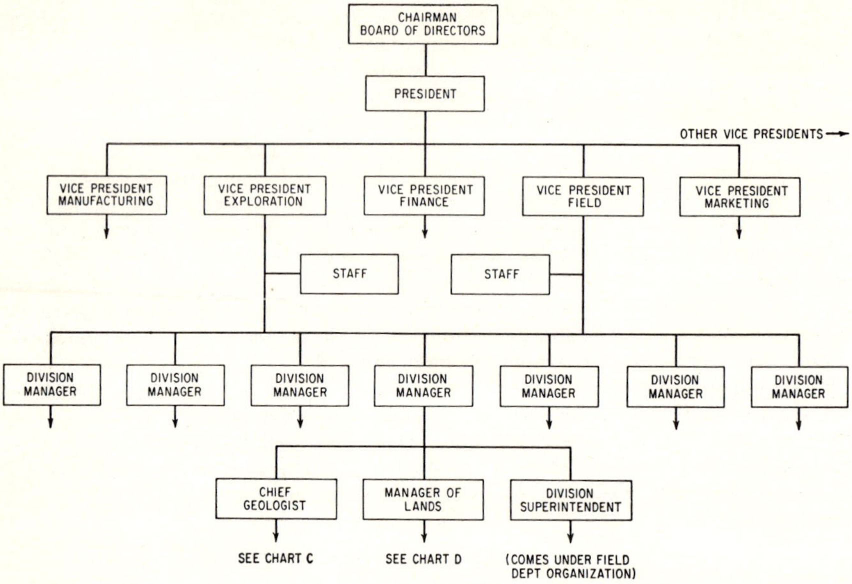

FIGURE 3-2. Organization chart of large company, domestic operations, executive personnel.

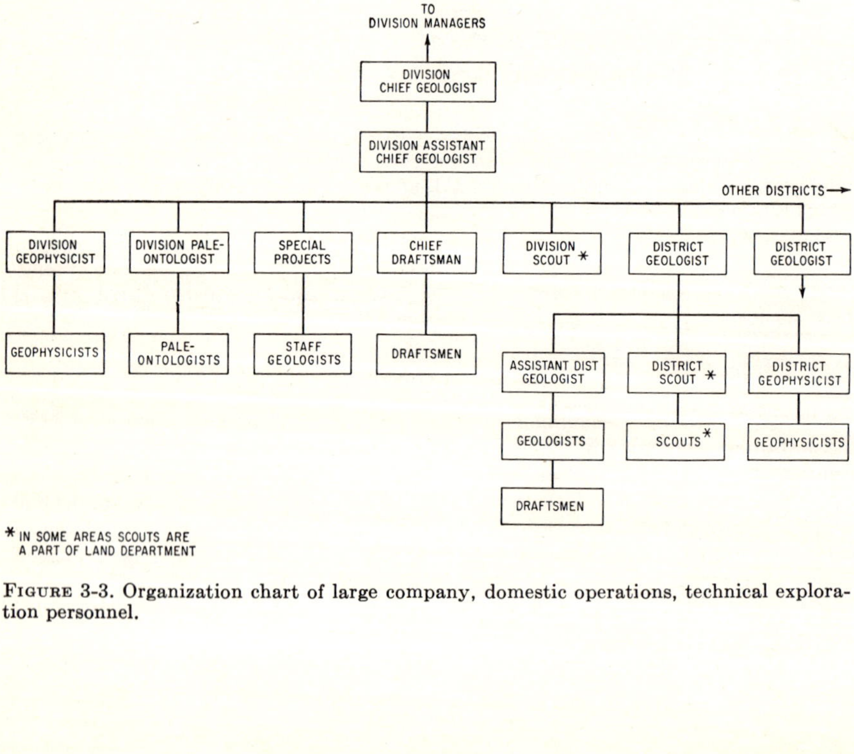

FIGURE 3-3. Organization chart of large company, domestic operations, technical exploration personnel.

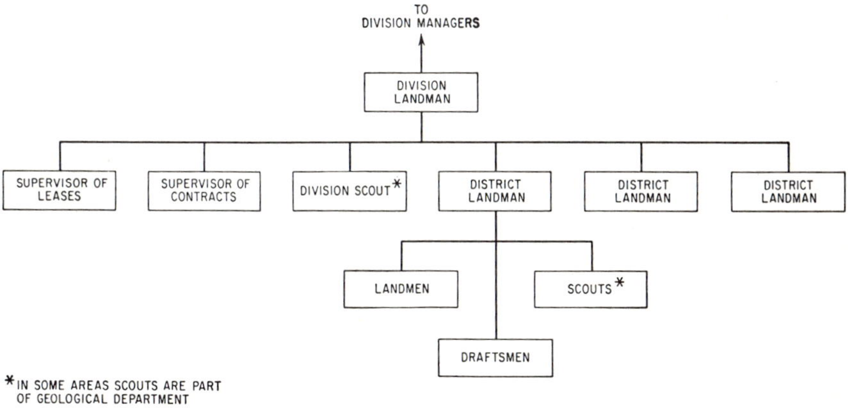

FIGURE 3-4. Organization chart of large company, domestic operations, land department personnel.

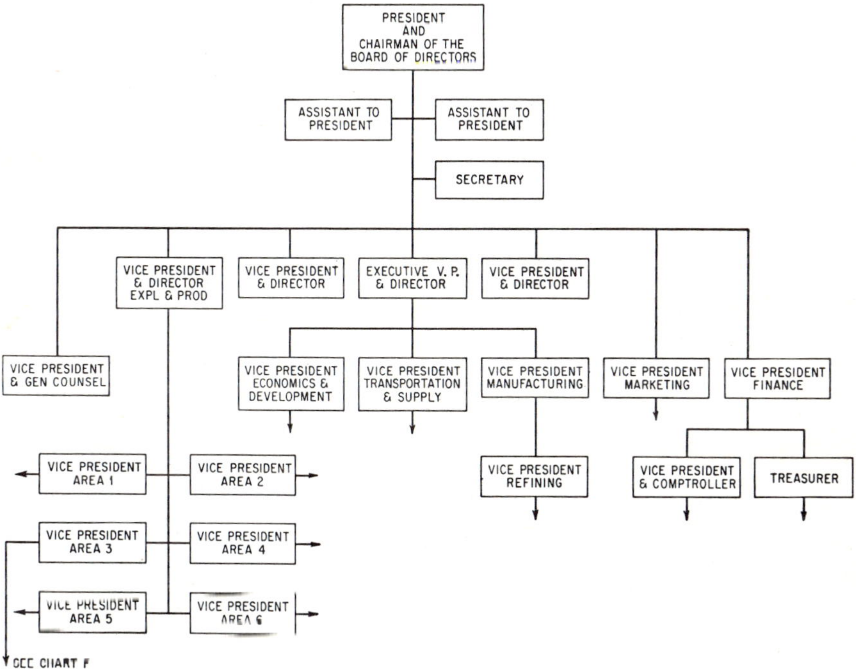

FIGURE 3-5. Organization chart of large company, domestic and foreign operations, executive personnel.

Branches or subdivisions of science used in petroleum exploration include (1) geology (structure, stratigraphy, photogeology, sedimentation, surface and subsurface studies), (2) geophysics (seismology, magnetics, and gravity), (3) chemistry and geochemistry, (4) electricity (with experts on the various types of electrical logs and interpretative techniques), (5) paleontology and paleobotany, (6) mineralogy (thin sections and

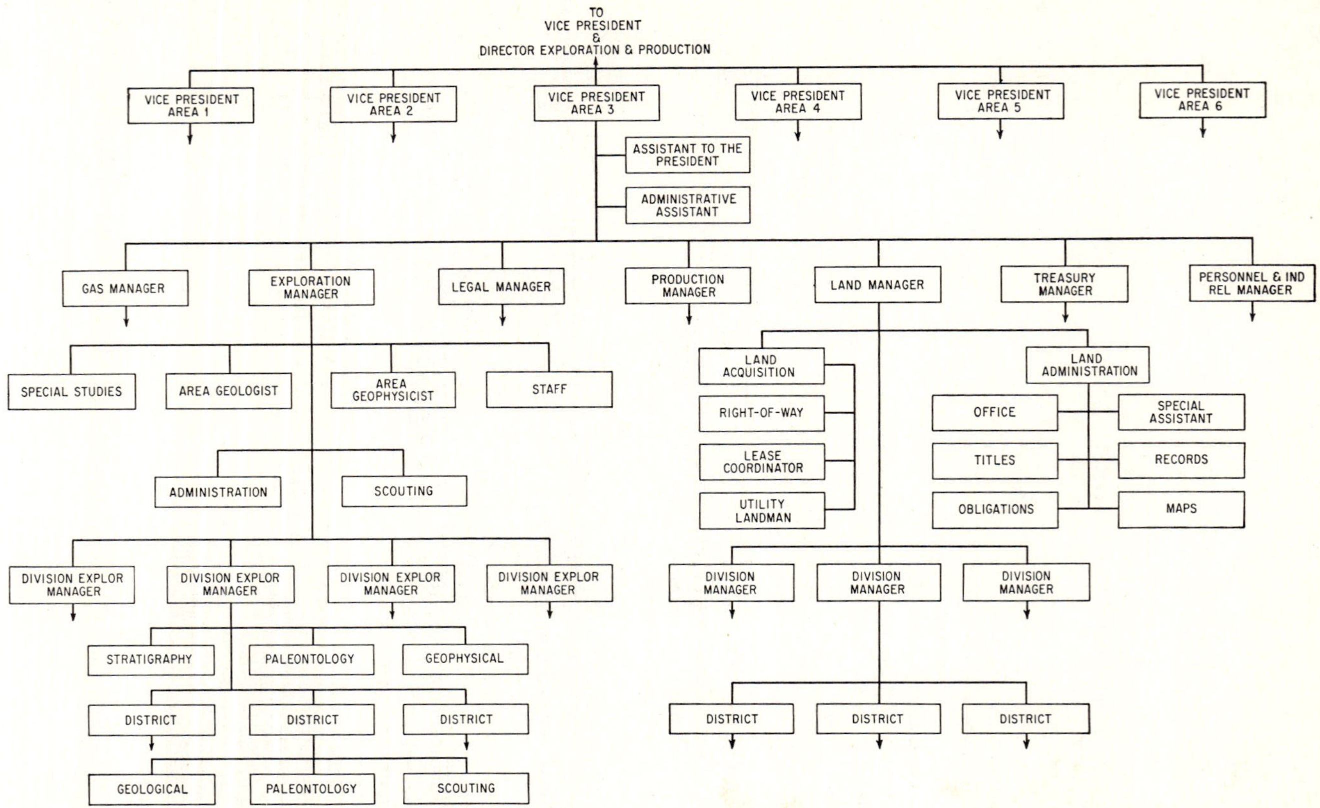

FIGURE 3-6. Organization chart of large company, domestic and foreign operations, technical exploration personnel.

mineral composition), (7) meteorology for weather forecasts and hurricane warnings, (8) hydrodynamics, the pressure and movement of fluids in rocks.

A wildcat "play" or idea could originate from an interpretation of geophysical data; surface, subsurface, or photogeologic studies; or any combination of these. When a general idea is conceived from one of these sources an attempt should be made to determine if cardinal points of the play can be substantiated by any of the other means. The assignment may originally have been a regional reconnaissance of an area to be supplemented later by additional work, in the field, office, and possibly the laboratory. It may be necessary to sample type sections in the field for lithology and stratigraphy or obtain cores from wells previously drilled in the area. The geologist will now enlist the aid of a paleontologist or paleobotanist to assist in the age determination or correlation. In instances it may be expedient to prepare thin sections of rocks or core samples to determine the nature of the mineral content under a microscope by a mineralogist. There may be some difficulty in the interpretation of data from the many types of electrical logs currently in use; consequently, he depends on an expert in logging technique.

Many of these services are carried out in elaborate research laboratories that are an integral part of a large company operation. Research departments of oil companies are primarily organized to upgrade commodities, modernize refinery techniques, and discover possible new products. Many of them are also equipped to assist in pure geological research. Some laboratories have small pilot-plant equipment to experiment with the recovery of oil from oil shales, waxes, or tar sands and for the development of new processes and techniques used in secondary recovery operations, all of which have a bearing on exploration.

Figure 3-1 depicts the organizational chart for a small company. Figures 3-2 through 3-6, inclusive, represent organization charts of two large oil companies, one in domestic operations only and the other in both domestic and foreign operations, and show the respective chains of command. Figures 3-2, 3-3, and 3-4, of a large company in domestic operations with one or two possible exceptions, have a vice-president heading up the different operating departments. This plan shows a definite trend toward decentralization of authority through a division manager. The exploration department in this case is responsible for the acquisition of potential oil and gas lands and the drilling of potential prospects, while the field department is responsible for development and production.

Figures 3-5 and 3-6 are examples of a large oil company in both domestic and foreign operations. The general plan of organization here is division by functions and by line and staff departments each headed by a vice-president. The exploration and production activity is further divided into geographical areas, also headed by a vice-president of lower rank.

EDITOR'S NOTE: Mr. Lee informs us that since he wrote this chapter the trend in oil-company organization has switched from decentralization to centralization; hence, his emphasis on the former may seem to be anachronistic. The Editor's comment is that this is a good example of how rapidly conditions change in the oil industry, that shortly the trend will revert again to decentralization, and that such changes keep the organization departments busy.

Chapter 4

PROBABILITY OF SUCCESS IN EXPLORATION

By FREDERIC H. LAHEE

Consulting Geologist. Formerly Geological Counselor, Sun Oil Company. Past President of AAPG

Dallas, Tex.

HISTORY OF EFFORT TO COLLECT DATA

As described in other chapters of this book, many methods have been applied in the search for accumulations of oil and natural gas underground; yet in spite of all the advances made in these methods and in spite of the ever greater accuracy of scientific techniques used for this purpose, there still remains an element of risk of failure. How great this risk is under various conditions of drilling is the theme of the present chapter.

Previous to 1935 no attempt had been made to ascertain the risk involved in wildcat drilling, to find out how many failures, on the average, were drilled for each successful venture. Beginning in that year, we gradually built up a procedure for compiling data on this subject. We soon realized that for reliable comparison of statistics, from year to year or from one region to another, a standard method of reporting would be essential and a standard classification of wells would be requisite. So we evolved a classification which is now used throughout the petroleum industry, and likewise we developed a standard procedure for compiling the necessary data, so that everywhere in the country members of the committee which was appointed to gather the facts would do so in the same way. This committee, known as the Committee on Statistics of Exploratory Drilling, was organized under the auspices of the American Association of Petroleum Geologists, with F. H. Lahee as chairman, in 1945. This committee was successor to a group of geologists who had assisted Lahee during the years 1935 to 1945. After Lahee's retirement from the committee in 1956, Graham B. Moody assumed the chairmanship for one year, to be followed by B. W. Blanpied. Each year, from 1937 to the present time, the chairman has compiled a report from the data furnished him by his committee, a report which has appeared annually in the June number of the *Bulletin of the American Association of Petroleum Geologists* since 1938.

CLASSIFICATION OF WELLS

For proper understanding of the significance of the important facts disclosed by the statistics gathered in this study, the reader must be reasonably familiar with our classification. All wells drilled specifically for oil or gas may be divided into two groups, or categories: (*A*) *field-development wells* which are drilled *within* the boundaries of known oil or gas reservoirs or which are short (usually one location) step-outs from wells already completed as producers and which are drilled to extend the productive area of the known "pool" or reservoir* and (*B*) wells which are *exploratory* in the

Table 4-1. Classification of Exploratory Wells

Drilling objective	Classification when drilling is started	Classification after completion or abandonment	
		Successful	Unsuccessful
For long extension of partly developed pool	Outpost	Extension well	Dry outpost
For a *new pool* on a structure or in a geological environment already productive—divided into tests drilled			
(1) Within limits of proved area of pool, for new pool			
(*a*) *Above* deepest producing pool	Shallower-pool test	Shallower-pool discovery well	Dry shallower-pool test
(*b*) *Below* deepest producing pool	Deeper-pool test	Deeper-pool discovery well	Dry deeper-pool test
(2) Outside limits of proved area of pool	New-pool wildcat	New-pool discovery wildcat	Dry new-pool wildcat
For a *new field* (on a structure or in an environment never before productive)	New-field wildcat	New-field discovery wildcat	Dry new-field wildcat

sense that they are drilled in search of oil or gas either above or below known reservoirs or at a considerable distance and entirely separate from known producing areas.

These exploratory wells are subdivided into (1) those which are drilled far from any oil-producing or gas-producing area and which, if oil or gas is found, will be the discovery wells of entirely new fields and (2) those which are drilled in association with known producing areas. The wells of group 1 are called *new-field wildcats.* Those of

* In this chapter we use the terms "pool" and "reservoir" as synonymous and as referring to a single interconnected porous rock medium which contains oil or gas in its interstices. A "field" may be one reservoir, or it may include two or more separate reservoirs, all of which, however, are related to the same geologic structure or to the same stratigraphic environment.

group 2 are further divided into (*a*) those located *outside* the known boundaries of partly developed pools and (*b*) those located *within* these known boundaries. The wells under (*a*) may be drilled to explore for new pools (reservoirs) laterally outside the confines of the known reservoir (called *new-pool* wildcats), or they may be wide step-outs intended to extend the known reservoirs (called *outposts* or *extension tests*),

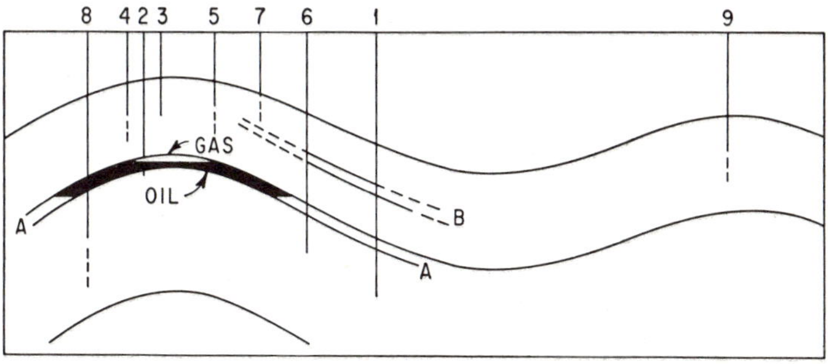

FIGURE 4-1. Diagrammatic illustration of different classes of hole drilled for oil and/or gas. 1. Dry new-field wildcat. 2. New-field wildcat producer. Discovery well of oil field in sand *A*. 3,4. Field-development wells, drilled after completion of 2. 5. Outpost, several locations from 3. 6. New-pool wildcat, outside of productive part of sand *A*, looking for a new pool on the same structure. Both 1 and 6 encountered a water-bearing sand *B* above the main pay *A*. 7. Shallower-pool test drilled in search of oil or gas near pinch-out edge of sand *B* which was absent in 5. 8. Deeper-pool test, drilled in search of possible production in unknown sand or sands below pay sand *A*. 9. New-field wildcat drilled on separate structure from that on which oil was discovered by well 2.

and the holes of subgroup (*b*) may be drilled to search for a new reservoir either stratigraphically *above* the known reservoir (called *shallower-pool tests*) or stratigraphically *below* the known reservoir (called *deeper-pool tests*). This classification is summarized in Table 4-1 and is diagrammatically illustrated in Figure 4-1.

RISKS OF FAILURE IN DRILLING

In the drilling of a well for oil or gas there is always risk of failure, and here we mean failure to find a permeable medium (sandstone, limestone, or other rock material) containing fluid hydrocarbons under sufficient pressure and in sufficient quantity to give promise that, when completed, the well will be a commercial success. The various methods, geological and geophysical, aimed at finding the more favorable geologic structures or environments on which to drill cannot yet surely determine whether or not these hydrocarbons are present or in what quantity they may be present. Only the drill can discover these facts and then usually only after the application of various technical tests made on the prospective reservoir after it has been penetrated by the drill. These statements apply especially to the new-field wildcats, far removed from known underground conditions, but they also apply to all other classes of exploratory hole and even, though to a considerably less degree, to field development wells.

Through the accumulation of analogous data from year to year, we are now able to ascertain approximately what the average risk of failure is in drilling the various classes of hole for oil or gas. Tables 4-2 to 4-5 record the number of failures and the number of successes as reported at the close of the year when the hole was abandoned as a failure (a dry hole) or was completed as a producer. This we call the "year of completion." Each table covers 15 years. The largest percentage of failures is recorded for new-field wildcats (Table 4-2) where, for the 15-year period, only 11.2 per cent of the holes drilled were called successful, or, stated in other words, there was only one success in every 8.9 holes drilled. Taking all five classes of exploratory hole into consideration (Table 4-3), in the 15-year period 19.7 per cent were successful, or

Table 4-2. New-field Wildcats Drilled

Year	Oil producers	Gas producers	Total producers	Dry holes	Total holes	Per cent success
1943	298	115	413	2,573	2,986	13.8
1944	247	94	341	2,752	3,093	11.0
1945	232	119	351	2,685	3,036	11.6
1946	236	97	333	2,800	3,133	10.6
1947	274	120	394	3,086	3,480	11.3
1948	366	135	501	3,795	4,296	11.7
1949	364	142	506	3,943	4,449	11.4
1950	462	130	592	4,698	5,290	11.2
1951	515	169	684	5,505	6,189	11.0
1952	555	185	740	5,957	6,697	11.1
1953	581	193	774	6,151	6,925	11.2
1954	639	263	902	6,478	7,380	12.2
1955	659	258	917	7,186	8,103	11.3
1956	628	240	868	7,841	8,709	10.0
1957	589	283	872	7,142	8,014	10.9
Total	6,645	2,543	9,188	72,592	81,780	11.23

Table 4-3. Total Exploratory Holes Drilled

Year	Oil producers	Gas producers	Total producers	Dry holes	Total holes	Per cent success
1943	529	178	707	3,291	3,998	17.7
1944	709	235	944	3,852	4,796	19.7
1945	840	374	1,214	4,399	5,613	21.6
1946	762	375	1,137	4,615	5,752	19.8
1947	981	397	1,378	5,397	6,775	20.4
1948	1,098	365	1,463	6,550	8,013	18.3
1949	1,406	424	1,830	7,228	9,058	20.2
1950	1,583	431	2,014	8,292	11,306	17.8
1951	1,763	454	2,217	9,539	11,756	18.8
1952	1,776	559	2,335	10,090	12,425	18.8
1953	1,981	699	2,680	10,633	13,313	20.1
1954	1,985	723	2,708	10,389	13,097	20.7
1955	2,236	869	3,105	11,832	14,937	20.8
1956	2,274	822	3,096	13,077	16,173	19.1
1957	1,945	865	2,810	11,897	14,707	19.1
Total	21,868	7,770	29,638	121,081	150,719	19.7

one success in every 5.09 drilled. Table 4-4 lists the data for *all* wells drilled for oil or gas, both field-development and exploratory. Here, in the 15 years, 64 per cent were successful, or one success in every 1.56 drilled. Finally, in Table 4-5, we show the data for field-development wells, excluding all types of exploratory hole. Even here, in the 15 years, there is shown to have been only 78.9 per cent success, or one success

Table 4-4. All Holes Drilled for Oil and/or Gas*

Year	Oil producers	Gas producers	Total producers	Dry holes	Total holes	Per cent success
1943	9,887	2,390	12,277	6,364	18,641	65.9
1944	13,502	3,078	16,580	7,153	23,733	69.9
1945	13,944	3,192	17,136	7,346	24,482	70.0
1946	16,087	3,562	19,649	8,496	28,145	69.8
1947	17,613	3,720	21,333	9,751	31,084	68.6
1948	22,197	3,312	25,509	11,939	37,448	68.1
1949	21,415	3,499	24,914	12,898	37,812	65.9
1950	23,775	3,480	27,255	14,918	42,173	64.6
1951	23,532	3,542	27,074	17,497	44,571	60.7
1952	23,371	3,693	27,064	18,211	45,275	59.8
1953	25,251	4,232	29,483	18,759	48,242	61.1
1954	28,063	4,219	32,282	19,137	51,419	62.8
1955	30,474	4,169	36,643	20,564	57,207	64.1
1956	30,641	4,495	35,136	22,254	57,390	61.2
1957	27,364	4,475	31,839	20,156	51,995	61.2
Total	327,116	55,058	382,174	215,443	597,617	64.0

* Excluding water-disposal wells and fluid-injection wells. These figures were taken from *World Oil*, Feb. 15, 1958, p. 139.

Table 4-5. Field-development Wells*

Year	Oil producers	Gas producers	Total producers	Dry holes	Total holes	Per cent success
1943	9,358	2,212	11,570	3,073	14,643	79.1
1944	12,793	2,843	15,636	3,301	18,937	82.5
1945	13,104	2,818	15,922	2,947	18,869	84.4
1946	15,325	3,187	18,512	3,881	22,393	82.8
1947	16,632	3,323	19,955	4,354	24,309	82.1
1948	21,099	2,947	24,046	5,389	29,435	81.6
1949	20,009	3,075	23,084	5,670	28,754	80.3
1950	22,192	3,049	25,241	6,626	31,867	79.2
1951	21,769	3,088	24,857	7,958	32,815	75.8
1952	21,595	3,134	24,729	8,121	32,850	75.2
1953	23,270	3,533	26,803	8,126	34,929	76.8
1954	26,078	3,496	29,574	8,748	38,322	77.2
1955	28,238	3,300	31,538	8,732	40,325	78.2
1956	28,367	3,673	32,040	9,177	41,217	77.9
1957	25,420	3,610	29,030	8,263	37,293	77.8
Total	305,249	47,288	352,537	94,366	440,903	78.9

* Excluding water-disposal wells and fluid-injection wells.

in every 1.27 drilled. These statistics impressively demonstrate how large is the factor of probable failure, or, to put it in another way, how small are the chances of success, when drilling for oil or gas.

One of the striking facts brought out by these tables is the relatively small variation in the indicated percentages of success from year to year when examined separately under each group. These variations are shown in Figure 4-2. This relative consistency through this period must mean only one thing, that improvements in the

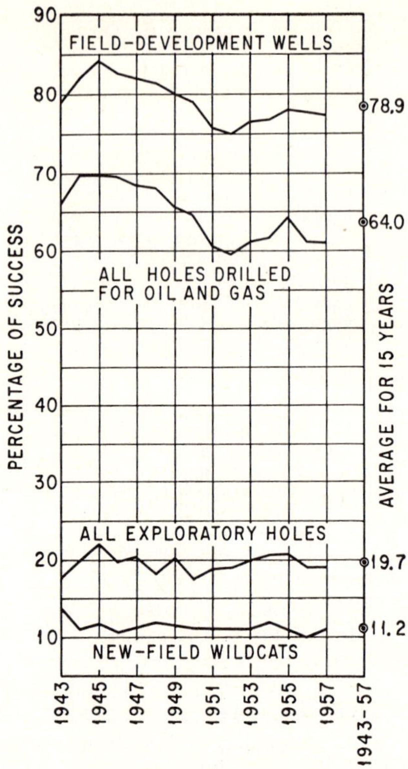

FIGURE 4-2. Curves showing percentage success of different classes of hole drilled for oil and/or gas during period from 1943 to 1957, inclusive.

Table 4-6. Average Depths of Exploratory Holes, Feet

Year	For all classes of exploratory hole	For new-field wildcats
1943	4,189	4,416
1944	4,217	4,217
1945	4,103	4,403
1946	3,854	4,007
1947	3,896	4,198
1948	4,086	4,362
1949	3,842	4,082
1950	3,898	4,181
1951	4,197	4,516
1952	4,476	4,852
1953	4,560	4,889
1954	4,549	4,808
1955	4,631	4,810
1956	4,574	4,823
1957	4,701	4,873
1943–1947	4,031	4,245
1948–1952	4,125	4,438
1953–1957	4,605	4,839
1943–1957	4,338	4,593

techniques of finding have been approximately balanced by the increasing difficulty of locating commercial underground accumulations of oil and gas, and this increasing difficulty is dependent mainly on the greater depths of drilling required to find new reserves and on the fact that the more obvious geologic structures have already been drilled. The increase in the average depth of drilling in exploratory holes is recorded in Table 4-6. Especially significant is this increase by 5-year periods.

Definition of Successful

When referring to a new-field wildcat as "successful" at the close of the year of completion, we mean that, as far as can be ascertained at that time, by examination of cores, by drill-stem tests, and by determinations of potential flow of the hydrocarbons, the well will prove to be commercial in the sense that its production when marketed will repay its costs of drilling, completing, and producing, although possibly with no additional returns to repay all the other costs which entered into the leasing, exploring, and overhead that preceded its drilling. However, if we wait a few weeks or a few months after the close of the year of completion, we find that some of these supposedly successful wells turn out to be failures, which are therefore abandoned. A careful study of these results reveals the fact that, on the average, not *one in nine*, but only *one in every ten* new-field wildcats can actually be rated as successful.

ESTIMATED RESERVES DISCOVERED BY WILDCAT DRILLING

Each successful new-field wildcat generally initiates the drilling of adjacent field-development wells, and these, after several years of development history, are usually sufficient to permit a fair estimate of the total ultimate recoverable reserves of oil and gas in the field. By a detailed investigation of such reserves, and by crediting them to the wells which discovered the fields, we can determine how many fields of certain selected sizes were discovered each year. For this purpose we use letter symbols, as follows:

Symbol	Barrels of liquid hydrocarbons	Cubic feet of natural gas
A	50 million or more	300 billion or more
B	25–50 million	150–300 billion
C	10–25 million	60–150 billion
D	1–10 million	6–60 billion
E	Productive, but less than 1 million	Productive, but less than 6 billion

When considering oil reserves and gas reserves together, it is useful, as well as scientifically correct, to relate oil and gas to one another on the basis of British thermal units, or Btu.* Roughly speaking, we may assume that a barrel of crude oil contains 6000 Btu, and that a thousand cubic feet (usually written 1 MCF) of natural gas contain 1000 Btu. In other words, 6,000 cubic feet of natural hydrocarbon gas are roughly equivalent to one barrel of crude oil, on this basis. Actually, individual samples of different crudes and different gases may vary somewhat from this value, but for most practical purposes it may be accepted as a reasonable approximation.

If we allow a period of 4 years or more for development of the fields discovered by

* A British thermal unit is defined as the amount of heat required to raise the temperature of one pound of water at its maximum density one degree Fahrenheit.

wildcat drilling in each of the 10 years, 1943 to 1952, inclusive,* and if we estimate the total ultimate recoverable reserves in each of these fields, adding together both the actual oil reserves and the gas reserves translated into their oil equivalent on the Btu basis, and if we attribute to each discovery new-field wildcat all these reserves, not only those in the first reservoir opened by the discovery well, but also those in other reservoirs opened subsequently in the same field, we learn by this procedure that:

1. To find a field containing total ultimate recoverable reserves of 50 million barrels or more (class A), 422 new-field wildcats were drilled.

2. To find a field with 25 million barrels or more (class A plus class B), 219 new-field wildcats were drilled.

3. To find a field with 10 million barrels or more (classes A plus B plus C), 84 new-field wildcats were drilled.

4. To find a field having 1 million barrels or more of total ultimate recoverable reserves (classes A plus B plus C plus D), 27 new-field wildcats were drilled.

These are very significant figures. They demonstrate how great is the risk of failure incurred in drilling for new reserves of oil and gas. To find a field which will pay the costs of drilling and producing (D fields or better), on the average 27 new-field wildcats must be drilled. But no company can stay in business and grow and pay all its additional expenses of leasing, geological and geophysical exploration, etc., unless it discovers some fields of the A and B categories. Yet to find one such field, as we have shown, on the average 219 new-field wildcats must be drilled.

SUMMARY AND CONCLUSIONS

We may summarize this chapter by citing its obvious conclusions together with a few additional thoughts, as follows:

1. Petroleum and natural gas, which were formed during long periods of geologic time, occur underground in accumulations which are exhaustible.

2. The number of these accumulations is limited, so that, as each reservoir is found and drained, there is one less accumulation to be discovered; there is no replacement of the hydrocarbons removed.

3. In those regions where the search for oil and gas has been in progress for many years, extensive geological and geophysical exploration has resulted in the discovery of the more obvious geological reservoirs for these hydrocarbons.

4. The search for new oil and gas fields is ever more difficult.

5. In this search drilling is being carried to greater and greater depths.

6. In spite of many technical improvements in exploring, drilling, completing, and producing, the quantity of reserves found per foot drilled is declining.

7. The costs involved in exploring, drilling, and producing are increasing.

All these important facts should be clearly borne in mind by all persons concerned in any way with this industry.

Oil and gas are of such tremendous and far-reaching importance, both in the nation's economy in times of peace and as an essential commodity in times of emergency, that nothing should ever be done to delay or impede the search for them. From every angle—industrial, political, and utilitarian—aggressive exploration for these hydrocarbons should be vigorously encouraged.

* This decade was selected because data prior to 1943 are inadequate and data for years after 1952, allowing at least 4 additional years for development history, were not yet available when this study was made (see Statistics on Natural-gas Discoveries, *Bull. AAPG*, vol. 42, pp. 2037–2047, September, 1958).

Chapter 5

ORIGIN, MIGRATION, AND OCCURRENCE OF PETROLEUM

By Lewis G. Weeks

Geological Consultant (Worldwide). Formerly Chief Geologist, Standard Oil Company of New Jersey. Past President of AAPG

Westport, Conn.

This chapter is devoted to an analytical discussion of the most common questions and problems concerning oil occurrence that confront the active petroleum geologist. Conflicting views and various facets of the problem are weighed in the light of the best and latest information available. Where diverse explanations might be applicable, the one presented is that which best fits the facts of oil occurrence as indicated by experience and judgment. Nature is not capriciously devious but always follows well-regulated principles. Apparent exceptions to the usual, such as, for instance, the time of oil migration and accumulation, generally disappear with adequate analysis when all facts are considered.

There is an extensive literature dealing with petroleum occurrence. Much that is new, however, has been learned by the vast amount of exploration experience embracing innumerable examples of each of many types of oil basin and oil occurrence over the world.* Theories which are not based sufficiently broadly and firmly on the great wealth of oil-occurrence facts that are now available may be more harmful than beneficial.

Petroleum Classification

The petroleum and petroleumlike substances may be roughly classified and defined for the purposes of the petroleum geologist as follows:

1. *Petroleum:* An almost infinitely complex mixture of saturated hydrocarbons, with relatively small amounts of sulfur, nitrogen, oxygen, and many other lesser constituents in combination. On account of its extreme variability of composition, the

* In the preparation of this chapter the author has drawn freely in several places on his editorial review in "Habitat of Oil," Special Symposium Volume published by the American Association of Petroleum Geologists, 1958. The reader is referred to that volume for further material on some of the subjects discussed here.

common reference to petroleum as a mineral or mineral fuel does not appear technically correct. It also does not meet Webster's Dictionary definition of a mineral as "any chemical element or compound occurring naturally as a product of inorganic processes." Petroleum is today generally considered to be of organic origin. Petroleum occurs in solid, liquid, and gaseous forms, as follows:

a. Asphalt or tar, paraffin waxes, and various forms of hard, brittle bitumen, such as uintaite or gilsonite, ozokerite, albertite, grahamite, manjak, wurtzilite, etc. (solids or semisolids). It also includes the solid hydrocarbon torbanite, usually classed with oil shales.

b. Crude oils (liquids).

c. Petroleum or natural gases.

2. *Tar sands:* Inspissated, semisolid petroleum-bearing sands.

3. *Oil shales:* These fine-grained sediments might better be termed kerogen shales or pyrobituminous shales, since oil can be obtained from them only by distillation at temperatures upward of about 650°F. The oil thus obtained is not true petroleum. The hydrocarbons are unsaturated. They generally contain nitrogen and sulfur and are particularly high in oxygen. Kerogen shales occur under basin environmental conditions that are sufficiently different from those that produce normal crude petroleum to suggest that such bitumens are not the primary source of crude petroleum. With increase in kerogen content the shales grade into:

4. Torbanites, boghead and cannel coals, and canneloid sediments, all very rich in bitumen.

The various forms of petroleum and tar sands (hydrocarbons) are associated in the sedimentary basin. Oil shales, bogheads, cannels, etc., may occur associated with the petroleums or separately.

Petroleum Reserves and Potential Resources

Petroleum is of widespread occurrence, a common constituent of sedimentary rocks. As in the case of all natural resources, the quantities of petroleum in the earth are far greater than man will ever use. What proportion of the total quantity will ultimately constitute *proved reserves* will depend on how deep into the total it will be economical to delve, for nothing can be classed as a reserve which cannot be produced usefully, that is, profitably. *Potential petroleum resources* comprise the accumulations of petroleum which may be found in the future and converted into reserves that can be produced and utilized at a profit at that time.

The incidence of petroleum occurrence varies widely over the earth as among the different continents, countries, and sedimentary basins and the different parts or elements of any basin. It varies from noncommercial quantities to very prolific occurrences, from an average of less than 5,000 to upward of a million barrels of oil per cubic mile of sediments.

Though oil occurs in sediments of all ages from Cambrian to Pleistocene, its incidence of occurrence varies widely both among the various geological periods and areally for all periods. For example, about 80 per cent of the oil found to date in the world occurs in what has been called the "oil axis poles"—the Gulf-Caribbean and the Mesopotamian–Persian Gulf areas. All the oil was reservoired in the Mesozoic-Tertiary. The upper half of the Tertiary has to date accounted for upward of 35 per cent of the oil found in the world. By contrast, the Silurian and Triassic, which together embrace twice as much of geologic time as the Upper Tertiary, have accounted for but a minor fraction of 1 per cent of the oil found.

It is axiomatic that one of the very best places to look for oil is in the belts or trends in which it has already been discovered in good quantity or under similar geologic situations elsewhere.

Ever since man became interested in the uses of petroleum and where and how to come by it, he has been intrigued by the manner and causes of its occurrence as guides to exploration. Whereas we can witness the origin of coal deposits today and see and measure the coal seams at the surface, the origin and occurrence of petroleum are for

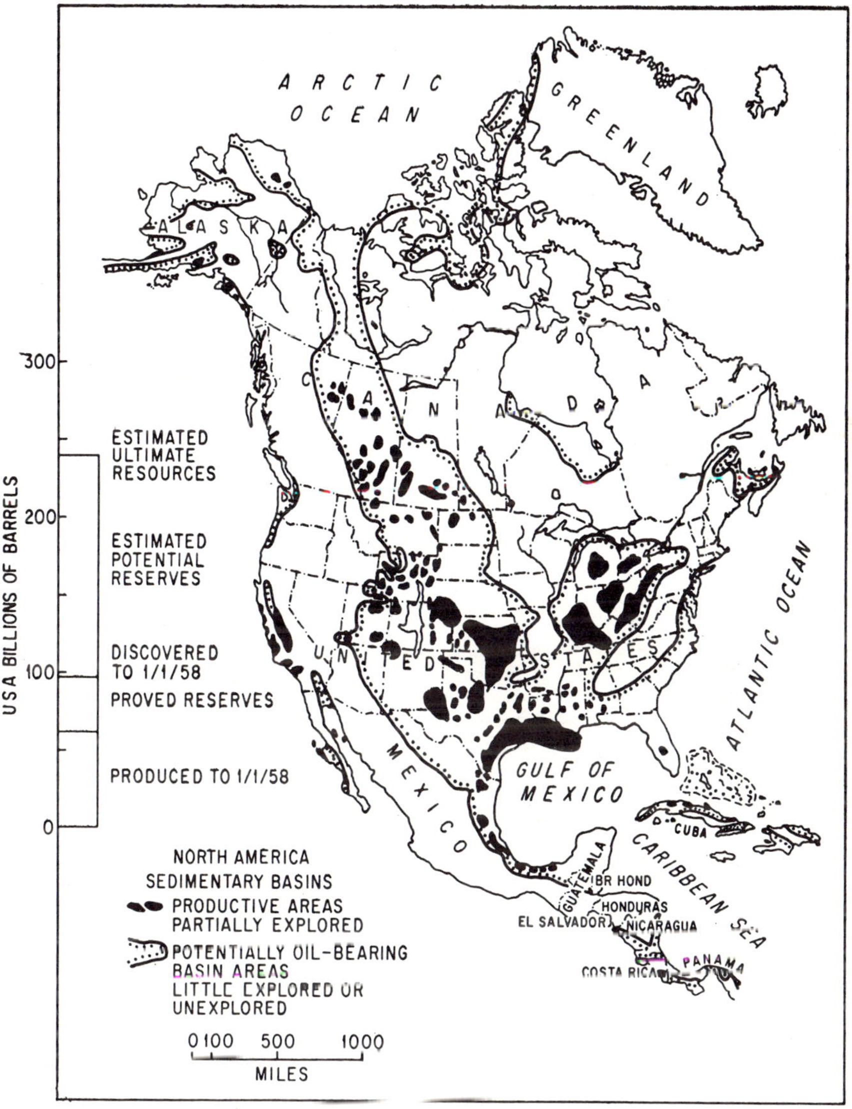

FIGURE 5-1. North America

the most part hidden from sight. It is partly for these reasons that there has been so much poorly founded thinking and guesswork with regard to the manner and time of petroleum origin and accumulation and the quantities available.

The history of petroleum reserves and total potential resource estimates has, by

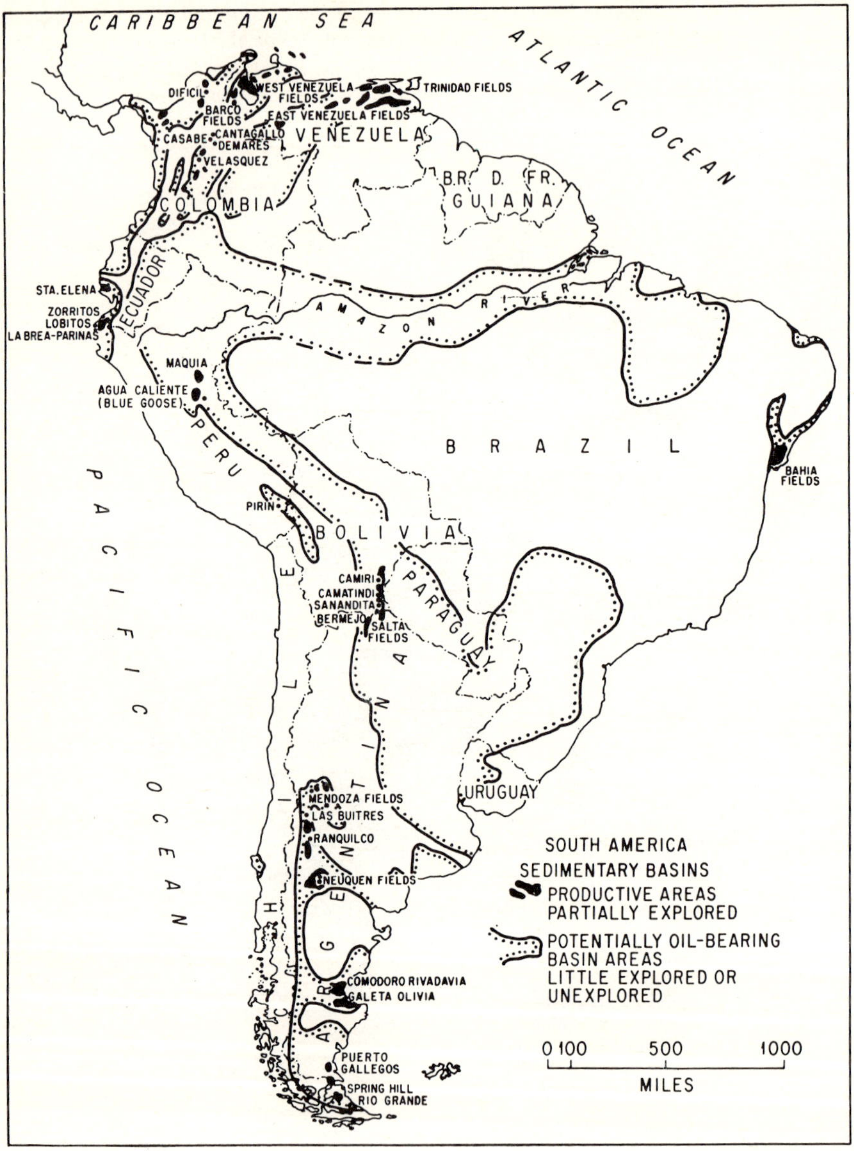

FIGURE 5-2. South America.

comparison with coal, been one of continuous conservatism and gross understatement. As a consequence, there may now be developing a tendency to go to the other extreme, based partly on wishful thinking and the fact that we have been wrong previously. Approaches that are not soundly based on geology are not applicable even to a country as highly developed as the United States; they are worthless when dealing with the prospects of undeveloped or little explored areas or regions.

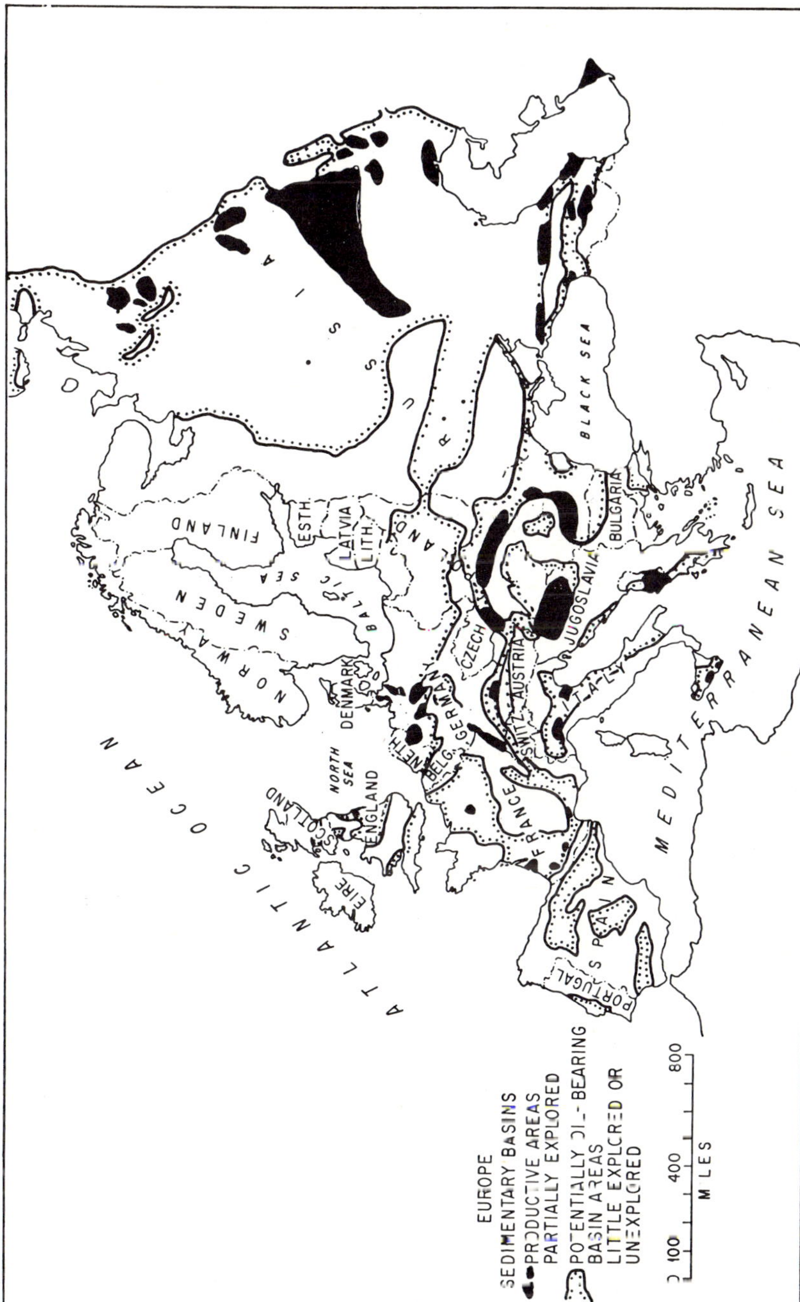

FIGURE 5-3. Europe.

Figures 5-1 to 5-7 show the sedimentary basins of the world. These represent the areas within which petroleum occurrence is possible. The shaded areas within the outlines of the basins are those within which petroleum has been discovered and where exploration has been mainly concentrated. For the most part, in those large areas of

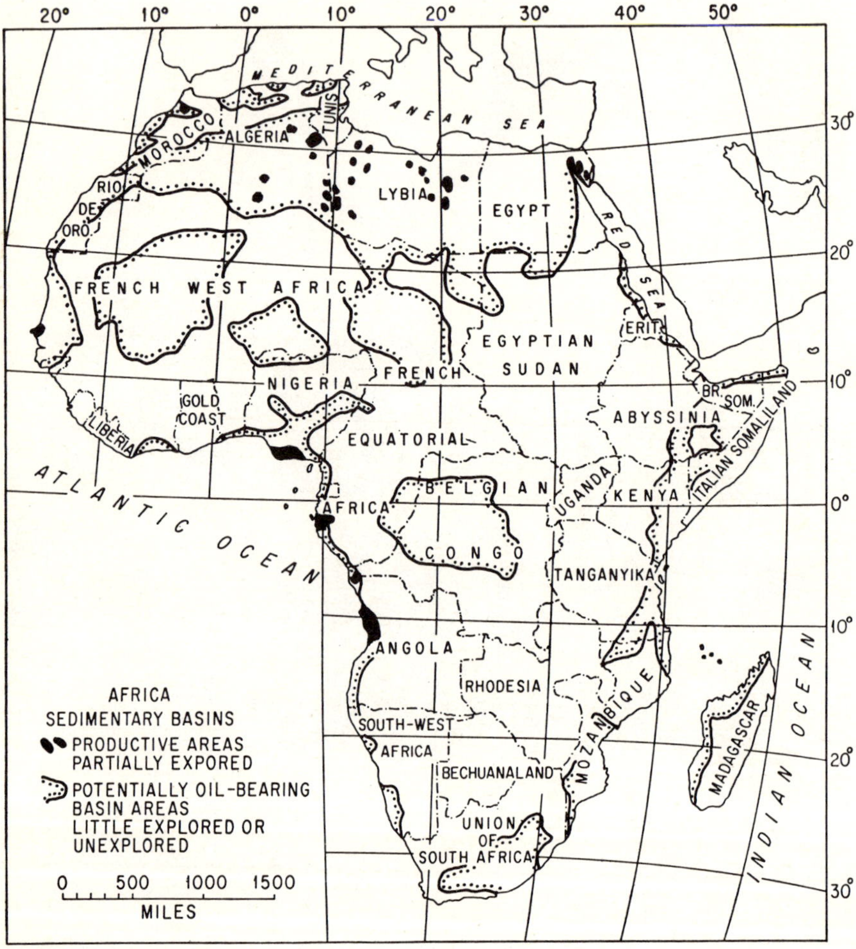

FIGURE 5-4. Africa.

the sedimentary basins outside the shaded areas there has been very limited or no exploration for petroleum.

Need for More Geologic Understanding

Much has been learned by the geological fraternity about oil occurrence from the vast and world-wide scope of exploration, encompassing many examples from each of many types and conditions of occurrence. At the same time there still exists widespread disagreement (which can only mean lack of understanding) on some of the most

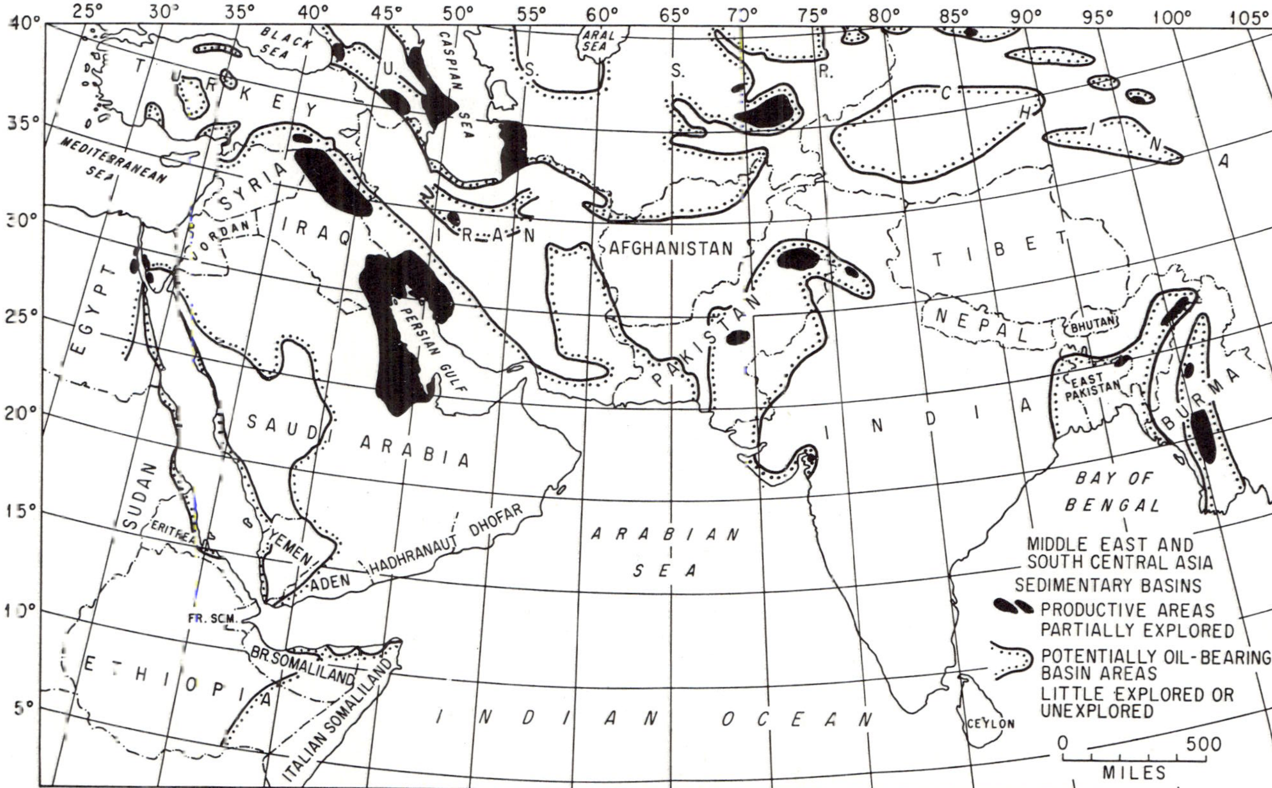

FIGURE 5-5. Middle East and South Central Asia.

fundamental of oil-occurrence principles. Archaic, completely unsound ideas still form part of the bible of many geologists. One reason for this situation is the fact that the average geologist does not have the opportunity or, more important, he commonly does not have the inclination for the empirical study and analysis in many places that are required to determine the facts. Some are primarily interested in getting into operations or management. When they go that route, they take whatever archaic ideas they may have with them, and that can be not only a bottleneck but a discouragement as well.

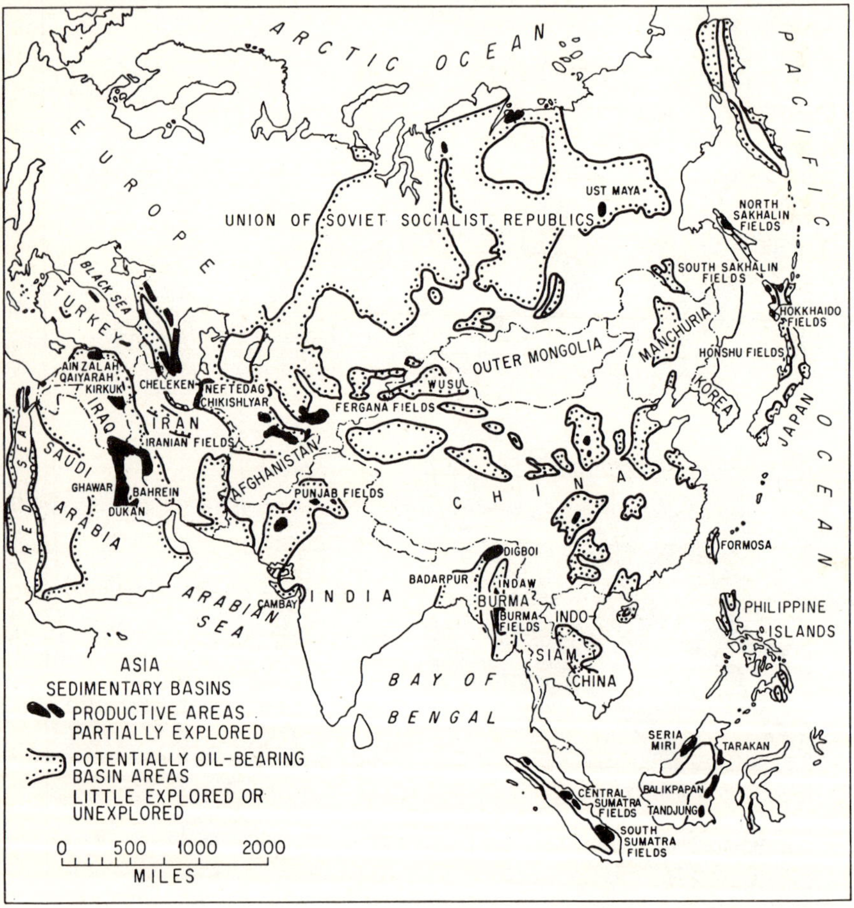

FIGURE 5-6. Asia.

Unfortunately, lack of understanding often leads to lack of confidence, and that commonly leads to blind conservatism. Petroleum is not generally found by the fainthearted, by the chronically pessimistic, or by those whose main interest is in making the current year's "profit" as high as possible, with little thought for the future. Nor is sound progress made by those who would proceed anyway, gambling on hunches. Some, many of them geologists, say that "it doesn't make too much difference really where you take acreage; the main thing is to take it." Needless to say, this is basically a mistaken notion, and it is very discouraging to geologic behavior.

Surface Petroleum Indications

Surface petroleum occurrences may be "live" or active, such as a seepage or exudation, or "dead" or "fossil," such as an asphalt deposit or a tarry or inspissated oil sand. An oil seepage may be directly associated with a commercial oil pool; much more often it is not. The reason why most seepages are not directly connected with an actual commercial-sized pool is that normally either petroleum has accumulated because it could not escape or it did not accumulate because it was not completely sealed against loss by leakage over a long period. Of course, there are numerous exceptions. For instance, a pool that has been caused to leak relatively recently geologically may still be seeping oil, e.g., the Kirkuk Pool in Iraq.

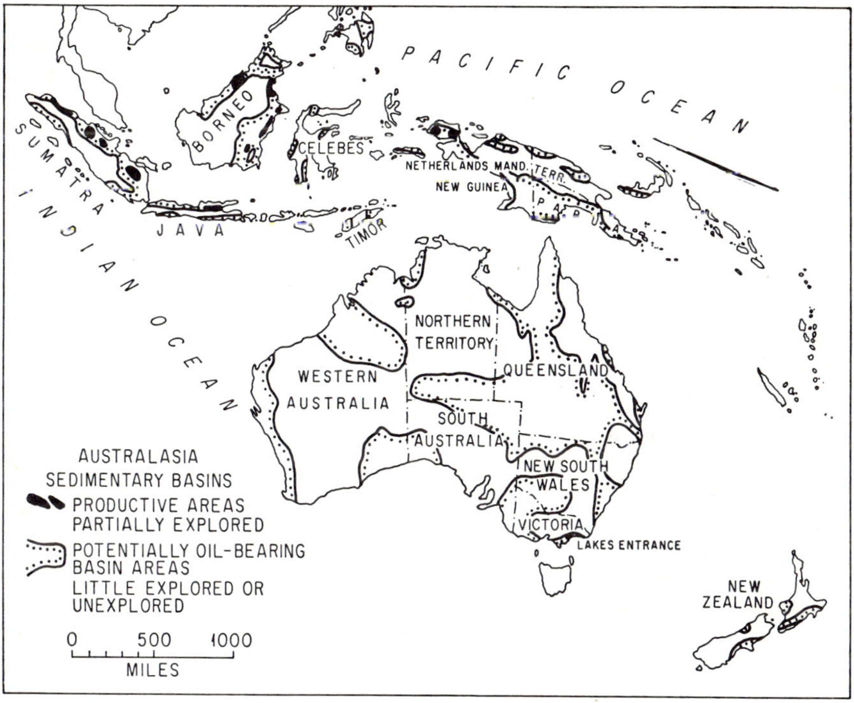

FIGURE 5-7. Australasia.

Heavy residues, such as asphalt or tar, are commonly found at seepages or in oil sands exposed by erosion. These residues result from the loss of the lighter fractions of asphaltic crudes through evaporation and, more particularly, bacterial oxidation. It is but a step from the surface type of heavy residue to that of the "tar mat" or tar belt, which is present along the shallow updip part of the shelf of many basins. This type of oil occurrence attained huge volumes in notable instances. As discussed later, its condition may have resulted, also in part, from bacterial invasion.

Most presently active oil seepages occur at or near the intersection of the water table with the earth's surface. They are found commonly in valleys near stream level or actually rising from the waters of a stream, lake, or sea. They may be found in association with water springs. Active oil seepages are practically unknown over the desert areas except where the water table meets the surface, as at an oasis or in a marginal gulley.

Another type of surface hydrocarbon occurrence is that of solid bitumen dikes, which not uncommonly reach the surface. They usually occur in and extend upward from very highly bituminous sediments where the quantities of the semisolid bitumen are so great that they are exuded and expressed under pressure of overburden. The bitumen rises plastically under pressure along any avenue of escape, such as that afforded by fracture systems of tectonic or simple adjustment to basin subsidence nature. Among many notable examples are those of the Uinta Basin, Utah; the Mesozoic deposition basins of Peru and of Neuquen–south Mendoza, Argentina; the occurrences in the Middle East; and many others. Inspissated, oil-saturated sandstone dikes are of a different nature, inasmuch as the fluid-saturated sand itself was the mobile intruding element. The oil is more or less incidental.

Still another type of heavy hydrocarbon occurrence is that of the semisolid residues or blebs not uncommonly found disseminated through sediments. They are particularly common in limestones. Where found within an oil pool they appear to be of a different time of emplacement from that of the main collection of oil. As explained later in another connection, increase in temperature and in gas saturation pressure with increase in depth of burial decreases the solubility of asphaltic fractions; this may cause them to be separated out. In cases where they are found isolated from pools, they appear to be inspissated residues of oil that were left behind during migration, caught in minor local porosity traps so common in sediments, particularly limestones.

While most petroleum-bearing basins have surface hydrocarbon indications, their relative abundance is no quantitative index of the prospects of the basin. Cuba and the Uinta Basin of Utah are good examples of situations where the prospects have not been found to be commensurate with the great abundance of surface hydrocarbon indications. On the other hand, in the Gulf Coast, in the mid-continent, in the African Sahara, or in the Russian plains surface indications of petroleum are relatively rare.

Surface oil indications are most abundant in the younger formations. They are more abundant in much disturbed structure within oil basins where the petroliferous formations crop out. Oil basins with a quiet history and in which the oil formations do not crop out have very few or no visible seepages. Petroleum seepages do not normally occur in formations, even though highly disturbed, that have been unduly metamorphosed.

Many of the world's oil fields had attention first attracted to them because of seepage or other petroleum indications. This is, of course, not at all peculiar to petroleum, for it is surface evidence that has attracted man to all minerals and other forms of natural resources. Surface oil indications were more of a guide to oil exploration in earlier years than they are today, but they are still a very important guide.

Surface petroleum indications vary in size from those which are invisible, the so-called microseeps, to tar accumulations like the Trinidad pitch lake, which originally contained perhaps as much as 200 million barrels of asphalt and tarry oil. Microseeps will be discussed later under a separate heading.

Some huge seepages or surface showings of oil or asphalt mark the unearthing by erosion of oil pools. Good examples include the Trinidad pitch lake; the Guanoco pitch lake in eastern Venezuela; the outcropping reservoirs at Quirequire, eastern Venezuela, and along the eastern margin of Lake Maracaibo and at Mene Grande, Venezuela; the outcropping La Brea sands in northwestern Peru and at the Cacheuta field in Mendoza, Argentina; and the outcropping reservoirs at Baku, Russia; in the Basra area of Iraq; along the southwestern side of the San Joaquin Valley, California; and the Athabaska oil sands of Alberta, Canada.

Metals Associated with Petroleum and Petroliferous Sediments

It has long been known that hydrocarbons occur incidentally associated in minute quantities with certain forms of metallic ores and mineralizations, whose occurrence is subject to similar basic environmental controls, such as pH, redox potential, etc. Similarly, many metals occur disseminated in trace quantities in bituminous, fine-grained, basinal sediments such as shales, marls, and dense shaly limestones. They are not commonly found in shallow, aerated water sands and limestones. Where they occur in such sediments, it is usually only in association with a local reducing environment that has been conditioned by some type of organic matter, including in some cases petroleum.

The association in the basinal and other sediments is thought to be related to an environment that is favorable both to the preservation of the hydrocarbons, bitumens, and other organic matter and to the deposition of the minerals from solutions carried into the basin of deposition by streams draining terrain which had been mineralized. Deposition was no doubt promoted by bacterial agencies.

Metals that occur associated with hydrocarbons in the sediments or in the ash of the petroleum include vanadium, nickel, uranium, lead, zinc, copper, chromium, iron, arsenic, antimony, magnesium, molybdenum, silver, aluminum, cadmium, and probably others. These are commonly spoken of as trace metals; they are of world-wide occurrence in basinal sediments. They are extracted from the sea water by organisms, including bacteria, which probably use them in their metabolism. They may possibly play an important role as catalysts in conversion of organic matter to petroleum.

Origin of Petroleum

There has been a great deal of study and speculation concerning the origin of petroleum, and much has been said and written about it. No one has knowingly witnessed the process in nature or convincingly demonstrated it in the laboratory. Yet today few, if any, petroleum geologists doubt that petroleum is of organic origin.

Earlier in the history of thinking about the origin of petroleum the inorganic theory had considerable following. Facts on which this theory was based include the demonstration in the laboratory that hydrocarbons can be formed by mixtures of alkali metal, carbon dioxide, and water and also by the action of water and acids on metallic carbides and the detection of hydrocarbons in the atmosphere of certain planets and in some meteorites.

The inorganic theory has been favored mainly by igneous, mining, or hard-rock geologists and by some chemists. Hard-rock geologists, in whose school the author began his career, are particularly assiduous in ascribing, wherever it is at all possible, direct high-temperature origins for the minerals and other substances with which they have to deal.

Following are some of the facts which negate a high-temperature origin and point to an organic source:

1. Composition as highly complex and varied and of such heavy molecular weight as that of the petroleums is found only in organic matter.

2. Many of the constituents and properties of petroleum could come only from an organic source. They include organic porphyrins and nitrogenous substances, both almost invariable constituents of petroleum; iodine; bromine; and the actual microscopic skeletal fragments of organisms. The various metallic trace elements, enumerated earlier, that are found in petroleum are utilized by organisms and occur in organic matter. Their abundance in marine sediments varies significantly with the organic-

matter content. Certain optical properties, such as optical rotation, possessed by petroleum are found only in the bodies of organisms.

3. The nitrogenous substances, organic porphyrins, and the property of optical rotation are very heat sensitive and are destroyed at even moderately high temperatures. Porphyrins are destroyed under an aerobic or oxidizing environment.

4. Many organisms, in fact most cellular material, contain hydrocarbons. Algae, particularly the abundant diatoms, are known to contain oil globules in the amount of several up to 50 per cent of their weight.

5. The view that adequate hydrogen to produce hydrocarbons is not present in organic matter and hence may have come from primary inorganic sources, such as volcanoes, is not supported by the fact that most petroleum occurrences lie far from volcanism. As pointed out later, a more likely hydrogen source is supplied by the abundant bacterial action which continues for a varying period after deposition in all sediments or, possibly, by radioactivity.

6. Finally, petroleum hydrocarbons have in recent years been demonstrated to be widespread in important amounts in recent sediments, where they are shown by C-14 studies to be of the same age as the sediments.

There are several schools of thought concerning the manner of organic origin. One postulates a transformation by a sort of temperature and pressure distillation at some depth of overburden or by tectonic stress. Another views the transformation as a biochemical process, the principal agency being bacteria. Another sees it as a more purely chemical process. A fourth, pointing to the widespread occurrence of hydrocarbons in living marine organisms, sees a direct origin. Still others have considered that radioactivity may have played a part. Many think that catalysts, such as certain clays, trace metals, etc., play an important role. Some have favored a combination of two or more of the several processes just listed.

The older thermal decomposition or distillation theory has few adherents today. This is largely owing to the very extensive evidence for both a low temperature and a very early origin subsequent to deposition. Residues of the type that should be left by thermal decomposition are not found in sediments. Experiments applying pressure to various types of rocks containing organic matter revealed no tendency to petroleum formation.

It has been demonstrated by extensive studies that bacteria of various types and the enzymes they produce operate to make organic matter more petroleumlike, and some hydrocarbons, mainly methane, have been actually produced from organic matter by bacterial action in the laboratory. Evidence of the production of hydrogen by bacteria in the laboratory has also been reported. This has been considered a possible source of the hydrogen required for transformation of organic matter into hydrocarbons. The biochemical theory has been regarded with some favor because it is concerned mainly with low temperatures and pressures and because of the ubiquitous nature and versatility of bacteria. However, to date, bacterial production of petroleum has never been demonstrated to occur in the sediments.

Hydrogen is one of the products of the action of alpha radiation on organic matter in the laboratory. But hydrogen is very rarely found in petroleum gases. A few instances only are known. It is possible that the absence of hydrogen resulting from any action, radioactive or bacterial, can be explained by its immediate combination with unsaturated molecules to form petroleum. Cracking of organic matter by alpha radiation, plus hydrogenation, has been suggested as a possible, though long-term, petroleum-formation process. However, many radioactive and highly bituminous sediments of great age occur in which there has been no noticeable formation of hydrocarbons; on the other hand, many of the world's most petroliferous sediments exhibit little radioactivity.

One of the chief objections to the radioactivity theory of petroleum formation stems from the fact that alpha bombardment of organic matter in the laboratory splits off hydrogen. This, it has been thought, would tend to increase the ratio of carbon to hydrogen in geologic time, resulting in progressively heavier oils. However, others have suggested this as a source of hydrogen needed by other processes.

The commonly posed question of the deficiency of hydrogen loses importance if our petroleum originates from the hydrocarbons formed directly by organisms, including bacteria. Moreover, this seems an adequate source as regards quantity. In fact most cells are said to contain hydrocarbons. The problem seems rather to account for what has happened to the vast aggregate amounts of such hydrocarbons as occur in the immeasurably great mass of organic matter that becomes entombed in the sediments.

The fact that, though hydrocarbons are widespread in recent sediments, no one has uncovered any processes or stages of transformation of organic matter into hydrocarbons may well indicate that the hydrocarbons were formed directly. One of the questions unanswered is whether all the great number of types of hydrocarbons occurring in petroleum are represented in organisms. Many different types of hydrocarbons, some of them very complex, have been found to be produced in the life cycle of certain present-day plants and animals. They include optically active sterols like those in crude oil and many forms of hydrocarbons, waxes, etc. Complex hydrocarbons found in marine kelp are said to range from 19 to 35 carbon atoms.

The laboratory of the living organism probably possesses greater versatility and potentialities than that in which the dead organic matter is found. However, if part or all of the hydrocarbons in crude oil are not formed directly by living organisms, we are faced with the problem of how they may have been derived from the dead organic matter. In this latter environment bacteria may supply the "living" energy either directly or in their capacity as a catalyst. The apparent versatility possessed by the infinite population of the bacterial world in mineralizations and transformations of many kinds may not be adequately appreciated. A limited amount of bacterial decay may at least have operated favorably to release hydrocarbons from the cells of organic matter.

It was not until recent years that critically diagnostic studies of organic matter and of hydrocarbons likely to meet with success have got under way in recent sediments. This appears today to be the most potentially remunerative field for study leading to the solution not only of the origin of oil but, as will be pointed out later, of its migration and accumulation as well.

Summary of Important Factors That Control Petroleum Occurrence

Before going into a detailed discussion of the factors that control petroleum occurrence in a sedimentary basin, it might be well to list some of the principal factors that tend to make a basin highly prospective or, conversely, the absence of which make it likely to produce little or no oil. They include:

1. A favorable basin form. This is a highly important factor.
2. An optimum degree of mobility.
3. Environments, as reflected in the litho- and biofacies, favorable for the deposition of both source and reservoir facies.
4. A favorable physical relationship, each to the other, of source and reservoir facies.
5. A favorable ratio of reservoir to source facies.
6. A high degree of sorting or screening of muds from the sands.
7. An adequate rate of deposition.
8. Variability in rate of deposition laterally.

9. Lateral lenticularity rather than uniformity of deposition across the basin.

10. The presence of unconformities, including diastems or lesser breaks.

11. Abundance, timeliness, and effective basin-wise distribution of traps, structural, compactional, or stratigraphic, or, as is so common with good oil occurrence, combinations of these.

12. The presence of deposition sinks* flanked by adequate porosity.

13. The presence of broad structural noses or bulges plunging alongside the basin or out into it.

14. Preservation from erosion of oil-bearing members where they lie near the top of a formation or series that is terminated upward by an erosional unconformity.

15. Lack of fresh-water flushing of preunconformity accumulations.

16. High percentage of the original deposition basin (particularly the more favorable or oil-bearing trend parts of the basin) that remains uneroded and otherwise intact in the present structural basin.

17. Favorable evidences of oil and gas. As stated on page 5–10, oil prospects are not necessarily proportionate to the abundance of oil and gas indications.

18. Favorable significance of any exploratory drilling.

19. Absence of regional or local metamorphism.

20. Optimum depth for prospective oil horizons.

21. Adequacy and timeliness of impermeable cover and of base and updip seal. These are among the most important of all factors.

22. Presence of evaporites, either below as a favorable structure former (example, salt domes or salt anticlines) or above as a cap rock (example, Middle East pools) or as an indicator of the probable existence of silled basin conditions (many striking illustrations are later listed).

23. A water system that displays the existence of good porosity and permeability. However, conditions of wide open artesian flow of wide basin dimensions are not generally favorable. Blanket sands require more closure and closure permanence than does lenticular porosity. Also lenticularity usually means more rapid deposition and closer, hence more favorable, lateral and vertical association of source and reservoir.

Deposition Environment and Facies

Study on a sufficiently broad scale clearly reveals that *the basic primary control of oil occurrence can be summed up in two words—deposition environment.* Environment is, in fact, the basic control for all natural phenomena, whether it be the occurrence of oil, coal, copper, bauxite, or any other mineral or the occurrence and quantitative distribution of any and all plants and animals, of peoples, of diseases, of the forms of weather, and so on without limit. Our problem, then, as oil finders is to establish what are the products of those all-important environmental factors and relationships in the original sedimentary basin and to recognize them in the present basin.

The only direct indicator of past basin environments that we have is the facies and the environment of similar facies in modern basins. We have the advantage in modern basins that the various elements that determine environment can actually be observed and measured. They include such things as the oxidation-reduction potential E_h, alkalinity-acidity range pH, the salinities, and other electrometric and physicochemical conditions; the bacterial and other biochemical activities; temperature, pressure, currents, and other physical factors.

* Throughout this chapter the author uses the term "deposition basin" or "deposition sink" rather than "depositional basin" or "depositional sink" because he has in mind a basin in which deposition was occurring rather than a basin that was produced by deposition.

The environment and the resulting facies are in turn controlled, more than anything else, by the basin-bottom form at the time of deposition and by the deposition rate and the type of sediments. The basin-bottom form varies with the type of basin and its own peculiar individual development. In résumé then, other things being equal, oil occurrence is related, in sequence, to:

1. The mechanics of basin subsidence, which varies in a general way with basin type.
2. The resulting basin form, which along with supply and currents determine:
3. The sediment distribution, which further modifies the basin form.
4. The environment, which is largely influenced by the basin-bottom form and the rate of deposition. All these control:
5. The facies distribution.

Finally, these factors, along with the adequacy of reservoir, timeliness of traps, and the efficacy of seal (to be discussed later) determine:

6. The incidence (volume and distribution) of the oil.

Relation of Oil Occurrence to Sinks of Relatively Rapid Subsidence

The incidence of oil occurrence is commonly highest around sinks of more rapid subsidence relative to deposition. Other things being equal, the incidence of oil occurrence over the world is definitely greatest in reservoir facies which are deposited over highs on the deposition bottom that lie within or adjacent to source facies depressions. The yield per unit of area and volume of sediments for each petroliferous unit decreases as we go away from these optimum trends, back over the forelands or interior stable areas. Stable, open, aerated sea basins or similar areas of basins that are remote from unaerated, stagnant deposition bottom sinks have contributed very little oil.

The facts just stated can be shown in various ways, such as by contouring the average yields per acre or per square mile over explored sedimentary areas. The principles can be applied to the exploration of basins generally. An outstanding, but by no means exceptional, example is that of the Middle East. The basin covers one million square miles, one-third the area of the 48 intracontinental states of the United States. Sediments of the same age as those which produce oil so very abundantly extend throughout the basin, yet 80 per cent of that large area will eventually yield but a very minor fraction of the huge total volume of oil. The oil is clustered around and near prominent deposition-bottom depressions filled with source-type sediments. Away from these centers, that is, over most of the one million square miles, deposition-bottom conditions were largely open and aerated and source facies are absent or sparsely scattered and of low quality.

There are, of course, other conditions which influence the incidence or volume distribution of petroleum than those just mentioned and which have operated importantly to give the Middle East accumulation trends such huge volumes of oil. These include exceptionally large reservoir volumes in intimate relation to the source and timeliness and adequacy of seal. A discussion of these features and the reason for their great importance will be discussed later.

There are many hundreds of known examples of the clustering of oil around the source facies of deposition-bottom subsidence areas. They range from major to minor in their proportions. A few examples are:

1. The San Joaquin Mio-Pliocene Basin, California
2. The Los Angeles Mio-Pliocene Basin of California
3. The Illinois Chester Basin

4. The southwest Texas, the Houston, and the New Orleans or delta sinks of the Gulf Coast, successively younger from west to east
5. The central sink of the Vienna Basin around which the oil is clustered
6. The Devonian Basin sink of Bolivia
7. The Permo-Carboniferous Tarija marine basin sink of southeast Bolivia, extending into adjacent Argentina
8. The Middle Magdalena Eocene basin sink in Colombia with its fringe of oil accumulations
9. The Jurassic basin sink of the petroliferous Neuquen Basin, west central Argentina
10. The very petroliferous little Rhaetic Basin sink of north Mendoza, Argentina
11. The Mesozoic sinks of eastern Peru
12. The central silled basin-sink part of the Comodoro Basin, Argentina
13. The various graben and trough sinks of north Germany
14. The trough sinks of Burma
15. The Williston Basin sink with its flanking oil occurrences wherever porosity is adequate
16. The Upper Devonian basin sink of Alberta, Canada
17. The various sinks of mid-continent United States, with rapidly decreasing incidence of petroleum away from source centers or belts
18. The local little Neocomian sink around which petroleum is clustered in the Parentis area of the Aquitaine Basin, France
19. The deposition sinks of both northwest and eastern Borneo, with the oil clustered along the immediate margins of the belts or areas of more rapid subsidence
20. The sink in central Sumatra, immediately south of which are huge oil accumulations
21. The trough sink of eastern Java, fringed immediately along the north by petroleum pools
22. The old classic fields of Russia and the newer ones of the Ural-Volga or "second Baku"
23. The new fields of the Sahara, North Africa, flanking the prominent Upper Paleozoic and Paleozoic-Mesozoic sinks of deposition
24. The new fields of Libya adjacent to the late Cretaceous–early Tertiary deposition sink
25. The Gulf of Suez, with its oil clustered around deep Miocene marl and shale sinks and occurring both in and over fault blocks, etc., overlapped unconformably by these same source sediments
26. The Bighorn Basin of Wyoming and a very long list of others

And so we could go on for pages and include scores of large and hundreds of minor local sinks and, also, embayments of basins with their source facies, such as in parts of north Texas, the Rocky Mountains, etc. These facts can be convincingly brought out only by the development of isopach-diagnostic-facies-oil-occurrence maps and with due respect for the great potentialities of a younger, overlapping, as well as contemporary source.

One of an infinite number of examples of the relation of oil occurrence to a deep deposition-sink source is that of the great accumulations along the south flank of the Anadarko Basin against the Amarillo ridge. This grand display of hydrocarbons contrasts sharply to the little evidence of their formation over the shallow, aerated water, platform environment spreading southward from the south side of that ridge.

The facts of oil occurrence reveal that the general run of wide open sea conditions of deposition, particularly over stable areas where deposition is slow and the bottoms are aerated, are scarcely more favorable to the genesis of oil than are continental sediments laid down in aerated fresh water. Incidentally, the open ocean bottoms are also well ventilated.

The effect of silling in producing a great increase in organic content is illustrated by the Channel Islands Basins off southern California; by Lake Maracaibo, Venezuela; by the Norwegian fiords, which are silled by their terminal moraines; by the Black Sea; and by literally thousands of other instances, ancient as well as modern. The deposition of source beds from which the great bulk of the world's oil originated, occurred in restricted, often silled basins or in basin areas below wave base and of adequate deposition rate where the favorable conditions similar to those of silled basins existed. Let us look further at what these favorable conditions are.

Environments and Basin Relations of Reservoir and Source Facies

A sedimentary facies is a product of some particular deposition-basin environment and is coextensive with that environment. Environments that are favorable for source facies are generally unfavorable for reservoir sediment accumulations, and vice versa. One of the most obvious facts of oil occurrence, but one whose significance is little appreciated, is that normally no oil can occur except where both source and reservoir facies coexist and where these exist in proper relationship, one to the other. The two facies groups are quite distinct and separate. With minor exceptions, concerned mainly with secondary porosity, such as that provided by a fractured shale, oil can accumulate only in the reservoir facies. Normally, neither source nor reservoir facies can result in oil accumulation by itself.

Principal effective porosity bodies of a primary reservoir environment are sand lenses that have been screened of their muds, carbonate reefs, oolitic limestones and dolomites, detrital and granular limestones and dolomites. Recrystallization, often with increase in porosity, is common in carbonate rock reservoir facies. Primary dolomites, of the evaporite type, are not normally a good reservoir and are not usually in a position to receive oil. However, "secondary" or diagenetic dolomites, formed by dolomitization along with recrystallization of limestones in waters of adequately favorable salinity, are commonly excellent reservoirs. Some of the best porosity in many carbonate reservoirs is where there has been minor to extensive dolomitization. This type of dolomitization occurs most commonly in organic limestones not far from the margins of source facies deposition sinks, such as in a belt of variable width along a flanking shelf or platform facing the sink or over highs rising above the source facies of the sink, etc. Dolomitization occurs commonly in reefs.

A primary reservoir facies of whatever nature can accumulate only in an aerated or oxidizing environment, which is usually current- or wave-agitated. This is not an environment in which petroleum can originate. Organic matter of a type favorable as a source of petroleum cannot come to rest permanently in such an environment. It could not be preserved from destruction by bacterial oxidation and scavenger organisms, both of which live up to the limit of food supply in an oxidizing environment. All forms of primary porosity normally accumulate over or along a basin's flanks or over highs of any origin, tectonic or static, where the bottom lies at or above wave or current base for muds and below that for sands.*

On the other hand, an environment favorable for the accumulation of source sediments, in which organic matter favorable as a source of oil can be preserved, can accumulate only in an environment of low oxidation-reduction potential. Thus, a source facies can accumulate normally only under the nonoxidizing or reducing conditions found in the stagnant, relatively deeper waters below the reach of wave or current agitation and outside the reach of important oxidation from any cause. Such

* Effective wave or current base for mud, sand, or other grade of sediment is the maximum depth at which waves and currents are effective in moving that sediment.

environments are inimical to the existence of aerobic bacteria and other scavenger organisms. Such bottoms carry only the remains of the surface plankton and certain free-swimming nekton. Only on such bottoms can the very light organic matter, dropping from above and washed away from the wave-agitated shallow areas, settle out and find lodging to be incorporated in the muds.

Few, if any, basins are filled entirely with nonporous sediments. The normal complement of clay-muds (shales) and limy muds (marls and dense shaly limestones) are, respectively, sands and crystalline, porous limestones and dolomites. Hence, if exploration turns up a section that is lacking in porosity, it only poses the problem of locating the areas or belts of the porosity complement.

Reefs, Source or Reservoir?

Reef-building organisms, because of their nature and the vast activity, which is nearly all in the reef surface, are an important potential source of organic matter. Based on previous discussion, it is doubtful, however, if those parts of the organisms which might produce petroleum could be preserved from rapid aerobic oxidation and the myriad of scavengers in or living off the reef as long as it stays in the reef, for reefs flourish only in the most turbulent and aerated of environments. Yet much of this same organic matter is no doubt washed from the reef, just as it is from shelf or any bottom-high area, to find rest and burial in the adjacent quiet deep-water muds outlying the reef front and its talus apron. There the environment is of adequately low oxidation-reduction potential for preservation of the dead organic matter and associated hydrocarbons.

With extinction and burial of the reef by muds to form a porosity trap hydrocarbons are fed into the dead reef from the adjacent muds. If petroleum could form *in situ* in the reef, most reefs should be oil-bearing instead of only the small percentage that are associated with an adjacent source facies environment *and have an adequate and timely seal.*

Environmental Significance of the Biofacies

One of the very best indicators of environment is the fauna. Basin bottoms and their sediments in which a benthonic fauna is abundant are not favorable as a source of oil. Bottom environments that are unfavorable for such organisms, receive organic matter of the plankton (surface-water plant and animal life), and protect such organic matter from destruction by bottom scavengers and bacteria are usually favorable sites for petroleum source sediment deposition. In the best oil basins it is normal to find the reservoir facies, commonly with a benthonic biofacies, in the sediments of the wave- or current-stirred and aerated bottom areas high on the deposition bottom. Sediments of source facies character in such basins, usually with planktonic but few or no benthonic biofacies remains, occur in the sinks or deeper areas of the bottom.

Ecology is the science which treats of the relations of organisms to one another and, particularly, to their environment. The paleontologist thus has at his command one of the most effective geological tools for oil finding but, unfortunately, one which is all too little put to use. Commonly all paleontological effort is devoted to fossil naming and correlation. Entirely too little attention is given to use of the biofacies as an indicator of environment. Many paleontologists take cognizance of only a few of the more widely occurring fossil forms, the so-called "key fossils." Too little attention is given to the differences in faunas and the reasons for these highly important differences, yet here are other key fossils which may be more important than those used.

Actually, most fossils do not faithfully represent a limited period of time. The faunas of similar environments of successive ages are commonly much more alike than are the faunas of different environments of the same age. Yet not uncommonly environment-indicator faunas are actually used for correlation, usually with uncertain, quite often with unfortunate, results.

The sediments of large areas of normal open sea bottoms have passed through the digestive tracts of scavenger animals, often many times. Instead of being preserved, the energy that was endothermically stored up by photosynthesis in the sunlighted surface waters, where the great bulk of all marine life exists, is exothermically released by such scavengers. This accounts for the low organic content and poor oil-source potentialities of sediments accumulated under such conditions.

It is not a case primarily of how much organic life exists in the waters as it is of how much unstable organic matter is preserved and in what state. Where conditions favorable for scavenger life exist, an increase in the supply of organic matter will only increase the number of scavengers. Every fisherman recognizes the effect of bottom topography on the distribution of fish and the food on which they prey. Fishermen supply houses have long known about the vastly greater demand for their products along some coasts than along others.

Importance of Rate of Deposition

Though the silled portion of silled basins is much more apt to be protected from oxidation than the open basin areas, on the borders of the latter organic matter may be preserved where the rate of sedimentation is adequate to bury it rapidly and thus protect it from bacterial destruction. This is an important point and requires further examination.

Bacteria act as activating agents, or "catalysts," to bring about oxidation and reduction processes. They get their energy from organic matter, including hydrocarbons. Where oxygen is absent, as in a stagnant sea-bottom area or at a depth of burial adequate to exhaust the free oxygen, aerobic bacteria cease to live. However, anaerobic bacteria are here to prey upon the organic matter. Anaerobes obtain their oxygen from sulfates and other oxygen-bearing compounds. They reduce sulfates to sulfides, and in the process the organic matter is oxidized to carbon dioxide and water. However, with increasing depth of burial, bacterial activity, even that of anaerobes, very rapidly decreases. Thus, rate of deposition is a highly important factor in preserving organic matter, including hydrocarbons, from bacterial destruction.

A common fallacy among geologists is that the percentage of organic matter in a sediment is a true measure of the areal incidence of organic-matter preservation or of the potential of the sediment as an oil source. This may be very far from the fact. Area A with the same percentage of organic matter as area B but with five times the rate of deposition carries five times the total preserved bio-mass. Thus basin flank areas of rapid deposition may contain as much or more total organic matter per unit of time as deep stagnant bottoms, where the deposition is very slow and where anaerobic bacteria have had an opportunity to destroy much of the organic accumulation before burial. Thus it is that such sea bottoms as that of the Black Sea, notably high in its percentage content of organic matter, actually has considerably less total bio-mass than the areas of comparatively rapid deposition adjacent to the deep.

Other facts emphasize the importance of the deposition rate. It appears that of all the constituents of organic matter it is the hydrocarbons that are the most easily destroyed. Investigations of recent sediments, such as along the Gulf Coast and in the Channel Islands Basins off California, show that hydrocarbons actually make up a greater percentage of the total organic matter in flank areas, where they are preserved by rapid burial, than in the highly organic sediments of the deeper bottom. In the near flank areas or belts of a basin, also, the sediments are more variably textured, both areally and vertically; hence the more abundant hydrocarbons not only are physically more free to migrate but are in closer proximity to most porosity and to the controlling influence of most traps.

It should be pointed out here, however, that the increase in the hydrocarbon percentage of the organic matter found in flank sediments immediately adjacent to the deep may not be a reflection of better conditions for preservation of the hydrocarbons in those positions but rather an indication that flankward migration is already in progress. Studies of recent sediments along the Gulf Coast and in the Pedernales area, Venezuela, adjoining the Orinoco delta, show that some concentration of hydrocarbons in small sand lenses has already begun in muds not over 10,000 years of age.

Incidentally, in large open seas and oceans the major current systems carry to great depths far from land. In the open ocean, also, the waters carry very little of the basic mineral nutrients and hence are sparsely inhabited by organisms. The great bulk of the life on the earth, both land and water, is in the nearer shore areas of the oceans and in the shallow seas. As shown by the studies of Parker Trask and also various others, the percentage of organic content in the deep and broad open sea-bottom sediments is normally much less than that of the upper continental slope, even though in the latter the deposition was much more rapid.

Partial Résumé

In partial résumé, it is not the amount of organic matter, including its hydrocarbons, formed in the seas that is the important criterion, but the amount and the condition in which it is preserved, for without the protection given only by the proper environment, scavenger animals live up to the limit of the available food supply. The conditions that are most favorable for oil are an abundant supply of organic matter and adequate preservation thereof from exothermic destruction, coupled with closely adjacent conditions of adequate porosity in timely, well-sealed traps. Such conditions are found adjacent to deposition-bottom areas of more rapid subsidence and deposition. They are seldom found on open aerated shelves remote from environments of source or on other stable open-sea areas.

Petroleum is one of the very few ingredients of matter in the earth's crust which, as a result of its several billions of years of history, has not been "shaken down" to its molecularly most stable state. The tremendous amount of sun-derived energy that was incorporated endothermically into the nonstable organic life in the surface sunlighted photosynthetic zone of the seas could be preserved for us still in its nonstable state only by special environmental conditions such as occur below the zone of aeration in basins where silling and/or rate of deposition is adequate to preserve the organic matter from biochemical destruction. In basin environments where aerobic bacteria and other scavenger organisms can abound, the energy derived endothermically from the sun is exothermically released by those organisms to form the stable resultant products carbon dioxide and water. Where such conditions alone prevail, commercial supplies of petroleum cannot result.

Basin Characteristics of Oil Trends

Oil trends result from the fact that the bathymetric form or contour of the basin controls the distribution of environments and their corresponding facies. Oil trends mark the belt where a favorable relationship exists between a source and a reservoir facies. It is strikingly common to find oil trends along the belt of maximum rate of change of thickness and of most intimate association of source and reservoir facies. The Eocene Wilcox trend of the Mississippi embayment is a good illustration, and those familiar with the Gulf Coast will immediately recall others. The Arabian Jurassic "Arab zone" illustrates a very common situation. The thickest isopachs mark the very prolifically oil-bearing carbonate section of the outer shelf and hinge belt. There is thinning up the shelf, with loss of porosity partly due to anhydrite in-filling. There is thinning also out into the basinal facies shales, marls, and shaly limestones where, in places, euxinic conditions existed. The four Arab zone reservoir lenses are thus enclosed within the isopach thick. Each zone is capped by anhydrite. Here is an ideal combination of an ample reservoir, fine source, and perfectly sealed and timely trap, a trap whose seal is general over the entire region. Other examples are so numerous as scarcely to require listing.

The highest sand/shale ratio or the greatest total thickness of sands along a basin is not usually the optimum situation for oil accumulation. In many basins the optimum belt is where there is some degree of total sand thinning with respect to shale. The optimum ratio for one basin or sedimentary unit may not be the same for another. Predominantly sandy formations tend to carry their oil lower in the basin than do predominantly shale sediments.

A study of world basins shows that many cycles of deposition, beginning above a major unconformity, receive their deepest water facies early, almost immediately or within a short distance above the unconformity. This indicates rapid early subsidence in relation to deposition. Later the rate of deposition speeds up, overtakes, and then surpasses that of subsidence, until more and more neritic and even nonmarine deposition takes over.

In a basin of gradually receding shore line a common arrangement of oil occurrence is a concentric one, with the oldest along the basin flank and successively younger basinward. This is illustrated by the Gulf Coast, where there exist no less than a dozen oil-age trends, from Jurassic to Pleistocene. It is common in the many instances of this sort for all the various trend oils to lie within certain maximum and minimum depth ranges. This does not mean that the various oils are in direct hydrostatic communication. Rather it means that in a simple basin, similar facies distribution and similar tectonic, compaction, and hydrostatic pressures left similar results at the time of migration. In a composite basin, made up, say, of two or more sedimentary basin sequences separated by angular erosion unconformities, there may be little or no such relation among the petroleum systems of the independent sequences. This is illustrated by the Carboniferous and Mesozoic basins of the Rocky Mountains.

The question is often asked, "Why is this physically good trap barren of oil?" To answer that "we cannot know why" is not usually excusable today. Particularly is this true after a few wells have been drilled and we acquire information about the basin form and facies, history of development, and timeliness of trap. Ample empirical evidence exists today of the factors that control the incidence of oil occurrence. The extent to which that knowledge is really understood and can be applied reveals itself whenever a new basin is opened to competition. The picture presented is not always complimentary, whether the selection was made by the geologist or by one of his superiors on the all too common theory that "it is not possible to discriminate; selection of acreage is largely a matter of luck."

Diagnostic Value of Isopach-facies-oil-occurrence Maps

One of the best keys to the relation of oil occurrence to the basin is the isopach-facies-oil-occurrence map. Isopach maps alone do not give a trustworthy picture of the deposition basin. All they show is the gross sediment distribution. The site of deposition of the thickest section may be that of the shallowest or the deepest water or of some intermediate water depth and environment. The true situation is relative and can be determined reliably only from the facies. An isopach-facies map speaks eloquently of where the oil trends should be and gives much information on the timeliness of traps. Add to this picture any existing data on oil occurrence, or from the relation of oil occurrence to facies generally, and the basic trinity for analysis is complete. Even in a little explored basin a large amount of indicative information is available from the surface geology and from an understanding of basin type and the regional geology. The picture should be coupled wherever possible with a little assistance from geophysical reconnaissance. For best analysis there is a fourth dimension, which may be more important than any of the other three; that is experience on the broadest possible scale in the meaning of the data.

Attempts at relating oil occurrence to the present structural basin have been grossly misleading. This is particularly true where there has been a history of varying mobility. A large portion of an original sedimentary unit, including the principal oil-occurrence facies, may have been eroded. Or large segments of the original deposit may have been incorporated into the rocks of an encroaching mountain. Many basins are composite; that is, they are made up of two or more successive series of marine invasions and sedimentations separated in whole or in part by unconformities. An unconformity may be quite angular over a considerable portion of the basin area. Thus a structural or highly deformed sedimentary basin that may have lost some of its original facies units, possibly along with some of its oil trends, may underlie a basin that still largely retains its original form.

Isopach maps show fully the depositional form only when used in combination with facies. In such combination they show the times of deformation, of folding, and of block movement or tilting and the relation of oil-occurrence trends to the basin of deposition. The environments and, in consequence, the oil-occurrence trends tend to follow the isopach strike for long distances and be bounded both up and down flank by limiting isopachs. Isopach-facies-oil-occurrence maps outline the "fairways," in contrast to the "rough," and help us to know where to look for them in undeveloped basins.

A paleogeologic map may cover an area ranging from that of a structure, a basin, a country, or a continent. Properly used, it can be one of the most informative tools available to the geologist. It seems well to remember that where petroleum-bearing beds were once exposed by erosion at the unconformity, the oil is likely to have been eroded, and it is usually more profitable to look for oil from an overlapping source. By comparison, the isopach-facies-oil-occurrence map is much more intimately and reliably associated in time and space with the basin environmental factors that control oil occurrence or its nonoccurrence.

A True Marine Environment Is Not Requisite for Oil Occurrence

In the consideration of the requirements for oil occurrence, the need for a marine environment is commonly overemphasized. While petroleum occurs widely in certain marine realms, many of our rich oil-bearing sediments are of restricted marine or brackish origin or even of still more continental aspect. In some of the latter type of occurrences there is a gradation basinward into brackish or marine facies, but in other places this is not the case, or at least it is not evident. It is commonly difficult to say with certainty that an environment was entirely that of fresh water or that it did not have brackish or even marine intercalations. Absence of marine fossils is in itself by no means an indicator of a nonmarine environment. Many marine bottom areas are too toxic for bottom-living organisms, and those of the surface plankton may have been devoid of hard parts. In some basins the formation waters of the so-called continental sediments are as saline as or more sc than those of the marine units. An example is the Wasatch of the Green River Basin, Wyoming. Petroleum has also been found to occur in sediments deposited in abnormally saline waters, though not in important amounts.

Illustrative of oil-bearing sediments that cannot be classed as marine under a true interpretation are:

1. The Maracaibo Basin Oligo-Miocene and Eocene.
2. The almost fresh to very restricted and at best never truly marine late Cretaceous and Lower Tertiary oil-bearing sediments of eastern Peru, Ecuador, and southeast Colombia.
3. Many of the oil-bearing Tertiary sediments of eastern Venezuela.
4. The richly oil-bearing brackish sediments of the small, nicely silled Comodoro Basin in southern Argentina.
5. The fresh to brackish, richly oil-bearing zone of Rhaetic sediments lying between basement and overlying rusty red beds in Mendoza, western Argentina.
6. The similarly situated, nonmarine, probably brackish water sediments of the large Kansu-Tarim Basin of western China in which one or more major oil pools have been discovered.
7. Part of the oil-producing section of the Punjab Basin of India in which a number of oil pools have been found.
8. The brackish-water, richly petroliferous Pliocene Meotic and Dacic formations of Rumania.
9. The oil-bearing Tertiary of the Magdalena Valley, Colombia.
10. The nonmarine Eocene Wasatch fields along the Colorado-Wyoming border and other oil-bearing parts of the Wasatch in the Green River and Uinta Basins of Wyoming, Colorado, and Utah.
11. The Eocene Green River formation of Utah's Uinta Basin. In this case the sites of deposition varied from possibly fresh water to abnormally saline.
12. The Oran Basin of northern Argentina. This was a deep, rapidly subsiding silled sink. It is margined by heavy oil seepages and bears supersaturated saline formation waters. Oil is found in such wells as were able to reach the oil horizon on the open flanks of the basin sink. The sink area forms part of a much larger region which carries only nonpetroliferous, open-basin, fresh-water-bearing sediments outside the limits of the Oran sink.
13. The oil-bearing Sespe of the Ventura Basin, California.
14. The Mio-Pliocene of the Russian Baku and Turkoman sections.

The important point to note in this connection is, that under the right kind of silled basin conditions, brackish- or even nearly fresh-water basins and also abnormally saline basins may be much more favorable for oil than the wrong kind of stable, open-sea, normal marine basin.

Significance of Evaporites

There are many examples of a cycle of deposition ending with evaporites. Most commonly the preevaporite sediments consist of carbonates and marls or muds, but there are numerous examples also of predominantly muds or marls and few sands. The evaporite may be calcium magnesium carbonates (primary dolomites), calcium sulfate (anhydrite), sodium chloride (common salt), or other salts or combinations of two or more of these.

The common significance of evaporite occurrence in a basin is this: The environments set up by the basin silling that led up to the more complete closure required for evaporite deposition were nearly always favorable for hydrocarbon development and preservation on a moderate to vast scale.

Though no two cases are exactly alike, all bear certain resemblances. A typical case is provided by the Mississippian of the Williston Basin or by the Permian of west Texas and southeast New Mexico. Commonly the series begins above an unconformity with a deep stagnant-water mud facies (shales or marls or muddy limestones). It continues up through somewhat less stagnant limy mud facies, with limestones or dolomites around the flanking shelves. Cherts may occur somewhat deeper in a part of the basin. Then the sequence may continue through a purer, partly recrystallized and dolomitized limestone facies. Isolated reefs and reeflike shelf fronts facing the deep basin, as well as patch reefs higher on the shelf, are common features of this and the preceding stage. The sequence commonly continues up with spreading evaporites—primary dolomites, anhydrites, and/or salt. Usually the evaporites begin in the back reef shelf areas, but with increasing restriction of the basin, sands and sandy shales, commonly red or brown in color, normally replace the evaporites on the shelf. Evaporites may now spread to the entire central area of the basin, where again sands and clays may ultimately terminate the cycle. While the foregoing is of common occurrence, there are, of course, variations. Not all sequences reach the stage of evaporites; also the sequence may end with a rapid fill-up by a clastic phase without going through an evaporite stage.

There are many examples of basin sequences in which petroleum occurs moderately to abundantly in the stage or stages preceding the evaporites. They include, in addition to the west Texas Permian and the Williston Basin of Canada, Montana, and North Dakota, just mentioned:

The Paradox Basin of southeast Utah and southwest Colorado
The Michigan Basin
The Gulf Coast Lower Cretaceous
The eastern Peru Lower Mesozoic (or Permian?) basin
The Amazon Paleozoic basin
The Neuquen, Argentina, Jurassic and Cretaceous basins
Possibly the central Parana Basin of Brazil
The north European Permian basin
The Ural Permian basin of Russia
The southern Rhine Oligocene graben basin
The Gulf of Suez Miocene basin
The Red Sea Miocene graben
The Iskenderun Miocene graben basin of Turkey
The Tuz Golu Tertiary basin of Turkey
The Upper Jurassic basin of Arabia, etc.
Local parts of the Cretaceous basins in the Middle East
The Oligo-Miocene basin of Iraq, Iran, etc.
The large modified half-graben basin of northern Siberia

Not uncommonly, evaporites, which are an excellent seal, form the cap rock for the oil. Outstanding examples are the huge Tertiary and Jurassic oil pools of the Middle East. Again, as in the producing trend of the Williston Basin Mississippian in Saskatchewan, oil may occur only where anhydrite infilling forms an updip seal. In cases where reservoir capacity is highly developed, as in the Middle East, huge volumes of oil occur. In other cases, as in the Williston Basin, accumulation is restricted by the relatively limited development of porosity or reservoir facies. Usually in evaporite-bearing basins of the type here described, petroleum source facies abound in the preevaporite stages of the sequence.

Unconformities and Oil Occurrence

One of the most important and interesting facets of oil occurrence is its relation to unconformities. A petroleum geologist fortunate enough to work in many places early recognizes how great are the volumes of oil that accumulate along or in the vicinity of unconformities. He notes such examples as that at the base of the Pennsylvanian in the mid-continent, at the base of the Permian Phosphoria in the Rocky Mountains, at the base of the Miocene in Lake Maracaibo, at the base of the Miocene La Pica in eastern Venezuela, within and below the overlap of the Permo-Pennsylvanian on the buried Amarillo ridge of the Texas Panhandle, etc.; the Kirkuk oil field at the Lower Miocene overlap in Iraq; the numerous Devonian limestone fields of Michigan, where the oil lies just below its shale source overlap; the oil fields of the Suez graben, where the oil is reservoired in sandstones and carbonate rocks ranging in age from Carboniferous to Eocene exposed in eroded fault block structures beneath a Miocene source shale and marl overlap. Here oil is reservoired also in early Miocene carbonate reefs that grew over the fault block-cored bottom highs. There are literally many hundreds of other similar cases.

But how common are lesser unconformities in the sediments, and what role do they play in oil occurrence? The author has often stated that if we were to define an unconformity as any break in deposition, then nearly all the world's oil is accumulated both with the aid of and along unconformities.

Much light is now being thrown on the depositional rate of recent sediments by the use of carbon 14 dating. The method indicates rates of 0.002 to 0.04 feet or more per year for short intervals in the last few thousand years of the earth's history datable by the method. These short period rates correspond to from 2,000 to upward of 40,000 feet per million years. Such rates, even allowing for compaction, are probably nowhere represented in basin sediments continuously over a hundred thousand, much less a million, years. Average rates for entire formations or series of formations of past geologic ages are of the order of a few hundred feet per million years; rates as high as 1,000 feet are not common.

Deposition, then, must occur in spurts, corresponding in part to episodic periods of subsidence, interspersed by long intervals of no deposition or even of subaqueous bottom abrasion during which the sediment is bypassed or moved out to areas still below the base level of deposition. Thus it is that our "conformable" sediments are just filled with minor unconformities.

The evidence from geologic history and from observations of the present day is that subsidence or other movements of the earth's crust are not gradual but rather sudden and episodic. They are interspersed by relatively long periods of quiet during which the stresses build up again. In many typical basins the belt where subsidence accommodates and receives the thickest sediment accumulation is commonly found along the outer shelf and what the author has called the hinge belt or hinge line of the basin. Following each subsidence, which may be but a few inches or feet, the interval is filled up quickly to wave or current base. Deposition then ceases over that part of the bottom while the sediments are moved farther out. Thus it is that with each subsidence sediments of a deeper basin environment having a lower redox potential and distinctly more favorable source characteristics may be spread far flankward over porosity-bearing sediments of the previous aerated environment. This intimate interfingering of source- and reservoir-type environments and their corresponding facies may be repeated many times along the outer shelf and hinge belt of a basin in a single formation. It is for this reason that this belt is the most common locus of oil trends along basin flanks all over the world.

These minor unconformities or deposition breaks, with which the sediments are filled, split and disappear out in the basin, and they tend to converge and merge on the flanks and shelves of the basin or over any bottom highs, such as are formed by simple compaction or by growing structures. Even a sand represents but partial deposition, as the fine components have been screened out and washed away to quieter waters. Sand belts and patches are no doubt filled with complete lapses of deposition of varying lateral duration.

Here is one of the most fertile fields for isotope study. It should tell us much about the true pattern or fabric of sedimentation, with its complex of permeability channels, the true nature and significance of bedding and lensing, the manner in which basins subside, and the response thereto of sediment accumulation. Finally, it will tell us much about the pattern of oil migration and its accumulation and occurrence. Turbidity currents spread coarser sediments well out into the central areas of basins. In a sense they are pseudounconformities, and they provide drainage channels of exceptionally wide reach.

There are three physical aspects of unconformities of probable very great importance for oil occurrence. The first is the condition of timeliness. Early existence of the trap is one of the cardinal requisites for oil occurrence. No other trap could be more timely than that associated with an unconformity where the source is from the overlap.

A second valuable aspect of an unconformity is its angularity. Even where slight, angularity greatly increases the thickness of sediments which may feed oil along the bedding to the unconformity and to porosity traps associated therewith. The number of strata, both source and reservoir, impinging on the common plane are increased many times. Pressures in an overlapping or capping shale source are those of the overburden; thus they closely approach rock pressure. Hence they are normally much greater than those of the underlying porous layer, such as a sand, a weathered limestone surface, etc. Obviously the easiest avenue of escape of fluids from the overlapping muds is into the porous channels below. Compaction of the overlapping series feeds vast volumes of hydrocarbon-bearing fluids tangentially into the unconformity. There the hydrocarbons will be "strained out," will collect in any of the many kinds of traps peculiar to such situations.

The third physical aspect of unconformities of great importance is the bringing into juxtaposition of two facies which may be moderately to greatly different. With renewed subsidence and deposition, even after a slight lapse of deposition, sediments of a lower redox potential and better source characteristics may be spread or intercalated far flankward to feed hydrocarbons into the progressively overlapped permeability planes beneath. Such ideal source-migration channel relationships may result along sedimentation breaks ranging in magnitude from the minutest lapse of deposition or basal part of a turbidity current deposit to that of a major unconformity.

The relating of oil along unconformities to a lower source is an old concept based on an early unsupported view that because oil and gas are lighter than water, they must rise up from below. In this respect there has long been a tendency not to appreciate or even to rule out unconformably overlapping sediments as a source. Though there has been a great change in viewpoint in recent years, some geologists still believe that a reservoir may be sourced from sediments that lie scores or even thousands of feet stratigraphically below. Physically and from the standpoint of position and timeliness there could be no better source-reservoir relationship than that which is overlapping or penecontemporaneous. By the same token there could be few, if any, poorer than a detached lower source. A quick mental survey will reveal a great number of oil occurrences adjacent to an unconformity overlapped by a good source.

One of the widespread types of unconformity oil occurrences is that of a buried limestone surface overlapped by source muds. Usually the oil lies below an erosion-weathered surface. Among a large number of good examples are the Devonian limestone pools of the Michigan Basin Traverse group, the Rogers City and Dundee formations, and the Detroit River group. Of course, a classic example is the Ordovician Trenton belt of pools in western Ohio. In England oil occurs in the top of the Carboniferous limestone for 25 miles along the strike on the east flank of the Derbyshire dome, unconformably below the overlapping marine shales of the petroliferous Millstone Grit. The Gulf of Suez, the Permian of the Ural Basin, the Devonian of western Canada, and the Golden Lane and other fields of Mexico are other good examples of shale or marl overlap of limestone surfaces, some of which are of reef character.

The porosity of limestones may not be greatest close to the unconformity surface. In some cases it extends 100 or 200 feet or more down from the unconformity, and it may take the form of openings from vug to channel or cavern dimensions. Solution weathering is known to have developed porosity to depths of as much as several hundred feet beneath old erosional surfaces, very much as we find beneath present surfaces in limestone areas. A notable example is that of the Dollarhide Pool in Andrews County, western Texas, where the drills found a cavern several hundred acres in area and up to 16 feet in height in the Silurian Fusselman limestone. The prominent erosion surface and Permian overlap, the probable source of the oil, lie about 1,000 feet above the cavern. There are a number of pools in the Central Platform limestone of west Texas that lie well below the Permian overlap but which are now widely considered to have been sourced by oil from the Wolfcamp, which overlaps from out of deeper basin levels than that of the oil pool.

At the other extreme, under other conditions, unconformable limestone surfaces exist beneath which there has been little penetration by solution weathering. Good examples are the pools near the top of the Mississippian on the north side of the Williston Basin in Saskatchewan, Canada, and along another trend off the foothill front in Alberta. In both cases the porosity is depositional or diagenetic and is sealed off updip by anhydrite infilling not far below the unconformable overlap of the Mesozoic.

The development of porosity in dolomitic limestones may result from the differential solution of calcite by acid-bearing waters. Calcite is twenty-four times as soluble as dolomite. Increase in jointing and fracturing as the present earth's erosion surface is approached was probably duplicated to some degree beneath ancient erosion surfaces, marked today by erosion unconformities.

Except for loss of light ends and inspissation by bacterial oxidation, petroleum accumulations are not necessarily lost by exposure at the erosion surface, any more than water will rise above the ground fluid table. This is attested by great numbers of examples over the world. Some of these run into scores or hundreds of millions or even billions of barrels of oil. Two outstanding examples are the young accumulations along the southwest side of the San Joaquin Basin in California and the great Miocene fields outcropping along the east side of the Maracaibo Basin, Venezuela.

As might be anticipated, instances are common over the world of finding two types of oil in a reservoir. One is an inspissated oil that accumulated from sediments of the age of the reservoir, and the other is a lighter gas bearing oil from the overlapping sediments which provided a good seal as well as a source.

Some shore sand wedges may contain one or more disconformities or lesser breaks in deposition, representing both regression and subsequent transgression or even just temporary static conditions. It is suspected that such may be the case with respect to the Woodbine sand under eastern Texas. This sand undoubtedly includes within it at least one and perhaps other breaks of significant time importance, represented by some additional as well as thicker section of sediments basinward.

Contemplation of the nature of an unconformity, whether major or very minor, must impress upon us the fact that every bed that was ever laid down ends or impinges somewhere at an unconformity of some sort. It is no wonder, therefore, that such a great bulk of the world's oil should be found today in traps along or in close relationship to unconformities. The innumerable bedding planes, diastems, and other larger depositional breaks converge and join toward the basin flanks and other high areas within the basin. They lead normally into porosity bodies, whether they be located within the wedge, within the base of the overlap, or beneath the overlap.

Migration and Accumulation Problems

The problems of origin, migration, and accumulation are interrelated; in fact they are interdependent. They include such factors as:

1. The time that each process occurs relative to the time of deposition, the rate of deposition and the weight of overburden, the type of sediment, the type of organic matter, the porosity and permeability, tectonic deformation, etc.
2. Long- or short-distance migration or origin *in situ*.
3. Manner of migration, whether dissolved in moving waters, carried physically by the waters or independently of the water; whether the migration was a part of the early compaction-induced fluid movement concomitant with burial or of some later fluid movement subsequent to lithification of the sediments.
4. The extent to which secondary migration may be operative and under what conditions.
5. The extent to which hydrodynamics may be a factor in migration and remigration.

Time and Manner of Migration and Accumulation

It is difficult to reconcile postlithification migration from and through shales with the fact that the same kind of shales normally serve as impenetrable cap rocks. It is likewise difficult to conceive of late migration along dips that are in many cases almost imperceptible. In an attempt to explain migration of the vast scale required, it has been suggested that oil will migrate along the hydraulic gradient at any time after complete lithification, even though the pressure gradients are inadequate to force dispersed oil droplets through the pore constrictions (the gradient pressure does not equal the displacement pressure) provided that a large enough patch of continuous phase is formed so that capillary effects are overcome therein. Others have suggested that the migration occurs in solution in the water, as a colloidal dispersion, or as droplets of oil.

Contrary to what was once widely supposed, there has been a notable swing among geologists in recent years, as evidenced by their writings and discussions, to belief in early formation, migration, and accumulation of oil. Overwhelming is the mass of evidence that nearly all migration and accumulation are both primary and early relative to the time of deposition of the contributing sediments. Included in this evidence are the following facts:

1. The world's youngest sediments, the Pliocene and Oligo-Miocene, have the highest incidence of proved oil occurrence per unit of volume of sediments and per

acre of all ages. The Pleistocene also might be included here if the great bulk of its sediments had not been deposited beyond the present land areas and continental shelves because of the high stand of the continental platforms during the Pleistocene.

2. There is direct evidence of early oil formation. Recent studies by advanced techniques are showing that hydrocarbons similar to those in crude oil are both present and widespread in sediments which, as shown by C-14 methods, are not over 10,000 to 20,000 years old. The amounts present appear to be far more than amply adequate, if they were similarly present in ancient sediments, as they seem to have been, to account for all the crude oil likely to have ever been accumulated. While many simple to complex types have been observed in the limited study to date, it has not been determined whether all the very many types of hydrocarbons found in crude oil are present among these recent hydrocarbons or to what extent they may result from organic matter by later evolution or from changes during and subsequent to migration.

3. The greatest and perhaps only effective migration mechanism operates early. Compaction takes place during basin subsidence and deposition adequate to reduce the porosity of muds from something like 70 to 90 per cent to 10 per cent or less in the resultant shales. Thus, something of the order of magnitude of 15 billion to 20 billion barrels of fluid is expelled from each cubic mile of the original mud. In addition to this, many times this amount of original fluid is forced through from the neighboring muds down basin. At no time in the history subsequent to its subsidence and deposition is there anything remotely approaching the magnitude of the migration and flushing action that takes place during that early subsidence and receiving of sediments. This point requires some discussion before we go on to point 4.

The hydrocarbons are no doubt expressed by the compaction along with the water from the muds of deposition sinks and other depressed areas of relatively greater overburden. They are forced outward and upward along bedding planes and along all types of depositional breaks, ranging from major unconformities to the innumerable diastems that occur in sedimentation. Owing to this differential compaction pressure and to other forces that are not too well understood but which are commonly referred to as "capillarity," the hydrocarbons selectively find their way into and follow along the paths of relatively larger porosity. These paths lead, naturally and with increasing development, to the porosity areas, belts, or trends, most of which are localized during deposition by highs within the basin and along the basin flanks. More specifically, these channels of sedimentary origin lead naturally into sand lenses and into reefs. They also lead into weathered surfaces or any form of porosity existent in older sediments or rocks of any kind that are overlapped tangentially updip and sealed off by the sediments in which the oil originated. It is in these porosity traps that oil is so widely found and where it must have been selectively screened from the migrating fluids.

In the very early stages some migration of hydrocarbons across the bedding is probable. Though most of these may be lost, some of them may be retained by the younger muds. The studies of Hedberg, Athy, and others show, however, that long before clays are buried to a depth of 200 feet, actually within a very few tens of feet, the clay particles are sufficiently in contact to be mechanically flattened to a point adequate to preclude important migration across the bedding. Very early in the deposition also, diagenesis effects extensive readjustment in the clays. The presence of sands, silts, bedding planes, turbidity flows, etc., greatly contributes to parallel permeability. Under compaction-induced expulsion of fluids, free hydrocarbons would probably be expelled as readily as or more than the water.

Many of the problems that perplex us when attempting to explain petroleum migration in very low-dipping, lithified sediments under, at best, extremely slight fluid movement against the considerable forces of capillarity in a complexly and variably textured medium seem to disappear when we consider early compaction-facilitated migration. They also tend to disappear when we consider that at the time of early migration, porosity and permeability were much greater than they are today.

4. A study of the distribution of hydrocarbons today suggests early migration. According to John M. Hunt and George W. Jamieson,* practically all shales and carbonate rocks contain, besides water, indigenous organic-matter residue disseminated in three forms: (1) soluble hydrocarbons that are similar in composition to the heavier fractions of crude oil found in reservoir rocks, (2) soluble asphalt similar to the asphaltic constituents of crude oil, and (3) insoluble organic matter (kerogen) which is pyrobituminous in nature. The presence of these materials in even young Tertiary sediments indicates that they are residues that were left behind early. They are not the source materials, and they are probably not likely to produce more petroleum by natural processes.

Another inadequately considered fact of petroleum occurrence seems to be the apparently overwhelming preponderance of gas in the total fluids formed. The quantities of gas that accumulated in reservoirs probably equaled in fuel value that of the oil. Examination of the lithified sediments today indicates that the porosity outside the petroleum traps contains principally water, in which are small amounts of dissolved gas. Studies carried on over the greater Gulf Coast region for over 10 years† indicate the general presence of petroleum gas dissolved in formation waters in concentrations up to 14 cubic feet per barrel. The gas is largely methane, but it contains much lesser concentrations of the heavier gas fractions.

Calculations by the author indicate that the total of such gas, if similarly present in the sediments of the world, as there is every reason to expect, may total in Btu value many times that of the oil accumulated in reservoirs.

In addition to these vast occurrences of petroleum gas, the total volume of this most fugitive of hydrocarbons that escaped to the earth's surface must have been many times that retained in the earth. Based on the limited studies to date, gaseous hydrocarbons appear to predominate among the total hydrocarbons found in recent sediments. Certainly if gas comprises such a preponderant part of the petroleum, it must greatly if not primarily determine the physical manner of migration.

It seems possible that oil may migrate on compaction-induced flow partly in solution in the gas and partly as films on gas bubbles in the water. Some but certainly not all of the hydrocarbons may have been carried in solution in the water. The formation waters studied by Buckley, Hocott, and Taggart were partially to substantially saturated at the formation pressures and temperatures. With the decreasing pressure and increasing porosity encountered in moving flankward the hydrocarbons would probably begin to segregate and aggregate by coalescence in the coarser porosity where they would be preferentially carried and in which they would be preferentially retained on entering a trap.

5. Traps that were nonexistent at the time of the deposition or very shortly thereafter are usually, perhaps always, barren of commercial oil. Adequate study seems invariably to eliminate the apparent "exceptions." This point calls for considerable discussion.

The oil-producing basins of the world and present deposition bottoms as well are replete with examples showing that the primary conditions for the localization of effective porosity must be present on the deposition bottom. Such conditions include first and foremost a bottom high. Bottom topography makes its own structure for localization of sand or carbonate porosity and accumulation of oil. It does this

* "Habitat of Oil," AAPG special symposium volume, 1958.

† Stuart E. Buckley, C. R. Hocott, and M. S. Taggart, Jr., in their paper, Distribution of Dissolved Hydrocarbons in Subsurface Waters, in *ibid.*

regardless of whether the high is of tectonic, compaction, or any other origin. The high does not need to be prominent, as a rule, in order to be adequate to accumulate a body or lens or successive lenses of effective porosity. The porosity may be a sand, a carbonate reef, or a shell, detrital, oolitic, or other form of potential limestone reservoir. Limestone reservoirs are usually recrystallized and are commonly dolomitized in part or entirely.

This same timeliness is a cardinal requisite for pooling the oil in its early migration. Late-formed structures, such as folds which have developed where the formations are at their thickest, are normally dry in those formations. Even though sands may be abundant, owing to shallow water maintained by relatively rapid deposition, they are of no avail if there was no closure there. Moreover, the effectiveness of porosity found in formations over late structures in sediments with this history is notably very inferior to that which occurs over structures which existed as highs on the deposition bottom. Incidentally, experience shows that porosity and permeability or "sand quality" is generally best in or near the thickest part of sand lenses that are accumulated over or alongside deposition-bottom highs.

The extent of sorting in basin sediments is an important criterion in the analysis of depositional conditions, timeliness of structure, and oil prospects. Where clastic sediments are dumped in a basin too rapidly for the capacity of the existing currents or of the waves or other forms of water movements to cope with their sorting, the resulting deposit will be an unsorted mixture of all sizes and shapes of grain. Thus we have clastic sediments ranging from silty and muddy ("dirty") sands to sandy and silty shales, depending on whichever grain size happened to predominate. Such sand may have little effective porosity or no more porosity than a shale. This is one of the difficulties in connection with certain large thicknesses of the Miocene section in the deep sink in the southeastern end of the San Joaquin Basin of California, where a large part of the sediments, mostly coming from the south at the time, were dumped into a rapidly subsiding sink. The same series carries large quantities of oil laterally on the adjacent, more favored flanks of the sink. Explorations of closed anticlines in the areas of thickest section do not normally find commercial oil even where they do find sand.

The extremely high percentage of Gulf Coast salt domes that are oil-bearing is significant. The domes were growing during deposition. The bottom highs, with their accompanying porosity facies, and the traps were in timely existence. Even the fact that the oil does not always occur all around the dome is significant. Naturally the structurally highest area at the time of oil migration and accumulation was not necessarily coextensive with that of today. Neither does the area of sand accumulation conform with the area and form of the present high, because it was controlled by the configuration of the deposition bottom at the time. Also, as so often happens over structures, some of the bottom highs may have been too elevated to receive any sediments, so that the sand may have accumulated only around the flank or only on a part of the flank, depending on currents, their direction, etc. Careful study is making these facts increasingly evident.

Many millions of dollars have been spent and are being spent each year to no avail on updip pinch-outs that were downdip at the time of deposition

Late formed structures, even in favorable basin position, are disappointing. Many of us can recall structures of large closure which are filled to but a minor fraction of capacity. Many of the deeply closed pre-Andean structures of Bolivia and northern Argentina are illustrative. Either there never was much oil to capture, or probably more often, as in the case just cited, the main development of closure was too late, and it captured only what oil may have been retained by the incipient structure or in its general vicinity.

And, of course, early or late structures in high shelf position, remote from areas of source sediments, are also notably disappointing. The structures in the greater part of the Near and Middle East occur in "remote-from-source" areas. They are representative of many examples over the world of little or no commercial oil accumulation in formations whose age equivalents in another part of the basin are very prolific. Certain areas of the United States are also illustrative.

Erroneous assumptions concerning the geologic history of a structure based on inadequately broad study have led some to use what superficially appears to be a late-formed structure as an example to prove late migration of oil. The same has led geologists to attempt explanations of where the oil was waiting in these "exceptions" to early oil migration. This statement can be briefly explained.

The principal locus of basin deformation over the world is around the margins of the basin sinks, that is, adjacent to the main depressions. This is true both on the so-called folded side of asymmetric basins and along the foreland hinge belt. In a petroliferous basin, the traps in these belts or trends are usually oil-bearing. In many basins or regions there was one period or episode in the history that tends to stand out as that of the main deformation. The so-called Laramide orogeny near the base of the Tertiary is one example, but there are innumerable other instances over the world. The common error is to look upon this one episode as the only one in the development of a structure. This is almost never true in the case of structures of this type.

In the Rocky Mountains province, for instance, early statements about a "Laramide orogeny" have been accepted as geologic gospel, to the extent that many other movements are overlooked. A number of investigators in more recent years have recognized that many of the so-called Laramide structures had their beginnings at least as far back as the Paleozoic. It is also being recognized as we widen our fields of observation that structures form over lines of weakness in the earth's crust, that these lines of weakness are real. Moreover, they are not ephemeral but persist over very long periods, scores or even hundreds of millions of years, during which time activity may be renewed, often many times. This is shown also and is being recognized more and more through study of the facies.

Conclusions concerning the time of deformation are commonly based on cross sections or diagrams made for other purposes. Even though they may be precise, they include only the immediate area. The problem commonly requires thickness, facies, and other data extending for miles off-structure, for the early structures did not look at all like the present.

It has also been stated that the Tertiary reservoirs in California, for instance, were not traps until the Pleistocene. This patently is not generally true, as is shown beyond all question by adequate examination not only of the structures but also of the sedimentary facies distribution. Sand lenses occur repeatedly through a considerable thickness of Mio-Pliocene and in places earlier, and they finger out off-structure into the deeper basin muds.

By the very nature of the earth's crust, lines of weakness are long existent and adjustments are many and episodic. A surprisingly large percentage of the world's oil-bearing anticlines are not primarily simple compressional folds. Rather the anticlinal form resulted from adjustments in the sediments over blocks in the subcrust. Even in the case of the more truly compression folds in sediments, if the fold carries to basement depth, it is almost certain to have been localized by persistent lines of weakness in the basement.

Surface anticlines and still more superficial wrinkles do, of course, exist. They commonly occur in the thick parts of sections. The structures are not directly controlled or activated down within the crust. Such structures have no "roots." They formed late and are as disappointing as they may be alluring. Numerous examples occur in northern Colombia, Venezuela, Papua, north Europe, and the Po Basin of Italy; in the Rocky Mountains; and in very many other places.

The fact that an early-formed trap is such a widely existent and usually obvious feature of oil pools should in itself make us suspicious of seeming exceptions and caution us concerning involved explanations. Even though nature may seem mysterious, we usually find with adequately wide and thorough study what perhaps we should expect, that she is not so capriciously devious as to go contrary on occasions to the laws which so generally, simply, and logically control her operation.

In conclusion, it must be admitted that the exact manner in which the hydrocarbons are carried has not been determined. It is far less clear than is the time of the important or major migration. This uncertainty regarding the manner of migration does not mean that the answer may not be a rather simple one. As pointed out later, the answer can probably best be sought from an area of investigation that has all too long been neglected—that of the recent and near recent sediments.

Secondary Migration

Much emphasis has been placed purposely on the well-documented fact of early migration and on the importance of timeliness of traps. At the same time it is recognized that cases of secondary migration do exist. They have been at most on a modest scale, varying with physical conditions. They have been more in the nature of local rearrangements in accommodation to physical changes within the vicinity of the present pool, rather than migrations *in toto* for distances of scores or even hundreds of miles across the basin. In the first place, in basin sedimentation of the type really favorable for oil, porosity pinches out rapidly basinward and does not spread across the basin. Second, when we consider that in primary migration and accumulation probably but a minor percentage of the vast total of hydrocarbons evolved actually ends up in a trap, then certainly the total trapped from any broad-scale remigration across a basin would indeed be insignificant.

Good examples exist of exceptionally huge oil accumulations in voluminous porosity bodies on the basin flank, oil that has migrated from a thick and excellent contemporaneous, penecontemporaneous, or, in many cases, overlapping source. They include the Lake Maracaibo Miocene wedge; eastern Venezuela Miocene wedges, one of which has a huge tar oil belt; the Athabaska oil sands; the Lower Cretaceous oil and tar belt of Kuwait, Basra, and bordering areas; and the Upper Jurassic Arab zone of Saudi Arabia, Qatar, etc. The total volume of oil that has moved into the several wedges just named probably runs into well upward of a trillion barrels. While the hydrocarbons are believed to have moved into the flank areas early, readjustments in varying degrees have no doubt occurred in accordance with the various physical changes that have occurred subsequently in the general areas of the traps.

Short- or Long-distance Migration

The question of short- or long-distance migration has been a subject of conflicting opinions. The fact that many billions of barrels of fluid are expelled from each cubic mile of sediments in the compaction process eloquently proclaims the magnitude of the migration. When one multiplies this large volume of fluid by the many other cubic miles of sediments farther basinward from which fluids are expelled, which must also pass through the sediments flankward, he realizes that migration is not just a local affair, however interpreted. Compaction studies show that the ratio of transverse to parallel permeability in muds very rapidly decreases with burial. To argue that hydrocarbons accumulate in a trap only from a closely adjacent source implies that they are formed only in the vicinity of traps; otherwise one has to explain why the hydrocarbons more distant basinward do not also migrate with the vast amount of fluids expelled on compaction or, if they do migrate, why they should avoid the traps.

No doubt the degree of long-distance migration of all fluids varies greatly, depending on the ratio of physical facility for egress parallel with the bedding to that transverse thereto. Long-distance migration would seem to be greatly facilitated by planes of porosity which may extend far into the basin. These include stratification planes, sand or silt tongues, innumerable diastems or lesser breaks, turbidity current deposits, etc. At the time of migration the porosity and more particularly the permeability were much greater than we find them today. This would favor a longer migration.

The relative importance of short- or long-distance migration will in a large measure vary with whether the hydrocarbons were mostly sourced near the porosity bodies or in more remote sediments. Here again, much depends on the distribution of supply of organic matter and environment favorable for its preservation. As already stated in another connection, rapid deposition is known to afford protection for organic matter, particularly hydrocarbons, from bacterial destruction. In most cases the belts adjacent to the porosity trends may thus be a much better source than the more remote and deeper water areas of very slow deposition. This may be true even though the latter had a more favorable bottom oxidation-reduction potential. Proximity to the reservoir facies, where there is also better parallel permeability, would in general also appear to make the nearer belts more favorable as a source.

Traces of oil are commonly found in drilling at the top of porosity layers outside commercial pools. These may be thought of as accumulations in microtraps due to the irregularities in porosity and permeability, slight changes of dip, microfault, or depositional-slump displacements, etc., normal in sediments. Such trace accumulation is certainly not valid evidence of either early or late migrations, nor is it valid evidence that petroleum will migrate independently of water updip along the top of a porosity layer. Oil is likely to be left there under any form of migration.

What Happens to All the Hydrocarbons Formed?

All studies that have been made of the organic matter or of the hydrocarbons present in sediments, recent or ancient, agree that the potential quantities of hydrocarbons are so vast that only a very minor percentage of them can have been trapped in a commercial pool. No doubt a considerable percentage of the originally formed hydrocarbons is permanently retained in the argillaceous muds, particularly where there may be physical difficulty of egress. Studies by John M. Hunt and George W. Jamieson* reveal that all shales and carbonate rocks contain indigenous organic matter in three forms: as soluble hydrocarbons similar to the heavier-than-kerosene fractions of crude oil; as soluble asphalt, like that in crude oil; and as insoluble pyrobitumens (kerogen). The amounts of the three forms vary widely with the facies. They probably also vary with the extent to which they were able to migrate away from the source. This facility of egress probably varies greatly with the texture and composition of the muds and with the presence or lack of permeable channels of egress.

Perhaps a great preponderance of all migrated hydrocarbons are lost in a disseminated form back to the sea bottom of both the offshore and shore areas. A certain part of this loss was transverse to the bedding in the very youngest sediments; probably more of the loss was by parallel migration. Some of the hydrocarbons might have reentered the cycle by being retained in the younger bottom muds; the balance probably escaped to be quickly destroyed by bacterial oxidation.

* "Habitat of Oil," *op. cit.*, p. 741.

Do Oil Pools Result from Erosion of Old Accumulations?

It has been suggested that some of our petroleum may have derived from resedimentation of hydrocarbons eroded from old sediments, particularly old oil pools, and carried back to the sea. Such a view takes little cognizance of the fact that most hydrocarbons are quickly destroyed by aerobic bacteria which live up to the limit of food supply in an oxidizing environment. Petroleum spilled on the ground soon disappears through the action of bacteria. The chances that eroded hydrocarbons may reach a favorable environment so as to reenter the cycle seem much less than in the case of recent hydrocarbons returned directly to the muds of the sea bottom. Likewise, the chances seem equally unlikely that the eroded hydrocarbons would reach the sea except on very short transport and there be deposited directly in the quantities and concentrations adequate to form some of our large commercial pools.

What Is a Source Rock?

It is probably safe to say that the efficacy of any sediment as a source of oil cannot be directly determined, on anything approaching a quantitative basis, by its appearance or even by its organic or bituminous content. We have no sure way of determining by inspection or analysis how much, if any, of the original organic matter has been converted into petroleum and has migrated from any sediment. In other words, we do not know how important any sediment may have been except by reasoning from analogy and widespread empirical study, and then only in a qualitative way.

Study of oil occurrence in many places and situations reveals that many prolific oil-bearing basins carry little or none of the dark bituminous sediments popularly thought of as a source. In those cases the geologist is content to accept the associated shales or marls as a source. Of course, the original organic content of the sediment must be diminished by the proportion thereof that went to furnish any hydrocarbons that may have been expelled from the sediment.

There appears to be a great deal of misunderstanding concerning the importance of color as well as bituminous content. The presence of only a moderate amount of calcium carbonate may appreciably mask the color of a shale, and many true marls have a higher organic content than is apparent. Even where sediments of an unfavorable environment are included in the series, a good source may still be present. Basin subsidence is episodic, and sediment environments vary as greatly with time as they do areally, and there are many repetitions. This may be particularly the case in a variegated series. The color of a reddish or brownish shale may have been brought in from the area of provenance rather than have resulted from the environment of deposition.

A highly bituminous black shale may still retain all its organic content because it was not in a form that could be readily given up or for some reason there may have been impedance of efflux of the bitumen. We know, for instance, that some fresh-water, highly bituminous beds, like the Permian Iraty black shales, which extend for over 800 miles through southern Brazil and neighboring countries, have furnished little or no free oil. Under an alkaline marine or brackish environment the preservation of the same or less amount of organic matter in the sediments might have supplied large quantities of oil.

A true source rock might be defined as one which originally contained adequately preserved organic matter that was high in hydrocarbons or was readily converted to hydrocarbons and from which the hydrocarbons were able to migrate—possibly to become trapped in a reservoir rock. To evaluate for source potential, then, it may be more important to ascertain the fact of development of hydrocarbons and the facility for egress rather than the quantity of organic matter now held as bitumen or as hydrocarbons. In true oil shales, for instance, most if not all of the bitumen appears to be still held in place. Also, in still highly hydrocarbonaceous shales, marls, and shaly limestones most of their hydrocarbons may still be held in place.

How Can We Solve Directly the Problems of Source and Early Migration?

From the above reasoning, it appears that the most logical procedure for evaluating for source potential is to go to the recent and near-recent sediments where the hydrocarbons are forming and where the environments of their formation can be observed at first hand. Our efforts at this type of research have not scratched the surface of the vast amount of available information of the very kind that the geologist has been groping for and speculating about for so many years.

Paraffin-naphthene and aromatic hydrocarbons similar to those in crude oil have been found in highly significant amounts in many recent sediments, including those of the Channel Islands Basins off California, along the Gulf Coast, and in an area neighboring the Orinoco delta of Venezuela. In the last area hydrocarbons were actually found to have accumulated in minor amounts in small sand lenses within muds less than 10,000 years old and but little over 100 feet deep. Similar evidence was obtained on the Gulf Coast.

To our recent sediments, then, we must go to demonstrate the answers which many cannot accept from empirical evidence or do not have the opportunity to seek for themselves. To the recent sediments, also, we need to go for the answers to other questions that perplex us with regard to source, migration, and accumulation. We now have powerful tools for this study.

Possible Errors in Correlating Oils with One Another and with Source

A common practice among geologists is to ascribe to an oil the same age as that of the reservoir rock. Thus we total up so many millions or billions of barrels of oil present in rocks of Ordovician or Devonian or Pennsylvanian or Eocene age in a pool or in a basin or in the world and speak of it all as being of the age of the reservoir rock. This may, of course, be the case, but the actual age of the oil in any pool may be anything from that of the reservoir rock to that of the overlapping sediments or that of sediments of any age subsequently brought into contiguity. The amount of oil that is pooled from a source in another, usually younger formation is by no means an inconsiderable part of the world total.

Another source of error and confusion arises from attempts at source correlations by means of hydrocarbon studies on the assumption that similar oils come from the same unit of sediments and different oils from some different formation. Such an assumption may be true within certain limits. However, adequate hydrocarbon studies in enough basins will impress upon one the importance of the word "limits." Two oils from different continents not uncommonly are more alike than oils from different areas of a single basin time unit or from different horizons of the same formation. One could hardly expect the facts to be otherwise when he considers that petroleums are products of environments, organic as well as inorganic, biochemical and physicochemical, all of which change greatly in both space and geologic time, and that they are products also of the varying vicissitudes of the history of the particular petroleum.

Possibilities for Source in Older Underlying Sediments

Still another early supposition persistent among some is that petroleum must have its source in older underlying beds. Two main considerations seem to motivate the geologists who look for source below the reservoir formation. One is an early and still persistent notion, possibly arising from the knowledge that oil is lighter than water, that the most likely way for oil to get to where it occurs is for it to rise from down below somewhere. It was once the author's opinion, taken for granted in common with perhaps most geologists at that time, that oil had to have its origin at depth in some older strata except, of course, where such was clearly impossible, such as where basement lies below. It had seemed that, somehow, oil could only move upward. That it had to cross the bedding did not concern us. At the same time we had no difficulty, nor has anyone today any difficulty, in recognizing the source of oil from the overlapping sediments when oil is found in fractured or weathered basement rocks or in the "wash" overlying the basement.

The second consideration that often causes geologists to look below for a source is the belief, generally unsupported, that the darker and more bituminous the formation, the more likely it is to have been a source. So, regardless of its geologic and physical relation to the reservoir, any dark and bituminous-appearing sediment, particularly if it lies somewhere below, is readily accepted as the most likely source. It has been interesting in the course of our review of world oil fields to note the many cases where, in search of some dark-colored underlying formation that came nearest to the geologist's idea of a "source bed," the source favored lay far below, even to thousands of feet below, the reservoir sands. In many cases the fact that porous horizons barren of any oil lay between the presumed source and the oil-bearing member seemed to make no difference. In the many prolific basins of the world that have no black bituminous shales, geologists are content to accept the intimately associated shales or marls without question as the source.

Studies of recent sediments show that the ratio of transverse to lateral permeability in the soft surface muds is approximately 1 but that it decreases very rapidly with burial. This is in line with the earlier studies of Hedberg and also of petroleum engineers and production geologists, which indicate the virtual absence of transverse permeability in shales, even for gas. The presence of sand partings in muds also greatly increases the lateral permeability. In many sands, particularly if there is a tendency to clay partings, transverse permeability may be poor or negligible. The basins of the world present a great number of examples of a succession of sands separated by thin shale zones in which the sands vary greatly in type and composition of fluids. Oil, gas, and water of varying compositions commonly alternate with completely independent relationships. The basins of the world are also replete with examples of excellent reservoir traps that are barren of oil but which should carry it if the oil in the pools above had risen from below. Why oil should wait until a certain period, long after deposition, and then move through normally impervious strata seems inexplicable. Particularly is this not understandable when penecontemporaneous or capping sediments are as good a source as any existing in many prolific oil-bearing basins.

In the author's experience there are two types of rock facies across the bedding of which oil may migrate under specially favorable conditions. Oil has been found to migrate upward into a young fresh-water series in which the clays are not overly abundant. It is thought that the clays were not diagenetically altered to platy forms in the same manner as those deposited in a marine or brackish environment but that they retained the vertical-permeability characteristic of soil clays.

A highly frangible limestone, shaly limestone, cherty limestone, or limy chert which is otherwise impervious may also permit a certain amount of vertical migration, and it may function as a reservoir under suitable cap. A modest amount of such migration seems to have taken place in the Cretaceous Ain Zallah anticline in northern Iraq. There oil occurs in both Upper and Middle Cretaceous limestones, which are separated by shaly limestones and marls, with bituminous marls, limes, and cherts in the lower part overlying the Middle Cretaceous limestone. Both limestones are unconformably overlapped by bituminous beds that might have been the oil source in each case. The beds which unconformably overlap the Middle Cretaceous limestone pool recall to mind the petroliferous cherty beds of the Santa Maria Basin in California, the oil-bearing La Luna of Venezuela and Colombia, and perhaps even the Phosphoria of Wyoming. At Ain Zallah the reservoir behavior is said to suggest that the upper oil, which is of limited occurrence and in fractures only, may have risen through fractures in the intervening beds. These beds are quite limy and frangible for the most part. The fact that the oils are undersaturated also suggests the leakage explanation.

Where a limestone sequence is more or less continuously porous vertically, there may also be no more of a seal or possibility of accumulation through the interval than would occur in sandstones. There appear to be intervals of that nature in the Paleozoic sequence in the Central Platform of west Texas.

The belief is held by some geologists in the Middle East that the huge volumes of middle Tertiary oil present there had their origin in dark shaly limestones and calcareous shales, of Middle Cretaceous or greater age, which occur thousands of feet deeper in the section. To reach the middle Tertiary reservoirs these great volumes of oil would have had to wait somewhere until after mid-Tertiary time and then migrate as a seepage across thousands of feet of strata. The intervening section includes beds such as form the caprock in many of the world's oilfields. Under some of the fields the section includes anhydrites such as form the very efficient caprock of a large proportion of the Middle East pools. The assumed deep-lying source sediments are said to be "polybituminous." This "polybitumen" is insoluble in the usual solvents, and it is the insoluble organic matter of the laboratory study discussed earlier in the chapter. In the author's opinion the Tertiary sediments of the Middle East carry more than ample facies of a good source type in highly favorable relationship to the reservoirs to supply the huge volumes of oil therein. In fact, the middle Tertiary sediments are much the most widely petroliferous of all those throughout the Middle and Near East areas. In the Qum area south of Teheran on the Iranian plateau, like middle Tertiary sediments are again highly petroliferous yet rest on what is essentially basement; thus they do not have a deep source for the oil such as has been assumed for the Tertiary fields of southwest Iran and Iraq.

Geologists have on occasion assumed a source of huge quantities of oil in old underlying strata separated from the oil-bearing series by an erosion interval of many millions of years duration. Such a source is seriously considered by some for the oil in the north Mendoza and Comodora Rivadavia Basins in Argentina. In both cases the oil-bearing beds lie on what the author considers to be basement. Based on his world-wide studies of oil occurrence there is nothing in the old, unconformably underlying rocks with anything like the source potentialities of the facies in which the oil occurs. Geologists have stated their belief that the oil in the Cretaceous-Tertiary of the major Langchunmiao field in Kansu Province, China, had its source in the unconformably underlying, rather metamorphosed rocks because they are probably of more truly marine origin than the sediments reservoiring the oil. It would appear to the author more likely that the oil had its source laterally in basin facies of the age of the reservoir beds.

Source of Athabaska Oil in Western Canada

Some geologists believe that the oil in the Cretaceous Athabaska oil sands of western Canada originated in the Devonian, which is separated from the oil sands by a major erosional unconformity. One of the urgently advocated views is that the Athabaska oil is a seepage, a seepage in the amount of several hundreds of billions of barrels! It is difficult for the author to visualize that the oil would wait about 200 million years to leave the Devonian and then all at one time seek a new home far away and above the long-existent erosional surface. The thinned Devonian limestones below the oil sands and for a long distance down the gentle dip constitute an open aerated water facies. World-wide oil-occurrence facts amply show that such limestone conditions are most unfavorable as a source. For the oil to come from the favorable source associated with the Alberta Basin Devonian fields it would have had to have been held somewhere for 200 million years and then migrate at one time for a distance of something like 200 miles.

One theory is that the Athabaska oil was supplied by the erosion of Devonian reefs in the Alberta Basin. This seems impossible to explain from a physical standpoint alone. Moreover, to supply 300 billion barrels of oil would require the long-distance migration and preservation intact of the equivalent of something of the order of 1,000 pools averaging 300 million barrels of oil each. No such remnant pools are known or likely. This also does not allow for the much greater volumes of oil that would be lost by inspissation and bacterial oxidation, plus that retained by the original reservoir.

The Athabaska oil is peculiarly different from any known in the Devonian. A simple, logical, indeed classic source for the oil is that of the great body of contemporaneous, penecontemporaneous, and overlapping sediments that fill the Mesozoic sink to the west of the sands and into which the great sand lens with its multitude of stratigraphic permeability planes interfingers across the basin. This Mesozoic basin with its great flanking oil-filled sand is one of many illustrations of a common principle of oil occurrence. The oil moves to the areas of greatest porosity and permeability and lowest pressure. Not only do the sedimentary channels of permeability subdivide and interfinger out into the basin, they also head back into the body of high porosity and low pressure. This kind of basin situation, in which the drainage channels of a prominent basin source are focused toward a great sand body, is a common setting for great accumulations over the world. Incidentally, the situation holds the answer to the commonly heard query, "If the source is in the same basin sediments, why don't we find more of the oil elsewhere in the basin?"

The now classic, often-referred-to Athabaska occurrence is by no means unique. The east-side Lake Maracaibo Miocene oil accumulation, which totals many scores of billions of barrels of oil, is one of many similar occurrences, and so is the much larger one in eastern Venezuela. Another similar classic as well as perfect setup for drainage out of a basin into a great body of abundant porous sand is that of Kuwait and adjacent areas in the Middle East. Here oil has accumulated in a volume fully equal to that of the Athabaska area. But there are many other cases similarly situated basin-wise.

Relation of Oil Gravity and Type to the Basin

The usual pattern of oil occurrence in a basin is that of increasing API gravity basinward in any one horizon or oil system. In some basins the amount of heavy oil high on the basin shelf totals many billions of barrels. The accumulation forms what has been termed a "tar mat." There is a tendency also for a decrease in API gravity flankward with decrease in chloride content. A decrease in API gravity commonly accompanies an increase in carbonate content of the formation waters.

It is not clear what the conditions are which control petroleum gravity and type distribution within the basin. The temperatures involved are low, for, as earlier recorded, certain of the usual constituents and properties of petroleum are destroyed well below temperatures of distillation. Oil-occurrence studies present an overwhelming mass of evidence that petroleum migrates early. Studies of petroleum hydrocarbons in recent sediments support the evidence. Space permits a list of only some of the possible reasons for the basin-wise type and gravity distribution of crude oil.

1. The heavier oils and tars are the less pure in that they contain much oxygen, nitrogen, and sulfur from the original organic matter. Perhaps they were the oils that formed first and hence migrated earliest and farthest toward the basin edge before compaction was completed. Some of the youngest oils are asphaltic. Perhaps paraffinic oils require greater depths of overburden and hence form later and do not migrate so far.

2. Just as there is a progressive metamorphism of sediments with increasing temperatures and pressures, perhaps there also are effected some physicochemical changes in petroleum with burial and time. On the other hand depth alone is not the controlling factor, as is seen in numerous instances. For example, the Permian and Carboniferous oils of the Rocky Mountains are heavier and darker than those of the Mesozoic.

3. Some think that certain active clays may have had a refining influence on hydrocarbons and that chromatographic changes should take place during migration. Actually, certain types of clays are utilized in refining. Clays may have an important catalytic influence in organic-matter transformation because of their extremely fine grain, the great surface area presented, and their intimate association with the organic matter. The effect may vary also with the temperatures and pressures of burial. Trace metals, such as vanadium, nickel, molybdenum, etc., which are extracted from sea water by organisms and occur in petroleum may serve well as catalysts in hydrocarbon synthesis. They may play an important role in conversion of organic matter to petroleum in the sediments.

4. Many geologists look upon an increase in gravity as due to "oxidation." Others look upon it as the simple result of loss of light ends at the shallow depths. Loss of the lighter fractions will result in a heavier crude, but it will not account for a change from a paraffinic- to an asphalt-base crude. The oxygen in heavy crude oils, like sulfur, nitrogen, etc., appears to be an "impurity" remaining from the original organic matter and not added later. Asphaltic oils are high in porphyrins which are rather readily destroyed by direct oxidation. Segregation of oil fractions may result from their differential solubility in migrating waters. The more oxygenated materials, which include the asphaltic fractions, are the most soluble and hence may be selectively taken out and moved farthest up the basin flanks.

5. Bacterial invasion of the shallower upflank belt may have caused an increase in gravity by selectively consuming the lighter, more paraffinic fractions. Most bacteria that prey on hydrocarbons seem to have a preference for the paraffins. On the surface of the ground bacteria rapidly destroy paraffinic oils. This may be a plausible reason why the paraffinic ends of paraffin-base crudes do not accumulate at seepages as does asphalt. Sulfate-reducing bacteria may account for the increase in gravity of oils occasionally occurring near the oil-water contact in pools where sulfate-bearing waters occur.

It is but a step from the surface type of asphalt occurrence to that of the so-called tar mat or tar belt found along the updip part of the shelf of many basins. Examples include the Athabaska oil sands, the great tar mat of the eastern Venezuela basin shelf, the tar mat of the Neuquen Basin in Argentina, that of the Middle East, and numerous others. Basinward in nearly all cases oils of progressively lighter gravity are found.

Though many of the oils in the tar mats are heavier than water, they occur above water in the same way as do the lighter oils of the deeper basin. They must have migrated in a somewhat lighter state and become inspissated by loss of light ends plus probable near-surface bacterial invasion and resultant oxidation. The facility with which bacteria remain with hydrocarbons, even adjust themselves permanently through geologic ages to underground conditions, makes invasion in force of the shallow flank hydrocarbon accumulations seem almost a certainty.

A rather common situation is to find lighter-gravity, less-dark oils protected in lenses near the deeper basin and varyingly heavier and darker oils in more exposed blanketlike sands farther up the shelf. The producing sands of the Venango series of northwest Pennsylvania may serve to illustrate. A belt of sand lenses faces northwest on the shaly deeper basin. The oil occupies the thickest, characteristically most porous parts of the sand lenses, and water tends to fill the tightest parts of the sands. Characteristically, too, the thickest part of the lenses generally faces the deeper water on the northwest, just as is so commonly the case with reefs and other carbonate oil-bearing bodies. A shale belt with few or no sands appears to follow on the southeast of the sand lenses, and that is succeeded in turn by a belt of blanket- or apron-type shore-front sands. In these more broadly porous sands water may be abundant in what appears to be hydrostatic instead of capillary adjustment with the oil, and the oil appears to be increasingly dark in color toward the southeastern shore.

Darkening of color shoreward in normally very light, green oils similar to those of Pennsylvania is notable in the south part of the Lomitas field of northern Argentina. California oils show a similar increase in gravity and asphaltic content with exposure, probably bacterial exposure.

6. Oils are a complex mixture of hydrocarbons. No two of them are entirely alike. The stamp of any particular crude may be controlled in large degree by the complexion of the marine life that supplied the organic matter. For instance, organic matter in which the structurally simple fats and carbohydrates predominate might be the source of the relatively pure paraffin oils. The complex, heavy-molecular-weight proteinaceous organic matter may give rise to the heavy-molecular-weight asphaltic-base crudes with their sulfur, nitrogen, oxygen, and other impurities present in heavy protein molecules.

Protein, which by definition contains nitrogen, is much the most abundant in animal matter. In the present seas copepods and other chitin-secreting animal scavenger organisms make up the bulk of the animal life in marine surface waters and on the shallow aerated bottoms. On the other hand the great bulk of all life in the seas consists of minute plants or phytoplankton of the sunlighted surface waters. Not only are the plants the most abundant, but their average life is much shorter than that of the animals. This multiplies the turnover as well as the proportion of plant substance in the organic matter that falls to the deeper bottom beyond the reach of the scavenger benthos. Thus, this natural distribution and incidence of occurrence and turnover may account, partially at least, for the characteristic oil-type distribution.

7. Many organisms, including bacteria, are known to produce hydrocarbons. Most cells, in fact, are said to contain hydrocarbons. Calculations suggest that the aggregate amount of hydrocarbons provided by the huge quantities of short-lived marine life is so vast as to be ample to supply all our petroleum very many times over. Plants and animals are known to produce large quantities of optically active sterols such as occur in crude oil. Many different forms of hydrocarbons, waxes, etc., are

formed in the life cycle of certain plants and animals. Hydrocarbons from marine kelp and from fresh-water algae, for instance, are said to be very complex, ranging from 19 to 35 carbon atoms. We need to know more about the extent to which the type of hydrocarbon varies with the kind of organism, if at all. Certainly we do not know whether all the forms of hydrocarbon found in crude oil have been produced directly by organisms.

8. Asphalt was found by Paul V. Smith to comprise the largest fraction of the soluble organic matter extracted from recent sediments.* The same finding was reported by John M. Hunt and George W. Jamieson in their study of ancient source sediments.† Thus this asphalt appears to be not only a normal and widespread but also an original constituent of the gross hydrocarbon content of sediments.

We have noted that the more oxygenated materials in petroleum, which include the asphaltic fractions, are the most soluble in the migrating waters. The amount of asphalt that would be carried in the migrating oil and hence be present in the crude accumulating in a reservoir would vary also with the solubility of the asphalt in the petroleum. Solubility, it appears, is very sensitive to changes in the surface tension of the solvent. Solubility, for instance, decreases rapidly with increasing surface tension. Surface tension has been shown by Katz, Monroe, and Trainer‡ to decrease with increasing temperature, and Swartz|| has shown that surface tension decreases with increasing saturation pressure. With increasing depth the content of gaseous hydrocarbons and the saturation pressure of the crude increases, and there is also an increase in temperature. Both operate to increase the surface tension markedly and greatly lower the solubility of the asphaltic fractions. Thus, other factors being equal, with increasing depth there should be a very pronounced and progressive decrease in the amount of asphaltic fractions carried away from the source by the migrating crude oil, and vice versa.

Accumulated oils high in asphalt might release some of their asphaltic content on being subsequently depressed beneath greater depths of overburden. However, the amount of such release would probably vary with the gas content and, of course, with the original asphalt content and thus may not be notable. However, this might account for some of the vestiges of asphalt found with lighter oil in some reservoirs.

Incidence of Oil Occurrence with Depth of Burial

There is an optimum depth of maximum burial for oil occurrence. Above and below such a depth or depth zone the incidence of oil occurrence decreases. This optimum depth varies with basin conditions and history, but in general it lies within 2,000 to 9,000 feet. There are various reasons for the gradual decrease in incidence at the greater depths. Perhaps the most important is the fact that petroleum tends to move out of the deeper basin to the shallower positions of increasing porosity up the flanks of the basin. The normal tendency is for the fluids to be forced flankward by the differential pressures of compaction, which approach 1 psi per foot of difference in depth of overburden. In addition to the great decrease in reservoir-type facies in the deep basin, even to its elimination, there is a varying amount of loss of porosity due to rearrangement, packing, and cementation of grains at the greater depths.

Increase in carbon ratio on approach toward a mountain front is well known. Less commonly do we consider the rapid increase that occurs with depth. This may be a

* *Bull. AAPG*, vol. 38, pp. 377–404, 1954.

† "Habitat of Oil," *op. cit.*

‡ *Petrol. Technol.*, vol. 6, p. 1, 1948.

|| *Physics*, vol. 1, p. 245, 1931.

critical factor in sediments that were once buried to greater depths than at present, as is indicated by the higher than normal densities and porosities in the shales. Probably much of the increase in the carbon ratio toward mountains is due to original depth of burial of rocks in deep geosynclinal positions that are now greatly unburdened by erosion.

Of course, oil once accumulated might later be depressed to greater depth. There is commonly some loss of closure due to tilt. Once oil has taken possession of a reservoir, the loss of porosity with increased burial is probably less than that outside the reservoir. The loss may little more than keep pace with the shrinkage of total volume of the hydrocarbons resulting from the increased pressure.

Another condition that affects the incidence of oil occurrence at depth is the tendency, under reservoir conditions of high temperature and pressure, lighter-gravity oil, and increased volumes of gas, for the gas and the oil to become physically indistinguishable. The light crude commonly goes into solution in the larger volumes of gas to produce what are generally known as condensates. Condensates may occur from depths as shallow as 6,000 feet.

Mud-flowage Phenomena

Gas is not uncommonly entrapped in plastic muds of the deeper basin and cut off from hydrostatic connection. Such gas occurs under pressures approximating that of the overburden. Should disturbance open up channels of egress for the gas and mud and contained water, mud volcanoes, mud flows, mud lumps, mud diapirs, etc., may be produced at the surface. Such phenomena are common. Notable occurrences include the mud lumps of the Mississippi delta front; the diapirs and mud volcanoes of Trinidad, of eastern Venezuela, of Barbados; in the Sinu in northwestern Colombia, the Baku area of Russia, and northwest Borneo; and the fantastically huge mud flows of the Makran coast of southern Iran.

Hydrodynamic Control of Petroleum Occurrence

There has been a recent renewal of interest in the effect of hydrodynamic forces on oil occurrence. It is contended by some that petroleum migration and entrapment are greatly influenced by hydrodynamic forces, even at any time geologically after lithification of the sediments. Some believe that oil and gas can be moved out of the porosity of traps in which they originally became fixed. These geologists believe that the oil and gas can be moved to new positions in the immediate area or even to other parts of the basin by a change in hydrodynamic forces, such as, they believe, would be brought about by a change in tilt of the basin. From the geologist's viewpoint the following are some of the many questions that arise with regard to these claims.

If hydrodynamic forces operate importantly in changing the position of oil and gas in a formation after the original accumulation, one might ask why it is that most oil-water tables are horizontal. One might wonder why the very much greater forces resulting from compaction-produced fluid expulsion should not have caused tilting of the oil-water table flankward or even flushed the oil out of the many slight closures in which it is reservoired. Or if we assume that the oil-water interface was originally horizontal, as it probably was and as it is in nearly all cases today, the subsequent subsidence and filling up with sediments more basinward from the reservoirs would certainly tend to give the accumulations a tilt toward the basin. As a matter of fact this is the direction of tilt occasionally found in basins that have had this type of history.

Hydrodynamic flow requires a pressure adequate to overcome the capillary forces. In addition it requires a continuously porous and permeable medium and an outlet as well as intake. Very few indeed are the basins of great value for petroleum in which porosity does not die out with a change to a more basinward facies.

It seems doubtful to the geologist that any slight movement of formation waters, which at best would be but an extremely slow creep, would be adequate to overcome the capillary forces holding oil in place in a reservoir. On the average less than 2 per cent of the pore spaces in a single horizon of a basin is oil-bearing. Therefore, the water would seem to have plenty of space to move through, should there be any movement, without having to overcome the not inconsiderable forces holding the oil in the porosity that was particularly suited in the first place for its retention. It would appear that in its almost infinitely slow creep past an oil accumulation, pressure equilibrium would be essentially maintained about the pool. It would seem that any forces tending to move the oil secondarily would be but a minute fraction of those which effected the original early emplacement of the oil in the trap.

Changes that sometimes occur in oil-water surfaces with production are not uncommonly cited to show the effects of hydrodynamics. Examples of the effects resulting from drastic man-produced changes are certainly not valid evidence of what occurs under the at best extremely slight forces normal in a basin. Assuming that the distribution of porosity across the basin were such as to permit it and that an extremely slight fluid movement would flush out the oil, in geologic time most accumulations would have been greatly disturbed if not flushed out. Quite the opposite is the case.

Experience indicates that in accumulation the oil shows preference for the larger pores and more porous parts of the sand. Moreover, once the oil is in place, the pores seem to be protected from cementation to a considerable degree. On the other hand, cementation in the water-bearing parts of the sand is quite common. These and other facts are not in harmony with the view that there is widespread remigration of oil.

Many cases of variable water table that are ascribed to hydrodynamic control have been found to be due to structural or stratigraphic causes or both. The author discussed examples of these in "Habitat of Oil."*

Millions of dollars are being spent annually in exploring updip pinch-outs that were downdip at the time of deposition and oil accumulation. World-wide study of oil occurrence shows that only under some very special circumstances are such situations commercially oil-bearing. More fantastic, however, seem the claims that with change of basin tilt (such as would make the erstwhile low rim of the basin the high side) the oil would move across the basin to the new high side. In any case, considering that in the primary migration and accumulation probably but a minor percentage of the vast total of hydrocarbons evolved actually ended in a trap, the total trapped from any broad-scale remigration, if such could take place, certainly would not be important.

A good illustration of points made in this discussion seems to be the Lake Maracaibo Basin, Venezuela. This basin is typical of many over the world. Like so many others, differential subsidence has left it crisscrossed with a network of faults. These are shown by the facts of oil occurrence, reservoir engineering, and repressuring to be perfect barriers to fluid migration. The basin changed its over-all slope from strong southwestward in the Cretaceous–Paleocene to strong northeastward throughout the Eocene to strong southwestward through the Oligocene–Lower Miocene to strong southeastward in the Upper Miocene–Pliocene. Very large volumes of oil have been found in all these units to date except the last, which is more continental in its facies and has been little explored. All oil occurrences present today in the Maracaibo

* *Op. cit.*, pp. 53–55.

Basin age sequences just cited are normally, indeed perfectly, related to their original depositional basin. Except for minor and very local adjustments in some of the very highly porous and permeable parts of the Eocene sands the oil is still where it originally accumulated.

The present is but an instant in the complete and ever-changing geologic history of a basin; hence the geologist may find himself somewhat nonplussed to discover that one of the basic measures for making a hydrodynamic study of a basin "to aid in oil exploration" is a series of levels on the stream crossings of outcrops around the basin. Too little thought, it sometimes seems, is given to the geology and complex history of the basin, the extent of faulting, the great changes in facies and in porosity basin-wise, or to the extent to which porosity determines where the oil will collect and cling in the kind of traps in which it accumulates.

The hydrodynamic-control-of-oil-occurrence theory is a fascinating one. The author believes that those who are working in hydrodynamics are developing an immense amount of fundamental data of great present and ultimate value to the oil industry. The questions raised above, of primary interest to the geologist, apply to only a part of the extensive work being done. As a geologist, the author feels that certain theories based on that part of the work should not be accepted blindly. Because new theories are intriguing, the occasional cases which might appear to support them should be very critically examined. In this case the theories should be so examined both in the subsurface and against the physical background of the basin and its geologic history.

Oil-formation Waters

Much has been learned about underground waters, both those of the shallow or vadose zone of normally fresh waters and of the deeper, usually saline waters. Then, of course, there are mixtures of both broad types. Much remains to be learned. Though there is a need for more data, there is a greater need for properly directed analysis of the data already available. This is not the place to discuss this big and very important subject or even to set forth the various classifications of underground waters given in many discussions of the subject.

The presence of carbonates, bicarbonates, and sulfates in oil-field waters suggests mixture with meteoric waters which picked these up in the vadose zone through oxygenation and carbonation or at ancient erosion surfaces marked by unconformities. So-called "connate" waters are high in chlorides. Some believe that the interstitial water of oil pools comes the closest to being true connate, the original water of deposition. Such water seldom occupies less than 10 per cent of the pore space in an oil pool, and it may range up to 50 per cent. The percentage of such water usually increases with decrease of permeability.

Two common differences between sea water and oil-field brines are, first, the presence of sulfate in sea water and its absence from some oil-field brines and, second, the presence of some alkaline earths (Ca and Mg) in sea waters and their absence in some oil-field waters. The reduction of sulfates may be the result of bacterial activity.

Silica is not normally found in reservoir waters. The source of silica, most common of cements, is probably in most cases primary. It appears to have been carried in solution by streams or more directly by volcanic activity and deposited on reaching certain salt-water environments. However, some of the silica may have been dissolved out of the clays on compaction, or it was derived from pressure solution of the sand grains themselves (Riecke principle).

Increasing temperatures and pressures with increasing burial increase the capacity of the waters to dissolve salts. Adsorption of ions on clay particles and their release later under high temperatures and pressures may well account for the common increase in salinity with depth and the changes that occur in the composition of the salt content in basins. There is much evidence of the precipitation of salts with the reduction of reservoir pressure with production.

Geochemistry

Geochemical map anomalies are believed by the author to be more likely an indicator of some kind of structure or other phenomenon in the near-surface sediments than of petroleum accumulations at depth. Hydrocarbons are normal, even though variable, constituents of all sediments, even those at and near the surface. Both laboratory research and exploration experience show that transverse migration of fluids, even gas, across shales, even minor shale partings, is unlikely, perhaps in most cases impossible. Because of the difficulty of migration across the bedding of the average lithified sedimentary section, any hydrocarbons emerging at the earth's surface would seem more likely to be those escaping from beds impinging on the erosion surface from out of the basin.

Studies of many basins show that deep-seated structure is commonly reflected, even though ever so slightly, to the surface beds. Even in basins like that of Alberta, western Canada, rigid structure or deposition features (such as a reef) as old as the Devonian are reflected in the younger beds through the Cretaceous. They are reflected not only in the structure of the younger beds but in the successive deposition-bottom topographies as revealed by the localization of sand and of petroleum accumulations over or around the very gentle deposition-bottom highs. To the extent that buried structures are reflected to the surface, geochemical anomaly maps may reveal them.

However, all sediments are affected by many other types of structure, depositional and tectonic, any of which may localize anomalous concentrations of hydrocarbons. Such minor structures include change of dip, even though slight; faulting, even to minute deposition settling faults; lensing; or changes in degree of porosity. Geochemical anomalies may result from simple transection by the surface of beds emitting or merely containing higher than background amounts of hydrocarbons. Superimposed on all this, though possibly in part resulting therefrom, there are, of course, the chemical and biochemical effects built up in the soil itself.

Perhaps it may be concluded that the value of geochemical maps depends both on their manner of preparation and on their correct interpretation and use. At the present stage of our knowledge these steps seem uncertain and the usefulness may be in large degree empirical, dependent on trial and error.

Adequacy and Timeliness of Seal

One of the most important and often not sufficiently considered facets of oil occurrence is that of the adequacy and timeliness of the seal. No doubt, on the average upward of 95 per cent of all hydrocarbons that have migrated miss being trapped and are lost to the sea bottom or land surface. In very few cases does the top, updip, and base seal of a basin flank exist on such a widespread or general scale as to capture a large proportion of the hydrocarbons. Two possible exceptions appear to be the Upper Jurassic Arab zone and the Tertiary Asmari of the Middle East. In both cases adequate, almost basinwide and direct seal was provided by anhydrites. Without aid from such widespread seal it is questionable whether the hydrocarbons would have been captured on such an unusually vast scale, vast enough to fill many huge structures, each to the extent of billions of barrels of recoverable oil. Where the seal is as widespread in a basin and the total of trapped porosity is as great as that just described, the gathering of oil and ultimate filling of structures may go on for longer than the average time after deposition. Numerous instances exist where an exceptionally effective seal seems logically to explain fair-sized oil accumulations in sediments of a type not normally important as an oil series. Certainly the matter of the adequacy and timeliness of seal is of first importance, not alone for an individual structure but on a more general scale.

Faults as Conduits or Barriers?

Probably most, if not all, oil geologists have at some time or other invoked the use of faults, now as barriers or closure, now as channels or conduits for migration of oil, as it seemed to suit their ideas at the time. Our studies indicate that faults far more commonly act as barriers to rather than as channels of migration. The studies show rather clearly that where timely faults occur in oil-bearing terrain, so disposed as to form a trap, the trap is about as sure to be oil-bearing as any other kind of structure. In fault trap occurrences, such as the many in eastern Venezuela, the Gulf Coast, and many other places, the oil accumulates on the sink or basinward side of the barrier. This is true in practically all cases where the faulting preceded and/or accompanied or very closely followed deposition. This occurs whether the faulting was up or down to the basin.

Most oil-bearing anticlines and salt domes are tension-faulted without prejudice to oil retention. If there is any one place where oil should not be retained, if faults or lesser breaks accounted for our oil accumulations through upward migration, it is in the average anticlinal or domal structure. This should be true whether a structure was formed tectonically or as the result of draping over a fault block or over any form of buried core. Production geologists know the barrier effect of faults, and so do the reservoir engineers, who look for and make use of the barriers in their repressuring operations, as is being done today on a vast scale so successfully in the Lake Maracaibo Eocene. Of course, there are cases where good porosity abuts both sides of a fault and the fault zone is not an adequate seal to migration across the fault. There are exceptions to many general rules in geology, even though not to principles.

The existence of an outcropping oil sand or an oil seep near a fault is often taken as evidence that the oil rose along the fault from depth. Just as it is natural for oil to accumulate against a fault at depth to form an oil pool, so it is natural for seepages to issue against faults on outcrop. The evidence that the seepage oil rose from depth along the fault is interpretive only. Study of many such cases in basins where numerous seepages exist has pretty well convinced us that the seeps are not localized at *any* horizon but are confined to particular stratigraphic horizons usually known to be petroliferous. The fault or fault vicinity is but one coordinate of the physical control for most such seepage phenomena. The other coordinate of the seepage is the bed which rises to or toward the fault. This coordinate is too often disregarded in considering the reason for the seepage. In the author's experience it is the near-surface bed which furnished the oil, just as in the case of most oil shows on the crest or flanks of anticlines or any other structures.

Deltas and Oil Occurrence

A common belief is that the delta is an optimum place for oil occurrence. This view seems to require serious qualification, both on theoretical grounds and on actual field evidence. Some lateral environments of the gross delta complex are undoubtedly very favorable, such as where well-screened sands have a chance to accumulate and there is quiet communion between them and frontal deeps.

In the physically active, growing part of the delta the picture changes. River currents carry light organic matter out beyond the active area, just as do waves and other currents of the sea. Also, the muddy waters of an active delta are a most unfavorable environment for marine flora and dependent fauna because of the very slight sunlight penetration.

A few have suggested that petroleum originates from organic matter, such as humic acid, which is carried to the sea in large amounts by such rivers as those which drain swamps and heavily vegetated regions. Such humic acid material is precipitated on reaching the colder salt waters of the ocean. It may be questioned whether organic matter carried from the land was originally of a type favorable as an oil source. It would be even less likely to be in a condition to source oil after its long journey to the sea in a highly oxidizing environment. This material also would be swept by the currents out beyond the delta proper.

REFERENCES

American Association of Petroleum Geologists, special volume, "Recent Marine Sediments," symposium, 1939.

American Association of Petroleum Geologists, special volume, "Stratigraphic Type Oilfields," symposium, 1941.

American Association of Petroleum Geologists, special volume, "Habitat of Oil," symposium, 1958.

Athy, L. F.: Density, Porosity, and Compaction of Sedimentary Rocks, *Bull. AAPG*, vol. 14, January, 1930.

Brooks, Benjamin T.: Origins of Petroleum: Chemical and Geochemical Aspects, *Bull. AAPG*, vol. 20, March, 1936.

Brooks, Benjamin T.: The Chemical and Geochemical Aspects of the Origin of Petroleum, in "The Science of Petroleum," vol. 1, Oxford University Press, New York, 1938.

Buckley, Stuart E., C. R. Hocott, and M. S. Taggart, Jr.: Distribution of Dissolved Hydrocarbons in Subsurface Waters, in "Habitat of Oil," special volume, pp. 850–882, American Association of Petroleum Geologists, 1958.

Clarke, F. W.: "The Data of Geochemistry," Bull. 770, 5th ed., U.S. Geological Survey, 1924.

Dunnington, H. V.: Generation, Migration, Accumulation and Dissipation of Oil in Northern Iraq, in "Habitat of Oil," special volume, pp. 1194–1251, American Association of Petroleum Geologists, 1958.

Dunstan, A. E. (ed.), et al.: "The Science of Petroleum," 6 vols., Oxford University Press, New York, 1938–1953.

Emmons, William H.: "Geology of Petroleum," 2d ed., McGraw-Hill Book Co., Inc., New York, 1931.

Forsman, James F., and John M. Hunt: Insoluble Organic Matter (Kerogen) in Sedimentary Rocks of Marine Origin, in "Habitat of Oil," special volume, pp. 747–778, American Association of Petroleum Geologists, 1958.

Frazer, H. J.: Experimental Study of the Porosity and Permeability of Clastic Sediments, *J. Geol.*, vol. 43, November–December, 1935.

Gussow, William C.: Differential Entrapment of Oil and Gas—A Fundamental Principle, *Bull. AAPG*, May, 1954.

Gussow, William C.: Time of Migration of Oil and Gas, *Bull. AAPG*, vol. 39, May, 1955.

Harrison, A. G., and H. G. Thode: Sulphur Isotope Abundances in Hydrocarbons and Source Rocks of Uinta Basin, Utah, *Bull. AAPG*, vol. 42, November, 1958.

Heald, K. C.: Essentials for Oil Pools, in "Elements of the Petroleum Industry," Sealy W. Mudd Series, American Institute of Mining and Metallurgical Engineers, 1940.

Hedberg, Hollis D.: Gravitational Compaction of Clays and Shales, *Am. J. Sci.*, fifth series, no. 184, April, 1936.

Heroy, William B.: Petroleum Geology, *Bull. Geol. Soc. Am.*, fiftieth anniversary volume, June, 1941.

Hobson, G. D.: Biochemical Aspects of the Origin of Petroleum, in "The Science of Petroleum," Oxford University Press, New York, 1938.
Hoots, Harold W.: Origin, Migration and Accumulation of Oil in California, *Calif. Div. Mines Bull.* 118, 1943. Reprinted 1948.
Hubbert, M. King: Entrapment of Petroleum under Hydrodynamic Conditions, *Bull. AAPG*, vol. 37, August, 1953.
Hunt, John M., and George W. Jamieson: Oil and Organic Matter in Source Rocks of Petroleum, in "Habitat of Oil," special volume, pp. 735-746, American Association of Petroleum Geologists, 1958.
Illing, V. C.: The Origin of Petroleum, in "The Science of Petroleum," vol. 1, Oxford University Press, New York, 1938; The Migration of Oil, *ibid.*
Katz, D. L., R. R. Monroe, and R. P. Trainer: Surface Tension of Crude Oils Containing Dissolved Gases, *Petrol. Technol.*, vol. 6, no. 5 (AIME Tech. Publ. 1624), 1943.
Kidwell, Albert L., and John M. Hunt: Migration of Oil in Recent Sediments of Pedernales, Venezuela, in "Habitat of Oil," special volume, pp. 790-817, American Association of Petroleum Geologists, 1958.
Knebel, G. M.: Progress Report of A.P.I. Research Project 43: The Transformation of Organic Material into Petroleum, *Bull. AAPG*, vol. 30, November, 1946.
Krumbein, W. C.: Principles of Facies Map Interpretation, *J. Sediment. Petrol.*, vol. 22, December, 1952.
Krumbein, W. C., and L. L. Sloss: "Stratigraphy and Sedimentation," W. H. Freeman and Co., San Francisco, 1950.
Krumbein, W. C., and R. M. Garrels: Origin and Classification of Chemical Sediments in Terms of pH and Oxidation-reduction Potentials, *J. Geol.*, vol. 60, January, 1952.
Kuenan, Philip H.: "Marine Geology," John Wiley & Sons, Inc., New York, 1950.
Landes, Kenneth K.: "Petroleum Geology," 2d ed., John Wiley & Sons, Inc., New York, 1959.
Levorsen, A. I.: Time of Oil Accumulation, *Bull. AAPG*, vol. 29, August, 1945.
Levorsen, A. I.: "Geology of Petroleum," W. H. Freeman and Co., San Francisco, 1954.
Muskat, Morris: "Flow of Homogeneous Fluids Through Porous Media," McGraw-Hill Book Co., Inc., New York, 1937.
Newell, Norman E., et al.: "The Permian Reef Complex of the Guadalupe Mountains Region, Texas and New Mexico," W. H. Freeman and Co., San Francisco, 1953.
Pettijohn, F. J.: "Sedimentary Rocks," Harper & Brothers, New York, 1949.
Pratt, Wallace E., and Dorothy Good (eds.): "World Geography of Petroleum," Special Publication 31, American Geographical Society, Princeton University Press, Princeton, N.J., 1950.
Sanders, J. McConnell: The Microscopical Examination of Crude Petroleum, *J. Inst. Petrol. Technologists*, vol. 23, 1937.
Smith, Paul V., Jr.: Studies on the Origin of Petroleum: Occurrence of Hydrocarbons in Recent Sediments, *Bull. AAPG*, vol. 38, March, 1954.
Sutton, F. A.: Geology of the Maracaibo Basin, Venezuela, *Bull. AAPG*, vol. 30, October, 1946.
Sverdrup, H. V., M. W. Johnson, and R. H. Fleming: "The Oceans: Their Physics, Chemistry and General Biology," Prentice-Hall, Inc., Englewood Cliffs, N.J., 1942.
Swartz, C. A.: The Variation in the Surface Tension of Gas-saturated Petroleum with Pressure of Saturation, *Physics*, vol. 1, pp. 245-253, 1931.
Thode, H. G., Jan Monster, and H. B. Dunford: Sulphur Isotope Abundances in Petroleum and Associated Materials, *Bull. AAPG*, vol. 42, November, 1958.
Trask, Parker D.: Organic Content of Recent Marine Sediments, in "Recent Marine Sediments," special volume, American Association of Petroleum Geologists, 1939.
Trask, Parker D., and H. Whitman Patnode: "Source Beds of Petroleum," American Association of Petroleum Geologists, 1942.
Trask, Parker D.: Dynamics of Sedimentation, in "Applied Sedimentation," John Wiley & Sons, Inc., New York, 1950.
Twenhofel, William H.: "Treatise on Sedimentation," 2d ed., The Williams & Wilkins Company, Baltimore, 1932.
Van Tuyl, F. M., and Ben H. Parker: The Time of Origin and Accumulation of Petroleum, *Quart. Colo. School Mines*, vol. 36, April, 1941.
Van Tuyl, F. M., Ben H. Parker, and W. W. Skeeters: The Migration and Accumulation of Petroleum and Natural Gas, *Quart. Colo. School Mines*, vol. 40, January, 1945.
Weeks, Lewis G.: "Principles of Basin Development, Sedimentation and Oil Occurrence," an unpublished "Red Book," 1948. Copies distributed by Standard Oil Co., N.J., to university, college, etc., libraries.
Weeks, Lewis G.: Concerning Estimates of Potential Petroleum Reserves, *Bull. AAPG*, vol. 34, October, 1950.

Weeks, Lewis G.: Factors of Sedimentary Basin Development That Control Oil Occurrence, *Bull. AAPG*, vol. 36, November, 1952.
Weeks, Lewis G.: Fuel Reserves of the Future, *Bull. AAPG*, vol. 42, February, 1958.
Weeks, Lewis G.: Habitat of Oil and Some Factors That Control It, in "Habitat of Oil," special volume, pp. 1–61, American Association of Petroleum Geologists, 1958.
Weeks, Lewis G.: World Mineral Fuel Areas, in "Encyclopedia of Science and Technology," 15 vols., McGraw-Hill Book Company, Inc., 1961.
ZoBell, Claude E.: The Role of Bacteria in the Formation and Transformation of Petroleum Hydrocarbons, *Science*, vol. 10, October, 1945.
ZoBell, Claude E.: Function of Bacteria in the Formation and Accumulation of Petroleum, *Oil Weekly*, Feb. 18, 1946.

Chapter 6

ENTRAPMENT OF PETROLEUM

By David G. Willis

Manager, Research Computer Systems, Lockheed Missiles and Space Division. Research Associate in Geophysics, Stanford University

Atherton, Calif.

In the hundred years which have elapsed since the drilling of the Drake well in Pennsylvania, there have been discovered throughout the world many thousands of commercial accumulations of oil and natural gas. They have been found at depths ranging from the surface of the ground to more than 15,000 feet and in rocks ranging in age from Precambrian to Pleistocene. In the case of oil they have varied in size from a few hundred to more than 20 billion barrels of liquid petroleum. Each of these accumulations is the result of a particular combination of geological and physical circumstances by which the oil or gas has been held in a stable configuration or trap underground.

The principal problem of petroleum exploration is to determine the location of undiscovered traps for oil and gas. Hence, among the most valuable tools which can be possessed by the petroleum exploration geologist are a basic understanding of the physical principles governing the entrapment of oil and gas and a wide familiarity with the various geological factors which have contributed to the formation of already discovered traps.

In this chapter are presented a review of our modern understanding of the physics of petroleum and natural gas entrapment and its historical development, and a detailed examination of the various geological factors which may contribute to the formation of traps together with examples demonstrating the wide variety of possible combinations of these factors which may occur.

The idea that oil and gas underground, being lighter than the surrounding water, should be moved by buoyant forces to the structurally highest position in a permeable rock has dominated the prevalent scientific thinking ever since underground petroleum accumulations were discovered. Its first expression was in the "anticlinal theory," which became clearly formulated (Howell, 1934) after the Drake discovery in 1859 and which found wide acceptance as an exploration tool prior to 1900. Since anticlines, with their concave downward geometry, are perhaps the most obvious of the possible geologic configurations which might form a trap, it is not surprising that the attention of petroleum explorers was first directed to them. Later, with the discovery of other kinds of traps in which geologic factors such as faults, permeability pinch-outs, and unconformities were responsible for producing an adequate trapping geometry, the anticlinal theory was extended to include a wider variety of possible geologic circumstances. In its expanded form this has been termed the "structural theory" or "trap theory" of petroleum accumulation and can be summarized by the statement that a trap will be formed wherever rocks permeable to oil and gas underlie rocks impermeable to oil and gas with their common boundary forming a surface which is concave downward.

Although the anticlinal and structural theories were widely accepted and applied ever since their formulation, an alternative minority viewpoint was propounded and discussed in the literature for almost 30 years commencing in about 1910 (Munn, 1909; Mills, 1920; Rich, 1921, 1934). This was the "hydraulic theory" of petroleum accumulation. Its basic tenets were that the flushing action of moving ground water was a major factor causing the migration of petroleum and that accumulation would occur wherever natural forces, either buoyancy or a decrease in permeability of the rocks, would impede the force of the water. Thus accumulations might be found on structural terraces or the sides of anticlines or associated with changes in permeability.

Recently, a comprehensive physical theory of the entrapment and accumulation of petroleum and natural gas has been developed by M. King Hubbert (1940, 1953). He has shown that the basic principle of the structural theory, whereby oil and gas migrate to the highest position of the reservoir rock, is a special case and is valid only under hydrostatic conditions, in which the environmental ground water is not in motion. In a hydrodynamic environment accumulations of oil or gas will occur in which the oil-water or gas-water contact is tilted in the direction of water motion. In some cases this tilt may be so great that a configuration which would trap oil or gas under hydrostatic conditions will now be incapable of holding those fluids. A review of already discovered petroleum and natural gas accumulations has revealed many examples in which these effects are strongly apparent. The following discussion of the physical principles of petroleum and natural gas accumulation is based primarily upon the work of Hubbert.

THE PHYSICAL PRINCIPLES GOVERNING ENTRAPMENT OF OIL AND GAS

Commercial accumulations of petroleum and gas invariably occur in or adjacent to sedimentary rocks, mostly of marine origin. These rocks are to a greater or lesser extent porous and permeable, and they are almost entirely saturated with water from very near the surface of the ground down to depths at which their porosity disappears. There exists strong evidence, based on the work of P. V. Smith, Jr. (1954), indicating that petroleum or natural gas has originated as a myriad of small particles or globules which were widely disseminated throughout certain of these water-saturated sedimentary rocks soon after their deposition. Migration of these initial small particles into large accumulations of oil and gas took place through the water-saturated environment. Although we do not know the exact process by which this migration took place, it is at least possible to describe the forces which were acting on the small globules and, what is of particular interest, the positions and configurations in which they approach equilibrium and become stable.

It is a fundamental physical principle implicit in the second law of thermodynamics that any mechanical system, no matter how complicated, will always evolve through a succession of configurations such that the sum of its kinetic and potential energies will become smaller and smaller by gradual conversion to heat. Furthermore, it will approach and eventually come to rest in an equilibrium configuration such that the kinetic energy is zero and the potential energy is at a minimum with respect to all small changes compatible with the constraints of the system. There exist many familiar examples of this process such as the slowing and stopping of a swinging pendulum, water flowing downhill into a lake, or a rolling ball coming to rest. Since underground fluids, water, oil, and natural gas in their environment constitute a mechanical system, they behave in this way. Although the motions of these fluids are so slow that kinetic energies are for all practical purposes zero, each element of fluid will possess an amount of mechanical potential energy which is principally a function of its position. It will in general be acted on by an unbalanced force, tending to move it from regions in which its potential energy is high to those in which its potential energy is low. It will cease moving and will become stable whenever it reaches a position in which its potential energy is at a minimum and it is surrounded by regions of higher potential energy. In particular, the original small particles of petroleum or natural gas will come to rest and accumulate in their respective positions of equilibrium where they are surrounded by regions in which their potential energies are higher. We can, therefore, define a trap for oil or gas as an underground locality in which the potential energy of those fluids is at a minimum with respect to their immediate surroundings. Such a configuration is shown in schematic form in Figure 6-1.

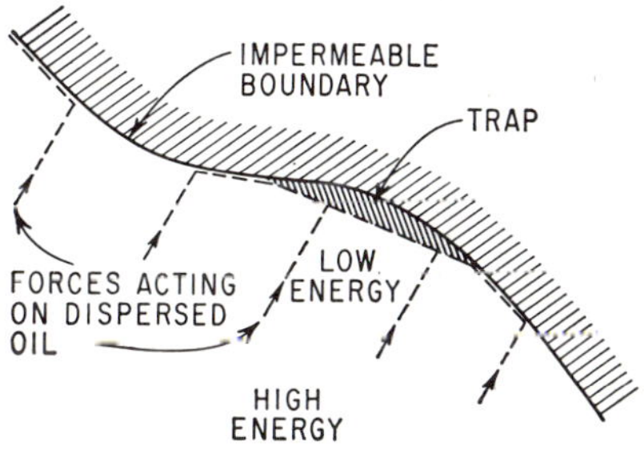

FIGURE 6-1. Schematic diagram of a trap showing the manner in which dispersed oil migrates from a high-energy region to a low-energy region.

To determine the location and configuration of traps it is first necessary to know the potential energies of the fluids of interest—water, oil, and gas in their underground environment. Inasmuch as the accumulation and migration of oil and natural gas take place in a principally water-saturated environment, the potential energy of water should be considered first.

Potential Energy of Water

Each small element of water underground will possess mechanical potential energy by virtue of both its elevation and the pressure existing within it. When it is referred to an element of unit mass, this potential energy is called the potential of the water. The value of the potential can be determined by calculating the work which would be required to bring a unit mass of water from some standard elevation and pressure to its present elevation and pressure. If the standard is taken as sea level and atmospheric pressure, it is first necessary to lift the unit mass to its present elevation z above sea level against a force equal to the acceleration of gravity g. It is then necessary to inject the unit mass of water, which has a volume equal to the reciprocal of its density ρ, against the local water pressure p. Therefore, the total work required for these two operations, and hence the potential ϕ of the water, is given by the expression

$$\phi = gz + \frac{p}{\rho}$$

A different interpretation of the potential can be obtained by consideration of the height to which water will rise in an open-topped manometer whose base is terminated at the point at which the potential is to be measured (Figure 6-2). When this height is referred to a standard datum of elevation which is again taken as sea level, it is called the "head" of the water and can be denoted by h. The value of the head will be given by the sum of the elevation of the point of interest z and the height of the static fluid $p/\rho g$ which can be supported by the local pressure p, giving

$$h = z + \frac{p}{\rho g}$$

If this expression is compared with the previous one, it is evident that the potential can be obtained simply by multiplying the head at any point by the acceleration of gravity, or

$$\phi = gh$$

Therefore, a manometer can be employed to measure the potential of water at any point. This is of particular value if we wish to determine the distribution of potential within a permeable subsurface stratum. Water wells terminated in the stratum can serve as manometers, and the static height h measured above sea level or other standard datum to which water will rise in the wells provides a direct measurement of the potential. Under hydrostatic conditions, when the water is not in motion, the head, and hence the potential, will have the same value wherever it is measured.

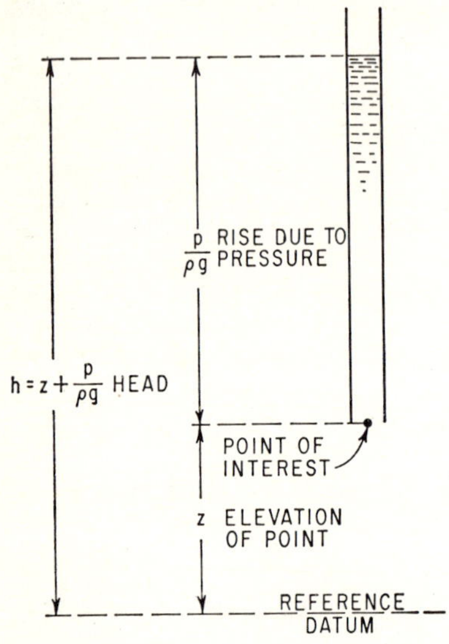

FIGURE 6-2. Relationship between head and pressure in an open-topped manometer.

Forces Acting on Water

The potential of water will be a scalar variable with a definite value at each point in an undergound space. Furthermore, since it is a function only of elevation and pressure and since the pressure will in general vary continuously, the potential itself will be either a constant or a continuous variable. If it is a continuous variable, surfaces will exist underground on each of which the potential assumes a constant value and which can be termed equipotential surfaces. The potential throughout a region can be described by a family of such equipotential surfaces. If the potential of water is known throughout some region of space, the forces which will act on a unit mass of water throughout that region can be determined. Each element of unit mass will be acted on by a force whose direction is that of the greatest rate of decrease of potential and whose magnitude is equal to the rate of decrease in that direction. Hence the force per unit mass $\mathbf{E}$ acting on the water will be a vector which is normal to the equipotential surfaces and which can be represented at any point by the negative of the gradient of the potential, or

$$\mathbf{E} = -\text{grad}\ \phi$$

The physical situation shown in Figure 6-3 depicts the force field acting on the water by lines of force normal to the equipotential surfaces. An alternative physical conception is obtained if we take the gradient of the expression for the potential for water, or

$$\mathbf{E} = -\text{grad}\,\phi = -\text{grad}\left(gz + \frac{p}{\rho}\right) = \mathbf{g} - \frac{1}{\rho}\text{grad}\,p$$

This is simply a vector equation which asserts that the force acting on a unit mass of water will be equal to the vector sum of the force per unit mass of gravity acting vertically downward and the negative gradient of the fluid pressure multiplied by the reciprocal of fluid density. There are thus two force components whose vector sum constitutes the net force acting on the fluid. The two components, as shown for two different cases in Figure 6-4, provide a simple physical conception of the situation. The first component **g** represents the force of gravity acting vertically downward on a unit mass of the water. The second component $(1/\rho)$ grad p represents the force produced by the negative gradient of the pressure acting on the volume $1/\rho$ occupied by the unit mass of water. In the hydrostatic case, illustrated in Figure 6-4*a*, the

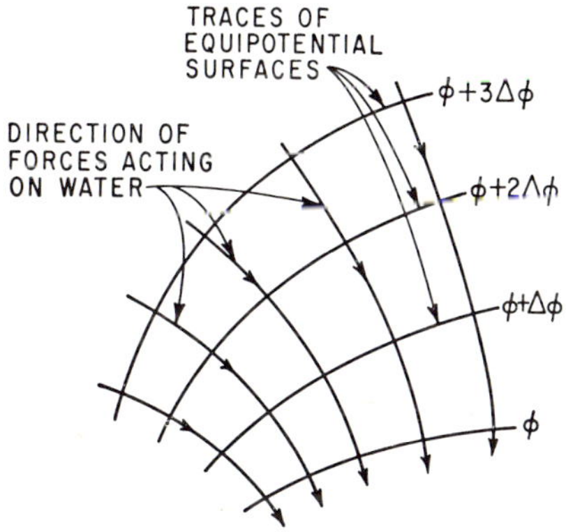

FIGURE 6-3. Section through an underground region showing traces of equipotential surfaces with lines of force normal to them and in the direction of decreasing potential.

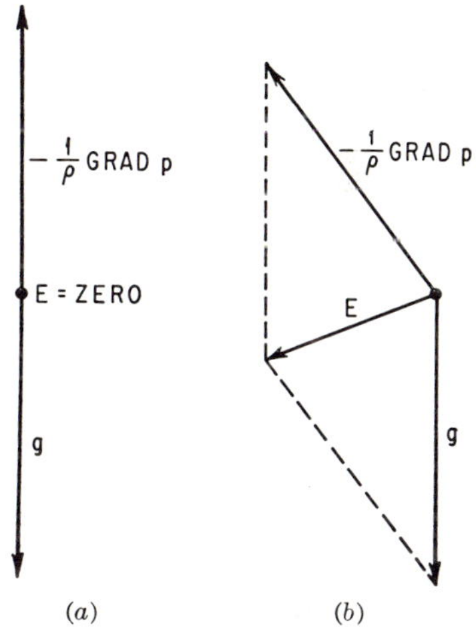

FIGURE 6-4. Forces acting on a unit mass of water in (*a*) hydrostatic and (*b*) hydrodynamic environments.

potential is everywhere constant and these two forces are exactly equal and opposite. The net force acting on the water is therefore zero. In a hydrodynamic situation, on the other hand, the pressure component will not exactly balance the gravitational component and a net unbalanced force will act on the unit mass of water as illustrated in Figure 6-4*b*.

The relationship between the actual motion of the water and the field of force acting on the water is provided by Darcy's law. This is not of direct concern here, and it is sufficient to note that the direction of motion of the water will virtually always coincide with the direction of the force acting on each element of water.

These principles provide a way of determining the state of motion of the water within a permeable subsurface stratum. In such a situation, if the strata immediately above and immediately below the stratum of interest are relatively impermeable to water, the fluid flow will be essentially two-dimensional and will be restricted to the stratum of interest. At each point in the stratum there will exist a value of the potential for the water and a corresponding value of h, or "head," defined by the elevation above sea level to which water would rise in a static well terminated within the stratum. The values of h measured with respect to a standard datum will define a surface which is called the potentiometric, or "potential-measuring," surface. The water will be flowing in a "downhill" direction on this surface from regions of higher

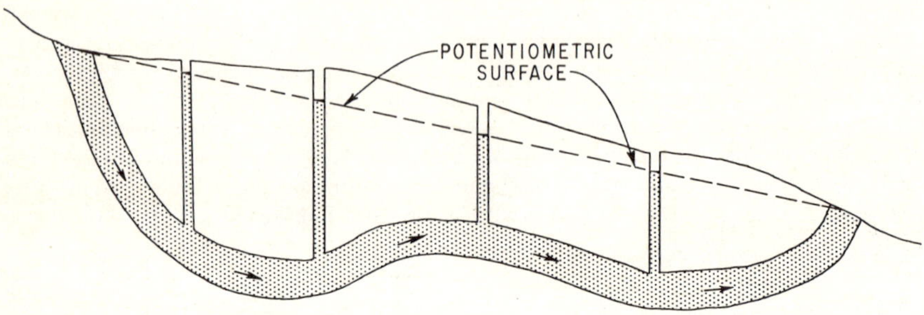

FIGURE 6-5. Schematic geologic section showing water entering the outcrop of a permeable stratum at a high elevation, flowing through it, and leaving it at a lower elevation, with the "head" decreasing in the direction of flow.

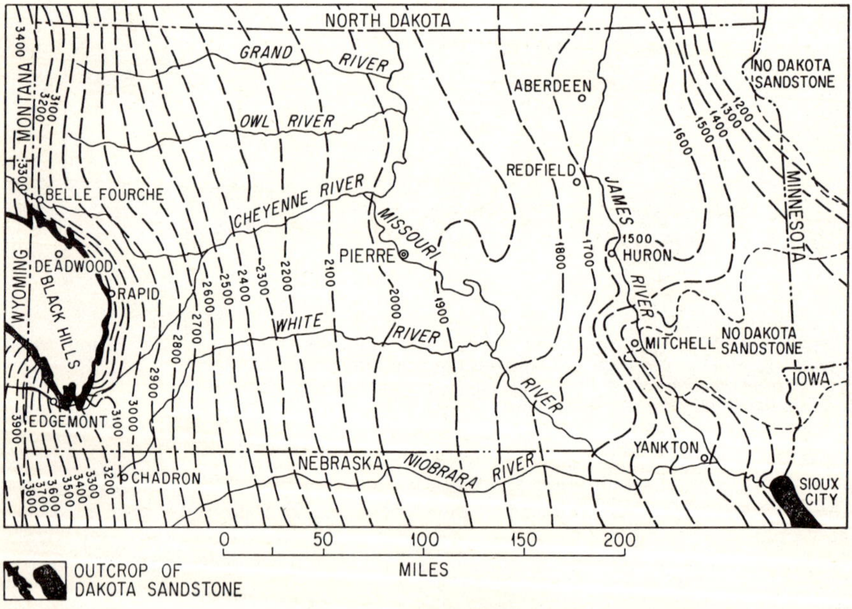

FIGURE 6-6. Regional map of South Dakota showing contours on the potentiometric surface for the Cretaceous Dakota sandstone. Water is entering the outcrop of the Dakota around the Black Hills and flowing in a "downhill" direction on the potentiometric surface. (*Redrawn after N. H. Darton, U.S. Geol. Survey Water Supply Paper 227, p. 61.*)

potential or h to regions of lower potential or h. The situation is depicted in Figure 6-5, which shows water entering the outcrop at a higher elevation, flowing through a permeable stratum, and leaving the surface at a lower elevation. The potentiometric surface is defined by wells terminated within the stratum. If sufficient measurements exist from wells within the same stratum over a large region, a contour map can be constructed which accurately depicts the regional motion of the ground water within that stratum. If, as is frequently the case, it is not convenient to allow the water to enter a well and rise to a static height which could be measured, the potential, or head, can still be determined by a measurement of the underground fluid pressure within the stratum of interest together with the elevation at which the pressure was measured.

Unfortunately few data of this kind exist. One of the few extensive compilations of potentiometric measurements is that performed by N. H. Darton (1905, 1909, 1918, 1918) in his classic investigations of ground water in the Cretaceous Dakota sandstone of the Great Plains region of the United States.

A map of a portion of this region is shown in Figure 6-6 on which are plotted equipotential lines as determined by Darton. It is clear from this map that water is entering the outcrop of the Dakota sandstone in the vicinity of the Black Hills and flowing eastward across the entire state of South Dakota.

Water of Variable Salinity

Implicit in the foregoing discussion is the assumption that the density of the water is constant throughout the region under consideration. In many situations, however, the ground water will contain varying amounts of dissolved solids which will, in turn, cause its density to vary from point to point. In such a case it is not possible to use a scalar potential to describe the forces acting on the water. The vector expression

$$\mathbf{E} = g - \frac{1}{\rho} \operatorname{grad} p$$

will always correctly describe the force acting on a unit mass of fluid under any conditions. Hence, if the pressure and density throughout a region are known, the force field can always be determined by its use.

However, a scalar potential of ϕ related to the force $\mathbf{E}$ by the expression

$$\mathbf{E} = -\operatorname{grad} \phi$$

will exist only if the density of the fluid is constant or, as a special case, if the density varies only as a function of pressure.

In a region of variable salinity and hence variable density the best procedure is to approximate the force field acting on the water by determining the potential for a reference water of constant density by the expression

$$\phi = gz + \frac{p}{\rho}$$

This potential will define the force which would act at any point on water having the reference density and will coincide with the actual force acting on the water only where the actual density is the same as the reference density. If the actual density is nearly the same as the reference density, the force field, as defined by the potential, will be approximately correct. If the actual density varies too widely throughout a region, it is possible to separate it into smaller regions and define a different potential for a different reference water within each of them and thereby obtain a closer approximation to the actual force field.

Potential Energies of Oil and Gas

The potentials of oil and gas can be determined in a fashion very similar to that of water. However, since these fluids in their initial dispersed state exist in a principally water-saturated environment, their energies will be modified by the effects of surface tension, which must therefore be taken into consideration for both oil and gas. In the case of gas, the additional effect due to the compressibility of that fluid must also be considered.

In the consideration of the effects of surface tension one of the most important factors is the wettability of the rocks containing the fluids of interest. In rocks which are hydrophilic or preferentially water wet, the water will tend to remain in contact with the rock grains, leaving the oil or gas as small globules or particles within the pore spaces. In rocks which are hydrophobic and preferentially oil wet, the oil or gas will tend to disperse itself along the surfaces of the rock, leaving the water as globules within the pore spaces. In almost every situation of interest the rocks are preferentially water wet and the oil exists as globules within the pore spaces.

In general, the fluid pressure within small particles or globules of oil or gas will be somewhat higher than that of the surrounding water because of the surface tension existing between the globules and the water. This difference is called the capillary pressure P_c and can be calculated from Plateau's equation

$$P_c = \gamma \left(\frac{1}{r_1} + \frac{1}{r_2} \right)$$

in which γ is the interfacial tension between the fluids and r_1 and r_2 are the principal radii of curvature of the surface.

For spherical globules this reduces to

$$P_c = \frac{2\lambda}{r}$$

in which r is the radius of the globule.

It is evident that small globules with small radii of curvature will have a high capillary pressure and hence a high potential energy per unit mass whereas larger globules with larger radii of curvature will have a lower capillary pressure and lower potential energy per unit mass. In fine-grained rocks with small pores the globules will, of necessity, assume geometric shapes having small radii of curvature and will therefore have a high potential energy per unit mass. In coarse-grained large-pore rocks, on the other hand, the globules will assume geometric shapes with larger radii of curvature and hence lower potential energy. Therefore, in a water-saturated environment the fine-grained rocks are regions of high potential energy for oil and gas globules whereas the coarse-grained rocks are regions of lower potential energy for those fluids. At boundaries between fine- and coarse-grained rocks, for example shale-sand interfaces, a sharp gradient will exist in the potential energy of those fluids and a consequent strong force tending to impel them from the fine-grained rock into the coarse-grained rock. Conversely, for oil or gas globules existing within the coarse-grained rock the interface will present an energy barrier which renders it effectively impermeable to those fluids. Aside from this effect and as long as there are no sharp changes in the grain size of the rocks through which movement of oil and gas takes place, the effects of capillary pressures can be ignored without serious error in determining potentials for oil and gas.

Therefore, in the case of oil we can determine its potential energy per unit mass by computing the amount of work required to bring it from standard elevation and pressure to the elevation z and pressure p at the point of interest, or

$$\phi_o = gz + \frac{p}{\rho_o}$$

in which ϕ_o denotes the potential of the oil and ρ_o its density.

Since gas is a compressible fluid, its potential differs from that of a liquid in that additional work must be done in compressing it from its initial standard pressure to the local pressure at the point of interest. If this process is carried out in such a way that the varying density ρ of the gas is a function of pressure alone, the expression for the potential of gas becomes

$$\phi_g = gz + \int_0^p \frac{1}{\rho}\, dp$$

By a procedure entirely analogous to that used for water, the forces acting per unit mass on oil or gas are determined by the negative gradients of their respective potentials at any particular point, giving the vector expressions

$$\mathbf{E}_o = -\operatorname{grad} \phi_o = \mathbf{g} - \frac{1}{\rho_o} \operatorname{grad} p$$

$$\mathbf{E}_g = -\operatorname{grad} \phi_g = \mathbf{g} - \frac{1}{\rho_g} \operatorname{grad} p$$

Again, in each case the force is composed of two components: one, the force of gravity acting vertically downward on the unit mass of fluid, and the other, the negative gradient of the fluid pressure acting on the volume occupied by the unit mass of

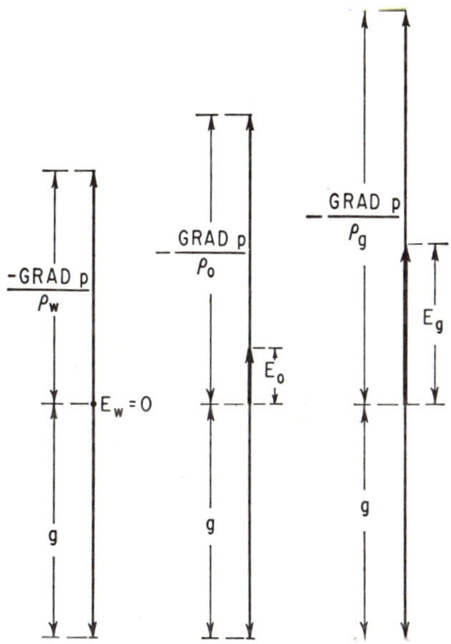

FIGURE 6-7 Forces acting on gas, oil, and water at the same point in a hydrostatic environment.

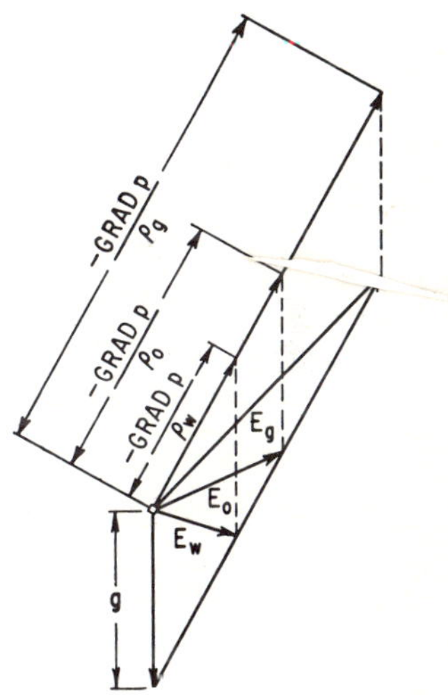

FIGURE 6-8. Forces acting on gas, oil, and water at the same point in a hydrodynamic environment.

fluid. It is of interest to examine the special case of hydrostatics, in which the environmental ground water is not in motion (Figure 6-7). In this case, there will be no net force acting on a unit mass of water, the vertically downward force of gravity being exactly balanced by an upward force in the direction of decreasing pressure acting on the volume occupied by a unit mass of water. However, for oil and gas, which both have less density than water and which therefore occupy a larger volume per unit mass, the upward force due to the negative pressure gradient will be greater and these fluids will tend to rise vertically, the gas much more strongly than the oil. This is the familiar situation of hydrostatics and represents a valid interpretation of the mechanics of oil and gas accumulation whenever the environmental ground water is not in motion.

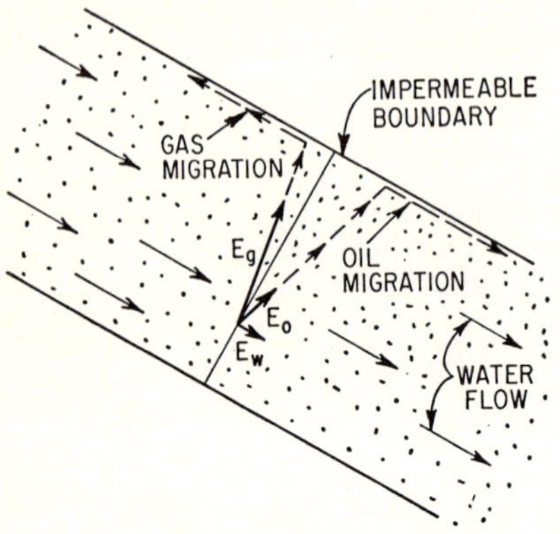

FIGURE 6-9. Schematic section showing conditions under which oil and gas could migrate in divergent directions in an underground permeable stratum.

The physical relationship between the three different forces which would act on the three fluids water, oil, and gas at the same point in a hydrodynamic situation is shown in Figure 6-8. It is clear that the first component **g** will be the same in each case while the second component, that due to the pressure gradient, will have the same direction in each case but will have a magnitude which is dependent upon the density of the particular fluid. Therefore, the resultant force vector will, in general, be different in both magnitude and direction for each of the three fluids, and they will tend to be driven in three separate directions.

It is easy to imagine a situation in which the oil and gas will move through a stratum in opposite directions and therefore end up in separate traps. This is illustrated in Figure 6-9 in which water is flowing down the dip in such a way that globules of oil will be moved down the dip whereas the gas will migrate up the dip.

Delineation of Traps

A trap will exist in any region in which the potential of oil or gas is at a minimum with respect to its surroundings. In order to delineate the positions and configurations of these regions, it is convenient to express the potential of oil or gas in terms of that of the environmental ground water. The potentials of oil and water are given by

$$\phi_w = gz + \frac{p}{\rho_w}$$

$$\phi_o = gz + \frac{p}{\rho_o}$$

in which the subscripts denote the particular fluid.

If the first of these equations is solved for p and substituted into the second,

$$\phi_o = \frac{\rho_w}{\rho_o}\phi_w - \frac{\rho_w - \rho_o}{\rho_o} gz$$

which gives the potential of oil in terms of that of water and the elevation.

A further simplification can be obtained by substituting the expressions

$$\phi_w = gh_w$$

$$\phi_o = gh_o$$

and dividing by g to give

$$h_o = \frac{\rho_w}{\rho_o} h_w - \frac{\rho_w - \rho_o}{\rho_o} z$$

which is the expression for the hydraulic head of oil in terms of the hydraulic head of water and the elevation at any point. In this expression, every term has the units of length; z is the elevation of the point, h_w is the height above datum to which water would stand in a manometer terminated at the point, and h_o the height that oil of density ρ_o would stand. We see that the value of h_o is equal to the sum of two terms. The first is h_w multiplied by a constant coefficient depending upon the densities of the fluids, and the second the elevation multiplied by another constant depending on the densities of the fluids. This relationship permits the determination of the equipotential surfaces for oil provided those for water are known. In order to do this it is convenient to divide the above equation by the factor

$$\frac{\rho_w - \rho_o}{\rho_o}$$

which gives the expression

$$\frac{\rho_o}{\rho_w - \rho_o} h_o = \frac{\rho_w}{\rho_w - \rho_o} h_w - z$$

If two quantities u and v are defined by

$$u = \frac{\rho_o}{\rho_w - \rho_o} h_o$$

$$v = \frac{\rho_w}{\rho_w - \rho_o} h_w$$

then by substitution,

$$u = v - z$$

The surfaces represented by constant values of u will coincide with the oil-equipotential surfaces, whereas those corresponding to constant v will coincide with water-equipotential surfaces. The surfaces along which z is constant are simply horizontal

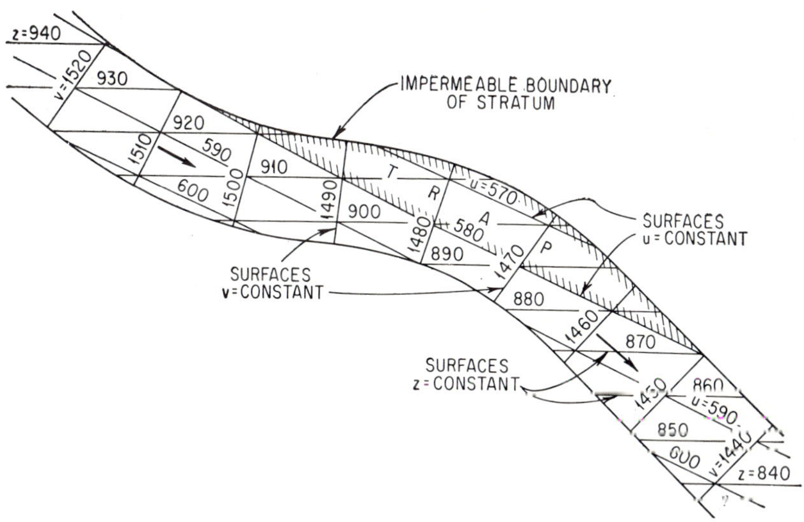

FIGURE 6-10. Section through an underground permeable stratum showing the relationship between surfaces of constant value u, v, and z. A trap is formed by the closed region of minimum value u.

surfaces with elevation z. The graphical expression of this relationship as it might exist in an underground permeable stratum is shown in Figure 6-10. It is clear that the oil-equipotential surfaces will be inclined and tilted at some angle to the horizontal in the direction in which the water is flowing. Since the oil will be impelled by a force perpendicular to the oil-equipotential surfaces and in the direction of decreasing potential, a trap will be formed wherever impermeable strata form a surface which is concave in the direction of increasing oil potential.

A straightforward method for determining the positions and configurations of traps underground is provided by the families of curves defined by the intersection of the surfaces of constant value u, v, and z with the upper surface of the permeable stratum. It is only necessary to have a structure-contour map of the region on the top of the stratum of interest and to know the configuration of the potentiometric surface referred to the upper surface of the stratum. The family of surfaces of constant z will form traces on the upper surface of the stratum which are simply the contour lines of the structure map. The lines of constant value v will coincide with the water-equipotential lines, and the values v on these lines can be obtained simply by multiplying the head by a suitable density factor. The difference between the values of v and z can be determined at the intersection of these two families of lines, and lines of constant value u along the upper surface of the reservoir can be drawn directly. These lines of constant value u will coincide with the oil-equipotential lines, and oil will be deflected along the upper boundary of the reservoir in the direction of decreasing u. Wherever the lines form a closed region of minimum u or oil-potential value, this area will represent a trap, since the oil cannot escape without moving from lower to higher potential. This procedure is illustrated in Figure 6-11.

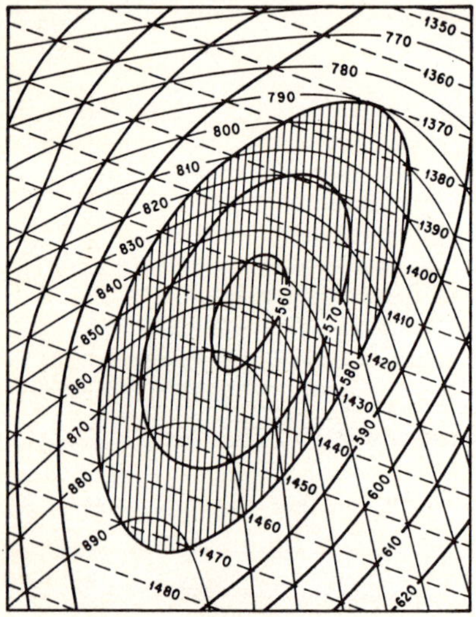

—— LINES OF CONSTANT VALUE z (STRUCTURE CONTOURS)
– – – LINES OF CONSTANT VALUE v (WATER EQUIPOTENTIALS)
—— LINES OF CONSTANT VALUE u (OIL EQUIPOTENTIALS)
▥ TRAP

FIGURE 6-11. Map showing relationship between traces of surfaces of constant value u, v, and z intersecting the upper surface of a permeable stratum. A trap is formed by closed regions of minimum value u.

This example has been worked out in terms of oil equipotentials for the delineation of oil traps. The problem of delineating gas traps is theoretically more complicated, by virtue of the fact that gas is not a constant-density fluid. The potential for gas strictly should depend upon the variations in density with local pressure. However, for practical purposes only slight error will result from considering gas as a constant-density fluid, treating it in exactly the same way as oil, and substituting the gas density ρ_g for the oil density ρ_o in all the expressions used. In this case the value h_o would represent the height to which a fluid with the density of gas would rise in a manometer terminated at the point of interest.

Configurations of Traps

By the foregoing procedure it is possible to delineate the traps for oil or gas in any region from structure-contour maps and a knowledge of the hydrodynamic situation within the strata. The oil- or gas-equipotential surfaces will be inclined or tilted in the direction of water motion, and a trap will be formed by any impermeable surface which is concave in the direction of increasing potential for oil or gas. It is of particular interest to determine the magnitude of the tilts of the oil- or gas-equipotential surfaces in terms of the slope of the potentiometric surface for water. There are several ways to obtain an expression for these quantities, but perhaps the most straightforward is that illustrated in Figure 6-12. A stable oil-water interface is shown in which the oil is at rest and the water is in motion below it. Since the oil is at rest, this surface itself will be an oil-equipotential surface. Two water manometers are separated by a horizontal distance Δx and are terminated in the water immediately below the interface, where the fluid pressures are p_1 and p_2, respectively. The difference in static level of the water in the manometers is given by Δh_w, and the difference of elevation of their bases is denoted by Δz. The difference in pressure Δp between p_1 and p_2 can be determined from the heights of static water supported by the manometers, or

$$p_2 - p_1 = \Delta p = \rho_w g(\Delta z - \Delta h_w)$$

Alternatively, except for minor differences caused by capillary effects, the fluid pressures within the oil will be very nearly equal to those in the adjacent water and Δp can be computed from the difference in elevation within the static oil. Hence,

$$\Delta p = \rho_o g \, \Delta z$$

Equating these two expressions, dividing by $\rho_w g \, \Delta x$, and combining terms give

$$\frac{\Delta z}{\Delta x} = \frac{\rho_w}{\rho_w - \rho_o} \frac{\Delta h_w}{\Delta x}$$

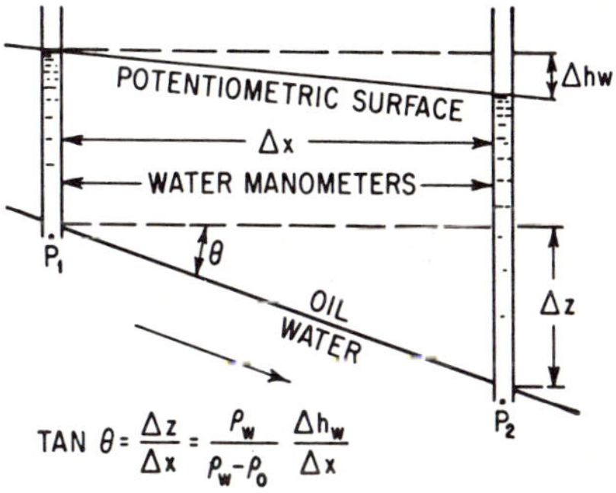

FIGURE 6-12. Diagram showing manner by which the equation describing the tilt of the oil-water interface in a hydrodynamic situation can be derived. Two water manometers are terminated in the water just below the interface.

which is the equation for the tilt of the oil-water interface in terms of the hydrodynamic gradient or slope of the potentiometric surface. The tilt of a gas-water interface will be given by exactly the same expression provided the density of the gas ρ_g is used in place of that of oil.*

In either case the interface and hence the equipotential surface for the particular fluid are equal to the slope of the water-equipotential surface multiplied by a dimensionless coefficient. This coefficient $\rho_w/(\rho_w - \rho_o)$ is therefore called the tilt-amplification factor. It may range from a value of very nearly 1 for some of the lighter gases to infinity (corresponding to complete instability for fluids the same density as water). In general the value will be between 1 and 2 for gases and between 7 and 15 for oils of various densities. For an average figure oil commonly has a density not far from 0.9 gram per cubic centimeter, giving a tilt-amplification factor of 10.

The magnitude of the hydrodynamic gradient or slope of the potentiometric surface is largely dependent upon the regional topography and the locations of outcrops of the permeable formations through which water may enter or leave. Thus in general, regions of low relief such as the Gulf Coast may have hydrodynamic gradients from zero to 2 or 3 feet per mile. In regions of great relief such as the sedimentary basins of the Rocky Mountains the gradients may be as high as 100 feet per mile. All gradations between these situations are possible.

* The foregoing method of deriving the Hubbert tilt equation is somewhat simpler than that given in either of the Hubbert (1940, 1953) papers. However, it was first derived jointly by Hubbert and his assistant Jerry Conner and is frequently used by Hubbert himself in private discussions and lectures.

Applying the tilt-amplification factors to these gradients, it is evident that oil-equipotential surfaces may be tilted as much as 800 to 1,000 feet per mile in extreme situations whereas tilts of gas-equipotential surfaces may be as much as 100 to 200 feet per mile in the same situation.

As a general consideration it is evident that if water is flowing through some stratum in a particular direction, the oil-equipotential surfaces will be tilted in that direction a certain amount and gas-equipotential surfaces will be tilted a lesser amount in the same direction. Only under hydrostatic conditions will the equipotential surfaces for gas and oil, and hence the traps for these fluids, exactly coincide.

Some of the configurations which may occur can be illustrated by considering the various positions of oil and gas traps associated with the same underground geometry

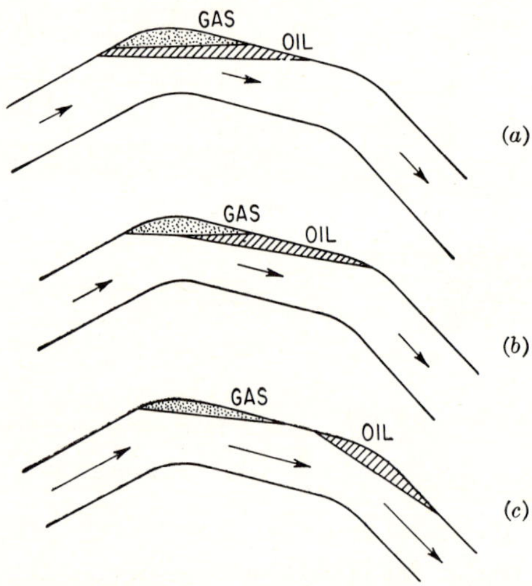

FIGURE 6-13. Schematic section showing several configurations of oil and gas traps which may occur under different hydrodynamic conditions with the same underground geometry.

but under different conditions of fluid flow. Such a situation is depicted in Figure 6-13. Under conditions of a relatively gentle hydrodynamic gradient in a permeable stratum the oil-water interface will be tilted slightly in the direction of flow. Since the oil and gas are both static, the gas-oil contact will be horizontal (Figure 6-13*a*). If the hydrodynamic gradient is somewhat stronger, the oil may be tilted sufficiently to cause a gas-water contact to occur, which will be tilted in the same direction as, but a lesser amount than, the oil-water contact (Figure 6-13*b*). Under a very strong hydrodynamic gradient the oil and gas traps may be entirely separated, with the gas on the crest of the structure and the oil in a structural terrace on its side (Figure 6-13*c*). With even stronger gradients the same structure may become incapable of holding either gas or oil.

A different possibility for a hydrodynamic trap is illustrated in Figure 6-14. In this case water is flowing downdip through a homoclinally dipping stratum containing a low-permeability region. In this local region the hydrodynamic gradient will become greater, causing the oil-equipotential surfaces to tilt more steeply. Where these surfaces resume their normal tilt, a region of upward concavity will occur. It is therefore possible for an equipotential surface to form a closed region against the flat upper surface of the stratum, forming a trap as shown.

Summary

It is apparent from the foregoing discussion that the locations and configurations of traps are determined jointly by the underground geometry of the rocks and the state of motion of the environmental ground water. Under hydrostatic conditions a trap will be formed wherever rocks permeable to oil or gas are overlaid by an impermeable surface which is concave downward. Under hydrodynamic conditions the impermeable surface must be concave in the direction of increasing potential for the particular fluid. As an additional possibility arising from permeability changes within

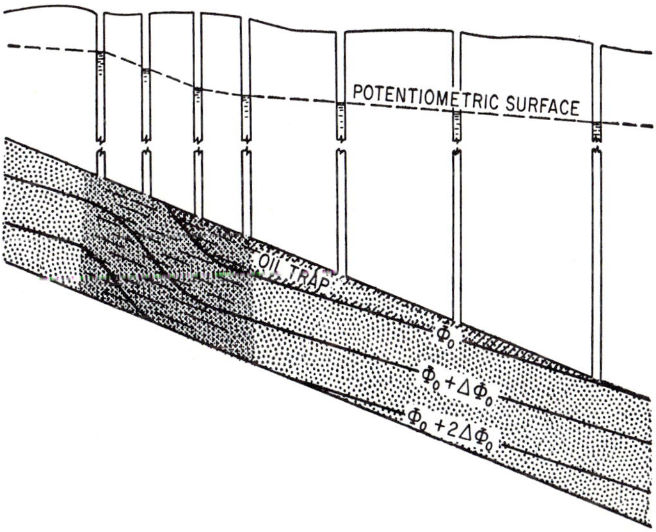

FIGURE 6-14. Schematic section showing the manner in which a trap may be formed by water flowing downdip in a homoclinal stratum through a region of low permeability. (*Redrawn from Hubbert, Bull. AAPG, vol. 37, p. 2001.*)

the rocks, the equipotential surfaces themselves may form a concave geometry terminating against the upper surface of a homoclinally dipping permeable stratum.

THE CLASSIFICATION OF TRAPS

In the previous section were described the physical principles governing the accumulation of oil and gas. It was shown that a trap is simply an underground region in which the potential energy of the oil or of the gas is at a minimum with respect to its surroundings. The geometry of the rocks together with the state of motion of the environmental groundwater jointly determine the location and configuration of these underground regions.

There are perhaps a dozen different geological factors which can contribute to making geometry favorable for a trap, and throughout the world traps have been found in which these factors have been combined in hundreds of different ways. With any group of phenomena as complex as this, schemes of classification are very useful in aiding our understanding. Many classifications of oil and gas traps based on many different kinds of criteria, ranging from age, size, and depth of the accumulation to extremely detailed and exhaustive classifications based on geologic conditions, have been described in the literature (Clapp, 1929; Wilhelm, 1945; Knebel and Rodriguez-Eraso, 1956).

To be of value to the petroleum exploration geologist, a scheme of classification should be based on those criteria which are useful in the exploration for new traps. By far the most widely accepted scheme of classification is of this kind and is based on the distinction between structural and stratigraphic factors, structural factors being those which have arisen by virtue of the deformation of the rocks and stratigraphic factors being those which have arisen by virtue of the conditions under which the rocks themselves were formed. Strictly speaking, almost every trap arises from a combination of both structural and stratigraphic factors. For example, in any sedimentary section, the stratified layers form many surfaces which are capable of acting as impermeable barriers to the movement of oil. One of these stratigraphic surfaces is an important element in almost every trap. Therefore, these traps are in part stratigraphic. Similarly, even a simple tilting of the strata into a homocline constitutes a structural deformation. Since there exist very few traps not containing at least this much of a structural element, almost every trap is also in part structural.

Although almost every trap is strictly both structural and stratigraphic, it has become customary to classify traps as structural if the major elements in their geometry have arisen through folding, faulting, or doming of uniformly stratified rocks and to classify traps as stratigraphic if the major elements in their geometry arise from lateral variations in the strata because of differences in sedimentation, the existence of unconformities, or the existence of anomalous organic deposits. Many traps exist in which structural and stratigraphic factors appear to be of equal importance, and these are commonly classed as combination traps. This three-way classification of structural, stratigraphic, and combination traps is widely used throughout the petroleum industry and is quite useful in obtaining a general understanding of the basic reasons for the existence of various accumulations. Yet in spite of its widespread use, the scheme is rather general and imprecise. In many cases the classification of a trap is to a large degree arbitrary and is a matter of personal opinion.

In the present discussion, therefore, there is no attempt to classify traps as such. Instead, the various possible geologic factors which can contribute to the formation of a trap are discussed in turn, and examples are presented of traps in which each of these factors is significant. At the same time the various examples will serve to show the multitude of ways in which the various factors may combine to form different traps. In this discussion the following breakdown of the various factors which may contribute to the formation of traps is used:

- Structural factors:
  - Folds
  - Faults
  - Intrusive structures—domes
- Stratigraphic factors:
  - Lateral variations in sedimentation
  - Unconformities
  - Organic deposits
- Miscellaneous factors:
  - Fractured reservoirs
  - Secondary chemical alteration
  - Asphaltic seals
- Hydrodynamic effects

Structural Factors

Traps which arise as the result of structural factors are characteristically much more easily discovered than those which arise from stratigraphic factors. Folds, faults,

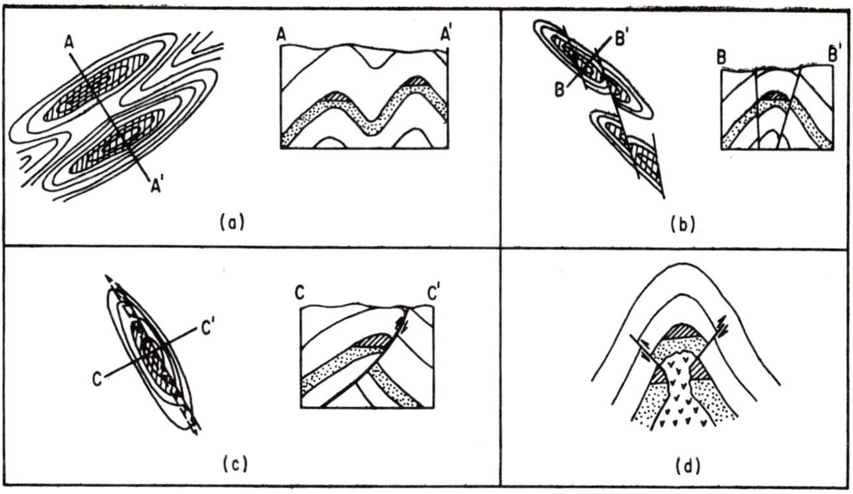

FIGURE 6-15. Schematic diagrams of traps which may be associated with anticlinal structures: (*a*) parallel folding, (*b*) echelon folding and characteristic transcurrent faulting, (*c*) more highly deformed fold with development of thrust faulting, (*d*) diapir fold.

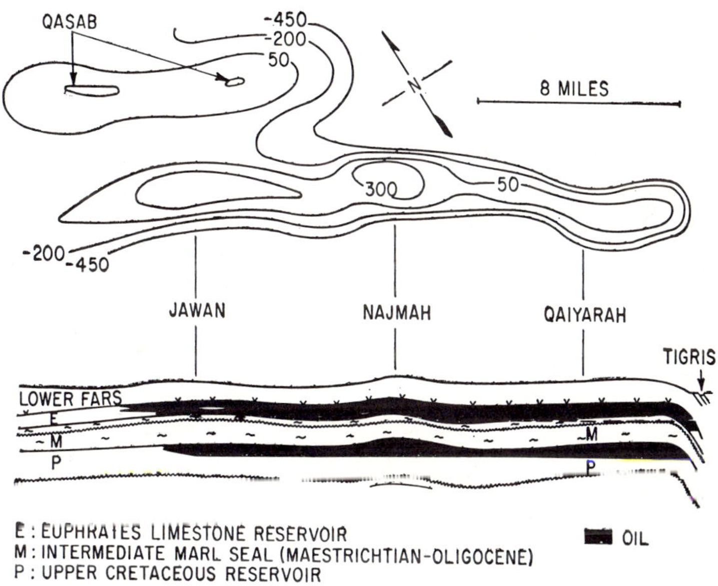

FIGURE 6-16. Qaiyarah oil fields of northern Iraq showing traps associated with parallel anticlinal folds. (*Redrawn from Dunnington, in "Habitat of Oil," p. 1238, AAPG*, 1958.)

and domes are often revealed in the surface geology. They represent strong anomalies in the distribution of subsurface rocks and are therefore easily revealed by geophysical methods. Consequently, when a new basin or province is explored, the structural traps are usually found earliest.

A single structural factor is also likely to produce more than one trap. It may cut across or affect the entire sedimentary section and thus form a trap in any potential

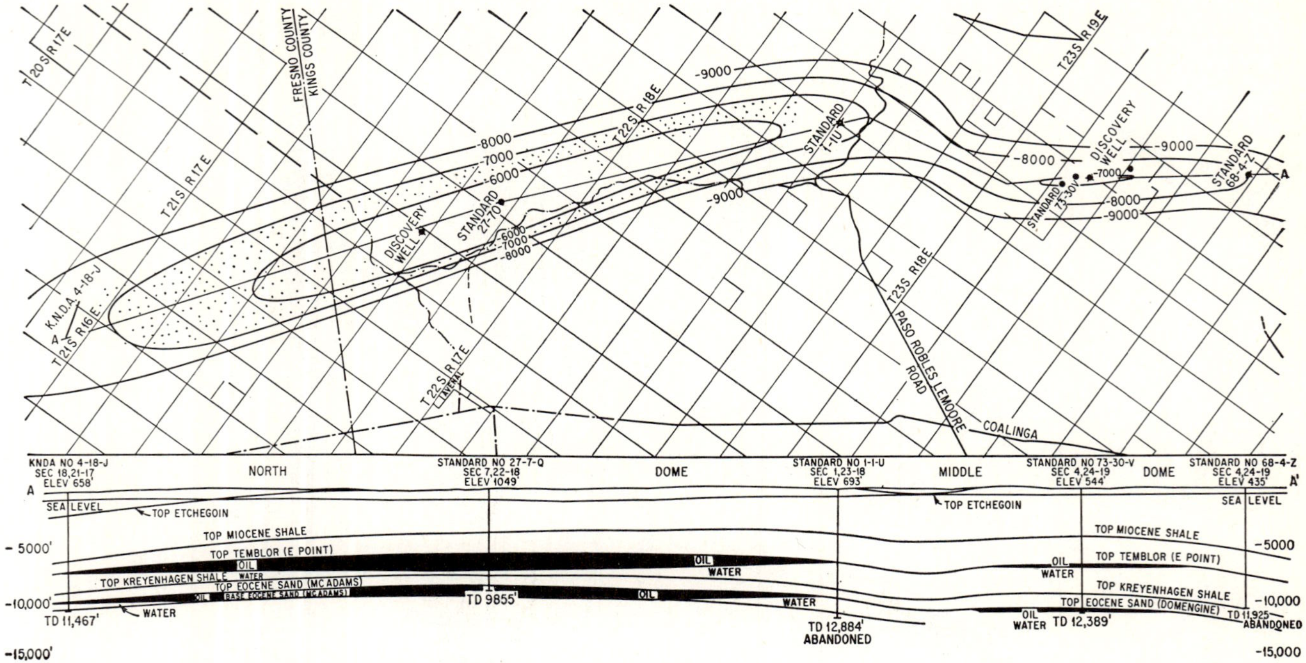

FIGURE 6-17. Structure contour map and longitudinal section of Kettleman Hills oil field, San Joaquin Valley, California, showing traps associated with unfaulted echelon folds. Structure contours are on the top of the Miocene Temblor shale. (*Redrawn from Porter, "1952 Guidebook," pp. 170–171, Pacific Section AAPG.*)

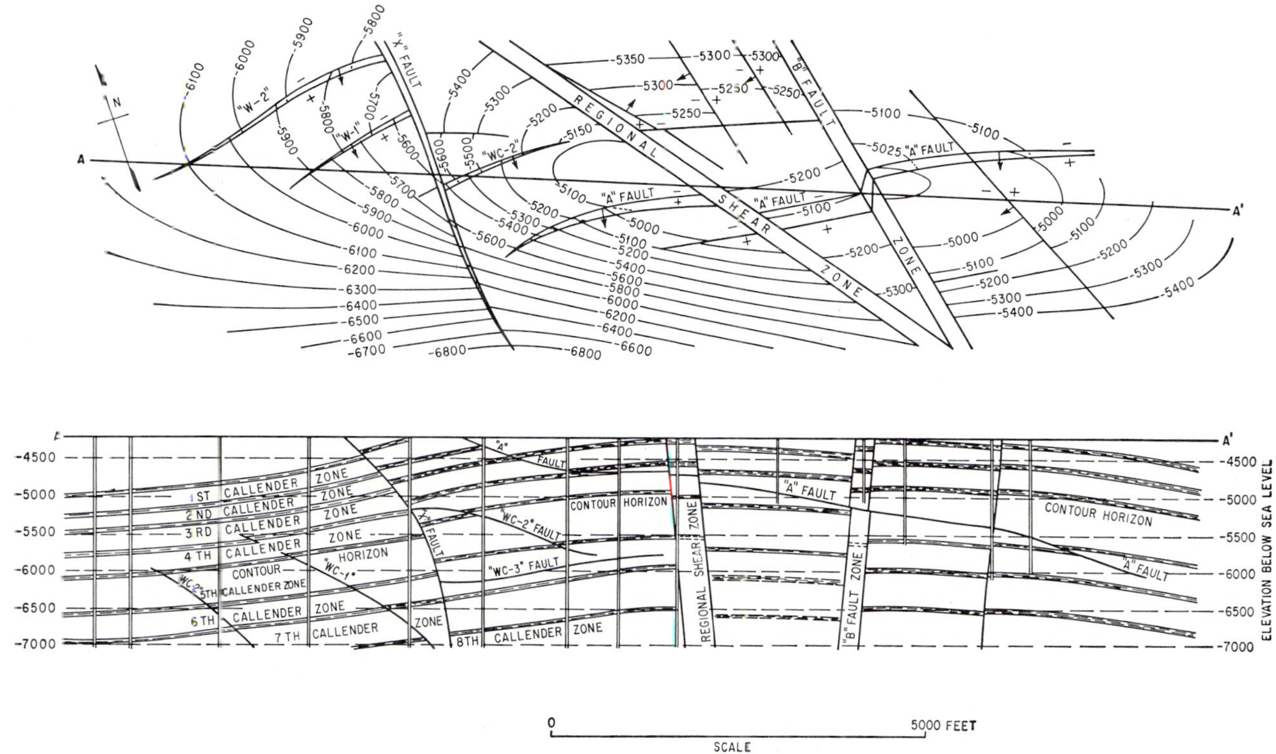

FIGURE 6-18. Structure map and section of the Dominguez oil field. This anticlinal structure lies in echelon relation to several others along the Newport-Inglewood trend in the Los Angeles Basin of California. Note the characteristic transcurrent faulting (regional shear zones). (*Redrawn from Graves, Map 32, Calif. Div. Mines Bull. 170, 1954.*)

reservoirs which may exist within the section. Examples are shown which illustrate dozens of separate traps associated with a single structural feature (Figures 6-18, 6-28, 6-30, and 6-32).

Folds. Of all the various kinds of traps which have been discovered by far the most common, and also the ones accounting for the largest proportion of the world's reserves of oil and gas, are those associated with anticlinal structures. Although they are comparatively easy to discover, their abundance does not arise solely because they have been found before other kinds of traps which are yet undiscovered. They appear to be truly much more common. Figure 6-15 depicts idealized sketches of some of the most common forms of traps associated with anticlines. These structures commonly arise as the result of lateral compression of the sedimentary section. Frequently, if the compression is not too great, a series of long, parallel, gentle folds will be formed. Wherever such folds plunge in opposite directions away from an axial

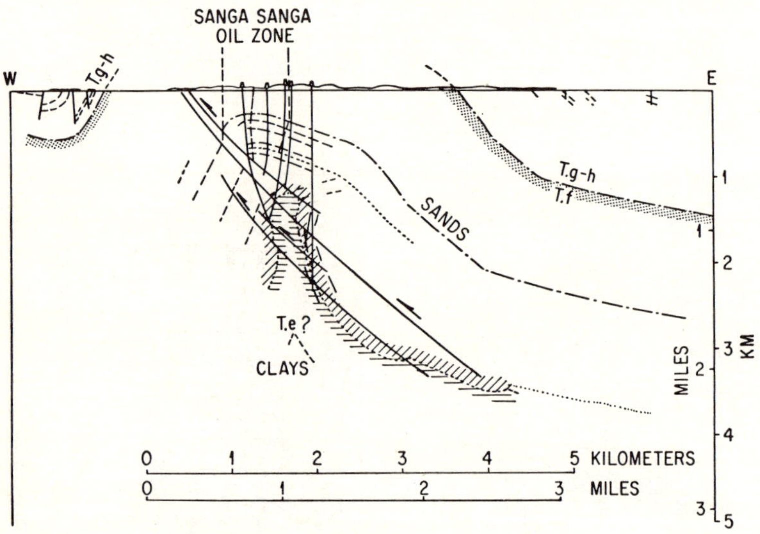

FIGURE 6-19. Structure section through the Sanga Sanga oil field on the Balikpapan trend in eastern Borneo. A number of minor thrust faults have developed near the crest of the fold. (*Redrawn from Weeda, in "Habitat of Oil," p. 1345, AAPG, 1958.*)

high, a trap may be formed as shown in Figure 6-15*a*. This kind of trap is typified by many of the huge oil fields of the Middle East. The Qaiyarah structures of northern Iraq depicted in Figure 6-16 are excellent examples (Dunnington, 1958). In these fields oil has been accumulated in Miocene and Upper Cretaceous limestone reservoirs. Within each of the reservoirs the oil-water contact is tilted to the southeast, giving indication of a hydrodynamic situation.

Often, anticlinal folds, instead of being adjacent, are formed in echelon arrangement. Such structures are frequently, but not always, cut by transcurrent faulting across their axes (Figure 6-15*b*).

An interesting example of traps associated with echelon folds in which transcurrent faulting has not developed is afforded by the Kettleman Hills oil field which is located in the San Joaquin Valley of California. This is depicted in Figure 6-17 (Porter, 1952; Woodring, 1940). There are three separate structures, only two of which are shown in the figure. Production is from multiple sands interstratified with shales of Miocene and Eocene age. The North Dome field is by far the largest and is expected to produce more than 500 million barrels of oil. The oil-water contacts in these reservoirs are also tilted, in this case toward the northwest, again giving evidence of hydrodynamic effects.

Another example of a series of echelon folds is afforded by the Newport-Inglewood trend of oil fields in the Los Angeles Basin of California. The Dominguez oil field, which is located on this trend, is depicted in Figure 6-18 (Graves, 1954). Here the

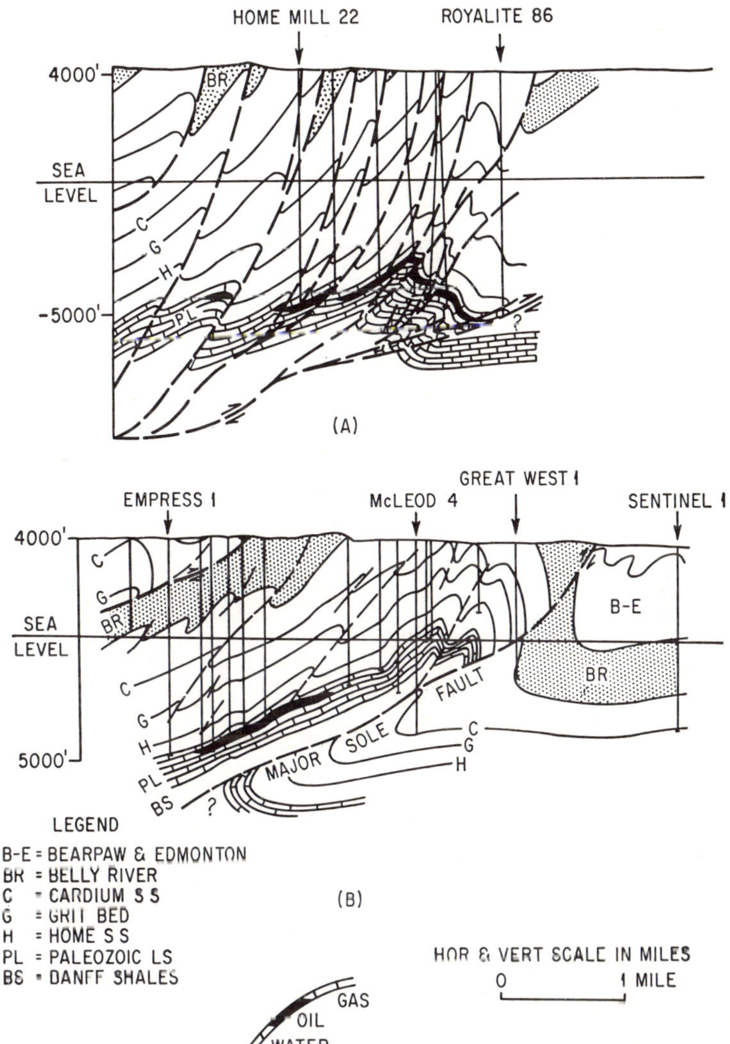

FIGURE 6-20. Structure sections through the Turner Valley oil field in the folded foothills belt of the Rocky Mountains of Western Canada: (*A*) section across the north end of the field, (*B*) section across the center of the field. (*Redrawn from Link, Bull. AAPG, vol. 33, pp. 1479, 1480.*)

characteristic transcurrent faulting across the axis is well developed together with minor thrust faulting parallel to the axis. In this case the faulting has, to some degree, influenced the accumulation of oil, and some of the fault blocks are much more productive than others. This field illustrates the way in which multiple traps may be formed by folding which extends through a large thickness of strata. In the

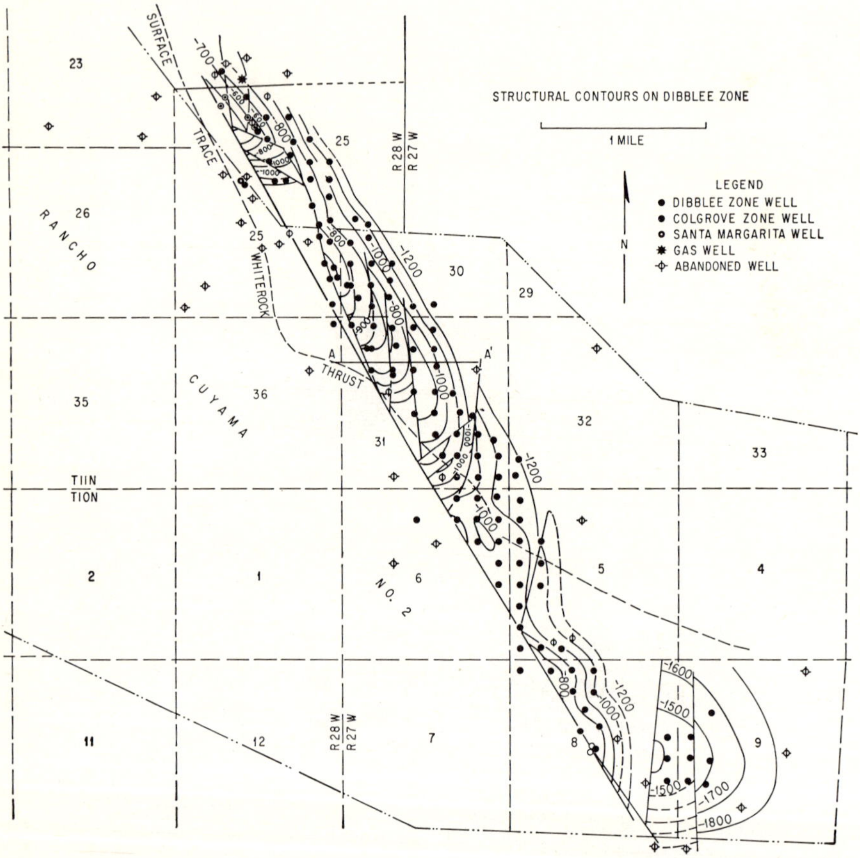

FIGURE 6-21. Structure contour map of the Russell Ranch oil field, Cuyama Valley, California. Contours are on the top of the Dibblee sand, which is the major reservoir in the field. This is an example of an anticlinal fold which has been strongly modified by faulting. (*Redrawn from Schwade, Carlson, and O'Flynn, in "Habitat of Oil," p. 94, AAPG, 1958.*)

Dominguez structure there are eight different pay zones. Production is from Miocene and Pliocene sands interspersed with shales.

In regions of more severe deformation, anticlinal structures may be overturned or cut by thrust faulting as shown in Figure 6-15*c*. An example of traps associated with an anticlinal structure in which some thrust faulting has taken place is afforded by the Sanga Sanga oil field located on the Balikpapan oil trend in east Borneo [Figure 6-19 (Weeda, 1958)]. Production here is from Tertiary sandstones.

The results of a much more intensive deformation are shown in Figure 6-20, which depicts structural sections through the Turner Valley oil field in the folded foothills of the Canadian Rocky Mountains (Link, 1949; Gallup, 1951). The accumulation is in a Paleozoic limestone reservoir which has been cut by many related thrust faults.

Another example of an anticlinal structure which has been strongly modified by several kinds of faulting is that of Russell Ranch oil field (Schwade et al., 1958) in the Cuyama Valley of California, which is depicted in Figures 6-21 and 6-22. The producing zones, which are sandstones of Lower Miocene age, have been cut by

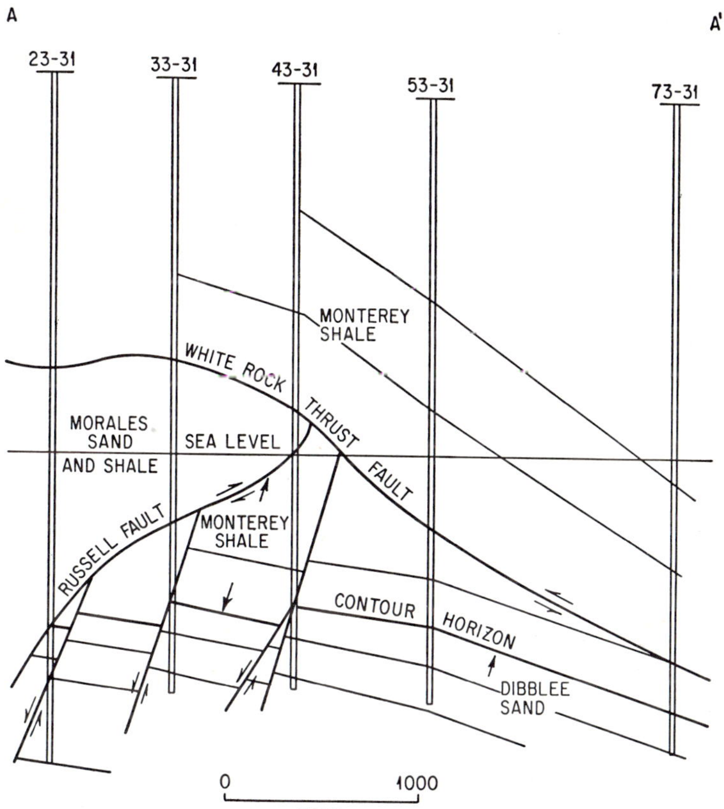

FIGURE 6-22. Structure section *A-A′* through the Russell Ranch oil field, Cuyama Valley, California. Note how the underlying structure has been covered and hidden by the White Rock thrust fault. *(Redrawn from Schwade, Carlson, and O'Flynn, in "Habitat of Oil," p. 95, AAPG, 1958.)*

multiple normal faults, and the entire structure has been covered by the White Rock thrust fault which probably occurred during a later period of deformation.

When folds are formed by compression of competent strata underlain by soft incompetent rocks such as salt or weak shales, diapiric anticlines may be formed by the soft rocks being squeezed up and intruded into the overlying strata along the axis of the structure. This situation is depicted in idealized form in Figure 6-15*d*. Diapiric structures are prevalent in the Aquitain Basin of southern France, the Carpathian regions of Eastern Europe, and the Caucasus region of Russia. In Figure 6-23 is depicted a section through the St. Marcet gas field in the southern Aquitain Basin

(Schneegans, 1948). Here plastic Triassic anhydrite and gypsum have been squeezed up into the overlying competent Cretaceous limestones and sands. One of the interesting characteristics of this kind of structure is the thrust faulting which develops on the edges of the diapiric mass. Figure 6-24 illustrates a section through the Moreni diapiric anticline (Walters, 1946; Small, 1959) in Rumania. Here plastic salt has been squeezed up and intruded along the axis of the long anticlinal structure. Accumulations occur in porous strata butting against the diapiric mass.

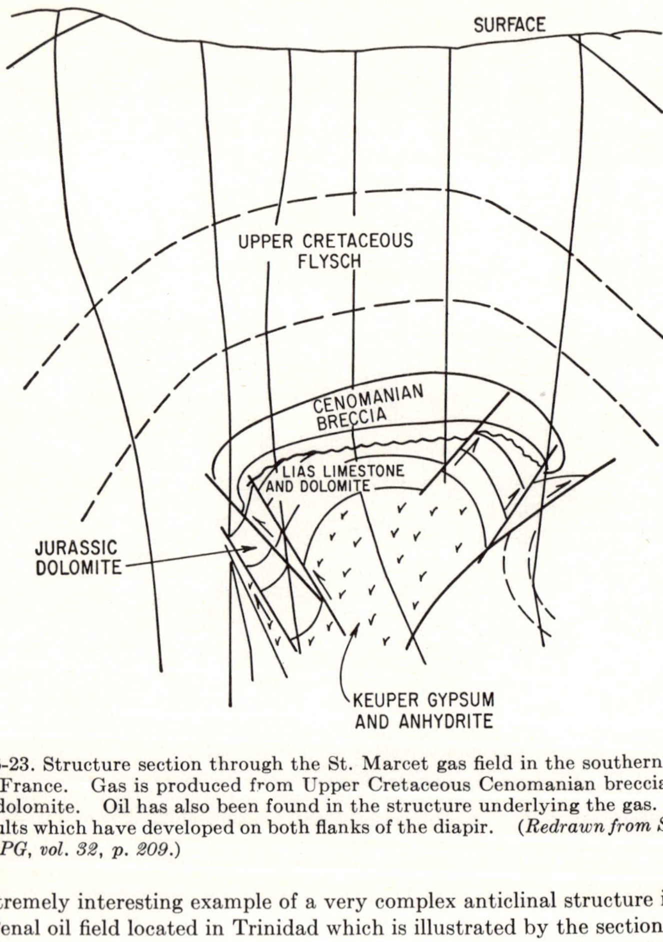

FIGURE 6-23. Structure section through the St. Marcet gas field in the southern Aquitaine Basin of France. Gas is produced from Upper Cretaceous Cenomanian breccia and from Jurassic dolomite. Oil has also been found in the structure underlying the gas. Note the thrust faults which have developed on both flanks of the diapir. (*Redrawn from Schneegans, Bull. AAPG, vol. 32, p. 209.*)

An extremely interesting example of a very complex anticlinal structure is afforded by the Penal oil field located in Trinidad which is illustrated by the section in Figure 6-25 (Bitterli, 1958). Here the structure has been overturned and two thrust faults have developed. Production is from the Herrera sandstone which, because of the overturn, is penetrated at three different depths in some of the wells. In addition to the overturning and faulting, a small diapiric intrusion occurs in the center of the structure.

Faults. Fault surfaces provide many different possibilities for the development of the proper geometry for a trap. Often there will be a zone of gouge, or broken rock, associated with the fault surface which is capable of acting as an impermeable barrier to oil. Frequently an impermeable stratum is displaced into a position where it can serve as a barrier to oil moving in a permeable stratum. However, faults do not always seem to be effective seals, and there is evidence in many cases that fault surfaces have served as conduits for the movement of oil up through the strata. Figure 6-26 depicts schematic diagrams of a number of possible traps associated with faults.

One of the most common types of fault trap is that formed by normal faulting. This may occur either by two or more normal faults intersecting homoclinally dipping strata or by a single normal fault intersecting a structural nose or terrace. These

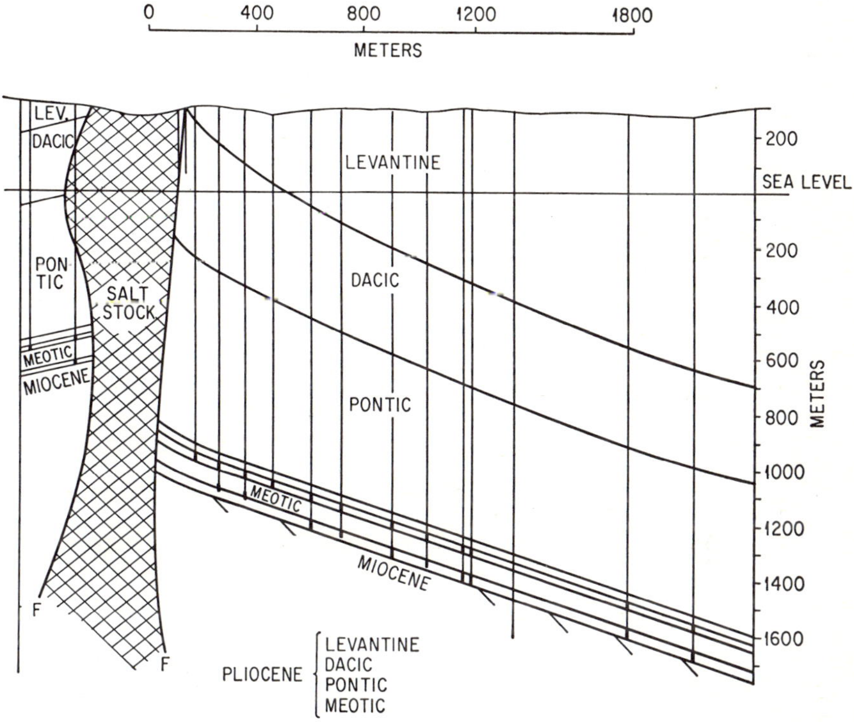

FIGURE 6-24. Section through the Moreni diapiric anticline in the Ploesti region of Rumania. Traps are formed by permeable strata butting against the diapiric mass. (*Redrawn from Walters, Bull. AAPG, vol. 30, p. 333.*)

situations are indicated in Figure 6-26*a* and *b*. In either case two possibilities exist for the relationship between the dip of the fault surface and the dip of the strata. If the strata dip in a direction opposite to that of the fault plane, the trap will be on the upthrown side of the fault, whereas if the dips are in the same direction, the trap will be on the downthrown side of the fault. For some reason which is not fully understood accumulations are much more frequently found on the upthrown side of normal faults. An excellent example of this kind of trap is afforded by the West Guara field (Hedberg et al., 1947) in the Greater Oficina area of eastern Venezuela (Figures 6-27 and 6-28.) This prolific field contains 47 separate productive traps in the same structural segment, illustrating again the probability of finding multiple traps associated with structural features. Production is from the Oficina sands of Oligocene and

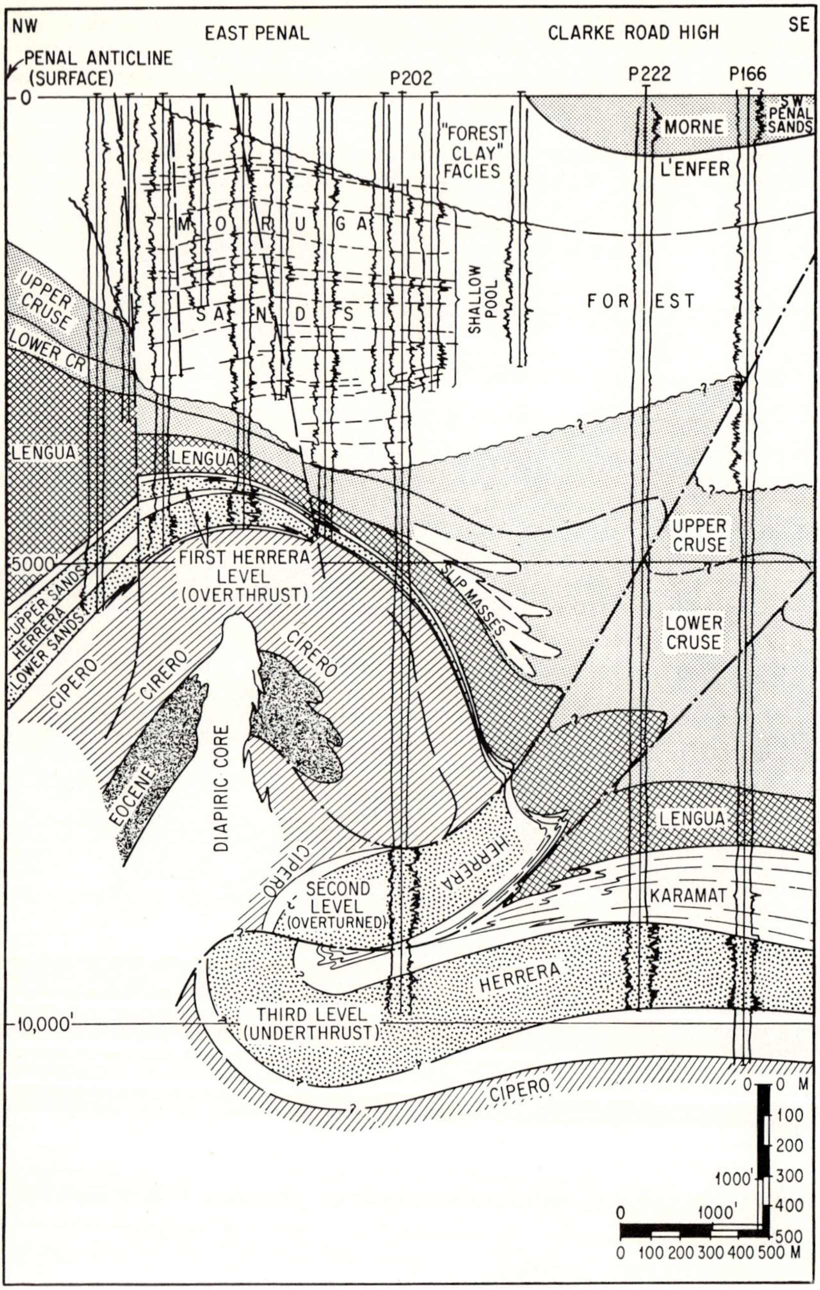

FIGURE 6-25. Section through the Penal oil field, Trinidad, B.W.I. Production is obtained from the shallow Moruga sands and the deeper Herrera sand, which is penetrated at three levels in some of the wells. Thrust faulting, overturning, and diapiric intrusion are all evident in this complex structure. (*Redrawn from Bitterli, Bull. AAPG, vol. 42, p. 145.*)

Miocene age. Samples taken from some of these sands show permeabilities as high as 10,000 millidarcys.

An example of multiple traps formed on the downthrown side of a series of normal faults is afforded by the Velasquez field (Morales, 1958) in the middle Magdalena Valley of Colombia which is depicted in Figures 6-29 and 6-30. Oil here is produced from early Tertiary sandstones.

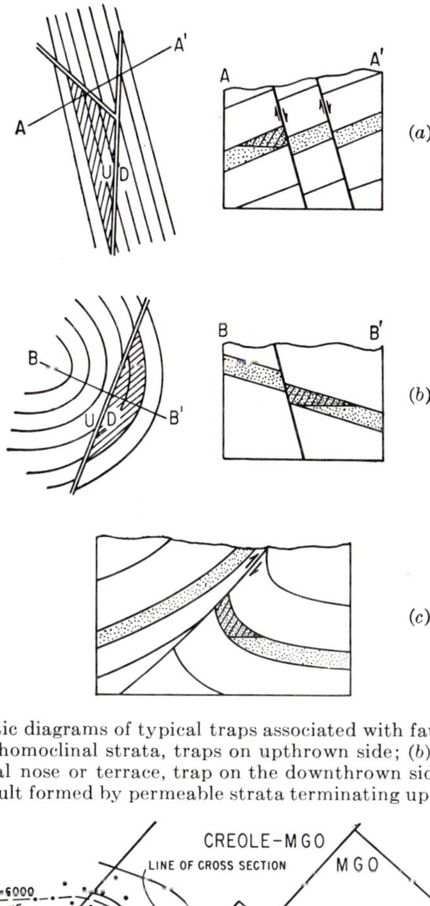

FIGURE 6-26. Schematic diagrams of typical traps associated with fault planes: (*a*) intersecting normal faults in homoclinal strata, traps on upthrown side; (*b*) normal fault intersecting a broad structural nose or terrace, trap on the downthrown side of the fault; (*c*) trap underlying a thrust fault formed by permeable strata terminating updip against fault plane.

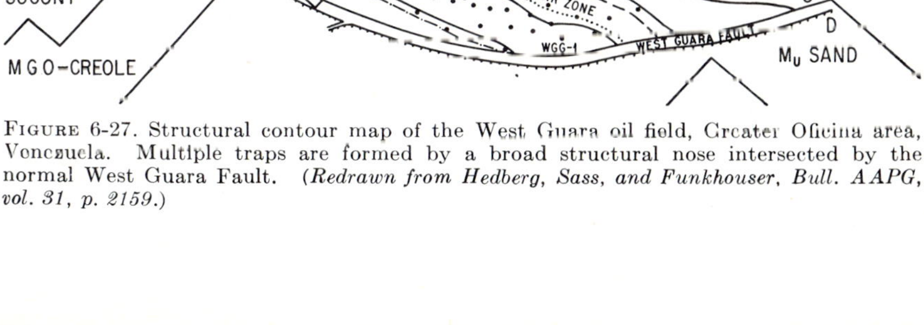

FIGURE 6-27. Structural contour map of the West Guara oil field, Greater Oficina area, Venezuela. Multiple traps are formed by a broad structural nose intersected by the normal West Guara Fault. (*Redrawn from Hedberg, Sass, and Funkhouser, Bull. AAPG, vol. 31, p. 2159.*)

Another example of multiple traps on the downthrown side of a normal fault is provided by the series of accumulations which occur for several miles along the Steinberg fault in the Inner-Alpine Vienna Basin of Austria and Czechoslovakia (Janoschek, 1958). Typical of these accumulations is the Gaiselberg field in which oil and gas are produced in Middle and Upper Miocene sandstones which have been faulted down

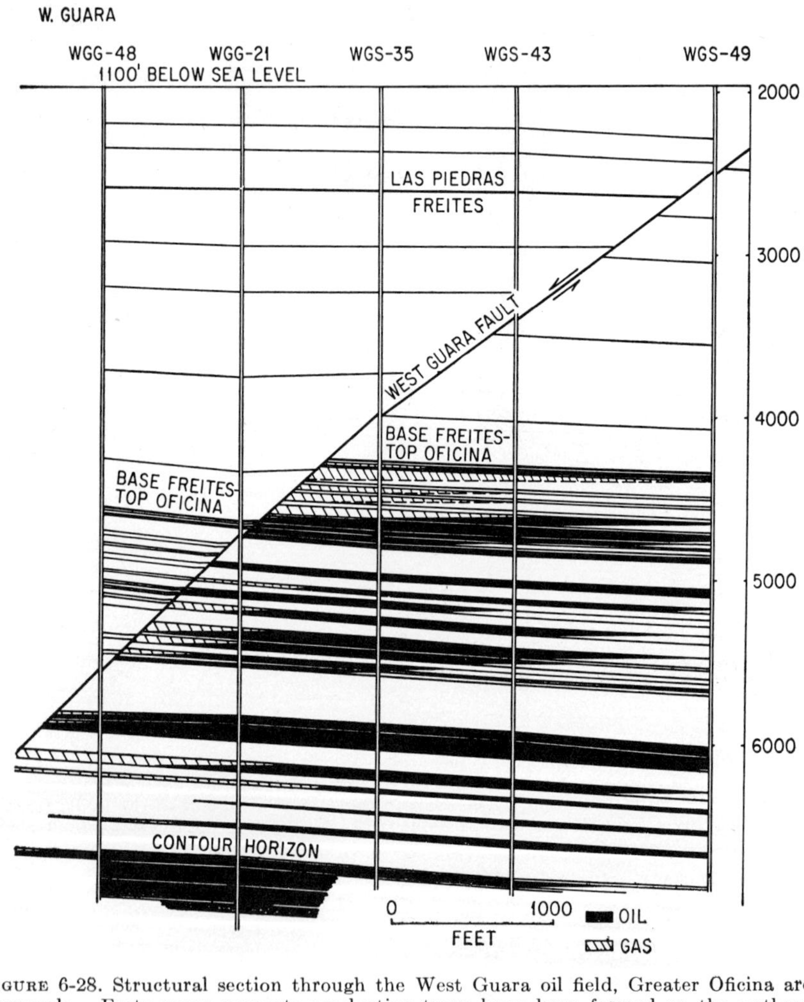

FIGURE 6-28. Structural section through the West Guara oil field, Greater Oficina area, Venezuela. Forty-seven separate productive traps have been formed on the upthrown side of the West Guara Fault. (*Redrawn from Hedberg, Sass, and Funkhouser, Bull. AAPG, vol. 31, p. 2160.*)

against the Eocene flysch basement. This field is illustrated in Figures 6-31 and 6-32. Again, multiple reservoir sands have formed several traps in a single structure. The map indicates a number of subsidiary normal faults which form a series of independent reservoir segments in each of the sands.

Traps are very frequently formed on the downthrown side of reverse or thrust faults as illustrated in Figure 6-26*c*. However, these traps are characteristically difficult to

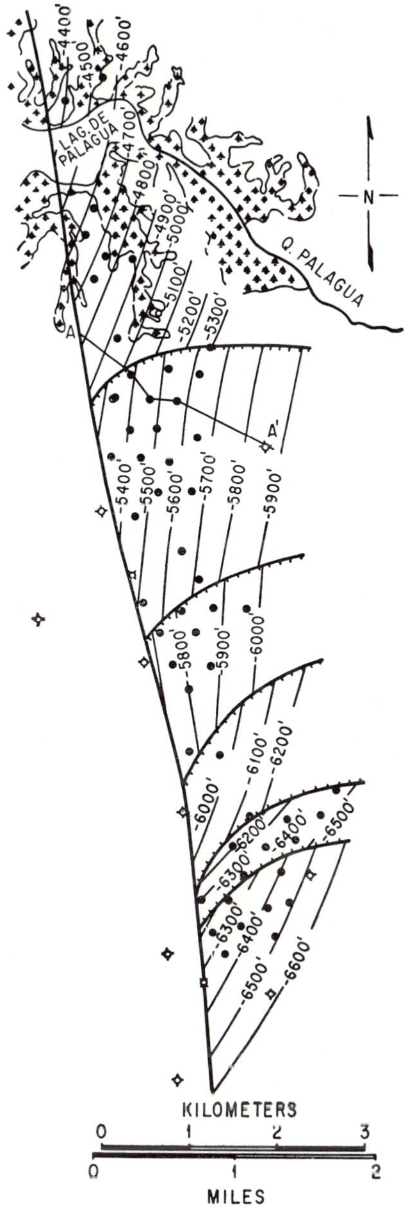

FIGURE 6-29. Structural contour map of the Velasquez oil field in the Middle Magdalena Valley of Colombia. Traps are formed by intersecting faults in approximately homoclinal strata. (*Redrawn from Morales and the Colombian Petroleum Industry, in "Habitat of Oil," p. 692, AAPG, 1958.*)

discover, since they are hidden by the structure and the deformation on the upthrown side of the fault. An excellent example of this kind of accumulation is afforded by a series of traps which have been formed underneath the Oakridge thrust fault in the eastern portion of the Ventura Basin of California. In this region Miocene, Oligocene, and Eocene rocks have been displaced over Pliocene sands and shales and oil has been trapped in the upturned edges of these porous Pliocene sands where they terminate against the fault surface (Ware and Stewart, 1958). Several large anticlinal struc-

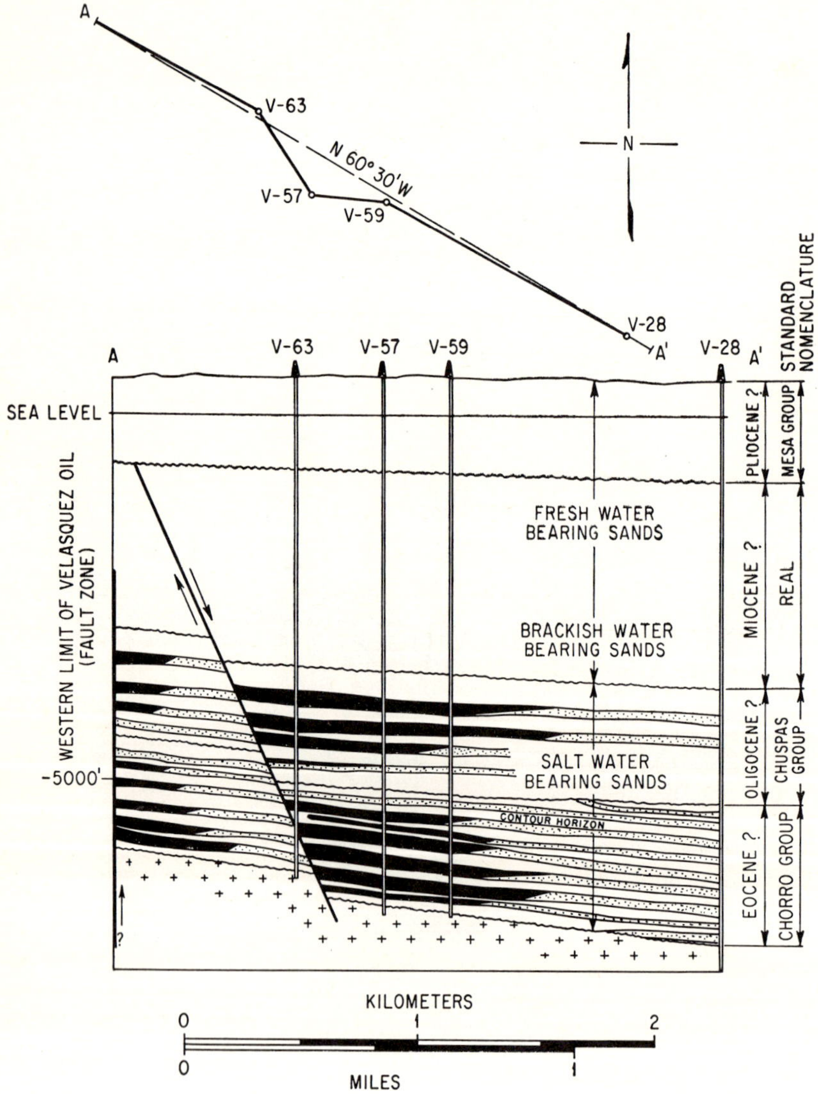

FIGURE 6-30. Structure section *A-A'* through the Velasquez oil field. Accumulation has occurred in traps on the downthrown side of the fault. (*Redrawn from Morales and the Columbian Petroleum Industry in " Habitat of Oil," p. 693, AAPG, 1958*).

tures have been formed in the overthrust block, and large accumulations are found within them. A north-south section through the South Mountain oil field and the Bridge Pool underlying the Oakridge thrust is shown in Figure 6-33.

An interesting prospecting technique has been developed in this region which greatly aids in the discovery of these traps. Wells are directionally drilled underneath

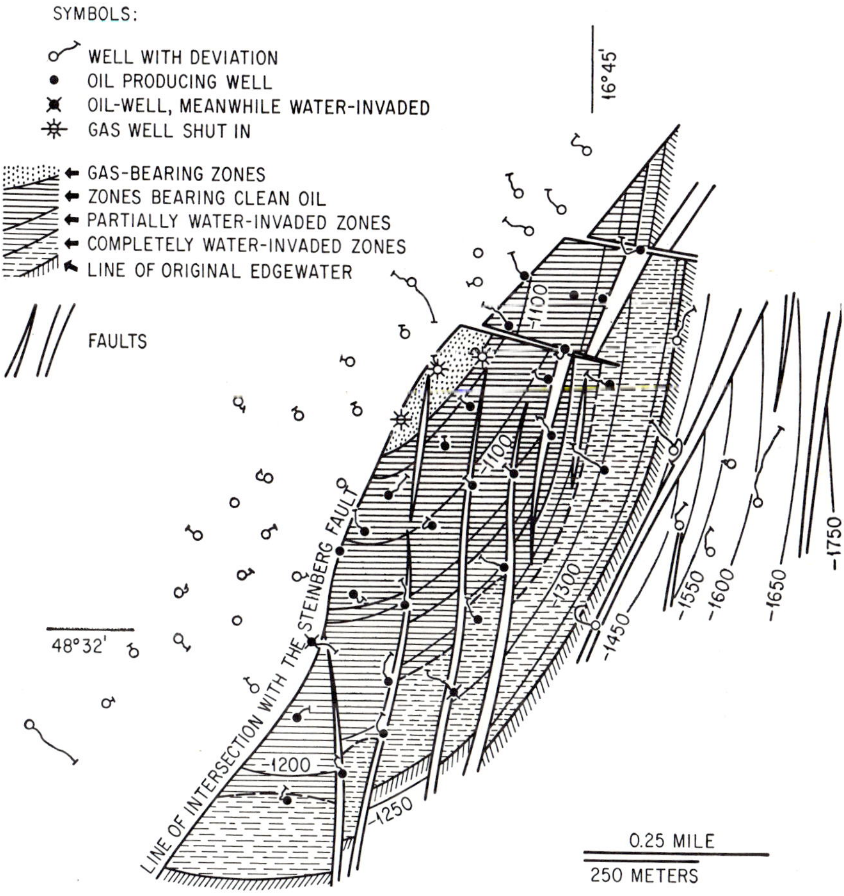

FIGURE 6-31. Structural contour map of the Gaiselberg oil field in the Inner Alpine Vienna Basin together with production data. Traps are formed in a broad structural nose on the downthrown side of the Steinberg Fault. (*Redrawn from Janoschek, in "Habitat of Oil," p. 1148, AAPG, 1958.*)

and parallel to the fault surface, so that a single test has a good probability of intersecting any accumulation that may be present. This technique has been successful in the discovery of several pools underneath the Oakridge thrust during the last few years.

An example of a series of traps in which both faulting and folding have played a significant part is shown by the Potrero oil field, which is another of the major structures along the Newport-Inglewood trend in the Los Angeles Basin of California (Willis and Ballantyne, 1943). This is illustrated in Figures 6-34 and 6-35. In this case, a series of multiple, independent traps is formed by several transcurrent faults cutting across an anticlinal structure. In particular, note the traps which are formed on the northwest plunging nose of the anticline. These traps have been formed in Miocene and Pliocene sandstones.

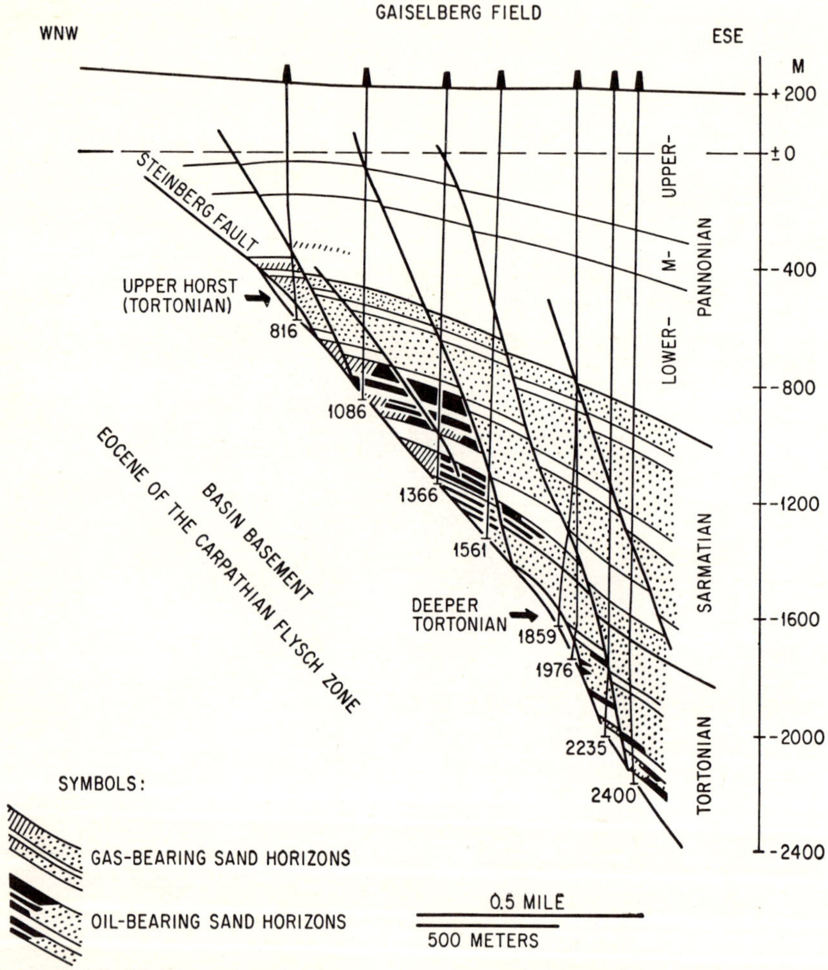

FIGURE 6-32. Structural section through the Gaiselberg oil field. Note the traces of the numerous normal faults subsidiary to the main Steinberg Fault. (*Redrawn from Janoschek, in "Habitat of Oil," p. 1147, AAPG, 1958.*)

Intrusive Structures. Deep-seated intrusion of the sedimentary section by plastic or fluid rocks provides the possibility for many different kinds of trapping configurations. Intrusions are often formed as the result of igneous activity in which the intruded material is a molten or semimolten rock. Although there undoubtedly exist many traps associated with igneous intrusions, they all appear to be unproductive, probably because of the high temperatures to which the rocks have been subjected.

A much more favorable situation arises in the case of intrusive salt structures or salt domes, which have originated at much lower temperatures and which have formed productive traps in many regions throughout the world.

The mechanism by which salt-dome structures are formed is fairly well understood for several reasons. A tremendous amount of subsurface information has been available for study and analysis, principally from well records in the Gulf Coast region of the United States, where these structures have been intensively exploited. In addition, the classic model experiments of Nettleton (1936, 1943) and of Parker and

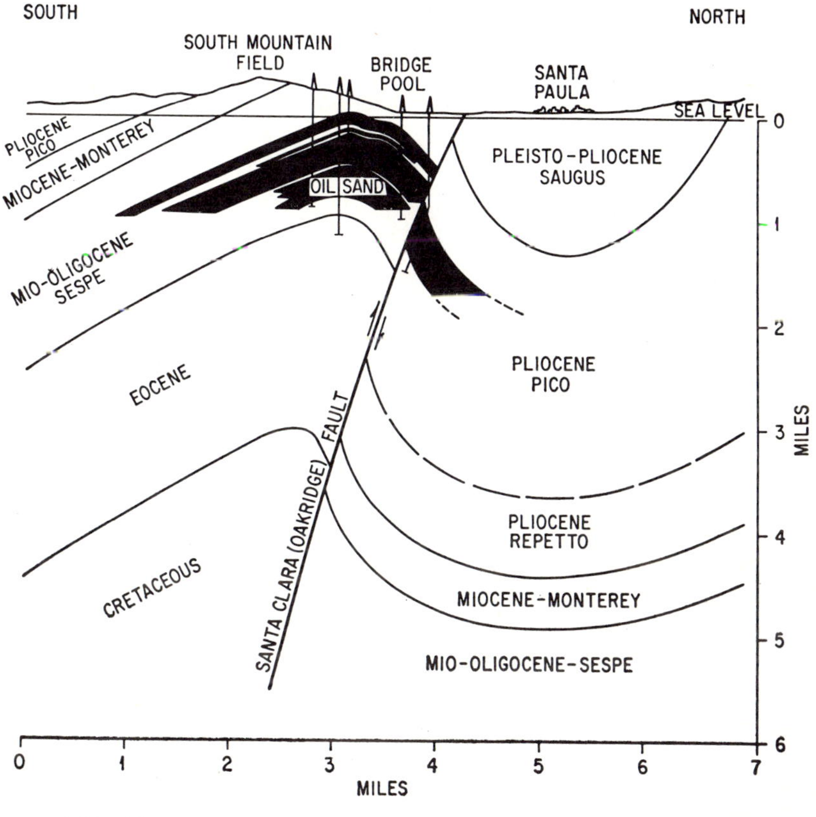

FIGURE 6-33. Schematic north-south structural section through the South Mountain oil field and the Bridge Pool in the eastern Ventura Basin, California. The trap in the Bridge pool is formed by Pliocene sands terminating updip against the Oakridge Thrust Fault. (*Redrawn from Ware and Stewart, "Guidebook," p. 180, AAPG Pacific Section, 1958.*)

McDowell (1951, 1955) have provided a convincing demonstration of the mechanism of salt-dome formation on a laboratory scale.

It is believed that a thick layer of sedimentary salt is overlain and buried by other sediments. The salt, being less dense than the overlying sediments, is in a mechanically unstable position with respect to them. Any small irregularities in the upper surface of the salt will have a tendency to allow the salt to penetrate and intrude upward into the overlying sediments. The salt, by virtue of its relative buoyancy, will then move upward as a plug or a dome, sometimes many thousands of feet, penetrating and piercing the sediments as it moves.

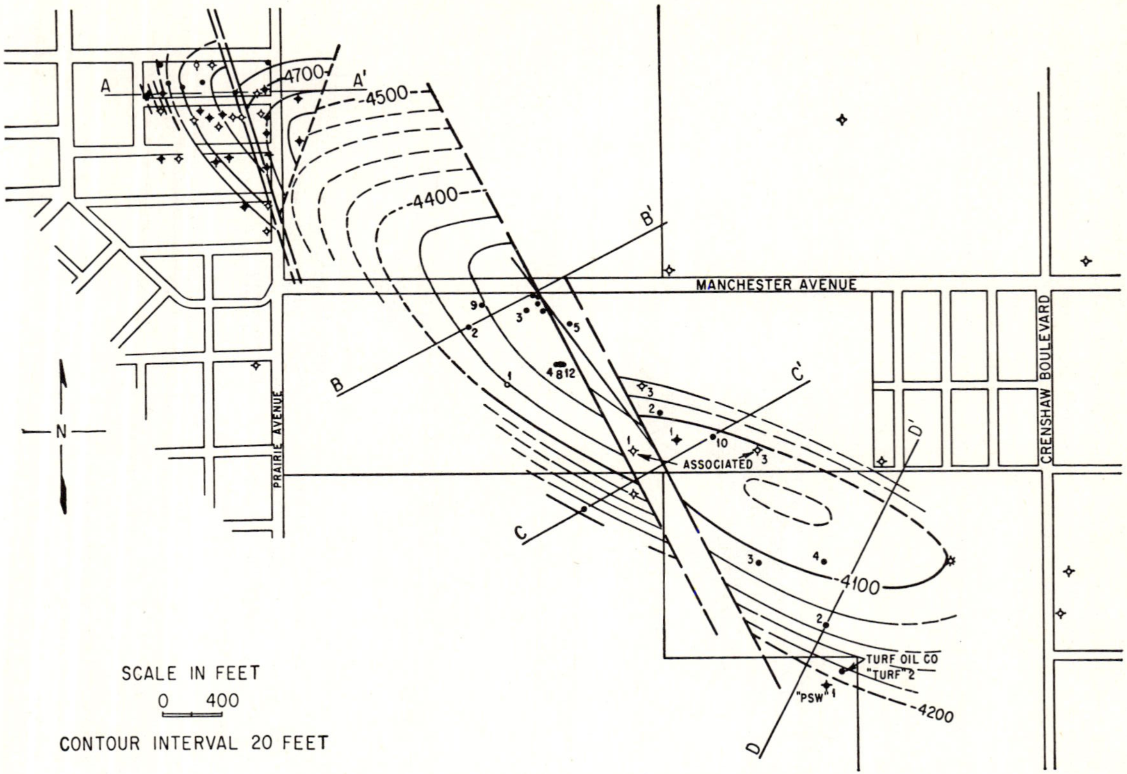

FIGURE 6-34. Structure contour map of the Potrero oil field in the Los Angeles Basin of California. This field is another of the echelon folds along the Newport-Inglewood trend in the Los Angeles Basin (Fig. 6-18). (*Redrawn from Willis and Ballantyne, Calif. Div. Mines Bull. 118, p. 312.*)

Structures of this kind are closely related to diapiric structures but with the distinction that in this case the energy is supplied primarily by the relative buoyancy of the salt whereas in diapiric structures the energy is supplied primarily by lateral compression of the sediments. All gradations between the two kinds of structures can be found. In many cases the rate at which the salt rises appears to be delicately balanced against the rate at which the overlying sediments are deposited, and as a result salt-dome structures are found in all stages of development, ranging from those which are too deep to be penetrated currently by the drill to structures in which the salt lies immediately below the surface of the ground.

Intrusive salt structures have been found in many regions throughout the world, and they have been responsible for the formation of many oil traps in the Gulf Coast region of the United States, the Isthmus of Tehuantepec in Mexico, the Hannover Basin of northern Germany, and the Emba region bordering the Caspian Sea in Russia. Intrusive salt structures have also been found in the Paradox Salt Basin of Utah and more recently in the Canadian Arctic islands (Fortier et al., 1954), but as yet no oil has been found associated with these. In these various regions the composition of the salt varies widely, with halite being the principal constituent associated with varying quantities of various chloride salts, anhydrites, carbonates, and clastic material. In the Gulf Coast region where salt intrusions, or domes, have been studied more intensively than anywhere else in the world, the salt is almost pure halite together with 5 to 10 per cent anhydrite and very minor quantities of other minerals (Hanna, 1934).

One of the extremely interesting characteristics of salt-dome structures which should be noted is the wide variation which can occur in the shape of the salt mass. One of the very common occurrences is the overhanging dome, which is particularly important from the exploration standpoint, as traps may exist below the overhang. Such a situation requires that wells must be drilled for varying distances through the salt before the traps can be reached.

Another very interesting characteristic of salt domes is the frequent occurrence of a cap rock. This is a layer of insoluble material covering the entire top surface of the salt mass. If present, it may be as much as 2,000 feet thick. The major constituent of the cap rock is usually anhydrite, but gypsum, limestone, dolomite, and free sulfur are frequently present. It was originally thought that cap-rock material represented a stratum which overlay the original salt and had been retained as it moved upward. However, the evidence available at present indicates that the cap rock is probably an insoluble residue which has accumulated as the salt mass was continuously dissolved and washed away by circulating ground water. This activity probably occurred more slowly than the concurrent upward movement of the salt plug, thereby precluding collapse and caving of the overlying strata.

Salt-dome structures are relatively easy to locate underground, especially if they are shallow. They represent such a strong anomaly in the physical properties of the rocks that geophysical methods are usually effective. In particular, gravitational measurements which indicate the distribution of deep-seated masses will often clearly reveal the less dense salt. However, in spite of the ease with which salt domes can be located, the multitude of various traps which can be associated with them is not nearly so easily discovered.

In the Gulf Coast region of Texas and Louisiana during the last 15 or 20 years large reserves of petroleum have been discovered in a great many relatively small fault and stratigraphic traps around the flanks of previously known piercement salt-dome structures. These discoveries have required extremely careful and detailed study of large numbers of well logs obtained from previous exploratory and development drilling.

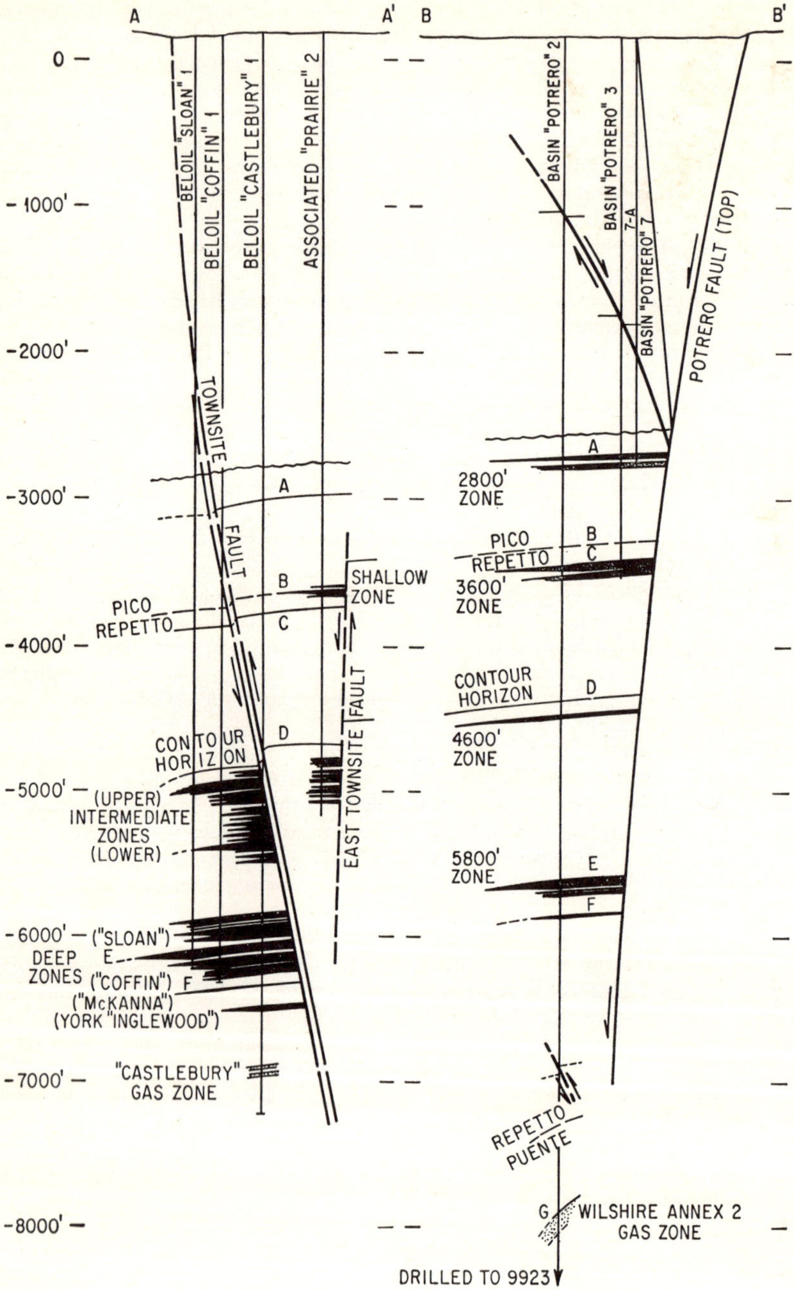

FIGURE 6-35. Structure sections through the Potrero oil field. Multiple traps are formed
118, p. 313.)

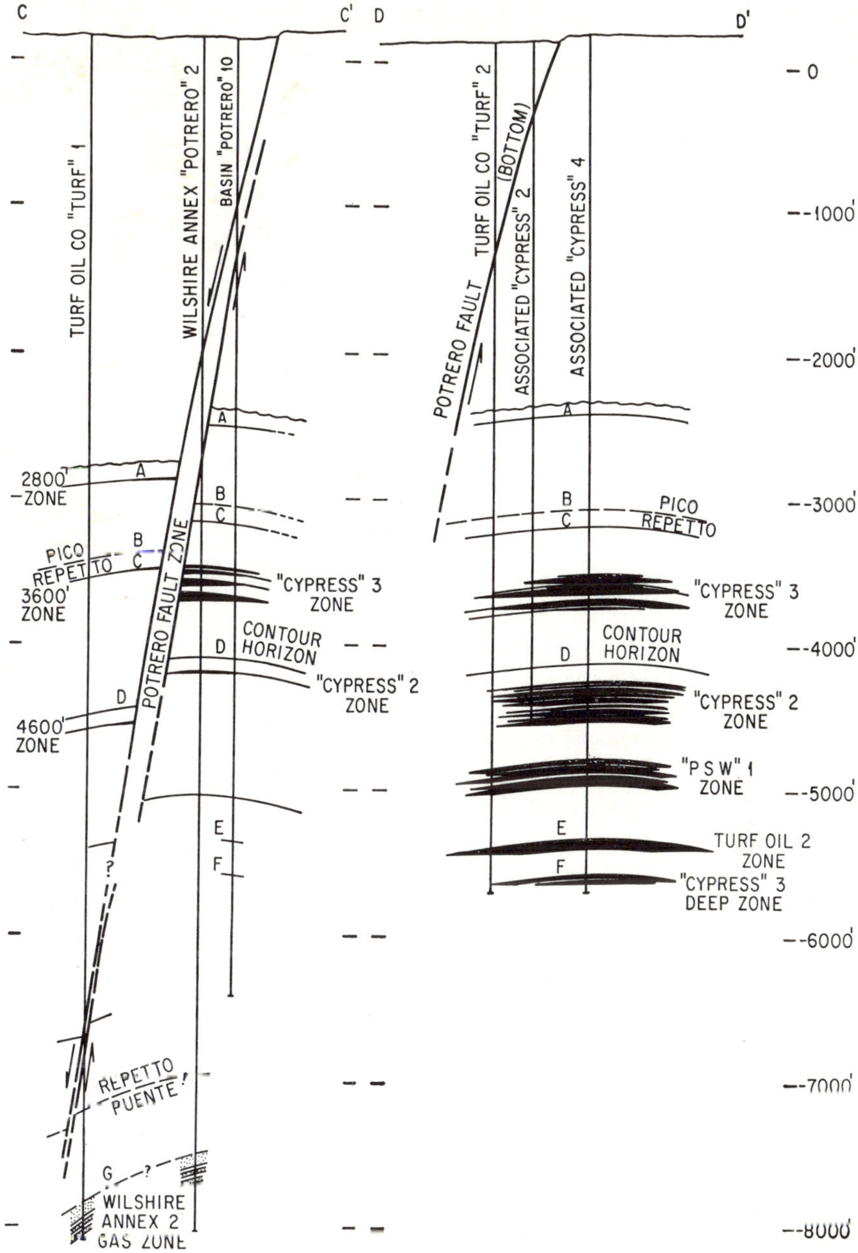

against the fault planes. (*Redrawn from Willis and Ballantyne, Calif. Div. Mines Bull.*

Idealized diagrams of traps which are commonly associated with intrusive salt-dome structures are shown in Figure 6-36. These may range from structural domes with little or no faulting which occur when the sediments likely to contain oil are far above the salt intrusion (Figure 6-36*a*) through more highly faulted structural domes

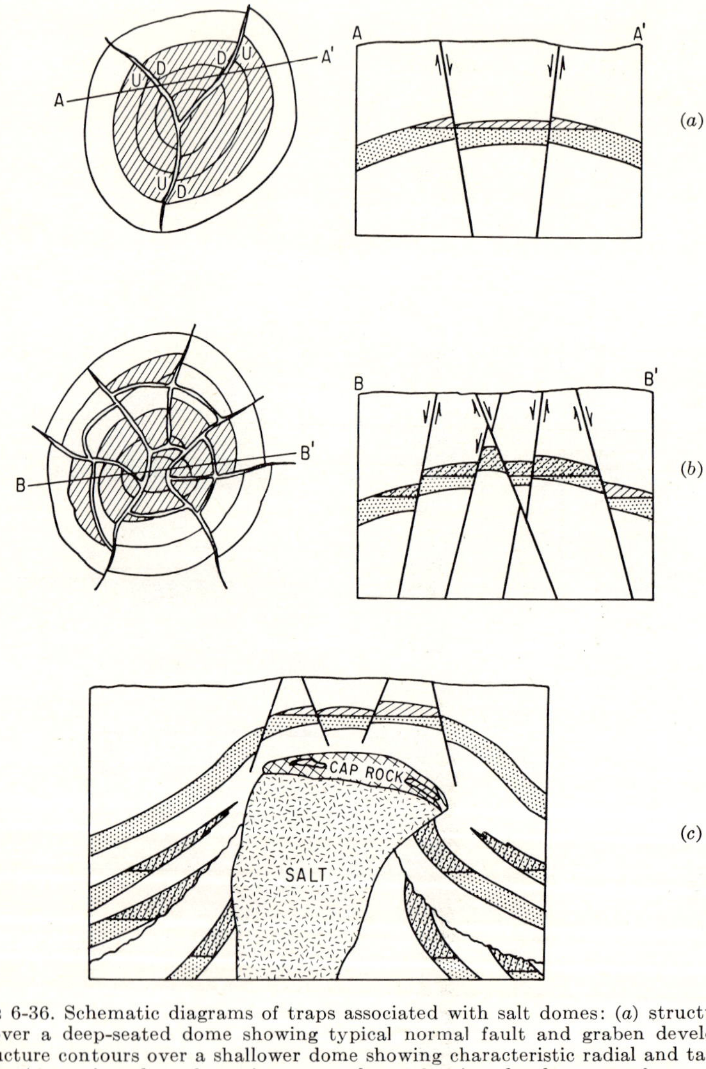

FIGURE 6-36. Schematic diagrams of traps associated with salt domes: (*a*) structure contours over a deep-seated dome showing typical normal fault and graben development, (*b*) structure contours over a shallower dome showing characteristic radial and tangential faulting, (*c*) section through a piercement dome showing development of traps on the flanks.

showing the characteristic radial and peripheral faulting which occurs nearer the salt intrusion (Figure 6-36*b*) to piercement structures in which the intrusion actually penetrates the sediments likely to contain oil and a great variety of traps are formed on the flanks of the dome (Figure 6-36*c*). The cap rock of the intrusion itself sometimes has developed excellent permeability and porosity, thereby forming a trap.

The examples which follow have been selected so as to show many of these various ways in which traps may be associated with salt-dome structures. In Figure 6-37 is depicted a structure contour map of the Brookhaven oil field located in the central portion of the interior salt basin of Mississippi (Womack, 1950). The contours are on

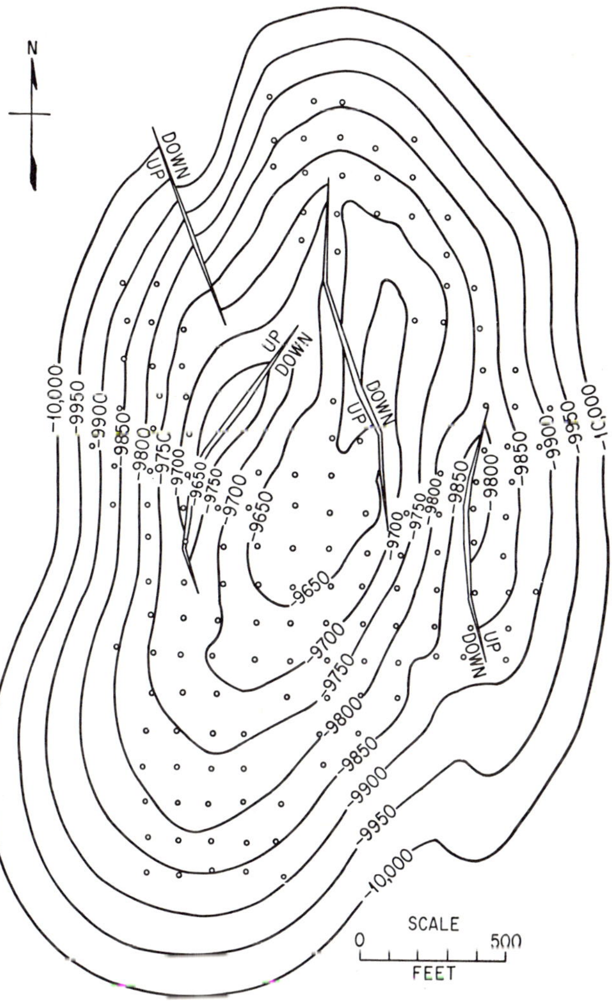

FIGURE 6-37. Structure contours on the basal Tuscaloosa sand in the Brookhaven oil field in Mississippi. This field overlies a deep-seated salt dome and shows the typical normal faulting (*Redrawn from Womack, Bull. AAPG, vol. 34, p. 1522.*)

the basal Tuscaloosa sandstone of Upper Cretaceous age, which is the reservoir rock of the field. Although the underlying salt has never been penetrated by the drill, its existence is inferred both from the structure and from a gravity minimum over the field shown by geophysical measurements. The three normal faults on the top of the structure form a central graben area which is typical of these deep-seated salt-dome structures.

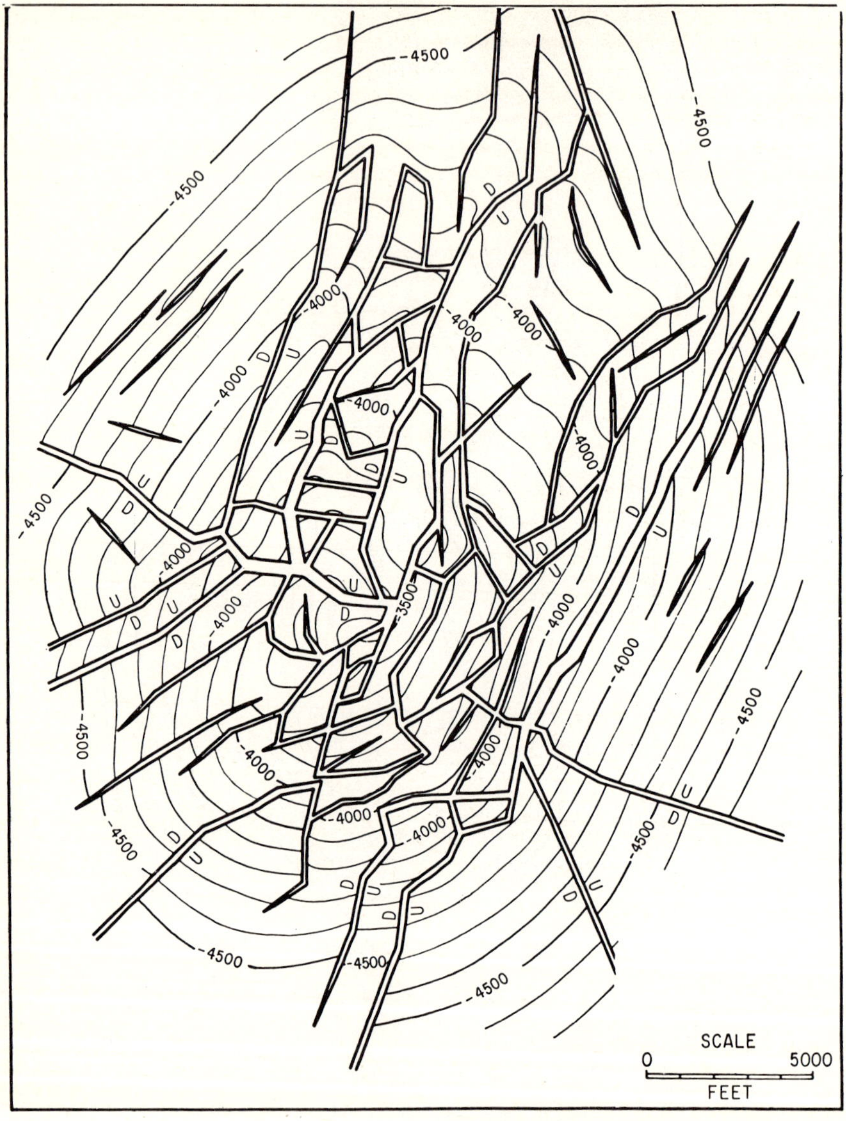

FIGURE 6-38. Structure contour map of the Hawkins oil field, Wood County, Texas. Contours are on the Woodbine sand, which is the producing zone of the field. The structure is formed over a deep-seated salt intrusion, and the characteristic radial and tangential faulting is apparent. (*Redrawn from Wendlandt, Shelby, and Bell, Bull. AAPG, vol. 30, and Parker and McDowell, Bull. AAPG, vol. 35, p. 2080.*)

A more intricate system of normal faulting is shown by the Hawkins oil field which is located in northeastern Texas (Wendlandt et al., 1946). Production is from the Woodbine sand of Cretaceous age. A structure map on the top of the Woodbine is depicted in Figure 6-38. The typical radial and tangential normal faults are evident.

Another example of the highly intricate system of normal faulting which may develop near the salt mass is afforded by the Reitbrook dome in the Hannover Basin of northern Germany (Reeves, 1946). A section through this structure is depicted in

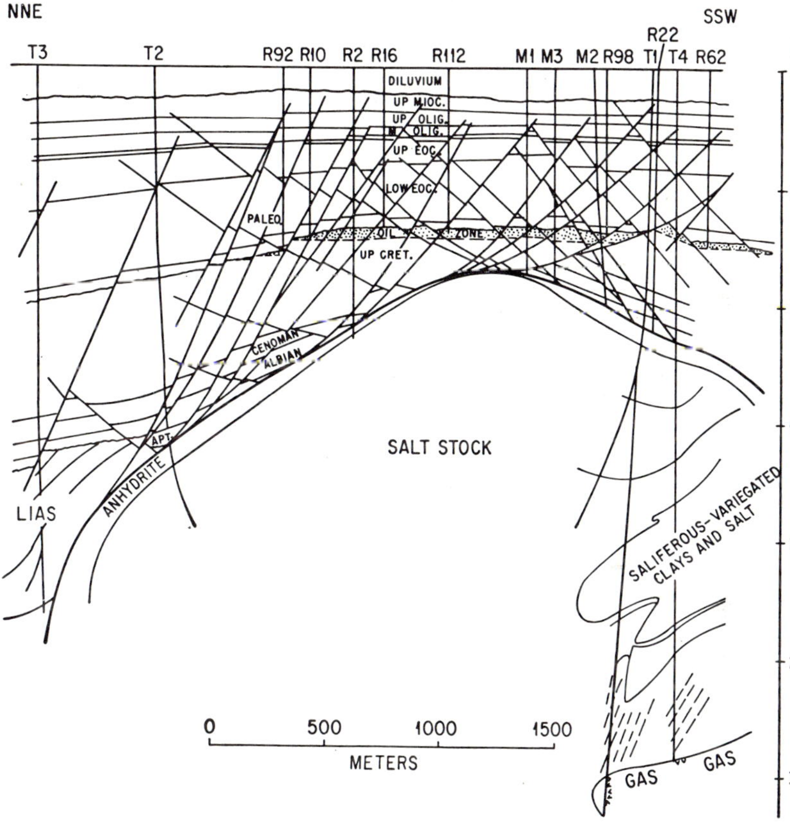

FIGURE 6-39. Structure section through the Reitbrook salt dome in the Hannover Basin of northern Germany, showing characteristic intricate normal faulting. *(Redrawn from Reeves, Bull. AAPG, vol. 30, p. 1564, and Behrmann, in British Intelligence Objectives Subcommittee, Oil Fields Investigation, Part III, Fig. 95.)*

Figure 6-39. Oil is produced principally from Marly limestones of Upper Cretaceous age in which the porosity is primarily due to fracturing. Some gas and oil are produced from the overlying Oligocene and Eocene sands. Note that the Cretaceous lies unconformably on truncated Lias beds, indicating that the salt structure had probably started to form prior to a period of pre-Cretaceous erosion.

When the salt actually pierces the sediments likely to contain oil, a great many possibilities exist for traps. Figure 6-40 depicts a structure map on the salt-dome material together with a number of structural cross sections in the vicinity of the South Liberty dome in the Gulf Coast region of Texas (Halbouty and Hardin, 1951).

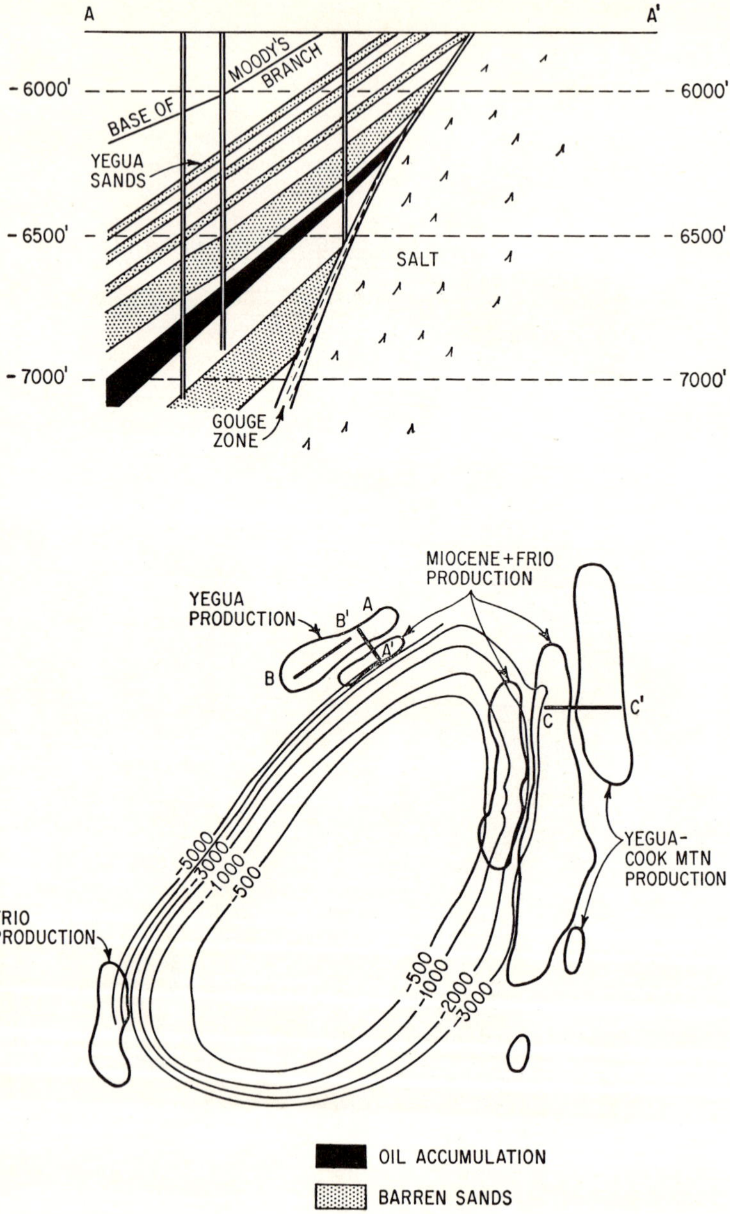

FIGURE 6-40. The South Liberty salt dome and associated oil accumulations, Liberty three sections illustrate some of the various traps which may be developed on the flanks *1939–1977.*)

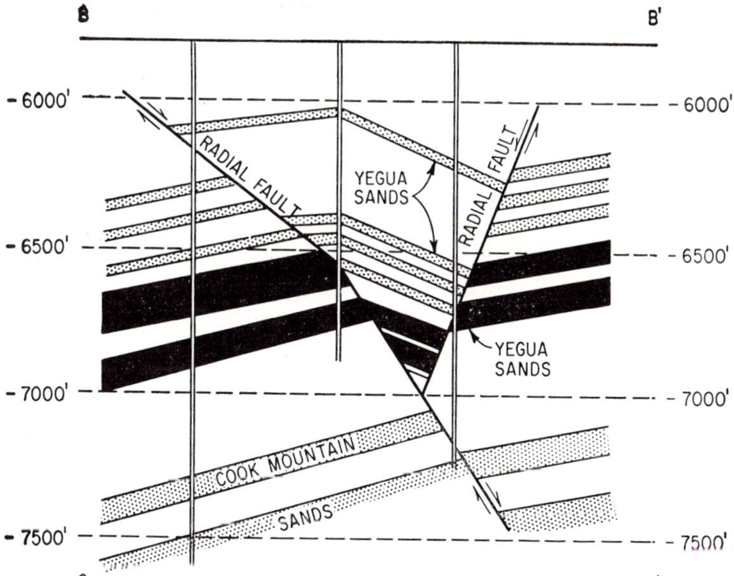

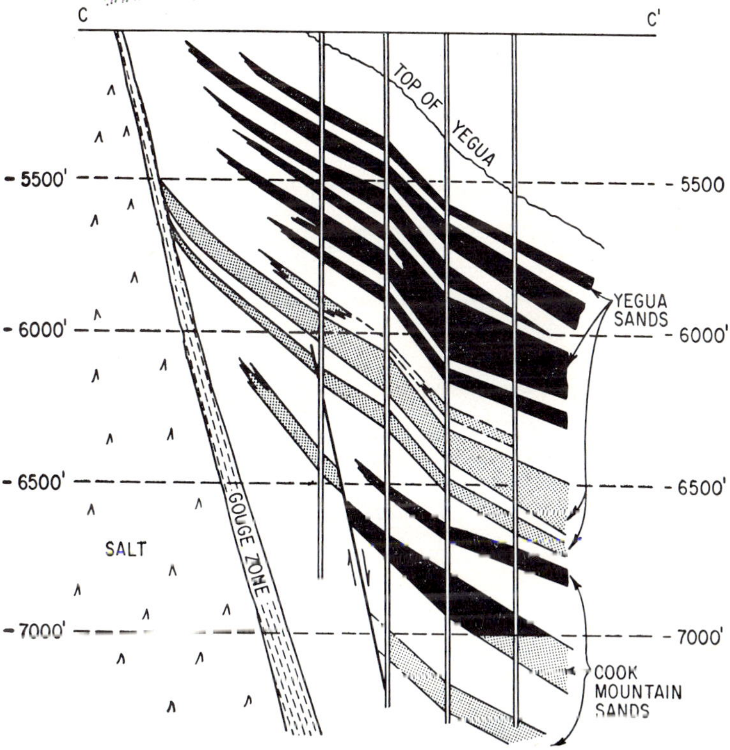

County, Texas. The structure contours are drawn on the top of the dome material. The of a piercement dome. (*Redrawn from Halbouty and Hardin, Bull. AAPG, vol. 35, pp.*

Over 50 different and separate oil traps associated with this structure have been discovered. Many of these are stratigraphic in part, being formed by the updip pinch-out of permeable sands. Some of them are formed by sandstones butting against the dome material, and yet others are developed because of faulting. The great variety and complexity of the traps which may occur on the flanks of a piercement salt-dome structure are evident in this example.

Stratigraphic Factors

There are, in general, three different kinds of stratigraphic factors which can contribute to the formation of traps. The first of these is lateral variation in the deposi-

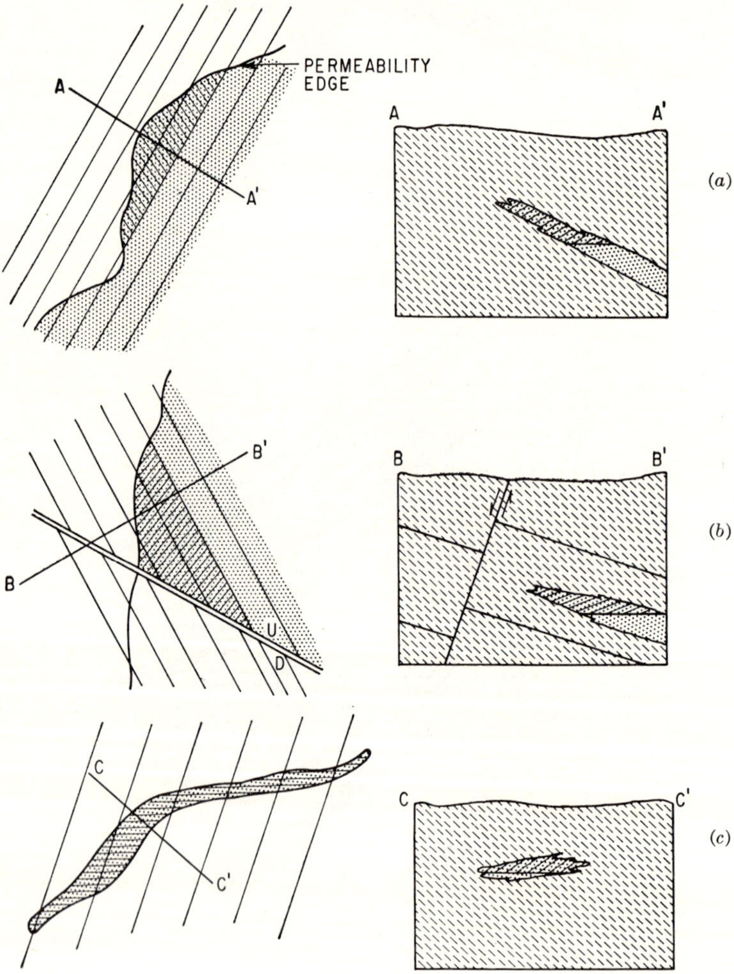

FIGURE 6-41. Schematic diagrams of traps associated with lateral sedimentary variations: (a) edge of a permeable stratum terminating updip in a homocline along a curving trend, (b) updip edge of a permeable stratum intersecting a fault, (c) shoestring sand entirely surrounded by impermeable rocks.

tion of sedimentary rocks. For example, a porous sandstone stratum which is deposited between layers of impermeable shale and which laterally pinches out or grades into impermeable shale will form a lateral boundary for the migration of oil or gas and may therefore contribute to the formation of a trap. A second class of stratigraphic factors are those which are associated with unconformities. In the

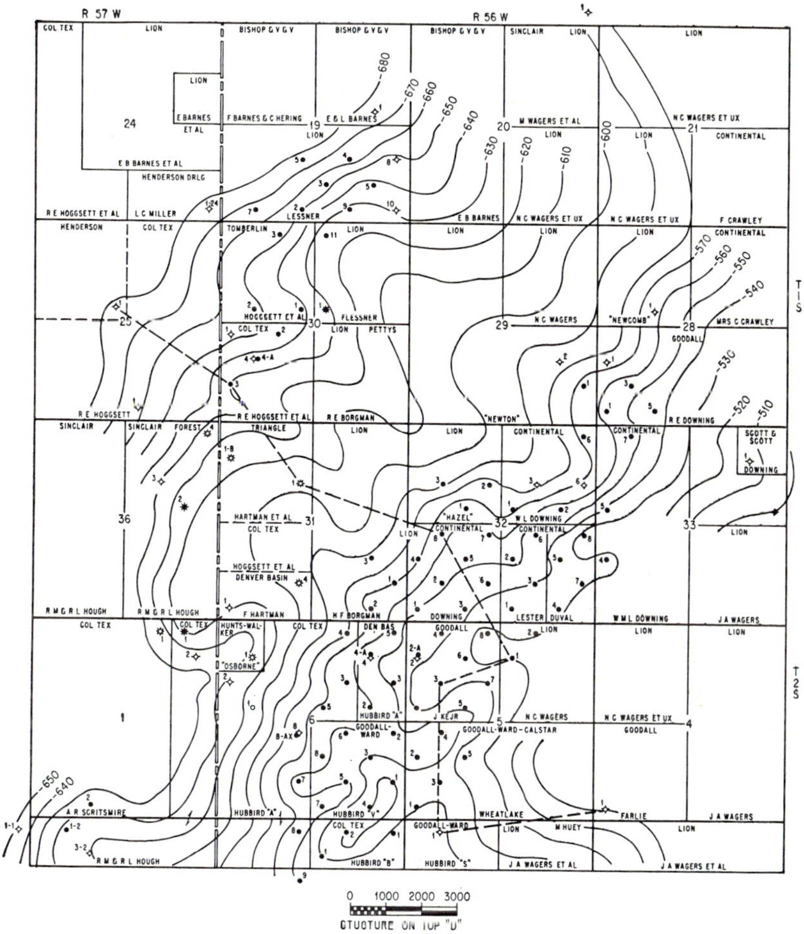

FIGURE 6-42. Structure contour map of the Little Beaver oil field in the Denver Basin, Colorado. Contours are on the top of the *D* sand body which terminates updip along a curving edge. (*Redrawn from Fentress, Bull. AAPG, vol. 39, p. 167.*)

third category are the organic sedimentary structures, biostromes and bioherms, or organic reefs. All these stratigraphic features are characteristically much more difficult to detect than structural features, principally because they constitute minor anomalies in the physical properties of the rocks in which they occur. They are seldom revealed in the surface geology and are frequently not reflected in geophysical measurements.

The most effective way of discovering traps associated with stratigraphic features is by the detailed analysis and comparison of the records obtained from the drilling of many wells in a region. Each stratum can be studied individually. Its lateral variations can be mapped, and unconformities can be detected. Since these kinds of data are available only when the exploration of a region is well underway, it is not surprising

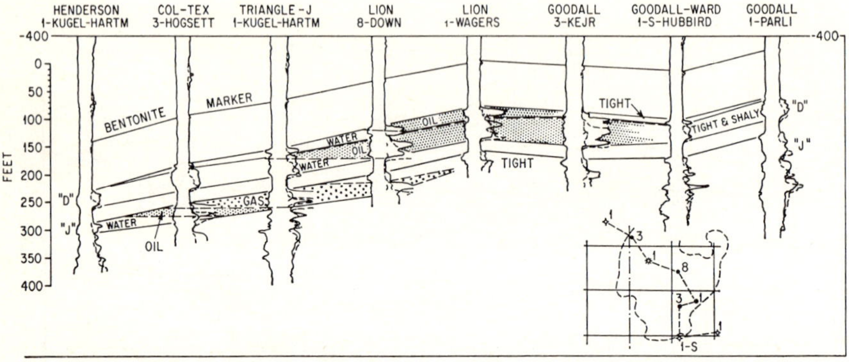

FIGURE 6-43. Structure section across the Little Beaver oil field showing accumulations in the *J* and *D* sands. (*Redrawn from Fentress, Bull. AAPG, vol. 39, p. 162.*)

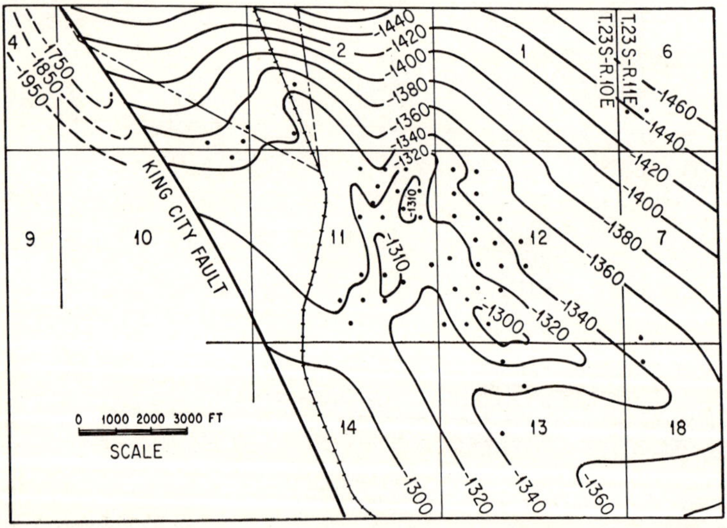

FIGURE 6-44. Structure contour map of the Lombardi sandstone in the San Ardo field in the Salinas Valley of California. The accumulation is controlled by the structural high and by the pinch-out of the Lombardi and Orradre sands along the King City Fault. (*Redrawn from Baldwin, Bull. AAPG, vol. 34, p. 1982.*)

that the structural traps in a new region are characteristically discovered before the stratigraphic traps.

Lateral Variations in Sedimentation. Several of the various ways in which lateral variations in sedimentation may contribute to the formation of traps are shown in idealized form in Figure 6-41. Perhaps the most common situation of this kind is the lateral "pinch-out" of a sand body. It is necessary that both the underlying and overlying strata be impermeable. If a homoclinally dipping sand body terminates updip along a curved edge, as shown in Figure 6-41*a*, a trap will be formed. If the updip edge is linear, it is necessary for some other factor such as a fault (Figure 6-41*b*) to be present before a trap exists.

A trap may also be formed by a lenticular sandstone being entirely surrounded by impermeable strata. The most common example of this situation is in "shoestring" sands in which the sand body is much longer than it is wide (Figure 6-41*c*). Deposits of this kind probably originate as stream channels or as offshore sand bars similar to those which occur today along the Atlantic and Gulf Coasts of the United States.

Although these kinds of traps are most commonly associated with sandstones and shales, they are not limited to these rocks. They might occur in any clastic sediments

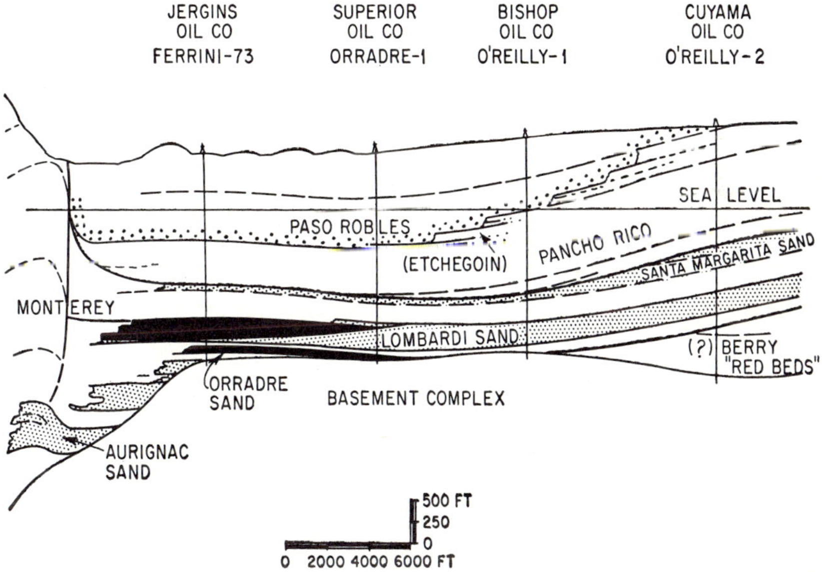

FIGURE 6-45. Southwest-northeast section through the San Ardo oil field. The Lombardi and Orradre sands grade laterally into Monterey shale. (*Redrawn from Baldwin, Bull. AAPG, vol. 34, p. 1986.*)

in which variations in permeability are possible, including organic clastic deposits, shell beds, breccias, arkoses, and many other kinds of rocks.

Some excellent examples of traps associated with sedimentary variations in sandstones and shales are depicted in some of the sections around the South Liberty salt-dome structure as shown in Figure 6-40. These sedimentary variations were undoubtedly caused by the presence of some topographic feature related to the dome itself at the time of deposition of the sediments.

Another good example of traps associated with lateral sedimentary variations is afforded by the Little Beaver oil field in the Denver Basin of Colorado which is depicted in Figures 6-42 and 6-43. Production in this field is obtained from two separate traps associated with updip pinch-outs in sandstones of the Dakota formation of Cretaceous age (Fentress, 1955).

A similar situation occurs in the San Ardo oil field located in the Salinas Valley of central California, which is depicted in Figures 6-44 and 6-45. Here, the Lombardi and Orradre sandstones of Miocene age grade laterally into siliceous Monterey shale

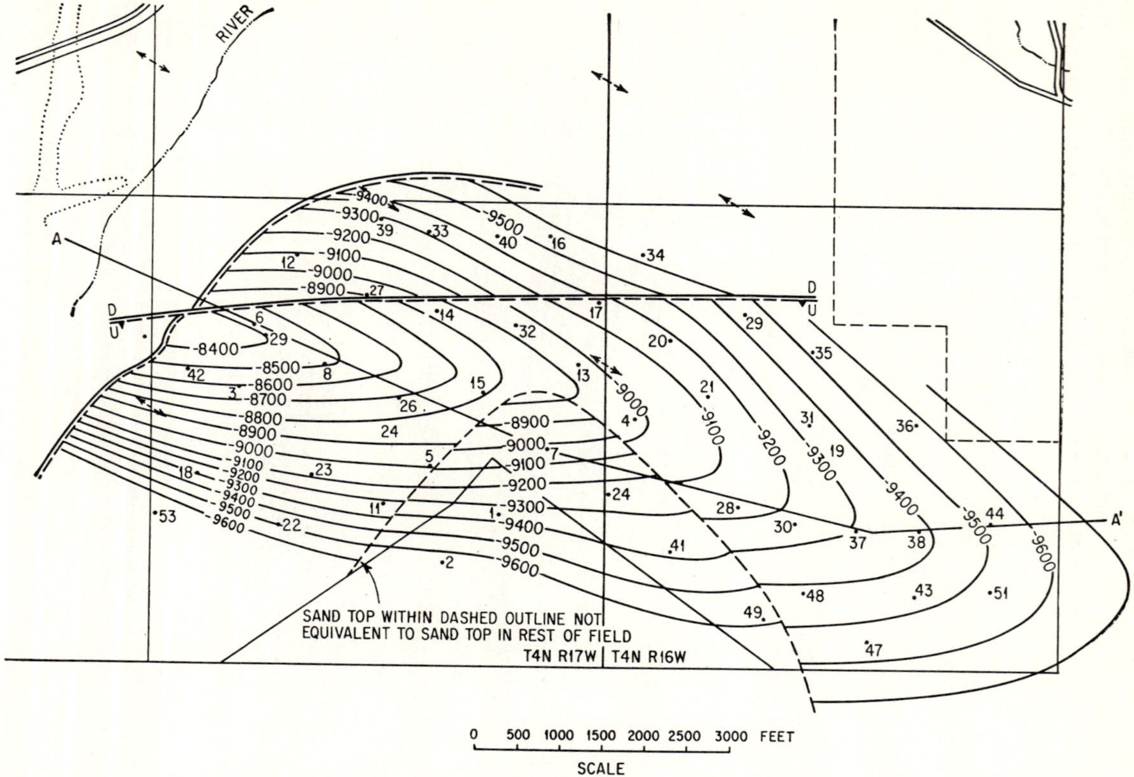

FIGURE 6-46. Contour map of the Castaic Junction oil field in the eastern Ventura Basin, California. Traps are formed in Miocene sands which pinch out updip to the northwest on a plunging anticlinal nose. (*Redrawn from Dudley, "Guide-book," p. 174, AAPG Pacific Section, 1958.*)

to the west. The gradation in lithology is apparently related to ancient topography associated with the King City fault, which is just to the west of the productive limits of the reservoirs. It is interesting to note that the oil-water contact in the Lombardi sand is tilted to the northwest, indicating a hydrodynamic situation (Baldwin, 1950).

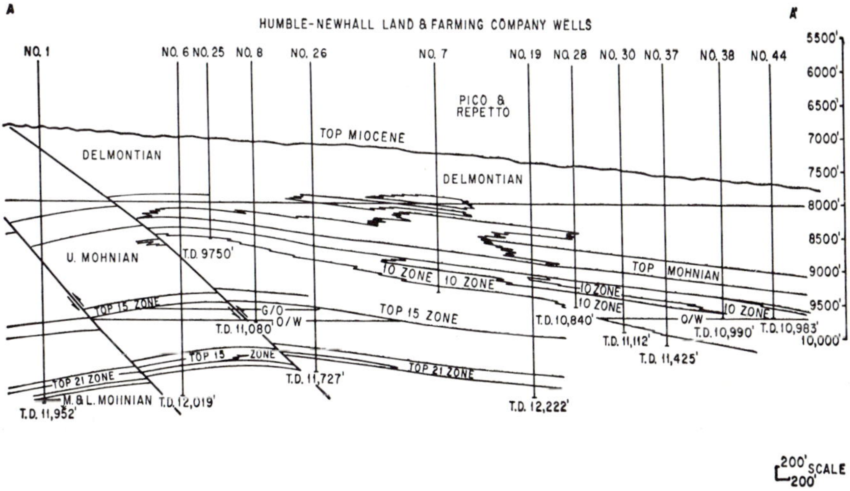

Figure 6-47. Structure section *A-A'* through the Castaic Junction oil field. (*Redrawn from Dudley, "Guidebook," p. 175, AAPG Pacific Section, 1958.*)

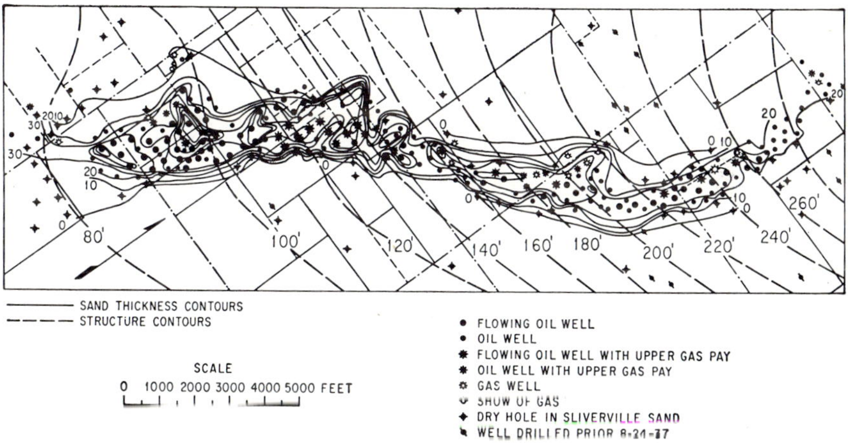

Figure 6-48. The Music Mountain oil field, McKean County, Pennsylvania. The solid lines are isopach contours showing the thickness of the producing Sliverville sand; the dashed lines are structure contours on the top of the underlying third Bradford sand. The producing sand body is apparently an ancient offshore sand bar. (*Redrawn from Fettke, in "Stratigraphic Type Oil Fields," pp. 493, 498, AAPG, 1941.*)

Another example of traps associated with lateral variation in sand deposition is found in the Castaic Junction oil field located at the eastern end of the Ventura Basin in California (Dudley, 1958) (Figures 6-46 and 6-47). Traps are formed by lensing sands of Miocene age which pinch out to the northwest across a southeastward plunging anticlinal nose.

An example of a trap formed by a shoestring sand entirely surrounded by shales is afforded by the Music Mountain oil field in northwestern Pennsylvania (Fettke, 1941), which is depicted in Figure 6-48. Isopach contours of the thickness of the producing Sliverville sandstone of Upper Devonian age are shown. Dashed structure contours are shown on the top of the Bradford third sand, which lies 240 feet below the Sliverville. It is apparent that the accumulation bears very little relation to the structure and is controlled primarily by the presence of the sand body.

Unconformities. In general, there are two ways in which unconformities may contribute to the formation of traps. These are depicted in idealized form in Figure 6-49. Perhaps the most common occurrence is the truncation of a permeable stratum by the unconformity with subsequent deposition of an impermeable stratum on the truncated edge. Of course, for a trap to be formed it is necessary that the permeable stratum be both overlain and underlain by impermeable strata (Figure 6-49*a*). The second way in which unconformities may contribute to the formation of traps arises when permeable strata are laid down over and lapping against a tilted unconformity surface as shown in Figure 6-49*b*. Again, an impermeable stratum is required to be deposited over the

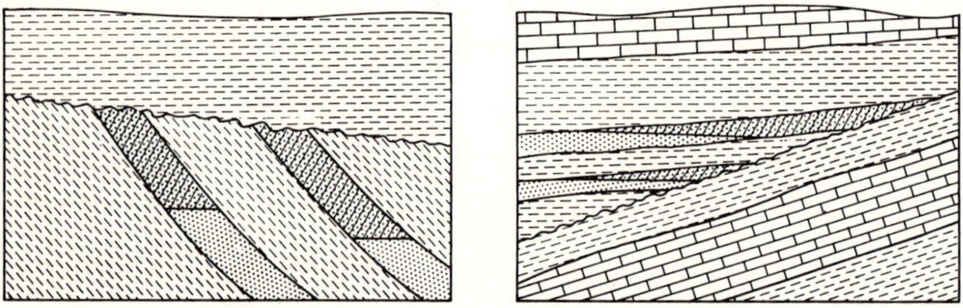

(*a*) (*b*)

FIGURE 6-49. Schematic diagrams of traps associated with unconformities: (*a*) truncation of permeable strata and subsequent deposition of impermeable rocks on truncated edge, (*b*) deposition of "buttress sands" lapping on to the unconformity surface.

permeable one. Sandstones which occur in this configuration are frequently termed "buttress sands."

Perhaps the classic example of a trap formed by unconformity surfaces is that of the east Texas oil field (Minor and Hanna, 1941). This is illustrated in Figures 6-50 and 6-51. It is a single trap in the prolific Woodbine sandstone of northeastern Texas. Both the upper and lower surfaces of the trap are represented by unconformities. The regional Sabine uplift which culminates some 80 miles to the southeast has created a broad structural nose which together with the two unconformity surfaces forms a very large trap. This field is the largest yet discovered in the Western Hemisphere and has an expected ultimate recovery of about 6 billion barrels of oil.

It is interesting to note that traps associated both with lateral variations in sedimentation and with unconformity surfaces have a tendency to occur along the same trend lines. It has been suggested that regions such as these are likely to occur along the edge of a basin shelf in the transition zone between thinly deposited strata on the shelf itself and the more thickly deposited strata toward the middle of the basin. This transition zone is therefore frequently termed a "hinge line" and is regarded as a favorable location for the occurrence of traps.

An example of a probable hinge-line region is afforded by the huge Pembina accumulation in western Canada (Patterson and Arneson, 1957) which is depicted in Figures 6-52 and 6-53. Multiple traps are present, and factors which have contributed

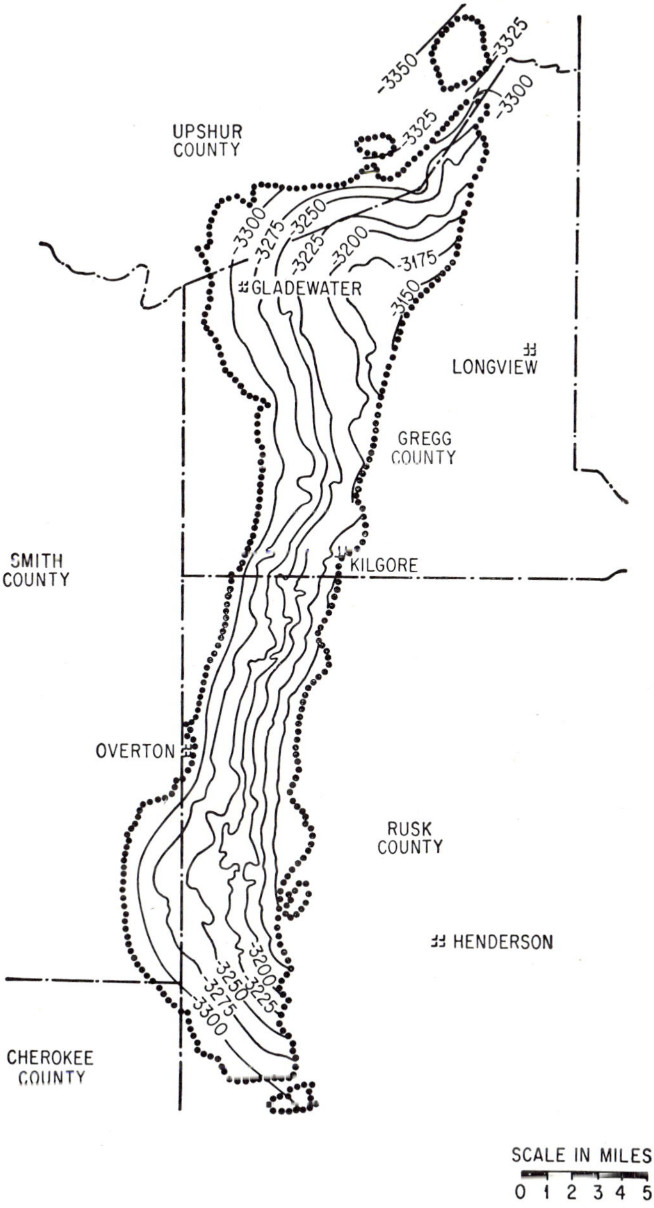

FIGURE 6-50. Structure contours of the producing Woodbine sand in the east Texas oil field. The trap is formed in the Woodbine by two bounding unconformity surfaces which have been gently arched by the Sabine uplift to the east. (*Redrawn from Minor and Hanna, in "Stratigraphic Type Oil Fields," p. 617, AAPG, 1941.*)

toward them include buttress sands, truncated sands, and lateral changes in lithology. The largest reservoir is represented by the Cardium sandstone, and although it is locally irregular and variable in thickness, it pinches out to the east. The entrapped oil covers an area of probably more than 800 square miles. The total recoverable reserves have been estimated to be in excess of 1 billion barrels of oil.

An excellent example of a field in which several traps have been formed by unconformity surfaces as well as other factors is afforded by the Midway-Sunset field of California,

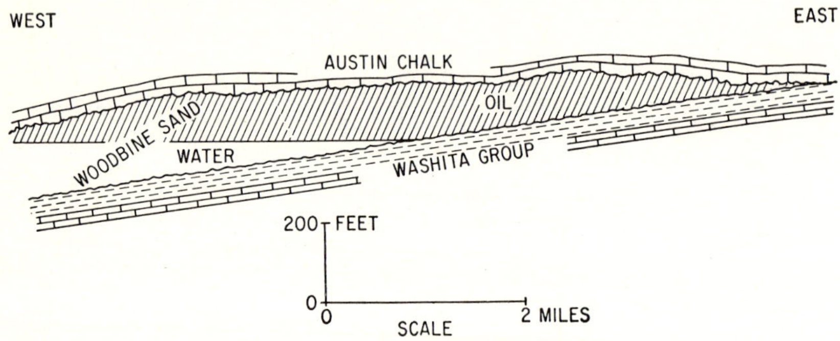

FIGURE 6-51. Schematic structure section across the east Texas oil field showing intersection of unconformity surfaces above and below the Woodbine reservoir.

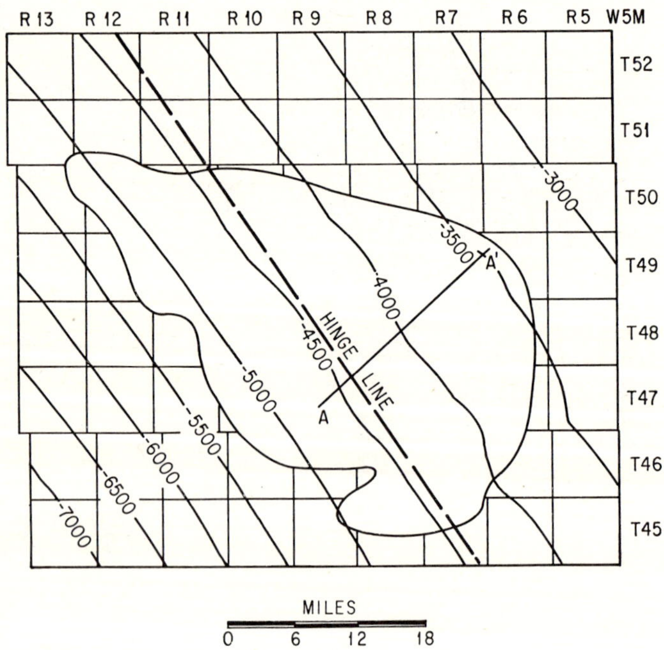

FIGURE 6-52. Map of the Pembina oil field, Alberta, Canada. The outline indicates the probable productive limits of the field. The contours are on the Mississippian Banff formation, and a "hinge-line" region is suggested by the steepening of the regional dip. (*Redrawn from Patterson and Arneson, Bull. AAPG, vol. 41, p. 947.*)

of which a generalized structure section is shown in Figure 6-54. In this field, anticlinal structure, truncations, buttress sands, and pinch-outs have all contributed to the formation of multiple traps in Pliocene and Miocene sandstones (Hoots et al., 1954).

Organic Deposits. Organic sedimentary structures have been responsible for forming the traps for many significant accumulations of petroleum. They are frequently

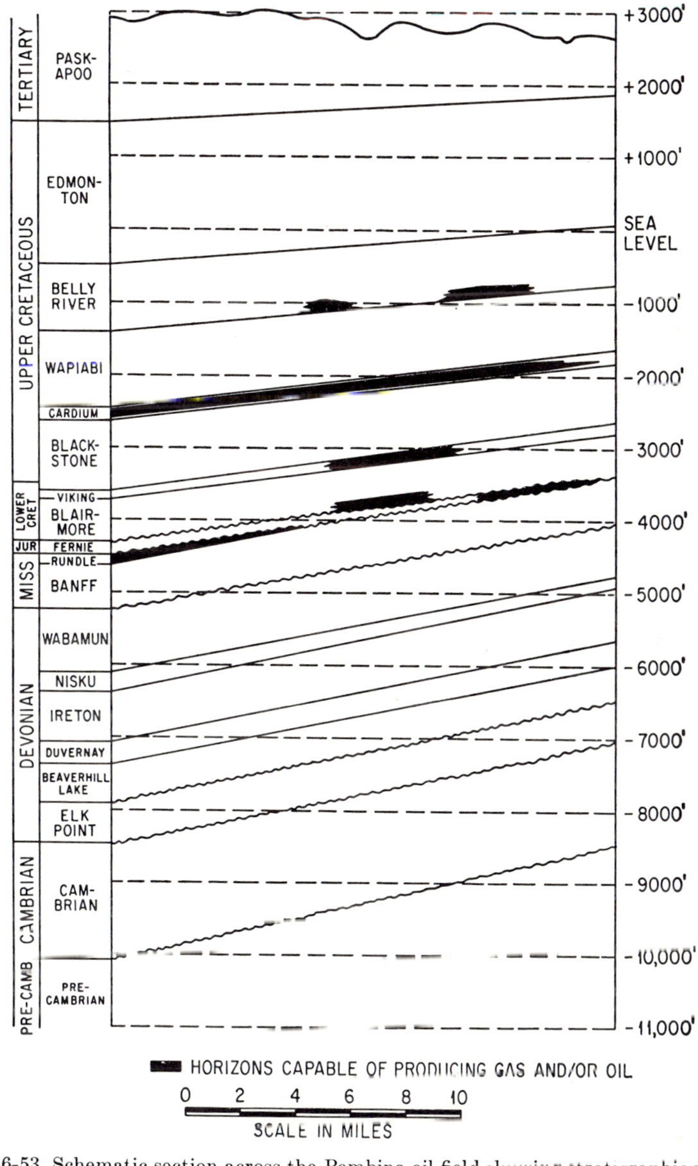

FIGURE 6-53. Schematic section across the Pembina oil field showing stratigraphic relations without structure. Horizons capable of producing oil or gas are indicated in black. (*Redrawn from Patterson and Arneson, Bull. AAPG, vol. 41, p. 940.*)

divided into two classes called biostromes and bioherms (Cumings and Shrock, 1928). Biostromes are simply bedded layers containing a high percentage of shells and other similar organic remains. If they are permeable, biostromes may form traps in the same way as other permeable sedimentary strata. On the other hand, bioherms, or reefs as they are commonly called, are moundlike structures which are the result of growth and accumulation of organic material in one localized area. These structures may form traps wherever they have developed sufficient permeability and are covered by an impermeable stratum.

Organic reefs are found today in the process of formation at many places in the world, and study of them has shed much light on the characteristics of ancient or fossil reefs which, in general, were formed in a similar fashion. Modern bioherms have popularly been termed "coral reefs." However, corals are only one of the organisms which contribute to their formation. Among others are algae, sponges, crinoids, mollusks, brachiopods, bryozoans, and foraminifera. In spite of the wide variety of different organisms and the consequent differences in the structure and form of modern

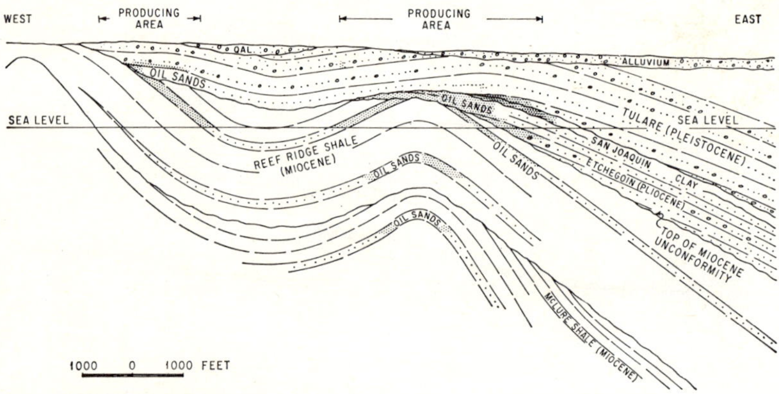

FIGURE 6-54. Schematic section through the Midway-Sunset oil field, San Joaquin Valley, California. Traps are formed by truncation, buttress sands, pinch-outs, and anticlinal structure in Pleistocene, Pliocene, and Miocene sands. (*Redrawn from Hoots, Bear, and Kleinpell, Calif. Div. Mines Bull. 170, chap. 9, p. 31.*)

reefs, there seem to be a number of common and essential factors for their formation. Reasonably warm water on the order of 65 to 70°F is required. The organisms also need some light, and as a result, reefs grow only in relatively shallow water, certainly not more than 50 fathoms. Because of this depth requirement reefs are frequently, although not always, associated with shore lines. Their size and shape may vary greatly. At one extreme are structures many miles in length and width such as the Great Barrier Reef along the northern coast of Australia. This extends for a length of over 1,000 miles and averages nearly 100 miles in width. At the other extreme are structures such as the typical South Pacific coral atoll with its sandy island and surrounding circular reef. Reefs may also occur in isolated moundlike masses.

Ancient reefs have been found throughout the world in sediments of every geologic age from the Cambrian onward. Although many of these were formed by different organisms from those found in modern reefs, their characteristics seem to be quite similar. Although they are not always productive, reef structures have accounted for some of the most prolific oil fields which have yet been discovered. This may perhaps be due to the nature of the permeability and porosity which are commonly found in

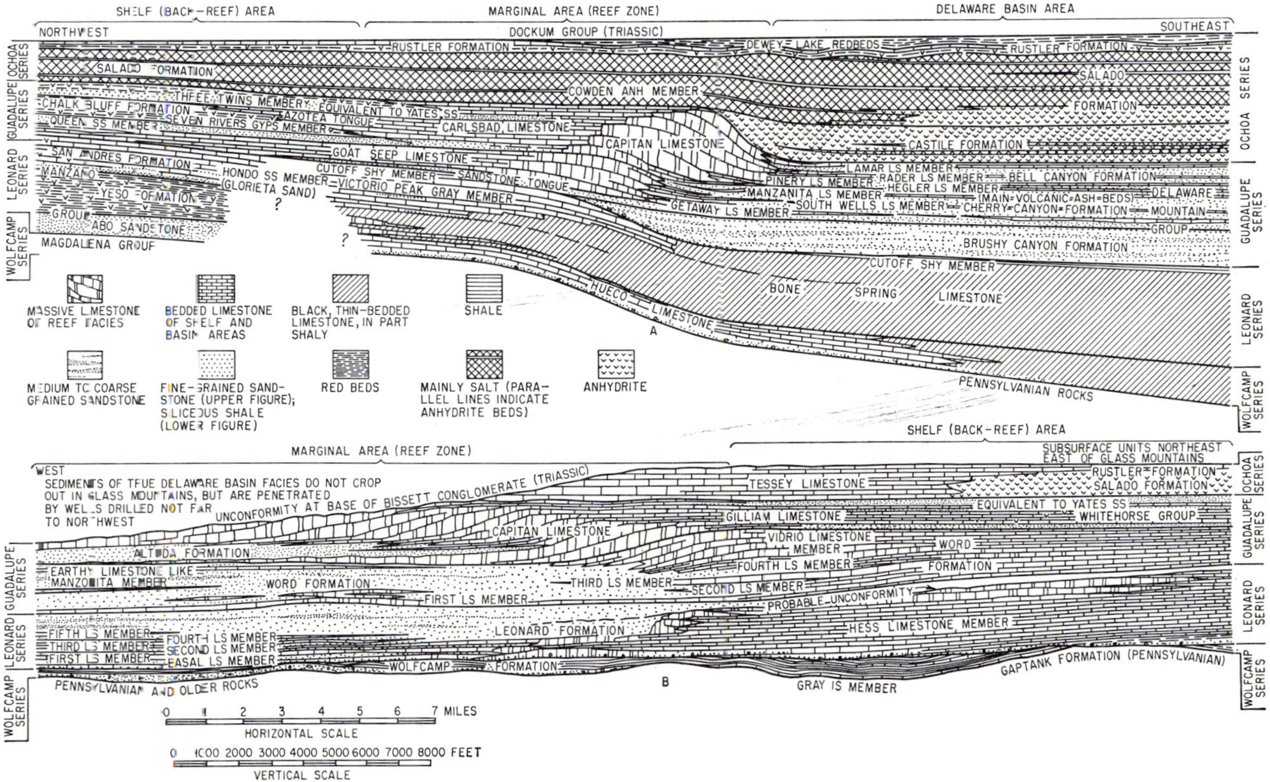

FIGURE 6-55. Stratigraphic diagrams of the Capitan Reef and adjacent formations in west Texas. The upper section is from the Guadalupe Mountains and is drawn normal to the strike of the reef trend. The lower section is from the Glass Mountains and is drawn obliquely to the strike of the reef trend. (*Redrawn from King, Bull. AAPG, vol. 26, p. 542.*)

them. In the development of the reef itself, many of the organisms involved grow in a branching structure which produces large numbers of cavities. Often much of the reef material is the result of deposition and accumulation of coarse permeable clastic material which has been broken off by wave action. Some of the organisms which live in the reef are of the kind which bore through the solid material, forming natural cavities with extremely high permeability. All these factors can be enhanced by secondary effects after the reef has been buried. For example, solution of the reef material may occur through the agency of circulating ground water, or as is frequently the case the reef may become dolomitized. Not all the factors involved contribute to the formation of high permeability and porosity, and there are a number of processes

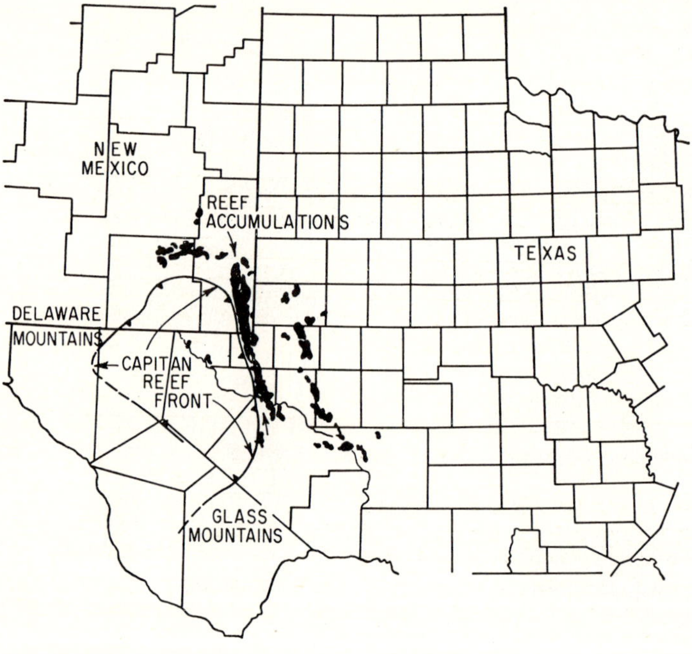

FIGURE 6-56. Regional map of west Texas and southeastern New Mexico showing oil and gas fields producing from Permian upper Guadalupe strata and the location of the Capitan Reef front. The line of fields adjacent to the reef front produces from the Capitan limestone. (*Redrawn from Galley, in "Habitat of Oil," p. 438, AAPG, 1958.*)

such as recrystallization of the reef material and the growth of other kinds of organisms which may tend to fill up cavities and voids which are present. In general, however, reefs are usually highly permeable and porous, giving rise to prolific accumulations.

On the other hand reef traps characteristically are very difficult to discover. They usually do not represent large or significant changes in the physical properties of the rocks in which they occur and are therefore not easily detected by geophysical or structural methods. Many of the known reef traps have probably been discovered by accident in the search for other kinds of traps. A possible tool which can be used in the exploration for reefs is based on the probability of their occurrence in association with shore lines. Analysis of already discovered reef traps has revealed that sediments of the same age on the basinward, or fore-reef, side of a bioherm are characteristically a deeper water marine deposit whereas those on the landward, or back-reef, side are frequently evaporites, red beds, or shallow-water or continental deposits. Many of these features can be correlated with those found in modern reef structures. The location of an undiscovered reef might well be revealed by very careful studies of the regional facies relationships of various strata (Cloud, 1952).

One of the most spectacular examples of a reef structure which has formed several prolific oil fields is afforded by the Capitan Reef in the Permian Basin of west Texas and southeastern New Mexico (Lloyd, 1929; King, 1938, 1942, 1948; Galley, 1958). Two sections through the reef drawn from its surface outcrops are depicted in Figure 6-55, and a map showing its regional trend is presented in Figure 6-56. As shown on the map, it forms a circular arc commencing in the Glass Mountains of west Texas, extending along the Pecos River and into southeastern New Mexico and thence back into west Texas, where it culminates at present in the massive exposed topographic feature known as El Capitan. The map shows oil accumulations in upper Guadalupe strata which include the Capitan Reef formation. The first line of oil fields immediately adjacent to the reef front occur within the reef itself. The Capitan Reef is particularly interesting, since it has been studied extensively and its outcrop constitutes one of the best known examples of an exposed fossil reef structure.

Another example of a series of prolific reef traps is the Devonian reef reservoirs of Alberta Province in western Canada (Downing and Cooke, 1955). Figure 6-57 shows structure contours on the surface of a portion of the Leduc trend, and Figure 6-58 a section indicating the relation of the reef to surrounding sediments. Some of the individual fields along this trend are expected to produce over 500 million barrels of oil.

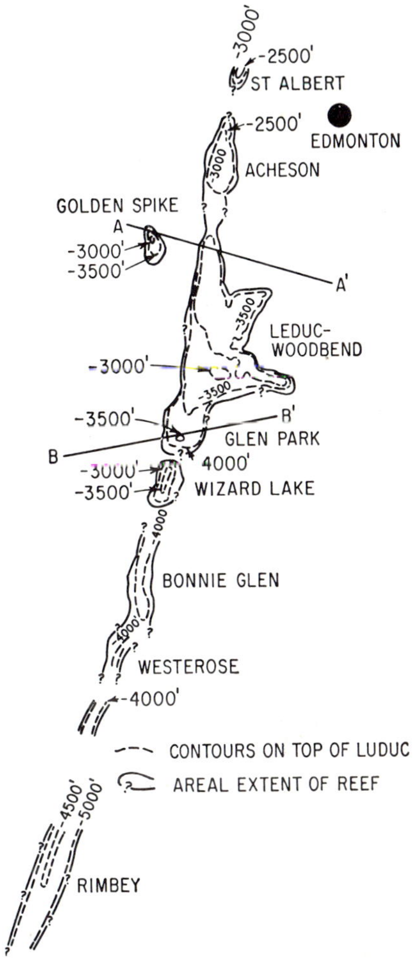

FIGURE 6-57. Map of the Rimbey-Leduc-Acheson Devonian reef trend in Alberta, Canada. The areal extent of the reefs together with structural contours on their surface are shown. Only the structurally highest portions of the reefs are productive. (*Redrawn from Downing and Cooke, Bull. AAPG, vol. 39, p. 198.*)

Another spectacular example of oil accumulation in a series of reef traps is afforded by the Golden Lane trend (Guzman and Mina U., 1956; Rockwell and Rojas, 1953) on the eastern coast of Mexico which is depicted in Figure 6-59. Here production is from the El Abra reef limestone of Cretaceous age and is expected to reach well over a billion barrels of oil. A similar reeflike structure accounts for the Poza Rica field shown several miles to the west of the Golden Lane trend. This field itself is expected to produce more than a billion barrels of oil.

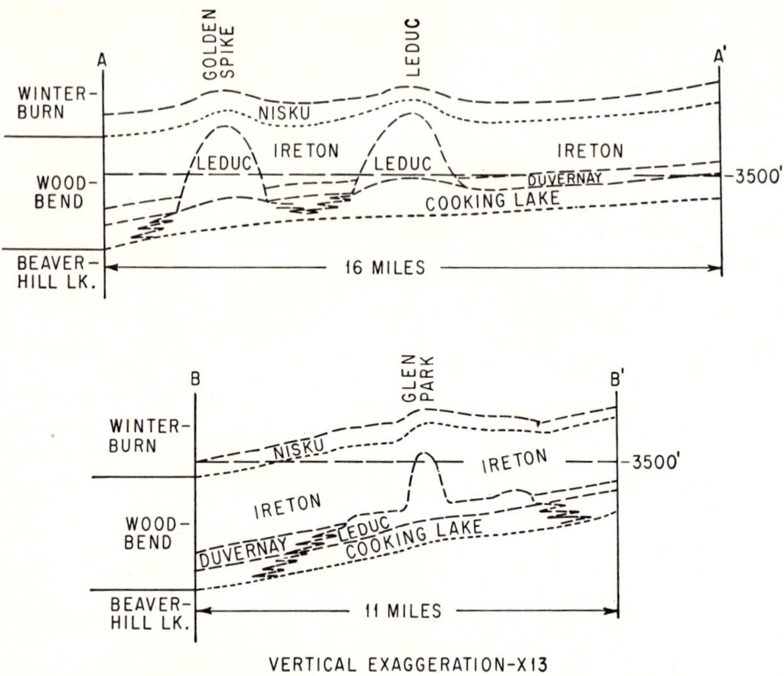

FIGURE 6-58. Schematic cross sections of Devonian reefs in the Leduc area. (*Redrawn from Downing and Cooke, Bull. AAPG, vol. 39, p. 200.*)

Miscellaneous Factors

There are a number of factors capable of producing traps which are of sufficiently unusual character that they can best be discussed separately from the structural and stratigraphic factors which have already been described. Specifically, these are the fractured reservoirs, traps formed by secondary chemical alteration of limestones to dolomites producing changes in permeability, and traps formed by the sealing of permeable strata by asphalt near the surface of the ground.

Fractured Reservoirs. A great many oil fields produce from fractured reservoirs. In many of these, the fracturing is only a secondary factor in the formation of the trap, whereas in a few it is a primary and essential feature.

An interesting example of a trap formed principally by a fractured reservoir is the huge Spraberry accumulation in west Texas (Figure 6-60). This field covers an area of nearly 1,000 square miles, and as shown in the figure, it bears no direct relationship to the structure. The oil is produced from the Spraberry sandstone of Lower Permian age. The average permeability of the unfractured reservoir rock is less than 1 millidarcy, and production is obtained principally by means of an extensive system of vertical fractures occurring within the formation. Most of the wells within the field have required artificial stimulation by means of hydraulic fracturing before significant production could be obtained.

Another interesting example of a fractured reservoir is afforded by the old Florence oil field which produces from fractures within the Upper Cretaceous Pierre shale at the southern end of the Denver Basin in Colorado (de Ford, 1929; McCoy et al., 1951).

A section through this old Florence oil field is depicted in Figure 6-61. Production here is entirely from fractures within the shale, and no water accompanies the oil. Apparently, solid unfractured shale surrounds the entire reservoir and forms the trap.

Secondary Chemical Alteration. Another factor sometimes contributing to the formation of traps is the increase in porosity and permeability which often occurs when

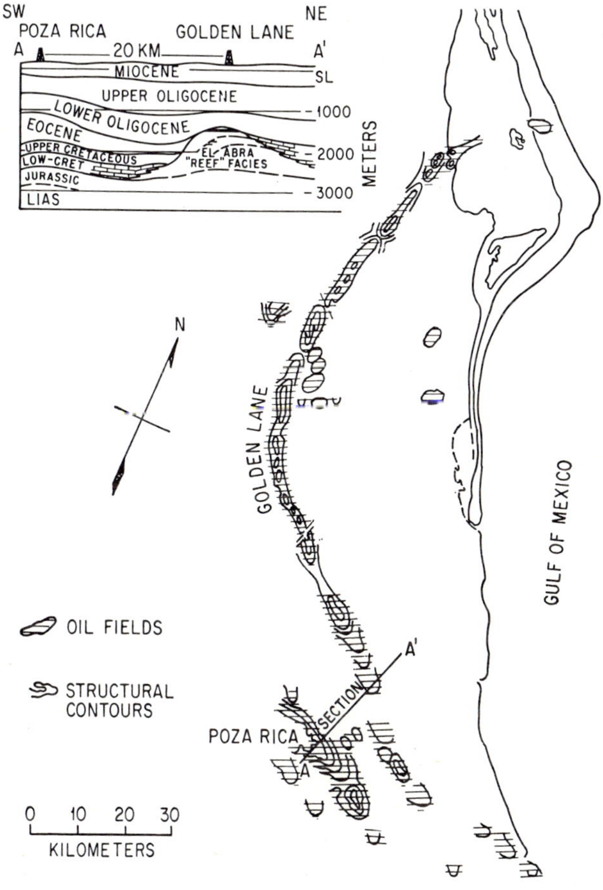

FIGURE 6-59. The Golden Lane trend on the eastern coast of Mexico. Schematic contours are shown on the top of the Cretaceous El Abra reef limestone. (*Redrawn from Guzman and Mina U., Bull. AAPG, vol. 40, p. 1492, and Rockwell and Rojas, ibid., vol. 37, p. 2555.*)

limestones undergo secondary alteration to dolomite (Landes, 1946). This process appears to occur by the replacement of some of the calcium within the limestone by magnesium carried by moving ground water. The exact mechanism and the manner in which permeability and porosity are thereby created are not well understood.

Excellent examples of traps occurring within patches of permeable dolomitized limestone are afforded by the individual pools of the Lima-Indiana oil field of Ohio and Indiana (Figure 6-64). The oil occurs in the Ordovician Trenton limestone and is almost entirely controlled by local dolomitization (Orton, 1888, 1889; Carman and

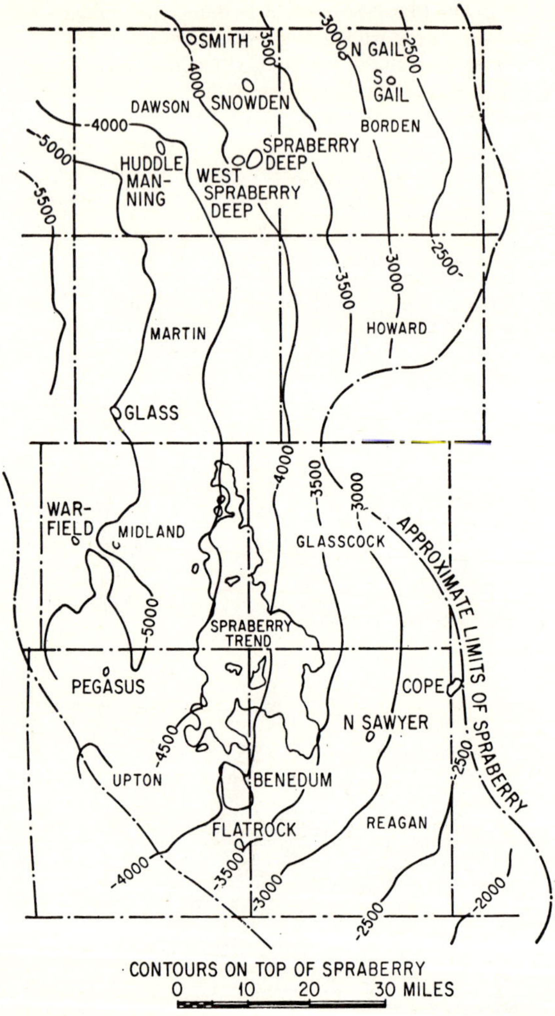

FIGURE 6-60. Structure contour map of the Spraberry sandstone in west Texas showing areas of oil accumulation. (*Redrawn from Hubbert and Willis, Proc. 4th World Petrol. Congr., Sect. I/A/1, p. 69.*)

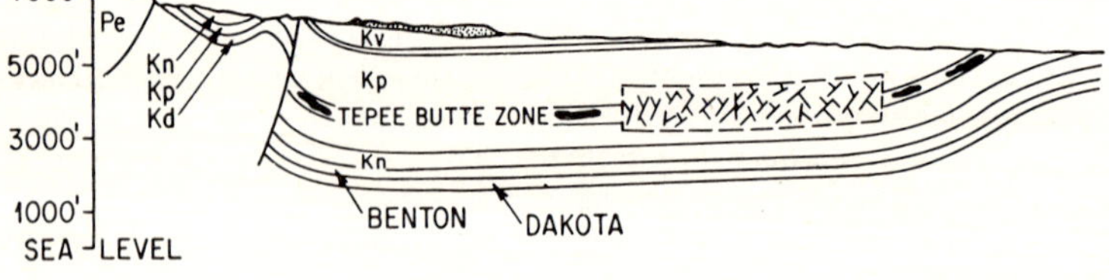

FIGURE 6-61. Section through the Florence oil field, Colorado. The trap is formed by fractures within the Cretaceous Pierre shale. No water is produced with the oil. (*Redrawn from McCoy et al., Bull. AAPG, vol. 35, p. 1008.*)

Stout, 1934). The accumulations are also interesting in that they appear to be influenced by hydrodynamic effects over a large region, as described below.

Asphaltic Seals. One unusual factor which can contribute to the formation of traps is the plugging of the pores of an otherwise permeable rock by asphaltic material. This situation is likely to occur where a sandstone or other impermeable rock outcrops at the surface and oil which has moved up the dip comes in contact with the atmosphere and is oxidized and thickened to form an asphaltic material. Some of the accumulations associated with the Midway-Sunset oil field in the San Joaquin Valley of California (Figure 6-54) have been formed in this way.

Hydrodynamic Effects

In the first section of this chapter the physical principles governing the accumulation of oil were reviewed and it was shown that under hydrostatic conditions in which the environmental ground water is not in motion, traps will be formed by any impermeable structural surface which is concave downward overlying a reservoir rock.

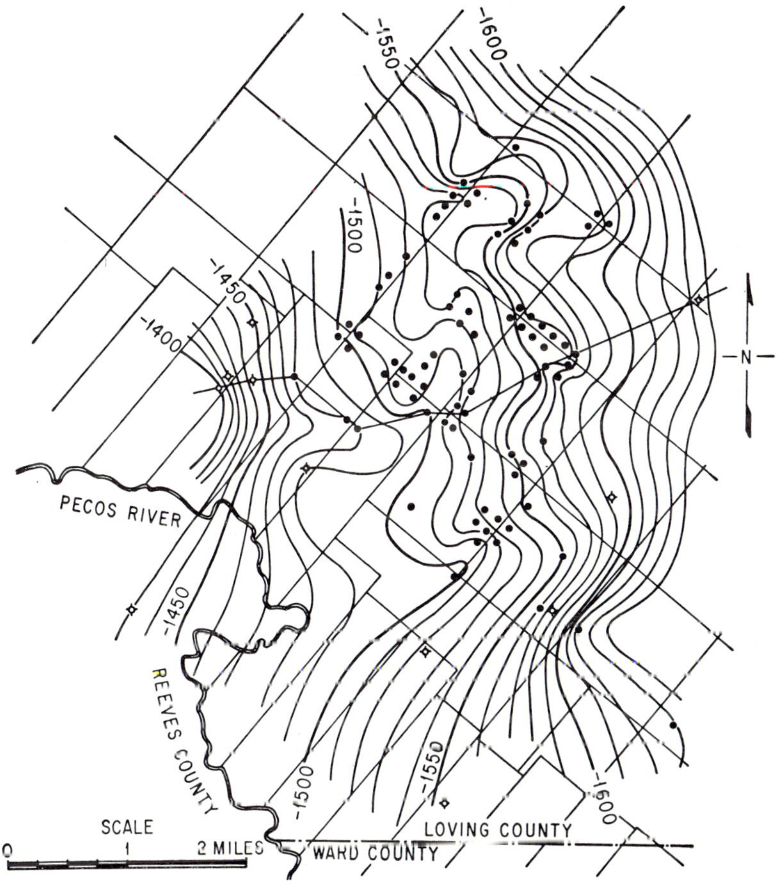

FIGURE 6-62. Structure contour map of the Wheat field in the Delaware Basin of west Texas. The trap is probably formed by a hydrodynamic gradient across the open structural terrace. (*Redrawn from Adams, Bull. AAPG, vol. 20, pp. 780–796.*)

However, under conditions of hydrodynamics in which the environmental ground water is in motion, traps will be formed by structural geometry which is concave in the direction of increasing potential for oil or gas. In general, the effects of hydrodynamics may range from a simple tilting of the oil-water or gas-water interface in the direction of water flow to a profound modification of the position of the trap to the extent that normally closed structures will not hold oil and that normally unclosed structures such as anticlinal noses will form excellent traps.

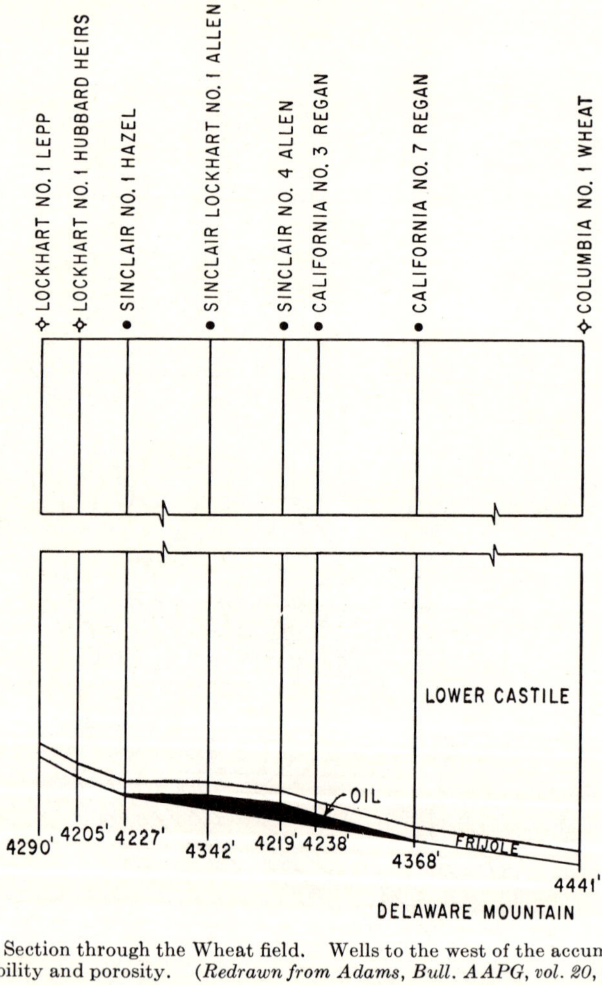

FIGURE 6-63. Section through the Wheat field. Wells to the west of the accumulation show good permeability and porosity. (*Redrawn from Adams, Bull. AAPG, vol. 20, pp. 780–796.*)

Some of the examples of traps already presented show the effects of hydrodynamic conditions. In the Kettleman Hills oil field (Figure 6-17), the oil-water contact shows a tilt of 50 feet per mile in a northwesterly direction, indicating that water is flowing through the reservoir sands in that direction. The water in these sands is of low salinity, providing further evidence of a hydrodynamic situation. The San Ardo oil field in the Salinas Valley of California (Figures 6-44, 6-45) also shows the effects of hydrodynamics. The oil-water contact in the Lombardi sand is tilted approximately

40 feet per mile to the northwest. A map showing contours on the oil-water contact has been presented by Baldwin (1950).

In these examples hydrodynamic effects are represented by relatively gentle tilts of the oil-water contact. However, in each case the difference in the area covering the

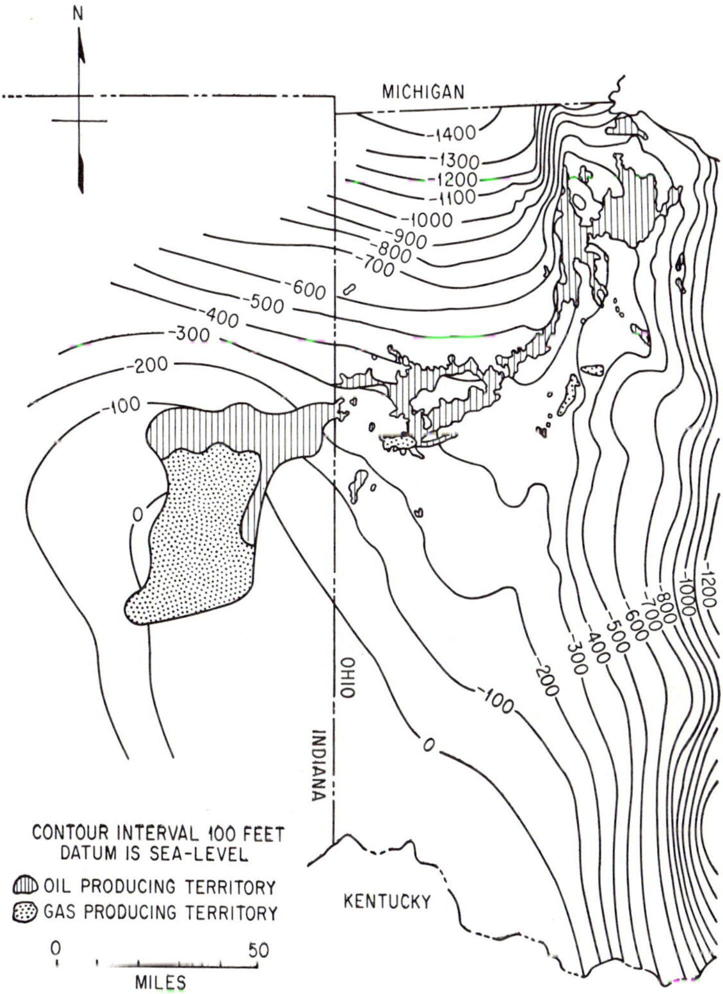

FIGURE 6-64. Map of the Lima-Indiana oil field of Indiana and Ohio showing structure contours on the surface of the Trenton limestone. Production is locally controlled by dolomitization and permeability of the Trenton but regionally appears to be affected by a hydrodynamic gradient. (*Redrawn from Carman and Stout, in "Problems of Petroleum Geology," p. 522, AAPG, 1934.*)

existing trap and that which would cover a hydrostatic trap containing the same amount of oil represents many thousands of acres and is therefore extremely significant in the exploitation of the accumulations.

There exist a number of more spectacular examples of hydrodynamic traps. One of these is the Wheat field in the Delaware Basin of west Texas (Adams, 1936). This is

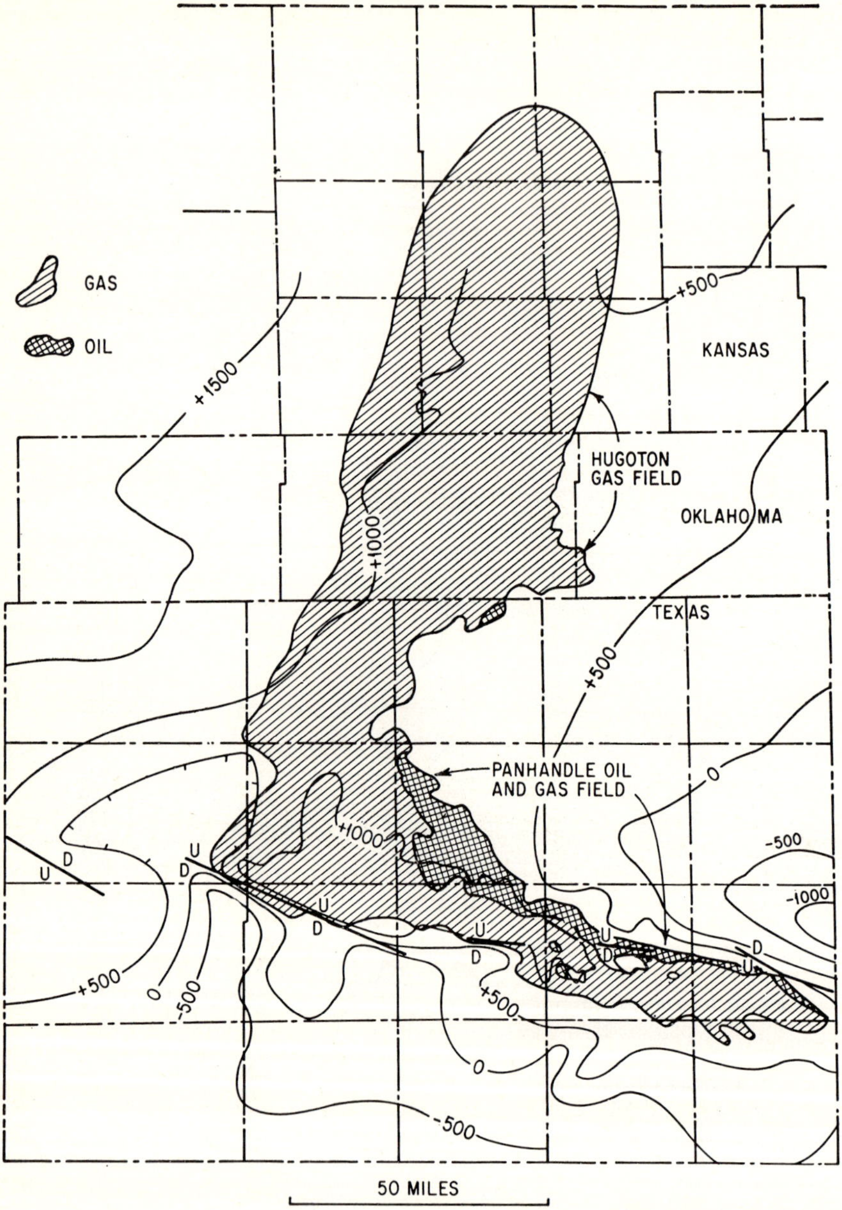

FIGURE 6-65. Map of the Hugoton gas field and the Panhandle oil and gas fields. Structure contours are shown on the top of the "Big Lime." The traps are apparently strongly modified by hydrodynamic effects with both the oil-water and gas-water contacts tilted eastward. The gas-oil contact is horizontal, and the oil is in the lowest structural position of both reservoirs. (*Compiled from Lilly, Bull. AAPG, vol. 42, p. 1235; Hemsell, ibid., vol. 23, p. 1055; and Cotner and Crum, in "Geology of Natural Gas," p. 387, AAPG, 1935.*)

illustrated in Figures 6-62 and 6-63 showing a structure map and a section through the accumulation. Oil is trapped in a sandstone stratum of the Delaware Mountain series of Permian age. The regional dip is to the east, and the field is located on a structural terrace having no structural closure to the west. There is no indication of faulting, and wells, as shown in the diagram, on the west and southwest of the field show excellent permeability within the reservoir rock. These facts have been pointed out by Adams in his discussion of the field, and it appears highly probable that the accumulation is a hydrodynamic one. This interpretation is consistent with the regional topography, since the Delaware Mountains to the west would be a natural source of water flowing eastward into the Delaware Basin.

Another very interesting example is afforded by the old Lima-Indiana oil field which extends for a distance of about 150 miles from central Ohio into eastern Indiana (Figure 6-64). It actually consists of a great many separated areas of production. The oil has accumulated in the 600-foot-thick Trenton limestone of Ordovician age. The field was discovered in 1885 and was described in detail by Edward Orton (1888, 1889). The oil-water contact occurs at about 650 feet below sea level in central Ohio and gradually rises to the southwest until it is about 50 feet above sea level in Indiana. This evidence, together with the presence of gas in the structurally highest portion of the field in eastern Indiana, indicates the strong possibility of hydrodynamic tilt. Another major factor in determining the location of the accumulation is the dolomitization of the Trenton limestone. The actual producing areas appear to be closely correlated with this chemical alteration. However, they appear to be interconnected as evidenced by the related oil-water contacts. Therefore the principal effect of the dolomitization is apparently the creation of sufficient local permeability to permit commercial production. The over-all tilt of the oil-water contact is on the order of 2 or 3 feet per mile. This effect could be produced by a hydrodynamic gradient of a few tenths of a foot per mile, a situation which could easily be present in this region.

Another interesting example of a probable hydrodynamic trap of very large dimensions is afforded by the Hugoton gas field and the adjacent Panhandle oil and gas field which occur in western Kansas and in the Panhandles of Oklahoma and Texas (Rogatz, 1939; Hemsell, 1939; Garlough and Taylor, 1941; Cotner and Crum, 1935; Lilly, 1958). A map of this region is depicted in Figure 6-65 showing the oil- and gas-producing areas and regional structural contours on the top of the Panhandle Big Lime formation of Permian age. Production is obtained from Permian and Pennsylvanian rocks. The initial gas pressure within both the Hugoton and the Panhandle fields was very nearly constant throughout, ranging from approximately 430 to 440 psi. No structural closure exists to the west of the Hugoton field, and this trap has been classed as a stratigraphic one on the basis that the limestone and dolomite reservoir rocks grade into red shales and silts some 20 miles west of the field. However, according to Garlough and Taylor (1941), the gas-water contact varies in elevation from about 500 feet above sea level on the west edge of the field to about 120 feet above sea level on the eastern edge of the field with an average tilt of 12 feet per mile. Since the gas, initially, was at nearly constant pressure, the inclined gas-water contact was an equipressure surface. Thus the equipressure surfaces in the water must have been tilted downward toward the east. This could happen only if water were flowing down the dip from the Rocky Mountain region to the west. This is strong evidence of a hydrodynamic condition which could arise from water flowing down the dip from the Rocky Mountain region to the west. A possible mechanism for the accumulation may be a lower permeability in the reservoir rocks just to the west of the field, creating a situation similar to that depicted in Figure 6-14. The associated Panhandle oil and gas field shows a tilt of the oil-water contact to the east on the order of 10 feet per mile. Considering the two fields together, the oil is found in the structurally lowest portion of both with an essentially horizontal contact between the gas and the oil, as should be expected if the two fields jointly form a single trap. It is interesting to note that in all probability any oil which ever entered the trap migrated in it to the lowest portion, ending up in Panhandle oil accumulation.

SUMMARY

At the beginning of this chapter, it was pointed out that among the most valuable tools which can be possessed by the petroleum exploration geologist are an understanding of the physical principles covering the accumulation of petroleum and natural gas together with a wide and thorough acquaintance with the geological conditions under which oil and gas have accumulated in already discovered traps.

It was shown that traps for oil and gas consist of underground regions in which the potential energies of these fluids are at a minimum with respect to their surroundings. Under hydrostatic conditions in which the environmental ground water is not in motion, traps will be formed in rocks permeable to oil and gas wherever they are bounded by impermeable surfaces which are concave downward. Under hydrodynamic conditions in which the environmental ground water is in motion, traps will be formed whenever the impermeable surface is concave in the direction of increasing potential for the particular fluid. The respective traps for oil and gas will not coincide except under hydrostatic conditions.

There are about a dozen different geologic factors which may contribute to the proper geometry for an oil or gas trap. There are literally hundreds of different ways in which combinations of these factors may occur, and examples have been presented of some of the most common of these. Undoubtedly traps are yet to be found which can combine these factors and perhaps new ones in new and unexpected ways.

REFERENCES

Adams, John Emery: Oil Pool of Open Reservoir Type, *Bull. AAPG*, vol. 20, pp. 780–796, 1936.

Baldwin, Thomas A.: San Ardo—A Stratigraphic Analysis of a California Oil Field, *Bull. AAPG*, vol. 34, pp. 1981–1989, 1950.

Bitterli, Peter: Herrera Subsurface Structure of Penal Field, Trinidad, B.W.I., *Bull. AAPG*, vol. 42, pp. 145–158, 1958.

Carman, J. Ernest, and Wilbur Stout: Relationship of Accumulation of Oil to Structure and Porosity in the Lima-Indiana Field, in "Problems of Petroleum Geology," pp. 521–529, AAPG, Tulsa, Okla., 1934.

Clapp, Frederick G.: Role of Geologic Structure in the Accumulation of Petroleum, in "Structure of Typical American Oil Fields," vol. 2, pp. 667–716, AAPG, Tulsa, Okla., 1929.

Cloud, Preston E.: Facies Relationships of Organic Reefs, *Bull. AAPG*, vol. 36, pp. 2125–2149, 1952.

Cotner, Victor, and H. E. Crum: Geology and Occurrence of Natural Gas in Amarillo District,. Texas, in "Geology of Natural Gas," pp. 385–415, AAPG, Tulsa, Okla., 1935.

Cumings, E. R.: Reefs or Bioherms? *Bull. Geol. Soc. Am.*, vol. 43, pp. 345–347, 1932.

Cumings, E. R., and Robert R. Shrock: Niagaran Coral Reefs of Indiana and Adjacent States and Their Stratigraphic Relations, *Bull. Geol. Soc. Am.*, vol. 39, p. 599, 1928.

Darton, N. H.: Preliminary Report of the Geology and Underground Water Resources of the Central Great Plains, *U.S. Geol. Survey Profess. Paper* 32, 433 pp., 1905.

Darton, N. H.: Geology and Underground Waters of South Dakota, *U.S. Geol. Survey Water Supply Paper* 227, 156 pp., 1909.

Darton, N. H.: Artesian Waters in the Vicinity of the Black Hills, South Dakota, *U.S. Geol. Survey Water Supply Paper* 428, 64 pp., 1918.

Darton, N. H.: The Structure of Parts of the Central Great Plains, *U.S. Geol. Survey Bull.* 691, pp. 1–26, 1918.

de Ford, Ronald K.: Surface Structure, Florence Oil Field, Fremont County, Colorado, in "Structure of Typical American Oil Fields," vol. 2, pp. 75–92, AAPG, Tulsa, Okla., 1929.

Downing, John A., and D. Y. Cooke: Distribution of Reefs of Woodbend Group in Alberta, Canada, *Bull. AAPG*, vol. 39, pp. 189–206, 1955.

Dudley, Jr., P. H.: The Castaic Junction Oil Field, in "A Guide to the Geology and Oil Fields of the Los Angeles and Ventura Regions," pp. 173–176, Pacific Section, AAPG, Los Angeles, 1958.

Dunnington, H. V.: Generation, Migration, Accumulation, and Dissipation of Oil in Northern Iraq, in "Habitat of Oil," pp. 1194–1251, AAPG, Tulsa, Okla., 1958.

Fentress, George H.: Little Beaver Field, Colorado, A Stratigraphic, Structural, and Sedimentation Problem, *Bull. AAPG*, vol. 39, pp. 155–188, 1955.

Fettke, Charles R.: Music Mountain Oil Pool, McKean County, Pennsylvania, in "Stratigraphic Type Oil Fields," pp. 492–506, AAPG, Tulsa, Okla., 1941.

Fortier, Y. O., A. H. McNair, and R. Thorsteinsson: Geology and Petroleum Possibilities in Canadian Arctic Islands, *Bull. AAPG*, vol. 38, pp. 2075–2109, 1954.

Galley, John E.: Oil and Geology in the Permian Basin of Texas and New Mexico, in "Habitat of Oil," pp. 395–446, AAPG, Tulsa, Okla., 1958.

Gallup, W. B.: Geology of Turner Valley Oil and Gas Field, Alberta, Canada, *Bull. AAPG*, vol. 35, no. 4, pp. 797–821, 1951.

Garlough, John L., and Garvin L. Taylor: Hugoton Gas Field, Grant, Haskell, Morton, Stevens, and Seward Counties, Kansas, and Texas County, Oklahoma, in "Stratigraphic Type Oil Fields," pp. 78–104, AAPG, Tulsa, Okla., 1941.

Graves, Doyle, T.: Geology of the Dominguez Oil Field, Los Angeles County, Map Sheet No. 33, in "Geology of Southern California," *Calif. Dept. Nat. Resources, Div. Mines Bull.* 170, San Francisco, 1954.

Guzman, Edwardo J., and Frederico Mina U.: Petroleum Developments in Mexico in 1955, *Bull. AAPG*, vol. 40, pp. 1485–1497, 1956.

Halbouty, Michel T., and George C. Hardin, Jr.: Types of Hydrocarbon Accumulation and Geology of South Liberty Salt Dome, Liberty County, Texas, *Bull. AAPG*, vol. 35, pp. 1939–1977, 1951.

Hanna, Marcus A.: Geology of the Gulf Coast Salt Domes, in "Problems of Petroleum Geology," pp. 629–693, AAPG, Tulsa, Okla., 1934.

Hedberg, H. D., L. C. Sass, and H. J. Funkhouser: Oil Fields of Greater Oficina Area, Central Anzoategui, Venezuela, *Bull. AAPG*, vol. 31, pp. 2089–2169, 1947.

Hemsell, Clenon C.: Geology of Hugoton Gas Field of Southwestern Kansas, *Bull AAPG*, vol. 23, pp. 1054–1067, 1939.

Hoots, Harold W., Ted L. Bear, and William D. Kleinpell: Stratigraphic Traps for Oil and Gas in the San Joaquin Valley, in "Geology of Southern California," chap. 9, Oil and Gas, pp. 29–32, *Calif. Dept. Nat. Resources, Div. Mines Bull.* 170, 1954.

Howell, J. V.: Historical Development of the Structural Theory of Accumulation of Oil and Gas, in "Problems of Petroleum Geology," pp. 1–23, AAPG, Tulsa, Okla., 1934.

Hubbert, M. King: The Theory of Ground-water Motion, *J. Geol.*, vol. 48, no. 8, pp. 785–944, 1940.

Hubbert, M. King: Entrapment of Petroleum under Hydrodynamic Conditions, *Bull. AAPG*, vol. 37, no. 8, pp. 1954–2026, 1953.

Hubbert, M. King, and David G. Willis: Important Fractured Reservoirs in the United States, *Proc.* 4th *World Petrol. Congr.*, Section I/A/1, Rome, 1955.

Janoschek, Robert: The Inner-Alpine Vienna Basin, in "Habitat of Oil," pp. 1134–1152, AAPG, Tulsa, Okla., 1958.

King, Philip B.: Geology of the Marathon Region, Texas, *U.S. Geol. Survey Profess. Paper* 187, 1938.

King, Philip B.: Permian of West Texas and Southeast New Mexico, *Bull. AAPG*, vol. 26, pp. 535–763, 1942.

King, Philip B.: Geology of the Southern Guadalupe Mountains, Texas, *U.S. Geol. Survey Profess. Paper* 215, 1948.

Knebel, G. M., and Guillermo Rodriguez-Eraso: Habitat of Some Oil, *Bull. AAPG*, vol. 40, pp. 547–561, 1956.

Landes, Kenneth K.: Porosity through Dolomitization, *Bull. AAPG*, vol. 30, pp. 305–318, 1946.

Lilly, Russell N.: Developments in Texas and Oklahoma Panhandle in 1957, *Bull. AAPG*, vol. 42, pp. 1234–1247, 1958.

Link, Theodore A.: Interpretations of Foothills Structures, Alberta, Canada, *Bull. AAPG*, vol. 33, no. 9, pp. 1475–1501, 1949.

Lloyd, E. Russell: Capitan Limestone and Associated Formations of New Mexico and Texas, *Bull. AAPG*, vol. 13, pp. 645–658, 1929.

McCoy, III, Alex W., Robert L. Sielaff, George R. Downs, N. Wood Bass, and John H. Maxson: Types of Oil and Gas Traps in the Rocky Mountain Region, *Bull. AAPG*, vol. 35, p. 1000, 1951.

Mills, R. Van A.: Experimental Studies of Subsurface Relationships in Oil and Gas Fields, *Econ. Geol.*, vol. 15, no. 5, pp. 398–421, 1920.

Minor, H. E., and Marcus A. Hanna: East Texas Oil Field, Rusk, Cherokee, Smith, Gregg, and Upshur Counties, Texas, in "Stratigraphic Type Oil Fields," pp. 600–640, AAPG, Tulsa, Okla., 1941.

Morales, Luis G., and the Colombian Petroleum Industry: General Geology and Oil Occurrences of Middle Magdalena Valley, Colombia, in "Habitat of Oil," pp. 641–695, AAPG, Tulsa, Okla., 1958.

Munn, Malcom J.: The Anticlinal and Hydraulic Theories of Oil and Gas Accumulation, *Econ. Geol.*, vol. 4, pp. 141–157, 1909.

Nettleton, L. L.: Fluid Mechanics of Salt Domes, in "Gulf Coast Oil Fields," pp. 79–108, AAPG, Tulsa, Okla., 1936.

Nettleton, L. L.: Recent Experimental and Geophysical Evidence of Mechanics of Salt-dome Formation, *Bull. AAPG*, vol. 27, pp. 51–63, 1943.

Orton, Edward: The Origin and Accumulation of Petroleum and Natural Gas, *Rept. Geol. Survey Ohio*, vol. 6, pp. 1–310, 1888.

Orton, Edward: The Trenton Limestone as a Source of Petroleum and Inflammable Gas in Ohio and Indiana, *U.S. Geol. Survey* 8th *Ann. Rept.*, part II, pp. 475–662, 1889.

Parker, Travis J., and A. N. McDowell: Scale Models as Guide to Interpretation of Salt-dome Faulting, *Bull. AAPG*, vol. 35, pp. 2076–2094, 1951.

Parker, Travis J., and A. N. McDowell: Model Studies of Salt Dome Tectonics, *Bull. AAPG*, vol. 39, p. 2384, 1955.

Patterson, A. M., and A. A. Arneson: Geology of Pembina Field, Alberta, *Bull. AAPG*, vol. 41, pp. 937–949, 1957.

Porter, Jr., Livingston: Kettleman Hills Oil Field, in "Guidebook," pp. 170–172, Pacific Section, AAPG, Los Angeles, 1952.

Reeves, Frank: Status of German Oil Fields, *Bull. AAPG*, vol. 30, pp. 1546–1584, 1946.

Rich, John L.: Moving Underground Water as a Primary Cause of the Migration and Accumulation of Oil and Gas, *Econ. Geol.*, vol. 16, no. 6, pp. 347–371, 1921.

Rich, John L.: Problems of the Origin, Migration and Accumulation of Oil, in "Problems of Petroleum Geology," pp. 337–345, AAPG, 1934.

Rockwell, D. W., and Antonio Garcia Rojas: Coordination of Seismic and Geologic Data in Poza Rica—Golden Lane Area, Mexico, *Bull. AAPG*, vol. 37, pp. 2551–2565, 1953.

Rogatz, Henry: Geology of the Texas Panhandle Oil and Gas Field, *Bull. AAPG*, vol. 23, pp. 983–1053, 1939.

Schneegans, Daniel: Gas-bearing Structures of Southern France, *Bull. AAPG*, vol. 32, no. 2, pp. 198–214, 1948.

Schwade, Irving T., Stanley A. Carlson, and James B. O'Flynn: Geologic Environment of Cuyama Valley Oil Fields, California, in "Habitat of Oil," pp. 78–98, AAPG, Tulsa, Okla., 1958.

Small, Walter M.: Thrust Faults and Ruptured Folds in Romanian Oil Fields, *Bull. AAPG*, vol. 43, pp. 455–471, 1959.

Smith, Jr., P. V.: Studies on Origin of Petroleum: Occurrence of Hydrocarbons in Recent Sediments, *Bull. AAPG*, vol. 38, no. 3, pp. 377–404, 1954.

Walters, Ray P.: Oil Fields of the Carpathian Region, *Bull. AAPG*, vol. 30, pp. 319–336, 1946.

Ware, Jr., G. C., and R. D. Stewart: Bridge Area, South Mountain Oil Field, in "A Guide to the Geology and Oil Fields of the Los Angeles and Ventura Regions," pp. 180–181, Pacific Section, AAPG, Los Angeles, 1958.

Weeda, Jan: Oil Basin of East Borneo, in "Habitat of Oil," pp. 1337–1346, AAPG, Tulsa, Okla., 1958.

Wendlandt, E. A., T. H. Shelby, Jr., and John S. Bell: Hawkins Field, Wood County, Texas, *Bull. AAPG*, vol. 30, pp. 1830–1856, 1946.

Wilhelm, O.: Classification of Petroleum Reservoirs, *Bull. AAPG*, vol. 29, pp. 1537–1579, 1945.

Willis, Robin, and Richard S. Ballantyne: Potrero Oil Fields, in "Geologic Formations and Economic Development of the Oil and Gas Fields of California," *Calif. Dept. Nat. Resources, Div. Mines Bull.* 118, pp. 310–317, 1943.

Womack, Jr., Robert: Brookhaven Oil Field, Lincoln County, Mississippi, *Bull. AAPG*, vol. 34, pp. 1517–1529, 1950.

Woodring, W. P., Ralph Stewart, and R. W. Richards: Geology of the Kettleman Hills Oil Fields, California, *U.S. Geol. Survey Profess. Paper* 195, 170 pp., 1940.

Chapter 7

PETROLEUM RESERVOIRS

By Dean H. Sheldon

Petroleum Engineer
South Pasadena, Calif.

The oil industry has reached the point where a substantial understanding of the physical principles related to the development and production of oil and gas accumulations is a prerequisite to the maximum recovery of hydrocarbons at a minimum cost.

The purpose of this chapter will be to review the more important physical principles related to the proper production of oil and gas. It is intended primarily to acquaint personnel in exploration with the basic concepts concerning the characteristics and behavior of reservoir fluids. The space allotted to the subject permits only qualitative considerations.

There are two primary aspects to the study of reservoir behavior and the proper use of the inherent characteristics thereof.

One encompasses all those features inherent in the reservoir at the time of discovery and during exploitation. These are the configuration of the gross reservoir and its dimensions, the initial fluid and/or gas content and the distribution thereof, the solubility of the gas in the oil, the reservoir temperature and pressure, the permeability variations of the pay zone, the permeability-saturation relationship, and the size of the aquifers bounding the hydrocarbon accumulation.

The other embraces those things that can be controlled by man to promote maximum economic results at minimum cost, such as the proper well spacing, completion techniques, and production rate control.

There are no formulas that will provide an infallible answer to a reservoir engineering problem. There is no substitute for experience that has been accompanied by a determined effort to understand the cause and effect of oil field phenomena.

Muskat (1949) states that "Oil producing reservoirs are not the product of a mass assembly line, designed and constructed according to specifications. They are objects of individual study and analysis."

Geologic events determine the lithologic character of the reservoir, its configuration, the amount of space available for the storage of hydrocarbons, and to some extent, the rate at which they can be recovered. These events probably control either in whole or in part the physical character of the liquids, oil and/or gas contained in the reservoir void space. This latter relationship is only suspected. It is not proved. Proper appreciation of the production performance of an oil and gas pool requires some understanding of the interrelationships of hydrocarbon systems, water, temperatures, and pressures.

An oil and gas reservoir, once tapped by the drill, becomes dynamic. It cannot be returned to its original state. Each barrel of oil or water or cubic foot of gas removed brings about some redistribution of the fluids and gases remaining in the reservoir and a change in their physical properties.

The chemical composition of hydrocarbon systems is limited to a few simple elements. In general, their percentage distribution by weight (Burcik, 1947) is as follows:

Carbon: 84 to 87 per cent
Sulfur: 0.6 to 2.0 per cent
Oxygen: 0.6 to 2.0 per cent
Hydrogen: 11 to 14 per cent
Nitrogen: 0.6 to 2.0 per cent

The similarity of the carbon and hydrogen weight ratios in various oil and gas systems does not imply simplicity in form. The hydrogen and carbon atoms combine in many molecular structures that exhibit a wide variety of physical properties.

The physical state of hydrocarbon compounds ranges from gaseous to liquid to a solid. Reservoir analysis is concerned with the behavior of the gaseous and liquid phases of hydrocarbon systems.

The thermodynamic state of a hydrocarbon system is specified in terms of its intensive properties, i.e., those properties which are independent of the amount of material present. Such properties are pressure, temperature, specific volume, density, etc. The various homogeneous physical forms of a pure substance are termed phases: for example, the gas phase, the liquid phase, and the solid phase. The term *phase* applies only to matter that is homogeneous. A gas is homogeneous, since all parts have identical properties. A mixture of pure gases is also homogeneous and can act as a single gaseous phase. The same statement applies to mixtures of miscible liquids. A change in the temperature of a system or the pressure applied to a system can produce change in the phase of any hydrocarbon system, either pure or mixed.

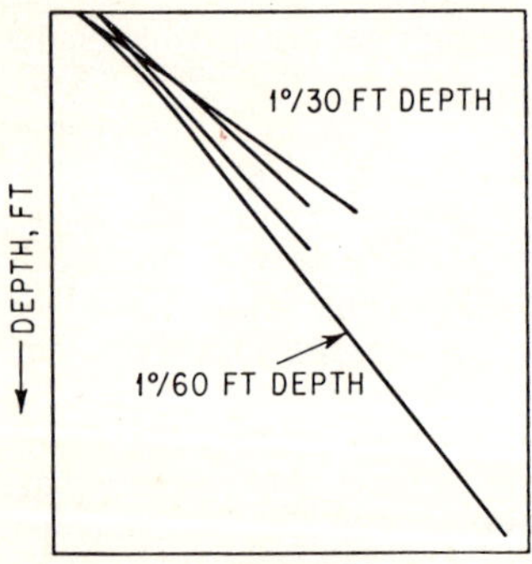

FIGURE 7-1. Average geothermal gradients.

The response of hydrocarbon systems to changes in temperature and pressure is of prime interest to anyone concerned with the exploitation of an oil and/or gas pool. The temperature of an underground reservoir is assumed to be constant while hydrocarbons are being extracted. Pressures in the reservoir may vary over a considerable range during the productive life of an oil and/or gas field.

Temperatures found in oil and/or gas reservoirs vary directly with the depth of the pool below the surface of the ground (see Figure 7-1). The geothermal gradient existent in the geographical location of the field determines the subsurface reservoir temperature for any given depth. If no measured geothermal gradient is available, an assumed average gradient of 1°F increase above surface (ground) temperature for each 60 feet of depth below the surface may be used. However, there can be large variations in the geothermal gradient within localized areas and even within the productive limits of oil pools that have substantial areal extent. If the reservoir containing oil and/or gas overlies and is in proximity to a large mass of igneous or metamorphic rock, abnormal temperature gradients may be anticipated. Temperature gradients of fields so situated in the Los Angeles Basin of California exhibit increases on the order of 1°F above the mean average surface temperature of the ground per each 35 to 40 feet of depth. Deep fields in the San Joaquin Valley have gradients of 50 to 60 feet per °F increase. The same values are common in fields that overlie the thick portions of the sedimentary basin in the Los Angeles area. Geothermal gradients in the Gulf Coast area of Louisiana increase 1°F for each 70 to 80 feet of depth. The Texas Gulf Coast has a slightly higher gradient, about 1°F for each 50 to 60 feet of depth. North central and northeast Texas fields have gradients that increase 1°F for each 50 to 60 feet of additional depth. Oklahoma fields generally exhibit low geothermal gradients. A rise of 1° for each 110 to 120 feet of depth is not unusual. The Fitts Pool has within the field itself gradients that vary from 60 to 117 feet per degree increase in subsurface temperature. Southeast New Mexico and west Texas districts have subnormal geothermal gradients. Here values range from 125 to 250 feet per degree increase in temperature. Many of the reported subsurface temperature data have been obtained from readings taken in connection with drilling or production operations. In such circumstances, complete temperature equilibrium between the sediments and the fluids in the borehole is probably not realized. Thus many of the gradients so reported may be lower than their true equilibrium value.

In areas of low regional dip, the geothermal gradient has an inverse relationship to the heat conductivity of the sedimentary section.

The pressures exerted on the fluids and gases are of prime interest in analyzing hydrocarbon reservoir behavior. The original pressures in a hydrocarbon reservoir affect that portion of the oil and/or gas originally in the reservoir that can be reduced to possession. Proper use of this inherent energy can increase substantially the hydrocarbons recovered from an oil or gas accumulation.

In general, it is found that the reservoir pressures existing in a virgin oil and/or gas pool are equal to the pressure exerted by a column of water equal in height to the distance from the ground-water table to the pay. There are, however, reservoirs in which the subsurface pressures exceed the normal hydrostatic pressure by a ratio of 2:1. A few pools have even greater pressure gradients. Some oil fields at the time of discovery have reservoir pressures that are less than the hydrostatic. Here movement of oil, gas, or water out of the trap is indicated.

Normal hydrostatic pressure gradients vary from 0.435 to 0.465 per foot of depth below the surface, the outcrop, or the water table. In California, Texas, Louisiana, and the mid-continent, continuity of the liquid and/or gas phases from the reservoir to the outcrop is uncommon. Some Rocky Mountain fields do exhibit this pressure relationship. The precise value of a normal pressure gradient is controlled by the composition of the water column that exerts or originally exerted the pressure on the interstitial fluids and gases. The use of 0.45 psi per foot of depth pressure gradient for estimating subsurface pressures in a trap at the time of discovery will usually produce results that approximate the true value within plus or minus 3 per cent. Many forms of reservoir analysis are quite sensitive to the values used for reservoir pressures. It is important therefore that adequate reservoir-pressure information be gathered immediately upon discovery and throughout the subsequent producing life of a field. The pressure history of an oil or gas pool coupled with accurate production data is of utmost importance in the proper planning of exploitation operations.

There are some oil-producing provinces, fields or pools, that have initial, i.e., virgin, pressures in excess of those imposed by a column of water. Reservoir pressures that range from several hundred pounds per square inch above the normal hydrostatic value to those that approximate the geostatic pressure, i.e., the pressure imposed on interstitial substances by the weight of sedimentary overburden, have been recorded. A geostatic pressure gradient is often assumed to be 1-psi increase per foot of depth. California, the Texas and Louisiana Gulf Coast areas, and southwest Texas have oil pools that upon discovery demonstrated abnormal pressures. Cannon and Craze (1938) reporting on pressures observed in a number of fields or pools that embraced nine major stratigraphic units show an average gradient of 0.465 psi per foot of depth. There were 34 petroleum reservoirs in the Gulf Coast and southwest Texas which had original pressure gradients reported to be equal to 0.735 psi per foot of depth at 10,800 feet to 0.56 psi per foot of depth at 8,000 feet. E. V. Watts (1948) reports original pressures in the oil-bearing measures of the Ventura field, California, of 8,300 psi at well depths of 9,200 feet or subsea depths of 9,000 feet. This is a pressure increase of 0.92 psi per foot of well depth. It is interesting to note that the literature reports only a few examples of the existence of abnormal pressures at depths of less than 7,000 feet. Some have been recorded at depths of 2,500 feet, more or less, from the Khaur field in India. Pressure gradients of 0.65 psi per foot of depth have been recorded at depths of 3,500 to 4,000 feet in the Cruse formation of the Guapo field, Trinidad (Meyers, 1959).

The failure of the granular structure of the overburden, compression of the rock and interstitial material by tectonic forces, deep burial of the reservoir with subsequent elevation, and erosion of part of the overburden of the host rock have all been used to explain the presence of abnormal hydrocarbon reservoir pressures. The existence of abnormal pressures prior to penetration of the reservoir by the drill does not so far appear to follow any predictable pattern.

There are three principal physical characteristics of a subsurface reservoir that interest the petroleum engineer. They are porosity, fluid and gas saturation of the interstitial void, and permeability.

Porosity is the percentage of the total unit volume of the rock that is available for the storage of liquids or gases. There are two types of porosity: absolute and effective. Work with hydrocarbon reservoirs concerns itself primarily with effective porosity or those pore spaces that are interconnected and permit the movement of liquid or gas from one pore space to another. Effective porosity is the interconnected pore space in a unit volume of rock expressed as a percentage of the bulk volume of the sample tested. There are a number of methods available for making porosity determinations on samples, either chip or cored, that have been taken from the pay zone. Porosity values measured in the laboratory, if sufficient in number and properly taken, are valid representations of the porosity of the rock in the reservoir. The pore space present in sedimentary rock reservoirs is the result of orderly geologic processes. This fact makes possible the application of porosity values derived from infinitesimal portions of the reservoir to studies of the gross reservoir. An understanding of the genesis of a pay zone is important when a reservoir sampling program is being considered. Such geologic knowledge can produce the maximum amount of representative information with the minimum amount of sampling.

The consideration of finite porosity values can, for present purposes, be limited to limestone, dolomite, and sandstone reservoirs. The classic example of maximum porosity, when the solid particles are in contact, is the cubicle packing of perfect spheres. In this case, the pore space occupies 47.64 per cent of the gross or bulk volume. Actually the sandstones and limestones of granular composition that are present in most reservoirs of commercial value have porosity values that range from 10 to 35 per cent. There are instances where the reservoir rock is unconsolidated and the porosity values may exceed 40 per cent, though these are not common. The intergranular porosity of limestones may be on the order of 3 to 5 per cent. The vugular porosity of limestones and dolomites covers such a wide range of values that general statements are meaningless. The essential problem, where vugular or fracture porosity is present, is that of getting truly representative samples of the reservoir material for porosity measurement. There is no direct method available for the determination of the capacity of such reservoirs to hold liquids and/or gases. The solution here must be sought through the analysis of production and pressure performance of the reservoir as the hydrocarbons are withdrawn.

The absolute permeability of a rock material containing oil or gas may be defined as the ease with which homogeneous fluids under pressure can be passed through the interconnected pore spaces. Absolute permeability values require that the pore space be filled with only the flowing fluid or gas. This condition is nonexistent in most hydrocarbon reservoirs, possibly excepting those that have large fractures or solution cavities. Permeability is independent of the dimensions of the porous medium, the pressure differential imposed upon the flowing material, and the viscosity thereof.

Permeabilities obtained by standard core-analysis procedures are usually referred to air. Dry air is passed through a sample that has been treated with a diluent to remove all hydrocarbons and subsequently dried to remove all moisture except the water of crystallization. Such information should be used with caution. The *in situ* condition of the reservoir sample may be altered by a rearrangement of the very fine particles in the test sample as a result of purging the residual hydrocarbons by washing with a solvent. Sometimes the drying process expels some water of crystallization or dehydrates the clay minerals. The montmorillonite groups are especially susceptible to this and other effects. Consequently, permeability to air values reported for the so-called dirty or argillaceous sands are often much higher than are the fluid permeabilities in the reservoir.

On many occasions, air permeabilities have indicated that satisfactory production rates should be anticipated. Subsequent testing of the wells shows little or no fluid entry. The disappointing performance is the result of the alteration of the *in situ* reservoir conditions when laboratory techniques are applied to the determination of air permeabilities.

Permeability is expressed in terms of darcys or millidarcys (1/1,000 darcy). A darcy is defined as the rate of flow in milliliters per second of a homogeneous fluid with a viscosity of one centipoise that passes through a porous medium of one square centimeter in cross section under a pressure differential of one atmosphere per centimeter of length. D'Arcy's law assumes conditions of viscous flow. There is a general relationship between porosity and permeability values of reservoir rock (see Figure 7-2). However, the application of such general correlations to specific pools or fields can result in very misleading conclusions.

S. M. Paine (1956) indicates a porosity-permeability relationship obtained from the analysis of a number of Wilcox sand samples as well as seven or eight productive horizons. The spread of data for even one formation in a particular field makes the usefulness of general correlations dubious. For example, at 20 per cent porosity, the permeability of the Upper Wilcox in the Mercy field ranged from 1 to 500 millidarcys. For eight reservoirs, a spread of porosity values at 150 millidarcys permeability varied from 14 to 36 per cent. The degree of sorting, i.e., grain-size distribution; the variation in grain shape; the amount of compaction; cementation; and in some cases, the chemical composition of the rock, all influence the porosity and permeability of reservoir rock.

Studies of the porosity permeability relationship indicate that permeability varies directly as a first power of porosity and inversely as a second power of the specific surface. As cementation usually takes place at the sharpest contact angles between grains, the specific surface decreases and permeability increases. At the same time, as porosity decreases, permeability decreases. The two factors porosity and specific surface tend to cancel each other. Where little cementation takes place, the effect of specific surface is greater than that of porosity. The relationship of these two parameters to permeability explains why comparable porosity values in the same formation may exhibit considerable differences in permeability.

The permeability of sediments that hold oil and gas is of importance to the reservoir engineer. It controls to a large degree the rate at which fluids and gases can move through the interconnected pore space and to some degree the quantity that can be recovered. Also there is a relationship between permeability and the percentage of pore space filled with interstitial or connate water (see Figure 7-3), thus affecting the quantity of hydrocarbons that may be stored in a given volume of reservoir material.

In all granular-type rocks, a portion of the pore space is filled with interstitial or connate water. The role played by connate water in vugular or fractured reservoir rock is not clear. It is considered to have little effect on the production behavior of these two types of oil and gas reservoirs.

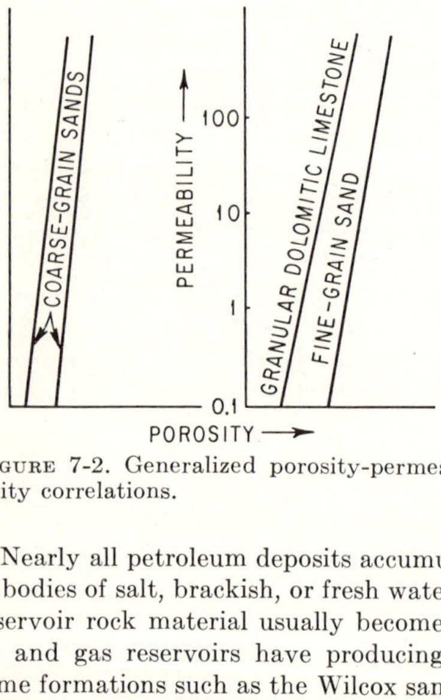

FIGURE 7-2. Generalized porosity-permeability correlations.

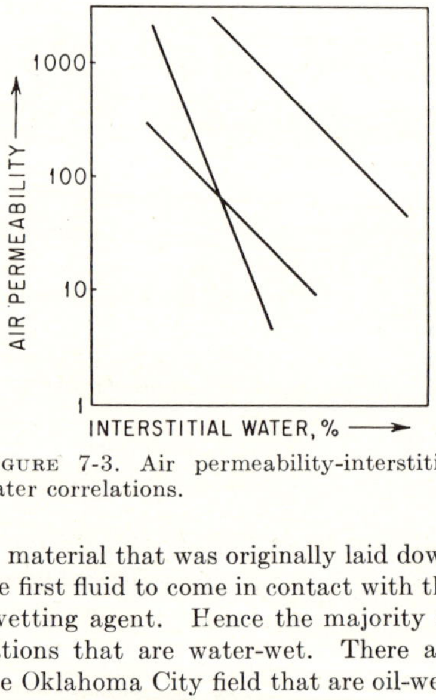

FIGURE 7-3. Air permeability-interstitial water correlations.

Nearly all petroleum deposits accumulate in material that was originally laid down in bodies of salt, brackish, or fresh water. The first fluid to come in contact with the reservoir rock material usually becomes the wetting agent. Hence the majority of oil and gas reservoirs have producing formations that are water-wet. There are some formations such as the Wilcox sand in the Oklahoma City field that are oil-wet.

The connate water forms a molecular-thin film that covers the surface of the granular material. It also occupies much of the angular-shaped space created by the points of contact between the individual sand grains.

Interstitial water in water-wet sands usually occupies 20 per cent or more of the pore space. When grain sizing is uniform and cementation has eliminated some contact angularity, connate water may drop to 8 or 10 per cent of the pore space.

Consider a water-wet sand that contains water and oil (see Figure 7-4). The connate water is considered to be immobile. The surface tension between the water and the sand grains exceeds the interfacial tension of the oil and water. The oil moves when sufficient pressure differential exists. This is the equilibrium saturation of the wetting phase. If the percentage of pore space filled with water increases, as water moves into the oil reservoir, the mobility of the water increases and it begins to move through the interstices. As the wetting-phase saturation increases, the nonwetting phase (oil) becomes discontinuous and its mobility is reduced (Jamin action). This is the equilibrium saturation of the nonwetting phase.

Any description of the movement of fluids or gases through a hydrocarbon reservoir entails consideration of absolute (effective) and relative permeability. Absolute permeability has already been defined. A measurement of permeability made with a liquid or a gas or both flowing through the interstitial space of a reservoir rock that

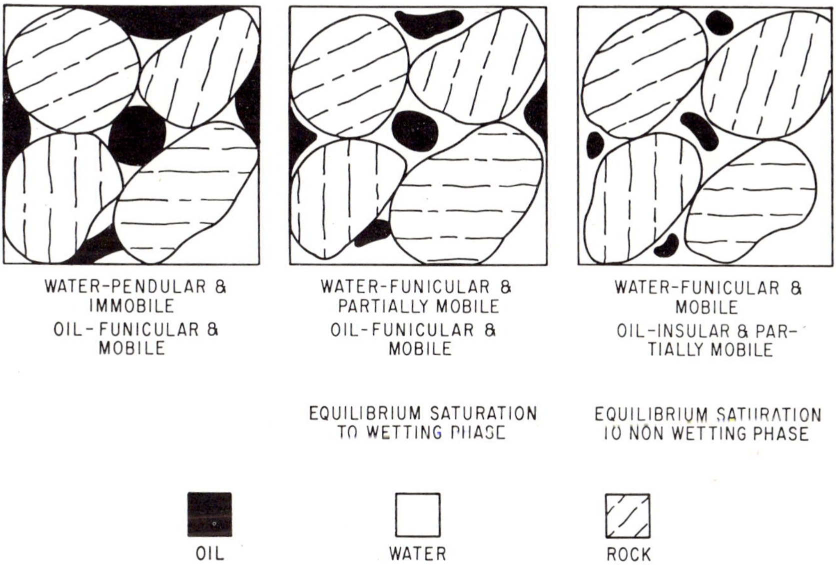

FIGURE 7-4. Principal stages of oil-water saturation.

contains connate water will be lower in value than that measured with only a single phase present and flowing. This is the effective permeability of the mobile phase. Relative permeability is the ratio of effective permeability to absolute permeability.

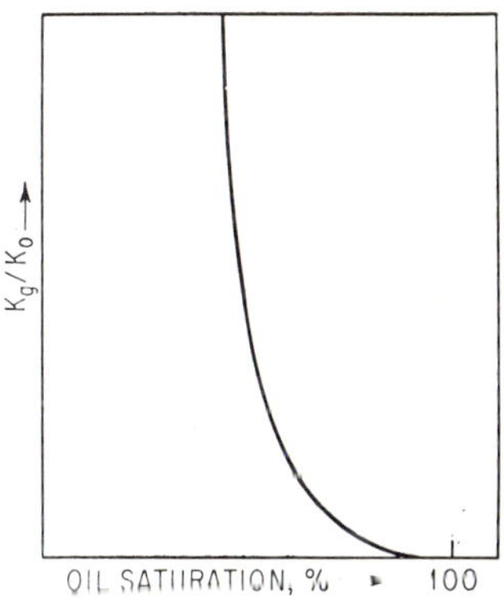

FIGURE 7-5. Relative permeability—gas to oil and per cent oil saturation.

Oil and/or gas reservoirs contain interstitial water and hydrocarbons in the liquid and/or gaseous states. When production begins, the latter (oil and gas) move through the pay to be recovered at the well bore. The reservoir analyst is interested in the relative mobility of these substances. This relationship, the relative permeability ratio (k_g/k_o), is a function of the percentage of pore space filled with liquid (see Figure 7-5). It is independent of pressure differentials, the value of the interfacial tension, and viscosity ratios. It cannot be calculated for a specific pay but must be determined experimentally. No two samples of reservoir rock will necessarily have the same relative permeability ratio of gas to oil for the same liquid-saturation value.

A system is a body of matter with finite boundaries which can be considered as isolated from its surrounding environment if such is desired. The boundaries may be physical or mathematical. The properties of a system are those characteristics which can be quantitatively measured experimentally. The properties of a system describe its condition at any particular time. The properties of a hydrocarbon system are usually related to the effect of temperature, pressure, and composition upon its specific volume, viscosity, etc. Extensive properties are directly proportional to the amount of material. Intensive properties are independent of the amount of material, i.e., pressure, temperature, composition, specific volume, specific gravity, viscosity, etc. A system is homogeneous when its intensive properties vary from point to point in a continuous manner. A heterogeneous system has two or more homogeneous parts at the surface contact of which there are abrupt changes in property. The homogeneous regions in heterogeneous systems are referred to as phases: the gas phase, the liquid phase, the solid phase of hydrocarbon mixtures. For example, ice, water, and steam are three phases of a system. A phase is a homogeneous body of material that is different in its properties from those of its surroundings.

The observation and examination of hydrocarbon reservoir behavior require study of the intensive and extensive properties of hydrocarbon systems. As a means of distinguishing between gaseous and liquid phases of the hydrocarbons, some definitions are pertinent.

A gas is a homogeneous fluid generally of low density and viscosity that has no definite volume but fills completely any vessel in which it is placed. Liquids, generally speaking, are homogeneous fluids that have higher densities and viscosities and greater intermolecular forces than do gases. The intermolecular force of a liquid is sufficient to maintain the material in one mass when it is placed in a container whose volume exceeds that of the material. There is, however, sufficient molecular freedom to allow liquids to flow.

The behavior of hydrocarbons in the gaseous phase are, for analytical purposes, usually related to the behavior of a perfect gas. A perfect gas follows the prediction of Boyle's law and Charles's law. The volume of a given weight of gas varies directly as the temperature and inversely as the pressure exerted upon it. Temperatures used for gas calculations are usually expressed as degrees Rankine (°F + 460). Pressures are expressed in terms of atmospheres (1 atm = psia/14.7) or pounds per square inch absolute (psia). Avogadro's law for perfect gases states that under identical conditions of temperature and pressure equal volumes of all perfect gases contain the same number of molecules. Hence one molecular weight in pounds (1 pound mole) of any perfect gas occupies 379 cubic feet at standard condition of pressure and temperature (14.7 psia and 60°F). Combining the perfect-gas laws and applying them to one molecular weight of gas gives $PV = RT$, in which R is a constant that has the same value (10.72) for all gases. For N moles of gas, the equation is $PV = NRT$, where N is the weight of the gas in pounds divided by the molecular weight (MW) of the gas. This is the general gas law when P is expressed in pounds per square inch absolute or in atmospheres, V is in cubic feet, temperature is in degrees Rankine, and N is pounds of material divided by MW. For example, molecular weight of air is 28.96. Therefore 28.96 pounds of air equals 1 pound mole.

The foregoing relationships apply to those gases that perform in accordance with Boyle's law, Charles's law, and Avogadro's law.

The hydrocarbon gases (see Table 7-1) in a subsurface reservoir do not perform in strict accordance with these laws. They are not simple single-component gases such as oxygen, hydrogen, and nitrogen but are mixtures of a complex group of gases. A deviation factor that compensates for the departure of such gaseous mixes from the behavior as predicted by the perfect-gas law has been developed by experimentation. Calculation of the deviation or gas-compressibility factor requires the computation of an average molecular weight for the gas mixture being studied. This is possible because each separate component in the mixture follows Avogadro's law. The volume of each component will be equal to the number of moles of that component in the mixture, or

Table 7-1. Summary of Physical Properties of Natural-gas Hydrocarbons

Compound	Molecular weight	Critical	
		Temperature, °R	Pressure, psia
Methane, CH_4	16.04	344	673
Ethane, C_2H_6	30.07	549	712
Propane, C_3H_8	44.09	666	617
n-Butane, C_4H_{10}	58.12	766	551
Isobutane, C_4H_{10}	58.12	732	544
n-Pentane, C_5H_{12}	72.15	846	485
Isopentane, C_5H_{12}	72.15	829	483
n-Hexane, C_6H_{14}	86.17	914	435

Table 7-2. Example of Calculated Molecular Weight of a Natural Gas

Component	Per cent molecular weight or gaseous volume	Molecular weight component	Molecular weight average
CH_4	83.19	16.04	13.31
C_2H_6	8.48	30.07	2.54
C_3H_8	4.37	44.09	1.92
C_4H_{10}	0.76	58.12	0.44
C_4H_{10}	1.68	58.12	0.97
C_5H_{12}	0.57	72.15	0.41
C_5H_{12}	0.32	72.15	0.23
C_6H_{14}	0.63	86.17	0.54
Av. mol. wt.			20.36

weight per cent equals mole per cent. The conversion from weight per cent to mole per cent can be calculated; an example is given in Table 7-2.

Gaseous material has weight. It therefore has density. The density scale for gases is the ratio of the weight of a given volume of gas at standard conditions of temperature and pressure to the weight of an equal volume of air at the same conditions of temperature and pressure. Hence a gas with a specific gravity of 0.60 has a molecular weight of (28.96 × 0.60) 17.27 pounds per pound mole.

Ideal- or perfect-gas laws are sufficiently accurate that their use is justified when pressures on the system are 50 pounds or less. Above this, the gaseous hydrocarbon mixtures begin to deviate in performance from the predictions obtained by using ideal-gas equations. Precise values for such deviations can be obtained only by direct experimentation. However, the practical application of the ideal-gas law to hydrocarbon gases can be made by introducing the deviation or compressibility or Z factor into the gas equation. $PV = ZNRT$, in which Z is the compressibility or deviation factor for the gas or gases under study.

The factor Z is expressed in terms of a ratio or percentage of deviation from the ideal gas law. It is dimensionless and intensive. It is established by the character of the gas, the existing temperature and pressure.

The critical temperature T_c of hydrocarbon components is that temperature above which they can exist only as a gas no matter what pressure is applied. The critical pressure P_c is the pressure required to liquefy the component at its critical temperature.

The critical volume is the volume occupied by one mole of the component at its critical temperature and pressure. The same statements can be applied to mixtures of hydrocarbon gases. The values of the critical temperatures and pressures for gaseous hydrocarbons are set forth in Table 7-1.

The theory of corresponding states proposes that certain gases including hydrocarbons will exhibit the same behavior if they are similar in terms of reduced temperatures, pressures, and volumes. Reduced temperatures, pressures, and volumes are expressed as the ratio of the absolute value to the critical value.

The critical values calculated for a mixture of hydrocarbon gas, i.e., average values, are called pseudocritical values. The pseudocritical temperature of a mixture of 30 per cent methane and 70 per cent ethane would be $(0.30 \times 344) + (0.70 \times 550)$, or 488.

Empirical correlations relating the gravity (air = 1) of a hydrocarbon gas to its pseudocritical temperature and pressure are available in the literature. The resulting values are satisfactory for most engineering computations.

The pseudoreduced pressure P_r or temperature T_r of a mixture of gases is the ratio of the pseudocritical value to the critical value. If the gas were at a temperature of 100°F, the pseudoreduced temperature would be $100 + 460/488$ or 1.15. The pseudoreduced pressures and temperatures of gaseous mixtures provide a means of calculating the deviation (Z factor) of the particular mixture from the ideal-gas laws. Charts have been developed for correlation of the pseudoreduced temperatures and pressures of natural gas mixtures with the value of the Z. These can be found in any text that deals with the behavior of gaseous hydrocarbon systems.

Pure hydrocarbon substances such as methane, ethane, propane, butane, etc., are of interest, since under reservoir conditions they may take either the form of a liquid or a gas or, under specific conditions of temperature and pressure, the form of the gaseous and liquid phases simultaneously. The vapor pressure of a liquid at a specific temperature is the pressure created by the vapor of the liquid when it is in equilibrium with its liquid. The vapor pressure of a liquid varies directly but not in linear fashion with the temperature of the liquid. Above the critical temperature and pressure of the component only one phase can exist.

The graph in Figure 7-6 indicates that at a temperature of 100°F two phases will be present when the pressure is 200 psia. At a temperature of 50°F, the two phases exist when the pressure is 90 psia. In either case, if the temperature were held constant and the samples were subjected to increased pressure, the substance would at equilibrium be all liquid, or if the pressures were reduced, the substances would at equilibrium be all gas. As the temperature and pressures of a two-phase system of a pure substance increase, the properties of the two phases become more alike. The temperature T_c and pressure P_c at which the two phases have identical properties is the critical state or point of the substance. Above the critical temperature, no amount of pressure can convert the gas to liquid. At the critical temperature, the least pressure that will liquefy the vapor is the critical pressure. As the state of a system passes from the gaseous to the liquid phase, the equilibrium condition in which an infinitesimal amount of liquid exists is called the dew point or the state of saturated gas. Conversely, as the state passes from a wholly liquid to a gaseous state, the condition at which an infinitesimal amount of gas appears is the bubble point or saturated liquid state.

The phase behavior of a pure substance is illustrated graphically in Figure 7-7.

Hydrocarbon systems in an oil field are not single pure substances. They are a mixture of many substances. Hence the phase-behavior diagram for a two-component system will in the two-phase region have a slightly different appearance from one constructed for a single pure substance (see Figure 7-8).

Whereas the vapor pressure, the dew point, and the bubble point are equal for a single pure substance, such is not the case for a mixture of hydrocarbons. Here with temperature held constant, the bubble-point pressure of the mixture exceeds the dew-point pressure. At the bubble point, the infinitesimal drop of gas present has a larger amount of the more volatile components than does the liquid or system as a whole. Conversely at the dew point, the infinitesimal drop of liquid present has a greater amount of the less volatile constituents than does the system as a whole.

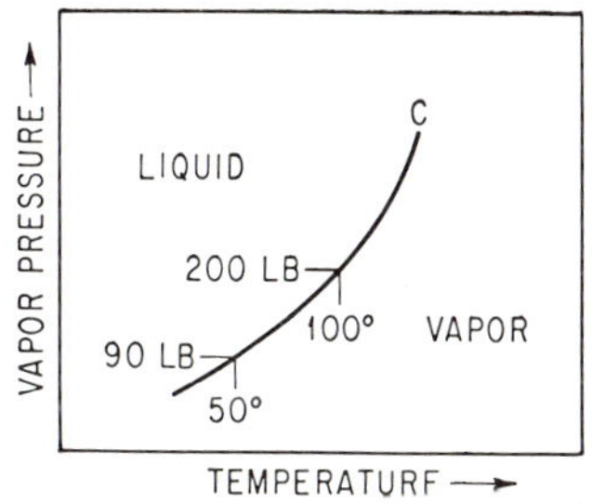

FIGURE 7-6. Vapor pressure–temperature relationship of a pure hydrocarbon substance.

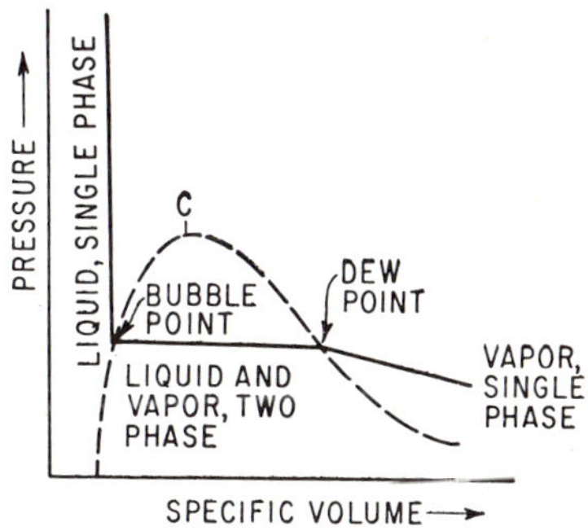

FIGURE 7-7 Single component system at a constant temperature below the critical.

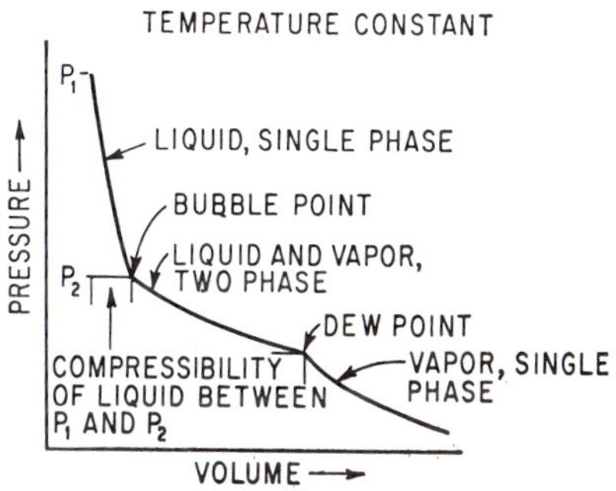

FIGURE 7-8. Pressure-volume isotherm for a two-component system.

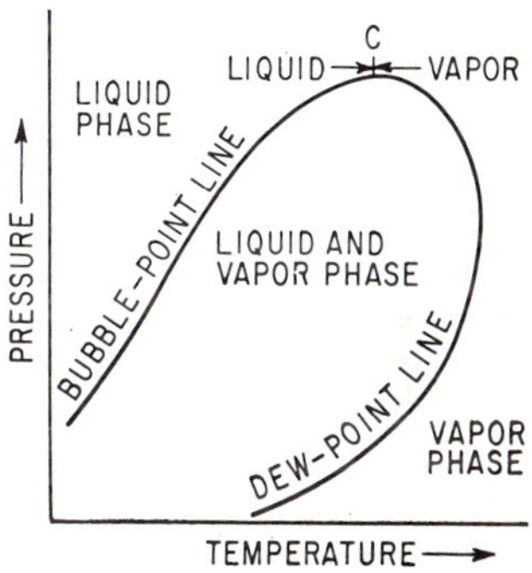

FIGURE 7-9. Pressure-temperature relationship for a two-component system.

For a single-component system, the critical point is that condition of temperature and pressure above which there is no discernible difference between the liquid and gaseous phases. In a multicomponent system, two phases may exist above the critical temperature. The critical point in this instance becomes the point of intersection of the dew-point line and the bubble-point line, in other words, the point at which the liquid and vapor phases have identical properties (see Figure 7-9).

This relationship of the dew point and bubble point can be determined for various hydrocarbon mixtures by laboratory analysis.

Some multicomponent systems have a dew-point line that doubles back; i.e., its critical temperature is less than its maximum temperature. Also the bubble-point line doubles back and the mixture has a critical pressure that is less than its maximum pressure. The characteristics of the two lines are determined by the composition of the mixture. Consider Figure 7-10: an isothermal increase in a pressure along the path P_1-P_2 would find a vapor phase only to the point P_1, a liquid and a vapor phase from P_1 to P_2, and a vapor phase at pressures above P_2. This illustrates the phenomenon of retrograde condensation whereby the hydrocarbons pass from a vapor to a liquid to a vapor phase as pressure increases.

Oil and gas reservoirs are considered to have constant temperatures throughout their productive life. Hence the performance of any one reservoir is considered to be isothermal. It is only when the behavior of hydrocarbons in different reservoirs or during their flow through the production equipment is compared that the temperature effect becomes important.

Retrograde reservoirs initially contain hydrocarbons in the vapor phase. If the pressure in the reservoir is not maintained above the maximum pressure on the bubble-point line, the hydrocarbons liquefy in the interstitial space of the pay. They become a disconnected phase within the interstitial space, and only a small portion can be recovered.

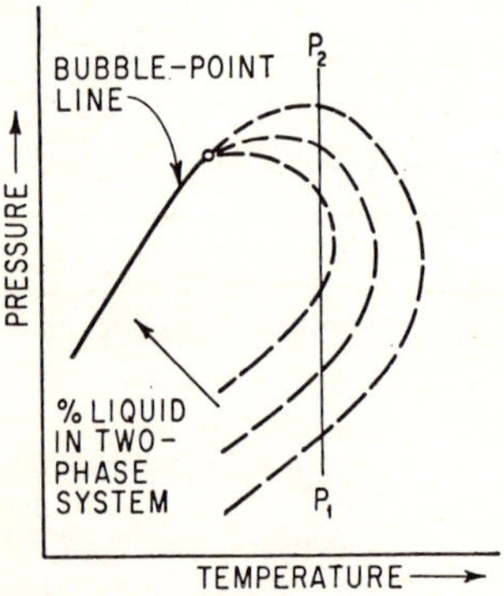

FIGURE 7-10. Pressure-temperature relationship—isothermal retrograde condensation.

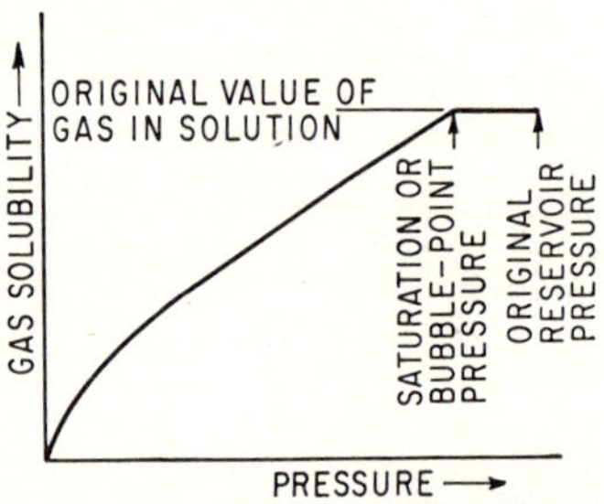

FIGURE 7-11. Gas solubility-pressure relationship—undersaturated crude oil.

Condensate or distillate is usually light straw or water-white in color. Its gravity is 50° or more. The gas/oil ratio is usually above 10,000 cubic feet per barrel. Condensate reservoirs may be associated with black oil deposits, or they may contain distillate only.

A precise description of a complex mixture of hydrocarbons is impractical when analyzing oil and gas reservoir behavior. The information necessary for such a description is not obtained in the ordinary course of developing and producing oil and gas. However, there are data obtained during the producing life of an oil and gas reservoir that permit the prediction of the future behavior of the system with a reasonable degree of accuracy. Composition of the gas phase can be determined by simple laboratory analysis, or its gross character can be obtained if the specific gravity or the average molecular weight of the gas is known. The compressibility factor can be computed from such information. Liquid fractions are more difficult to describe in gross terms.

The gravity of the liquid hydrocarbon is one of the most diagnostic features of oil composition and its behavior. It is indicative of the viscosity of the liquid, the solubility of gas in the liquid, and the specific volume of the liquid. Changes in the produced gas/oil ratio (cubic feet/barrel) as reservoir pressures decline are descriptive of particular hydrocarbon systems.

Oil reservoirs may have one or several sources of energy for the movement of hydrocarbons to and through the well bores. It is seldom that any single source remains predominant throughout the productive life of a pool or field. The principal producing mechanism of an oil and gas reservoir is determined by many factors that vary from field to field and within the confines of a single pool or reservoir. The structural configuration of a reservoir, the lithologic character of the rock, the subsurface temperature, the original reservoir pressures, the viscosity of the entrapped oil, the quantity of gas dissolved in the oil and in a free state, the amount of connate water in the interstitial space, and the relative volume of the hydrocarbons and the water in the continguous aquifers, all influence the producing mechanism of oil and gas fields.

Energy sources may be (1) gravitational forces proportional to the density of the various liquid and gas phases; (2) energy stored through the compression of the liquid and gaseous hydrocarbons, both dissolved and free, compression of the interstitial water in the hydrocarbon reservoir and that in the interconnected aquifers; (3) capillary forces; (4) compression of the reservoir rock. The two last forces are not usually considered to be of primary interest. However, their existence should be noted.

The most important sources of energy for the production of hydrocarbons are (1) the expansion of the compressed gas, both dissolved and/or free; (2) the expansion of the oil and water present in the hydrocarbon reservoir and the aquifers bounding it; (3) the force of gravity acting on the gaseous and liquid systems present in the reservoir. All crude oil in subsurface reservoirs contains some dissolved gas. The amount of gas in solution may vary from a few to many cubic feet per barrel of oil. Some hydrocarbon accumulations have more gas than can be dissolved in the oil at the existing reservoir temperature and pressure. Here the free gas gathers in the high structural positions as a result of gravitational forces that segregate the various hydrocarbon mixtures according to their density. The energy available in a unit of such free gas is proportional to the volume of the free gas unit at standard conditions of temperature and pressure and the logarithm of the pressure.

Although most hydrocarbon reservoirs have only that gas which can be held in solution at the existing reservoir temperature and pressure, i.e., no free gas, other reservoirs contain oil that is undersaturated or oil that contains less gas than it could hold in solution at existing subsurface temperature and pressure. Most crude-oil–gas mixtures are, at original reservoir conditions of pressure and temperature, in the liquid phase and contain the maximum amount of gas soluble at the existing temperature and pressure. The bubble-point pressure is the same as or very close to the original subsurface pressure. As pressures in the field decline, the oil can hold less gas in solution; the liberated gas accompanies the oil produced and fills the space voided by the oil and gas produced. Thus produced gas/oil ratios increase until the supply of gas is near exhaustion, at which time they begin to decline. Figure 7-11 illustrates an average isothermal-pressure–gas-solubility relationship.

Hydrocarbon reservoirs produced by gas expansion are classified as gas drives. There are two types of gas drive reservoirs: a solution gas drive that has available for expansion only the gas dissolved in the oil and a gas-cap expansion drive that has the gas dissolved in the oil and free gas as an additional source of energy.

The behavior of a solution gas drive mechanism operating in a single reservoir with homogeneous characteristics will show good agreement with that expected from theoretical considerations. Unfortunately most oil pools developed and produced as a unit are a combination of many separate oil accumulations. The homogeneity of the permeability of the pay zones will determine to a large extent the similarity between field history and analytical predictions of behavior.

The performance of solution gas drive reservoirs, where no water encroachment takes place, gravity drainage is not effective, and the oil is saturated at original reservoir pressures, is characterized by (1) initial oil production at original solution gas/oil ratios, (2) a slight decline in gas/oil ratios as the gas saturation in the interstitial spaces of the pay builds up to its equilibrium value (10 to 20 per cent of the pore space), (3) rising gas/oil ratios as the relative permeability of gas to oil increases, (4) subsequent decline in gas/oil ratio as the supply of gas in the reservoir diminishes (see Figure 7-12). Isogor and isobaric maps of a pool controlled by solution gas drive do not necessarily show correlation with structural features. They are indicative of the permeability distribution in the area surrounding the individual well bores and the quantity of oil taken from the individual wells. A reservoir being produced by solution gas drive, where permeability variation is nominal, is more closely related to the cumulative reservoir voidage or cumulative oil produced than it is to the rate at which the hydrocarbons are recovered. Practically this is not to say that from the viewpoint of the gross reservoir, production rate has no effect on recovery, since seldom are the well bores open to a single homogeneous pay.

The ultimate oil recovery from a reservoir subjected to a solution gas drive only will be in the range of 10 to 35 per cent of the tank oil originally in place. There are no average recovery figures that can be applied with confidence to a specific pool. An

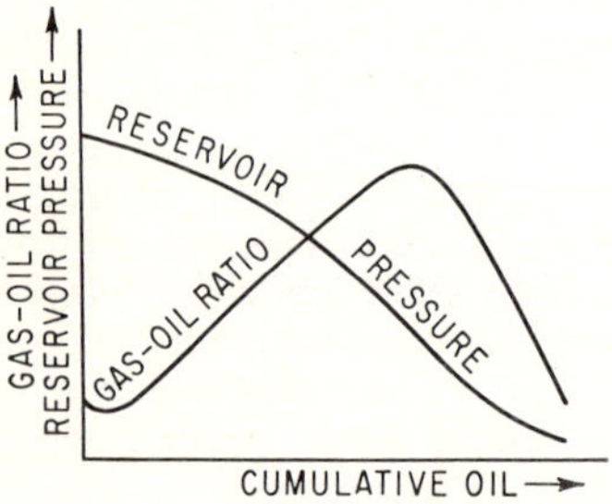

FIGURE 7-12. Illustration of production history of a gas-depletion-type reservoir.

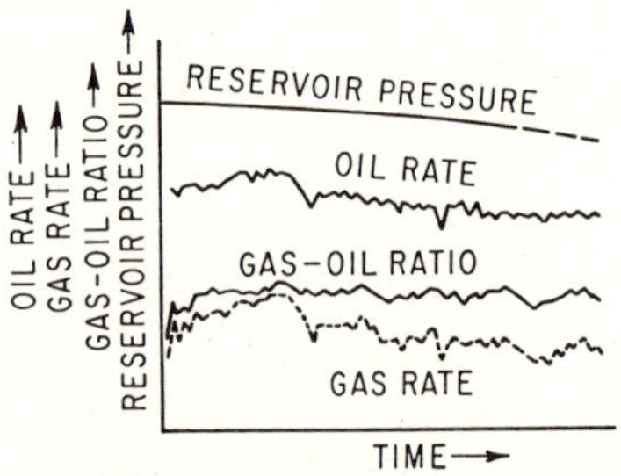

FIGURE 7-13. Illustration of production performance of a property subject to gravity drainage.

examination of the case history of oil reservoirs indicates that ultimately other production mechanisms come into play. The viscosity and the gravity of the oil, the permeability variation of the pay, the original solution gas/oil ratio, and the original reservoir pressure, all influence the effectiveness of the solution gas drive as a recovery mechanism. For example, the ultimate recovery of oil from a solution gas drive reservoir can range from 12 to 32 per cent of the tank oil originally in place as a result of variation in oil gravity alone.

The presence of free gas (gas cap) can provide a source of energy for the recovery of additional oil from a reservoir that would otherwise produce as a solution gas drive. Realization of increased recovery depends upon controlled production of the individual wells in a manner that permits gravity drainage to operate by minimizing reservoir oil viscosity and maintains maximum fluid saturation at the flank positions on structure. When the depletion of pressure in the gas-cap area exceeds that in the oil belt, subnormal oil recoveries may be expected. Here the oil invades the upstructure (gas-cap) portion of the reservoir, the liquid saturates the dry gas portion of the pay and becomes nonrecoverable. Proper control of gas production (minimum gas/oil ratios) reduces the rate of reservoir-pressure decline and retards the increase in oil viscosity. This aids subsurface segregation of the oil and gas phases. In turn, the percentage of pore space filled with oil saturation at downdip structural positions ahead of the advancing gas front is maintained at a maximum value. Gas bypassing of this oil is reduced or eliminated, and oil recovery from the pool is increased. Structural relief, permeability of the pay, reservoir viscosity of the oil, and ratio of the volume of the gas cap to the volume of the oil belt, all influence the amount of oil which can be obtained by exercising proper control of the gas cap as it expands. Theoretically the reservoir should at depletion contain only low-pressure gas with a minimum residual oil saturation in the original oil belt. When the volume of the gas cap is equal to the volume of the oil belt, proper control of gas-cap expansion can produce recoveries that exceed those obtained from solution gas drive by as much as 50 per cent. Field performance, however, is not apt to realize this calculated amount of recoverable oil.

Gravity drainage as a producing mechanism will ultimately result in minimum residual oil saturation of the producing measures. The presence of an initial gas cap or the return of gas to the reservoir (gas injection) at high structural positions increases the rate of oil recovery by aiding gravity drainage through maintenance of low reservoir oil viscosity and high fluid saturation at the flank positions on structure. Characteristically performance histories of controlled gas-cap expansion mechanisms have the following features (see Figure 7-13). (1) High gas/oil ratio wells appear first at top structural positions. (2) Structural position and individual well gas/oil ratio trends can be related to cumulative oil produced. (3) Complete control of the expanding gas cap finds the produced gas/oil ratio declining as reservoir pressures drop.

The performance of one oil pool indicates that the production of oil by gravity drainage will ultimately yield 65 per cent of the tank oil originally in place. Such recovery efficiency is unusual. It cannot be anticipated for very many pools producing by gravity drainage. Other reservoirs with low permeability, low-gravity oil (high reservoir oil viscosity), little gas in solution, poor structural relief, and no free gas cap will recover by gravity drainage only a very small percentage of the tank oil originally in place. Some fields with 30 to 40 years of productive life are now and have been for some time controlled by gravity drainage. Estimates of recovery indicate an ultimate realization of 10 to 14 per cent of the tank oil in place.

Water plays a substantial role in the recovery of oil. It may contribute a part or all of the reservoir energy required to deplete an oil or gas reserve. It may be effective only in the initial stages of production through its expansion within a limited aquifer when the reservoir pressure is reduced, or it may continue to be active throughout the productive life of a field. In such situations, either the adjoining aquifer is replenished at outcrop or the ratio of the aquifer to the hydrocarbon reservoir is exceptionally large.

Liquids have a coefficient of expansion that relates the increase in specific volume for a given weight of material to temperature and a coefficient of compression that relates the decrease in specific volume to an increase in pressure. These values vary with each liquid and are numerically quite small. The average compressibility of the water at reservoir temperature is 0.000469 cubic foot per atmosphere. Assume a hydrocarbon reservoir filled with oil and no gas and which is open to a 100 per cent water-saturated pay with ten times the volume of the oil reservoir. Assume, for simplification, that the oil and water have the same compressibility factor and the reservoir pressure is 3,000 psia. Upon reduction of reservoir pressure to atmospheric, the liquids will expand and replace 10 per cent of the hydrocarbons originally in place. As the ratio of the volume of the adjacent aquifer to the volume of hydrocarbons increases, the recovery due to fluid expansion alone becomes more and more significant. Thus some hydrocarbon reservoirs can be depleted with a water-drive mechanism even though the supporting aquifer is not open to replenishment at outcrop.

The efficiency of water as a recovery mechanism is related to the permeability variation of the pay or pays, relative permeability of oil to water at reservoir conditions, and the pore volumes displaced.

In some oil fields, the effect of water encroachment is not discernible until reservoir pressures have experienced considerable decline. In other reservoirs, particularly those composed of highly fractured dolomite, limestone, or siliceous shale, very little drop in reservoir pressure is required to initiate the movement of water. Encroachment may be updip from flank positions (edge water), or it may underlie all the hydrocarbon reservoir (bottom water) and move vertically. The presence of high water cuts or volumes does not determine the presence or absence of an active water drive. Rather it is the movement of water as reflected by individual well performance and changes in volumes produced that with reservoir-pressure data is diagnostic of active water drives.

Recovery of oil by natural or artificial water drive as a rule will result in high ultimate yields. However, the field experience indicates that when instituted after primary depletion is well advanced, the added recovery may range from zero to 100 and 200 per cent of the ultimate primary recovery. As with all other production techniques and reservoir controls, the use of a water-drive mechanism does not guarantee added oil recovery.

In those instances where the natural drive can contribute large volumes of water, the residual oil saturation can be reduced to a minimum, hence maximum oil recovery realized. The degree to which such a program can be carried depends on the economics of producing and disposing of large volumes of fluid.

It is difficult to decide what material should be covered in a brief presentation of the behavior of oil and gas reservoirs. However, if this chapter leads to a better understanding of the relationship of geology to the performance of oil and gas fields, its purpose will have been served.

REFERENCES

Brown, G. G.: "Deviations of Natural Gas from Ideal Gas Laws," Clarke Bros. Co., Inc., 1940.

Bulnes, A. C., and R. U. Fitting, Jr.: An Introductory Discussion of the Reservoir Performance of Limestone Formations, *Trans. AIME*, vol. 160, 1945.

Burcik, Emil J.: "Properties of Petroleum Reservoir Fluids," John Wiley & Sons, Inc., New York, 1947.

Calhoun, Jr., John C.: "Fundamentals of Reservoir Engineering," University of Oklahoma Press, Norman, Okla., 1953.

Cannon, G. E., and R. C. Craze: Excessive Pressures and Pressure Variations with Depth of Petroleum Reservoirs in the Gulf Coast Region of Texas and Louisiana, *Trans. AIME*, vol. 127, 1938.

French, R. W.: Geothermal Gradients in California Oil Wells, *API Drilling and Production Practice*, 1939.

Hawkins, J. E.: "Thermodynamics," John Wiley & Sons, Inc., New York, 1947.

Meyers, Willis: Personal letter, Jan. 12, 1959,

Muskat, Morris, "Flow of Homogeneous Fluids," J. W. Edwards, Publisher, Inc., Ann Arbor, Mich., 1946.

Muskat, Morris: "Physical Principles of Oil Production," McGraw-Hill Book Company, Inc., New York, 1949.

Muskat, Morris, and M. O. Taylor: Effect of Reservoir Fluid and Rock Characteristics on Production Histories of Gas Drive Reservoirs, *Trans. AIME Petrol. Div.*, vol. 165, 1942.

Nichols, Earl A.: Geothermal Gradients in Mid Continent and Gulf Coast Oil Fields, *AIME Petrol. Technol.*, November, 1946.

Paine, S. M.: Petrophysical Analysis of Some Wilcox Wells, *AIME Petrol. Technol.*, October, 1956.

Pirson, S. J.: "Oil Reservoir Engineering," 2d ed., McGraw-Hill Book Company, Inc., New York, 1958.

Sage, B. H., and W. N. Lacy: "Volumetric and Phase Behavior of Hydrocarbons," Stanford University Press, Stanford, Calif., 1939.

Standing, M. B.: "Volumetric and Phase Behavior of Oil Field Hydrocarbon Systems," Reinhold Publishing Corporation, New York, 1952.

"Secondary Recovery of Oil in the United States," American Petroleum Institute, New York, 1950.

Watts, E. V.: Some Aspects of High Pressures in the D-7 Zone of the Ventura Avenue Field, *Trans. AIME Petrol. Div.*, 1948.

Chapter 8

PLANNING EXPLORATION PROJECT

By Ed J. Hamner
Consulting Geologist

and

John L. Loftis, Jr.
Regional Exploration Manager, Southwest Region, Humble Division, Humble Oil & Refining Company

Houston, Tex.

An exploratory project designed to search for oil and gas may be defined as an organized endeavor intended to result in finding hydrocarbons in commercial quantities. The project itself may vary in scope from an effort to explore by one or several of the many tools at the disposal of exploration geologists in provinces where established oil and gas fields exist to exploration of the more remote regions where there are no known fields and where perhaps no wells have ever been drilled.

Since the regional setting for the project may vary, so may the basis for its conception. It may have been inspired from a desire to participate in lucrative oil production along a trend or in a province where competitors are enjoying success. Or it may have been the result of studied geologic deliberation in new and probably more remote areas of no known oil. This latter type could be the more rewarding for an individual pioneering company if it meets with success. However, the former type, lower risk by nature, is much more likely to result in early economic gain to the operator.

In the case of an exploratory project in a known oil province, most oil companies and individual operators would ordinarily have current subsurface geologic maps available. The oil companies would probably have geologic maps currently kept up to date by geologists in their own organization. Individual operators in most cases have geologic information available to them through the services of consulting geologists. The project itself, in all probability, would be restricted to the utilization of one or more specific tools particularly adaptable to the problem, such as core drill and reflection seismograph. These newly acquired data would then be integrated with the previous geologic interpretation of the region.

This chapter will attempt to deal primarily with the exploratory type of project undertaken in a geologic province that at the outset has no commercial oil and gas fields within its confines. It will be conducted in a region where there has been very sparse drilling and where the participating company or individual has performed no previous geologic work. It will encompass utilization of the majority of known exploration techniques in orderly succession. In actual practice one or several of these techniques may often be omitted from the project. Oil and gas normally occur in the sedimentary basins of the world, and no two geologic basins are alike. They may be similar in size or in basin configuration, yet many differences exist which will vary the tools and techniques utilized in exploration. For example, some sedimentary basins may be covered by recent sediments in a manner which precludes the use of surface geology. Others, from the standpoint of regional setting, might even preclude the use of geophysics.

The origin of this pioneering type of exploration project must always be conceived in the minds of men and normally has as its impetus the desire on the part of the company or an individual to expand. The beginning may even be motivated by an educated hunch on the part of one or several geologists. History tells us that many of the world's more important oil fields were found by almost random drilling by people who thought those spots might be a good place to find oil. The discovery of the great East Texas oil field is a classic example. This newly considered geologic province may be similarly located geographically and topographically as is some known oil-producing province. On the other hand, the inspiration for the project may be only following competition after it has been developed through scouting information that other companies or individuals have explored a remote area and are following up their exploration with lease acquisitions. This might excite oil operators to enter into an exploration project simply out of respect for the ability of its competitors to explore. Classic examples to bear out this statement have occurred in the last decade, i.e., the rush into the waters of the Gulf of Mexico by many companies because of their respect for the first company who entered this venture and their fear that they might be left out in the finding of lucrative new reserves. The other example is the tremendous rush into Alaska by many people and companies following the discovery of oil there by a highly respected company in 1957.

Once it has been decided to embark upon an exploratory project, a considerable amount of preliminary work and research must be done. All available files and all published literature must be thoroughly examined to determine whether or not geologic reports or maps are already available for this or similar areas. Maps showing the area and the land ownership should be secured immediately. Aerial photographs, which show topography and often reveal structural geologic information, are a necessity. They should be of high quality, and the scale should be large enough to permit detailed examination. These photographs may be available through commercial firms or governmental agencies. It is imperative that all data from wells drilled previously in the province be studied in the early stages. If this information, including logs and other surveys of these wells, is not already a part of the operator's files, then every effort must be made to acquire these data. Should a review of the available literature indicate that surface geology in this area reflects the possible presence of reservoir rocks in the subsurface and that this information can generally be correlated with the logs of the wells which have been drilled, then certainly the necessary ingredients are known to be present to warrant pursuit of the project from a geologic point of view.

At this point, very little money has been expended on the project because most of the work has been performed by geologists in file and literature research. Now the decision must be made whether or not to send personnel and equipment into the area at a considerable outlay of money. It is essential that all the factors influencing the practicability of going forward with the exploration program be critically examined.

First, and most important, it must be determined that leases, concessions, or exploration permits will be available and can be acquired. Probable costs of lease acquisitions must be determined. Acreage limitations, such as are imposed by the United States on Federally owned lands, must be investigated. In some cases these restrictions limit participation in an area to the extent that a company may decide, for this reason alone, not to enter a new area. If the new region to be explored lies in foreign lands where all minerals are owned by governments, the possibility of obtaining reasonable lease concessions must be explored. The royalty that must be paid to the mineral owner must also be checked. Here, again, unless a reasonable percentage for royalty can be agreed upon, this factor alone may cause an operator not to enter into the new project. Should competition already exist in the area, it is highly necessary to ascertain how much acreage is already under lease to other operators, and its accurate location should be placed on maps of the area. If it should be learned that all the favorable acreage is already under lease, enthusiasm for the area would certainly diminish. The effective term of the acquisition agreement or lease is also of great importance, particularly since the complete exploration program with subsequent drilling is not only costly but time consuming. A prospective company must be assured that it will have adequate time to perform the work required by the project in a manner consistent with both good business and the geologic factors involved. Any other unusual terms required in available leases or other acquisition contracts must be studied thoroughly. It must be remembered here that the type of lease form or contract made with the early property owners usually sets a pattern that will endure for some time. Extreme care should be taken in negotiating with the early property owners to secure contracts reasonably from not only the royalty aspect but also performance requirements, producing terms, damage clauses, etc.

Following the determination of the likelihood of obtaining reasonable acquisition agreements, an estimate of logistics for the contemplated operation must be made. The accessibility of the region to both men and equipment, direct factors influencing cost, is also highly important. The cost of drilling wells and transporting oil and gas to market must be considered.

It has previously been determined that the bare essentials of a geologic setting favorable for the accumulation of oil and gas are present. Now that the other factors of leasing and logistics appear satisfactory, the operator may proceed to plan the on-the-ground exploratory program. The exploration manager, chief geologist, and chief geophysicist, with the help of their staff of trained men, will now begin to plan this program. They will look at it in much the same way that consultants and contract personnel will consider the problem when individuals or independent operators are concerned. Depending on the size and accessibility of the area, the number of surface geologic parties will be determined. If aerial photographs are not available for immediate purchase, it will be necessary to contract with some commercial aerial photograph service to fly the area. High priority should be assigned to this work. If gravity and magnetic surveys are not already available, then this work should be done either by company or contract personnel.

In the initial stages of the utilization of these geophysical tools, they will probably be used only as an aid in determining basin configuration. The initial plan might well include the employment of surface geology, magnetics, and gravity simultaneously, commencing at the very start of the on-the-ground project. Any two or all three of these techniques may be used at the same time without conflict, and this will speed up the initial stages of the project. The importance of aerial photographs cannot be overemphasized because they are an invaluable aid to the surface geologist. They can be interpreted directly for revelation of geologic structure. They can also be used as topographic maps and to trace rock outcrops. Later on in the project they will prove of material aid to civil engineering survey parties.

After this initial phase has progressed sufficiently to indicate that the area is favorable for the accumulation of oil and gas, then core drill equipment and seismograph may be deployed. Sometimes rough terrain will preclude the use of either of these tools, but in most cases, rough terrain provides opportunities for excellent surface geologic work which may be sufficient to outline geologic structures for drilling. The time of deployment of the core drill or seismograph usually depends on the findings of areal surface geology, magnetic or gravity meter surveys. Consequently, specific planning will have to be flexible. Closely following this stage of planning, or even simultaneously with it, acquisition of leases, concessions, or permits will probably be accomplished. Lease acquisition time varies considerably and is influenced by factors of governmental attitude, development of the land, urban population, and many other things. In foreign countries it is usually necessary to obtain concessions before any exploration is done. Usually, the initial information at hand is very sketchy. If the area contains large ranches or other ownerships which can be obtained cheaply, it is generally a wise policy to acquire these ownerships immediately following the decision to conduct the exploratory project. In cases where ownerships consist of only a few acres, it is usually wise to conduct the exploration and then lease the small ownerships. If the initial phase of the exploratory project dictates that small ownerships be acquired, much time-consuming work will be necessary.

Once the plans for land acquisition and the exploratory program have been completed, it is time to begin the more costly phases of the project. Men and equipment must move to the area. Suitable living quarters must be established for the surface geologists and geophysicists. The problem of living for the men will vary from residence in populated areas with modern conveniences to living in tents. In some areas of forbidding terrain and in arctic conditions, food as well as other types of supplies and equipment must be furnished to the men by air.

After the geological field parties and the geophysical crews have located themselves as centrally as possible to the working area, reconnaissance by both groups will commence. A magnetic survey, if decided on, should be made from the air because this is both a faster and more accurate method of obtaining the coverage. After studying the aerial photographs and the available maps, the field men should know the road net for the area and, in the case of the gravity meter crew, can soon thereafter start their survey. Since little is known of the surface geology, the field parties will do considerable reconnaissance, based on information gleaned from the aerial photographs or driving the roads, or in difficult terrain, they may even reconnoiter by helicopter or boat.

After the preliminary reconnaissance, the surface parties will commence to map the areal geology of the project area, using paleontology and lithology in connection with the aerial photographs. Utmost care should be exercised to note the character of all rocks cropping out. Of particular importance in this phase of the work are the physical properties of the rocks, such as porosity and permeability. If no probable reservoir rocks are mapped at the surface, the project may look much less inviting. Also of prime importance is the observation and mapping of all oil and gas seeps in the area. These seeps are usually considered as direct evidence of the presence of source rocks for hydrocarbons. Key beds, which are easily recognized throughout the area, must be found, and the thickness of section between these formations must be measured accurately. This information will be of vital importance in determining the structural attitude of the surface beds and, more specifically, in locating local structure. Elevations can be determined by aneroid barometers or by plane table. Structural information can also be obtained from the attitude of the beds by measuring and recording strike and dip. All of the surface geology should then be reduced to as few maps as possible, preferably to one. Once the magnetic or gravity survey and surface work has progressed sufficiently to recognize the probable existence of a sufficiently deep sedimentary basin with local structures present, then additional land acquisitions may be made.

Some, or all, of the area may have been found to be covered by alluvium or some other recent deposit which masks the bed rock. The decision will probably then be made to move a seismograph crew into the area. Core-drill exploration may also be used to supplement surface geology in these "blind areas." It must be necessary for good key beds to exist at shallow depths before the core drill will be effective. Once the area has been covered by combined surface geology, the seismograph, and any other necessary geophysical work, an integrated subsurface map of the area can be constructed, using a potential reservoir rock mapped at the surface and projected into the subsurface as a key bed.

Based on this integrated map, final lease acquisitions will be attempted. It is likely that competitors have moved in by now, and additional acreage may be unavailable. It is also probable that much of the previously acquired acreage may be relinquished because of unfavorable position. This will be particularly true where legal limits as to the amount of acreage held are imposed. It is possible that oil and gas in a new province may be reservoired in stratigraphic traps, and judgment will usually dictate that all acreage be held under lease until wells can be drilled to evaluate this possibility. Should lease rentals be exorbitant, it is sometimes impossible to follow this policy.

It is now time to decide upon the location for the first wildcat test well. The operator now has the integrated information mapped and all the available acreage in the apparently favorable areas. This first well will probably be located on a structure mapped by surface geology and confirmed by the seismograph and in an area where gravity and magnetics have indicated a thick subsurface sequence of sedimentary rocks. Civil engineers will do the necessary surveying for the location and determine ownership boundaries within reasonable limits of the proposed well. They will supervise the preparation of the location and any road building necessary. Land and title men will cure the titles to the properties immediately involved. In the meantime, the drilling rig and accessory equipment will be moved to the location. An important part of the accessory equipment on a wildcat well in an unknown province is the mud logger. This instrument will detect the presence of hydrocarbons in the drilling fluid and the bit cuttings. If the area is remotely located, other well-logging equipment including electric equipment, radioactivity, and dipmeter must be made readily available. It is also highly important to include drill-stem testing equipment. Should the area be located reasonably near service company bases, this special equipment can be ordered as needed.

A well-site geologist should be on location at all times excepting down time while fishing or repairing the rig or when the well is waiting for cement to set after running casing. This is a very confining responsibility, and it may prove necessary to assign a minimum of two and preferably three geologists to the well. These men should be sufficiently trained to allow intermittent rotation to other types of geologic work in the area. After the well has been spudded, the well-site geologist in charge should exercise extreme care to ensure complete sample gathering and logging from the grass roots down. He should work closely with the men operating the mud-logging unit in order to observe all drilling breaks and take care to see that the drilling fluid is circulated up to the surface before drilling deeper. He will also check the cutting samples for possible shows of hydrocarbons both visually and with ultraviolet light for fluorescence. On a wildcat well in a region almost completely unknown every show of hydrocarbons should be drill-stem-tested. All information obtained should be kept confidential. If the operator is sufficiently encouraged by either oil or gas shows or fluid recovery on drill-stem test, it is likely that a more determined attempt will be made to acquire additional favorably located land. It will be necessary to take cores periodically in an attempt to correlate subsurface rocks with others which may have been found updip at the surface. Porosity and permeability data, as well as fluid saturation, should be determined from these cored sections. Well logs should be run during the course of drilling both for stratigraphic information and to ensure the obtaining of this information in the event the hole is lost. Below approximately 10,000 feet well logs should be made every 500 to 1,000 feet because the risk of losing the hole is much greater with depth.

If the first wildcat is dry, the information gained must be integrated into the over-all geologic picture and the decision as to whether to continue prospecting must be made. If this well failed to make a commercial producer but the dipmeter indicated the possibility of a higher structural position on the same fold, then a second well will probably be drilled at this higher position. If the second well is a failure, it is probable that this structure should be abandoned at least temporarily and another structure selected for drilling. If the second prospect, too, fails to yield an oil or gas discovery, again the information gained will be integrated into the over-all picture.

It may now appear that the chances for success of the project are slim. The risk of success may be considered so high that the operator will wish to promote the drilling by others on a portion of the holdings retained. This can be done either by the farm-out route or by contributing either money or acreage or both to other operators for wildcat drilling on their own acreage adjacent to the company's land. Eventually, the project may have to be written off as a complete failure, and many suffer this fate.

Should the wildcat be completed as a producer and the reservoir contain enough oil or gas pay to ensure commercial success, then the project can be regarded as successful. Other prospects can then be drilled and upon discovery can be developed in an orderly fashion. The degree of success of each individual prospect must be determined through confirmation drilling. More well-site geologists will move in accompanied by petroleum engineers who will study the new reservoir in an effort to determine optimum producing rates and well spacing patterns. The engineers and geologists should work very closely in arriving at a proper development program. Exploration geologists will continue to work in the area as new well information is developed. How long they will continue working is difficult to determine, since it is doubtful that an exploration project is ever really complete. It is impossible in any geologic basin ever to turn over all the rocks in order to learn what may lie underneath.

Chapter 9

PHOTOGEOLOGY

By Norman C. Smith

Consulting Photogeologist
Dallas, Tex.

and

Henry L. Buthman

Regional Photogeologist, California Oil Company (Western Division)
Denver, Colo.

Definition

Photogeology is the method of determining surface geology from the study and interpretation of air photographs. It is applied both as an independent reconnaissance method and, when coordinated with field mapping, for detailed surface exploration. While normal practice in petroleum exploration ordinarily calls for the collation of all available types of data, photogeology is treated here as a special method for the delineation of structural features as potential traps for the accumulation of hydrocarbons. This recognizes the unique attribute of built-in perspective which the air photograph brings to surface mapping, by which the surface of the earth is reduced to convenient scale for comprehensive geological analysis. With greater perspective than can be obtained by field observation alone, the geologist using air photographs can quickly and accurately piece together scattered fragments of ground detail to reconstruct the areal distribution and structural disposition of surface rocks.

Limitations

All gross geological features present on the earth's surface are to some degree and in some manner recorded on the air photograph. Such features may be recognizable by the appearance and configuration of the outcrop, but they are more commonly represented photographically as zones of contrasting tone resulting from lithologic control of soil or vegetation or relief patterns or as linear elements indicative of fracture. In spite of difficulties presented by natural camouflage, the principal limitations controlling the caliber of photographic interpretation are generally human rather than mechanical. Given photographic coverage of adequate quality, the most important factor is the interpreter's experience as a geologist; less important is the geologist's experience as a photointerpreter.

Advantages

Photographic interpretation is the only exploration method that, under ordinary circumstances, cannot be scouted by a competitor. Without the need for urgency required of most exploration methods open to observation by rivals, reconnaissance can proceed at a more deliberate pace, allowing a reasonable, unhurried period for collating the results of photographic interpretation with data from other sources to obtain a more reliable picture of oil possibilities.

In competitive oil operations, saving time is often as important as achieving accuracy, since being right is small consolation to the operator who is too late. The development of photogeological techniques brings to surface mapping the element of speed, without sacrifice of accuracy, which the economics of modern exploration requires. Photographic interpretation can be used to cover vast areas in a fraction of the time required for field surveys, especially in regions of poor accessibility.

In oil exploration, where cost is a critical factor in successful operations, photogeology cannot be excelled for low cost per unit area. Both high-quality photographic coverage and experienced interpretation can be obtained for a fraction of the cost per area of any other exploration method. This does not mean that photogeology can be used to supplant other techniques, but it certainly suggests that photogeology can profitably be used as an inexpensive and rapid method of obtaining reliable reconnaissance data to precede and supplement other more detailed methods.

Training

A photogeologist is first and fundamentally a geologist. He should also have training in geomorphology, since the rocks in most of the oil basins of the world are at least partially covered and major dependence will invariably be placed upon physiographic interpretation. Any geologist is capable of interpreting geology from the aerial view if he has sufficient interest and flexibility to adapt more conventional mapping techniques. Photogeology remained a specialty as long as relatively few geologists took advantage of air photography as a source of basic geological data. However, photogeology now enjoys an increasingly widespread application which is resulting in a revival of surface mapping in oil exploration.

THE PHOTOGEOLOGICAL PROJECT

Methods of photogeological interpretation are presented within a general outline of procedure. Each interpreter will, of course, develop personal techniques in the effort to gather as many accurate geological data as possible.

Problem

Scope is the first consideration of a project. During reconnaissance study an attempt is made to produce a generalized interpretation of major structural axes and of the bedrock types present. Such data can usually be acquired through a study of the mosaics and will support a decision of whether to explore further in a new province.

On the other hand, in a known favorable province, a more deliberate analysis is directed toward isolating favorable prospects. Either type of project must be adapted to terrain conditions.

Approach

1. In areas of exposed strata: by stereoscopic study of contact prints, map the thickness, lithology, and attitude (dip and strike) of the beds; where possible, select map datum and elevation control to produce a structural contour map and an areal geologic map.

2. In areas of scattered rock exposures: by stereoscopic study of contact prints and study of mosaics, map and interpret structural relationships from the distribution of rocks and residual soils as reflected by differences in vegetation, cultivation, and relief to produce an areal geologic map and structural anomaly map.

3. In areas blanketed by transported soils: usually by mosaic study, map the vegetation zones, drainage and relief patterns, and, in brief, integrate all surficial physiographic and geographic evidence that might indicate structural influence beneath the surface to produce a physiographic anomaly map.

4. In some areas, where stratigraphic control of accumulation is expected: map the areal geology with particular attention to convergence in surface beds to aid in localizing convergence and facies change of possible subsurface significance.

EQUIPMENT AND MATERIALS

Depending upon the amount of detail required and the nature of the specific geologic answers sought, the essential equipment can vary from the simplest stereoscope to extremely complex and correspondingly expensive photogrammetric instruments for plotting and making measurements. Readers interested in such specialized equipment and quantitative photogrammetric techniques, which are not directly involved in geological interpretation, are referred to the selected references at the end of the chapter.

Stereoscope

Many makes of this instrument exist; however, there are basically just three types: the lens, mirror, and prism. Each has its advantages, and the choice of stereoscope will depend upon the scope of the study being made, the quality of the photography in use, and the surface characteristics of the area involved. In areas of varying surface characteristics it will often be expedient to augment one with another.

For general office and field study of the standard 9- by 9-inch prints, the folding lens stereoscope with 2× magnification, commonly referred to as the pocket stereoscope, is adequate (see Figure 9-1). It has the advantage of being small and compact, making it particularly adaptable for use in the field, and its size and weight offset the somewhat awkward necessity for curling the edge of the overlapping print to expose the complete field of stereoscopic coverage. When studying areas of minute outcrops or when using small-scale photography, the interpreter may desire added magnification. Suited to this purpose, at least one model of the folding, lens stereoscope is adjustable for either 2× or 4× magnification.

Mirror and prism stereoscopes are similar in that they permit full separation of the photographs and allow the viewer to see all portions of stereo coverage without the curling action mentioned above (see Figure 9-2). This results in greater ease in handling and better over-all perspective. Binocular attachments are usually available for added magnification.

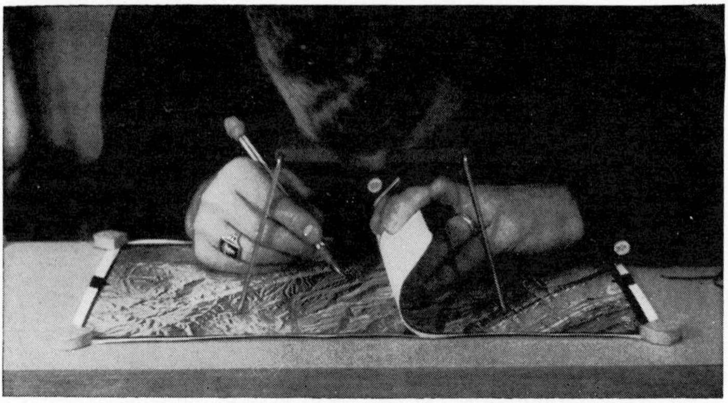

FIGURE 9-1. Folding lens stereoscope, Abrams CF-8.

FIGURE 9-2. Mirror stereoscope, Ryker M-11, with binocular attachment.

Photography

Within practical limits the photographic coverage is selected with reference to the geological problem; i.e., small-scale photography is commonly used for regional studies and large-scale for local area studies; mosaic coverage is better adapted to reconnaissance physiographic studies and stereo coverage to detailed outcrop studies. However, time and cost factors often dictate the use of whatever coverage is at hand, regardless of scale, type, or quality. Ideally, detailed stereoscopic analysis is done with relatively large-scale (about 1:20,000) vertical contact prints, although improvements in photographic materials and equipment have produced high-altitude photography of relatively small scale (about 1:60,000) which is quite satisfactory for reconnaissance and even semidetailed studies.

Color photography is continually being improved in quality and applicability. Variations in photographic tone, which are essential to recognizing rock and soil boundaries, are rendered with heightened contrast in color. However, such advantage must be weighed against the added cost of photographic coverage. Little color coverage has been flown, whereas widespread coverage in black and white is available (approximately 90 per cent of the United States) from various governmental agencies and commercial firms at a cost per unit area far less than it would be to have any given area photographed specially. Therefore, the cost differential between color and black-and-white coverage, though much less where initial flying is required, would be appreciable for most areas. Unlimited and universal substitution of color for black and white cannot be justified at present either from the technical or economic point of view.

Contact Prints

Contact prints of both vertical and oblique aerial photographs are made directly from the film negative without enlargement or reduction. The most common print size is 9 by 9 inches and is preferably printed on a semimatte paper for an optimum combination of photographic detail and tolerable reflection.

Mosaics

Mosaics are constructed by matching ground detail in overlapping continguous contact prints of vertical aerial photographs. Controlled or accurately scaled mosaics require reconstituted prints with radial distortion and tilt removed by corrective photographic means. Wet mounts, which use only the central, least distorted portions of prints, are the most satisfactory. However, lacking controlled mosaics, the geologist can construct his own temporary stapled mount or use uncontrolled stapled mosaic index sheets issued by various governmental agencies and commercial distributers of air photography.*

Trimetrogon. In the initial stages of exploring inaccessible regions trimetrogon coverage, which provides oblique views of the area between widely spaced vertical flight strips, can be used for reconnaissance mapping at minimum cost. Data annotated on the oblique prints are transferred to the base map by the use of a perspective grid (Canadian Topographical Survey, 1928).

Lighting

For stereoscopic work good lighting may be provided by a 75- or 100-watt frosted bulb. However, since the lamp must be situated close to the head, the cooler fluorescent fixture is preferred. Mosaics should also be lighted with an even, bright, but not brilliant source. Two or more floodlight fixtures fitted with 300-watt bulbs will generally be adequate to illuminate a wall or floor area.

Mounting

Since floor space is usually at a premium, wall mounting of mosaics is most practical. A pressed composition (upson) board (1/4 inch minimum) is the most satisfactory material for portable or permanent mounts. The unfinished appearance of this material on an office wall can be improved by covering it with untreated burlap.

Annotation

The contact prints are marked with rough annotations for later transfer to the base map. Both the colored china marker (wax pencil) and a recently developed, all-

* Aerial Photography of the United States; index map showing photographic coverage and holding agencies, issued periodically by USGS, Washington 25, D.C.

purpose (omnichrome) lead pencil are most satisfactory for this purpose, since the mark can be easily removed by an artgum eraser or by a cloth dampened with acetone or other similar solvent without injury to the print. When semipermanent annotation is desired, certain ball-point pens have been found to be ideal. They are available in numerous colors and make a vivid annotation which will not smear and is easily eradicated. Delicate line work and permanency are best achieved by the use of crow-quill pen and india inks.

Overlays

When reporting interpretation data, it is advantageous to present it in conjunction with the photography to demonstrate areal relationships. To eliminate the necessity of marking directly on the photoexhibit, the use of an acetate overlay of the mosaic is recommended. The interpretation detail can be transferred from the contact prints by inspection with pencil and redrafting on clean acetate with ink. Although frosted film acetate (thickness 0.001 inch) is adaptable, an ideal overlay is a matte film positive of a base map constructed from the photomosaic in use. This permits direct printing of the completed overlay for work sheets and ultimately for printing as the final map. The mosaics with overlay are invaluable for use in field checking.

PROCEDURE

Preparation

Before actually initiating work on the photographs it is important that the geologist prepare himself as well as possible to make the study.

Literature Review

Because photogeology is ideally adapted to exploring inaccessible, unknown provinces, the interpreter will not infrequently experience the stimulation and frustration of being a geological pioneer. Although he becomes adept at initiating projects with little or no prior work to guide him, the interpreter should routinely check all possible sources of information, since it would be foolhardy to ignore the existence of helpful data. This is especially emphasized because the use of photogeology as an independent tool to supplant other applicable exploration techniques or reducing its effectiveness by withholding the potential benefit of basic geological information may defeat or at least diminish the successful outcome of petroleum exploration.

Therefore, the literature should be reviewed just as thoroughly and purposefully as the geologist would prepare himself for geologic mapping under any circumstances. Perhaps more than other approaches, photogeology can profit in the initial stages of a project by information from other sources because of the added hazard of working over varied terrains often with unfamiliar climatic, cultural, and other, possibly unknown, surface influences. As his experience throughout the world increases, these hazards are proportionately reduced, but they are never actually dissolved; the problems of each new project must be faced squarely and resolved within the context of its purpose and requirements.

Preliminary Photo Review

Before any project can proceed far, reference must be made to the photography. Much time can be lost by incorrectly assuming that complete coverage is at hand. When the photographs are delivered, they should be checked immediately against

Table 9-1. Map Scales and Equivalents
(Courtesy of Transmares Corporation, New York, N.Y., and Arlington, Va., representing Zeiss-Aerotopograph, Munich, Germany)

Fractional scale	Feet per inch	Inch per 1,000 feet	Inches per mile	Miles per inch	Meters per inch	Acres per square inch	Square inch per acre	Square mile per square inch
1:500	41.667	24.00	126.72	0.008	12.700	0.0399	25.091	0.00006
1:600	50.00	20.00	105.60	0.009	15.240	0.0574	17.424	0.00009
1:1,000	83.333	12.00	63.36	0.016	25.400	0.1594	6.273	0.00025
1:1,200	100.00	10.00	52.80	0.019	30.480	0.2296	4.356	0.00036
1:1,500	125.00	8.00	42.24	0.024	38.100	0.3587	2.788	0.00056
1:2,000	166.667	6.00	31.68	0.032	50.800	0.6377	1.568	0.00100
1:2,400	200.00	5.00	26.40	0.038	60.960	0.9183	1.089	0.0014
1:2,500	208.333	4.80	25.344	0.039	63.500	0.9964	1.004	0.0016
1:3,000	250.00	4.00	21.12	0.047	76.200	1.4348	0.697	0.0022
1:3,600	300.00	3.333	17.60	0.057	91.440	2.0661	0.484	0.0032
1:4,000	333.333	3.00	15.84	0.063	101.600	2.5508	0.392	0.0040
1:4,800	400.00	2.50	13.20	0.076	121.920	3.6731	0.272	0.0057
1:5,000	416.667	2.40	12.672	0.079	127.000	3.9856	0.251	0.0062
1:6,000	500.00	2.00	10.56	0.095	152.400	5.7392	0.174	0.0090
1:7,000	583.333	1.714	9.051	0.110	177.800	7.8117	0.128	0.0122
1:7,200	600.00	1.667	8.80	0.114	182.880	8.2645	0.121	0.0129
1:7,920	660.00	1.515	8.00	0.125	201.168	10.00	0.100	0.0156
1:8,000	666.667	1.500	7.92	0.126	203.200	10.203	0.098	0.0159
1:8,400	700.00	1.429	7.543	0.133	213.360	11.249	0.089	0.0176
1:9,000	750.00	1.333	7.041	0.142	228.600	12.913	0.077	0.0202
1:9,600	800.00	1.250	6.60	0.152	243.840	14.692	0.068	0.0230
1:10,000	833.333	1.200	6.336	0.158	254.000	15.942	0.063	0.0249
1:10,800	900.00	1.111	5.867	0.170	274.321	18.595	0.054	0.0291
1:12,000	1,000.00	1.0	5.280	0.189	304.801	22.957	0.044	0.0359
1:13,200	1,100.00	0.909	4.800	0.208	335.281	27.778	0.036	0.0434
1:14,400	1,200.00	0.833	4.400	0.227	365.761	33.058	0.030	0.0517
1:15,000	1,250.00	0.80	4.224	0.237	381.001	35.870	0.028	0.0560
1:15,600	1,300.00	0.769	4.062	0.246	396.241	38.797	0.026	0.0606
1:15,840	1,320.00	0.758	4.00	0.250	402.337	40.000	0.025	0.0625
1:16,000	1,333.333	0.750	3.96	0.253	406.400	40.812	0.024	0.0638
1:16,800	1,400.00	0.714	3.771	0.265	426.721	44.995	0.022	0.0703
1:18,000	1,500.00	0.667	3.52	0.284	457.201	51.653	0.019	0.0807
1:19,200	1,600.00	0.625	3.30	0.303	487.681	58.770	0.017	0.0918
1:20,000	1,666.667	0.60	3.168	0.316	508.002	63.769	0.016	0.0996
1:20,400	1,700.00	0.588	3.106	0.322	518.161	66.345	0.015	0.1037
1:21,120	1,760.00	0.568	3.00	0.333	536.449	71.111	0.014	0.1111
1:21,600	1,800.00	0.556	2.933	0.341	548.641	74.380	0.013	0.1162
1:22,800	1,900.00	0.526	2.779	0.360	579.121	82.874	0.012	0.1295
1:24,000	2,000.00	0.50	2.640	0.379	609.601	91.827	0.011	0.1435
1:25,000	2,083.333	0.480	2.534	0.395	635.001	99.639	0.010	0.1557
1:31,680	2,640.00	0.370	2.000	0.500	804.674	160.000	0.006	0.2500
1:48,000	4,000.00	0.250	1.320	0.758	1,219.202	367.309	0.003	0.5739
1:62,500	5,208.333	0.192	1.014	0.986	1,587.503	622.744	0.0016	0.9730
1:63,360	5,280.00	0.189	1.000	1.000	1,609.347	640.00	0.0016	1.0000
1:96,000	8,000.00	0.125	0.660	1.515	2,438.405	1,469.24	0.0007	2.2957
1:125,000	10,416.667	0.096	0.507	1.073	3,175.006	2,490.98	0.0004	3.8922
1:126,720	10,560.00	0.095	0.500	2.00	3,218.694	2,560.00	0.0004	4.00
1:250,000	20,833.333	0.048	0.253	3.946	6,350.012	10,063.907	0.0001	15.5686
1:253,440	21,120.00	0.047	0.250	4.00	6,437.389	10,244.202	0.0001	16.00
1:500,000	41,666.667	0.024	0.127	7.891	12,700.025	39,855.627	$0.\underline{_4}25$	62.2744
1:1,000,000	83,333.333	0.012	0.063	15.783	25,400.050	159,422.507	$0.\underline{_5}62$	249.0977
Formulas	$\frac{\text{Scale}}{12}$	$\frac{12{,}000}{\text{Scale}}$	$\frac{63{,}360}{\text{Scale}}$	$\frac{\text{Scale}}{63{,}360}$	Ft per in. × 0.3048006	$\frac{(\text{Scale})^2}{43{,}560 \times 144}$	$\frac{43{,}560 \times 144}{(\text{Scale})^2}$	$\frac{(\text{Ft per in.})^2}{(5{,}280)^2}$

inventory lists and flight index maps. Any blank areas or missing photographs can then be replaced with a minimum loss of time.

As with any map, scale is a fundamental property of the air photograph which must be known before reliable areal or linear measurements or estimates can be made. Scale can be derived from the formula $S = f/h$, where S equals the scale, f the focal length of the camera, and h the altitude of the plane above the ground. Scale can be determined from unlabeled photography by using known ground or map distances. Such scale computations are approximate, for the actual scale changes from print to print as the altitude of the photoplane, and perhaps within a single print with topographic relief. However, approximate scale is adequate for most qualitative interpretation purposes.*

The mosaic coverage of the project is then studied, with attention directed to the prevalence of outcrop, soil and alluvial cover, prominent fold or fault axes, and the physiographic grain of the terrain expressed by the drainage and associated land forms. If rock outcrops dominate the terrain, stereoscopic study of contact prints is indicated. If, however, the area is generally blanketed by unconsolidated materials, greater emphasis is placed on the use of mosaics to analyze physiographic relationships. Mosaics of differing scales (e.g., 1 inch = 2,000 feet and 1 inch = 6,000 feet) are often compared effectively to bring out features of subtle expressions in tone or topography.

Area

Large areas can be worked more efficiently than small areas. The necessary elements of control are more likely to be present in the larger project, whereas a relatively small area, such as a single township, would only fortuitously contain a structural feature in its entirety. Therefore, ordinarily a large area is examined for regional information before concentrating on the local features which may be present in a restricted area.

Photo Study

Fundamentally the only real difference between mapping surface geology in the field and on air photographs is in the point of view. The field man deals with profiles of partial sections and integrates them into the plan view with his surface map. The photogeologist, starting ordinarily with the plan view, must learn recognition of objects seen from above and translate them to familiar geologic entities. Once the geologist has adjusted to this viewpoint, the procedure is not basically different from that of any surface analysis.

However, the elevated view of the earth's surface, although introducing the advantage of perspective, necessitates mapping the gross characteristics and general relations of rocks, i.e., composition, bedding, dip, and strike. In areas of outcrop these rock properties and relationships are easily mapped from the photographs through the stereoscopic study of vertical contact prints. In blanketed areas, however, the geologist must deal with their physiographic expression only, often best seen on the mosaic, to determine the areal distribution and structural relationships of the bedrock.

APPLICATION AND INTERPRETATION

Outcrop Mapping

In areas of outcrop, such as mountainous regions of considerable relief and their associated basins, photomapping is for the most part as straightforward as field map-

* For general reference see Table 9-1. For detailed application of scale data, such as might be needed in flight planning, see "Manual of Photogrammetry" listed in References at end of chapter.

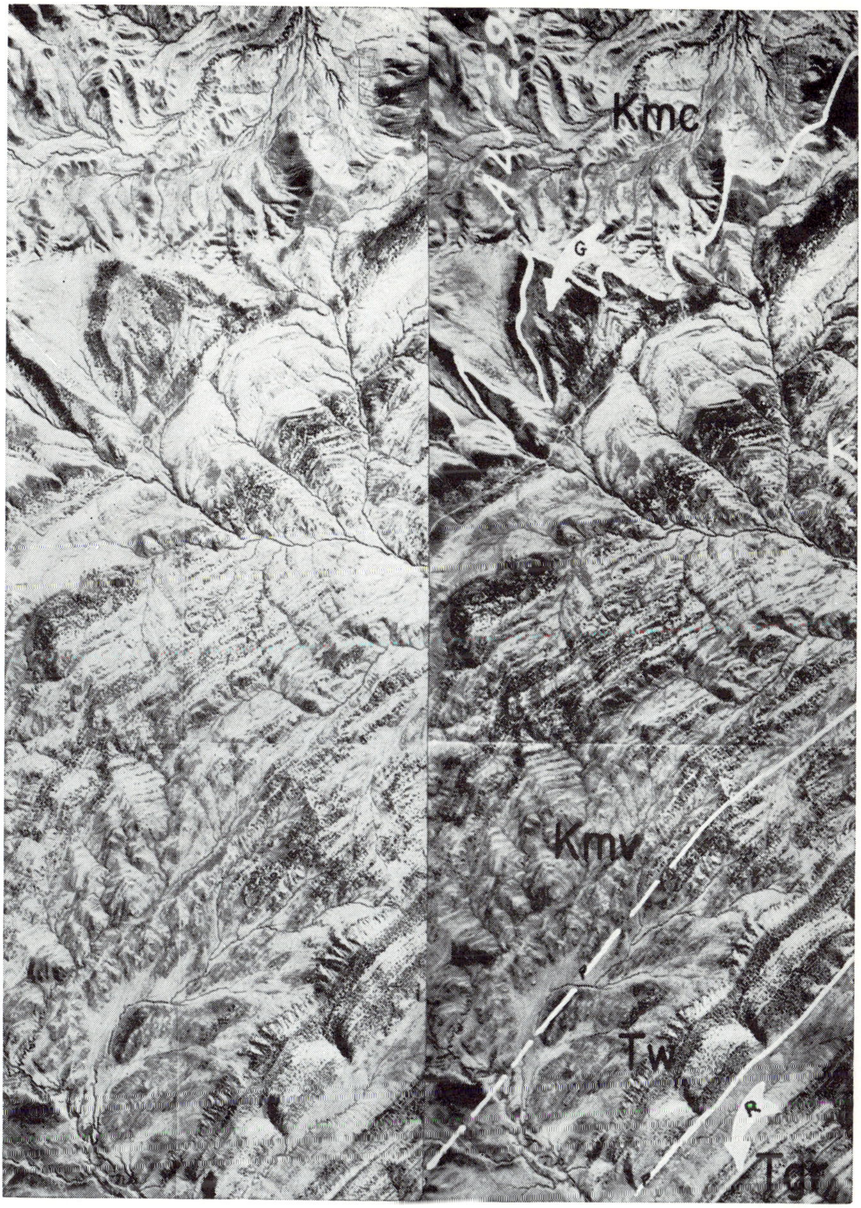

FIGURE 9-3. Stereogram showing hogback ridges in the Cretaceous and Tertiary sediments of the Rangely area in northwestern Colorado. The hogback ridge is of particular interest because of its conspicuous representation of surface structure. Note the ease with which dip can be read, particularly in the vicinity of the arrow marked R. (*Soil Conservation Service, U.S. Department of Agriculture.*)

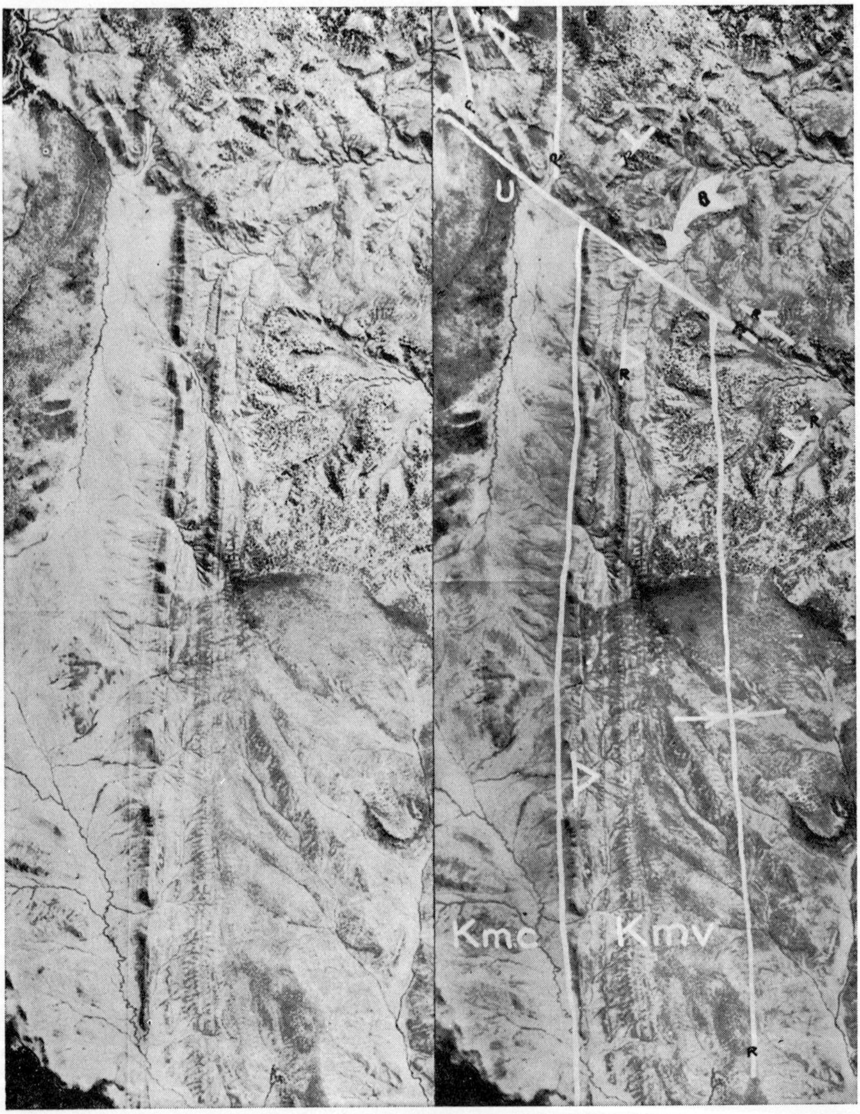

FIGURE 9-4. Stereogram showing a high-angle fault with a horizontal component in the Rangely area of northwestern Colorado. Structural attitudes are annotated on right photograph to demonstrate the effects of this kind of faulting on structure. Note displacement of synclinal axis in vicinity of arrow marked B. The fault trace, due to clarity of topographic expression, is marked only where necessary to show its effect on surface bedding. Note the prominent drainage alignment which marks the fault trace on the unannotated photograph. (*Soil Conservation Service, U.S. Department of Agriculture.*)

ping (see Figures 9-3 and 9-4). Key beds and formation contacts, recognized by their persistent lithology and erosional habit, are traced and annotated directly on the photographs. The identification of bedrock lithology is based upon the comparative study of contiguous outcrops and the contrast observed in the erosional behavior of different rocks within the same climatic environment. For example, whereas well-cemented sandstones are commonly found to be resistant in a variety of erosional environments, tending to be the ledging and capping elements of topography, shales are usually less resistant and are typically found in valley floors and in the lower slopes of valley walls. However, under certain climatic conditions the topographic expression could be reversed if the lithologies involved were friable sandstones and indurated claystones. Similarly, a limestone tends to be resistant in an arid climate, but in a humid climate, where solution effects tend to dominate physical erosion, the same limestone may weather deeply. The erosional behavior of metamorphic and igneous rocks likewise varies with climate and mineralogical composition.

Therefore, it may be very difficult to reach a definite conclusion regarding specific lithology without reference to the field or literature. However, general categories of rocks may at least be recognized and the relative distribution of contrasting lithologies established.

Estimating Dip and Slope

The true angle of inclination of any surface is distorted by vertical exaggeration inherent in stereoscopic viewing (Fichter, 1953–1954). Therefore, when estimating dip or slope, the observer must give proper attention to the factors causing exaggeration (focal length, overlap, and type of stereoscope) and correct accordingly. Slope conversion charts to simplify this procedure are included.* It will be noted that the greatest exaggeration occurs in low-order dips, 1 to 20 degrees, and progressively diminishes as the dip approaches 90 degrees. Fortuitously, this factor assists the interpreter to recognize low dip where the true dip may be exaggerated as much as 3½ times. For example, a true dip of 10 degrees may appear to be as much as 35 degrees and a dip of 1 degree as much as 3½ degrees. Experience will generally equip the interpreter to estimate dips with reasonable accuracy. If greater accuracy is required, photogrammetric devices can be employed. In reconnaissance mapping, however, the amount of dip is not so important as the direction.

Radial displacement resulting from surface relief is inherent in all aerial photography and must also be considered when estimating dip and slope. Because of its effect a surface dipping away from the center of the model (area of stereoscopic coverage) appears to be steeper than it is, and a surface dipping toward the center of the model appears to be gentler than it is. This distortion is not significant when estimating true dips of 10 degrees and less (estimated as 35 degrees or less stereoscopically); however, when working with true dips of over 10 degrees (estimated as over 35 degrees stereoscopically), it is significant, and the appropriate correction should be made. A radial-displacement correction chart (see Charts) is included to simplify this procedure. Note that this correction is made to the dip as estimated stereoscopically before it has been referred to the slope conversion chart.

A third distortion, inherent to some extent in almost every aerial photograph, is commonly called *tip-and-tilt*. This refers to the orientation of the airplane, and therefore of the camera, relative to the horizontal at the time the photograph was taken. Tip is introduced into the photography if one wing is lower than the other. Tilt results if the nose is either high or low. Tip-and-tilt is usually the resulting vector of both. Present-day camera mounts eliminate all but a minor amount of

* See Slope Conversion Charts 9-1 to 9-4 at the end of the chapter.

this distortion, and it is seldom that more than 2 degrees of tip-and-tilt is introduced. In dealing with relatively high dips (true dip of 10 degrees and over), this is not significant. However, in dealing with low dips, particularly those of less than 2 degrees, it can have a great effect on both the dip and the strike, and an adjustment

FIGURE 9-5*A*. Figures 9-5*A*, *B*, and *C* illustrate the influence of bedrock buried beneath an unconsolidated cover. Figure 9-5*A* shows widely cultivated cover of Lisbon (sandstone and shale) formation in upper left contrasted to uncultivated, densely forested Tallahatta (indurated shale) formation, Clarke County, Ala. (*Edgar Tobin Aerial Surveys, San Antonio, Tex.*)

should be made. Because of the fact that numerous references to the true horizontal plane are present in the model in the form of drainage, alluvial planes, lakes, etc., tip-and-tilt is usually readily recognized. Adjustment can be made by slightly rotating one or both of the photographs until the reference area appears normal. This requires some practice but is not difficult once it is understood.

In estimating dip and slope, the photographs must first be correctly oriented for stereoscopic viewing. This is accomplished by placing the photographs in such a way that the center points and the conjugate points of both photographs fall in a straight line. Photo separation (distance between like ground points) should be approximately 2¼ inches when using a pocket stereoscope. Separation varies slightly with

FIGURE 9-5*B*. Relatively rugged relief of the mantled Jackson (marl) formation reflected in irregular field patterns where cultivation is confined to sharp divides, Clarke County, Ala. (*Edgar Tobin Aerial Surveys, San Antonio, Tex.*)

the various types of mirror stereoscope in use, being governed by the distance between the side mirrors. The correct separation when using the Ryker M-11 stereoscope and other mirror stereoscopes of similar dimensions is from 8 to 8½ inches. Small magnets and a metal plate are helpful in keeping the photographs in proper orientation.

After the photographs have been oriented, the procedure involved in estimating

FIGURE 9-5*C*. Broad divides and relatively gentle relief of the blanketed Marianna (limestone) formation permit the development of rectangular field outlines (sectionized survey), Clarke County, Ala. (*Edgar Tobin Aerial Surveys, San Antonio, Tex.*)

dip and slope is relatively simple. The interpreter first looks for evidences of tip-and-tilt and corrects for it if necessary, using a reference area as close to the dip to be estimated as possible. He then estimates the dip. If the dip appears to be greater than 35 degrees, he corrects it for radial distortion before referring it to the slope-conversion chart. If the dip appears to be less than 35 degrees, it can be referred directly to the slope-conversion chart and converted to true dip.

Outcrop Patterns

Under the normal processes of denudation, surface outcrop patterns will reflect the structural relief of surface and near-surface beds. As erosion cycles develop one after the other, following regional uplift and local deformation, older beds are exposed over structurally high areas, and vice versa. Although the local exposure of older rocks can often be attributed to drainage dissection alone, it more often reflects some degree of positive structural movement. For this reason a careful study of formation contacts, key beds, and their stratigraphic relationships can be of much assistance in evaluating structural warping in low dip areas.

In areas of relatively young, unconsolidated cover, the areal geology at shallow depth is often reflected upon a blanket surface in such subtle ways as relief and drainage texture, fracture intensity, vegetational differences, and tone due to moisture content (see Figure 9-5). Recognition of the outcrop pattern beneath the unlithified cover will aid the geological evaluation of an area in the same manner as does a study of the outcrop pattern on an uncovered surface. Some of the more significant outcrop patterns and their structural implications are listed below (see Figure 9-6):

Inliers and partial inliers (older beds surrounded by younger) indicative of uplift by folding or faulting

Outliers and partial outliers (younger beds surrounded by older) indicative of down-warping or downfaulting

Downdip reentrants indicative of a plunging anticline or nose

Updip reentrant indicative of a syncline

The V pattern [rule of V's (Dake and Brown, 1925) in drainage cuts] point of V usually indicates direction of dip, with certain important exceptions and modifications (Smith and Wengerd, 1947)

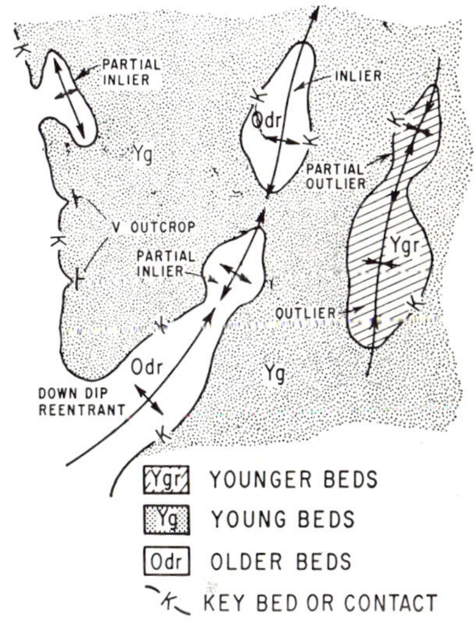

FIGURE 9-6. Significant outcrop anomalies.

Physiographic Mapping

The term "physiographic mapping" as used here refers to the mapping and interpretation of the various physiographic and geomorphic criteria which reflect bedrock structure and distribution in areas of negligible surface relief complicated by the presence of overburden. This phase of photointerpretation requires experience and imagination. Physiographic expression often reflects the composition and disposition of the buried bedrock as well as the structural attitude. That a blanket of unconsolidated materials does not necessarily obscure such expression may be due to a winnowing and sorting adjustment of the overburden to an erosional topography during deposition or by differential settling due to postdepositional movement. Or inherited characteristics may be apparent in relatively undisturbed residual soils overlying the parent bedrock so that even the vestiges of bedding may still be recognizable in zones of contrasting tone or differing erosional behavior. The chemical composition of some transported soils, although not genetically related to the underlying rocks, may support a distinctive vegetation or characteristic jointing to reflect the influence of an underlying lithology (see Figure 9-7). By similar reasoning the geologist can develop the clues from which a reconnaissance picture of the areal distribution of buried rocks can be reconstructed.

FIGURE 9-7*A*. Shows linear fracture in caliche cover of the Goliad formation in Duval County, Tex. (*Agricultural Adjustment Administration, U.S. Department of Agriculture.*)

Structural Interpretation

Doming or folding is recognized in blanketed areas by the isolated occurrence of an area of distinctive soil or relief type, indicating the possible presence of a hidden inlier or outlier. Nosing axes are observed where physiographic zones trend with or across the regional dip without obvious topographic reason. Faults are indicated where a physiographic zone or pattern of zones terminates abruptly against an area of con-

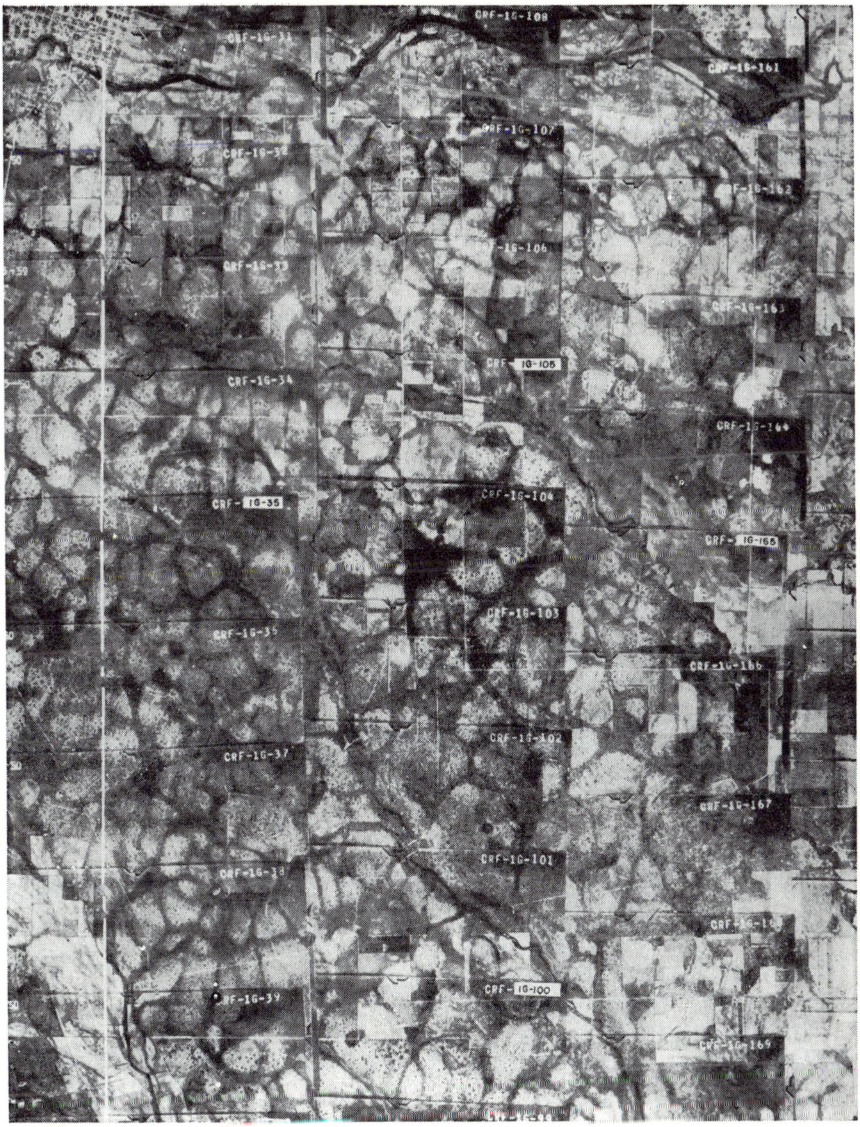

FIGURE 9-7B. Shows polygonal fracture in caliche cover of the Gueydan formation in Jim Wells County, Tex. (*Production and Marketing Administration, U.S. Department of Agriculture.*)

trasting pattern. Faults can also be recognized by drainage elements which are directly controlled by the surface trace or by jointing developed in adjustment to faulting at depth. Drainage pattern disarrangement may occur in the overburden which has been adjusted to structural movement in the buried bedrock along either fault or fold axes.

Land Forms

An understanding of the origin of land forms and their structural implications is fundamental to a thorough geological interpretation of air photographs. This includes an appreciation of the reaction of the many rock types to erosional processes, the various topographic expressions of structure, and especially the manner in which obscure surface features may reflect underlying structure and lithology.

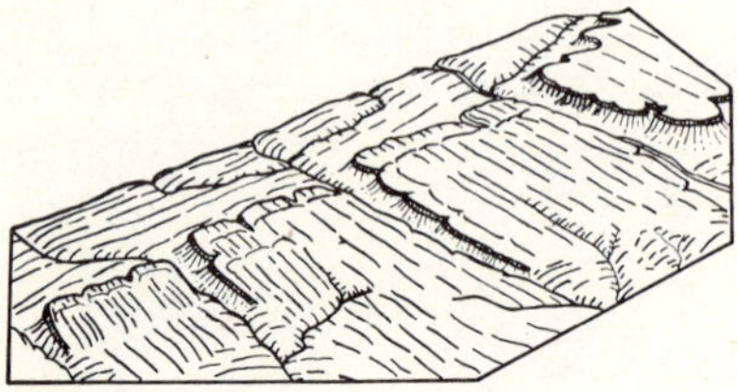

FIGURE 9-8. Diagram showing the transition from a hogback through a homoclinal ridge and cuesta to a mesa as controlled by varying dip of the rocks. (*Modified after W. M. Davis.*)

Some of the more common land forms and their structural implications are listed below (see Figure 9-8).

Cuesta—indicative of gentle dip in a direction away from the scarp face

Hogback—indicative of steep dip in a direction away from the scarp face

Mesa—(sedimentary cap) indicative of horizontal or nearly horizontal beds

Dip slopes—direction of dip usually identified with ease

Valley and canyon walls—usually helpful in determining direction of dip

Local outcrop—usually helpful in determining strike and frequently dip

Drainage Patterns

Because drainage is controlled almost entirely by the physical character and attitude of the surface rocks, a study of drainage patterns represents one of the most practical approaches to an understanding of surface structure and lithology. Dendritic drainage patterns are "normal," usually reflecting homogeneity of lithology and horizontal beds. Any marked departure from normal should be carefully considered and, if possible, explained. In studying any particular area it may be found that some pattern other than dendritic predominates. The dominating pattern should be considered normal, and any local deviations anomalous. Before assuming, however, that anomalous drainage indicates structural effects, the influence of topography and the lithology of surface rocks must be carefully taken into account. Only after all other nonstructural conditions are discounted should the anomalous pattern be designated as a probable indication of structural influence.

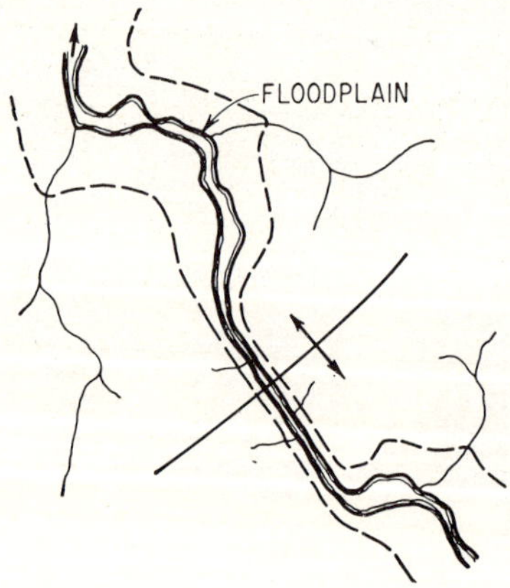

FIGURE 9-9. Deep stream incision and narrowed floodplain developed transverse to structural axis.

Actually, the relating of drainage behavior to structural influence is based on recognizing the occurrence and trend of local base levels, variations in gradient, or adjustments in maintaining grade. Some of the more common drainage anomalies are listed and illustrated with suggested structural controls briefly described and annotated.

Anomalous Trunk Drainage

Incised drainage suggests superimposition or contemporaneous structural uplift (see Figure 9-9).

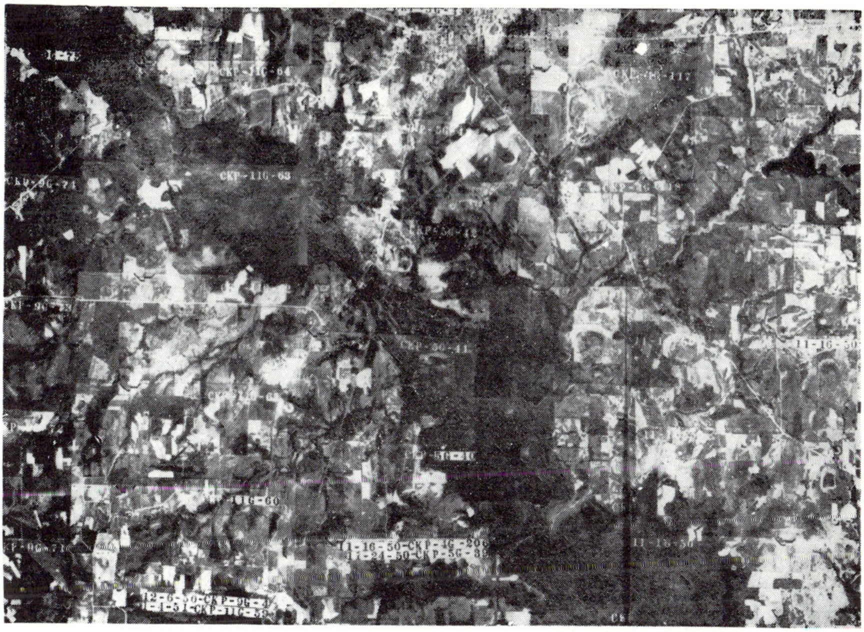

FIGURE 9-10*A*. Northeast-southwest graben (controls production at Merigale-Paul field), Wood County, Tex., identified by narrowed floodplain of trunk stream and right-angle convergence of tributary streams controlled by main faults. (*Production and Marketing Administration, U.S. Department of Agriculture.*)

FIGURE 9-10*B*. Same as Figure 9-10*A* with edges of graben and floodplain of trunk stream indicated.

Narrowed floodplain suggests change of lithology due to folding or faulting at the site of the anomaly (see Figures 9-9 and 9-10).

Broadened floodplain suggests possible structural influence at the downstream side of the anomaly.

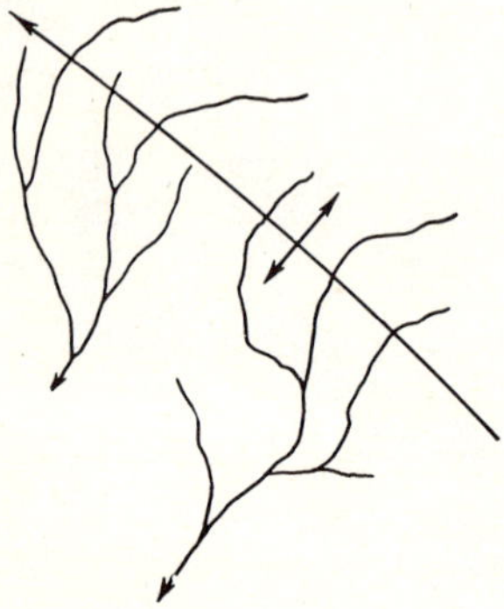

FIGURE 9-11. Parallel arcs suggest plunging anticline.

FIGURE 9-12. Opposed arcs suggest closed anticline.

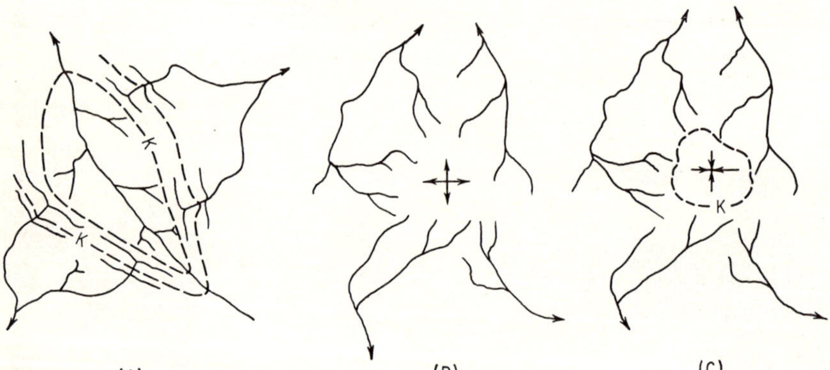

FIGURE 9-13. (*A*) Composite drainage pattern showing possible relations of radial, trellis, and centripetal drainage to a closed anticline. (*B*) Radial drainage on young erosional surface suggests doming. (*C*) Radial drainage and erosional remnant on old erosional surface suggest horizontal beds or structural low.

Abrupt change in direction may reflect local uplift by folding or faulting.

Alignment of bends may reflect fault or fold axes.

Abrupt change in meander frequency suggests local fold or fault zone normal to drainage course.

Anomalous Tributary Drainage

Parallel arcs suggest plunging anticlinal axis (may be synclinal) (see Figure 9-11).

Opposed arcs suggest closed anticline (may be synclinal) (see Figure 9-12).

Trellis suggests adjustment to truncated folded or steeply dipping beds, commonly developed on anticlinal flanks or tilted fault blocks (see Figure 9-13*A*).

Radial (see Figure 9-13*A*) on young erosional surface suggests doming (see Figure 9-13*B*); associated with capping remnant on old erosional surface suggests horizontal bed or structural low (see Figure 9-13*C*).

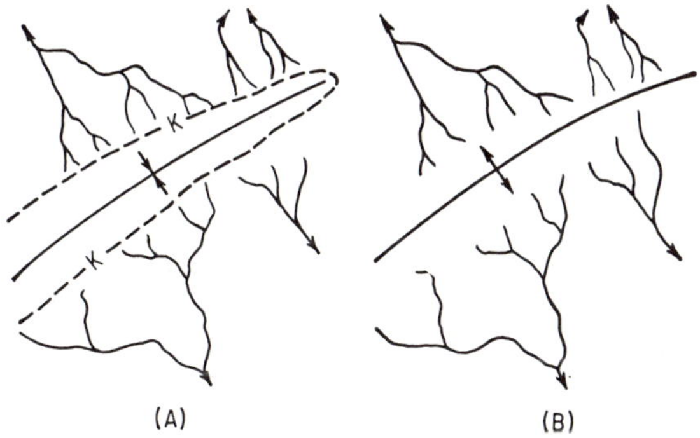

FIGURE 9-14. (*A*) Drainage divide and erosional remnant on old erosion surface suggest synclinal axis. (*B*) Drainage divide on young erosional surface suggests anticlinal axis.

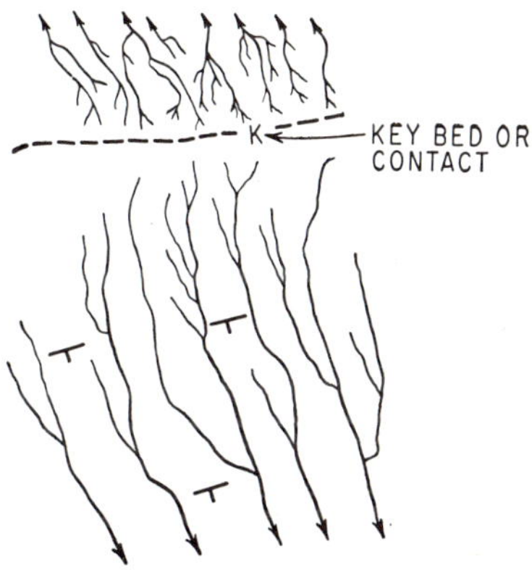

FIGURE 9-15. Parallel linears associated with dip slope.

FIGURE 9-16. Alignment of linears suggests fault or fracture zone.

Persistent divide on sedimentary surface, commonly associated with a cuesta or hog-back, suggests anticlinal or homoclinal dips; associated with capping remnant on old erosional surface, suggests synclinal axis (see Figure 9-14*A*); on young erosional surface, suggests anticlinal axis or fault scarp (see Figure 9-14*B*).

Parallel linears are usually controlled by structure, possibly dip slope (see Figure 9-15).

Alignment of linears suggests fault or fracture zone (see Figure 9-16).

Rectangularity usually reflects fault or fracture system (see Figure 9-17).

Local angularity usually reflects local fault or fracture (see Figure 9-18).

Abnormal convergence suggests fault or fracture control (see Figure 9-10) when angle of convergence is greater or less than regional norm.

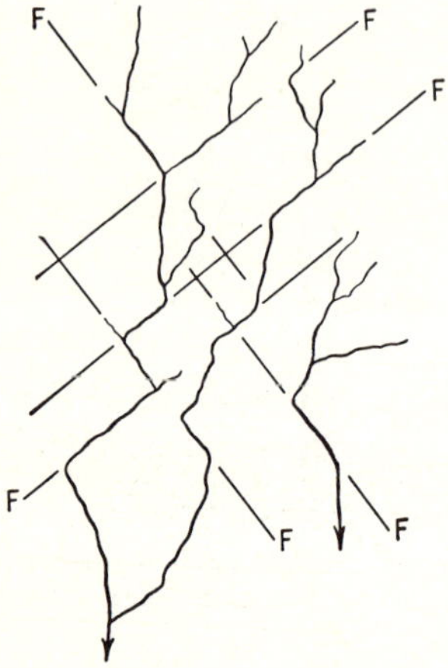

FIGURE 9-17. Rectangular drainage pattern suggests fault or fracture system.

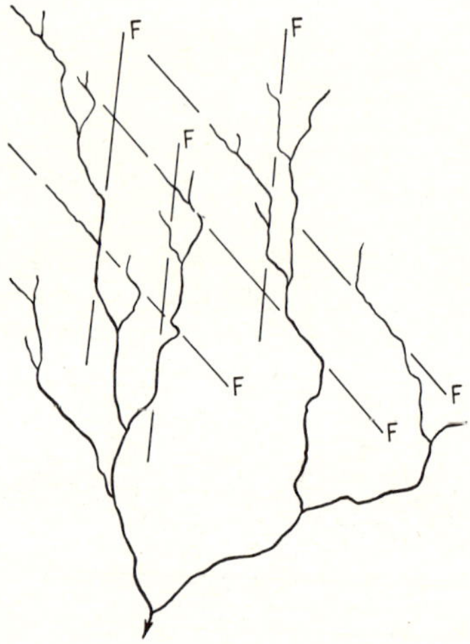

FIGURE 9-18. Angulate drainage pattern suggests fault or fracture.

Abnormal branching (fanning) suggests fault or fold axis.

Ponding suggests structural obstacle on downstream side of anomaly or collapse over salt dome.

Alignment of falls or rapids indicates probable fault axis.

Vegetational Criteria

Vegetational changes play an important part in the delineation of the different lithologies exposed at or near the surface. They may constitute the principal evidence of lithologic control by the outcrop in covered areas, and their interpretation may establish structure. In addition to the factors of climate and elevation, vegetation is controlled by the moisture and mineral composition of the soil. The movement of mineral-bearing waters may bring critical types or amounts of precipitates into transported soils, permitting a distinctive vegetation type or density to duplicate the areal pattern of the source rocks. Ground waters may percolate to the surface through a fault and by the concentration of moisture encourage a more luxuriant growth of vegetation along a linear zone (see Figure 9-19).

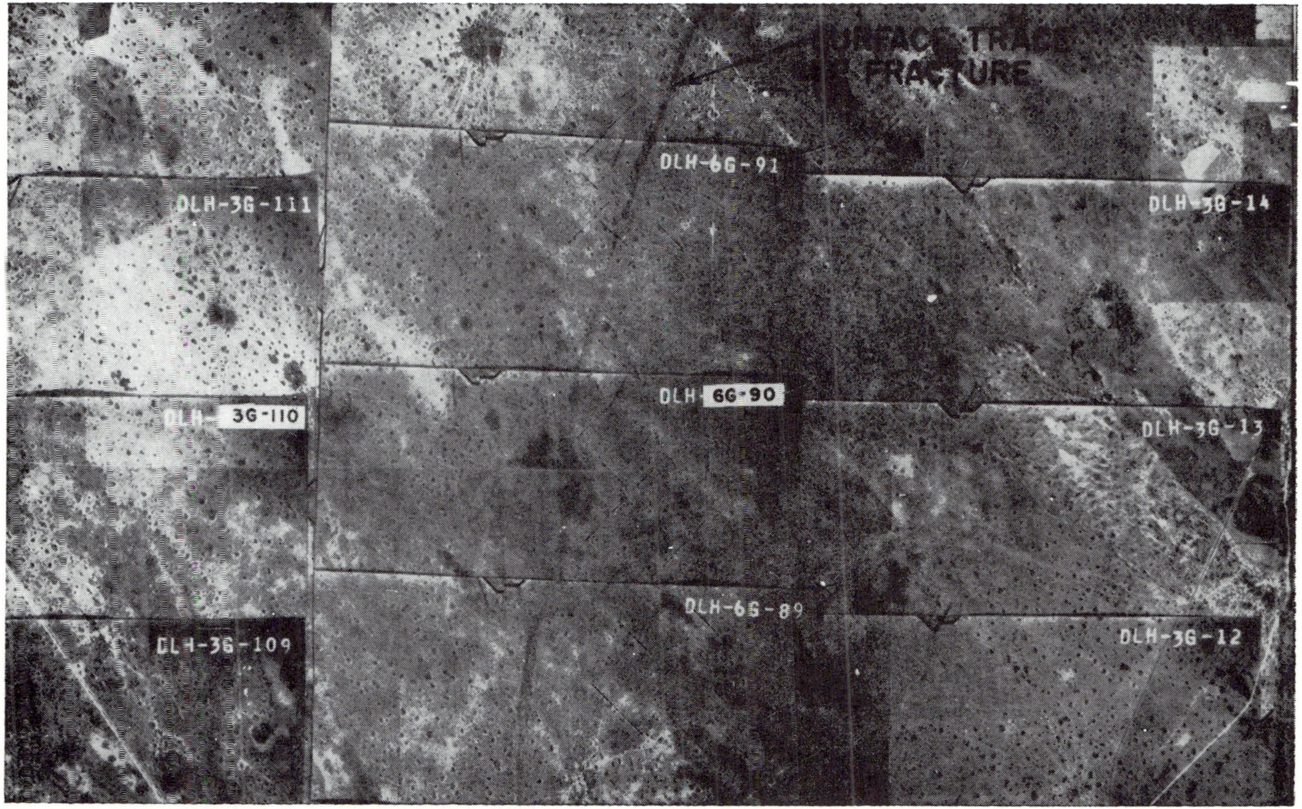

FIGURE 9-19. Surface trace of fracture (just left of center of photograph) localized by concentration of vegetation, Jim Hogg County, Tex. (*Production and Marketing Administration, U.S. Department of Agriculture.*)

Soils

Residual soils may reflect the contrast in lithology between contiguous beds of decomposing rock by erosional behavior or relief texture, as well as by tonal differences due to inherited mineral content or to moisture retention properties. Sandy soils, being more porous and better drained, tend to be lighter in tone than clay soils in the same physical environment.

Linear Anomalies

Lineations, lineaments, and alignments are all terms commonly given to the straight-line relationships observed in air photographs. They are recognized in the form and texture of drainage, in relief and vegetation patterns, and often in subtle changes of shade or tone in soils. When they cannot be related to cultural influence, they are

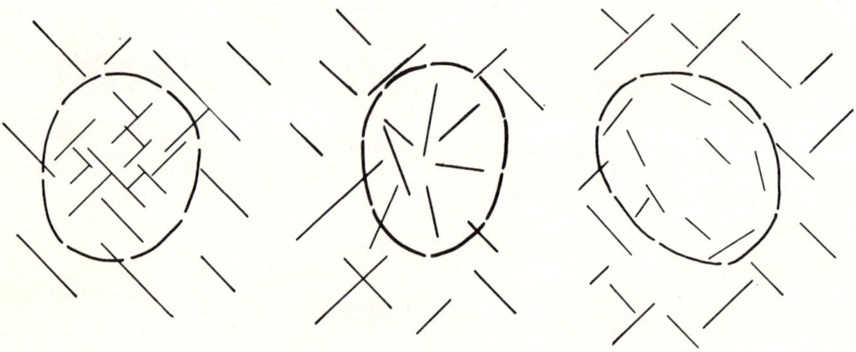

FIGURE 9-20. (*A*) Local increase in density of fractures is anomaly that suggests anticlinal structure. (*B*) Local complication in fracture pattern is anomaly that suggests anticlinal structure. (*C*) Local changes in direction of fractures suggest anticlinal structure.

usually attributed to rock fracture. Termination or actual displacement of drainage patterns or soil zones or outcrops may reflect faulting. Distinct and persistent drainage lineation may be related to subsurface faulting, particularly if orientation is consistent with the tectonic framework. Less prominent fractures are designated as joints with possible relations to local uplift.

It should be noted that physiographic evidence of doming is frequently difficult to discern in association with faulting, perhaps because the effects of faulting tend to dominate the surface of the disturbed area. Hence, lineal anomalies should be further explored for domal possibilities. However, it is also observed that minor fracturing is often found in direct association with a domal area, either in radial pattern or notably divergent from the regional trend. Therefore, in contrast to the areas dominated by faulting, where indications of doming may be overshadowed, the reflection of doming may be emphasized by the presence of subordinate fracturing.

This direct evidence of local fracturing is, of course, fundamental to the purpose of geologic mapping. However, the broader structural implications of fracture pattern and density can also be of great importance in the study of regional structure. Local changes in orientation and density of fracture elements are anomalous and should be considered possible indications of structure. Some of the more common fracture anomalies are listed below.

Local increase in density (see Figure 9-20*A*) indicates possible anticline. Owing to greater structural relief of uplift area, fractures inherent to the surface formation are locally subjected to increased erosional action, and therefore, a greater number are more clearly revealed and a greater number may actually occur.

Local complication (see Figure 9-20*B*) indicates possible anticline. A greater number of fractures may occur in incompetent beds over crestal or domal areas during structural growth. These would not necessarily conform to the regional orientation and might, therefore, result in a significant anomaly.

Local deviation (see Figure 9-20*C*) indicates a relatively stable area usually associated with positive structure. The presence of an area tending to be resistant to stress during the development of a regional or large-scale fracture system may cause slight variations in orientation of fractures in and around a local high.

Areal Anomalies

The outline of the anomaly is ordinarily drawn to enclose all physiographic elements apparently related to an arbitrary center. It seems probable that structurally negative portions of some features might be included. For logical structural reasons or simply owing to technical limitations, the position of the surface anomaly may be offset with respect to a structural feature in the subsurface. In other words, to be realistic in the application of physiographic interpretation, whereas the areal limits and amount of closure are not necessarily definable, the presence of uplift can ordinarily be inferred and supported with geomorphic and geologic evidence.

Physiographic Prospecting

The most elementary approach to prospecting for oil-bearing structures is empirical; that is, one may recognize the association between a known producing structure and a physiographic anomaly and therefore consider another similar anomaly prospective. However, this attack can at best be very limited where photointerpretation is concerned, for it incorrectly presupposes that all structural features will influence the surface in the same manner. Actually there are no definitive criteria in physiographic mapping, since the same physiographic adjustment may be associated with either anticlinal or synclinal folding and doming and both folding and doming may be associated with a variety of different physiographic features. This is why physiographic mapping requires some training in land-form genetics and why a geologic background is essential to the intelligent and reliable application of the observed data. Used with discretion and full awareness of its limitations and potentialities, photogeologic mapping has become one of the most effective exploration tools ever developed.

Pocket Stereoscope

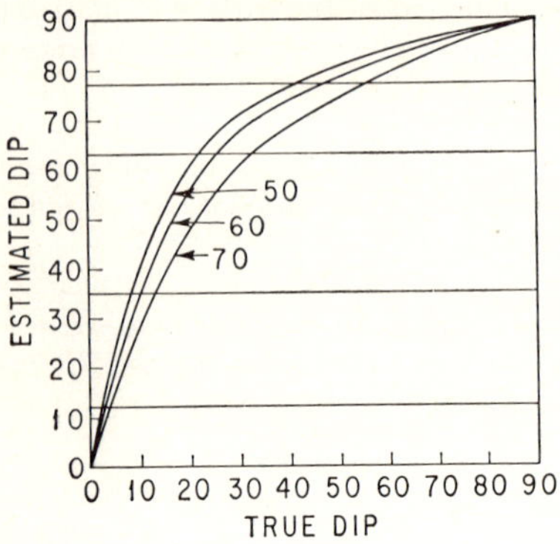

PHOTOGRAPHIC CONDITIONS

SIZE = 9 x 9
FOCAL LENGTH = 6"
OVERLAP = 50, 60 OR 70%

INSTRUMENTATION

VIEWING DISTANCE = 4.5"
MAGNIFICATION = 2 x

CHART 9-1

Pocket Stereoscope

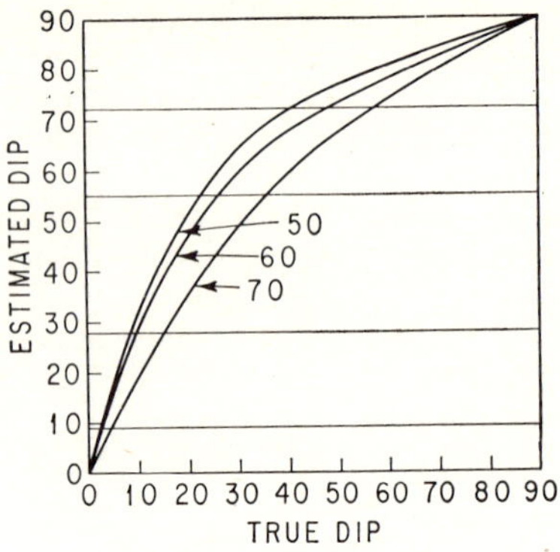

PHOTOGRAPHIC CONDITIONS

SIZE = 9 x 9
FOCAL LENGTH = 8.25"
OVERLAP = 50, 60 & 70%

INSTRUMENTATION

VIEWING DISTANCE = 4.5"
MAGNIFICATION = 2 x

CHART 9-2

Ryker Mirror Stereoscope

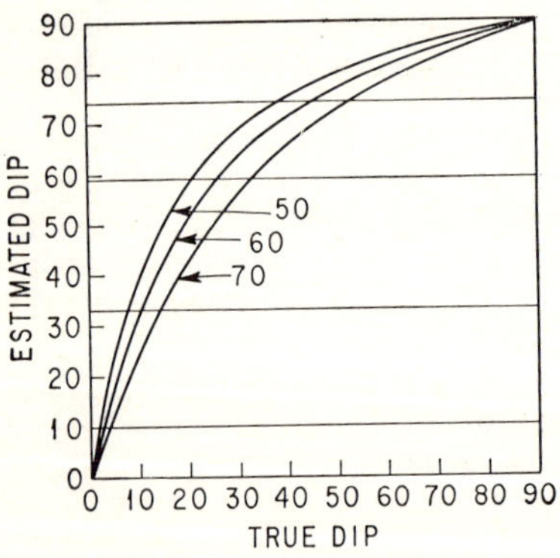

PHOTOGRAPHIC CONDITIONS

SIZE = 9 x 9
FOCAL LENGTH = 6"
OVERLAP = 50, 60 & 70%

INSTRUMENTATION

VIEWING DISTANCE = 11.3"
MAGNIFICATION = 1.1 x
PHOTO SEPARATION = 8" TO 8.5" BETWEEN LIKE POINTS

CHART 9-3

Ryker Mirror Stereoscope

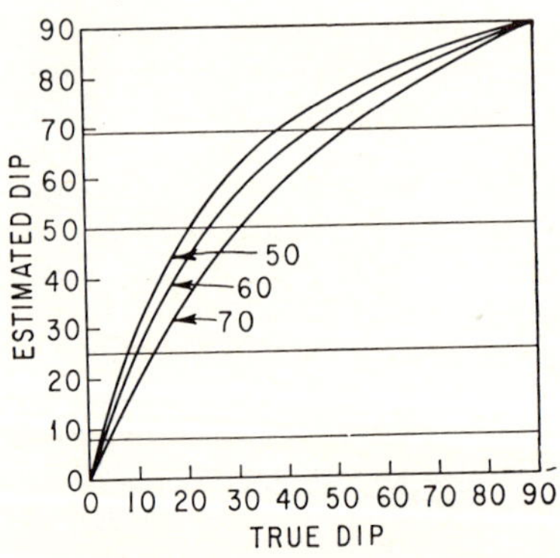

PHOTOGRAPHIC CONDITIONS

SIZE = 9 x 9
FOCAL LENGTH = 8.25"
OVERLAP = 50, 60 & 70%

INSTRUMENTATION

VIEWING DISTANCE = 11.3"
MAGNIFICATION = 1.1 x
PHOTO SEPARATION = 8" TO 8.5" BETWEEN LIKE POINTS

CHART 9-4

CHARTS 9-1 to 9-4. Slope-conversion charts to accommodate the most common combinations of photography and stereoscope now in use.

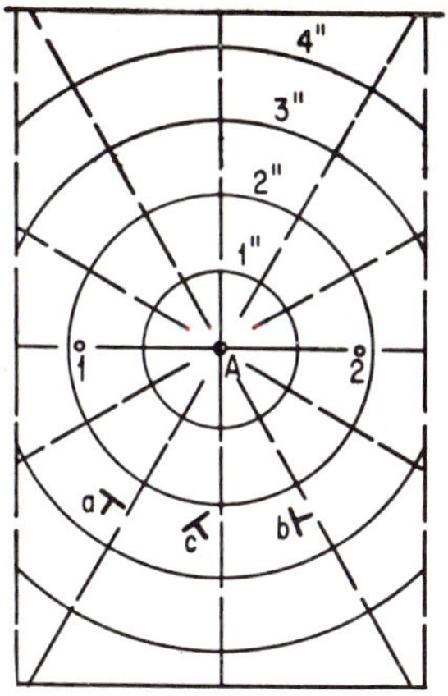

RADIAL DISPLACEMENT CORRECTION CHART

Figure at left represents area covered by model. Points 1 and 2 represent consecutive photo centers, with *A* being the model center. Concentric lines represent distance from the model center in inches. The correction for radial displacement varies with the distance of the object from the model center as indicated in chart below. Corrections also vary with dip orientation, as follows: (1) When the dip is radial from *A*, as in dip *a*, maximum distortion is encountered and 100% correction is applied. (2) When the dip is perpendicular to a radial from *A*, as in dip *b*, no distortion is encountered and therefore no correction is applied. (3) When the dip orientation falls somewhere between the above, correction is applied according to per cent of variation. For example, dip *c* is oriented 45° from the radial. Therefore 50% correction is applied.

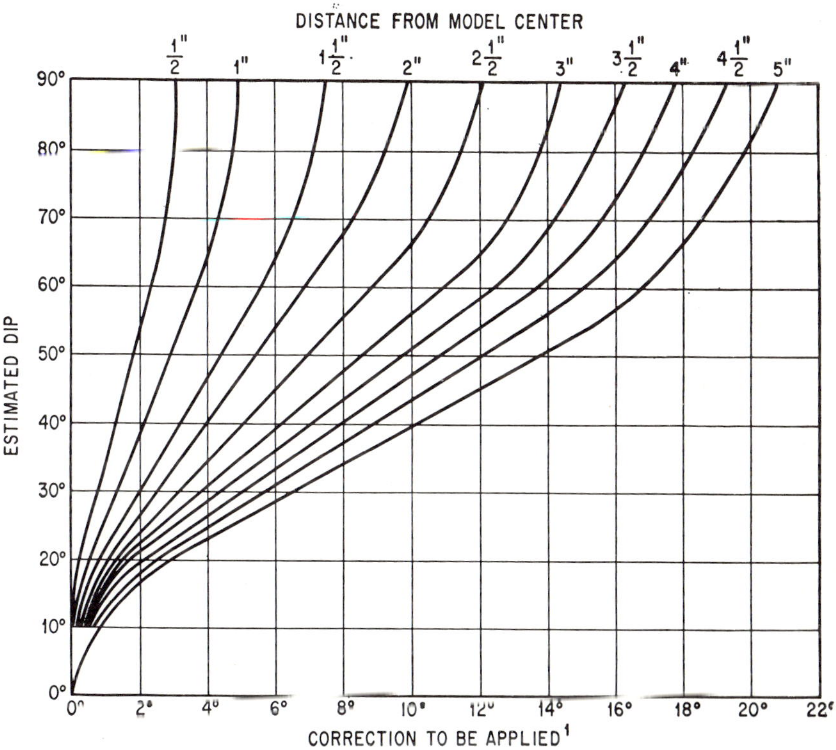

[1]ADD WHEN BEDS DIP TOWARD CENTER, SUBTRACT WHEN BEDS DIP AWAY FROM CENTER

CHART 9-5. Radial-displacement correction chart.

REFERENCES

Abrams, Talbert: "The Essentials of Aerial Surveying and Photo Interpretation," 289 pp., 210 figs., McGraw-Hill Book Company, Inc., New York, 1944.

American Society of Photogrammetry: "Manual of Photogrammetry," 2d ed., 876 pp., photoplates, diagrams, and tables, Washington, D.C., 1952.

American Society of Photogrammetry: "Manual of Photographic Interpretation," 859 pp., 600 + plates, bibliography, Washington, D.C., 1960.

Canadian Topographical Survey: A Graphical Method of Plotting Oblique Aerial Photographs, *Can. Topographical Survey Bull.*, Department of Interior, Dominion of Canada, 1928.

Dake, C. L., and J. S. Brown: "Interpretation of Topographic and Geologic Maps," pp. 221–230, McGraw-Hill Book Company, Inc., 1925.

Fichter, H. J.: Geometry of the Imaginary Stereoscopic Model, *Photogrammetria*, vol. 10, no. 4, pp. 134–139, 1953–1954.

Lueder, Donald R.: "Aerial Photographic Interpretation—Principles and Applications," 445 pp., photographs, sketches, maps, and bibliography, McGraw-Hill Book Company, Inc., New York, 1959.

Miller, Victor C.: "Photogeology," McGraw-Hill Book Company, Inc., New York, 1961.

Parvis, Merle: Drainage Pattern Significance in Airphoto Identification of Soils and Bedrock, *Photogrammetric Eng.*, vol. 16, no. 3, pp. 387–409, June, 1950. Sketches and photoplates, bibliography.

Ray, Richard G.: Procedures and Studies in Photogeology, *USGS Bull.* 1043–A, 21 pp., instruments illustrated, Government Printing Office, Washington, D.C., 1956.

Smith, Norman C., and Sherman A. Wengerd: Photogeology Aids Naval Petroleum Exploration, *Bull. AAPG*, vol. 31, no. 5, pp. 824–828, May, 1947.

Taverne, N. J. M.: Aerial Photography and Exploration for Oil, *J. Inst. Petrol.*, vol. 34, no. 396, pp. 515–540, August, 1948. Sketches and photographs.

Thornbury, William D.: "Principles of Geomorphology," pp. 120–140, illustrated discussion of significance of drainage pattern and texture; pp. 208–275, topography on folded and faulted structures with airphoto illustrations; John Wiley & Sons, Inc., New York; Chapman & Hall, Ltd., London, 1957.

Chapter 10

SURFACE GEOLOGY

By Kenneth G. Smith

Continental Oil Company
Durango, Colo.

and

Sherman A. Wengerd

Petroleum Consultant
Professor of Geology, University of New Mexico
Albuquerque, N. Mex.

An interpretation of subsurface geological conditions from geophysical data and studies of well logs and paleontological determinations is predicated to a large degree on comparison of the subsurface data with analogous conditions that have been observed on the surface, either in the immediate area or in other regions.

It is of paramount importance, therefore, that the explorer for oil make a comprehensive examination of geological phenomena exposed on the surface in the area under consideration and in contiguous areas. Only through a thorough understanding of exposed geology can he hope to interpret satisfactorily any subsurface data developed.

Surface geology becomes of prime importance in new areas of exploration, such as in a large, undrilled structural or depositional basin. In such an area, stratigraphic conditions within the basin must be postulated, based on stratigraphic sections exposed around the perimeter of the basin, while good surface structural data within the basin will establish trends and will serve as a guide for planning future seismograph work in the area without first shooting long, costly seismic reconnaissance lines. Rapid, accurate work by the surface geologist can reduce costs of exploration for his company in a new area, as well as give the company a slight lead in a dynamic and highly competitive part of the oil industry.

The rapid development and expansion of the oil industry in recent years have led to an emphasis on subsurface rather than surface geology, and opportunities to undertake a large field project are becoming rarer within the continental United States. Alaska and many foreign countries offer new areas for oil exploration which must first be observed by the field geologist in order to plan a well-coordinated exploration program which will ultimately lead to the drilling of prospects which are favorable to the accumulation of oil both structurally and stratigraphically.

Basic plans for the schedule and equipment for a field party are generally the responsibility of the geologist in charge (party chief), and it is therefore necessary for him to familiarize himself as thoroughly as possible with the terrain, climate, and accessibility of the area in which the field work is planned. These factors will govern his selection of transportation, camping equipment (if necessary), and timing of the work, depending on climatic conditions which might render certain seasons of the year unfavorable for geological field work.

In remote areas, where it is necessary to operate from a camp which is moved periodically as work progresses, timing is of utmost importance, and the party chief should previously have completed a set schedule for camp moves and areas to be covered from each camp, dependent upon the total area outlined for the field project as well as its major objectives. Generally, 10 to 20 per cent additional time should be added over and above the total estimated to finish the project in order to allow for lost time due to bad weather conditions and general climate of the area in which the field work is planned.

Selection of the geological equipment to be used is also the responsibility of the party chief and will be dependent on the type of field work planned. Geological equipment required will vary depending upon the objectives of the field project, that is, if the project is structural, stratigraphic, or areal geology or a combination of two or all of these. Equipment will also vary with the amount of detail required. Reconnaissance geology, detailed reconnaissance geology, or detailed geology will require progressively more elaborate field equipment.

The major purpose of this chapter is to set forth use and selection of equipment and methods for field mapping which will obtain maximum results in reasonable time. In the highly competitive petroleum industry, time is always the important element, as each company desires to keep up with or ahead of its competitors in all fields of geological endeavor. The following discussion is not designed to serve as a complete text on field methods but is more a review of common field methods with added short cuts which have been gained through experience by field geologists.

EQUIPMENT

Transportation

Field Vehicles. The automobile has become the major transportation means for the geologist in the United States; many of the areas to be covered by field projects can be traversed in part or entirely with this vehicle. If the area to be mapped has many secondary roads and no extensive work is planned off these roads, an ordinary short-wheel-base model of most low-priced cars, without alteration, is generally satisfactory. If, on the other hand, considerable work is planned where cross-country driving is necessary but the terrain is not too rough, any one of these cars can be armor-plated to protect the oil pan, heavy-duty springs can be placed in the rear, and the car can be equipped with 15-inch wheels. A positraction rear end may also be necessary. An ordinary automobile with this extra equipment will serve satisfactorily as a field vehicle in the greater part of the continental United States.

If the field project is to be in an area of rugged topography and very rough roads, with considerable cross-country driving, the universal jeep is the best all-round field vehicle. Its short wheel base enables it to cross deep gullies with ease, and the four-wheel drive makes it possible to "get that last inch" up a steep slope, cutting walking time to a minimum to observe an outcrop and thus conserving valuable time.

The universal jeep is not exactly the best highway vehicle, for it is slow and rough riding. The best arrangement when a jeep is used for field work is to equip it with a tow bar and to tow it to the area of operation with a conventional automobile. Tow bars can be obtained that apply the brakes in the jeep when the brakes on the forward car are applied, thus reducing wear on the towing car and increasing the safety factor. The jeep should also be equipped with free-wheeling hubs, which, when properly adjusted, disengage the front axle, thus reducing wear on the forward transmission when the vehicle is either towed or driven on a highway.

If it is possible to have only one vehicle for field use and work in rugged terrain is contemplated, a six-cylinder, four-wheel-drive jeep station wagon equipped with free-wheeling hubs is suitable. This vehicle will go almost any place that the universal jeep will go and is faster and more comfortable for highway driving. Although the jeep station wagon can also be towed, it is slightly top-heavy and bulky, making it difficult to control with the towing car, especially in high winds or on mountainous roads. It is also considerably heavier than the universal jeep, and towing is not recommended for the station wagon.

Some companies are now using four-wheel-drive pickup trucks or carryalls for field work, with the same type of disengaging front-axle arrangement as on the jeeps, for highway driving. These vehicles are much more powerful than the jeeps and are especially useful when considerable field equipment is constantly needed or when the field party is camping out in the field and moving at intervals. However, they are not recommended for normal field work because of the longer wheel base, bulkiness, and hard steering as compared with the universal jeep.

If the party chief has a choice of vehicles to use, final selection should be governed by the scope of the field project, its accessibility, and the needs of the field party.

Fixed-wing Aircraft and Helicopters. Since the end of World War II fixed-wing aircraft and helicopters have taken the place of automotive vehicles as field transportation in many instances. This is especially true in remote, inaccessible foreign areas and Alaska and in poorly accessible areas within the continental United States.

Small planes have long been used for reconnaissance work by the field geologist and even in accessible areas of the United States can still be put to good use. For example, if a field project is planned and photomosaics are available, the entire area can first be flown by the party chief and all roads observed from the air that are not shown on existing county road maps can be sketched in on the mosaic. Then having seen the entire area from the air and with a good knowledge of accessibility and disposition of roads, the party chief can better plan and schedule his field project.

In remote foreign areas where field camps are maintained, the small plane becomes the major means of supplying and moving the camp. When the party chief is faced with the problem of moving and supplying his camp by small plane, selection of the type of plane will depend upon the climate and terrain of the field project area. As an example, in northern Canada, the Canadian Geological Survey supplies its field parties with a single-engine float plane because there are thousands of lakes upon which they can land.

In Alaska, where the fixed-wing aircraft is the major mode of transportation and where there are many lakes and rivers, selection of the plane to be used is governed by the size of the bodies of water from which it will have to operate. If they are of considerable size, a twin-engined amphibian is the preferable plane because of its greater legal load limit and the fact that it can make use of airstrips at the nearest source of supply. For areas with smaller and shallower lakes a high-horsepower single-engine float plane is the preferable plane, as its legal load limit may almost equal that of the twin-engine amphibian; but its use requires that the source of supply be near a body of water, which is true of all float planes and which can be inconvenient at times. A lower horsepower single-engine float plane can also be used for this type of operation, but as it is much smaller, with a lesser load capacity, it requires a great number of trips to accomplish the same amount of work as the other larger planes mentioned.

If desert camps are planned, with moving and supplying by small plane, any of several types of small plane can be used that can land on short runways, as generally a campsite can be selected within or near an area of relatively flat, clear ground. If the ground is too rough for ordinary planes, a light plane with tandem wheel gear can be used.

Small planes can also be used to fly over areas in which photomosaics are not available in order to gain a general knowledge of the prominent outcrop geology and general structural trends. Small planes, though they vary in cost depending on size, are still more costly to operate than automotive vehicles. However, when wisely used, they can ultimately save valuable time for the field geologist, as well as cut over-all expenses for the company in terms of cost per square mile of area covered.

The helicopter, on the other hand, is becoming the main type of field transportation for the field geologist at work in remote areas. It can be used for reconnaissance work, running traverses across the structural grain of the area, or, for detailed work, it can accommodate two or three two-man teams in the field, transporting the teams to and from the place of work. The helicopter can be used for plane-table mapping in rugged areas, alternately moving rod man and instrument man or moving the rod man around the instrument man to locations where instrument shots are desired. The latter method has been used by some companies in remote, rugged areas of the continental United States, while other companies doing detailed surface stratigraphy have also used helicopters as a means of transportation for the field party.

When the cost of operation of the helicopter, which is quite high, is weighed against the increase in area covered by a field party using a helicopter, the comparison with cost when automotive vehicles are used is quite surprising. In Alaska, it is not uncommon for a field party to cover 40,000 to 50,000 square miles in a 4-month season, doing reconnaissance, structure, and stratigraphy, at a cost of approximately $100,000, including helicopter, fixed-wing aircraft, and maintenance of a well-equipped field camp. This type of reconnaissance thus costs about $2 to $2.50 per square mile.

The most economical use of the helicopter for field work is dependent upon careful planning of traverses and areas to be covered each day. The best method of approaching the problem of planning the daily work is to fly the complete area and observe the outcrops and landing places from the air, then plot traverses on the topographic map or air mosaic, set up a daily schedule, and attempt to maintain this schedule. Contrary to belief, the helicopter cannot land just anywhere, and for this reason the advice of the pilot can be very helpful in planning the traverses.

Another problem that arises when using a helicopter in the field is the need for fuel. On the average, a three-place helicopter uses from 12 to 15 gallons of gasoline per hour with a usable fuel-tank capacity of 40 to 45 gallons. If long traverses are planned or the work planned is a considerable distance from camp, gasoline caches must be put out, either with a fixed-wing aircraft if suitable landing places are available or with the helicopter. It is preferable to use a fixed-wing aircraft for this, if possible, as it is faster, can carry more gasoline in one load (probably sufficient for two or three caches), and its use leaves the helicopter free to continue geological work; hence no time is lost.

Fixed-wing aircraft and helicopters, when used as transportation for a field party, will generally justify their higher cost of operation if the field work is well organized and scheduled. The responsibility of obtaining maximum benefit from their use rests directly on the party chief and his planning.

Water Transportation. Although water transportation is not commonly required by a field party, occasionally a need for this type of transportation may arise. The nature of the body of water (river, lake, etc.) will govern the type of craft to be used, with safety of the party personnel being a governing factor in its selection.

In northern Canada, the Canadian Geological Survey uses canoes for its field parties. A two-man field party, with complete camping and field equipment, is constantly on the move doing geological work during the field season and is supplied periodically by float planes. This is probably the maximum use of water transportation by geological field parties at the present time, as there are not many areas that lend themselves to it so well.

Horses and Mules. Occasionally a situation arises where it is necessary to use horses or mules for field work. With saddle bags, considerable equipment can be carried on horseback, such as the plane table and alidade or a transit, with the tripod tied to the back of the saddle.

If long traverses or several days' work are planned using horses or mules, with possibly a field camp, it is usually best to take along a wrangler to serve as camp cook and handy man and to tend the horses. The wrangler can then go ahead of the field party with pack horses to a suitable location and set up camp each day, thus freeing the geologists from time-consuming unfruitful work and giving them a longer day in the field.

Although this form of transportation may seem outmoded, it should never be overlooked in an emergency. Many field geologists have had the experience of having all the roads into the project area washed out by torrential rains, and under such conditions the horse or mule becomes the main means of moving into and around the area being mapped.

No method of transportation should ever be classed as obsolete, as conditions vary considerably in the many vastly different areas in which a petroleum exploration geologist may find himself, and he will have to use whatever will best serve his needs, including walking.

Camps

Geological field work within the continental United States may require a field camp only occasionally, whereas outside the United States, as in Alaska, northern Canada, or remote foreign areas, the establishment of field camps is the only manner in which a field project can be completed.

While camps in the United States may just be temporary establishments to facilitate work for a short time in some rather inaccessible area, the other areas mentioned generally require several months in the field with periodic moves of the camp. For the long-term major operation of camps for field parties, not only equipment necessary but also character of personnel must be taken into consideration. This is especially true if there are to be other personnel besides geologists, such as pilots, mechanics, cooks, or camp helpers. Once the party is in the field in some remote area, if there is a misfit in camp, he should be replaced, and this will disrupt the operation if it becomes necessary. Even the most genial of men, furthermore, can become disgruntled in the field after several months of the same type of food, discomforts due to inclement weather, and the loneliness of being away from their families. Under these circumstances, differences will arise from time to time among members of the party, and here the party chief must act as a mediator and attempt to prevent friction from going too far. The best approach is to talk to the individuals separately, show each of them where perhaps he might be wrong, and then try to ease the tension among them. By no means should a man be reprimanded sharply by the party chief in the presence of other party members, particularly in an extended field camp in remote areas. A complete review of personnel problems which might arise would require a lengthy discussion, but needless to say, the harmony, and thus the success, of a major field operation is the responsibility of the party chief, and his ability to cope with personnel problems, as well as to direct the required field work, is the major factor to be considered in this selection.

Camping equipment required will vary with the nature of the area in which the field project is planned and its remoteness. Whether it is within the continental United States or in foreign areas, certain basic data should be compiled before final arrangements are made to equip the camp. The following list includes most of the major items which should be investigated, and additions might be made, depending upon the scope of the field project and its needs:

1. Distance from points of supply and type of transportation required.
2. Topography and drainage, including average relief, vegetation, size of lakes and rivers, water supply, roads and trails, etc.
3. Climate, including weather characteristics, mean and extreme temperatures, average precipitation and dates of maximum occurrence, etc.
4. Animal and insect life, such as large predatory animals, poisonous snakes, rodents which might bother supplies, insect pests. Choice of firearms and medical supplies will be governed in part by these data.
5. Medical facilities, including location of nearest hospitals or available doctors and source of medical supplies.
6. Clothing and camping requirements, including tents, sleeping bags and beds, cooking and eating equipment, tent heaters, etc.
7. Communications equipment, such as two-way radio or nearest available telephones, for emergency and routine contacts.

Many other items might be included on the above list, most of which can be added as an investigation of each item is made. In new areas, especially where an extended camp is planned in a remote area, local people in the nearest town or city may be quite helpful in adding items needed to the above list, and their help should be sought whenever possible. If other companies have worked in the same general area, they also might be very helpful with suggestions based on their previous experience.

Subarctic Camps. The use of plane and helicopter for transportation and field vehicles, respectively, was a major contributing factor in bringing about extensive field investigations by oil companies in northwestern Canada and Alaska. Geological operations in these areas require, in most instances, more elaborate camping equipment than would normally be used, with an accompanying higher cost. Therefore, everything selected for use in such a camp should be chosen with consideration of the efficiency of operation, which will include adequate transportation for supplying and moving the camp and a minimum of weight in the camp in order to facilitate rapid camp moves.

Transportation for the camp should include a fixed-wing airplane with a load capacity sufficient to move the entire camp in three or four trips and a three-place helicopter for the actual field work. Necessary personnel for transportation facilities would include a fixed-wing pilot, a helicopter pilot, and a helicopter mechanic. For a group as large as this, which would amount to five, including the party chief and his assistant, a camp cook is also necessary. Thus, a minimum number of personnel for a normal bush operation would be six men.

Assuming that the camp was established for six men, if 12- by 12-foot pyramidal tents are used, one can serve as a cook and mess tent, the other three for living quarters, two men to each tent. The floored tents should preferably be white, with zippered entrances. Although these tents may appear to be extremely large to place only two men in each, it must be remembered that the operation may be planned for several months, and ample room to live in and store personal belongings is required to maintain high morale.

Wood supplies in subarctic areas are generally poor, and such as is available is generally wet. Hence, for cooking facilities for six men, one three-burner and one two-burner portable pressure-fuel stove will best fill the requirements. Other cook-tent equipment should include two 8-foot metal folding tables and such pots and pans, dishes, silverware, and cutlery as can adequately serve the needs of six men, yet maintain a minimum of weight.

Clothing and sleeping equipment should be selected also with considerable thought given to weight and bulk. A 4-pound down-filled sleeping bag will be adequate for most weather conditions encountered, since subarctic field projects are limited to 3 or 4 months in the summer. If colder temperatures are expected, a lightweight surplus army blanket, which can be rolled into the sleeping bag, might be included. Army-type canvas folding cots should be used, and for increased comfort and warmth, some type of rubber air mattress will be necessary. Clothing should include heavy-duty rain gear, rubber boots, shoe pacs, regular laced field boots, both lightweight cotton and heavy wool shirts, and some type of lined jacket (preferably with removable lining) with a hood. Additional minor items of clothing should be governed by the wants and needs of the individual, but numbers of items should be limited so that all field clothing can be packed in one duffle bag for camp moves.

Pressure-fuel tent heaters can be obtained which will greatly increase the comfort of party personnel on cold nights and mornings. If outside temperatures are such that it is necessary to use the heaters while sleeping, however, then some type of venting should be arranged for the heater in the tent. A Coleman-type two-element lantern should also be furnished for each tent and gives adequate light for limited map work at night without eyestrain.

If the camp is a considerable distance from any type of communication lines, a two-way radio is essential for its operation, for contact both with a base station and with aircraft serving the camp. The most satisfactory communications equipment is a 50- or 100-watt radio with an all-wave receiver and six to eight fixed crystals for transmitting, frequencies of which will be governed by those in greatest use in the area. Power for the radio can be furnished by a small portable 110-volt electric generator. The generator can also be used for light for the geologist's tent if considerable night-work is necessary on maps and sections. The cook can also serve as radio man in camp to maintain contact on schedule with camp aircraft when they are operating and to order supplies and report to the base station each day on a regular schedule.

Field operation of this size camp requires the services of a good logistics man at the main base of operations who, on short notice, can furnish items needed and who will help in maintaining lower costs by careful filling of supply orders given to him over the radio by the cook. Quantity buying of food supplies, gasoline, and pressure fuel can result in considerable savings for this type of operation, and the large supplies can be stored at some centrally located airfield around which the camp will be moved and where the fixed-wing plane attached to the camp can pick up gasoline and food supplies every 2 or 3 days as required. Camp moves should be anticipated well in advance so that all gasoline hauled into camp will be used before or during the move and there will be a minimum of fuel and food supplies to transport. Previous to the move, one or two loads of gasoline can be placed at the new campsite by the camp plane. While the logistics man can cut costs by quantity buying, the party chief can eliminate waste in the field by good supervision in the supplying of the camp, such as working with the cook on determining future needs of the camp up to the time of a planned move and by anticipating amounts of gasoline required for daily use, based on previous experience, so that there will be no surplus at the time of the camp move. An extra round trip with the plane on the day of a camp move should not be required, and careful supervision can prevent its necessity.

Foreign Operations—Desert Camps. This information was compiled and furnished by Robert W. Blair,* with the permission of Continental Oil Company, based on maintenance of field parties in Egypt.

* Mr. Blair was in charge of surface geology operations for the Sahara Petroleum Company in Egypt, where field parties were maintained in large desert camps.

For the desert work two field parties operated out of each camp. Each field party consisted of a party chief, an assistant geologist, and a geological trainee. Other personnel in camp included a camp supervisor, four drivers for hauling water and supplies, an automobile mechanic, a cook, a cook's helper, and four to eight laborers. When a helicopter was used, pilot and mechanic both stayed in camp. Pilots of the small fixed-wing aircraft often stayed overnight in camp also, and hence quarters had to be maintained for them. Thus living facilities (cooking and dining accommodations and sleeping quarters) were required for 22 to 26 men, as a minimum expected figure, for each camp.

Shelters used for the desert camp were of two general types: huts and tents. The huts used were 20 by 20 feet in size, with one-piece walls and floor and a two-piece slanted roof, all parts held together by bolts, requiring a minimum effort to put up or take down. One hut was used for cooking, while the other was used as a combination office and mess hall.

Two sizes of tents were used, 8 by 10 feet and 10 by 12 feet. All tents had 30-inch walls and 6-inch sand flaps around the base. Each tent had a single entrance at one end with an air vent and flap at the other end, with netting at the entrance so that flaps could be opened and insects kept out. Over each tent was a separate canvas fly that shaded the tent and allowed free circulation of air between the fly and tent. These flies are essential in hot climates and are stretched across the tents and about 2 feet above them, supported by the tent poles. Geologists, pilots, and camp supervisor were quartered in the 8- by 10-foot tents (7 feet high), one man to a tent. The 10- by 12-foot tents (8 feet high) were used to quarter laborers and drivers, four men to a tent, and to store supplies.

Motor transportation included three jeeps and three power wagons in each camp, one of each for each field party operating out of the camp and an extra one of each in camp. Other vehicles used included Ford 500, Ford 800, and Kenworth trucks, which varied in number and use depending on the needs of each camp.

Water and supplies were usually brought into camp by truck, but it was found to be more satisfactory to fly in perishable foods with a DC-3, servicing several camps with one flight. The best method found to haul and store water was with a two-wheeled trailer with a 250- to 500-gallon tank mounted on it. Other methods of water hauling used included the building of permanent tanks on power wagons and the use of specially made galvanized 50- to 60-gallon drums. Small single-engine fixed-wing airplanes were also used to transport personnel and to fly in emergency supplies. Gasoline was transported in the Kenworth trucks, which could carry from fifty to sixty 55-gallon drums at one time.

Electric power for the camp was furnished by diesel generators on skids. Gasoline generators were found to be easily damaged by sand and were too costly to service. Each camp was supplied with a two-way radio powerful enough to reach the base of operations. In addition, each jeep and some of the power wagons were equipped with two-way radios to maintain contact with camp when used in the field.

Sleeping equipment included folding metal beds, mattresses, blankets, and sheets. Light sleeping bags were also available in camp for overnight trips. The large amount of equipment used in this operation required several trucks for a camp move, but the nature of the work and life on the desert for long periods of time can become quite a strain on camp personnel, and the extra comforts provided were considered necessary to maintain their morale.

The hazards and problems of work on the desert cannot be overemphasized. It is easy to lose a man on the desert, even if he is in a jeep, truck, or helicopter, and these means of transportation should be adequately supplied with emergency rations and water. Thus, if the vehicle or helicopter fails mechanically while in the field, there is less danger of losing a man because of starvation or dehydration before aid comes from the camp.

Foreign Operations—Jungle Camps.* These data on two types of jungle operations were furnished by the Conorada Petroleum Corporation. In presenting this information it should be pointed out that for each type of operation the requirements can vary depending on the magnitude of the survey, remoteness of the area, season of the year, terrain, and perhaps the preferences of the individual geologists.

(1) *Requirements for conducting a geological reconnaissance of jungle areas using mules for transportation.*

A two-geologist reconnaissance party requires additional personnel such as a head man and guide, two trail cutters, and a cook. Additional camp personnel will include men to pack and care for the mules. These men will be furnished by the contractor who supplies the mules.

Each geologist should be furnished with a 10- by 12-foot lightweight fly, a native-style hammock, and a mosquito net constructed for use with a hammock. If desired, these three items can be replaced by an army jungle hammock. The geologist should also have a cotton blanket or sleeping bag, a pocket first-aid kit, a pocket snake-bite kit, a hunting knife, a machete, a revolver (if allowed by local law), a canteen, and a flashlight.

General camp equipment should include a first-aid kit complete with such medicines as antibiotics, antimalaria, antidysentery, antivenom serum, morphine, salt tablets, and fungicidal soap. A local doctor should be consulted for advice on the specific needs of the area, which might add to the above list.

Other necessary general equipment includes two water filters, a length of ⅜- or ¼-inch rope, wire, and twine. Saddles would probably be supplied by the mule contractor, but when working from mules for an extended period of time, it is suggested that a new saddle be purchased for each geologist at the start of the project. Saddle bags and sufficient canvas bags to pack all equipment are also necessary.

Cooking equipment should be kept to a minimum and should be lightweight. Two nested cooking kits, enameled cups and plates, assorted cutting and cooking utensils, and waterproof containers for matches and food are essential items. Food supplies will depend on personal taste, but a supply of beans, rice, canned meats, and vitamin pills will provide a nutritious, though monotonous, diet.

If more detailed work is to be done, or if it is otherwise practical, comforts such as tents, cots, camp chairs, and tables may be added. Laborers hired locally usually prefer to furnish their own sleeping and food supplies, and therefore this discussion includes only the needs of the geologists.

(2) *Requirements for conducting a geological reconnaissance of jungle areas using river transportation.*

A two-geologist field party using rivers for transportation in jungle areas would require as supporting personnel a mechanic, a cook, and native helpers. One and preferably two of the native helpers should be capable of acting as guides.

General and personal equipment requirements are basically the same as for jungle camps using mules for transportation, with the exception of a radio transmitter and receiver. Radio equipment, although quite heavy, is readily transported by boat, but a radio powerful enough to reach great distances would be too heavy to transport on mules.

For river transportation two 2-ton canoes are required, as well as one small aluminum boat or light canoe to be carried for entering smaller streams. Three outboard

* Information compiled by Dr. B. F. Uhl, Jr., Mr. D. Anderson, and Mr. G. H. McDonald, of the Conorada Petroleum Corporation, has been incorporated into a brief discussion which includes only what are considered the minimum operational requirements.

motors are also required, of sufficient power to move the larger canoes. Extra equipment necessary includes spare propellers, shear pins, emergency repair kits, and tools. The amount of extra gasoline, oil, and lubricants carried depends upon the availability of stocks from previously established depots. Tarpaulins should also be included for cargo coverage and to provide shade for personnel in the boats.

Rifles, shotguns, or pistols should be carried for protection and to obtain fresh meat. Fishing tackle may also be desirable for obtaining fresh food, since only canned foods can be carried into the field in jungle areas. All foods should be packed in tins or sealed glass, even dry cereals, powdered coffee, crackers, or powdered milk, as those not packed in this manner will be found worthless owing to mildew and insects in a humid jungle operation.

Geological equipment has not been mentioned in the preceding discussion of two types of jungle operation, since it is generally similar for all geological work, depending upon the magnitude of the project and nature of the desired results. Geological equipment required should be well determined before initiating the project, however, and an ample supply of needs be included in camp equipment, including extra material. The loss or damage of a Brunton compass or other material might cause considerable delay while being replaced unless extra equipment were available in camp.

Two other types of jungle operations might be mentioned, one using helicopters for transporting men and equipment, the other using porters to transport equipment over jungle trails. Basic camp requirements remain the same, except that when porters are used, the weight of equipment must be kept to a minimum, as food must be carried for both geologists and porters.

MAPS

Preexisting Maps

First and foremost in the needs of a field party are the available maps of an area, no matter how old they may happen to be or whether or not they are presently considered adequate or accurate. Maps of great value to a field project include older geologic maps, topographic maps, county road maps, and land survey plats.

Geologic Maps. Many areal geologic maps are available covering most of the continental United States, Canada, and Alaska. Some are the result only of reconnaissance work, whereas others are detailed. In most areas to be worked by a field party there is probably some previous work available from the U.S. Geological Survey, from the Canadian Geological Survey, or in one of the many geological publications which will be an aid in determining in advance the type and age of formations that will be encountered in the field. If air photographs are to be used, any available detailed geologic map, even if covering only a small part of the field project area, will give a general idea of the outcrop expression on the air photographs and will ultimately increase the efficiency of the field party if this is known in advance of the actual field work.

Topographic Maps. Topographic mapping by the U.S. Geological Survey is quite extensive throughout the United States and Alaska, although there are still some unmapped areas in a few of the western states. Scales of the maps vary, depending on age of the maps or accessibility of the area and the needs for smaller or larger scales. Typical scales of available topographic maps are 1:250,000, 1:125,000, 1:62,500, 1:31,250, and 1:24,000. Although the smaller-scale maps are useful, scales of 1:62,500 or larger are preferable when they can be obtained. Topographic maps are valuable to a field party for preparatory work, as the newer ones indicate roads and other culture in the area, location of bench marks and triangulation stations, and in some instances include the township and section lines. All these items are essential to the field party, in addition to a knowledge of the topography of the area to be mapped and the general accessibility.

County Road Maps. Most states have available, through the state highway department or some associated agency, maps of the major Federal highways, state highways, county roads, and many secondary roads. These are usually available by counties, sometimes with several sheets to cover the area of a larger county. Scale of these maps is generally ½ inch to the mile, with some occasionally available at 1 inch to the mile. In addition to the roads, these maps also indicate culture, including ranch and farm houses, water tanks, towns, and railroads, all of which may be of use to the field party. Major drainage may also be shown on these maps, and if the land is surveyed, township and section lines will be included.

Land Plats—General Land Office Surveys. Individual land plats of townships and sections in lands surveyed by the General Land Office are available through

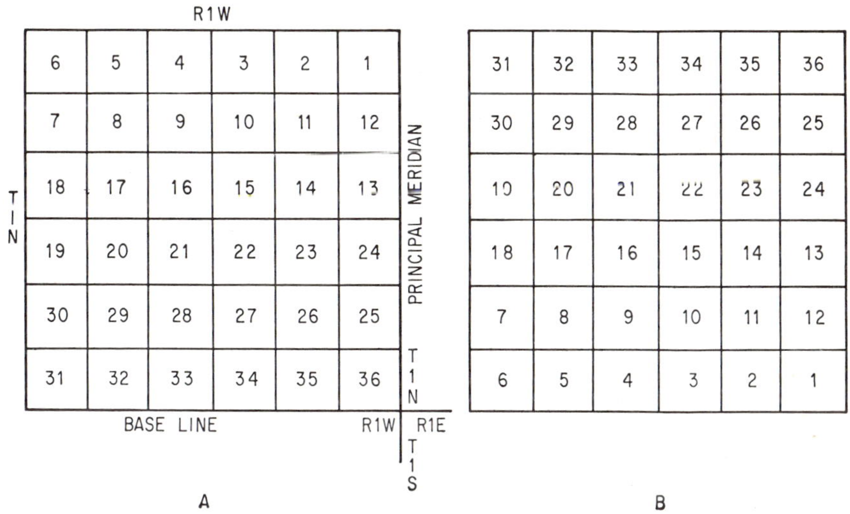

FIGURE 10-1. (*A*) Land subdivision in the United States. (*B*) Canadian township plat showing reverse order of section numbering

county and Federal agencies. These maps, in addition to the township and section lines, generally include sketches of local major topographic features, major drainage lines, and much of the culture. They can be very useful for locating section corners in the field and, if available, should be obtained for field party use.

Within the continental United States and also in Alaska, the normal township is 6 miles square and contains 36 sections each 1 mile square; the sections are numbered as shown in Figure 10-1*A*. Variations which occur in these, with short or long sections along one margin, are generally due to the meeting of two separate surveys or the northward convergence of the lines of longitude.

The Canadian survey also uses a township 6 miles square with 36 sections, each one 1 mile square, but the numbering starts at the southeast corner of the township, i.e., Section 1 is in the southeast corner and Section 36 is in the northeast corner of the township (see Figure 10-1*B*).

Texas, New Mexico, and other states which were originally under Spanish and then Mexican control before becoming part of the United States are characterized by numerous irregularly shaped old land grants which replace the usual township and section divisions. In many of the states these land grants follow a pattern of having one margin along a major stream, while in other more humid areas there is no regular pattern to the disposition of land grants. The land grants are usually designated by the name of the original owner followed by the number of acres in the grant, e.g.,

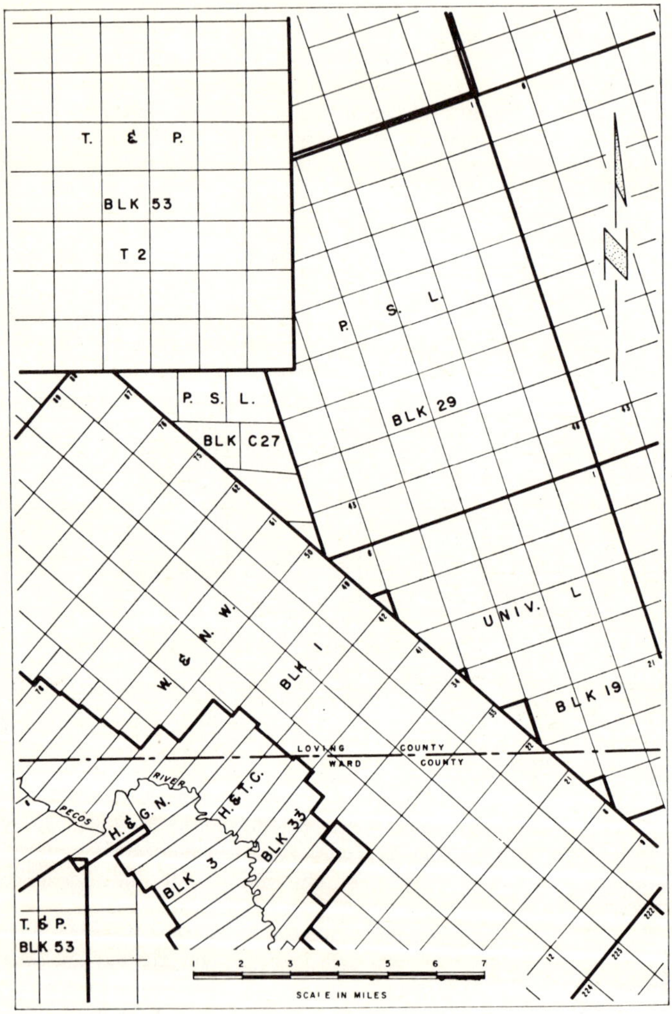

FIGURE 10-2. Example of land surveys in West Texas. (*Courtesy of Muldrow Aerial Surveys.*)

Valdez 247 acres, because parts of land were granted to men having the same last name. Texas is also noted for having several outsized "townships" containing as many as 100 or more sections in addition to overlaps of surveys and vacancy strips due to vagaries of surveys (see Figure 10-2).

Base Maps. Most large companies presently have sets of base maps or assembly maps compiled for each district within the continental United States, prepared by one of the several aerial survey companies furnishing this service. These maps are generally prepared photogrammetically and include townships and sections where surveyed, major drainage, bench marks, triangulation stations, highways and secondary roads, and railroads. If this type of map is available in a field project area, its use as a base map onto which field data are transferred from mosaics or plane-table sheets can save considerable drafting time and speed up the finalizing of the maps and report. A good sepia of the base map, if carried in the field and kept up to date every evening, will be ready for printing at any time in case an early progress report is required by the company or if some structural feature is observed that the field party chief feels should be brought to the immediate attention of his superiors for possible acreage acquisition.

Air Photograph Maps. Contact prints and controlled photomosaics are just as important to the field geologist as to the photogeologist. When carried in the field, contact air photographs, which are usually of larger scale than accompanying mosaics, are relatively accurate maps of the ground surface and are excellent for determining exact location of a point in the field, which can then be more easily located on the mosaic for a permanent record. If a pocket stereoscope is also carried for the field, contact prints can be checked on the station location if there is some doubt as to geologic interpretation, in which case an air view of the area is helpful. The scale of contact air photographs varies, depending upon the altitude of the plane taking the photographs, extreme variations in topography, or possibly tilt of the camera due to weather conditions at the time the area was flown. If the contact prints are to be used for actual mapping and data transferred to a smaller-scale base map, a constant check must be kept on the scale of the print by means of exact location of section corners or other points between which the distance is constant, known, or easily measured in the field.

Controlled photomosaics are useful for any type of geologic field work reconnaissance, detailed reconnaissance, or detailed geology. A 15-minute quadrangle on a scale of 1 inch to the mile will fit directly on the regular-size plane-table board and can then be used as the actual plane-table sheet for more detailed work, or the board can serve as a field desk on which the photomosaic is mounted to sketch in local geology. If it is preferable not to mark up the mosaic, it can be overlain with a sheet of thin, frosted acetate or, if damp weather conditions are expected, a sheet of stabilene. The overlay can be taped onto or over the mosaic on the plane-table board, and to facilitate its removal and correct replacement when the overlay must be used for transfer of data, the lines of longitude and latitude on the mosaic should first be traced onto the overlain transparent sheet.

If more detailed work is required than can adequately be shown on a 1-inch-to-the mile mosaic, in some areas 7½-minute quadrangles are available on a scale of 2 inches to the mile or the mosaics can be enlarged to this scale by a commercial photographer.

A base map of the available photography in the continental United States and Alaska can be obtained from the U.S. Geological Survey. This map indicates areas flown and government agencies from which photographs can be obtained. If mosaics are not available from a government agency, they can often be obtained from private companies, but at a considerably higher cost. For air-photograph coverage in Canada, the Canadian Geological Survey offers similar information.

Foreign photography is also available, generally from government agencies in the country in which field work is planned, but in most instances it may be found that private companies are better sources for photographs of remote foreign areas.

INSTRUMENTS

Brunton Compass

The most important field tool of the geologist is the Brunton compass or Brunton pocket transit. Its major uses include determination of strike and dip, triangulation for location on photographs or topographic maps by resection, hand leveling, and taking of bearings for pace and compass traverses.

Brunton compasses can be obtained with the face graduated from 0 to 90 degrees in each of four quadrants or graduated from 0 to 360 degrees. The Brunton divided into 90-degree quadrants is preferable with bearings taken and recorded with reference to degrees east or west of north (or south). Amount of magnetic declination can be adjusted by a screw on the side of the Brunton. The parts of a Brunton pocket transit are shown in Figure 10-3.

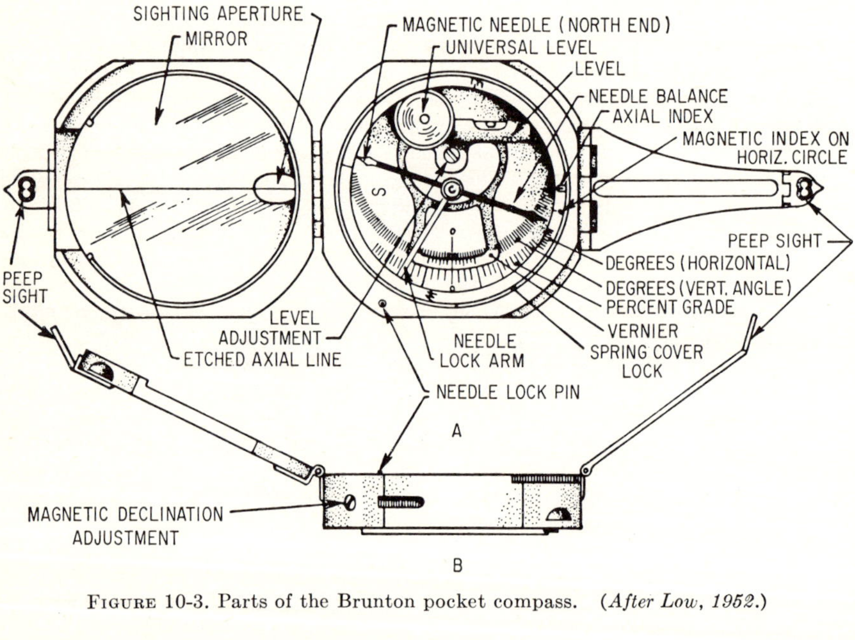

FIGURE 10-3. Parts of the Brunton pocket compass. (*After Low, 1952.*)

Strike-and-dip Determination with the Brunton Compass. The determination of strike and dip of beds with a Brunton requires its use as a compass to determine the direction of strike and as a clinometer to determine the angle of dip.

By definition, the strike of a bed is the direction in space of the line of intersection of the bedding plane with a horizontal plane and the dip of the bed is the angle of inclination from the horizontal plane at right angles to the line of strike (Figure 10-4). Strike of the bed should be read on the compass dial with respect to the direction faced when it is taken. If facing northwest, the bearing would read N45°W, or S45°E if facing southeast. The dip is recorded in degrees as read on the vernier of the compass level, and the general direction recorded to indicate on which side of the strike line the dip symbol should appear. For example, if the strike is N45°W, there are two possible directions of dip, southwest or northeast, and care should be taken to assure the recording of the correct direction. This may appear to be very simple and not worth consideration, but it would be difficult to find a field geologist who could honestly say that he had never recorded the wrong direction of dip as a result of improper orientation or carelessness.

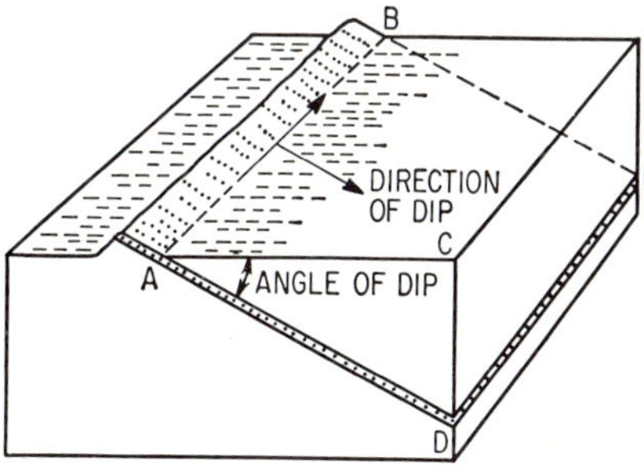

FIGURE 10-4. Diagram illustrating strike and dip. Line *AB* is the strike, representing the position in space of the line of intersection of the bedding plane *ABED* and a horizontal plane.

The method of strike-and-dip determination in most common use is the direct-contact method, in which a bedding plane is exposed and the strike and dip read directly from the rock exposure. If this method is to be used, the bedding surface should first be cleaned of all minor debris and erosional material over a sufficient area to allow the placing of a flat board or field notebook on the surface. The arc level should then be set on 0 degrees on the vernier, and the compass turned on its side extended, with the bubble of the arc level on top and the lower edge of the Brunton placed on the surface of the board or notebook. The Brunton is then rotated slowly until the bubble of the arc level is centered, and then a line is drawn along the lower edge of the compass. The bearing of this line is the strike of the bed,

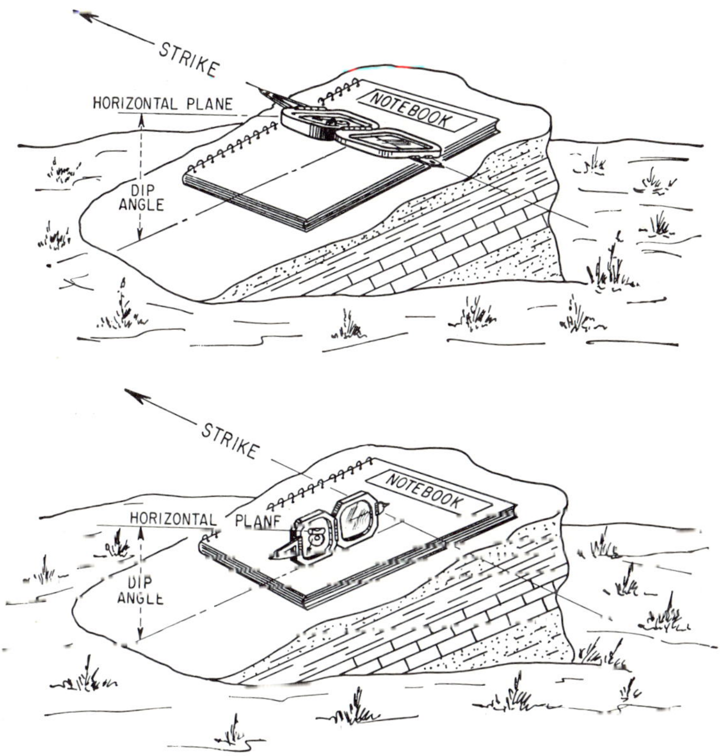

FIGURE 10-5. (*Top*) Direct-contact method for measuring strike of rock bedding surface. (*Bottom*) Measuring dip by direct-contact method.

and the angle of inclination, measured at right angles to it with the clinometer, is the dip (Figure 10-5). If this method is used, caution is necessary to prevent disturbance of the flat board or notebook once the strike line is drawn. Although this method is widely used, it requires the recording of more data to determine average strike in an area, since very few beds are characterized by continuously flat and regular bedding planes.

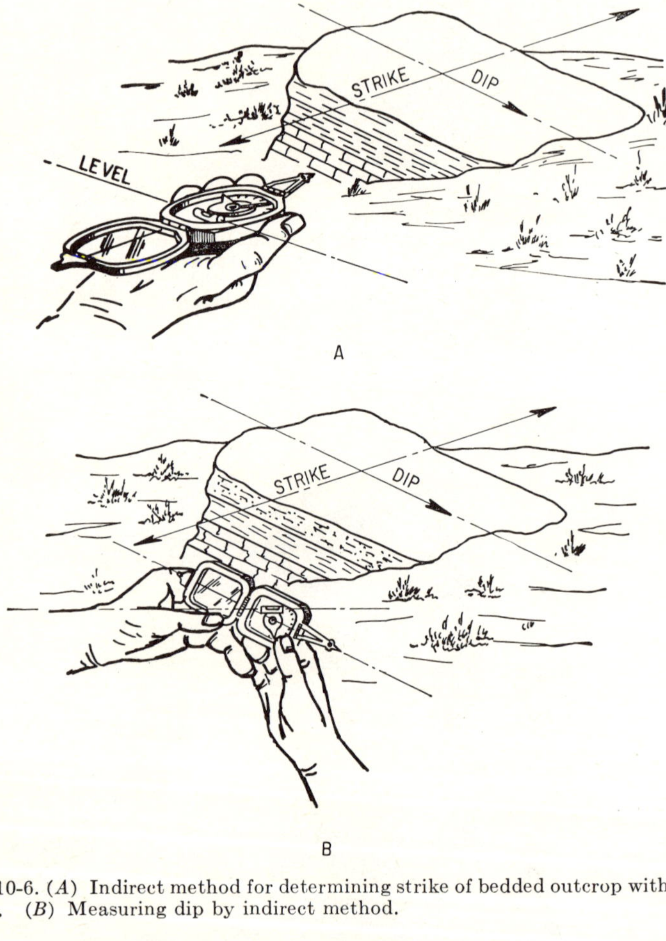

FIGURE 10-6. (*A*) Indirect method for determining strike of bedded outcrop with a Brunton compass. (*B*) Measuring dip by indirect method.

As an aid for structural contouring, the hand-level strike and dip is preferable to one obtained by the contact method. In narrow canyons a prominent bed may be well exposed on both sides of the canyon, dipping either upstream or downstream. If the geologist places himself in a position where his eyes are level with the top or the base of the prominent bed and then, using the Brunton as a hand level, sights across the canyon and finds the similar position on the bed where it is level with him, a bearing toward this point will be the true strike and the dip is measured at right angles to it along the canyon walls (Figure 10-6). The Brunton should be held at arm's length when this method is used in order to assure proper eye focus on its dial and to avoid parallax.

Another method of hand leveling a strike and dip, used on a dip slope, is to pick an eye-level point on a tree or on the field assistant, then proceed along the dip slope 50 or 100 feet away approximately along strike, and then, using the Brunton as a hand level, move up or down slope until the eyes are level with the original point. The bearing between the two points is the strike of the bed, and the dip of the bed can then be determined by moving down or up slope from the original eye-level point, at right angles to the strike line, and reading the vertical angle between the eye-level points (Figure 10-7).

Location by Resection with the Brunton Compass. For work on air photographs, the Brunton can become very useful when a station location occurs in a featureless area on the photograph and is difficult to locate but where it is surrounded by prominent features that stand out clearly on the photograph. Bearings can be taken on two or more of these prominent features, lines of resection drawn on the photograph, and the general area of the station indicated by triangulation. Although the lines of resection may not occur at the exact location of the station, owing to photographic distortion, they will generally come close enough so that some nearby minor feature can be seen on the photograph and the point more accurately located with respect to it.

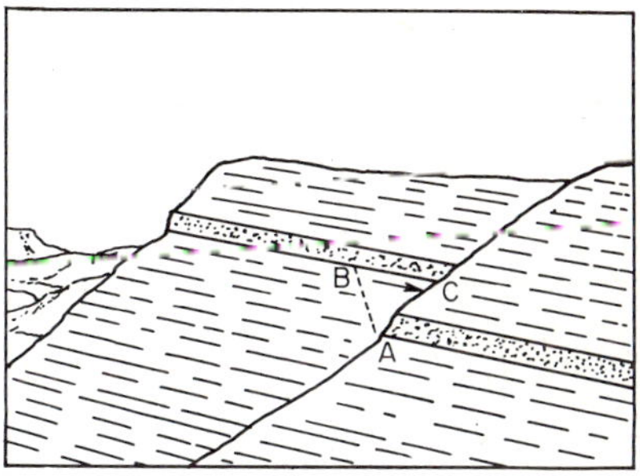

FIGURE 10-7. Hand-level method for determining strike and dip. Standing at point *A*, a geologist using a Brunton compass or hand level can determine the position of point *B* which is level with *A*; the line of sight between *A* and *B* is the strike. Angle of dip is measured in the face of the cliff between points *B* and *C* with a clinometer.

Traversing with the Brunton Compass. With modern mapping methods and transportation, the pace-and-compass traverse may appear obsolete when taught to students in a beginning field course, but it is doubtful that any geologist who devotes much of his time to field work will complete his career without using it numerous times.

The first requirement for this type of traverse is that the field geologist should know the length of his pace to convert to feet and should have established a constant manner of pacing and counting. The second requirement is to be completely familiar with the dial of the Brunton compass so that bearings used are correctly read with respect to north and south. Bearings should be read sighting with the compass held approximately level with the belt and using the mirror of the compass in which to view the object being used as a sighting point.

A pace-and-compass traverse is generally used in the field for work on photographs or mosaics, for work on some type of base map without benefit of photographs and upon which certain known points can be located in the field, and when more detailed information concerning the local geology is required than can be shown on the scale of the photographs or base maps.

A pace-and-compass traverse is started from a known point on the map or photographs which can readily be located in the field. This point is established as station 1. With the general direction in which the traverse is to be run determined, a bearing is taken toward some conspicuous object or outcrop along the line of traverse, and the distance paced off to that point, which can then be established as station 2. At this point, perhaps the direction of traverse may change slightly, another bearing is taken, and distance paced off to station 3. This process is followed until either the end of a linear traverse is reached or perhaps a closed circular traverse is made, tying back to station 1. At each station geological notes are taken of required data, and between sighted stations, if geological phenomena are observed that are important enough to include in the notes, substations can be established.

Notes for a pace-and-compass traverse should be recorded in some manner similar to the following example, which is a set of notes for the illustration of a pace-and-compass traverse map (Figure 10-8).

Closed traverse starts at station 1, a bench mark on contact of massive gray sandstone and underlying gray-green shale, strike of contact N74°W, 6°S.

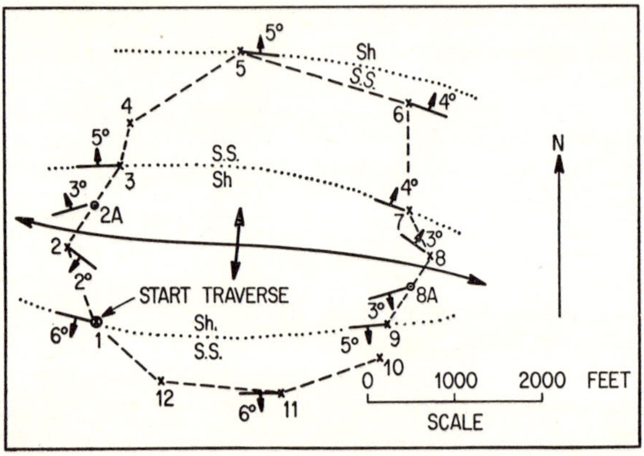

FIGURE 10-8. Sketch map of pace-and-compass traverse showing data that can be worked up in the field.

From station 1: N20°W, 950 feet to station 2. Thin sandstone bed in gray-green shale, strike N51°W, 2°SW.

From station 2: N34°E, 575 feet, to station 2A, thin sandstone bed in shale, strike N74°E, 3°NW.

From station 2A: N34°E, 525 feet to station 3, contact of massive gray sandstone bed with underlying gray-green shale, strike E-W, 5°N.

From station 3: N13°E, 500 feet, to station 4, turning point, weathered exposures of massive gray sandstone, no good surfaces for strike and dip data.

From station 4: N58°E, 1,500 feet to station 5, contact of massive gray sandstone with overlying dark gray shale. Strike N84°W, 5°N.

From station 5: S73°E, 2,000 feet to station 6, outcrops of massive gray sandstone exposed along line of traverse. At station 6, thin shale bed in massive sandstone, strike N71°W, 4°NE.

From station 6: S1°E, 1,225 feet to station 7, contact of massive sandstone with underlying gray-green shale. Strike N69°W, 4°NE.

From station 7: S23°E, 550 feet to station 8, thin sandstone bed in gray-green shale. Strike N54°W, 3°NE.

From station 8: S35°W, 440 feet to station 8*A*, thin sandstone bed in shale, strike N75°E, 3°SE.

From station 8A: S35°W, 460 feet to station 9, contact of massive gray sandstone and underlying gray-green shale. Strike N86°E, 5°S.

From station 9: S9°W, 400 feet to station 10, turning point, poor exposure of massive gray sandstone, badly weathered, no strike and dip obtained.

From station 10: S71°W, 1,200 feet to station 11, thinly bedded sandstone zone with recognizable bedding planes, strike E-W, 6°S.

From station 11: N84°W, 1,400 feet to station 12, turning point, an outcrop of massive gray sandstone, badly weathered, no strike or dip obtained.

From station 12: N46°W, 1,000 feet to station 1, end of traverse.

Only a minimum of data has been recorded in the example of a pace-and-compass traverse. Additional data which should usually be included are remarks on the nature of bedrock exposed along line of traverse between two stations, quality of strikes and dips at each station, and perhaps methods used to obtain the strike and dip. Substations along lines of traverse have been given a letter designation following the number of the station from which the line started. The method of numbering stations and substations is arbitrary but should be done uniformly if more than one such traverse is needed for a field project. The sketch map may be plotted up in the field if so desired, so that if a structural or stratigraphic feature of interest is found, additional data might be sought out before leaving the area of the traverse. Topographic observations should also be made if pertinent to the problem.

Hand Level

The hand level has varied uses in field work for both structural and stratigraphic work. It can be used to determine the difference in elevation between two points on a slope (Figure 10-9) in order to measure the thickness of section or to obtain a structural datum. A hand level may be used rather than a Brunton compass for a more accurate determination of strike for a hand-level strike-and-dip determination.

There are two hand levels in common use, the *Locke* and the *Abney* hand level. Another type for more detailed and accurate work is the *telescopic,* or *stadia,* hand level.

The Locke hand level is the simplest of the three, containing only a small bubble tube, a horizontal sighting line, and a prism which allows a simultaneous view of the leveling bubble, the horizontal line, and the object on which the sight is taken. No angles can be determined with the Locke hand level, and its use is limited to horizontal sighting for interval determinations.

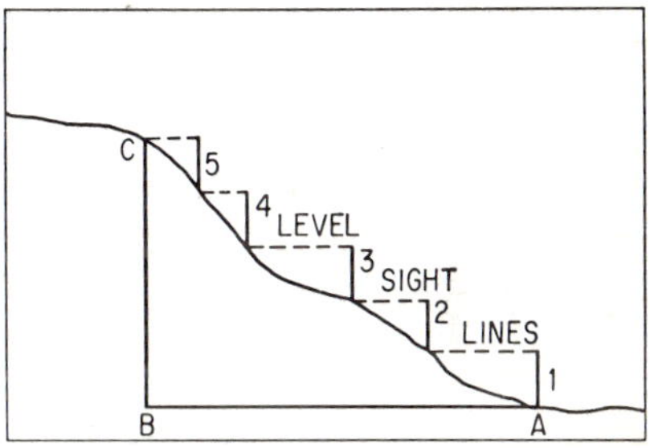

FIGURE 10-9. The process of hand leveling. Difference in elevation *BC*, is determined by a series of hand-level steps from point *A* to point *C*. The sum of equal steps 1 through 5 is the differential elevation between *A* and *C*.

The Abney hand level is more versatile in that it is equipped with a vertical arc and an arc level. In addition to horizontal sighting, angles of slope or dip can also be measured with this hand level. If it is to be used for direct leveling, the vertical-angle arc is set at zero, as is done when the Brunton compass is used for hand leveling.

The *telescopic*, or *stadia*, hand level is designed for use with a stadia rod. It is equipped with a set of stadia wires with the same effective spacing as in an alidade or transit. The level is low precision, and it has a system of lenses with magnifications as high as 10 diameters. A view of the level bubble adjacent to the cross wires is provided by prisms. When the level is used with a stadia rod, the distance determinations are the same as with an alidade or transit; at 500 feet the stadia intercept (between lower and upper wires) is 5 feet.

Jacob's Staff

A common accessory used with the hand level is the *Jacob's staff*, a light wood pole the length of which depends upon the height of the geologist using it, since the hand level is placed on the top and should be at a comfortable reading height. The Jacob's staff should be used when more accuracy is required than can be attained by the more common methods of hand leveling, as it can be graduated in feet and tenths of feet for measuring fractions of steps up a slope and it also provides a steady rest for the hand level. The Jacob's staff is most commonly used for the measurement of stratigraphic sections.

Aneroid Barometer and Altimeter

The aneroid barometer and altimeter are instruments which can be used for the determination of elevations by variations in atmospheric pressure at given locations.

The aneroid barometer is an instrument containing a metal vacuum chamber which is affected by and responds to changes in atmospheric pressure. The face of the aneroid barometer has a circular dial on which a needle rotates to indicate the changes in atmospheric pressure affecting the vacuum chamber. The dial of the aneroid barometer has an outer scale calibrated in feet of elevation and an inner scale graduated in inches of mercury. Smallest divisions of the elevation scale are usually 10 or 20 feet, and values between must be estimated.

The altimeter is a type of aneroid barometer but with an adjustable dial needle which requires that the instrument be leveled before it can be properly and accurately read. The scale on the dial is graduated in feet, with smallest divisions usually of 5 or 10 feet. Accurate reading of the altimeter requires the use of a magnifying glass.

Both instruments require corrections for extreme variations in temperature, local storms and weather conditions, and the diurnal variation in atmospheric pressure. Exceedingly severe local storms limit the use of the barometer, and it is best to restrict its use to times of favorable weather conditions.

Assuming the aneroid barometer or altimeter to be calibrated to record elevations correctly at 50°F only (most are calibrated at this temperature, but they may vary to 60°F), at temperatures above the calibration temperature, the difference in elevation between two stations as indicated by the instrument will be smaller than the time difference. If the temperatures recorded at the two stations are below the calibration temperature of the instrument, the indicated difference in elevation will be larger than actual difference. The correction is approximately $\frac{2}{10}$ of 1 per cent of the elevation difference for each degree of temperature above or below 50°F (Lahee, 1961, p. 506).

For example, if the elevation difference recorded between two stations is 130 feet, the temperature at station 1 is 65°F and at station 2 is 75°F, the corrected difference in elevation would be

$$130 + \frac{(65 - 50) + (75 - 50)}{2} \times 0.002 \times 130 = 135.20 \text{ feet}$$

or (130 + 5.20) feet.

This amount of correction, 5.20 feet, is less than the limits of accuracy in reading the barometer or altimeter, and the use of this correction would be justified only if extremely detailed work was being attempted with the barometer or when extreme temperature differences were recorded.

The most important correction to be made when using the barometer is that for the diurnal variation in atmospheric pressure. No standard curve can be drawn and used for this variation, and when traverses are being made with a barometer, a curve of the variations should be kept each day.

A typical curve illustrating diurnal variations from 8:00 A.M. to 5:00 P.M. on a day with steady conditions is shown in Figure 10-10. Any correction on the curve would be added to the recorded elevation where the curve is below the zero line and would be subtracted if in that part of the curve above the zero line.

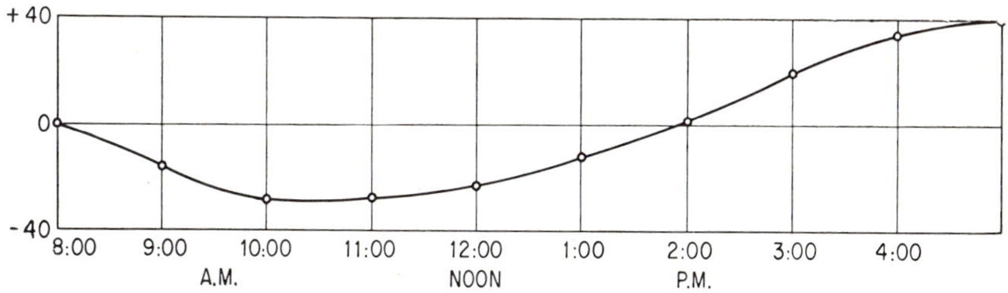

FIGURE 10-10. Curve illustrating diurnal variations in atmospherical pressures on a day with steady conditions. (*After Lahee, 1961.*)

To show differences in diurnal variations which can occur, two additional illustrations of curves recorded in the field have been included. Both are in the same general area in southeastern Utah, and it should be noted that whereas one (Figure 10-11) has a rather erratic pattern, its tendency is to follow the form of the typical curve, with no variation exceeding more than 40 feet above or below the correct elevation. On the other hand, the second curve (Figure 10-12) shows a continuous rise, with slight variation, up to almost a 200-foot correction before it drops off between 5:00 and 6:00 P.M. When one realizes that such differences can occur in diurnal curves, the

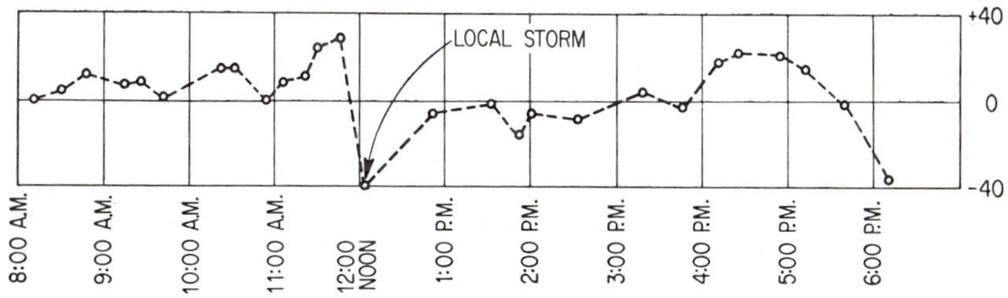

FIGURE 10-11. Barometer curve recorded on a step traverse near Monticello, Utah, on May 16, 1957. (*Courtesy of Continental Oil Company.*)

necessity of keeping an accurate record at points of known elevation for each day in the field can readily be seen.

It is generally recommended in the plotting of the data for the diurnal curve that the points should be connected by an arcuate curve, as shown (Figure 10-10), rather than by a series of straight lines. However, with a fairly large scale and with fixed elevations read at regular intervals of time fairly closely spaced, the use of straight lines to connect the data points should give reasonably accurate data for reconnaissance or detailed reconnaissance work when a structure contour interval of 100 feet is used. This curve will serve well enough even for a 50-foot contour interval if the diurnal data time interval is short. The two field curves have been drawn in this manner so that a comparison can be made between the two methods of establishing the diurnal curve.

Field work with the barometer may be carried out using two or more barometers or with only one barometer, depending on the size of the field party and area to be covered.

If several barometers are to be used for a large field party, one or two barometers can be kept at fixed stations of known elevations in the area and readings taken at regular intervals to establish a diurnal curve. At the same time men in the field may be collecting geological data and recording apparent elevations for points on the map, which can be corrected later using the diurnal curves established at the fixed points of known elevation. This type of work or any type using more than one barometer requires that all barometers used be checked and set at one known elevation point each day before starting the day's field work. Before any such large-scale project is started, if several barometers are to be used, one day should be allowed to plot diurnal curves for each barometer to be used, all at the same point of known elevation, to make sure that they all maintain the same rate of variation. If a barometer has been

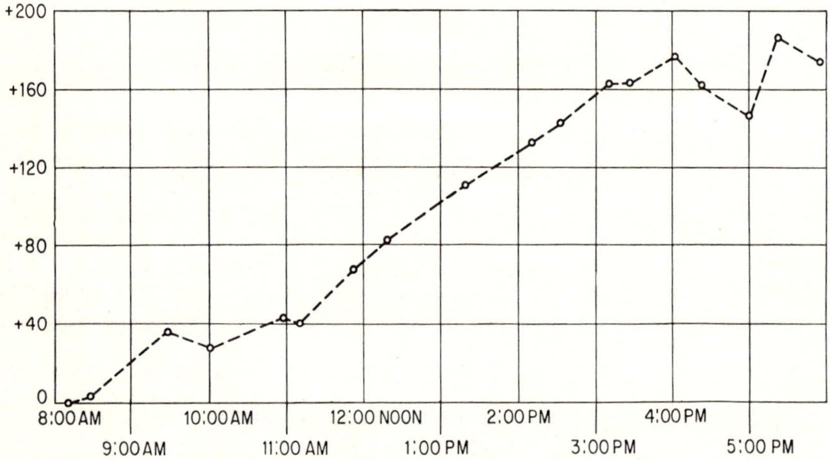

FIGURE 10-12. Barometer curve recorded on a step traverse near Blanding, Utah, on Sept. 25, 1956. (*Courtesy of Continental Oil Company.*)

mishandled, it may have been damaged sufficiently to affect its sensitivity, and a check of all barometers in advance will eliminate the possibility of taking a defective one into the field.

If only one barometer is to be used, a diurnal curve can still be plotted in the field, and there are two methods of operation which will maintain a reasonable amount of accuracy. The first method would be employed in an area in which there were many elevation control points such as U.S. Geological Survey or Coast and Geodetic Survey bench marks. The barometer can be set for elevation on a known point, and the time recorded. The geologist can then proceed to a point of unknown elevation where geologic data are required, record the apparent elevation and time, and then return to the original point to record the barometer reading and time. Several of unknown elevation points can be recorded between known points providing the time between readings at known points of elevation is kept at about 30 to 40 minutes. The diurnal change can be plotted in the field and corrections made before leaving the area so that if any apparent elevation, when corrected, appears out of line with the others, it can be rechecked.

The second method using only one barometer might be described as a "step traverse," in which the barometer is first set at a point of known elevation (Table 10-1). The geologist then proceeds to his second station along a traverse, records the apparent elevation and time, and then returns to the first station to record elevation and time again. The second station is then used as a point of known elevation, and the third station is corrected to true elevation by points on the diurnal curve as established at

Table 10-1. Field Data Sheet for Part of Barometer Curve for Figure 10-11*
(8:10 A.M. to 1:34 P.M.)

Map area________________

Observer________________ Date May 16, 1957

Station	Time	Barometer reading (apparent elevation)	Correction	Elevations		Remarks
				Corrected	Control	
V21	08:10 A.M.	7066	0	....	7066	U.S.C. & G.S. BM V21 Barometer set
B22	08:30	6929	−4		6925	U.S.C. & G.S. BM B22 Control point
(S464)	8:48	6645	−12	6633	6633	Temporary BM—see notes for 5/15/57
S477	9:03	5680	−9	5671	5671	Temporary BM—to establish control
(S464)	9:14	6640	−7	....	6633	
(S477)	9:27	5680	−9	....	5671	
S478	9:35	5610	−5	5605	5605	Instrument station
(S477)	9:42	5672	−1	....	5671	
(S478)	10:23	5620	−15	....	5605	
S470(D)	10:28	5665	−15	5650	....	Datum
(S478)	10:32	5619	−14	....	5605	
S481	10:49	5545	−4	5541	5541	Temporary BM
(S478)	10:55	5605	0	....	5605	
(S481)	11:05	5550	−9	....	5541	
S481(D)	11:12	5680	−10	5670	....	Datum
(S481)	11:20	5552	−11	....	5541	
S482	11:26	5550	−19	5531	5531	Temporary BM
(S481)	11:30	5565	−24	....	5541	
(S482)	11:47	5560	−29	....	5531	
S482(D)	11:57	5720	+16	5736	....	Datum
(S482)	12:05 P.M.	5490	+41	....	5531	
(S482)	12:52	5525	+6	....	5531	
S484	1:29	5480	+1	5481	5481	Temporary BM and instrument station
(S482)	1:34	5530	+1	....	5531	

* Courtesy of Continental Oil Company.

the second station. In this manner a long traverse can be made recording elevations and geological data using a type of "backsight" and "foresight" with the barometer. When possible, the traverse should be tied in to points of known elevation for a check of accuracy during the day. If this is not possible, the traverse should at least be tied to the original starting point before the party leaves the field at the end of the day, determining its elevation from the last field station and seeing how the result agrees with the known elevation. If the barometer work has been done carefully throughout the day, the error in vertical closure should not be more than 10 feet. Although this method may seem slow for one man in the field, if it is a two-man party, work progresses quite rapidly, as one man returns to preceding stations to read the barometer while the other is recording pertinent geological data and collecting samples at the new station.

One other method which might be used and can be applied to both methods just discussed is one in which a diurnal curve is not plotted but the apparent elevation is read at a new station and the geologist immediately returns to a point of known elevation and again reads the barometer. The difference in time of readings is then known, and the amount of change at the point of known elevation is known. The apparent elevation is then corrected in the following manner:

Station A:		
Known elevation	5,350	
Barometer reading	5,372	at 10:10 A.M.
Elevation correction	22	
Station B:		
Barometer reading	5,420	at 10:25 A.M.
Station A:		
Barometer reading	5,362	at 10:40 A.M.
Elevation correction		

As the barometer is falling, the total change in correction at station A is 10 feet in 30 minutes. The corrected elevation at station B, which was read halfway between the time elapse between readings at station A, would be

$$5{,}420 + (-22 + {}^{10}\!/_{2})$$

or

$$5{,}420 - 17 = 5{,}403$$

This method assumes a straight line connecting points on the diurnal curve, and to maintain a reasonable degree of accuracy, time intervals between readings at points of known elevation should be kept as short as possible.

Plane Table and Telescopic Alidade

The plane table and telescopic alidade are two of the most important basic tools used by petroleum geologists. They are the best instruments for combined topographic and geologic mapping or for geologic mapping only, as with them the map can be drawn in the field, where observed data can be placed accurately on the map and the amount of data required can be determined by observation and obtained before leaving an instrument station.

Amount of field data mapped with the plane table and alidade will vary depending upon the objectives of a field project. However, it is important to realize that plane-table work does not necessarily mean detailed geology, since it can also be used to great advantage for reconnaissance or detailed reconnaissance field geology.

The Plane Table. The plane table can be simply described as a drawing board which may vary in size. The size most commonly used in petroleum exploration is a board 18 inches wide and 24 inches long, although some use is made of boards 15 by 15 inches and 24 by 31 inches in size.

The plane-table board is mounted on a tripod, which may be of two general types, one having rigid legs, the other with adjustable, or telescope, legs. The rigid-leg type is preferable for petroleum exploration, as it is more durable and provides better support for the plane-table board. The telescope-leg tripod has certain advantages, however, if space is at a premium in transporting field instruments, because when it is collapsed, it is considerably shorter than the rigid-leg type of tripod. Also, when an instrument station must be located on a steep slope, the legs can be adjusted to varying lengths to permit a more stable setup and greater ease in final leveling of the plane-table board.

The plane-table board is generally mounted on the tripod by a device known as a Johnson head (Figure 10-13), a ball-and-socket arrangement with two separate thumb-screw-type lock nuts, commonly called wing nuts, which permit the board to be tightened in a level position with one nut while it is still free to turn in a horizontal plane for orientation and then can be set in a fixed position by tightening the lock nut.

The Telescopic Alidade. The telescopic alidade consists of a transit telescope, with varying adjustments, mounted on a flat base or straight edge (Figure 10-14*A* and *B*). The telescope contains stadia hairs and has a vertical movement in a plane that is parallel with the straight edge on which it is mounted. Other features of the telescopic alidade include a vertical-angle arc with universal level, a Beaman stadia arc, a striding level, a circular bubble on the straight edge for leveling the plane-table board, and a compass rigidly attached to the base so that when oriented on magnetic north the compass needle is parallel with the straight edge.

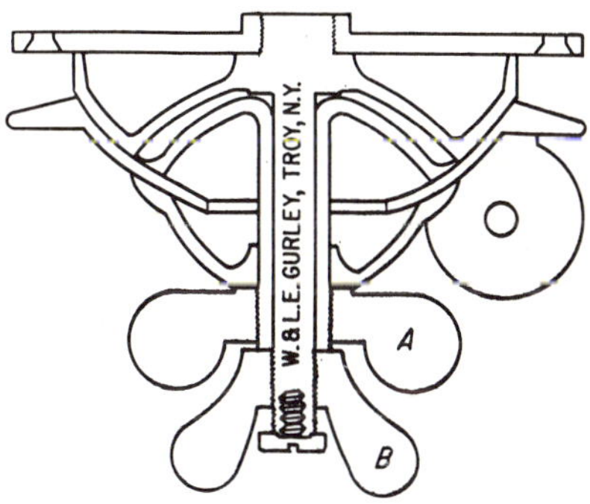

FIGURE 10-13. Sectioned view of Johnson tripod head assembly. When wing nuts *A* and *B* are loosened, the table can be rotated or tilted in any direction. Nut *A* is tightened when the board is leveled to prevent tilt of the table while being rotated in horizontal plane for orientation. Nut *B* locks rotational movement when the board is properly oriented. (*Courtesy of W. & L. E. Gurley.*)

The larger and more accurate of available alidades are the *standard telescopic alidades*, which have a base 18 to 24 inches long and 3 inches wide, with the telescope mounted on a standard or pillar 3½ to 4½ inches high. Telescope magnification can vary from 13 to 32 diameters. These larger instruments are delicate, heavy, and difficult to carry in the field in some instances, and therefore, the smaller *explorers' model* alidade is preferable for petroleum exploration field geology.

The explorers' model telescopic alidade has an 11½-inch base on which a smaller telescope is mounted on a standard about 2 inches high. Magnification of the telescope may vary from 10½ to 18 diameters' but all other features are basically the same as on the *standard alidade*. The explorers, model alidade is only about 3½ inches high and weighs only 4¾ pounds, complete in leather case with sling for carrying ease in the field.

The Stadia Rod

The principal method of determining distances with the alidade and plane table is by means of the stadia rod. The telescopic alidade has two cross hairs on the lens, one vertical, one horizontal, that are equal in length to the diameter of the lens and intersect at its center. Two additional horizontal hairs, called stadia hairs, are placed above and below the center horizontal hair and equidistant from it. Most alidades have a fourth horizontal hair between the middle and upper hairs and equidistant between them, called a quarter hair.

The stadia rod is usually 12 to 15 feet in length and may be jointed in the middle for folding, or sometimes, in the case of a 15-foot rod, it may be divided into three sections

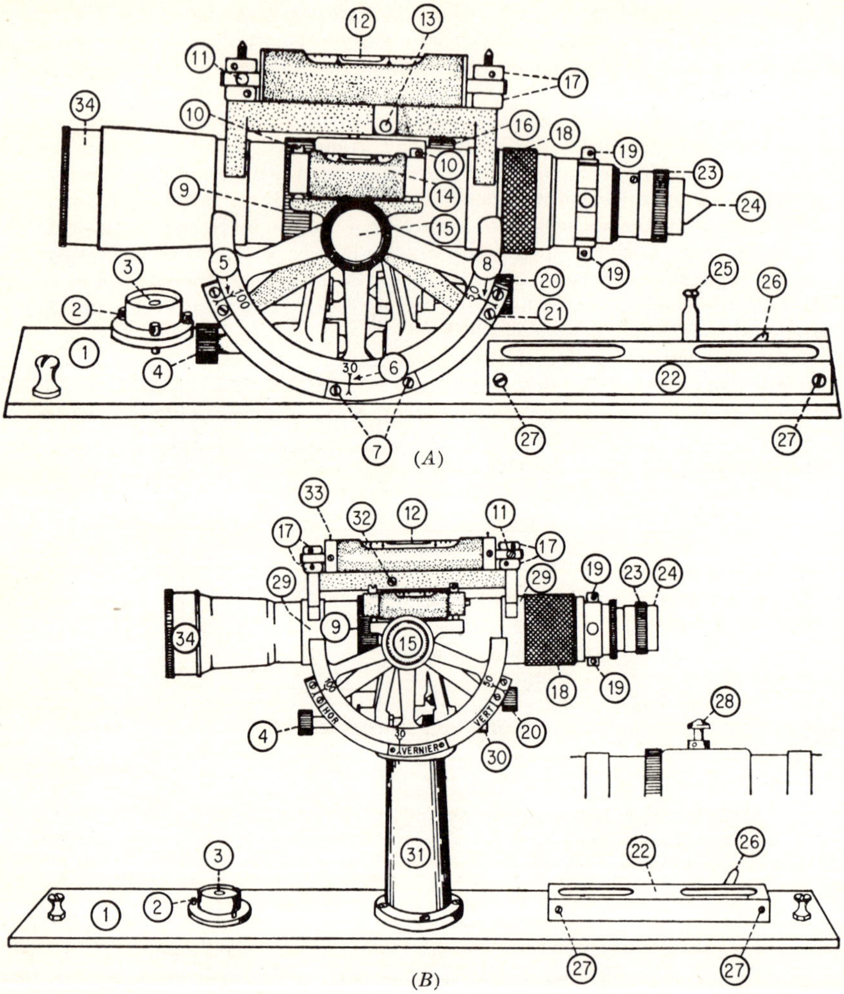

FIGURE 10-14. Principal parts of the alidade. (*A*) Explorers' model. (*B*) High standard alidade: (1) Blade or base. (2) Adjustment screw for universal level. (3) Universal level center mark. (4) Vertical-angle and Beaman arc index tangent screw. (5) Horizontal-distance arc index. (6) Vertical-angle arc index. (7) Vertical-angle arc index adjustment plate. (8) Beaman arc index. (9) Knurled telescope holding collar. (10) Control-level adjustment screw. (11) Lateral adjustment screw for striding level. (12) Striding-level bubble vial. (13) Striding-level spring lock. (14) Control-level sleeve shield. (15) Telescope axis bearing. (16) Axis clamp (on right side of instrument). (17) Vertical adjustment capstan nuts of striding level. (18) Knurled focusing ring. (19) Capstan adjustment screws for adjusting collimation. (20) Telescope tangent screw (with or without gradienter attachment). (21) Beaman arc index adjustment screws. (22) Magnetic needle box or trough. (23) Cross wire or eyepiece focusing ring. (24) Prismatic eyepiece on *A*, inverting eyepiece on *B*. (25) Keeper stud for striding level, used when instrument is laid in case. (26) Needle-lifting lever. (27) Compass box assembly screws. (28) Striding lifting lever. (29) Ring foot bearing for striding level. (30) Plunger opposing Beaman tangent screw. (31) Pedestal or standard of topographic-type alidade. (32) Striding level spring lock releasing screw. (33) Finder sights on striding level (not on all levels). (34) Dust protector cap for objective lens. (*From Low, 1952.*)

so it can be folded up to measure 5 feet in length for convenient carrying. The rod should be 4 inches in width for good visibility when it is viewed through the alidade and not over ¾ inch thick to keep the weight down to a minimum. One side of the rod is divided into feet, painted red, white, and black to increase visibility when viewed through the alidade. Some rods have one foot near the base and one near the top, divided into tenths of a foot, while other rods may have every foot divided into tenths.

The uppermost and lowermost cross hairs of the alidade are spaced so that, on level readings with the alidade, the number of feet between the points at which they intersect the rod is approximately 1 foot on the rod for each 100 feet of distance from the instrument station (Figure 10-15). This is called the stadia ratio. Hence, if the reading, called the stadia intercept or stadia interval, is 8 feet on the rod, the distance is 800 feet; if the stadia intercept is 4.7 feet, the distance is 470 feet, the partial foot being determined by placing one stadia hair in a foot divided in tenths, the other on an even foot. If the distance to the rod is so great that when the lower horizontal cross hair is at the base of the rod, the uppermost cross hair is above the rod, a half-hair reading can be made on the intercept between the middle cross hair and the lower or upper cross hair. Thus, if a half-hair intercept of 9 feet is read, the distance is equal to 2 times 9 times 100 feet, or 1,800 feet. Even longer distances can be determined with a quarter-hair reading, multiplying the product of the intercept and the stadia ratio by 4 to obtain the distance in hundreds of feet. Theoretically, with a 15-foot rod, a distance of 6,000 feet could be determined with a quarter-hair reading, provided visibility of the rod permitted this long a sight with the alidade. Examples of stadia intercepts are given in Figure 10-16.

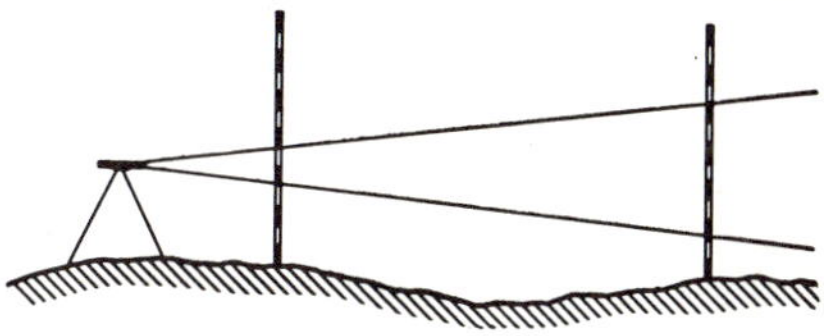

FIGURE 10-15. Diagram illustrating the stadia principle. (*From Mather and Washburn, 1938.*)

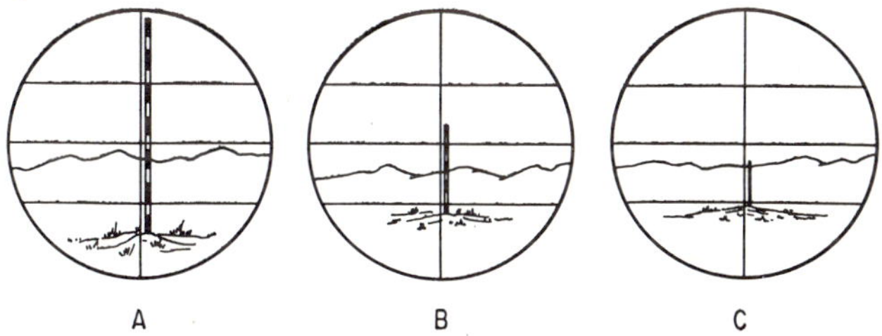

FIGURE 10-16. Examples of stadia intercepts. (*A*) Stadia rod in field of view at 720 feet. (*B*) Stadia rod in field of view at 1,740 feet. (*C*) Stadia rod in field of view at 3,360 feet. (*From Mather and Washburn, 1938.*)

When plane-table work requires a high degree of accuracy, two important factors should be taken into consideration and corrections made on distances determined with the alidade. A minor correction is for the *stadia constant*, which is the distance from a vertical line at the center of the alidade to the point of outer focus (Gurley, 1941). For distances greater than 50 feet, a correction of 1 foot, added to the distance obtained by stadia-rod measurement, is necessary. The *stadia ratio*, which is equal to 100 feet for every foot of stadia intercept, may vary slightly in some instruments and should be checked in the field before a project is started. Starting from a point in front of the instrument equal to the stadia constant, a line should be accurately measured with a tape and divided into 100-foot intervals, up to several hundred feet from the instrument. With the telescope level, the stadia intercept can then be read on the rod at each 100-foot interval. Then divide the distance measured by the observed stadia distance, and multiply by 100 to give the exact stadia ratio. The ratios can then be averaged, and if a stadia ratio of 97 or 98 is obtained, it should be used to multiply the stadia intercept when the project is started rather than a stadia ratio of 100 (Figure 10-17).

Long Sights. Occasionally, when using an alidade equipped with only three stadia hairs, it is necessary to determine a distance that is greater than can be obtained with a half-hair intercept. There are several methods by which this distance can be determined, of which two lend themselves favorably to geological work with the plane table.

1. First rotate the telescope 90 degrees in its sleeve so that the stadia hairs are vertical instead of horizontal. Then, after placing one hair on the station, signal the rodman to mark his station and then to move laterally in the direction of the other stadia hair until the rod, held vertically, is in line with the second hair. The rodman can then measure the horizontal distance between the two points on the ground and record it, to be used later as the stadia intercept or interval by the instrument man. This method requires a prearranged set of arm signals between the rodman and instrument man.

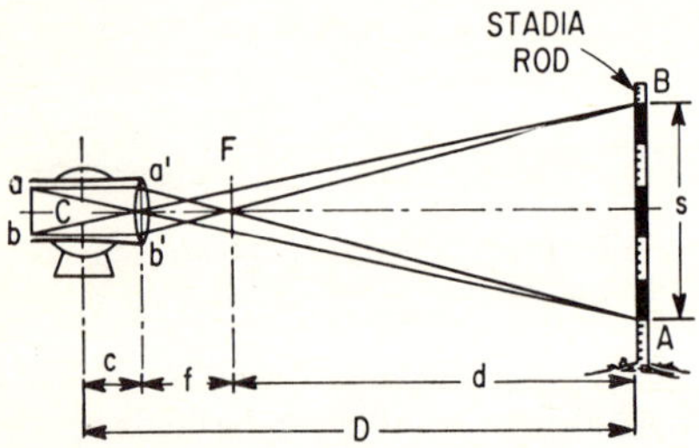

FIGURE 10-17. Diagram showing paths of light rays entering telescope with external focus. *a* and *b*, top and bottom stadia hairs; *C*, vertical axis of telescope; *a'* and *b'*, points in object lens at which rays projected from *a* and *b* are deflected; *F*, principal focus of telescope; *A* and *B*, points on stadia rod to which rays from *a* and *b* are projected through object lens; *S*, stadia intercept; *c*, distance from vertical axis to center of object lens; *f*, distance from center of object lens to principal focus; *d*, distance from principal focus to stadia rod; *D*, true distance from alidade to stadia rod. This figure illustrates a stadia ratio of 100. (*From Mather and Washburn, 1938.*)

2. If the alidade is equipped with a *gradienter screw* or *Stebinger drum*, there are several available methods for determining distances on long sights, of which one is most commonly used for geologic mapping. The Stebinger drum, which is attached to the tangent screw of an alidade, is divided into 100 parts and so graduated that one complete revolution of the screw will raise (or lower) the telescope through one stadia intercept. A graduated horizontal bar is placed just above the drum and parallel with the axis (or screw) on which it turns to indicate the number of complete revolutions.

When the rod station is at a distance where the half-hair intercept is greater than the length of the rod, the Stebinger drum can then be used with the following procedure: First place the middle cross hair on the top of rod, read and record the Stebinger, denoting it as *A*. Then lower the telescope until the middle hair intersects the base of rod, and read and record the Stebinger as *B*. Finally continue to lower the telescope until the top stadia hair rests on the top of the rod, reading and recording the Stebinger as *C*. The distance can then be computed:

$$D = 200r\frac{C - A}{B - A}$$

in which D represents the distance, r the length of the rod, and A, B, and C the three readings of the Stebinger drum (Lahee, 1961).

Inclined Sights. The preceding paragraphs have discussed distance determinations with the alidade and stadia rod, assuming a level line of sight. When the line of sight is inclined to the vertical, the distance read on the rod should be corrected for the angle of variation from the horizontal as read on the vertical-angle arc. Either stadia tables or a stadia slide rule can be used to determine the correct horizontal distance on an inclined sight and are both based on the stadia formula

$$D = d \cos^2 A + (f + c) \cos A$$

in which d = observed stadia distance from rod reading
A = measured vertical angle
$f + c$ = stadia constant
D = correct horizontal distance to rod station on a level line of sight (Figure 10-18) (after Gurley, 1941, p. 62)

Use of this correction will depend upon the scale of the map, the distance of the rod shot, and the magnitude of the vertical angle. For example, at 1,000 feet, with a vertical angle reading of 4 degrees, the correction is only 4.9 feet. If a scale of 1 inch to 2,000 feet is being used, it would be difficult to plot this small difference correctly on the plane-table sheet. On the other hand, if a scale of 1 inch to 100 feet is being

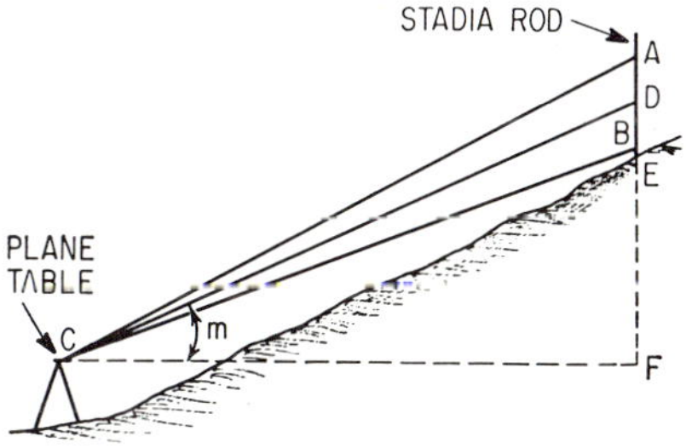

FIGURE 10-18. Inclined sights. AB is the stadia intercept that gives the slope distance CD. When corrected, using angle m, it will give the true horizontal distance CF. Elevation difference EF is also determined using angle m and the stadia intercept AB. The correction factor times the stadia intercept will give the elevation difference DF, from which DE (middle hair reading) must be subtracted to obtain EF.

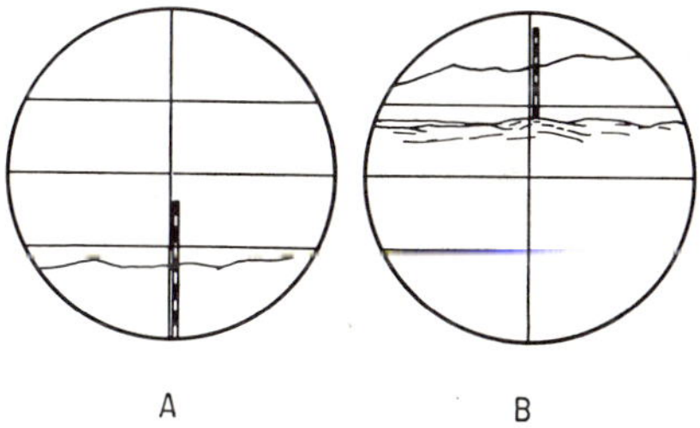

FIGURE 10-19. Differential leveling with the plane table and alidade. (*A*) Stadia rod in field of view with level telescope at a distance of 1,050 feet. The base of the 13-foot rod is 14.8 feet below the elevation of the alidade. (*B*) Stadia rod in field of view with level telescope at a distance of 2,140 feet. The base of the rod is 8.9 feet above the elevation of the alidade. (*From Mather and Washburn, 1938.*)

used, the correction would be almost 1/20 inch on the map and could easily be plotted. The correction should be determined for every inclined rod shot in order to maintain accuracy of data even if it is too small to plot correctly on the plane-table sheet.

Determination of Elevation Differences

The height of the center of the alidade above the ground should be measured at each instrument station once the board is leveled and oriented. This measurement is added to the known ground elevation of the instrument station, and the sum is known as the HI (height of instrument). Or if the instrument station is located by a backsight on a point of known elevation and the elevation established by a level shot or with vertical angles, the figure obtained is the HI, and the distance from the center of the instrument to the ground must be subtracted to obtain the ground elevation. The HI is always used for determination of elevations of other points sighted in from the instrument station, whether by rod shot or triangulation.

Differential Leveling. The simplest method to determine elevations is by *differential leveling,* which is limited by the length of the rod, and only points that permit intersection of the rod by one of the cross hairs when the alidade is level can be determined in this manner (Figure 10-19). For example, if the HI is 4,550 feet and the middle cross hair intersects the rod at 9 feet when the alidade is level, the elevation of the point sighted in is 9 feet lower than the HI, or 4,541 feet. If a 15-foot stadia rod is used, the maximum elevation difference that could be determined by differential leveling is 30 feet, with the top or bottom cross hair intersecting the top or the base of the rod with a full stadia interval reading, 60 feet with a half-hair reading and 120 feet with a quarter-hair reading.

Most of the readings of the stadia rod on an extensive field project will be on inclined sights, for which there are several methods to determine elevation differences with the alidade, the most common of which are by means of vertical angles and the stepping method and by using the *Stebinger drum* or *gradienter screw.*

The Vertical-angle Method. The vertical-angle arc is divided into degrees numbered from 0 to 60 degrees, with ½-degree divisions. The arc is so designed that when the instrument is leveled, the vertical-angle arc can be adjusted so that the zero line of the lower vernier scale bisects the upper arc at 30°00′. Minutes of arc are read on the vernier, which is graduated from 0 to 30 minutes. The vertical angle is the difference between the arc reading and 30°00′. When the reading is less than 30°00′, it is a negative angle and the sign is *minus;* when the reading is greater than 30°00′, the angle is positive and the sign is *plus.*

The vertical correction, using angles, is expressed by the formula

$$h = d(\tfrac{1}{2} \sin 2A)$$

in which d = observed stadia distance from rod reading
A = measured vertical angle
h = elevation difference between instrument and point sighted on rod (Gurley, 1941)

Stadia tables give the value of $\frac{1}{2} \sin 2A$ which can be used in the formula, or a stadia slide rule has one scale graduated in these values. If precise work is required, the correction $(f + c) \cos A$ can be added to the elevation difference computed.

The most favorable procedure for reading the vertical angles is first to level the instrument and adjust the control level so that the zero line of the vernier is at 30°00′ on the vertical-angle arc, then raise or lower the telescope to the rod, read the stadia distance, and record it. Following this, the middle cross hair should be placed on a foot marker on the rod, preferably the same one on every rod shot, such as the 5-foot marker, and then the vertical angle read in degrees and minutes on the vernier and recorded in the notes. The data can then be incorporated into the formula, using the value of $\frac{1}{2} \sin 2A$ as obtained from the stadia tables. The cross-hair reading, at the 5-foot mark, known as the *rod correction,* is subtracted from the elevation difference when the angle is positive, added when the angle is negative.

If only the striding level is used to level the instrument, the procedure can be reversed, first reading the vertical angle on the rod, then leveling the instrument and reading it again. In most instances when this method is used, the level angle will vary above or below a value of 30°00′ and the difference between the two will be the angle used to obtain the elevation difference. This method can lead to errors by the instrument man, and it is preferable to level to 30°00′ on each rod shot with the control level to avoid possible human errors.

The method discussed in the preceding paragraphs has assumed that the rod readings were on *foresights* from the instrument station, with the elevation difference added for a positive vertical angle, subtracted for a negative angle. When a rod reading is made on a backsight and the elevation of the instrument station is to be determined, the reverse is true, in that the elevation difference is subtracted for a *plus* angle, added for a *minus* angle.

The Stepping Method. This method depends upon the determination of the number of stadia intervals required to raise or lower the telescope from a level position to one in which either the top or bottom cross hair intersects the rod, respectively. The procedure is described in Lahee (1961, pp. 538–539) as a series of steps, as follows:

"Level and orient the board at Station A, place the alidade in proper position, sight at the rod held at B, and draw the alignment, *ab*. Determine the distance, *AB*. Let us assume that *B* is higher than *A*. Level the telescope by the striding level. Looking through the telescope, choose some recognizable object, such as a rock, bush, etc., which is crossed by the horizontal cross-hair. Revolve the telescope on its horizontal axis till the lower stadia hair is on this object. This is a *half step*, or a *half interval*. Note a new object crossed by the upper stadia hair. Revolve the telescope till the lower stadia hair is on this object. Repeat, recording the number of *steps* until the upper stadia hair falls on the rod at *B*. Read the rod and wave the rodman to proceed. The elevation at *B* is the total number of *steps* times the vertical stadia interval for the distance, *AB*, minus the rod reading. If *E* is the elevation at Station B, H. I. is the height of the instrument above sea level at Station A; *N* is the number of steps, including the last one which brought the upper stadia hair on to the rod; *S* is the vertical stadia interval for the distance *AB*; and *R* is the reading of the rod where crossed by the upper stadia hair; then,

$$E = (\text{H. I.}) + (N \times S) - R$$

"If *B* is lower than *A*, the process consists of *stepping down*, and the formula becomes:

$$E = (\text{H. I.}) - (N \times S) + R."$$

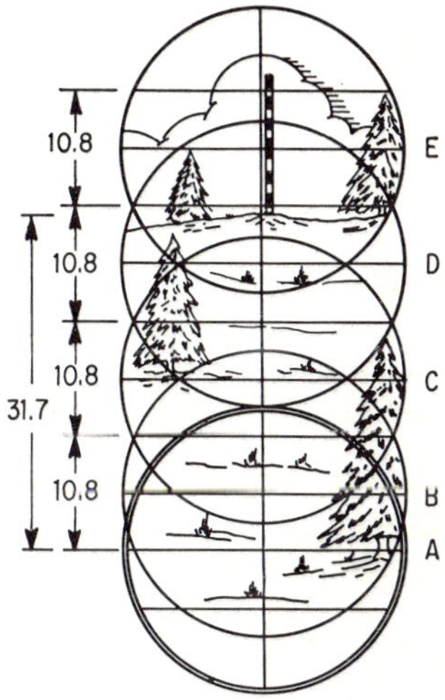

FIGURE 10-20. Stepping method of determining elevation differential. Circle *A* encloses the field of view as observed through the level telescope. Circle *B* is the field after the telescope has been raised so that the bottom cross hair occupies the position occupied in *A* by the middle hair. *C* and *D* represent the second and third steps, in which the bottom cross hair rests in the position previously occupied by the top hair. In *D*, the third step, the top hair cuts the rod 0.7 foot above the base. The rod intercept, 10.8 feet, is indicated in *E*. The base of the rod is therefore 3 × 10.8 − 0.7 = 31.7 feet above the instrument. (*From Mather and Washburn, 1938.*)

Mather and Washburn (1938, p. 43) describe the *stepping method* in detail, and their discussion emphasizes the fact that though the method is fast and fairly accurate when used by an experienced instrument man, it is not sufficiently accurate for important work, as there is a wide margin of error involved in the placing of one hair in the position previously occupied by another (Figure 10-20).

The Stebinger Drum or Gradienter Screw. The *Stebinger drum* or *gradienter screw*, which is a part of the right tangent screw on the alidade, is graduated on the periphery from 1 to 100. One full revolution of the drum is equal to one stadia interval, and each of the divisions in which it is graduated is equal to 1⁄100 revolution, or stadia interval, when the screw is properly adjusted.

The procedure for determining elevations with the gradienter screw first requires the orientation and leveling of the board and then sighting in on the new station. The stadia interval is then read, and for maximum accuracy the telescope should then be leveled with the striding level. The reading of the Stebinger drum should then be recorded, and the telescope then raised to a point where the middle cross hair intersects the rod at the desired point. The difference in elevation can be expressed by the following formula:

$$E = (S \times G) - R$$

in which E = elevation difference to be determined
S = stadia interval
G = number of revolutions of gradienter screw
R = rod reading where intersected by middle cross hair

For a station higher than the instrument station, the following data might be obtained to substitute in the formula:

Stadia interval	= 10.0 feet
Gradienter drum (level)	= 0.45
Gradienter drum (middle cross hair on 5-foot mark)	= 4.95
Difference in gradienter readings	= 4.50

Therefore

$$E = (10.0 \times 4.50) - 5 = 445 \text{ feet}$$

Use of this method assumes that one revolution of the drum moves the telescope through one stadia interval. In his discussion of the use of the gradienter screw, Low (1952, p. 48) points out that though the gradienter is theoretically a more precise measurement than the vertical angle (the smallest division on the vernier, 1 minute of arc, being approximately 0.03 interval, while the gradienter reads 0.01 interval directly), the actual precision depends upon the amount of mechanical wear in the instrument, exact adjustment of the screw, the precision with which the threads of the tangent screw are cut, and extreme care in operation. Accuracy will decrease with increased elevation or depression of the telescope, and the method is not recommended when the inclination of the telescope exceeds 4 revolutions of the drum.

The Beaman Stadia Arc

An alidade equipped with a Beaman stadia arc (Figure 10-21) facilitates rapid field calculation of horizontal distances and elevation differences for inclined sights in plane-table mapping with a stadia rod. Since all computations are those of multiplication, addition, and subtraction, they can be made in the field at the time of taking

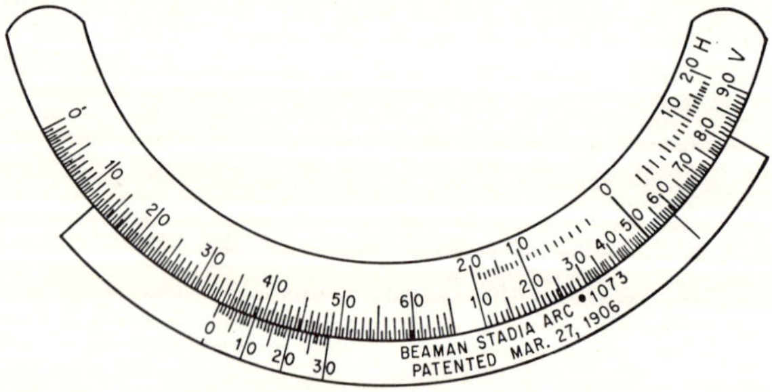

FIGURE 10-21. Combined vertical-angle arc with vernier and Beaman stadia arc. Beaman arc set to show a positive difference in elevation of eight times the observed stadia interval. (*After Mather and Washburn, 1938.*)

the notes, and the reduction of stadia measurements by means of tables or slide rules is entirely eliminated (Gurley, 1941, pp. 64–65). The Beaman stadia arc was devised and patented by W. M. Beaman, a topographer with the U.S. Geological Survey. It provided the original method of mechanical stadia reduction (Gurley, 1941, p. 67).

The Beaman arc is incorporated into the vertical-angle arc of an alidade and has selected angles laid off in two special scales, marked *H* (horizontal) and *V* (vertical), with both scales read by a common index (Gurley, 1941, p. 67).

For horizontal distance corrections, the *H*, or horizontal, scale of the Beaman arc levels at zero and is numbered consecutively from 0 to 20, each division representing

Table 10-2. Tabulation Based on Scaling Error of 1/100 Inch*

Scale	Probable scaling error	Maximum vertical angle, degrees	Probable error in difference of elevation, feet
1:96,000	80 feet	1	1.4
(1″ = 8,000 feet)		5	7.0
		10	14.1
		20	29.1
1:361,680	25 feet	1	0.4
(2″ = 1 mile)		5	2.2
		10	4.4
		20	9.1
1:12,000	10 feet	1	0.2
(1″ = 1,000 feet)		5	0.9
		10	1.8
		20	3.6

* Distances scaled from plane-table sheets, based on resection, will yield the vertical elevation errors indicated for each vertical angle at that map scale, irrespective of length of line scaled. (From J. Low, "Plane Table Mapping," p. 109, Harper & Brothers, New York, 1952.)

a 1-foot-per-100-feet correction on an inclined sight. The horizontal distance correction can be determined by substituting in the following formulas (Gurley, 1941, p. 68):

Horizontal corrections = rod reading × *H* scale reading
Horizontal distance = (rod reading − horizontal correction) × 100

In the above formulas the *H* reading is actually a percentage value, so that a value of 3 read on the *H* scale represents a correction of 3 feet per 100 feet of horizontal distance and is substituted in the formula as 0.03.

For elevation differences, the *V* scale zero point is numbered 50. Inclined sights to points of higher elevation will fall in the range of the Beaman arc from 50 to 95, while points of lower elevation, or minus angles, will be read from 50 down to 5. The difference between 50 and the *V* scale reading times the stadia interval will give the elevation difference between the instrument station and the point sighted on and will be added or subtracted to the HI, depending upon whether the difference is positive or negative. The procedure can be expressed by the formula

$$E = (V - 50) \times S - R$$

in which *E* is the elevation difference to be determined, *V* is the Beaman arc reading, *S* the stadia interval, and *R* the selected point on the rod where the middle cross hair was sighted for the reading. If *V* is less than 50, that part of the formula $(V - 50) \times S$ will be negative and the difference in elevation must be subtracted from the HI; if *V* is more than 50, $(V - 50) \times S$ will be positive and the elevation difference is added to the HI. As in the use of vertical angles, the reverse is true for backsights, with negative values added and positive values subtracted to obtain the elevation of the instrument.

Gurley (1941, p. 69) lists the following advantages in using the Beaman stadia arc:

1. The use of stadia tables, slide rules, or diagrams is entirely obviated.

2. There is no vernier or similar contrivance to be read.

3. Final results are obtained in less than one-third the time required by ordinary methods.

4. The accuracy of results is identical with formulas or table computations, regardless of the angle or distance.

5. The simplicity of the process practically eliminates the chances of error incidental to the use of other methods.

These advantages should be given serious consideration when a detailed plane-table mapping project is planned. If the alidade available is equipped with a Beaman stadia arc, its use can save valuable time without loss of accuracy.

Suggestions for Stadia Work

When plane-table mapping is carried out using the stadia rod, the geologist generally serves as rodman, walking the outcrops, recording the geological data at desirable points, and giving the instrument man rod shots at these points to locate them on the plane-table sheet. The instrument man, in most instances, is also a geologist who does not as yet have sufficient experience to direct and carry out the field work but is capable of learning the entire procedure through his experience as instrument man with the help of a competent geologist and rodman. Both men should be thoroughly familiar with plane-table procedure, and the following suggestions, abstracted from Gurley (1941, p. 64), are of importance to both members of a plane-table field party:

1. Accuracy of stadia work is directly proportional to the ease, care, and accuracy with which the stadia interval is read. If focus is poor, the rod indistinct because of too great a distance, or the rod is not held plumb, the observed distance is uncertain and good results cannot be expected.

2. The greater the vertical angle, the greater the error due to the rod not being held plumb.

3. Focus of the cross hairs should be such that it will give no apparent displacement of objects when the stadia hairs are in sharp focus for the observer.

4. When the objective is focused on the object and the eye moved up and down in front of the eyepiece, there should be no parallax. Parallax is the apparent movement of the image with respect to the cross hairs with movement of the eye and is the result of the image formed by the object glass not being in the same plane with the cross hairs. Sharp focus of the cross hairs will eliminate parallax.

5. When the air is unsteady, with occurrence of heat waves, it should be remembered that the greatest refraction in the line of sight occurs in the lower air stratum, and distances should not be read in which the lower wire is nearer the ground than 2 or 3 feet.

6. For refined work the stadia ratio should be determined for the alidade being used under conditions similar to average field conditions. Midmorning or afternoon is the preferable time of day to make this determination.

7. For elevation work with inclined sights, extreme accuracy in the observed distance is required when the vertical angle is large and a carefully determined vertical angle is necessary when the distance is great.

8. Distance determination will be more accurate if the rod is held plumb, especially on inclined sights, for which either a folding-type rod level can be carried in the field or the Brunton compass can be used to plumb the rod.

Orientation of the Plane-table Board

Plane tabling is a surveying method which is best adapted for geological field mapping. By definition, surveying is the method of locating points on or near the surface of the earth and the measurement of horizontal distances and differences in elevation between these points. Although a line connecting any two points on or near the surface of the earth is a curve, the radius of curvature is so large that whenever the points are close together, the earth can be considered a plane and the line between the points as straight. Surveying projects which encompass relatively small areas and consider the earth as a plane are classified under that branch of surveying designated as *plane surveying* (Bouchard, 1940, p. 1). Geological field work by plane-table methods, generally, can be designated as a type of plane surveying.

Since all surveying methods are subject to errors, a carefully set up and oriented plane table at each instrument station will reduce the chance for errors in points located from that station. The first step in orientation of the plane table requires that the legs of the tripod be placed so that the tripod head is approximately level. Assuming that the tripod used is equipped with a Johnson head, after the plane-table board is attached to the head, the two wing nuts at the base of the Johnson head are loosened, the alidade placed on the board, and the board can then be moved up or down by pressure on the sides until it is level. Then the upper wing nut of the Johnson head is tightened to hold the board firmly in the level position. (Leveling is accomplished by use of the circular level attached to the base of the alidade.) This leaves the lower wing nut still loose, and it is possible to rotate the board in a horizontal plane to obtain the correct orientation.

Proper orientation of the board requires that, once the board is leveled, it should then be rotated to and tightened in such a position that any line drawn on the board and intersecting the instrument station will correspond in direction and be parallel with the line on the ground it represents. It also requires a magnetic north line as well as a true north line on the plane-table sheet, the declination of magnetic north being determined from a recent isogonic chart.

Initial Instrument Station. Every plane-table traverse will generally be started from a known point on the map, which can be plotted arbitrarily on the plane-table sheet in a position favorable for allotting a maximum amount of field data by sight lines from this station to unknown points on the map and for locating additional instrument stations on the plane-table sheet along the traverse line or lines. Elevation of the initial station may be known or assumed. Once the initial instrument station has been plotted on the map, all additional instrument stations will require careful orientation of the plane-table board.

Orientation by Backsighting. Orientation of the plane table by backsighting is the most common method used in detailed plane-table mapping. To orient by this method, the instrument station A must already have been plotted on the plane-table sheet and another instrument station B must also have been previously plotted on the map and must be visible from station A. The straight edge of the alidade can then be placed so that it lies on the pin-point markings of both stations A and B, and the board can then be rotated until the vertical cross hair in the telescope is centered on the marker or flag at station B. The lower wing nut can then be tightened, and the board is correctly oriented. Orientation can be checked then either by sighting in on another previously plotted station or by placing the straight edge along the magnetic north line to see if the compass on the alidade will substantiate a correct orientation.

Another method of backsight orientation is to sight in a point to be occupied as an instrument station and draw a "foresight" ray toward the point emanating from the sighting station *A*. Then, when the point, station *B*, is occupied with the instrument, the board can be oriented by placing the alidade along the ray drawn from station *A* and the leveled board rotated until the vertical hair of the alidade is brought to bear on the marker or flag on station *A*, since station *B* is known to occur along the ray, or line. Once the board has been oriented, either instrument station *B* can be located on the plane-table sheet by a backsight rod shot to station *A* for distance and elevation difference between the two stations or it can be located by resection from other visible points or stations previously plotted on the map. Orientation can be checked by magnetic north and sights on other visible stations once the new instrument station is plotted on the plane-table sheet.

Magnetic Orientation. The box compass attached to the base of the alidade offers a simple and reasonably accurate method for orienting the plane-table board at a new, unplotted instrument station. A magnetic north line should previously have been plotted on the plane-table sheet, generally marked with an arrow on the north end, and the line should be as long as or longer than the base of the alidade being used. After the board has been leveled, the fiducial edge of the alidade is placed along the magnetic north line on the map, and the board then is rotated until the compass needle indicates the magnetic north line to be correctly oriented. The instrument station can then be located by a backsight rod shot to a previously occupied station or by resection.

Location of Stations

Field stations can be located on the plane-table sheet by *radiation*, by *plane-table traverse* (foresight and backsight), and by *triangulation* (intersection and resection). Distances to stations located by *radiation* and *plane-table traverse* are determined with the stadia rod. When stations are located by triangulation, distances can be scaled on the map to determine elevation differences at the intersected or resected points, using the vertical angles recorded when the points are sighted in from the instrument station.

Radiation. If only a small area is to be mapped with the plane table, it may be possible to sight in all necessary points from a centrally located instrument station. This procedure is known as radiation, and for convenience, it is preferable to establish the instrument station on a high topographic point in the center of the area, which allows clear visibility of stations to be occupied by the rodman.

In Figure 10-22 the plane table has been set up at station *A*, which was arbitrarily plotted in the center of the plane-table sheet to allow adequate space to plot additional stations sighted in around it. Stations *B*, *C*, *D*, *E*, and *F* are occupied by the rodman and sighted in from station *A* with the alidade, and rays drawn toward each station, emanating from *A*. Horizontal distance is determined with the stadia rod, and these distances are then scaled and plotted on the corresponding rays, thus locating points *b*, *c*, *d*, *e*, and *f* on the plane-table sheet.

Plane-table Traverse. This method of location, also known as the *method of progression* or the *stadia traverse*, enables the field party to map a series of stations, moving from one station to another with the plane table and finally returning to the original station to close the traverse.

Figure 10-23 illustrates a closed traverse, starting with the original instrument station at *A*, represented by *a* on the map. At the starting point, if no data have yet been plotted on the plane-table sheet, instrument station *a* can be arbitrarily placed on the sheet in a favorable position to allow a maximum amount of data to be mapped on a single sheet and, in this instance, was placed near the left margin of the plane-table board, so that the entire traverse could be mapped on a single plane-table sheet.

For accurate, detailed work, once the plane table has been set up at point *A*, the station arbitrarily located on the plane-table sheet, and a magnetic north line drawn on the map, the following procedure is recommended:

1. If geological data are required in the vicinity of the instrument station, the necessary number of points should be mapped by radiation.
2. Sight in the next desirable instrument location at *B*, draw a ray toward it from *a* on the map, and either read the rod there on a foresight to obtain the distance or set up the plane table at *B*, orient by backsight to *A*, and read the rod at *A* to determine distance.
3. Scale the distance as determined from the stadia rod reading on a ray emanating from station *a* and plot station *b* on the map.
4. Map any geological stations near *B* by radiation, and then proceed to location *C*, following the same procedure.
5. Continue the traverse, and finally return to location *A* from location *D* to close the traverse. Orientation by backsight, along with careful stadia rod readings and

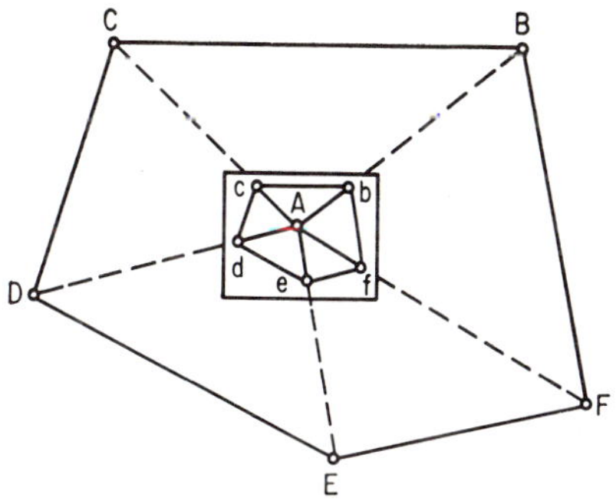

FIGURE 10-22. Radiation with the plane table.

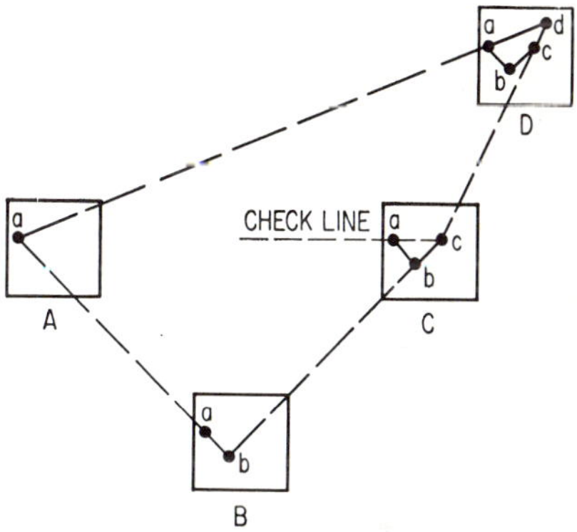

FIGURE 10-23. Mapping by plane-table traverse or progression method. Squares represent locations of plane table.

vertical-angle determinations, will contribute considerably to a minimum error in horizontal and vertical closure of the traverse.

A more rapid method, however, if no instrument station is required at station *B*, is to read the stadia rod at *B* and plot the station on the plane-table sheet, designating it as a *turning point*. This is known as a foresight reading. The instrument man then proceeds to location *C*, sets up the plane table there, orients a magnetic north, and again reads the rod at location *B* on a backsight reading. Distance is scaled on a ray from *B* toward *C* (*b* and *c* on map, Figure 10-23) drawn on the plane table sheet when the rod is read, and *c* is then plotted on the map. Although this procedure will speed up the progress of the traverse, magnetic orientation may increase the possibility of errors in closing the traverse unless extreme care is exercised when leveling and orienting the board.

Occasionally a check line may be drawn on the map, as shown at location *C* (Figure 10-23), whenever a previously occupied station is visible in order to substantiate correct orientation and location.

Triangulation. This method permits the location of stations on the plane-table sheet with the alidade without the necessity of a rod reading at the point being located. In general, triangulation is divided into two separate methods:

1. *Intersection,* in which unknown points on the map are sighted in from a previously located instrument station

2. *Resection,* in which an unknown instrument station location is sighted in from known points previously located on the map.

Bouchard (1941, p. 287) points out that the ease with which points can be located by *intersection* is one of the decided advantages of the plane-table method. A diagram illustrating, in simplified form, the method of intersection is shown in Figure 10-24. If the board is set up and oriented at location *A*, with the station arbitrarily plotted or previously located on the plane-table sheet as *a*, the alidade is pivoted at *a* to sight in and draw rays of indefinite length toward *C*, *D*, and *E*. Then the plane table is moved and set up at location *B*, which either is located by a foresight and rod reading from *A* or was previously located and plotted on the map as *b*. Locations *C*, *D*, and *E* are then sighted in, and rays drawn toward them on the map from *b*. The intersections of the rays with those drawn from *a*, on the plane-table sheet, determine the positions of *c*, *d*, and *e* on the map, as shown at the plane-table location *B* in Figure 10-24.

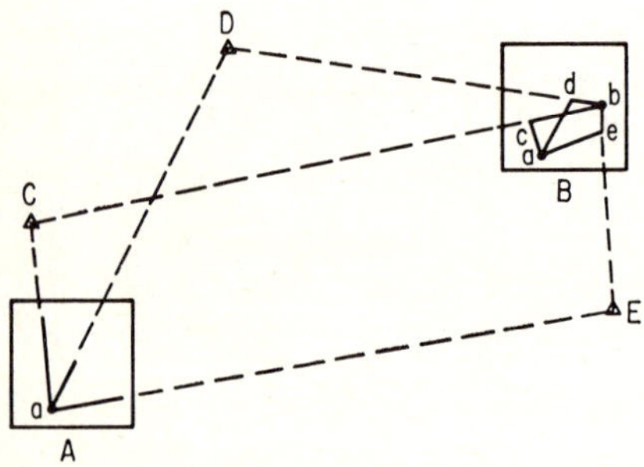

FIGURE 10-24. Mapping by intersection. Squares represent plane-table locations.

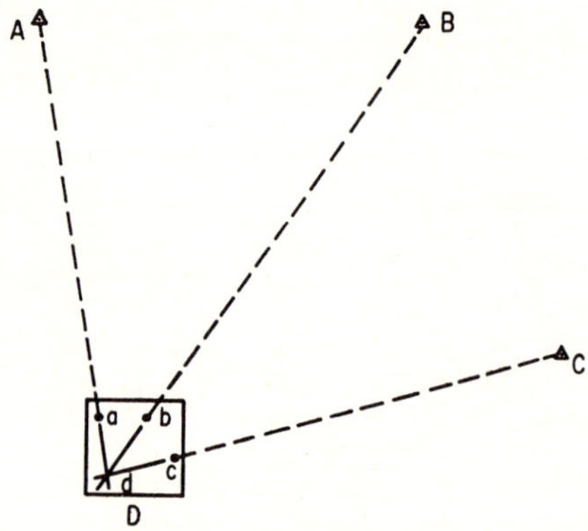

FIGURE 10-25. Location by resection. Squares represent plane-table locations.

If the points located by intersection are on inclined lines of sights, no horizontal correction is required and the elevation difference can be determined by the formula

$$E = \tan AD$$

in which *A* is the vertical angle, *D* is the distance scaled on the map once the point is located by intersection, and *E* is the elevation difference obtained. Since the location of a point by intersection requires a line of sight from two known instrument station locations, the elevation difference should be computed from both instrument stations as a check for accuracy.

The *method of resection* permits the occupying of a station with the plane table when that point has not been previously located on the map and is of great advantage in an extensive survey (Bouchard, 1941, p. 288). To locate an instrument station by resection, the plane table is set up at a convenient point which allows visibility of two or more previously located and plotted triangulation or rod stations. The board is leveled and oriented on magnetic north on the new instrument station, and the known points are then sighted in with the alidade and rays drawn from them toward the unknown point. The intersection of rays on the plane-table sheet determines the location of the instrument station.

Figure 10-25 illustrates the location of an instrument station by resection. The board, at location *D*, was first oriented on magnetic north; then locations *A*, *B*, and *C* are sighted in; and rays drawn on the plane-table sheets from *a*, *b*, and *c*, along the lines of sight, determine the location of *D* to be in the lower left corner of the plane-table sheet where they intersect at *d*.

When determining elevations by intersection and resection, it should be remembered that lines of sight for intersection are *foresights* while those for resection are *backsights*. Therefore, plus angles give higher elevations and minus angles lower elevations when points are located by intersection, whereas the reverse holds true for resection.

The Three-point Problem. When the method of resection is used to locate plane-table stations, if the points are located by only two intersecting rays on the plane-table sheet, no correction can be made and it must be assumed that the locations are reasonably accurate. For precise work locations by triangulation should be made with three intersecting rays wherever possible. If the plane table has been accurately oriented on magnetic north or by backsight at each instrument station, the three rays should intersect at a common point.

If the board has not been properly oriented, the three rays will form a *triangle of error* the size of which will be governed by the accuracy of the plane-table orientation. A slight rotation of the table to the left or right may resolve the triangle to a point when sighted in again, and no additional correction is needed. If the triangle of error cannot be resolved by a slight rotation of the board, it requires a solution of the *three-point problem*, which will resolve the three points of the triangle of error to a common point.

In a discussion of the three-point problem, Mather and Washburn (1938, p. 17) point out that the solution of the problem is more difficult for an inexperienced surveyor but that it can always be accurately solved if the following four axioms are kept in mind.

1. If the triangle of error lies *inside* the triangle which can be drawn by connecting the three control points *A*, *B*, and *C*, the *position of P is inside the triangle of error.*

2. If the triangle of error lies *outside* the triangle formed by lines joining *A*, *B*, and *C*, the *true position of P lies outside the triangle of error* and in one of the two sectors to the left or to the right of all the rays. (The left or right of the ray is determined when one looks along this ray toward the point from which it emanates.)

3. In either case 1 or case 2 the final position of *P* is found by the following law: The distances measured perpendicularly to *P* from the three rays which form the triangle of error are exactly proportional to the lengths of these rays.

4. There is no possible solution for the three-point problem if the three known points *A*, *B*, and *C* and the observer's true position *P* lie on or even approach the circumference of a circle. This is known as the *revolver case.*

Circumstances described in cases 1, 2, and 4 above are illustrated in Figure 10-26. Diagram I shows the triangle of error which might occur when the plane table is within the triangle formed by joining the three stations *A*, *B*, and *C* with lines to form a triangle and the true location *P* is within the *triangle of error*, as described in the first axiom. Diagram II illustrates a circumstance where the plane table lies outside the triangle formed by joining the three stations *A*, *B*, and *C* and the true location *P* is outside the triangle of error, as in the second axiom. Diagram III is a *revolver case,* where the true position of *P* as well as stations *A*, *B*, and *C* lie on the circumference of a circle, as discussed in the fourth axiom, for which there is no solution (Mather and Washburn, 1938, p. 18).

There are several methods for solving the three-point problem, one of which will work fairly well for diagram I, Figure 10-26. In this instance, knowing that the true location of *P* lies inside the triangle of error, an arbitrary point *P* can be chosen inside the triangle of error and plotted on the map. A line is then drawn to *B*, which is the greatest distance from *P*; the edge of the alidade can then be placed along the line *PB*; and the board rotated until the sights are in line with *B*. The point can then be checked by sighting along the lines from *P* to *A* and from *P* to *C*, and if there is still a triangle of error, the procedure can be repeated (after Lahee, 1961, p. 534).

The three-point problem can also be solved graphically, directly on the plane-table sheet, by rotating the board slightly and plotting three more resection lines from points *A*, *B*, and *C* to obtain another triangle of error. If this triangle is larger than the original triangle of error, the board should be rotated in the opposite direction; if the triangle is smaller, the process should continue until the triangle of error will reach a minimum size, after which it will become larger again with further rotation of the board in the same direction and will be *inverted*. To determine the correct position for *P*, straight lines can then be drawn connecting the corresponding apices of the

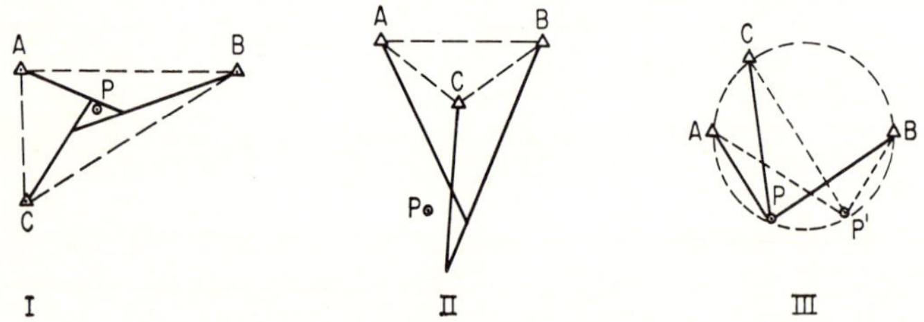

FIGURE 10-26. Diagrams illustrating the *three-point* problem. (I) Plane table is within the triangle formed by joining the three foresight stations *A*, *B*, and *C*; the true location *P* is within the *triangle of error*. (II) Plane table is outside the triangle formed by joining three foresight stations *A*, *B*, and *C*; the true location *P* is outside the triangle of error. (III) The *revolver case*. (*From Mather and Washburn, 1938.*)

normal and the inverted triangles of error, and the intersection of these three lines will be the correct location for point *P*. The board can then be oriented by sighting from point *P* to the most distant control point and then checked by sighting at the other two stations used for the resection location (Lahee, 1961, p. 535).

The most frequently used method for solving the three-point problem is the tracing-paper method. If the board is first oriented with the compass, a piece of tracing paper can then be placed over the plane-table sheet and secured with drafting tape or scotch tape. An arbitrary point can then be selected on the tracing paper to plot point *P*, from which points *A*, *B*, and *C* can be sighted in with the alidade and rays drawn toward the three points from *P*. The tracing paper is then loosened and can be moved around on the top of the plane table until a position is found where all the rays drawn from point *P* pass on the underlying plane-table sheet by a pin prick through the tracing paper. The table can then be reoriented by sighting on the most distant station from point *P*, and the orientation and location checked by backsight on the other two stations.

SUPPLIES

It is nearly impossible to compile a complete list of the supplies required by a geologic field party, inasmuch as the items needed vary considerably depending upon the nature of the field work and its objectives, the climate of the area to be worked, character of the topography and vegetal cover, and the accessibility and means of transportation. These and other factors, when taken into consideration, may limit the amount of material supplies that can be carried or may dictate certain requirements that make it necessary to eliminate other desirable items that could be used but are not essential to the project. Hence, the best approach is to compile a check list of field supplies from which the individual geologist can select those which the field party will require and then may include additional items that will be helpful to the field party and that will not be a burden when carried in the field. Therefore, the following list is included, from which selected items are discussed in detail in the following paragraphs.

Camera, exposure meter, tripod
50- or 100-foot tape
Bunting or strip cloth
Hatchet or axe
Geologic hammer
Hand lens
Notebook and pencil case to carry on belt
Rulers and scales
Protractor and triangles
Drafting set, inks
First-aid and snake-bite kits
Sample bags
Rucksacks and pack boards
Pocket stereoscope
Clip board
Field notebooks
Engineering tables
Sketching pads
Tracing paper, fade-out transparent cross-sectioned paper, graph paper
Plane-table sheets
Air photographs and mosaics
Topographic maps
Planimetric maps or base maps
Pencils, 2H to 9H
Colored pencil set
Pocket knife
Sandpaper
Erasers
Needles and pins
Drafting tape and scotch tape

Most of the items on the check list are standard items that do not vary in specifications, such as the 50- or 100-foot tape, geologic hammer, hatchet, hand lens, etc., while others may vary according to what is preferred by the geologist in such items as rules and scales, protractors, triangles, drafting sets, watch, camera, etc. The need for other items such as engineering tables, maps, air photographs, and mosaics, as well as instruments discussed in previous sections, may be dictated by the objectives of the field project.

Although many of the items on the check list may vary in type, depending upon the geologist's preference, certain suggestions for notebooks and equipment, plane-table sheets, tracing paper, and sketching pads are offered as an aid in their selection.

Dependent upon the inaccessibility of the country and absence of geodetic control, initial geologic work must of necessity be preceded by geodetic location mapping of high-order precision. In such completely uncontrolled areas latitude and longitude may be determined by star sights with the French astrolabe or the wild-type first-order theodolite. After base-line measurement and sun-line orientation, a survey net may be erected and expanded with theodolites by triangulation, or trilateration nets may be expanded through use of the tellurometer, geodimeter, or terrameter (Marussi, 1960).

Notebooks and Related Equipment

The standard-size field notebook used by most geologists is approximately 5 to 5½ by 8 to 8½ inches in size, bound, with lined or cross-section divided pages. The notebook may be bound so that it opens lengthwise, which is convenient if the book is to be used for plane-table work, or it may be bound along the side, which is more convenient for general reconnaissance geology.

Since one of the problems of field geology is loss of notes or damage in the field, a bound notebook requires that each day's field notes be copied into another notebook each evening, which can be tedious and use valuable time that could be better spent working on maps or geological data. Therefore, one of the most favorable types of notebook is a 4- by 7-inch six-ring loose-leaf notebook, for which lined, blank, and cross-sectioned pages can be obtained, as well as plane-table notebook pages which will fit in this size loose-leaf notebook. If this type is used, an extra cover can be kept in camp or wherever the party is based, and each evening, those pages of notes which are no longer required in the field can be transferred to the other notebook. Loose-leaf notebooks are also available in the larger, standard field notebook size, if preferred, from standard manufacturers of engineering field notebooks.

If either size notebook discussed is used, a leather carrying case, also designed to hold pencils, scales, protractors, and perhaps a small circular slide rule, is very convenient. The notebook case should have a loop for fastening to the belt for carrying convenience. If an abundance of extra items is needed in the field, a surveyor's vest might be preferable, as in addition to places for notebooks, pencils, and scales, this vest has extra pockets in which other equipment can be carried. If the work is to be in a warm climate, a normal-size surveyor's vest can be used; if in a cold climate, or if cold weather is expected, the vest should be large enough to fit over the heaviest clothing one expects to wear.

For barometer surveys, the most convenient method for keeping notes is to use a clip board and a stenographer's 8½- by 6-inch spiral-type dictation notebook, which can be carried on the clip board. In addition, a few sheets of 8½- by 11-inch transparent graph paper can also be carried on the clip board if the barometer curve is to be plotted in the field. Then each evening the barometer data notes can be torn out of the spiral notebook and placed in a separate folder with the barometer curve for the day. If this type of equipment is used, it can be conveniently carried in the field, with other items, in a canvas musette bag or some similar canvas carrying case that can be slung over one's shoulder. The use of transparent graph paper in the field permits a ready comparison with the other barometer curves, and any daily curve can be reproduced without additional drafting if required for a progress report.

Pencils, considered part of notebook equipment, vary in requirement, 8H to 9H preferred for work on plane-table sheets, 3H or 4H for work on air photographs or mosaics, 2H or 3H for recording notes and plotting barometer curves. A pocket knife and sandpaper should be carried to maintain proper sharpness on the pencils to be used. For work on air photographs, mosaics, or topographic maps, a rapidograph-type pen can be carried in the field to ink stations and numbers, as pencil points may rub off during the day's work if numerous stations are plotted on the map.

Standard plane-table notebooks are available from several sources, the choice of which rests with the party chief. Forms for the notes vary slightly with different manufacturers, but in general any standardized plane-table notebook will be satisfactory, depending on the preference of the instrument man. Plane-table notebooks are also available in a bound or a loose-leaf type.

Plane-table Sheets

Standard plane-table sheets are generally made with a good grade of drawing paper mounted on a strong cloth. *Single-mounted* sheets have the drawing paper on one side only, while *double-mounted* sheets have the drawing paper mounted on both sides of the cloth.

The cloth-mounted paper may be subject to shrinkage or expansion due to local atmospheric conditions. Although this may be avoided, in part, by exposing them to the air for a few days before using the sheets, there is no way to adjust to variations in local weather conditions, and heavy summer thundershowers in desert areas may cause considerable unexpected shrinkage.

A new material now available for plane-table sheets is *stabilene,* a heavy, transparent plastic, which is not subject to shrinkage or expansion and will not be damaged if not covered immediately in a sudden rainfall. This type of plane-table sheet is highly valuable in areas of extreme weather and temperature variations, especially when exact detail work is required. Another advantage is the transparency of the stabilene, which permits reproduction of the original field plane-table sheet if required for a progress or final report

Tracing Paper and Sketching Pads

A roll of plain white tracing paper should be carried in the field also, especially if reproducible base maps are not available. The tracing paper can be used in solving three-point problems or for compiling field data onto a single map from plane-table sheets or air photomosaics.

A type of tracing paper that may prove most useful for the field project is a 36- or 48-inch roll of transparent, fade-out, cross-sectioned paper onto which maps can be traced or cross sections can be readily constructed in the field. When reproduced, the cross-section lines fade out, giving a clear print of the map or cross section drawn. The paper, divided into square inches and tenths of inches, enables the field geologist to construct cross sections with a minimum of effort spent on scaling in points, and the finished field drawing, if done with care, can be used as an illustration for the final report without complete redrafting, which can save valuable time in finalizing a field project.

Sketching pads, 8½ by 11 inches in size, can be carried on a clip board in the field and in some cases should be a light tracing paper rather than heavy drawing paper so that field sketches can also be reproduced for progress or final reports without redrafting if time is essential. Fade-out, cross-sectioned paper is also available in 8½- by 11-inch pads, and in many instances it is preferable to clear tracing paper for field sketching, as it allows some estimate of scale for the drawing if required.

PRE-FIELD EXPLORATION REPORT

Major field projects are usually proposed by a district or division office within an oil company and submitted to the main office for approval before work is commenced. In many instances the proposed project is the result of recommendations by a district-office geologist who is familiar with the project area and who may be the party chief, since the field project arose through his work and knowledge of the area and his

Proposed Field Project

I. *Name of Project*

II. *Area Involved*

(1) Location, with approximate townships and ranges, county and state, and estimate of number of square miles.

III. *Land Situation*

(1) Type of lease available: Federal, State, Fee, or Indian tribal or allotted lands.

(2) Approximate percentage of land under lease.

(3) Availability of land already under lease for drilling deals or purchase.

IV. *Time Required*

(1) Estimated number of days or weeks to complete project.

V. *Number of Field Personnel Required*

(1) Availability of field party personnel in local office.

VI. *Supporting Discussion*

(1) Geological data supporting need of project.

(2) Project objectives.

(3) Field methods to be used.

FIGURE 10-27. Example of usable short outline form for preparation of a field project by a district office.

realization that the project was required to further the company's search for oil in the district.

Whether or not this is the circumstance or perhaps a new man is moved into the district to act as party chief for the field project, it is essential that the geologist supervising the field party should have a good outline of the project objectives so that he understands the problems involved, the type of information to be recorded, the types of map needed, and the amount of area to be covered. Size of the area is very important in determining the scale of the final map, as it is always best to have only one scale map, if possible, to submit with the final report, providing the area covered is not so large as to require too small a scale to include it all. Simplicity in final presentation is the best procedure if the field work results in data which the party chief wishes to emphasize and perhaps use as a basis for recommending additional field work, geophysical work, or acreage purchase (Figure 10-27).

Examination of Prior Work

Before field work is commenced, a search of the company files should first be made to determine if any previous work, perhaps only a very rapid reconnaissance, has been done by company geologists. Perhaps, if the district office has been in the location for some length of time, an earlier geologist spent only a short time going through the project area but, while doing so, took the time to stop and record data on geological phenomena observed which later became the basis for the recommendation of the field project, or he may have noted on a map the location of a well-exposed stratigraphic section which might serve as a good starting point for the field project.

Literature Search

A search of the geological literature on the area should also be made, and a bibliography compiled. Some of the literature may be helpful in locating type sections which should be included in the project or may contain geological maps of small parts of the project area, with possibly some structural data. All published articles on the area will usually include important data which will be useful for geological orientation in the field, especially if the party chief is unfamiliar with the project area.

Aim of Project

It is also necessary to understand the scope of the project thoroughly in order to determine the amount of materials needed and the expense warranted. For example, if the project is to be a very rapid reconnaissance of a large area, covering only major stratigraphic sequences and breaks and major structural elements, the immediate purchase of contact prints of air photographs may not be warranted and photomosaics would suffice. On the other hand, if the project is to include detailed stratigraphy and structure, contact stereographic air photographs are necessary and should be included as field party needs, as well as photomosaics.

Photographs and Maps

Previous to leaving for the field, if contact prints are to be used in conjunction with mosaics, these should be examined under a stereoscope and strikes and dips of beds determined to be transferred to the mosaics, along with major structural axes and faults that can be readily identified. If mapping is planned directly on the mosaics in the field, only a sufficient number of dips to establish regional trends and structures should be placed on them as guides for the work unless an overlay of acetate is to be used. If only mosaics are available for field use, it is still possible to determine some of the major structures and faults on these and to estimate directions of strike and dip from outcrop and drainage patterns.

It is also possible before leaving for the field to select several potential mappable units from observation of the contact prints or mosaics. Although some may not prove satisfactory as mapping units when observed in the field, if they are already outlined on the mosaics when the party first goes into the field, selection of a mappable unit for structural work is much easier and intervals can be checked for "jumping" units and reducing them to a common datum.

Contact prints, mosaics, and topographic maps are also very helpful for checking accessibility of the field project area and should be studied in the office prior to the field work. If county road maps are available for the area, a thorough knowledge of existing roads and nature of the topography will facilitate a well-coordinated plan for the entire field project, including means of transportation to be used, field methods and equipment, and the best bases of operation. The last, of course, includes the consideration of whether the field party is to be based in town or perhaps will have to operate from a field camp.

Reconnaissance Flight

Accessibility of the project area should also be checked from the air by small plane if possible. In a flight over a new area with the mosaics as a map and guide, many small jeep trails and roads not shown on existing maps can be observed and should be traced out on the mosaics to facilitate finding them when working on the ground. The importance of a preliminary air reconnaissance of a field project area cannot be overemphasized, as it may be the party chief's first view of the area. He will see for the first time many of the things he has already observed on contact prints and mosaics, and he may well change or resolve his project plans based on what is seen on the flight.

Land Net and Elevations

If the project area is in surveyed land, with section, township, and range, land plats should be obtained for the area and, if possible, the township lines should be traced out on the mosaics in some temporary form until some section corners are located in the field and accurately placed on the mosaic. If the township lines are carefully placed on a controlled photomosaic while still in the office, it will usually be found that the degree of error will be slight once horizontal control is established in the field.

If controlled elevations are needed for the field project, available data on U.S. Coast and Geodetic Survey or U.S. Geological Survey bench marks should be obtained and the approximate locations scaled in and designated on the mosaics. A line of bench marks is generally along a secondary or main highway with approximate locations given with respect to distance from a known point and are usually easy to find by driving the highway and carefully checking the mileage as indicated on the data sheets. Where U.S. Coast and Geodetic Survey triangulation stations are present in the area, data should also be obtained on the locations and elevations to serve as both horizontal and vertical control check points if plane table and alidade are to be used in the field.

The preceding paragraphs have been written with the assumption that the objectives of the field project are only a general reconnaissance or detailed reconnaissance of the area, contact prints and photomosaics being used on which to map and record data. If the project is to be detailed in nature, with the use of plane table and alidade with closed traverses for structural geology, areal geology, or stratigraphy, the same preliminary work should be done, as the same basic problems will arise with respect to previous work, accessibility, mappable units, bases for operation, horizontal control, and vertical control.

Once these items have been covered in the office, the party chief can then select the field equipment and type of transportation needed, based on his findings from the preliminary work and on his knowledge of the needs to fulfill the objectives of the field project.

Time Schedule

A final important part of any field project which should be discussed is planning and timing with respect to amount of area to be covered and time to be allowed. This is especially important in areas with short field seasons because of heavy rains or snow, while in some other areas where field work can be done on a year-round basis, timing might still be an important factor to retain a lead over competing companies.

No set rules for planning and timing can be set forth, since there are so many variable factors to be considered, the two most important of which are (1) the party chief's ability and confidence in his ability and (2) the nature and objectives of the field project.

Assuming that the field project is in an area in which the company has done no previous geological work but that the nature and objectives of the project are along the lines of specialization of the party chief, he might set up his planning in this manner:

1. Divide the project area into smaller areas on a base map, based on means of access or possibly changes in bases of operations.
2. Record all known geological data on each smaller area, and outline what field data might be expected in the area that must be evaluated by the field party.
3. Based on previous experience, estimate the amount of time necessary for each small area, taking into consideration travel time, accessibility, and amount of detail that can be obtained.

Party No. ____________ Date ________________

WEATHER CONDITIONS

Temperature ______________________________

Atmosphere ______________________________

____

GEOLOGICAL FIELD WORK

Location ______________________________

Description ______________________________

____

Personnel ______________________________

Hours Worked ______________________________ | |

GEOLOGICAL OFFICE WORK

Description ______________________________

____

Personnel ______________________________

Hours Worked ______________________________ | |

MISCELLANEOUS WORK

Location ______________________________

Description ______________________________

____

____

____

Personnel ______________________________

Hours Worked ______________________________ | |

REMARKS

____

____

____

____

____

____

Signed

FIGURE 10-28. Sample sheet of field diary. (*Courtesy of Continental Oil Company.*)

If all these items are summed up in a short outline form, a tentative schedule for the entire field project can be established, and every possible effort should be made to remain on schedule. If weather conditions present a problem in the area, an allowance of 10 to 20 per cent additional time may be justified.

For the benefit of geologists who might do additional work in the area or similar areas in the future, a well-kept diary, separate from field notes, should be maintained. Important items that should be recorded in the diary each day are such things as weather conditions, amount of time spent in the field and nature of work, amount of time spent on maps in evening, any time spent on miscellaneous work not pertinent to geology but necessary to the project, as well as general remarks as to minor or major problems that arise which have been solved in the field or require solution. Recording of this amount of information may appear tedious at first, but if the party chief had had access to a similar diary before starting the project, perhaps many of his problems would have been readily solved, and realization of this will tend to make the task of keeping a field diary a little less tedious. Figure 10-28 shows one type of diary sheet.

AIR PHOTOGRAPHS IN SURFACE MAPPING

Advantages of Air Photographs

Separation of engineering and geological duties in many oil companies has resulted in application of engineering staffs to varied and diverse problems of producing economically the oil found by geologists, whereas surface geologists at times do not utilize necessary engineering principles applicable to air photographs. This gradual process has led engineers away from those features of exploration which require air photogrammetric engineering, and it has led geologists away from the continued increase of engineering knowledge which would enable them to apply basic air mapping principles to the geologic interpretation of air photographs in preparation for surface exploration. Many geologists have become photogrammetric engineers of high caliber, and a byproduct of this engineering use of air photographs has been a keener interpretative sense for what was recorded on the photographs. New engineering methods will be invented for making better maps from a set of photographs, but the quality of final geologic interpretation in preparation for surface mapping is directly dependent upon the experience of the geologist who examines the photographs.

There are four basic advantages which air photographs present to the surface geologist, namely, speed of coverage, accuracy of the maps, continuous reference, and accessibility of the air surveying technique to all terrains. In spite of these advantages air photographs will never completely replace ground surveying and detailed surface geology in problem areas. A geologist's interpretation and detailed delineation of geologic contacts on vertical air photographs at a large scale with subsequent transfer by eye to a small-scale base map usually result in considerable inaccuracy. Adequate time for engineering compilation of photographs is very often considered unnecessary. Maps are generally wanted "day before yesterday"—usually owing to poor planning. The compilation time for an air photographic project is often used as an argument against air mapping. However, vertical air photographs compiled into accurate base maps and used in conjunction with field examinations can save surface parties and geophysical parties much time and money in the field. Accuracy of maps constructed from air photographs with an optimum of field control is greater than plane-table surveying under any given set of conditions and specifications.

Identification of geologic features and accurate geologic delineation of those features on base maps were formerly two widely separated functions. The geologist of today who deals with surface geologic features in attempts to bring virgin territory into oil production must know how to do both. In certain terrains, the surface geologist can do both so much more rapidly and accurately and thus minimize field work.

Air photographs are a continuous source of reference. If an area has been photographed from the air at a reasonable scale, those photographs are a continuous source of both geologic and engineering data. A final advantage is the single one of accessibility, for photographs can be taken of terrain where horse and man cannot go and many disheartening months of bush-whacking can be avoided in field work if the problem areas are found by prior photographic interpretation. Many areas can easily be interpreted and mapped by photographic evidence without field examination but must be checked by the surface geologist.

Application of photogrammetric techniques is vital to many phases of the oil industry which require good maps. On the other hand, few new techniques have been discovered in the field of geologic interpretation of air photographs. Some surface geologists, who see in the air photograph a serious contender to detailed field mapping, are not so sure that there is much on a photograph that will help in solving the vast problems of geologic mapping. This type of geologist uses the photograph as a direct map to help him get where he wants to go and fails to recognize the wealth of leg-saving detail which the air photograph affords. There is no substitute for photogrammetric training combined with field mapping experience for the modern surface geologist. His interpretative technique grows as his field experience widens; hence the development of new interpretative techniques is a personal factor based on reconnaissance and detailed geologic mapping.

The end result of a carefully planned photogrammetric program is a series of maps at a suitable scale showing natural features in their true relation to one another. If the map shows these features without relief, it is generally known as a *planimetric base map.* If relief is shown by contours or form lines, it is a *topographic base map.* The geologist, satisfied with fewer elevation data than a topographic map provides, can construct a good structural contour map by combining his geologic contact interpretation with data gained by precise photogrammetric measurements from air photographs. Accurate base maps constructed from air photographs are valuable to geologists as base maps for plane-table mapping of geology in areas which cannot be interpreted accurately from photographs or as base maps for plotting contact elevations in those areas where interpretation is possible with a minimum of field examination.

Air-Photogrammetric Maps

The Vertical Photograph. The truly vertical photograph is somewhat of an accident, for seldom can the camera be leveled exactly during air photography. Knowing that all photographs are tilted to some degree, it is then important for the surface geologist to realize the errors that such a photograph produces and how much tilt can be rectified by simple radial-line triangulation before detrimental inaccuracies are incorporated in a map. Definition of some of the common terms in orienting an air photograph is necessary for those not acquainted with the photogrammetric processes.

Principal point: the center of the photograph, found by drawing lines between special collimation remarks recorded on each air photograph

Plumb point: the point on the photograph over which the plane was flying when the photograph was taken

If the photograph was taken with the optical axis of the camera truly vertical, the principal point and the plumb point are the same and the photograph is a true vertical.

Tilt: a resultant of longitudinal tip and lateral list and a measure in degrees of the departure of the photograph from true horizontality

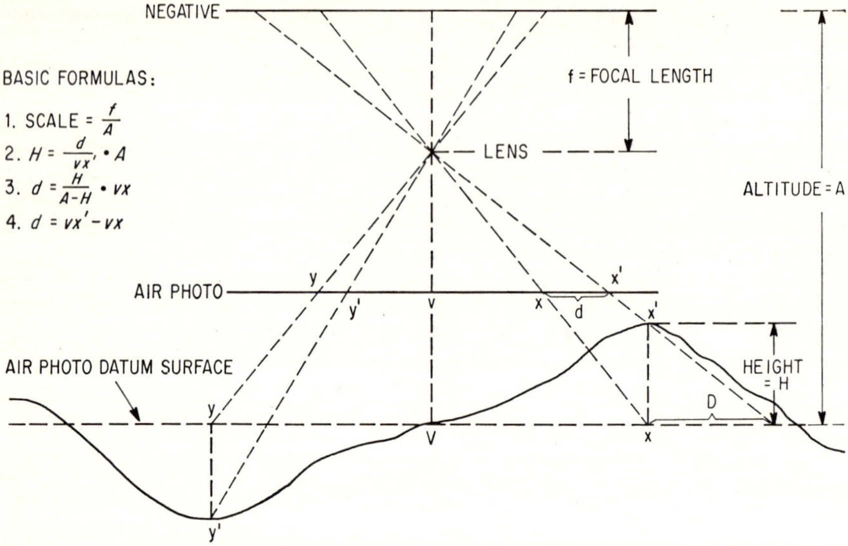

FIGURE 10-29. Displacement of air-photographic images due to topographic relief are radial from the plumb point *V* which is the principal point of a vertical photograph. (*After Smith, 1943.*)

Isocenter: a point lying on the junction of the axis of tilt and the principal line

Principal line: a line passing at right angles to the axis of tilt and extending through the plumb point and the principal point

Isoline: a line passing through the isocenter at right angles to the principal line

Focal Length: distance from the rear nodal point in the lens to the film in the camera

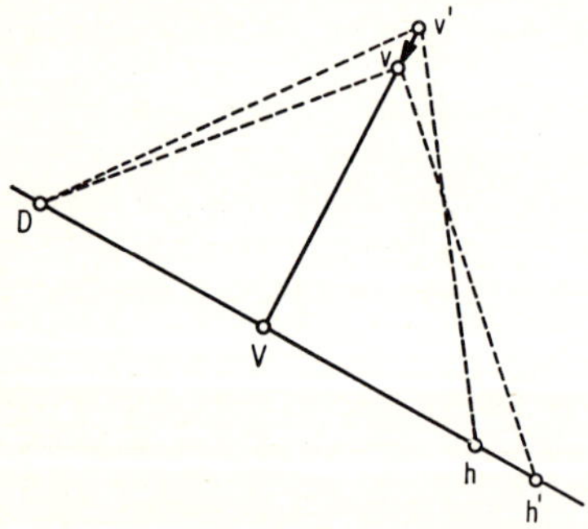

FIGURE 10-30. Displacement of air-photographic images in relation to air-photographic datum. *V* is principal point; *v'* is photographic position of bottom of valley; *v* is true position of valley bottom; *h'* is photographic position of top of hill; *h* is true position of hill top. Note that *Dv'* is not a true direction in space. (*After Smith, 1943.*)

The isocenter is found by dropping the bisector of the angle between the optical axis and the plumb line to the photograph. That point is one of the most important ones in air photogrammetry. An important rule to remember is that *tilt causes image distortions which extend radially from the isocenter* whereas *distortions due to relief extend radially from the plumb point.* It is then simple to imagine how difficult it is to find out exactly where the top of a hill is in a tilted photograph. In an untilted photograph the relief is radial from the center of the photograph (principal point), as shown in Figure 10-29, and the displacement can be computed by simple trigonometry, using focal length of the camera, height of the hill, and altitude of the airplane. Simple rectifying projectors have been designed to eliminate tilt from air photographs. Distortions due to relief may be eliminated in the radial-line process but are always present to a degree in all air photographic mosaics.

Simpler photogrammetric instruments include small refracting stereoscopes and mirror stereoscopes. Their use is world-wide, and though they are very necessary to many stages in radial-line triangulation and photographic interpretation, few new techniques have been developed for their use. Radial-line triangulation is a well-known process which has wide application. Simple geometric principles led early workers to realize that "coupled" stereoscopic air photographs have survey "strength." In other words, the photograph covered enough country to enable an interpreter to pick strong intersection points falling on surrounding photographs. With 60 per cent forward overlap and 30 per cent side lap, a maximum of strength was gained for picking points common to successive photographs. The principle of three-point resection was applied with the result that many photographs could be related in scale

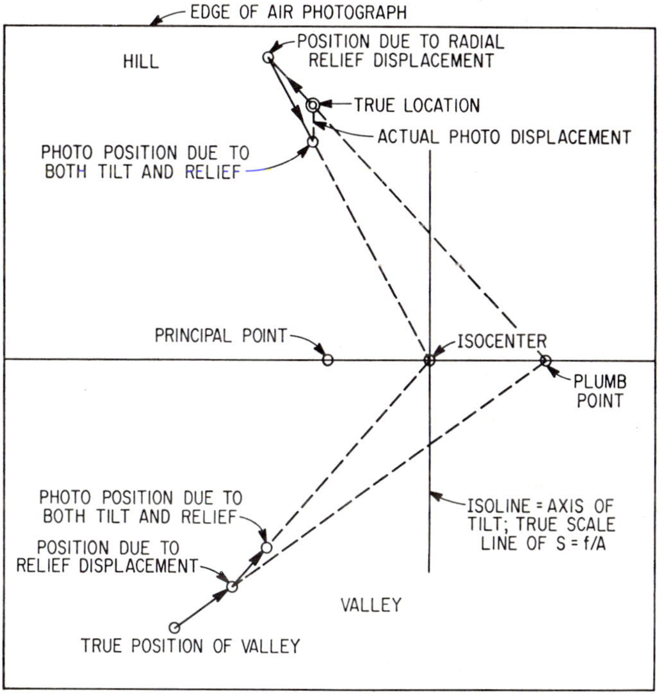

FIGURE 10-31. Positions of relief features on tilted photograph. (*After Smith, 1943.*)

despite varying scales of flight. Because the radial-line triangulation system assumes that the principal point is very nearly the plumb point, it must also assume that all displacement of photographic image due to relief and tilt radiates from some point very near center of the photograph. Air photographs may be overlaid by acetate, and lines drawn radially from the principal point to each chosen point; slots radiating from the center to those points may be cut in cardboard; or mechanical slotted guides may be laid to form a template right over the photograph. These techniques for getting an even scale all over a photographic project depend upon at least one control point for each eight pictures before the whole template assembly is laid down on a Polyconic or Lambert conformal projection. Pegs representing field control are fastened to the projection board, and all the templates are laid surrounding the control. The radially intersected points are then pegged firmly to the layout board, where they constitute accurate secondary control located on the projection of that section of the earth's crust. Radial-line triangulation is a graphic means of rectifying tilts below 3 degrees and radial distortion caused by relief. After templates are all correctly related to one another, they are taken up and all radially intersected points marked. Acetate overlays are made onto which all planimetric data desired from the photographs can be drawn by use of additional detail point intersections, the vertical sketchmaster, or the reflecting projector. The end result is a planimetric map showing everything but contours.

In construction of air mosaics, the scale changes of the photographs can be computed from radially intersected points and the photographs enlarged or reduced (ratioed) to fit the projection scale. The mosaic is then laid picture by picture, great care being used to match for both scale and image coincidence. If sufficient control is available, this can rightfully be called a semicontrolled mosaic. This type of mosaic is much more accurate than the simple stapled-down mosaic which is constructed by image matching without control.

Elementary elevation-determination processes are not nearly so accurate for vertical control as is the radial-line process for horizontal control. Two reasons for this are immediately apparent. *First,* the person using the map often does not care if detail is horizontally mislocated by 100 feet, but he is particularly anxious that a certain contour interval be used and that all points be within half a contour interval of their true vertical position. *Second,* horizontal positions are gained by broad intersection angles at the ends of lines several thousand feet long, whereas vertical differences are on the order of tens of feet measured by minute parallax differences and differential radial displacements between two photographs. Inasmuch as these small differences represent many feet vertically, a small error or even variations within the plotting error of a pencil-line width will throw a parallax-measured point off by two or three contour intervals. Hence no simple accurate graphic method exists for gaining data to locate geologic contacts to a vertical accuracy of 10 feet unless the photographs are flown too low to be practical in most other respects. The *Contour-finder* and *Stereocomparagraph* can be used for form-lining the relief of a photographed area or in spot measurements of the elevations of geologic contacts. Because the photographs used are not corrected for tilt, and because the tables now widely extant to facilitate these elevation-measuring methods are unreliable, the methods can be relied upon to be accurate only to a footage interval no less than $\frac{1}{250}$ the flight altitude of the photography in areas where abundant vertical ground control is available. However, structural form lines can be plotted on a radial-line triangulation planimetric base sheet if the geologic contacts are transferred from the photographs to the base sheets. Stereocomparagraph operators can become proficient in handling these simpler elevation-measuring instruments in about 2 weeks. A great disadvantage to this type of form-line plotting is the fact that the form lines are on a perspective scale and hundreds of added detail points must be radially intersected to scale both the geologic contacts and the structural form lines to an accurate planimetric base map. Tilt of 3 degrees in a 12-inch-focal-length 9- by 9-inch photograph taken at 10,000 feet, if uncorrected before plotting or not compensated for by a correction graph during contouring, can cause vertical errors of 200 feet in drawing geologic contacts and structural form lines near the edge of the photographs analyzed by means of Stereocomparagraph or Contour-finder. Tilt is the problem which led to development of highly refined instruments for more accurate mapping from air photographs.

The controlling factors of flight altitude, terrain conditions, and map accuracy desired regulate the type of air photography used. A 500 to 1,500 multiple of the specified elevation accuracy is now being used to compute flight altitudes, dependent upon the reconnaissance or detailed nature of final results. Thus, if a 10- to 30-foot accuracy is desired, the plane may fly at altitudes of about 15,000 feet above terrain.

In detailed topographic mapping, the Multiplex process has already been proved as the speediest, least expensive, highly accurate technique for producing contour maps for use as plotting bases in geologic mapping. Application of the instrument to exploration for oil without drawing contour maps will be shown later. The Multiplex consists of a mounting bar with small projectors which project transparent diapositives of air negatives in complementary colors to form a spatial model of terrain to be viewed with colored glasses. The diapositives are printed in special Multiplex-reducing printers designed to handle various standard-focal-length air negatives. The bar-mounted projectors, through use of adequate control, reproduce the attitude of the air camera in space, and the distance between projectors represents the distance between air camera stations at the scale of the plot. Each projector thus reproduces the altitude, tilt of the camera, and position in space of the air camera at the time of exposure. With alternate projectors projecting the images in red and green and the Multiplex operator wearing glasses having one red ocular and one green ocular, two projectors correctly oriented to plotted control lying on smooth sheets at table level will produce a stereoscopic model in space. Working in a darkened room, the operator sees hills and valleys stand on the table. A Multiplex tracing table set at the elevation to be contoured is now checked with known elevations gained from ground control present on the spatial model. The tracing table has a small dot directly over a tracing pencil. As the table is moved over the topography, keeping the dot on the ground, the pencil draws an orthographically projected contour line. No further photogrammetric correction of this contour line is needed for scale, tilt, or orientation. The tracing table is lowered or raised by a scaled amount representing the contour interval, and the next contour is traced out on the spatial model standing in space on the table in front of the operator. Adequate tests have proved the technique to be more accurate and much speedier than plane-tabling in the field.

Detailed topographic contouring is unnecessary for structural contouring of geologically interpreted aerial photographs. Photogrammetric measurement of spot elevations on interpreted contacts can be made at optimum locations with the tracing table. Dips can be computed by three-point geometrical solutions, and a structural contour map can be derived at an accuracy greater than that obtainable in field surveying. Problem areas must, of course, be examined in the field where complex faulting and folding make difficult the tracing of contacts, but these data entered on air prints in the field and supplied to the geologically trained Multiplex operator will enable him to trace out contacts directly and get spot elevations at *all* critical points. In operations of this type, a keen geologic interpreter working with a skilled Multiplex operator over unexplored terrain could produce maps of existing structures faster than any known surface mapping techniques. This application may never be used in areas as accessible as Oklahoma and Texas, where surface rocks in many places show relationships which can be projected into the subsurface. Geologic interpretation of the Cretaceous terranes of Utah and Alaska, where extensive surface structures can easily be traced in photographs, is definitely an easier way of doing geologic field mapping. Accuracy of Multiplex in contouring approximates 1 in 1,200 if adequate vertical control is present. In other words, for 90 per cent of the contour lines to be within one-half contour interval in accuracy, it is necessary to work with photographs flown below 12,000 feet for 10-foot contours or below 24,000 feet for 20-foot contours. Spot elevations, however, can be gained at an accuracy of 1:4,000, which means that clearly recognizable geologic contacts could be spot-controlled to within 3 feet vertically on photographs flown below 12,000 feet. This falls in the plane-table range of vertical accuracy and makes Multiplex applicable to structural contouring. The training of a skilled operator takes between 6 months and 1 year.

The price range and accuracy of photogrammetric instruments which produce phototopographic maps are breached by a wide zone. The Multiplex is an expensive instrument which is accurate with a minimum of control. The Stereocomparagraph instruments are inexpensive and inaccurate and necessitate much control. The need for an instrument lying between these two types which would not be far above the Stereocomparagraph price range and would result in map accuracy near that produced by the Multiplex was well known. Other types of stereoscopic plotters are designed to gain planimetric and topographic data from overlapping vertical air photographs and plot them directly upon a map projection at any chosen scale between 1:15,840 (25-foot contours) and 1:31,680 (50-foot contours), 1 inch = ¼ mile to 1 inch = ½ mile, respectively. These plotters differ from other precise stereoscopic plotting instruments in that they plot contours by taking advantage of known change in scales in a photograph for each contour interval. Scale of the photographs is changed by raising and lowering the picture plates in relation to the floating marks, whereas most other stereoplotters depend upon measuring the differential parallax for each contour.

Table 10-3. Line Accuracy on Maps and Photographs*

Scale	*Feet covered by 1/100-inch line*† *(nearest foot)*
1:2,000	2
1:4,000	3
1:4,800	4
1:10,000	8
1:12,000	10
1:15,000	13
1:20,000	17
1:24,000	20
1:31,680	25
1:40,000	33
1:48,000	40
1:50,000	42
1:62,500	52
1:63,360	53
1:96,000	80
1:125,000	104
1:250,000	208

* Tolerances in the location of points and lines in surface mapping may be double the distance shown at these various scales excepting in first-order plane-table triangulation.

† Resolution of the normal human eye is at a minimum line width of 1/200 inch, which is the maximum accuracy attained in careful photogrammetric compilations in the laboratory.

Pantograph attachments allow plotting to other scales which might be applicable to oil exploration. Design and operation of these plotters are very simple, and anyone having stereoscopic ability and a desire to learn can be trained to turn out accurate work after one month of intensive training. Accuracy obtained is in direct proportion to the amount of good vertical control available. Using the generally applied specifications of having 90 per cent of the contours within one-half contour interval of their true positions, these plotters can produce contours at an interval one five-hundredths of the flight altitude. Some operators can produce contours at intervals ranging up to one seven-hundredths of flight altitude. Skilled operators can contour from 1 to 2 square miles a day. The use of these plotters by photogeologists would no doubt result in the 1:700 accuracy contour interval because of their greater experience in topographic interpretation. If it can be assumed that the higher accuracy can be obtained easily, 20-foot contours could be produced from photographs taken at 14,000 feet and 10-foot contours from 7,000-foot photographs. Since four times more area is present in photographs taken with the 6-inch lens at 7,000 feet than with the 12-inch lens at 7,000 feet, only one-fourth as many 6-inch focal-length photographs would need to be interpreted and contoured to get the same contour coverage. When continuous contouring is done away with and only spot heights measured on interpreted geologic contacts (for contouring the geologic structure), a much greater area can be covered per day by the experienced photogeologist using photographs over terrain where geologic contacts can be picked on the photographs. These instruments are applicable to many exploration problems in less accessible parts of the Rocky Mountain area or in foreign areas. The capital outlay and short period of training needed to get good results make these techniques of definite value to the exploration departments of oil companies who have photogeologists with a knowledge of the closer applications of practical photogrammetry to field geology and the geologic interpretation of air photographs.

The Oblique Photograph. Photogrammetric measurements on oblique photographs are generally more difficult to convert to map data despite the greater ease with which data can be interpreted on an oblique aerial view. The same terms defining points on vertical photographs apply to obliques except in special cases.

The Canadian grid process developed in Canada prior to 1928 led many investigators to adapt this simple graphic means of handling oblique photographs to other sciences besides ordinary mapping. In this process, three cameras are mounted in the nose of an airplane in such a manner that one camera points forward, one 45 degrees to the right, and the third 45 degrees to the left. By tilting the cameras so that the horizon shows in each photograph and by shooting with sufficient overlap in the line of flight, the Canadian photogrammetric engineers have been able to map thousands of square miles of lake country on the Canadian shield in a very short time. Originally developed as a reconnaissance method, care in handling at somewhat larger scales plus more field control would allow adaptation to reconnaissance geologic mapping. Graphic methods have also been devised to relate two oblique photographs in space in such a manner that rough topographic mapping can be done over high-relief terrain. For the geologist working on small areas where no other maps are available, the oblique photograph handled graphically (without complex instruments) will provide data for good planimetric maps if the terrain is essentially level and of low relief. In high-relief terrain where a few control points and several photographs of the same area are available, graphic intersection techniques serve to get abundant spot elevations and many planimetric data. Application of these graphic techniques is not warranted over wide areas of reconnaissance when constant scale is important; hence these techniques, except for expanding control for later precise vertical flying, do not offer a wide field to large-scale detailed exploration by the oil industry.

Two instruments were developed between 1930 and 1940 for gaining a wealth of vertical and horizontal data from oblique photographs. The Miller Single-Eyepiece Plotter has not found wide use except in certain special noncommercial projects for production of 50-meter contour maps at small scales. Obliques for this plotter should be flown with 50 per cent of the photographs at right angles to each other and 50 per cent of the photographs with parallel axes. Height of the photographic plane should be at least as high above the highest point in the area as the highest topographic point is above the lowest topographic point. Sparse triangulation and some vertical control are necessary to relate the oblique data from one photograph to the next on the base sheet. This method is not recommended for detailed oil exploration but may prove a great reconnaissance aid when coupled with geologic interpretation of the oblique photographs.

A second instrument, the Wilson photo-alidade, measures horizontal and vertical angles directly from a horizon oblique. The instrumental method is faster and more accurate than graphic resection techniques employed on obliques. The photo alidade is the air photogrammetric equivalent of the plane table. From the perspective of the airplane, much more country can be covered with an accuracy equal to reconnaissance plane-tabling for the scale of mapping and amount of vertical control available. It has been found that points surveyed in the field by ground methods can be measured to a vertical accuracy of 9 feet (plus or minus) with the photo-alidade at a scale of 1:24,000. This accuracy decreases as a horizontal line through the principal point is approached because air obliques show rapid scale change from the lower part of the picture toward the horizon. This instrument could well be adapted for use in the oil industry. Heretofore it has been used for reconnaissance contouring of small-scale geologic base maps and aeronautical charts. In areas where outcrops are prominent and good horizon obliques are available with some vertical and horizontal control, a structural map of reconnaissance accuracy could be produced if easily recognizable outcrops had dips greater than 5 degrees. These geologic conditions are found widespread in many potentially petroliferous but unexplored terranes of the Rocky Mountain Province, on the Arctic slope of Alaska, and in North Africa.

First application of Trimetrogon photographs to geological reconnaissance was made under special specifications over U.S. Naval Petroleum Reserve No. 4 in Alaska during 1944 and 1945. A description of its first geological application may prove a guide to the value of the process in regional reconnaissance for petroleum-bearing structures. Trimetrogon photography is normally flown with three 6-inch-focal-length K-17 cameras in trimounts at an elevation of 20,000 feet with flight lines 20 miles apart. Compilation is done at a scale of 1:80,000, and final aeronautical strip charts are issued at scales of 1:500,000 and 1:1,000,000 with 500-foot contours. These wartime compilation scales were too small for geologic investigation of an area. Specifications were designed to fly a large part of the Arctic slope with Trimetrogon from an altitude of 12,000 feet with flight lines spaced at 10 miles. Size of the geologic structures interpreted from older 20,000-foot Trimetrogon photography led to adaptation of flights at this lower scale to enable geologists to make geologic interpretation of the resulting vertical and the wing obliques. The qualitative interpretation done by geologists in preparation for field mapping allowed accurate delineation of structural axes and generalized structural analyses from the vertical and obliques. Compilation of the Trimetrogon photography by the U.S. Geological Survey provided a base upon which the geologic interpretation was entered. When the position of likely looking structures was known, a planning map was set up for covering certain areas with precise vertical photography for final Multiplex compilation and construction of structural contour maps by field mapping. Correlation of measured field sections, photo-alidade control, plane-table mapping, and geologic interpretation on Multiplex base maps resulted in adequate oil-exploration surface geologic maps.

In a survey of all processes now available, the apt handling by a geologist–photogrammetric engineer of the combination Trimetrogon-Multiplex process for compilation, coupled with geologic interpretation from photographs and field work, will yield maps many times faster and much more accurately than ordinary ground surveying coupled with surface geology.

Geologic Interpretation

Interpretation of geology from air photographs is based on a critically descriptive approach backed by an excellent background in structural and geomorphologic analysis based on wide experience in field geology. Factors which make interpretation of air photographs successful are present in geologists to varying degrees. The photogrammetric ability to construct accurate base maps from air photographs and to place thereon the interpreted geologic data with precision is a necessity in preparation for speedy field mapping. There are few new techniques in geologic interpretation inasmuch as new techniques arise in the application of engineering precision with which interpreted data are represented in map form. There is no better technique than hard work and experimentation with air photographs in the field and laboratory. The gamut of uses to which geologists put air photographs runs from the preliminary one of finding one's way around an area to a final examination and mapping from photographs alone. Detailed geologic interpretation can readily be applied in areas where sedimentary contacts and structural features are plainly visible or reflected by form, vegetation, or soil patterns, but the surface geologist must examine and check critical areas in the field. The factor of visibility eliminates most heavily wooded areas, grassy lowlands, and swamps where rocks are not clearly exposed and coastal plain or alluvium-filled valleys where bedrock is not visible, excepting for the mapping of generalized anomalous-looking areas. Detailed geologic interpretation is successful only in areas where the geomorphic expression of underlying rocks is directly indicative of their structural and sedimentational or lithologic characteristics. It can thus be seen that the exploration of a block or concession from the air falls into several rather definite steps, namely, flight check or reconnaissance photography of the whole area, photographic interpretation to delineate areas where detailed geologic interpretation from precise vertical photographs is possible, delineation of other areas where a combination of office interpretation and surface mapping must be used to do the geology, choice of those areas where the photographs are of no interpretative aid and may be used merely to aid geologists to cover the area while detailed surface work is done.

Qualitative Interpretation. The present status of most geological interpretation from air photographs is mainly qualitative. Oblique and vertical photographs are amenable to plotting of contact strikes and estimation of dip. Contacts can be traced for transfer to ordinary base maps, either in the office from photographs or in the field with the aid of photographs. If the relief of terrain is low, the flight altitude and focal length can be used for scale (scale = focal length/flight altitude) and simple orthographic maps prepared showing geologic contacts, strike-and-dip symbols, and the structural axes, or these features can be interpreted directly onto vertical photographs and a crude matched mosaic constructed. Overlays (tracings) of these crude mosaics are often called geological maps and published as such. If detail is sufficient, an interpretative geologic column can be built up, thicknesses estimated, dips estimated, and structural axes plotted with enough accuracy to serve as field guides to problem areas and locations for field measurement of dip and strike and measurement of stratigraphic sections. Obliques can be treated in the same manner by computation and construction of a Canadian grid which enables the interpreter to trace the detail that he interprets directly onto an orthographic projection. By plotting from only a few well-chosen horizon obliques, a qualitative picture of the geologic terrane can be built up over a wide area. Interpretation of obliques is generally easier but reduction of data to the map plane more difficult by simple graphic methods than similar work on vertical photographs. High-relief terrain does not lend itself to simple graphic plotting with the Canadian grid. More complex graphical plotting methods using instrumental techniques are available for work on obliques covering mountainous areas.

Quantitative Interpretation. The quantitative interpretation of air photographs completes what qualitative reconnaissance interpretation must leave undone. It is in quantitative measurements that the full part of photogrammetric engineering must be utilized by the photogeologist and the field geologist. Graphic methods and trigonometric solutions will yield remarkably accurate dip-and-strike measurements from both vertical and oblique photographs if some ground control is available to provide exterior orientation of the photographs and scale for thickness, slope, and dip computations. The outstanding problems in an interpretative job are accurate elevation determination and geographic position of points on the earth's surface. Vertical field control is necessary for the former, and horizontal field control is necessary for the latter. By bridging techniques discovered in radial-line triangulation and Multiplex mapping, both problems can be solved for control extension. Oblique photographs often provide better vertical-control data for bridging than do vertical photographs because they cover more area and relate in one perspective a greater number of topographic elevations which can be measured by the photo-alidade. Applications of the photogrammetric instruments have stressed topographic contouring, whereas the geologist needs far fewer elevation data than a detailed topographic map presents for constructing structural geologic maps and detailed cross sections and for making strike-dip measurements. Selection of contact elevations, measurements of strike and dip, construction of a columnar section, and location of rock outcrops can be performed by use of spot data obtained from Multiplex, Kelsh stereoscopic plotter, or the Stereocomparagraph as applied with geologic interpretation. Use of these modern photogrammetric techniques by geologists to save needless field work in easily interpreted areas adds up to one main point: Geologists should study photogrammetric engineering in order to gain precision in quantitative applications of their qualitative geologic interpretations, and the field geologist and photogeologist should be one and the same man.

Program Planning

Concessions and large lease blocks in unexplored areas are often taken on the scanty evidence supplied by examinations over an area. Any geologist who has flown over many different terranes is quick to recognize structures outlined by stream characteristics, vegetation distribution, and visible contacts of rocks. Knowledge of the existence of such structures in potentially petroliferous areas has led to check flying of the area and direct examination from the air. Very often an inaccurate small-scale map is taken along, and the structural axes marked thereon. A recent experience of this type led to delineation of four anticlines over an unexplored area and an average inaccurate placement of 2 miles for their axes as well as serious errors in defining likely structures. Corrections were made later from Trimetrogon photography. Spotting geologic structures on small-scale maps by direct observation without photography is both inaccurate and dangerous. Planning based on such structural distribution patterns has many times cost enough to have paid for precise photographic coverage. Flight checking by a geologist results in one man's interpretation on an often inaccurate base map. No other reference material is available after the flight is over, whereas a reconnaissance Trimetrogon or vertical coverage provides continous reference for technical study by the experienced geologist. No added data are gained for construction of better reconnaissance base maps if check flying only is applied in planning for later surveys on the ground. Preliminary geological reconnaissance maps made from small-scale photography are far more accurate and indicative of areas valuable for concentrated effort than are structural axes drawn on a check flight map. Preliminary reconnaissance photography provides data for precise flight maps in the form of planimetric maps or mosaics if vertical photographs are flown. The cost of small-scale photography plus subsequent selective large-scale photography of a concession is small compared with the amount of ground survey and control which must be provided if only poor large-scale photography is made available owing to the lack of good flight planning. This same small-scale photography provides an excellent planning base for field survey operations in gaining vertical and horizontal control for Multiplex or Kelsh stereoscopic plotter compilations and in accessibility studies prior to surface geological mapping.

Oblique photographic reconnaissance of an area provides many data of aid to planning of field mapping. Type of vegetal cover, distribution of rock outcrops, strike of structural axes, and distance of the area from airfields control the type of oblique photography which must be utilized. Scattered horizon obliques, if taken over geologic structures, are excellent for small-scale interpretation and construction of simple planimetric base maps. If the obliques are taken as pairs, stereoscopic examination is facilitated in interpretation, and the stereo pairs can be used in the oblique height finder or the photo-alidade to gain spot elevations for construction of preliminary cross sections and determination of maximum closure in anticlines with bold dips. If the area warrants controlled oblique photography, regular oblique flight lines can be laid out to cover an area thoroughly. This type of reconnaissance photography, though it furnishes the geologist with excellent initial interpretative data, is difficult to compile onto maps owing to the individual attention necessary for each oblique photograph. If an anticline is found on one or two obliques and it is desired to cover the anticline with precise vertical photography, flight lines can be computed and laid out directly on the oblique photographs through application of the Canadian oblique grids. Aviators flying with such flight aids have proved that precise flight lines can be followed more closely on several oblique photographs covering an area than can be followed on a detailed map at a scale of 1:100,000. Distortions due to relief require that flight check points in the obliques be picked in valleys which are crossed by the flight lines. If the area to be photographed with obliques has low relief and many water features, a Canadian grid can be constructed quickly for each oblique, the planimetry traced at the flight scale, and the resulting orthographic templates matched to construct a reasonably accurate reconnaissance map (Wengerd, 1945). This has been done for thousands of square miles in Alaska and the Canadian Arctic and has proved successful in producing maps of value as flight guides for later precise vertical photography as well as plot maps for geologic data and the guidance of geological parties in the field.

Should better maps, and even mosaics, constructed from obliques be necessary prior to more complete vertical coverage of related areas, the utilization of oblique rectifiers or oblique detilters will result in vertical photographic templates which can be controlled by radial-line methods or matched to make up crude but highly valuable reconnaissance mosaics. The mosaics constructed from such detailed obliques could provide abundant reconnaissance aid for the geologist and field control parties. These amazing detilted photographs will even provide good stereoscopic models for examination and interpretation by the geologist prior to field work.

Trimetrogon reconnaissance coverage carries far more weight as a unified system of producing accurate maps than do the rather hit-or-miss reconnaissance oblique techniques mentioned above. Two horizon obliques flanking a vertical photograph provide enough data, when flown in regular flight lines and controlled by several ground survey stations, to give excellent interpretative coverage and control for photogrammetric compilation of planimetric base maps which can be used as plane-table sheets.

Semiarid and arid terrain, mountain terrain without tree cover, and subarctic and arctic treeless plateaus not masked by glacial and alluvial debris lend themselves to Trimetrogon reconnaissance photography which can be used for geologic interpretation. An area 85 miles long and 60 miles wide flown with Trimetrogon photography can be covered with six flight lines 10 miles apart if the plane carries a 6-inch-focal-length camera at an altitude of 12,000 feet. Small structural geologic anomalies can be delineated and easily interpreted from photography at this scale. In photographing Naval Petroleum Reserve No. 4, it was found advisable to fly Trimetrogon lines magnetic north-south directly across the axes of the visible synclines and anticlines. In coastal plain areas of Alaska where geologic interpretation of structure is almost impossible, the direction of flight was entirely controlled by weather conditions.

These photographs allow initial geologic examination to determine areas which require precise vertical coverage for Multiplex or Kelsh stereoscopic plotter compilation. Compilation of this Trimetrogon photography under the direction of a competent photogeologist results in direct choice of areas for intense air mapping, allocation of geologic field parties to areas not amenable to photographic interpretation, planning of field control, selection of good triangulation stations, and distribution of ground control parties to gain horizontal and vertical control for use in detailed photogrammetric compilations of precise vertical photography over areas chosen for intense study.

Although vertical photography is often more difficult to interpret, the experienced geologist can do far more valuable reconnaissance geology from a set of verticals than he can from a set of Trimetrogon photographs. Photographic compilation of verticals by radial-line technique is also easier for fewer personnel than the involved compilations necessary to convert Trimetrogon photography into planimetric maps. Vertical photographs lend themselves far more readily to rapid reconnaissance phototopography than do obliques. All the advantages of oblique reconnaissance photography are present in vertical reconnaissance photography except a more easily interpreted angle of view and the necessity for more ground control to complete vertical photographic compilation. Several features of small-scale verticals make them far more desirable than obliques for reconnaissance work, namely, ease of mosaic construction, simple application of radial-line triangulation, production of better flight maps for gaining precise photography, better initial geologic interpretation and compilation of geologic information on the mosaic base, rapid construction of controlled mosaics if enough ground control is available, and more suitable base maps for planning precise field mapping. The cost of both oblique and vertical reconnaissance photography is indeterminate unless many variables are known for the particular area to be covered.

The type of ground control necessary to compile a map successfully from air photographs depends directly upon the terrain, the photography available over the area, and the uses to which photogrammetric maps will be put. For Trimetrogon coverage, an astronomic position for each 600 square miles of an oblong or square area is sufficient for control of radial-line triangulation in the compilation process. If only one astronomic position is available, a base line, expansion net, and a long-distance triangulation net can be laid out to tie all parts of the area together. Astronomic points scattered about 75 miles apart provide a net which is by no means so accurate as triangulation but has been proved adequate for reconnaissance Trimetrogon coverage and the later construction of small-scale reconnaissance maps.

Choice of areas for precise vertical coverage requires application of second-order triangulation and trigonometric leveling within that area for control of the vertical photographs to be compiled by Multiplex and Kelsh stereoscopic plotter. Internally these smaller areas are generally surveyed by traverse, plane-table triangulation, or second-order triangulation expanded from local base lines measured from an astronomic position or a triangulation point of the long-distance triangulation net. Precision altimetry, using surveying aneroids in accordance with the two-base technique, provides excellent vertical control at minimum cost and cuts down plane-table control necessary for accurate phototopographic mapping. If the small areas are widely separated and spirit-level traverses or trigonometric leveling prove too costly, use should be made of the terrain clearance indicator and two aneroids in a traversing plane. Small areas can thus be tied together vertically, and base elevations provided from which precise altimetry can be carried out within the smaller areas of precise mapping.

When all major control tying together smaller areas is surveyed and computed for latitudes and longitudes, the points are accurately marked on the available air photographs for the guidance of the photogrammetric engineer. A projection of the whole area at one-half the flight scale (if covered with reconnaissance Trimetrogon) or full flight scale (if covered with vertical reconnaissance photography) is constructed on a large lay-down board. The photographs are prepared for radial-line triangulation, templates are laid, and resulting photocontrol is plotted on the projection. Acetate overlays can be employed to gain all photo control and field control on sheets easily handled in the final process of delineating the planimetry and drawing the contours or reconnaissance geologic contacts. If Trimetrogon photographs are used, planimetry is delineated from the verticals with vertical sketch masters and from the obliques with oblique sketch masters. Topography is gained by photo-alidade work or through use of the oblique height finder. Interpretative contouring is then done of the area where spot elevations have been measured for reconnaissance coverage, and a mosaic can be laid directly on the lay-down board containing the field control and photocontrol. Direct overlays provide planimetry, and the Multiplex or Kelsh plotter can provide rapid reconnaissance contouring at large contour intervals for geologic contact maps.

Areas of great interest, such as specific structures and special terranes where seismic or gravimetric coverage must be gained later, can be chosen directly with the aid of the reconnaissance photographic coverage. Whichever type of coverage is planned, over-all precise vertical coverage or reconnaissance photography with later precise vertical coverage of areas of geologic or other interest, the precise coverage should be planned to provide accurate plane-table sheets for the surface geologist.

Experience has shown that geologic interpretation of areas coupled with phototopographic techniques to gain vertical data over structures requires flying of a precision in direct ratio to the precision desired in the final maps. The two operations are in direct opposition in the matter of scale. The higher a plane flies, the fewer stereoscopic models need be handled to cover an area with contours by Multiplex or Kelsh plotter. The higher the flights, the less the vertical accuracy and the less detail can be discerned by the geologist. Obviously a happy medium must be found. Since it is the height of photography and not the focal length of the cameras which determines the accuracy of a chosen contour interval and the height of photography coupled with focal length controls scale, which in turn controls the amount of detail visible in photographs of an area, a short-focal-length camera flown at a medium height will satisfy both the phototopographer and the geologist. For Multiplex topographic contouring at 10-foot contour intervals, a flight height of 10,000 feet is recommended with either 6- or 12-inch lens. Use of the 6-inch lens, however, results in processing approximately one-fourth the number of stereoscopic models necessary to cover the same area with a 12-inch lens, even though the 12-inch lens gives double the scale, a desirable feature for the geologist. The height of photography to suit the geologist must then be controlled by the types of terrane he has to interpret. It is believed that 6-inch-focal-length photographs flown at 7,500 feet should provide the necessary scale over most areas for accurate interpretation by the geologist and still cover most smaller areas of intense study with few enough stereomodels to keep down the time-consuming instrumental work by the photogrammetrist. If contouring is not needed and the geologist is a capable Kelsh or Multiplex operator, much time can be saved by spot elevation work on geologic contacts for construction of structural maps rather than complete topographic contouring of the area.

The aim of much reconnaissance in oil exploration is production of a final geologic map to guide the planning of a deep-drilling or core-drilling campaign. If the concession area is amenable to geologic interpretation as a whole or in part, those parts should be photographed at a scale which will offer fine detail and the areas should be contoured with a small contour interval. Final geologic interpretation then results in the plotting of all structural axes, attitude of beds, faults, joint systems, plus structural contouring from photographic key beds, and construction of detailed cross sections. Much of this can be done with a minimum of field examination over many areas, but the final interpretation must be done in the field by checking the stratigraphic sections and faulted or strongly folded areas.

Simple Aids for Field Mapping

Techniques in the utilization of air photographs for oil exploration include many simpler aids developed within the last few years. These simple aids have not found wide use because the oil industry is not given to using halfway measures in photographic mapping. An oil company either depends heavily upon the excellent aids provided by photogrammetric processes or uses air photographs only for pictorial examination of terranes. Between these two extremes lies a vast field of methods which can save companies much expense in planning and executing ground exploration. It is axiomatic that the more time spent on office preparation of photographic coverages prior to surface geology, the faster, more efficiently, more accurately, and less expensively geologic field work can be carried out. Two simple timesaving functions can make field work much more effective, namely, choice of areas for intensive study and preparation of working base sheets. Whether the geologic features can be interpreted directly from air photographs or not, these two aids can be used over any area. Any geological interpretation which can be done is merely a greater application of the photography over an area amenable to photographic interpretation. In some instances, air photographs are belatedly furnished as an afterthought. The coverage may consist of mounted unrectified photographs at a scale of 1 inch to 1,000 feet. Such nonstereoscopic coverage may be used merely to aid the party in getting around in the field. Half the time a party spends in the field may be used in the search for outcrops. Strict application of stereoscopic photographic technique in the office would result in accurate location of nearly all outcrops and contacts necessary for structural contouring of the area by plane table. Additional time in preparation of controlled mosaics by radial-line triangulation utilizing the abundant control available would yield accurate plane-table base sheets, use of which would reduce field work to getting elevations by plane table. Further projection of techniques using the precise single short-loop or two-base altimetric technique may reduce field work to what it was originally intended to be, a study of surface geology with a minimum of survey engineering in preparation of plane-table planimetric base sheets.

Areas of low relief can be mapped quickly with plane-table base sheets consisting of photographs enlarged to a definite scale ratioed to abundant field control available from land-ownership plats and United States topographic quadrangle maps. Horizontal closing errors up to 500 feet are allowed in some types of detailed surface geologic mapping, yet the vertical closing error may be limited to 1½ feet. The best graphic realignment and correction of large-loop plane-table traverses having horizontal closing errors of 500 feet may often put certain geologic contact elevations over 100 feet from their true horizontal positions. In prairie grasslands containing outcrops, these measured elevation points may be over 200 feet from their true positions. Horizontal position errors of 200 feet in terranes having a formation dip of 4 degrees will result in a vertical error of 14 feet should the direction of closing error coincide with direction of 4-degree dip. Allowance of a closed-traverse error which cannot be graphically corrected so that each measured elevation point is within 50 feet of its true geographic position (in relation to all other elevation points measured) may lead to serious structural misinterpretations of plane-table mapping. The errors are not compensating, and their seriousness depends upon the direction of error in relation to the direction of dip. Photogrammetric standards for planimetric mapping at scales less than 1:20,000 in which one geographic control point is available for every series of eight photographs radially triangulated (approximately 6½ square miles of stereoscopic coverage at a photographic scale of 1 inch to 1,000 feet) require that 90 per cent of all detail points on that planimetric base sheet be within 1/50 inch of their true position; i.e., 90 per cent of *all* detail must be within 20 feet of its true horizontal position on a map having a scale of 1 inch to 1,000 feet. The accuracy of good radial-line triangulated or Multiplex-compiled planimetric base sheets is much greater than the plane-table mapping resulting from ordinary surface work.

If 9- by 9-inch vertical photographs flown at a scale of 1:24,000 (6-inch lens at 12,000 feet) are enlarged to 1:12,000 (1 inch equals approximately 1,000 feet), plane-table base sheets 18 inches square result. These base photographs can then be placed on the plane table with an acetate overlay. If alidade shots involving only level determination on the rod are kept below 2,000 feet in horizontal distance, an area can be covered without recourse to any intercept distance plotting. Approximately 50 per cent of the instrument man's time is thus saved, and, with experience, three rodmen can be kept busy running out contacts and doing surface geology. If relief is low, these base sheets can be matched directly and final structural contouring done on overlay acetate sheets or the elevation points transferred to a radially triangulated controlled mosaic to gain a more accurate geographic relation of all points. If relief is moderate, the points whose elevations are measured can be transferred to the 9- by 9-inch photograph in the office and an accurate radial-line plot made of all photographs covering the surface-examined area. Radial-line triangulation corrects distortions of point positions due to relief and moderate tilt in the photographs. In high-relief terrain where the Beaman stadia arc, the gradienter, or the vertical circle must be used to compute elevations using the true horizontal distances, the radial-line planimetric base should be used as a plane-table sheet in preference to the enlarged photograph because of relief and possible tilt distortions present in most "vertical" photography.

A simpler technique has been devised for use of enlarged photographs with the plane table in cases when only one man is available to do the surface geology of an area. The method will work in any type of terrain but requires the added step of completely contouring the photographs in the field by taking rounds of level shots to all points identifiable on the photograph from many different instrument positions. Instrument stations are picked on points readily identifiable in the photographs. Subsequent radial-line triangulation of all photographs so used in the field results in geographic rectification of all measured points and instrument stations. The resulting rectified base sheets can then be used for plotting geology without further alidade work in the field. In cases where two men are available in rough country, geology and level shooting can be done at the same time with subsequent geologic interpretation from stereoscopic photographs to fill in detail. The end result in all these technique variations is a structural contour map of accuracy comparable to and often greater than straight plane-table work with far less time spent bushwhacking through the country, 50 per cent of which, for lack of outcrops, should never have been tramped. The ideal simple technique for aiding surface geology to be done with the plane table is to prepare complete detailed planimetric plane-table sheets in the office. These can then be sent out to the field with air photographs to be interpreted directly in the field. A combination of plane-tabling and geologic interpretation in the field and on photographs is the fastest and most accurate method of doing surface geology over terranes which cannot be interpreted from photographs alone.

STRUCTURAL FIELD METHODS

Objectives of Projects

Methods of field work will vary, depending upon the major objectives of the project, the amount of time allowed for the project, and the availability of base maps, air photographs, and other basic tools essential to the successful completion of a geologic field project.

Objectives of a field project might include all or one of three phases of geological field work: *areal geology, structural geology,* or *stratigraphy.* Areal geology and structural geology are generally done simultaneously if maximum benefit is to be gained from a project. In the process of doing the work the field geologist must be a competent enough stratigrapher to work out sequences of beds in the project area to aid in

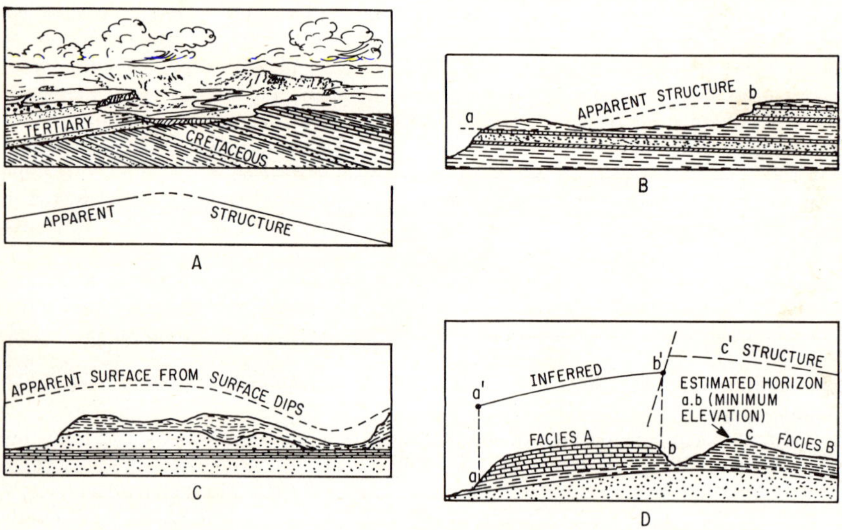

FIGURE 10-32. Stratigraphic anomalies that can cause errors in structural field work. (*A*) Unconformity producing dips simulating an anticline. (*B*) Repetition of similar sequences of beds erroneously correlated, giving false impression of structure. (*C*) False dips caused by solution of near-surface gypsum. (*D*) Abrupt changes in facies suggesting presence of a fault. (*From Low, 1957.*)

the mapping of units or groups of units, and he must also be able to recognize facies changes, unconformities, and disconformities where they occur, as they might otherwise affect the accuracy of his map if not observed correctly in the field. Examples of stratigraphic anomalies are shown in Figure 10-32.

The areal and structural geology field project is mainly concerned with accurate recording of surface data with respect to distribution of outcrops and structural configurations which might carry to depth in an area underlain by what are believed to be desirable drilling objectives. The final map and report should lead to the outlining and recommendation of specific drillable prospects to be considered for acreage acquisition or seismograph work if it is believed the structure may shift with depth.

The *stratigraphic project,* on the other hand, requires the work of a well-trained and experienced stratigrapher. His work may take him around the margins of a large basin where formations are exposed which are or may be beneath the surface in the center of the basin and which are to be evaluated as potential source or reservoir beds. The stratigrapher must also be a competent enough structural geologist to allow for changes in dip and other structural features which must be recognized in order to arrive at a reasonably correct thickness when measuring a section. His work might also indicate sufficient differences in section on opposite sides of a basin to suggest the possibility of stratigraphic as well as structural traps in the basin.

A major field project may include all three (areal, structural, and stratigraphic work), or it may be divided between two or more parties, each doing a separate phase of the work. In this case, one party might be doing the areal and structural geology while the other does the marginal stratigraphy, and the final report is written as an

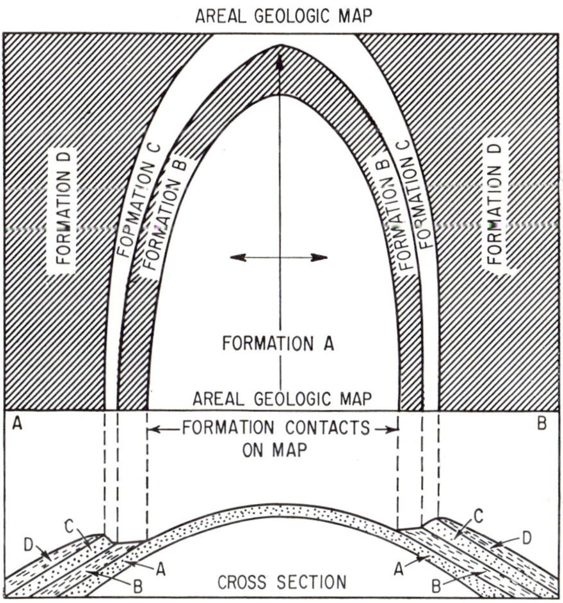

FIGURE 10-33. Topography and areal geology. Cross section of a northerly plunging anticline on which a more resistant bed forms a hogback on the limbs of the structure. Contacts are projected upward to illustrate their expression when reduced to the common plane of an areal geologic map.

evaluation of the entire area relative to potential structural prospects, potential stratigraphic traps, and major drilling objectives.

The type of field work planned then might be classified in one of three groups, depending upon amount of field data desired and time allowed for the project: *reconnaissance*, *detailed reconnaissance*, or *detailed* field work. Methods of carrying out actual field work may differ somewhat between *areal* and *structural geology* and *stratigraphic geology*, and therefore a discussion of methods has been divided into two sections. Although the methods are treated separately, the fact should not be overlooked that the types of geology discussed are interrelated and the well-trained geologist should be familiar with all the methods he can possibly use in the field, since his work might require that he assume responsibility for all phases of geology in an assigned field project.

Reconnaissance Areal and Structural Geology

Reconnaissance geology is just what the name implies, a rapid field study of an area that is done by whatever method is most suitable. A field reconnaissance of an area should be done in such a manner that when the project is completed, it will give a reasonable picture of outcropping formations, possibly with contacts of major units shown. It should establish regional dip and local variations; structures, such as noses, anticlines, and faults; and regional trends or structural grain should be well defined on

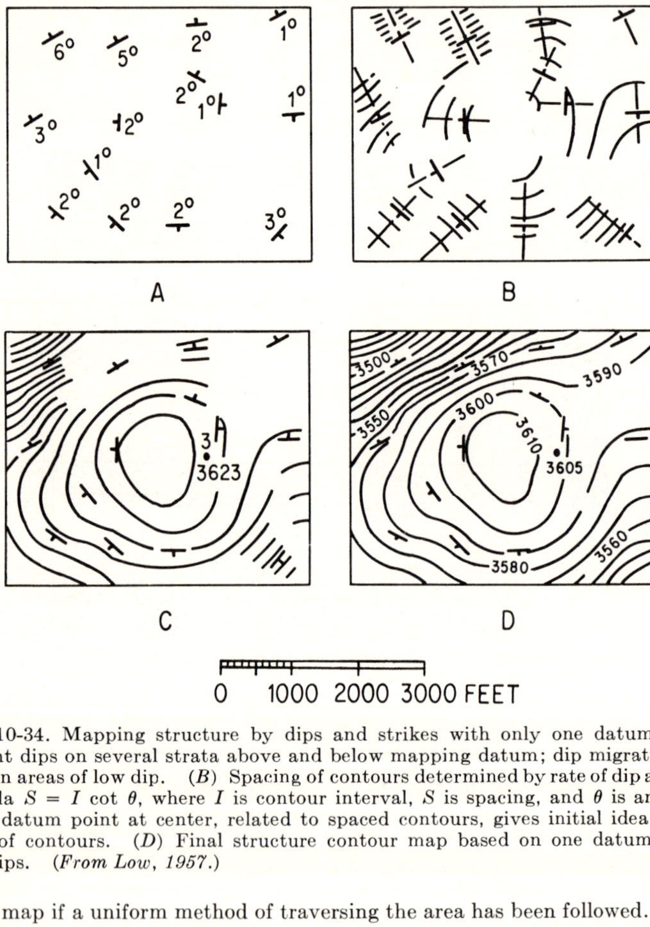

FIGURE 10-34. Mapping structure by dips and strikes with only one datum elevation. (*A*) Slight dips on several strata above and below mapping datum; dip migration is not a problem in areas of low dip. (*B*) Spacing of contours determined by rate of dip as expressed in formula $S = I \cot \theta$, where I is contour interval, S is spacing, and θ is angle of dip. (*C*) One datum point at center, related to spaced contours, gives initial idea of relative position of contours. (*D*) Final structure contour map based on one datum point and several dips. (*From Low, 1957.*)

the final map if a uniform method of traversing the area has been followed. Detailed work should be included only to the extent that, in the case of a large anticline being observed in the field, for instance, sufficient strike-and-dip data should be obtained to indicate reversal, if present, and thus establish possible closure. The resulting dip diagram can then be plotted on a map, the axis of the structure shown, and perhaps a form-line contour constructed to indicate possible closure (Figure 10-35).

Photomosaics are best to work on for reconnaissance work, but if they are not available, topographic maps, county road maps, or good base maps with sufficient known points shown can also be used with equal success. If none of these is available, compass traverses from known points can also be run, on foot or with a vehicle, and data can be plotted on a map tied to some known location outside the project area.

The most rapid results can be obtained in an area if traverses are run across the structural grain. If photomosaics are used, prominent structural features can generally be outlined on them, and with this information traverses can be planned accordingly. If air photographs are not available to the field party and work is to be done on base maps or road maps, an air reconnaissance of the area in a small plane, as previously mentioned, can be very helpful for observing the structural pattern, and the lines of traverse can be intelligently planned from these observations.

The field geologist should bear in mind that although the final results of a reconnaissance field project do not include a controlled structure contour map, a careful record of all pertinent observed data can result in structures outlined sufficiently well to warrant their consideration as prospects without additional work. This might become necessary if the area becomes one of increased activity by competing oil companies and there is insufficient time to do detailed structural work on prospects outlined by the project. Therefore, it is important to emphasize that the field geologist selected to undertake a reconnaissance field project should be an experienced one with complete confidence in his ability to map and evaluate an area and to be selective in his choice of data and field analysis so as not to be slowed down by indecision in the field.

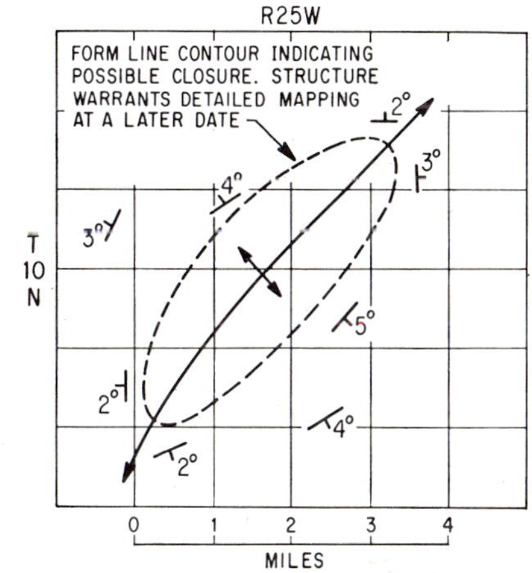

FIGURE 10-35. Reconnaissance geology. Diagram illustrates method of emphasizing possible closure by use of one form-line structure contour.

Use of Air Photographs (Mosaics and Contact Prints) for Reconnaissance. A complete set of photomosaics covering the area to be studied is the most satisfactory base on which to do reconnaissance field work. A well-planned reconnaissance geology field project can be completed very rapidly using photomosaics because not only do they give an air view of the area to be worked but many of the larger structural features and major outcrop patterns can be observed in advance.

If a definite structural pattern for the area can be defined on the photomosaics, then traverses can be planned across the structural grain at regular intervals, preferably along good access roads if present. If strike and dip, contacts, and other geological features are recorded along each line of traverse, the areas between lines of traverse can be filled in by tracing the contacts and faults on the mosaic, providing the geology is not too complex. If a complex area is found, side traverses can be run to help work the structure into the regional pattern. If contact prints are also available to the party, they are a great aid for location of points in the field when the scale of the mosaic makes this difficult and should be used whenever possible for more accurate location of stations where data are observed and recorded.

If no general regional structural pattern can be defined on the photomosaics, then some type of parallel traverses should be planned across the field project area in order to obtain an initial idea as to the type of structure present, if any, and then additional traverses can be run at right angles to the original ones to fill in the areas between them. Random, unplanned traverses can be very time consuming, and though they may add considerably to the amount of geological data obtained, their actual value may be small when compared with what could have been obtained with better planning.

Topography may prevent a complete adherence to this method of approach, but there may still be a sufficient number of roads and stream valleys in the area to allow an initial effort in this manner, and an additional plan of traverses can be worked out depending upon the results.

The accompanying illustration (Figure 10-36) is a preliminary map of reconnaissance work done in an area where the structural pattern was not readily seen on the mosaics. First the Cretaceous-Jurassic boundary was located in the field, and its general outline traced out on the mosaic. Following this, an east-west traverse was run along the contact area at the south end of the map, followed by a traverse to the north along the western boundary of the Cretaceous outcrop. Owing to inaccessibility and poor exposures, additional points were located within the area of Cretaceous outcrops, and sufficient data were obtained to indicate a possible anticline in the northeast corner of the map and a southeast-plunging nose in the southwest corner of the map. Axes of the structural features may be shown by dashed lines on the map, after which the party chief must evaluate his field data as to reliability and decide whether additional, more detailed work is warranted. As no plunge can be indicated on the anticline in the northeast corner of the map, based on the dips obtained, additional work on this structure might find reversal, and it is probably the more favorable structure on which to plan additional work.

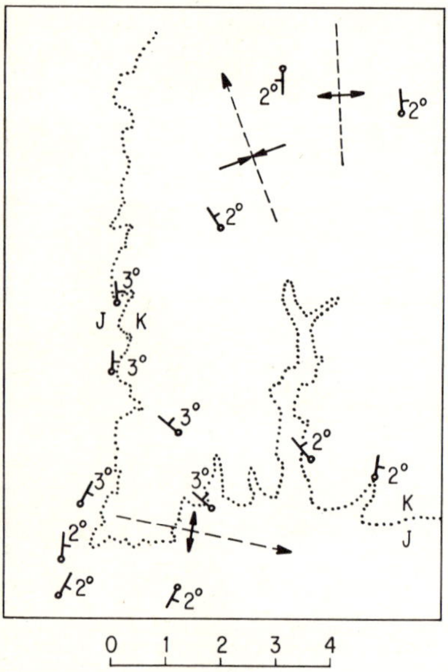

FIGURE 10-36. Reconnaissance geologic map of a small area illustrating structural features located and perhaps requiring additional, more detailed field work. (*Courtesy of Continental Oil, Company.*)

While in the process of obtaining additional strike-and-dip data for a structure such as this, the party chief should bear in mind the possibility of returning to construct a detailed structure contour map of the area, and hence, if any persistent marker bed is observed, it should be recorded in the notes as a potential mapping horizon. If possible, it can be sketched in on the final map as an aid in visualizing closure if established, rather than using a form-line contour. These data might be sufficient to evaluate the structure without benefit of a detailed contour map. The value of good dip diagrams cannot be overemphasized, and therefore, when field data are recorded, it is important to note the quality of the outcrop and apparent reliability of the strike and dip.

Reconnaissance Work on Topographic Maps. If photomosaics are not available for the field party but the area is one in which good detailed topographic maps are available on a scale of 1 inch to the mile or larger, reconnaissance field work can be done in a manner similar to that used with mosaics.

Persistent, more resistant beds will be cliff formers and can be readily observed on topographic maps. Once the formation forming this cliff is observed in the field, its lower contact can be traced on the topographic map along the base of the cliff. An

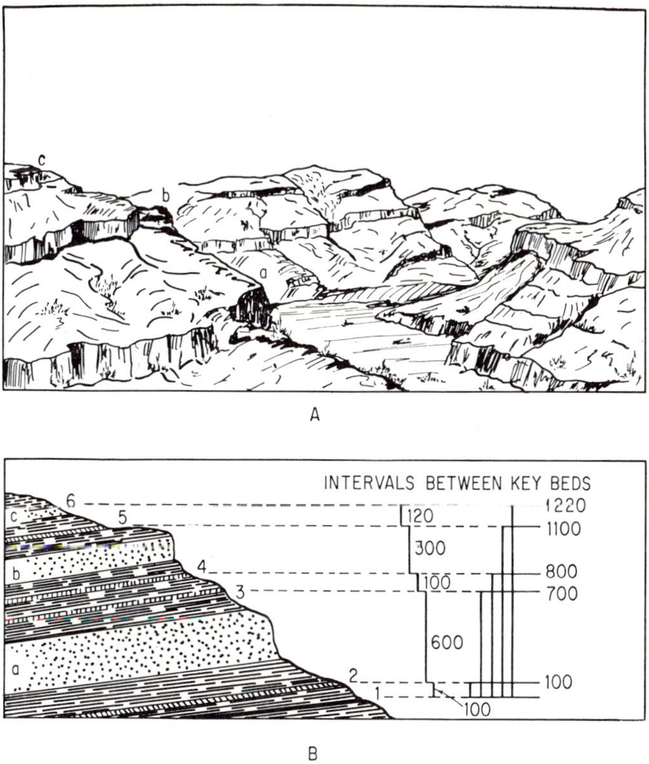

FIGURE 10-37. Identification of key beds and intervals between them for structural mapping. (*A*) Sketch of exposure in cliff showing key beds *a*, *b*, and *c* that can be used for obtaining structural data. (*B*) Cross section shows stratigraphic positions of key beds *a*, *b*, and *c*. (*From Low, 1957.*)

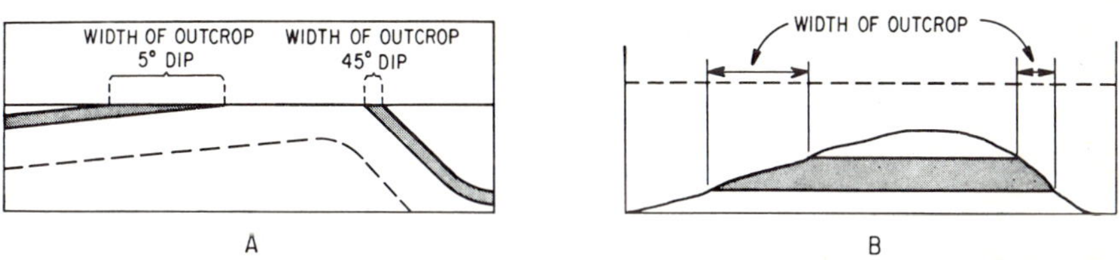

FIGURE 10-38. Topographic expression of outcrop patterns. (*A*) Variation in width of outcrop for a dip of 5 degrees compared with a dip of 45 degrees where ground is essentially flat. (*B*) Relation of slope of topography to width of outcrop in area underlain by horizontal beds. (*Modified from Duke and Brown, 1925.*)

excellent example of this can be seen on the topographic map (Figure 10-39), on which the approximate contact between two units in the Cretaceous is outlined: the Mancos shale and the overlying cliff-forming Mesa Verde sandstone. This area is in northwestern New Mexico in the Shiprock quadrangle, a 15-minute quadrangle mapped on a scale of 1:62,500 by the Topographic Branch of the U.S. Geological Survey.

Although location of stations on the map illustrated may not be so easily accom-

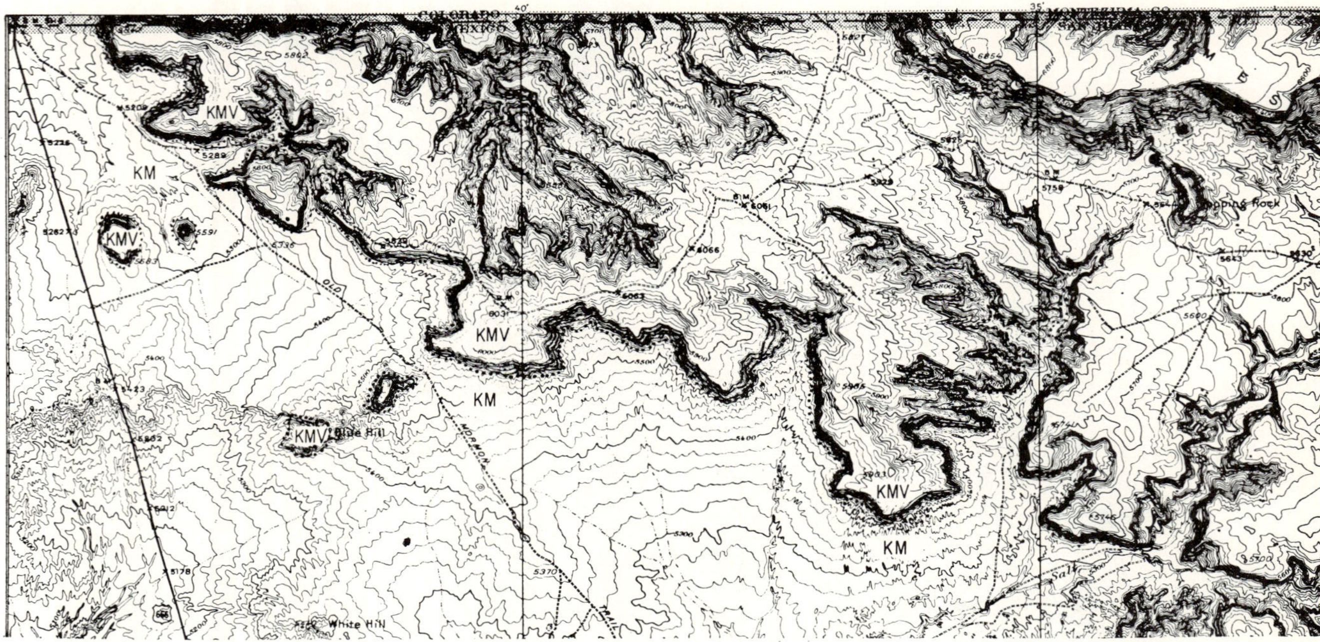

FIGURE 10-39. Tracing approximate contacts on topographic maps. U.S. Geological Survey topographic map of the Chimney Rock quadrangle, San Juan County, New Mexico, offers a good example of the topographic expression of the Mesa Verde–Mancos contact along the base of a steep cliff.

plished as on photomosaics, stream junctions and isolated buttes, as well as small branches off major canyons, can usually be readily recognized in the field. In the event it is desirable to locate a station on a traverse out onto the featureless area shown on the map, there is a sufficient number of widespread buttes and prominent high points to allow the use of two or more as triangulation points and to locate the station by resection with a Brunton compass.

This particular example is in an area of relatively low dips, and any structures present are not readily apparent on the map. However, in areas of strong folding, the major structural features can be easily seen on a topographic map, in many instances, by the outline of the cliff-forming ledges as indicated by topographic contours.

Reconnaissance Work on County Road Maps or Base Maps. In both the previous sections the topographic maps and photomosaics discussed offer many ways to locate each station in the field as to position on the map or mosaic. Sometimes, however, it becomes imperative to obtain some type of geological information before either of these items can be obtained, and perhaps only county road maps or a base map are available to the field party, on which only a few prominent features are shown that can be readily located, such as road junctions, stream crossings, railroad crossings, schoolhouses, etc. The important thing is that none of the features shown can be used to locate a field station by resection with the Brunton compass. Under these circumstances, there are at least two methods of approaching the problem that will obtain a reasonably accurate picture of the major geological features for a reconnaissance field project. The two methods assume that a field vehicle, either jeep or car, will be used and that distances can be measured by recording mileage differences on the odometer, using all existing roads first and filling in important areas between the roads by pace-and-compass traverse if it becomes necessary.

1. If only a base map can be obtained showing township and range but on which roads are not accurately mapped except for junctions with secondary roads, the field vehicle can be utilized for a compass survey with distances recorded in miles or tenths of miles. Thus, starting from a known point on the base map, a copy of which should be carried in the field, a bearing can be taken with the Brunton compass to determine the direction of line of traverse, and then the field party can proceed down the road to a station where geological data are to be recorded or where it is necessary to take another bearing due to a turn in the road. At this point the mileage is recorded, another bearing taken for direction of traverse, and then the party can proceed down the road in the vehicle. Stations and data can be plotted on the base map in the field as the traverse progresses. Whenever a known point shown on the map is reached and recognized in the field, it should also be recorded as a station, mileage should be corrected, and the traverse continued.

All secondary roads or any major stream beds crossed on the traverse should also be recorded and plotted on the base map and marked or flagged in the field. If additional traverses are warranted at right angles to the original traverse, these locations may be helpful in planning them and knowing the possibility of access to the area. If none is present and a secondary traverse is required, then it must be negotiated as a pace-and-compass traverse from the nearest previous station, thus emphasizing the importance of marking all stations along a traverse, with either a rock cairn or some brightly colored flagging.

2. If good county road maps are available on which roads are accurately shown with all turns, river crossings, railroad crossings, as well as schoolhouses, ranches, etc., correctly mapped, the same procedure can be followed as previously discussed, except that it may not be necessary to take bearings with the Brunton compass. Mileage can be recorded at a known starting point, such as a junction or bridge, and the traverse can then be run down the road to the first station where geological data can be obtained,

the mileage recorded there, and the distance scaled on the map along the road from the starting point. Mileage should be checked and adjusted at regular intervals when identifiable side roads or other features shown on the map are encountered by the traverse. As new minor roads may have been constructed that are not shown on the county road map, the position of any encountered should be plotted on the map, and if it should be necessary to use one for a side traverse, it will require the use of a compass traverse.

The accompanying illustration (Figure 10-40) is a map of a traverse run along a main highway in an area over which only one formation is exposed. Starting from station *A*, at the top of the map, the traverse proceeded southward, and where geological data were obtained, mileage was recorded and the distance measured on the map so that the point could be plotted. It should be noted that stations *B*, *G*, and *I* were identified on the map and are indicated as mileage check points. The traverse crossed three anticlinal structures from north to south, one showing no plunge from data obtained, one showing northwest plunge, and one showing southeast plunge. The scale of the map indicates these structures to be broad enough, and hence large enough, to warrant additional side traverses to determine possible closure by means of a dip diagram.

This type of traverse is recommended only for reconnaissance geology in which the major objective of the project is the delineation of major structural features which can be readily defined by a dip diagram and outlined for recommendation of future detailed structural mapping. Sufficient data could be obtained, however, to warrant acreage acquisition in an area of high interest, provided the area was quite active and it was necessary to act rapidly without benefit of detailed geology.

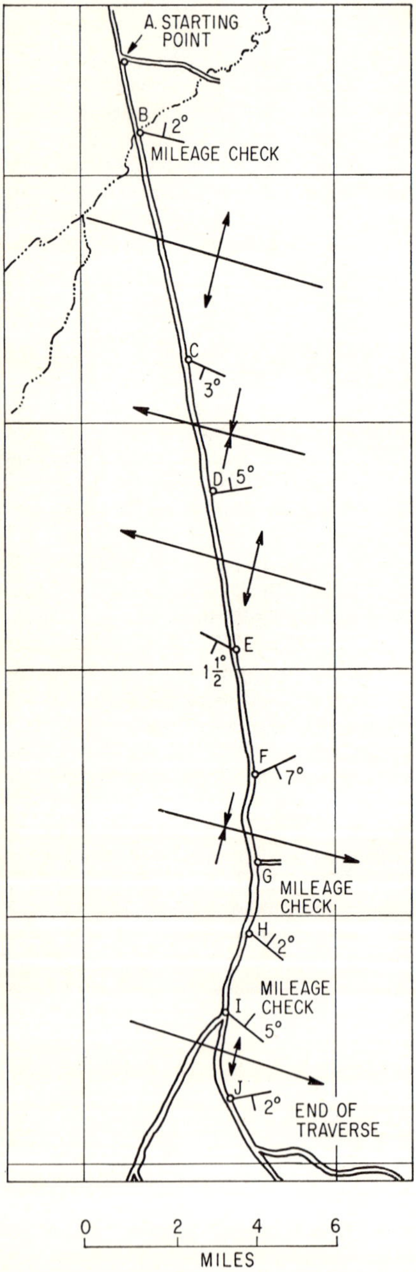

FIGURE 10-40. Reconnaissance geology on a county road map. Traverse starts at *A* and ends at *J*. Distances between points measured with odometer in automobile, using mileage checks to correct errors at recognizable features shown on map. (*Courtesy of Continental Oil Company.*)

Detailed Reconnaissance: Areal and Structural Geology

The term *detailed reconnaissance* is introduced to differentiate field work of the type in which, in addition to strike-and-dip data and outcrop patterns, a varying amount of vertical control is obtained in the field so that the end product includes a structure-contour map, as opposed to the final map of a reconnaissance field project, which will indicate only structure by dip diagrams and form-line contours.

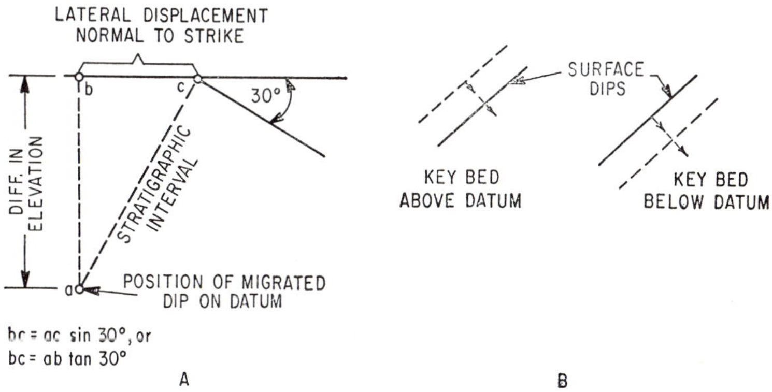

FIGURE 10-41. Migration of dip in structural field work. (*A*) Trigonometric relation of dip migration. (*B*) Relative direction of dip migration with respect to key bed. If key bed is above datum, migration is updip; if key bed is below datum, migration is downdip. (*After Low, 1957.*)

Detailed reconnaissance differs from detailed structural work in that, although vertical- and horizontal-control errors are kept to a minimum by closing on known points as often as possible, the project is not limited to strict closures by plane-table methods. A barometer can be used in conjunction with a plane table and alidade, so that long traverses with the plane table are not necessary to carry vertical control through an area. When a good station is found from which many triangulation or rod shots can be made with the alidade and which can be readily located by triangulation or on a photomosaic, the elevation can be determined with the barometer and the surrounding area then shot in with the alidade.

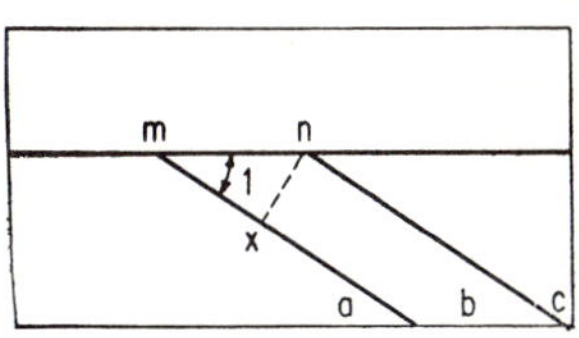

FIGURE 10-42. Solution for dip, thickness known, horizontal topography. Sine angle $1 = nx/mn$, in which nx is thickness of formation b and mn is width of outcrop. (*Modified from Dake and Brown, 1925.*)

For this type of work, if long triangulation shots are contemplated with the alidade, it is best to use a 100-foot contour interval for the structure map, although if extreme care is exercised in the barometer work and with the alidade, a 50-foot contour interval might be satisfactory.

Detailed reconnaissance, like regular reconnaissance geology, is done with the objective of outlining the larger structures of an area but with the added advantage of a fairly well-controlled structure-contour map. In some instances the barometer may be the only instrument used to obtain elevation data for structural control, but if it is combined with the use of the alidade and plane table, an area can be covered much more rapidly.

The most satisfactory maps on which to do detailed reconnaissance geology are

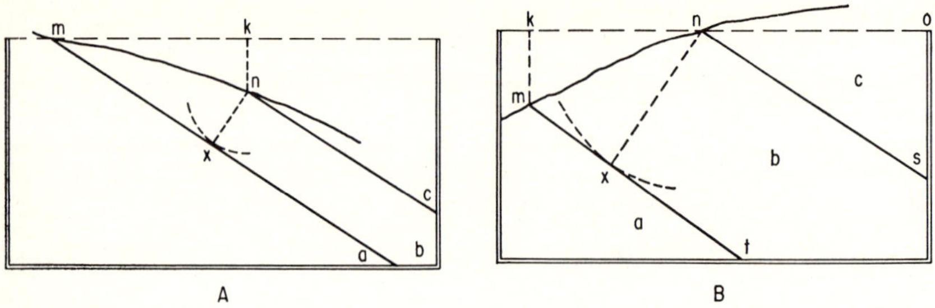

FIGURE 10-43. (*A*) Solution for dip, thickness given, beds dipping same direction as topography.

$$mn = \sqrt{mk^2 + kn^2}$$

$$\tan \angle kmn = \frac{kh}{mk}$$

$$\sin \angle xmn = \frac{xn}{mn}$$

$$\text{Dip} = \angle kmn + \angle xmn$$

(*B*) Solution for dip, thickness given, beds dipping opposite to topographic slope.

$$mn = \sqrt{mk^2 + kn^2}$$

$$\tan \angle knm = \frac{mk}{kn}$$

$$\cos \angle xnm = \frac{xn}{mn}$$

$$\text{Dip angle } sno = 180° - (90° + \angle xmn + \angle knm)$$

(*From Dake and Brown, 1925.*)

photomosaics, topographic maps, county road maps, and regular base maps, in descending order of preference. However, if a good photocontrolled base map is available, with U.S. Coast and Geodetic Survey and U.S. Geological Survey triangulation stations and bench marks accurately located and with a correct land grid, it may be preferable to a topographic map or a county road map.

Earth Curvature and Refraction. This subject is usually discussed under plane-table mapping, but as its primary use is generally in detailed reconnaissance geology, where long triangulation shots are made with the alidade, it has been included in this section. Its importance cannot be overemphasized in this type of field work, where triangulation shots of 30,000 to 40,000 feet may be made and for which the elevation correction for curvature and refraction is 18 and 32 feet, respectively.

The elevation correction for curvature and refraction can be expressed by the equation

$$C = 0.02044 \left(\frac{d}{1{,}000}\right)^2$$

in which C = elevation correction, feet
d = distance, feet

The correction is always added algebraically to the elevation as determined by distance and angle recorded with the alidade. The equation given includes the correction for both curvature and refraction, the basis of which is the fact that the level datum to which elevations are referred is a surface having the curvature of the earth, whereas the line of sight of the alidade is tangential to this surface, and hence distant objects appear lower than they really are (see Figure 10-44). The refraction correc-

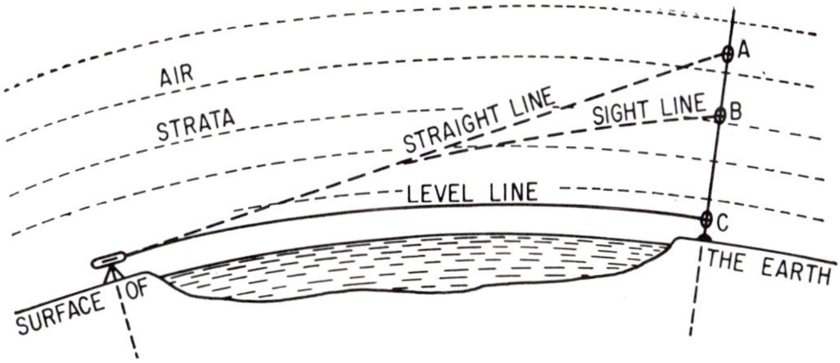

FIGURE 10-44. Exaggerated diagram showing influence of curvature and refraction upon observations for determining differences in elevation between two points. (*From Mather and Washburn, 1938.*)

tion, a minor correction applied to the equation, is based on the density stratification of air above the surface of the earth which causes a downward bending or refraction of light rays passing through them, and hence this causes a reduction of the curvature correction, as opposed to what it would be without the effect of refraction (Lahee, 1961, p. 528–529).

Table 10-4. Curvature and Refraction Corrections for Field Use

Distance	Correction*	Distance	Correction*	Distance	Correction*
1,000	0.02	25,000	12.73	49,000	49.10
2,000	0.08	26,000	13.84	50,000	51.10
3,000	0.18	27,000	14.90	51,000	53.20
4,000	0.32	28,000	16.04	52,000	55.30
5,000	0.51	29,000	17.20	53,000	57.40
6,000	0.74	30,000	18.40	54,000	59.60
7,000	1.00	31,000	19.65	55,000	61.70
8,000	1.31	32,000	20.92	56,000	64.10
9,000	1.65	33,000	22.22	57,000	66.41
10,000	2.04	34,000	23.68	58,000	68.76
11,000	2.47	35,000	25.05	59,000	71.15
12,000	2.94	36,000	26.52	60,000	73.58
13,000	3.46	37,000	28.00	61,000	76.05
14,000	4.01	38,000	29.52	62,000	78.56
15,000	4.60	39,000	30.90	63,000	81.13
16,000	5.23	40,000	32.70	64,000	83.73
17,000	5.92	41,000	34.40	65,000	86.36
18,000	6.63	42,000	36.20	66,000	89.04
19,000	7.38	43,000	37.80	67,000	91.75
20,000	8.18	44,000	39.60	68,000	94.51
21,000	9.02	45,000	41.40	69,000	97.31
22,000	9.90	46,000	43.30	70,000	100.15
23,000	10.80	47,000	45.20	71,000	103.2
24,000	11.78	48,000	47.20	72,000	106.2

SOURCE: Continental Oil Company.

* Elevation correction per 1,000 feet of distance.

$$C = 0.02044 \left(\frac{d}{1{,}000}\right)^2$$

where C = correction (in feet)
d = distance (in feet)
Correction is always positive.

A table of corrections for curvature and refraction (Table 10-4), for distances given in 1,000-foot intervals has been included in the text and can be used for field reference when required. However, the equation

$$C = 0.02044 \left(\frac{d}{1,000}\right)^2$$

will plot as a straight line on log-log paper, as shown in Figure 10-45, with distance, in 1,000-foot units, plotted as ordinates and elevation corrections as abscissas. If field work is planned in which several curvature corrections are contemplated, data from Table 10-4 can be plotted on four-cycle log-log paper, thus enabling one to obtain the correction for distances between the 1,000-foot units. The distance obtained can be picked on the ordinate line, and a horizontal line extended to intersect the straight line of the graph. From this point a vertical line can be extended downward to the abscissa, where the elevation correction for that distance can be read (Figure 10-45).

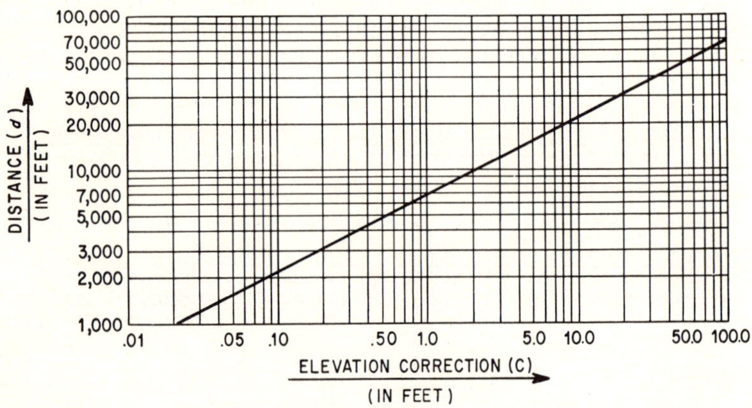

FIGURE 10-45. Curvature and refraction correction chart. $C = 0.02044(d/1,000)^2$. Correction is always positive. Compiled and prepared by K. G. Smith.

Detailed Reconnaissance on Photomosaics. Several methods can be used to obtain a fairly accurate and detailed areal and structural geology map of an area when mapping is done with photomosaics. The simplest of these is a series of traverses across the area with a barometer, obtaining elevations on the base or top of the bed being used as a mapping horizon. The step traverse with the barometer, as discussed in a previous section of this chapter, enables a two-man field party to move rapidly across an area and to obtain sufficient data to draw a structure contour map, as well as obtaining strike-and-dip data and mapping rock units or formations.

Although this method or one similar to it with one barometer read regularly at one station during the day to establish the diurnal curve is often used, the actual data obtained in the field require that the barometer be read on the contact being mapped or on some other bed from which the interval is known to the mapping horizon. A much more rapid method of structural mapping on air photographs combines the use of the barometer with a plane table and alidade and enables the geologist to shoot in several reference datum points on the mosaic from one instrument station. There are three ways in which this type of work can be carried out, in all of which the elevations for instrument stations are obtained with the barometer.

1. Low-angle shots with alidade to mapping horizon, using distance as scaled on the mosaic
2. High-angle shots with alidade to mapping horizon, using distance as scaled on mosaics
3. Level-angle shots with alidade to mapping horizon or to other beds of known interval above or below the mapping horizon

All three methods require the presence of one or more good mappable units in the project area that are prominent enough to stand out on the mosaics used. Contacts used as mapping horizons should be traced out on the mosaic and transferred to the final map. In this type of work a contour interval of not less than 50 feet is recommended, with a 100-foot contour interval being preferable. If necessary, a combination of the three methods can be used, depending upon field conditions which might require low-angle, high-angle, and level sights with the alidade.

This method of doing detailed reconnaissance geology requires a controlled photomosaic for field use, preferably on a scale of 1 inch to the mile for maximum benefit of detail shown and areal coverage on each sheet used. A magnetic north line should be drawn on the mosaic for orientation in the field, and then the mosaic to be used can be taped to the plane-table board with ordinary drafting tape. Contact prints for each mosaic may be helpful for location of stations in the field and for tracing out minor geological features that should be included on the map and, if available, should be carried in the field.

The method can best be described by use of the map (Figure 10-46) as an illustration of the results that can be obtained. The two instrument stations *I*-1 and *I*-2 were located on the mosaic by means of resection from known points and with the help of the contact prints. Elevations of the instrument stations were obtained by a step traverse with the barometer. The horizon being mapped is the contact between the younger bed *A* and the older bed *B*. This contact was drawn in on the mosaic before any shots were taken with the alidade.

Stations *S*1 through *S*4 were shot in from instrument station *I*-1, while stations *S*5 through *S*15 were shot in from instrument station *I*-2. In each instance, the stations where structural data were obtained are prominent outcrops of the beds with the contact well exposed. The middle cross hair of the alidade is sighted in on the contact, and the vertical angle read and recorded. For this type of work it is best to level the alidade before each shot, then the point being sighted in is marked with a needle or pin on the mosaic and further marked and given a station number once the alidade is removed. The distance is then measured from instrument station to datum station and also recorded in the notes. Elevation of the contact is then calculated from the vertical angle, the distance recorded, and a structural datum is obtained for the horizon being mapped. Curvature and refraction corrections are applied to the longer shots, such as stations *S*6 and *S*7, which are over 28,000 feet from instrument station *I*-2. The illustration (Figure 10-46), though only part of a structural map of a large anticline, illustrates the large area that can be covered rapidly by this method, thus increasing the efficiency of the entire project and saving valuable time. The area of structural control represents in excess of 20 square miles covered from the two instrument stations and does not include several more points that were shot in from instrument station *I*-1. In this instance the outcrop pattern does not indicate the presence of the structure, and it would not be possible to identify it on the photomosaic without the aid of contact prints.

An objection that might be raised to this method of obtaining structural data is that distances measured might not always be accurate on the photomosaics. Two items previously mentioned, the use of a 50- or 100-foot contour interval and the use of

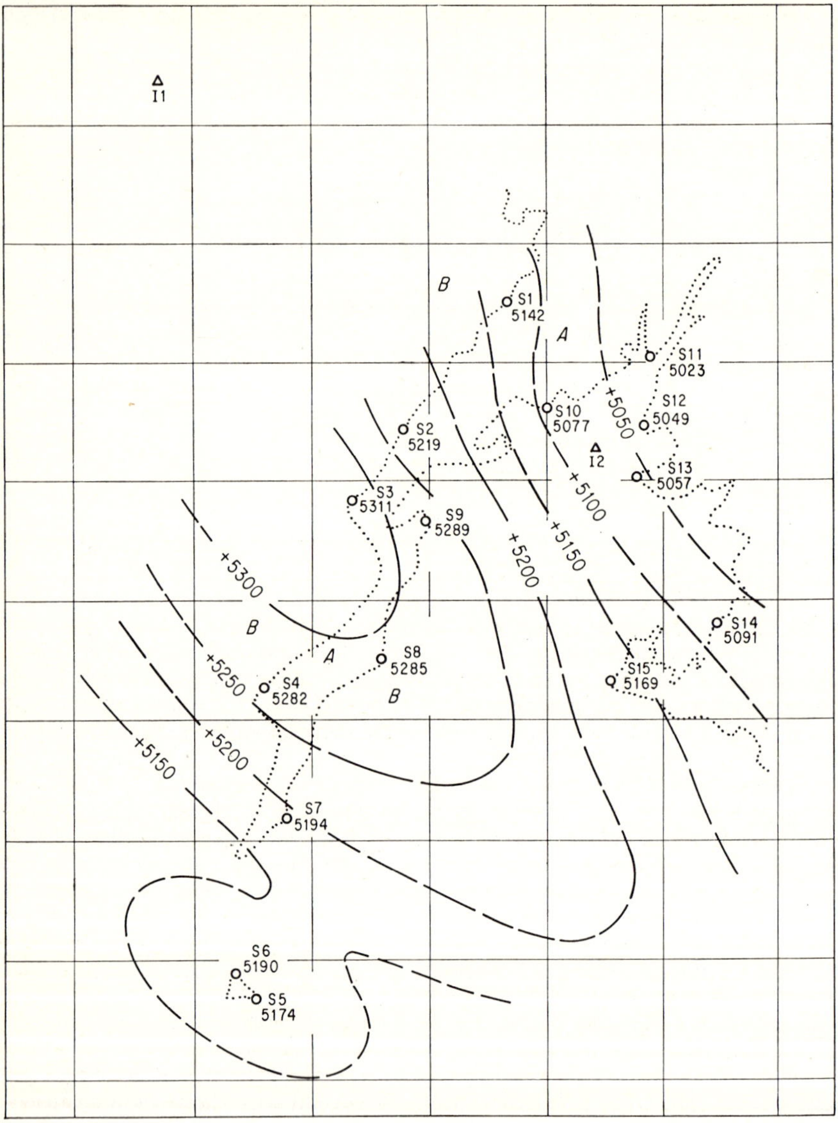

FIGURE 10-46. Detailed reconnaissance geologic map. Structural data obtained by combined use of barometer, plane table, and alidade on photomosaic. Datum is contact between younger sandstone bed *A* and underlying shale bed *B*. (*Courtesy of Continental Oil Company.*)

low-angle shots, should keep errors to a minimum. A vertical angle of 2 degrees or less is considered best for this type of work, although a higher angle can be used occasionally when distance is not too great. For example, if the vertical angle read was 2 degrees and there was an error of 200 feet in the distance scaled due to distortion on the photomosaic, the resulting error in elevation difference would amount to 7 feet;

for a vertical angle of 1 degree, an error in distance of 500 feet would result in an elevation difference error of about 9 feet. Since the objective of detailed reconnaissance structural work is the outlining of major structures in the area, these errors would not greatly affect the picture if a 100-foot contour interval were used. In any plane-table work, including a stadia traverse, the greater angles of inclined sights increase the possibilities for errors in both distance and elevation determinations.

FIGURE 10-47. Dip-and-strike determination with the plane table. (*A*) Rod is held at a point level with the instrument station and strike is drawn. (*B*) Rod is placed downdip and at right angles to the strike line, and amount of dip is read on the vertical arc. (*From Low, 1957.*)

Since this type of work requires the determination of elevations by barometer traverse for instrument stations, whenever this traverse happens to cross the contact being used as a mapping horizon, its elevation should also be recorded and the point located on the mosaic and given a station number but designated as a barometer datum.

Structural data should be supplemented with strikes and dips, where possible, as an aid for final contouring on the map. However, in low-dip areas this method can be extremely helpful in working out structure when it may be difficult to obtain a strike and dip in the field.

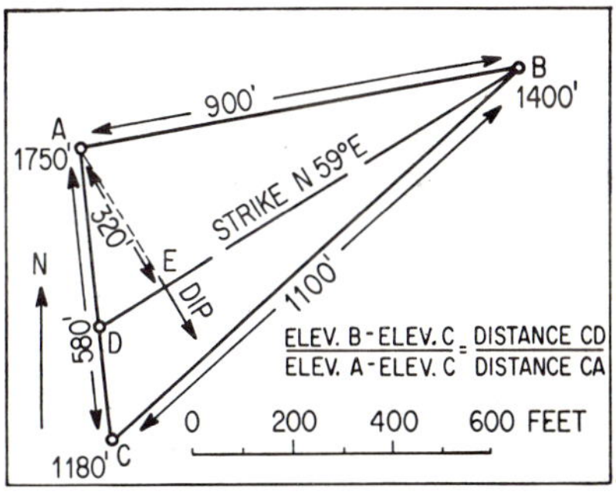

FIGURE 10-48. Three-point method for determining strike and dip. Given three points of known elevation on top of a bed, location of point *D* can be determined. Elevation at point *D* is 1,400 feet, and line *BD* is the strike. Angle of dip can be determined from line *AE*, along which the rate of dip from point *A* is 350 feet in 320 feet and the angle of dip is 47.5 degrees.

Detailed Reconnaissance on Topographic Maps. Well-controlled U.S. Geological Survey topographic maps on a scale of 1 inch to the mile are readily adaptable to a method of detailed reconnaissance mapping similar to that described for work on photomosaics. However, when work is done on topographic maps, it is necessary that the geologists be completely familiar with the stratigraphic section in the area to be mapped and know the topographic expression of mappable units. This can be attained by detailed measurement of sections before starting the major part of the project, followed by reconnaissance trips across the area to be mapped either by car or in a plane.

Once the stratigraphic section and topographic expression of units are known, then approximate contacts (tops or bases) of units can be sketched in on the topographic map from high points in the field. Elevations can be determined for instrument stations by means of barometer traverses, and the topographic contours will aid in assuring correct determinations. Sightings from the instrument stations can be made with the alidade on the topographic map, mounted on the plane-table board, using the contacts as sketched in on the map. Vertical angles can be recorded, and with the distance as scaled on the map, elevation differences can be determined. Instrument stations can be located by resection from recognizable high peaks or triangulation stations which are shown on the topographic map.

An illustration of the results of this method is shown in Figure 10-49, a structural contour map of the Muley Creek anticline in Garfield County, Utah. Structure contours are drawn on the base of the Dakota sandstone based on data obtained by sighting in on contacts as sketched on the topographic map, recording vertical angles, and scaling distance on the map, with 8 shots from instrument station Number 1 (points sighted in from this station indicated with a small cross) and 28 shots from instrument station Number 2 (indicated by small circle). In the total of 36 shots from instrument stations, 29 were on the horizon mapped while 7 were on a bed 250 feet lower than the mapping horizon. A complete set of notes as recorded for the field work is shown in Figure 10-50, and the manner in which the notes are recorded is recommended for use in this type of work, even though it requires the construction of notebook pages in the field, since this arrangement is not shown in standard field notebooks available. The same arrangement of notes can also be used for detailed reconnaissance on mosaics.

Datum points which required conversion to the mapping horizon are underlined on the map and are indicated in the field notes. Curvature and refraction corrections, though of minor consequence in most instances, have been included in the field notes to illustrate their use. In the vertical-angle column, it can be noted that most of the angles used were less than 2 degrees, with only two datum points requiring the use of a vertical angle in excess of 2½ degrees. Occasional use of greater angles is permissible, but their general use is not recommended.

To realize fully the possible value of this method, the time element should be taken into consideration. The data necessary to complete the map as drawn (Figure 10-49) require less than a full day to obtain. The end results can be further evaluated by a comparison with a U.S. Geological Survey map of the same structure, contoured on the base of the Ferron sandstone (Figure 10-51) (Geology and Geography of the Henry Mountains Region, Utah, by Charles B. Hunt, Paul Averitt, and Ralph L. Miller, *U.S. Geol. Survey Profess. Paper* 228, 1953). Mapping for this project was completed by standard plane-table methods. The base of the Ferron sandstone is approximately 550 feet stratigraphically higher than the base of the Dakota sandstone in this area, and the Dakota structure map can be rapidly converted to a comparable map by adding 550 feet to each datum point. Although the rapid-method map does not have the refinements of the more detailed map, carefully selected instrument stations and datum points, accompanied by drawing of structure contours in the field, have resulted in the outlining of a major structure in a very short period of time.

Detailed Reconnaissance Using Well-controlled Base Maps. Well-controlled base maps, on which both U.S. Coast and Geodetic Survey and U.S. Geological Survey bench marks are accurately shown, can be used to excellent advantage for detailed structural reconnaissance. If the land has been surveyed by the General Land Office, it facilitates location of instrument stations if section corners can be found, but this is not essential.

A rapid method of doing reconnaissance on controlled base maps is to run barometer traverses across the area and establish a series of temporary bench marks to use for

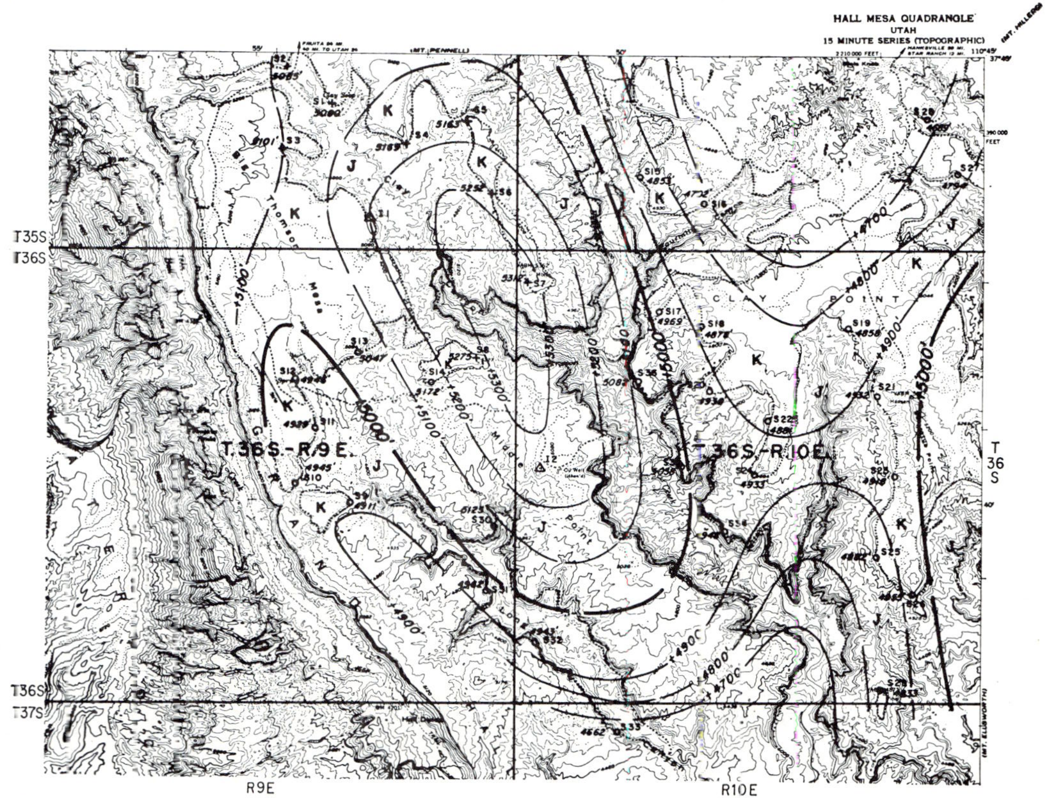

FIGURE 10-49. Detailed reconnaissance geologic structure map of the Muley Creek anticline, Garfield County, Utah, mapped on U.S.G.S. topographic map using barometer, plane table, and alidade. Structure contours on base of Cretaceous. (*Courtesy of Continental Oil Company.*)

Date________________________

	No. inst. sta.	No. sight sta.	Angle			Vert. corr. factor	Distance	Diff. elev.	H. I.	Curvature correction	Elevation	Diff. to datum	Datum	Geological column
			Level	+ or −	Diff.									
1	I-1	S1	30°00′	29°03′	−0°57′	1.66	8200	−137.0	5216.0	+1.5	5080.5		+5081	Base Dakota sandstone
2		S2	30°00′	29°21′	−0°39′	1.135	11,800	−134.0	5216.0	+2.5	5084.5		+5085	" " "
3		S3	30°00′	29°14′	−0°46′	1.34	7800	−104.5	5216.0	......	5101.5		+5102	" " "
4		S4	30°00′	29°31′	−0°29′	0.84	5600	−47.0	5216.0	......	5169.0		+5169	" " "
5		S5	30°00′	29°40′	−0°20′	0.58	9400	−54.5	5216.0	+1.8	5163.3		+5163	" " "
6		S6	30°00′	30°14′	+0°14′	0.41	8400	+34.4	5216.0	+1.5	5251.9		+5252	" " "
7		S7	30°00′	30°28′	+0°28′	0.81	11,500	+93.1	5216.0	+2.5	5311.6		+5312	" " "
8		S8	30°00′	30°16′	+0°16′	0.47	11,950	+56.2	5216.0	+2.9	5275.1		+5275	" " "
9	I-2	S9	30°00′	28°55′	−1°05′	1.89	13,100	−263.0	5170.0	+3.5	4910.5		+4911	" " "
10		S10	30°00′	29°12′	−0°48′	1.40	16,600	−232.0	5170.0	+5.5	4943.5		+4944	" " "
11		S11	30°00′	29°05′	−0°55′	1.50	15,400	−246.0	5170.0	+4.8	4928.8		+4929	" " "
12		S12	30°00′	29°15′	−0°45′	1.31	17,600	−230.5	5170.0	+6.2	4945.7		+4946	" " "
13		S13	30°00′	29°30′	−0°30′	0.87	14,700	−128.0	5170.0	+4.5	5046.5		+5047	" " "
14		S14	30°00′	30°00′	0°00′	.....	9500	Level	5170.0	+1.8	5171.8		+5172	" " "
15		S15	30°00′	29°07′	−0°53′	1.54	21,200	−326.0	5170.0	+9.1	4853.1		+4853	" " "
16		S16	30°00′	28°43′	−1°17′	2.24	21,300	−477.0	5170.0	+9.2	4702.2		+4702	" " "
17		S17	30°00′	29°07′	−0°53′	1.54	13.300	−205.0	5170.0	+3.6	4968.6		+4969	" " "
18		S18	30°00′	28°50′	−1°10′	2.04	14,500	−296.0	5170.0	+4.3	4878.3		+4878	" " "
19		S19	30°00′	29°11′	−0°49′	1.425	22,600	−322.0	5170.0	+10.4	4858.4		+4858	" " "
20		S20	30°00′	28°53′	−1°07′	1.89	12,450	−235.0	5170.0	+3.0	4938.0		+4938	" " "
21		S21	30°00′	29°23′	−0°37′	1.08	23,000	−249.0	5170.0	+10.8	4931.8		+4932	" " "
22		S22	30°00′	28°53′	−1°07′	1.89	15,600	−294.0	5170.0	+4.9	4880.9		+4881	" " "
23		S23	30°00′	29°22′	−0°38′	1.11	23,700	−263.0	5170.0	+11.2	4918.2		+4918	" " "
24		S24	30°00′	29°02′	−0°58′	1.69	14,300	−241.0	5170.0	+4.2	4933.2		+4933	" " "
25		S25	30°00′	29°16′	−0°44′	1.28	23,400	−299.0	5170.0	+11.0	4882.0		+4882	" " "
26		S26	30°00′	29°30′	−0°30′	0.87	26,400	−229.0	5170.0	+14.0	4955.0		+4955	" " "
27		S27	30°00′	29°20′	−0°40′	1.16	34,500	−400.0	5170.0	+24.0	4794.0		+4794	" " "
28		S28	30°00′	29°11′	−0°49′	1.425	35,300	−507.0	5170.0	+25.5	4688.5		+4689	" " "
29		S29	30°00′	29°16′	−0°44′	1.28	27,500	−352.0	5170.0	+15.0	4833.0		+4833	" " "
30		S30	30°00′	26°32′	−3°28′	6.04	4900	−295.0	5170.0	......	4875.0	+250.0	+5125	Sandstone bed 250 ft below base Dakota sandstone
31		S31	30°00′	27°06′	−2°54′	5.05	9100	−460.0	5170.0	+1.7	4711.7	+250.0	+4962	" " " " " " " "
32		S32	30°00′	27°41′	−2°19′	4.04	11,900	−480.0	5170.0	+2.8	4692.8	+250.0	+4943	" " " " " " " "
33		S33	30°00′	27°40′	−2°20′	4.07	18,800	−765.0	5170.0	+7.0	4412.0	+250.0	+4662	" " " " " " " "
34		S34	30°00′	27°55′	−2°05′	3.63	13,100	−476.0	5170.0	+3.5	4697.5	+250.0	+4948	" " " " " " " "
35		S35	30°00′	27°40′	−2°20′	4.07	9500	−384.0	5710.0	+1.8	4787.8	+250.0	+5038	" " " " " " " "
36		S36	30°00′	27°47′	−2°13′	3.87	8800	−340.0	5170.0	+1.5	4831.5	+250.0	+5082	" " " " " " " "

FIGURE 10-50. Complete set of field notes for structural data on detailed reconnaissance map of Muley Creek structure. (*Courtesy of Continental Oil Company.*)

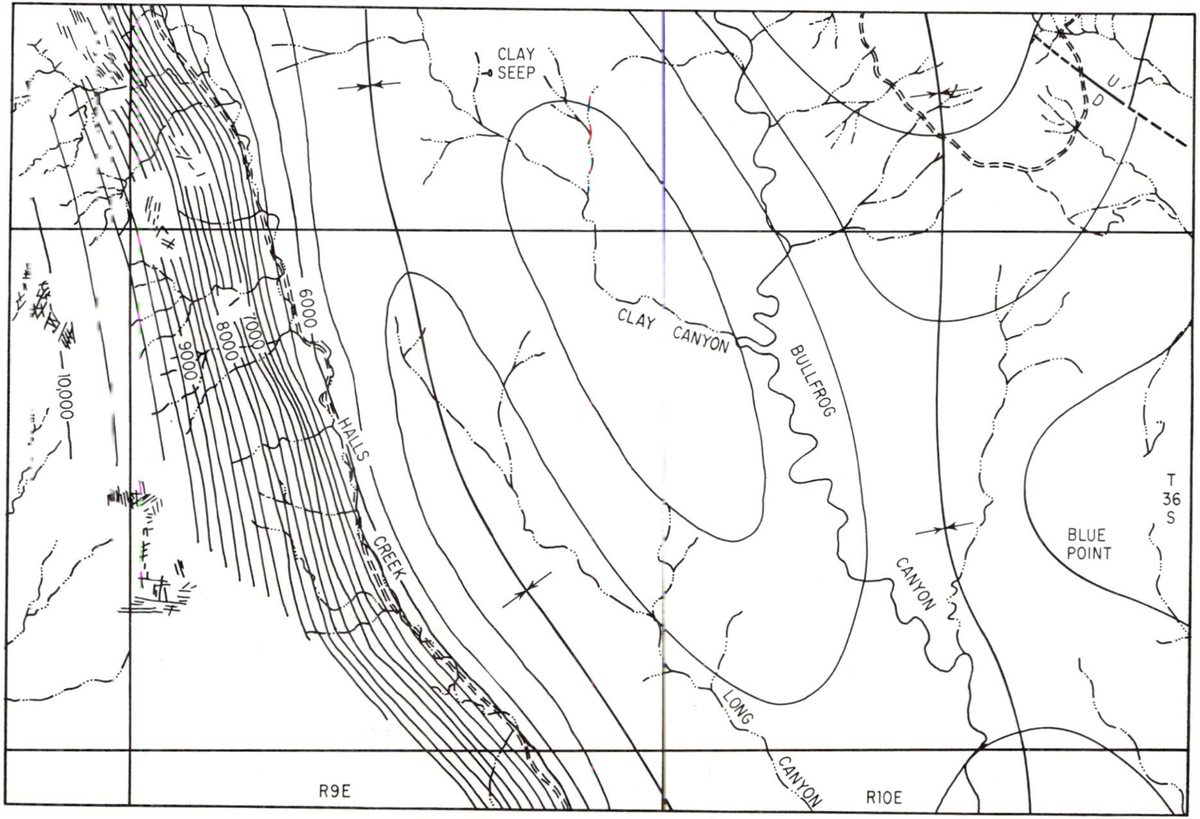

FIGURE 10-51. Detailed structure map of the Muley Creek anticline obtained by regular plane-table methods, contoured on base Cretaceous Ferron sandstone. (*After C. B. Hunt et al., U.S. Geol. Survey Profess. Paper 228, 1953.*)

elevation reference. At any point where structural data can be obtained, an instrument station can be established, a barometer elevation used, and the contact or horizon to be mapped can then be shot in using the alidade, plane table, and stadia rod. The station can be located easily if its selection has been made so that two or more triangulation stations can be seen from it and can be used to locate by resection.

This method requires a series of overlapping plane-table sheets to which the triangulation stations and bench marks can be transferred accurately from the base map. An excellent type of plane-table sheet to be used for this is a regular stabilene base. This material is very transparent and does not require a light table when tracing data from the base map onto it, it will not shrink or wrinkle, and it is very durable and will not be damaged by rain.

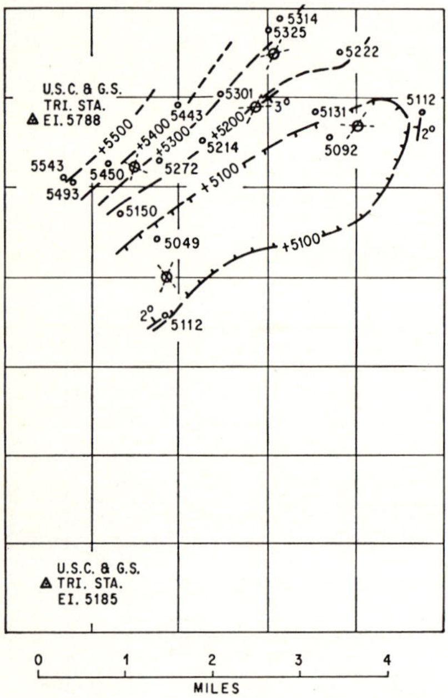

FIGURE 10-52. Detailed reconnaissance on base map using two triangulation stations for horizontal control, barometer for vertical control. Instrument stations (circle crossed by two dashed lines) located by two-point resection. Datum points obtained by radiation with plane table and stadia rod. (*Courtesy of Continental Oil Company.*)

The small structure map (Figure 10-52) is part of a plane-table sheet used for this method of detailed reconnaissance work. Magnetic north was plotted on the original map, elevations were carried by barometer traverse, and whenever a desirable instrument station location was found, the board was first oriented on magnetic north and the elevation established. The station was then located by resection from the two triangulation stations on the west side of the map. The angles of resection for the two lines were considered sufficiently large to give an accurate map location, and any instrument station that was on a high point was flagged to be used as a check point from other instrument stations after they were located by resection from the same two triangulation stations.

Once the location of the instrument station and its elevation were established, the structural data were shot in on the map (Figure 10-52) using the stadia rod. Although this method is slower than in the use of photomosaics, owing to the necessity of the rod shots, once a geologist and instrument man become accustomed to an area and are thoroughly familiar with the surface stratigraphy and intervals of marker beds above and below the horizon being mapped, work can progress quite rapidly. The use of two field vehicles will greatly increase the speed of the project. If a step traverse is being made with the barometer, considerable time can be saved if, while one man is returning to the last known point of elevation to read the barometer, the other orients the plane table and locates the station by resection. It is preferable that the geologist stay to orient the board and locate the station on the plane-table sheet, as then he can be scouting the area and have his rod shots planned out by the time the other returns from recording the necessary barometer readings. Both men, assuming a two-man party, should be more than competent in the use of the alidade and the barometer, so that there can be an interchange of work when necessary.

Although fewer actual field data can be recorded by this method in a given amount of time, as compared with work on mosaics, a series of traverses across an area using only base maps can furnish sufficient data to determine regional trends and locally large structures. In areas where favorable structures are found, additional rod shots might be taken for more detailed contouring.

Detailed Mapping

When field projects with objectives of obtaining areal and structural geologic data are graded in amount of detail to be recorded as reconnaissance, detailed reconnaissance, and detailed mapping projects, there is no sharp line of demarcation between each type of project.

Detailed mapping includes the more precise methods of recording field data, location of stations, and determination of datum elevations. The more precise methods would also imply a greater number of control points to determine the configuration of structural contours in the project area. Detailed reconnaissance mapping may require only one datum control point per square mile to obtain the desired results, whereas, for detailed mapping, 10 or more points per square mile may be necessary. Detailed mapping, with these requirements, will progress more slowly and hence require more time to cover a given area as compared with detailed reconnaissance.

Detailed mapping can be carried out with the plane table and alidade on base maps, on air photomosaics, or with a barometer survey on contact prints or air photomosaics. The greater number of control points requires a larger scale of mapping; 2 to 4 inches to the mile is preferable.

Permissible limits of error are smaller for detailed mapping than for reconnaissance and detailed reconnaissance work. For stadia traversing with plane table and telescopic alidade, accurate results within 1 in 200 for horizontal closure should be attained with reasonable care and 3 to 5 feet of closure in elevation (Lahee, 1961, p. 550). For barometer traverses, closures should not be more than half a contour interval. Hence if a 20-foot contour interval is used in detailed mapping, permissible limits of error are considerably smaller than for detailed reconnaissance using a 50- or 100-foot contour interval.

The decision to do detailed mapping in an area may originate within the district office of a company planning a field project based on previous experiences in the area and on the type of information needed to evaluate it, or the party chief, while doing reconnaissance or detailed reconnaissance mapping, may find certain smaller parts of the project area that require detailed mapping to emphasize certain structural features, and the decision will rest with him.

Broad structural features with low dips and a small amount of vertical closure sometimes require a detailed structure contour map with a small contour interval to outline and evaluate them properly as prospects. Contour intervals of 10 or 20 feet can be used in this instance, with a scale of 2 to 4 inches to the mile. A good example of a low-relief structure requiring detailed mapping is the Table Mesa anticline in San Juan County, New Mexico. A detailed surface map of this structure, on an original scale of 4 inches to the mile and with a contour interval of 10 feet, indicates the structure to have approximately 50 feet of vertical closure. The number of control points required to draw the map and determine amount of closure is shown on the reduced scale map of the Table Mesa anticline (Figure 10-53). The time and effort expended to complete a detailed map of this structure were well worthwhile, as the Table Mesa Field, located on the crest of the structure, has produced in excess of 1 million barrels of high-gravity oil since its discovery in 1925 from a shallow pay horizon at an average depth of 1,300 to 1,400 feet.

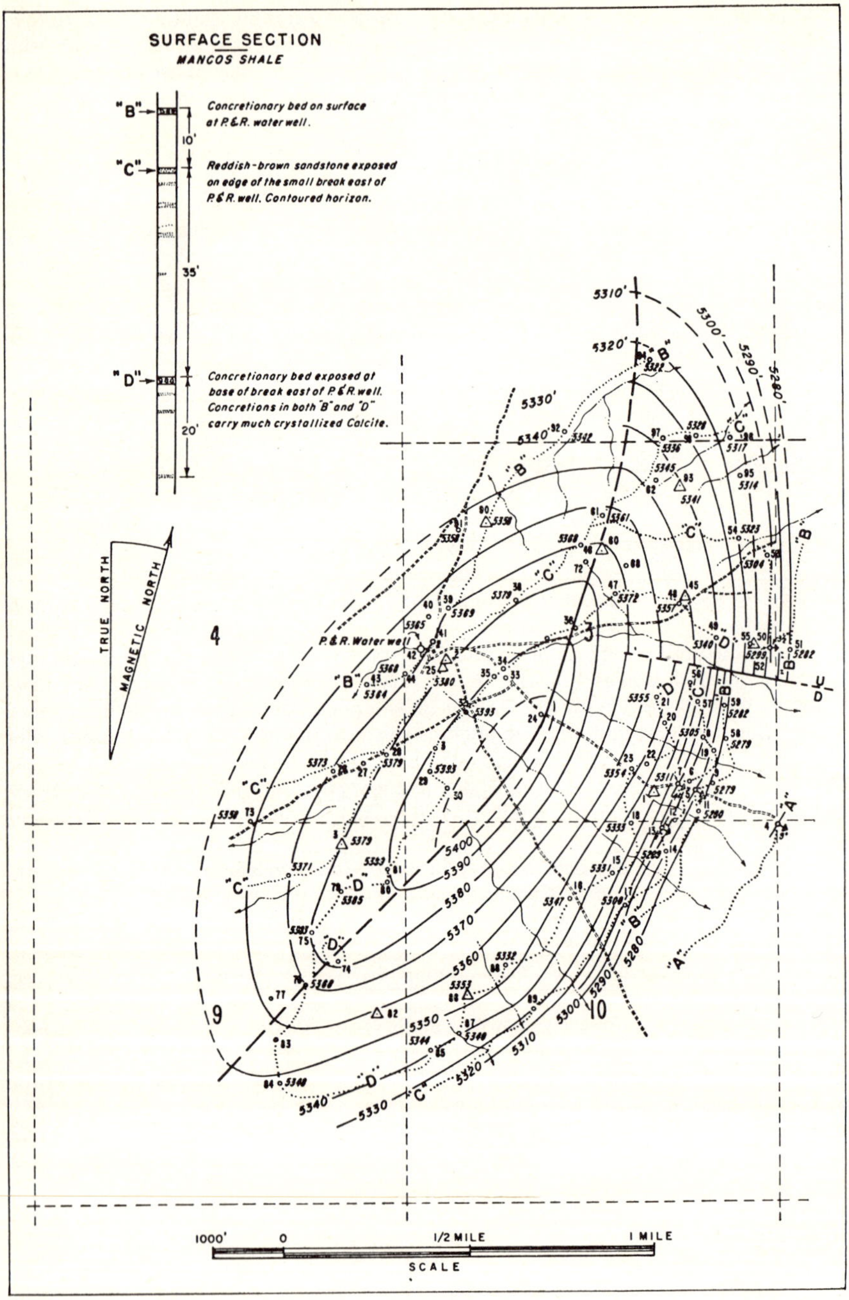

FIGURE 10-53. Detailed plane-table structure contour map of the Table Mesa anticline, San Juan County, New Mexico, illustrating amount of control necessary for detailed maps. (*Courtesy of Continential Oil Company.*)

Factors which will influence a decision to map an area in detail are distribution of surface exposures, type of geologic structure, vegetal cover, topography and accessibility, and the amount of time and number of men available for the project. Low-dip areas may necessitate detailed mapping to outline structural prospects. Poor outcrops and accessibility, in addition to a limited amount of time to complete the project, may render detailed mapping untenable. These as well as other factors, such as project objectives, available funds, and accuracy of final results required, should be given careful consideration before a decision is reached as to the amount of detail to be mapped on a field project.

STRATIGRAPHIC RECONNAISSANCE

Pre-Field Work

The reconnaissance measurement and collection of stratigraphic data in the field should be preceded by assembly of all published literature, previous company data, air mosaics, stereoscopic air photographic coverage, and available topographic and geologic maps.

Sections are measured by reconnaissance techniques in the field for two main reasons: (1) to gain stratal intervals of mappable units for reduction to a common datum in structural mapping and (2) for general analysis of the sedimentary rocks as to lithology, paleontology, sedimentary structure, and thickness variation as related to facies gradation. In the former, the strata provide a control for the projection of thickness of exposed strata into the subsurface, hence interval measurement is stressed rather than lithologic or paleontologic analysis. In the latter, outcrops are analyzed to give lithologic control on facies variation which can be collated with subsurface samples to yield a regional picture of the stratal and sedimentologic history. In regions such as the Rocky Mountains where regional structural gradients are high, Cretaceous strata measured on the surface for structural control on Paleozoic strata must also be sampled with care, for these same Cretaceous strata may be found, in remarkably short distances, to lie beneath Tertiary strata in the subsurface where they are potentially petroliferous.

It is not possible to detail every combination of stratal and structural variation which will control the technique utilized in section measurement, but a summary of all available previous work collated with vertical air photographs and controlled air mosaics allows the planning of a careful field attack with a minimum of lost motion (Low, 1957, pp. 267–331).

Pre-field work includes stereoscopic analysis of contact air photographs, erection of generalized stratigraphic columns utilizing estimated dips, plotting of contacts from air photographs onto air mosaics or onto topographic maps, delineation of areas where continuous sections or composite sections can be measured, and the assembly of all field gear for a logical attack on the area.

Field Tools

The field tools necessary to reconnaissance section measurement are topographic maps; air mosaics; contact air photographs; Brunton, Abney, or Locke hand level; Jacob's staff; plane table; alidade; stadia rod; long steel tape; short steel tape; yardstick or folding carpenter's rule; sample bags; hand lens; acid bottle; rock-identification charts; dip and slope correction charts; field notebook; and rapid graphic forms for construction of field logs as measurement proceeds.

Techniques

In rapid reconnaissance of stratigraphic sections, the topographic map coupled with office-annotated air mosaics are invaluable, particularly in the estimation of major unit thicknesses of recognizable formations from helicopter or reconnaissance flights in fixed-wing aircraft. On the ground, thick stratigraphic sections can be measured quickly for stratigraphic interval control in structural mapping by the following means.

1. Pace-and-compass measurements across strike (parallel to dip) with trigonometric corrections for slopes and stratal dip
2. Long-distance horizontal taping across strike with corrections for stratal dip
3. Plane-table traversing or central-location plane-table setups to gain stadia distances between contacts or use of the controlled mosaic as plane-table sheets with stadia-distance checks and dip measurement with the Brunton at convenient intervals dictated by obvious dip changes
4. Rapid hand-level, Jacob's staff, or Brunton interval sequence measurements of near-horizontal strata in areas of good outcrops

Combination of all these techniques may be necessary in long-section measurement where dip and strike as well as mappable unit thickness are highly variable in short distances.

In most detailed and reconnaissance plane-table work in low-dip terranes, structural mapping and interval measurements proceed simultaneously and structural elevation differences approach true stratal thicknesses.

Most reconnaissance stratigraphic work is done on scales of the contact air photographs which may range from 1 inch equals 1,666 feet to 1 inch equals 4,400 feet. At times broad regional stratigraphic measurements may be done in areas of thick stratigraphic units at air mosaic scales ranging from 1 inch equals ½ mile to 1 inch equals 1 mile. Much of this type of measurement is done while doing reconnaissance structural analysis with the plane table.

Mappable Units

Regardless of names, fossil content, or thickness of the total visible stratigraphic section, most stratal sequences are divisible based on gross aspect. Inasmuch as subdivision must be done for any type of structural mapping and most strata are either tilted or cut through to reveal sequences of lithology, the following criteria for subdivision are useful: (1) breaks in slope, (2) vegetational changes, (3) soil changes, particularly in residual (nontransported) soils on low slopes, (4) lithologic changes in well-exposed strata, (5) continuity of prominent ledges in areas where most of the stratal section is covered with soil or alluvium.

Inasmuch as correlation and subdivision must go hand in hand during stratigraphic reconnaissance and it must be recognized that fossil zones ("time surfaces") usually transgress lithologic surfaces, the field geologist should construct field logs with as nearly complete sections in small areas as the structural deformation allows. Widely scattered segments of composite sections may result in the crossing of hidden faults, angular unconformities, and areas of major lateral facies gradation and thickness change. Subsequent structural deformation and consequent erosion on the present surface section may occur parallel to facies or isopachous strikes; hence it must be with some trepidation that the field geologist will select a structural datum near the bottom of the entire exposed section for the mapping of younger strata from mountainous areas far out into the central parts of undrilled basins. Thus several stratigraphic "datums"

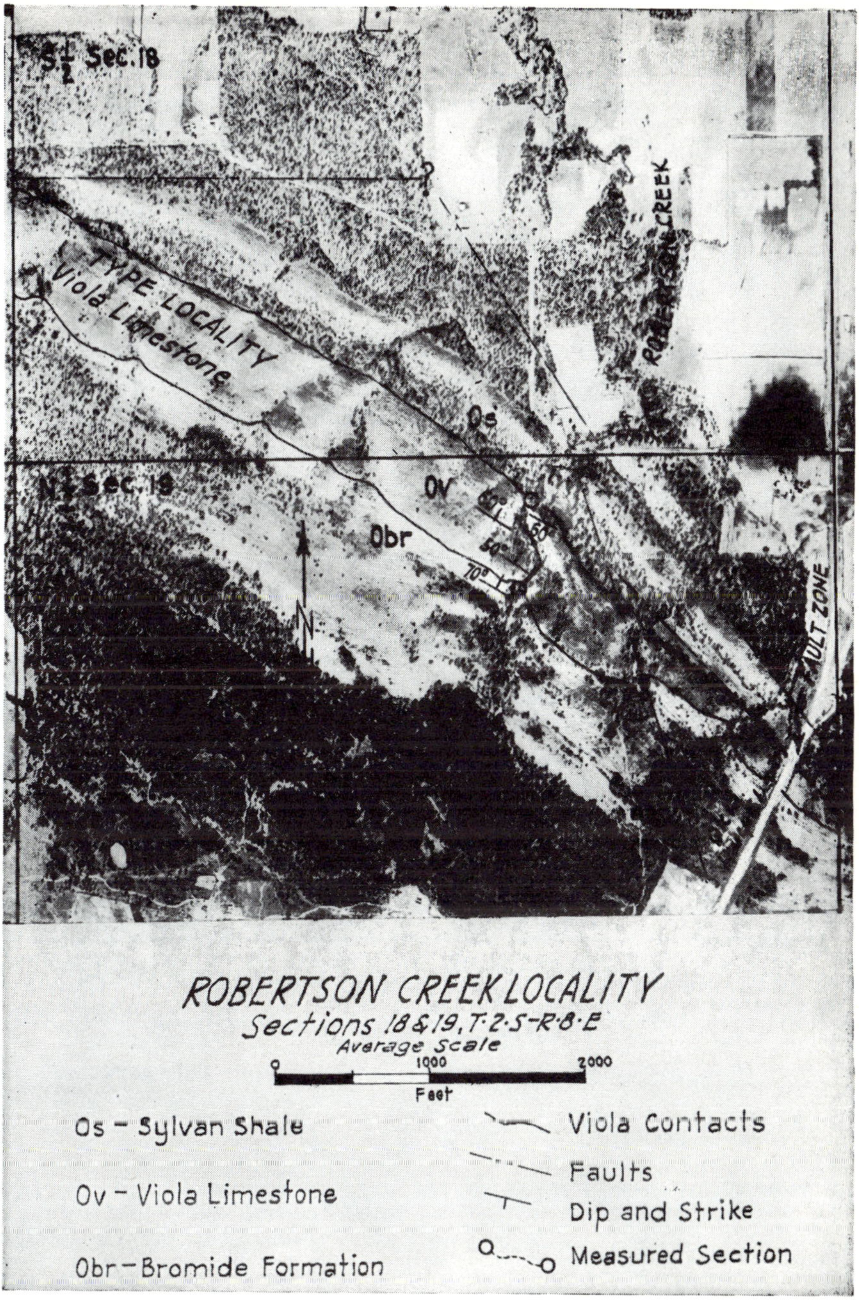

FIGURE 10-54. Robertson Creek locality in Arbuckle Mountains, Oklahoma. Example of use of air photographs to locate stratal sections in steeply dipping Viola limestone in creek bottom on north limb of Belton anticline south of Wapanucka graben. Measured thickness of 434 feet corrects earlier 195-foot measurements complicated by faulting along the highway in corner of vertical air photograph.

FIGURE 10-55. Murray Lake locality, Arbuckle Mountains, Okla. Example of repetition of section by longitudinal faulting in Viola strata of moderate dip along highway on north limb of Hunton anticline, south of Franks graben. These strata are productive of oil in Fitts field only a few miles to the north; thus this section provided an excellent comparison with subsurface cuttings in deep tests as well as a stratal interval for surface mapping.

may be selected in broad regional work for the mapping of surface closures in different parts of the stratigraphic section.

Within logical limits controlled by recognized stratal changes, identification and correlation are done by utilizing lithology, interval, sequence, fossils, sedimentary structure, recognized direction of facies gradation, weathering characteristics, soil and vegetal zones, and geomorphic form.

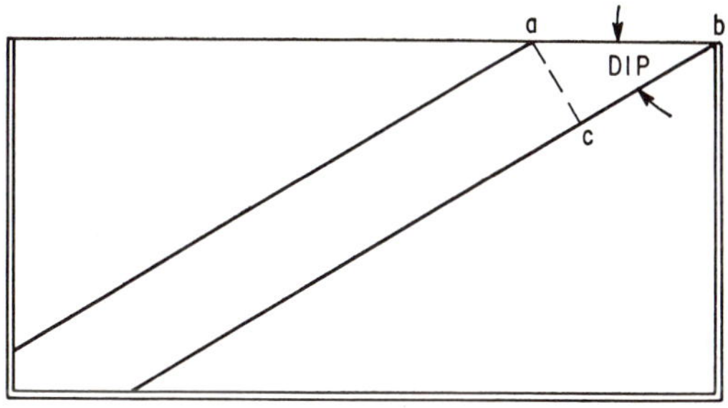

FIGURE 10-56. Solution for thickness of bed *ac*, dip given, topography horizontal.

$$\sin \angle abc = \frac{ac}{ab} \qquad \text{hence } ac = ab \sin \angle abc$$

(*From Dake and Brown, 1925.*)

Utilizing these parameters, collated with field photography, sketching, and the construction of generalized field logs, whether in notebook, as hinge-card logs, or as sketch logs on the sides of plane-table sheets and mosaics, the astute field geologist engaged in structural and stratigraphic reconnaissance seldom gets lost in the section. Memory is a poor substitute for graphic reference, and the validity of a man's field

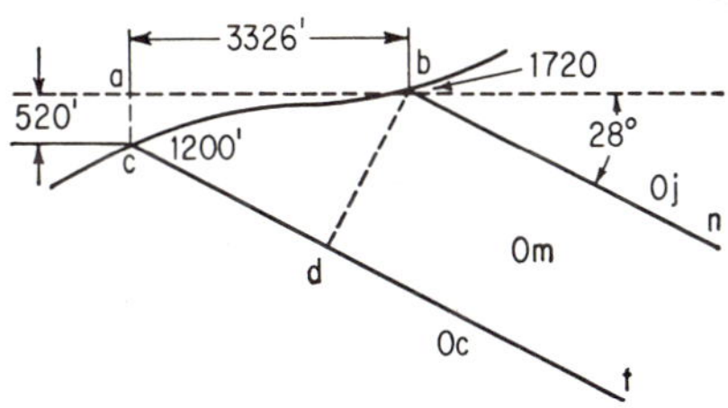

FIGURE 10-57. Solution for thickness, dip given, beds dipping opposite to topographic slope.

ab = the distance across the formation, normal to the strike = 0.63 mile, or about 3,326 feet

a = the west or basal contact, at a 1,200 foot elevation

b = the east or top contact, at a 1,720-foot elevation

$$ac = 1{,}720 \text{ feet} - 1{,}200 \text{ feet} = 520 \text{ feet}$$

$$bc = \sqrt{520^2 + 3{,}326^2} = 3{,}366 \text{ feet}$$

$$\tan \angle abc = \frac{520}{3{,}326} = 0.1564$$

$$\angle abc = 8°53'$$

$$\angle cbd = 180 - (90° + 28° + 8°53') = 53°7'$$

$$\cos \angle cbd = \frac{bd}{bc}$$

$$\text{Whence } bd = 3{,}366 \times 0.6002 = 2{,}020 \text{ feet}$$

(*From Dake and Brown, 1925.*)

work is the ease with which his work can be taken back into the field at a later date to do more detailed work. In areas of poor accessibility, where reentry is uneconomical whether foreign or in the United States, the entire job must be done to a degree of accuracy which will not necessitate reexamination prior to the detailed surface work required for staking drill locations or planning detailed core-drill or geophysical investigation.

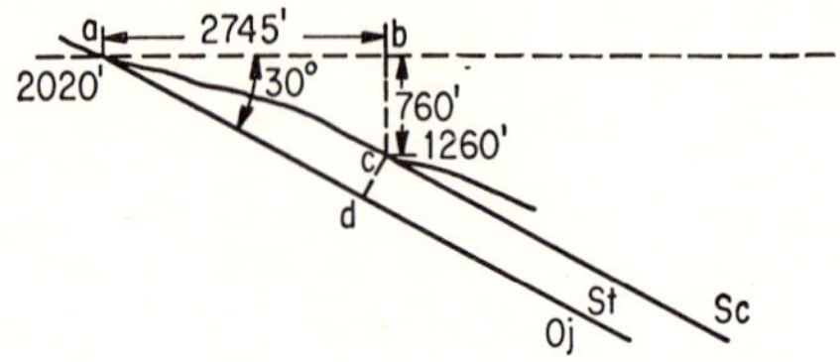

FIGURE 10-58. Solution for thickness, dip given, beds dipping in direction of topographic slope.

ab = the distance across the formation, normal to its strike, about 0.52 mile, or 2,745 feet

a = the west (basal) contact, at a 2,020-foot elevation

c = the east (top) contact, at a 1,260-foot elevation

bc = 2,020 feet − 1,260 feet = 760 feet

$$ac = \sqrt{760^2 + 2{,}745^2} = 2{,}848 \text{ feet}$$

$$\tan \angle bac = \frac{760}{2{,}745} = 0.2768$$

$$\angle bac = 15°28'$$

$$\angle cad = 30° - 15°28' = 14°32'$$

$$\sin 14°32' = \frac{dc}{2{,}848}$$

$$\text{Whence } dc = 2{,}848 \times 0.2509 = 715 \text{ feet}$$

(*From Dake and Brown, 1925.*)

Sample Collection

In reconnaissance structural analysis, sample collection is not often undertaken unless it is recognized that the surface sections in one sector are present in depth basinward where they may be petroliferous. In such cases, general sampling may be carried on by spot sampling of each thin mappable unit, sampling at a regular footage commensurate with the thickness and facies gradation of each mappable unit, or sampling at each plane-table and stadia rod location.

Rapid reconnaissance sampling in stratigraphic analysis is usually done in areas where little has been published on the stratal section. If sampling is not contemplated in such small-scale work, the field description of lithology must be complete yet the detail should be commensurate with the chosen sample interval. The technique of sampling or field lithologic description must thus be fitted to the purpose and scale of the reconnaissance project.

Above all, the exact stratal interval and, in most cases, the exact geographic location of all collected samples or field-described lithologies must be recorded in field notebook, on plane-table sheets, on air mosaics or contact air photographs, or on rapidly constructed field logs and keyed directly with sample bag numbers for ease of office construction of the stratal section.

In areas of good exposures a combination of field sketch showing sample location and a medium-distance photograph will aid in determining sequence and the extension of loggable section above and below the sampled interval. The concept of photostratigraphy developed by Leslie LeRoy is invaluable in logging thick sections rapidly

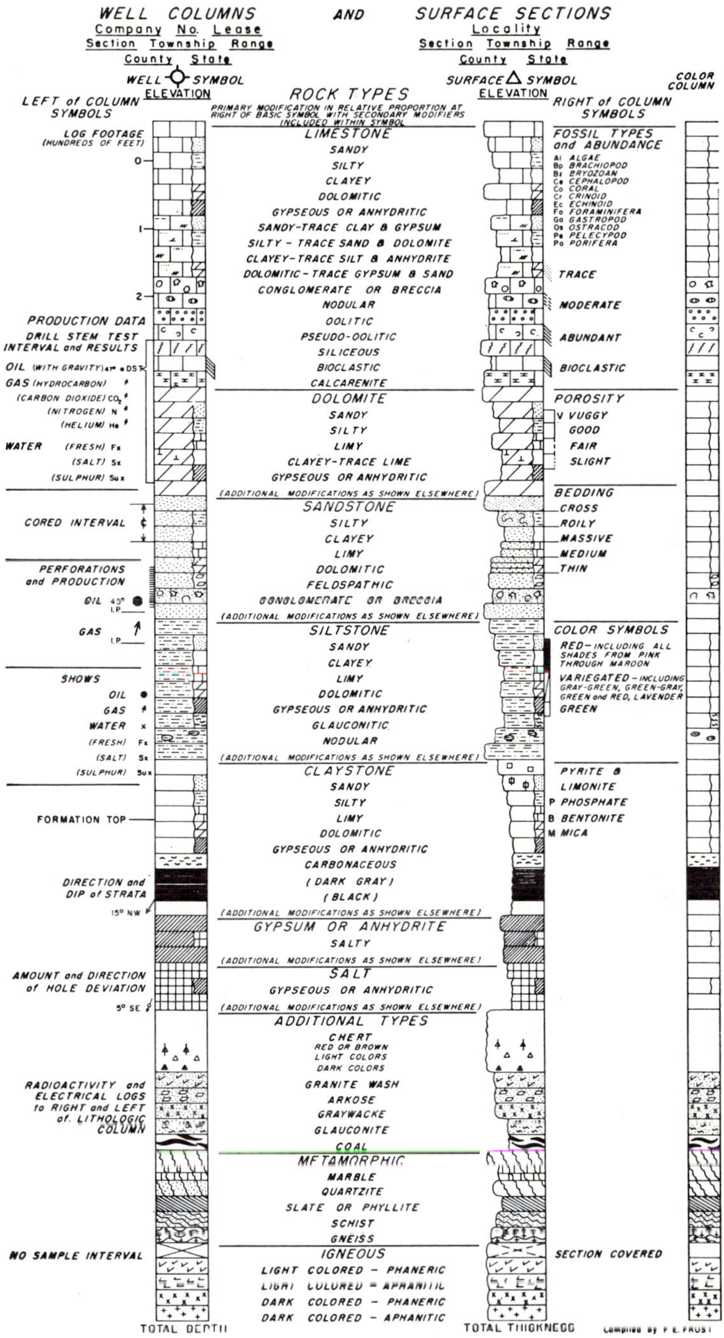

FIGURE 10-59. Graphic representation of lithologic-stratigraphic data for well columns and surface sections. A simplified lithologic symbology for surface sections which allows direct comparisons with subsurface logs.

in deep canyons. If thick, easily recognized lentils are measured in the shade behind the observer and a photograph taken of the steep, possibly inaccessible, opposite canyon wall in the sunlight at the sampled interval, a general sequence log can be built up for structural interval control inasmuch as each spot-measured interval yields the scale of a sizable segment of the photograph. Corrections can be made in tilted strata provided there is little change in strike and dip across the canyon. This technique was used in reconnaissance measurement of Pennsylvanian strata in the deep San Juan Canyon of Utah prior to detailed measurement and sampling.

Correlation with Subsurface Sections

Three prime reasons for the reconnaissance measurement of stratal sections on the outcrops in a region with sparse subsurface control are:

1. Construction of mappable-unit convergence maps to control the projection of mapped surface structure into the subsurface
2. Appraisal of regional facies gradation and facies strike variations
3. Projection of regional dip and the analysis of the relations between present regional dips and ancient sedimentational dips to reconstruct basinal, shelf-slope, and shelf distribution

The more dense the subsurface control, the more carefully and more closely together surface sections must be measured and collated.

Sources of Error

Common sources of error have been noted by Low (1957, p. 299) as follows:

1. Stratigraphic thinning or wedging of units within the area where measurements are made
2. Small exposures, necessitating short sight lines in establishing dip and strike
3. Near-surface weathering, sagging or slumping, incipient faulting
4. Irregular bedding "planes," resulting in inaccurate or unrepresentative dips
5. Attenuation of strata on the flanks of sharp folds and mashing and thickening in crestal regions
6. Sparse dip-and-strike control, especially important where there is appreciable variation
7. Inaccuracies in reading (or recording) all instruments
8. Careless plotting of control points and inaccurate drafting of graphic constructions
9. The use of scales too small for accurate plotting or reading
10. Gross errors in determination of distance or elevation

DETAILED STRATIGRAPHY

The Approach

Geologists have long utilized the field notebook and the planimetric or topographic map in recording all field observations. In basic geologic mapping, the measurement of stratigraphic sections may be considered secondary to the mapping of contacts and geologic structure. For the sedimentologist, sedimentary petrologist, stratigrapher, and paleontologist, however, spatial location, important as it is, is secondary to careful measurement, field description, and collection of typical samples from outcrop sections.

Each geologist, after several years of field experience, devises his own special techniques for the rapid measurement, collection, and field notation of stratigraphic sections. These techniques vary not only because of differences in lithology, variations in lentil and formation thickness, accessibility, and attitude of rocks but also because of variables in the uses to which field lithologic data are put, length of the field project, size of the area, density of section distribution, and the temperament of the geologist examining the sections.

The value of techniques for rapid and efficient completion of field work in some graphic form which can be utilized with a minimum of replotting in the laboratory is recognized. The end point of section measurement is not the final log but the adequacy of graphic representation for correlation, computation of lithofacies, determination of lithologic associations, appraisal of sedimentary environments, and paleontologic-paleogeologic studies. Laboratory time should not be cluttered with the replotting of field data. Valuable time is saved by adding lithologic and paleontologic data to an already well-planned field log through careful field and laboratory regimentation.

After recognition of the problem, choice of the region whose stratal history is to be studied, reading of all background materials, and assembly of previously published maps, sections, and stratigraphic descriptions, the geologist should study air photographs to plot logical distribution, density, and accessibility of the outcrops to be measured and described. Study of each section can be facilitated greatly by graphic forms easily handled in the field and on which all field data can be plotted. Immediate collation of samples and sample interval measured and described must be possible, yet the field log should be so complete that fault-repeated sections, gaps, and field correlations can be recognized as the work advances. The ideal technique, then, is one in which detailed laboratory data such as color, sorting, texture, matrix, composition, chert, insoluble residue-heavy mineral descriptions, and positive fossil identification can simply be added to the completed field log so that replotting of field data is unnecessary during laboratory examination.

Despite the fact that the end point of field or laboratory work is not the log itself, the theory of the geogram as elucidated by Marr in 1924 and the technique of collection as outlined by Otto (1938) should be kept in mind. The desired end result, via correlated lithologic cross sections, quantitative lithofacies-isolith-isopleth maps, and schematic environmental reconstruction, is to discover the sedimentational and tectonic history of a region. Although it must be remembered that the field work should be entirely complete, any time saved in grubbing for data in the hills is additional time that can be utilized in appraising basic data to arrive at the truth (Wengerd, 1956).

Field Methods

The best summary of the many field techniques usable under different conditions of geography, slope, attitude, and accessibility has been prepared by Low (1957, pp. 267–331).

Both pre-field and field reconnaissance utilizing all available published materials, topographic maps, stereoscopic air photographs, semicontrolled or controlled mosaics should result in answers to the following questions (adapted from Low, 1957, p. 260):

1. Are outcrops extensive?
2. Are exposures badly weathered?
3. Is the region accessible by means of available transportation?
4. How complex is the structure?
5. What are the topographic slopes?
6. Is there a wide choice of instruments and techniques for measurement and collection?

7. What are the time and expense limits in completion of the stratigraphic work?
8. Are lodging or campsites near or in the area?
9. Are supplies readily available?
10. Are the personnel adequate to do an optimum job for the desired results?

The equipment necessary for detailed stratigraphic measurement includes means of transportation (afoot, horses, automobiles, helicopters, boats, etc.), instruments (Brunton, hand lens, watch, stereoscope, hand level, aneroid barometer, plane table and alidade, stadia rod, Jacob's staff, binoculars, camera, etc.), accessory equipment (notebooks, geologic hammer, acid bottle, tapes, field tables, charts, air photographs, topographic maps, graphic log blanks, etc.), camp equipment.

Field reconnaissance can be lessened by careful air photographic interpretation of well-exposed sections, but there is no substitute for general reconnaissance with helicopter, fixed-wing aircraft, or automobile in the location of suitably distributed stratal sections. The major effort in reconnaissance is to locate continuous exposures in accessible places that can be measured and collected with a minimum of field effort. Every stratal section, complete as it may be, whether in high mountains, deep canyons, road cuts, along ephemeral stream bottoms, on mature hill slopes, in mines, or on peneplaned surfaces cut across folded and faulted structure, requires its own peculiar combination of measurement and collection techniques. In particular, the attitude of strata in combination with topographic slope, depth of weathering, and depth of soil or alluvial cover will dictate the methods used to get the most data in stratal continuity.

Teamwork in the field is a practical necessity in detailed measurement and collection of stratal sections. Whether the plane table and alidade are used or sections are measured directly with the tape or Jacob's staff, a three-man field party is most efficient in measuring long, continuous sections. The chief of such a party measures the section by hand level, Brunton, tape, or Jacob's staff and calls out the footage and lithologic description; a second member collects chip samples of the measured interval and labels the sample bags; the third member records the sample number and the lithology described by the party chief and constructs the graphic field log as the party moves along the section. Stratal sections are most easily measured and collected from the bottom to the top in canyon country where access is from the top. Reconnaissance of such a section is done while going down over the section. In mountain country where access is from the bottom, the section can be measured from top to bottom, thus obviating carrying the great number of samples upward across the section and then bringing the completed collection back down to the access point.

In plane-table measurement, the party chief is the rodman, a second member handles the plane table and alidade, and the third member collects and records the samples in sequence and referenced to the rod positions.

All combinations of techniques are usually utilized in measuring long sections, but it is the party chief who chooses the measured and sampled intervals based on his judgment.

Mappable Units

Identification and Recognition. The adequate field identification of all types of rocks is prerequisite to both structural geologic mapping utilizing recognizable stratal units and detailed stratal measurement and collection. The field geologist must be well grounded enough in petrology, petrography, sedimentology, stratigraphy, paleontology, and structural geology so that he is "at home" in the section. The early recognition of a correlation of weathering characteristics and lithology for all units to be measured, regardless of thickness, structural attitude, slope, facing direction of outcrop, altitude, etc., enhances the speed with which lentils can be identified as mappable subdivisions and possible sedimentational units. Tables for the identification and naming of commonly encountered rock types will be found in the Appendix.

Subdivision and Correlation. Studies of previous reconnaissance, whether published, or from company files, or made as an immediate prelude to the detailed stratigraphic work, by field or office methods, result in the recognition of large mappable units that can be traced into accessible areas of good outcrops. Mappability on a

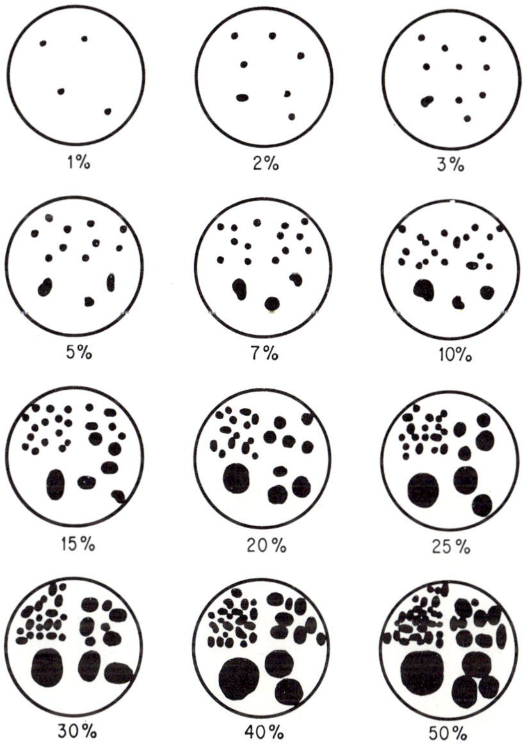

FIGURE 10-60. Visual estimation of percentages in sediments and rocks.

small scale is based on ease of subdivision related to weathering and structural conditions. Good base maps showing the geologic contacts, numerous topographic features, and structural attitude are a practical necessity to the planning of a grid control on the distribution of stratal sections. Whether the location of the stratal section to be measured and collected is completely at random in relation to other sections or the outcrop accessibility is such as to allow a statistical pattern, the density of measured sections should be collated with the rates of facies gradation. Thick basinal sections with low gradients of facies gradation are notably difficult to subdivide but easy to measure and collect in mountain country; unstable shelf sections with high gradients of facies gradation, thin strata, and great vertical alternation of lentils may be easy to subdivide but time-consuming to collect and measure. In these two situations, adequate field reconnaisance leads the geologist to determine the density of stratal sections necessary to determine the stratigraphic, sedimentational, and paleontologic history of the area. Where drilled tests are present in structural basins whose strata are exposed in the ringing uplifts or mountains, the density of these tests may exert some control on the density of sections measured. In some instances subsurface data will direct attention to specific outcrop areas where stratal problems can be solved by surface measurement and collection.

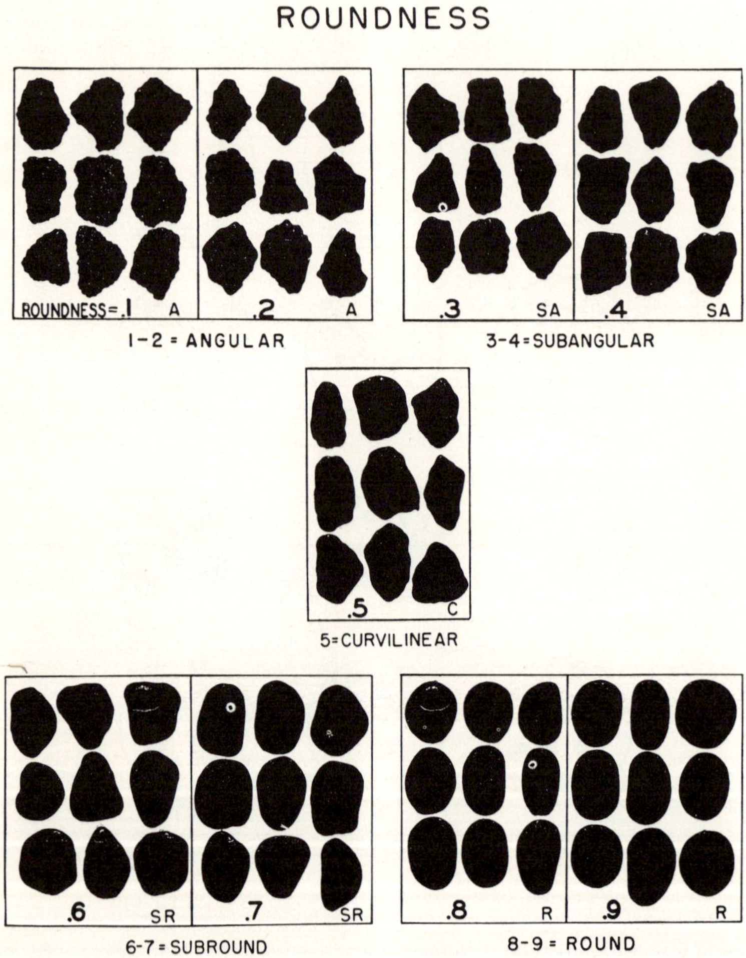

FIGURE 10-61. A visual chart for determining roundness of grains in sedimentary rocks. (*Modified from Krumbein, 1941.*)

Correlation is classically done via (1) lithologic characteristics, (2) thickness, (3) sequence, (4) recognizable direction and rates of facies gradation, (5) direct tracing on air photographs, and, in the field, (6) use of fossils. Inasmuch as correlation cannot be done without subdivision, the many variables inherent to detailed stratigraphic work are overshadowed by accessibility of outcrops, excellence of exposure, complexity of structure, and density of control necessary. Each stratigraphic problem has its own peculiar solution which may sometimes be gained in the field, at other times only after detailed and laborious laboratory work. The early recognition of disconformities, critical sedimentary structure, and directions of shore lines and basinal areas for all mappable units is at times not possible, yet such prior knowledge based on reconnaissance or published materials will inevitably lead to careful planning of the stratigraphic attack.

Detailed Sampling and Measurement

Essential Factors in Collecting Samples. Regardless of the attitude of strata, it is imperative that continuous exposures be available for direct measurement and collection. Differential erosion due to differing resistances of argillaceous, siliceous, and calcareous strata may yield prominent hills which are excellent localities on which to collect spot samples used in generalized study of the formation during contact mapping. These hills, however, may be inadequate exposures for collection of samples to be compared with samples from drilled tests. Not only must the exposure be continuous, but it must be relatively unweathered, for that is the condition of samples from the subsurface. The outcrop should be in a readily accessible locality owing to the number of samples which must be collected. The section must not be so fractured and faulted that more time is spent in working out the local structure than is spent in studying the sedimentary features of the outcrops. Needless to say, all these essential factors do not exist in most localities.

Methods of Measurement. Techniques of measurement vary dependent upon the type of exposure available. Every effort should be made to measure the strata directly instead of using slope measurements and trigonometric computations. Correlation of direct measurement and sample collection allows the construction of a field log for field comparison with other measured sections. Direct measurement of steeply dipping beds with the tape is more accurate and less difficult than similar measurement of strata with low dip; however, strata with low dip are more easily measured with the hand level or the plane table. By exercising care and checking the final measurements with thickness gained from air photographs, it is possible to obtain accurate information on the thickness of any well-exposed section.

Direct measurements can be made with a semirigid 6-foot steel tape or Jacob's staff, and samples collected, description recorded, and the field log constructed as measurement proceeds.

Indirect measurements may be resorted to across gaps in the sections and across areas where only field exposures are available. Tables are constructed for conversion of these indirect distances to log thickness. All indirect measurements should be done directly across strike with the tape held horizontally, thus obviating an extra trigonometric correction for slope variation.

Subsequent to measurement and collection of each section, the section should be chained directly from contact to contact as an over-all check on the direct measurement and to check the local scale of the air photograph or air mosaic on which the field locality is shown.

Methods of Collection. Otto (1938, p. 575) states: "The sedimentation unit at any sampling point is that thickness of sediment which was deposited under essentially constant physical conditions." However, samples for sedimentation units may be difficult to select because determination of sedimentary unit boundaries presupposes a degree of knowledge not existing when collection is made. The problems of adequate sampling are difficult, and two extremes may be noted. The first is to collect the section at 50-foot intervals and attempt subdivision directly from field observations; the second is to collect a 100-gram sample from every bed regardless of its thickness. The first method yields only a crude stratigraphic subdivision and virtually no lithologic

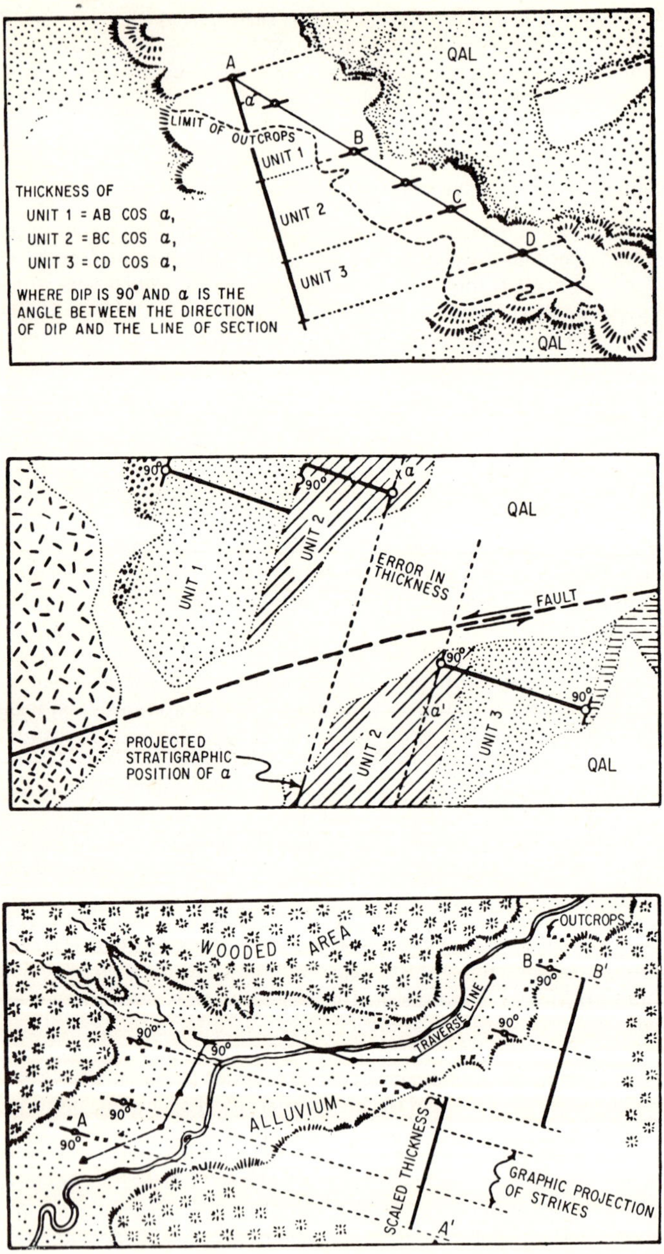

FIGURE 10-62. Field conditions affecting approach to thickness determinations. (*Top*) Computation of thickness where line of section is oblique to the direction of dip. (*Center*) Errors introduced by obscure faulting when offsetting along strike. (*Bottom*) Projecting strike lines to reconstruct section where outcrops are widely scattered. (*From Low, 1957.*)

data to aid in ascertaining the origin of the formation. The second method would require thousands of samples from each section collected depending upon the number of sandstone, shale, claystone, siliceous limestone, and chert layers in the section. Obviously the latter method is too time-consuming for the results gained, and a median course must be chosen which can be varied as different lithologies are noted on the field log.

Four types of sampling may be used by the geologist, namely, *engineering sampling* in which specimens are carefully chosen for porosity, permeability, absolute content, and crushing-strength determinations; *descriptive sampling* in which only sufficient samples are taken to aid in recalling the general rock types noted in the field descriptions; *environmental sampling* in which the rocks are collected solely for the purpose of determining their history; and *correlation sampling* in which a continuous series of samples is taken to determine correlative horizons.

Sampling of any outcrop should be carried out for three main purposes: to gain representative lithologic types comprising the formation; to get enough samples which, combined with field estimates of relative proportions in each sample interval, yield an accurate computation of the insoluble residue, heavy mineral, or fossil content of the formation and its members; to compare lithologic intermember boundaries gained by field description and measurement with those established after laboratory analysis of the collected samples.

Three methods of sampling may be applied to the shortest cogenetic stratigraphic interval during the collection of representative specimens. *Channel sampling* requires a constant-width trenching of the formation to a specific depth. Such a specimen would contain hundreds of small chips representing the chosen interval regardless of the lithologic boundaries traversed. The siliceous character of many strata, distribution of nodular and layered cherts, and the length of time necessary to collect such samples generally militate against the use of channel sampling. *Spot sampling* consists of identifying lithologic units in the field and the collection of one specimen per chosen lithologic interval. The interval is measured, and the specimen is broken out and sacked with a label which applies to both the sample and the interval. The variable percentage of limestone and chert or clastic components renders infeasible a strict application of this method. *Chip sampling* is a logical combination of channel and spot sampling and can be applied to a formation containing several equally important and variable constituents. For example, a limestone section is examined and an interval chosen which has a constant percentage of chert; this interval is measured, the percentage of chert estimated, and a specimen of the chert is collected. If two types of chert are represented, they are collected separately and labeled according to their position in the interval. If hand lens inspection reveals that the interval contains limestone showing no appreciable gradation as to either content or texture, *one* specimen is chosen which is considered representative of the specific chert interval measured. If two different limestones are present, two specimens are collected. If the limestone is highly variable but homogeneous in its heterogeneity, a closely spaced series of chips is taken and sacked as one specimen representative of the interval—without chipping the chert into the limestone sample. As is so often the case, other bedded ingredients are present in the section. These, other than chert, comprise layers of silica or plastic claystone, shale, and beds of highly siliceous limestone, sandstone, and shale. One sample of each type is collected separately (not chipped in with the limestone), and the percentage of each rock type for the measured interval is estimated. It is therefore necessary to collect four or more samples to represent the measured interval in some parts of the section.

Field and Laboratory Logs. Study of any section in the field is facilitated by construction of a field log which shows the sample numbers in their correct scaled position on the left side of the log and the descriptive field notes on the right side of the log. Noted sedimentary structure and lithologic content visible with 10-power hand lens are entered in the log column for comparison with logs from other field localities as the field project proceeds.

Samples collected in the field can be described in the laboratory on a special log consisting of hinged 8- by 5-inch cards or on ordinary log strips. Field data of aid to laboratory descriptions were entered on these logs to facilitate description of individual samples. Similar logs can be prepared for the sections measured previously by other workers. The color system generally adopted by commercial stratigraphers is used to describe the color of samples under blue light. No effort need be made to apply the more accurate color terms in the field; accurate color designations cannot ordinarily be made easily under the binocular microscope without the personal equation that characterizes all color descriptions which are made without the aid of a colorimeter. Until a colorimeter is designed which will permit rapid color evaluation at the same time lithologic descriptions are made, the personal equation may as well be allowed to exert its influence utilizing the color chart as a guide for color appraisal of samples. The binocular microscope, with magnifications of 9X to 12X and 24X, should be used for all lithologic study in the laboratory.

Field Notes

Written Description. Complete word descriptions of strata may be made at the fancy of the investigator or may follow a schedule of observations such as that devised by Goldman and Hewitt (Krumbein and Pettijohn, 1938, pp. 8–10). A simple, lined, tabular form is used by many field men. At best, such written descriptions, with or without simple graphic additions in the field, are difficult and time-consuming to transcribe in preparation for laboratory work and usually result in omission of pertinent field data necessary to the sedimentologist.

Verbal Check-list Description. Simple printed forms can be devised so that a single sheet is utilized for each sample interval (Figures 10-63 and 10-64). Presumably the field investigator bases his sample interval on some logical lithologic subdivision such as Otto's "sedimentational unit."

Despite the fact that these forms are vast improvements in field notation, these data must be transcribed completely into a log system in the laboratory prior to detailed lithologic work. A second detrimental factor is the bulk of the paper that must be carried in the field in order to measure just one complex section.

Graphic Check-list Description. Graphic check lists have been devised leading to completeness in stratigraphic descriptions rarely utilized by field investigators (Figures 10-65 and 10-66). Generally they are vast improvements over the simple verbal check-list forms, since any number of lentils can be recorded on one sheet of paper if size of the sheet is no consideration. Reduced to a printed form in a loose-leaf field book, a graphic check list could prove useful to most sedimentologists; however, the data must be transcribed into log form manually. Although such transcription can be done by nongeologists, there are numerous opportunities for error. Another detrimental factor is the lack of proportions for the lentils, resulting in a serious lack of opportunity for graphic correlations with other field sections measured utilizing the same system.

Graphic-spatial Description. Scaled thickness for measured sections recorded in a bound geologic notebook results in graphic portrayal of lithology as determined in the field and records the type and position of the sample collected. Word descriptions are used with appropriate lithologic symbols and abbreviations.

Transcription of field data to hinged-card strips is tedious and overly time-consum-

FIELD FORM FOR USE IN DESCRIBING SECTIONS

Typ: **Col-** w f **Col Distri:** home-grad-band-mottl
Top Exp: Cl rnd-ver-ir Sl reg-ir-rl **Exp:** gd-fr-pr-cov **Jnt:** **Prdt:** rub-sol
Att: 90-60-30-15-0 **Form:** tab-wedge-lens-ir **Base:** sh-grad, ev-ir, par-ang, cov
Bed Type: **Bed Thickness:** **Bed Succ:** rhy-nrhy-misc **Bed Strut:** lam-gra hom-het **Lam Type:** **Lam Thickness:** **Lam Strct:** grad-hom-het
Gr Size: 256-64-4-2-1-.5-.25-.125-.062-— **Gr Sh:** tab-eq-prol-bld
Gr Rnd: ang-sang-srnd-rnd **St:** dul-sm-rgh-pol **Sort:** **Comp:** qtz mica femag $CaCO_3$ cl feld rxfrag C misc
Cem: fe-si-ca-cl-o **Deg of Cem:** **Ct:** xln-gran-aph-ph-xbl-col-bio
Chert: layrd-nod-lens-ir **Chert Col:** **Concr:** cal-fe-si-o
Foss: abd-rare **Types:** **Condition:**
Fab: **Sm:**
Remarks:

FIGURE 10-63. Field form for use in describing sections; devised by Fuels Branch, U.S. Geological Survey. This form requires the use of one sheet per mappable unit, subdivisible lentil, or sedimentational unit. This technique is especially valuable in rapid work where no samples are collected.

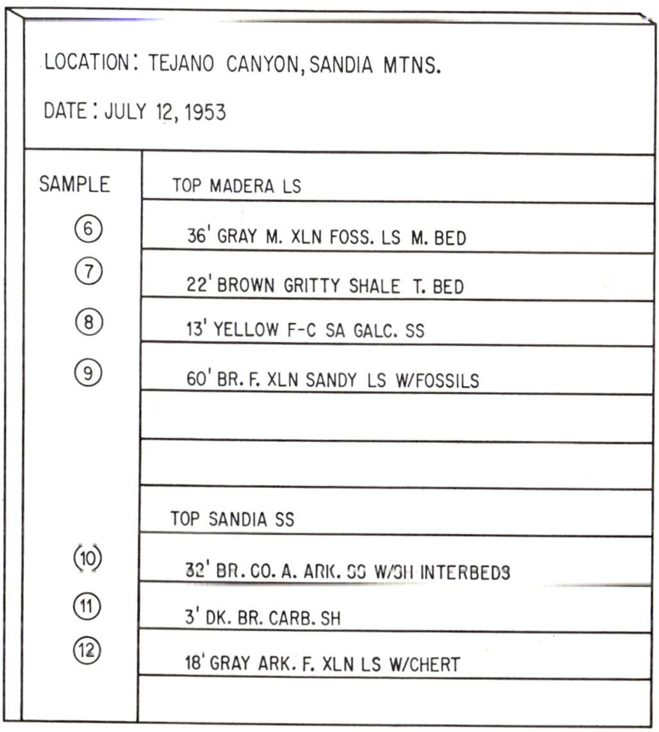

FIGURE 10-64. Tablet-outcrop description. A simple verbal technique showing a minimum of field data usually gathered during structural or areal mapping. Samples must be assumed to be representative of interval measured but may be valueless for detailed sedimentational, stratigraphic, or paleontologic analyses. If sample tags are annotated with short graphic columns, this difficulty is obviated.

ing, but transcription is necessary in order to compare the field data with similar logs devised for the study of well cuttings and cores. Approximate correlation can be done in the field by doubling the field notebook pages of later sections forward to previously logged sections; however, only the left-hand pages of the notebook could be used for the graphic log.

Loose-leaf forms can be devised for use on a clip board, and they provide a quick means of taping pages together for field correlations (Figures 10-67 and 10-68). However, a complete replotting of field data is necessary before lithologic study can be done in the laboratory.

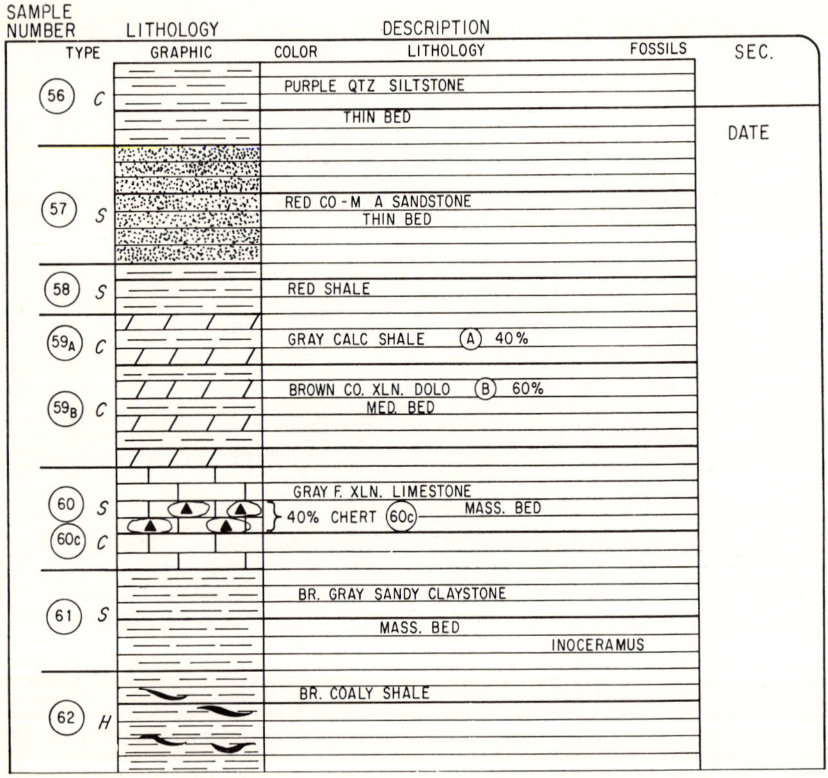

FIGURE 10-65. Bound notebook—outcrop description. *C* is channel sample, *S* is spot sample, and *H* is mixed sample.

A hinged-card (Figure 10-69) log was devised which could be made up prior to field work and on which one could record the following data: footage, chert, sample number and interval, color, texture, outcrop form, beddedness, composition, fossil abundance and types, and other field observations. Immediately after field work each day, these cards could be hinged with opaque scotch tape and utilized for correlation with previously measured sections.

Figure 10-70 shows the form "Description of Measured Stratigraphic Section" from the New Mexico Bureau of Mines and Mineral Resources.

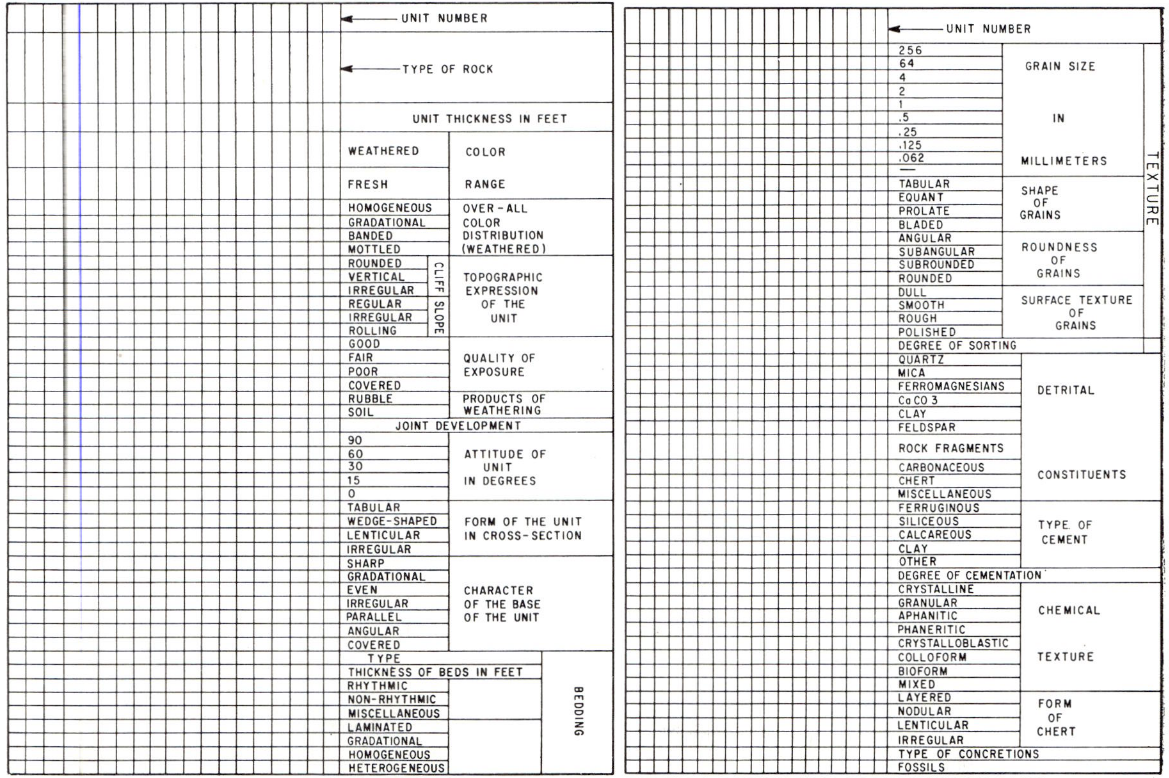

GRAPHIC CHECKLIST FOR FIELD DESCRIPTION OF SEDIMENTARY ROCKS

FIGURE 10-66. Graphic check list for field description of sedimentary rocks; devised by the Fuels Branch, U.S. Geological Survey. A distinct improvement over verbal forms but similar to verbal forms, data on this form must be converted to graphic logs for correlation and publication.

Final Field-laboratory Log

Because of variations in scale on field logs employed by sedimentologists working different types of sections, uses to which the final logs will be put, or plain personal differences, it is recognized that no truly final field-laboratory log system can be devised to fit all situations. Nevertheless, whether the field scale is 1 inch to 5, 10, 20, or 50 feet, if time of transcription is to be saved, the field and laboratory logs must be the same scale. The length of section measured, finiteness of stratigraphic detail, and amount of lithologic detail will always control this scale choice, but the scale of 1

AGE ________
LOC. ________
THICK ________
SAMP. ________

CLASSIFICATION

CONGL. OZITE SS SILT CLAY LS DOLO. GYP. ARK GRWKE

COLOR

RED MAROON GREEN PURPLE YELLOW BUFF TAN BROWN BLACK GRAY

STRUCTURE (BEDDING)

MASSIVE MEDIUM THIN SHALY FISSILE

GRAIN SIZE

COARSE MEDIUM FINE VERY FINE CRYPTO

ROUNDNESS

ANGULAR SUB-ANGULAR SUB-ROUND ROUND

SORTING

GOOD FAIR POOR

FOSSILS TYPES

ABUNDANT MODERATE TRACE BRACH. PELECY. GASTRO. TRILO. CRIN. FUS. COR.

MODIFICATIONS

CONG SANDY SILTY CLAYEY LIMY SIL. QZITIC ARK FELD OOLITIC

CHERTY MICACEOUS NODULAR GLAUC BRECC. FRAG. CALCAR.

WEATHERING AND OTHER DESCRIPTION

FIGURE 10-67. Loose sheets—outcrop description. A verbal form for each lentil measured and collected, this field note system requires plotting to a graphic system in continuity. Reductions of this form could be used on large sample tag or enclosed with each field sample.

inch to 20 feet is, in most cases, wholly practical, utilizing hinged logs to represent stratigraphic sections as thick as 1,000 feet.

The attempt to combine on one form card all the columns necessary for the entry of all field and laboratory data led to three major difficulties: (1) The card was too wide for a clip board which could fit into a field notebook case; (2) if the clip board was carried in the open, one hand was lost for climbing (a factor the flatlander would not consider in measuring sections along road cuts); (3) the columns into which laboratory data were to be entered became inordinately smudged and scarred during field work.

Figure 10-71 illustrates a graphic-spatial field card *on the left,* which makes up into a final hinged field log for preliminary correlations. This card is a 7- by

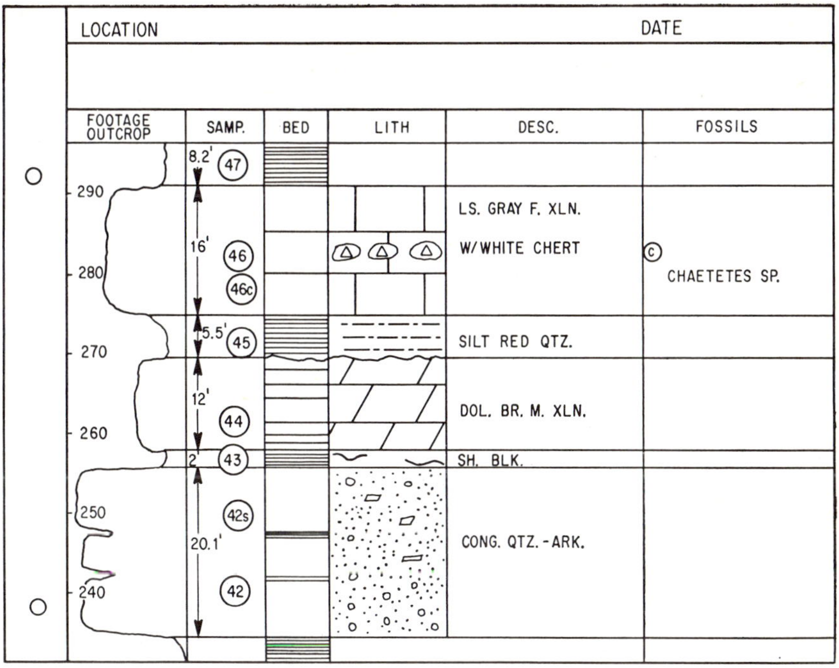

Figure 10-68. Loose-leaf form—outcrop description. Graphic form that can be used in a clip board or in three-ring notebook during field work but can be taped to other sheets to yield a temporary field log. *c* is chert, *s* is shale. This form has sufficient space for the entry of simple laboratory data.

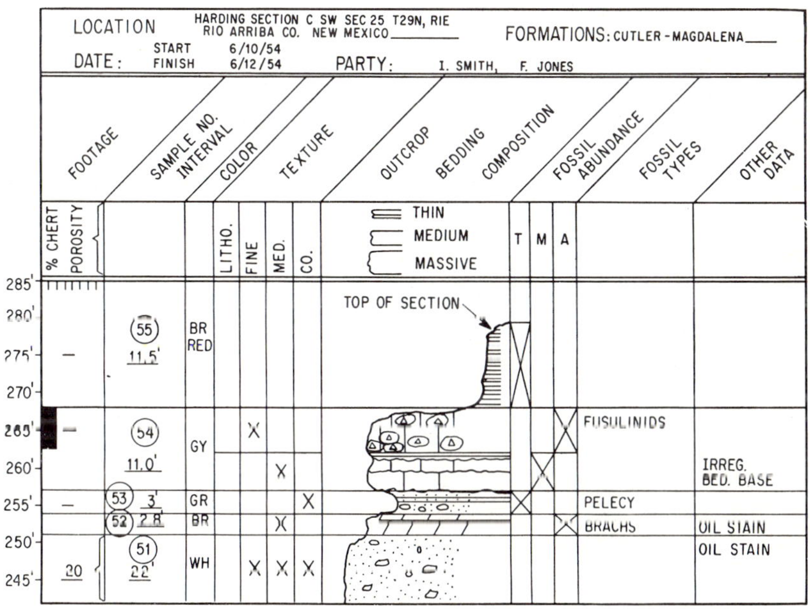

Figure 10-69. Hinged-card log—outcrop description. A simple field log heading card that results in a continuous graphic log of correlative elements for comparison with other field sections. Ideal for rapid stratigraphic measurement and collection.

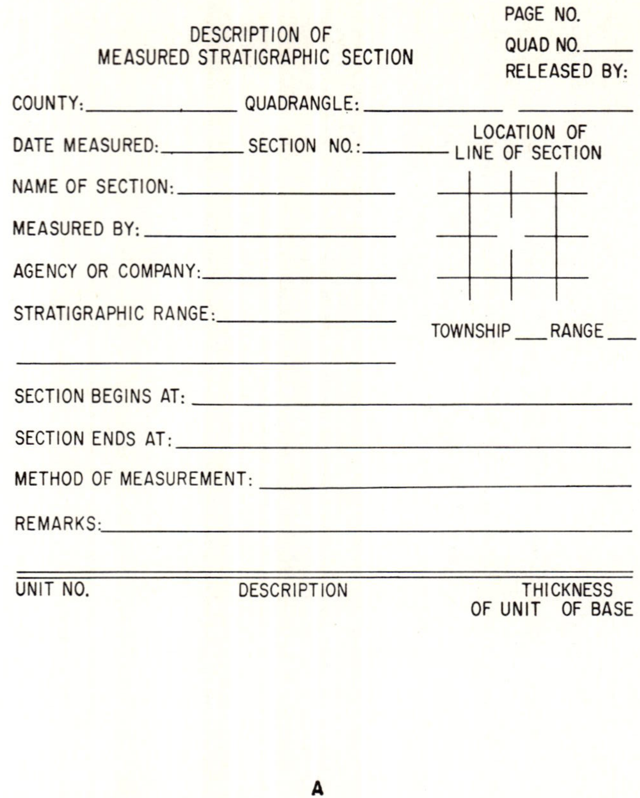
DESCRIPTION OF
MEASURED STRATIGRAPHIC SECTION

PAGE NO.
QUAD NO. ____
RELEASED BY:

COUNTY: ________ QUADRANGLE: ________ ________

DATE MEASURED: ________ SECTION NO.: ________ LOCATION OF LINE OF SECTION

NAME OF SECTION: ________

MEASURED BY: ________

AGENCY OR COMPANY: ________

STRATIGRAPHIC RANGE: ________

TOWNSHIP ___ RANGE ___

SECTION BEGINS AT: ________

SECTION ENDS AT: ________

METHOD OF MEASUREMENT: ________

REMARKS: ________

UNIT NO.	DESCRIPTION	THICKNESS OF UNIT	OF BASE

A

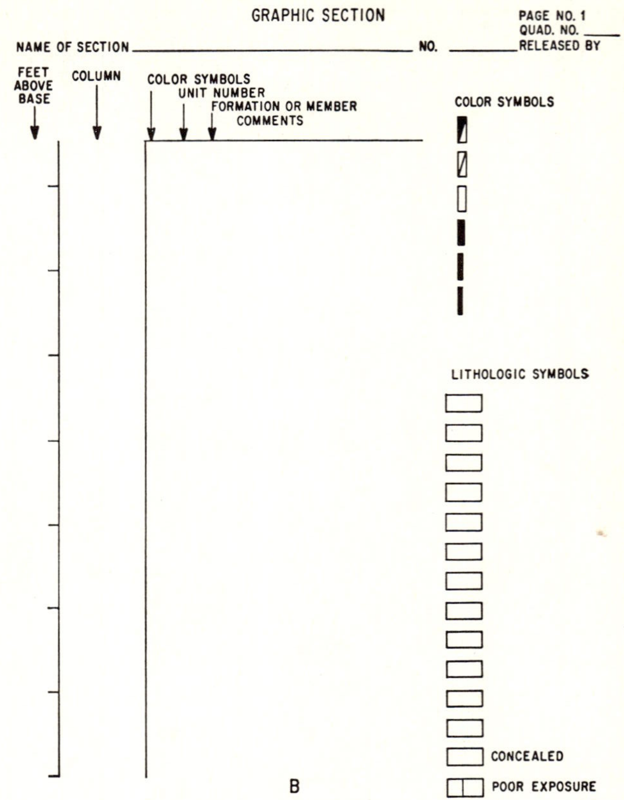

FIGURE 10-70. Stratigraphic sections as measured and described (*A*) and graphic representation (*B*) in New Mexico. (*After New Mexico Bureau of Mines and Mineral Resources.*)

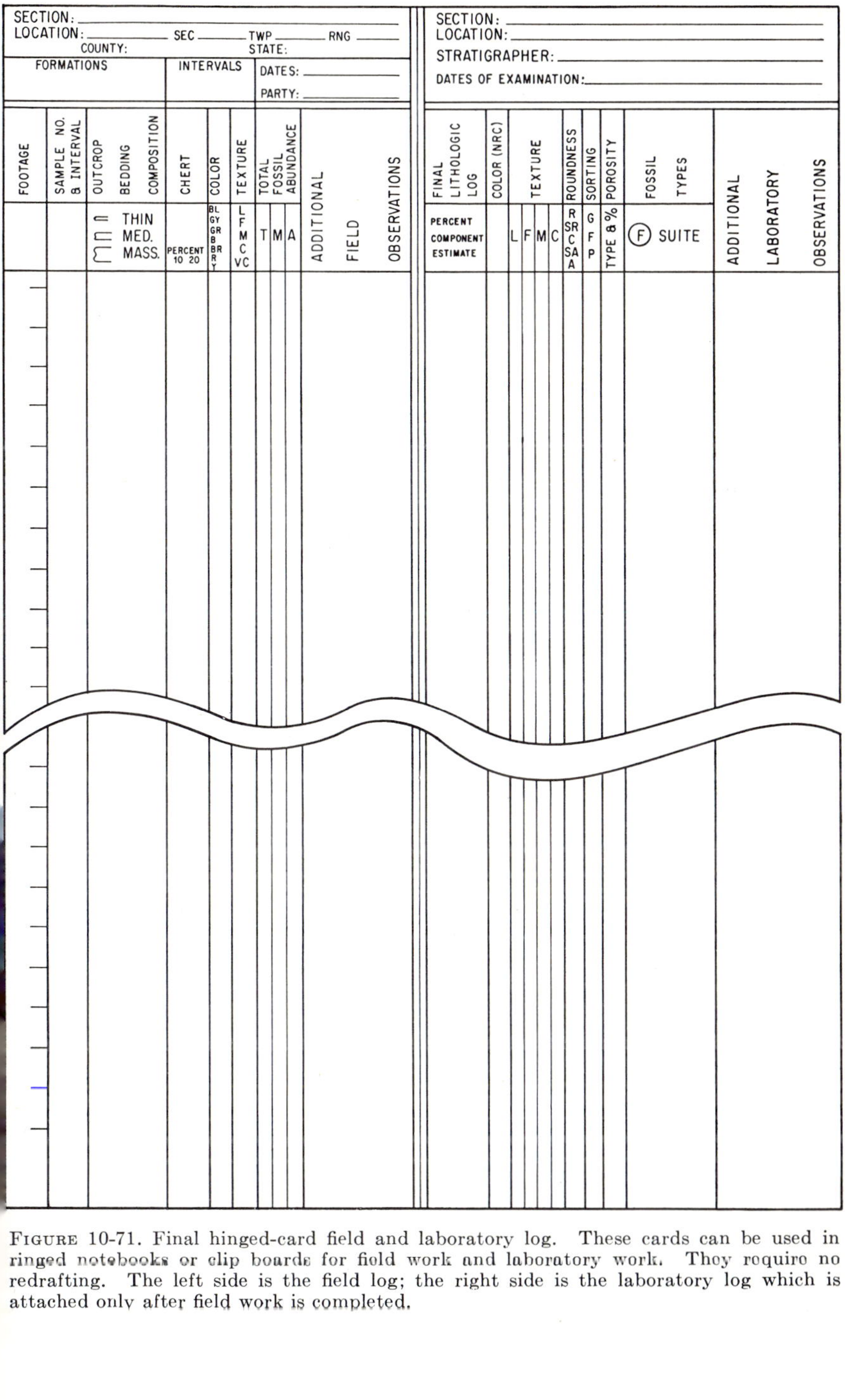

FIGURE 10-71. Final hinged-card field and laboratory log. These cards can be used in ringed notebooks or clip boards for field work and laboratory work. They require no redrafting. The left side is the field log; the right side is the laboratory log which is attached only after field work is completed.

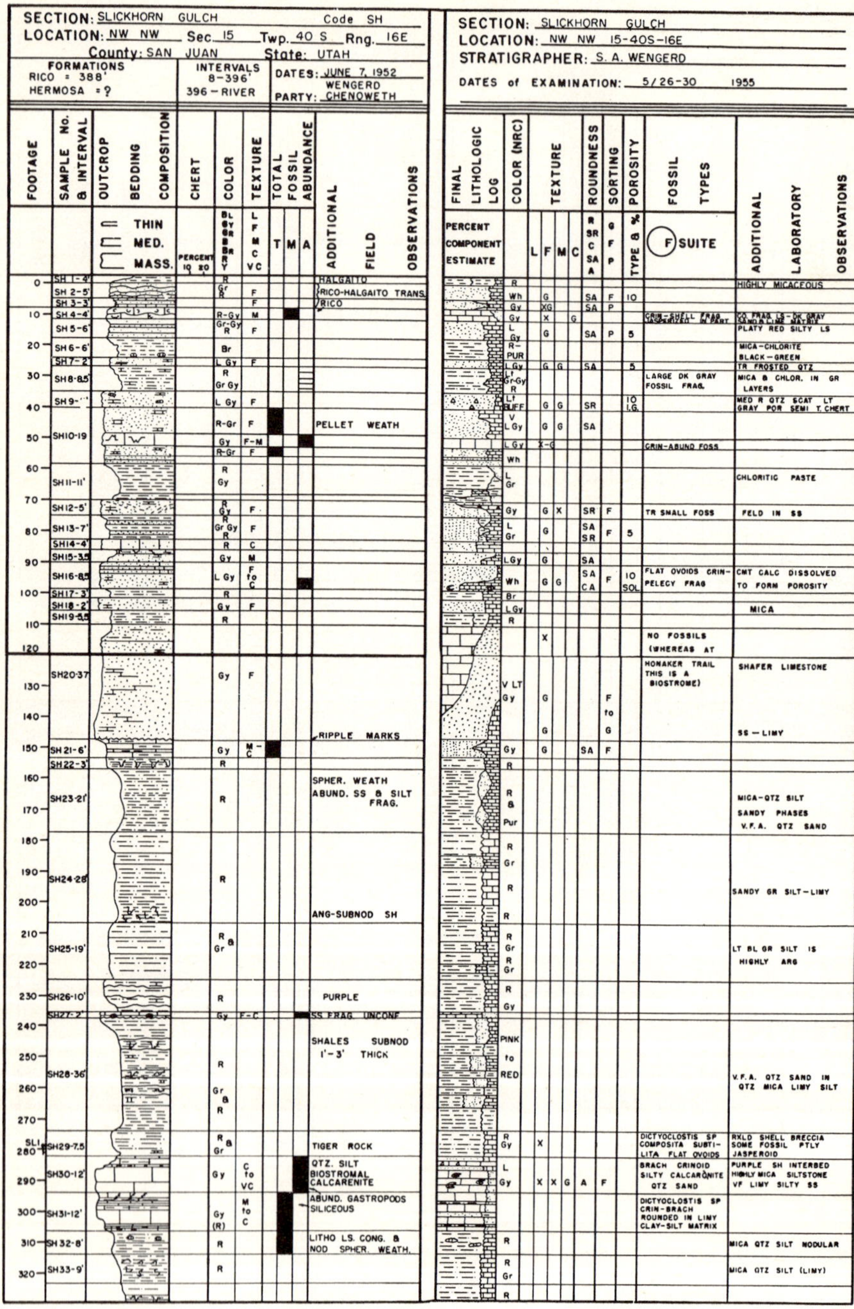

FIGURE 10-72. Completed field-laboratory log. This is a section of an actual log.

10½-inch form or can be photographically rescaled to approximately 5- by 8-inch size, printed on heavy filing-card stock, with one heading card per stratigraphic section and the remainder of the log made up of card-stock forms lined as shown on the bottom part of the left half. Utilizing the same scale, a heading card was devised on which all laboratory data are entered after the field work is completed.

Upon completion of the full study of a stratigraphic section, the field log and laboratory log are taped together lengthwise. A third narrow laboratory-entry log could also be designed on which to record microfossil, heavy mineral, and insoluble residue data.

Summary. A summary of written, check-list, graphic, and combination logs prepared in the field as measurement and collection of stratigraphic section proceed suggests that the ideal field log is one which

1. Can be easily handled in the field and will fit into the field kit on a clip board
2. Is graphic and to a scale which will allow direct comparison with other measured sections in the field
3. Can be hinged so that any length of log can be opened for inspection to determine gaps and overlaps of measured section
4. Can be taped to a laboratory log for direct comparison as the laboratory log is prepared
5. Can be reproduced economically without transcription of field data
6. Can be redrafted to any smaller scale, after photographic reduction, by the overlay method utilizing tracing cloth over a light table

It is recognized that every field geologist has personal preferences as to scale, data to be recorded, graphic rendition of final results, and interpretation of final results. However, any technique which speeds up the process to its final conclusions yet loses few of the vital facts must be considered an advance in field and laboratory procedure.

FIELD PHOTOGRAPHY

Field photography is carried out for several purposes: (1) general views to illustrate topography, access, and geomorphic sequences in an area; (2) outcrop photographs to show stratal section, sequence, and stratal detail such as sedimentary structure and lithologic detail; (3) sequence photographs to show wide-angle coverage in the horizontal plane; (4) vertical sequence photographs to show stratal succession in near-horizontal strata.

Ground photographs may be used (1) for study and collation with field observation, (2) as controlled views from which survey measurements can be made, (3) as stereoscopic coverage for both of the above, (4) as illustrations to support field observations.

Because of the many uses for good, clear black-white photographs and color photographs, the field man must be an accomplished photographer, familiar with his equipment, its limitations, and the value of adequate illustration of field data difficult to describe.

Cameras

In reconnaissance field work where adequate transportation is available and weight is of little consequence, cameras of the film-pack type with ground-glass back and with fully corrected lenses of both short- and long-focal length are invaluable for office-study photographs and for illustrations. Where weight and volume of equipment are considerations, the many types of small reflex or 35-millimeter cameras are invaluable, particularly where work must be done on foot or on horseback.

In most field work, both a black-white camera of the film-pack or large-roll type and a 35-millimeter color camera are necessary, whether the surface geologist is doing fast reconnaissance for structure, detailed plane-table mapping, or measuring sections for stratigraphic and sedimentational data. Adequate, rugged but light cameras are fully as essential as the Brunton, the geologic hammer, and the hand lens.

Accessory Equipment

Tripod. A light telescopic tripod with universal tilt head, a rugged exposure meter, a tape or folding ruler, extra lenses, and proper filters are all necessary items for adequate field photography, particularly in the measurement of stratigraphic sections.

Films. Adequate film, as to both supply and type, must be carried in the field, particularly in areas far from supply centers. Color film resulting in slides or in color negatives and black-white film of moderate contrast, fine grain, large size, and sufficient speed result in photographic coverage which can be used for anything from precise measurements to pictorial illustration of geologic phenomena.

Techniques

Distant Views. Distant views of major geomorphic or structural features should be taken showing from one-third to one-fourth of the sky in the top part of the photograph. Most illustrative photographs, except in steep canyon country, are taken with the long axis of the photograph horizontalized. Effort must be made to shoot in the 180-degree semicircle away from the sun. Photographs taken in full sunlight without large cloud shadows and with nearby objects in focus to give a clear scale result in highly usable records of the terrain. Roads should not lead out of the lower corners of photographs, and any feature which divides the photograph into a left and right half should be avoided.

Foreground Views. This type of photograph is used to show single objects such as a slope, an outcrop, or a structure. A minimum-distant background and an uncluttered foreground are necessary. Photographs from the tops of field vehicles, tops of trees, or tops of scarps result in pleasing composition and meaningful coverage as compared with photographs from low ground positions. A part of the sky should show for orientation, but steep obliques, either up or down, contain considerable distortion in the perspective view. A scale, such as a man, a field vehicle, a horse, or other familiar object, should be stationed on or very near the prime object of the photograph.

Outcrop Shots. Outcrop photographs (Figure 10-73) are basically of three types: (1) the whole outcrop to show geomorphic expression; (2) a vital part of the outcrop to show stratal pattern, important contacts, sedimentary structure of large scale, and erosional unconformity; (3) close-up detail shot to show contact phenomena, small-scale sedimentary structure, and sedimentational detail. Only in the first can the sky be showing in most instances.

In all these types, single small-scale shots can be augmented with sequence shots to yield semipanoramas or long stratal successions. Scale must be provided for the viewer, ranging from a man on whole outcrop shots to an easily read ruler or scale on detail shots. Particularly in the shooting of oriented specimens should an adequate scale be present, as well as the annotated chalk, crayon, or paint orientation marks. After photography, the orientation marks should be sprayed with a heavy acrylic spray, shellac, or varnish to ensure their preservation when placed in sample bags.

Orientation. Regardless of the type of photograph, the direction of the principal line (the line passing from the plumb point beneath the observer through the center of the picture to the top of the photograph) must be recorded in the photographic notes. If the camera is tilted from the horizontal, the tilt should be recorded in degrees. Both the direction and the tilt can easily be measured with the Brunton if a tripod is used. In all instances, the lateral axis of the camera (parallel to one side of the photograph) must be exactly horizontal, particularly if sedimentary and deformational structure is being photographed.

Scale. The scale of any photograph changes greatly with distance from the rear optic node of the camera lens, except where photographs are taken normal to planar surfaces. In attempting to show both major and minor features in large outcrops or

FIGURE 10-73. Outcrop photographs. Chosen to show stratal detail of Viola limestone. (*A*) Medium distance (50 feet) showing variable massiveness. (*B*) Close-up (2 feet) to show chert in limestone. (*C*) Interbeds of chert and limestone from distance of 20 feet. (*D*) 10-foot shot to show sedimentary structure.

nearby structures such as oblique faults or folds oblique to the optic axis, the field geologist will do well to place several large stadia rods from the foreground into the background, thus providing criteria for distance measurements directly in the finished photograph. The scale of any point on a photograph is

$$\text{Scale} = \frac{\text{focal length (inches)}}{\text{distance (inches)}}$$

The field geologist, therefore, will do well to measure the exact focal length of his various lenses in order to utilize the photographs for relatively precise measurements. When the focal length, the distance from photographed stadia rods, and the elevation of the observation point are known and very thin ink lines are drawn from diagonal corners of the photograph to determine the principal point on a photograph which shows a distinct, relatively flat horizon, generalized elevation measurements can be made by trigonometric computations. Two such oriented ground photographs from plane-table stations whose locations are known by traversing or resection will enable measurements to be made of hundreds of common points well beyond the principal point and allow resection location of these points (Bagley, 1917). This technique is particularly applicable in fast reconnaissance in rugged country where few plane-table stations can be occupied. An added parameter which should be known for each line-up of the field camera is the exact angle of coverage, both horizontally and vertically, particularly if the photographs are to be used to make precise measurements.

Stereoscopic Photography

Stereoscopic ground photographs for both illustration and measurement are today too seldom utilized because ground surveying cameras have not generally been designed to the precision found in air cameras. Although the bulk of most surveying air cameras is too great for reconnaissance and detailed mapping on the ground, they have been so utilized, and with the wide range of optically precise enlargers now available in photographic laboratories, smaller precise cameras can now be used for crude ground surveying and the location of major features on reconnaissance maps. Where field notes are many times inadequate owing to limited range in field vehicles or to short field sessions, stereoscopic ground photography gives a permanent record from which many types of planimetric, topographic, and structural control can be obtained in the office.

Photographic "records" from each plane-table station in a reconnaissance triangulation net yield almost limitless stereoscopic points provided the exact orientation of the camera is indicated on the plane-table sheet.

In photographic illustrations of nearby objects such as outcrops (Prusok and Ege, 1960), a photographic base line of as little as 1 foot gives adequate stereoscopic depth when the two camera stations are located on top of a flat surface (plane table) and the shots are taken with the camera optic axes parallel. The more distant the view, the longer must be the base line to get good stereopsis of specific distant objects.

Photostratigraphy*

Photostratigraphy involves that phase of geologic study whereby the distribution, characteristics, and relationships of stratigraphic units of an area are *systematically* and *sequentially* shown by prints, projected transparencies (two or three dimensional), or motion pictures or by a combination of all three.

Photographic data frequently demonstrate stratal variables and details which have been disregarded or minimized during field observations of outcrops. Some of these photographed features may suggest reexamination of an exposure. Stratigraphic information recorded on film only supplements field investigations.

A well-selected series of stratigraphic pictures, properly captioned and organized, enables those individuals who are unable to witness field conditions to visualize, understand, and become acquainted with the stratigrapher's problems. They improve understanding and appreciation of the (1) topographic relief, (2) relation of topography to stratigraphic units, (3) degree and nature of exposure, (4) type and density of vegetation, and (5) structural aspects of an area. This information assists the geologist to organize a field party more efficiently and outline stratigraphic programs. For example, 12 appropriate, well-toned, colored transparencies of the Grand Canyon section studied by an individual never before having observed this sequence

* Prepared by L. W. LeRoy, Colorado School of Mines, Golden, Colo.

would familiarize him with the structural and stratigraphic fabric of the region prior to arrival on location. Photostratigraphic data of an area permit the geologist to review rapidly his accumulated field information, which eventually will be compiled into report form. It is possible that a geologist, after years of absence from an area, may return to reevaluate it. The stratigraphy and structure of the area could be quickly reanalyzed if proper photographs were available. It is mandatory that the stratigraphic photographs be coordinated with a definite field and subsequent presentation procedure.

Control Map and Stratigraphic Sections. During photographic studies of a stratigraphic section, the field geologist should have available a control base map and a generalized lithic column. In areas where stratal units are lithologically and stratigraphically variable, a correlation chart serves as an additional aid. These drawings should be arranged and integrated for orientation both areally and stratigraphically as the photographs are examined. *It is imperative not to lose continuity among the photograph, map, and graphic section.*

The base control map (air photograph, topographic, geologic) should be of such size as to minimize concentration of plotted photographic data. It should include only the locations of the pictures being studied. These locations are shown by a number enclosed within a circle. An arrow drawn from the circle edge indicates the direction in which the photographs are taken. The ground location numbers and corresponding picture numbers can be recorded on the columnar section and the correlation chart to improve clarity and coordination during study of the photographs.

Photograph Sequence. Each stratigraphic unit of an area should be photographed as completely as possible, and the photographs then arranged and presented in sequence. It is recommended that photographs of the oldest stratigraphic unit be viewed first. Pictures of successively younger units may then follow.

Stratigraphic Unit Discussion. The field geologist should have in his possession summary text data pertaining to age subdivision, formation terminology, lithologic description, thickness variation of the stratigraphic units, and the position of unconformities. Before the details of the individual stratigraphic units are shown, a distant photograph giving a panoramic view of the section should be introduced. This photograph acquaints the observer with the general structural and stratigraphic setting of the specific unit. Photographs of this type may be an aerial vertical or oblique or a distant "shot" from a ground vantage point. Once such views have been studied, near-distant photographs showing the details of the unit then follow. Photographs of specific stratigraphic units should illustrate such features as outcrop tendencies, vegetation pattern, color, bedding, jointing, and boundary characteristics. If multiple stratigraphic relationships and facies intergradations exist, they should also be included. If the unit is characterized by several distinct lithofacies, photographs of these should likewise be included.

In order that a particular feature characteristic of the unit be emphasized, several photographs taken at different angles may be required. These special features may be indicated on the photograph by a colored dot, arrow, or line. The chroma of the designations placed on colored transparencies is controlled by shades of the transparency.

For comparative purposes a recognizable scale—geologic pick, Brunton compass, or an individual should always be included in photographs. An illuminated, large-scale columnar section prepared on plyboard and arranged to flank the projection screen has proved emendable for constant orientation.

It may be of interest to those teaching stratigraphic geology in universities that the procedure as herein discussed is followed at the Colorado School of Mines and has proved extremely beneficial to both instructor and student.

Photographic Field Notes

Complete photographic field notes should include roll number, exposure number, subject, exact direction of shot in degrees, tilt of camera, exposure time and f/value, type of film used, exact location by number keyed to the map, and weather conditions. An entry form in a loose-leaf notebook should be printed up the same size as the field notebook and clipped, bound, or taped right into suitable places in the project field notebook. Another field technique utilizes a code sequence to allow entry of all these data with geological observations in proper places in the notebook. There are as many ways to annotate field photographs as there are geologists who use this valuable adjunct to field geological mapping.

Final Photographic Quality

Given a precise field camera (without chromatic or geometric aberrations) whose focal length and angular coverage are known, the final negative must be one exposed on film of fine grain and wide photolatitude. These negatives should be on a film base with very low but equidimensional shrinkage and expansion. For survey measurements, the photographic positive should be printed unenlarged on double-weight semimatte aero film; for purposes of illustration, the photograph must be single-weight glossy. Most ground photographs used in making precise horizontal and vertical measurements are enlarged two to four diameters in precise enlargers dependent upon the original negative size and the degree of accuracy required for the scale of geologic mapping. Photographs of good quality are easily examined for geologic detail under the mirror stereoscope. Both horizontal and vertical angles to similar points on stereoscopic coverage can be computed trigonometrically by collating the focal length and the measured x and y coordinates of each point on the photograph from the principal line and a line parallel to the horizon (parallel to the top line of the photograph if the camera is precisely horizontalized in the lateral axis upon exposure). Distances can be joined by resection of rays to similar points provided the orientation of the optic axis is measured in the field. The vertical angles plus the resected map distances will yield data that allow computation of elevations to reconnaissance accuracy.

UNDERWATER GEOLOGIC MAPPING AND OBSERVATION*

The development of reliable self-contained underwater breathing apparatus (SCUBA) by the French (Cousteau-Gagnan) in 1943 made it possible and practical for geologists trained in its use to make underwater geological observations in the shallow areas of the continental shelves. The first organized underwater field mapping was done along the coasts of southern California in 1953 (Menard et al., 1954). Since then, it has been used extensively on the marine shelves of southern California and in many other areas of the world where bedrock crops out on the sea bottom. SCUBA diving has proved to be a most useful tool for the geologist in the study of recent sedimentary processes because it allows him to make direct observations *in situ* as phenomena occur.

Equipment for Diving

For the most elementary investigations in water less than about 5 meters deep, the basic equipment for simple observations consists of a faceplate and a pair of fins. For

* Prepared by George Shumway and Robert F. Dill, U.S. Navy Electronics Laboratory, San Diego 52, Calif.

routine work in deeper water, a complete set of SCUBA equipment for at least two divers is needed. Diving is always done in pairs both for the sake of safety and to ensure the correctness of data collected under the sometimes difficult conditions existing in the underwater environment. A team of two divers requires, as a minimum, the equipment listed below, although efficiency of diving can be increased by the addition of the auxiliary equipment:

Minimum equipment for a scuba diving team (two-man):

4 40-cubic-foot high-pressure air tanks
2 air regulators
2 exposure suits (wet-suit type)
2 sets of fins
2 faceplates
2 weight belts including extra weights

Auxiliary diving equipment:

2 underwater watches
2 depth gauges
1 pressure gauge for determining pressure of air tanks
1 diving ladder
Decompression tables

Equipment for Making Geological Observations

For measuring attitudes of strata, several instruments have been combined to make the underwater geologist's "Brunton" (Figure 10-74). A small lucite pad with an attached well-damped compass in a watertight case, a clinometer, and a writing surface are the basic tools for measuring the dip and strike of outcrops underwater. The writing surface is necessary because it has been found that a man breathing compressed air forgets easily while underwater. Unless geologic observations are recorded immediately and checked by the buddy diver, they are not dependable. In addition to his compass-inclinometer, the diving geologist carries a geological hammer for rock sampling, a knife, a depth gauge, an underwater watch, and small marker buoys to be sent to the surface if exact locations of bottom features are desired. The diver wears an exposure suit to protect him from the cold water in which he must work most of the time. Fortunately, the problem of working in cold water has been almost eliminated with the development of efficient exposure suits. Temperature is not a problem for divers, even in waters which are covered with ice. Geological observations have been made in both the Arctic and Antarctic.

Geological equipment:

2 compass-inclinometers
1 echo sounder (fathometer)
1 sextant
1 navigation kit:
  Large-scale map of area
  Dividers
  Protractor (three-arm)
  Protractor (regular)
  Ruler
  Writing pad
50 sample bags (cloth)
6 diver-carried marker buoys

6 surface location buoys and anchors
1,000 feet of nylon line (parachute shroud or equivalent)

Auxiliary geological equipment:

Underwater camera (a wide-angle lens is a necessity)
Underwater tape recorder (for dictating detailed notes under certain unfavorable conditions or while being towed at relatively high speeds over a bottom)
Underwater sled for towing divers on long traverses
Underwater high-pressure water jet equipment for determining thickness through 20 feet of sediment

FIGURE 10-74. Underwater geologist's Brunton. (*Courtesy of United States Navy.*)

Techniques for Underwater Geological Mapping

A three-man diving team utilizes its personnel better and is able to accomplish more work per man than a two-man team because of the physiological limitations imposed by the breathing of compressed air at depth. The experienced underwater geologist can work to a depth as great as 50 meters with relative ease. The number of dives he is capable of accomplishing in a given day depends on the depth, the characteristics of the water in which he works, and the amount of time he stays there, inasmuch as the deeper the diver works, the shorter the period of time he can spend underwater. The practical lower limit of geological work underwater is about 50 meters because of nitrogen narcosis and submerged time limitation with greater depths and light availability. A three-man team diving alternately in pairs and operating in waters less than 40 meters deep can complete about nine dives of 10 to 20 minutes' duration in a working day provided they start in deep water and gradually move into shallower water as the day's work progresses. New repetitive-dive decompression tables established by the U.S. Navy are used to determine the length of time a geologist can work at any given depth (Figure 10-75).

The usual procedure followed is for two divers to descend and swim on a compass course examining the geology as they move along the bottom. The tending boat, a necessity for any type of extensive mapping, follows the air bubbles of the divers;

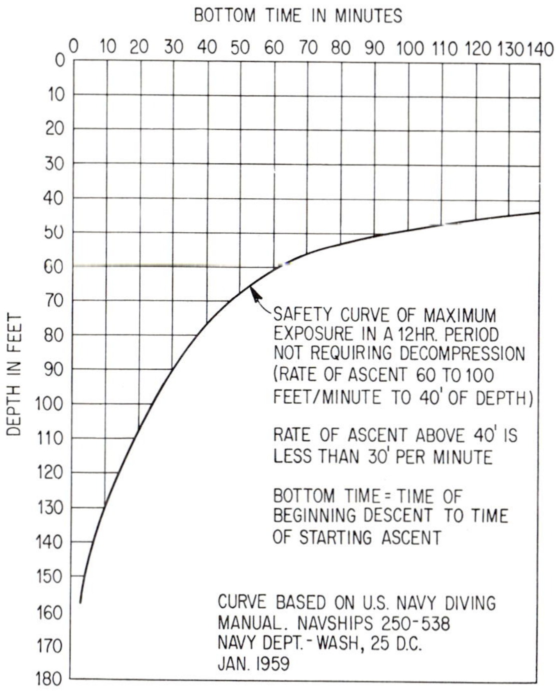

FIGURE 10-75. Safety curve of maximum exposure in a 12-hour period for underwater operations not requiring decompression. (*Courtesy of United States Navy.*)

the third man, using the sextant held horizontally, measures two angles between three points on the shore and plots their position at both down and up locations of the divers. In addition, small floats are released by the divers as a signal for the recorder in the tending boat to locate their position while submerged. For faster reconnaissance, where distance traveled across the bottom is important, divers can be towed along the bottom at speeds up to 2 or 3 knots.

In addition to his basic knowledge of geology, the diving geologist also must have a working knowledge of diving physiology to be able to cope with the problems that can occur during his work underwater. His knowledge of diving physiology and his limitations are necessary to plan efficiently and determine the techniques he will use while mapping underwater. Terms like *bends, oxygen poisoning, nitrogen narcosis, air embolism,* and *decompression* are as common in his vocabulary as *dip* and *strike* and *contact*. In addition to the problems of diving physiology the diving geologist finds that he is also forced to become a practicing biologist and marine ecologist, inasmuch as the general appearance of bottom life or the presence of a certain type of fish will indicate where rock outcrops may lie in an area predominantly covered by sediment. As on land, the finding of rock outcrops is, of course, a necessity for underwater structural mapping. The geologist who ventures beneath the sea finds himself acting in much the same role as an early naturalist, for he is in a new environment and he is forced to take advantage of every bit of information available in order to determine the geological picture of the area.

FINAL RESULTS

Realization of maximum benefits from a field project depends upon the ability of the party chief not only to map and record field data correctly but then to be able to summarize the data in an orderly report accompanied by illustrative material. Illustrations may include maps, stratigraphic columnar sections, stratigraphic cross sections, structural cross sections, field sketches, and photographs.

Reports

The final written report may sometimes be a difficult task for the field geologist, as it requires the assembly of all field notes and geologic data into a summary of the field project results, followed by conclusions with respect to the petroleum potential of the project area, including an evaluation of both the stratigraphic section and structural prospects. Hence, though many companies require weekly or periodic progress reports, if progress reports are not required, a periodic review of field data by the party chief will be of considerable benefit when the time comes to incorporate into a written report the ideas and opinions that have been developed over a period of several weeks or months, based on observed field data. Therefore, even though no progress report is required, a periodic written review of data, kept in a file for later reference, will be a reminder of ideas formulated in the field which might otherwise have been forgotten.

The Progress Report. This report should be true to its title, in that it is a record, weekly, biweekly, or monthly, of the progress of the field party. Prints of a small-scale base map of the project area can be carried in the field so that each progress report can be accompanied by a map indicating the amount of area mapped during the period covered by the report.

If certain pertinent data have been found, structural or stratigraphic, the report should contain a tentative interpretation of these data, with some type of illustrative material, such as a map or cross section, accompanying it. If the material submitted to the district office, for example, is considered important enough for immediate action, the progress report may be forwarded to higher management for additional evaluation, hence the importance of a well-written report with illustrations drawn in ink on tracing paper so that they can readily be reproduced.

The greatest value of progress reports is that they represent the changing ideas or theories of the party chief as the field work proceeds. Hence, at the end of the field season, the party chief is able to review his field notes in conjunction with a review of the progress reports, which will facilitate a rapid conclusion of the final report without the necessity of drawing ideas from memory. The field party chief should keep carbon copies of the field reports so that they can be reviewed in the field if necessary.

The Final Report. Most companies generally have a certain style of preparation for the report to be submitted at the end of the field season if one is required for the project. Many of the details should already have been recorded in progress reports or periodic summaries, and a good portion of the final report can be outlined or preparatory work started toward the close of the field season.

Although the main part of the text may vary in emphasis depending upon the objectives of the field project, stratigraphic or structural evaluation, the same general outline should be followed for every field project whenever possible. The following outline is suggested as a possible form of presentation for the material in a final report; other headings may be added or some of those headings suggested may be deleted.

I. Introduction
   1. Location
      a. Geographical setting
      b. Accessibility
   2. History of previous exploration
   3. Field project objectives
   4. Field methods

II. Stratigraphy
   1. Surface outcrops
   2. Precambrian rocks
   3. Paleozoic
   4. Mesozoic
   5. Cenozoic
   6. Tertiary
   7. Igneous intrusive and extrusive rocks

III. Geologic structure
   1. Regional setting
   2. Major structural trends
   3. Local structure

IV. Geologic history

V. Petroleum potential
   1. Surface indications of petroleum
      a. Oil and gas seeps
      b. Petroliferous rocks
   2. Stratigraphy favorable for accumulation of petroleum
      a. Potential source beds
      b. Potential reservoir beds
   3. Structural prospects
      a. Nature of potential traps
      b. Potential drilling objectives
   4. Potential areas for stratigraphic traps

VI. Conclusions

VII. Recommendations

The introduction of the report should include the exact location of the project area, as well as a general discussion of its location with respect to major geographical features, such as large lakes and rivers, mountain ranges, and major towns. If there has been any previous petroleum exploration in the project area, including old wells, a brief discussion should be included, possibly with formation tops, oil and gas shows, and tests if the information is available. There should also be a brief discussion in the introduction of the objectives of the field project, followed by a few remarks on the general procedure, or field methods, used to complete the field work.

If the major objective of the field project was a detailed surface study of the stratigraphy as exposed around the margins of a structural basin to be correlated with subsurface data from old wells drilled within the basin, the section of the outline on stratigraphy might well be expanded to include such items as subsurface stratigraphy, formational relationships, facies changes, regional aspects, or other related features observed in the field. If the field project was predominantly for the purpose of outlining structural trends and drillable prospects, the section on stratigraphy might be quite short, especially if none of the potential drilling objectives is exposed in the project area.

A geologic history of the area can be very helpful in grading structural prospects. This is especially true in an area having two or three structural trends, as expressed by local structure. If the geologic history of the area is worked out, the multiple trends may be found to represent different periods of deformation. In turn, this may tend to support downgrading of some structural prospects, upgrading of others, based on the respective age of folding as postulated from the geologic or tectonic history of the area.

A section on the petroleum potential of the area should be included near the end of the report, in which possible structural and stratigraphic traps are discussed, as well as lithologic characteristics of exposed strata with respect to their potential as reservoir or source beds. Although many areas may have no surface indications of the presence of oil or gas, if oil and gas seeps or petroliferous rocks are observed in the project area, the locations should be indicated on the map and their significance should be interpreted in the report.

Conclusions and recommendations can be grouped under one heading or can be written in separate divisions of the report. Whether the results of the field project are negative or positive with respect to the oil potential of the area and its prospects, the party chief should summarize the conclusions at the end of the report, followed with his recommendations as to future exploration activities to be followed if the area warrants additional consideration. It is desirable generally to include a summary of conclusions and recommendations on the first page of the final report. This summary is for the benefit of management personnel who are interested mainly in the answers to the problem and, furthermore, lack time to digest the entire report.

Illustrations. The key to good report writing is the preparation of clear illustrations, which can reduce substantially the number of words required to describe the observed field relationships. Maps, cross sections, field sketches, and field photographs contribute considerably to the completion of a well-integrated report, and every effort should be made to incorporate such illustrations into the final report.

Final Maps. The major result of a field project is some type of map on which field stations are located and the data, stratigraphic, areal, or structural, are illustrated as to distribution of outcrops, geologic structure, or interpretation of distribution of stratigraphic units.

If sufficient field data are obtained, isopach maps and lithofacies maps will adequately illustrate the results of a stratigraphic field project. The isopach map, showing changing thickness of stratigraphic units, or formations, is useful both in reducing elevations on key beds to datum values for structure contours and in defining thinning and possible stratigraphic traps where units may pinch out updip on plunging noses or in structural reentrants. Lithofacies maps, illustrating lateral changes in lithology as caused by changes in environment of deposition, can also lead to the location of potential stratigraphic traps, as well as to determining the presence of favorable drilling objectives on structural prospects.

The areal geologic map is used to illustrate both structural and stratigraphic field projects. On it, major stratigraphic units and their outcrop patterns are shown, with a sufficient number of strike-and-dip symbols to outline local structures and regional dip. A complete areal geologic map should include outcrop patterns of all exposed formations, all faults observed, axes of observed anticlines and synclines, and a complete legend designating the formational age, sequence, and the map symbols used.

Structure contours can be superimposed onto an areal geologic map to illustrate better the pattern or trend of local and regional structures. Structural maps are generally considered those on which the geologic structure is shown by means of contours. In some areas, however, the use of symbols to indicate structure may be sufficient, especially if the area is one of fairly high dips. In this instance, however, if datum elevations were not obtained, a form-line closing contour on any structure believed to have reversal, as indicated by strike-and-dip symbols, will be helpful in emphasizing the structure, making it stand out on the map.

The structure contour map is of greatest importance in the search for petroleum, as it has been the basic map of the oil companies since the origin of the anticlinal theory of petroleum accumulation. Structure maps may also prove essential to resolve other problems in a project area, such as changing environments of deposition as related to regional tectonics, patterns of formation distribution as related to local or regional structures, depth to drilling objectives on a structural or stratigraphic prospect, and many other more basic problems which might arise. Hence, even though a field project was planned as a stratigraphic study of an area, the inclusion of a structure contour map with the final report will permit a more thorough evaluation of the area as to potential stratigraphic prospects as well as structural prospects.

Although a structure contour map superimposed on a map of the areal geology of a project area presents the best illustration of actual field relationships, for detailed reconnaissance work, especially when time is important, a map showing only those formational contacts used to construct the structure contour map may be satisfactory. The party chief is the best judge of what type of maps will satisfactorily illustrate the final report, and he should bear in mind that the report may be read only once by those who will later evaluate the work but the maps will be used over and over again for reference to the project area. Hence, the maps should present a detailed graphic picture of the accumulated field data, regardless of original objectives, containing sufficient information to be useful references for future evaluation of the project area without referring back to the report.

Cross Sections. Two general types of cross section can be used to illustrate a final report for the field project—structural and stratigraphic. The structural cross section is drawn with reference to a *datum*, generally sea level, which is a level line and from which formation boundaries are scaled and plotted in a vertical plane. The *line of section*, which should be designated on an accompanying map, represents the line of intersection of the plane of the cross section and the land surface.

The stratigraphic cross section is generally drawn with reference to the top or base of a stratigraphic unit. Other stratigraphic units are then plotted up or downward, to scale, from the line of reference, or datum, which represents some stratigraphic horizon and is drawn as a straight horizontal line on the section. The reference datum should be clearly designated on the section to avoid confusion with structure sections.

Whenever possible, structure sections should be drawn true-scale, that is, with the same horizontal and vertical scale. Although exaggeration of the vertical scale tends to emphasize the structure at a glance, it can also lead to false impressions of amount of reversal on a structural prospect and should be avoided.

On the other hand, detailed stratigraphic cross sections generally require considerable exaggeration of the vertical scale, especially if detailed changes in lithology are to be illustrated along the line of section. The amount of vertical exaggeration should be kept at a minimum if the section is drawn to illustrate both stratigraphic and structural relations in an area. Otherwise, the amount of vertical exaggeration of scale for a stratigraphic cross section should be sufficient to allow presentation of the details required.

Sketches and Photographs. Field sketches and photographs are often used to illustrate minor features of interest observed in the field, as well as general views of major features and topography. Although the camera offers the best method of obtaining such illustrations, light conditions or camera failure may prevent its use, and therefore the field geologist should exert every effort to become competent in making accurate field sketches (Figure 10-76).

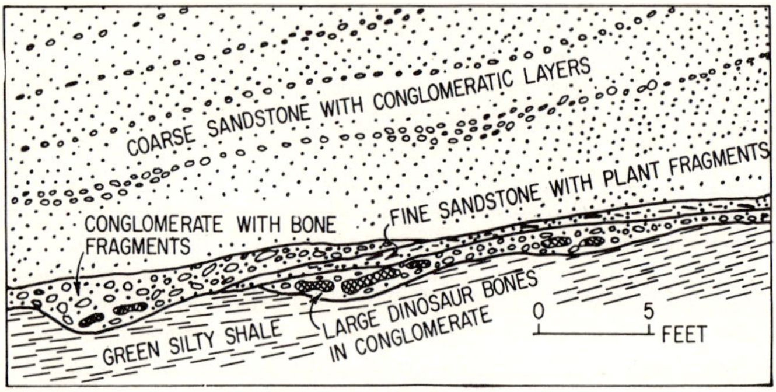

FIGURE 10-76. Field sketch of exposure of Morrison formation in Wyoming, showing details of depositional variations as observed in the field. (*From Smith and Bradley, 1954.*)

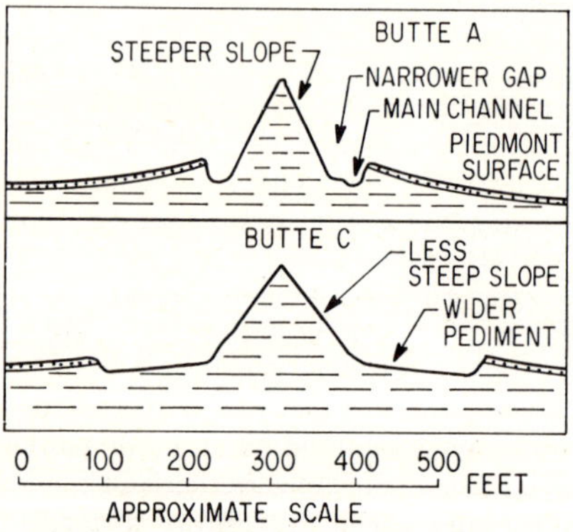

FIGURE 10-77. Generalized field-sketch diagrams to show general relationships of piedmont slopes to remnant buttes in Badlands National Monument, South Dakota. Note that details of stratigraphy are absent. (*From Smith, 1958.*)

Field sketches may be of two types: an accurate drawing of observed field conditions or a schematic diagram of observed relationships (Figure 10-77). Although scale of the field sketch may only be approximate, it should be estimated and given on the final drawing.

Photographs of minor geologic phenomena should include a hammer, Brunton compass, or some other familiar item for an estimate of scale. For photographs of large-scale features, distance should be estimated by the field geologist and approximate scale shown.

Field sketches and photographs can greatly improve the presentation of subject matter in the final report on a field project, and their inclusion in the report will, in many instances, result in a reduction of the number of words required to describe the observed phenomena.

Summary. Other illustrations which may prove useful for illustrating the final report include columnar sections, block diagrams, and panel diagrams. If the final result of a project is only a structure contour map, the report should include at least one columnar section showing exposed stratigraphy in the area and the mapping horizon, and intervals used to convert to datum should be indicated. Block diagrams can be useful for illustrating complicated geology, and panel diagrams for complex stratigraphy.

A well-written, well-illustrated final report on a field project will compensate any organization for the time and money allowed, whether the findings are positive or negative with respect to the oil potential of the project area. Hence, the importance of careful selection by the field geologist of illustrations and data to be discussed in the final report cannot be overemphasized, since his observations as presented in the report will guide his company's future activities in the project area (see Chapter 15 for detailed discussion of report writing).

Supplementary Tables

Tabulations of particular interest to the geologist in the field as well as other tabulations covering material of more general interest (not applicable primarily to field geology alone) will be found in the Appendix. Appendix Tables A-1 to A-15, B-1 to B-7, and C-1 (Part C) were prepared originally for this chapter.

REFERENCES

Alling, Harold: A Metric Grade Scale for Sedimentary Rocks, *J. Geol.*, vol. 51, pp. 259–269, 1943.

Badgley, Peter C.: "Structural Methods for the Exploration Geologist," 280 pp., Harper & Brothers, New York, 1959.

Bagley, James W.: The Use of the Panoramic Camera in Topographic Surveying, *U.S. Geol. Survey Bull.* 657, 88 pp., 1917.

Bagley, James W.: "Aerophotography and Aerosurveying," 324 pp., McGraw-Hill Book Company, Inc., New York, 1941.

Bagley, James W.: Oblique Photographs in Mapping, *Photogram. Eng.*, vol. 10, no. 4, pp. 240–244, 1944.

Bashaw, W. O.: New Geophotometric Mapping Technique, *World Oil*, pp. 63–64, January, 1948.

Bergstrom, J. R.: Jacob Staff Refined, *Bull. AAPG*, vol. 42, pp. 2251–2254, 1958.

Birdseye, C. H.: Topographic Instructions of the United States Geological Survey, *U.S. Geol. Survey Bull.* 788, 1928.

Bishop, Margaret S.: "Subsurface Mapping," 198 pp., John Wiley & Sons, Inc., New York, 1960.

Bouchard, Harry: "Surveying," 2d ed., 625 pp., International Textbook Company, Scranton, Pa., 1940.

Bureau of Land Management: "Standard Field Tables and Trigonometric Formulas," 7th ed., 245 pp., Department of Interior, 1950.

Campbell, M. R.: Rapid Section Work in Horizontal Rocks, *Trans. AIMME*, vol. 26, pp. 298–315, 1896.

Canadian Topographical Survey: "A Graphical Method of Plotting Oblique Aerial Photographs," 21 pp., Department of Interior, Dominion of Canada, 1928.

Cox, G. H., C. L. Dake, and G. A. Muilenburg: "Field Methods in Petroleum Geology," 305 pp., McGraw-Hill Book Company, Inc., New York, 1921.

Dake, C. L., and J. S. Brown: "Interpretation of Topographic and Geologic Maps," 355 pp., McGraw-Hill Book Company, Inc., New York, 1925.

Davis, W. M.: Nomenclature of Surface Forms on Faulted Structures, *Bull. Geol. Soc. Am.*, vol. 24, pp. 187–216, 1913.

DeFord, Ronald K.: Rock Colors, *Bull. AAPG*, vol. 28, pp. 128–137, 1944.

DeFord, Ronald K.: Grain Size in Carbonate Rocks, *Bull. AAPG*, vol. 30, no. 11, pp. 1921–1928, 1946.

Desjardin, L., and S. G. Hower: "A Technique for Geologic, Topographic and Structural Mapping," Aero Exploration Co. (private publication), 1938.

Dill, Robert F., and George Shumway: Geologic Use of Self-contained Diving Apparatus, *Bull. AAPG*, vol. 38, pp. 148–157, 1954.

Eardley, A. J.: "Aerial Photographs: Their Use and Interpretation," 203 pp., Harper & Brothers, New York, 1941.

English, W. A.: Some Plane Table Methods, *Bull. AAPG*, vol. 8, pp. 47–54, 1924.

English, W. A.: Use of Airplane Photographs in Geologic Mapping, *Bull. AAPG*, vol. 14, pp. 1049–1058, 1930.

Erfle, M. E.: Surveys Made from Air Aid Remote Exploration, *World Oil*, pp. 210–212, October, 1958.

Folk, Robert L.: Practical Petrographic Classification of Limestones, *Bull. AAPG*, vol. 43, pp. 1–38, 1959.

Forgetson, James M., Jr.: Review and Classification of Quantitative Mapping Techniques, *Bull. AAPG*, vol. 44, pp. 83–100, 1960.

Fuller, M. E.: New Tool Measures Distance Fast, *Oil Gas J.*, vol. 57, no. 38, pp. 106–110, 1959.

Fuller, M. L.: Quick Method of Reconnaissance Mapping, *Econ. Geol.*, vol. 14, pp. 411–423, 1919.

Geologic Division, U.S. Geological Survey: Interpreting Ground Conditions from Geologic Maps, *Geol. Survey Circ.* 46, 10 pp., 1949.

Goldman, M. I., and D. F. Hewett: "Schedule for Field Description of Sedimentary Rocks," Committee on Sedimentation, National Research Council, Washington, D.C.

Gurley, W., and L. E. Gurley: "Manual of Surveying Instruments," 49th ed., 208 pp., W. and L. E. Gurley, New York, 1941.

Hager, Dorsey: "Practical Oil Geology," 6th ed., 589 pp., McGraw-Hill Book Company, Inc., New York, 1951.

Hanson, Wallace R.: Improved Jacob Staff for Measuring Inclined Stratigraphic Intervals, *Bull. AAPG*, vol. 44, pp. 252–255, 1960.

Hayes, C. W.: "Handbook for Field Geologists," 2d ed., 159 pp., John Wiley & Sons, Inc., New York, 1909.

Hayes, C. W.: "Handbook for Field Geologists," 3d ed. (Sidney Paige, ed.), 166 pp., John Wiley & Sons, Inc., New York, 1921.

Hewett, D. F.: Measurements of Folded Beds, *Econ. Geol.*, vol. 15, pp. 367–385, 1920.

Hunt, C. B.: Geologic Mapping by Helicopter, *GeoTimes*, vol. 4, no. 7, April, 1960.

Irving, J. D.: Field and Office Methods in Preparation of Geological Reports, *Econ. Geol.*, vol. 8, pp. 66–69, 177–181, 264–297, 373–397, 489–499, 578–597, 691–721, 759–797, 1913, and vol. 9, pp. 67–72, 184–189, 1914.

Jones, E. J.: An Electronic Break-through in Surveying, *Civil Eng.*, 3 pp., July, 1959.

King, J. E.: KEK Stereoscopic Plotter, *Photogram. Eng.*, vol. 10, no. 4, pp. 252–263, 1944.

Kissam, Philip: Precision Altimetry, *Photogram. Eng.*, vol. 10, no. 1, pp. 25–45, 1944.

Krumbein, W. C.: Measurement and Geological Significance of Shape and Roundness of Sedimentary Particles, *J. Sediment. Petrol.*, vol. 11, pp. 64–72, 1941.

Krumbein, W. C.: Principles of Facies Map Interpretation, *J. Sediment. Petrol.*, vol. 22, no. 4, pp. 200–211, 1952.

Krumbein, W. C., and F. J. Pettijohn: "Manual of Sedimentary Petrology," 549 pp., Appleton-Century-Crofts, Inc., New York, 1938.

Krynine, Paul D.: The Megascopic Study and Field Classification of Sedimentary Rocks, *J. Geol.*, vol. 56, no. 2, pp. 130–165, 1948.

Kummel, Jr., Bernhard: New Techniques for Measurement of Stratigraphic Units, *Bull. AAPG*, vol. 27, pp. 220–223, 1943.

Lahee, Frederic H.: The Barometric Method of Geologic Surveying for Petroleum Mapping, *Econ. Geol.*, vol. 25, pp. 150–169, 1920.

Lahee, Frederic H.: "Field Geology," 6th ed., 926 pp., McGraw-Hill Book Company, Inc., New York, 1961.

Landes, Kenneth K.: "Plane Table Notes," George Wahr Publishing Co., Ann Arbor, Mich., 1947.

LeRoy, L. W., and Julian W. Low: "Graphic Problems in Petroleum Geology," 238 pp., Harper & Brothers, New York, 1954.

Littleton, James W.: Detail Mapping by Use of Enlarged Aerial Photographs, *Photogram. Eng.*, vol. 10, no. 3, pp. 214–216, 1944.

Loel, Wayne: Use of Aerial Photographs in Geologic Mapping, *AIMME Tech. Publ.* 890, 54 pp., 1938.

Low, Julian: "Plane Table Mapping," 365 pp., Harper & Brothers, New York, 1952.
Low, Julian: "Geologic Field Methods," 489 pp., Harper & Brothers, New York, 1957.
Lueder, D. R.: "Aerial Photographic Interpretation," 452 pp., McGraw-Hill Book Company, Inc., New York, 1959.
McCurdy, P. G.: "Manual of Photogrammetry," Hydrographic Office of U.S. Navy, Washington, D.C., 1940.
McNeal, R. P.: Lithologic Analysis of Sedimentary Rocks, *Bull. AAPG*, vol. 43, pp. 854–879, 1959.
Marr, J. E.: "Deposition of Sedimentary Rocks," Cambridge University Press, New York, 1924.
Marussi, Antonio, Chairman: International Symposium on Distance-measuring Techniques, *J. Geophys. Research*, vol. 65, pp. 385–528 (25 papers), 1960.
Mather, Kirtley, F.: The Manipulations of the Telescopic Alidade in Geologic Mapping, *Bull. Sci. Lab. Denison Univ.*, vol. 19, pp. 97–142, 1919.
Mather, Kirtley, F., and Bradford Washburn: The Telescopic Alidade and the Plane Table as Used in Topographic and Geologic Surveys, *J. Sci. Lab. Denison Univ.*, vol. 33, pp. 1–60, 1938.
Melton, Frank A.: Preliminary Observations on Geological Use of Aerial Photographs, *Bull. AAPG*, vol. 29, no. 12, pp. 1756–1765, 1945.
Menard, H. W., et al.: Underwater Mapping by Diving Geologists, *Bull. AAPG*, vol. 38, pp. 129–147, 1954.
Meyer, Jr., William H.: Laying a Controlled Photo Mosaic, "Manual of Photogrammetry," pp. 436–448, American Society of Photogrammetry, 1944.
Mitchell, J. G., and J. C. Maher: Suggested Abbreviations for Lithologic Descriptions, *Bull. AAPG*, vol. 41, pp. 2103–2107, 1957.
Moffitt, F. H.: "Photogrammetry," 455 pp., International Textbook Company, Scranton, Pa., 1959.
Otto, George: The Sedimentation Unit and Its Use in Field Sampling, *J. Geol.*, vol. 46, pp. 569–582, 1938.
Paine, Paul: " Oil Property Valuation," 204 pp., John Wiley & Sons, Inc., New York, 1942.
Payne, Thomas: Stratigraphic Analysis and Environmental Reconstruction, *Bull. AAPG*, vol. 26, pp. 1697–1770, 1942.
Powers, M. C.: A New Roundness Scale for Sedimentary Particles, *J. Sediment Petrol.*, vol. 23, pp. 117–119, 1953.
Prusok, R. A., and J. R. Ege: A Simple Stereophotographic Field Method of Outcrop Description for the Geologist, *Photogram. Eng.*, vol. 26, no. 1, pp. 98–100, 1960.
Reid, H. F.: Geometry of Faults, *Bull. AAPG*, vol. 20, pp. 171–196, 1909.
Reid, H. F., et al.: Report of the Committee on the Nomenclature of Faults, *Bull. Geol. Soc. Am.*, vol. 24, pp. 163–186, 1913.
Rosenfeld, G. H.: The Accuracy of Mosaics, *Photogram. Eng.*, vol. 21, pp. 670–673, 1955.
Russell, William L.: "Structural Geology for Petroleum Geologists," 427 pp., McGraw-Hill Book Company, Inc., New York, 1955.
Salisbury, R. D., and W. W. Atwood: Interpretation of Topo-Maps, *U.S. Geol. Survey Profess. Paper* 60, 1908.
Sharp, H. O., and R. K. Palmer: Faster and Cheaper Field Surveys, *Eng. News-Record*, Sept. 30, 1948.
Shrock, Robert R.: "Sequence in Layered Rocks," 507 pp., McGraw-Hill Book Company, Inc., New York, 1948.
Shumway, George: Compass-inclinometer for Underwater Outcrop Mapping, *Bull. AAPG*, vol. 39, pp. 1403–1404, 1955.
Smith, H. T. U.: "Aerial Photographs and Their Application," 372 pp., Appleton-Century-Crofts, Inc., New York, 1943.
Smith, Kenneth G.: Erosional Processes and Land Forms in Badlands National Monument, South Dakota, *Bull. Geol. Soc. Am.*, vol. 69, pp. 975–1008, 1958.
Smith, Kenneth G., and Daniel A. Bradley: Radioactive Fossil Bones in Teton County, Wyoming, *Geol. Survey Wyo., Rept. Invest.* 4, 12 pp. 1954.
U.S. Army Air Forces: Reconnaissance Mapping with Trimetrogon Photography, "Manual of Photogrammetry," pp. 645–712, American Society of Photogrammetry, 1944.
U.S. Geological Survey: Geologic Map Compilation, "Technical Workbook," part 1, Fuels Branch, Albuquerque, N.M., 1948.
Van Nouhuys, J. J.: Geological Interpretation of Aerial Photographs, *AIMME Tech. Publ.* 825, 18 pp., 1937.
Warner, C. A.: "Field Mapping for the Oil Geologist," 145 pp., John Wiley & Sons, Inc., New York, 1921.

Wengerd, Sherman A.: 'Logarithmic Form for Rapid Computation of the Oblique Grid, *Photogram. Eng.*, vol. 11, no. 5, pp. 258–261, 1945.

Wengerd, Sherman A.: Newer Techniques in Aerial Surveying, *World Oil*, vol. 127; no. 3, pp. 37–42; no. 4, pp. 46–54; no. 5, pp. 49–54; no. 6, pp. 136–142; no. 8, pp. 131–136, 1947.

Wengerd, Sherman A.: Visibility of Signal Cloth under Tropical Conditions, *Trans. Am. Geophys. Union*, vol. 31, pp. 165–168, 1950.

Wengerd, Sherman A.: Graphic Aids in Study of Strata, *J. Sediment. Petrol.*, vol. 26, pp. 42–48, 1956.

Wentworth, C. K.: A Scale of Grade and Class Terms for Clastic Sediments, *J. Geol.*, vol. 30, pp. 377–392, 1922.

Wheeler, R. R., and N. C. Smith: Finding Faded Structures, *World Oil*, vol. 135, nos. 1 and 3, 8 pp., 1952.

Wilson, R. M.: Oblique Photographs for the Surveyor, *Photogram. Eng.*, vol. 8, no. 1, pp. 36–54, 1942.

Chapter 11

GEOPHYSICAL SURVEYS*

By KARL DYK

Chief Geophysicist, Pan American Petroleum Corporation. A Past Vice-President of the Society of Exploration Geophysicists Tulsa, Okla.

Geophysical surveys are used in petroleum exploration to gain information concerning the possible location of subsurface petroleum accumulations from physical measurements made at or near the earth's surface. All generally accepted methods are indirect in that they do not determine the presence or absence of petroleum but measure some parameter that may be related to favorable conditions. Seismic data may define the attitude and persistence of (reflecting) beds and in some cases yield evidence of lithologic changes, gravity data can be interpreted in terms of subsurface density variations and their causative factors, and "the detailed mapping of the earth's magnetic field should be helpful in guiding our speculations with regard to the structure of that portion of the earth's rock mantle which contains appreciable amounts of ferromagnetic minerals" (*A*,† Vacquier, 1951). The great value of geophysical surveys lies in their ability to reduce the over-all hazards of exploration drilling by a factor well out of proportion to their costs.

The basic limitations of both gravity and magnetic surveys reside not in precision of measurement but in the multiplicity of geologically sound solutions that can satisfy the reduced data; seismic reflection data are plagued by similar uncertainties but to a much lesser degree. Unfortunately it is not always possible to obtain seismic reflection data, while gravity and magnetic readings can always be obtained.

Because of their limited resolving power and relatively low cost, gravity and magnetic prospecting are primarily reconnaissance tools. Seismic refraction surveying today has only a limited usefulness in special problems. The principal geophysical petroleum prospecting tool as of now is the reflection seismograph.

The initial use of geophysics in the search for oil appears to date from World War I and consisted of a torsion balance survey made in Czechoslovakia. The first gravity prospect to be proved in the United States was the Nash Dome, in Fort Bend County,

* The material under Gravity Surveys and under Magnetic Surveys would not have been possible without the generous collaboration of Mr. William R. Sype, Staff Geologist with Pan American Petroleum Corporation, Casper, Wyoming. His assistance is gratefully acknowledged.

† The letter *A* refers to part *A*, Seismic Surveys, of the reference list at the end of the chapter.

Texas. "Oil was discovered on the flank of this dome January 3, 1926, and this was probably the first oil pool to be discovered by geophysical methods in the entire world" (*A*, DeGolyer, 1935). At about this same time the first seismic refraction survey discoveries were being made, to be followed in a few years by the successful evolution of reflection shooting. Reflection techniques and instrumentation improved significantly for some years. There ensued a period during which great stress was placed on instrumentation, which did not proportionately improve the effectiveness of the seismic tool. About 10 years ago, under pressure of necessity, today's techniques involving multiple holes and multiple phones began to emerge, opening to exploration many new areas and extending the depth range of exploration in others. These techniques appear to be approaching a limit, one factor being the size of the surface area over which data can be usefully composited in the effort to reduce the noise that obscures the desired reflections. In order for the data to be useful, the structural and lithologic changes over the composited subsurface area must not be significant.

Another instrumental phase is now in progress, which Clifford (*A*, 1959) has termed a part of "The Revolution in Petroleum Exploration 1955—." Seismic information is being transcribed to continuous record sections with the data on each trace adjusted for differences in surface elevation, weathering, normal move-out, and velocity distribution. Seismic reflection records are being simulated with the help of continuous acoustic velocity logs and analogue or digital computers.

The search for stratigraphic traps has led to a more systematic study of the relation of seismic reflection character to stratigraphy and lithology (*A*, Rummerfield, 1954). Many theoretical papers touching on various aspects of this approach have appeared, but empirical data are not being published. It is likely that interpretation of reflection character changes will always require detailed geologic information. In other words, character study may help to extend *known* production if it happens that variations in the productive zone produce recognizable changes in reflection character.

In what follows we shall assume some familiarity with the elements and terminology of the geophysical methods discussed. No descriptions of computational procedures will be made, as they are largely standardized and well covered in a voluminous literature. Discussions of theory will be general and free of mathematics. It is also felt that description of crew organization and equipment is not pertinent in a book of this type.

SEISMIC SURVEYS

Seismic Reflection Surveys

General Description

The significant features of a conventional land operation are depicted in Figure 11-1. For clarity of representation, only a limited number of seismic-ray paths are shown, but it should be noted that for each seismometer there is a corresponding reflecting point at each reflecting interface as well as at least one refracted ray path.

A wide variety of specialized instruments and equipment has been developed for particular operations, such as swamp, marsh, muskeg, sand, shallow water, and open gulf water. Monthly coverage may vary from 20 miles for a swamp operation to 500 miles for a gulf boat operation.

The basic seismic field instrumentation can be considered as the seismometers, the seismic amplifiers, and the magnetic tape recording system. It is not uncommon for operators to have strong opinions concerning all three. During the early history of seismic exploration, instruments were not commercially available. Each operator set up a shop to design and build his own instruments and equipment, and this background and "organization in being" often influence present-day thought. However, the field of modern seismic instrumentation manufacture is now highly competitive, and instrumentation with any desired performance characteristics is available commercially.

The generally useful range of seismic frequencies for reflection work lies between 25 and 100 cycles. Seismic instrumentation needs to be designed so this range of

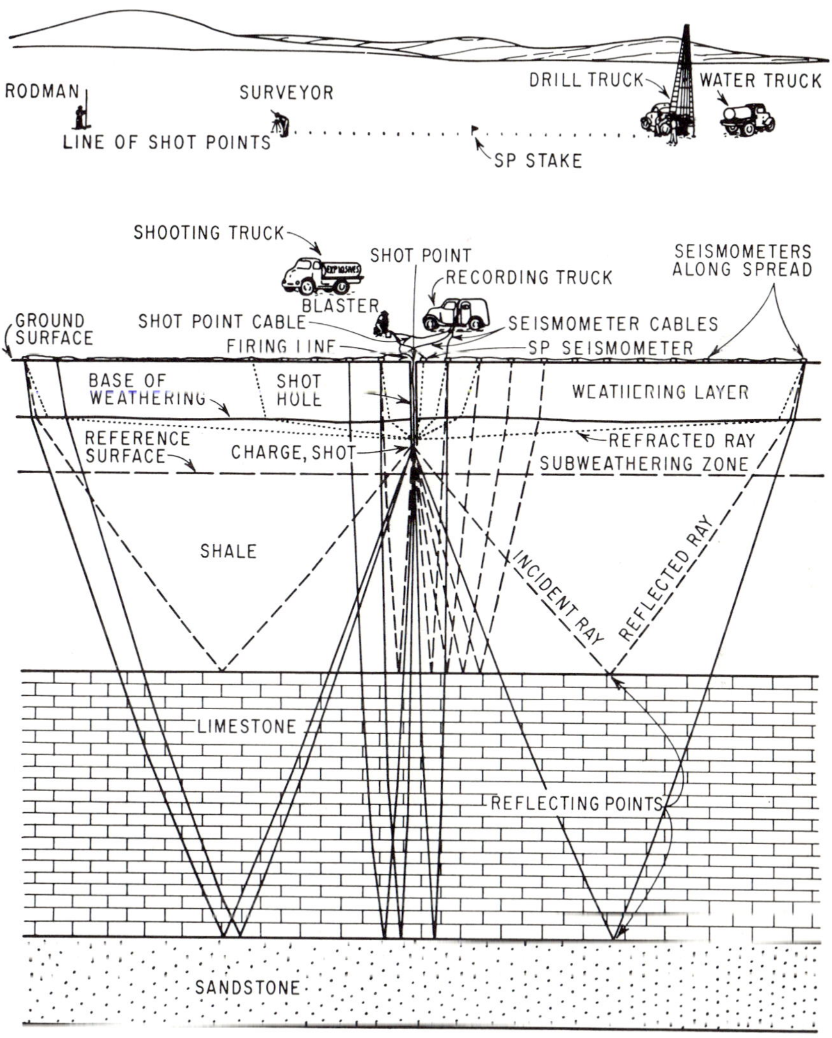

FIGURE 11-1. Schematic diagram of reflection shooting.

frequencies can be explored and specific subranges accentuated or attenuated. Seismometers with peak frequencies ranging from approximately 10 to 30 cycles are in general use. With low-frequency seismometers it is often necessary to prefilter ahead of the field seismic amplifier in order to minimize ground roll. Seismic amplifiers should provide a wide range of filtering options, including choice of cutoff rates. The

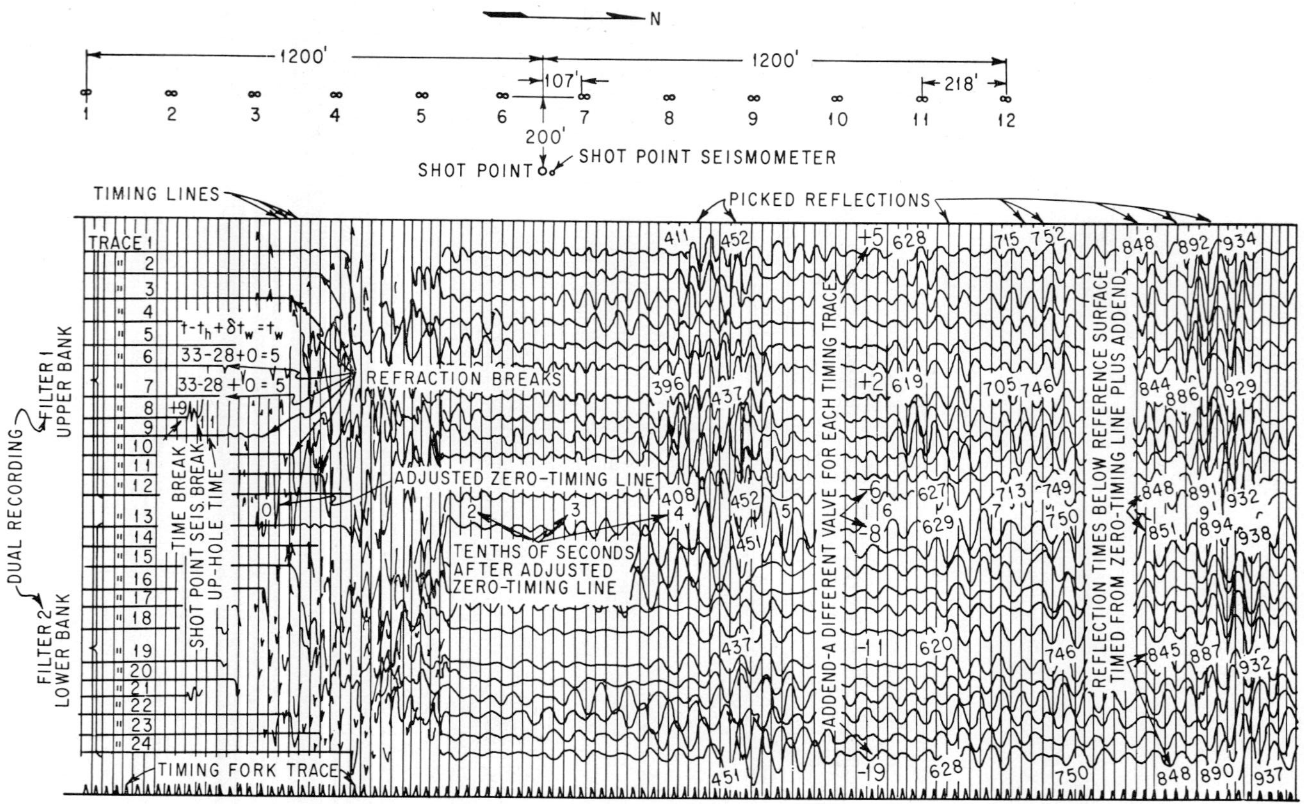

FIGURE 11-2. Typical reflection record.

filter phase-shift curves should be linear with phase axis intercepts of zero or an integral multiple of pi (*A*, Anstey, 1956).

Tape recording of field seismic data is now generally accepted, and probably over 90 per cent of the field crews are so equipped. The overwhelming majority of the field systems in use are either direct recording or frequency modulated. (There is one system employing pulse-width modulation presently being marketed.) The relative merits of the two systems have been discussed objectively by Erath (*A*, 1957), but many operators appear to have violent opinions on the matter. Both systems can readily be made to operate well within the required fidelity limits and are at present equally versatile in playback centers.

A considerable variety exists in purely dimensional elements in tape recorders of different manufacture, involving track width and spacing as well as tape length. A number of manufacturers have adopted one or the other of two of the more popular specifications, resulting in a degree of standardization.

Full utilization of tape recording depends on the operation of replay or data-processing centers where the corrected record section, to be discussed later, is produced. These centers may also permit the utilization of certain data-improvement techniques such as selective compositing of traces and/or shots as well as the selection of presentation modes (*A*, Kramer and Peterson, 1956; Reid, 1957). It will be several years before the data-processing centers have their full impact on interpretation.

The Reflection Record

The early portion of a typical good-quality reflection record appears in Figure 11-2, with the various elements identified. This is an "uncorrected" record, as taken in the field at the time the shot is fired—uncorrected in that normal move-out and weathering patterns are present. The shot was recorded in two filters simultaneously, which was a frequent practice before the days of magnetic recording.

Reflection Identification

Travel paths of seismic events are deduced from their time-distance relationships. For this reason such events must be identifiable over a distance sufficient to establish the relationship to an appropriate degree of accuracy. The minimum number of traces over which the line-up must appear in order to be used often depends on the degree of desperation of the interpreter, but it is a function of the coverage between traces and the probability of chance line-ups. A two-trace "line-up" is clearly meaningless, and on a statistical basis there will be many three-trace line-ups in random data. The probability of chance line-ups decreases rapidly as the number of traces involved is increased, and it is relatively low for six traces.

After correction for normal move-out and weathering, primary reflections should lie on a straight line if the subsurface dips are relatively uniform. When reflector depths are large in comparison with spread length, the normal move out becomes too small to be diagnostic and it becomes very difficult to differentiate between primary reflections and other events except when the apparent velocity is unusually large. After normal move-out and weathering corrections are applied, the limits of apparent velocity for reflections are infinite for flat dips, with the lower limit of apparent velocity being the first-break velocity for 90-degree dips.

Timing of Reflections

In order to make structural maps it is necessary to determine travel times to the degree of precision required for the particular problem. The raw times on the records measure time from the shot to reflecting horizon and back to the seismometer at the earth's surface. To relate conveniently travel times between shot points, the raw times are corrected to a reference surface as well as to vertical travel paths. The end result of this process corresponds to the familiar use of a sea-level datum for geologic markers, but the seismic reference is not necessarily level.

Weathering and Elevation Corrections. A correction is usually made to take care of both elevation and near-surface velocity anomalies including the weathered layer (see *A*, Dobrin, 1952, pp. 265–271; Widess, 1946; and Thralls et al., 1952). The particular method employed to correct for weathering and elevation changes depends on the nature of the near surface as well as the specific exploration objective. It is customary to obtain absolute corrections at selected groups and differential corrections for the remaining groups. Three methods of obtaining absolute corrections follow:

1. Use of uphole times at the shot point and assuming this applies to the near groups. This depends on the existence of gradual and systematic weathering changes and placing the shot below the depth of lateral subweathering velocity changes (*A*, Woods, 1952).

2. Use of the first-break times from the near seismometers of the spread. This yields the correction at the timing group for adequate shot depth as defined above.

3. Use of the first breaks at the most distant seismometer groups in the spread. These refraction arrivals may penetrate appreciably below the shot level and, of course, provide weathering data to the depth of penetration. Because of path uncertainties (discussed later under refraction surveys), these data often do not provide the accuracy necessary for structural mapping, regardless of mathematical manipulation.

In practice it is well to check the several methods against one another. It should be clear that in continuous profiling, method 3 out of one shot hole can be checked against method 2 out of the adjacent shot hole. Where subweathering lateral velocity variations are known to exist, "weathering" computations should be checked at strategic locations by uphole measurements from below the depth of change.

Isochron Mapping. In some areas exploration objectives can be satisfied by mapping the time difference between specific reflections, effectively eliminating near-surface correction problems. This use of isochroning is distinct from the more conventional application of isopaching in subsurface mapping.

Normal Move-out Correction. The determination of normal move-out is best done on a statistical basis from the records (*A*, Swan and Becker, 1952) rather than by computation from velocity surveys. Careful screening of the statistical data is necessary to minimize the influence of interbedded multiples.

Cavity Delay. Reflection times for successive shots in a hole do not necessarily change by the same amount as the uphole times. Often successive shooting of a shot hole at the same depth produces marked changes in reflection times (*A*, McCready, 1940), in some cases as much as 10 milliseconds. This is undoubtedly related to the effects of shattering and aeration and has been termed cavity delay. The usual procedure to minimize such timing discrepancies is to relate reflections to the first shot in the hole through differential uphole times, the effect of cavity delay on uphole times being relatively small.

Filter Correction. Reflection time (and character) changes are introduced by instrumental filtering (*A*, Anstey, 1956). The delay is a function of both the instrumental filter characteristics and the frequency content of the recorded energies. Average delays among filters are readily determined by recording the identical signal through several filters (see Figure 11-2). For convenience, one can select some filter in common use as a standard and correct to this standard.

The Corrected Record Section

The field seismic recording, whether it be photographic or on magnetic tape, places the several elements or traces of data in a time relationship determined in part by near-surface conditions at the geophone sites, in part by spread geometry, shot characteristics, instrumental factors, and, of course, the attitude of and average velocity to the reflecting beds. Each shot is usually displayed on a separate piece of paper with its own set of timing lines. If the records from a line of points are laid side by side in a proper time relationship, the timing lines are not continuous from one record to those adjacent. Furthermore, if corresponding events are followed from trace to trace, the pattern so developed may be affected by near-surface and geometric effects to the extent that the significant relationships are somewhat obscured. The historic solution has been to transfer selected data to a cross section, with a resulting loss of many useful data. The overwhelming trend today is to display the seismic recordings themselves in section form with all traces corrected to a common reference and displaying reflection time perturbations caused solely by subsurface changes (see pages 11-5 to 11-6).

Much credit is certainly due V. T. Reynolds (*A*, 1955) for hastening this trend by development of his mechanical plotter. Reynolds' plotter has been largely supplanted by magnetic recording and playback.

Corrected record sections can be presented in a number of different visual representation modes. The commonest modes are the conventional oscillograph trace, variable density (first used by Frank Rieber in 1935), and variable area (see Figure 11-3).

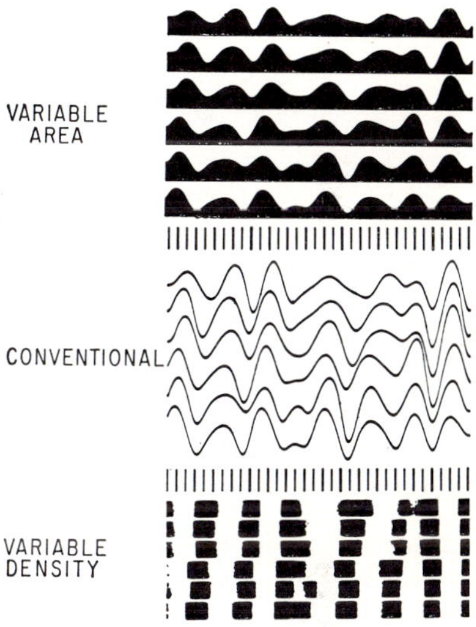

FIGURE 11-3. Comparison of presentation modes. (*United Geophysical Corporation.*)

Of these, the conventional oscillograph trace is capable of providing the greatest amount of detail, but variable-intensity displays can be by far the most arresting (see Figure 11-5). Reflections show up as continuous black-and-white striations, so that a corrected seismic record section gives the appearance of a geologic section. On variable-intensity sections the horizontal scale (trace width) can be greatly compressed without a further loss of detail on the individual trace, so that the eye can readily encompass structural relations over a large area (*A*, Palmer, 1957). It is an extremely valuable mode of presentation, both from the interpretive and from the "selling" viewpoint. Variable-area displays involve considerable loss of data, as the troughs are usually completely blocked off and with poor data quality present an unnecessary obstacle to the interpreter.

Record sections can be made using either a uniform time scale or a uniform depth scale. Such depth scales are necessarily only approximate, first, because the velocity function at any point is generally known only approximately and, second, because it is impractical to introduce velocity changes into the sections. Thus if depth sections are made in areas of velocity change, depth values on maps and record sections will not correspond.

Record sections, regardless of whether they are made to a depth or to a time scale, place the data below the point of observation and not in a position corresponding to the points of reflection. Dipping reflections on a record section must be migrated if they are to appear in the plane of the section at the *projected* position of the beds that gave rise to the reflections. The only practical way of doing this is by use of conventional seismic sections.

The preparation of corrected record sections is simplified if field conditions permit the use of a uniform seismometer group interval. This makes it easier to maintain horizontal scale on the sections.

The interpretive procedure on record sections is an extension of that mentioned under reflection identification. Individual reflections are carried as far as possible, and every effort is made to correlate persistent reflections across gaps where they are not developed. Identification of diffractions and multiple reflections is facilitated as is the picking of poor-quality data in areas of erratic weathering changes.

The Seismic Cross Section

If a record section is available, there is little need for a conventional seismic section except in cases of faulting or steep dip, where reflection migration is necessary to "untangle" the data. Reflections on a conventional section are represented by lines, the length of the lines often indicating subsurface coverage. With unmigrated sections the preparation of a cross section involves merely the transfer of selected data from the individual records.

No seismic section is complete without the posting of the known geologic data. These data may be from wells in the line of the section or may be projected from other control. A sound structural interpretation depends on full utilization of all pertinent available data. In general, the more subsurface control that is available, the better the interpretation is likely to be. Seismic record interpretations depend on the same kinds of skills and judgments that are involved in the correlation of well logs and are subject to the same degree of hazards. In this connection the matter of the relationship between the geologist and geophysicist in the exploration program is often raised. This question is rooted in the origins of exploration geophysics, and we shall consider it specifically as it concerns the seismic tool. The primary concern of the early geophysicists was with the instrumentation and techniques for making measurements, along with crew management. More often than not their maps represented picks on reflecting interfaces rather than geologic structure. Indeed, the geophysicists were all too frequently carefully shielded from geologic data so their record interpretation would not be influenced by a knowledge of the "facts." Today's geophysicists are trained structural geologists and are well qualified to combine geological and geophysical data. It is primarily a matter of policy and organization whether the interpreter spends his time managing a crew and picking records or makes structural interpretations using available geology, leaving crew operations in the hands of a crew foreman.

Migration of Data

Seismic reflection data involve wave travel in three dimensions. Record sections and seismic cross sections are two-dimensional representations. For dip lines this presents no problem, but it may for other line directions.

As mentioned previously, the points of reflection from dipping beds are not located vertically beneath the mid-point between shot and geophone but are displaced updip an amount depending on the dip and depth of the reflecting horizon as well as on the velocities in the intervening section. With complex subsurface conditions this displacement causes a tangling in the data that must be resolved by migrating the data to their approximate position in the section. It would be impractical to do this precisely even if it were possible, and a number of practical solutions have been advanced. One of the simpler in application is that proposed by Rice (*A*, 1953, 1955). As ordinarily done, migration is confined to the plane of the section, but the cross component may also be appreciable; in its migrated position the data thus represent a projection into the plane of the section.

Mapping Control Types

The spacing of observation points and the interrelationship of the data determine mapping control. The most efficient shot-point pattern from many points of view is undoubtedly a uniform two-dimensional (spot) pattern, but interpretational difficulties and/or field expediency usually dictate continuous line shooting (*A*, Salvatori, 1937) with a pattern of intersecting lines of shot points forming closed traverses or loops.

Spot Control

This type of control can be used where the character, as exhibited by recorded wave form, of one or more specific reflections can be recognized from shot point to shot point. At best only a limited number of reflections from prominent acoustic boundaries are in practice correlatable. Spot control can also be used to prepare dip projection maps if "four-way" dips are obtained at each location or a combination of correlation and dip projection is possible. Maps based on spot control suffer from general lack of reliability. When it does work, this type of shooting can be used for rapid coverage and field location of points can be more readily selected for ease of accessibility and uniform shooting conditions.

Line Control

As the name implies, shot points and the corresponding spreads are disposed linearly to form a network of lines, intersecting ordinarily at right angles. This provides a succession of closely spaced depth points which, if unbroken, is termed "continuous profiling" and if interrupted periodically is termed "discontinuous profiling." The distance between depth points or more specifically between seismometer groups is determined by reflection quality and continuity, dip of the beds, mapping depths, and the geologic problems and objectives. The distance between shot points is determined by the same factors and is, of course, also limited by the available seismic instruments and equipment. Economics dictates the use of the maximum spacing between shot points consistent with the objectives of the survey; the normal limits are between ¼ and 1 mile. The spacing between lines is determined by the objectives. Line control with continuous profiling is now in almost universal use in reflection work.

Continuity in Continuous Profiling

An important advantage of continuous profiling is that certain reflection time relations are available which can be utilized as a check on the inherent consistency of record interpretation. In general, two problems are involved in the continuity process: that of following events across a record and that of tying events on interlocking profiles. We shall not consider carrying continuity across the near traces, as this need involve no serious problems.

It will be assumed that records have been counted to datum so that timing mis-ties are immediately evident. If the weathering and initiation errors are less than one-fourth leg, the correct tie between interlocking records is established by time alone without regard to reflection character. When differential timing errors between the interlocking traces are of the order of a half cycle or half leg, there may be serious doubt concerning the correct tie in the absence of good character correlation.

Figure 11-4 shows the nature of the problems arising from weathering errors alone and initiation errors alone. The correct section is represented by a uniformly dipping bed. A dashed line indicates the section as it would appear with a 10-millisecond error at shot point 2 using the correct continuity. One possible result of using incorrect continuity is shown by a dotted line.

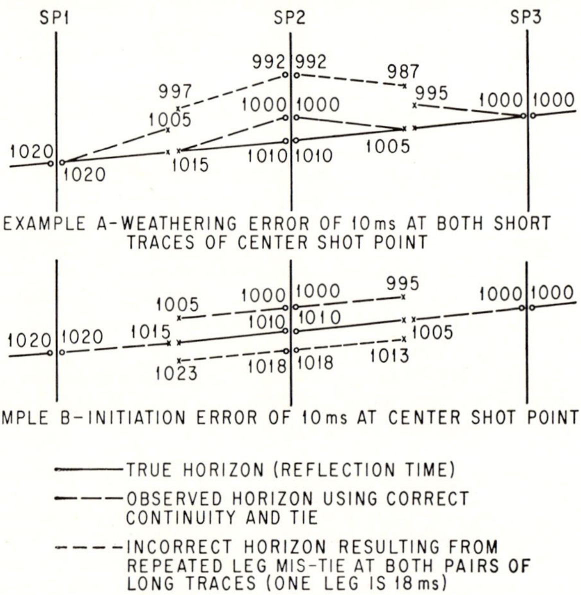

FIGURE 11-4. Cross section illustrating nonrandom sources of error.

Some Useful Relations. The following useful relations can be observed from the figure:

1. The difference in long trace times between shot points 1 and 2 is opposite to that between shot points 2 and 3, whether the error is in the weathering or initiation. The difference between long trace times should therefore be determined and tabulated at the top of the cross section in order that a ready check can be made to assure that these differences are canceling each other rather than cumulating down the continuous line. Care in this regard should generally confine a leg error, if it should be committed, to a single profile.

2. In the case of weathering as the source of error, if a wrong tie by a leg is made at the long times (dotted line on Example A), it is significant that the departure of the resulting horizon with respect to the true bed is equal to the magnitude of weathering error plus a full leg. Since the weathering error in itself must be in the neighborhood of half a leg in order to cause a mis-tie, a very abrupt rise or drop in the cross section would result, which would immediately arouse the suspicion of the interpreter.

3. The fact that initiation and weathering errors appear in different ways on the

cross section results from the fact that the two are related differently at long traces and at short traces. This difference therefore makes possible the identification of misties as due either to weathering or to initiation, merely by comparing the particular mis-tie appearing on a section with Examples A and B. Such a comparison is based on the assumption that extraneous errors are not predominant; combinations of weathering, initiation, and normal move-out errors may confuse the comparison.

Except as modified by normal move-out and dip, practically the same trace-to-trace weathering pattern should be shown by all reflections on a record. The same weathering pattern should appear on the interlocking portions of two adjacent profiles, the interlocks being shot from adjacent shot holes into the seismometer spread between the holes. This feature of continuous profiling provides a very powerful analytic tool in poor record areas having pronounced weathering changes.

Running Horizons

The separation of incidence points for adjacent seismic traces is of the order of 100 feet. This close spacing is made necessary by the difficulty of correlating most events from trace to trace as the spacing is increased. Even so, few reflections can be identified continuously over many spreads, necessitating the running of phantom horizons. Frequently no reflections originate in a zone of particular interest, so no seismic map for that zone can be prepared. Phantom horizons as mapped should be shown on record sections and cross sections in a manner to permit ready verification of the supporting data.

Closed traverses provide a powerful tool in seismic interpretation. For our purposes we shall consider a closed traverse as either one between two points of known subsurface control or one forming a loop with or without subsurface control. If a reflection, continuous between two wells, mis-ties the well logs, we may look for miscorrelation of either the seismic or well log data, a velocity gradient, or severe cross-line dips which would throw the seismic data out of the plane of the section. One-leg faults (faults which cause a reflection displacement of one cycle on the records) can be particularly difficult to pin down. Misclosure in carrying a continuous reflection around a loop is, of course, indicative of error in picking, which can be resolved by study of several filter presentations of the same data.

It is normally necessary or desirable to run horizons that are not supported by continuous reflections. This is accomplished by utilizing discontinuous data in a depth zone encompassing the horizon of interest. It is of prime importance to limit this zone as much as practical, and it should not transgress unconformities. The phantom horizon is run paralleling the selected dip information. The individual segments of dip that are used can be in error because of uncertainties in weathering corrections, variations in shot conditions, normal timing errors, and the uncertainties incident to the use of marginal data. These possible sources of loop misclosure must be critically analyzed and minimized as a first step in the running of phantom horizons. The remaining misclosure in a loop results either from error or from faulting. Remaining misclosures resulting from error are adjusted, usually on some sort of percentage basis taking due account of data quality. The system used depends on individual preference. It is common practice not to reverse critical dip in the adjustment process even though such a practice can be questioned on purely statistical grounds.

Adjustment of Errors

Most seismic prospects involve a network of interlocking loops. When phantom horizons are run in such a network, each individual loop has its own error of closure.

An accepted approach in such a circumstance is to start adjustments in lead areas with a view to putting the least possible adjustment in the areas of interest. Considerable judgment is required in order to avoid introducing undue adjustments into some seismic lines as a result of adjustments made in lines that are common to two loops. Any adjustment introduced into one loop necessarily affects the misclosure in the remaining legs of all loops with sides common to it.

Fault Identification

We shall concern ourselves here with a consideration of seismic evidence for faulting as it may manifest itself in continuous line shooting. The principal criteria are:

Direct correlation of reflections across the fault zone
Extinction of reflections that have been continuous for some distance
Anomalous dips near the fault zone
Absence of reflections or diminution in their number, particularly from below the fault zone
Structural deformation in beds above the zone of faulting
Overlapping of reflections

Direct Correlation. Direct correlation depends on the presence of one or more reflections whose persistence can be established on both sides of a fault zone. The reliability of such correlations depends in large measure on reflection prominence (or the degree to which it stands out amplitude-wise from the preceding and following seismic signals) as well as on the reflection character or wave form. It is, of course, necessary to establish the degree of stability of both character and prominence. If there are two or more correlatable reflections, the time intervals between can also be a useful guide, although these may change abruptly in crossing a fault.

Extinction of Reflections. The extinction of reflections is, of itself, not diagnostic of faulting, as it usually results from other causes. When there is other evidence of faulting, the disappearance of a reflection that has carried for thousands of feet or perhaps a matter of miles constitutes additional evidence even though the reflection is so lacking in character that it cannot be identified beyond the suspected zone of faulting.

Anomalous Dips. Anomalous dips from above the fault zone are generally representative of the structural deformations of the beds resulting from drag or from slumping. Dips from below the fault zone are further distorted by velocity effects, and the actual attitude of the beds may be indeterminate except in a gross way (*A*, Laubsher, 1956), even when identifiable reflections from below the zone of faulting are present.

Diminution in Number of Reflections. Distortion of beds in the vicinity of faults is often so mild that reflection picks appear to cross without interruption, especially in the case of minor faults with small throw. However, with increased throw and distortion of the formations, conditions become unfavorable for the generation of reflections. The net result often is a diminution in the number of reflections from above the fault, absence of reflections from the fault zone proper (it is not common to record reflections that can definitely be related to the fault surface), as well as virtual absence of reflections from below the fault. As stated earlier, recognizable reflections that do make a round trip through the fault zone are apt to be so distorted as to make them of little value other than to suggest faulting. An interruption in the normal pattern of reflection density is often the most diagnostic seismic evidence for faulting and provides the best control for the hade. The value of this criterion diminishes with diminishing over-all reflection density and as the amount of section involved is reduced. If only a few thousand feet at the base of a seismic section is involved in actual faulting, the interpretation problem can be enormously difficult.

Structural Deformations. Continued movement along a fault during or following deposition without rupture of the overlying formations produces characteristic structural deformations which can serve as useful supporting evidence for fault interpretation on seismic sections.

Low-angle thrust faulting presents particularly difficult problems for seismic resolution, especially where imbrication is well developed. The individual thrust terminations give rise to major velocity anomalies for reflections from beneath, with other

FIGURE 11-5. Seismic reflection section across a fault zone.

evidences of faulting usually only poorly developed. A knowledge of the stratigraphy that may be involved is needed for intelligent evaluation of such data.

Overlapping of Reflections. Seismic reflection records taken across faults frequently show an overlapping of reflections from the displaced blocks (*A*, Krey, 1952) as a result of diffraction from the reflector terminations. As the source of such diffracted waves is the terminus of the bed, cross-section migration of such events should bring them to a point in the fault zone, but inaccuracies in the data limit actual application. (The development of diffractions in model experiments is admirably shown by *A*, Rieber, 1937.)

An extensive selection of seismic cross sections across normal faults is shown by Quarles (*A*, 1950) and Laubsher (*A*, 1956). Faulted record sections can be seen on pages 95, 97, and 503 of "Geophysical Case Histories" (*C*, 1956). A variable-intensity record section across a faulted area is shown in Figure 11-5.

Identification of Reflections with Geologic Markers

Offhand it would appear to be a rather simple matter to relate seismic reflections to the generating interface. It should only be necessary to measure reflection travel time, and with the use of a nearby velocity survey, the depth of the reflector should be determined. Unfortunately this simplicity does not develop in practice if we wish to make such a correlation to the nearest 100 feet. Apart from some of the basic problems of timing, discussed under Timing of Reflection, and those of migration, mentioned under Migration of Data, there is an even more fundamental problem in attempting to reconstruct the earth layering from a reflection record. These are inherent in the limitations in resolving power. A wide variety of earth models will produce reflection records that are not recognizably different, especially when the presence of noise is considered.

The inverse process of synthesizing a reflection record from an assumed earth model also involves serious problems. The earth models are ordinarily based on smoothed, continuous velocity logs, and density data are usually lacking. Reflection coefficients depend on both density and velocity contrasts. Published information does not appear to indicate that a close correlation between synthetic records and associated shot recordings is to be expected. Both digital and analogue computers have been used in the synthesis of reflection records from well data (see *A*, Peterson et al., 1955; Berryman et al., 1958; and Woods, 1956). The problems of accounting for the presence of multiple reflections and earth filtering are formidable.

In dealing with prominent reflections, it is often possible to estimate the onset to within a half cycle. In such cases it is possible to determine the depth within estimable limits.

Reflection Character

The character or shape of reflections can be adequately described for our purposes by the amplitude relationships of successive legs and the time interval between successive peaks and troughs. An idealized single reflection, free of background noise and other disturbances, as it may appear on a conventional seismogram is shown in Figure 11-6. The wave envelope is a useful concept and can be considered as a pair of smooth lines, symmetric to the time axis, drawn tangent to the reflection near the peaks and troughs as illustrated. Changes in individual reflection character are concomitant with changes in the position of the envelope maximum with respect to the wave peaks and troughs.

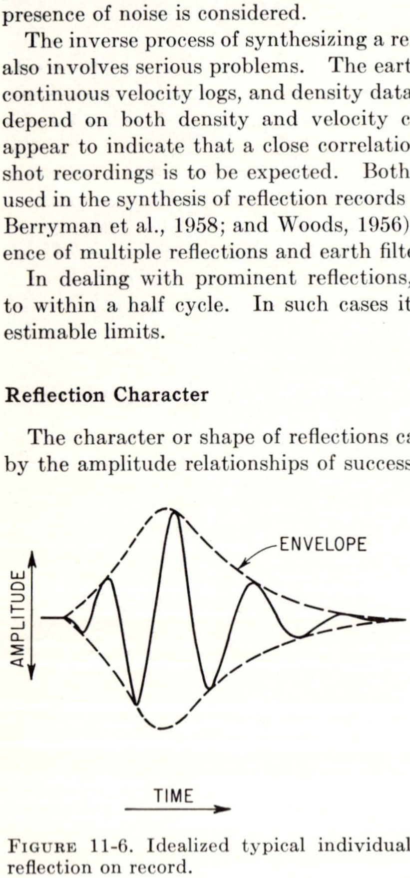

FIGURE 11-6. Idealized typical individual reflection on record.

Description of Character. A qualitative description of reflection character is very useful in correlation studies. A notation developed by M. B. Widess (unpublished) is explained in Figure 11-7. This shows very clearly how character is related to the envelope maximum indicated by the vertical arrow.

The character of reflections on seismograms is ordinarily not as clear cut as shown in the figure, and it may vary considerably over the several traces of an individual record. Amplitude relationships and/or period measurements can be used to put character on a quantitative basis, and average values for a number of traces are readily determinable.

Effect of Bedding Changes. Innumerable factors, including initiation, transmission, and reception, may affect reflection character. Let us consider the character changes that may be useful from an exploration viewpoint. They result from interference between reflections from beds separated by less than a wavelength,

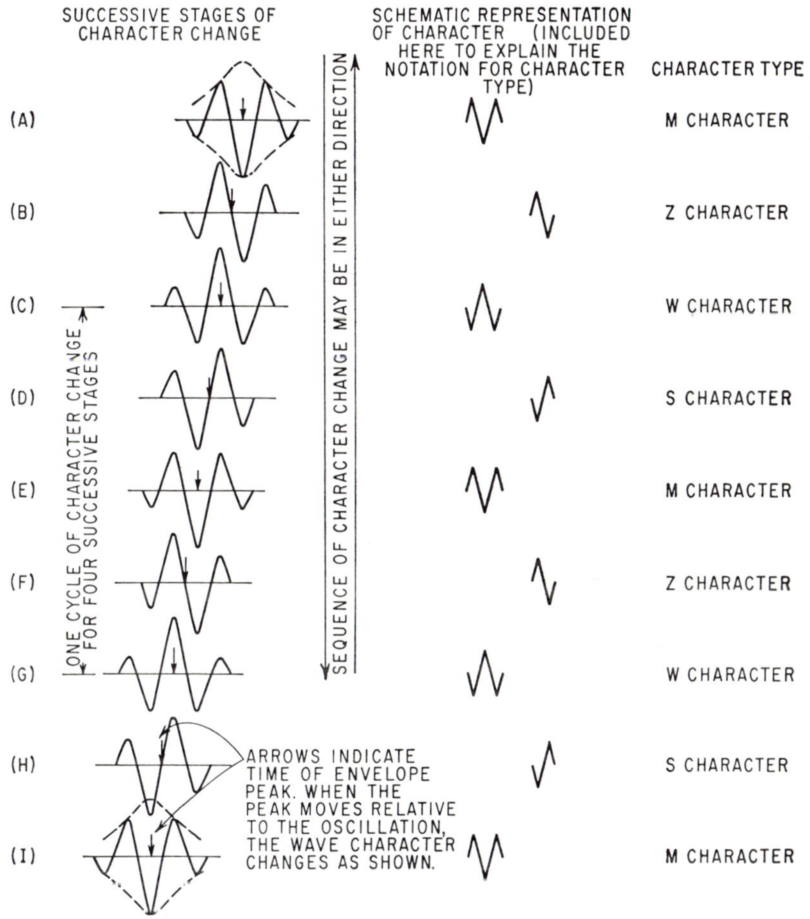

FIGURE 11-7. Theoretical successive stages of character change. (*After Widess.*)

because of either velocity changes in the intervening medium or changes in bed separation. Reflections from beds separated by more than a wavelength can usually be identified as separate reflections. Velocity changes may be due to lithologic changes, such as shaling of sands, or to fluid changes in the rock matrix, oil to water. Computed change of reflection character with bed thickness increasing from $\frac{1}{40}$ wavelength to a wavelength are shown in Figure 11-8, taken from Widess (*A*, 1957). This is a rather simple case of wavelet reflection at normal incidence from a single bed with rather high velocity contrasts and no density contrasts. It would appear from this example that isochron analysis may be very useful in the study of the behavior of thin beds. However, there is ambiguity in that practically identical character changes can result from velocity changes in the reflecting bed or a combination of velocity and bed-thickness change. For example, the character change occurring between bed

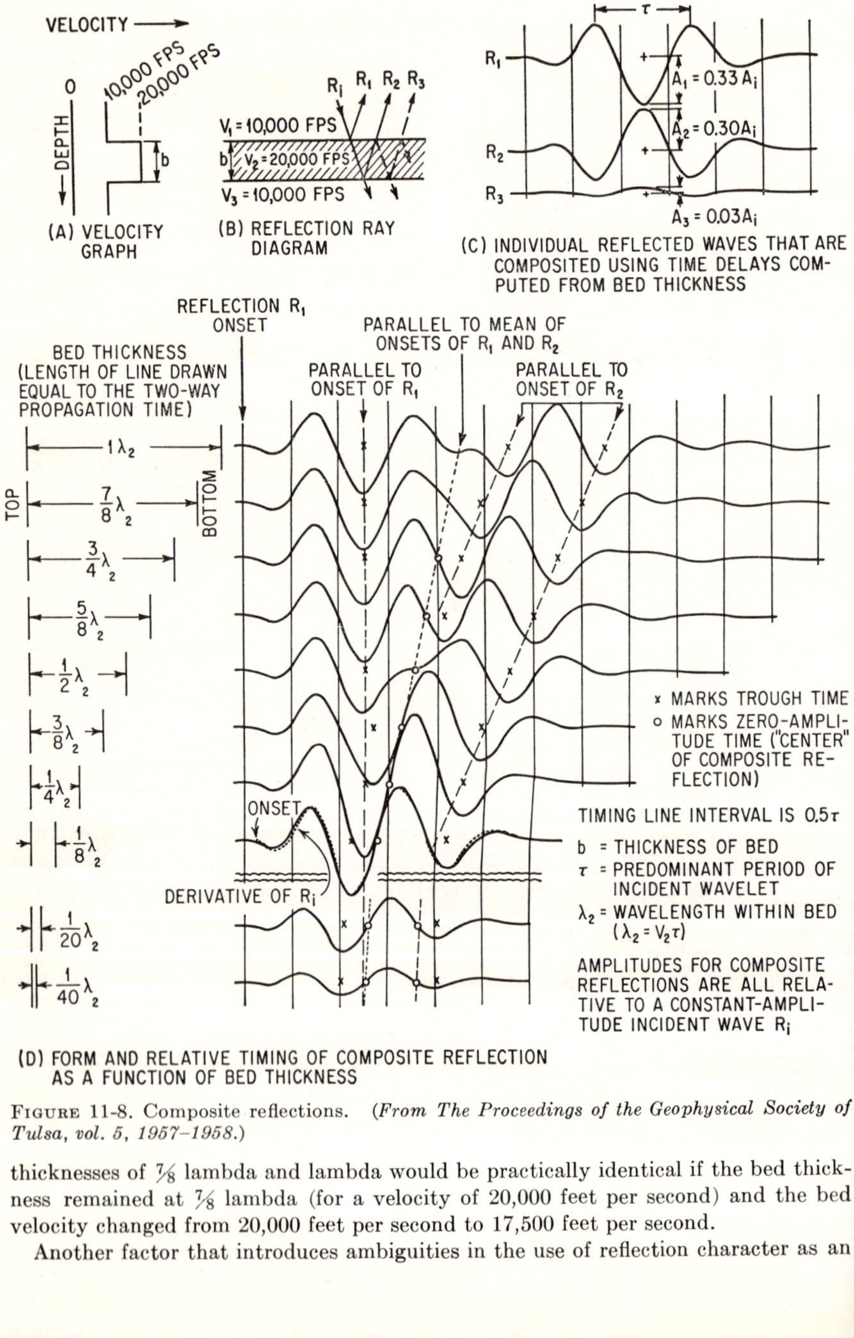

FIGURE 11-8. Composite reflections. (*From The Proceedings of the Geophysical Society of Tulsa, vol. 5, 1957–1958.*)

thicknesses of ⅞ lambda and lambda would be practically identical if the bed thickness remained at ⅞ lambda (for a velocity of 20,000 feet per second) and the bed velocity changed from 20,000 feet per second to 17,500 feet per second.

Another factor that introduces ambiguities in the use of reflection character as an exploration tool is the effect of lithology above the reflecting bed. This is well illustrated in Figure 11-9, taken from Berryman et al. (*A*, 1958). The full line shows a synthetic reflection record for about 2,000 feet of section, while the dashed line shows the synthetic record for the bottom 1,000 feet of the same section with the effect of transmission through the overlying 1,000 feet removed. The earth acts as a filter, each overlying bed having transmission characteristics depending on the relation between bed thickness and wavelength (*A*, Osborne and Hart, 1946).

It is apparent that there are formidable difficulties in the use of Rummerfield's (*A*, 1954) fourth dimension (reflection quality or character) as an exploration tool, even when one is so fortunate as to have a persistent reflection associated with a

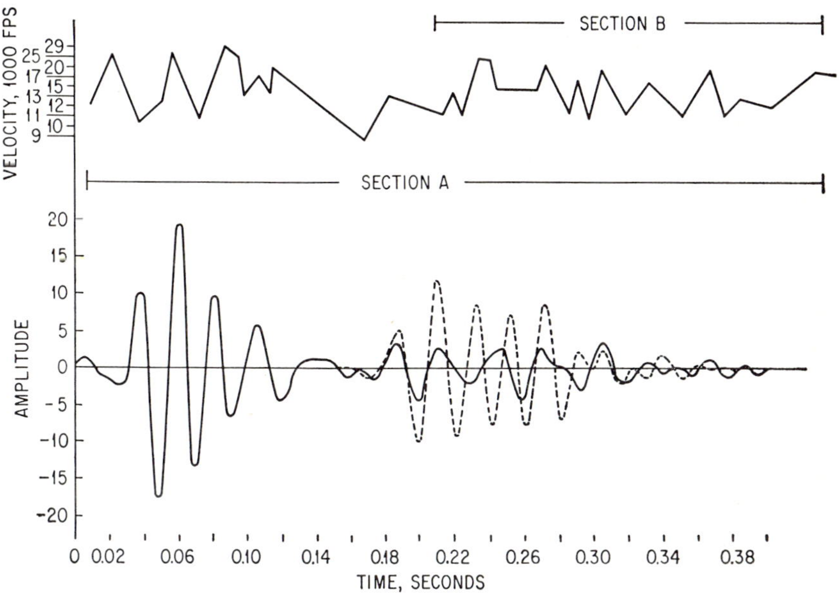

FIGURE 11-9. Simulated reflection record. (*Reproduced from Geophysics, vol. 23, by permission.*)

producing horizon. However, the industry is accumulating large numbers of acoustic logs, and the detailed study of synthetic records made from them in the light of known geology will bring a better understanding of the geologic significance of reflection character changes.

Other Factors. Mention was made earlier of additional factors affecting reflection character. They may cause character changes more drastic than those caused by changes within the reflecting beds, and such factors must be meticulously controlled even for the usual types of seismic interpretation. Some of the more important of these factors will be mentioned.

Charge weight: Light charge sizes tend to generate a greater proportion of high frequencies than do heavier charges.

Shooting medium: Shot-generated frequencies tend to increase with increase in velocity of medium in which the detonation occurs.

Charge characteristics: Elongated charges with detonation velocities matching the formation velocities generate relatively more high-frequency energy than do equal weights of concentrated charges; such charges, when capped at the top also minimize secondary or ghost reflections (*A*, Musgrave et al., 1958).

Charge depth: In addition to the effect of the shooting medium, charge depth affects the formation of secondary reflections with their resulting interference with the primary reflections.

Energy attenuation: Higher frequencies are attenuated during transmission more rapidly than lower frequencies.

Shot-generated noise: The effect on character, of course, depends on the relative strengths of reflection and noise and their phase relationships.

Field techniques: In addition to those relating to charge weight, depth, and placement, such items as spread configuration and even seismometer placement may affect reflection character.

Instrumentation: Filtering, of necessity, changes the amplitude and phase relationships of the different frequency components of the recorded energy—in other words, changes the character. (Some filtering is almost always necessary to minimize noise.) Filter distortion can be minimized by appropriate design and use of minimal filtering necessary for an optimum signal-to-noise ratio. Improper automatic-gain-control characteristics may also cause appreciable character distortion, including timing errors (*A*, Anstey, 1956).

Multiple Reflections

It is convenient to define a multiple reflection as one arriving at the earth's surface after an odd number (>1) of reflections. This definition excludes secondary reflections, which are first reflected from an interface above shot level. For simplicity in the discussion that follows, we shall assume that both shot and receiver are at the earth's surface.

Perhaps the simplest type of multiple is one that makes two identical round trips, but with a large number of reflecting interfaces, the number of possible multiple-reflection paths becomes fantastic. Thus with a normal geologic section we might expect an almost continuous "rain" of multiple reflections to arrive following the advent of the first few primary reflections, adding to the background of noise through which the primary reflections must be identified. The relative strengths of simultaneously arriving primaries and multiples determine which may be recognizable as an event.

A massive high-velocity section near the surface may effectively shield the beds beneath, the energy reverberating between the bed boundaries to produce a rather oscillatory recording in which primary reflections from below are completely submerged. Deeper beds having high-velocity contrast are more insidious in that they may be instrumental in the generation of multiples that are difficult or impossible to identify as such. Referring to Figure 11-8*C*, we see that R_3 is an interbedded multiple reflection with an amplitude about one-tenth that of the primary reflections R_1 and R_2. A primary reflection from the first interface below the V_3 bed (not shown in Figure 11-8*B*) would have practically the same amplitude at the surface as would R_3 if the velocity contrast at the V_3 to V_4 interface were 8 per cent. Thus it would require a velocity contrast considerably in excess of 8 per cent to produce a recognizable primary reflection that happened to arrive at the same time as R_3.

Interbedded multiple reflections having identical path elements which differ in sequence as in Figure 11-10 will arrive at the surface simultaneously and reinforce (*A*, Bortfeld, 1956). The number of such equal time paths increases in geometric progression as the number of beds involved is increased. This is another important factor which may raise the noise level on seismic reflection records. We must regard multiple reflections as noise, as when present they interfere with the primary data and provide no useful subsurface information which cannot be better obtained from primaries.

The identification of multiples as such is usually extremely difficult, although at times it is definite. The principal criteria are normal move-out, time relationships with primary reflections, and pattern repetitions. Any reflection with a consistently large normal move-out for its time of arrival is suspect. If persistent, strong reflections are recorded, times for simple as well as interbedded multiples should be computed and plotted for comparison with the other recorded data. A good working rule is to regard any reflection as a primary unless it can be proved to be a multiple.

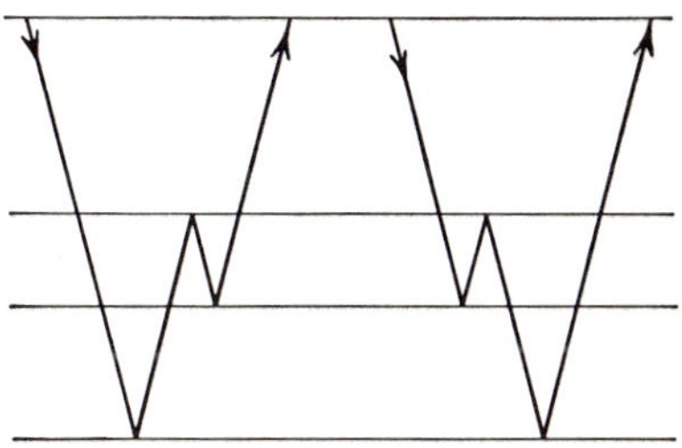

FIGURE 11-10. Interbedded multiple reflections.

A comprehensive discussion of multiple reflections appeared in the January, 1948, issue of *Geophysics*. The paper by Bortfeld (*A*, 1956) is recommended for those with a reading knowledge of German. Richard's (*A*, 1953) discussion in French is also illuminating. Conklin (*A*, 1953) shows some examples of multiple reflections in the Texas Panhandle and southern Mississippi.

There is no known method of eliminating multiple reflections in the initiation process without also eliminating primary reflections. There has been some effort to use trace compositing techniques to enhance primary reflections at the expense of multiples. The process does not appear to be practical at present, although it does hold hope for the reduction of simple surface-to-surface multiples. For most interbedded multiples the normal move-out differential is probably too small to be useful.

Secondary Reflections

Energy emanates in all directions from an explosive charge detonated in a bore hole. The uptraveling energy is reflected downward at acoustic interfaces and then may still have sufficient energy to produce observable reflections from below.

Secondary reflections are usually generated either at the ground surface or at the base of weathering because of the large acoustic contrasts (*A*, Van Melle and Weatherburn, 1953; Giulio, 1954). They follow the primary reflections at a time equal to twice the time from shot to the reflecting interface above. It is apparent that their travel time increases with increasing shot depth, which identifies them when the shot depth is sufficient to permit resolution between the primary and secondary. For shallower shots an interference pattern between primary and secondary develops which may cause severe character and timing changes in the primary reflection.

The effect of secondaries can often be held to a safe level by control of shot depth. This method is not applicable if secondaries are troublesome at the best shooting depth as developed by other criteria. It is then desirable to use elongated charges capped at the top, with the detonation velocity approximating the velocity of compressional waves in the formations in which the detonation occurs (*A*, Van Melle and Weatherburn, 1953). Such a charge is highly directional, with a much greater proportion of the energy traveling downward than travels upward, with consequent reduction of energy in secondary reflections. The same thing can be done by delayed detonation in a system of vertically disposed charges (*A*, Shock, 1950) or by the addition in appropriate time relationships of a series of separate magnetically recorded shots at different depths.

Other Shot-generated Noise

Multiple and secondary reflections are principal sources of vertically propagating "noise" on seismic records. In addition, large numbers of primary reflections that persist over no more than two or three traces are undoubtedly recorded from most shots. As they are not identifiable as reflections, they must be regarded as part of the noise.

Horizontally traveling shot-generated noise is of two types: surface waves (or ground roll) and body waves.

Surface Waves. Ground roll has an appreciable vertical component and consequently is recorded along with reflections. It travels along the surface with speeds ranging from 300 to 3,000 feet per second. Ground roll can usually be satisfactorily controlled by shooting below the weathering base (*A*, Howell et al., 1953) and by instrumental filtering. Some operators use seismometers peaked at 25 to 30 cycles per second to help reduce the effects of ground roll. The use of multiple detectors in a linear array can also reduce the effects of ground roll; however, in some areas weathering changes limit the use of this technique. (Sharp weathering changes may change the direction of travel of ground roll by refraction, may act as secondary sources of ground roll, and, of course, cause reflection time changes between seismometers in a group.) Elongated charges should also reduce ground roll. In some areas, ground roll cannot be satisfactorily controlled by shot placement, and filtering is ineffective because reflection and ground-roll frequencies tend to overlap. It is then necessary to offset the reflection spread from the shot hole a distance sufficient that the ground roll arrives at the spread after all desired reflection information has been recorded. Offsets in excess of a mile are at times necessary.

Reflected Refractions. Horizontally traveling body waves (so named because they travel through the body of a material rather than along surfaces) are generated in great profusion by seismic shots. The direction and speed of travel of such waves can often be determined by special techniques, for example, closely spaced seismometers recording independently. It would appear that in noisy areas, body-wave energy arrives at a seismometer simultaneously from many directions, and it is almost impossible to trace distinct arrivals to their sources even if there were practical reasons for trying in individual instances. The near-surface materials of the earth are often far from homogeneous. Any sudden change of properties will cause reflection or diffraction of seismic waves (*A*, Tatel and Tuve, 1954). It is well known that caliche and surface limestones with their multitudinous vertical fractures are particularly noisy seismically. The near-random nature of the resulting shot-generated noises is the basis of the multiple seismometer and multiple-hole techniques for their suppression. (See, for example, the report in Vol. 52, No. 50, of the *Oil and Gas Journal* of a symposium by the Geophysical Society of Tulsa. Also see *A*, Pommier and Richard, 1957.)

Noise Control Using Multiple Holes and Seismometers

Depending on the theories held or general operating practices or conveniences, multiple seismometer and/or multiple-hole arrays are laid out in radial, spiral, star-shaped, or rectangular patterns. The most efficient suppression of noise occurs when, on the average, noise between adjacent seismometers is 180 degrees out of phase. A similar statement can be made for pattern shot-hole spacing. The number of seismometers and shot holes required to obtain useful data is determined empirically in each area. The distribution of "power" between shot holes and seismometers depends on economics and physical conditions in the field. Regardless of other factors, the maximum dimensions of both the shot pattern and the seismometer group are controlled by the requirement that the maximum time difference for a primary reflection arriving at different seismometers in an individual group should not cause destructive interference. This time difference results from elevation and weathering differences at the seismometers and also at the shots (the shots may be in the weathering layer) as well as differences caused by normal move-out and dip of the beds.

To a very rough approximation, "power" can be interchanged between seismometers and shot holes, keeping the product of the number of seismometers per group and the number of shot holes per pattern constant. Changing the depth of shots usually changes the product, the most significant change occurring in going from holes shot in the weathered layer to holes shot below the base of weathering.

A good discussion (with a rather complete bibliography) of the use of "power" to improve record quality is given by Heinitz (*A*, 1959).

Seismic Refraction Surveys

In refraction surveying, the distance between shot and receivers is large with respect to the depths to be mapped, so that considerable horizontal travel of the energy is involved. The name "refraction" is technically incorrect, but it has become accepted. The method has its origin in observations made by Mohorovicic (*A*, 1909). In plotting the time-distance data for the first arrivals of a near earthquake, he noted that the data could be fitted by two straight lines. He interpreted this to mean (near) horizontal travel in two earth layers, the lower layer having the higher speed. The top of his lower layer has come to bear his name: the Mohorovicic discontinuity. The application to seismic prospecting was patented by Mintrop (German Patent, 1919).

The Refraction Process

O. V. Schmidt (*A*, 1938) has photographed these "refracted" waves in model experiments and called them "headwaves." Their existence depends on the presence of layered media. A medium with velocity *V* effectively shields the development of the headwaves in those layers farther from the energy source and having longitudinal wave velocities less than *V*. Simple optical theory is not adequate to explain the development of the headwaves, the problem being to explain the observed energies. Present theoretical treatments appear to be inadequate.

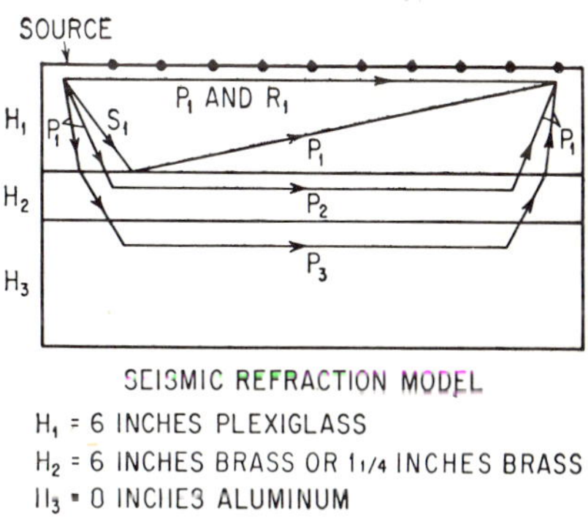

FIGURE 11-11. Types of (refraction) waves observed in triple-layer model. (*Reproduced from Geophysics, vol. 19, by permission.*)

Press et al. (*A*, 1954) have reported on some results of seismic model studies of refractions. A diagram of their model showing observed wave paths is given in Figure 11-11. The resulting travel-time plots appear in Figure 11-12. These models correspond to the simplified earth structures used in conventional refraction-technique interpretations, and the range of ratio of wavelength to bed thickness covers that encountered in oil exploration.

A very significant observation is that the velocity of the wave through the high-speed refractor depends on the thickness of the overlying beds. Lacking a theoretical explanation, the question of dependence on other overburden parameters is open. The implication with respect to mapping is clear.

A further significant result is the observation of strong "composite" refractions which because of their long wavelength have a velocity determined by the properties of two (or more?) adjacent layers, as indicated by the phases $P_1P_{32}P_1$ and $P_1P_{23}P_1$ in

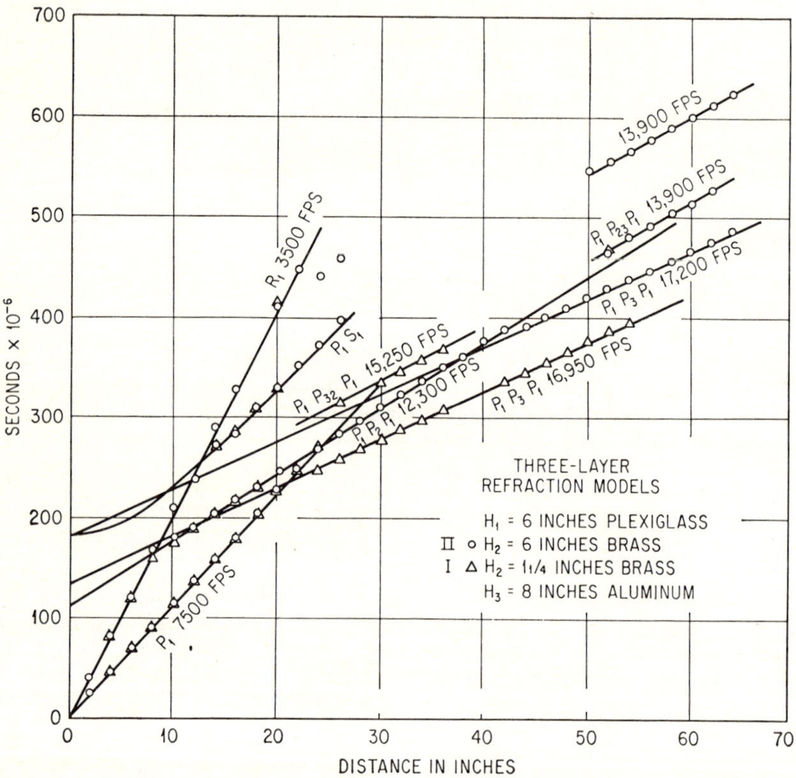

FIGURE 11-12. Combined travel-time curves of three-layer model. *(Reproduced from Geophysics, vol. 19, by permission.)*

Figure 11-12. These phases appear as the most prominent events on the recording and could therefore well be the only pickable events. It would appear reasonable that the observation mentioned in the previous paragraph and the one here noted are manifestations of the same mechanism of propagation as affected by wavelength. Instead of propagation along an interface at a predictable velocity, there is involved the layering in the overburden. Thus the changes in perhaps 1,000 feet of section above a refractor may determine the arrival times of refracted waves as well as the computed relief.

Earth structure is enormously complicated in comparison with the simple model of Figure 11-11. This leads to a further problem of echeloning in the time-distance plots caused by the apparent loss of early legs either with or without changes in the apparent velocity.

Early History

It should be apparent from the foregoing that refraction shooting is not a particularly reliable exploration tool. It was used with considerable success in the early days of seismic prospecting for the location of shallow piercement-type salt domes in the Gulf Coastal areas, employing the well-known fan techniques, which will not be described here. The observable effect of a large salt mass embedded in low-velocity strata on seismic travel times is so large that little ambiguity is introduced by path uncertainties. The virtue in this application lies in the large area that can be explored in one fan, since shot-point-to-receiver ranges of 5 to 10 miles are used. This range requirement is also one of the elements that limit the resolving power of conventional refraction shooting, as all effects along the entire path are represented in the total travel time.

Present Applications

More recently, refraction surveys have been used in areas where reflection methods were unsuccessful and, less frequently, as an auxiliary tool to reflection prospecting. As an auxiliary tool, they are often used to verify faulting and to obtain approximate depths to a high-speed bed—say a massive carbonate section or basement. If reflection work is unsuccessful because of complexities in the subsurface or highly erratic near-surface conditions, it is unlikely that refraction shooting would be useful.

Refraction shooting is to a great extent limited to the use of first arrivals and the mapping of one interface. The use of later arrivals involves greater risk of creeping from one event to another through interference patterns.

In setting up a program it is necessary to determine experimentally the shot-point-to-spread range for which a suitable mapping horizon provides a first arrival. This can be done by laying out a series of spreads end to end and shooting into each spread in turn from the same shot point until the desired penetration is obtained. In practice, this operation can be considerably shortened if data are available from which the range can be estimated. The process is repeated in the opposite direction, duplicating subsurface coverage to eliminate dip in the computation of the horizontal velocity in the mapping horizon. The computational procedure is described by Dobrin (*A*, 1952). See also Mota (*A*, 1954) and Wyrobek (*A*, 1956). The vertical velocity distribution can be approximated from the refraction-line data or from what is known of the section if velocity-survey data are not available.

With the range determined, the next problem is to select a shooting pattern. Some form of continuous profiling should be employed. Considering that required ranges may be 7,500 feet or greater, the shooting pattern may be dependent on the availability of shooting permits. If no such constraint exists, the most practical pattern is one that most simplifies the land-surveying problem. One such technique puts shots and detectors on (two systems of) parallel lines separated by the range distance, one system being at right angles to the other to provide closed traverses. The spacing between adjacent lines of shot holes is governed by the exploration objectives, but it should be some integral fraction of the range for field convenience. Considering the resolving power of the refraction method, 2 miles is a reasonable spacing for mapping depths of 8,000 to 10,000 feet. The corresponding seismometer spacing may be about 500 feet. This would require the use of 40 seismometers on the ground to achieve a 10,000-foot spacing between shot holes with continuous coverage. Shots are occasionally taken with the seismometer spread in line to check the refractor velocity and ensure that the picked arrivals are still associated with the section of interest. Shot and detector lines are interchanged to provide continuity, and as a by-product, this furnishes up-hole weathering data at key seismometer locations.

Continuity in coverage can also be just as well achieved by an in-line technique, as illustrated by Dobrin (*A*, 1952). A convenient computing method is described by Tarrant (*A*, 1956). This technique is adaptable to prospects with areal limitations in one dimension. A variety of refraction field patterns has been used for particular purposes, some involving multiple-shooting crews and/or multiple-recording crews, plus the necessary land-survey crews. It is well to gear the program to modest levels so that technique parameters can be readily adjusted to meet changing requirements.

As with reflection data, refraction times must be corrected for weathering and elevation and, for convenience, to an appropriate reference surface. A further convenience is to reduce these corrected times to that for a standard range or shot-to-seismometer distance. These net times can then be readily converted to mid-point (between shot and receiver) depths, making the computations on the assumption of a two-layer problem with no dip. A more sophisticated approach, "exact" for in-line shooting and approximate for broadside shooting, is to assign the differences in net times from a shot point to two seismometers entirely to the depth points at the seismometer end of the spread. In a closed traverse these data can be handled in much the same way as reflection data as regards errors of closure. A further theoretical refinement is to redistribute the incremental times between the two sets of depth points (shot end and receiver end) after the lines have been shot in both directions.

Faulting of a refracting bed will appear as a displacement in the refraction arrivals on the record sections. With a fault throw in excess of 200 feet or thereabouts, record evidence becomes rather clear cut. Some excellent examples are reproduced by Ledoux (*A*, 1957).

One of the specialized uses of refraction shooting is to outline the boundaries of salt domes, particularly overhang. A seismometer is lowered in a deep well which penetrates or flanks a dome, and shots are recorded for a number of seismometer depths. Shallow shots fired above the dome into a flank well are used to outline the salt mass adjacent to the well; shots from distant points opposite the well then are used to delineate the far profile of the dome. The method was described by Gardner (*A*, 1949). An article by Clasen (*A*, 1958) indicates that the method gives qualitative information adjacent to the well but provides no useful data at other points, including the offset location. Clasen's evaluation was conducted under conditions most favorable for the success of the method.

Blundun (*A*, 1956) has described "the most valuable use in the Alberta foothills of the refraction seismograph for the definition of overthrust faulting in the Mississippian limestone which is overlain by a faulted, overthrust, and overturned Cretaceous section." Using ranges of 20,000 to 30,000 feet, a refraction from the Mississippian at depths of from 9,000 to 12,000 feet was recorded as a second arrival. Both in-line and broadside spreads were used. Some of the problems previously mentioned can be seen in his excellent specimen seismograms, but they do not appear to impair seriously the prime objective of fault definition.

Velocity Surveys

The empirical relation between time and depth is fundamental to the seismic interpretive process. Even in the absence of marked lateral velocity changes, it is necessary to know the depth to a seismic time anomaly and desirable to translate milliseconds of relief into feet or meters. Rapid lateral velocity changes may produce reflection time changes that are not representative of structural relief (*A*, Swan, 1942; Conklin, 1953), the general effect being to increase time (and computed depth) in the direction of decreasing velocity. Velocity determinations may be made from surface observations or from observations made in wells.

Types of Velocity Surveys

Well observations are of two general types. The first, or conventional, measures travel time from a shallow shot to a seismometer lowered in a well. The seismometer occupies a series of selected depths, with one or more shots taken at each position. From this, a time-depth curve and velocity-depth curves can be drawn. The shooting procedures have become so standardized that they need not be described here. Most United States velocity surveys, including continuous velocity logs, are run by contract operators for organized groups according to standards adopted by the group. A high percentage of surveys already made are available through purchase or trade.

The second type of velocity survey measures interval velocity continuously as a function of depth by means of a probe lowered in a well (*A*, Vogel, 1952; Summers and Broding, 1952). These readings can be integrated to obtain a time-depth curve with the aid of control points obtained by the method described in the preceding paragraph.

The techniques involved in surface-to-surface measurements, termed "velocity profiling," are outlined in excellent fashion by Widess (*A*, 1952). In simple concept, the times of a reflection to seismometers at two different distances from the shot point determine the depth and average velocity to the reflector. In practice, many such observations are averaged. The method yields approximate velocity values but does not have the accuracy required for most standard mapping purposes (*A*, Widess, 1952; Swan, 1942). This is unfortunate, as velocity profiling can be done before a prospect is drilled.

Application to Structural Mapping

We shall now discuss the application of velocity surveys to structural mapping, that is, the time-depth relationship. A time-depth determination at a well does not describe the time-depth relationship for a given reflection or geologic marker. To illustrate the point, let us assume that we run a velocity survey through a 1,000-foot section of limestone from a depth of 5,000 feet to a depth of 6,000 feet, the limestone having a velocity of 20,000 feet per second. The survey will show a (one-way) time difference of 50 milliseconds between the two depths, equal to 100 milliseconds of reflection time difference. If 5 miles away the top of the limestone is at 6,000 feet because of thickening of overlying shale beds with a 12,500 feet per second velocity, the increase in reflection time to the top of the limestone will be 160 milliseconds. It should be clear that the time-depth data from one well cannot be used "as is" for structural mapping. It is preferable to derive average velocities to mapping horizons and use these in the absence of information on lateral variations.

The determination of lateral velocity variations requires areal distribution of surveys. The average velocity to correlatable marker horizons used for seismic mapping can then be contoured. The contoured value at each shot point is used to convert time to depth. The reference surface must be the same for both the seismic map and the contoured velocity map. Contouring of the average velocities should be guided by the lithologic information available between well surveys as well as by seismic ties to other wells in an effort to minimize recontouring as additional surveys are made. If the distribution of velocity surveys is inadequate for contouring, it is at times possible to work out special solutions. For example, it may happen that the time-depth or the time-average velocity data to a particular marker fall on a smooth curve. Relationships of this type usually break down as the volume of data increases.

A velocity survey is made after an area is shot and mapped and a well is drilled. Surveys can thus be considered as a second-stage correction following which a new survey may necessitate further changes in the contoured seismic depth maps. Each stage reduces the probable error in a seismic map. This sequence follows in large measure the changes in exploration objectives in an area from major structures to successively smaller features.

Considerable progress has been made in the interpretation of resistivity logs in terms of velocity (*A*, Faust, 1953). This approach is frequently of considerable aid in the preparation of contoured average velocity maps.

Continuous Velocity Logs

Continuous velocity logs serve an important exploitation function in the determination of formation characteristics such as porosity and fluid content (*A*, Wyllie et al., 1956, 1958). Their prime use in seismic work is to identify prominent reflecting horizons and in the production of synthetic seismograms. They are helping to provide a much-needed insight into reflection formation and the transmission-reflection characteristics of the sedimentary section.

Specialized Reflection Techniques

A number of unconventional methods of energy generation have been used in reflection shooting, and we shall give a brief description of several.

The Poulter Method

In this method (*A*, Poulter, 1950) a plurality of charges are supported above the surface of the ground. The detonation in air imparts seismic energy to the ground which travels downward to reflecting interfaces. The charge, charge pattern, and manner of detonation (simultaneous or sequential) are varied to optimize the seismic pulse, particularly as to frequency content. It is claimed that an improved signal-to-noise ratio can be obtained as well as the almost complete elimination of multiple reflections. The explosion in air generates a large air wave, and the impact on the ground surface generates surface waves, both of which can be controlled by long offsets. Recording instruments may be conventional.

It is not known whether the method is currently in use.

The Vibroseis

Crawford and Doty (*A*, 1958) recently described the use of a vibrator as an energy source. The vibrator consists of a pair of unbalanced flywheels rotating in opposition and arranged so that horizontal forces are canceled and vertical forces are added. A maximum force of 14,000 pounds weight is said to be generated at 60 cycles. The vibrator is mounted on two massive rollers.

A coded signal is generated, the frequency being allowed to decrease during the operating cycle. The duration of the signal is stated to be longer than the time of the deepest reflection to be mapped. Two in-line offset seismometer groups are used on opposite sides of the shot point. The vibrator occupies successive positions for each location of the seismometer groups to obtain coverage. At each vibrator position a number of "shots" are taken and combined to minimize the effect of ambient noise. The "shots" are tape-recorded by means of a pickup mounted on the flywheel case, and the seismograms are also tape-recorded.

By means of electronic data-analysis machinery, the outgoing signal and each seismogram are multiplied and the product integrated over an appropriate range of phase relationships. The presence of a reflected signal on the seismogram gives rise to a maximum in the integrated product. The resulting record bears considerable resemblance to a conventional seismic recording. It is claimed that useful information from depths of 7,000 to 8,000 feet has been obtained.

The method is still in the experimental stage and will require considerable further work for evaluation. The ability to control the transmitted signal may lead to some novel uses of the tool.

Weight-dropping Techniques

In this technique the seismic signal is generated by the ground impact of a weight of several tons falling a distance of some 8 to 10 feet. The weight is truck-mounted, the truck being stationary during a drop and then being moved to the next drop location 10 to 50 feet distant. Each drop is tape-recorded from two seismometer groups separated by the normal shot-hole spacing used in conventional shooting. The method may be considered the inverse of the conventional process, with the positions of the shots (now drops) and seismometers being interchanged. "Magnetic recordings at a seismometer station from successive drops are corrected in time to a common datum . . . and to a zero or common drop distance. . . . Successive recordings then are composited ordinarily in groups ranging from 12 to as many as 96 per trace" (*A*, Domenico, 1958). Linear, rectangular, circular, and spiral drop patterns for a group of drops have been used.

The use of falling weights to generate seismic waves is old (*A*, Hubert, 1925). The practical application to seismic exploration rests on the development of magnetic recording and allied instrumentation which makes it possible to combine the recordings of many drops in their proper time relationships. Reflected energy from one drop is often well below the ambient noise level.

Neitzel (*A*, 1958) reports a case in which the efficiency of a weight drop in producing a reflection was 1,000 times that of a (36-hole shallow pattern) dynamite shot. This figure would hardly hold for a shot below the base of weathering. It is certain that a falling weight is a much more efficient generator of low-frequency body waves than are explosives as now used; these waves are less attenuated than the higher frequencies in transmission through earth materials.

Reflection data to depths of 10,000 to 15,000 feet have been obtained by means of weight drops. Published data indicate that the optimum weight-drop records in an area are generally inferior to optimum conventional records. This is due in large measure to the high level of surface waves generated by a weight drop. The frequency range of the surface waves overlaps to a high degree the frequency range of the reflected energies, which limits discrimination by filtering. Furthermore, the absence of higher frequency components makes the weight-drop recordings rather characterless. Generally speaking, coverage costs with weight-dropping techniques can be appreciably lower than for conventional shooting, as the cost of explosives is eliminated and the cost of operating a weight-drop truck should be less than drilling costs. It should be mentioned here that no satisfactory solution to the problem of determining time through weathering when using weight-dropping techniques has emerged.

In its present state of development, the method undoubtedly has applications in reconnaissance programs where the objectives are comparatively large features with in excess of several hundred feet of relief. In view of the large percentage of overlap frequently employed in this work, it may be advisable to check features with conventional methods before drilling an expensive test.

Offshore Seismic Operations

Offshore seismic (deep) surveys do not differ fundamentally from conventional land surveys. Tremendous coverage is achieved as the spreads are towed, and shot-hole drilling is unnecessary. Rapid and precise positioning is achieved through electronic surveying techniques. One significant instrumental departure from land operations is the use of pressure-sensitive detectors in place of velocity type. The design is such as to minimize the disturbances due to water wave motion, and considerable improvement in record quality is achieved. Pressure detectors are also improving record quality in shallow-water and marsh work.

Subbottom Sediment Exploration

The use of pulsed oscillators and spark discharges as sound sources for the delineation of subbottom geologic structures in water-covered areas has become operational. Earlier efforts (*A*, Hough, 1952) were successful in obtaining information concerning the nature of the bottom materials. Later work (*A*, Smith, 1958; Bechmann et al., 1958) has succeeded in mapping to depths of 500 feet in 50 feet of water. It seems probable that the depth range may be substantially increased through use of lower frequencies and "shot" compositing techniques.

GRAVITY SURVEYS

Introduction

In petroleum exploration the principal use of gravity surveys is as a reconnaissance tool for the guidance of lease plays and to localize the slower and more expensive seismic work and in determining an order of priority in the selection of local areas for processing by seismic means.

In the seismic method, observations from different depths are separated in time on the recording. In gravity (as in magnetic) surveys, the observation at a point is the integrated effect of a physical property of rocks at all depths and distances. The physical property is density, and the effect measured is gravitational attraction. The changes in gravitational attraction over an area must be interpreted in terms of changes in structure and lithology.

Interpretive effort rarely produces a final, positive solution of a geologic problem. Every gravity anomaly can be explained by an infinite number of density distributions, a circumstance known as the ambiguity of gravity interpretation. It is therefore necessary to consider all available geological and geophysical data in arriving at the most reasonable interpretation of gravity data. In common with other geophysical methods, gravity is not uniformly successful in all areas and should not be used without due consideration of the possible limitations.

Principles and Units

Gravity measurements derive from the mutual mass attractive force between the centers of mass of two bodies as expressed by Newton's law

$$F = \gamma \frac{m_1 m_2}{r^2}$$

F being the force, m_1 and m_2 the respective masses of the two bodies, r the distance between their centers of mass, and γ the universal gravitational constant. The force on m_1 per unit of its mass, due to the attractive force of m_2, is

$$\frac{F}{m_1} = \gamma \frac{m_2}{r^2} = a$$

It is equal to the acceleration a of the mass m_1, due to the attractive force of the mass m_2. (Force equals mass times acceleration.) A unit force acting on a unit mass will cause unit acceleration. In the cgs system, a force of one dyne acting on a mass of one gram causes an acceleration (velocity increase) of one centimeter per second per second. This unit rate of acceleration is termed the *gal* after Galileo, who discovered the laws of free fall. The dimensional units of the gal are those of acceleration, or dynes per gram. Some early gravity maps show the dyne or millidyne as the mapping unit, which is technically incorrect. At the surface of the earth, the force of gravity on a unit mass (the acceleration of gravity) is approximately 980 centimeters per second per second or, in geophysical units, 980 gals. Since very small variations in the gravitational field are usually of interest, the gal is an inconveniently large unit of measurement. Therefore, the *milligal* (one one-thousandth of a gal) is a common measuring unit. The *gravity unit,* which is one-tenth of a milligal, is sometimes used.

Gravity-measuring Instruments

Gravity surveys have been made with three types of instruments: torsion balance, pendulum apparatus, and gravity meter (sometimes called "gravimeter"). Pendulum observations are reduced to absolute-gravity values; these have been determined at many places over the earth to provide standard reference data. Torsion-balance and gravity-meter observations, on the other hand, furnish measurements of variations in gravity but not of absolute values. In recent years virtually all commercial and many geodetic surveys have employed the gravity meter because of its operational simplicity, reliability, and speed. For purposes of petroleum exploration, the torsion balance and the pendulum are obsolete instruments, except for specialized surveys, and are not described here. A brief account of the gravity meter, however, may contribute to clearer understanding of surveys and maps.

The Gravity Meter, or Gravimeter

Any gravity meter is essentially a highly sensitive weighing device. It consists of a small mass hung on a spring or a torsion fiber within a sealed thermally insulated case. The force of gravity varies from place to place, and the force on the mass varies accordingly and stretches the spring to greater or lesser degree. The amount by which the spring is stretched can be read on a scale. In practice, many gravity meters are designed as null instruments and the mass is restored to its null position by means of a second spring attached to a screw and calibrated reading dial. The restoring force is then a measure of the change in the force of gravity with reference to a previous observation.

Gravity-meter observations give figures in reading-dial scale divisions. In order to determine corresponding gravity values, the instrument is calibrated in milligals per scale division. Calibration can be accomplished by taking readings at two or more stations between which the gravity difference is known or by other methods. Instruments now used have meter constants approximating 0.1 milligal per scale division;

the observer estimates to the tenth of a division, or approximately to 0.01 milligal. Some special geodetic instruments are designed for even higher sensitivities. Since anomalies of interest in exploration are often of the order of 1 milligal or less, the gravity meter must respond to variations of less than one-millionth of the total gravitational field. It is a tribute to the ingenuity of instrument makers that such sensitivity can be achieved with simplicity of design.

Field Methods and Data Reduction

Gravity surveys consist of observations at selected points (stations) on a series of interlocking closed traverses. Each traverse includes an initial observation at a point (base-station), successive observations at stations on the traverse, and a final observation at the original station to tie the traverse. Later traverses may utilize portions of existing traverses to effect closure. Gravity-meter readings are then reduced to equivalent gravity values in milligals or gravity units on a reference datum, plotted on a map, and contoured. As previously stated, such a survey does not develop absolute-gravity measurements. It is comparable to a topographic survey made on an arbitrary datum without reference to sea level. The gravity survey may be tied to a previously established datum such as an earlier survey or one of the absolute-value pendulum stations (as of the U.S. Coast and Geodetic Survey), just as topographic data are tied to elevation bench marks. Locations and elevations of gravity stations are usually determined by transit or alidade-stadia traverses run concurrently with the meter survey; error limits are generally of the order of 1 to 1½ feet for elevations and up to a maximum of 400 feet for horizontal position. These tolerances are necessary because the elevation and geographic location of each gravity station determine corrections to the observed data, as described below.

Gravity-station locations are chosen to minimize disturbing effects of topographic relief and so reduce the amount of correction for rough topography. Station spacing depends on exploration objectives and may range from ¼ mile to smaller intervals along traverses. The distance between traverses may range from 1 mile in sectionized regions to wider or closer spacing depending on objectives and field conditions.

The high sensitivity of the gravity meter causes response to conditions other than the local and regional geologic anomalies of interest in exploration. If not recognized and as far as possible eliminated, these extraneous responses may mask the gravity effects of subsurface geologic features. Two types of instrument response vary with time: the tidal effect and the drift.

The Tidal Effect

Motions of the sun and moon cause gravitational effects, which are variable in time due to the relative motions of the earth and those bodies. This variation, which can be calculated from astronomical data for any point on the earth, is the tidal effect, and it contributes small but measurable components to gravity-meter readings. Maximum amplitude of the tidal effect is about 0.3 milligal during certain 24-hour periods. A universal table for calculating the effect is given by Goguel (*B*, 1954).

Instrumental Drift

Spring systems sensitive enough to register small changes in gravity are not perfectly stable; physically, they approach but do not quite attain a condition of perfect elasticity. Therefore, when a gravity meter is returned to its starting point or base station to complete a traverse, the instrument may fail to repeat the original base-station reading. After any necessary correction for the tidal effect, the remaining component of misclosure due to mechanical characteristics of the spring system is called the drift. If the length of time between base-station readings is limited to a few hours, the time depending on the precision requirements of the survey, the amount of drift is reduced to a small value, which may be prorated as a linear time function among all stations on the traverse, and the tidal effect is also minimized.

Normal Corrections to Field-survey Data

The purpose of normal corrections is to remove the effects of elevation, terrain or topographic conditions, and latitude, producing gravity values on a standard reference surface. Variations of gravity may then be ascribed to subsurface conditions on a local, regional, or continental scale. Occasionally an isostatic correction is applied, although this is usually restricted to geodetic survey data.

Latitude Correction

The latitude correction consists of two parts, one due to the shape of the earth and the other resulting from the earth's rotation. The earth's equatorial radius is about 13 miles greater than its polar radius. The gravitational attraction on a body is thus less at the equator than at the poles, and the centrifugal force of rotation further lessens the force; the total change due to the latitude effect is about 5,000 milligals. The *rate* of change varies from zero at the equator and poles to about $1\frac{1}{3}$ milligals per mile at mid-latitudes. Since the latitude effect may account for gravity differences exceeding 1 milligal per mile in a north-south direction, station locations are determined within limits permitting latitude-correction errors of less than 0.1 milligal.

Free-air Correction

If two gravity meters are so situated that one rests on the surface while the other is suspended, as it were, in the "free air" some distance above, then the second meter is farther away from the earth's center of mass and therefore responds to a correspondingly smaller gravitational force. The vertical gradient of gravity above the surface of the earth is −0.09406 milligal per foot or −0.3086 milligal per meter.

Bouguer Correction

Gravity stations at different elevations are separated not only by free air but also by earth material which contributes the "Bouguer effect" to the gravity measured at the higher elevation. The additional material under the elevated meter, if considered to be a slab of infinite horizontal extent, causes a gravitational attraction directly proportional to its density and thickness. The attraction of such a slab is $0.01276 \times \text{density} \times \text{thickness}$ (milligals per foot), or $0.04185 \times \text{density} \times \text{thickness}$ (milligals per meter), where density is measured in grams per cubic centimeter, thickness (or elevation above reference surface) in feet or meters, respectively.

With reference to a datum plane, the free-air correction is positive (added to observed gravity) and the Bouguer correction negative (subtracted from observed gravity) for stations at higher than datum elevations. The two corrections are usually combined into a single elevation-correction factor, which depends on the density of surface rocks within the topographic range of the survey. Methods for field determination are given by Nettleton (*B*, 1940), Domzalski (*B*, 1955), and Whetton et al. (*B*, 1957). Elevation corrections are made to sea level or some other convenient datum. A few representative surface densities and corresponding elevation factors are listed:

Density	*Milligals per meter*	*Milligals per foot*
1.6	0.2416	0.07364
1.8	0.2333	0.07109
2.0	0.2249	0.06854
2.2	0.2165	0.06599

It is apparent that gravity-station elevation errors will not greatly exceed 1 foot if the elevation correction is held within 0.1 milligal limits.

Terrain Correction

Extreme or abrupt variations in topography develop "terrain effects" which are not compensated by the elevation correction. Suppose a gravity observation is made on top of a steep-sided hill or mesa. Applying the Bouguer correction assumes that this hill is a slab of infinite horizontal extent. But there is a "mass deficiency," since the earth materials are not present beyond the limits of the hill and an overcorrection results. A *terrain correction* must be added to approximate the effect of the missing mass. If the meter is situated at a topographically low point, then the upward components of gravitational forces due to adjacent masses—hills, mountains—reduce the measured gravity, and a terrain correction must also be added in this case. Terrain corrections are always positive (added to observed gravity). In rough terrain, the estimation of terrain corrections, which depend on the density, dimensions, and distances of the topographic features, is an essential and often laborious element of gravity surveys. Careful selection of meter sites can minimize the labor involved. Nettleton (*B*, 1940) gives a terrain-correction table.

Isostatic Correction

Gravity measurements in and near the high mountains of India, about the middle of the nineteenth century, disclosed differences between the measured values, after corrections for elevation and latitude, and those calculated for the international ellipsoid. Measured gravity values in the mountains were much smaller than calculated values. This circumstance led to the proposal that the material of the earth's crust is less dense under large land masses than it is under large depressions. Thus, mountains and continents are evidently associated with relatively light crust, ocean areas with relatively heavy crust. The association of major variations in topographic relief with corresponding variations in density of the earth's crust led to the concept of isostatic compensation, the hypothesis of the approximate uniformity of mass per unit area over the earth's surface as measured from some local reference surface in the interior. For geodetic purposes, *isostatic corrections* are applied to gravity data, since geodetic investigations embrace large areas within which the effects of isostasy, or *isostatic compensation*, are significant. For purposes of petroleum exploration, however, isostatic corrections can be omitted, since they are relatively constant within the smaller areas of interest to such exploration. Heiskanen and Vening Meinesz (*B*, 1958) give extended discussion to isostasy. See also Martin (*B*, 1954).

Gravity Maps

After the proper corrections have been applied to all field data, gravity-station locations and values are plotted on a map and contoured. Contour intervals of

0.2 milligal, and occasionally 1 milligal for reconnaissance data, are generally used. This is sometimes called the *total gravity map*, because it represents variations in the total gravitational field and includes the effects of all subsurface elements which contribute to the field. Total gravity maps may be considered as representing a combination of two types of components: a trend, or *regional gradient*, due to very large or deep-seated geologic features and *anomalies* associated with local or relatively shallow density contrasts in the subsurface, which may have significance as regards petroleum accumulation.

Interpretation

The interpretation of gravity data may be considered in two phases, the identification of anomalies on a gravity map and the determination of their probable geologic causes.

Areas of interest will appear as perturbations superimposed on broad features, the complete outlines of the broad or regional features often not appearing in the limits of the survey. Departures from regional are manifested on a contoured total gravity map as changes in curvature of the contours and changes in their spacing. Recognition of such an anomaly is often fairly easy, but some experience is required to delimit the areal extent and approximate magnitude in milligals on a total gravity map. As a basis for a quantitative approach to gravity interpretation, it is often customary to prepare a derived map from the total gravity determinations. Most derived maps are of two types, the residual gravity map and the second-derivative map.

Residual Gravity Map

We can consider the residual map as being constructed by drawing a smoothed surface through the average values of total gravity and plotting the difference between the actual value of total gravity at each point and the smoothed value. The mechanics of computation can be handled in a number of ways. Griffin (*B*, 1949) describes a method based on grid networks. Such methods require a careful selection of grid dimensions to yield optimum results. More sophisticated calculation methods depending on modern computers are now also in use and provide a sounder basis for quantitative analysis than do those based on grid computations. See Oldham and Sutherland (*B*, 1955), and Grant (*B*, 1957). Residuals can also be determined graphically from total gravity profiles.

Second-derivative Maps

These are curvature maps, and the rate of change of gradient of total gravity is contoured. As gravity observations are not continuous, the maps depend in some measure on the particular computational procedure (see, for example, *B*, Elkins, 1951). Zietz and Henderson (*B*, 1955) have pointed out that "The residual map using the same grid system may be substituted for the second derivative map since it can be shown that the maximum, minimum and zero curvature contours appear at exactly the same respective positions on both maps."

There are often sharp differences of opinion regarding the relative merits of residual and second-derivative maps. See, for example, Elkins (*B*, 1951); also see *Geophysics*, vol. 23, no. 4, p. 860, 1958, for a letter by Amalendu Roy and a reply by L. L. Nettleton. The physical basis for residual anomalies is much easier to see than is the case for second-derivative maps, which is a distinct advantage to the interpreter. Certainly the degree of refinement justified in the preparation of derived maps is dependent on the quality and density of control; in order to achieve high resolving power, both factors must be commensurate with the objectives.

Many of the papers cited show total and derived gravity maps. Additional examples may be found in "Geophysical Case Histories," vols. 1 and 2, published by the Society of Exploration Geophysicists. A very interesting example of the resolving power of gravity observations is given by Hammer (*B*, 1953).

Geologic Deductions from Gravity Data

Skeels (*B*, 1947) has presented an excellent discussion concerning the fundamental ambiguity in gravity (and magnetic) interpretations: that gravity data cannot, of themselves, be interpreted uniquely. One of his figures, reproduced in Figure 11-13,

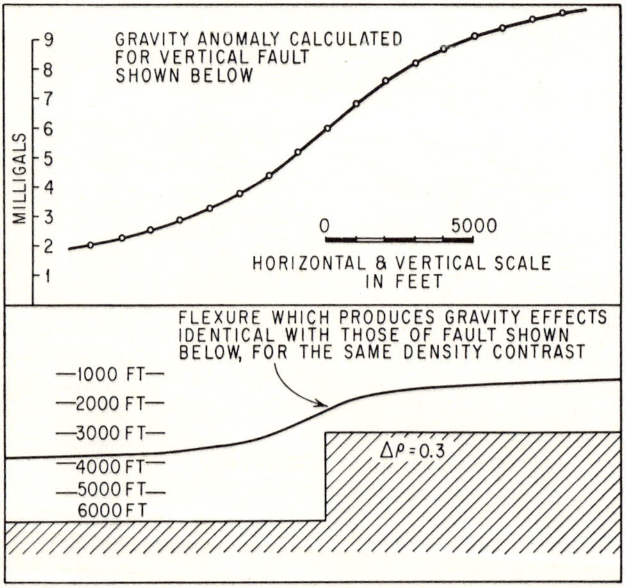

FIGURE 11-13. Gravity profile for a vertical step fault. (*Reproduced from Geophysics, vol. 12, no. 1, by permission.*)

shows a flexure and a vertical step fault as two models which satisfy a given set of data equally well. Other models could be constructed that would also fit the data. Despite this fundamental weakness, a careful evaluation of the geologic probabilities can serve to limit the solutions to a useful degree.

One of the more useful aids in the geologic interpretation of gravity anomalies is to compare the computed effect of models based on simple geometric forms with observed anomalies. Nettleton (*B*, 1942) has published a number of charts and tables, and most interpreters have extensive additional tabulations. Once the geologic possibilities have been outlined, anomalies for models with contrasts at various depths and different structural dimensions are compared with observations in an effort to determine models that would produce the observed data. Some experience in this

Table 11-1. Table of Densities and Magnetic Susceptibilities*

Location	Organization company	Farm	Description	Depth, ft	Material	Density, g per cc	Magnetic susceptibility $\times 10^6$, cgs units
Alberta	Imperial	Claremont	16-25-72-5W6	...	...	2.57	10
	Imperial	Normandville	1-16-79-22W5	...	...	2.77	17
	Imperial	Spirits River	12-20-78-6W6	...	...	2.75	14
Alaska	USGS	S.Barrow No. 3	...	203	Sedimentary	2.44	15
	USGS	S.Barrow No. 3	...	2,413	Precambrian	2.53	21
	USGS	S.Barrow No. 3	...	2,879	Precambrian	2.60	24
	USGS	Simpson No. 1	...	6,545½	...	2.70	34
	USGS	Umiat No. 1	...	258–290	Sedimentary	2.45 avg	16 avg
	USGS	Oumalik No. 1	...	920–1,609	...	2.43 avg	17 avg
Arizona	...	...	...	78	Escobrosa Limestone	2.71	18
Colorado:							
Archuleta	Phillips	No. 1 Crowley	...	1,607–1,609	Sedimentary	2.58	8
Las Animas	Skelly	No. 1 Tercio	...	5,975–6,158	Sedimentary	2.60 avg	25 avg
	Shell	Green A-16	30-9S-53W	7,802	Granite	2.58	31
Florida:							
Jackson	Humble	No. 1 Tindal	...	8,881–8,891	Very hard gray granite	2.80	1,519
Mississippi:							
Warren	Magnolia	No. 1 Ragsdale	24-16N-4E	7,721–7,726	Reworked igneous	2.55	1,392
Warren	Magnolia	No. 1 Ragsdale	24-16N-4E	8,599–8,609	Tuscaloosa sandstone	2.53	344
Newton	Magnolia	...	30-7N-13E	Surface	Winona fm Linonite	3.46	125
Newton	Magnolia	...	20-7N-13E	Surface	Winona fm	2.47	150
Nebraska	Shell	Phillips No. 1 Simmons	21-10N-22W	3,365–3,371	Granite	2.63	48
New Mexico:							
Chaves	Magnolia	No. 1 Turney Federal	23-14S-22E	3,780–3,790	Red shale Permian	2.60	32
Chaves	Pure	RR.MTN.Dike	...	Surface	...	2.93	3,655
Oklahoma:							
Beckham	Shell	No. 1 Oklahoma	8-10N-21W	9,388–9,409	Conglomerate Wash	2.58 avg	25 avg
Texas:							
Presidio	Argo	No. 1 State	40-1	1,500–2,000	Tertiary intrusive	2.58 avg	2,320 avg
Presidio	Argo	Mitchell State	40-1	3,110–3,120	Cretaceous	2.32	35
	Gulf	C. F. Blackman	Sur. 3 ABS 175-B W.A. McKinney Sur.	3,548	Basement	2.55	265
Wyoming:							
Goshen	Gen. Petr.	No. 1 Van Tussel	32-30N-60W	575–585	Tertiary sandstone	2.26	167

*Selected from J. A. Sharpe, *Proc. Geophys. Soc. Tulsa*, 1953.

trial-and-error approach provides an appreciation of the value and limitations of reliable gravity data.

Gravity-model computations lend themselves to handling by digital and analogue computers. Specialized analogue computers for this application are on the market.

Density Distributions

Some knowledge of formation densities is required to help delimit the range of possible solutions in gravity interpretation, as gravity anomalies result from lateral density variations. Horizontal changes may result from either structural or lithologic changes, and it is necessary to evaluate the geologic probabilities of each. It has been rather common practice to make density determinations from well samples to provide background for such an evaluation. Continuous density log surveys in wells are now available as a commercial service and provide a much greater measure of control for gravity interpretation. A selected list of densities from sample determinations taken from Sharpe (*B*, 1953) is given in Table 11-1.

MAGNETIC SURVEYS

Magnetic maps, like gravity maps, represent variations in a potential field of the earth which are not directly analogous to geologic structure. These maps are useful in petroleum exploration because they provide data which can be interpreted in terms of (1) approximate depths of basement rocks, and hence approximate thickness of sedimentary section, and (2) anomalies of possible geologic and economic interest when these are related to basement structure. The interest in this exploration method, particularly for reconnaissance purposes, is evident from the numerous commercial surveys over large areas in recent years. Most of these surveys have been made with air-borne equipment, which includes a magnetometer designed to measure and record variations in the earth's total magnetic field and suitable positioning devices, photographic or electronic, to record geographic locations of the magnetic observations. Prior to about 1946, when the total-field equipment became generally available, most magnetic surveys were made with instruments designed to operate on the ground and to measure either the vertical or horizontal components of the earth's field.

Magnetic surveys are often the first geophysical surveys to be made in an area, particularly in large, unexplored sedimentary basins for which few other data are available. They are then used to focus attention on possible trends for further exploration with more definitive tools. They are seldom used "on their own," although J. W. Peters (*C*, 1949) lists six United States fields as discovered on the basis of magnetic information, alone.

Principles and Definitions

Any magnetized body, including the earth or its geologic components, may be thought of as a large number of small magnets. Each small magnet, or dipole, has a north pole and a south pole, and the assemblage of these small magnets is a group of dipoles. The magnetic force F between poles depends on the respective pole strengths P_1 and P_2 and on the distance r between the poles. This force is defined as

$$F = \frac{CP_1P_2}{r^2}$$

where C is a constant of the material in which the poles are situated.

Each of two equal poles, P_1 and P_2, separated by unit distance (1 centimeter) in a nonmagnetic medium ($C = 1$) has unit strength when F is 1 dyne. The strength or intensity of a magnetic field at any point is defined as the force in dynes on a unit magnet pole at that point. The unit of field strength, 1 dyne per unit pole, is defined as the *oersted*. The earth's field intensity is usually expressed in geophysical literature in terms of *gauss*, although the correct term is the numerically equivalent oersted. The *gamma* (γ), which is 10^{-5} oersted, is a more convenient unit for the designations of the small changes in field intensity usually mapped.

In the case of a uniformly polarized spherical earth with no external disturbances, the field would be horizontal at the magnetic equator, vertical at the magnetic poles; field strength would vary at a predictable rate over its surface, reaching a maximum at each pole; and field strength would be constant at any point. This is not true for two reasons: (1) the magnetically heterogeneous nature of earth materials, which modifies the field spatially, and (2) the occurrence of magnetic field disturbances over varying periods of time. These conditions bear on both the objectives and problems of magnetic surveys.

Any spatial perturbation in the earth's field is a magnetic anomaly. Anomalies of interest in petroleum exploration are caused by variations in the magnetic character and distribution of rocks composing the upper part of the earth's crust. Rock magnetism depends almost entirely on the amount of magnetite present, magnetite being the most important ferromagnetic mineral because of its relatively high susceptibility and wide distribution. Very small percentages of magnetite occur in most sedimentary rocks, and most sediments are only weakly magnetic. Magnetite occurs more generally and in greater concentrations in igneous rocks, which therefore usually develop greater field strength. These circumstances account for the correlation of prominent magnetic anomalies with the distribution of igneous rock types, whether these types occur as the basement-complex floor of a sedimentary basin or as intrusives or flows within or overlying the sediments.

The disturbing effects of rocks on the earth's magnetic field are determined in part by the intensity of magnetization, a property having both magnitude and direction. Magnetic intensity consists of two components, remanent (residual, or permanent) magnetism induced by previously existing fields and the portion induced by the presently existing external field. Although many rocks are known to have remanent magnetization, there is little information presently available to evaluate this factor adequately in petroleum surveys. In what follows, the possible effects of remanent magnetization will not be considered.

The intensity of magnetization I induced in magnetic materials by an external magnetic field H is proportional to strength of the field, the constant of proportionality being called the volume susceptibility k. This relation is

$$I = kH$$

The units of I are "magnetic moment" per unit of volume. Magnetic moment is the product of the pole strength of one of two equal poles of opposite polarity multiplied by their distance of separation. A unit north pole and south pole 1 centimeter apart have a magnetic moment of 1 cgs unit. The direction of the moment is along the line between them and toward the north-seeking pole.

Preceding paragraphs outline some of the factors which contribute to the earth's magnetic field and define some of the measuring units. The field is not uniform; the principal magnetic poles are not symmetrically situated, and they are some 1,200 miles distant from corresponding geographic poles. The north magnetic pole, visited as early as 1811, is in northern Canada; the south magnetic pole, reached in 1909, is in South Victoria Land of the Antarctic continent. Early interest in determining deviations of "magnetic north" from "true north" (declination) for navigation purposes led to development of instruments for observing other components of the field. These observations revealed small daily changes in the magnetic field, called "diurnal variations," which range from a few to 100 gamma or more, depending on the place of observation. Larger time-related effects are called magnetic disturbances, and violent disturbances of the field, ranging in duration from a few hours to a few days and in magnitude to some thousands of gamma, are called magnetic storms. Both diurnal variations and the larger disturbances are thought to be caused by ionization changes in the upper atmosphere due to variations of solar radiation.

Magnetic Field Measurements

At most points on the earth's surface, the magnetic field is inclined at an angle to the surface. The field may be defined by its horizontal component H, its vertical component V, and its inclination i. Magnetic measuring instruments used in petroleum exploration are the vertical magnetometer, which measures changes in the vertical component; the horizontal magnetometer, which measures changes in the horizontal component; and the total-field magnetometer.

Magnetometers

The first two of those mentioned are small, lightweight instruments set up on a tripod on the ground; a bar magnet assembly balanced on knife-edges within a thermally insulated case is deflected by the H or V component of the total field. The amount of deflection, read on a scale, is a measure of the H or V intensity at that point (*C*, Joyce, 1937). A series of such measurements, usually of the vertical component, made at intervals of ½ to ¼ mile on mile-spaced traverses, provided the data for most of the earlier magnetic surveys. While the vertical magnetometer can be designed for a sensitivity of about 10 gamma per scale division or better with a reading accuracy of 1 or 2 gamma, a survey of this precision presents difficulties in maintaining instrument adjustment; many surveys have been run only with sufficient precision to warrant contouring the data on a 25-gamma interval. Among the requirements of "ground" magnetic surveys are avoidance of nearby magnetically or electrically disturbing elements, such as wire fences and power lines; repeat observations at base stations to check instrumental error or presence of magnetic disturbances; and temperature observations which affect temperature corrections of the magnetic data. Elevation control is not so critical as in gravity surveys.

Since about 1946 virtually all large-scale magnetic surveys have been made with instruments which respond to variations in the total field. There are two types of instruments for this purpose.

Flux-gate Magnetometer

The flux-gate magnetometer includes an element sensitive to small changes in the earth's total field, mounted on a platform with an orienting mechanism that automatically adjusts parallel to the direction of total field. So oriented and provided with a recording device, the instrument when mounted in a moving vehicle, as an aircraft, provides a continuous record of variations in the total field. Measurements represent relative values of total intensity; an accuracy of 1 gamma is obtainable. One embodiment of an air-borne flux-gate magnetometer and associated equipment is described by Wyckoff (*C*, 1948).

Nuclear Precession Magnetometer

Packard and Varian (*C*, 1954) have described a method of measuring the earth's total magnetic field by a determination of the frequency of free precession around this field of the protons in a small volume of water. The water in a bottle is magnetized in a direction approximately normal to the earth's field by passing a current through a coil surrounding the bottle. The current is abruptly terminated. The protons act as miniature gyroscopes, and the precession of the magnetic-moment vectors around the earth's field direction induces a measurable alternating voltage in the coil. The frequency of the induced alternating voltage is a measure of the earth's total field. Waters and Phillips (*C*, 1956) report that accuracies of ±2 gamma can be expected if readings are made at 1-second intervals; with readings at 10- to 15-second intervals, an accuracy of ½ gamma can be achieved. The method has been adapted to operation from aircraft and appears to have some advantages over flux-gate types. A disadvantage is the intermittency of the observations and perhaps the limitations in accuracy.

Field Procedure in Aeromagnetic Surveying

The magnetometer with its associated orientation instrumentation may be fixed on the aircraft, as in a wing-mount, or it may be housed in a "bird" and towed at the end of a 100-foot cable. Magnetic observations are continuously recorded on a moving strip with provision for automatically shifting the recording range to keep the pen on the scale. The position as well as the time of the observations needs to be known.

Position Control

Horizontal control can be obtained either photographically or by a radio-location method. Continuous-strip or intermittent photography is used, the photographs being matched with ground-controlled mosaics. Photographic methods are inapplicable over water or wherever ground identification of position fails. Radio methods, such as Shoran, are limited to line-of-sight distances. Elevation control is maintained by barometric altimeter, and elevation above ground by radio altimeter.

Flight Pattern

This usually consists of a series of parallel lines at 1- to 2-mile intervals, depending on objectives. Tie lines at right angles to the pattern are spaced at intervals of usually from 15 to 30 miles. The flight-loop pattern is designed to minimize errors of adjustment at intersection points.

Corrections to Aeromagnetic Data

These include corrections for the diurnal effect, for instrumental drift, and sometimes for the normal change of total-field intensity over the survey area. Usually a continuously recording total-field magnetometer is set up as a ground monitor, preferably at a location near the center of the survey area, to record time-related variations in the total field. This permits correction of the flight data for the observed effects of diurnal variations; minor loop misclosures remaining after this correction can then be adjusted as instrumental drift corrections or as due to acceptable errors of position.

Large or abrupt variations of the field intensity recorded by the ground monitor during a survey flight necessitate reflying traverses flown during the distrubances. The irregular variation of the earth's total intensity from minimum values along the magnetic equator to maxima at the magnetic poles is called the "normal change." The total-field intensity ranges from 25,000 gamma, in a region along the magnetic equator in Brazil, to about 61,000 and 72,000 gamma, respectively, at the north and south magnetic poles. The rate of change is not constant in magnitude or direction; it ranges from zero to several gamma per mile. Most commercial magnetic surveys include a normal correction to the data in order to approximately remove this regional gradient. Published maps (U.S. Coast and Geodetic Survey, Hydrographic Office of the United States Navy) provide data for this correction.

A detailed account of operating procedures and problems is given in Wyckoff (*C*, 1948) and Domzalski (*C*, 1957).

Interpretation

It has already been stated that magnetic anomalies are in large part due to changes in the elevation or composition of basement rocks. The interpretive problem can then be considered as being to determine the generalized basement structure and make estimates of the depth to basement, since these provide a measure of the thickness of the sedimentary section. The magnetic effects of susceptibility changes in basement rocks are usually large in comparison with those from changes in basement relief, so most magnetic computations are based on susceptibility changes in an assumed flat basement.

Derived Maps

As with gravity, the interpretation of magnetic data is greatly facilitated by the removal of trends of a dimensional scope that transcends the survey objectives. This handling of the data is a necessary part of some interpretive processes, and the resulting maps define the anomalies to the uninitiated more clearly than do the total-intensity maps. Either residual or second-derivative maps are used. Methods of preparing such maps have been described by L. J. Peters (*C*, 1949), Henderson and Zietz (*C*, 1949), Grant (*C*, 1957), and others. In an actual large-scale application, Henderson and Zietz (*C*, 1958) chose to compute residuals because of the saving in time involved. One can at times question the degree to which refinement of computational procedures should be carried in view of the uncertainties involved in interpretation.

Comparison with Models

One method widely used in the structural interpretation of magnetic data is described in considerable detail by Vacquier et al. (*C*, 1951). It consists of comparing magnetic features of actual surveys with those computed for models of idealized basement units. The principal assumptions made in the model computations are: (1) The basement surface is horizontal; (2) the basement complex consists of rock prisms that are rectangular in plan, with horizontal upper surface and vertical sides, and infinite in downward vertical extent; (3) the magnetic susceptibility is constant throughout a prism and contrasts to adjacent prisms; (4) the polarization is in the direction of and entirely due to the earth's present field; i.e., there is no remanent polarization. Total-intensity and curvature contours for a wide variety of model dimensions and orientations with respect to the magnetic meridian are presented in a series of 83 figures.

The features on both the total-intensity and the second-derivative field maps are compared with the computed model anomalies. Comparisons of the second-derivative maps are the more significant, as the outlines of the anomalies are more clearly defined. For anomalies of large areal extent, the zero-curvature contour roughly outlines the shape of the model to be selected for comparison. A rough estimate of the depth below flight elevation is made from the horizontal extent of the steepest gradient on an anomaly. The estimate is made from the second-derivative map from the distance between the maxima and minima and the zero contour. A graphical method of estimating depth of basement from second-derivative contours is given by L. J. Peters (*C*, 1949). Determination of anomaly gradients for depth calculations entails examination of the original magnetic profile records as well as the total-intensity and curvature maps.

Accuracy of Depth Estimates

J. W. Peters (*C*, 1949) states that "Over the past 12 or 15 years some 90 per cent of our depth estimates are probably accurate to better than ±20 per cent, and some 75 per cent are probably accurate to within ±10 per cent. The word 'probably' is used above because only a few of our estimates have been checked by wells which reached the igneous basement. The accuracy of the others can only be inferred from other information." Vacquier et al. (*C*, 1951) report computations on six surveys, in three of which the depth to crystalline rock is known, the crystalline rocks being at the surface. In these cases the computed average "depth of basement," which was the flight level, differed from the actual by 0, 12, and 14 per cent, with computations from individual anomalies differing from the actual by as much as 50 per cent. Vacquier et al. (*C*, 1951) do not consider any of these surveys a good test of the interpretational procedures. More recently, Henderson and Zietz (*C*, 1958) reported on their interpretation of an aeromagnetic survey of Indiana. Where wells to or near basement provided checks, the calculated depths to basement are correct, on the average, to 10 per cent.

Delineation of Structural Trends

A rather significant feature of Henderson's and Zietz's work is that for the entire state of Indiana, only 79 anomalies, one in about 13 townships, were selected for depth computation. To be suitable for depth computation, an anomaly must be well defined, well shaped, and isolated. A conclusion of great import is that the " . . . topographic trends in the Precambrian basement rocks are *in general* similar to the structural trends on top of the Trenton limestone." This can perhaps be considered as a clear statement of the principal contribution to be expected from aeromagnetic mapping in petroleum exploration.

As with all geophysical methods, magnetic data need to be constantly reviewed in the light of development. This frequently suggests empirical relationships between structure and anomalies that would be difficult to predict. Such relationships need to be viewed in their proper statistical framework, and leads should be checked with seismic surveys before drilling.

Magnetic maps of 13 different areas are reproduced in "Geophysical Case Histories" (*C*, Society of Exploration Geophysicists, 1948, 1956), most of them with associated seismic, gravity, and subsurface structural maps. In addition, many of the references cited have magnetic maps of local areas. Probably the most instructive magnetic maps that have been published are in the report by Henderson and Zietz (*C*, 1958).

A very enlightening evaluation of basement-depth determinations from air-borne magnetic data was presented at the March 14, 1960, meeting of the Asociación Venezolana de Geofísica in Caracas by P. Jacobsen, Jr. He compared two independent interpretations of the same data for a portion of the eastern Venezuela Basin for which seismograph and well information was available. He concluded:

1. In proper hands, careful analysis of intrabasement-type magnetic anomalies may yield a reliable and useful generalized basement-depth map.

2. Local basement relief based on a study of the usually smaller, suprabasement magnetic anomalies cannot be reliably interpreted from magnetometer work alone.

OTHER METHODS OF GEOPHYSICAL EXPLORATION

Over the years there has been an increasing effort both within and outside the industry to develop new exploration methods either to supplant or supplement conventional methods, not to mention the hundreds of ideas proposed by crackpots and charlatans. Most major oil companies have examined proposed methods, as have private organizations (*D*, Cook, 1959), often at considerable expense. Every plausible idea has undoubtedly been explored by several independent groups, each of which hoped to come up with a successful embodiment. It would appear highly probable that any method proposed more than 5 years ago and not in common use today is not competitive, because of either economics or results or both.

Patrick (*D*, 1958), in his report of geophysical activity for 1957, shows 8,556 seismic crew months, 1,266 gravity crew months, and 136 magnetic crew months in the Western Hemisphere with no crew months for "other" types of geophysical exploration. This does not mean that there may not have been some isolated activity in the "other" category but indicates that no other method has attained a status to merit discussion in this work.

GEOPHYSICS IN THE EXPLORATION PROGRAM

It is frustrating that the most rapid and least expensive geophysical tool for petroleum exploration, the magnetometer, is at the same time the least definitive. Many large, undeveloped areas are in the initial exploratory phases, and it is essential that by some means the most promising areas in large regions be identified. Perhaps the first requirement may be to make some approximation of the thickness of sediments. Such estimates can be made within reasonable limits from magnetic surveys. At times, the nature of the data warrants a generalized contouring of basement topography. In this happy circumstance, the magnetic data can localize further exploration effort to areas whose structural elements are thus determined. Under less favorable circumstances, only scattered basement-depth estimates can reasonably be made. One can then only assume that structural trends are related to magnetic trends and base the next geophysical phase on that premise.

The role of gravity in the exploration program is not so clearly defined. This is due in part to the fact that while there can be grave ambiguities in the interpretation, local gravity changes over sedimentary basins statistically have close association with changes in the sedimentary section whereas magnetic variations, in large measure, are related to crystalline rocks. In passable terrain, the relatively low cost often makes gravity surveys the second geophysical phase in the search for hydrocarbons. Certainly if air-borne gravimeter surveys with adequate accuracy were possible, there would be no question concerning the role of gravity measurements in the initial evaluation of petroleum prospects (see *C*, L. G. D. Thompson, 1960). The high dependence of gravity-data interpretations on geological data frequently creates a paradoxical situation in that when the geological data are available, it may not be economical to conduct a gravity survey while it certainly is important to reexamine existing gravity data in the light of new subsurface information.

Seismic efforts in a virgin area leave many questions unanswered. There is no way of recognizing basement and no sure way of differentiating structure from velocity effects. Statistically, however, we can place a great measure of reliance on reflection-time anomalies of some magnitude when based on data of reasonable quality and when there is reason to believe that the reflection data are from sedimentary beds. (Intrabasement reflections have been reported in the literature; see *D*, Widess and Taylor, 1959.) Initial seismic control in a new province is often rather loose, the effort being to pick up large features. After production is established, the gradual tightening of control in the search for smaller and still smaller features is historic. In many of the more developed areas, as well as in less developed areas where structure is not a primary factor, the reflection method is being forced beyond its present resolving power in the effort to delineate stratigraphic factors.

The function of all the geophysical aids to petroleum exploration is to help determine the selection of drill sites. It is an interesting speculation to consider at what point a reduction in drilling costs would render geophysical surveys uneconomical on a routine basis. Such a point has already been reached in selected areas where production is at comparatively shallow depths and the success ratio for random drilling is rather high. When this practice becomes general, the geophysicist can concentrate his talents on developing a direct method of determining when the drill bit has penetrated oil-bearing formations.

REFERENCES

A. Seismic Surveys

Anstey, N. A.: *Geophys. Prospecting*, vol. 4, no. 1, 1956.

Beckmann, W. C., et al.: Paper presented at the American Geophysical Union meeting, May 5, 1958.

Berryman, L. H., et al.: *Geophysics*, vol. 23, no. 2, 1958.

Blundun, G. J.: *Geophysics*, vol. 21, no. 3, 1956.

Bortfeld, R.: *Geophys. Prospecting*, vol. 4, no. 4, 1956.

Clasen, Gerhard: *Erdöl u. Kohle*, 11 jahrg. nr. 1, 1958.

Clifford, O. C.: *Geophysics*, vol. 24, no. 1, 1959.

Conklin, Glenn M.: *Geophysics*, vol. 18, no. 4, 1953.

Crawford, John M., and W. E. N. Doty: *Proc. Geophys. Soc. Tulsa*, vol. 5, 1958.

DeGolyer, E.: *J. Soc. Petrol. Geophys.*, vol. 6, no. 1, 1935.

Dobrin, M.: "Introduction to Geophysical Prospecting," McGraw-Hill Book Company, Inc., New York, 1952.

Domenico, S. N.: *Geophysics*, vol. 23, no. 4, 1958.

Erath, Louis W.: *Geophys. Prospecting*, vol. 5, no. 2, 1957.

Faust, L. Y.: *Geophysics*, vol. 18, no. 2, 1953.

Gardner, L. W.: *Geophysics*, vol. 14, no. 1, 1949.

Giulio, S. E.: *Proc. Geophys. Soc. Tulsa*, 1954.

Heinitz, K.: Das Problem der Bündelung von Geophonen, *Freiberger Forschungshefte*, C66, *Geophysik*, Berlin, 1959.

Hough, J. L.: *J. Sediment. Petrol.*, vol. 22, no. 3, 1952.

Howell, L. G., et al.: *Geophysics*, vol. 18, no. 1, 1953.

Hubert, F.: *Z. Geophysik*, vol. 1, 1925.

Kramer, S. S., and R. A. Peterson: *Oil in Canada*, Aug. 6, 1956.

Krey, T.: *Geophysics*, vol. 17, no. 4, 1952.

Laubsher, H. P.: *Geophysics*, vol. 21, no. 2, 1956.

Ledoux, Y.: *Geophys. Prospecting*, vol. 5, no. 4, 1957.

McCready, H. J.: *Geophysics*, vol. 4, no. 4, 1940.

Mohorovicic, A.: *Jahrb. Meteorol. Observatoriums Zagreb für 1909*, 1909.

Mota, Lindonor: *Geophysics*, vol. 19, no. 2, 1954.

Musgrave, A. W., et al.: *Geophysics*, vol. 23, no. 1, 1958.

Neitzel, Edwin B.: *Geophysics*, vol. 23, no. 1, 1958.

Osborne, M. F. M., and S. D. Hart: *J. Acoust. Soc. Am.*, vol. 18, no. 1, 1946.
Palmer, E. M.: The Gulf Seismic Profile Printer, *Geophysics*, vol. 22, no. 2, 1957.
Peterson, R. A., et al.: *Geophysics*, vol. 20, no. 3, 1955.
Pommier, G., and H. Richard: *Geophys. Prospecting*, vol. 5, no. 3, 1957.
Poulter, T. C.: *Geophysics*, vol. 15, no. 2, 1950.
Press, Frank, et al.: *Geophysics*, vol. 19, no. 3, 1954.
Quarles, Jr., M.: *Geophysics*, vol. 15, no. 3, 1950.
Reid, A. C.: *Proc. Geophys. Soc. Tulsa*, vol. 5, 1957–1958.
Reynolds, V. T.: *Proc. Geophys. Soc. Tulsa*, 1955.
Rice, R. B.: *World Oil*, vol. 137, no. 2, 1953.
Rice, R. B.: *Geophysics*, vol. 20, no. 1, 1955.
Richard, H.: *Geophys. Prospecting*, vol. 1, no. 1, 1953.
Rieber, F.: *Geophysics*, vol. 2, no. 2, 1937.
Rummerfield, Ben F.: *Geophysics*, vol. 19, no. 4, 1954.
Salvatori, H.: *Geophysics*, vol. 2, no. 4, 1937.
Schmidt, O. V.: *Z. tech. Physik*, no. 12, 1938.
Shock, Lorenz: *Geophysics*, vol. 15, no. 2, 1950.
Smith, W. O.: *Bull. Geol. Soc. Am.*, vol. 69, no. 1, 1958.
Summers, G. C., and R. A. Broding: *Geophysics*, vol. 17, no. 3, 1952.
Swan, B. G.: *Geophysics*, vol. 7, no. 4, 1942.
Swan, B. G., and Adrian Becker: *Geophysics*, vol. 17, no. 3, 1952.
Tarrant, L. H.: *Geophys. Prospecting*, vol. 4, no. 2, 1956.
Tatel, H. E., and Merle A. Tuve: *J. Geophys. Research*, vol. 59, no. 2, 1954.
Thralls, H. M., et al.: *Geophysics*, vol. 17, no. 2, 1952.
Vacquier, Victor, et al.: *Geol. Soc. Am. Mem.* 47, 1951.
Van Melle, F. A., and K. R. Weatherburn: *Geophysics*, vol. 18, no. 4, 1953.
Vogel, C. B.: *Geophysics*, vol. 17, no. 3, 1952.
Widess, M. B.: *Geophysics*, vol. 11, no. 3, 1946.
Widess, M. B.: *Geophysics*, vol. 17, no. 3, 1952.
Widess, M. B.: *Proc. Geophys. Soc. Tulsa*, 1957–1958.
Widess, M. B., and G. L. Taylor: *Geophysics*, vol. 24, no. 3, 1959.
Woods, J. P.: *Geophysics*, vol. 17, no. 2, 1952.
Woods, J. P.: *Geophysics*, vol 21, no. 2, 1956.
Wyllie, M. R. J., et al.: *Geophysics*, vol. 21, no. 1, 1956.
Wyllie, M. R. J., et al.: *Geophysics*, vol. 23, no. 3, 1958.
Wyrobek, S. M.: *Geophys. Prospecting*, vol. 4, no. 2, 1956.

B. Gravity Surveys

Domzalski, W.: *Geophys. Prospecting*, vol. 3, no. 3, 1955.
Elkins, T. A.: *Geophysics*, vol. 16, no. 1, 1951.
Goguel, Jean: *Geophys. Prospecting*, vol. 2, Supplement, March, 1954.
Grant, F.: *Geophysics*, vol. 22, no. 2, 1957.
Griffin, W. R.: *Geophysics*, vol. 14, no. 1, 1949.
Hammer, S.: *World Oil*, vol. 52, no. 21, 1953.
Heiskanen, W., and F. A. Vening Meinesz: "The Earth and Its Gravity Field," McGraw-Hill Book Company, Inc., New York, 1958.
Martin, Rodolfo: *Geophysics*, vol. 19, no. 1, 1954.
Nettleton, L. L.: "Geophysical Prospecting for Oil," McGraw-Hill Book Company, Inc., New York, 1940.
Nettleton, L. L.: *Geophysics*, vol. 12, no. 3, 1942.
Neumann, R.: *Geophys. Prospecting*, vol. 2, no. 4, 1954.
Oldham, C. H. G., and D. B. Sutherland: *Geophysics*, vol. 20, no. 2, 1955.
Sharpe, J. A.: *Proc. Geophys. Soc. Tulsa*, 1953.
Skeels, D. C.: *Geophysics*, vol. 12, no. 1, 1947.
Whetton, J. T., et al.: *Geophys. Prospecting*, vol. 5, no. 1, 1957.
Zietz, I., and R. G. Henderson: *Geophysics*, vol. 20, no. 2, 1955.

C. Magnetic Surveys

Agocs, W. B., and K. Isaacs: *Geophys. Prospecting*, vol. 4, no. 1, 1956.
Deel, Samuel A., and H. Herbert Howe: United States Magnetic Tables and Magnetic Charts for 1945, *U.S. Coast Geodetic Survey Serial* 667.

Domzalski, W.: *Geophys. Prospecting*, vol. 4, 1957.
Grant, F.: *Geophysics*, vol. 22, no. 2, 1957.
Henderson, J. R., and I. Zietz: *U.S. Geol. Survey Profess. Paper* 316–B, 1958.
Henderson, R. G., and I. Zietz: *Geophysics*, vol. 14, no. 4, 1949.
Hydrographic Office, U.S. Navy: "Total Intensity Map of the Earth's Magnetic Force, for the Year 1955."
Joyce, J. Wallace: "Manual on Geophysical Prospecting with the Magnetometer," U.S. Bur. Mines Publ. printed by American Askania Corp., Houston, 1937.
Packard, Martin, and Russell Varian: *Phys. Rev.*, vol. 93, 1954.
Peters, J. W.: *Mines Mag.*, vol. 39, no. 7, 1949.
Peters, L. J.: *Geophysics*, vol. 14, no. 3, 1949.
Society of Exploration Geophysicists: "Geophysical Case Histories," vol. 1, 1948; vol. 2, 1956.
Thompson, Lloyd G. D.: *J. Geophys. Research*, vol. 65, no. 1, 1960.
Vacquier, V., et al.: *Geol. Soc. Am. Mem.* 47, 1951.
Waters, G. S., and G. Phillips: *Geophys. Prospecting*, vol. 4, no. 1, 1956.
Wyckoff, R. D.: *Geophysics*, vol. 13, no. 2, 1948.

D. Other Methods of Geophysical Exploration

Cook, John C.: *Geophysics*, vol. 14, no. 1, 1959.
Patrick, H. G.: *Geophysics*, vol. 23, no. 5, 1958.
Widess, M. B., and G. L. Taylor: *Geophysics*, vol. 24, no. 3, 1959.

Chapter 12

GEOCHEMICAL SURVEYS

By John A. S. Adams

Professor of Geology, Rice University
Houston, Tex.

Since the end of World War II many completely new geochemical approaches and techniques have been developed for the solution of geologic and exploration problems. These new developments have led to a great burst of research activity. In response to this increase in geochemical research a new journal, *Geochimica et Cosmochimica Acta*, was founded in 1950; several surveys and special volumes have appeared (Rankama and Sahama, 1950; Faul, 1954; Rankama, 1954; Abelson, 1959; and several volumes of the "Physics and Chemistry of the Earth"); and in 1955 the Geochemical Society was organized on an international basis. In addition, a large and increasingly available Russian literature on geochemistry continues to accumulate.

This greatly expanded activity in geochemical research continues, and it threatens early obsolescence for any discussion of the applications of geochemistry to the problems of petroleum exploration. For example, there are serious doubts about the accuracy of the currently accepted absolute time scale for the Paleozoic (Adams et al., 1958; Faul, 1959; Mayne et al., 1959; Holmes, 1959; Adams et al., 1960). A major revision of the Paleozoic time scale would have implications for many major geologic processes, including the formation, accumulation, and stability of petroleum.

A variety of geochemical techniques have been applied to a central problem in petroleum exploration, namely, the delineation of those situations favorable for the accumulation of petroleum. For convenience, these techniques will be discussed under the following headings: (1) trace-element analysis, (2) detailed analysis of hydrocarbons, (3) isotopic analysis, (4) absolute-age determinations.

TRACE-ELEMENT ANALYSIS

The determination of the abundance and distribution of the chemical elements is one of the basic studies of geochemistry. Green (1953) has summarized the present knowledge about the concentrations of elements in various rocks. The observed distribution of the elements, together with the geochemical mechanisms and geologic processes responsible for the distribution, are of interest to petroleum exploration to the extent that they provide clues about the accumulation of petroleum. The following examples illustrate a few of the ways in which this type of information can be used.

Keith and Degens (1959) and Degens, Williams, and Keith (1958) have indicated how "trace-element analyses of sedimentary rocks may provide a number of criteria for differentiating rocks of marine origin from those of fresh-water origin. The most promising are elements of the marine chemofacies, such as boron, lithium, sulfur, and fluorine." Marine shales tend to have more boron and lithium than fresh-water shales, and this observation may be helpful in delineating basins in those cases where more familiar criteria such as fossils are not available. The evaluation and applicability of such "geochemical indicators" must await more extensive studies. It is fortunate that newer techniques of analysis such as X-ray fluorescence permit more rapid and accurate determinations of many of the trace elements of interest in this connection.

Adams and Weaver (1958) studied the ways in which thorium and uranium are distributed by sedimentary processes. During weathering, uranium tends to be oxidized and brought into solution whereas thorium is not. Thus thorium tends to be concentrated relative to uranium in continental deposits; the uranium, on the other hand, tends to be concentrated relative to thorium in carbonate rocks and certain black shales. These two elements form a "geochemical facies" that can be used to distinguish marine from continental sediments and as a weathering index.

In many carbonate rocks and sandstones the thorium content is indicative of the shale or clay content (Baranov, Ronov, and Kunashova, 1956; Adams and Weaver, 1958). Rabe (1957) found a relationship between natural gamma radiation and permeability in the Denver-Julesburg basin. This relationship held because the matrix, or cementing material, in these reservoir sands was mainly clay, which is generally higher in thorium, potassium, and uranium relative to quartz sands. Thorium, uranium, and potassium are the only sources of natural gamma-ray activity, and studies of their distribution in sedimentary rocks may provide still other opportunities for more detailed interpretation of gamma-ray logs.

Another type of geochemical study with broad implications for petroleum exploration is that recently completed by Ronov (1958). Ronov reports the determination of organic carbon in 26,000 Russian platform sediments of all types and ages. He reports "an alternation of epochs of intensive and of insignificant accumulation of organic matter," which was also found by Trask and Patnode (1942) in their study. He concludes that "with the aid of quantitative litho-geochemical maps, the organic carbon distribution in clayish, carbonate, and sandy upper Devonian beds of the Frasnian stage of the Russian platform has shown that in approaching the Volga-Ural oil basin the organic carbon content in rocks increases, reaching its maximum value in the ranges of the latter." The trend is most pronounced in the clay beds, which also show that the percentage of iron in the ferrous state increases as the amount of organic carbon increases.

DETAILED ANALYSIS OF HYDROCARBONS

The great advances that have been made recently in inorganic analysis have been paralleled in the analysis of hydrocarbons. It is now possible to detect and characterize many hydrocarbons that are present in sediments in concentrations of a part per million or less (Smith, 1954; Meinschein, 1959; Abelson, 1959). This great increase in sensitivity has been achieved with a whole battery of new techniques, including absorption spectrometry, chromatography, fluorescence, and mass spectrometry. Hanson (1959) and Meinschein (1959) have recently discussed the problem of petroleum formation and accumulation in the light of this new evidence. An even more direct implication for petroleum exploration lies in Abelson's (1959) suggestion regarding the Precambrian: "A diligent search around the stable shields might uncover old undisturbed sediments containing quantities of extractable organic substances." Swain et al. (1958) succeeded in determining the concentration and general composition of bituminous substances in Precambrian rocks from Minnesota.

The application of detailed analysis of hydrocarbons to petroleum exploration is made very difficult and complicated by the fact that modern techniques are so sensitive to so many different types of organic compounds. On one hand, the great wealth of possibilities has plagued attempts to locate seeps from reservoirs by means of soil analysis for hydrocarbons or bacteria. On the other hand, who can say that it is not possible to find empirically or deduce theoretically a type or suite of organic materials that will distinguish hydrocarbons escaping from a petroleum reservoir from those present in modern soil? Already important systematics in these complex systems are known. For example, the hydrocarbons in all known petroleum reservoirs have been out of any contact with the atmosphere for so long that they should not have any radioactive carbon 14 left in them. Thus, any hydrocarbon leaking directly from a petroleum reservoir should be devoid of the carbon 14 with which all modern carbon is tagged (Libby, 1955). Another process that takes place in geologic time is the differential breakdown of the members of a particular class of organic compounds. Thus, the relative stabilities of many amino acids have been established by both laboratory experiments and analysis of ancient rocks (Abelson, 1959; Erdman, Marlett, and Hanson, 1956). Finally note should be made of the fact that certain features of biochemical or metabolic evolution are beginning to emerge. Thus, the relative abundances and types of organic compounds made by organisms in the geologic past have changed just as the relative abundances and morphology of their hard parts have changed (Baldwin, 1949; Florkin, 1949).

ISOTOPIC ANALYSIS

The fractionation in nature of the stable isotopes of elements such as hydrogen, carbon, oxygen, nitrogen, and sulfur has been studied for only a few years (Rankama, 1954). Yet enough is already known from the analysis of rocks, laboratory experiments, and theoretical considerations to enable the study of isotope fractionations to contribute importantly to our understanding of geologic processes. An outstanding example is the isotopic analysis of the free sulfur and the carbon in the limestone cap rock of the Gulf Coast salt domes (Thode et al., 1954; Ault, 1959). The enrichment of the lighter carbon and sulfur isotopes in this situation points clearly to bacterial reduction of the anhydrite in the salt and bacterial oxidation of the petroleum in the caprock.

Silverman and Epstein (1958) made a preliminary study of the carbon isotopic compositions of petroleum. Their data and conclusions provide several important insights into petroleum formation and accumulation: "The fact that the magnitude of C-13/C-12 ratio differences between marine and terrestrial organisms is similar to the differences between petroleums geologically related to marine deposits and those related to non-marine sediments provides additional confirmation that petroleum can be derived from non-marine as well as marine organic sources." They also observed that the carbon in natural gas is isotopically "lighter" than the carbon in the petroleums studied. This led to the conclusion that it was unlikely that any of the petroleums analyzed had released or been converted to natural gas to any great extent. This conclusion is supported by the studies of the isotopic composition of the nitrogen in natural gas and petroleum made by Hoering and Moore (1958). The nitrogen in natural gas is also isotopically "lighter" than the nitrogen in the accompanying petroleum. To date, the main contribution of these and other isotopic studies has been to limit further the possible hypotheses for the formation and accumulation of petroleum.

ABSOLUTE-AGE DETERMINATIONS

Tilton and Davis (1959) have recently reviewed the present status of geochronology. As regards the absolute dating of the fossil time scale they have come to the conclusion that "at present the situation is far from satisfactory." However, the refined measurements sought are not so vital for many applications in petroleum exploration, and it should be pointed out that approximate, but very useful absolute dates can be obtained from rather small amounts of such minerals as glauconite, authigenic potassium feldspars, and biotites. In particular, in situations where it is important to know whether or not an igneous rock is actually Precambrian, the refined methods of modern mass spectrometry may be most helpful in identifying those subsurface situations where an intrusion or a buried lava flow is overlying sediments that may contain petroleum reservoirs.

REFERENCES

Abelson, P. H.: Geochemistry of Organic Substances, pp. 79–103 in "Researches in Geochemistry," P. H. Abelson (ed.), John Wiley & Sons, Inc., New York, 1959.

Adams, J. A. S., and C. E. Weaver: Thorium-to-Uranium Ratios as Indicators of Sedimentary Processes: Example of concept of geochemical facies, *Bull. AAPG*, vol. 42, pp. 387–430, 1958.

Adams, J. A. S., G. Edwards, W. Henle, and J. K. Osmond: Absolute Dating of Bentonites by Strontium-Rubidium Isotopes, *Bull. Geol. Soc. Am.*, vol. 69, p. 1527, 1958.

Adams, J. A. S., J. K. Osmond, G. Edwards, and W. Henle: Absolute Dating of the Middle Ordovician, *Nature*, vol. 188, pp. 836–838, 1960.

Ault, W. U.: Isotopic Fractionation of Sulfur in Geo-Chemical Processes, pp. 241–259 in "Researches in Geochemistry," P. H. Abelson (ed.), John Wiley & Sons, Inc., New York, 1959.

Baldwin, E.: "An Introduction to Comparative Biochemistry," Cambridge University Press, New York, 1949.

Baranov, V. I., A. B. Ronov, and K. G. Kunashova: On the Geochemistry of Thorium and Uranium in Clays and Carbonaceous Rocks of the Russian Platform, *Geokhimiya* (Geochemistry), no. 3, pp. 3–8, 1956. (See review, *Bull. AAPG*, vol. 41, pp. 146–148, 1957).

Degens, E. T., E. G. Williams, and M. L. Keith: Environmental Studies of Carboniferous Sediments, II, Application of Geochemical Criteria, *Bull. AAPG*, vol. 42, pp. 981–997, 1958.

Erdman, J. G., E. M. Marlett, and W. E. Hanson: Survival of Amino Acids in Marine Sediments, *Science*, vol. 124, p. 1026, 1956.

Faul, H. (ed.): "Nuclear Geology," John Wiley & Sons, Inc., New York, 1954.

Faul, H.: "Geologic Time Scale, *Bull. Geol. Soc. Am.*, vol. 71, pp. 637–644, 1960.

Florkin, M.: "Biochemical Evolution," Academic Press, Inc., New York, 1949.

Green, J.: Geochemical Table of the Elements for 1953, *Bull. Geol. Soc. Am.*, vol. 64, pp. 1001–1012, 1953.

Hanson, W. E.: Some Chemical Aspects of Petroleum Genesis, pp. 104–117 in "Researches in Geochemistry," P. H. Abelson (ed.), John Wiley & Sons, Inc., New York, 1959.

Hoering, T. C., and H. E. Moore: The Isotopic Composition of the Nitrogen in Natural Gases and Associated Crude Oils, *Geochim. et Cosmochim. Acta*, vol. 13, pp. 225–232, 1958.

Holmes, A.: A Revised Geological Time-Scale, *Trans. Edin. Geol. Soc.*, vol. 17, part 3, pp. 183–216, 1959.

Keith, M. L., and E. T. Degens: Geochemical Indicators of Marine and Fresh-water Sediments, pp. 38–61 in "Researches in Geochemistry," P. H. Abelson (ed.), John Wiley & Sons, Inc., New York, 1959.

Libby, W. F.: "Radiocarbon Dating," 2d ed., University of Chicago Press, Chicago, 1955.

Mayne, K. I., R. St. J. Lambert, and D. York: The Geological Time-Scale, *Nature*, vol. 183, pp. 212–214, 1959.

Meinschein, W. G.: The Origin of Petroleum, *Bull. AAPG*, vol. 43, pp. 925–943, 1959.

"Physics and Chemistry of the Earth," vol. 1, 1956, vol. 2, 1957, Pergamon Press, Ltd., London.

Rabe, C. L.: A Relation between Gamma Radiation and Permeability, Denver-Julesburg Basin, *Petrol. Trans. AIME, Tech. Note* 403, vol. 210, pp. 358–360, 1957.

Rankama, K.: "Isotope Geology," Pergamon Press, Ltd., London, 1954.

Rankama, K., and T. G. Sahama: "Geochemistry," University of Chicago Press, Chicago, 1950.

Ronov, A. B.: Organic Carbon in Sedimentary Rocks (in Connection with Source Sediments of Petroleum), *Geokhimiya* (Geochemistry), no. 5, pp. 409–423, 1958.

Silverman, S. R., and S. Epstein: Carbon Isotopic Compositions of Petroleums and Other Sedimentary Organic Materials, *Bull. AAPG*, vol. 42, pp. 998–1012, 1958.

Smith, Jr., P. V.: Studies on the Origin of Petroleum: Occurrence of Hydrocarbons in Recent Sediments, *Bull. AAPG*, vol. 38, pp. 377–404, 1954.

Swain, F. M., A. Blumentals, and N. Prokopovich: Bituminous and Other Organic Substances in Precambrian of Minnesota, *Bull. AAPG*, vol. 42, pp. 173–189, 1958.

Thode, H. G., R. K. Wanless, and R. Wallouch: The Origin of Native Sulfur Deposits from Isotope Fractionation Studies, *Geochim. et Cosmochim. Acta*, vol. 5, pp. 286–298, 1954.

Tilton, G. R., and G. L. Davis: "Geochronology," pp. 190–216 in "Researches in Geochemistry," P. H. Abelson (ed.), John Wiley & Sons, Inc., New York, 1959.

Trask, P. D., and H. W. Patnode: "Source Beds of Petroleum," AAPG, Tulsa, Okla., 1942.

Chapter 13

SUBSURFACE GEOLOGY

By F. B. Loomis

Chief Geologist, Clark Oil & Refining Corporation
Denver, Colo.

ROLE OF SUBSURFACE GEOLOGY IN PETROLEUM EXPLORATION

Petroleum exploration in any region usually follows a logical sequence in that it proceeds from the known and obvious to the less known and less obvious, that is, from a study of the exposed surface geological features to the unexposed, buried subsurface features. In a sense, subsurface geology may be defined as the study of geology in three dimensions in that it extends the observations and interpretations made at the surface into the underground. In petroleum geology the term includes also the interpretation of data furnished by geophysical surveys of various types and by geochemical investigations. The subsurface geologist is concerned, then, with the distribution, thickness, character, attitude, elevation, sequence, correlation, fossil content, and geologic age of the underground or subsurface rock layers. From the study of logs, cuttings, and cores from wells; physical measurements made within the well bores; and data from geophysical and geochemical surveys, information on the stratigraphy and structure of the underground rock formations is obtained. When properly interpreted and presented on appropriate maps, cross sections, and other types of diagrams, this information, as we shall see, has wide application in both exploration and exploitation.

Subsurface geologic exploration may be roughly subdivided into three distinct but overlapping phases, each distinguished by the amount and quality of data available upon which to base an interpretation of the geologic merits of the region or area of interest under investigation. Phase 1 may be thought of as the pioneer or reconnaissance phase in which only sparse, scattered data may be available on which to base an evaluation of an area of interest. There are occasions when a geologist must make recommendations to management or client even though the nearest outcrops to the area under consideration are many miles away, the nearest well previously drilled is equally distant, and geophysical surveys have yielded only indefinite indications of subsurface structural conditions. Such circumstances require the geologist to use lively but controlled imagination in order to evaluate the possibilities of the area. Study of distant outcrops, wells (perhaps only one), and inconclusive geophysical findings can result in tentative conclusions as to regional structural trends, source and distribution of sediments, and possible presence of source and reservoir rocks. The final and most imaginative step in this study is projection of the tentative conclusions on regional conditions to the area of interest and the formulation of a definite recommendation for positive action on the project. This is true pioneering exploration.

Phase 2 of subsurface exploration is the advanced or detailed phase. After a region has been explored more thoroughly, considerably more data are available. Perhaps one or more oil or gas fields have been discovered and developed. Data control from numerous wells and from detailed geophysical surveys permits structural and stratigraphic analysis of the area of interest and construction of structure contour, isopach, facies, paleogeologic, and paleogeographic maps and sections, both regional and local in extent. This is the stage in which subsurface geology is most rewarding, for careful analysis of the increased amount of dependable data, together with clear imaginative geologic thinking, will inevitably result in the development of drillable prospects in any petroliferous province.

Phase 3 of subsurface exploration is the application of subsurface geology to the development and possible extension of an existing field. This is known as production or exploitation geology, and as our discussion is primarily concerned with exploration for new accumulations of oil and gas, we shall limit our discussion of subsurface geology principally to phases 1 and 2.

The purpose of petroleum exploration is to find traps that contain accumulations of oil and gas, a major objective that is often overlooked. In pursuing this objective subsurface geologic mapping is approached in the same manner as in surface geologic reconnaissance. As previously pointed out, subsurface geology may be thought of as the projection of surface geology into the underground. All the information obtainable at the surface is also available to the subsurface geologist. The subsurface geologist must know the regional stratigraphic section. Formations are usually measured, described, and named from surface observations; the names are in most cases applied to analogous buried rocks wherever possible. The structural geology at the surface is of utmost importance to the three-dimensional interpretation. Regional and local formation dips and strikes can be extrapolated downward. The location and attitude of faults at the surface may be projected into the subsurface and used to predict structural conditions in the prospect area.

It cannot be emphasized too strongly that solution of local subsurface problems repeatedly requires regional geologic knowledge. The more a geologist knows of the geology of an area, the more complete will be his interpretation of the bits of evidence he is able to acquire from subsurface data. For example, the stratigraphic section penetrated by a well may not be comparable to that measured and described elsewhere in the region either on the surface or in other wells. The subsurface geologist must immediately determine the possibilities of faulting and, perhaps of even greater importance, the possibilities of stratigraphic traps in strata pinching out along the flanks of structural uplifts or wedging out along the basin margins. Again, regional information makes it possible to determine the deeper prospects in a field or unexplored area. Regional patterns of folding and faulting may provide important clues to the solution of local patterns. A knowledge and understanding of the regional geology of an area should be acquired from the published literature, from unpublished reports, and from frequent consultation with geologists familiar with the stratigraphy and structure exposed on the surface. Nor should attention to the surface geology be relaxed after current information is obtained. New exposures are continually being provided by road cuts, surface excavations, etc. These data should be integrated with the subsurface studies in progress.

Geophysical mapping is an integral part of the systematic exploration program. Geophysical surveying is, in fact, a form of subsurface mapping performed at the surface without drilling. The mapping is done from measurements of various physical properties of the rocks, such as rock density, energy transmissibility, radioactivity, magnetic susceptibility, etc. These measurements can be translated into geologic data giving information on structure, stratigraphy, depth, and attitude. Geophysical data must be combined with other subsurface information, and all the information synthesized into a complete picture. This synthesis becomes more important as the prospect becomes more detailed or its size more limited, for less reliability can be placed on details or on minor structural features, particularly when the measurements are near or below the limits of accuracy of the instruments. A subsurface geologist should be familiar with the degree of accuracy and the limitations of the data produced by the different geophysical methods of surveying, and he should also be able to judge where additional geophysical work will aid in the interpretation of subsurface structure.

Although we are not concerned here with the role of subsurface geology in the exploitation or development of the producing oil or gas field, much information of value to subsurface exploration is obtained from careful and detailed studies of producing areas. From such studies of areas with a concentration of good well logs can be learned such things as the stratigraphic relationships of the formations, the position and nature of unconformities, the facies changes in the formations being studied, the changes in folding with depth, the time of folding and of faulting, and the nature of the potential reservoir rocks. Other valuable data include the nature of the hydrocarbons present; the effects of the reservoir fluids on electric, radioactive, and acoustical logs; the probable recovery factors; etc. After making one or more detailed case history studies, the subsurface geologist proceeds from the known toward the unknown with more assurance that he will be able to correlate the scattered bits of information he is able to acquire and fit them together into logical and more nearly accurate regional or local geologic interpretations.

SOURCES OF SUBSURFACE GEOLOGIC DATA

The principal source of the basic data of subsurface geology, excepting the data obtained from geophysical surveying, is from well drilling—both wildcat and development. Examination of actual formation samples obtained during drilling—drill cuttings, cores, and formation fluids—together with measurements by indirect means of certain physical properties of the rocks penetrated, yields such information as formation lithology, formation boundaries, elevation of formation boundaries, formation age, dip and strike of formations, location of unconformities, the presence of faults, porosity, permeability, fluid content of porous formations, fluid pressure, formation temperatures, etc. The techniques for obtaining these data are described in Chapter 17, Drilling the Exploratory Well; Chapter 18, Coring; Chapter 19, Electrical Logging; Chapter 20, Radioactivity Well Logging; Chapter 21, Miscellaneous Well Logs; and Chapter 22, Laboratory Procedures in Exploration. We shall discuss in this chapter, therefore, only some of the more important applications to subsurface geology of the data obtained from those sources.

Drill Cuttings and Cores

Sample study is basic geologic work. It requires concentration, patience, well-developed powers of observation, and a good understanding of basic drilling and sampling procedures on wells. The principal piece of equipment needed for sample examination is a good binocular microscope. The lens magnification should range from approximately 6 to 45 diameters. Magnification should be great enough to see the essential structure and texture of the sample. It should be low enough to reduce eyestrain and to provide a sufficiently wide field of view for estimating percentages of rock constituents. A magnification of approximately 10 to 12 diameters is used for routine sample examination, and 30 to 45 diameters for more detailed study. Illumination of the sample is provided by use of a suitable microscope lamp. Any lamp with a blue bulb or blue filter plate can be used. A lamp that produces a "rainbow" of colors should not be used, as it will tend to mask or distort the true colors of the sample and will produce eyestrain. Not only should sufficient light be applied to the sample itself, but the work area around the microscope should be well illuminated to prevent excessive eyestrain. Other miscellaneous items of equipment useful in sample work include black sample trays, spot plates, tweezers, pick (probe or needle), small magnet, hydrochloric acid (10 per cent solution), carbon tetrachloride, pens, colored pencils, and other drafting supplies as needed.

Methods of Interpretation

There are three general methods of sample description and logging: the percentage system, the interpretive system, and the combination percentage-interpretive system. In the percentage system, the geologist describes and logs all material in the sample, disregarding only *obvious* foreign material and cavings, and estimates by eye the various percentages. This system is most useful for inexperienced geomicroscopists, in areas of rapid stratigraphic change, or where the detailed stratigraphy is not well known. The disadvantages of the system are that formation boundaries do not show up clearly on the sample log, and if the samples are badly contaminated with cavings, the log may become so congested as to be useless.

The interpretive system depends essentially on the ability of the geomicroscopist to recognize cavings, and the entire sample is described and logged as being composed of the rock type found in the cuttings believed to be representative. This system is dangerous for the inexperienced geologist and may also be misleading if the formation penetrated is thinly interbedded with several rock types. Auxiliary logs, such as electric, radioactive, or penetration logs, are often used to help build the interpretive log. The interpretive system has the advantage of sharply defining formational boundaries on the sample log, but unless it is used with discretion, many of the valuable minor details in the rock section may be omitted from the log.

The combination percentage-interpretive system shows the percentages of rock types interpreted as freshly cut in the sample interval. The material interpreted as cavings does not enter into the percentage calculation. This system combines the advantages of both of the above systems, has the advantage of flexibility, and can be varied by the geologist according to the conditions involved, the area concerned, or the experience of the individual.

It is common practice to run all samples wet for lithology, color, texture, accessory minerals, fossils, etc. This procedure tends to remove some of the personal element involved in sample examination, makes for uniformity in technique, and often brings out important features that may be missed when examining dry samples. Features such as porosity and hydrocarbon showings, however, are generally easier to evaluate in dry samples. The usual procedure is to run through a tray of samples dry for porosity and oil showings, then dip the tray in water, drain off the excess, and examine the wet samples for other features.

Description of Samples

Sample descriptions should be made in the following order:

1. Rock type, e.g., sandstone, shale, limestone, etc.
2. Color

3. Texture and porosity, including grain size, angularity, sorting of sandstones, degree of lamination and hardness of shales, texture of matrix and character of pore structure of carbonate rocks, etc.
4. Cement, including modifying adjectives, such as calcareous, siliceous, argillaceous, etc.
5. Accessory minerals and fossils
6. Hydrocarbon showings, including estimate of intensity and remarks regarding cut or cut fluorescence

All samples should be checked with dilute hydrochloric acid. Observation of the degree of effervescence will often help distinguish between a limestone and a dolomite; limestone reacts rapidly, dolomite slowly on the surface of the rock chip. The reaction will be more rapid if the chip is scratched. This, of course, is a field test and is not satisfactory for precise determination. It is often difficult to distinguish between calcareous and dolomitic shales. If the shale is relatively soft, the calcareous type will react much more strongly. If the shale is consolidated, a reasonable estimate can often be made by observing the acid reaction on a crushed chip.

The nature of the residue after the acid reaction may be important. In distinguishing between shaly limestone and limy shale, for example, it is useful to crush a small chip and estimate the percentage of residue remaining after the material is digested. If more than half the material remains, the rock is a calcareous shale, and vice versa, etc. The residue should also be examined for accessory minerals such as anhydrite, chert, etc.

A third important acid reaction is observed when oil is present in a carbonate rock. The oil often forms a film encasing the effervescing carbon dioxide gas, allowing the bubbles to grow larger than usual. When this happens, the rock chips may bob up and down in the acid or float to the surface. This test gives a clue to the possible presence of oil but should not be taken as conclusive. Further tests must be made to determine accurately the presence or absence of hydrocarbons.

One of the most important observations to be made in sample logging is the detection of porosity. If porosity is observed, it should be thoroughly described, and the sample should always be checked with carbon tetrachloride for cut fluorescence regardless of whether or not staining is observed on the porous rock chips. A fresh surface should be exposed by breaking the chip before or after it is immersed in the carbon tetrachloride, particularly if the samples have been in storage for some time.

The original sample log should be plotted in color for maximum usefulness. Color photographic reproductions are available, although presently rather expensive. If a number of reproductions are required, the log can be plotted using black-and-white symbols and the copies later colored. Suggestions for suitable color and black-and-white symbols to be used in plotting sample logs can be found in the references at the end of the chapter.

Sources of Error

The principal sources of error in sample logging are the contamination of samples from upper beds, termed caving, and the recirculation of cuttings. The amount of caving in a sample varies considerably, depending on hole condition, nature of the drilling mud, and the current drilling activity. One of the main problems of the geomicroscopist is to detect the appearance of any *new* rock type or feature in the sample. The detection and identification at a glance of all previously penetrated material (caving) versus a small amount of new material requires well-developed powers of observation, a good memory, and careful attention to detail, as well as experience. These are the reasons why sample examination is very exacting work.

Bentonitic shales, thin-bedded brittle shales, coal, finely fractured brittle rocks, and loose sand cave more readily than do well-consolidated or thick-bedded sediments. Rig shutdowns such as for drill-stem testing, coring, bit changes, etc., cause an increase in caving when drilling is resumed. Good drilling mud will line the hole with a tough mud cake and reduce caving to a minimum. Nevertheless, the average sample contains at least 50 per cent caved material.

Small rock particles, particularly fine sand grains, will pass through shale shaker screens and may not settle out in the mud tanks or baffle boxes. Such cuttings may not appear at all in the samples but are continuously recirculated, or they may appear in later samples if the mud thins down or changes in gel strength.

Occasionally cuttings may be lost altogether. If lost circulation occurs, cuttings may be carried into the formation or fractured interval which is taking the mud. Bentonite and soft bentonitic shales may break down and become so thoroughly disseminated in the mud that they do not appear in the samples. These materials can also be lost by too thorough washing of the samples. Anhydrite and salt are soluble in drilling mud to a greater or lesser extent and tend to be at least partially lost. Rock salt cuttings are often completely dissolved.

The preceding discussion of well sample examination and logging has emphasized the importance of describing and recording the data observed in the formation samples. From these data the imaginative geologist may now with more or less assurance, depending on the quality of the material available and the care with which it has been analyzed, reconstruct the sedimentary environment in which the rock was deposited and sometimes trace the ensuing history of the region by determining the nature and extent of secondary alteration affecting the formations. The major depositional environments—continental, fluvial, lacustrine, and marine—are not difficult to recognize. At least one company, employing a large staff of sedimentologists, has subdivided these environments into 29 units which can be determined from well cuttings and cores.

Figure 13-1 shows a method of recording basic subsurface data on a modern sample log. Figure 13-2 shows a generalized sample log using the percentage-interpretive system.

Fossils

The identification of fossils present in drill cuttings and cores is the field of the micropaleontologist. The subsurface geologist generally must rely on the identifications and paleoecological interpretations determined by the micropaleontologist, for he seldom has time to become familiar with the descriptions of the thousands of species, their geologic ranges, and the particular environments under which each was deposited. However, he must intelligently integrate the paleontological data with all other available information in selecting favorable areas and prospects for oil exploration.

Microfossils are useful in subsurface work because they are found in practically all types of sedimentary rocks ranging in age from Precambrian to Recent. Thus, they are extremely valuable for determining the age of sediments. In many instances, through identification of index or guide fossils or taxonomic groups, world-wide age correlations within the epoch level can be made. Formations can often be zoned through the use of microfossils, permitting more accurate correlations to be made. This can be accomplished by noting the first occurrence of a new form or a change in the assemblage; by population counts, counting a given number of individuals (200

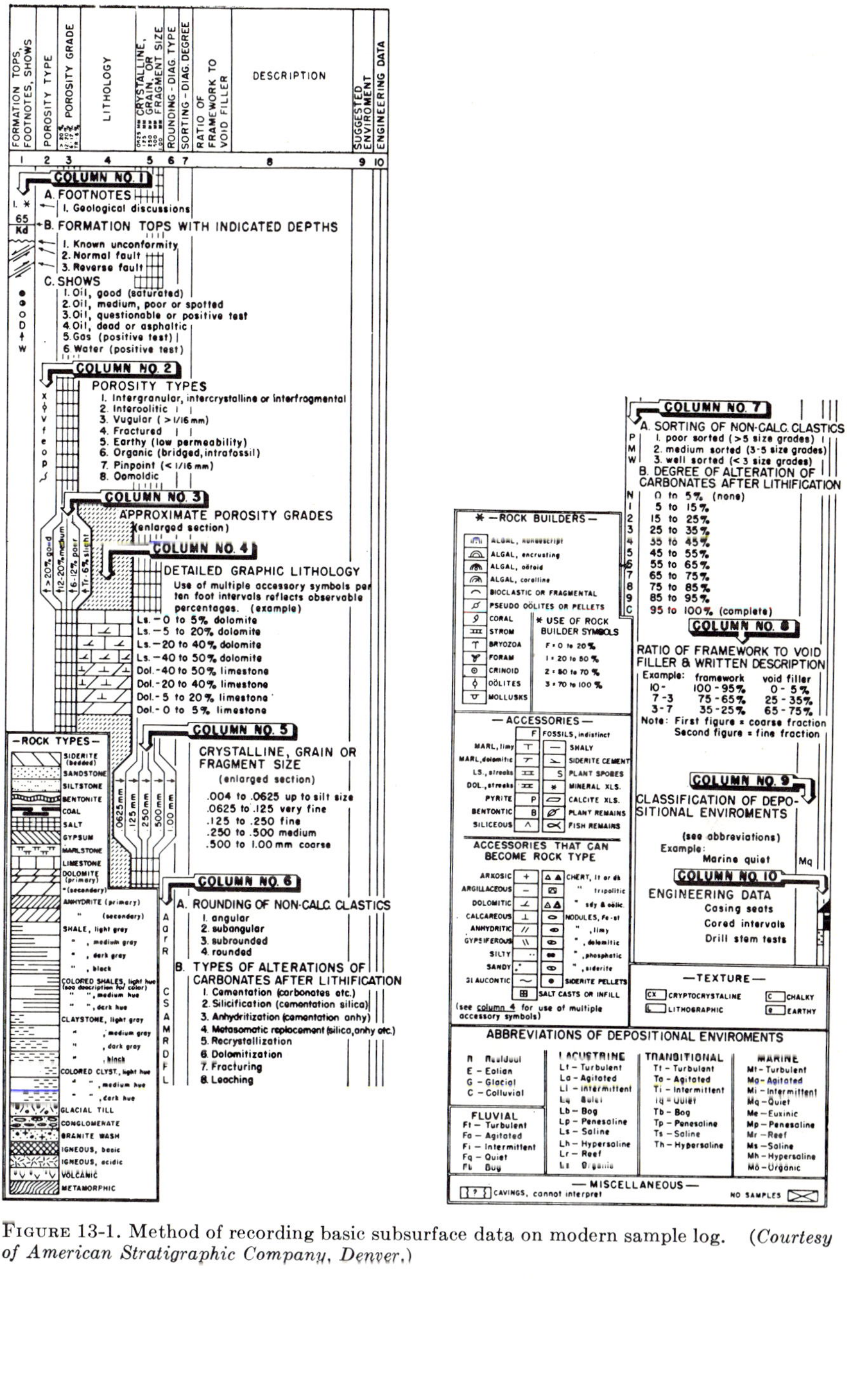

FIGURE 13-1. Method of recording basic subsurface data on modern sample log. (*Courtesy of American Stratigraphic Company, Denver.*)

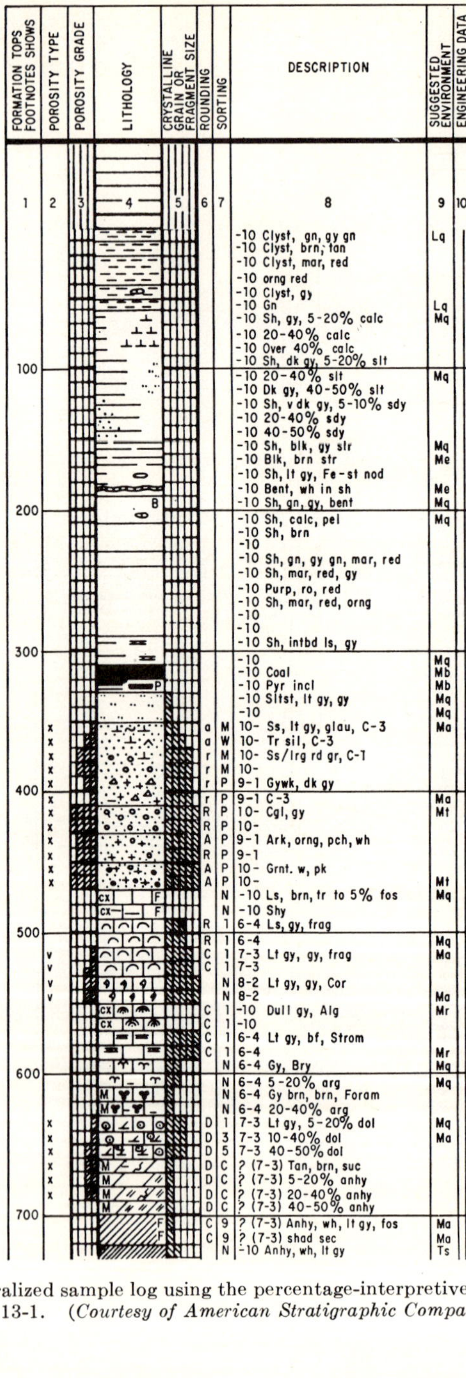

FIGURE 13-2. Generalized sample log using the percentage-interpretive system, and plotted as shown in Figure 13-1. (*Courtesy of American Stratigraphic Company, Denver.*)

to 300) of a certain group of microorganisms and plotting the percentage of each species in histogram form; or by determining the ratio of plant microfossils to brackish or marine animal microfossils. Microfossils are good environmental indicators and can be used to evaluate areas for petroleum exploration. They are particularly useful in subsurface exploration because, owing to their small size, they escape destruction from the drilling operation and only a small amount of sample, requiring little time and no elaborate equipment to prepare, is needed for examination.

Three major types of microfossils are distinguished: animal microfossils, plant microfossils, and microfossils of uncertain affinity. The commoner animal microfossils include foraminifera—small, one-celled animals, most of which have calcareous shells (tests); radiolaria—marine spheroid, asteroid, or bell-shaped organisms having tests of silica or strontium sulfate; ostracods—small bivalved crustacea, especially useful in nonmarine beds; and conodonts—minute, toothlike objects, particularly useful as index markers in Paleozoic rocks. The useful plant microfossils include coccoliths—disk-shaped planktonic forms particularly abundant in Mesozoic rocks; charophytes—a form of algae ranging in age from Devonian to Recent; diatoms—algal forms with siliceous cell walls consisting of two closely fitting or overlapping halves; and plant spores and pollens. The principal microfossils of uncertain affinity are the chitons—organisms composed of chitinous material. Of all these microfossils the most useful have been the foraminifera, which have been used by the oil industry for more than 40 years. At the present time plant spores and pollens are becoming very useful in stratigraphic geology. Unlike foraminifera, which are confined principally to marine sediments, pollen and spores are found in nonmarine as well as marine formations. They are widely distributed by wind and water and by insects. Foraminifera have different forms in each type of environment, such as near-shore, offshore, shallow water, deeper water, and abyssal. Correlation by foraminifera across facies boundaries is often difficult. In contrast, the same species of pollen or spores are found in sediments laid down in all environments, thereby permitting correlation across differing depositional facies.

Drill cuttings and cores have a number of other uses in subsurface geologic analysis. Correlations can be established, formation sources traced, depositional environments postulated, or hydrocarbon content analyzed by such varied techniques as preparation and comparison of insoluble residues, identification of heavy minerals, grain-size logging, differential-thermal analysis, thermoluminescence measurements, and X-ray analysis. These techniques are all laboratory methods requiring equipment and procedures not readily adaptable to rapid analysis in the field. These methods are described elsewhere in this volume.

Logs

Although the drill cuttings and cores recovered while a well is being drilled are the most important sources of subsurface information, the fluids present in the formations penetrated also provide valuable data for geologic analysis. Much information concerning the formation fluids is obtained indirectly from electric, radioactive, and sonic logs and from mud logging, but study of the fluids themselves and their behavior yields important data regarding the nature of traps to be expected in a formation, the direction and extent of possible migration of oil and gas, entrapment under hydrodynamic conditions, and correlation by means of formation fluid analysis. Samples of formation fluids are commonly recovered from laboratory analyses of well cores and from drill-stem or production tests. Usable samples can sometimes be obtained from the fill-up in cable-tool holes. Fluid-pressure data are obtained from the pressure bomb used in drill-stem and production tests and indirectly from such information as the height of fluid fill-up in an open hole, such as cable-tool well tubing and casing; lost circulation observations; and blowout data. The study of ground-water flow and its effects on petroleum entrapment constitutes the field of hydrodynamic analysis (see Chapter 6). The application of hydrodynamic analysis in recent years has provided a new tool for mapping stratigraphy and subsurface faults, as well as providing data for more accurate prospect evaluation.

In rotary well drilling, drilling fluid or mud is pumped down the hole inside the drill stem, through holes in the bit, and back to the surface between the drillpipe and the wall of the hole. The bit cuts and dislodges chips of the formation being penetrated which are carried upward by the circulating mud. As these cuttings travel upward, the hydrostatic pressure in the mud column decreases, permitting any gas occupying pore space in the cuttings to expand. A portion of this gas will escape into the drilling mud, and a portion will remain in the cutting chip. In oil-bearing formations, the associated gas will expand similarly and, in addition, will force a portion of both the gas and the oil into the mud column. When the mud and cuttings reach the surface, these hydrocarbons can be detected by specialized equipment. If the mud and cuttings are continuously analyzed and suitable allowance is made for the time necessary for the mud to reach the surface, the oil and gas-bearing formations can be determined and their depth ascertained.

Mud Logs. In general, reliable showings of oil and gas are obtained in both the drilling mud and the cuttings. In unconsolidated sands or where mud circulating rate, hole conditions, or mud properties are such as to prevent the cuttings from reaching the surface promptly, analysis of the cuttings may not be reliable. The amounts of oil or gas in either the drilling mud or the cuttings are a qualitative indication only of the formation content. Too many factors affect the relative amounts of hydrocarbon released to permit a quantitative interpretation to be drawn from the size of the showing. In practice, the drilling mud and cutting analyses are generally combined with a drilling-time log and a graphic log of the lithology (Figure 13-3) of the formations penetrated as determined from examination of the cuttings. Such a combination of data is extremely useful in determining the need for immediate drill-stem testing of promising formations and for rough correlation of the well. The drilling mud log is particularly valuable where soft, unconsolidated formations are being drilled, permitting a very fast drilling rate; where there is a low pressure differential between the mud column and the formation; in formations with medium to low permeability; and when using mud of low filtration rate.

Drilling-time Logs. Drilling-time, or rate-of-penetration, logs consist of a graph or curve in which drilling time or penetration rate is plotted against depth. The shape of the curve is determined by the speed of penetration. Abrupt breaks or changes in slope of the curve indicate contacts between rocks of different penetrability. The principal factors influencing the drilling rate of a formation include the strength of the rock being penetrated and its toughness. Toughness depends on the mineral content, the kind of cement and the degree of cementation, the texture of the rock, and its porosity. Rock strength varies with depth of burial and with fluid pressures in the pore spaces. Drilling rate, of course, also varies with such factors as design and sharpness of the bit, depth and weight on the bit, drilling mud characteristics and pump velocity, rotation speed, skill of the driller, etc. However, most of these variables remain more or less constant for a given hole or portion of a hole, so that the drilling-time log is ordinarily a reliable index to the formations being penetrated. Normally the log is made automatically by devices such as the Geolograph or the Log-O-Graph. However, the rate of penetration can be logged manually by timing the downward progress of the Kelly with a stop watch.

Drilling-time logs are particularly valuable during drilling of a well because of their

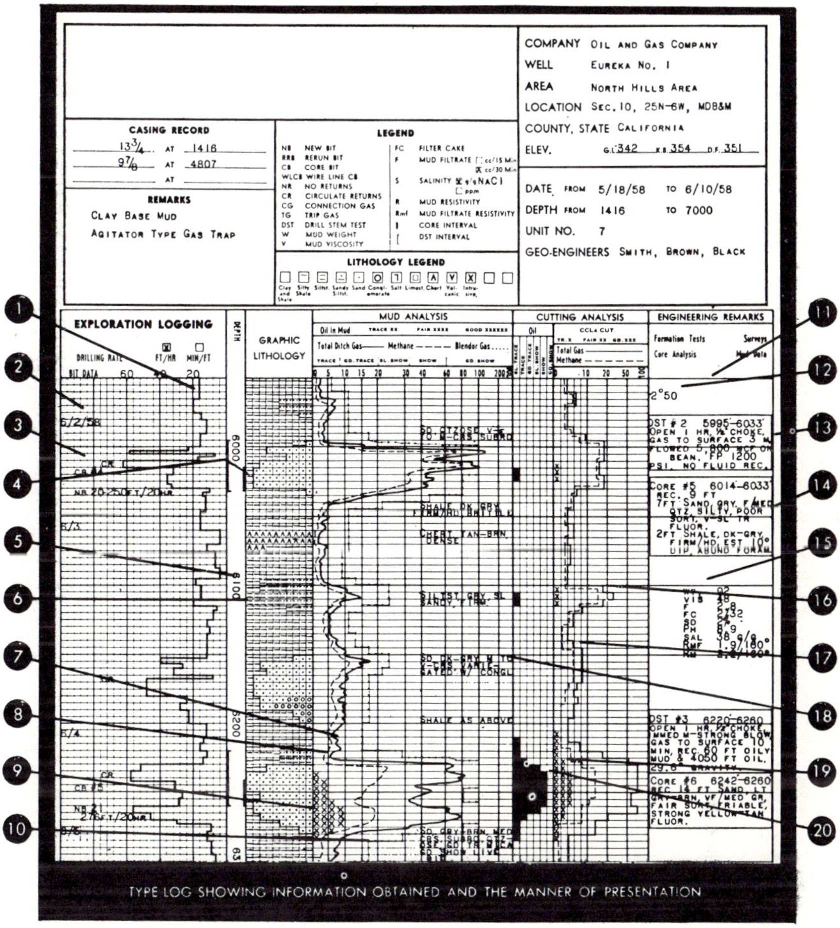

1 DRILLING RATE CURVE
2 DATE
3 BIT AND CORE BIT DATA
4 CORE AND DST INTERVALS
5 DEPTH
6 PERCENTAGE LITHOLOGY
7 DITCH GAS CURVE, TOTAL
8 DITCH GAS CURVE, METHANE
9 OIL IN MUD CURVE
10 MUD BLENDOR GAS, TOTAL
11 REMARKS COLUMN
12 DEVIATION SURVEYS
13 DST DESCRIPTION
14 CORE DESCRIPTION
15 MUD DATA
16 CUTTINGS GAS, TOTAL
17 CUTTING GAS, METHANE
18 LITHOLOGIC DESCRIPTION
19 CARBON TETRACHLORIDE
20 OIL IN CUTTINGS CURVE

FIGURE 13-3. Typical drilling mud log showing information obtained and manner of presenting the data. (*Courtesy of Exploration Logging Company, Sacramento, Calif.*)

immediate availability. When plotted on a time scale selected so that the amplitude between fast and slow peaks corresponds roughly to the range between the shale base line and the maximum peak of an average electric-log SP curve for the general area, the drilling-time log provides the geologist with data which can be used, within reasonable limits, for much the same purposes as the electric log (Figure 13-4). Many times the stratigraphic position of any portion of a drilling well can be established in advance of electric logging by observing the sequence of beds penetrated as shown by the drilling-time log. This may become especially valuable if for some reason it later becomes impossible or impractical to run an electric or other mechanical log in the hole. The passing of a formation contact can be indicated by the drilling-time log before samples arrive at the surface. Coring points can thus be selected as drilling progresses. Also, the drilling-time log is often used to interpret gaps in the core record due to incomplete recovery. Other important uses of the drilling-time data include the recognition of porous and permeable intervals in a thick carbonate rock section and the identification of thin shale breaks or tight calcareous streaks interbedded in a sandstone formation.

Electrical Logs. Electrical well logging is, essentially, the recording of the resistivities (or their reciprocals, the conductivities) of the subsurface formations and the spontaneous potentials generated in the boreholes. The parameters are measured *in situ* by means of appropriate down-hole instruments, called "sondes," and are continuously recorded at the surface. The measurements are performed in the uncased portions of the boreholes only. Several different kinds of resistivity curves

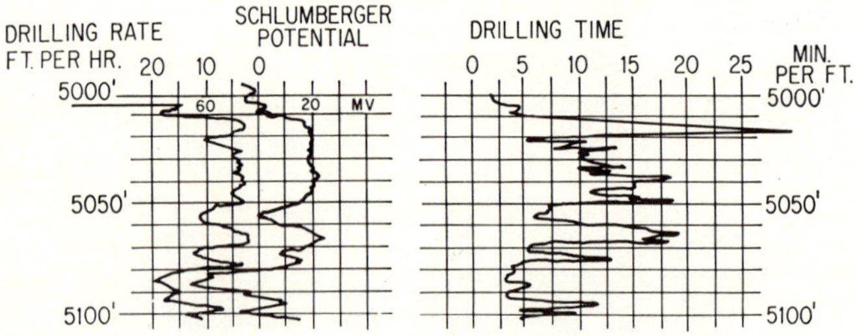

FIGURE 13-4. Correlation of the electrical (SP) curve with a drilling-time log and its reciprocal log, a drilling-rate log. (*After J. A. Simons, from G. F. Shepherd, Drilling Time Logging, p. 378, in Haun and LeRoy, 1958.*)

or logs can be recorded in addition to the spontaneous potential, or SP, curve. These correspond to the use of different conductivity or resistivity measuring devices and are designated as induction logs, conventional logs (normal and lateral), Laterologs, Micrologs, etc. The typical appearance of a modern electrical log is illustrated in Figure 13-5. The left-hand track contains the SP curve; the right-hand tracks contain the conductivity and resistivity curves for the induction and short-normal devices. Each curve shows a sequence of peaks, deflections, and flat zones corresponding to the different formations penetrated by the borehole. The logs recorded with other types of resistivity-measuring devices have a similar appearance.

The general information provided by the logs is essentially the following: (1) differentiation between shales, hard rocks, and permeable beds; definition of boundaries between beds; and correlation (these data make possible the delineation of structural features and the determination of the thickness and lateral extension of potential reservoirs) and (2) in most cases qualitative discrimination between oil- or gas-bearing and water-bearing beds, as well as location of oil-water contacts; in favorable cases quantitative evaluation of porosity and water saturation.

Electrical well logging was introduced more than 25 years ago. The vast majority of wells drilled within the last 10 years have been logged by electrical devices, and the logs of these boreholes constitute one of the principal sources of subsurface geological information.

Radioactivity Logs. Radiation well logging measures the natural radioactivity of the subsurface formations (gamma-ray logging) and the secondary effects due to the bombardment of the formations by neutrons emitted from a source contained in the sonde (neutron logging). As with electrical logging, the parameters are measured *in situ* and are continuously recorded at the surface (see Figure 13-6). The measurements can be made in both open and cased holes filled with any fluid whatsoever (including air, i.e., empty).

Of all formations encountered in drilling wells, shales contain the greatest concentrations of radioactive salts. Practically without exception, shales are more radioactive than sands, limestones, anhydrites, etc. Therefore, the gamma-ray log essentially distinguishes shales from other formations. The recording equipment is so designed that the curve deflects toward the right when the radioactivity increases. Therefore the gamma-ray log is somewhat similar to the SP curve for which deflections to the right also indicate the presence of shales, and the two curves can sometimes be used interchangeably. The gamma-ray log is used principally for bed definition, determination of interfaces, and correlation.

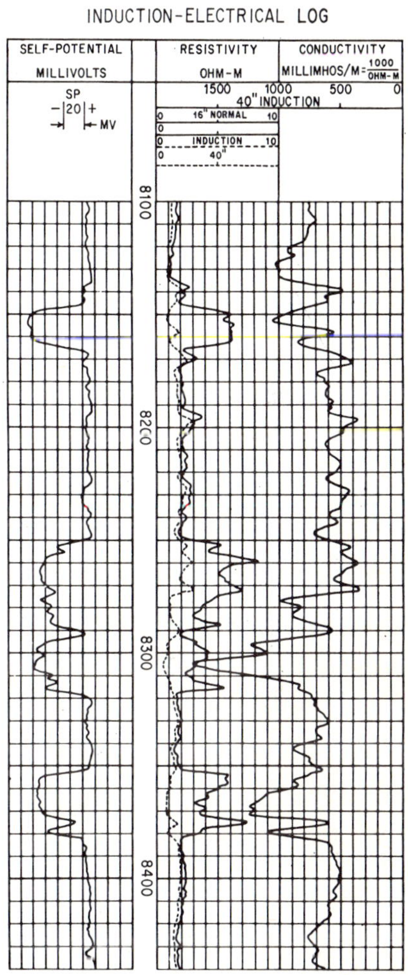

FIGURE 13-5. A typical electric log. Left-hand track contains the SP curve; right-hand tracks contain the conductivity and resistivity curves for the induction and short-normal devices. (*From Schlumberger, 1958.*)

The neutron curve is produced by an artificially induced neutron bombardment of the wall rocks and is an inverse measurement of the amount of hydrogen present. Inasmuch as most hydrogen in rock is in either interstitial water or oil, the neutron curve indicates (by subnormal counting rates) the presence of fluids and hence is a porosity log. It is a natural companion to the gamma-ray curve in that it makes possible the distinction between porous and nonporous formations. Figure 13-6 illustrates the qualitative interpretation of a typical gamma-ray–neutron log.

Sonic Logs. Sonic or acoustical well logging basically involves the recording of the time required for a sound wave to travel through a definite length of formation. These travel times are recorded continuously, versus depth, as the sonic sonde is pulled up the borehole and are inversely proportional to the speed of sound in the various formations. Table 13-1 gives the sonic velocities and corresponding travel times (in microseconds per foot) for various fluids and solid materials. Some of the values are averages and may vary with depth in subsurface formations.

Table 13-1. Sonic Velocities and Corresponding Travel Times (in Microseconds per Foot) for Various Fluids and Solid Materials

Material	Sonic velocity V, ft per sec	Travel time Δt ($= 10^6/V$), μsec per ft
Air (STP)	1,088	919
Methane (STP)	1,417	706
Oil	4,300	232
Water (mud)	5,000–5,300	200
Neoprene (typical)	5,300	189
Shales	6,000–16,000	167–62.5
Rock salt	15,000	66.7
Sandstones	Up to 18,000	55.6
Anhydrite	20,000	50.0
Limestones	Up to 23,000	43.5
Dolomite	24,500	40.8

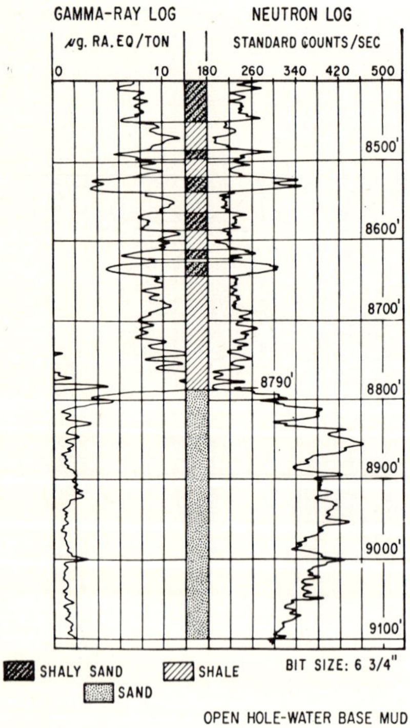

FIGURE 13-6. A typical gamma-ray–neutron log together with a qualitative interpretation of the curves. (*From Schlumberger, 1958.*)

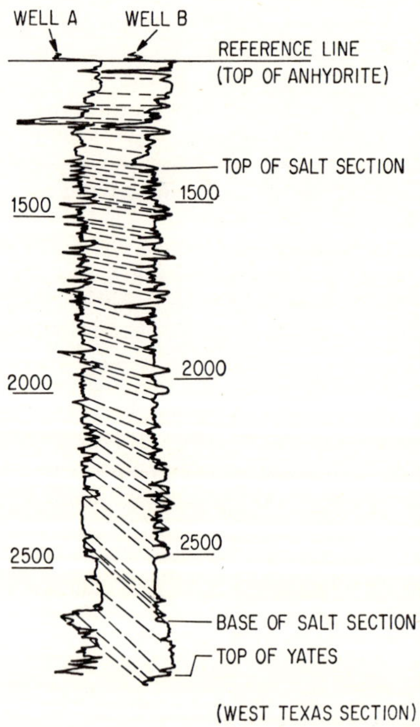

FIGURE 13-7. Two sonic logs illustrating the amount of detail revealed by the logs and showing degree of correlation possible. (*From Schlumberger, 1958.*)

Speed of sound in subsurface formations depends upon the elastic properties of the rock matrix, the porosity of the formations, and their fluid content and pressure. In moderate to hard formations, where porosities are low, the sonic log is considerably affected by the amount of fluid in the formations; hence, it gives reliable indications of their porosities. In soft formations of high porosity, it is more responsive to the *nature* of the fluids contained within the pores (sound velocity in oil is approximately nine-tenths that in water), and so it may sometimes give indications of oil saturation. In all types of formations, it is of value to the geophysicist in interpreting seismic charts. Figure 13-7 illustrates the amount of detail revealed by the sonic log.

For the subsurface geologist the acoustical well log provides another correlation parameter, inasmuch as the log is derived from a physical property not otherwise measured in boreholes. The sonic well survey can provide (1) a log revealing outstanding correlation data; (2) a log showing small changes in lithologic character of the formations penetrated; (3) a log differentiating stratigraphy in thick limestone, dolomite, sand, and shale sections; (4) a log which, under proper conditions, will reveal fault contacts; and (5) an effective log in highly saline drilling fluid.

Sonic logging is relatively new. As experience is gained and more data become available, it is reasonable to suppose that interpretative techniques will be expanded and refined and that sonic logs will become even more valuable and versatile than limited experience now demonstrates.

While the electrical, radiation, and acoustical well logs are by far the most important sources of data which can be recorded in boreholes, other logs can be run in conjunction with these and will yield additional information of

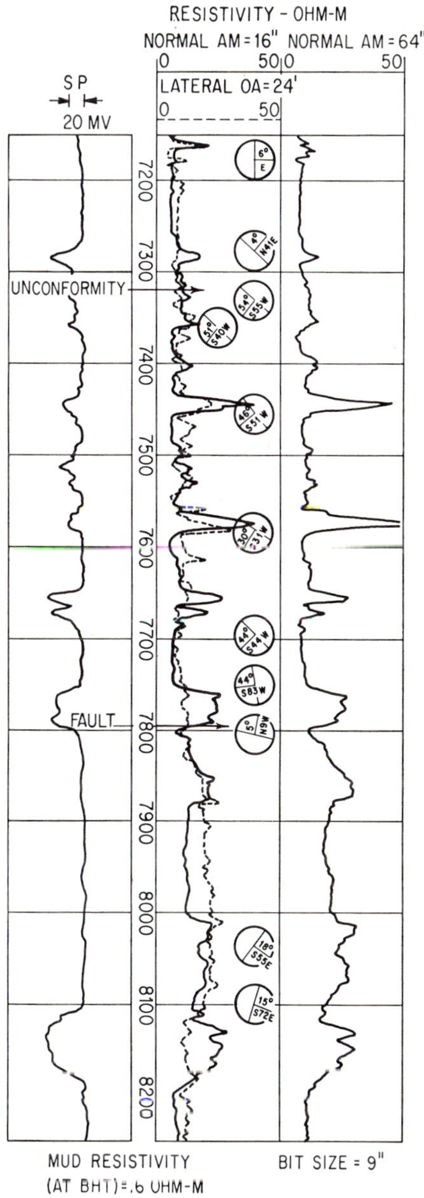

FIGURE 13-8. Electrical log and results of dipmeter measurements, showing changes in angle and direction of dip. Based in part on these data, the presence of an unconformity and a fault has been inferred. (*From Schlumberger, 1958.*)

value to the subsurface geologist. These include dipmeter surveys, caliper logs, and temperature logs.

Dipmeter Logs. The dipmeter is a survey device run in the hole to determine amount and direction of dip of formations penetrated by the drill. Dipmeter surveys provide data assisting in the solution of many structural and stratigraphic problems encountered in exploratory and development wells. For example, when the first well of a prospective producing area turns out to be dry, the problem is to determine in which direction to sidetrack, or to drill the next well. The dipmeter will provide the solution. If the dip is great enough, it can be computed with precision and restrict to a small sector the zone in which to drill the next well. If the dip is shallow, so that accurate determination is difficult, only the general direction of the next location can be given. Such information is precious in wildcat territory.

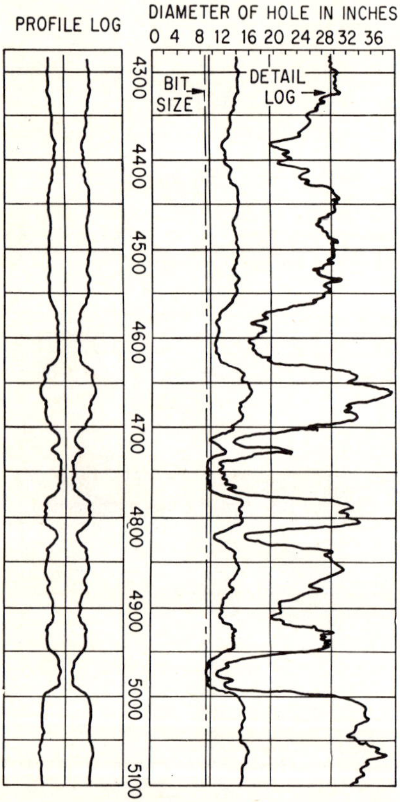

FIGURE 13-9. Portion of a linear section-gauge log. (*From Schlumberger, 1958.*)

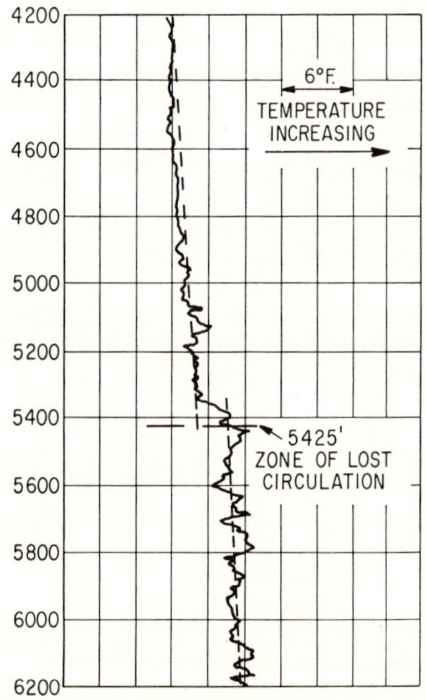

FIGURE 13-10. Temperature survey showing location of a zone of lost circulation. (*From Schlumberger, 1958.*)

In producing areas, an oil reservoir may be overlain by gas or underlain by water. A knowledge of the dip of the formations assists in the location of offset wells to avoid drilling into the productive horizon where it is so low as to produce water or so high as to produce gas. Dipmeter surveys have also been of considerable help in unraveling structural conditions adjacent to piercement salt domes.

As illustrated in Figure 13-8, dip measurements made at relatively short intervals in the hole will often show that the dip angle or direction changes with depth. Sometimes marked dip reversals are observed in a single well. Such data may indicate the presence of unconformity surfaces or of faults.

Caliper Logs. The caliper, or section gauge, is designed to measure the diameter and cross-sectional area of a borehole. As it is moved along in the hole, the actual size of the hole at the depth of the instrument is recorded continuously on film by a recorder at the surface (Figure 13-9). Primarily the size of a drill hole is of interest for the following purposes: (1) It makes possible the computation of hole volume and of the quantities of cement necessary in cementing operations, (2) it is useful in the selection of packer seats, and (3) it provides an accurate value for hole diameter, which is essential for the more accurate interpretation of the various electrical and radioactive logs. However, a study of numerous caliper logs leads one to the conclusion that caving patterns exist. Certain generalizations can be made as to the relative ability of rocks to stand up to bit size. Although similar drilling conditions and similar muds tend to standardize caliper logs, some astonishing long-range correlations can be made using them. In some areas caliper-log correlations are more trustworthy than those made with electric logs.

Temperature Logs. Temperature in the earth increases with depth. The rate of increase (under conditions of thermal equilibrium known as the *geothermal gradient*) depends on locality considered and on heat conductivity of the geological formations. Because of the dependence of geothermal gradients on the heat conductivity of the formations, measurements of the temperature gradients, at various depths of wells in which thermal equilibrium has been established, have been used for correlation purposes. The temperature gradients are in general low in formations having high heat conductivities and high if heat conductivity is low. However, the temperature log will not soon replace the electric log, and its use in subsurface correlation is limited. Figure 13-10 shows a temperature survey record.

Figure 13-11 presents a comparison of seven fundamental ways of logging subsurface phenomena.

CORRELATION

The collection of subsurface data has no geologic value whatsoever until the data have been recorded, classified, studied, and interpreted. The first task for the exploration subsurface geologist is to work out the stratigraphic section in detail. This includes the correlation of sedimentary units and the recognition of unconformities. Correlation is the process of determining mutual relations. Correlation of stratigraphic units serves to indicate which rocks in different regions are of the same age and arranges rocks of different ages in their proper sequence. Until this is done, the primary objective of finding hitherto hidden traps, either structural or stratigraphic, or of increasing the yield of hydrocarbons within a producing area cannot be commenced.

Correlation may be based on three types of stratigraphic units: *lithogenetic* or *rock units*, *biostratigraphic* or *faunal units*, or *time-stratigraphic* (*time-rock*) *units*. In other words, the demonstrated equivalency may be in terms of lithologic or biologic continuity or in terms of the geologic time scale. The distinction among these three approaches must be kept in mind, as nothing but confusion will result from attempting to correlate rock units by strictly biologic methods or time-rock units by reference to lithology alone.

Lithogenetic Units

Lithogenetic or sedimentary-rock units are specific bodies of rock identified and recognized by observable objective criteria. The fundamental rock unit is the

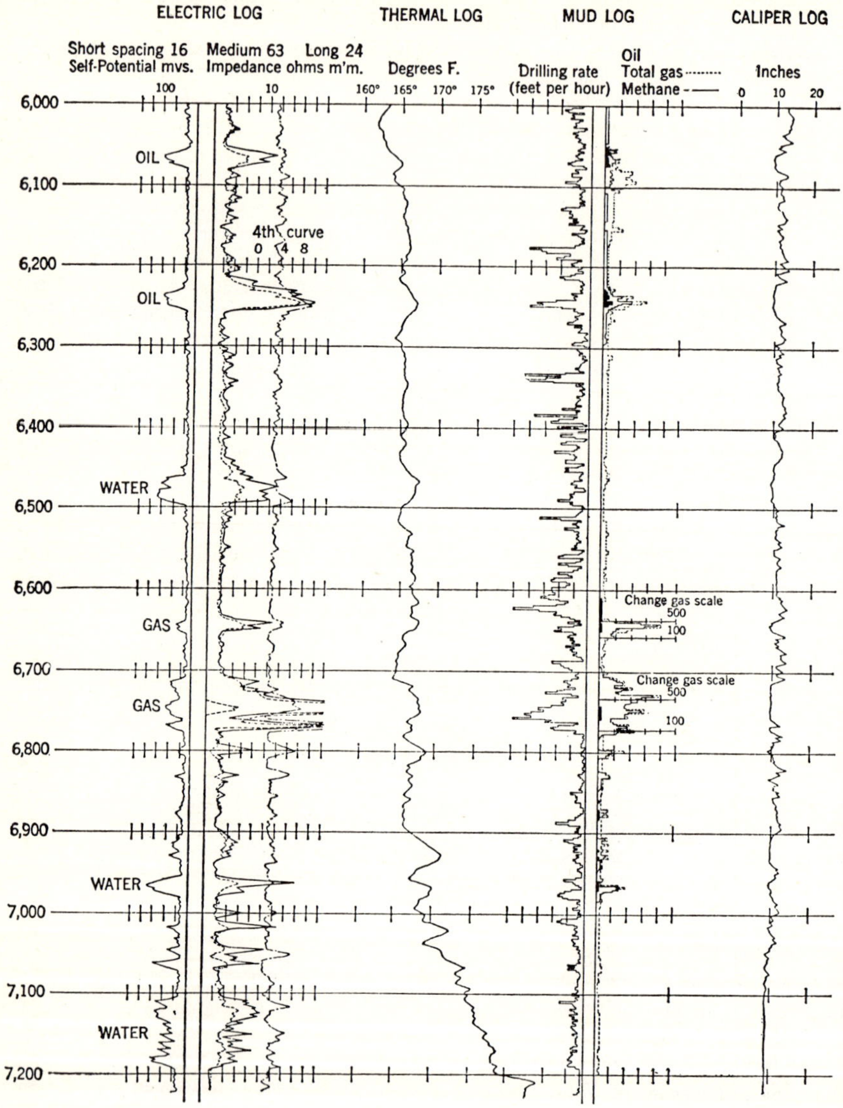

FIGURE 13-11. Comparison of seven fundamental ways in which subsurface data are logged, either in a borehole or from rock materials brought to the surface. Direct comparisons can be made as these logs are all from the same well. (*After Humble Oil & Refining Com-*

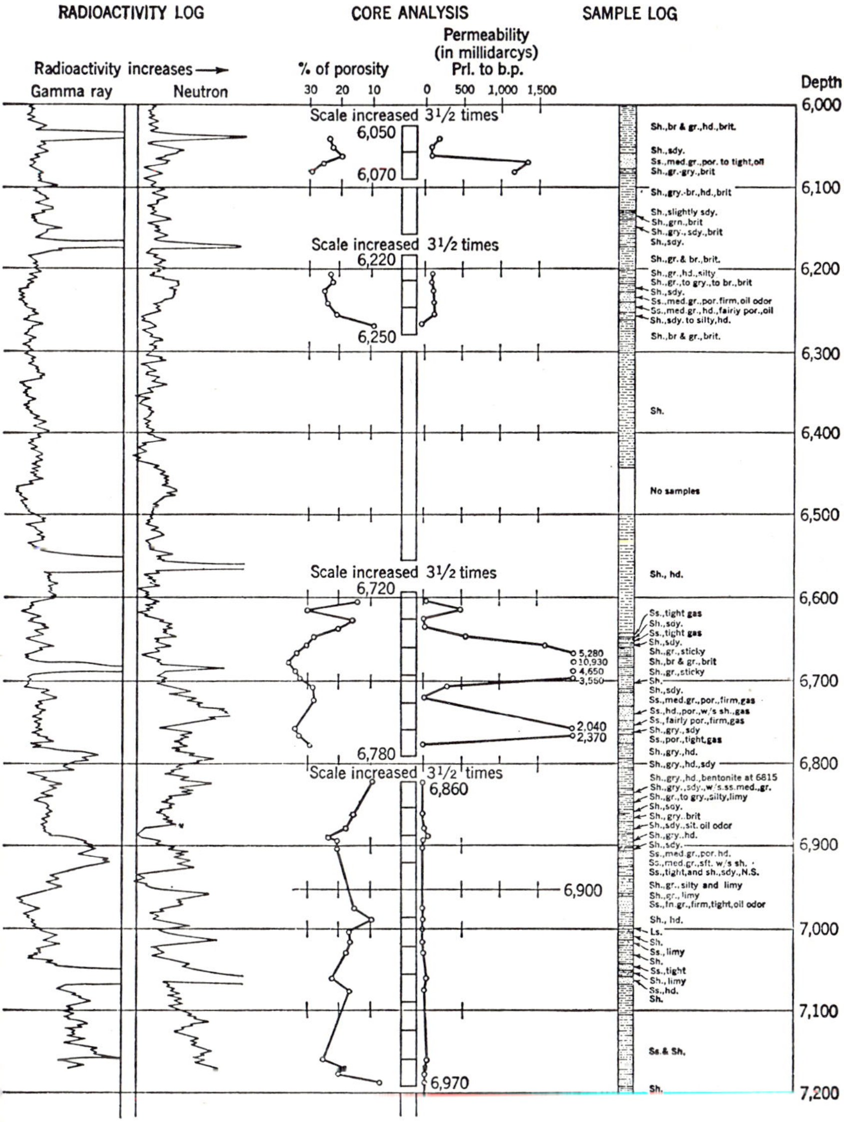

pany in D. A. Busch, Subsurface Techniques in "Applied Sedimentation," P. D. Trask (ed.), John Wiley & Sons, Inc., New York, 1950.)

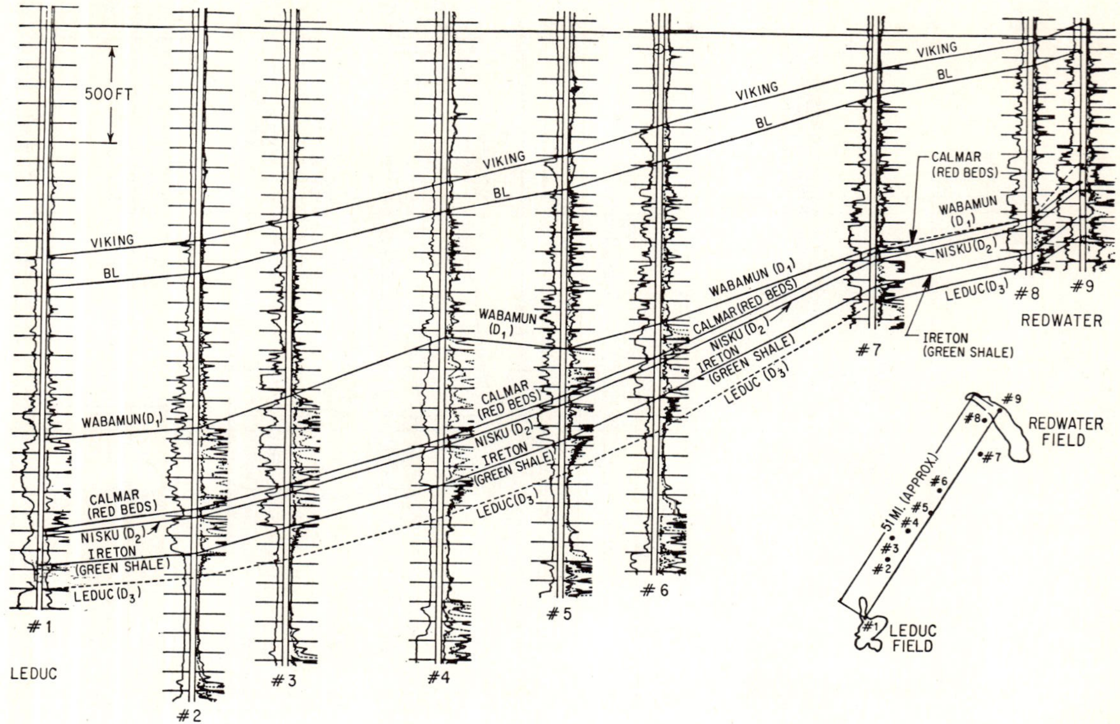

FIGURE 13-12. Long-distance correlation between Leduc and Redwater fields (distance about 90 miles) using electric logs. (*From Schlumberger, 1958.*)

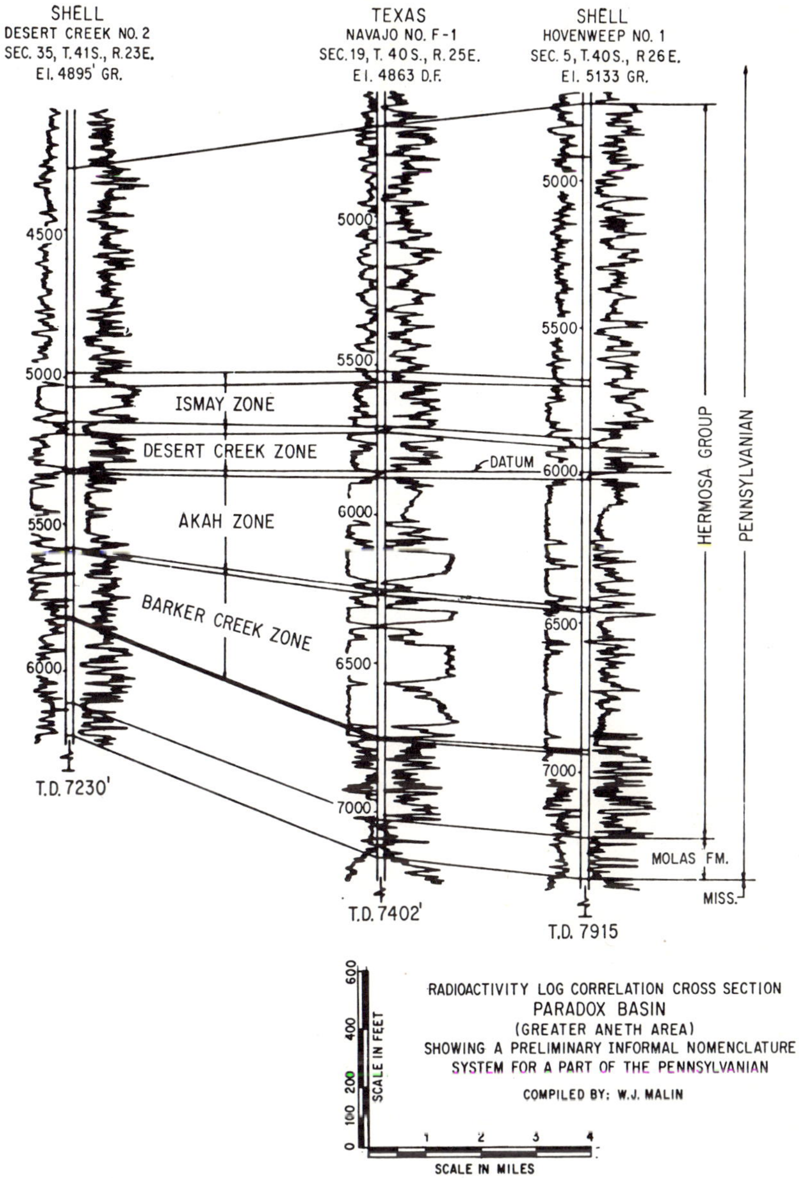

FIGURE 13-13. Example of correlation using radioactivity logs. (*From W. J. Malin, A Preliminary Informal System of Nomenclature for a Part of the Pennsylvanian of the Paradox Basin, "Guidebook to the Geology of the Paradox Basin," p. 136, Intermountain Association of Petroleum Geologists, 1958.*)

formation. Formations may be assembled into *groups* or differentiated into *members*, *tongues*, *beds*, etc. All the features useful in the recognition of formations are also useful in correlation. The most important and most commonly employed types of evidence include:

Physical evidence:
- Lithologic similarity
- Continuity of strata
- Position in stratigraphic sequence
- Orderly variation in lithology
- Electrical characteristics
- Radioactivity
- Acoustical characteristics
- Unconformable relations
- Structural development

Paleontologic evidence:
- Index fossils
- Paleontologic sequences
- Paleontologic similarity

Figure 13-12 is an example of long-distance correlation using electrical characteristics. Figure 13-13 illustrates correlation of radioactivity logs.

Correlation of a lithogenetic unit is limited to the areal extent of the unit in ques-

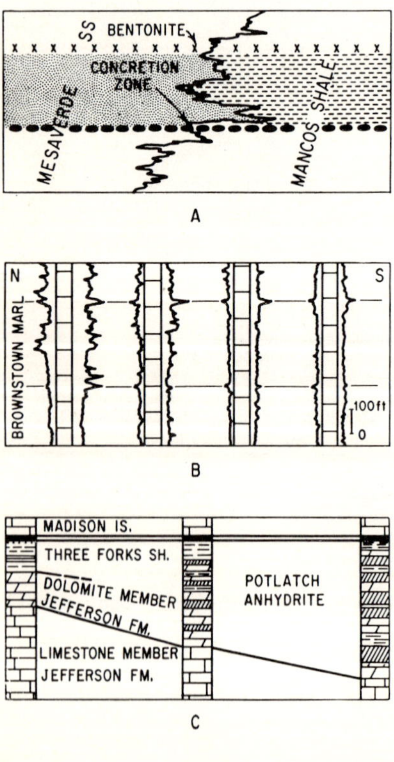

FIGURE 13-14. Three methods of correlation by identity of interval. In *A* a bentonite layer and concretion zone establish the identity of an interval which crosses a lithotope boundary in the Upper Cretaceous of Utah. In *B* an interval in the Brownstown marl (Cretaceous) of Louisiana is traced by means of distinctive electric log "kicks" across an area in which the interval grades from interbedded limestone and shale to nearly all shale (adapted from *Shreveport Geol. Soc.*, 1945). *C* illustrates the correlation of an interval in the Upper Devonian of Montana by means of persistent recognizable units above and below. In this case the black shale at the base of the Madison limestone and the limestone member of the Jefferson formation maintain their character and establish correlation of the interval composed of the Three Forks and the dolomite member of the Jefferson, even though these units grade into an evaporite sequence. (*From Krumbein and Sloss, p. 294, 1955.*)

tion. Where the extent of the unit is limited by nondeposition or by erosion, this area over which correlation is possible is easily defined. However, problems arise where the unit changes laterally into varying lithofacies. Sometimes such problems can be side-stepped by using different names for the varying aspects of the unit without attempting to correlate the individual unit over a number of facies. Correlation by identity of interval (Figure 13-14) also yields data of less obvious but equal importance, and it may be useful in the construction of facies maps.

Biostratigraphic Units

Biostratigraphic units are bodies of strata characterized by their contained fossil assemblages. The fundamental unit is the *zone* or *biozone.* Biozones, like rock units, are of limited areal extent and generally cannot be correlated beyond a single depositional basin or province. Biostratigraphic zones are not necessarily related to rock units. However, environmental factors commonly affect *lithotopes* and *biotopes* simultaneously, and therefore, biozones and rock units may coincide over broad areas. Characteristically, the two types of units tend to be more parallel or coincident around the rim of a sedimentary basin and more widely divergent downdip into the basin.

Subsurface correlation of biozones is based on the identification of micropaleontologic zonal guide fossils. Normally, consideration of assemblages of species rather than individual forms is required. Emphasis is placed on rapid recognition of biostratigraphic horizons in the routine analysis of subsurface samples. The data are of importance in determining structural position and in detecting changes in thickness of stratigraphic intervals from well to well. The first appearances of zonal guide fossils in the samples are used to establish the position of "tops" of successive zones penetrated. These tops may be correlated like key beds in a lithologic section (Figure 13-15).

Time-Stratigraphic Units

Time-stratigraphic (time-rock) units are bodies of strata deposited during a specific interval of geologic time. No name has been assigned to rocks deposited during an era, these being commonly referred to as "Paleozoic strata," "beds of Cenozoic age," etc. Strata deposited during an epoch form a *series,* and those laid down during an age are designated a *stage.* The boundaries of time-rock units are invisible planes or surfaces of synchronous time, although sometimes these surfaces may be delineated by unconformities. A time-stratigraphic unit frequently contains within its boundaries various types of lithology, such as sandstone, reef limestone, and shale. The time-rock units may transect lithogenetic units.

The aims of time-rock correlation are generally interprovincial, interregional, or even intercontinental and are somewhat less likely to be of more than passing concern to the petroleum exploration geologist. The methods of correlation include recognition of index fossils and biozone limits, absolute-age determinations from radioactive decay rates, observation of eustatic changes in sea level, and comparison of gross lithologic character of the units.

Some of the more common difficulties encountered in all correlation work include discontinuity of stratigraphic units; structural complexity; lateral variations in thickness, lithology, and paleontology; poor development or absence of marker beds; presence of unrecognized unconformities; limited amount of control; multiplicity of nomenclature; and erroneously compiled data.

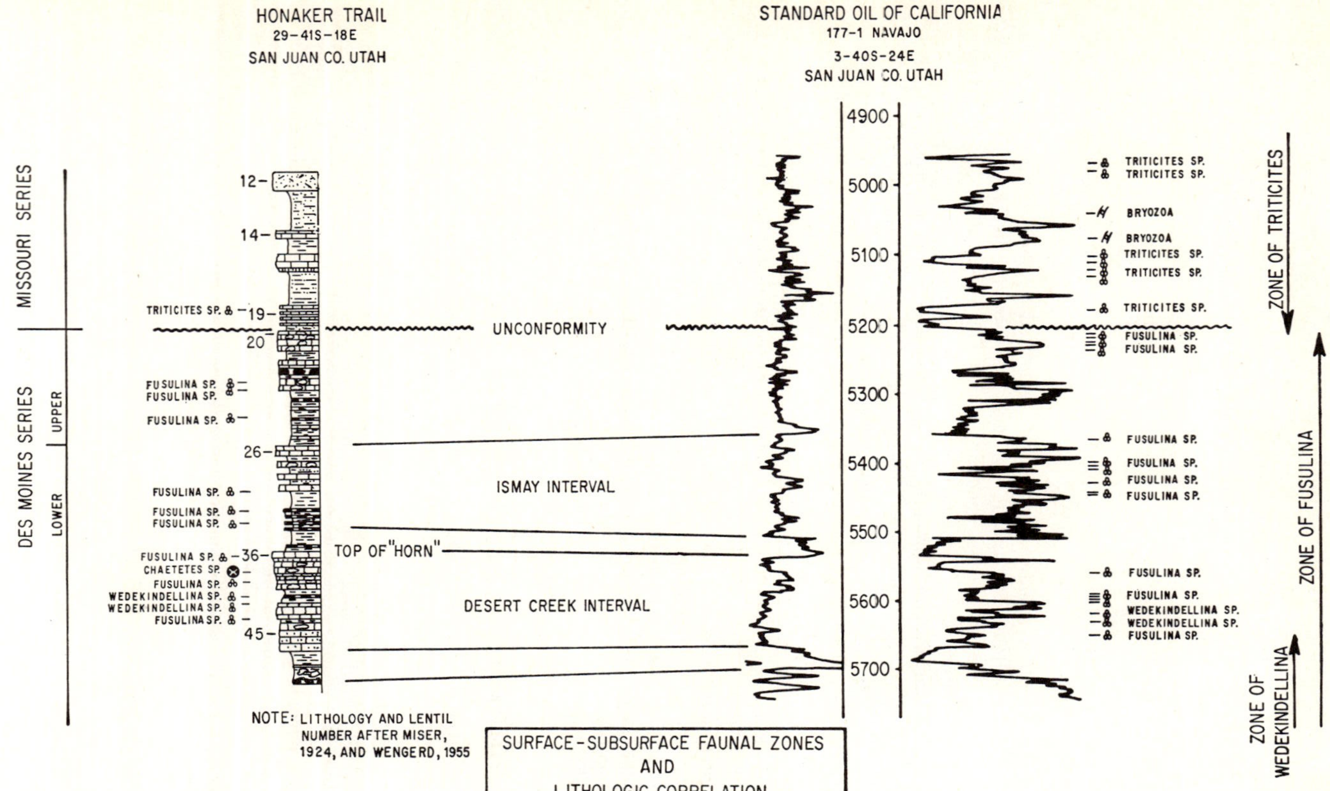

FIGURE 13-15. Example of correlation of biozones by identification of guide fossils. (*From J. E. Walsh, Faunizones of the Pennsylvanian and Permian Rocks in the Paradox Basin, "Guidebook to the Geology of the Paradox Basin," p. 155, Intermountain Association of Petroleum Geologists, 1958.*)

INTERPRETATION AND PRESENTATION OF SUBSURFACE GEOLOGIC DATA

Undoubtedly, the most important technique that a petroleum geologist must acquire is the ability to express structural and stratigraphic concepts graphically and to be able to interpret the graphic representations of others. These representations, generally having the form of maps or sections, present the nature of structural or stratigraphic features much more clearly than any amount of verbal description. The widespread applicability of maps for delineating geologic relationships early led to their almost universal use for depicting structural, stratigraphic, and sedimentary data. Among the first were structure maps to which contour lines were added to show the position of a rock surface, usually a coal bed or some other unit of economic importance, as it existed below the surface. Combinations of measured outcrop sections and carefully compiled well data make possible complete, three-dimensional studies of structural and stratigraphic units. It is impossible to exaggerate the importance of maps in subsurface geology. They often summarize the results of geological and geophysical investigations costing millions of dollars. More than 50 varieties of maps are presently used in oil exploration, reservoir engineering, regional stratigraphic analysis, and related fields.

Maps

The purpose of any map is to present data in an interpretable form. Topographic maps, geologic structure maps, isopach maps, and some lithofacies maps are quantitative. Each is a graphic integration of specific information presented in an understandable form. The subsurface geologist must weigh carefully all possible interpretations of the data at hand; many times his map may be merely a suggestion of probable relationships. It is never a mathematical solution of a problem. Usually subsurface data can be interpreted in a number of ways, and although only one solution is selected as preferable for the map, all possibilities must be considered and evaluated. The amount of error in subsurface interpretations may be reduced if values are made to fit known structural and stratigraphic patterns. The more geology the interpreter knows, the more successful his interpretation will be. The exploration geologist must understand sedimentation, stratigraphy, ecology, and structural concepts in order to interpret the rock data. However, he can seldom find an answer to subsurface problems without making geological imagination an important part of his interpretation.

Marker Beds

The fundamental element for structural and stratigraphic mapping is a stratigraphic horizon, commonly termed a "key horizon," "key bed," or "marker bed." The horizon is used for preparing subsurface structure maps, paleogeologic maps, and others which are based on the attitude or nature of a plane or surface. Many stratigraphic maps, however, provide information regarding a three-dimensional body of rock and require a stratigraphic unit of finite thickness. Definition of a stratigraphic unit requires the selection of two horizons which form the top and bottom of the three dimensional body.

Some commonly used kinds of marker beds are listed below. These are divided into three classes and include lithologic characteristics, biologic properties, and structural discontinuities which may be present.

Examples of lithologic key beds:
- Thin limestone bed
- Thin sandstone or conglomerate
- Coal or lignite bed
- Bentonite (volcanic ash) bed
- Insoluble residue zone
- Heavy mineral zone
- Characteristic electric, radioactive, or sonic log "kick"
- Seismic reflection surface

Examples of biologic markers:
- Teilzone (occurrence of an index or guide fossil)
- Biostratigraphic horizon (occurrence of a typical faunal assemblage)

Examples of structural discontinuities:
- Angular unconformity
- Disconformity
- Diastem

It is of the utmost importance that the selected key beds or markers be correctly correlated. The purpose of most subsurface maps is to show conditions within specific intervals of time or at some geologic "instant." Miscorrelation of the key horizons will result in a map cutting across the interval being studied. Correlation of the key bed is generally based on such criteria as its lithologic identity, position in sequence, identity of interval, etc., as discussed on preceding pages. It goes without saying that the accuracy of the map and of the interpretation presented therein depends in major part on the care and accuracy with which the data on which it is based are treated.

Structure Contour Maps

Accurate representation of structure is the objective of a large part of the subsurface geologist's work. The map generally used to show the structure of areas which may have important oil prospects is the *structure contour map*. Structure maps show the geometric shape of a stratigraphic unit, and the structure contour map is drawn to depict the configuration of a rock surface by means of contour lines passed through points on the same stratigraphic horizon which have the same elevation above or below a selected datum or reference plane. The commonly used datum is sea level, and the elevation of the key bed above or below sea level is indicated at each control point on the map. The elevation of the key bed with respect to the sea-level datum is the algebraic difference between the surface elevation of the well head and the drilled depth to the key bed. Thus, if the surface elevation at the well is 3,000 feet and the depth to the marker bed is 2,000 feet, the elevation of the key bed with respect to the datum is 1,000 feet. Some confusion regarding use of the word "datum" may be experienced. The terms "datum" and "datum horizon" are used by many geologists with the same meaning as key horizon or marker bed. However, as applied to structure maps, datum usually means the base to which the elevations shown on the structure map are referred.

If the drill hole is crooked, the apparent depth to the key bed will be either too great or too small. Figure 13-16 shows two examples of holes deviated from the vertical. In *A*, the hole is inclined downdip and has penetrated the marker bed at *a*. If one is unaware that the hole is crooked, the marker bed would appear to have been penetrated at *b* and the apparent dip would be the reverse of the actual dip. In *B*, the hole is inclined updip, and the resulting apparent dip is much steeper than the real dip. Unless a directional survey has been made, it is impossible to adjust the log of a crooked hole to obtain a correct control point. However, in general, it is wise to use with caution data from wells in areas of steeply dipping beds whose inclinometer surveys show deviations from the vertical exceeding 2 or 3 degrees.

The first step in making a structure contour map is selecting the scale and the contour interval. The scale is determined partly in order to make the map of convenient size, partly in order to bring out important details of the features to be presented, and partly as a result of the accuracy of the surveying methods. The choice of the structure contour interval depends on several factors; the size of the errors in elevation, the types of structures to be mapped, and the purpose for which the structure map will be used, all have a bearing. Where the dips are steep and the scale of the map fairly small, a large interval should be used to avoid close spacing of the contours. The size of the structures is also an important factor. If the crests of anticlines are only 25 to 35 feet above the adjacent synclinal axes, a 50-foot structure contour obviously would show little structure and a 10-foot interval should be selected. On the other hand, if the anticlines rise thousands of feet above the adjacent synclines, an interval of 10 or 50 feet would clutter the map with meaningless lines and a contour interval of 100, 200, or even 500 feet would be more satisfactory. The structure contour interval should be at least two or three times the probable error in the elevations of the control points. This is necessary to eliminate the appearance of apparent structures due merely to the errors.

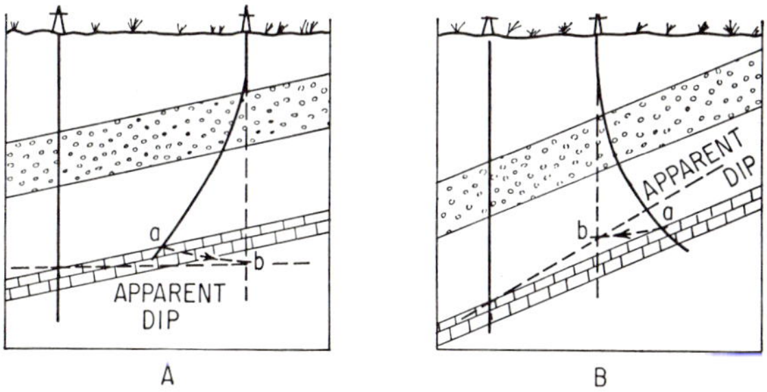

FIGURE 13-16. Effects of crooked or deviated holes on apparent datum elevations and interpreted dips.

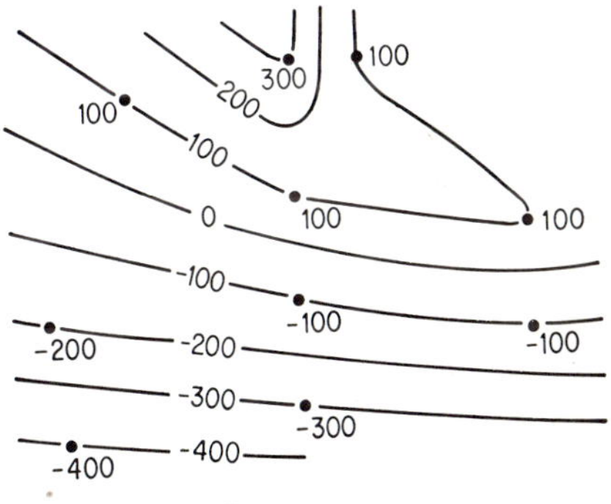

FIGURE 13-17. Structure contour map contoured by mechanical spacing method.

Data available for subsurface structure mapping are rarely, if ever, sufficient for accurate contouring; therefore, structure contour maps are subject to the factor of individual interpretation. Contouring can be done mechanically (Figure 13-17). By this method the vertical difference in elevation between two control points is distributed evenly according to the number of units required by the chosen contour interval. The structural pattern develops as the contours are drawn connecting the points of equal elevation. If the control points are situated on opposite limbs of a fold, or if local anomalies cause strike-and-dip variations, the resulting patterns can be very misleading. Purely mechanical contouring is feasible where sufficient information is available to determine with reasonable accuracy the highest and lowest points as well as all changes in the rate of dip for a given area. Sufficient data of this sort are commonly available only in drilled-up oil or gas fields.

Another method of contouring is to space the contours evenly over all or broad areas of the map (Figure 13-18). If a few wells are close together, a rate of dip is established from their relationship and used for spacing the contours. Through use of this method high and low areas can be drawn that are not based on any established points but that result from fitting the number of contours required to maintain the established dip into the available space. Equal spacing of contours is based on the assumption that if the rate of dip can be determined from three known points, it will continue over broad areas without change. Inasmuch as the rate of dip may have

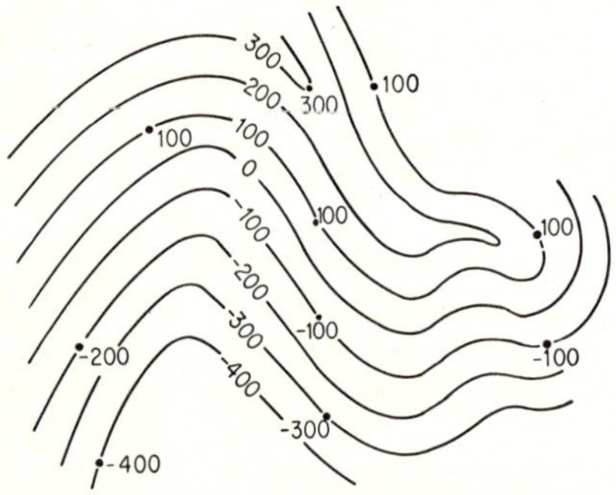

Figure 13-18. Structure contour map based on same data as Figure 13-17 but contoured by equal spacing method.

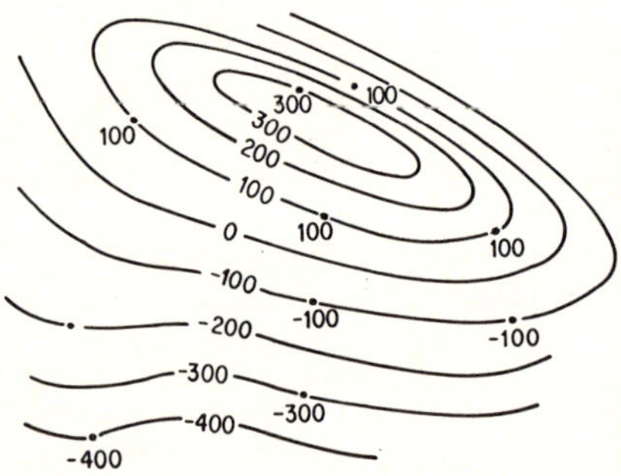

Figure 13-19. Structure contour map based on same data as Figures 13-17 and 13-18 but contoured by interpretive contouring technique.

been established on a nose or from wells on opposite limbs of a fold, it is apparent that this method should be used with caution.

A knowledge of the general character and form of structures in the region aids greatly in correctly interpreting the subsurface structure where the well control is sparse. When the character of folding is known, an attempt should be made to contour the data so that the structural pattern will bear out regional trends or tendencies. This technique may be termed interpretive contouring (Figure 13-19). If the control is widely spaced, it is obvious that there may be variations in the interpretation of the data. In general, the more sparse the data, the more varied may be the possible interpretations.

A few rules for contouring numerical data on a map must be observed; they are specified in the following five paragraphs.

1. A contour line of given value must everywhere pass between points whose numerical values are higher and lower, respectively, than that of the contour. For example, two points having elevations of 60 to 40 feet cannot be on the same side of the 50-foot contour.

2. A contour line cannot cross itself or any other contour. Contours represent definite values, and inasmuch as each line represents a different value, the lines cannot cross. There are two apparent exceptions to this rule: overturned or recumbent anticlines and reverse faults. However, in space these lines would not be in contact but would be one above the other. In practice, the underside of a recumbent anticline

and that part of the datum bed lying below a thrust sheet are omitted on a structure contour map in order to avoid the confusion of lines which would result. Occasionally these portions are contoured using dashed or dotted lines.

3. Contour lines may not merge with contours of the same or of different values. Where the contoured bed is vertical or where faulting has displaced it along strike by an amount equal to or exceeding the contour interval, the contours may appear to be merging. In space, however, the lines are not in contact but are one above the other.

4. Contours must close or end at the edge of the map. Small hills within the boundaries of a map will always appear as a series of closed contours. If only part of a hill is shown, the direction of slope may be all in one direction, and the contours will end at the edge of the map.

5. Contours of the same value must be repeated to indicate a reversal of direction of slope. For example, groups of high values surrounded by low values are segregated within a closed contour. However, if the high values are flanked by low values on

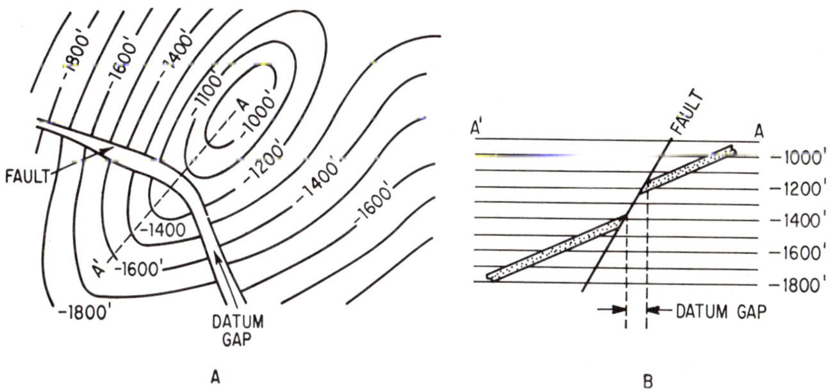

FIGURE 13-20. Structure contour map *A* and cross section *B* of a faulted dome, showing datum gap.

two sides only, separation of the groups is indicated by a repetition of contours on each side of the high values.

Contouring faulted beds presents additional problems. On contour maps a fault is commonly drawn as a single heavy line. However, a normal fault with a low-dipping plane results in a zone where the key horizon is absent. This zone is known as a "datum gap." The width of the gap depends on the degree of dip of the marker bed surface, the dip of the fault plane, the amount of throw, and the angle between the strike of the marker bed and of the fault plane (Figure 13-20). The datum gap can be worked out on a structure contour map if there are sufficient control points to determine the general contouring of the structure and at least three wells not in a straight line that have penetrated the fault plane.

For reverse faults, no datum gap is present and the contours on the upthrown block overlie the contours on the downthrown side (Figure 13-21). To avoid confusion, it is sufficient to draw the fault trace as a heavy line on the upthrown side of the fault.

Faults in the subsurface usually are recognized by inference only unless a well actually penetrates the fault plane and establishes a cut point. On a structure contour map faulting may be suggested by an abrupt increase in the rate of dip, by an abrupt change in the direction of strike, or sometimes by an elongated closed low.

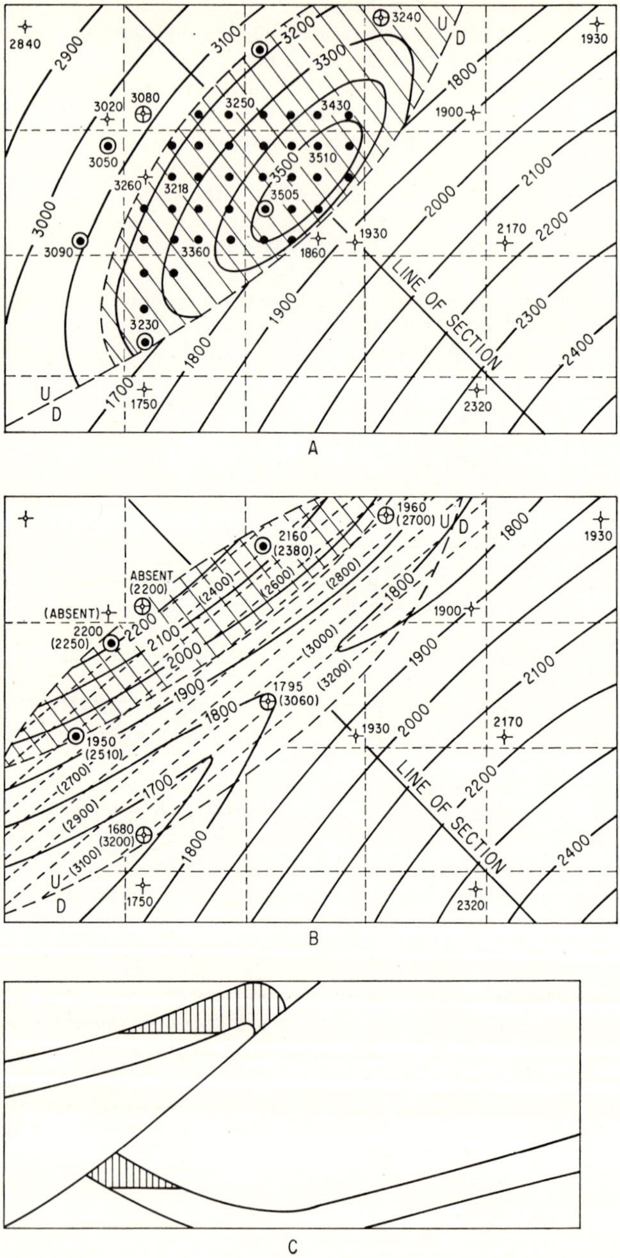

FIGURE 13-21. Structure contour maps and cross section of a feature cut by a high-angle reverse fault. *A*, Contours above the fault plane. *B*, Contours on and below the fault plane. *C*, Cross section showing dip of datum plane and fault. (*From J. W. Low, Subsurface Maps and Illustrations, p. 464, in Haun and LeRoy, 1958.*)

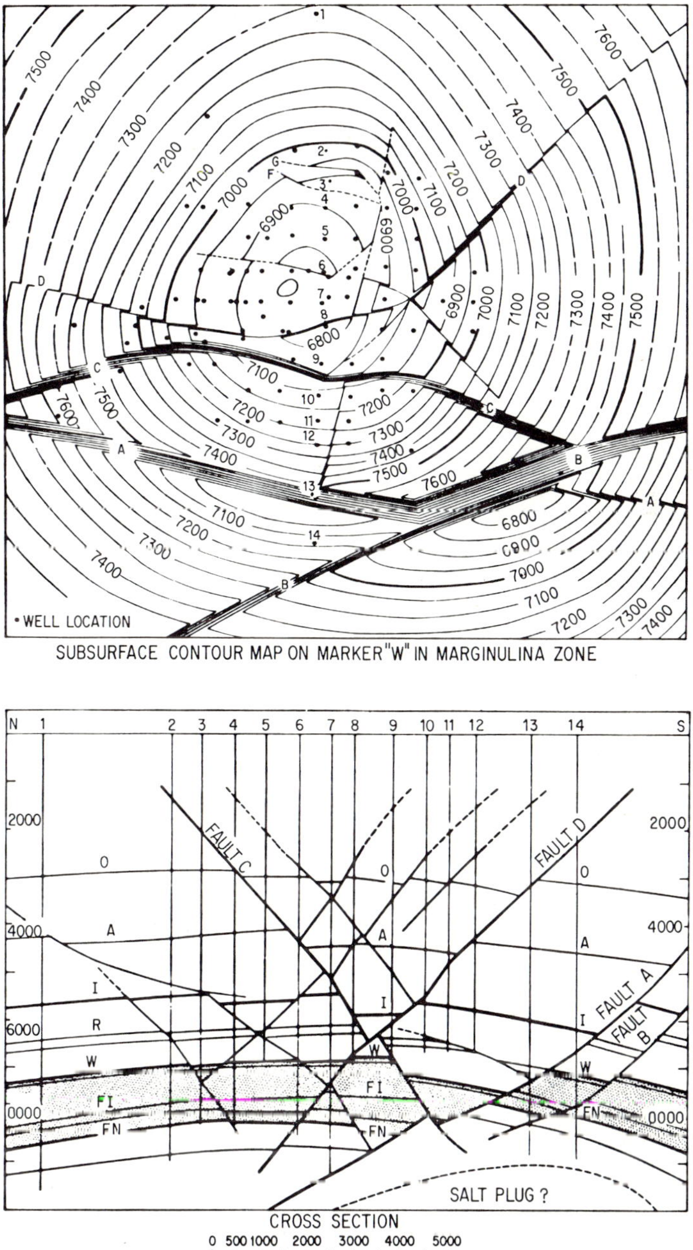

FIGURE 13-22. Structure contour map and cross section of Iowa Dome, Calcasieu and Jefferson Davis Parishes, Louisiana.

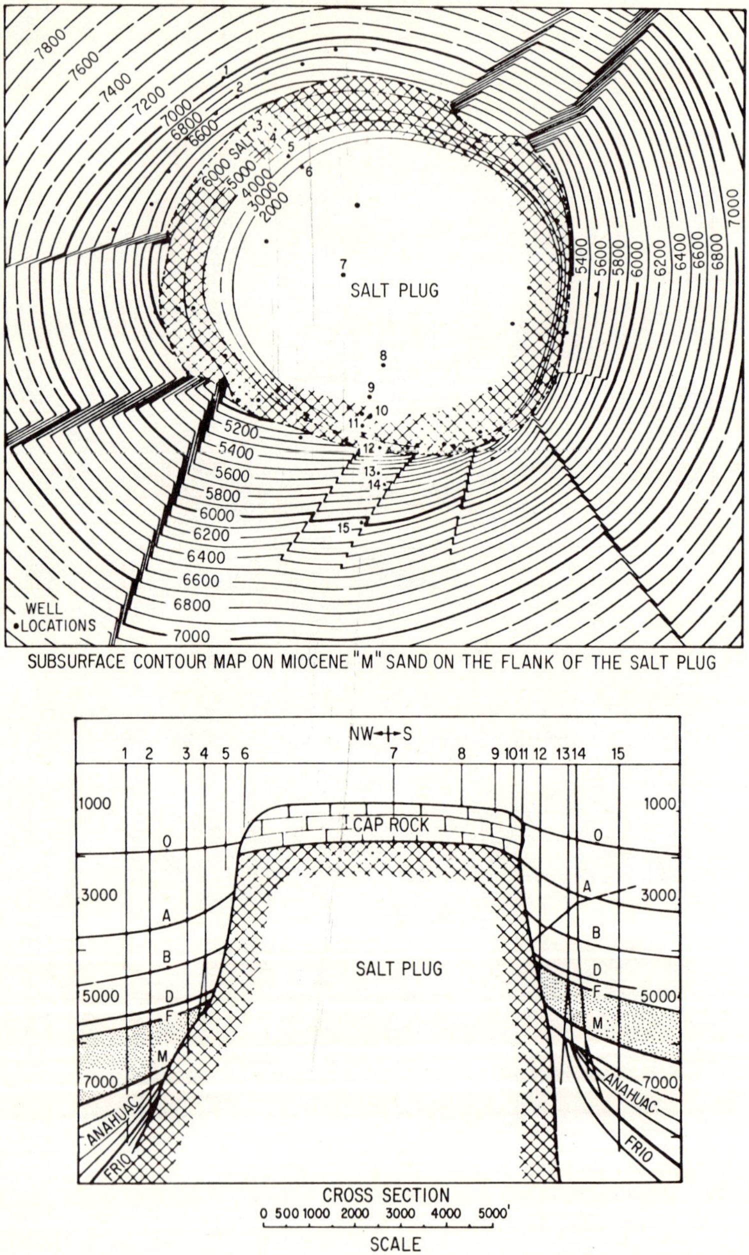

FIGURE 13-23. Structure contour map and cross section of Black Bayou salt dome, Cameron Parish, Louisiana.

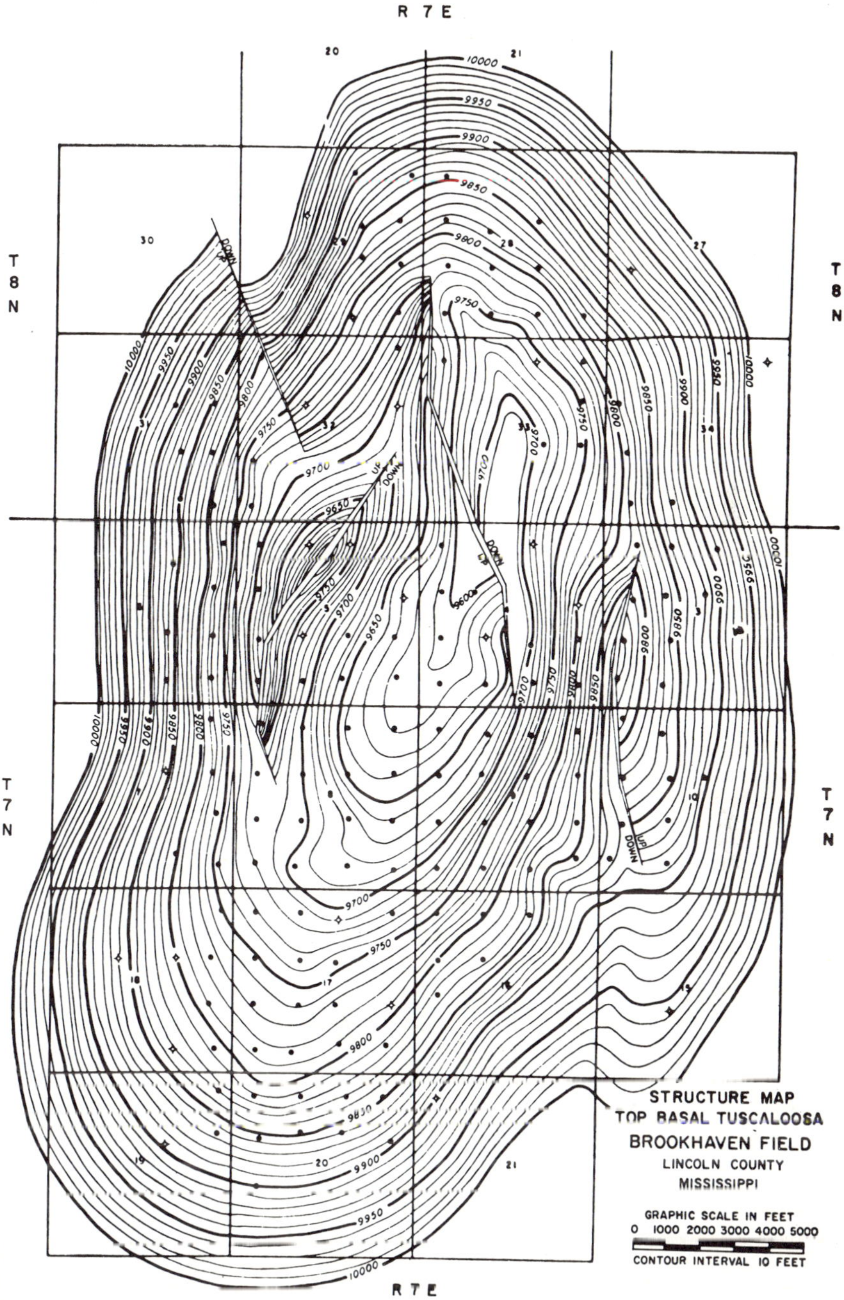

FIGURE 13-24. Detailed structure contour map of Brookhaven field, Lincoln County, Miss. (*From R. Womack, Jr., Brookhaven Oil Field, Lincoln County, Mississippi, Bull. AAPG, vol. 34, p. 1522, 1950.*)

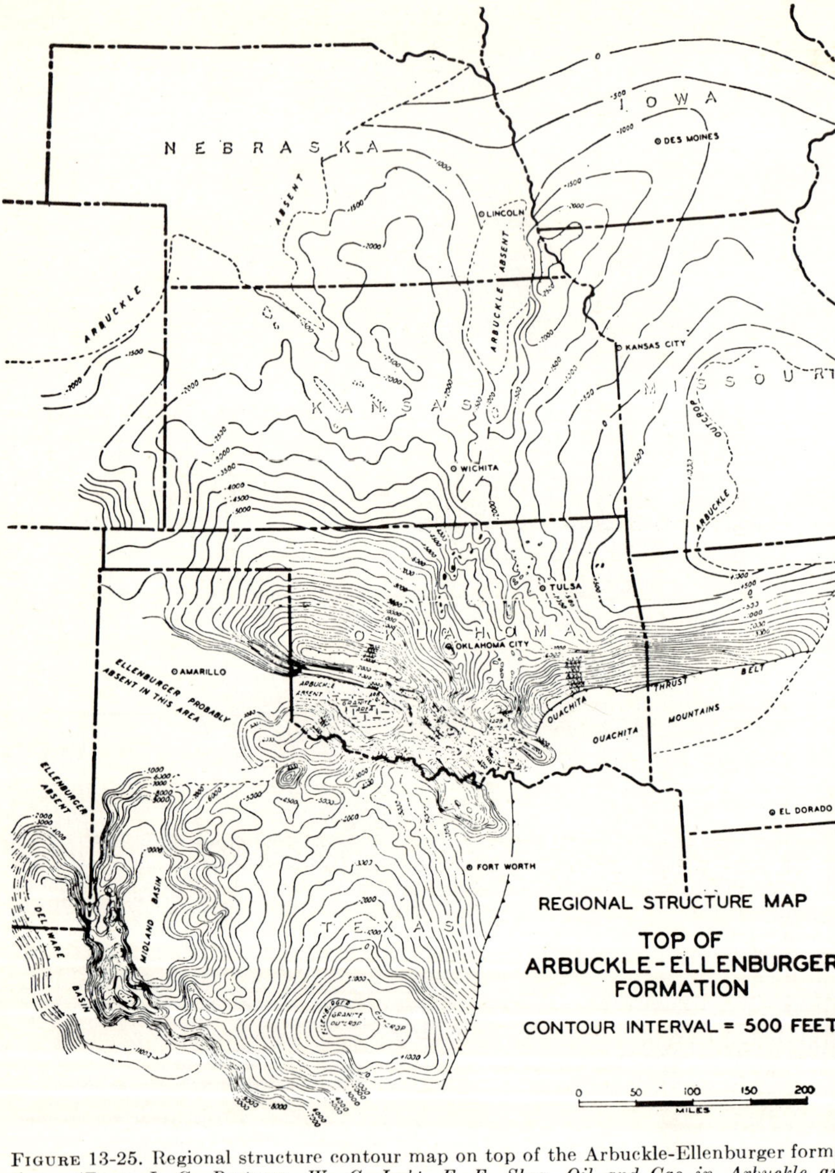

FIGURE 13-25. Regional structure contour map on top of the Arbuckle-Ellenburger formation. *(From J. G. Bartram, W. C. Imbt, E. F. Shea, Oil and Gas in Arbuckle and Ellenburger Formations, Mid-continent Region, Bull. AAPG, vol. 34, pp. 690–691, 1950.)*

Such interruptions in normal contouring are regarded as evidence of the probable presence of faulting; however, they are not proof, and the data on which the faulting is based should be carefully considered.

Every geologist should be able to make a structure contour map from strike-and-dip determinations, and he should also be able to determine the strike and dip from structure contour maps. It is quite simple to determine the dip and strike at any point on a structure contour map. The strike is parallel to the structure contours, and the dip is determined by the spacing. The rate of dip between two points is found by dividing the difference in elevation between the two points given by the structure contours by the map distance between them. The result is the tangent of the dip angle. This tangent, multiplied by 100, gives the percentage slope or grade.

The first step in making a structure contour map based on strike-and-dip data is to draw in the anticlinal and synclinal axes. These show where the contours must pass around folds. Faults, if present, should also be indicated. Next, a series of cross sections at right angles to the strike can be constructed. Elevations of key horizons at points where they intersect the structure contour levels on the sections are transferred to the map. Between these cross sections the structure contours are sketched in. It is helpful to remember that the contours must everywhere be parallel to the strike and that the distance between the structure contours can be determined for each dip by dividing the structure contour interval by the tangent of the dip angle.

A common example of the need to change a map in which the structure is shown by local dip measurements to a structure contour map is found in seismic dip surveys which commonly yield a large number of strike and dip measurements. It is generally desirable to change these to structure contours drawn on a phantom horizon. Phantom horizons are imaginary key beds used in making structure contour maps where it is not possible to draw the structure contours on real key horizons. It may be impossible to use real key horizons either because none is present or because the relation of the strike and dip determinations to the real key horizons is not known. The phantom horizon is assumed to have a definite elevation at some point at depth, and structure contours are drawn as explained in the preceding paragraph. Although the key horizon is imaginary, the structure it shows is real except for errors made during the survey. The structure contour map on the phantom horizon shows the same structure as would a real key horizon at the same depth.

Figures 13-22, 13-23, and 13-24 are included as excellent examples of subsurface structure contouring where sufficient control is available. Figure 13-25 is an example of regional structure contouring.

Strike lines and form lines are sometimes used to indicate the structure of an area where something is known about the structure but not enough to justify drawing ordinary structure contours. Strike lines and form lines are similar to structure contours. They are parallel to the strike, and they are closer together where the dip is steeper. They differ from structure contours in that they have no elevations, they may not be uniformly spaced, and they need not obey rules of contouring except that they should not cross themselves or other strike or form lines. Sometimes strike lines or form lines are used to indicate the structure in areas where the principal source of information consists of determinations of local dip. If the dips were shown without strike lines, the map would be difficult to interpret. Often they are of great help in locating structural axes in areas of confused dip data.

Isopach Maps

An *isopach* is a contour line drawn through points of equal stratigraphic thickness, and an *isopach* (or *isopachous*) *map* is one which shows by means of isopachs variations in the thickness of some stratigraphic unit from place to place. Isopach maps may show variations in the thickness of a formation, a group of formations, the strata belonging to a given geologic period or epoch, a lithologic unit such as a coal seam or a sandstone or limestone bed, or they may show merely variations in the interval between any two key horizons or beds that can be correlated. The area covered by an isopach

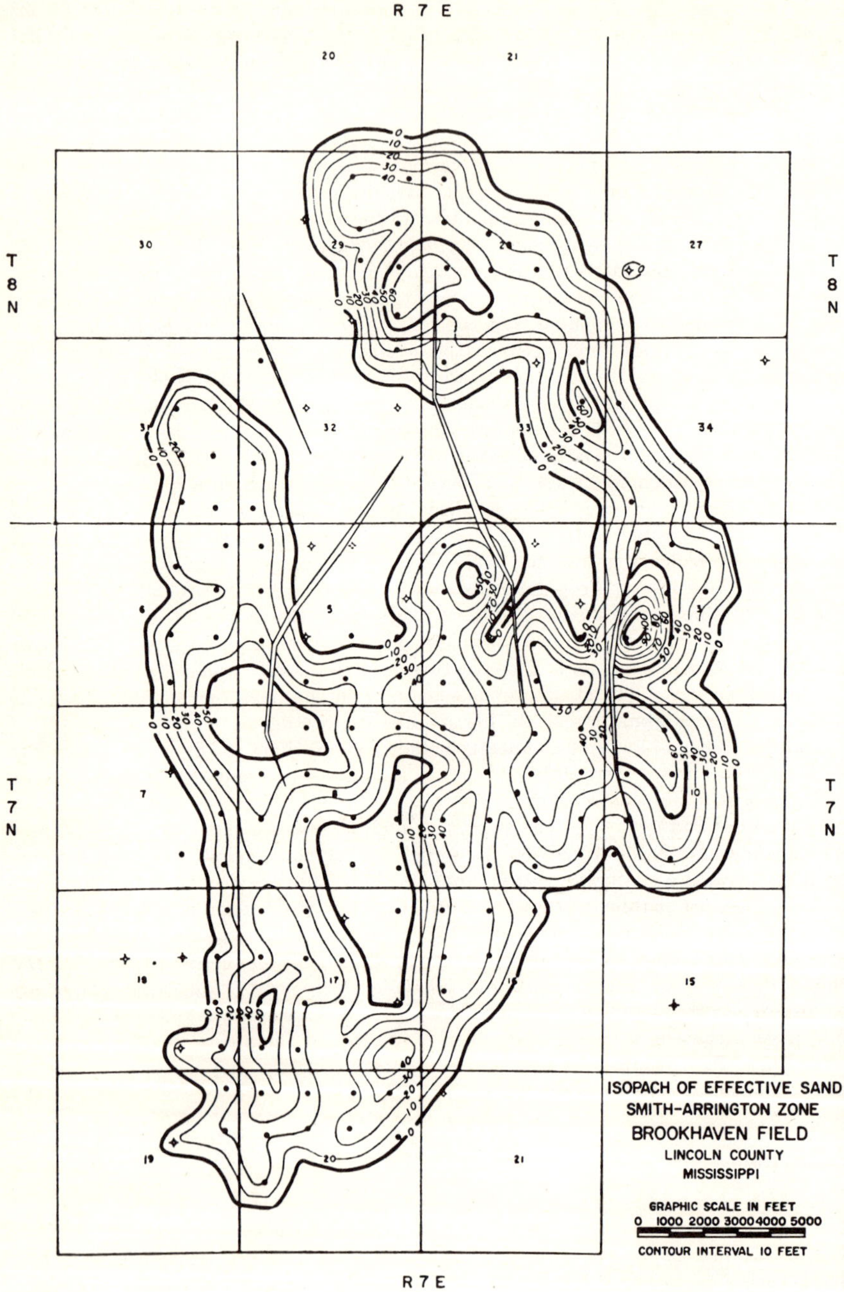

FIGURE 13-26. Example of an isopach map covering a limited area. Isopach map of effective sand, Smith and Arrington zones, Brookhaven field. (*From R. Womack, Jr., Brookhaven Oil Field, Lincoln County, Mississippi, Bull. AAPG, vol. 34, p. 1526, 1950.*)

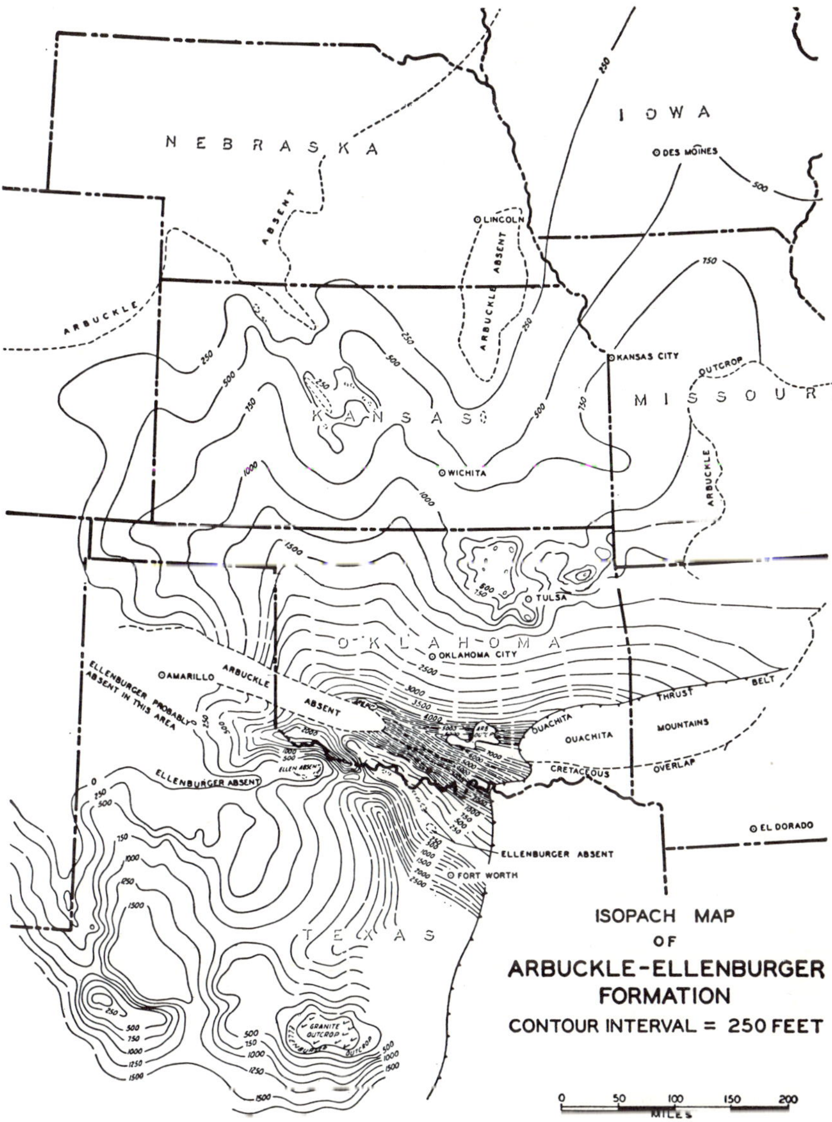

FIGURE 13-27. Regional isopach map of the Arbuckle-Ellenberger formation. See Figure 13-25 for structure contour map covering the same area. (*From J. G. Bartram, W. C. Imbt, and E. F. Shea, Oil and Gas in Arbuckle and Ellenburger Formations, Mid-continent Region, Bull. AAPG, vol. 34, pp. 686–687, 1950.*)

map may be small, such as a single pool or field (Figure 13-26), or it may be very large for regional studies (Figure 13-27) covering structural basins or provinces including thousands of square miles.

Isopach maps are constructed in much the same way as structure contour maps. Once the stratigraphic unit for isopaching is selected, the observed thicknesses are plotted at the control points on the map and the isopach lines drawn through points of equal thickness. The vertical interval between isopachs may be selected from several hundred feet on a regional map to intervals as small as 10 feet for detailed studies of thin stratigraphic units.

A common source of error in subsurface isopach maps results from steeply dipping strata at the point where the well is drilled. Obviously, a correct stratigraphic interval will be measured in a straight hole only where the beds are flat-lying. If the beds are dipping, the well records will show a drilled thickness that is greater than the correct stratigraphic section. Since most wells are drilled on structures, appreciable dips are encountered quite commonly. The apparent thinning of the section found on the tops of many structures is probably often only a result of this condition. If beds are dipping more than 450 feet per mile (about 5 degrees), the data used in detail maps should be corrected. The true stratigraphic thickness is equal to the drilled thickness multiplied by the cosine of the dip angle.

A subsurface map showing by contours the drilled thickness of a stratigraphic interval, without regard to true stratigraphic thicknesses, is correctly termed an *isochore map* (Figure 13-28). The term is not commonly used, but occasionally the distinction is made when data of regional extent are being presented in areas where little is known about local structural conditions. Where the dip is low-angle, the difference between isopach and isochore maps is insignificant.

An isopach map differs from other types of contour maps in that its datum is at the top of the stratigraphic unit. A model of an isopach map would have a flat top with the varying stratigraphic thicknesses extending down from it. By this convention the isopach map shows the structural attitude of the lower marker with respect to the upper horizon taken as a level plane. This feature is of inestimable value in reconstructing the structural history of a region. An isopach map based on the interval between two angular unconformities shows the structure of the lower unconformity at the time the upper unconformity was developed. Similarly, an isopach map based on an interval with a conformable top shows the structure of the lower surface during the time of deposition of the unit. A series of isopach maps using successive stratigraphic units makes it possible to trace the structural changes which took place in the area.

Isopach maps are useful in the evaluation of areas of maximum subsidence during an interval of sedimentation. A map prepared on a conformable interval with no unconformities in the section will indicate the relative subsidence of the depositional site during the stratigraphic interval selected. The upper and lower key beds or horizons selected must be assumed to have been horizontal depositional surfaces at the time of their formation.

Isopach maps can help in restoring the original depositional edge of a stratigraphic unit. In some cases the present-day zero ispoach may represent the ancient shore line. More often, however, the shoreward parts of the formation have been eroded away. Successful extrapolation of the isopach map unit to its original depositional edge depends on understanding the sedimentary facies now present near the zero line and their probable placement with respect to the ancient shore line. The isopach map combined with facies maps will provide integrated data permitting more critical evaluation of the evidence.

Isopach data are used to locate buried structures in regions where deposition was less, resulting in thinner formations, over the crests of structures. They can also be used to locate the highest portions of areas which were positive at the time of original deposition. In many areas, particularly biohermal reef areas, deposition was thinnest in the immediate postreef sediments over the top of the ancient structures. However, later tilting caused the crest of the structure to migrate up the regional dip from the original high point.

Where formations or groups of formations thin or converge more or less uniformly in one direction, the isopach maps are often called *convergence maps.* The thinning may be due to one of several causes, such as offlap, onlap, erosion and truncation, bypassing, or a lesser supply of sediments in the direction of thinning. One of the important

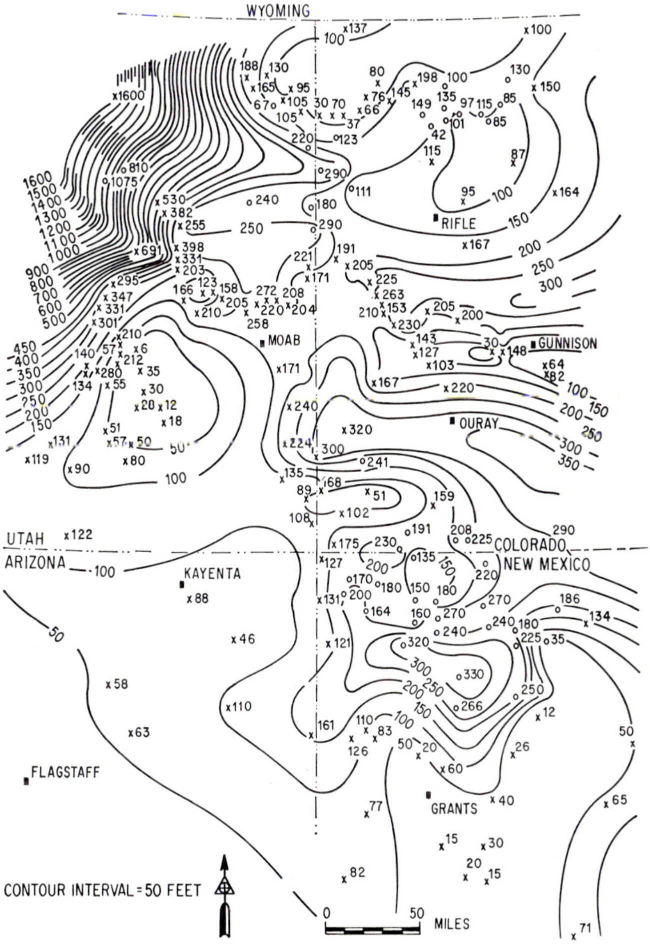

FIGURE 13-28. An isochore map, showing by contours the drilled thickness of a stratigraphic interval. (*From R. G. Young, Dakota Group of the Colorado Plateau, Bull. AAPG, vol. 44, p. 101, 1960.*)

effects of converging strata is the shifting of the position of folding with depth and increasing or decreasing its intensity.

Isopach maps are of assistance in preparing prognoses for wildcat wells by helping to determine in areas of limited deep control the depth to specific formations and to predict the structural attitude of such formations by contouring a shallower bed for which adequate control is available. If the upper and lower horizons are parallel, the structure of the deeper bed has the same attitude as the upper one, and it is necessary only to add the thickness of the intervening section to determine the elevation of the deeper unit. If, however, the horizons are not conformable, an isopach map of the interval between the two is necessary to show that the deeper bed is affected by variations in thickness. First, a structure contour map is drawn on the upper key bed (Figure 13-29*A*). Next, an isopach map showing the thickness of the section between the upper and the lower beds at the same scale as the structure map is superimposed over the structure map (Figure 13-29*B*). The position of each point of intersection of the two sets of contours is plotted on a third overlay sheet. The elevation of the upper bed is then added to the thickness figure, and the sum entered at each point of intersection. These data represent the estimated elevations for the deeper horizon, and they are contoured to produce a structure contour map (Figure 13-29*C*) of the probable attitude of the bed.

Isopach data are also used in conjunction with other information such as porosity, water saturation, etc., to calculate the volume of oil or gas in a reservoir, the size of the reservoir, and other factors of importance to reservoir engineering.

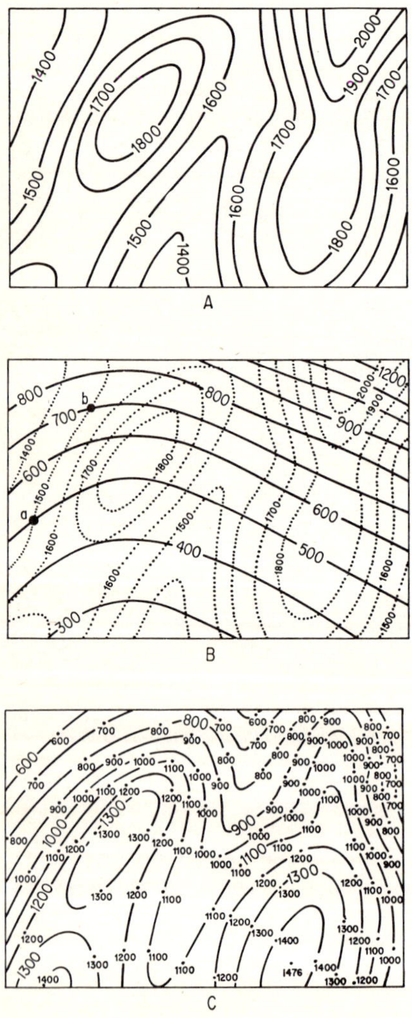

FIGURE 13-29. Steps in constructing a structure contour map of a deep horizon from a shallower bed for which adequate control is available. *A* is a subsurface contour map of the shallower bed. *B* shows isopach data of the shallower bed superimposed on the structure map. Point *a* is the intersection of the 1,500-foot structure contour and the 500-foot isopach contour. At this point the top of the deeper bed is 500 feet below the shallower datum at an elevation of 1,000 feet. At point *b* the deeper bed is 700 feet below the structure datum and its elevation, therefore, is 800 feet. All intersections are reduced to the deeper elevations in this manner. *C* is the subsurface structure contour map on the top of the deep bed resulting from contouring the intersections in *B*. (*From J. W. Low, Subsurface Maps and Illustrations, pp. 470–472, in Haun and LeRoy, 1958.*)

Facies Maps

The rock characteristics in practically all stratigraphic units change in some manner from one part of a basin to another. The degree of variation and rate of change may be great or small, depending upon such factors as the size and configuration of the basin and adjacent terrane, the climate, the chemistry of the waters, etc. A single formation may be a conglomerate at one locality, a sandstone or shale at another, a limestone or anhydrite at a third, and perhaps a combination of three or more lithologies at still another place. The changing patterns of sedimentation within a genetically related stratigraphic unit reflect differences in both the tectonic framework and the local environment, and it is, therefore, of utmost importance to the subsurface geologist in working out geologic history to determine the nature of these variations and where they occur and to be able to show the changes on maps.

A *facies map* shows the areal variation in the over-all aspect of a stratigraphic unit. The aspect may be expressed in terms of lithology (lithofacies maps), faunal content (biofacies maps), or tectonic implication (tectofacies maps). Facies can be distinguished and mapped qualitatively or quantitatively. Qualitative maps commonly employ patterns to identify more or less indefinite facies. They are used when the section contains a great number of variables or if the area being mapped is large. The simplest qualitative map is made by plotting the dominant lithology of a stratigraphic unit at a number of control points and segregating the areas of similar aspect by drawing boundary lines around them (Figure 13-30). This type of map indicates the major lithologic, biologic, or tectonic changes which took place during the deposition of a stratigraphic unit, but it does not show the rate at which the changes occur.

If an unconformity is used for the upper boundary of the mapping unit, errors in interpretation of the qualitative map may result, because the distribution of the lithology is an erosional effect and does not indicate the depositional environment. Another source of error in the interpretation of a qualitative map is caused by intertongued or interbedded members of a thick section which may cancel the characteristics of one another. The average lithology of such a rock body often leads to erroneous conclusions.

The quantitative facies map shows by means of contour lines the areal distribution of some measurable characteristic of the mapping unit. The number of variables that can be shown is limited, but the contour method has the advantage of showing the rate at which change occurs regardless of the variable being analyzed. The choice of whether to map the thickness, the percentage, or the ratio of one component to another generally depends on the nature of the stratigraphic unit and the size of the area being studied. Ratio maps are commonly used in regional work, thickness maps in detailed analysis, and percentage maps for either scale study. Because of the close relationship between facies and thickness, the most suitable base map for any one of the quantitative analyses is the isopach map of the same stratigraphic unit, permitting comparison of the thickness, percentage, or ratio of individual components to the thickness of the entire unit.

Lithofacies maps, as the name suggests, are maps of facies differentiated on the basis of lithologic characteristics. The simplest lithofacies map is the *isolith map*. An isolith is a line denoting the aggregate thickness of beds of only one lithology in a stratigraphic succession composed of one or more lithologies. Where only one lithology is present, an isolith and an isopach of the same stratigraphic interval are identical. An isolith map is a contour map depicting the thickness of an exclusive lithology over the map area. In a section composed of sandstones and shales, two isolith maps are required to show the total thickness distribution, whereas only one isopach map

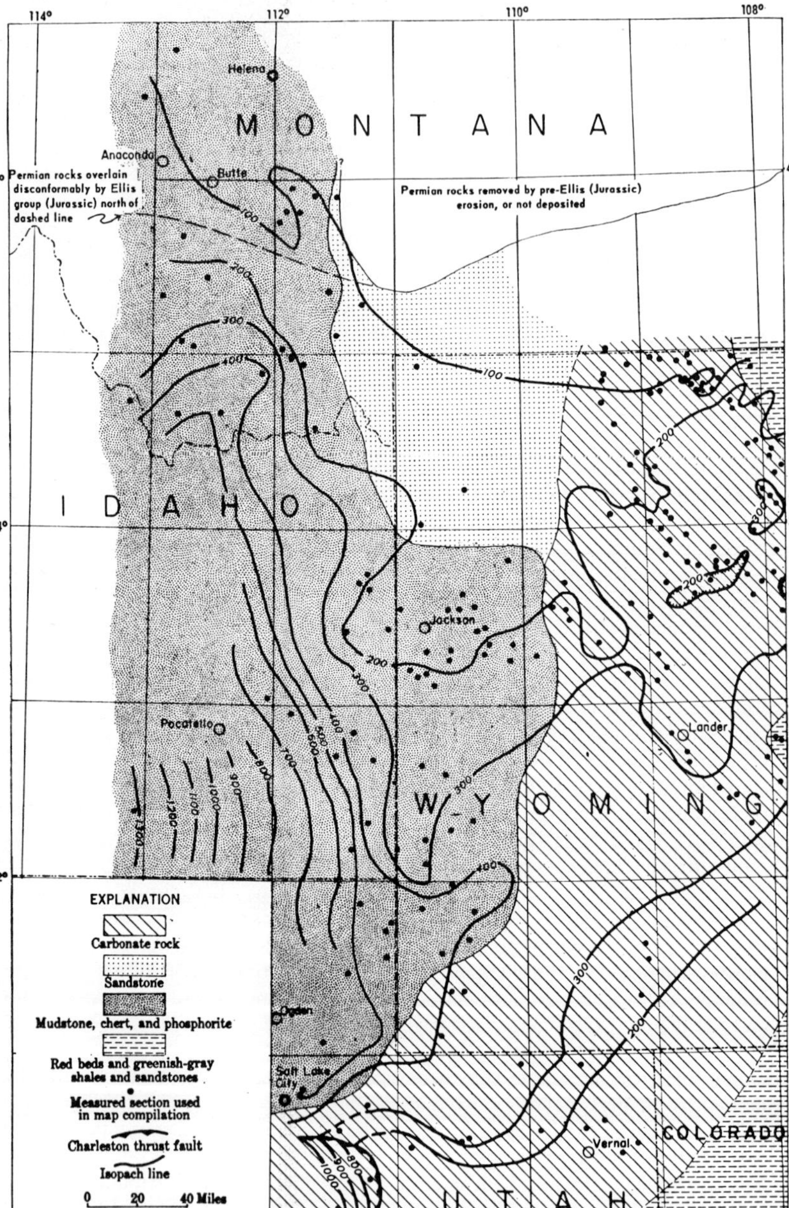

FIGURE 13-30. Dominant lithology of the Phosphoria formation and its stratigraphic equivalents. A simple qualitative facies map made by plotting the dominant lithology of the stratigraphic unit and segregating the areas of similar aspect. (*From V. E. McKelvey et al., The Phosphoria, Park City, and Shedhorn Formations in the Western Phosphate Field, U.S. Geol. Survey Profess. Paper 313-A, 1959.*)

is needed. The sum of all the thicknesses shown on a complete set of isolith maps of a given stratigraphic interval equals the isopach map of the same interval.

Figures 13-31 through 13-35 (from Low in Haun and LeRoy, 1958) present an isopach map of a basin of sedimentation together with a cross section illustrating the complex relationships of the sediments occupying the basin, followed by isolith maps of the sandstone, shale, limestone, and evaporite components comprising the sediments. Note that in each isolith map qualitative data have been added to the quantitative isolith aspect. There are many other qualitative aspects which might be differentiated within each of the lithologic groups depending, perhaps, on local conditions.

A special adaptation of the isolith map is the *isofacies map*. This map is designed to show the change in lithologic character within a facies. Construction of such a map requires first delimiting the area and stratigraphic interval of the facies, then treating this portion of the formation as if it alone were a formation and constructing the required isopach and isolith maps.

A common type of facies map is made by drawing contour lines to connect points at which the ratio between two variables is the same. In these *ratio maps* two variables are thus delineated by one set of contour lines, and other sets can be used if it is desirable to show more components. The ratio of sandstone to all other rock types in the section provides a map showing the relative importance of sandstone in the interval. The ratio values range from infinity (a section composed entirely of sandstone) to zero (no sandstone). The value 1.0 means that the amount of sandstone in the section is equal to the sum of the other sedimentary types.

Combined lithologic ratios provide methods for contrasting detrital and nondetrital rocks in the section. The clastic ratio and the sand-shale ratio are examples.

$$\text{Clastic ratio} = \frac{\text{thickness of clastics}}{\text{thickness of nonclastics}}$$

$$\text{Sand-shale ratio} = \frac{\text{conglomerate} + \text{sandstone}}{\text{siltstone} + \text{shale}}$$

The clastic ratio shows the proportion of detrital clastic rocks and locally formed nonclastic rocks which occurs in the unit and thus differentiates between areas of chemical activity and those of mechanical deposition. The sand-shale ratio indicates the proportion of coarse to fine detrital components, without regard to the nonclastic rocks in the section, and thus differentiates areas of turbulent near-shore deposition from those of quiet- or deep-water environment. In an evaporite basin comparison among limestone, dolomite, and anhydrite is more significant than a comparison between clastics and nonclastics; hence, the ratio of anhydrite to carbonates and of limestone to dolomite is substituted for the clastic and sand-shale ratios.

Both the clastic and sand-shale ratios range from zero to infinity. The clastic ratio is infinity when there are no nonclastic rocks in the section, and its value is zero when there are no detrital rocks. The value 1.0 represents equal abundance of clastic and nonclastic rocks. Similarly, the sand-shale ratio is infinity when no shale is present and zero when there is no sandstone. The value 1.0 represents equal amounts of coarse and fine clastics.

Clastic ratio and sand-shale ratio maps are prepared by drawing lines of equal ratio value between control points at which are plotted the values of the ratios. Because of the wide spread from zero to infinity, a convenient method of contouring is to draw the line of value 1.0, followed by values 2, 4, 8, etc., on one side and ½, ¼, ⅛, etc., on the other (Figure 13-36). This provides a geometric contour interval with each succeeding contour doubling or halving the ratio.

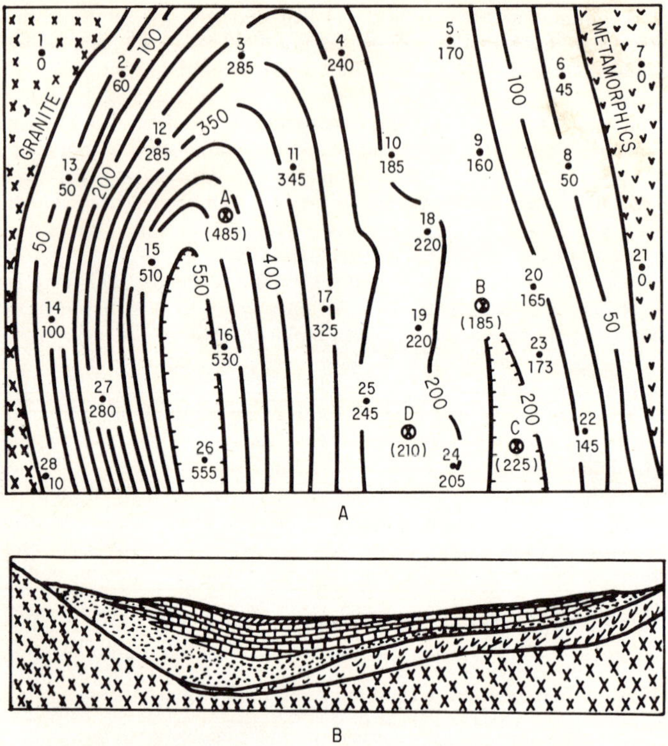

FIGURE 13-31. *A*, Isopach map of a basin of sedimentation. *B*, Cross section showing the complex relationships of the sediments. *(From J. W. Low, Subsurface Maps and Illustrations, p. 487, in Haun and LeRoy, 1958.)*

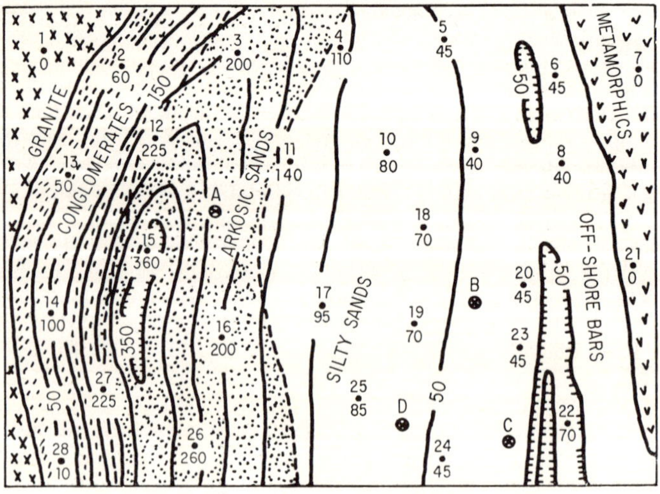

FIGURE 13-32. Sandstone isolith map of the basin shown in Figure 13-31. *(From J. W. Low, Subsurface Maps and Illustrations, p. 488, in Haun and LeRoy, 1958.)*

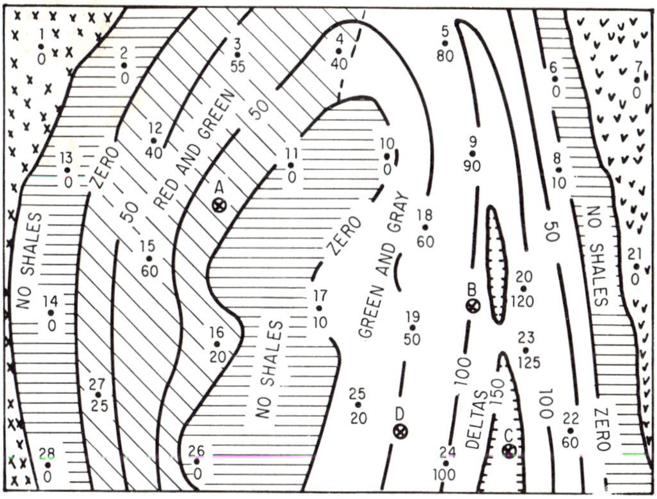

FIGURE 13-33. Shale isolith map of the basin shown in Figure 13-31. (*From J. W. Low, Subsurface Maps and Illustrations, p. 489, in Haun and LeRoy, 1958.*)

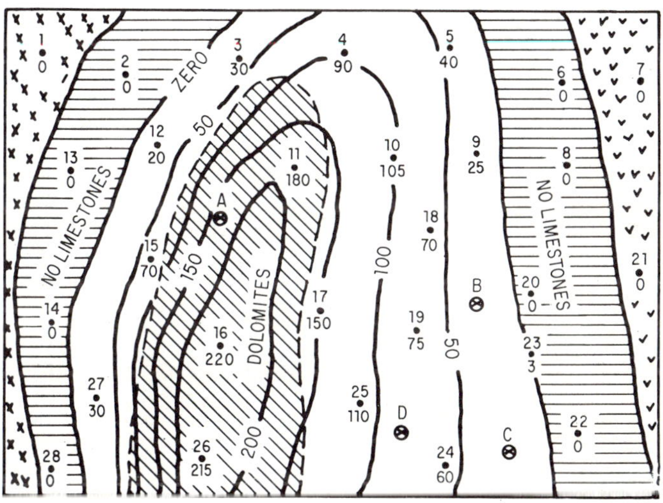

FIGURE 13-34. Limestone isolith map of the basin shown in Figure 13-31. (*From J. W. Low, Subsurface Maps and Illustrations, p. 490, in Haun and LeRoy, 1958.*)

Figure 13-37 is a map that shows both sand-shale ratio and thickness of total clastics.

The two sets of contour lines of the clastic and the sand-shale ratio can be combined into a single lithofacies map by segregating areas limited by selected values of the ratios. Figure 13-38 depicts a lithologic triangle designed for this purpose. The triangle is divided into nine lithologic groups, named for the dominant sediments in each. Table 13-2 lists the dominant rock names, the ratio limits, and a summary statement of the lithologic characteristics of each group. Although this particular

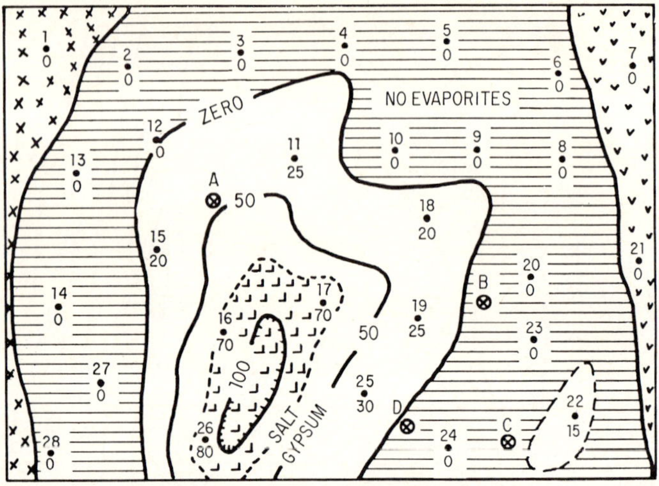

FIGURE 13-35. Evaporite isolith map of the basin shown in Figure 13-31. (*From J. W. Low, Subsurface Maps and Illustrations, p. 491, in Haun and LeRoy, 1958.*)

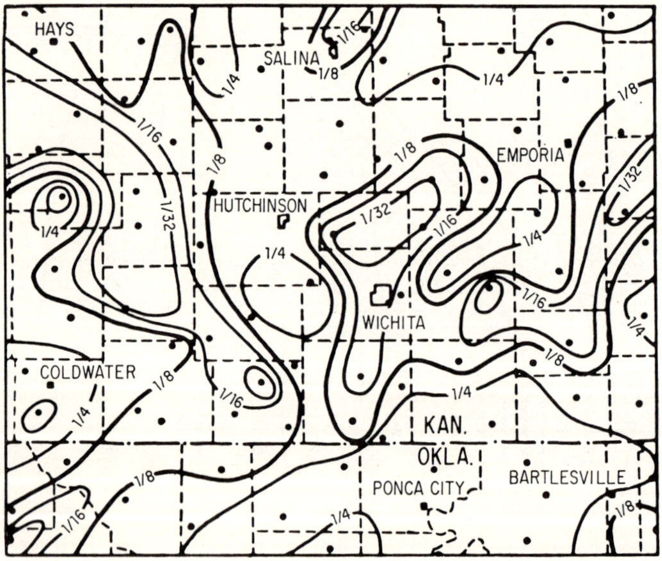

FIGURE 13-36. Sand-shale ratio map of Pennsylvanian strata in part of south central Kansas and adjacent Oklahoma. [*After Krumbein (J. Geol., 1955), from Weller, p. 529, 1960.*]

segregation has had rather widespread use, other systems for expressing these ratios have been attempted.

Figure 13-39 shows the steps involved in compiling a lithofacies map from isopach, clastic ratio, and sand-shale ratio maps. Figure 13-40 is a ratio-type lithofacies map.

Percentage maps are contoured on the percentage of one component compared with the total unit. They are similar to ratio maps except that the comparison is between

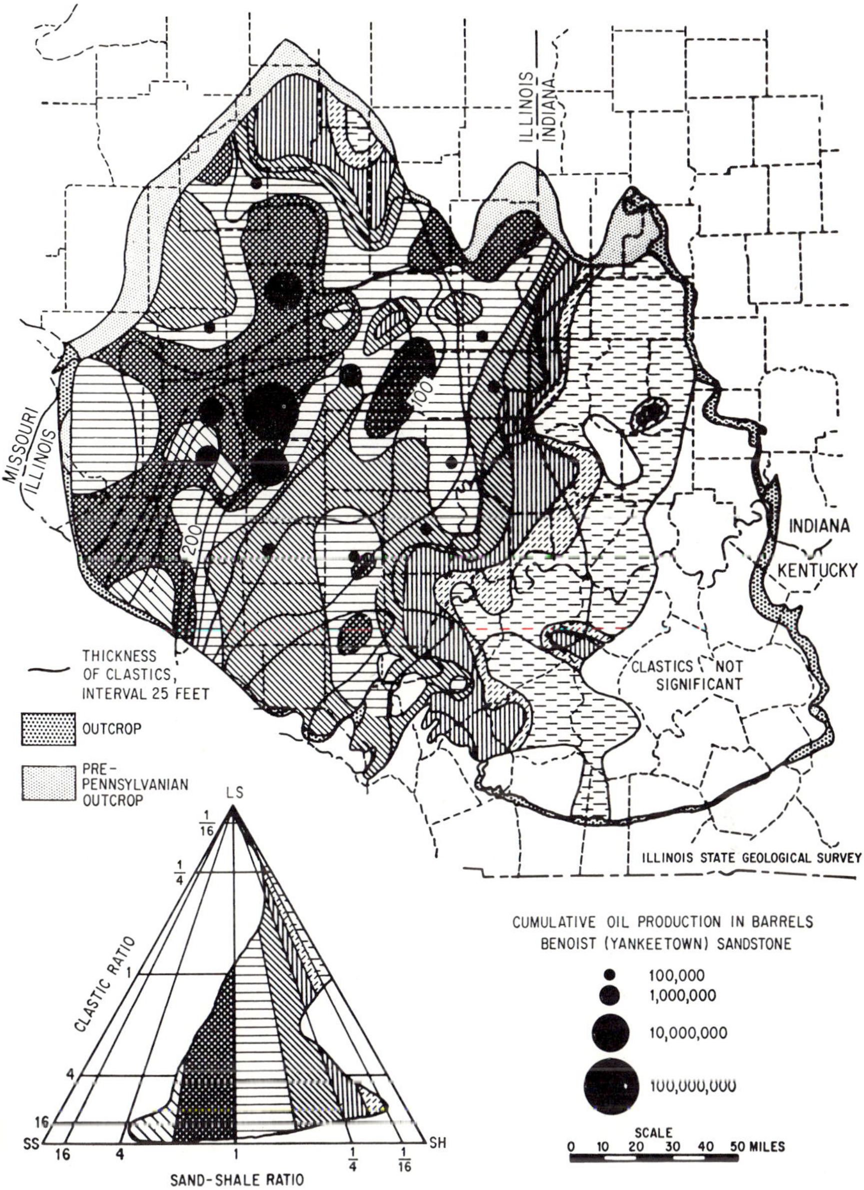

FIGURE 13-37. Map showing sand-shale ratio and thickness of total clastics of the Benoist Aux Vases section, Illinois Basin. (*From D. H. Swann and A. H. Bell, Habitat of Oil in the Illinois Basin, p. 463, in "Habitat of Oil," AAPG, 1957.*)

Table 13-2. Limits and Characteristics of Lithologic Groups

Group name	Clastic ratio limit	Sand-shale limit	General features
Sandstone	>8	>8	More than 79% sandstone
Sand-shale	>8	8–1	More sand than shale; less than 11% lime
Shale-sand	>8	1–⅛	More shale than sand; less than 11% lime
Shale	>8	<⅛	More than 79% shale
Sand-lime	1–8	>1	More sand than shale; 11–50% lime
Shale-lime	1–8	<1	More shale than sand; 11–50% lime
Lime-sand	¼–1	>1	50 to 80% lime; more sand than shale
Lime-shale	¼–1	<1	50–80% lime; more shale than sand
Limestone	<¼	Any value	More than 80% lime

the variable and the whole instead of between two variables (Figure 13-41). Separate contours are required for each component unless the mapping unit consists of only two components, in which case each can be expressed in terms of the other.

Lithofacies maps generally indicate the relative or absolute quantities of the principal lithologies in a section, but they do not reveal the manner in which these lithologies occur; a section consisting of alternating thin lithologic units is not distinguished from one composed of a few thick units. Lithofacies maps which attempt to show the arrangement of rocks in a vertical sense are termed *vertical variability maps.* They are much more difficult to prepare because of the almost infinite variations in the arrangement of sediments in any stratigraphic unit. There are several methods used to indicate vertical variability on a map. In one method the number of alternating lithologic units in the section studied is determined for each well, and these numbers are plotted at well locations and contoured (Figure 13-42). These contours, when superimposed on an isopach map of the same stratigraphic interval, yield the average thickness of the lithologic units at any location.

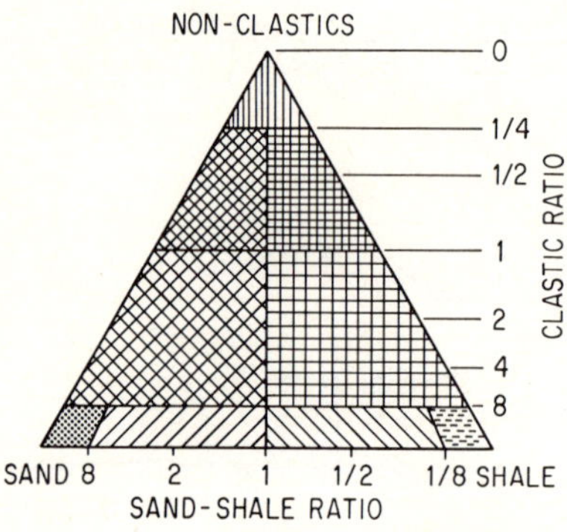

FIGURE 13-38. A lithologic triangle used for mapping the composite lithologic aspects. The ratios are used as contour intervals, and the patterns for areal maps.

A more practical application of this technique is to determine the number of changes in lithology per 100, 500, or 1,000 feet. The advantage of this variation is that the rate of alteration is independent of the thickness of any one section. If a well does not penetrate the entire interval, the alternation rate can still be determined for the portion of the section drilled.

Another vertical variability map can be made by superimposing contours drawn on the number of discrete sands within the boundaries of the mapping unit upon an isolith base map of the total sand thickness. When the map is used in conjunction with an isopach map or a sand percentage map of the mapping unit, a decrease in the number of discrete sand bodies when accompanied by a decrease in the total thickness of the mapping unit or by an increase in the sand percentage suggests that several sands have merged into a single unit in the shelf zone. A decrease in the number of sands when accompanied by an increase in the total thickness of the whole unit or by an increase in the shale percentage suggests an approach to the deep basin. An increase in the number of sands when accompanied by a decrease in the total sand content or by an increase in the total unit thickness indicates the offshore zone where the sand bodies tend to feather out, etc.

The placement of beds in the section can be shown by ratios based on the amount of sandstone in the upper half as against the lower half of the section, and the zone in which thin sand bodies merge to form a thick, continuous body can be determined by a map showing the ratio between the number of sands of minimum thickness and those of maximum thickness.

Vertical variability maps are illustrated in Figures 13-43 and 13-44.

Biofacies maps show the areal variation in biologic aspect of a stratigraphic unit. The preparation of biofacies maps presents many of the same problems encountered in

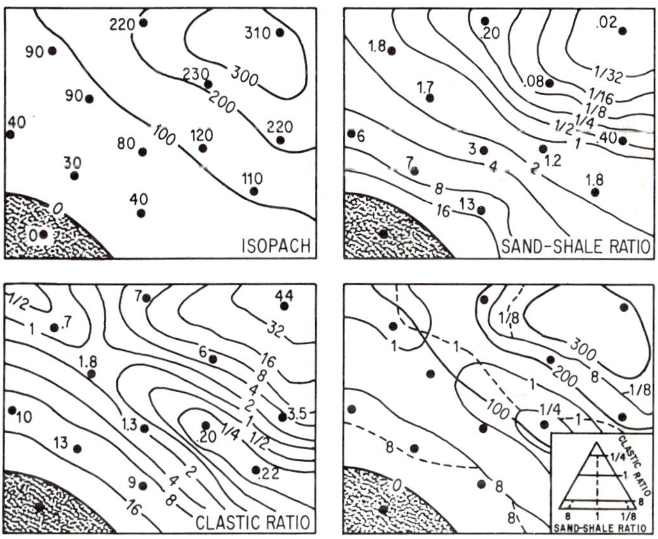

FIGURE 13-39. Steps in compiling a lithofacies map from isopach, clastic ratio, and sand-shale ratio maps of an area. (*From Sloss, Dapples, and Krumbein, 1960.*)

lithofacies mapping; however, biologic features present a much more complicated problem because they are less easy to classify or express in quantitative terms and their meanings are much less certain. The measurements to be used in biofacies differentiation are particularly difficult. Some of those that have been suggested and used to some extent include geographic distribution of species, genera, or other taxonomic groups; number of species representing selected taxonomic groups; number of specimens representing selected taxonomic groups; and weight of specimens belonging to such groups. The first makes no distinction between very rare and very common fossils. Counts of individuals are not satisfactory for bryozoans and crinoids, seldom found whole; for trilobites and ostracods that molted; or for colonial corals that might be counted as one or many individuals. The relative weights of such disparate fossils as, say, brachiopods or pelecypods and foraminifera are without meaning. Despite these difficulties, however, much can be done with the fauna if allowances are made for the limitations.

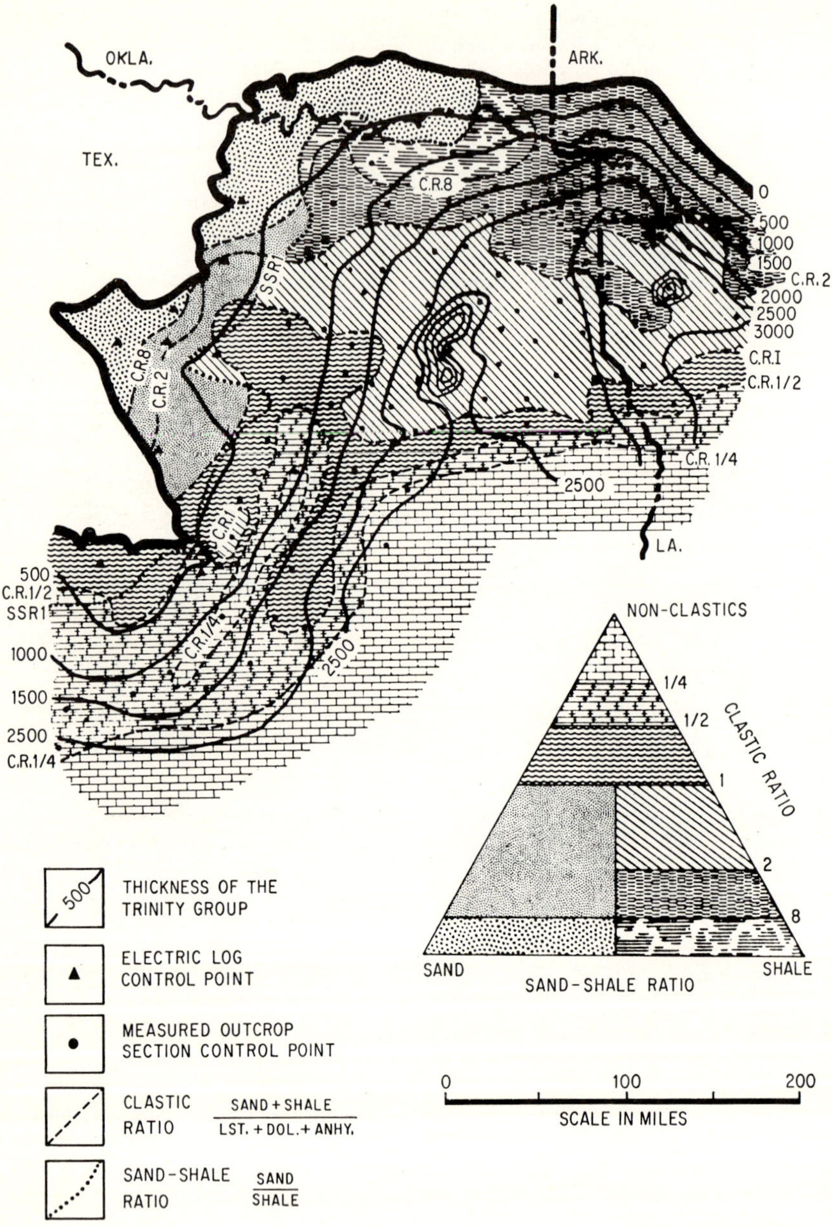

FIGURE 13-40. Ratio-type lithofacies map of the Commanchean Cretaceous Trinity group. (*From J. M. Forgotson, Jr., Review and Classification of Quantitative Mapping Techniques, Bull. AAPG, vol. 44, p. 87, 1960.*)

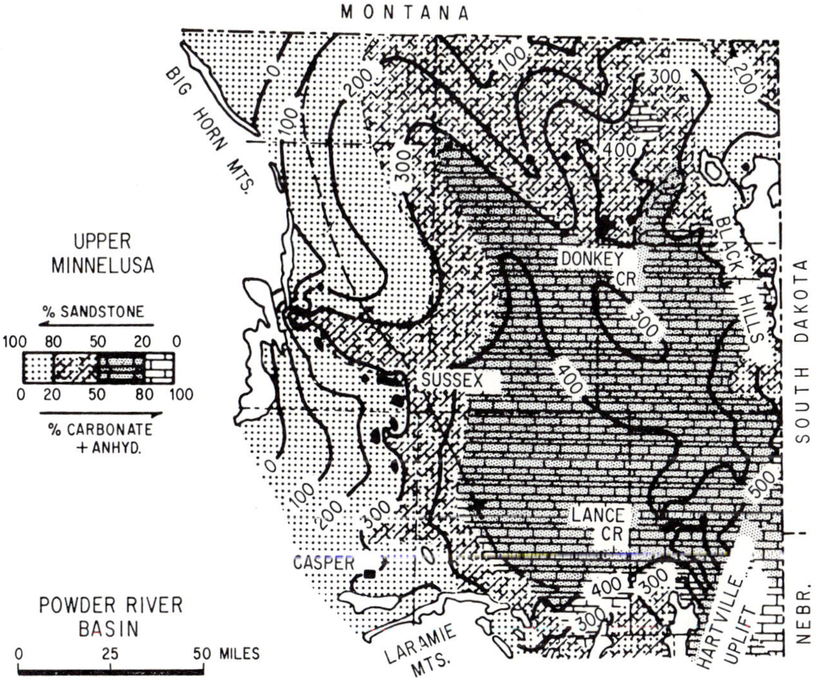

FIGURE 13-41. Sandstone percentage map of the upper Minnelusa formation, Powder River Basin. (*From D. I. Foster, "Oil Potential of the Minnelusa Formation, Powder River Basin, Wyoming," 1959 Geological Record, published by Petroleum Information, Denver, 1959.*)

Biofacies maps are most valuable for interpretations related to the nature of the environment of accumulation. Bathymetric, thermal, and other indicators present among the fossils, as well as distinctions between clear-water and muddy-bottom dwellers, etc., make possible environmental evaluations. The interpretation of biofacies maps is greatly strengthened when they are used with lithofacies maps of the same stratigraphic interval. Each type of map supplements the other and adds independent lines of evidence which strengthen inferences by the convergence of data. Figures 13-45 and 13-46 are examples of biofacies maps which contribute significant information, particularly when used with lithofacies data.

Tectofacies maps show the areal variation in tectonic aspect of a stratigraphic unit. They differ from lithofacies and biofacies maps in being interpretive rather than

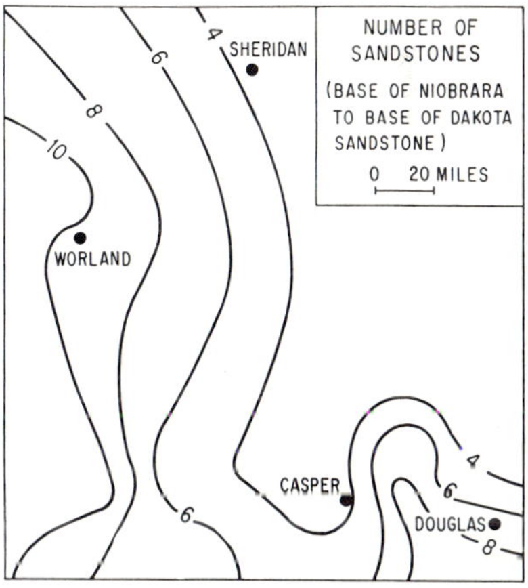

FIGURE 13-42. Vertical variability map: number of sandstones between the base of the Niobrara formation and the base of the Dakota sandstone, Central Wyoming. (*From W. C. Krumbein and W. G. Libby, Application of Moments to Vertical Variability Maps of Stratigraphic Units, Bull. AAPG, vol. 41, p. 207, 1957.*)

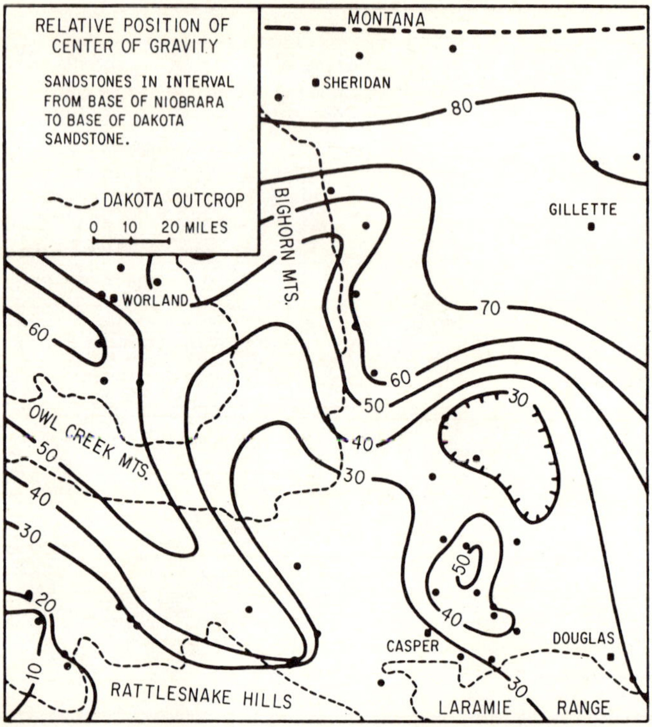

FIGURE 13-43. Vertical variability map: relative position of center of gravity of sandstones in same interval as Figure 13-42. Contours indicate weighted mean position of discrete sands expressed as percentage of distance down from top of unit. (*From W. C. Krumbein and W. G. Libby, Application of Moments to Vertical Variability Maps of Stratigraphic Units, Bull. AAPG, vol. 41, p. 207, 1957.*)

factual. The purpose of tectofacies maps is to show the distribution of tectonic elements over the area of accumulation of the stratigraphic interval and the degree of tectonism in each element. The basic data for the maps are *tectotopes*, tectonic subdivisions representing particular tectonic environments (Figure 13-47) (cf. *lithotopes*, distinctive lithologies characterizing particular depositional environments, and *biotopes*, distinctive faunal assemblages characterizing particular biologic environments). The identification of tectotopes depends on evaluation of the lithologic associations and thicknesses of outcrops and well sections in a stratigraphic unit.

No standard methods have been developed for presentation of tectofacies data. Tectonic framework maps, showing the distribution of tectonic elements in the depositional area, are so far the nearest approach to over-all tectofacies maps. These are usually prepared from isopach and one or more lithofacies maps which enable the geologist to determine the shelf, basin, and geosynclinal tectotopes which may occur within the region studied as defined by the thicknesses and lithologic associations present in the section.

Paleogeologic Maps

Paleogeologic maps show the areal geology of the surface as it was at some time in the geologic past or as it would now appear if all the strata were stripped off above

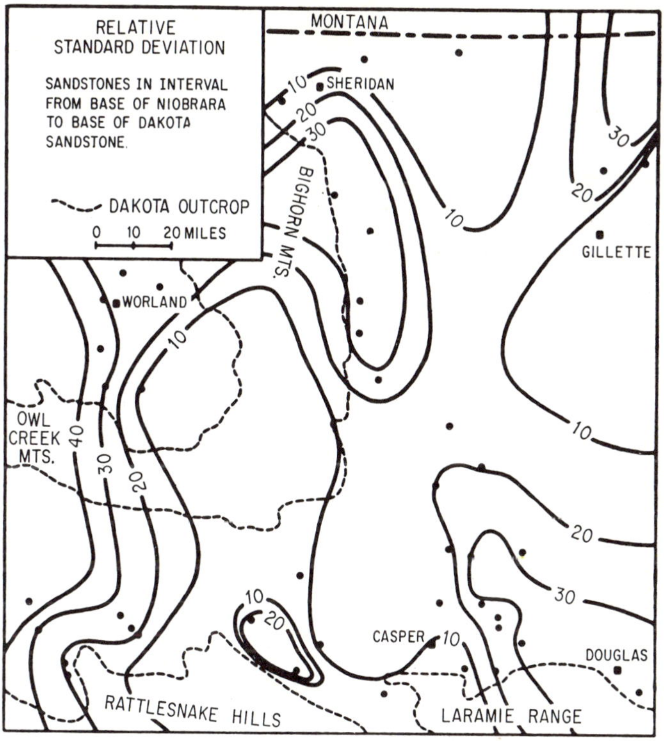

FIGURE 13-44. Vertical variability map: relative standard deviation of sandstone occurrences in same interval as Figure 13-42. Contours indicate percentage of total interval thickness occupied by bulk of sand occurrences. *(From W. C. Krumbein and W. G. Libby, Application of Moments to Vertical Variability Maps of Stratigraphic Units, Bull. AAPG, vol. 41, p. 206, 1957.)*

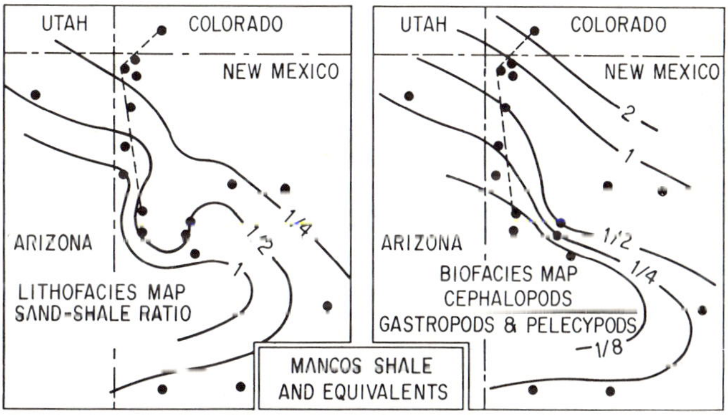

FIGURE 13-45. Lithofacies and biofacies maps of the Mancos shale and equivalents in the southwest. Reciprocal relationships between the sand-shale ratio and the faunal ratio are shown by the maps. *(After W. S. Pike (Geol. Soc. Am., 1947), from Krumbein and Sloss, p. 414, 1955.)*

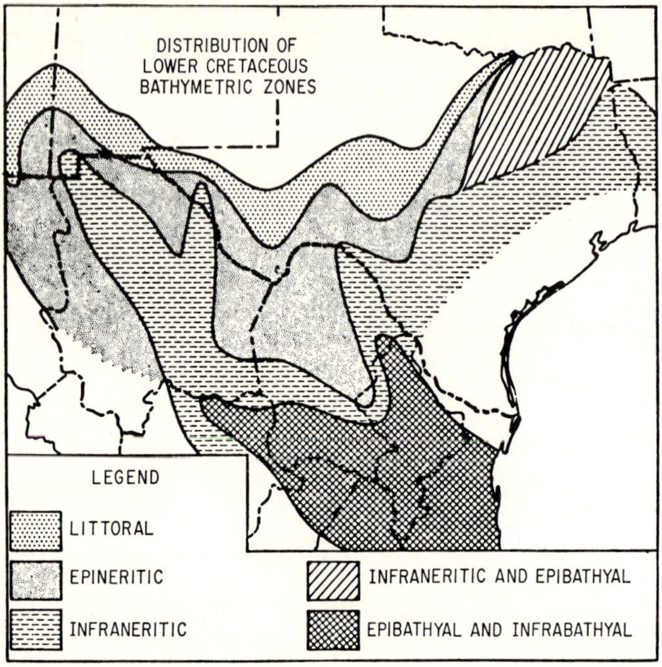

FIGURE 13-46. Bathymetric biofacies of the Lower Cretaceous of the Texas-Mexican area based on ammonoid assemblages. The areas or bodies of strata characterized by each assemblage are *biotopes*. (*After Scott (J. Paleon., 1940), from Krumbein and Sloss, p. 282, 1955.*)

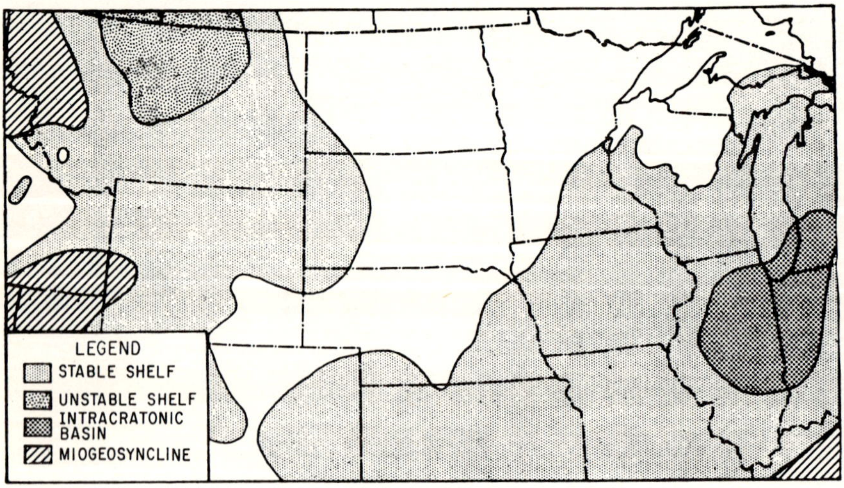

FIGURE 13-47. Tectofacies map of the Croixan Series. The areas distinguished are *tectotopes*, tectonic subdivisions representing particular tectonic environments. (*From Krumbein and Sloss, p. 440, 1955.*)

some particular horizon (Figure 13-48). Such maps are also called *subcrop* maps because they show where the rocks previously cropped out at a surface that is now covered. Paleogeologic maps are most commonly drawn on major unconformities where several rock units are exposed. However, local unconformities are also important mapping horizons if beds are tilted and folded and a number of formations are truncated.

In some areas a single stratigraphic unit overlies the erosion surface and is in contact with the truncated edges of a number of different formations below the unconformity. In other areas, for example where transgression of the sea has been relatively slow, there may be several beds overlapping the beveled edges of the different units below the erosion surface. The precise correlation of an erosion surface is the most critical problem in the construction of a paleogeologic map. The maps are based principally on subsurface data, mainly well logs, and their detail is dependent upon the abundance

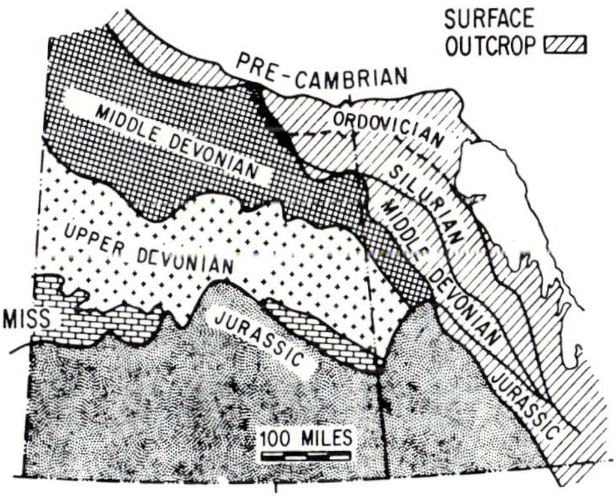

FIGURE 13-48. Precretaceous paleogeologic map of the Canadian portion of the Williston Basin. (*From G. B. Darling and P. W. J. Wood, Habitat of Oil in the Canadian Portion of the Williston Basin, p. 136, in "Habitat of Oil," AAPG, 1957.*)

and distribution of the data. Control is also provided by the stratigraphic relationships observed along the outcrop of an unconformity. The maps cannot be extended with much confidence beyond such outcrops.

Subgeologic Maps

A *subgeologic* or *worm's-eye* map is another type of map that can be made at an unconformable surface. The subgeologic map shows the areal relations of strata that immediately overlie the unconformity. Its pattern may indicate overstepping formations or other units deposited in a transgressing sea. More often, however, the stratigraphic relations suggest transgression interrupted by one or more regressions indicated by corresponding younger unconformities. Like paleogeologic maps, subgeologic maps are based principally on data provided by well records. Figure 13-49 shows diagrammatically the relationship between a paleogeologic map and a subgeologic map constructed at the unconformity indicated in the cross section.

Paleogeologic and subgeologic maps have a number of uses in subsurface exploration. Stratigraphic traps of various types, buried zones of weathering, and outcrop trends of productive formations can be mapped with accuracy in areas where a considerable number of wells have been drilled. Depositional history can be determined in some instances. When the rock distribution on the two maps is compared, the size and

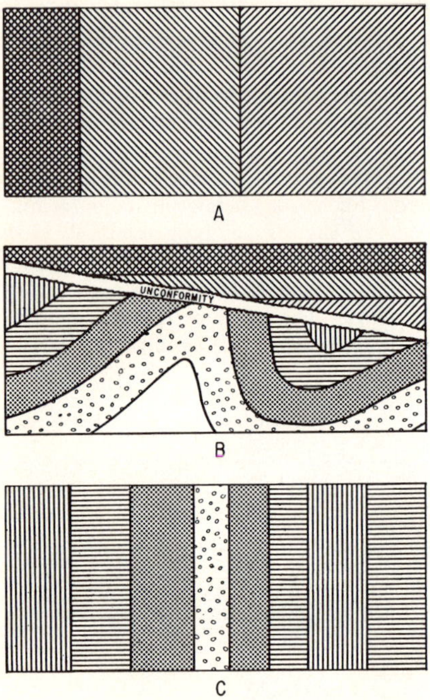

FIGURE 13-49. Subgeologic map *A* showing areal relations of strata immediately overlying unconformity in cross section *B*, which shows attitude of strata above and below an unconformity. Paleogeologic map *C* showing areal geology at the surface immediately below the unconformity in the cross section.

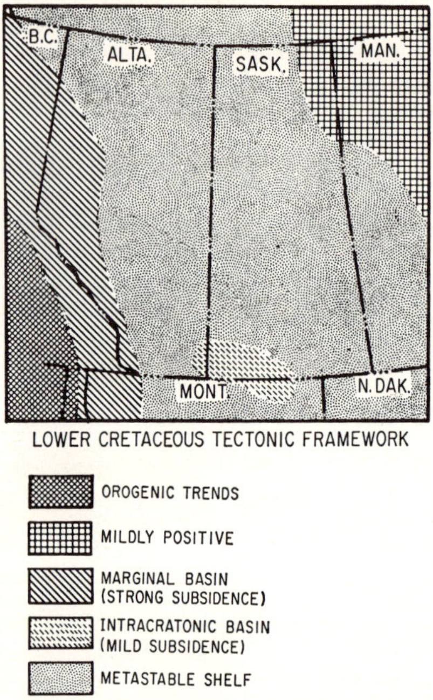

FIGURE 13-50. Paleotectonic map of the Lower Cretaceous in western Canada showing structural pattern existing during that time. (*From R. P. Glaister, Lower Cretaceous of Southern Alberta and Adjoining Areas, Bull. AAPG, vol. 43, p. 630, 1959.*)

shape of ancient positive areas as well as the time and rate of their burial can be deduced. Layer by layer the tectonic history can be unraveled by constructing paleogeologic maps at each unconformity. The effects of more recent events can then be evaluated and removed from the interpretation of older deformation patterns.

Paleotectonic Maps

Paleotectonic maps show the structural patterns that existed during a chosen division of geologic time. The maps are generally regional in scope, and their principal use is to illustrate the distribution of various components of the tectonic framework (Figure 13-50). Areas of volcanism, the position of orogenic belts, basin areas, and positive areas can be illustrated. The data upon which these maps are based are derived from sources such as lithofacies, paleogeologic, and paleogeographic maps. Their accuracy, therefore, is generally quite good in subsiding basin areas, relatively less so in rising portions of any region, and poor in uplifted areas.

Paleogeographic Maps

Paleogeographic maps attempt to show the distribution of land and sea as these existed at some moment in the geologic past (Figure 13-51). Unlike the previously

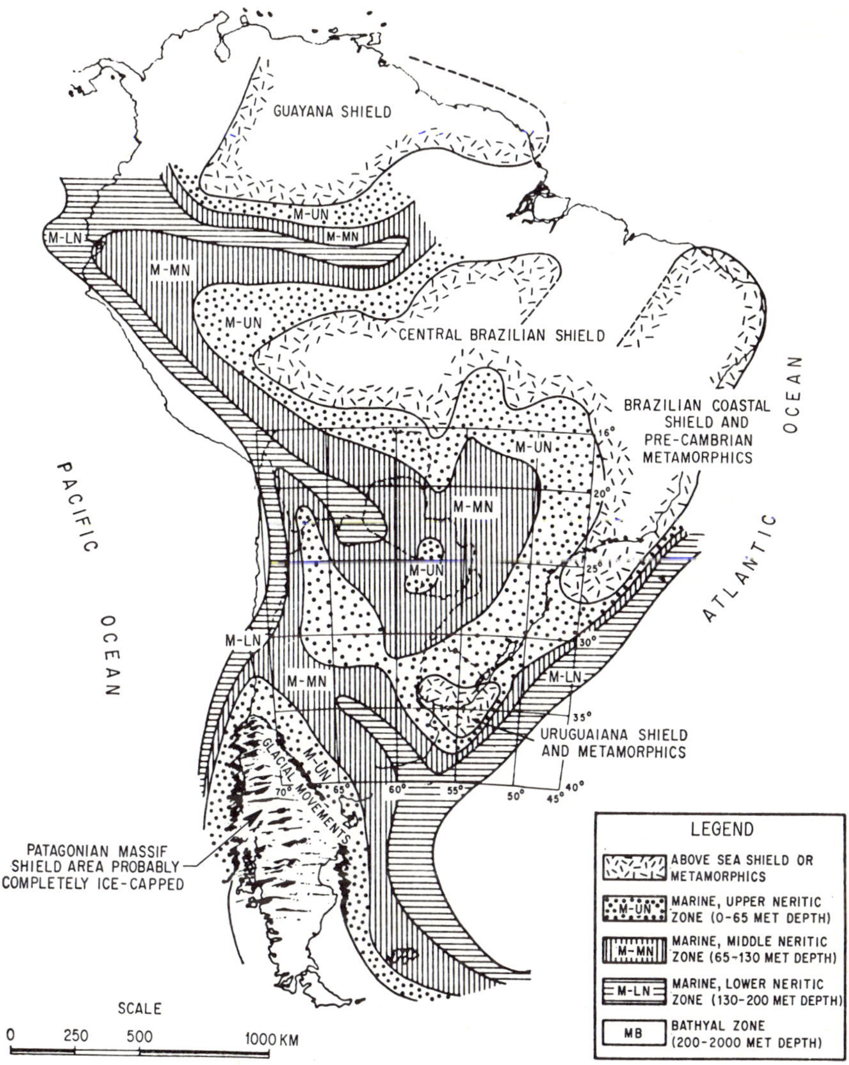

FIGURE 13 51. Paleogeographic map of South America during Lower Devonian time. (*From R. M. Sanford and F. W. Lange, Basin-study Approach to Oil Evaluation of Parana Miogeosyncline, Bull. AAPG, vol. 44, p. 1353, 1960.*)

described geologic maps, which are predominantly factual, paleogeographic maps are principally interpretations. Paleogeographic maps can be elaborated in various ways by adding areas of nonmarine deposition, evaporite basins, highlands subject to severe erosion, etc. The maps are based on geologic and paleontologic data supplemented by considerable imagination. They actually represent conditions and relations that existed for an appreciable interval of time, hence are generalized. The uncertainties inherent in paleogeographic interpretation are such that no two persons would be likely to construct similar maps of any large region, even though the maps were based

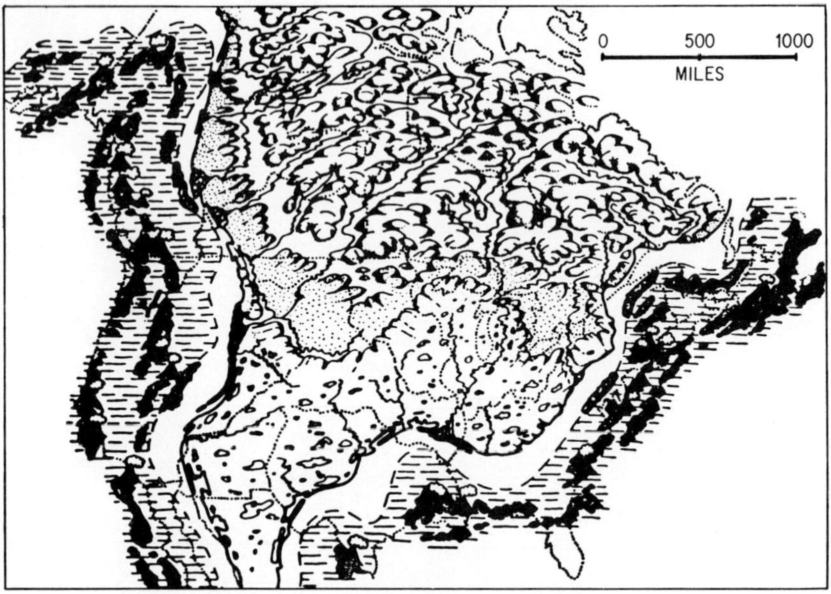

FIGURE 13-52. Paleogeographic map of the United States and Canada during Middle Ordovician time. The base is palinspastic and the map is drawn out laterally to compensate for later thrust faulting. (*After Kay (Geol. Soc. Am., 1951), from Weller, p. 590, 1960.*)

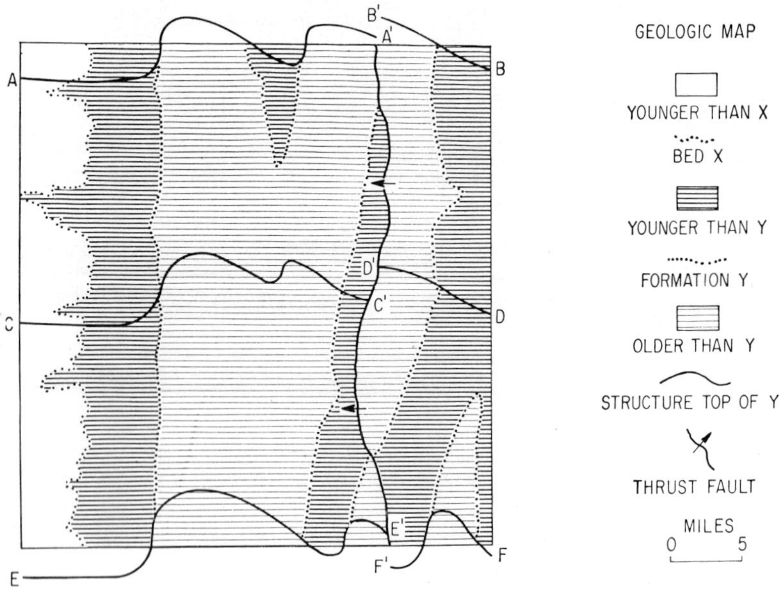

FIGURE 13-53. Geological map of an hypothetical area of folded and faulted sediments. See Figure 13-54. (*From M. Kay, Paleogeographic and Palinspastic Maps, Bull. AAPG, vol. 29, p. 440, 1945.*)

on the same data. These maps are most useful for illustrating major events in geologic history and least usable for showing details of local stratigraphy.

A *palinspastic map* is a special type of paleogeographic map. Folding and overthrusting, chiefly in geosynclinal zones, have resulted in compression of the rocks and shortening of segments of the earth's surface that may be measured in many miles.

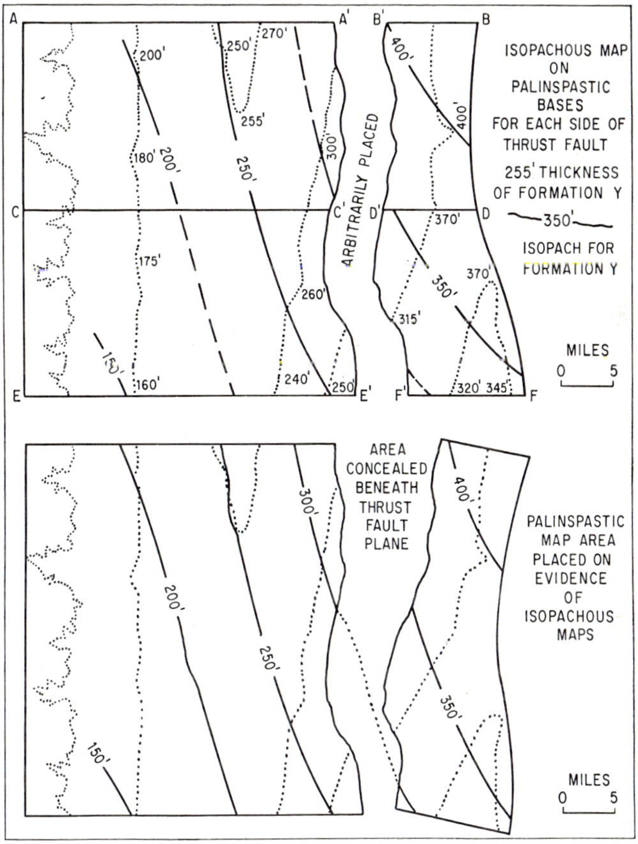

FIGURE 13-54. Steps in constructing a palinspastic map of the hypothetical area of Figure 13-53. (*From M. Kay, Paleogeographic and Palinspastic Maps, Bull. AAPG, vol. 29, p. 441, 1945.*)

Points that are now close together may originally have been far apart. If it is desired to return geographic localities or geologic features to their former relative positions, expansion of the map showing present relations is required. The resulting expanded map is a palinspastic map (Figure 13-52). These maps are ordinarily used to remove crustal shortening accompanying orogenic movements and are important in reconstructing geologic history (Figures 13-53 and 13-54). They are also useful in local problems where the environment of individual thrust sheets, folds, or restored shore lines may be important.

Sections, Diagrams, and Models

The subsurface maps discussed above depict geologic conditions in more or less horizontal planes. In order to show the vertical as well as the horizontal relationships among the rock units, cross sections usually are constructed. Neither maps nor cross sections alone present the whole picture of a geologic situation, and any complete subsurface investigation of an area must utilize them both. Cross sections are not difficult to construct when adequate subsurface data are available from boreholes or from geophysical surveys, but they may be extremely difficult to draw when only surface data are available.

Cross Sections. In the construction of cross sections two considerations are particularly important, choice of scale and degree of vertical exaggeration. The scale should be selected so that the feature or features illustrated can be readily encompassed by the human eye and should not be so large that drafting problems render construction impractical. Cross sections drawn with the same horizontal and vertical scale are called normal. In such sections there is no distortion of the dip of the beds or of the thickness of the units, and the point of intersection between faults and bedding planes, the angle of dip of faults, dikes, and stratigraphic units, and the thickness of beds can be measured directly from the section. However, in many cases significant variations in the vertical direction are too small to be shown unless the scale in this direction is exaggerated. This vertical exaggeration may amount to as much as 100 times, although an exaggeration of 10 times is more common. Exaggeration of the vertical scale introduces certain distortions that may suggest geologic conditions which do not exist. Sections with greatly exaggerated vertical scales tend to show thinning of the formations where dips are steep, giving the effect of attenuation on the flanks of structures and thickening on the tops of anticlines, on the bottoms of synclines, and at other places where the dips are flat. Likewise, rates of dip become increased in accordance with the amount of exaggeration (see Figure 13-55).

Cross sections are divided into two general groups, structural and stratigraphic, although both stratigraphic and structural features may be combined to a greater or lesser extent on sections in each of the groups. Accurate cross sections of folded and faulted structures are of great value in petroleum exploration. The ability to predict the position of structural features at depth results in substantial savings in drilling expense and in a much higher success ratio. Sections that are constructed primarily to show structural relationships commonly use sea level as the plane of reference. Each control point is selected along a line perpendicular to the strike or projected into it. The lithology of the stratigraphic units may be shown, but often only the elevation of each bedding plane is plotted. Lines connecting the elevations show the attitude of the bedding planes with respect to sea level. A uniform unit thickness is used unless otherwise indicated by the data. An understanding of the fundamental processes involved in folding and faulting is an essential requirement for good cross-section construction (see Figure 13-56).

Stratigraphic cross sections are drawn to illustrate the correlation among units. The plane of reference is one continuous stratigraphic horizon, either a bedding plane or an unconformity. Stratigraphic sections emphasize variations in the thickness of units but do not show structural attitudes. In this respect they are comparable to isopach maps. The complete lithology of the stratigraphic units is commonly shown (Figure 13-57) so that the manner in which changes in the character of the deposits take place can be understood. Variations in thickness are most apparent in the units immediately below or above the reference plane. These variations are less obvious with increased distance from the datum plane, an important consideration

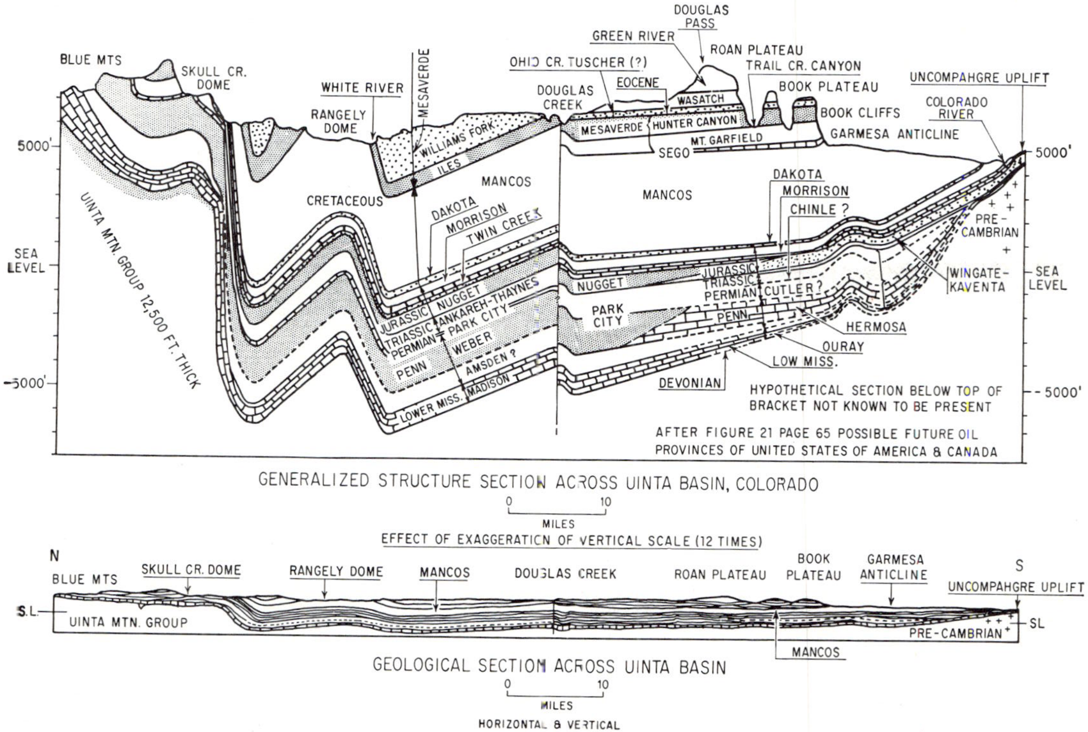

FIGURE 13-55. Geologic cross sections across Uinta Basin, Colorado, showing effect of vertical exaggeration amounting to twelve times. (*From H. H. Suter, Exaggeration of Vertical Scale of Geologic Sections, Bull. AAPG, vol. 31, p. 330, 1947.*)

in selecting the plane of reference when constructing the section. A paleontologic cross section is a special type of stratigraphic section. In a paleontologic section correlation of fossil genera and species is emphasized rather than lithology (Figure 13-58).

Columnar Sections. Columnar sections illustrate the exact measurements and lithology of a measured outcrop or well log. They have no relationship to a datum plane and are not really cross sections, although several columnar sections can be shown together to demonstrate changes in lithology within a given stratigraphic

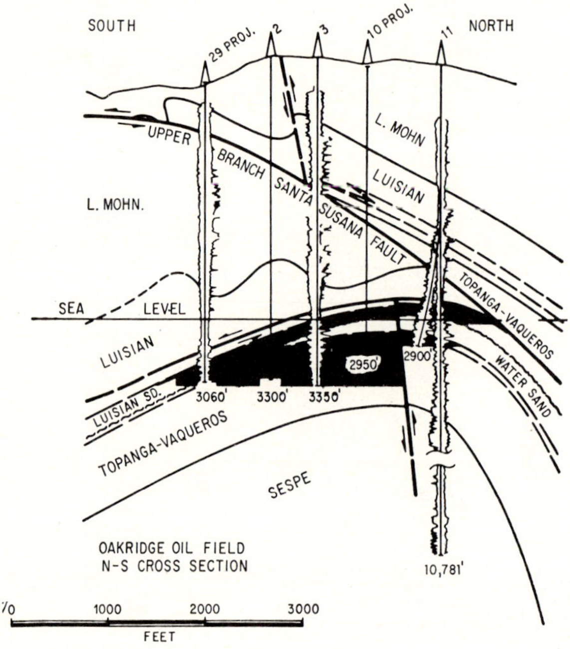

FIGURE 13-56. Structural cross section of Oakridge oil field, Ventura County, Calif. *(From E. A. Hall, G. H. Rickels, and W. O. Plant, Oakridge Oil Field, in "A Guide to the Geology of the Los Angeles and Ventura Regions," p. 187, Pacific Section, AAPG, 1958.)*

sequence. Electric or mechanical logs can be used for columnar sections where correlations are dependent on time. The lithology associated with each change in self-potential or resistivity usually is plotted on the log, and the position of the various units within the geologic time scale is indicated.

Log Segment Maps. The desire to combine both the horizontal and the vertical geologic aspects of an area has led to the development of several three-dimensional diagrams which illustrate the areal relationships among wells not located along a single line. This type of illustration seldom can be constructed with a common horizontal and vertical scale. The simplest, and often the most effective, of these is the *log segment map.* It consists of segments of the well logs plotted to any adaptable scale in their respective locations on the map. Either the tops or the bases of the selected log intervals may be plotted at the map control points. The logs may be joined by formational correlation lines, or if the map base is an isopach or facies map, the logs can be used to demonstrate variations in thickness or facies changes. If the

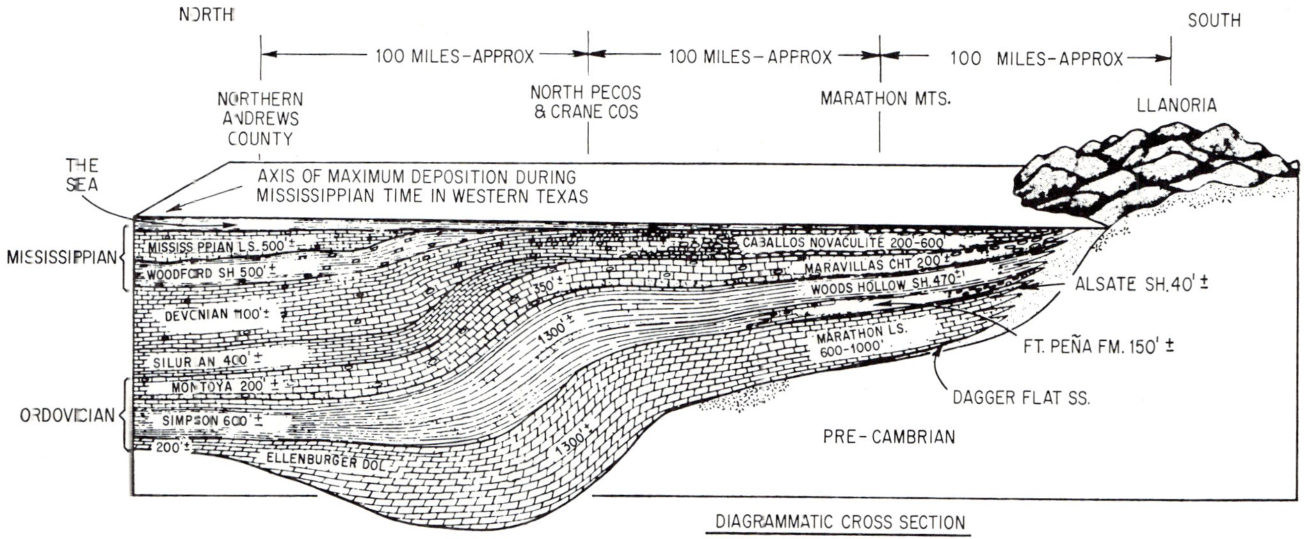

PRE-PENNSYLVANIAN ROCKS BEFORE LATE MISSISSIPPIAN DEFORMATION

FIGURE 13-57. Stratigraphic cross section showing Prepennsylvanian rocks before late Mississippian deformation. This section, in addition to demonstrating stratigraphic relationships, shows conditions at a definite time in the geologic past. (*From J. M. Barton, Pre-Permian Axes of Maximum Deposition in West Texas, Bull. AAPG, vol. 29, p. 1345, 1945.*)

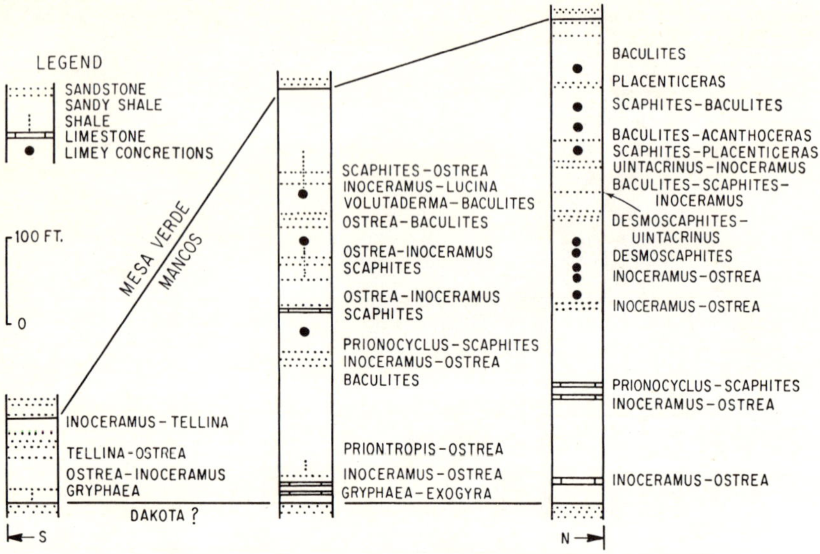

FIGURE 13-58. Cross section illustrating components of successive faunas in Mancos shale of the southwest. See Figure 13-45 for location of this section. *(After W. S. Pike (Geol. Soc. Am., 1947), from Krumbein and Sloss, p. 413, 1955.)*

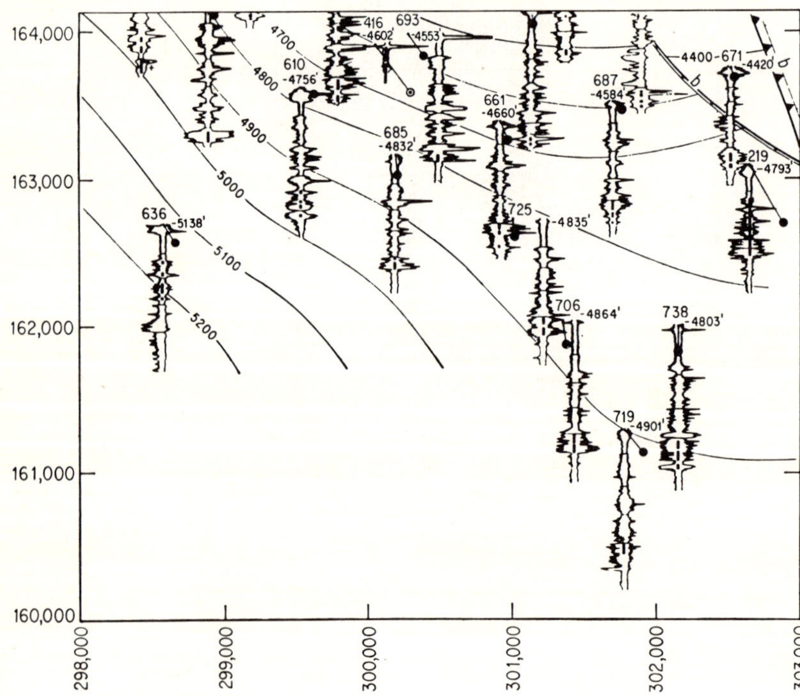

FIGURE 13-59. Portion of a log segment structure contour map, utilizing electric log segments of the zone being studied. *(From T. H. Bower, Log Map, New Type of Subsurface Map, Bull. AAPG, vol. 31, p. 343, 1947.)*

base is a structure map, the log segments will illustrate the correlation of the contour datum used in constructing the map (see Figures 13-59 and 13-60).

Panel Diagrams. If the formational correlation lines connecting the log segments in the log segment map described above are filled in to show the lithologic facies, pinch-outs, and other stratigraphic conditions occurring between the wells, a *panel diagram* results (Figure 13-61). In the construction of this type of diagram only the

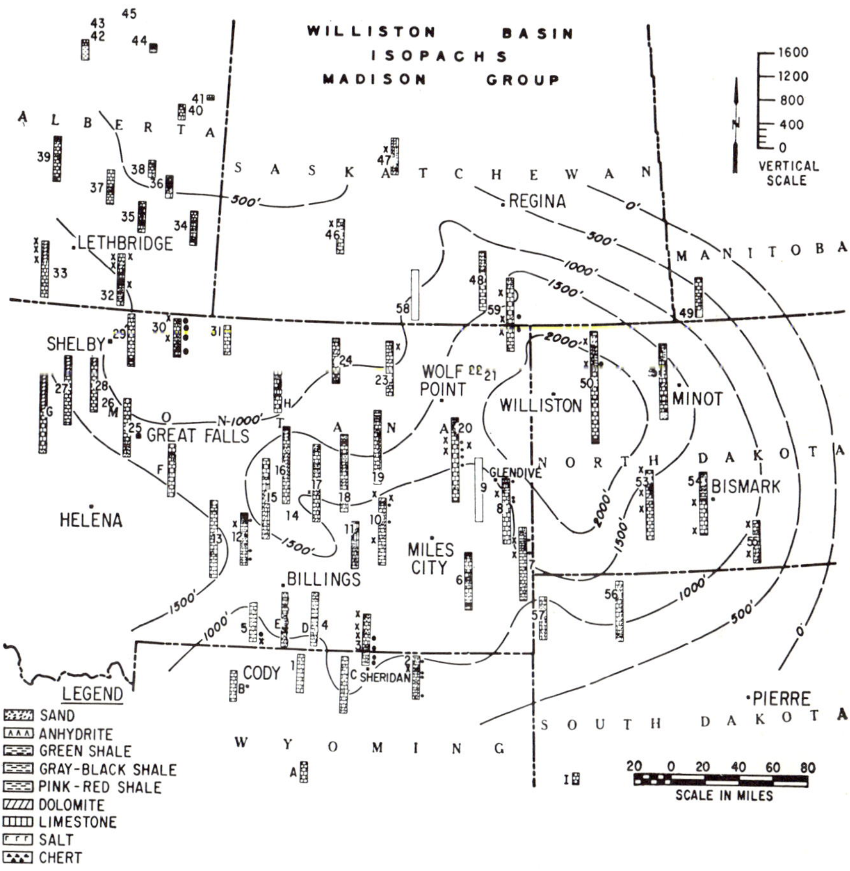

FIGURE 13-60. Log segment isopach map illustrating the lithology of the unit being studied. (*From T. R. Barnes, A New Province for Oil Exploration, Bull. AAPG, vol. 37, p. 346, 1953.*)

front or lowermost panels are shown in their entirety and panels joining wells along north-south lines are omitted, as they would appear only as single lines on the map.

In the construction of a structural panel diagram the plane of the map is sea level for each location and the surface elevation is plotted upward from this point. The topographic profile is found by connecting these surface elevations at each well. The depth to each bedding plane is plotted directly from the log, using the surface elevation as zero point. Elevations may be referred to some plane selected a chosen distance consistent with the scale of the map above or below sea level.

The stratigraphic *isometric projection* is a special adaptation of the panel diagram, differing from it in the treatment of the base map. It is redrawn in such a way that regularly arranged grid lines, intersecting at right angles on the plan map, meet at angles of about 60 and 120 degrees. This produces foreshortening and makes the map appear to lie at an angle to the plane of the diagram. An isometric panel diagram is superior to a normal one in pictorial quality because perspective is accentuated (Figure 13-62). The vertical scale of all sections is uniform, but the horizontal scale is variable.

Block Diagrams. The most realistic of all techniques for illustrating geologic concepts is the *block diagram.* Block diagrams are perspective drawings of solid bodies that represent portions of the earth. They are particularly well adapted to showing the generalized relations of surface geology and topographic relief to stratigraphy and structure (Figure 13-63). The simplest block diagram shows a surface area and two geologic cross sections on the forward vertical faces of the block. Many variations,

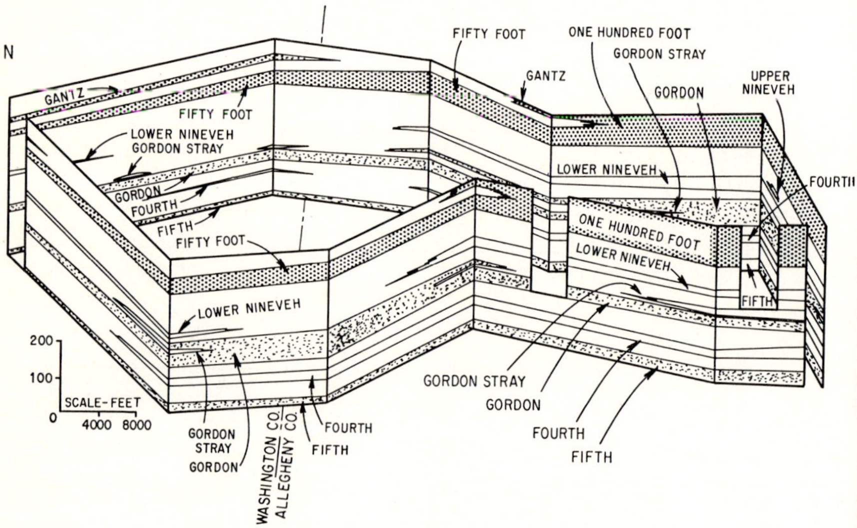

FIGURE 13-61. Panel diagram of a productive area in Washington and Allegheny Counties, Pa. [*After Ingham (Prod. Monthly, 1949), from Levorsen, p. 573, 1954.*]

however, can be produced by altering the shape of the block so that other sections are presented or by dividing the block into parts and offsetting them.

Space in this chapter does not permit describing the details of block diagram construction. They can be constructed by either isometric or stereographic projection. The latter method is more complex, and the blocks are more difficult to draw, but the better perspective that stereographic blocks provide makes them more effective. The reader is referred to the volume "Block Diagrams" by A. K. Lobeck (1958) for detailed data regarding this subject.

Peg Models. Models are sometimes used by subsurface geologists as a means to illustrate and study stratigraphic and structural conditions in three dimensions. The easiest model to construct and a most useful tool in solving complex structural problems is a *peg model.* Peg models are not adapted to broad regional work where wells are widely separated but are ordinarily used in detailed subsurface studies of local areas where well control is abundant. The principal objections to peg models are that they are time-consuming to construct, occupy much floor space, and cannot be transported readily.

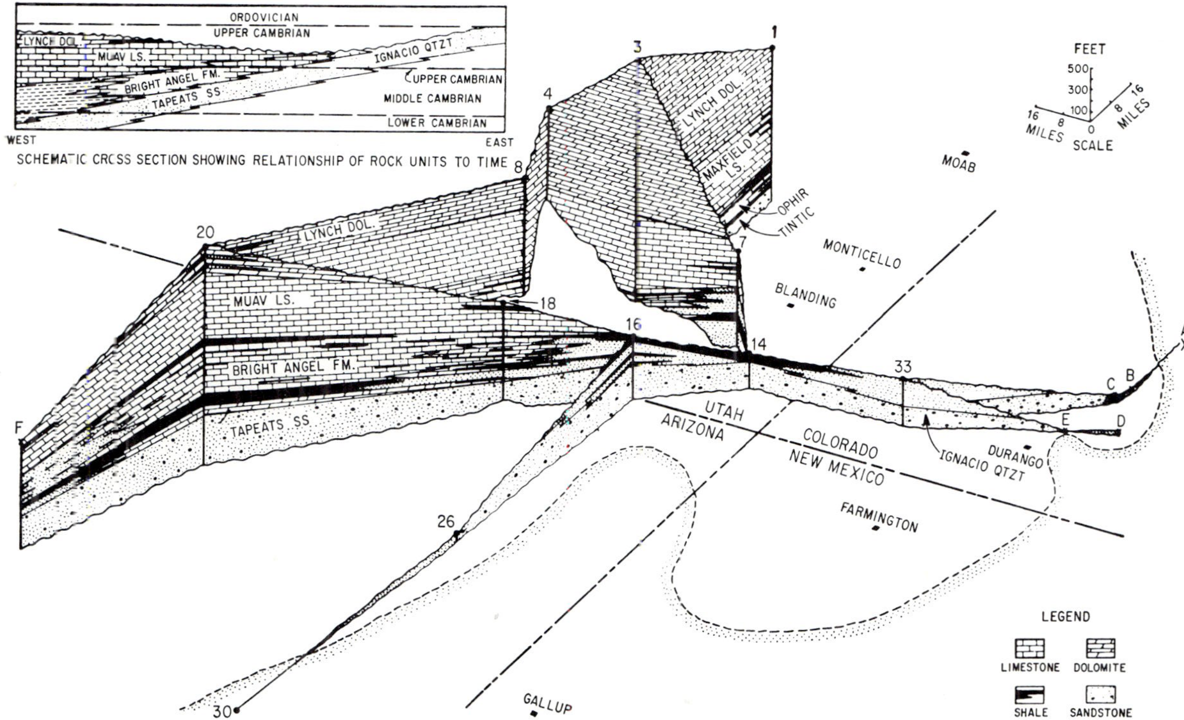

FIGURE 13-C2. Isometric projection illustrating stratigraphic relationships of the Cambrian system in the Paradox Basin. (*From L. L. Baars, Cambrian Stratigraphy of the Paradox Basin Region, "Guidebook to the Paradox Basin," p. 99, Intermountain Association of Petroleum Geologists, 1958.*)

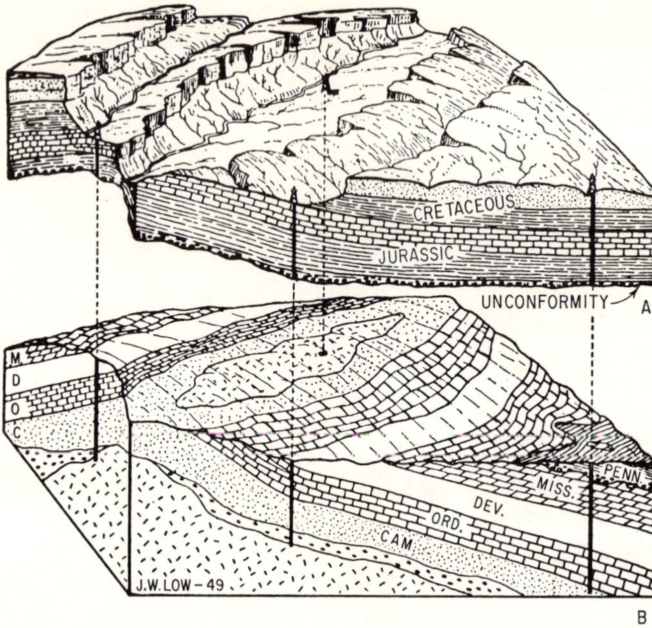

FIGURE 13-63. Block diagram illustrating an eroded anticline involving a major unconformity in the subsurface. (*From J. W. Low, Subsurface Maps and Illustrations, p. 458, in Haun and LeRoy, 1958.*)

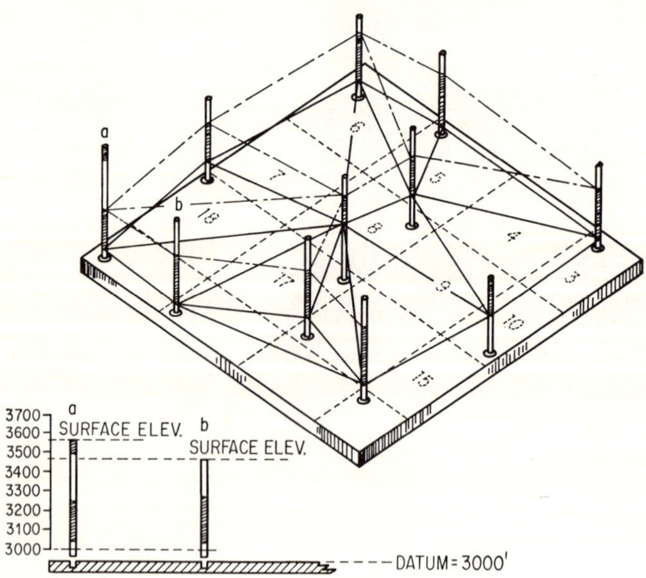

FIGURE 13-64. Peg model of a simple dome. (*From J. W. Low, Subsurface Maps and Illustrations, p. 506, in Haun and LeRoy, 1958.*)

The base for the peg model is made of wood, about 1 inch thick. This should be painted white or some neutral color, following which section and township lines and other necessary surface features are drawn on it at the scale selected for the model. At each well location a ¼-inch hole is drilled to a depth sufficient to receive the ¼-inch peg which will represent the well. The plane of the board is the elevation datum for the model. This assumed elevation should be below that of the lowest bottom-hole elevation in the area mapped. Rods are cut from ¼-inch doweling to lengths equal to the surface elevations of the wells on the vertical scale selected, plus an amount equal to the depth of the holes into which they will be inserted. The formations penetrated by the wells are represented on the rods by painted bands of different colors. After the painted rods are each set in their respective locations, correlations of the formation tops may be indicated by stretching strings colored the same as the formations they represent from well to well (Figure 13-64). A great advantage possessed by the peg model is that it can be readily altered to incorporate structural data becoming available at a later date, through either additional drilling or reinterpretation of older information.

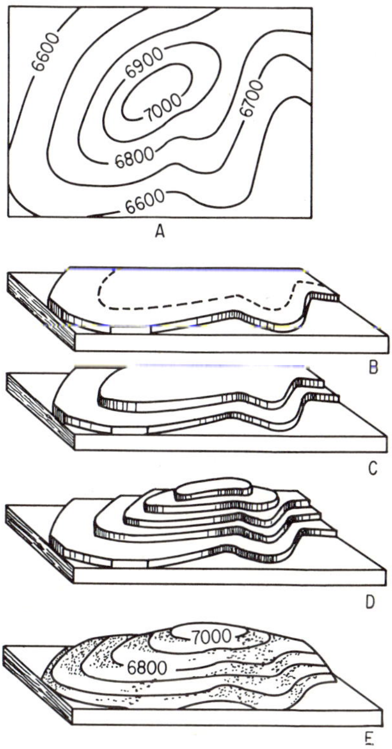

FIGURE 13-65. Steps in the construction of a solid relief model of a structure. (*From J. W. Low, Subsurface Maps and Illustrations, p. 507, in Haun and LeRoy, 1958.*)

Solid Relief Models. A *solid relief model* of a structure can be constructed in much the same way as a topographic relief model provided a structure contour map of the area is available. Steps involved in making such a model are shown in Figure 13-65. The contour map is used as a pattern. Sheets of fiberboard or masonite are cut with a coping saw along each contour and mounted one above the other. When all the contours have been cut and the sheets firmly mounted, the resulting skeleton is covered with papier-mâché or plaster of paris, care being taken to use a thickness only sufficient to cover the edges of the fiberboard contours. Structure contours can be shown on the finished model by drawing lines along the edges of the covered blocks. The finished model is useful only to illustrate some special structural feature. It has the disadvantage that new structural data cannot easily be incorporated in the model.

Section Models. A *section model* is constructed by mounting a series of cross sections drawn or painted on any thin, rigid, boardlike material in properly spaced slots in a solid base. Use of a transparent material such as cellulose acetate or plexiglass for the sections has the advantage of making it possible to view one cross section through another (Figure 13-66). The section model is more practicable than the solid model in that any one of the sections can be revised without disturbing the others.

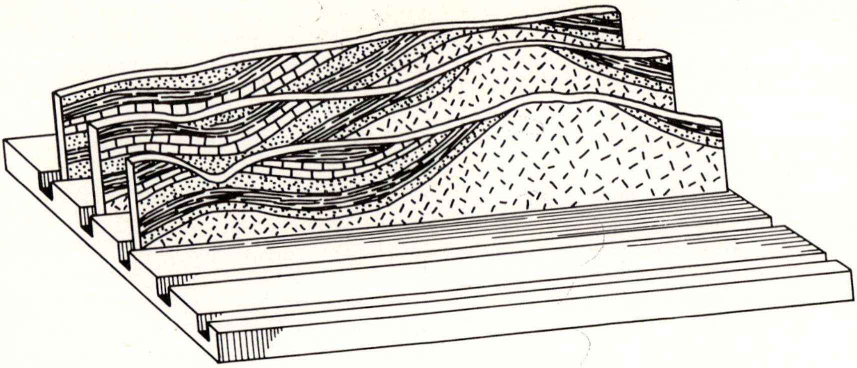

FIGURE 13-66. Section model built up from models of parallel structural cross sections. (*From J. W. Low, Subsurface Maps and Illustrations, p. 507, in Haun and LeRoy, 1958.*)

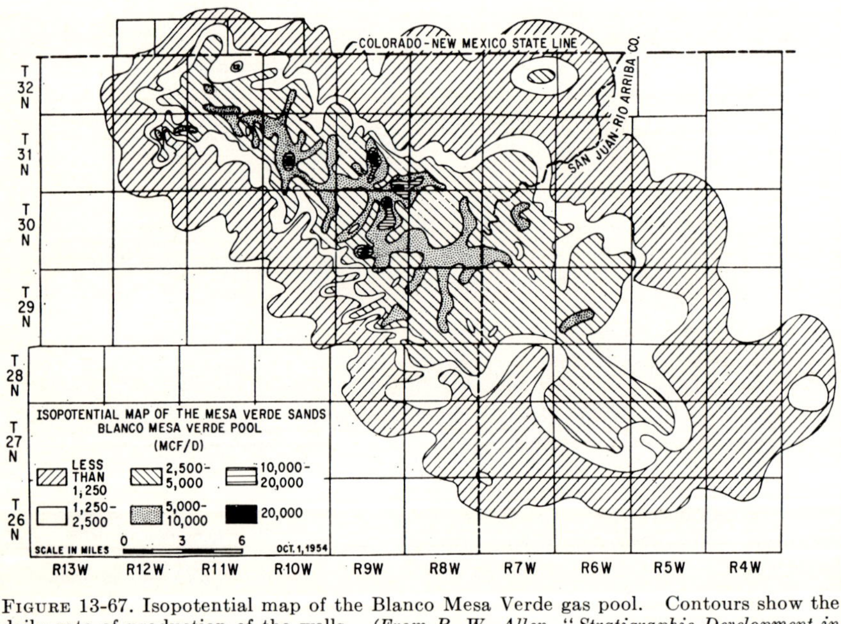

FIGURE 13-67. Isopotential map of the Blanco Mesa Verde gas pool. Contours show the daily rate of production of the wells. (*From R. W. Allen, "Stratigraphic Development in the Blanco Mesa Verde Pool of the San Juan Basin, 1955 Geological Record, published by Petroleum Information, Denver, 1955.*)

Geophysical Anomalies

Geophysical mapping may be thought of as a kind of subsurface mapping done in advance of drilling, although enough drilling to calibrate the instruments may be required beforehand. The mapping consists of measuring various physical properties of the rocks, such as the travel time of energy waves for reflection and refraction seismic surveying, magnetic susceptibility for magnetic mapping, rock density for gravity mapping, radioactivity for radioactivity surveying, electrical conductivity for self-potential mapping, etc. These geophysical measurements are then translated into geologic data, such as structural attitude, stratigraphy, depth, etc. This information, combined with data obtained from other types of subsurface studies, becomes a part of the total geologic interpretation. A local variation or deviation from the normal pattern developed over the area mapped is known as an *anomaly* or, specifically, a *seismic anomaly, gravity anomaly, magnetic anomaly,* etc. After all pertinent corrections have been applied to the data, the residual anomaly is considered due to some geologic effect, the correct interpretation of which may depend on the regional geology, the local geology, and the ability and experience of the geologist and the geophysicist. The translation of geophysical measurements into geologic concepts requires basic assumptions with regard to the rock characteristics or a definite knowledge of the lithology of the units being mapped. If the area is unexplored, interpretation of the geophysical data is based on assumed physical properties for an average lithology. Such interpretations are of value in a regional study, although they tend to be inaccurate in detail, as it is impossible to predict the variables that may be present in an unknown rock section. The better the geologic background of the analyst, the more accurate will be the interpretation of the geologic phenomena from geophysical data.

Miscellaneous Maps

All the maps, cross sections, and diagrams discussed in this chapter have been normal, orthographic, perspective, or isometric projections, and the geologic structures

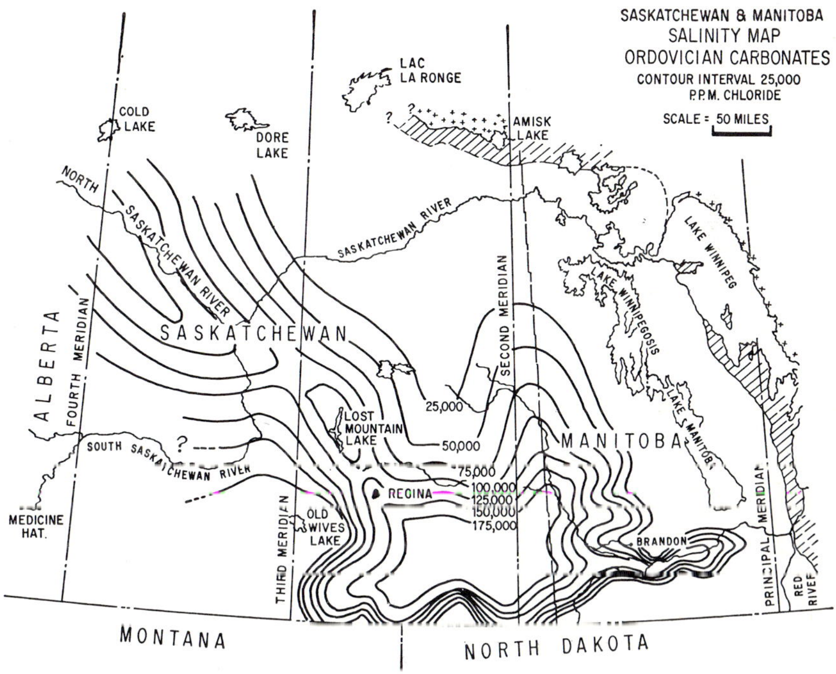

FIGURE 13-68. Isoconcentration map of Ordovician carbonates in Saskatchewan and Manitoba. Contours show salinity in parts per million. (*From A. L. Evans, The Red River Formation in Saskatchewan and Manitoba: A Structural and Stratigraphic Interpretation, 1960 Geological Record, published by Petroleum Information, Denver, 1960.*)

have been represented on them in either vertical or plan view. Actually, geologic structures are mapped on the spherical surface of the earth, and it is sometimes desirable to study structures or petrofabric relations in their true spherical relationship. *Stereographic projections* are used to represent spherical projections on a plane surface. They are geometric projections of spherical coordinates onto a horizontal plane. Stereographic projections are particularly useful in solving structural problems such as the determination of strike and dip from two apparent dips, the attitude of the lines of intersection of intersecting plane surfaces, the original configuration of geologic structures prior to tilting, etc. Other uses of these projections include the

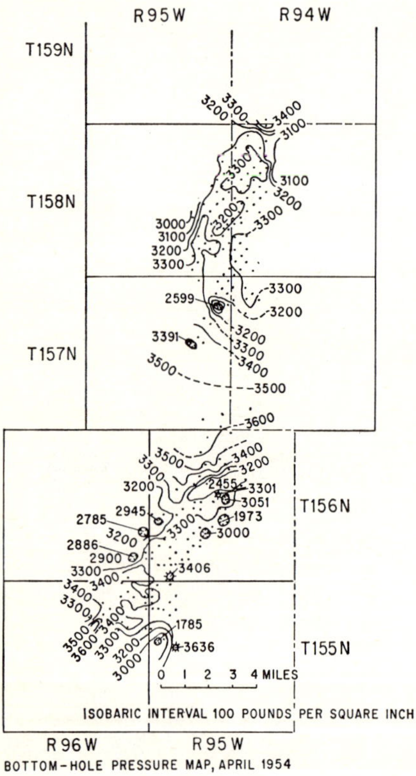

FIGURE 13-69. Isobar map of Beaver Lodge and Tioga fields, North Dakota, showing Madison formation pressures in April, 1954. (*From W. M. Laird et al., The Beaver Lodge and Tioga Fields, Mountrail and Williams Counties, North Dakota, 1955 Geological Record, published by Petroleum Information, Denver, 1955.*)

determination of stress orientations and the statistical analysis of structural features such as joints and lineations and of petrographic relations such as the orientation of sand grains, fossils, etc. Construction of stereographic projections is rather difficult, and for more complete information regarding them, the reader is referred to the list of references.

Many kinds of subsurface maps, cross sections, and diagrams other than those described in this chapter can be prepared to illustrate or help visualize specific phenomena. For example, an *isopotential map* can be prepared to show by contours the initial or calculated daily rate of production of wells in a field (Figure 13-67). An *isoconcentration map* can be made to show the concentration of salts in oil field waters (Figure 13-68), and an *isobar map* can be drawn to show by contours the lines of equal reservoir pressure of a pool (Figure 13-69). Figure 13-70 is a map showing by means of rose diagrams the orientation of plant stem fragments in a sandstone, a feature which is indicative of current direction at the time of deposition. Figure 13-71 is an example of another *iso* map—isogravity of crude. Figure 13-72 is an isosulfur contour map.

The number of variables in subsurface geology is large, and the number and type of maps that the subsurface geologist may find helpful in discovering a potential trap for the accumulation of oil or gas are limited only by the quantity of reliable data available, his ability to analyze these data, and his imagination.

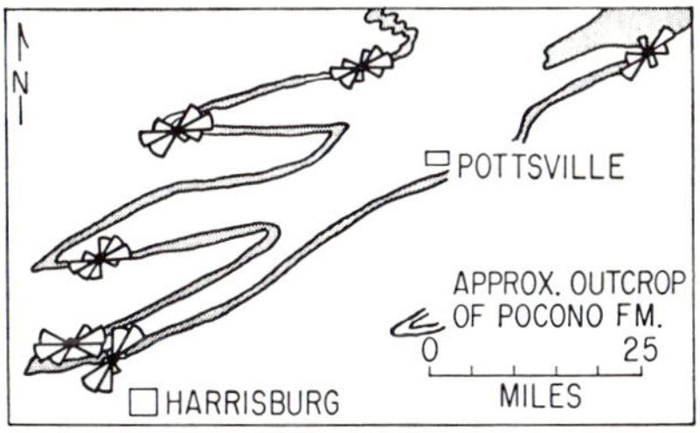

FIG. 13-70. Map showing by means of rose diagrams the orientation of plant remains in the Pocono formation in Pennsylvania. (*After B. R. Pelletier, Pocono Paleocurrents in Pennsylvania and Maryland, Bull. Geol. Soc. Am., vol. 69, p. 1052, 1958.*)

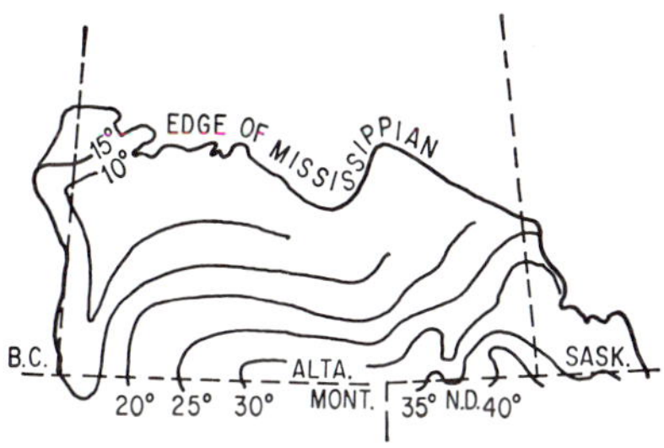

FIGURE 13-71. Isogravity contour map of crude oil in Mississippian rocks in the Canadian portion of the Williston Basin. (*After G. B. Darling and P. W. J. Wood, Habitat of Oil in the Canadian Portion of the Williston Basin, in "Habitat of Oil," p. 141, AAPG, 1957.*)

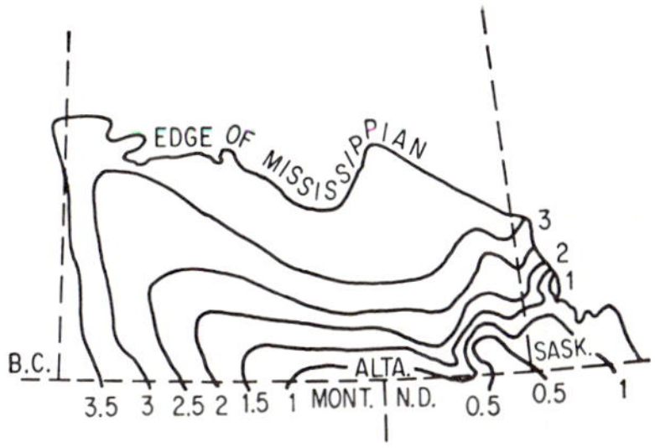

FIGURE 13-72. Isosulfur contour map showing percentage of sulfur in crude oil in Mississippian rocks in the Canadian portion of the Williston Basin. (*After G. B. Darling and P. W. J. Wood, Habitat of Oil in the Canadian Portion of the Williston Basin, in "Habitat of Oil," p. 141, AAPG, 1957.*)

CONCLUSION

It must be emphasized that the work of the subsurface geologist is never finished Subsurface maps are progress maps, only as complete as the data used when they are made. When additional information becomes available, as new wells are drilled or old wells reexamined, the maps must be corrected, revised, perhaps entirely reinterpreted. Also, better techniques are continually being developed to obtain new or more precise subsurface data or to interpret data more rigorously. The subsurface geologist must, therefore, be alert and imaginative, ready to use new kinds of data, to reanalyze old data, and perhaps to discard earlier interpretations in favor of a more realistic way of putting his information together. When one realizes that literally hundreds of geologists are working with the same data, seeking some indication that will suggest a drilling prospect, it becomes evident that the successful subsurface geologist will be the one using dependable data, painstaking and careful thought, and, above all, the imagination to fill in the gaps left by data that are all too frequently widely scattered, insufficient, or inconclusive.

REFERENCES

Badgley, P. C.: "Structural Methods for the Exploration Geologist," 280 pp., Harper & Brothers, New York, 1959.

Bishop, M. S.: "Subsurface Mapping," 198 pp., John Wiley & Sons, Inc., New York, 1960.

Dake, C. L., and J. S. Brown: "Interpretation of Topographic and Geologic Maps," 355 pp., McGraw-Hill Book Company, Inc., 1925.

Haun, J. D., and L. W. LeRoy (eds.): "Subsurface Geology," a symposium, 887 pp., Colorado School of Mines, Golden, Colo., 1958.

Kay, M.: Paleogeographic and Palinspastic Maps, *Bull. AAPG*, vol. 29, pp. 426–450, 1945.

Krumbein, W. C., and L. L. Sloss: "Stratigraphy and Sedimentation," 497 pp., W. H. Freeman and Company, San Francisco, 1955.

Lahee, F. H.: "Field Geology," 6th ed., 926 pp., McGraw-Hill Book Company, Inc., New York, 1961.

Landes, K. K.: "Petroleum Geology," 2d ed., 443 pp., John Wiley & Sons, Inc., New York, 1959.

LeRoy, L. W., and J. W. Low: "Graphic Problems in Petroleum Geology," 238 pp., Harper & Brothers, New York, 1954.

Levorsen, A. I.: "Geology of Petroleum," 703 pp., W. H. Freeman and Company, San Francisco, 1954.

Levorsen, A. I.: "Paleogeologic Maps," 174 pp., W. H. Freeman and Company, San Francisco, 1960.

Lobeck, A. K.: "Block Diagrams and Other Graphic Methods Used in Geology and Geography," 2d ed., 212 pp., Emerson-Trussell Book Company, Amherst, Mass., 1958.

McKee, E. D., et al.: "Paleotectonic Maps of the Jurassic System," Miscellaneous Geologic Investigations Map I-175, U.S. Geological Survey, Washington, D.C., 1956.

McKee, E. D., et al.: "Paleotectonic Maps of the Triassic System," Miscellaneous Geologic Investigations Map I-300, U.S. Geological Survey, Washington, D.C., 1959.

Russell, W. L.: "Structural Geology for Petroleum Geologists," 427 pp., McGraw-Hill Book Company, Inc., New York, 1955.

Schlumberger Well Surveying Corporation: "Introduction to Schlumberger Well Logging," Document 8, 176 pp., Schlumberger Well Surveying Corporation, Houston, Tex., 1958.

Sloss, L. L., E. C. Dapples, and W. C. Krumbein: "Lithofacies Maps, An Atlas of the United States and Southern Canada," 108 pp., John Wiley & Sons, Inc., New York, 1960.

Weller, J. M.: "Stratigraphic Principles and Practice," 725 pp., Harper & Brothers, New York, 1960.

Chapter 14

ECONOMIC ANALYSIS OF EXPLORATORY PROJECT

By Frank B. Carter

Consulting Petroleum Geologist. Formerly Senior Administrative Geologist, General Petroleum Corporation
Glendale, Calif.

and

Milton T. Whitaker

Staff Reserves and Evaluation Engineer, Mobil Oil Company, Los Angeles Division
Santa Fe Springs, Calif.

After an apparently favorable petroleum prospect has been located by the various geological and geophysical methods available, there are numerous factors which must be considered before the prospect can be recommended as economically attractive for drilling.

LAND SITUATION

The first of these factors must necessarily be the cost of acquiring the mineral rights (concession, lease, or fee) on the land covering the prospect. Leasing problems are relatively simple in areas in which land holdings are in large tracts but tend to become difficult and costly when the holdings are in small parcels and in town-lot areas. Large land-holding companies usually have fixed rental and royalty rates from which they will not deviate, and these rates are sometimes very high. Many states have statutory regulations, which require leases on state-owned lands to be obtained only by competitive bids. Leases on Indian tribal lands are likewise nearly always issued only on the basis of cash bonus bids plus a fixed royalty. Many cities have ordinances which prohibit drilling within their limit except by special permission. Waterways, school lands, and even cemeteries can cause a great deal of trouble and expense when they occur within an oil-prospect area.

City and county planning commissions frequently must approve all drill sites within their area of jurisdiction. State and/or Federal agencies, such as the California

State Division of Oil and Gas and the Texas Railroad Commission, must usually be consulted prior to drilling, and their approval obtained for the drilling program. The U.S. Geological Survey must be consulted, and their approval obtained when operating on government leases.

In most countries, other than the United States and to some extent in Canada, all minerals are government owned and leases or concessions to prospect must be obtained from the government (see Chapter 16).

GEOLOGICAL FACTORS

Before actual drilling begins on an exploratory project, it is advisable to consider certain geological factors, viz.:

1. *The probable depth* to the objective zone is an important consideration that affects the cost of drilling. Drilling costs usually can be estimated with considerable accuracy, particularly by experienced drilling contractors, if the probable total depth can be predicted.

2. *The thickness and reservoir properties of the objective zone or zones* are important factors in the ultimate pay-out of an exploratory project. The thicker the pay zone, the longer lived the well will be, is a general rule applicable to most wells. Similarly, the better the reservoir conditions, i.e., good permeability and porosity that exist in the oil- and gas-bearing zones, the more prolific will be the production.

3. *The geologic age of the producing zone* is of less importance than the reservoir characteristics except by analogy. In a given area or region it may be evident from the history of nearby oil or gas fields that certain zones or formations are more productive than others. Thus, an exploratory prospect whose objective zone is one which historically has been found to be a good producer is more attractive than one which does not possess this characteristic.

4. *The probable extent of the productive area* must be considered in the event oil or gas is found by the exploratory well. The structural conditions which have created the trap in which the accumulation has occurred usually are the most important factors governing the extent of the productive area. Sedimentation may also be important, particularly in stratigraphic-type fields. From an economic point of view, the difference between a small-size prospect at considerable depth and a larger one at a moderate or shallower depth is readily recognized.

ACCESSIBILITY AND OUTLET

The operator of an exploratory project must also consider a means of access to the site selected for the prospect well and an outlet for the oil and/or gas which he may discover. Gas discovered in a remote area, for example, must be in sufficient quantity and quality to justify the building of a pipeline to the field. Oil can be moved by tank truck to a market point in limited quantities, but this is a costly and laborious means of transportation. Road building, particularly in mountainous areas, is costly, and the roads are difficult to maintain. Ultimately, pipelines will have to be built if the reserves of the new find are sufficient to justify the expense of construction.

OPERATIONAL PROBLEMS

After drilling operations have started, operating difficulties can often spell the difference between success and failure in the development of an exploratory project. The distance from roads and other facilities, such as fuel, water, and power, is an important factor to consider. In recent years, electric, diesel, or other internal-combustion engines have almost completely supplanted the time-honored steam engines for power in the oil fields. This has eliminated the need for the large supplies of water and fuel that were needed by the steam rigs. However, reliable sources of water and fuel are essential. The availability of supplies, tools, drilling mud, bits, and other equipment is also important.

Certain areas are frequently characterized by their own peculiar drilling problems, and the operator should look at the records of any wells previously drilled in the area before making too many assumptions based on experience elsewhere. A search of the drilling records in a given area will often reveal the presence of such troublesome obstacles as very hard formations, extremely high-pressure flowing waters, heaving shales, running sands, and cavernous or vugular formations which cause circulation losses. Any of these obstacles may cause the loss of a hole after many weeks of costly effort. Directional, or "slant-hole," drilling, frequently employed in offshore areas or where surface sites directly above the desired bottom-hole position are not available, will add appreciably to the drilling costs. In the final analysis, the probable cost of the well can be most accurately determined by estimating the time to drill, which is based on the anticipated total depth, and applying a per diem rate, to which must be added the cost of casing, plus special services, such as cementing, mud logging, electrical logging, coring, hole surveying, etc. The actual cost of drilling the hole can be estimated by drilling contractors on either a footage or daily basis. These costs will vary widely with the depth and area in which the wells are drilled. In California, for example, where depths commonly range from a few hundreds of feet to more than 15,000 feet, the cost of drilling currently ranges from $2.50 to more than $12 per foot. On a time basis, these costs might range from $875 to more than $1,400 per day.

Offshore drilling costs are normally far in excess of those for comparable depths onshore. This is principally owing to the expense involved in preparing a suitable drill site and the problems of logistics in connection with drilling operations. Experience in the offshore areas of the Gulf of Mexico and California indicates that costs will be in the order of four times as much as onshore drilling to similar depths. These costs added to the high bonuses and royalty rates paid in competitive bids for state-owned tideland leases indicate a long period before pay-out in most offshore operations (Cram, 1957).

WELL SPACING

Well spacing ordinarily is not important prior to discovery. It is, however, pertinent to economic studies and for that reason is discussed briefly here. Reservoir characteristics largely control the well-spacing pattern during exploitation. Fields with thick or multiple zones will require more wells to effect maximum drainage than those with a single or thin zone. Porous reservoirs will produce more oil per unit of measurement than will tight reservoirs. Thus, the economics of the anticipated recoveries, based on thickness and saturation of the pay zone, enters the picture. Obviously, the greater the number of wells drilled to a single reservoir, the greater the ultimate recovery, per surface acre, of oil and/or gas will result. There is, however, a practical limit to the number of wells that can be drilled, which is controlled by the cost of drilling and operating and is based on the estimated ultimate recovery from each well. Since depth is the principal factor governing drilling costs, this again has a bearing on the problem of well spacing. For example, with an anticipated ultimate recovery of 10,000 barrels per surface acre and a drilling cost of $100,000 per well, plus the cost of production, it is evident that a spacing of less than 10 acres per well would probably be uneconomic. There is no rule of thumb in the matter of well spacing. Each field or pool must be considered separately.

Factors of a technical nature to be considered in well spacing are as follows:

1. Thickness of productive zones.
2. Depth to productive zones.
3. Viscosity of the oil. Viscous oils do not possess the mobility which permits their ready passage through the reservoir as do the lighter, less viscous oils and will require a closer spacing of wells to effect maximum efficient drainage.
4. Gravity of the oil. Lighter gravity oils usually contain more dissolved gases, possess more mobility, and are less viscous than the lower gravity oils and will require fewer wells to effect maximum efficient drainage.
5. Reservoir pressures. High-pressure reservoirs, particularly if maintained by recycling operations, will result in higher recovery per well, and therefore, a wider spacing can be employed.
6. Reservoir properties. Porosity, the ability to contain fluids, and permeability, the ability to allow fluids to flow through the reservoir to the well bore, are very important factors in well spacing. Porous and permeable reservoirs can be effectively drained with fewer wells than tight reservoirs.

The following non-technical factors affect well spacing:

1. The spacing pattern adopted by neighboring or offsetting properties
2. Rate of production desired, due to lease expiration, market price, market demand, etc.
3. Proration laws, in effect in some states, which dictate the amount of oil or gas an operator may produce
4. Rate of pay-out desired
5. Tax problems, e.g., desire to defer income and distribute over a longer period of time or to take income rapidly over a short period of time

Well spacing in gas-field development is generally considerably wider than in oil fields. This is owing to the fact that hydrocarbons in a gaseous state possess more mobility than do liquid hydrocarbons. Consequently, a much larger area can be effectively drained of its gas by a single well than if the product being produced is a liquid hydrocarbon. Basically, however, the same principles, both economic and geologic, which govern the spacing of oil wells apply also to gas development. Thickness and multiplicity of zones, depths, pressures, and reservoir properties, which determine the ultimate recovery per surface acre, will also control the well-spacing pattern in gas fields. As a general rule, well spacing in the principal gas fields, in the United States, ranges from 160 to 640 acres per well. Many states have regulatory agencies which determine the spacing pattern for gas wells within their area of jurisdiction.

ESTIMATING PROSPECTIVE RESERVES

An exploratory project is generally on unproved lands, i.e., lands not known to contain oil or gas but which, based on geological or geophysical data, have good prospects of becoming productive. However, the project might be on a known producing structure and is programmed with the objective of finding new pools either above or below producing formations or separated from known pools by faults.

The geologist should be able to assign "prospective or possible" reserves to lands upon which an exploratory well is to be drilled. This estimate of reserves should enable the geologist to evaluate, or "grade," his prospective projects. The evaluation should also serve as the basis to justify to management the acquisition of lands or the drilling of an exploratory well.

Prospective or possible reserves represent volumes of oil or gas that might be produced from a reservoir when and if it is drilled. On a recommended project the geologist can make an estimate or "educated guess" as to the ultimate recovery based on analogy or on solution of the volumetric equation using assumed values for the terms involved. Methods of calculating reserves after discovery are discussed in Chapter 25.

Volumetric Method for Oil

The volumetric method basically represents a study of the space in the productive zone and the proportion of that space that contains hydrocarbons that can be produced economically, i.e., yield a profit after costs. This method is important because it is the only analytical approach that can be used to estimate the quantities of hydrocarbons originally in an undiscovered reservoir and the possible ultimate recovery therefrom.

The equation for a volumetric calculation is

$$V = \frac{At \times 7{,}758 \times \phi(1 - S_w)}{FVF} R \tag{14-1}$$

where V = ultimate recovery of stock tank oil, bbl of 42 U.S. gal
A = productive area, acres
t = average net thickness of pay, ft
7,758 = number of 42-gal barrels in 1 acre-ft
ϕ = porosity of pay zone as decimal fraction
S_w = interstitial water as decimal fraction of pore space
FVF = formation volume factor which converts barrels of oil and its dissolved gas in the formation to stock-tank barrels (the reciprocal, the shrinkage factor, may be used in lieu of FVF; if used, it appears as a multiplier in the numerator)
R = recovery factor as decimal fraction of the oil in place that is recoverable

Values assigned to the terms in Equation (14-1) must be based largely on assumptions until a well actually penetrates the objective zone and gets production or at least showings of oil therein. Subsurface structural maps constructed from geophysical data and/or from information furnished by wells located not too far away may suggest a reasonable value for the productive acreage and perhaps other parameters. In the majority of instances, particularly in pioneering exploratory projects, values for the terms are selected by analogy with the stratigraphic, structural, and reservoir characteristics of nearby or even distant producing areas.

It will be noted in Equation (14-1) that the first three terms in the numerator give the bulk volume of the reservoir in barrels, multiplication of the bulk volume by the fourth term gives the total pore space available for all fluids, multiplication of total pore space by the fifth term gives the pore space available for hydrocarbons, and division of pore space for hydrocarbons by FVF gives barrels of stock-tank oil in place. The final step in the calculation is application of the recovery factor R, the value of which *must* be chosen by analogy.

Example of Calculation. *Assumptions:*

Productive area A = 1,000 acres
Average net thickness of pay t = 25 ft
Porosity ϕ = 20 per cent (0.20)
Interstitial water S_w = 30 per cent (0.30)
Formation volume factor FVF = 1.11
Recovery factor R = 30 per cent (0.30)

Solution: Substitution of the assumed values in Equation (14-1) gives

$$V = \frac{1{,}000 \times 25 \times 7{,}758 \times 0.20 \times (1 - 0.30)}{1.11} \times 0.30$$
$$= 7{,}339{,}000 \text{ bbl (rounded)}$$

This calculated ultimate recovery amounts to approximately 294 barrels per acre-foot from the 25,000 acre-feet, the bulk volume of the reservoir.

The assumed value of 1.11 for *FVF* indicates that each barrel of oil in the reservoir contained about 150 cubic feet of dissolved gas (see Chapter 25). It is sufficiently accurate to assume, prior to actual production, that the ultimate recovery factor for the dissolved gas will be the same as for the oil. In the problem as stated, the ultimate production of gas would be 1,100,850,000 cubic feet or, as customarily written in thousands of cubic feet, would be 1,100,850 MCF. Assuming 1 gallon of natural-gas liquids from each thousand cubic feet of gas gives 1,100,850 gallons of recoverable natural-gas liquids, the ultimate recovery of gas calculation is

$$7{,}339{,}000 \times 150 = 1{,}100{,}850{,}000 \text{ cubic feet}$$

Data Used in Volumetric Calculations

An oil reservoir is made up of porous and permeable rock, the interstices of which contain oil, gas, and water under pressure. The reservoir rocks are generally sands, limestones, or fractured shales which have the ability to contain and transmit fluids.

Porosity ϕ. This represents the void space in the bulk reservoir rock that is capable of containing fluids. If one-fourth of the reservoir volume consists of pores, then the reservoir has a porosity of 25 per cent. Most sand reservoirs have porosities that range from 10 to 35 per cent, although some may be as low as 5 per cent.

Interstitial Water S_w. Oil reservoirs, particularly sands, contain in their pores interstitial or connate water (ranging from 10 per cent upward) held by capillary forces. This interstitial water is an important factor, as it reduces the amount of pore space available for hydrocarbons. It has been established that a relationship exists between the interstitial water content and the permeability of the rock. Generally, the lower the permeability or the "tighter" the reservoir, the greater the water saturation of the pore volume.

Formation Volume Factor *FVF*. This represents the number of barrels of reservoir space occupied by one barrel of tank oil plus its dissolved gas. Oil, under reservoir pressure, normally contains gas in solution, and oil expands in volume when gas is dissolved in it. Therefore, each barrel of stock-tank oil measured at 60°F, which contains dissolved gas, will occupy more space in the reservoir where it contains dissolved gas.

The formation volume factor is sometimes expressed as a shrinkage factor. The shrinkage factor *SF* is the reciprocal of the formation volume factor and represents the percentage reduction in the volume of a barrel of reservoir oil as the dissolved gas is removed from it. It should be noted that the gas dissolved in the reservoir oil has important beneficial effects on the physical properties of the oil, the most important being the reduction of viscosity. Gas in solution thus renders the oil more mobile than it is after the release of gas from solution which occurs with pressure decline.

Recovery Factor *R*. This represents that percentage of the tank oil in place that is commercially recoverable, i.e., economic. The percentage recovery is a function of the producing mechanism, or "drive." There are three general classifications of producing mechanisms: dissolved gas, gas-cap expansion, and water drive. Actually, in some reservoirs because there exists a favorable combination of oil mobility and dip in the producing beds, gravity drainage or segregation takes place. These reservoirs will yield a greater portion of their oil than those where gas alone is the only source of energy. Most reservoirs will produce some oil as a result of gravity drainage. Many reservoirs will have a combination of two or more of these producing mechanisms.

Dissolved Gas Drive. This is probably the most common producing mechanism and is sometimes referred to as the "solution gas" or "depletion" type drive. The dissolved gas drive derives its energy from the gas which is in solution in the oil. When the pressure in the reservoir is reduced, the gas comes out of solution and expands. This expansion serves as an expulsive force causing the oil to be driven to the well bore. This mechanism generally has the lowest efficiency in percentage recovery of the original tank oil in place of the three basic drives. Range of recoveries depends upon the amount of gas in solution and the viscosity of the oil. Lower gravity oils will range from 10 to 15 per cent; higher gravities up to 25 per cent.

Gas-cap Expansion Drive. In this type of drive, the oil is in direct contact with "free gas" or gas cap. The external energy of the expanding gas cap maintains pressure on the oil sand and thus retards the gas from coming out of solution. Where an original gas cap does not exist, it is sometimes advisable to create one by injecting gas at the crest of the structure. Gas can also be injected to maintain or enlarge an original gas cap. Such an operation is commonly known as "pressure maintenance." Depending upon the extent of the free gas and the viscosity of the reservoir, recoveries may range up to 35 per cent and higher.

Water Drive. In this mechanism, water advances into the oil sand, replacing oil and gas as they are withdrawn. This displacing action of the advancing water tends to maintain pressure in the reservoir. Water-drive pools are generally believed to be the most efficient natural oil-recovery mechanism. Recoveries may range up to 50 per cent and higher, particularly if the sands are homogenous and possess good porosity and permeability.

Volumetric Method for Nonassociated and Associated Gas

The equation for the calculation of recoverable gas by the volumetric method is

$$V = \frac{43{,}560}{1{,}000} At\phi(1 - S_w) \times \frac{460 + 60}{460 + T_r} \frac{P_r}{P_b} \frac{1}{Z_r} R \tag{14-2}$$

where V = volume of recoverable gas, MCF
43,560 = number of cubic feet in 1 acre-ft
A = productive area, acres
t = average net thickness of pay, ft
ϕ = porosity of pay zone expressed as decimal fraction
S_w = interstitial water expressed as decimal fraction
460 = number added to °F to get °R
T_r = temperature at mid-point of reservoir, °F
P_r = average initial reservoir pressure, psia
P_b = standard gas pressure measurement base, psia
Z_r = compressibility factor at original reservoir conditions
R = recovery factor expressed as a decimal fraction

The first five terms in Equation (14-2) give the reservoir volume (in thousands of cubic feet) filled with gas. The sixth, seventh, and eighth terms convert that volume of gas to volume at standard conditions of temperature and pressure at the surface and adjust it for the deviation of hydrocarbon gas from the ideal-gas laws. The final term R is the very important recovery factor.

The standard gas-measurement temperature base is 60°F. There is no universally accepted gas-measurement pressure base. Some states, for example, establish the base pressure by law. The Federal Power Commission uses 14.73 psia, whereas the American Gas Association uses 14.65 psia.

The terms A, t, ϕ, S_w, and R were explained in the discussion of Equation (14-1). The only further comment necessary is that the percentage recovery of the gas in place in nonassociated and associated gas reservoirs is generally much higher than the like figure for oil from oil reservoirs. Gas recoveries of 60 to 90 per cent are common.

Charles's law states: At constant pressure the volume of a gas will be directly proportional to its absolute temperature. Multiplication by the fraction $(460 + 60)/(460 + T_r)$ converts the volume of gas at reservoir temperature to its volume at base measurement temperature. Boyle's law can be stated as follows: At constant temperature the volume of a gas is inversely proportional to its pressure. Multiplication by the fraction P_r/P_b converts the volume of gas at reservoir pressure to its volume at some base measurement pressure. Hydrocarbon gases are not perfect gases; they deviate from the ideal-gas laws. Multiplication by $1/Z_r$ corrects the calculated volume for deviation from the ideal-gas laws. There is fuller discussion in Chapter 25 of the three terms briefly treated in this paragraph.

Example of Gas Reserves Calculation. *Assumptions:*

$$
\begin{aligned}
\text{Productive area } A &= 5{,}000 \text{ acres} \\
\text{Average net thickness of pay } t &= 100 \text{ ft} \\
\text{Porosity } \phi &= 20 \text{ per cent } (0.20) \\
\text{Interstitial water } S_w &= 30 \text{ per cent } (0.30) \\
\text{Temperature at mid-point of reservoir } T_r &= 120°\text{F} \\
\text{Average initial reservoir pressure } P_r &= 3{,}015 \text{ psia} \\
\text{Standard pressure } P_b &= 14.65 \text{ psia} \\
\text{Compressibility factor } Z_r &= 0.94 \\
\text{Recovery factor } R &= 0.90
\end{aligned}
$$

Solution: Substituting the assumed values in Equation (14-2) gives

$$
\begin{aligned}
V &= \frac{43{,}560}{1{,}000} \times 5{,}000 \times 100 \times 0.20 \times (1 - 0.30) \\
&\qquad \times \frac{460 + 60}{460 + 120} \times \frac{3{,}015}{14.65} \times \frac{1}{0.94} \times 0.90 \\
&= 540{,}000{,}000 \text{ MCF (rounded)}
\end{aligned}
$$

Analogy

In estimating undiscovered reserves, analogy based on actual recoveries in other fields probably will be of more value than technical knowledge of reservoir mechanics or the mathematics of volumetric calculations.

Production statistics and reservoir data are available on older fields, thus enabling the geologist or engineer to calculate actual cumulative recoveries in barrels per acre or barrels per acre-foot for any given field or reservoir. Nothing is of more value to the estimator than historical knowledge of a similar reservoir. However, these statistical yardsticks should not be used as a substitute for judgment but as tools to make judgment more accurate.

Basically, either the geologist or his fellow engineer can go only so far with mathematics, for in the final analysis the judgment and the experience of the estimator determine to a large degree the realism of any reserves estimate of undiscovered petroleum.

EVALUATION OF VENTURE

The final step in the economic analysis of an exploratory project, prior to recommending or rejecting acquisition of acreage and drilling, is to compare anticipated costs with anticipated revenue, i.e., to put a dollar sign on the venture. It is expedient to do this in order to compare the relative economic merits of different projects available for exploitation. It should be kept in mind when considering such an evaluation that all figures used in the calculation are merely estimates, with the exception of the actual costs incurred prior to time of calculation.

Costs to Be Considered

Included in costs are all the expenses of the exploration department and the land department that are rightfully chargeable to the project under consideration. The other important item of cost is that for drilling the exploratory holes to prove the prospect; this includes dry-hole contributions to other operators' wells the results of which should give pertinent information on the possibilities of the prospect.

Source of Revenue

Anticipated revenue will come from sale of the operator's share (working interest) of the oil, natural gas, and natural-gas liquids that are estimated to be recoverable from the prospect. It is customary to use current posted prices for oil, gas, and liquids in these calculations.

Example of Evaluation

The values assumed for the different terms in the volumetric calculation on pages 14-5 and 14-6 are used in this example. Other factors that must be included in the evaluation were discussed in the first pages of this chapter. The evaluation outlined below is known also as an *exploratory project prospectus*.

Logistics. Located in broad valley; no topographic relief; readily accessible by good roads; 10 miles from Exburg, population 22,500; 3 miles from trunk pipeline.

Geology. No surface outcrops; nearby dry holes penetrated formations that are productive in Empty field, 20 miles away; one of dry holes had show of oil in objective zone; favorable structure outlined by geophysics and confirmed by structural interpretation based on data from dry holes; probable closure 100 feet.

Objective Zone. The Waterless zone at about 4,000 feet; this zone in the Empty field has produced 10 million barrels to date and has current estimated reserves of 5 million barrels.

Estimates for Volumetric Calculation. Productive area, 1,000 acres; net thickness pay, 25 feet; porosity, 20 per cent; interstitial water, 30 per cent; formation volume factor, 1.11, recovery factor, 30 per cent; gas/oil ratio, 150 cubic feet per barrel; natural-gas liquids yield, 1 gallon per MCF.

Exploration Costs

	Past	Estimated future	Total
Geology and geophysics	$100,000	$ 10,000	$110,000
Land and lease	5,000	50,000	55,000
3 exploratory wells (dry)	....	210,000	210,000
Total risk capital in case of failure	....	....	$375,000

Cost of Development if Project Is Successful. Drilling and equipping 50 producing wells with lease lines and tankage cost $3,000,000.

NOTE: One of these producers, the one that discovers production, will be charged back to exploration in calculating the *finding* cost of oil.

Prospective Value. Estimated ultimate recoveries are 7,339,000 barrels of oil, 1,100,850 MCF of gas, and 1,100,850 gallons of natural-gas liquids. No shrinkage in gas due to extraction of natural-gas liquids is shown because the sale value of those liquids is included in sale value of the gas.

Estimated revenue from the oil, including the natural gas and natural-gas liquids produced with it, is $2.40 per barrel, after taxes except income taxes. Estimated royalty is one-eighth; hence, working interest ultimate recovery is 6,421,625 barrels. Estimated total operating revenue is 6,421,625 × $2.40 = $15,411,900, or, in round numbers, $15,400,000.

Ratio of estimated operating revenue to risk capital is 15,400,000 ÷ 375,000, or about 41:1.

Ratio of estimating operating revenue to total exploration and development costs is 15,400,000 ÷ 3,375,000, or about 4.5:1.

Summary. The risk factor in this project is low due to better than average data on stratigraphy and structure; chance of success is probably 35 per cent. The project rates high also on accessibility, estimated ultimate recovery, and ratio of anticipated operating revenue to risk capital and to total cost of exploration and development. Acreage should be acquired and a test well drilled.

It will be noted that the preceding evaluation applies to a project that is successful and carries it through the development stage. Many times a prospectus will be submitted for a single exploratory well, which is simpler but involves the same factors. Many companies have prescribed prospectus forms. The refining of calculations beyond the point reached in the preceding evaluation is not warranted. All the calculations are based on estimates some of which may be fairly realistic but others just pulled out of the air. It is enough for the exploration man to give his best estimate of the reservoir volume, anticipated ultimate recovery of petroleum, and the hoped-for money return. Sharp-pencil calculations on the pay-out time, annual rate of production, ultimate net profit after income taxes, etc., are better postponed until some oil is found.

COST OF FINDING OIL

It is self-evident that an exploratory company must find oil at a cost per barrel less than the posted price. Actually the cost of finding must be much lower than the posted price so that the company can develop and produce the oil found and still sell it at a profit. Operators should know what it costs them to find oil and should compare their costs with the so-called average cost for the industry. Finding costs have increased rapidly over the past years; recent estimates suggest that it costs more to find a barrel of oil than the market value of that barrel. If such is the case, the industry would run out of funds for developing and producing the oil, which leads to the thought that oil producers would be liquidating themselves. The wide variance in published figures on the cost of finding oil is due to the way the "authorities" determine the number of barrels discovered and the dollars spent in discovering them.

There is need for a simple, realistic method of arriving at the cost of finding oil. Obviously the cost of finding will vary widely from year to year with the degree of success or lack of success in finding new supplies of oil which the individual operator or the industry as a whole experiences. For this reason, an average of several years will more truly represent the probable cost of finding oil in any given area. A method of arriving at a realistic figure, which will represent the cost of finding oil, is outlined below.

Costs to Be Included

1. All expenses in connection with operating an exploration department, including those attributable to:
   a. Geological operations
   b. Geophysical operations
   c. Research costs
   d. Consultants
2. All expenses in connection with operating a land department, including:
   a. Bonuses and rentals paid for leases and in lieu of drilling
   b. Taxes and other carrying charges attributable to these lands, less any revenue such as crops, sale of water, grazing rights, etc., from lands held in fee
   c. Dry-hole contributions to other operators, who drill wells affecting lands under lease or owned in fee
3. All expenses in connection with drilling exploratory wells including:
   a. New-field wildcat wells drilled in search of distinct new fields
   b. New-pool wildcat wells drilled in search of new pools in or near existing fields
   c. Deep tests drilled to explore below the deepest known productive zone in a field
   d. Outpost wells drilled a substantial distance from known production with the express purpose of extending a known productive zone but which involve sufficient risk to be classified as exploratory

A more complete definition and classification of exploratory wells is given in Chapter 4.

The total cost to be considered having been arrived at (items 1, 2, and 3 above), the next problem is to establish the quantity of recoverable oil against which to apply these costs. To attempt to arrive at a figure which would represent a "current" cost of finding oil would be a wasted effort, since the factors which enter into such a calculation are constantly changing and subject to much future revision. It is possible, however, to ascertain the cost of finding oil within a given period of time, such as a year or, with more accuracy, over a span of several years.

The ultimate recoveries of oil from all discoveries, which enter into the cost of finding oil, are evaluated in gross barrels, including condensate, on the basis of current information. Discoveries under this definition include any successes resulting from the drilling of any wells in the exploratory-well category listed above. The volumetric method of calculating ultimate recoveries from measurable physical data has been covered earlier in this chapter (see Chapter 25 also).

The ultimate recovery initially assigned to any given discovery is nearly always subject to revision (usually upward by addition) in the course of subsequent development. For the purpose of computing the cost of finding oil all revisions in such cases are credited back to the year of original discovery. In other words, a discovery made a number of years past is credited with all the oil or gas which it will ultimately produce, as estimated in the light of present knowledge, after several years of productive history have elapsed. Thus, the ultimate recoverable oil or gas discovered in any given year is subject to revision periodically. Obviously, the more productive history available, the more accurate the estimate of ultimate recovery will be. The greatest difficulty is experienced in evaluating the more recent discoveries in terms of ultimate recovery. It should be kept in mind, however, that new, partially developed discoveries are usually underestimated with respect to ultimate recovery, since the tendency is to evaluate on the basis of *proved* productive area rather than on the *probable* future size of the new pool or field. To arrive at some idea of the probable extent of the newly discovered field or pool requires a knowledge of the geology of the area in which the discovery has been made. From this knowledge, the geologist can make an educated guess as to the probable future extent of the field to which he may justifiably apply certain factors of reservoir thickness, saturation, porosity, etc., to arrive at an estimate of ultimate recovery.

Oil discoveries are evaluated in terms of ultimate recoveries of the number of barrels of liquid oil. Gas discoveries are evaluated in terms of ultimate recoveries of cubic feet of gas. To put these gas discoveries on a comparable basis to oil in order to compute the cost of finding oil, the ultimate recoveries of gas are converted into barrels of oil on the basis of the dollar value of the gas and its recoverable liquids equated to the average price of crude oil in the area under consideration. For example, if the gas has a wellhead value of 20 cents per MCF and the average price of crude oil is $2 per barrel, the conversion ratio would be 10,000 cubic feet of gas equals 1 barrel of oil. Gas which is produced with crude oil is similarly converted to an equivalent quantity of oil.

The gross costs accumulated under items 1, 2, and 3 above, which include all exploration, land, and drilling costs over a span of several years, divided by the ultimate recoveries of all discoveries of both oil and converted gas made during the same span of years will give what the authors consider to be a realistic figure for the "cost of finding oil."

REFERENCES

Brown, G. G., D. L. Katz, G. G. Oberfell, and R. C. Alden: "Natural Gasoline and the Volatile Hydrocarbons," 92 pp., sponsored by Natural Gasoline Association of America, 1948.

Buckley, Stuart E. (ed.): Petroleum Conservation, *AIMME Trans.*, 304 pp., 1951.

Calhoun, Jr., John C.: "Fundamentals of Reservoir Engineering," 417 pp., University of Oklahoma Press, Norman, Okla., 1958.

Cram, Ira H.: Some Offshore Exploratory Problems, *Bull. AAPG*, vol. 41, no. 7, pp. 1422–1428, July, 1957.

Dodge, John F., Howard C. Pyle, and Everett G. Trostel: Estimation by Volumetric Methods of Recoverable Oil and Gas from Sands, *Bull. AAPG*, vol. 25, no. 7, pp. 1302–1326, July, 1941.

Hager, Dorsey: "Practical Oil Geology," 6th ed., 589 pp., McGraw-Hill Book Company, Inc., New York, 1951.

Lahee, Frederic H.: Our Oil and Gas Reserves—Their Meaning and Limitations, *Oil Gas J.*, Nov. 30, 1950.

Lahee, Frederic H.: "Field Geology," 6th ed., 326 pp., McGraw-Hill Book Company, Inc., New York, 1961.

Paine, Paul: "Oil Property Evaluation," 204 pp., John Wiley & Sons, Inc., New York, 1942.

Vance, Harold: "Appraisal of Oil and Gas Properties from the Viewpoint of a Banker and an Oil Man," 80 pp., Bank of the Southwest, Houston, Tex., 1959.

Chapter 15

WRITING THE REPORT

By KARL A. MYGDAL

Coordinator, Producing Property Acquisitions, The Pure Oil Company
Denver, Colo.

Other People
Ending the Investigation

Evaluation

Appraising the Gathered Material
Reaching Conclusions

ORGANIZING THE REPORT

The Content and Its Arrangement

Selecting from the Gathered Material
Organizing to Serve the Reader
- The Summary
- The Introduction
- The Discussion
- The Conclusions and Recommendations
- The Acknowledgments
- The References and Bibliography
- The Appendix
- Examples of Report Organization

Illustrations

Elements of Effectiveness
Specific Suggestions for Geological Material

Format

Neatness
Spacing
Headings
Optional Features

WRITING THE REPORT

The First Draft
Checking and Modifying
Preparing the Master Copy
Acknowledgments
References and/or Bibliography

CONCLUSIONS

REFERENCES

INTRODUCTION

The report is the culmination of the geologist's* investigation of an exploratory project. It is the medium by which he transmits his ideas, conclusions, and recommendations to management or client. It must be clear, concise, and readily understandable by those not familiar with technical terms. The conclusions and recommendations must be devoid of ambiguity, must be summarized in the opening statement of the report, and must convey the geologist's confidence in their validity. The geologist's success in securing acceptance of his recommendations depends largely on his report. It is evident, therefore, that the ability to write a convincing report is one of the most valuable skills that a geologist or any other investigator in petroleum exploration can possess. On it may depend the course of his career.

The writing of reports that attain desired results is an art which can be learned. Those whose formal education did not include such instruction can gain much by study of the literature dealing with effective writing, by discussions with associates, and from thoughtful appraisal of good reports written by others. All these approaches are of immense value, but facility in writing will be achieved mainly by practice. "Practice makes perfect" is especially applicable to the writing of understandable and convincing reports.

This chapter summarizes the principles of effective report writing and the practical steps by which geological reports are planned and written.

Nature of Geological Reports

Many kinds of reports are prepared by exploration personnel. These range from comparatively simple ones, such as prescribed forms with blank spaces for information wanted or the well-sitter's description of samples recovered from a drilling well, through those which require much preparatory work prior to writing, such as interpretation of geophysical surveys, surface and subsurface geological investigations, or laboratory examinations of cuttings and cores, to the final geological report that correlates the results of all phases of endeavor on an exploratory project. Each report is necessary and helpful in exploration; the last-mentioned is the most important because it integrates the results attained by everyone who worked on the project.

The purpose of a report may be to inform, to recommend, to initiate action, or merely to record data. It may be directed to a district geologist, an exploration manager, the president of a company, a client, or a government agency. The readers may not be geologists or even familiar with geological terms. The writer may be of any rank and length of experience and may work in any phase of exploration. Whatever the nature of the required report, its proper design and completion depend on principles that apply to all report writing.

* EDITOR'S NOTE: *Geologist* is a generic term that includes all those actively engaged in exploration: scouts, landmen, geophysicists, photogeologists, surface and subsurface geologists, paleontologists, and those in laboratory and research work related to exploration.

Geological reporting is expository—a form of writing characterized by objectivity, honesty, clear reasoning, adequate information, and simple unadorned style. It is free from personal bias. The facts are permitted to speak for themselves; none is distorted or suppressed. Opinions are not expressed without supporting evidence.

A report that is published in a technical journal is called an article or a paper. The geological report discussed in this chapter is the unpublished private report prepared for management or client. While it must have the same clarity and objectivity as the published article, it has one marked difference: It is more purposeful because its chief intent is to stimulate action. The answers are more desired than the machinery by which they were captured. Most executives and clients want the answers early and directly and may not be interested in or even understand the data and reasoning that led to them. Consequently in geological reports the conclusions and recommendations are given in a summary on the first page. The rest of the report itemizes the steps performed, the evidence gathered, and the deductions made to arrive at the conclusions. A report should not create suspense but understanding—the sooner the better.

A comparison of the main divisions of the geological report and the published article is given in the following tabulation.

Geological report	*Published article*
Summary	Abstract
Introduction	Introduction
Discussion	Discussion
Conclusions and recommendations	Summary and conclusions
Acknowledgments	Acknowledgments
References and/or bibliography	References and/or bibliography
Appendix	Appendix

These divisions will be discussed in a later section of this chapter.

Problems and Procedures

Principal Difficulties

Geologists, in company with other report writers, usually face three principal difficulties:

1. Designing their reports to do the particular job required in a specified situation
2. Knowing where to start and what steps to follow in writing the report
3. Knowing when to stop

The geologist's difficulties usually do not stem from the fact that the report deals with geology. Most of them result from violation of the elementary rules of grammar, coherence, or emphasis, and from lack of knowledge of the basic principles of report design and organization. These nongeological aspects will be treated at some length in this chapter.

Four Stages in Report Preparation

The actual writing of a report is the final step in a four-stage process: (1) analyzing the requirements of the report, (2) investigation, (3) design and organization, and (4) writing.

Each of these four main steps will be treated in a corresponding section of this chapter. While they are distinct as concepts, in actual practice they fuse into one another. It will be necessary to retrace some of them from time to time, but despite this they are useful subdivisions which aid in visualizing the entire process of report preparation.

This chapter is intended to be fairly comprehensive, and not all readers will want or need all that is presented here. Not all reports will require the same attention to analysis and design; some are of standardized form, routine within an organization. More has been provided in this chapter than will be needed in all situations but it is hoped that the geologist will consider all of it. If the basic principles and practices described herein are made part of his thinking about report writing, he will be able to select and to adopt with some confidence that which is appropriate to his special writing problem.

Getting Started

Many persons find it difficult to begin work on a report. They are easily distracted and shrink from the mental effort needed to carry the project forward. This is especially true for those who write reports only occasionally; those who write often feel much less concern. For the novice the best advice is to plunge in, to roll up his sleeves and get started with enthusiasm and interest.

It is necessary to concentrate on the task. The greatest aid to concentration is interest (it is difficult *not* to concentrate on something of real interest), and the initial step toward concentration is the development of interest. Everything under the sun has something of interest, and once a reason for interest, however slight, has been discovered, it can be developed. The habit of conviction that interest is present and that it can be found will lead to the asking of self-directed questions on which any investigation can be launched or enlarged.

A great aid to concentration is the pressure of time, as last-minute toilers on income tax reports can testify. It is harder to concentrate when one is underworked than when there is too much to do. The habit of always attempting to complete a report in the shortest possible time, whether or not there is a deadline, will generate at least the interest of a race and set the stage for natural concentration. Sufficient time must be allowed, however, to ensure that the report will be comprehensive and convincing.

ANALYZING REQUIREMENTS OF THE REPORT

The initial point of attack in report writing is analysis of the problem. What is the purpose of the report? How will it be used? Who will read it? How can the needs of the reader be met with the least expenditure of his time and effort? The geologist must know what he wants to achieve, what is wanted by those for whom he is writing, and what is needed to satisfy that want. He must understand the function of his report. When he knows clearly what he has to say and why, he can proceed to design a report which will be appropriate to the need.

Purpose

Most industrial reports and correspondence, including those written by geologists, have as their purpose to inform, to advise or recommend, to persuade, to explain, or to provide a record. Whatever the purpose may be, the geologist must be aware of it. If he fails to make a recommendation when one is wanted, his report has failed. An unwanted recommendation in a report intended only to inform may be embarrassing. A report intended to provide a record may require more detail and different organization from reports for the other purposes listed.

How the Report Will Be Used

The geologist should determine the specific use to be made of his report, its time of usefulness, and the number of readers. The use made of reports varies considerably. A report may be read only once, by an immediate superior, or it may be read through several years by many people. It may travel vertically and horizontally within an organization or even outside it. In joint-interest operations some reports are read by other companies. A consultant's report may not go beyond his client's filing cabinet, but in some instances it may be forwarded to a bank or to prospective investors. These factors will influence not only what is put into the report but also the choice of the report form, the organization of the report, and perhaps even the format and choice of binding.

Considering the Reader

Giving the reader what he wants is fundamental to any writing, and the report that does not accomplish this is a failure. Nevertheless, technical writers often become so concerned with what *they* want to say that they fail to consider adequately the wants of the reader. The geologist shares with his supervisor or client the responsibility for establishing clearly what is wanted. If he finds that the instructions are not clear, he should seek clarification through discussion.

Three questions concerning the reader demand answers: (1) What does the reader want? (2) What does he need to know? (3) What is his background? Whatever the geologist decides, the answers to these three questions will affect all that he does and writes.

The report must present ideas so that the reader will understand them; that is its purpose. The transmission of ideas to the reader is the writer's responsibility, not the reader's. He cannot assume that the reader will struggle through highly technical language with which he is not familiar or that he will look up supporting material which has not been summarized. It is much easier not to read the report. Worse, antagonism to the geologist may be generated. The experienced geologist anticipates the reader's problem and provides clear answers and appropriate emphasis, all arranged so they can be readily recognized. Such consideration earns the gratitude of the reader and leads to acceptance of the report's recommendations.

The written word is easily misunderstood, and this fact should be recognized. Particular words and phrases may have special meanings for certain persons, and where the geologist is aware of these, he must be sure that he writes so those persons cannot misunderstand him.

Correspondence and Report Forms

The professional writing done by geologists can be grouped into the general categories of reports and correspondence. The difference between these is not so much in the nature of the contents as it is in the amount and complexity of the material itself and in the purpose and use of the writing. The geologist must be familiar with both, and the principles which apply to the preparation of reports apply also to the writing of correspondence. A letter is usually a small report.

Correspondence Forms

These consist of letters and memoranda. The letter form may be used for correspondence circulated within a company and must be used for any correspondence going outside the company. The memorandum is restricted to correspondence within a company.

Correspondence is particularly marked by two characteristics: It is short, and it is confined to one topic. Confining the subject to one topic simplifies consideration, decision, action, and filing. When more than one topic is considered, the effectiveness of the messages is reduced and filing is immensely complicated.

Report Forms

Four basic report forms are commonly employed in geological exploration. The choice of form depends on the use to be made of the report and on the length, complexity, and nature of the material to be presented. Within a company it will also be governed by established procedures.

Form Reports. These are usually employed in routine reporting situations and often consist of printed forms which are to be filled in. The design of the form report develops from a full knowledge of the specific information needed and of the use to be made of it. When filling in the blanks of a form report the geologist must never be general or vague, for this type of report challenges him to state his answers concisely and directly.

Letter Reports. The letter report is well suited to reporting situations that call for a short report of six to eight pages to be addressed to a specific reader, especially if it is likely that the report will not be read by any number of individuals. It is particularly suited to reports that are part of company correspondence. It is a modification of the regular business letter and, of all the report forms, allows the closest contact between writer and reader. It is approved practice to give the conclusions at the beginning of such a letter report and to let the supporting evidence follow in the body of the letter.

Article Reports. Short reports addressed to several readers or a group of readers are usually presented in article report form similar to a magazine article. This form is useful for reports not over 12 to 15 pages in length. A summary and conclusions section is provided as the first item.

Formal Reports. The more comprehensive reports are best handled in the formal report form, which resembles a book in method of presentation. A table of contents and a summary are always included to make the large amount of material easily accessible to its readers. Like the article report the formal report allows the geologist to shape his material for a wide range of readers.

The Report Outline

When the requirements of the report have been determined, the geologist should make a tentative outline. Outlining is planning, and it is inescapable in the long run. The first outline may consist only of principal topics and subtopics. Changes will almost certainly be needed as the design of the report progresses, but the preliminary outline gives a generalized picture of the task ahead which allows the geologist to proceed with confidence. Not only will all his work be easier once the outline is made, but for all but the briefest reports time will be saved and the final product will be better. By means of the outline, the major contours of the job can be formed on an experimental and flexible basis.

Two types of outlines are used by report writers: the topic and sentence types. The first consists of brief nonsentence entries arranged in divisions and subdivisions of rank. In the sentence type the major ideas are stated in sentences, and these can usually serve as the topic sentences of the paragraphs of the report.

The topic type of outline is most common and is essentially identical with a list of the headings and subheadings used in the body of the report. It serves well in geological reports of conventional area-evaluation type, as will be seen in the later section on Organization. It is less suitable in more complex reports and correspondence which deal with abstract matters, since the absence of sentence discipline makes it possible to jot down ill-digested fragments of ideas so that the result is more a record of confusion than an orderly guide to a finished report.

Whatever its type the outline should be built from a list. The geologist should list all the topics which occur to him without any thought to their importance or order. Any idea which would take up a paragraph in the finished report is large enough to list. Each topic should be written on a separate card or sheet. The principal topics, those which clearly identify an entire section of the report rather than just a paragraph, are then marked, and the remaining topics are segregated under the section topic in groups and subgroups in logical sequence. When a satisfactory sequence has been devised, the topics within the section can be transferred to section cards or sheets and finally to a single sheet which will form the outline, provide the headings of the report and, if desired, the table of contents. The alternative plan of trying to work out the entire outline at once on a single sheet of paper hinders the experimentation which is necessary to find the best arrangement.

Much of the difficulty in outlining lies in the selection of the divisions and subdivisions of the subject material. In most geological reports the principal topics, such as stratigraphy and structure, and their subdivisions related to time of deposition or folding are straightforward. Examples of suitable outlines can be constructed from the headings of the articles published in the *Bulletin of the AAPG*. The Bulletin uses only four ranks of headings, and this is ordinarily sufficient for an outline. Appropriate divisions must be selected on the basis of logic and relevance. Headings equal in rank must be equal in form, and one must not overlap the other in concept.

One purpose of the outline is to isolate topics so they can be treated separately by the geologist and by the reader. The geologist must be sure that each portion of his material is located in the appropriate division and in one division only. If a major topic is subdivided, there must be two subdivisions, since dividing the larger topic cannot logically result in a single smaller one.

When all the tentative subject material is accounted for in appropriate divisions and subdivisions without overlap and in logical sequence, the outline will constitute a model of the final report from which the geologist can estimate how large each section will be. After he has gathered and analyzed his material and has formed his conclusions, he will return to his tentative outline to polish and balance it, to insert new topics, and perhaps to eliminate others. A constant appraisal of the outline and a readiness to modify it when possibilities for improvement appear are necessary to achieve that orderly organization which distinguishes the good report.

The outline from which this chapter was written has been presented at the beginning of the chapter. It is hoped that it will illustrate the principles of outlining and provide a check list of the elements of report writing.

INVESTIGATION

The report may be written by any of the various qualified persons who have taken part in an exploratory project. Whoever does the actual writing will be thoroughly "saturated" with information on the phase of activity in which he participated—geophysics, geology, or other. His main sources of information for the report are his personal field or research work and that done by others engaged in the same investigation. The correlation of all data from original investigations related to the project is, of course, the vital procedure in preparing the report. There are other sources of information, however, which must not be overlooked. Earlier company reports and published articles may contain valuable information applicable to the specific problem in hand.

The geologist prefaces his own original investigations, if time permits, by a careful search of company reports and published articles that may pertain to his proposed project. Sufficient time for such examination of previous work is often lacking at the start of a venture; hence, the literature search must be prosecuted after completion of original investigation and before writing the report. The system of finding and recording pertinent information is discussed in this section.

When the requirements of the report have been determined, a program of gathering and evaluating information must be set up. The immediate problem is to determine how much information is needed. Obviously the geologist must be better informed than his reader, and he must gather sufficient data to allow him to reach and support pertinent conclusions. On the other hand all extraneous material must be excluded from consideration. There must be a proper regard for time, and the geologist must curtail his investigation early enough to allow himself time to review the data and to reach his conclusions. The criterion is always to relate the investigation to the requirements of the report.

Gathering Material

Gathering material is a process of discovering sources, using them, and recording their contents. Each source may suggest several others, and the approach is constantly modified as material is located.

The examination of large quantities of written material can be very time-consuming if not done with purposeful attention to what is needed. It is seldom necessary to read a source from end to end. The geologist should master the technique of scanning the entire source item, deciding what need *not* be read carefully, and marking the pertinent portions for study. Irrelevant material, no matter how intriguing, should be passed over. The techniques of rapid and efficient reading are important; several useful references to this field are provided at the end of this chapter.

For most geologists the principal sources of information are (1) original research or field work; (2) company (or other organization) files; (3) publications, technical and industry; (4) other people.

Original Investigation

For our consideration this important activity can be defined as the gathering of new data in the field, in the laboratory, or at the drilling well. It may form the entire subject of a report or be part of a more comprehensive investigation. Each type of original investigation will have its own useful tricks of the trade, but these often can be improved by the alert and ingenious worker. In all cases the data must be recorded accurately and completely in notes and journals understandable to others who may wish to check them. Specific suggestions for procedures are provided by Lahee (1961) and Sebring (1958).

Company Files

These include geological and geophysical reports, correspondence, scout information, maps and cross sections, and a great variety of data on individual wells. Every geologist should be familiar with the organization and content of his company's files and should know who is in charge of them and how to determine whether they contain material pertinent to his problem. If the writers of earlier reports are still employed, they can be consulted to advantage. Checking the files is a first step.

Geologic Literature

The library habit is essential to sound scientific research. Serious errors can be made in appraising local situations if the geologist is unaware of the geology of the region and of the conclusions of earlier workers. To dig out by original investigation what has already been discovered is an intolerable waste of time and energy.

While these views are generally accepted, some geologists are not sufficiently familiar with the procedures of a literature search. In this, as in all things, practice brings ease and what seems complex at first becomes second nature through use. There is nothing essentially difficult about library techniques. Success depends more on orderly systematic procedures, the mechanics, than on brilliance.

Indexes and Sources. Howell and Levorsen (1957) list each of the libraries in the United States and the Provinces of Canada, in Mexico, and in Central America. Pearl (1951) discusses geological literature in general, how literature searches are conducted, the organization of library indexes, library and abstract services, and the most important index guides, bibliographies, periodicals, and government publications.

The "Bibliography of North American Geology" is published by the U.S. Geological Survey and is complemented by "Bibliography and Index of Geology Exclusive of North America," published by the Geological Society of America. Since these bibliographies are published at intervals of 2 to 3 years and are thus always somewhat out of date, it is often helpful to refer to the monthly *Industrial Arts Index*, which covers technical and business periodicals, including 14 professional and trade periodicals dealing with oil and gas.

The periodical most important to petroleum exploration is the *Bulletin of the AAPG*, available in all libraries which cover the field of petroleum geology. Two comprehensive indexes have been published, 1917 through 1945 and 1946 through 1955. Annual indexes are printed in the December issues.

Excellent journals and guidebooks to field conferences are published by some of the local geological societies affiliated with the AAPG. The geologist should become acquainted with these and with the publications of the state or provincial geological surveys in his region. In some areas comprehensive bibliographies have been prepared by local geological societies or governments. Since geologic knowledge is expanding rapidly, most of it is recorded in periodicals and government bulletins rather than in books. Nevertheless books should not be ignored. They often provide a good general review and, if they are recent, a helpful list of references.

Finally the geologist should not overlook the most simple approach of all—the library card catalogue. Since catalogue practices vary he should make sure that he understands the system of the particular library he is using. If his company has its own reference library, the initial search for pertinent references can be conducted by the librarian.

In a literature search two separate problems are involved: how to find material of value and how to record what is found and to keep a record of its exact source. The recording problem is simplified by the use of two separate systems of cards, one for the bibliography and the other for notes.

Bibliography Cards. A bibliography should be started before the literature search begins and should be added to as the search progresses, using a separate note card for each item. There are many advantages in using cards instead of sheets of paper. The card can be filed conveniently. It is easier to use during the investigation; a rubber band can be flipped around a group of bibliography cards but not around sheets of paper.

Complete bibliographic details are recorded so there can be no question later about identification. The author, date, title, periodical, volume, page, publisher, and city are listed in exactly the same way as they would be if included in the bibliography section of the report. A consistent style should be adopted and adhered to. If the employer has not specified how bibliographic references shall be listed, the style used by the *Bulletin of the AAPG* (Wengerd, 1959, p. 209) is recommended. Examples follow:

Articles:

Gussow, William Carruthers, 1954, "Differential Entrapment of Oil and Gas: A Fundamental Principle," *Bull. Amer. Assoc. Petrol. Geol.*, Vol. 38, No. 5 (May), pp. 816-853.

Books:

Levorsen, A. I., 1956, *Geology of Petroleum,* W. H. Freeman and Company, San Francisco, 703 pp.

The bibliography cards may be typed and become part of a permanent file after the report is completed, with the cards filed on a geographic basis where possible. Fortunate is the geologist who can draw on such a file already established. If one is not available in his office, it is never too late or too early to start one. It may be profitable to annotate the entry, giving a brief description of its content and significance for later reference or for inclusion in the reference section of the report.

Note Cards. Notes should be taken on cards, never on sheets of paper. A separate card should be used for each entry, no matter how short. The topic or topic sentence to which the note applies, taken from the initial outline, should be written at the top of the note card. At the bottom should appear an adequate description of the source, including page number, so that weeks later there will be no question as to exactly where the note originated. Quoted passages should be copied exactly, enclosed by quotation marks, and be checked immediately for accuracy in order that no later verification will be needed. Any omission should be shown by marks of ellipsis: . . . . The geologist's own comments or paraphrasing, if any, should be distinguished from the quoted material by brackets. A sample note-card entry follows:

> Pre-Devonian history:
>
> "Relatively positive movements within central and southern Alberta prior to Devonian sedimentation are indicated by the regional relationships of the Cambrian, Ordovician and Silurian sediments. [Broad uplift may have removed Silurian and Ordovician; possibly no deposition occurred during part of time] . . . . Southern Alberta and Montana behaved tectonically as a shelf area during Devonian deposition, in some respects coinciding with their pre-Devonian positive tendencies." Andrichuk, J. M., and Wonfor, J. S., 1954, p. 2504.

Conducting the Literature Search. With the bibliography and note-card systems clearly visualized and with library and index sources located, the literature search can begin. Many geological reports deal with a structural or geographic entity, and a good starting point is to examine the listings under the appropriate heading (the Sweetgrass arch of Montana, for example) in the indexes of the *Bulletin of the AAPG*, in Eardley (1951), or in the regional index prepared by the local geological society, government agency, or library if one is available.

The most promising recent item should be examined first. In most cases the references or bibliography given in one publication will lead to items in other publications, so that the investigator is led into ever-widening areas of material.

Other People

The knowledge and experience of other geologists can be extremely valuable; this source is often overlooked. Where employers' interests are not affected and where this inquiry is skillfully presented, most geologists are quite willing to give information. In requesting it, the questioner must determine in his own mind what he wishes to learn. He must be prepared with specific questions. This preparation will aim the inquiry at the objective and will lessen the time which the giver of the information must sacrifice. Such prior preparation is an essential courtesy. Saving time and getting to the point are fundamental to cooperation.

Information can be gathered through a personal or a telephone interview, a questionnaire, or group discussions. Of these the interview is the most common and by far the most useful. It is important that the notes taken during the interview be "cleaned up" and recorded as soon as possible. Participants, place, time, date, and the preliminary questions should be listed, since often it is impossible to recall these details later.

Ending the Investigation

In the competitive world of petroleum exploration time is valuable. Investigation, no matter how fascinating, must not be extended for its own sake into areas not significant to the purpose of the report. There is always a limit of usefulness as well as of time. Thus when the search begins to yield significantly diminishing returns and when the data already gathered seem to have made valid conclusions possible, it is time to stop. The geologist must then leave the library and move on to the evaluating, organizing, and writing stages.

Evaluation

Once the material has been gathered, the important and difficult task of evaluation follows. It has two objectives: (1) to appraise the significance, reliability, and completeness of each item of the gathered material and (2) to reach conclusions which are based on it. Both these objectives must be pursued with the function of the report in mind: What does the reader want? How much does he need to know?

Appraising the Gathered Material

The geologist has naturally been evaluating each item as it came to him. Now he needs to review these tentative evaluations in the light of the whole and particularly in relation to the purpose and audience of the report.

The first step is to study the facts and data critically, to determine whether valid conclusions have been reached by the authors of the publications referred to and by the geologist himself in his tentative appraisal. Are these conclusions supported by the gathered material? Is the information accurate, complete? Do all the parts of the picture fit into a reasonable whole?

In the second step the main facts must be separated from the supporting facts. The certain must be distinguished from what is suggested but not fully supported.

Though both may be important, the true nature of each must be recognized in order to reach valid conclusions.

Reaching Conclusions

With the gathered material appraised, the geologist must consider the problem and the evidence and must reach his principal conclusion, at least tentatively. This must be done before the final organization of the report is begun and before the writing starts. When the geologist can write down these tentative conclusions, his investigation will have terminated. He can then proceed to the stages of organization and writing. In them his conclusions will almost certainly be modified and refined as he presents his findings in orderly fashion and states them in accurate language.

ORGANIZING THE REPORT

Organizing a report involves determination of the following:

1. The content and its arrangement in a logical sequence
2. The illustrative material required to make the report more understandable
3. The form and appearance

The general pattern of organization began to form with the preparation of the initial outline. It was modified during the investigation; now it must be formalized.

The Content and Its Arrangement

Selecting from the Gathered Material

This is a most important step, one that many writers find extremely difficult. Their tendency is to include too much, sometimes because they are reluctant to discard hard-won information, more often because they prefer to shirk the task of selection of what is pertinent. The result is a hodgepodge of material, without preparation and without purpose. So much of the reader's time is required to determine what is important that he is certain to become exasperated with the report and with the person who wrote it. The report writer shows his ability as much by what he omits as by what he includes.

When the investigation has been completed and the geologist knows what position he should take in his report, he faces the problem of communicating that position and the support for it to the reader. The two fundamental requirements of successful communication are:

1. That the writer knows clearly and specifically what he wants to say
2. That he states it in such a way that the reader will fully understand him

Selection of material should be approached with these two requirements in mind. The report writer need not include everything uncovered in his search for his conclusions. He does, however, need to state those conclusions and to support them with the most important evidence at hand.

Clear thinking is the basis of clear writing. The first step in organizing a report should be to write down in concise and specific terms the "core idea." This core idea becomes the basis for the structure of the report. All the material which the investigation has yielded should then be reviewed, and all material which is not relevant to the core idea should be discarded.

Organizing to Serve the Reader

Once the material for a report has been selected, it must be organized into a clear, concise, and purposeful unit. This step is the most important in the entire writing process. There is no one way to organize any report. The geologist's task is to determine which of the many possible organizations will prove most effective *for the reader*. His object is to inform the reader, as quickly and lucidly as possible, of the conclusions and of the principal supporting facts and reasoning.

An earlier section of this chapter dealing with the nature of geological reports referred briefly to the main divisions of such reports. Each of them will be discussed now more fully.

The Summary. This section should answer the questions which are most important to the reader: What has been learned from all this work? What is important about it? What should be done about it? The geologist must provide these answers or his report will have failed. If a recommendation was called for, it must be given. Even though the geologist might find it safer to let the reader determine the significance of the report, he must present his own conclusions. This section represents his professional judgment, and this judgment is his most valuable contribution to his employer. Appraisals and decisions are unavoidable.

The summary pertains to the entire report and includes a brief description of the problem which prompted the report, the methods used to solve it, and the principal findings. The conclusions which stem from the latter are given, as are the recommendations which grow out of them if such were requested.

This section should not exceed one page and is placed at the beginning of the report. The geologist must remember that this may be the only part of the report which will be read. Clarity, conciseness, logic, and firm conviction are more important here than in anything else that the geologist may write professionally. Whatever may be the content of the body of the report, and whether or not the latter is read, the geologist will be judged on the quality of this first page. He should strive to make it so effective that the reader will find it unnecessary to read further.

The title of the report precedes the summary and should be chosen carefully to attract immediate attention and to disclose directly and concisely what the report is about.

It will be noted that the *summary* of the geological report is replaced by the *abstract* in the published article. The abstract is a very brief statement of the purpose of the work done, its implementation, and the conclusions reached. It should give just enough to excite the reader's further interest or, conversely, to indicate that the article is not what the reader wants. It rarely includes a recommendation except perhaps to remark that the problem involved is not completely solved and merits further investigation. It should be even shorter than a summary.

The Introduction. This part of the report or article gives more detail than does summary or abstract on the reason for the investigation, the location of the project being considered, when the work was done, what branches of exploratory effort were involved, culture of the region or country, logistics of operating in the region or country, and the political situation. It is true that in some instances a few of the items mentioned might be of sufficient significance to merit individual sections in the report rather than inclusion as a small part of an introduction.

The Discussion. This is the body of the report or article. It describes and interprets the data accumulated during the investigation, i.e., the evidence on which the conclusions are based. The evidence and interpretation must be presented in a meaningful order. Man thinks consecutively, proceeding naturally from then to now and from here to there. He responds naturally to material which is presented in a time sequence or in an orderly grouping in space. If the mind must skip around and is made to consider information in an unconsecutive manner understanding becomes much more difficult.

Common bases for presentation of geological material are:

1. Event-time sequence, as in historical geology
2. Space relation, as in describing from top to bottom, east to west
3. Cause and effect, as in discussion of tectonics and sedimentation processes
4. Function, as in a report on the organization of a mapping program
5. Order of importance as in describing possible reservoir beds

Of these the one must be selected which best fits the reader's needs—the one which provides the clearest answer to the problem. If, for example, the purpose of the report were to determine the most important prospective reservoir beds in an area, these might better be described in the order of excellence based on past yield, thickness, porosity, and other criteria rather than by age.

Within these bases two patterns, applied concurrently, will help the reader.

1. The familiar-to-unfamiliar pattern in which each step necessary to understand that which follows is explained. The reader is led by the report writer, who answers his questions as he goes along. If the reader is forced to jump to a new idea unprepared, he usually becomes lost, confused, or at least disturbed.

2. The presentation of the main idea first, followed by a detailed and developed statement. This principle runs through several levels of the organization plan. The summary of the report precedes the details. Each main division has an opening general statement followed by detailed discussion, and each paragraph of the report has a topic sentence which is the main idea to be developed in the paragraph itself.

The practical approach to the establishment of a logical and consecutive order is to group the note cards covering the selected material under the headings of the modified outline. If note cards are left over, the material on these cards is superfluous or the outline needs further revision. The final arrangement of the central body of the report should be accomplished through this reconciliation of note cards and outline before the writing of the report is begun.

The effectiveness of a report depends on emphasis of the most significant items; this requirement is often overlooked. The geologist himself must first recognize the significant in his material, and he must know how to give it emphasis. Emphasis is provided by (1) position—placing material first or last in the report, paragraph, or sentence, (2) quantity—devoting more space to important material, (3) style—expressing important ideas with emphasis, and (4) layout—reinforcing important ideas with the mechanical devices of the layout; style and layout are considered chiefly during the writing stages.

If the investigation indicates that acreage should be acquired and a test well drilled, a rough estimate of costs vs. possible returns from the venture should be made (see Chapter 14).

The Conclusions and Recommendations. This section weighs the evidence presented under Discussion and states the conclusions developed therefrom. Recommendations should stem so smoothly and logically from the conclusions as to be almost self-evident. Both conclusions and recommendations must be forthright and decisive.

The Acknowledgments. Appropriate recognition of assistance received from others in conducting the investigation or writing the report is given in this section. This is required as a matter of professional ethics and, furthermore, is common courtesy.

Credit for quotations is given as a footnote at the bottom of the page on which the quotation starts. Credit for borrowed illustrations is given in a line below the illustration or as a part of the caption. Permission to use copyrighted material in a published article must be obtained from the original copyright holder, who may specify the exact form and position of acknowledgment required.

The References and Bibliography. Books or published articles specifically referred to in the text are listed under References. These are numbered and may be arranged chronologically by year of issue or by order of reference in the text or alphabetically by author. If there are only a few references, they may be entered parenthetically in the text or as footnotes.

The Bibliography cites other sources consulted in preparing the report or article. It is particularly useful for the specialist who wants to delve deeper into some subject treated and to the student.

The Appendix. Material that complements the text but is not immediately essential to an understanding of it is carried in an appendix. Conversion tables, for example, belong in this section.

Examples of Report Organization. The geologist should study the reports of others in order to become proficient in effective report organization. When the headings and subheadings of these are listed in outline form, the organization becomes clearly visible. Good models are provided by Heylmun, by Wengerd and Matheny, and by Sprigg, all in 1958.

A typical area-evaluation report outline follows.

SUMMARY

State the problem, the method of solution thereof, the conclusions, and the recommendations.

INTRODUCTION

Outline reasons for the work, authorization, location, types of exploratory effort utilized, culture of area, logistics, etc.

DISCUSSION

Technical Investigations

Geophysical Surveys
Surface and Subsurface Geology

Base Maps. Land net, air photographs, topographic.

Stratigraphy. Description and classification of formations, measured sections (from surface exposures and well information), areal distribution of formations maps (surface and subsurface), facies maps, history of sedimentation.

Structure. Structural contour maps, isopachous maps, stereograms, diagrams, structural sections, periods of deformation, surface and subsurface structure coordinated.

Regional. Correlation of local geological features with regional features as described in the literature or other reports.

Petroleum Possibilities

Currently Producing Fields

Location. Distance from area being considered.

Vital Statistics. Productive formations; productive areas; net thickness of pays; production cumulative and current; number of producing wells; gravity of oil, etc.

Estimated Reserves.

Economic Analysis

Relate objective zones to productive zones in nearby fields; summarize significance of results in nearby dry holes; make rough estimate of possible net revenue from production which may be found.

CONCLUSIONS AND RECOMMENDATIONS

State clearly and convincingly the conclusions that are derived from a review and evaluation of the evidence developed under Discussion. It is emphasized that the conclusions must be in conformity with the evidence. Recommendations proceed directly from the conclusions and should be equally unambiguous and persuasive.

REFERENCES AND/OR BIBLIOGRAPHY

APPENDIX

Illustrations

The importance of illustrations is apparent to geologists, since they deal constantly with maps, sections, and well logs and with tables of numerical and stratigraphic data. The report writer must select the most appropriate and expressive material from all that submitted by geophysicists, geologists, and others involved in the project.

Elements of Effectiveness

To be effective an illustration must be (1) useful, (2) properly integrated with the text (names used in the text must be shown on the supporting map), (3) located at point of greatest convenience (it should never precede the discussion), (4) clear and self-explanatory, (5) accurate, (6) grasped quickly (this depends on clarity of drafting and lettering and elimination of all that is not essential), (7) be remembered easily (this depends on design).

Simplicity is the key to effectiveness. A good rule is "the simplest figure that does the job is the best." All material not pertinent to the particular point being illustrated should be omitted; unnecessary detail distracts. Levorsen (1956) provides many examples of simple and effective illustrations.

Specific Suggestions for Geological Material

The geologist should develop an appreciation of what is effective and pleasing in maps, sections, charts, diagrams, and other illustrations by critical observation and emulation of suitable materials in the files of his company and in use by other companies and geologists. Maps should be neat and attractive. The drafting and reproduction should be of good quality, consistent with reasonable expenditures of time and money. If contours on maps are subject to later revision, they should be penciled on the prints rather than on the transparencies. The control data on contour maps should be clearly indicated. Contour values should be shown abundantly, particularly near important features. Oil fields should be outlined, named, and colored. Township and range or latitude and longitude data must not be forgotten, and key cities and topographic features should be shown so the reader can place himself. It is helpful to show a type electric log or lithologic log on maps.

Illustrations should be no larger physically than is necessary for clarity. A busy reader resents having to unfold large sheets on a desk which may already be crowded, and material which requires this chore may go unexamined. Waste margin space should be eliminated, and map coverage should be restricted to the area of pertinent data. If possible the sheet dimensions should be multiples of the text-page size. Where folding is necessary, it should be exactly to the text-page size, using an accordion-pleat bottom-to-top folding system. The title should be placed in the lower right-hand or lower left-hand corner, and the sheet folded so that this title will appear on the outside. Large maps should not be bound as sheets within the printed material. All maps, sections, and graphs should be numbered as "figures." Tables of data are called "tables" and numbered separately.

The use of color to emphasize important features is helpful. Colors should be applied lightly and should be those which are conventional in oil exploration: green for oil fields, red for gas fields, yellow for sand, blue for limestone, and so forth. If it is suspected that the report will be presented orally and that the maps will be used as wall illustrations, the titles and labels should be of sufficient size to allow reading across the room.

Tables of formations are important, since they will be referred to often. The U.S. Geological Survey uses the following form:

System	Group	Formation	Thickness (feet)	Character and distribution
Jurassic	San Rafael	Carmel formation	0–600	Red earthy sandstone in Colorado & eastern Utah; thickens westward & becomes gray & red shale, limestone & gypsum

Unconformity

A guiding rule in preparing illustrations is, as with all other parts of the report, to *consider the reader*. Will he, perhaps far away and unfamiliar with the problem (if that be the case), understand the illustrations easily? How can the geologist facilitate his understanding? Has he emphasized what is important and eliminated what is not pertinent?

Format

Neatness

Pleasing appearance gives a tremendous advantage to any typed or printed material. Meaning may be modified by the physical appearance of the sentence which conveys it. A penciled scrawl does not convey the same message as crisp typewriting, though the words may be the same. It follows therefore that the first requirement is neatness—absence of erasures, insertions, overwritings, or smudges.

Spacing

The second requirement is proper spacing. The eye associates spacing with significance. Key thoughts should be apparent physically as well as mentally. For the typist the chief means of giving visible extension to mental patterns are indenting, paragraphing, numbering, and underlining.

Headings

The third requirement of appearance concerns the use of proper headings. These will display graphically to the reader the structure of the report and will make the various parts easily accessible to him. By conveying a sense of order and purpose they will increase the reader's receptiveness and increase his respect for the writer.

It is seldom desirable to have more than four ranks in headings, and three are often enough. Center headings are of greatest importance, those at the left-hand margin above the first line of the paragraph are next, and those at the left-hand margin run into the paragraph are lowest in importance.

A consistent system of headings must be adopted. The following commonly used style is suitable for the typewriter:

FIRST RANK: CAPITALIZED, CENTERED, UNDERLINED

SECOND RANK: CAPITALIZED AND CENTERED

Third Rank: Separate Line, Flush with Margin, Underlined, Main Words Capitalized

Fourth Rank: Paragraphed, Underlined, Main Words Capitalized

A four-rank system is used in this chapter. When only three ranks are needed, the first or second rank shown above is generally eliminated.

Optional Features

In long reports a title page, a table of contents, and a letter of transmittal may be desirable. In a formal report an index may also be required. The choice of bindings and cover or their omission will depend on the customs within an organization and on the purpose and intended use of the report, the likes or dislikes of the reader, and the filing methods. The report writer should be aware of the choices which are possible and should select the report form and make-up which seem most appropriate to the purpose of the report and *to its reader.*

WRITING THE REPORT

If the geologist has followed the steps described so far in this chapter, he is ready to write his report. He knows what he needs to say; he has only to say it. He can follow his modified outline with assurance and clear purpose. His remaining task has three steps: writing the first draft, checking and modifying it, and preparing the final copy.

The First Draft

With a detailed outline before him and his note cards in proper sequence the geologist can begin to write. It is not such a formidable task. He should be full of his subject, and writing should come easily. The main thing is to start writing—with the introduction, with the section on stratigraphy or structure—anywhere fancy chooses, except with the summary or conclusions and recommendations, which are always written last. The idea is to get going, to write.

Whether the geologist works by section or on the report as a single unit, he should write as rapidly as possible. He should not be overconcerned with exact expression, for if he stops to polish and rephrase he is likely to interrupt the flow of ideas. Writing and rewriting will produce a polished final draft faster than striving for perfection from the beginning. As he writes, the geologist may have doubt about whether to include an item. It is usually best to include it and to mark it in the margin, since it is easier to cut a report than to add material later. If new material suggests itself, it should be included. Though he may follow his determined organization, he should not become a slave to it. In writing, plans are always tentative and open to alteration. The very act of writing produces new ideas which should be utilized if they are relevant and helpful.

In most technical writing, the impersonal viewpoint is preferable to the personal, since personal pronouns deprive the data and ideas of emphasis. Nevertheless, the personal approach may be suitable or preferable in business correspondence, provided the "you-attitude," the emphasis of the reader's interest, prevails.

The problems of effective expression in writing are too broad to allow adequate treatment here. Competence in this field can be earned only through attentive study and much practice. It is essential to use plain English that the expected reader can understand. Smith, director of the United States Geological Survey, wrote in 1922: "The geology I plead for is that which states facts in plain words—in language understood by the many rather than only by the few. Plain geology needs little defining. There is really somewhat of an obligation upon us, both as scientists and as partners in the world's business, to show the world that geology is not mystery or magic but only common sense. I have told practical men of business that they should give little credence to the geologist who cannot tell his story in common language. The world has a right to discount our usefulness and even to distrust our honesty if we persist in concealing our thoughts, or lack of thoughts, behind a mask of professional jargon."

Checking and Modifying

This is the stage where the report is polished. During it the geologist must again assume the reader's point of view. The report should be checked in four main areas: content, organization, expression, and layout. It is helpful to put the report aside for a day or two before checking it—or at least overnight. Review by others is helpful. If constructive criticism is requested, it should be welcomed when it is given and should be considered objectively and impersonally. The geologist and his report are not the same.

Every sentence and every paragraph should be carefully examined for clarity, meaning, emphasis, and logic. Paragraphing should be checked to make sure that each represents a logical thought division. The ideas must follow a logical and obvious pattern. Each statement should add to the meaning of the discussion, and the intended relationship should be made evident by the use of proper transitions.

The content and the conclusions which grow out of them must be reexamined. The geologist must read his first draft as if he had never seen it before and ask himself of every conclusion "Is this fully substantiated?" and of every fact "Is this really important?" The tendency is to include too much.

Preparing the Master Copy

When the report has been thoroughly checked and modified, it is ready for the preparation of the master copy, which will serve as the basis for reproduction of the required number of final copies. Where the report writer is experienced and where the report is not complex, the first draft can often be modified to serve as the master copy from which the stenographer or typesetter can prepare the reproductions. More often the modified draft copy is retyped into the master copy.

In all cases the geologist is responsible for giving the typist, draftsman, and reproduction workers clear and complete instructions as to what shall be done. There must be no doubts or uncertainties. Producing a clean and accurate master copy depends on a readable modified first draft and on complete and detailed instructions. The geologist is the only one who can be responsible for these. He will conserve the time of all concerned by anticipating all questions which the typist may ask.

The master copy must be proofread and all errors eliminated before final reproduction. The final report should be neat, attractive, and free from error.

Acknowledgments

All material borrowed from published sources must be acknowledged. Apart from the obvious ethical responsibility involved acknowledgment permits the reader to investigate the material further should he desire to do so, provides substantiation for points of view, and indicates the extensiveness of the study being reported. Often the list of references is among the most valuable parts of the report.

Acknowledgment can be made in three ways:

1. Where only a few entries are to be made, by reference within the body of the text: for example, Mr. William H. White, in the January, 1959, issue of the *Bulletin of the AAPG,* page 62, states in his article entitled "Cordilleran Tectonics in British Columbia" that "The Purcell rocks. . . . "

2. By listing all sources in the References section at the end of the report. Reference can then be made by a number assigned to the source (7, p. 16) or by the surname of the author, in which case the authors are listed alphabetically. This last system is common in petroleum exploration and is used by the *Bulletin of the AAPG.* (Wengerd, 1958, p. 206.)

3. By footnoting at the bottom of the pages. Each footnote is given a number, and the corresponding number is placed in the text as a superscript.

Each of these systems has its advantages and disadvantages. If a documentation system has not been prescribed by the employer, that used in the *Bulletin of the AAPG* is suggested. The techniques of documentation often seem awkward to novice report writers, but once mastered they become automatic and easy. They are an essential skill in all technical writing.

References and/or Bibliography

In a major report the references cited may be supplemented by the listing of additional published works of general background interest which the geologist believes the reader might wish to examine. The cited references and the background articles may be combined in a single list called Selected References (Wengerd, 1958, p. 208), or may be separated as was done by Levorsen (1956), who for each chapter listed pertinent background material under the heading Selected General Readings and the cited references under Reference Notes.

CONCLUSIONS

The geological report is best organized in the following sections: Summary, Introduction, Discussion, Conclusions and Recommendations, Acknowledgments, References and/or Bibliography, Appendix. It differs from many other technical and industrial reports in that it states briefly the conclusions and recommendations in the opening section—the summary. The report must be designed to fulfill a specific purpose and should be designed and written with *the needs of the reader always in mind.* The four stages of reporting are the analysis of the requirements of the report, the investigation, the organization of the gathered material, and the actual writing. The marks of a good report are clarity, brevity, accuracy, objectivity, adequate information, and elimination of irrelevant material.

The principles of good reporting are available in the literature on writing, and it is important to the individual geologist and to those whom he serves that these principles be learned and that they be perfected by attentive practice.

REFERENCES

American Geological Institute: Glossary of Geology and Related Sciences, *Natl. Acad. Sci. Publ.* 501, National Research Council, Washington, D.C., 1957.

Anderson, J. Ansel, and M. W. Thistle: On Writing Scientific Papers, 28 pp., *Can. J. Research*, National Research Council No. 1691, Ottawa, 1947.

Beard, Jr., J. C.: Plain Talk, *Petrol. Engr., Reference Ann.*, pp. A-35, 1954.

Bell, R. W.: "Write What You Mean," 109 pp., George Allen & Unwin, Ltd., London, 1954. Presents first principles of writing business documents and of clear thinking, as seen by a lawyer.

Blickle, Margaret D., and Kenneth W. Houp: "Reports for Science and Industry," 310 pp., Henry Hall and Company, New York, 1958. Contains many examples of actual reports and correspondence gathered from industry and specific suggestions regarding methods.

Cairnes, C. E., and H. M. A. Rice: "Guide for the Preparation of Geological Maps and Reports," Geological Survey of Canada, Ottawa, 1957.

Dix, C. Hewett: "Seismic Prospecting for Oil," 414 pp., Harper & Brothers, New York, 1952. Seismic reports are discussed on pp. 195–200.

Eardley, A. J.: "Structural Geology of North America," 624 pp., Harper & Brothers, New York, 1951. The bibliography contains 701 references.

Flesch, Rudolf: "The Art of Readable Writing," 228 pp., Harper & Brothers, New York, 1949.

Gunning, Robert: How to Get Your Ideas Across on Paper, *Oil Gas J.*, pp. 78–81, Oct. 31, 1955. This is a condensation of his book "The Technique of Clear Writing," McGraw-Hill Book Company, Inc., New York, 1952.

Heylmun, Edgar B.: Paleozoic Stratigraphy and Oil Possibilities of Kaiparowits Region, Utah, *Bull. AAPG.*, vol. 42, no. 8, pp. 1781–1811, 1958.

Howell, J. V., and A. I. Levorsen, with Robert H. Dott and Jane W. Wilds of the AAPG: "Directory of Geological Material in North America," 300 pp., American Geological Institute, Washington, D.C., 1957.

Judson and Baldridge: "The Technique of Reading," Harcourt, Brace & World, Inc., New York, 1954.

Kent, John L.: Improve Your Writing Skill, *World Oil*, pp. 116–118, Oct., 1957.

Lahee, Frederic H.: "Field Geology," 6th ed., chap. 22, McGraw-Hill Book Company, Inc., New York, 1961.

Landes, Kenneth K.: A Scrutiny of the Abstract, *Bull. AAPG*, vol. 35, no. 7, p. 1660, 1951.

Leedy, Paul: "Reading Improvement for Adults," McGraw-Hill Book Company, Inc., New York, 1956.

Levorsen, A. I.: "Geology of Petroleum," 703 pp., W. H. Freeman and Company, San Francisco, 1956. Each chapter includes a comprehensive bibliography.

Linton, Calvin D.: "How to Write Reports," 240 pp., Harper & Brothers, New York, 1954. General and philosophical treatment. Specific suggestions on how to proceed.

Lobeck, A. K.: "Geomorphology," 731 pp., McGraw-Hill Book Company, Inc., New York. 1939. On pp. 22–23 the deductive, leading sentence presentation is compared with inductive details-first writing.

"McGraw-Hill Author's Book," McGraw-Hill Book Company, Inc., 1955, New York.

Merrill, Paul W.: The Principles of Poor Writing, *Sci. Monthly*, vol. 64, pp. 72–74, 1947.

Paine, Paul: "Oil Property Valuation," John Wiley & Sons, Inc., New York, 1942. Chapter VII, The Examination and the Report, discusses evaluation reports.

Patrick, W. W.: English? *Bull. AAPG*, vol. 38, pp. 2558–2561, 1954. Concerns up-to-date language. Written with snap and humor and is a good reminder to avoid stuffy, cumbersome, and outdated usage.

Pearl, Richard M.: "Guide to Geologic Literature," 299 pp., McGraw-Hill Book Company, Inc., New York, 1951.

Sebring, Jr., Louis: The Subsurface Structural Map, *Bull. AAPG*, vol. 42, pp. 561–587, 1958.

Shaw, Phillip B.: "Effective Reading and Learning," Thomas Y. Crowell Company, New York, 1955.

Shurter, Robert L.: "Effective Letters in Business," 2d ed., McGraw-Hill Book Company, Inc., New York, 1954.

Souther, James W.: "Technical Report Writing," 70 pp., John Wiley & Sons, Inc., New York, 1957.

Sprigg, Reg E.: Petroleum Prospects of Western Part of Great Australian Artesian Basin, *Bull. AAPG*, vol. 42, no. 10, pp. 2465–2491, 1958.

U.S. Geological Survey: "Suggestions to Authors of the Reports of the United States Geological Survey," 5th ed., 255 pp., 1958.

U.S. Government Printing Office: "Style Manual," rev. ed., 292 pp., Washington, D.C., 1953. Contains specific guides to capitalization, spelling, compounding, punctuation, and abbreviation.

Wengerd, Sherman A.: Guide to the Preparation and Distribution of Manuscripts for Publication in the Bulletin, *Bull. AAPG*, vol. 43, no. 1, pp. 202–216, 1959.

Wengerd, Sherman A., and Marvin L. Matheny: Pennsylvanian System of Four Corners Region, *Bull. AAPG*, vol. 42, no. 9, pp. 2048–2106, 1958.

Wood, G. M., and B. H. Lane: "Suggestions to Authors of the Reports of the United States Geological Survey," 4th ed., 126 pp., U.S. Geological Survey, 1935. Pages 49–102 cover suggestions as to expression.

Wrenn, C. Gilbert, and Luella Cole: "How to Read Rapidly and Well," 16 pp., Stanford University Press, Stanford, Calif., 1935. A manual of silent reading. 15th printing 1951.

Chapter 16

LEASES, CONCESSIONS, OPTIONS, PERMITS

By EDWIN M. BEDFORD

Independent specialist in leasing. Formerly Manager, Land Department, Western Gulf Oil Company
Los Angeles, Calif.

Conditions differ widely between the United States and other countries in regard to securing rights to explore for and to develop petroleum prospects. This chapter outlines some of the major points and variations in permits, options, leases, and concessions in the United States and some other countries, with more detail being given to typical leasing examples in the oil-producing districts of the United States and some of the more prolific of the foreign producing countries.

Just as land is said to be the basis for all wealth throughout the world, so is land the basis for the success of an oil operation. Therefore, before detailed exploratory or any development work can be undertaken, the operator must have the required land safely in his possession, as shown by the proper documents, supported by adequate evidence of good title. It is the function of the landman to get the land.

The ramifications and work of a complete land department are, however, so extensive that covering all the necessary details would be beyond the scope of a single chapter. Conditions vary, furthermore, from state to state and are quite different in all foreign countries. A typical wildcat "play" in practically any county in the oil country of the United States will be considered first and will be followed by more detailed discussion of various phases of a landman's work.

SUMMARY OF WILDCAT PLAY

Land work is nearly always preceded by a considerable amount of geological work in one or more of its various phases, and if a play is approved by the geological department, it is then presented to management for approval. This must be considered in light of the availability of funds in the company budget for both lease purchase and drilling purposes. Assuming the play is acceptable geologically and approved by management, it is ready to be turned over to the landman for acquisition.

At this point it should be emphasized that for complete security of the position of the company or operator, great care should be taken by each and every member of the group or company, extending from those engaged in geological and geophysical work in the field to all personnel in the offices and, subsequently, to the land and title men, to keep the project confidential so that news of the play reaches the general public only when the operator moves in equipment to start drilling operations.

Prices to be paid for the leases are a very important item and can be decided upon only by a trained landman. The area should be spot-checked to ascertain whether or not any other landmen are active in the area and, if so, whether any leases have been acquired and at what prices. He tries to ascertain, furthermore, if any earlier leasing has been done in the area and at what prices. It is important that these first contacts in the field be made quietly and without divulging the identity of the operator. A trained landman can generally find someone in the neighborhood who can supply the desired information.

The next step is to go to the county seat and search the records for the existence of possible leases in the desired area. There are several ways to obtain the information, but here again, it is most important to go about the job quietly and in utmost secrecy. In states where title companies have a good "plant" (a duplication of the county records), it is easier and faster to get the information from the so-called lot books of the title company. It is advisable for the landman or title searcher first to contact the manager of the title company, explain his mission, and request permission to use the books for the search. It has been the practice generally of title companies to extend this courtesy gratis, in expectation of securing subsequent title business in the form of preliminary reports and policies of title insurance on leases acquired. More recently, however, title companies have been making a charge at a fixed fee per hour or day for the time that its books are used.

For a preliminary survey, the foregoing is usually sufficient, but there are many cases of multitudinous ownerships in a specific area where it is better to employ the title company to make the lot book search, as in the case of a large town-lot play. The title company is requested to furnish the operator with a report of ownership, giving the correct name of the owner and his marital status. If the owner is married, the name of the spouse is given. The address, percentage of ownership, and any significant matters affecting the title, if any, such as mortgages, trust deeds, or outstanding leases, are listed. It is also very important to know if the property is being purchased under a contract of sale, in which event the title company must supply the name and address of the seller, who is still the record owner, so that he, too, can be contacted for joining in the execution of the lease. In this matter the title company manager is requested to keep strictly confidential the name of the operator and the area under search.

The benefit of using title company plant lies in the fact that the search can be made as to an area, whereas if the landman goes directly to the county offices, he first must ascertain the names of owners in his area and then make a detailed search of each owner, since all filing in the county recorder's office is alphabetical by grantor-grantee, and not by parcels. In order to find the names in his area, he first must go to the county assessor's office and take the names from the assessor's ownership plats. This latter procedure is followed in states where title and/or abstract companies are not available or where it is deemed inadvisable to work through a title or abstract company.

When the play is turned over to the landman and he has been furnished with an outline of the area of interest on a map or he has been given an adequate description of it, he has a property map prepared for use in the leasing of the desired lands from the landowners. Where the area has already been covered by a property map, it is merely necessary to use the correct report of ownerships and descriptions received from the title company to bring the map up to date for use in the leasing campaign. The landmen in the field, the field office, and headquarters are all supplied with an adequate number of the up-to-date maps.

The next step is preparation of the actual lease to be used. In most states a standard form of printed lease is recognized as acceptable, and most companies have their own lease form printed with name of the lessee and many of the other blanks in the ordinary form already filled in to aid the landman in the field. An oil and gas lease should be fair to both lessor and lessee, and most forms used by oil companies and printed and sold by stationery houses are fair and workable. It is now considered best procedure to have the name, address, and description of each owner completely filled in when the landman presents it to the owner for signature, for the sake of neatness and accuracy. If any changes have to be made in the printed form, they must be initialed by all parties signing as lessor and lessee or one of the officers signing for a company as lessee.

When the landman has his leases all prepared, price to be paid per acre (or per lot) has been agreed upon, and final approval has been given to begin leasing, he moves into the area or reasonably near the county seat. The leasing crew in the field will usually consist of a landman in charge, sufficient clerical help to process leases, and enough landmen to cover the project with dispatch. An attorney or title man is also of great help in many problems that arise from time to time.

It is obvious that leasing should start at the center of the play and work outward. It is also most important to acquire leases on all the larger tracts in the play as quickly as possible, thus protecting the operator against raids by competitive landmen. If any parcels are lost to competitors, it is best they be small or out on the edge of the play.

In order to close each lease transaction as promptly as possible, it is now the general practice to pay the landowner immediately on signing the lease, providing the preliminary title search has indicated he is the true owner and his title acceptable. Generally these payments in the field are limited to reasonably small ownerships or small outlay of money. In dealing for a much larger tract or one that will be quite costly, it is best to put the signed lease in escrow along with funds to pay for the lease and to authorize closing of the escrow and payment for the lease only after the lessee's attorney has given final approval of the lease form and title.

In states where title insurance is available, it is the usual practice to obtain policies of title insurance. These are often for the minimum of $1,000, but it is the practice of some operators to order a much larger policy on the parcel of land upon which the first well is to be drilled.

In starting to lease a new block, it is best to use a force of landmen large enough to cover the entire play as rapidly as possible. As leasing progresses and toward completion of the play, the number of landmen can be gradually reduced until only those who are necessary for cleanup work remain on the job.

When the operator is ready to start drilling operations, it is strongly recommended that a landman call on the owner of the land to be drilled to advise him of the plans, inform him of the proposed location of the well and get his friendly cooperation for the mutual benefit of both parties. There is quite often a question of reimbursement to the owner for crop damages which should be agreed upon or embodied in the lease itself before they occur. Also, rights of way may be needed for pipelines for water, gas, etc., to the drill site and other operations, and later for pipelines for gathering systems, marketing, etc. The landman who did the leasing is the best representative of the operator to continue this job, having in mind at all times the best of public relations.

ECONOMIC INTERESTS

There are several kinds of economic interests in real property which the landman must deal with from time to time, before and after the first lease is executed. Basically there is the owner of land in fee simple, which means he owns the land from the heavens above to the center of the earth, all surface rights and all substances which may be found in the ground below the surface, such as all minerals, including hydrocarbons. Title search, however, may disclose that one party owns the surface and another the subsurface, or the ownership of these may be divided fractionally. It may be revealed that a party owns merely an interest in the mineral rights or a royalty or a portion thereof merely as to oil and gas and sometimes merely for a limited period of time unless said substances are discovered prior to the time his ownership interest expires.

When an oil and gas lease is executed, the lessee may create an overriding royalty, often as payment of consideration or commission for services rendered in negotiations for the lease. After a lease is signed, the lessor may sell a part of his royalty. If the instrument of transfer is a mineral deed, the purchaser acquires an interest in the minerals of the current lease, or if said lease expires, he participates in bonuses, rentals, and royalties of any future leases; if merely a royalty interest under the current lease, his interest would expire with the lease. Likewise the owner of an overriding royalty created by the lessee loses his rights upon expiration of the lease. Often when leases are transferred by assignment from one party to another, the seller may retain an overriding royalty, a share in the net profits, a full working interest in which he pays his share of costs of operations, or a carried interest in which the operator pays all costs, carries the seller, and reimburses himself out of proceeds from production. Or the assignor and assignee may be in full partnership. A lease may be assigned to the operator of a group of leases called a unit, and the assignor participates in the costs and proceeds of the operations in the proper proportions. This chapter is devoted mainly to a discussion of straight leasing and merely the landowner's royalty.

LEASES

Forms

There are many and varied forms of oil and gas leases, but they must all contain certain basic terms. Oil operators, in general, prefer a reasonably short form of lease with the terms and conditions briefly but specifically stated. There are many of these forms of leases printed by publishing or stationery houses. One of the oldest and best known is the Producers 88 form, used in Oklahoma for nearly a half century with its variations in neighboring states throughout the mid-continent and Gulf Coast States. In the Pacific Coast States, the most popular of the earlier forms of printed lease was the Oil Age 86 lease.

Experience has shown that operators can use the printed form of lease, generally without changes, quite easily throughout the Rocky Mountain, mid-continent, Gulf Coast, and Pacific Coast States. There has been some evolution of the older forms of the lease, however, and most oil companies have their own forms printed, but usually these foll w the general basic terms of the old 88.

In the Pacific Coast States, California in particular, where real estate values are considerably higher than in many parts of the country, especially in certain highly developed areas, even more evolution of the old Oil Age 86 form has transpired. In the sparsely settled areas little difficulty is experienced with the usual form of printed lease. In the highly developed areas along the coast and in southern California, where land is sometimes worth more for real estate development than for any but high flush production of oil and gas, leasing is a far more difficult undertaking. Here the land is so valuable that the lease form is scrutinized most carefully by the owner and his attorney. The latter will often reject the lease form submitted and draw a lengthy typewritten form to cover all the basic terms agreed to and, in addition, many other clauses for the protection he believes needed for his client. Some large land-holding companies have their own form of lease and even stipulate the terms they will accept.

In California the printed lease forms furnished by publishing houses as well as those of the major companies have been modernized to protect the landowner adequately against damages and to assure him of periodic payments of rentals, exploration, or development as the case warrants. Although countless numbers of leases have been executed on the 1-page form, one large California landowner prints its lease form on 20 pages. Therefore, it would be difficult to list all of the many terms and variations of terms that are used. One example of a satisfactory lease (Oil Age Form 86-C-R—Revised) is given at end of this chapter.

Provisions

A lease is a grant or conveyance by the owner of an estate to another of a portion of his interest therein for a term less than his own (Chandler v. Hart, 161 Cal. 405, 413). The maker of the conveyance is called the "lessor," and the recipient of the interest conveyed is called the "lessee." The terms *lessor* and *lessee* are frequently used interchangeably with *landlord* and *tenant*. In practical oil-industry language these terms are also used interchangeably with the *landowner* and the *oil company*.

The *lessor* must sign the lease, but the *lessee* need not sign provided he accepts the lease and acts thereunder; however, it is better practice to require his signature in order to be sure that he is bound under the covenants made by him. Acknowledgment of the signature of a lease by the *lessor* is a prerequisite to recordation.

An oil and gas lease differs greatly from an ordinary lease and in many ways. Whereas a surface lease generally passes the right to *use* the property, an oil lease has, in addition, the right to take something *from* the property, the oil, gas, and other hydrocarbons. A surface lease generally runs for a definite period, while the usual oil and gas lease runs for a period of years and for an indefinite period *thereafter*, such as: "and so long thereafter as oil or gas is produced."

Oil and gas leases may be signed by individuals; corporations; executors and administrators of estates; guardians of minors or incompetents; national banks; state banks and trust companies; attorneys in fact; trustees; city, county, and state agencies; Indian agencies; cotenants; a lessor with no interest in oil but only a reserved right to make a future lease; cemetery corporations; railroad companies; and the United States government.

Details. If a typical oil and gas lease were abstracted, it would show the following important features:

Date
Lessor's name
Lessee's name
Granting clause
County and state
Legal description of land leased
Number of acres

Habendum clause or the primary term of the lease. This is the usual clause which states that the lease is to run for a term of "20 years from and after the date hereof and as long thereafter as oil or gas, or other hydrocarbon substances . . . is produced therefrom . . . ."

Royalty clauses for oil, gas, and casinghead gasoline

Clause providing for rate and method of paying rentals, varying from periods of monthly, quarterly, or annually

Term within which drilling must be started or the lease terminated

A provision for starting the next well in the event production has not been obtained in the first and provision for further exploration

Next is the clause governing development of the lease after production is obtained. This usually provides for starting operations on the second well within 3 months after completion of the first and continuing orderly development of the lease, allowing the same period of 3 months between the completion of a well and the starting of the next. This clause also stipulates the well spacing on the lease by providing that one well shall be drilled to a certain number of acres. The expected depth and/or quality or quantity of products usually dictate the spacing. For the average oil well in west Texas, it might be one well to each 40 acres, and for gas, one well to each 640 acres. This varies to closer spacing in the Gulf Coast and also in other mid-continent States. In California the spacing also varies widely. In the San Joaquin Valley, the average varies from one well to 10 acres to one well to 20 acres for oil and one well to 80 acres to one well to 160 acres for gas. In areas of very deep production, the operator provides for the more economic wider spacing of one oil well to each 40 or 80 acres. In areas of shallow, low-gravity production, 10-acre spacing is not unusual. The spacing problem is essentially an economic one, and the lessee must have a reasonable provision to protect him from marginal or subcommercial situations. Even where the lease may call for 20-acre spacing, if economical, the operator often conducts "5-spot" drilling, that is, a new well drilled in the center of a square previously formed by a group of four older wells, reducing the spacing to 10, 5, and even, in some cases, as low as 2½ acres per well. Much depends on the depth, rate of recovery, price per barrel of oil, and in some cases even the term of the lease may induce the lessee to attempt to recover as much oil as possible within a limited time. Some leases are only for 20 years, others 25, etc., and may not contain a "so long thereafter" clause. There is also a protective provision in this clause to provide that drilling and producing operations may be suspended while the price is a stipulated amount or less per barrel or when there is no available market for same at the well.

Lease provisions for development of natural gas are somewhat different from those for oil, as mentioned above, owing to the larger areas that can be depleted by each gas well. The basic terms are similar. Development after production, if established, is much less complex.

All modern leases contain an offset clause, generally from 330 feet for oil and gas to 1,320 feet for gas only, with the offsetting well to be started within a specified time after the completion of the well to be offset, generally 3 months. An "offset" is described as a well drilled opposite or reasonably near another well on the same or on adjoining property. A lease also contains the usual *force majeure* clause where the lessee is permitted suspension of drilling and production operations when he is prevented from complying with lease provisions by strikes; lockouts; the elements; accidents; rules and regulations of any Federal, state, municipal, or other governmental agency; inability to obtain materials or supplies in the open market; or other matters or conditions beyond the control of the lessee.

The lease also provides that the lessee shall pay all taxes on its improvements and its share (total less royalty) of mineral taxes and the lessor pays all taxes levied against the land and his share (same percentage as royalty, such as one-eighth) of mineral taxes. The same division of tax payments is provided for in the event of production or severance taxes. The lessee agrees to pay all damages caused by its operations, and the lessor may at all reasonable times examine the land and work in progress and inspect the books of the lessee in relation to production. The lessee usually agrees, on written request, to furnish to the lessor copies of logs of all wells drilled on the lessor's land by the lessee.

Most leases now provide for treatment of oil, where necessary, to make it merchantable, and lessor is charged a proportionate part of the cost of such treatment.

A very important clause in the lease is the default and forfeiture provision. Upon violation of any of the terms or conditions of the lease by the lessee and failure to begin to remedy within a specified time, usually 3 months after written notice from the lessor, the lease shall terminate and the lessee's rights come to an end, save and excepting certain acreage, usually 10 to 20 acres, around each well producing or being drilled where the lessee is not in default.

The modern provision for the method of paying royalties and rentals under a lease provides for mailing checks for same to the lessor's bank or to his address. It is safer to handle payments through a bank, but in many cases banks make an extra charge for handling these transactions, especially if payments are small, in which event the lessor may insist on direct delivery by mail.

Most leases carry the right to convey, transfer, or assign by both the lessor and lessee, but many lessors insist on a provision that the lessee shall not assign in whole or in part without written consent of the lessor. Usually the lease provides that the lessee may surrender and quitclaim the lease in whole or in part. If the lease is only partially surrendered, rentals and drilling requirements are reduced pro rata, the lessee retaining, however, easements and rights of way to operate the retained part.

Contrary to the older leases, all modern leases, unless community leases are used, should contain a "pooling clause." This provides that the lessee may combine all or any part of the lessor's land into an operating unit with any other lands (whether held by the lessee or others). The clause generally limits the *size of the unit,* such as from 40 to 160 acres for oil to 640 acres or more or less for gas. Therefore production from the unit is allocated to each owner therein in the proportion that his land in the unit bears to total acreage in the unit.

The unitizing of several small leases into an operating unit by one operator for the economy and convenience of operations within a field is not to be confused with the grand scale of unitization where all companies, operators, and landowners unitize and pool all leases on a structure to cover an entire oil or gas field. In the latter case one of the companies is named operator and is responsible for development of the entire field.

Federal Leases

Leases for oil and gas from the United States may be issued under several different acts, according to the type of lands to be leased. All are issued and administered by the Bureau of Land Management from the several district land offices in 10 western states and at Washington, D.C., for states that have no local land office.

Except for the Right of Way Leasing Act of 1930 and the Mineral Leasing Act for Acquired Lands of 1947, all public lands of the United States are leased for oil and gas under the Leasing Act of 1920, as amended. By this leasing act the old practice of acquiring United States land under government prospecting permits was terminated.

By evolution of documentary form the procedure of acquiring a Federal lease has recently been changed from the old method of filing an application and waiting a considerable time for issuance of lease to the modern form of filing in quintuplicate an Offer to Lease. This form is also the *lease* when properly filled out, signed by the lessee and a signing officer of the United States.

Accompanying the offer to lease is the offeror's check for $10 filing fee plus 50 cents per acre. The acreage limit under any one lease is 2,560 acres, subject to the rule of approximation. All lands in a single lease must be within a 6-mile square. An offer for a lease must not be for less than 640 acres, except when the lands are within a unit or cooperative agreement, or where the land is surrounded by lands not available under the Act or already covered by an existing lease. Only a native-born citizen, one naturalized, or a corporation or other legal entity qualifying may apply for a lease. The offeror's interests direct and indirect in oil and gas leases and applications or offers may not exceed 46,080 acres in each state or 100,000 acres in Alaska. Acreage included in unit plans and certain leases stemming from claims under laws prior to the Leasing Act of 1920 are not chargeable, nor is acreage chargeable that is held under a nonrenewable option, valid only for 3 years or such longer period as authorized by the Secretary of the Interior. However, no optionee may hold options at any one time for more than 200,000 acres in any one state. The offer to lease shall apply only to lands not within a known geologic structure of a producing oil or gas field.

The current granting clause leases the described lands for a period of 5 years and provides for an application for an extension of 5 additional years, thus amounting under certain conditions to a 10-year lease. Federal leases differ from ordinary fee leases in the government's requirements for the filing of bonds and for the lessee to subscribe to a pooling or unit plan approved by the Secretary of the Interior. Instead of a definite spacing program, the government requires the lessee to drill and produce all wells necessary to protect the leased land from drainage.

Rentals on lands outside a known geologic structure are 50 cents per acre for the first year, no rental for the second and third years, 25 cents per acre per year for the fourth and fifth years, and where renewed by extension, 50 cents per acre per year for the sixth and each succeeding year.

Rentals of $1 per acre per year are required if on a known geologic structure producing oil or gas. This, however, is treated as a minimum royalty, and the lessee may credit himself with this payment against earned royalty at the end of each lease year.

Royalty on production is 12½ per cent. Provided, however, if a first discovery on any geologic structure is made on a government lease in Alaska, the royalty is 5 per cent for 10 years and thereafter 12½ per cent. As in most fee leases, the government also reserves the right to take its royalty production *in kind*. If there is no well on the leased land capable of producing oil or gas in paying quantities, the failure to pay rental on or before the anniversary date shall automatically terminate the lease.

Any instrument or assignment or transfer, in whole or in part, must be filed with the government for approval within 90 days from date of execution, the same being effective only upon final approval by the Bureau of Land Management.

Overriding royalty in excess of 5 per cent may not be created except as otherwise authorized by the regulations.

FOREIGN REGULATIONS

The acquisition of foreign concessions is quite different from any type of leasing in the United States and also varies from country to country. Although the procedure is different in each country, there is some similarity in that a certain series of steps must be taken in making the application. Almost everywhere, except in the United States, the minerals, including oil, are the property of the state.

Since each foreign state has its own laws governing concessions, the applicant must either be thoroughly familiar with the laws or employ the aid of experienced personnel, more often local men in the country to be prospected. In some countries one may obtain what may be called an exploration permit on large acreage, often without bonus, on which geological and geophysical work, structural and stratigraphic drilling, or sometimes even deep drilling for oil must or may be done to keep the permit alive. Usually rentals must be paid during this stage, but the amount of the rentals is ordinarily small and often may be deducted from the required expenditures made on the permit. Some countries restrict the type of operation permissible during this first stage and require the oil company to convert the exploration permit to what may be called a petroleum license, covering a smaller block of acreage on which higher rentals are paid, before drilling is commenced. Like the permit, its duration may be in the neighborhood of 5 years. In most countries the final stage, whether it was preceded by one or two stages, is called the concession (or lease), which is almost always granted only after a commercial discovery has been made. In almost no countries is it permissible to produce and sell oil without first having been granted a concession. The concession will ordinarily run for 20 to 50 years, usually with a provision for extension for the same or a lesser period. However, the extension, when granted, will carry new provisions based on the country's recent laws and regulations. There have also been occasions where new provisions are enforced before the expiration date of a concession.

In general, the acreage figures for concessions are very large in comparison with land holdings in the United States. Thus, one might wind up with one or more concessions of 100 square miles each, after having first conducted geological exploration on a much larger area. Some random practices are outlined in the following pages.

Venezuela

In Venezuela the state now puts up lands in certain areas for bids, whereupon eligible and qualified companies may bid as to both the bonus and royalty. There are four types of concessions: exploration-exploitation, exploitation, manufacturing-refining, and transportation. The first are granted for 3 years on "lots" not to exceed 24,710 acres, of which up to half may be converted to exploitation, the remainder going into the national reserve.

Exploitation concessions may be granted over areas held in the national reserves (or, theoretically, over any other area), after publication of a resolution by the Ministry specifying the areas involved, upon the receipt of sealed bids. Such concessions have a life of 40 years with a 40-year extension possible.

Venezuela has many taxes which cannot be listed here. However, the basic production tax (royalty) is $16\frac{2}{3}$ per cent and there is a 50-50 split (very recently increased by the imposition of new nondeductible taxes) of the net profits.

Libya

In Libya rights for exploration are divided into the permit stage, having a duration of 1 year without the right to drill, and the exploration and exploitation concession, having a duration of 50 years with a 10-year extension. The oil companies have bypassed the permit stage and applied directly for concessions, since the permit holder has no preferential right to a concession.

Libya is divided into four zones in which the maximum area per concession runs from 11,583 square miles in zones I and II to 30,888 square miles in zones III and IV. There is no limit as to the number of concessions a company may hold. However, the concessions must be reduced in size by 25 per cent in the first 5 years, by 50 per cent within 8 years, and to $33\frac{1}{3}$ per cent of the original area within 10 years in zones I and II and to 25 per cent of the original area in zones III and IV. Minimum annual investments are required, which run from $4.20 to $16.80 per square kilometer, depending upon the zone and the age of the concession. Royalty is $12\frac{1}{2}$ per cent with the customary 50-50 split of the net profits. Libya grants a depletion provision of 25 per cent. No bonus is payable.

Algeria and France

In Algeria, as in France, permission to explore and produce is granted to a qualified company by negotiation, principally on the basis of the amount of money it will expend. At present, no foreign group is allowed to acquire more than a 50 per cent interest in any concession. The other 50 per cent or more must be controlled by the French. The first stage is the Prospection Authorization, but since it is nonexclusive and gives no priority rights to later permits, it is seldom applied for. The second stage, the Exploration Permit, is initially grant d for 5 years or less. It has two possible extensions of 5 years or less each; however, at the end of the initial period a 50 per cent reduction in acreage must be made and, upon the second extension, the area must be reduced to 25 per cent. Working and other obligations are written into each agreement, and if the minimum expenditures have not been made, the unspent balance is payable to the government. The Exploration Permit allows for all prospecting and exploration operations, with the right to dispose of hydrocarbons extracted during exploration. There is no bonus or rental, and there is no legal limit on the size of the permit.

The final stage, the Exploitation Concession, may be obtained only by the titleholder of an exploration permit. It is limited to the area covered by the exploitable deposit and runs for 50 years with no extension possible. There are no bonuses or rentals payable. Royalties are 12½ per cent on crude and 5 per cent on natural gas. There is the usual 50-50 split of the net profits, and a 27½ per cent depletion provision is granted, but actually the latter is merely a 5-year deferment of the tax on 27½ per cent of the gross. The concession is granted by decree of the Council of State on the recommendation of the Minister of Mines and the Minister of Saharan Affairs.

Turkey

In Turkey a nonexclusive Preliminary Reconnaissance Permit constitutes the first stage. Following it, the permittee will apply for an Exploration License, which gives him exclusive rights to explore and to drill for oil within a maximum area of 50,000 hectares (123,500 acres). However, a total of eight such licenses may be held by an individual in each of the nine petroleum districts into which the country has been divided. The duration of the license is 6 years, with two 2-year normal extensions possible, plus a 5-year extension in case of discovery. Annual rentals are on a sliding scale and run from U.S. 18 cents per hectare for the first year to $1.10 per hectare for the eleventh and following years. Exploration expense may be credited against rentals. Drilling must be commenced on a license within 5 years. The final stage, the Lease, may be applied for whether or not a previous license has been acquired. Drilling must commence on it within 6 months, and commercial production must begin within 3 years in order to retain a lease. The lease runs for 40 years with a 20-year extension; it carries sliding-scale rentals of from $1.10 to $3.40 per hectare. Royalty is 12½ per cent. After recovery of invested capital, a 50 per cent tax is levied on net profits. A depletion provision of 27½ per cent is granted.

Australia

In Australia, although the minerals are the property of the Crown, the oil laws are promulgated and enforced by the individual states. Only the Australian territories fall under Federal oil laws. The laws of the various states are alike in some respects and quite different in others. Each state provides for three stages, but few companies apply under the second stage, since it offers practically no advantages over the first stage though it costs more and restricts the amount of acreage obtainable. Unfortunately, the names of the permits (or *titles* as they are called in Australia) are different in different states. Thus, in Queensland, the first stage is called the Authority to Prospect whereas in New South Wales, it is the Petroleum Exploration License. The second stage is, in Queensland, the Prospecting Permit and in New South Wales the Petroleum Prospecting License. The final stage seems always to carry the term *lease* in its designated name. Universally, the terms under which a title is granted are written "at the discretion of the Minister," but the law provides for certain maximum and minimum figures. In Queensland, there is no limit on the size of Authority to Prospect that the Minister may grant, and therefore many such authorities are very large—as much as 100,000 square miles or even more. No bonus is payable, and rentals are only U.S. 3 cents per square mile in Queensland, but 22 cents per square mile in New South Wales where the Petroleum Exploration License cannot exceed 5,000 square miles in area. The length of time permitted under these titles is again "at the discretion of the Minister." Also the Minister may and does require certain minimum expenditures on the title each year, the amount generally varying with the area granted. The Lease which is granted after a commercial discovery has been made is more consistently the same throughout Australia. It usually has a maximum area of 100 square miles, runs for 20 or 21 years with a similar extension possible, and carries a 10 per cent royalty to the state. Rentals are U.S. $22.40 per square mile. The maximum area of a lease in the Territory of Papua and New Guinea is 500 square miles. Rentals are £15 to £50 or $33 to $112. All taxation in Australia is by the Federal government, and for the ordinary American company, the income tax is 37½ per cent. There is no depletion provision. Since Australia is a "have-not" country with no appreciable amount of production of oil or gas yet established, it is doing all it can to attract capital for oil exploration. At this writing, the Federal government is offering a subsidy of 50 per cent of the cost of approved wildcat wells and other exploration costs.

Canada

Canada has come rapidly to the fore in recent years as a producer of oil and gas; its land and leasing situation merits brief comment.

Most of the production is in the western provinces, but the whole of Canada has profited by this development. The people of the western provinces have benefited because of the considerable amount of public ownership of minerals, although the Crown is still the principal owner. Aside from the railroad lands, about 95 per cent of the minerals is estimated as being "Crown minerals."

In contrast to the Federal lands leasing system in the United States, where the rules are all the same regardless of the state in which the lands are located, in Canada the Crown lands are now owned and administered by each province. The Prairie Provinces (Alberta, Saskatchewan, Manitoba) oil and gas regulations are somewhat similar, but each province has its own regulations, and in some matters the Minister has full authority to issue leases under certain terms not covered by the regulations. In these Prairie Provinces there is a Minister of Mines, except in Saskatchewan, where he is designated Minister of Mineral Resources.

Prior to 1930, all Crown lands, minerals, etc., were owned and administered by the Dominion government for the Prairie Provinces. After 1930, the provinces administered the affairs of their own lands with the exception of Indian reserves, national parks, and soldiers' settlement lands. As in the United States, original homesteaders owned minerals under their lands, but after October 31, 1887, Canada reserved its minerals in lands west of the third meridian, which divides Saskatchewan about in half and therefore includes all of Alberta.

Since 1670 Hudson's Bay Company has been a large landowner in Canada. About 200 years later its holdings were considerably reduced and in the Prairie Provinces it was granted all of Section 8 and the west half and the southeast quarter of Section 26 in each township, except townships divisible by five, where all of Section 26 was retained. Title carried mineral rights, but the grant was only in the area bounded on the south by the United States and on the north by the North Saskatchewan River. Another division of minerals in these lands, however, was that prior to 1908 homesteaders of Hudson's Bay Company lands received title to minerals, whereas after that date, Hudson's Bay Company reserved them.

Again, as in the United States, where railroads were given large grants for building rail lines into undeveloped areas, the Canadian Pacific Railway Company was granted large acreage for building its line from eastern Canada to the Pacific. In 1881 Canada granted the Canadian Pacific all odd-numbered sections, except 11 and 29, within a distance of 25 miles of the railway. Also, as in the United States, the railway company made trades of lands to make larger blocks in different areas to change the pattern. In general, purchasers of lands from the Canadian Pacific Railway up to 1903 also received the minerals. Generally purchasers from 1903 to 1912 got all minerals except natural gas, owing to the Court's interpretation of the reservations. After 1912 the Canadian Pacific reserved all minerals including natural gas. The deeds issued by the railroad were not all identical. Some had ambiguous reservations of minerals and required a court decision as to exact definition of minerals reserved.

The entire leasing situation in Canada appears complex, so, instead of looking at the laws of all the Prairie Provinces, let us merely have a brief review of those of its leading producer—Alberta.

Royalties under Alberta leases are on a graduated scale for oil, starting at 5 per cent of the oil from each well having a production of 0 to 600 barrels per month, a barrel being 35 imperial gallons. Oil royalties are graduated upward to a top of 16⅔ per cent on wells producing 4,050 barrels per month, or about a 130- to 135-barrel daily well.

Royalty on natural gas is 15 per cent, but not less than 0.75 cent per MCF, except that the Minister may alter this under certain circumstances.

When an applicant files for a reservation for an oil and gas lease, he pays a fee of $250 and a deposit of $2,500 for each 2,000 acres or portion thereof. No application shall be for more than 100,000 acres, and the length of the tract shall not be greater than three times its width, except by consent of the Minister. A reservation is for 4 months, and the applicant must submit a plan of the proposed exploration. When the plan is approved, the reservation is extended 4 months and further variations are only by consent of the Minister. After 8 months, if the progress report is satisfactory, the reservation may be extended for further periods at 7 cents per acre for 3 months, 7 cents per acre for a second additional 3 months, and 8 cents per acre for the next two 3-month periods of renewal, to bring the reservation up to 2 years' duration. By submitting a summary of operations, showing delays due to inaccessibility and an estimate of cost of additional work to be done, the applicant may be granted by the Minister further periods of renewal at whatever fee the Minister may set. A reservation may be surrendered in whole or in part from time to time.

Providing he has met with all requirements, a reservation holder may apply for a lease, and where oil has been discovered in a well, the holder shall, within 3 months, apply for a lease. No other well may be started within 4½ miles until leases have been applied for. Under the lease the rental is $1 per acre per year, but a credit for first-year rental may be allowed for not more than 50 per cent of the sum spent on exploration with no credits for subsequent years. The term of the lease is 21 years or as long thereafter as production is obtained. Maximum area of a lease block shall be 9 sections (5,760 acres) if in form of a square or 8 sections (5,120 acres) if a rectangle. Maximum length is 4 miles, and the length shall not exceed twice the width. There must be a corrider 1 mile wide between lease blocks, although lease blocks may have a common owner. Leases, however, shall not exceed 50 per cent of Crown lands in any one township.

If no oil is found on a petroleum and natural gas reservation but natural gas is discovered, the holder applies for a natural gas license and he must surrender his oil rights on lands where a gas license is applied for. The fee for application for a gas license is $250, plus rental of 5 cents per acre for the first 6-month period. Upon approval of satisfactory progress of exploration and exploitation, the license may be renewed for 5 cents per acre for each subsequent 6-month period to a maximum of 3 years. The first well must be started within 3 months from issuance of the gas license, and the second well is to be spudded within 3 months of completion or abandonment of the first. Six-month extensions are possible, however, upon showing that weather or inaccessibility prevented compliance. Gas licenses are issued to certain zones, but where other gas zones are found, the Minister may permit the new zone to be included in the original license. Like an oil and gas reservation, a gas license may be surrendered in whole or in part, providing it is in good standing.

In the event of gas production under a license, a lease may be granted with acreage depending upon depth of production, 6 sections if production is above 3,000 feet, 8 sections if between 3,000 and 6,000 feet, 10 sections for each gas well below 6,000 feet deep, and an additional section for each additional 1,000 feet drilled if gas is found. If more than one zone is productive, acreage is calculated on basis of the deepest productive zone. The lease shall be for 21 years and as long thereafter as commercial gas is produced. Annual rental under the gas lease is 33⅓ cents per acre but may be reduced to 10 cents per acre if there is no market for the gas. Drilling expense may be applied as a credit only against the rental for the first year. Well spacing for a gas lease is one to a section unless otherwise modified by the Minister.

If oil is found in any zone under a gas lease, the operator is entitled to a Petroleum and Natural Gas Lease on the quarter section upon which the oil discovery well is located, but he must surrender three times the area or three quarter sections back to the government as a Crown reserve. These three quarter sections would later be put up for bid as proved Crown reserves. The Minister may put up the relinquished Crown reserve gas rights for competitive bidding, the terms of such license and subsequent lease being similar to those described above for original gas license and lease.

Likewise, any person may request the Minister to post Crown reserve Lands for a drilling reservation. In this case the applicant indicates the depth or formation he intends to penetrate and the Minister may act accordingly, putting up more acreage for deeper drilling and less for shallow. The fee is the same as other reservations, $250, but rental is 25 cents for each 6 months with renewals allowed, if satisfactory work is done, up to 3 years. Drilling must start, however, within 1 year of the date of reservation and be carried to the predetermined depth. Within 3 months after the discovery of oil, the holder shall apply for a lease, the amount of acreage being generally about one-quarter of the amount of the reservation, but there is no rigid rule in this case. Other terms are similar to those outlined above for original leases.

Proved Crown reserves usually go at competitive bidding in quarter, half, or full sections at the discretion of the Minister. On these parcels the rental is $1 per acre per annum; the term is 21 years and as long thereafter as production continues. Offsets must be started 90 days after production is found on adjoining land.

Summary

Oil companies are more or less welcomed in the "have-not" countries, and as would be expected, laws are quite favorable. In contrast, in countries having a considerable amount of oil-production and oil-industry history, the applicant must "pay to bid," "pay to lease," and "pay to drill." In some of the "have" countries, the trend now is to attempt to acquire an interest in the integrated oil company's operations or a share in the profits.

To illustrate the inducements put out by the have-not countries, for instance, the royalty for a concession in Australia is 10 per cent, Libya 12½ per cent. In contrast Egyptian royalties may run from 22½ to 27½ per cent and on up in the have countries to 50 per cent or more.

About 50 countries have oil and gas laws, many of which are more flexible than in the United States.

France and Spain have their mineral laws and numerous (20 to 30) companies are now actively in the process of considering and evaluating the possibilities in Spain and Spanish Morocco.

In order to keep abreast of the industry in foreign countries, one must follow the trends in oil laws and concession contracts. World Petroleum Report* is to be complimented for having made a systematic effort over a number of years to assemble the text of most laws and concession contracts of the world.

To illustrate changing conditions in regulations, the following may be mentioned for the past 2 years. Government lands were opened in Alaska for exploration before it became a state. In Canada new laws were passed in Manitoba, British Columbia, and other areas. Although a considerable amount of the new legislation was more favorable, some of the unfavorable changes were the increases in royalties, such as Saskatchewan raising its Crown-land royalties from a minimum of 5 to a maximum of 16 per cent on a sliding scale. In Central America, several countries amended their laws to provide for more liberal treatment of foreign capital, including Nicaragua and Honduras. Several favorable laws have been passed in the Eastern Hemisphere, including Turkey, France, Denmark, and Sicily. The Italian law passed in 1957 is considered unfavorable. There were few changes in the Near East laws except for Iran, whose laws now encourage the government to participate in oil company profits.

In 1958, new oil laws were passed in Algeria, Spain, Nicaragua, Argentina, Morocco, Cuba, Venezuela, Alaska, and Syria. In 1959 some changes were made or proposed in Nigeria, Chile, Paraguay, Colombia, Panama, Costa Rica, Honduras, El Salvador, Haiti, Somaliland, Italy, Saudi Arabia, Pakistan, Afghanistan, Burma, India, the Philippines, Indonesia, and Australia.

The unfavorable trend, at present, is for foreign countries to participate in the entire integrated operations of the operating company, especially in the have countries, which might have the effect of going beyond 50 per cent of the profits, up to possibly 60 per cent.

CONCLUSION

From the foregoing it should be apparent to the exploration man that the process of acquiring leases on prospective oil- or gas-bearing land, with the right to drill thereon, is not a simple matter. The services of trained and experienced landmen are essential at an early stage when embarking upon a petroleum exploration program. The results of the landman's efforts represent that very important link in the chain of events that starts with the exploration geologist in the field and ends with the drilling of a wildcat well which may lead to the discovery of an important oil field.

* World Petroleum Legislation Report, 604 Fifth Avenue, New York 20, N.Y.

Oil Age Form 86-C-R—Revised

OIL AND GAS LEASE

THIS AGREEMENT, made and entered into this____day of________________, 19____, by and between____

____

______________________________________party of the first part, herein styled "Lessor," and____ party of the second part, herein styled "Lessee."

WITNESSETH: That for and in consideration of Ten Dollars ($10.00) lawful money of the United States of America, to the Lessor paid, and of other valuable considerations, the receipt of all of which is hereby acknowledged, and in consideration of the covenants and agreements hereinafter contained by the Lessee to be kept and performed, the Lessor has granted, leased, let and demised, and by these presents does grant lease, let and demise unto the Lessee, its grantees, successors and assigns, the land and premises hereinafter described, with the sole and exclusive right to the Lessee to explore (by geological, geophysical or other methods, whether similar to those herein specified or not, and whether now known or not, including the drilling of shallow holes for the purpose of determining subsurface geological conditions), to drill for, produce, extract, take and remove oil, gas, asphaltum and other hydrocarbons (and water without cost for its operations) from, and to store the same upon, said land during the term hereinafter provided, with the right of entry thereon at all times for said purposes, and to construct, use, maintain, erect, repair and replace thereon and to remove therefrom all pipe lines, telephone and telegraph lines, tanks, machinery, buildings and other structures which the Lessee may desire in carrying on its business and operations on said land, or adjoining or neighboring premises operated by Lessee, with the further right to the Lessee or any of its subsidiaries to erect, maintain, operate and remove a plant with all necessary appurtenances, for the extraction of gasoline from gas produced from said land and/or other premises in the vicinity of said land, including all rights necessary or convenient thereto, together with rights-of-way for passage over, upon and across, and ingress and egress to and from, said land, for any or all of the above mentioned purposes. The possession by the Lessee of said land shall be sole and exclusive, excepting only that the Lessor reserves the right to occupy said land or to lease the same for agricultural, horticultural, or grazing uses, which uses shall be carried on subject to, and with no interference with, the rights or operations of the Lessee hereunder. The land which is the subject of this lease is situated in the County of________________________, State of California, and is described as follows, to-wit:

and contains____________acres, more or less, together with all right, title and interest which the undersigned may now have or hereafter acquire in the real property included in any avenues, alleys, highways, roads, streets or streams adjacent to the above-described premises.

TO HAVE AND TO HOLD the same for a term of Twenty (20) years from and after the date hereof and so long thereafter as oil or gas, or casinghead gas, or other hydrocarbon substances, or either or any of them, is produced therefrom or from lands pooled therewith, or drilling operations (including but not limited to drilling, deepening, plugging back, redrilling, repairing, cleaning out and similar operations) are conducted thereon (with no more than six (6) months between the cessation and commencement of operations or between the cessation of production and the commencement of such operations), or Lessee is excused therefrom, under the terms hereof.

In consideration of the premises it is hereby mutually agreed as follows:

1. Lessee shall pay Lessor as royalty on oil the equal One-Eighth (1/8) part of the proceeds of all oil produced, saved and sold from the leased premises, after making the customary deductions for temperature, water and b.s. at the market price as posted by__________________________or any of its present or future subsidiaries, in the vicinity in which the premises are located for oil of like quality and gravity the day the oil is run into purchaser's pipe line or storage tank, and settlement shall be made by Lessee on or before the 25th day of each month for accrued royalties for the preceding calendar month. At Lessor's option, exercised not oftener than once in any one calendar year upon sixty (60) days' previous written notice, Lessee shall deliver into Lessor's tanks on the leased premises, or at mouth of well to pipe line designated by Lessor free of cost, Lessor's royalty oil, provided that Lessee may at any time purchase and take Lessor's royalty oil at said posted available

market price. No royalty shall be due to the Lessor for or on account of oil lost through evaporation, leakage or otherwise prior to the marketing of the same or delivery to Lessor if royalty oil is being taken in kind.

2. For all gas produced, saved and sold from said land by Lessee, the Lessee shall pay as royalty the One-Eighth (1/8) part of the net proceeds from the sale of such gas, but nothing herein contained shall be deemed to obligate the Lessee to produce, save, sell or otherwise dispose of gas from said land. For the purpose of having gasoline extracted from gas produced from said land, the Lessee may transport, or cause to be transported, to a gasoline extraction plant located either on said land or on other lands, all or any portion of such gas where it may be commingled with gas from other properties. Lessee shall meter such gas so transported and such meter readings, together with the results of content tests by recognized methods made at approximately regular intervals, at least once every month, shall furnish the basis for computation of the amounts of gasoline and residue gas to be credited to this lease. Gas used or consumed, or lost in the operations of any such plant, shall be free of charge, and Lessee shall not be held accountable to the Lessor for the same or for any royalty thereon. Lessee shall not be required to pay royalty for or on account of any gas used for repressuring any oil-bearing formation which is being produced from by a well or wells on the leased premises, even though such repressuring is done by injecting such gas into wells not situated on the leased premises.

3. In the event Lessee shall extract or cause to be extracted gasoline from gas produced from said land, Lessee shall pay to Lessor as royalty and rental the "sale value," as hereinafter defined, of six per cent (6%) of the gasoline credited to this lease as herein provided. The "sale value" of such gasoline shall be deemed to be the value of such gasoline at the prices currently offered or paid by ____________________ or any of its present or future subsidiaries to producers for gasoline of like specifications and quality in the same vicinity. For the purpose of having gasoline extracted from gas produced from said land, Lessee may transport, or cause to be transported to a gasoline extraction plant located either on said land or other lands, all or any portion of such gas where it may be commingled with gas from other properties. Lessee shall meter such gas so transported and such meter readings, together with analyses of the gasoline content of such gas made at approximately regular intervals, at least once in every month, shall furnish the basis for computation of the amounts of gasoline and residue gas to be credited to this lease. Gas actually and reasonably used or consumed or lost in the operation of any such plant shall be free of charge, and Lessee shall not be held accountable to Lessor for such proportion thereof as the total amount of gas produced from the land covered by this lease and so transported to such plant bears to the total amount of gas treated at such plant.

4. The Lessee shall not be required to account to the Lessor for, or pay royalty on, oil, gas or water produced by the Lessee from said land and used by it in its operations hereunder, but it may use such oil, gas and water free of charge.

5. Commencing with the _______________ of the term hereof, if the Lessee has not theretofore commenced drilling operations on said land or terminated this lease as herein provided, the Lessee shall pay or tender to the Lessor _______________ in advance, as rental, the sum of ____________________ Dollars per acre per _______________ for so much of said land as may then still be held under this lease, until drilling operations are commenced or this lease terminated as herein provided.

6. The Lessee agrees to commence drilling operations on said land within Ten (10) years from the date hereof (which period is hereinafter referred to as "primary term") and to prosecute the same with reasonable diligence until oil or gas is found in paying quantities, or to a depth at which further drilling would, in the judgment of the Lessee, be unprofitable; or it may at any time within said period terminate this lease and surrender said land as hereinafter provided. No implied covenant shall be read into this lease requiring the Lessee to drill or to continue drilling on said land, or fixing the measure of diligence therefor. The Lessee may elect not to commence or prosecute the drilling of a well on said land as above provided, and thereupon this lease shall terminate.

7. If the Lessee shall elect to drill on said land, as aforesaid, and oil or gas shall not be obtained in paying quantities in first well drilled, the Lessee shall, within six (6) months after the completion or abandonment of the first well, commence on said land drilling operations for a second well, and shall prosecute the same with reasonable diligence until oil or gas is found in paying quantities, or until the well is drilled to a depth at which further drilling would, in the judgment of the Lessee, be unprofitable; and the Lessee shall in like manner continue its operations until oil or gas in paying quantities is found, but subject

always to the right of Lessee to quitclaim as provided in Paragraph 27 hereof and provided that Lessee shall not be obligated to commence drilling operations for any such well or to terminate this lease during any period for which rental has been paid pursuant to the provisions of this lease; and provided, further, that Lessee may at any time and from time to time during the primary term hereof, defer, suspend or resume drilling operations provided for under this paragraph; but, whenever Lessee shall so defer or suspend such drilling operations during any rental period for which rental has not been paid, Lessee shall pay or tender to Lessor rental at the rate specified in paragraph 5 hereof for the rental period or portion or portions thereof during which such drilling operations shall be so deferred or suspended. When such deferment or suspension occurs during only a portion of a rental period for which rental has not been paid, the rental payable for that particular rental period shall be such proportion of the total rental for that rental period as the number of days during which drilling operations are deferred or suspended during that rental period shall bear to the entire number of days included in that rental period. The aforesaid interval of six (6) months allowed between completion or abandonment of any well and commencement of drilling operations upon the next succeeding well shall be deducted in computing any such period of deferment or suspension and no rental shall be payable with respect to any such interval of six (6) months. The rentals provided for in this paragraph with respect to any rental period or portion thereof shall be paid or tendered to Lessor on or before the expiration date of such rental period or at the end of the period of such deferment or suspension, whichever shall first occur. All rental provided for in this paragraph shall cease when this lease shall be quitclaimed, surrendered or terminated as herein provided.

8. If oil is found in paying quantities in any well so drilled by the Lessee on said land, the Lessee, subject to the provisions hereof and to the suspension privileges hereinafter set forth, shall continue to drill additional wells on said land as rapidly as one string of tools working with reasonable diligence can complete the same, until there shall have been completed on said land as many wells as shall equal the total acreage then held under this lease divided by forty; whereupon the Lessee shall hold all of the land free of further drilling obligations; provided, that the Lessee may defer the commencement of drilling operations for the second or any subsequent well for a period not to exceed six (6) months from the date of completion of the well last preceding it. Except as herein otherwise provided, it is agreed that the Lessee shall drill such wells and operate each completed oil well with reasonable diligence and in accordance with good oil field practice so long as such wells shall produce oil in paying quantities while this lease is in force as to the portion of said land on which such well or wells are situated; but in conformity with any reasonable conservation or curtailment program affecting the drilling of wells or the production of all oil and/or gas from said land, which the Lessee may either voluntarily or by order of any authorized governmental agency subscribe to or be subject to. Drilling and producing operations hereunder may also be suspended while the price offered generally to producers in the same vicinity for oil of the quality produced from said land is seventy-five (75) cents or less per barrel at the well, or when there is no available market for the same at the well.

9. If oil in quantities deemed paying quantities by Lessee shall not be found in said land and Lessee shall complete a well or wells on said land which shall be capable, in the opinion of Lessee, of producing gas of commercial quality and in paying quantities, Lessee shall hold all of said land free of further drilling obligations subject, however, to the provisions of Paragraph 12 hereof, unless and until in its judgment the drilling of any additional well is warranted in view of its existing or anticipated market requirements for gas produced from the field in which said land is situated. In the event oil in such paying quantities is not found in said land and Lessee shall complete on said land as many gas wells as shall equal the total acreage hereby leased divided by six hundred forty, Lessee shall hold all of said land free of further drilling obligations, except as provided in Paragraph 12 hereof, subject to this lease and for the full term hereof. If Lessee shall complete a well or wells on said land which shall produce oil in such paying quantities and shall also complete on said land a well or wells which shall not produce oil in such paying quantities but shall be capable in the opinion of Lessee of producing gas of commercial quality and in paying quantities and Lessee shall drill the number of oil wells in Paragraph 8 hereof provided for with respect to the portions of said land which are capable of producing oil in such paying quantities, Lessee shall be entitled to retain all of the land hereby leased, subject to the provisions of this lease and for the full term hereof.

10. If it should hereafter appear that the Lessor at the time of making this lease owns a less interest in the leased land than the fee simple estate or the entire interest in the oil and gas under said land, then the rentals and royalties accruing hereunder shall be paid to the

Lessor in the proportion which his interest bears to the entire fee simple estate or to the entire estate in said oil and gas.

11. There is hereby expressly reserved to the Lessor, and as well to the Lessee, the right and privilege to convey, transfer or assign in whole or in part its interest in this lease or in the leased premises or in the oil and/or gas therein or produced therefrom, but if the Lessor shall sell or transfer any part or parts of the leased premises or any interest in the oil and/or gas under any part or parts thereof the Lessee's drilling obligations shall not thereby be altered, increased or enlarged, but the Lessee may continue to operate the leased premises and pay and settle rents and royalties as an entirety.

12. In the event a well is drilled on adjoining property within three hundred thirty (330) feet of the exterior limits of any land at the time embraced in this lease and oil or gas is produced therefrom in paying quantities and the drilling requirements as specified in paragraphs 8 and 9 hereof are not fully complied with, and the owner of such well shall operate the same and market the oil or gas produced therefrom, then the Lessee agrees to offset such well by the commencement of drilling operations within ninety days after it is ascertained that the production of oil or gas from such well is in paying quantities and that the operator thereof is then producing and marketing oil or gas therefrom. For the purpose of satisfying obligations hereunder such offset well or wells shall be considered as other wells required to be drilled hereunder.

13. The obligations of the Lessee hereunder shall be suspended while the Lessee is prevented from complying therewith, in whole or in part, by strikes, lockouts, actions of the elements, accidents, rules and regulations of any Federal, State, Municipal or other governmental agency, or other matters or conditions beyond the control of the Lessee, whether similar to the matters or conditions herein specifically enumerated or not.

14. The Lessee shall pay all taxes on its improvements and all taxes on its oil stored on the leased premises on the first Monday of March in each year, and seven-eighths (7/8) of the taxes levied and assessed against the petroleum mineral rights. Lessor agrees to pay all taxes levied and assessed against the land as such and one-eighth (1/8) of the taxes levied and assessed against the petroleum mineral rights. In the event the State, United States or any municipality levies a license, severance, production or other tax on the oil produced hereunder, or on the Lessee's right to operate, then and in that event the Lessee shall pay seven-eighths (7/8) of said tax and Lessor shall pay one-eighth (1/8) of said tax.

15. The Lessee agrees not to drill any well on said land within one hundred (100) feet of the now existing building thereon without the written consent of the Lessor. The Lessee agrees to pay all damages directly occasioned by its operations to crops on said land.

16. The Lessor may at all reasonable times examine said land, the work done and in progress thereon, and the production therefrom, and may inspect the books kept by the Lessee in relation to the production from said land, to ascertain the production and the amounts saved and sold therefrom. The Lessee agrees, on written request, to furnish to the Lessor copies of logs of all wells drilled by the Lessee on said land.

17. All the labor to be performed and material to be furnished in the operations of the Lessee hereunder shall be at the cost and expense of the Lessee, and the Lessor shall not be chargeable with, or liable for, any part thereof; and the Lessee shall protect said land against liens of every character arising from its operations thereon.

18. Upon the written request of the Lessor, the Lessee agrees to lay all pipe lines which it constructs through cultivated fields, below plow depth, and upon similar request agrees to fence all sump holes or other excavations to safeguard livestock on said land.

19. The Lessee shall have the right at any time to remove from said land all machinery, rigs, piping, casing, pumping stations and other property and improvements belonging to or furnished by the Lessee, provided that such removal shall be completed within a reasonable time after the termination of this lease. Lessee agrees after termination of this lease to fill all sump holes and other excavations made by it.

20. If royalty oil is payable in cash, Lessee may deduct therefrom a proportionate part of the cost of treating unmerchantable oil produced from said premises to render same merchantable. In the event such oil is not treated on the leased premises, Lessor's cash royalty shall also bear a corresponding proportionate part of the cost of transporting the oil to the treating plant. Nothing herein contained shall be construed as obligating Lessee to treat oil produced from the herein described premises. If Lessor shall elect to receive royalty

oil in kind, such royalty oil shall be of the same quality as that removed from the leased premises for Lessee's own account, and if Lessee's own oil shall be treated before such removal, Lessor's oil will be treated therewith before delivery to Lessor and Lessor in such event will pay a proportionate part of the cost of treatment.

21. Upon the violation of any of the terms or conditions of this lease by the Lessee and the failure to begin to remedy the same within sixty days after written notice from the Lessor so to do, then, at the option of the Lessor, this lease shall forthwith cease and terminate, and all rights of the Lessee in and to said land be at an end, save and excepting forty (40) acres surrounding each well producing or being drilled and in respect to which Lessee shall not be in default, and saving and excepting rights-of-way necessary for Lessee's operations; provided, however, that the Lessee may, at any time after such default, and upon payment of the sum of Ten Dollars ($10.00) to the Lessor as and for fixed and liquidated damages, quitclaim to the Lessor all of the right, title and interest of Lessee in and to the leased lands in respect to which it has made default, and thereupon all rights and obligations of the parties hereto one to the other shall thereupon cease and terminate as to the premises quitclaimed.

22. All royalties and rents payable in money hereunder may be paid to the Lessor by mailing or delivering a check therefore to____________________________________ Bank at____________________its successors and assigns, herein designated by the Lessor as depositary, the Lessor hereby granting to said depositary full power and authority on behalf of the Lessor, his heirs, executors, administrators, successors and assigns, to collect and receipt for all sums of money due and payable from the Lessee to the Lessor hereunder. No change in the ownership of the land or minerals covered by this lease, and no assignment of rents or royalties shall be binding on the Lessee until it has been furnished with satisfactory written evidence thereof.

23. Lessor hereby warrants and agrees to defend title to the land herein described, and agrees that the Lessee, at its option, may pay and discharge any taxes, mortgages, or other liens existing, levied or assessed on or against the above described land; and, in the event it exercises such option, it shall be subrogated to the rights of any holder or holders thereof and may reimburse itself by applying to the discharge of any such mortgage, tax, or other lien, any royalty or rentals accruing hereunder.

24. If and when any oil produced from the demised premises shall for any reason be unmarketable at the well at the price mentioned in paragraph 8 hereof, the Lessor agrees in such case to take and receive his royalty in kind, and should he fail or refuse so to do, then the Lessee may sell the same at the best price obtainable, but not less than the price which the Lessee may be receiving for its own oil of the same quality.

25. The words "drilling operations" as used herein shall be held to mean any work or actual operations undertaken or commenced in good faith for the purpose of carrying out any of the rights, privileges or duties of the Lessee under this lease, followed diligently and in due course by the construction of a derrick and other necessary structures for the drilling of an oil or gas well, and by the actual operation of drilling in the ground, and shall also include drilling, deepening, plugging back, redrilling, repairing, cleaning out, and similar operations.

26. On the expiration or sooner termination of this lease, Lessee shall quietly and peaceably surrender possession of the premises to Lessor and deliver to him a good sufficient quitclaim deed, and so far as practicable cover all sump holes and excavations made by Lessee, and shall abandon all wells in accordance with the applicable laws, regulations and orders of any governmental agency having jurisdiction of the lands herein leased.

27. Lessee may at any time quitclaim this lease in its entirety or as to part of the acreage covered thereby, with the privilege of retaining forty (40) acres surrounding each producing or drilling well, and thereupon Lessee shall be released from all further obligations and duties as to the area so quitclaimed, and all rentals and drilling requirements shall be reduced pro rata. All lands quitclaimed shall remain subject to the easements and rights-of-way herein-above provided for. Except as so provided, full right to the land so quitclaimed shall revest in Lessor, free and clear of all claims of Lessee, except that Lessor, his successors or assigns, shall not drill any well on the land quitclaimed within three hundred thirty (330) feet of any producing or drilling well retained by Lessee.

28. If this lease shall be assigned as to a particular part or as to particular parts of the leased premises, such division or severance of the lease shall constitute and create separate

and distinct holding under the lease of and according to the several portions of the leased premises as thus divided, and the holder or owner of each such portion of the leased premises shall be required to comply with and perform the Lessee's obligations under this lease for, and only to the extent of, his portion of the leased area, provided that nothing herein shall be construed to enlarge or multiply the drilling or rental obligations, and provided further that the commencement of the drilling operations and the prosecution thereof, as provided in paragraph 6 hereof, either by the Lessee or any assignee hereunder, shall protect the lease as a whole.

29. Lessee is hereby given the right at its sole option by written declaration of pooling or unitizing, at any time or from time to time during the term of this lease but in no event after expiration of twenty (20) years from the date hereof, to combine or pool or unitize all or any part of this lease and all or any part of said land with all or any part of any other land lying within one-half mile of the exterior boundary of any land included in this lease, so as to create by such combining or pooling or unitizing one or more operating units for the production of oil, gas or other hydrocarbon substances; provided, however, that any such operating unit or units shall each contain not more than one hundred sixty (160) acres if combined or pooled or unitized for the development or production of oil and not more than six hundred forty (640) acres if combined or pooled or unitized for the development or production of gas; such designation of such operating unit may be made at any time but in no event later than ninety (90) days after the completion of a well for oil or gas on the lands to be pooled or unitized and shall define the area which shall constitute the operating unit. In the event and as soon as any such oil or gas operating unit is so created, Lessee shall promptly record in the office of the County Recorder in the county in which said land is situated such written declaration of pooling or unitizing; such written declaration shall become effective when recorded as above provided and shall identify and describe the pooled or unitized acreage. Lessee shall give written notice of such pooling or unitizing to Lessor. In the event there shall be lands within the boundaries of the operating unit as to which operating rights are not owned by Lessee or as to which a lease has not been made to Lessee, but Lessee should thereafter acquire an oil and gas lease or operating rights on such lands, then Lessee may, at its option, within the term of this lease but in no event after expiration of twenty (20) years from the date hereof, by a supplemental declaration of pooling or unitizing recorded in the office of the County Recorder in the county in which said land is situated include such additional land and lease in such operating unit; such supplemental declaration shall become effective when recorded as above provided and shall identify and describe the lands included therein, but no retroactive adjustment of royalties shall be made. Lessee shall give written notice of such supplemental declaration of pooling or unitizing to the Lessor affected thereby. Lessors agree that in the event production of oil, gas or other hydrocarbons is obtained from any lands included within any such operating unit, whether or not from lands covered by this lease, there shall be allocated to and deemed to have been produced from the lands covered by this lease and included in such operating unit, only that proportion of the entire production from such operating unit that the amount of surface acreage within or overlying the land herein leased and included in such operating unit bears to the total surface acreage within or overlying the land in such operating unit, and royalty payable under this lease with respect to leased land included in such operating unit shall be computed only on that portion of such production so allocated to such leased land. Drilling obligations in such operating unit shall be at least equal to the drilling obligations above set forth in this lease as applied to the acreage in the operating unit. The entire acreage so pooled or unitized shall be treated as if it were covered by one lease and the drilling of a well or performance of any other obligations in any part of such operating unit, whether or not on land covered by this lease, shall fulfill Lessee's drilling and other obligations under this lease to the same extent as if such well were drilled and other obligations performed on land covered by this lease. Lessee shall not be required to drill any offset well on that portion of the land hereby leased which shall be within such unit to offset a well drilled on other acreage pooled or unitized into such unit. Upon the pooling or unitizing of less than all of the land hereby leased as hereinabove provided, this lease shall be severed and separately pooled or unitized acreage and any unpooled or ununitized acreage shall be segregated and constitute separate and distinct leaseholds, so that performance of the obligations of one shall validate that one regardless of any default in the other. Commencement of drilling operations on any operating unit created hereunder shall not relieve Lessee of its obligation hereunder to pay rental with respect to any unpooled or ununitized acreage then covered by this lease, subject to the other provisions of this lease including the right to quitclaim. Lessee may at any time quitclaim to the persons entitled thereto all or any part of the land in any such operating unit by filing a quitclaim in the office of the County Recorder in the county in which

said land is situated, and thereupon Lessee shall be released from all further obligations with respect to land so quitclaimed and all drilling requirements thereunder shall be reduced pro rata and no owner of the land in the operating unit not owning any interest in quitclaimed land except by virtue of such pooling or unitizing shall have any interest in such quitclaimed land after the quitclaim is delivered or recorded. The payment of royalty hereunder, and the payment of royalty under any other lease covering land which was originally included within the operating unit, shall not be affected or reduced by any such quitclaim unless and until a well for oil or gas shall be commenced on or under the land so quitclaimed, whereupon the Lessor herein having any interest in the production from such well or the proceeds thereof, whether as royalty or otherwise, and all of the quitclaimed lands of such Lessor shall be excluded from said apportionment of benefits and from royalty payment hereunder. Lessee may, at its sole option, at any time when there is no production in such operating unit of oil or gas in quantities deemed paying by Lessee, terminate such operating unit by a written declaration thereof, in the same manner in which it was created. Upon the termination of any such operating unit, if only a part of the leased land was included therein, this lease shall no longer be considered as a separate and distinct lease and the effect of such termination shall be to return this lease to the status in which it was immediately prior to the creation of such operating unit.

30. This lease and all its terms, conditions and stipulations shall extend to and be binding upon the heirs, executors, administrators, grantees, successors and assigns of the parties hereto.

31. Any notice from the Lessor to the Lessee must be given by sending the same by registered mail addressed to the Lessee at____________________, and any notice from the Lessee to the Lessor must be given by sending the same by registered mail, addressed to the Lessor at____________________

IN WITNESS WHEREOF, the parties hereto have caused this agreement to be duly executed as of the date first hereinabove written.

LESSOR	LESSEE
____________________	____________________
____________________	____________________
____________________	____________________
WITNESS: ________________	____________________

[*Required notarial acknowledgment*]

Chapter 17

DRILLING THE EXPLORATORY WELL

By B. W. Beebe

President, Exploration, Inc., and Petroleum Exploration Consultant. Formerly Vice-president of Production and Director of Keating Drilling Company. A Past Vice-president of AAPG

Boulder, Colo.

The drilling of the exploratory well is the culmination of the entire exploratory effort. Are the theories correct, and will there be oil at this location? Prodigious effort and time have been spent to reach this point. An exploratory well is usually termed a "wildcat" in oil-field parlance, and no matter how sophisticated the geologist, there is always an undercurrent of excitement in commencing a wildcat. Exploratory test wells may be drilled in an area in which there are nearby oil and gas fields, or they may be drilled hundreds or thousands of miles from the nearest production. Although the primary objective of any exploratory test is to find oil, nevertheless, even though the test fails to produce, geological information of immense value is secured and must be properly observed and recorded. Each bit of information from numerous dry exploratory test wells may finally lead to the discovery of a new field. The responsibility for observing and recording this information rests with the well-site geologist, the man who is "sitting on the well." The success or failure of the entire venture, into which thousands of dollars and countless hours of effort have been poured, may depend entirely on his skill in recognizing obscure but telltale signs of oil and gas.

His actual responsibility will vary considerably with the size and nature of the organization he is representing. In any instance, he guides the sampling, coring, testing, and logging and evaluates the results for casing or plugging. If he is employed by a large company, his capacity in many phases may be advisory, with much of the responsibility shared with or assumed by the drilling superintendent, the drilling engineer, and the petroleum engineer. If he is employed by a small independent operator, all phases of the operation, aside from the actual mechanical cutting of the hole, may be his responsibility. Practices differ among companies; no manual on geological responsibility in the drilling operation can hope to cover more than generalities. Furthermore, over a period of years, each geologist develops his own special techniques and methods.

This chapter is not a technical description of equipment and drilling operations. However, the geologist "sitting" on a well must have a thorough familiarity with and working knowledge of the equipment, techniques, and methods used in order to discharge his responsibilities efficiently. He must know enough about the procedure, equipment, and the problems to avoid costly, uneconomical operations and to do everything in his power to assist in completing a clean, well-logged, well-tested hole. No matter what his responsibility in detail, he is the key man in the success or failure of the venture. Obviously he cannot place hydrocarbons where they are not, but he must find them if they are there.

Today the geologist is an accepted and respected member of the drilling team. This has not always been the case. Less than three decades ago a geologist around a drilling well was a pariah, unwanted, unloved. The drilling crew delighted in making him feel unwelcome. Numerous "accidents" around the rotary hose connections sprayed him with mud, and occasionally he was nudged into the slush pit. The average driller, his crew, the tool pusher, the superintendent, and the contractor had no use for that "college feller." Their feelings were often heartily reciprocated, but over a period of years, the contractor and the crew have learned that the geologist is a valuable asset on tour (pronounced tower). The geologist, in turn, has learned that, while the members of the crew are not learned scientists, they are nevertheless skilled, earnest technicians and mechanics, craftsmen in their own right. Thus a mutual bond of respect has developed, with consequent better results for exploration drilling.

Duties and Responsibilities

There are two maxims a geologist sitting on a well must constantly bear in mind. First, there is no substitute for his being on the rig floor and observing events with all his senses. A drilling well cannot be watched from a hotel lobby. Second, he can always order an additional hole cut, but once cut it can never be retrieved. Circulating time is cheap when the well-site geologist is in doubt. A third maxim might be added: When in doubt, seek expert advice. A geologist walking the catwalk in the watches of the night can be as lonely as the captain on the bridge of a great ocean liner cutting its way through the fog.

The equipment used by the well-site geologist is not elaborate. It usually consists of:

1. A good binocular microscope, with magnification giving a large field and higher magnification for careful examination of individual cuttings
2. A light suitable for illuminating sample trays, preferably 6-volt
3. An ultraviolet light
4. Metal or plastic trays for samples
5. 3 × 5 sample envelopes
6. Dishes for testing with acid or petroleum solvent
7. A set of tweezers
8. A small pick or probe
9. A supply of dilute hydrochloric acid
10. Several sharp pencils
11. A tablet or a pocket-sized notebook
12. A good set of colored pencils
13. Well-log strips, preferably gridded, so that the time log can be graphed with the sample descriptions
14. Logs of nearby wells and measured sections
15. Test tubes and petroleum solvent
16. A strong constitution

There are times when sleep is hard to come by for the geologist sitting on a well, for, like births, important events usually happen on wells in the small hours of morning.

Prognosis

The first step in drilling the exploratory well, as far as the geologist is concerned, is the prognosis. What is likely to be encountered? A sample of a typical form is

PROGNOSIS

________________ District ________________ 19___

Well No. ________________

Location ________________

FILE ________________

TRACT NAME

AREA

COUNTY AND STATE

Electric Log: ________________

Cores ________________

Estimated Elevation ________________ G. L.

Drill Stem Tests: ________________

FORMATION	TOP	MINUS	BASE

Probable Producing Horizon ________________ T. D. if Dry Hole ________________

Method of Drilling (Rotary or Cable Tools) ________________

Estimated Casing Program (Size and Setting Depth)

Surface ________________ Intermediate ________________ Production ________________

Set through or on top ________________ Pump or Flow ________________

Method of Completion ________________

Remarks: (Include possible gas or blow-out zones, and artesian or corrosive water zones.)

FIGURE 17-1. Prognosis form.

shown in Figure 17-1. The prognosis furnishes the guide for everyone concerned in the drilling of the well; it is analogous to blueprints for building a house. The authority for expenditures (AFE), the casing program, the preliminary mud program, and the drilling contract all depend on the accuracy of the prognosis estimates. The geologist can be most helpful in this regard and save the operator considerable capital and many headaches by attempting to anticipate possible zones where circulation may be lost and where blowouts or artesian flows of water may occur and possible corrosive water zones where it may be desirable to protect the production string of casing. Zones of extremely hard drilling caused by cherts, orthoquartzites, igneous rock, or other unusually hard formations should be anticipated.

The casing program is highly important to the geologist. He is the only one who can anticipate fresh-water sands which must be protected from contamination. If properly prescribed, the program assures him of a good set of drill cuttings and a good electric log. He should attempt to estimate where he expects to core and what zones may merit drill-stem testing.

At first consideration, some of this information may appear almost impossible to determine. However, if other test wells have been drilled in the area and information is available on them, most of the estimates can be made after careful scrutiny of these records. The geologist's knowledge of the section to be penetrated, even if based on surface exposures, is most helpful. Surface, subsurface, isopachous, paleogeographics, and topographic maps all furnish information. Examination of aerial photographs in the vicinity of the location is of value. The prognosis furnishes a preliminary pattern, always subject to alteration, but absolutely necessary in the operation. This form may or may not be prepared by the well-site geologist, but he must have all the information required on the form.

Drilling Methods

There are two basic methods for drilling a test well: cable tool and hydraulic rotary. The former method, which for years was predominant, decreases in importance year by year. Most wells, either exploratory or development, are now drilled using the rotary method. Cable tools are used principally in the older, shallower producing areas or areas in which rotary drilling presents unique problems. The following tabulation gives some comparative statistics ("Primer," 1957, pp. 91–92):

RIGS OPERATING

Year	Rotary rigs	Cable-tool rigs
1951	2,554	1,462
1956	2,622	1,105

TOTAL WELLS DRILLED

Year	Rotary	Cable	Total
1947	23,203	9,944	33,147
1952	38,080	7,799	45,879
1956	51,278	6,993	58,271

It is obvious that both the number of cable-tool rigs and wells drilled by cable tools are steadily declining while the total number of wells drilled each year has been generally steadily increasing.

Once a method of drilling has been selected, will the well be drilled with company-owned tools or by a contractor? The percentage of wells drilled by company-owned tools has also steadily declined, as shown by the following table ("Primer," 1957, p. 93):

WELLS DRILLED BY CONTRACTORS

Year	Wells	Per cent
1947	26,849	81
1952	41,750	91
1956	54,192	93

Instructions by Geologists

Once the decision is made to drill the exploratory well and the method of drilling has been decided upon, the geologist has a number of duties preliminary to the commencement of the hole. He must decide on the sampling program and how he wants drilling time recorded, if the operator employs rotary tools. Again, his actual responsibilities and duties will depend on the size and type of organization drilling the well. In some cases, instructions will be given to the production, engineering, or drilling departments. In others, instructions will be given directly to the contractor. In any case, all instructions at all times should be simple, concise, and written. Reporting procedure must be arranged. Where the geologist assumes most of the responsibility, as in a small organization, he will write the contractor a letter, attaching several copies of the instructions, requesting that one copy be given to the tool pusher and one posted permanently in the doghouse; a copy should also be sent to the production department. It is also good practice to fasten a copy of the instructions securely to the driller's daily report book which is sent to the rig. With the use of these precautions, there is never any doubt.

A typical set of instructions is shown in Figure 17-2. Note that instructions are given regarding the depth at which the crew will commence saving cuttings and the intervals at which samples are to be taken, and if a mechanical device for keeping record of the drilling time is not available, the depth and intervals at which drilling time is to be recorded are noted. Names, addresses, and telephone numbers of persons to be called at certain times and under certain conditions are shown. If there is a call point at which the well-site geologist is to be notified, the depth at which he is to be called, his name, address, and telephone number, and those of his alternates and superiors must be given. Most important is the shutdown point, where, if the geologist is not present, drilling operations are to cease and the operator notified so that further orders can be obtained. Instructions should be given regarding procedure to be followed in case of unexpected showings of oil and gas or unforeseen emergencies. Finally, possible depths of artesian-water flows, blowouts, lost circulation, or extremely hard zones should be noted. Once the well-site geologist has arrived on the scene, any of these instructions may be modified, but again, modification should be in writing so that there can be no possible misunderstanding.

There are a definite procedure and chain of command in the drilling operation. In the large organizations, any requests or instructions for the crew are given to the operating or drilling departments. Where the well-site geologist assumes most of the responsibility, suggestions, instructions, orders, or complaints should be made to the contractor, tool pusher, or, if on location, to the driller on tour. He is the man responsible. An inexperienced well-site geologist may find himself in difficulties by indiscriminately ordering crew members around. The respect and good will of the driller are mandatory. He can be very helpful to the geologist. Both have a common goal:

to drill a clean, thoroughly tested hole. The geologist can be helpful to the driller by anticipating types of formations to be encountered, as well as possible hazards, but he must remember that the driller is responsible for cutting the hole. The importance of mutual respect, confidence, and cooperation between the driller and the well-site geologist cannot be overemphasized.

An experienced well-site geologist can easily determine the type of contractor and crew with which he must deal when he approaches the location. If the grounds,

OPTIMISTIC OIL COMPANY
P.O. Box 77
McCook, Nebraska

Telephone: Murray 2-3262
Saturday and Sunday:
Murray 2-5120

Producer No. 1
SE NW SW Section 7, T1N, R33W
Hitchcock County, Nebraska

Morning tour driller to report to district office by telephone collect, and mail copy of daily drilling report to district office each morning.

10′ samples	200–2,500 feet
5′ samples	2,500–T.D.

All samples to be thoroughly washed, carefully dried, sacked, labeled, tied in bunches of 20, and shipped collect to 110 Second St., McCook, Nebraska, by bus each Monday.

10′ drilling time	200–2,500 feet
5′ drilling time	2,500–3,250 feet
1′ drilling time	3,250–T.D.

Call geologist at 3,150 feet if he is not on location.
Shut down at 3,250 feet and telephone collect for orders if geologist is not on location.

Geologist	*Home Telephone*
Richard Roe (well-site geologist)	Murray 2-5120
John Doe (alternate)	" 2-4361
I. M. Rockhound (District Geologist)	" 2-6290
U. R. Drillingin (Prod. Supt.)	" 2-5031

Mud up at 2,500 feet, weight 9.5#, viscosity 33, maximum water loss 10. Check condition of drilling mud at least once each tour.

If oil or gas showing or odor noted, shut down and telephone collect for orders.

cc—Contractor, tool pusher, log book; well-site, alternate, district geologist; prod. supt., file.

FIGURE 17-2. Example of instructions to drilling crew.

equipment, and rig floor are neat and clean, he knows that the operation is in good hands. However, if there are junk, dirt, and grease all around the location, rig floor, and equipment, and roughnecks "snoozing" or playing cards, he should prepare for trouble. One final word of caution: The crew may not be college trained, but these men are not stupid. Do not try to bluff!

Cable-tool Methods

The well-site geologist of today will rarely be called on to sit on a cable-tool exploratory well. The cable-tool drilling method, also known as the percussion method, is rarely used for exploratory operations. It was the first drilling method and was supposedly originated by the Chinese, who used a spring bamboo pole and manpower.

Early cable-tool wells in this country were drilled using a supple sapling, powered by human muscle as shown in Figure 17-3. Modern cable-tool techniques are a far cry from these primitive operations. Cable tools are of two types: standard and machine. The machine is used for drilling only very shallow wells. Standard cable tools have drilled to depths of over 11,000 feet. Figure 17-4 is a drawing of a standard cable-tool unit.

The modern standard unit consists of a derrick; a rig, which is fundamentally the hoisting and running equipment; the power unit; drilling tools and accessories; and casing and accessories. The power unit may be steam or a diesel-, gas-, or gasoline-fueled internal-combustion engine. The crew consists of a driller and a tool dresser. Hole is made by the up-and-down motion of a bit, which pulverizes and penetrates the rock layers. Tools are attached to the bull-wheel wire-line cable to which the walking beam, which imparts the up-and-down motion, is attached by clamps on the drilling line. As the bit penetrates the formation, additional line is played out through the temper screw which controls the clamps. The tools are withdrawn from the hole from time to time, and a bailer is lowered on a line from the calf wheel to remove the cuttings which accumulate as the bit pulverizes the rock. The tools, which consist from the bottom up of a bit, stem, jars, and socket and which are on the end of the drilling line at the bottom of the hole, must swing so that the bit strikes the bottom of the hole at the proper moment. This motion depends on the skill of the driller and the elasticity of the drilling line. Some water is usually put into the hole to emulsify the cuttings and assist in drilling. Too much water impedes the action of the tools.

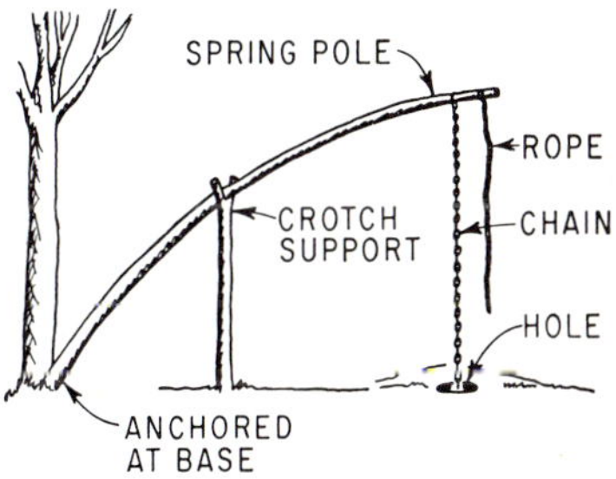

FIGURE 17-3. Primitive spring-pole method of churn (cable-tool) drilling.

A spudding machine imparts the up-and-down motion to the tools by means of a sheave, which rolls up and down on the drilling line, rather than through a pitman transmitting power from the eccentric to the walking beam.

Drilling in hard or clayey formations, particularly where fluid in quantity is constantly entering the hole from porous strata, is a slow and tedious process. When water in quantity is encountered, it is necessary to run a string of casing to dry up the hole or underream and lower casing which has previously been set. When a new string of casing is run, the hole must be reduced and a smaller bit used. There is always danger of "running out of hole" before reaching the objective, because the hole size becomes so small it is impossible to drill ahead. In times past, cable-tool holes with as many as six or more diminishing-diameter strings of casing were not uncommon. The deeper the well and the smaller the hole, the slower the progress.

There are certain definite advantages to drilling with cable tools. The source of the cuttings can be located with precision, and the geologist knows within a few feet the precise nature of the formation being drilled. Furthermore, drilling can cease at any time and an accurate, representative sample can be obtained from the bottom of the hole. The bailer will also reveal oil which may be entering the hole if the column of water in the hole above the formation showing oil is not too great. Gas is also readily detectable unless a column of water in the hole exerts too much pressure on the formation. It is therefore of the utmost importance to penetrate any possible producing formation with a minimum of fluid in the hole. Where formations are firm, the walls of the hole stand up, and few formations which yield large quantities of water are anticipated, cable-tool drilling is often less expensive and more efficient.

Highly fractured or porous and cavernous formations, which cause serious lost circulation in rotary operations, in many cases can also be drilled more efficiently with cable tools.

However, there are numerous disadvantages to the cable-tool method, particularly

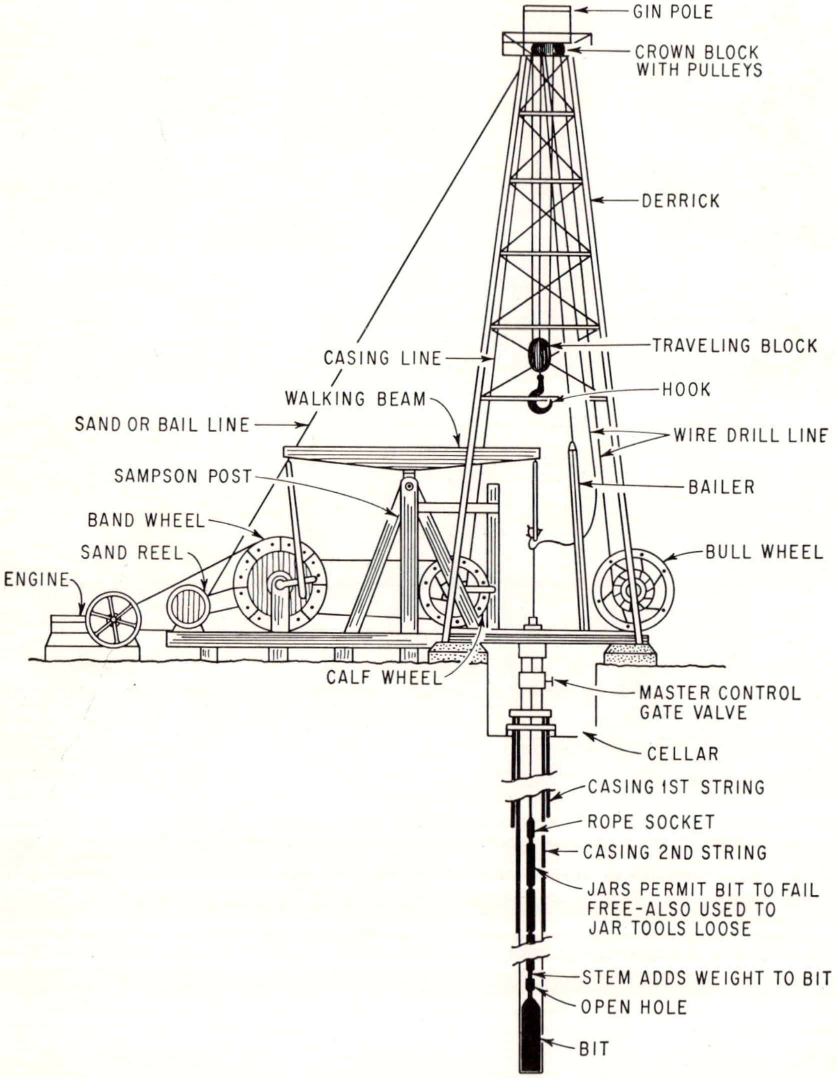

FIGURE 17-4. Cable-tool drilling rig.

where the nature of the strata to be penetrated is not well known. First, it is slow and particularly inefficient on deep holes which can be drilled by the rotary method. Second, if frequent water-bearing strata are encountered, casing must be lowered by underreaming to shut off water or a new string of casing run. This is also true if many soft or caving formations are encountered. Third, if unexpected oil or gas under high pressure is found, it is difficult to proceed with drilling. Fourth, if drilling does proceed with a large quantity of water in the hole, it is often difficult to detect low-pressure oil and gas zones. Fifth, if there is a large amount of fluid in the hole, drill cuttings are pounded so fine that the character of the formation is difficult to determine, and if the well is flowing, there are usually no cuttings from which to determine the character of the formations. Finally, a conventional electric log of the entire borehole cannot be secured without undue expense.

The geologist's duties on a cable-tool rig are not arduous. He collects, examines, describes, and preserves samples whenever the bailer is run. If an oil or gas showing is encountered, he attempts to determine the amount and evaluate the showing by bailing, then drilling a short interval, and bailing again. Finally, he attempts to assist the driller in anticipating zones which carry large amounts of water and offers advice on whether to underream or run another string of casing.

Hydraulic Rotary Drilling

Drilling by the hydraulic rotary technique has largely superseded the historic cable-tool method. Although attempts to drill shallow holes using rotary tools had been made previously, the general acceptance of this method actually dates from its use in the Spindletop field in 1901. Except for rare and unusual situations, it is the customary method of drilling an exploratory well.

The mechanics of rotary drilling are simple. Three operations are carried on simultaneously. A string of drillpipe with a cutting bit is rotated. The bit is lowered as the formation drills out from under the bit. Finally, the bit is cooled and lubricated, and the debris ground up by the bit and the cuttings are cleared away so that the bit can continuously cut new formation.

The drilling equipment is collectively known as a drilling rig, or *rig*. Its essential component parts are the prime movers or engines; the derrick or mast; the hoisting equipment, including the draw works, crown block, the traveling block, and hook; the rotating equipment, including a swivel, kelly, drillpipe, drill collars, and bit; and finally the mud system which circulates the drilling fluid from pits through the drillpipe and bit, lubricating and cooling, collecting the cuttings, and returning to the pits through the annulus between the drillpipe and the walls of the hole. A typical rotary rig with derrick can be seen in Figure 17-5.

While on location, the drilling rig operates continuously 24 hours per day. Each 24-hour period is divided into three shifts, or tours. Each tour is composed of four or five men, depending on the size and complexity of the operation. The driller is in charge. His first assistant is the derrick man, whose principal responsibility is the proper racking of the pipe on the monkey board, located high in the mast. If the rig is larger, number three man is the cathead man, or motorman. Then there are two roughnecks, who are floormen. The derrick man is usually in charge of the mud. One of the roughnecks samples the cuttings at specified intervals. If there is no motorman, one of the roughnecks is responsible for engine maintenance. The entire crew and operation are under the supervision of a tool pusher, the foreman, on 24-hour call supervising one or more rigs.

The power, or prime mover, for a rig may be of several types. The earliest type used was steam power, which is the most efficient, simplest, and most flexible and with the most efficient torque characteristics, but it lacks portability. Steam is now rarely used except in areas where continuous and large quantities of water and fuel are available and where moves are short. Steam power has largely been superseded by other types of power as they have become available and necessary improvements have been made.

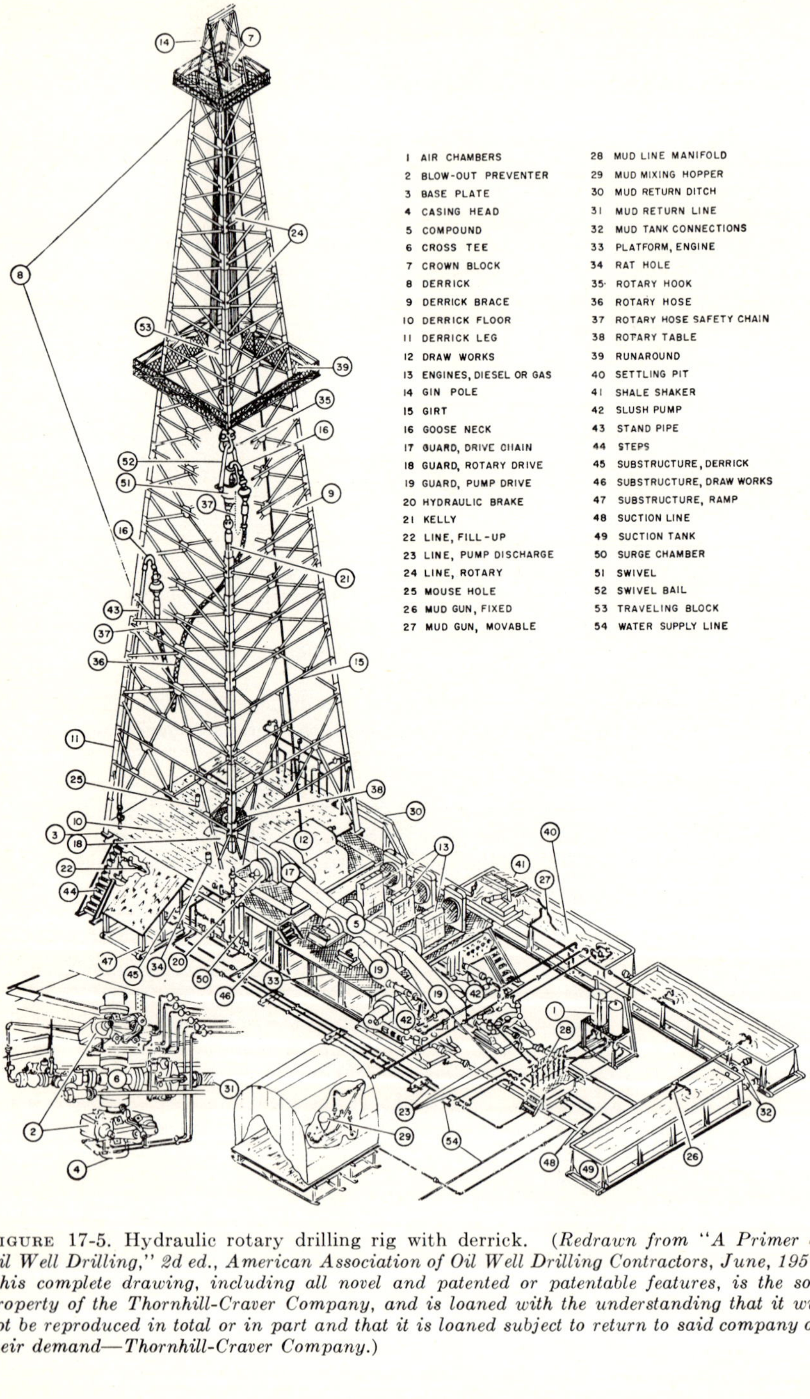

FIGURE 17-5. Hydraulic rotary drilling rig with derrick. (*Redrawn from "A Primer of Oil Well Drilling," 2d ed., American Association of Oil Well Drilling Contractors, June, 1957. This complete drawing, including all novel and patented or patentable features, is the sole property of the Thornhill-Craver Company, and is loaned with the understanding that it will not be reproduced in total or in part and that it is loaned subject to return to said company on their demand—Thornhill-Craver Company.*)

Many large rigs are now powered by electricity, the diesel-electric prime mover being the most common. Diesel engines drive generators which provide electric current to power electric motors. These prime movers are quite as satisfactory as steam. They are smooth and flexible with good torque characteristics. They are somewhat cumbersome and expensive and are generally used only on large operations for deep holes. A few experiments have been made with rigs powered directly with alternating current from public utilities.

Most rotary rigs with which a geologist will come in contact are powered by internal-combustion engines which burn natural gas, butane, or, occasionally, gasoline. Power-transmission devices are necessary so that the equipment is not excessively strained by jerks or jars characteristic of this type of power. A number of drilling rigs are powered with diesel engines. These are most economical to operate but present certain maintenance problems. Torque convertors are a necessity.

The second essential piece of equipment of a rotary drilling rig is a derrick or mast. A conventional derrick may be used on deeper holes. Generally, portable derricks or masts are used. The purpose of the derrick is to hold the traveling block assembly which holds the drillpipe and is, in turn, held by the rotary line feeding over the crown block from the draw works. When a trip is made, the derrick furnishes a most convenient, efficient place to stack the drillpipe vertically when it is temporarily out of the hole. The depth capacity of the drilling rigs depends primarily on the size of the draw works, which provides the hoisting and rotating action. The draw works allows the bit and drilling string to be lowered or pulled from the hole and furnishes the medium for rotating the drill string and bit. Power from the engines is transmitted to the draw works, then to the rotary table, which rotates the kelly, a square or hexagonal hollow shaft, the bottom of which is screwed into the topmost section of the drillpipe. The top of the kelly is attached to the swivel. The swivel is the connection between the slush pump and the drill string and allows the latter to rotate while drilling fluid passes through into the drill string. The swivel is attached by a hook to the traveling block on which the drilling line passes through to the crown block down to the draw works. As the bit cuts formation, the driller feeds drilling line from the drum on the draw works. The drum is controlled by a brake, permitting the drilling string and the bit to descend slowly.

The fourth component of the drilling rig is the mud system which is activated by the slush pumps. The pump sucks drilling fluid from the pits and pumps the fluid through the rotary hose, swivel, kelly, drillpipe, drill collars, and finally the bit. The fluid passes through the bit with a jetting action which assists the bit in cutting the hole. This is depicted in Figure 17-6. The fluid cleans and cools the bit, carrying the cuttings out through the annulus, from which it passes over a shale shaker which removes the cuttings, then flows into the pits. The cycle is continuously repeated, as can be seen in Figure 17-7.

To bore a hole with rotary tools, or, in oil-field parlance, "to make hole," there must be a bit on the bottom. Above the bit are drill collars, which furnish weight. These are hollow shafts of the toughest alloy steel, weighing from ½ to more than 2 tons each and generally are 30 feet in length. As many as 20 or more may be used, depending on the weight required to make the bit cut. The drillpipe extends from the top of the drill collars to the surface. It ranges from 2⅜ to 6⅝ inches in diameter and weighs from 6 to 25 pounds per foot. Joints of drillpipe are in lengths of about 20, 30, or 40 feet, known, respectively, as range 1, 2, or 3. The lengths of drillpipe are attached to each other by tool joints. The string of drillpipe performs three functions: setting the bit on the bottom, turning it, and providing an avenue for the drilling fluid to descend to clear away cuttings, lubricate, and cool the bit. As the

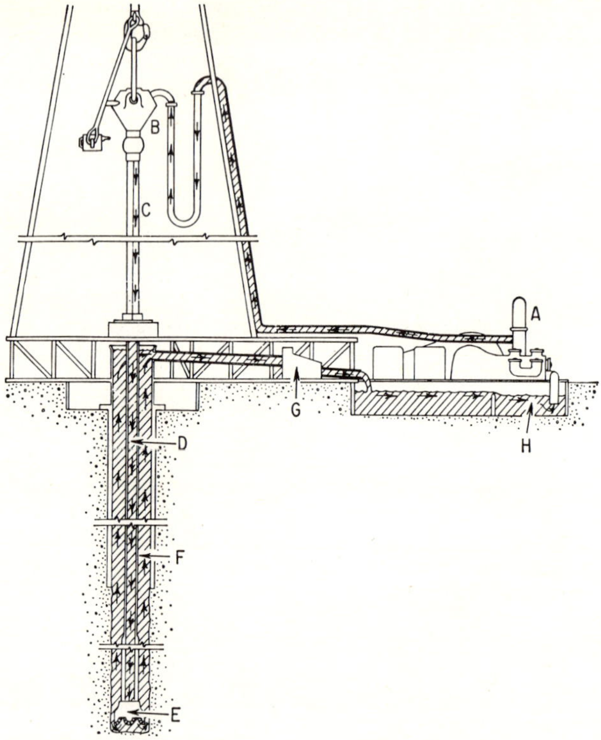

FIGURE 17-6. Diagram of rotary mud system.

hole is drilled deeper, additional lengths of drillpipe are added at the derrick floor. A diagram of the operations below the derrick floor is shown in Figure 17-8.

The drillpipe is rotated at the surface by the kelly, which fits into recesses in the drive bushings of the rotary table. When the rotary table turns, the kelly, the drill string, and the bit are turned. During this time, the continuous cyclic flow of fluid passes through the drillpipe to the bottom of the hole and returns. When enough formation is drilled, so that the kelly is "down," the entire string is lifted enough so that the kelly can be unscrewed and slipped into the "rathole," a new joint of drillpipe is pulled from the "mousehole," the connection is made, the kelly is reconnected, and once again hole is being made. When the bit becomes dull, it must be changed, a process known as "making a trip." All the drillpipe, drill collars, and bit are removed from the hole and stacked vertically in the derrick and resting on the derrick floor in lengths of two joints, doubles; three joints, thribbles; or four joints, fourbles, depending on the height of the derrick or mast. The bit is replaced, the drillpipe is connected again

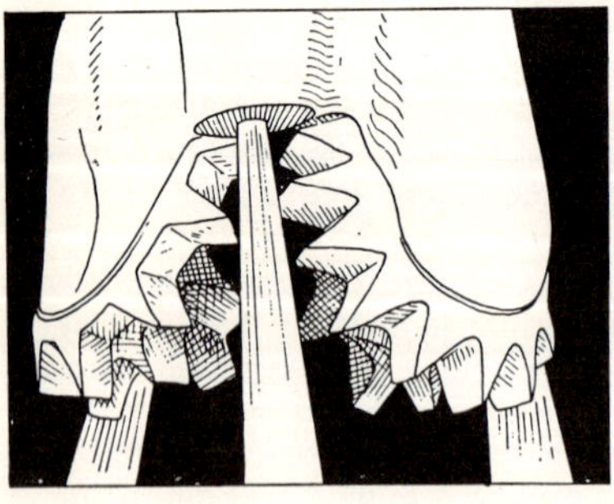

FIGURE 17-7. Jetting action of fluid through rotary drilling bit.

and lowered into the hole until the bit is on the bottom, and once again hole is made.

Slim-hole drilling is an adaptation of conventional rotary drilling which has become an important exploration technique. A slim-hole rig is a highly mobile, portable drilling rig, usually mounted on a truck and trailer which can be moved and rigged up rapidly. Small drillpipe is used to cut a hole ranging in size from 4¾ to 6¾ inches. This hole is presumably less expensive to cut, often is too small for conventional production purposes, but is ideal for obtaining stratigraphic information and determining the presence of oil or gas in porous zones. The slim hole, often termed a "strat

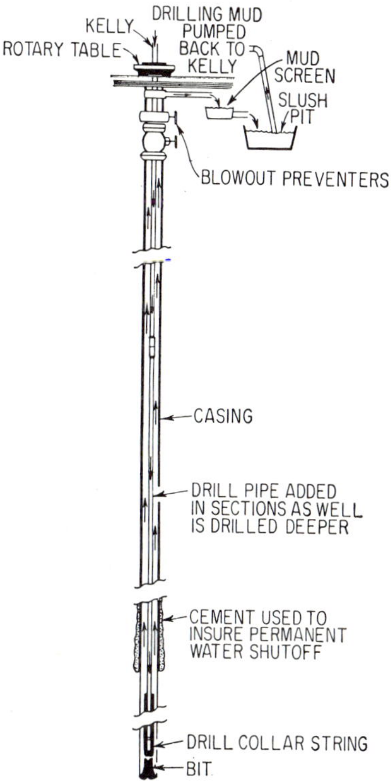

FIGURE 17-8. Cutaway diagram of rotary drilling operation below derrick floor.

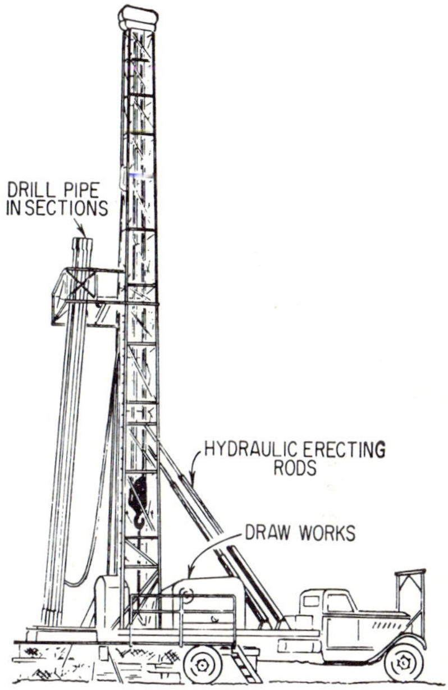

FIGURE 17-9. Diagram of truck-mounted portable (slim-hole) rotary rig.

hole," can be reamed out, cased, and put on production. Figure 17-9 is a diagram of a truck-mounted slim-hole rotary rig.

Bits and their proper choice are of prime importance in rotary drilling. They are basically of four types. The fishtail bit is a slab of tough steel hammered out in a blacksmith's shop and used primarily in very soft formations. For consolidated formations that are not too hard, a soft-formation rock bit is used. The teeth are longer and more widely spaced so that the cuttings will not clog them as the formation is torn. The hard-formation rock bit has stubby, closely set teeth which are designed to have a maximum rolling and crushing motion. For the very hardest formations, a tungsten carbide or diamond drilling bit is used. As many as 900 carats of industrial diamonds may be set in the alloy steel of a diamond drill bit. These types of bits are shown in Figure 17-10. All the bits have orifices along the sides or in the bottoms which act as jets for the drilling fluid to wash away the cuttings and assist in penetration. The choice of bit, which is the driller's responsibility, is extremely important in obtaining the maximum penetration rate. The geologist can often be of

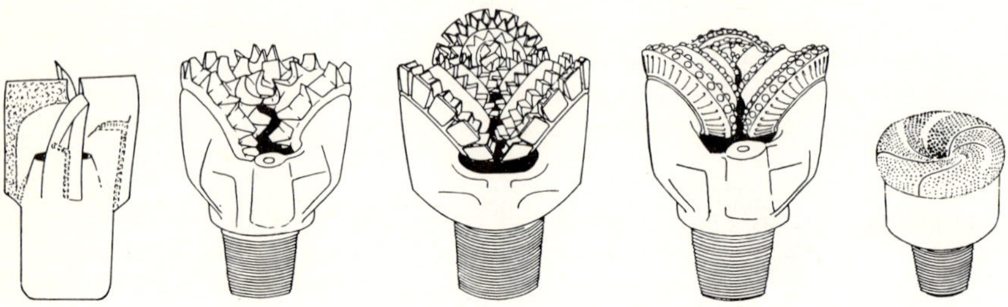

FIGURE 17-10. Five types of rotary drilling bits: (*A*) Fishtail bit; (*B*) soft-formation bit; (*C*) hard-formation bit; (*D*) tungsten carbide bit; (*E*) diamond bit.

considerable assistance to the driller in predicting the type of formation to be penetrated. The speed at which the formation can be drilled depends to a large extent on the type of bit, weight on the bit, rotating speed of the bit, and the pump pressure.

Fluid System and Fluids

The most important phases of a drilling operation to the geologist are the circulating system and the character and quality of the fluid in the system. The fluid and the circulating system perform a number of very important functions. The fluid clears the cuttings from the path of the bit, cooling it in the process; the jetting action aids the bit in penetrating the formation. An equally important function of the fluid is to coat and plaster the walls of the hole and prevent the hole from caving.

The important phases of the process to the geologist are the return of representative cuttings and the plastering of the walls of the hole. The latter prevents the hole from sloughing and caving, which would contaminate the cuttings. If the geologist wants representative cuttings from the bottom of the hole, the rotation of the bit is stopped, the bit is raised a few feet from the bottom, and the pumps continue to circulate the fluid. If drilling is not stopped, the bit is always ahead of the cuttings which the geologist sees at the surface. This time lag depends on the rate of penetration and the time that it takes for the fluid to return the cuttings to the surface. When the geologist wishes to know how long it takes for the fluid to make a complete cycle, dye can be introduced into the suction pipe, or a more handy substance, such as corn, wheat, or oats, often is used. The time is noted, and when the dye or foreign substance reappears on the surface, the time of the complete circulating cycle is known. The time required to bring the cuttings from the bottom of the hole is somewhat greater than half the time observed.

The character of the drilling fluid is obviously highly important. In addition to the functions previously cited, it reduces friction on the drill string and assists in keeping the well under control in case of an unexpected oil, gas, or artesian-water flow. If the only consideration were speed in cutting hole, clear water would be the ideal drilling fluid. However, drilling fluids of good quality and character assure good samples of the formation drilled, reduce invasion of potential porous producing zones, and materially reduce sloughing and caving of the hole.

A drilling fluid has a number of primary characteristics: viscosity, gel strength, density and colloidity, fluid loss, salt and sand content. Viscosity is a measure of the body of the mud, technically of its internal resistance to flow, which assures that the mud can carry the cuttings and that a sheath of mud cake will be deposited on the walls of the hole. Bentonite, commonly known as gel, which swells in water, gives body to the mud. It also is quite slippery and reduces friction of the drill string. Clay or shale beds which are penetrated often make mud and increase viscosity to an undesirable level. The viscosity is controlled by adding water or chemical thinners, including tannin extracts and common lye. Viscosity is measured several times daily by a member of the crew or a mud engineer using a Marsh funnel or a Stormer or Fann viscometer.

Control of the fluid loss (commonly termed water loss in the field in water-base mud) is necessary to reduce filtration of water from the mud into porous formations. This filtration might contaminate oil and gas zones and leave a thick, undesirable mud cake on the walls of the hole. Loss of water also increases viscosity. Water loss is controlled by adding cornstarch, vegetable gum, CMC, and similar materials. Preservative may be required to prevent mud thus treated from fermenting.

If additional weight is needed in the mud to control gas pressure or artesian-water flow, barium sulfate, commonly known as barite, is added. In a deep well, it is very important that the mud not be too heavy, because a column of drilling fluid can exert tremendous pressure on the formation and actually retard penetration by the bit. Fluid with too much weight may also lead to lost circulation. It is a very neat trick, when porous formations carrying gas or fluids are drilled under high pressure, to secure a weight sufficient to prevent a blowout and yet light enough so that fluid will not enter a formation, causing lost circulation. Mud can be aerated by jetting it with air. If there are continuous gas pockets in the mud from penetrating gas-bearing zones under high pressures, a degasser may be necessary to keep the mud in condition.

Although blowouts are relatively infrequent, they are a serious hazard in drilling. The most serious hazard, however, is lost circulation or loss of returns, which is frequently encountered in some areas and adds materially to the cost of drilling. Circulation is lost when a very porous, cavernous, or highly fractured formation is penetrated. Coarse material susceptible to swelling in water, usually cellophane flakes, ground leather, nutshells, cottonseed hulls, redwood or cane fiber, even hay and feathers, are put in the mud system to plug up fractures, fissures, and cavities and to seal off the formation. It is sometimes necessary to drill these zones for short intervals without circulation and cement them off before normal drilling operations can be resumed.

Oil or diesel fluid can be added to the mud under certain conditions to assist the bit in penetration and to lubricate the drillpipe. Oil also assists in keeping the teeth of the bit clean and helps prevent shales from balling up on the bits and drill collars. Oil-emulsion fluid or oil base fluid can be used to reduce fluid loss when drilling into potential producing zones subject to contamination by water or characterized by low pressures.

Drilling fluid with the proper characteristics is essential, and its maintenance is a complex problem. Close supervision of the mud program by a trained mud engineer is now routine on exploratory wells. If the drilling fluid is not in condition and will not yield good samples or shows excessive water loss, or if the hole sloughs and caves, the well-site geologist should shut down the operation, call for assistance, and drilling should not be resumed until the mud is properly conditioned. The geologist must insist that the weight, viscosity, gel strength, and fluid loss be checked at least once a day.

Drilling Control Instruments

The driller controls the various operations by a series of instruments on a conventional internal-combustion-powered rig. These are usually five in number, shown in Figure 17-11. The weight indicator shows how much weight in thousands of pounds is resting on the bit and how much is supported by the drilling line. The pump pressure gauge shows the pressure on the slush pump in pounds per square inch. The

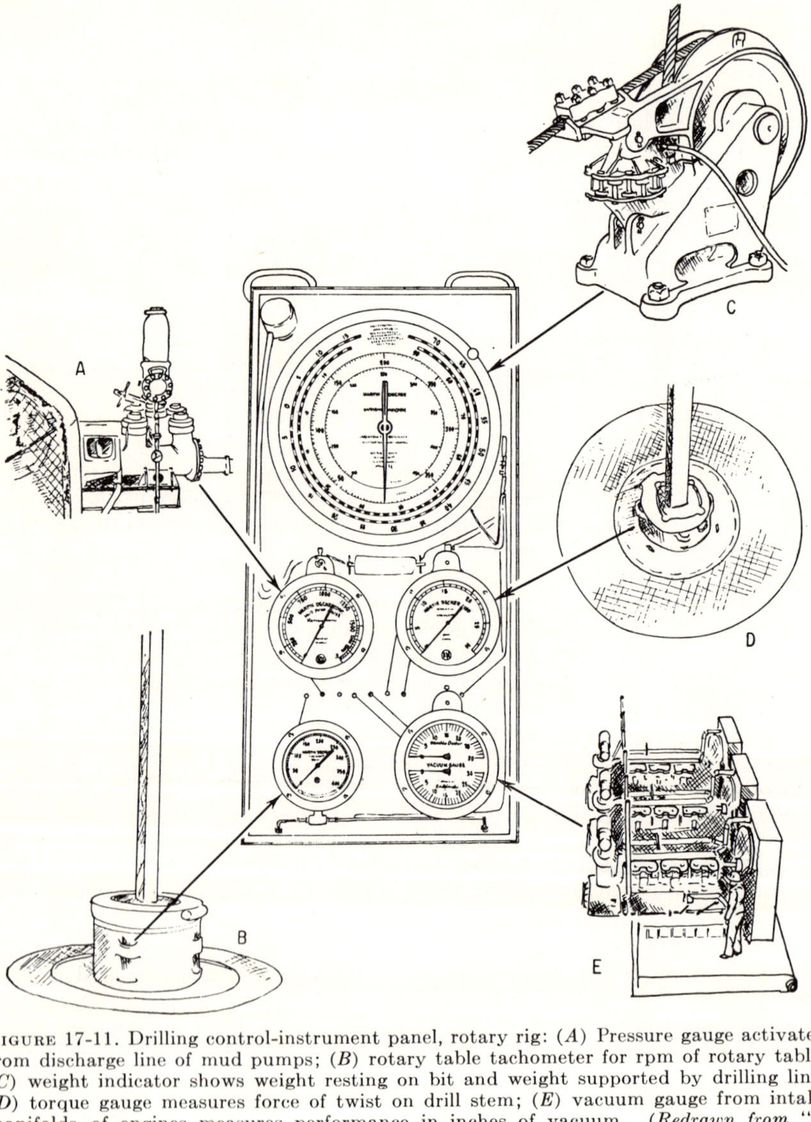

FIGURE 17-11. Drilling control-instrument panel, rotary rig: (*A*) Pressure gauge activated from discharge line of mud pumps; (*B*) rotary table tachometer for rpm of rotary table; (*C*) weight indicator shows weight resting on bit and weight supported by drilling line; (*D*) torque gauge measures force of twist on drill stem; (*E*) vacuum gauge from intake manifolds of engines measures performance in inches of vacuum. (*Redrawn from "A Primer of Oil Well Drilling," 2d ed., p. 40, American Association of Oil Well Drilling Contractors, June, 1957.*)

rotary table tachometer gives the number of revolutions per minute of the rotary table. The torque gauge measures the twisting force on the drillpipe, an important measurement to prevent twisting off the drillpipe. Performance of the engines is measured by a gauge showing the number of inches of vacuum on the intake manifold of each engine. If the rig is powered by steam or electricity, other types of instruments are necessary to show the performance of the power plant. The driller has at his disposal a variety of instruments which are run at least once daily to determine the deviation of the hole from vertical or both the angle and the direction of the deviation, which are expressed in degrees. The mud weight is measured in pounds per gallon or cubic foot and is measured several times daily by a balance which weighs small samples. Viscosity is usually measured by a Marsh funnel or Stormer or Fann viscometer several times daily. The Marsh funnel is the simplest, and the measurement is made by checking the number of seconds required for a given quantity of fluid to pass through the funnel. Water loss is recorded in cubic centimeters by a standard test. Salt content of the drilling fluid is determined in parts per million. There may also be gauges showing the speed and volume of the slush pumps.

A recording device may measure rate of penetration of the bit. Each length of pipe added to the drilling string is measured and recorded to show the length of each joint and the total amount of pipe in the hole. These measurements are made with a steel tape, either in feet and inches or feet and tenths of a foot. The latter is preferable. An accurate check on the measurement of the depth of the hole can be obtained by strapping the pipe stacked in the derrick with a similar tape, or for a rough check, a calibrated steel line with a weight is lowered into the hole, and a record shows the number of feet to the bottom of the hole.

Drill Cuttings

The well-site geologist is completely dependent on the quality of samples of the cuttings. From a microscopic study of the cuttings, he identifies the type of strata penetrated and examines them minutely for showings of oil or gas. If the cuttings are not representative of the strata penetrated, he is completely lost and can unknowingly drill through potential producing zones. At regular intervals, usually 5 to 10 feet, samples of the cuttings are caught at the shale shaker, screen, or trough by a member of the crew. He washes them with great care. The geologist takes a part of the wet sample for examination; the remainder of the sample may be meticulously dried without scorching by the crew member or sacked wet in cloth bags provided for the purpose. The name of the well and the depth are accurately noted on a label on the sample sack. The geologist will usually rewash a part of his sample, examine it wet and again after it has been carefully dried. The well-site geologist must check frequently to see that sampling is methodically and properly handled. Irregular large chunks or slivers of formation, usually shale, previously drilled are indications that the hole is caving, and caving materials tend to obscure cuttings of formation being drilled. Smooth rounded fragments of formation previously drilled are recirculated material. This is a warning that the cuttings are not dropping out of the fluid. Both must be kept at a minimum to assure good representative cuttings. Both conditions are primarily a result of drilling fluid of poor quality. Powdering of the cuttings often results from this condition. The utmost care must be used continuously to secure representative samples.

Logging the Well

The detailed discussion of various methods and techniques for logging the exploratory well is in Chapters 19, 20, and 21. Therefore only a few comments need be made at this time. The geologist has two primary responsibilities on an exploratory well. The first of these, of course, is the detection of oil and gas. The second, equally important, is the preparation of an accurate, descriptive, lithologic log from the drill cuttings. This can be accomplished only by insisting on drilling fluid which will produce good representative drill cuttings and carefully describing each sample in detail. It is good practice to examine each sample of the cuttings immediately after it has been caught, when actions of the bit and instruments are well in mind. Proper placement of samples at their true depth is assisted by a drilling time log, either manual or mechanical, and by first-hand observation of the action of the rotary table and the instruments. If drilling time is kept manually, the geologist must insist that it be kept with precision by the driller, and changes in action of the bit, new bits, and other pertinent data must be recorded. The time log should be plotted graphically on a log strip to assist in locating the actual material observed in the drill cuttings. A good lithologic and time log not only shows the material being drilled, accurately placed as to depth, but allows the geologist to compare structural position of various markers with the records of other test wells in the general area and compare lithology and thickness of units with these well records and/or measured surface sections. Each sample of the drill cuttings should be tested wet under ultraviolet light for fluorescence from oil or gas. This is of particular importance when there is a change in the rate of penetration of the bit.

There are many types of logging devices at the geologist's disposal, but unless there is a continuous mud logging unit on location, the only means he has of determining the character of formations being drilled and their content is through the observation of the action of the tools, the continuous examination of the cuttings, and the rate of penetration.

The driller also maintains a log of the type of material drilled, which varies greatly in quality, depending on the experience, interest, and sense of responsibility of the driller. He usually examines a handful of the cuttings each time a sample is taken, and he can often determine the type of material being drilled by the action of the tools and the rate of penetration. Obviously, a driller's log cannot compare in accuracy and detail with that made by a geologist with a microscope.

One of the greatest aids to the well-site geologist is the mud and cutting logging device, which may vary from a simple "iron man" mud logger recording gas in the drilling fluid and sounding an alarm when the gas units exceed a certain predetermined value to a complex, mobile laboratory with operators, which continuously tests both the drilling fluid and the cuttings for traces of hydrocarbons.

There are many other logging devices which are of material assistance, but all of them find their use after the fact. The event has already taken place, and they aid only in evaluation and post-mortems.

The principal devices are the electrical and radiation logs of which there are many types from which to choose. Each type performs a specific function or functions and is best adapted to certain areas. On an exploratory well, it is common practice to run such surveys at intervals to assist in positioning the well stratigraphically. In addition, there are other types of logs: the sonic log, which is relatively new but finding wide acceptance, and the caliper log, which records the size of the hole throughout its depth, showing where it is in gauge and where it is badly washed out. Finally, there is the dip meter, which is run in steeply dipping formations or where faulting is suspected to determine direction and the amount of dip at various intervals in the hole.

Coring the Well

In addition to cuttings, driller's log, time log, mud analysis, and many types of mechanical, electric, radiation, and sonic logs at the disposal of the geologist, a formation core is often either necessary or most helpful. Coring, however, should not be done indiscriminately. Coring adds materially to the cost of the well. In some instances, a program may be devised before the well ever commences for cores to be taken at specific intervals. In other instances, a decision to core may be reached through telephone reports from the well-site geologist to his supervisor. In some cases, the matter of coring may be left entirely to the discretion of the well-site geologist. There are actually two schools of thought on the subject: One group believes that cores should be taken at specific intervals and frequently; the other, that cores are a last resort on an exploratory well. It is the author's opinion that a carefully supervised well with good samples requires a minimum of coring unless formations are unusually soft, as, for example, in the Tertiary of California or the Gulf Coast.

The principal purposes of coring are specifically to determine the character of formation and the position of certain marker beds and contacts, for paleontological information, and to assist in the evaluation of showings of oil and gas. There are several types of cores which may be taken. The conventional core barrel may be used, although it is frequently most unsatisfactory because of the loss of soft, porous formation. Use of the diamond core bit reduces this possibility but is usually more expensive. Both are slow and require special equipment and round tripping of the drillpipe. Of the two methods in relatively indurated formations, diamond coring is preferable. In soft formations, the use of a wire-line core barrel obviates the necessity for round tripping the drillpipe, is satisfactory and much less expensive. Finally, after the well has been drilled and logged by electrical or radioactive methods, side-wall cores may be taken on the same line. Small-diameter cores can be recovered by tubes driven into the formation and retrieved by the logging line. In the author's opinion, the principal reason for coring is to obtain reservoir information, and an exploratory well, the principal purpose of which is to find oil and gas in commercial quantities, is a poor place to experiment. Cores for reservoir information may be taken on confirmation wells.

The entire subject of coring, the various techniques, and the responsibility of the well-site geologist in the operation will be thoroughly explored in Chapter 18.

Testing for Oil and Gas

The principal duty of an exploratory well-site geologist is to detect the presence of showings of oil and gas and to evaluate such showings. There are numerous methods at his disposal for both qualitative and quantitative evaluation. The most obvious is the appearance of oil or gas showings in the drilling fluid, which may be quite apparent, or the well may commence to unload mud, or blow out. There is no doubt that something unusual has been encountered in the latter case. Gas may break out of the mud, and/or bubbles and rainbows of oil appear. If this occurs, the geologist may not have been on his toes. Showings of oil or gas on the pits are not uncommon, but the neophyte should not become confused by the oily-appearing scum which often appears on the pits after a long shale section has been drilled or from oil that has entered the mud through lubrication of equipment. Such "showings" result in the "farmer tales" so often heard about wildcat operations.

When an abrupt change in the rate of penetration of the bit is noted, particularly unusually rapid drilling in soft formations, the well-site geologist is forewarned. The safest procedure is to circulate for samples after a few feet of unusually soft drilling. The geologist should be at the shale shaker when samples are taken. His first step is to grab a handful of the muddy samples in which he buries his nose and sniffs. If there is no odor of hydrocarbons, he should briskly rub a blob of the cuttings between his palms and immediately bury his nose in the rubbed-up mass. If there is still no odor, he should carefully wash a sample and look at it under the microscope for porosity and any indications of bubbling or staining. Porous fragments should be crunched between his front teeth. A pungent taste of petroleum may result. Hence the names "shale-eater," "sample-sniffer," and "mud-smeller" for the geologist. These are the practical, less scientific tests. After careful examination of the wet sample, including an examination of the sample under water, the wet sample should be examined under ultraviolet light. Petroleum in the samples will fluoresce with light yellow to blue color.

After the samples have been carefully dried, petroleum often leaves a brownish stain on the porous fragments. However, very light oils leave very little stain, and it is often difficult to differentiate stain from the actual color of the rock. In the case of coarse sands or sandstones where the material is carried up in individual grains rather than in clusters, the detection of petroleum may become quite difficult. The greatest caution must be used in drying samples to keep them from burning and changing the natural color.

When porous formation is drilled and petroleum cannot be determined visually or by taste or odor, it may be necessary to make a test using a petroleum solvent. Ether and carbon tetrachloride are most frequently used. A small amount of the crushed sample should be put in the test tube with sufficient solvent to more than cover the sample, and the tube should be briskly shaken. The liquid will show various shades of brown if petroleum is present, and if the sample is poured on filter paper, a dark ring will indicate the presence of petroleum. If high-gravity light oil is suspected, the same procedure is followed with acetone after the mixture has been shaken, with the tube corked so that there will be no contamination of the liquid by oil from the skin. A little distilled water should be added. If oil is present, the liquid will become milky and cloudy. This test can be considered conclusive if result is negative but cannot be considered absolutely conclusive as to the presence of oil if the result is positive.

If there is a mud-logging device or a composite mud-cutting analysis logger on location constantly testing and analyzing the mud and cuttings, the evaluation of showings of oil or gas is considerably less difficult.

When soft drilling has been encountered, cuttings from the bottom indicate porosity, and if there is any evidence of hydrocarbons, a decision regarding procedure must be made. It must be constantly borne in mind, however, that the rate of penetration depends on the condition of the bit and that some porous formations, particularly those where oil and gas occur in fractures, may drill very slowly. There are several procedures which can now be adopted. First, the hole can be deepened a few feet at a time, drilling suspended, and cuttings circulated from the bottom. Each sample can then be carefully checked for visual porosity and evidence of oil and/or gas.

The second procedure is the cutting of a core. The character of the suspect formation may be much better determined by a core, and evidence of oil, gas, and salt water are much easier to detect. However, there is still no proof of production, despite obvious indications. The core can be analyzed on location for porosity, permeability, oil and water saturation by an expert engineer from a service company or the operator. Special equipment is necessary. This on-location analysis rapidly supplies quantitative evidence of the potentialities of the zone, and it may be desirable to suspend drilling operations until the results of the analysis are known. If the on-location method is not used, samples of the core can be quick-frozen by a representative of the core laboratory or canned by the geologist and delivered to the laboratory.

Although such laboratory tests are much more conclusive, they are still not proof positive. The best means of establishing the productive possibilities of a porous zone where the presence of hydrocarbons has been noted is a formation test.

These tests are known as drill-stem tests. To take a conventional drill-stem test, it is necessary to remove the drillpipe from the hole, remove the bit and the collars, and attach the testing tool. Figure 17-12 shows the action of the testing tool and pressure charts. The essential components of the tool are, from bottom to top, a perforated anchor with a shoe, containing the pressure recorders and through which the formation fluid enters; a packer, which collapses to seal the interval to be tested

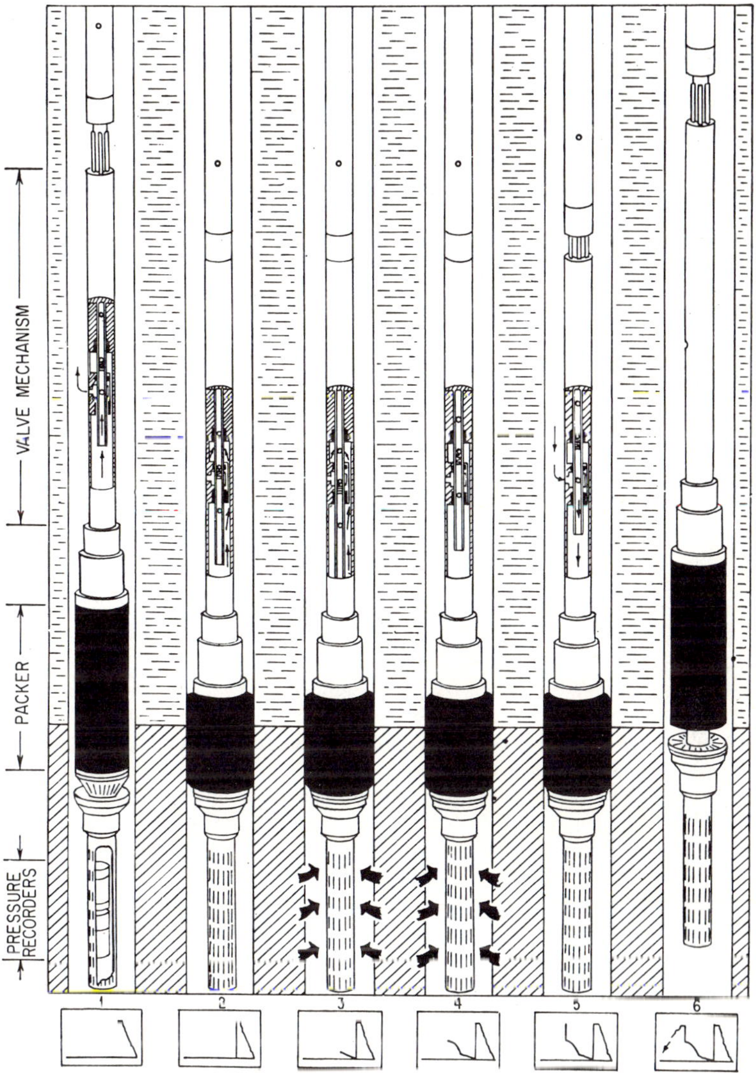

FIGURE 17-12. Action of the drill-stem tester with pressure charts.

from the hole above; and the valve mechanism immediately above the packer but below the drill stem. The conventional drill-stem test is known as an open-hole test, using one packer. However, at times it is desirable to isolate the zone to be tested from the hole both above and below, and a straddle-packer test can be run with a packer both above and below the perforated section of pipe and with a blank anchor below the bottom packer. It is possible to test in a reduced-size hole after coring by using either a regular or cone packer. Occasionally an additional packer will be set in the open hole above the cone packer to make sure of a proper shutoff of the open hole above. Finally, it is often expedient to drill-stem-test through perforations in casing, after casing has been cemented, using a hook wall packer.

The drill-stem test furnishes important information for both the geologist and the engineer. A successful drill-stem test will yield the following information:

1. The ability of the formation to produce fluid or gas naturally and an index of its productive capacity
2. A sample of the formation fluid or gas
3. Initial shut-in, flowing, and final shut-in bottom-hole pressures which are an index of the relative permeability of the formation
4. Hydrostatic pressure exerted by the mud column
5. After casing has been run, the effectiveness of the cementing of a string of casing

Once the hole is in condition for testing, the special tool is lowered into the hole on the drillpipe with the bypass valve open and the tester valve closed. When the anchor shoe reaches the bottom of the hole, the annulus should be full of drilling fluid and the packer collapsed, thus sealing off the zone to be tested. A weight then is dropped to puncture the metal-alloy disk at the top of the test tool to obtain the initial shut-in pressure. When this measurement has been obtained, another disk is ruptured to open the tool. The characteristics of the area and specific zone being tested and the condition of the hole will determine the time of the test. At the surface there is usually a control head with a pipe leading to the slush pit or storage tank. A small outlet valve to which a rubber hose is attached can be used to determine light blows, with one end of the rubber hose in a container of water. When the testing has been completed, the tester valve is closed and the tool is left in position to allow the pressure below the packer to build up to the true formation pressure. The packer is unseated, and the drillpipe and tester are removed from the hole.

Recently, a wire-line formation tester and fluid sampler has been developed which can be lowered and retrieved on a wire line, thus obviating the necessity for making two round trips of the drillpipe. Formation pressure is recorded electrically on the surface, and a small sample of the formation fluid is recovered. For obvious reasons, this device may not be so satisfactory as the conventional test, but it results in a substantial saving of money and has been highly satisfactory in many instances. A normal drill-stem test may often require as much as 24 hours of rig time, with an expenditure of $500 to $1,500, depending on the size of the rig, whereas the wire-line tester can be run in a fraction of the time.

Although a drill-stem test is the most conclusive method of evaluating oil and gas showings and potential producing zones, a negative test, that is, where no formation fluid or gas is recovered, may not be conclusive. Many formations are found to be productive where there is no recovery of fluid from the formation. Conversely, the recovery of oil or gas is not conclusive that production will be obtained. However, if formation water is recovered in quantity, with no showings of oil or gas, this test is conclusive and the formation will not produce hydrocarbons.

The geologist's responsibilities in drill-stem testing are manifold, depending on the responsibility delegated to him. He recommends or makes the decision to take the test. He selects the packer seat, being certain that the packer is set in a formation where an effective seal can be effected to isolate the zone to be tested. When pipe is being removed from the hole preparatory to taking a drill-stem test or coring, the geologist must insist that the pipe be measured so that errors in measurements can be found and corrected. This is known as "strapping the pipe." He carefully observes and records the action of the tool during the test and makes certain that the initial shut-in bottom-hole pressure is obtained and that the tool is left open long enough to secure a representative test. There is an all too frequent tendency not to leave the tool open long enough. He makes certain that flow pressures are recorded and a final shut-in bottom-hole pressure test of sufficient duration is made. Two pressure recorders should be required to be certain a good test has been made. In formations with low permeability, it is mandatory that the tool be open long enough to secure a satisfactory test and that the bottom-hole pressure be given ample time to reach maximum build-up after the tool is closed.

If the responsibility is entirely the well-site geologist's, he must be sure that the hole is in proper condition for testing. He must make certain there is an adequate water blanket in the drillpipe to prevent its collapsing in a deep hole. He also carefully records and evaluates the recovery of fluid, samples the fluid and gases, and attempts to gauge any flow. After the test has been completed and the tester recovered, he evaluates the results of the test and, with other information obtained from samples, cores, drilling times, and logs, makes an estimate of the potential productive capacity of the formation tested. Each factor must be carefully weighed, for all too often a test is not conclusive. The shorter the section tested, the more conclusive the test.

Deviated and Crooked Holes

A normal hole should be drilled nearly vertically. The conventional drilling contract limits the deviation of the hole from true vertical. The usual limits are from 3 to 5 degrees. Consequently, surveys are made by instruments at frequent intervals during the drilling to determine the deviation from vertical and often the direction. Crooked holes are usually a result of carelessness on the part of the driller. The "hole-crazy" driller often finds himself in trouble. Steeply dipping formations add to the hazard of crooked holes. In these formations, the hole tends to migrate updip, and the driller must use the utmost care not to have too much weight on the bit. Crooked holes are the cause of many problems in drilling and producing. They are harder to drill, more difficult to case, and a plethora of production problems results. As far as the geologist is concerned, a crooked hole causes the illusion of abnormal

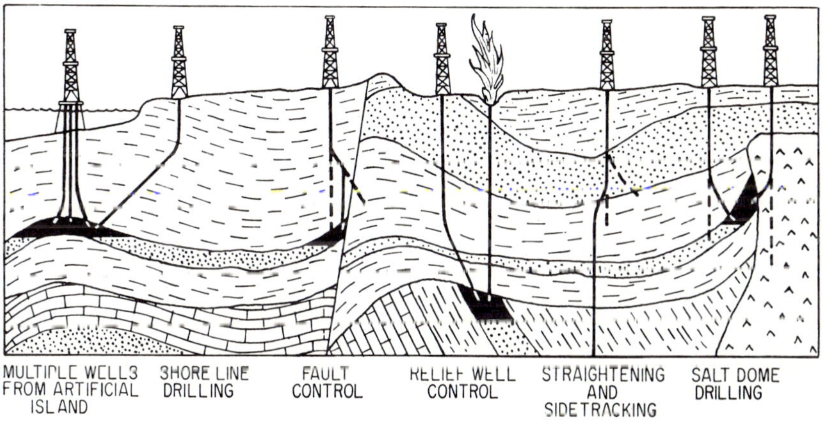

FIGURE 17-13. Applications of directional drilling.

thicknesses of beds and false structural position. When a crooked hole is drilled in proximity to faults, the bottom of the hole may be in an unfavorable structural position. In the early days of rotary drilling, holes were so crooked they actually bottomed across lease boundaries, and in the days of the Seminole boom, wells actually drilled into the bores of offsetting wells.

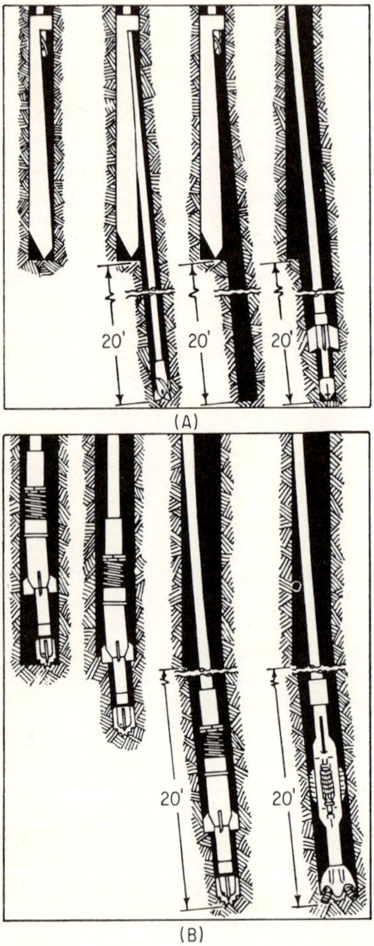

FIGURE 17-14. Action of directional drilling tools.

If the hole exceeds the normal tolerance from vertical, it must be straightened. If it is not too far off vertical, it is possible to reduce weight on the bit, reduce the rate of penetration, and bring the hole back to near vertical. In extreme cases, the crooked portion of the hole is cemented off and, by the use of whipstocks and knuckle joints, a new hole is drilled below the corrective point.

Not to be confused with a crooked hole is the hole which is intentionally deviated from vertical. This may be done to drill around drillpipe which has been lost in the hole and cannot be recovered or to drill to an oil deposit located where it is not possible to make the necessary surface installations vertically above it. In the Oklahoma City field, wells were purposely deviated to capture oil under the capitol building and grounds. In areas in southern California, holes were purposely deviated from a surface location on the beach to capture offshore oil. On the Gulf Coast, holes may be deviated from vertical to obtain production from sands under the overhang in a salt dome. A hole which is commenced on the downthrown side of a fault may be deviated to find production on the upthrown side. Finally, should a drilling well blow out and catch fire, a relief hole may be drilled off vertical to intersect the original borehole, shut off production, and stop the fire. Figure 17-13 shows various reasons for deviating a borehole. The drilling of an intentionally deviated hole is an extremely complicated process, requiring constant surveying. The same tools are used in drilling a purposely deviated hole as are used in straightening a crooked hole. These are the knuckle joint, whipstock, and the spudding bit. The first two of these and their action can be seen in Figure 17-14.

Offshore Drilling

Although there are many problems in conventional rotary drilling on land, some of which occasionally may be unique, an entirely new set of problems is encountered when the drilling location is under water. These problems are not primarily geological, with the exception of those of transportation and communication, which affect the geologist as well as all others connected with the operation. It is advisable for the geologist to be generally conversant with the various methods of setting up to drill when the location is under water. Figure 17-15 shows several types of above- and under-water installation for drilling offshore.

In shallow water, a submergible barge can be used. The rig is assembled on the barge, and the entire unit is towed to the location. The barge is sunk into position, and when drilling is completed, the water is pumped from the barge and it can be moved and again be used. Conventional submergible barges are not generally used where the water is more than 12 feet in depth. There is a submergible barge, however, designed to be used in water up to 40 feet in depth. This barge is, in essence, a mobile drilling platform where the barge is sunk and the platform raised above water by

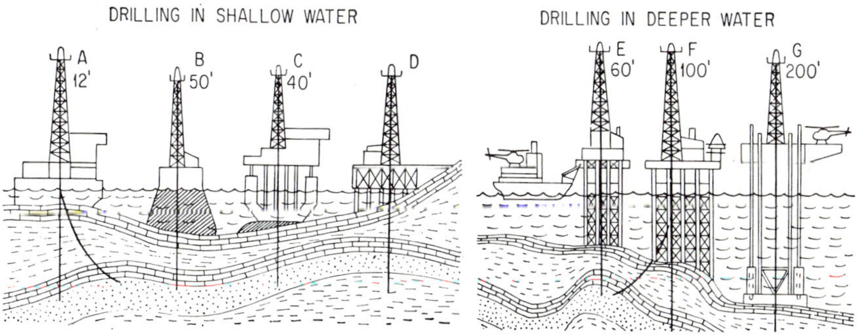

FIGURE 17-15. Surface and underwater installations for offshore drilling (two sheets).

hydraulic jacks. In some instances, an actual island is constructed in water up to 50 feet in depth, and the rig is assembled on this man-made island. It is also possible to drive piers and erect the rig on a platform supported by the piers. All these are generally operations where water is no more than 50 feet deep.

The mobile drilling platform with supporting legs is the most widely used method of offshore drilling at any depth. When on location, the legs are lowered to the ocean floor and the floating platform is elevated above water. The platform may be self-contained and most elaborate, with living quarters, storage space, and the complete drilling unit. Other mobile platforms require a floating barge or tender which contains the mud system, storage space, and living quarters. Drilling in water over 100 feet deep is now possible, and platforms for drilling in water 200 feet deep are currently being designed.

The platform drilling tender unit is generally preferred because space required on the platform is less, which results in a smaller, more economical platform. Transportation to and from offshore units is generally by boat, but the helicopter has won an important place. Landing decks are familiar sights on tenders and drilling platforms.

Offshore drilling presents many unique problems besides transportation and communication, all of which add to the expense. Maintenance problems are much more serious in salt water. A far greater amount of supplies must be kept on hand. Storms present a major problem and often shut down the operations until they cease. Precautions must be taken against pollution. Many types of permits must be obtained from state and Federal agencies. Communications are difficult and are carried on by radio-marine telephone or two-way FM radio. All these problems, coupled with the cost of normal and special equipment, result in an astronomical increase in costs. All types of offshore drilling, particularly the drilling of an exploratory test well, are reckoned in terms of millions of dollars per project.

Air and Gas Drilling

Drilling a hole using air or gas in lieu of conventional fluid as the circulating medium is a successful adaptation of rotary drilling. This method is known as the rotary pneumatic drilling method. Figure 17-16 shows a typical surface installation. The

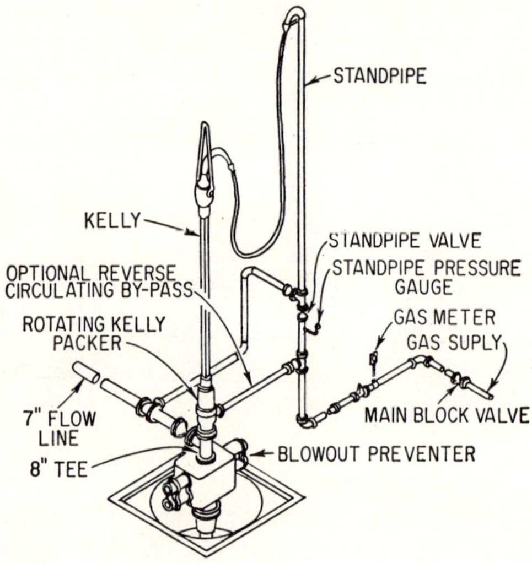

FIGURE 17-16. Surface installation for rotary drilling with gas or air.

technique has its greatest application where any or all of the following three circumstances may be encountered, but only in areas where formations yielding quantities of water are not found:

1. Producing formations are susceptible to injury from invasion of drilling fluids.
2. Severe lost circulation zones are likely to be encountered.
3. Drilling fluid is unusually expensive or not available.

Rotary pneumatic drilling has these advantages over conventional hydraulic drilling:

1. Less possibility of damage to potential producing zones by fluid
2. Less susceptibility to lost circulation
3. Less possibility of drilling through potentially productive formations
4. Less contamination of cores and cuttings
5. Higher rates of penetration
6. Fewer bits needed

Mechanical and operating disadvantages are:

1. Difficult, if not impossible, to cope with formation water in quantity. Water is deadly. Samples become doughy and clog up the bit. Experiments are under way to solve this problem.
2. Little support for formations which sluff badly.
3. Poor control of formation pressure.
4. Great fire hazard.

The principal disadvantages to the geologist are that the cuttings are so badly powdered and so difficult to sample that it may be impossible to determine the stratigraphic sequence. It is impossible to run a conventional electric log without filling the hole with fluid, and furthermore, conventional mud logging to assist in detection of hydrocarbons cannot be used. However, since there is no fluid in the hole, tests can be made of potential producing formations at any time.

The method has been used extensively in the San Juan Basin of New Mexico, where Mesa Verde strata are particularly susceptible to plugging action by conventional drilling fluid. Gas drilling materially reduced the time required for completion and resulted in better completions. Both coring and directional drilling operations were successfully conducted using gas. The technique is also used in other areas to combat severe problems encountered in conventional rotary drilling. It will undoubtedly assume increasing importance, as improvements are made, in areas where conventional operations encounter continuous problems.

Other Drilling Methods

Several methods of rotary drilling are in the experimental stage. The principal of these is known as turbodrilling, first developed in this country but greatly improved by the Russians. This method uses much of the same basic equipment used by conventional hydraulic rotary drilling. However, the kelly and drillpipe remain stationary. A section of the drillpipe immediately above the bit is a hydraulic turbine, and the force of the circulating fluid is used to turn the bit while the drill string remains stationary. Heavy-duty pumps are the principal modification of equipment. Slow rotation of the rotary table and drill stem assures straighter holes and fewer mechanical problems. It is still a moot question how universally this technique can be used.

A second development of great interest is the percussion, or hammer, drill, several types of which are still in the experimental stage. This method was devised to assist in keeping holes straight and penetrating hard rock formations which wear out bits and slow penetration. Normal drilling fluid powers a reciprocating piston and cylinder which deliver percussion blows to the bit. The drill string is slowly rotated, and the bit is driven much like a star drill penetrating masonry, rather than by chewing or boring the rock. Principal advantages are the increase in bit life and higher rate of penetration.

Experiments are being conducted with various types of vibration and sonic drills.

A final deviation from conventional rotary drilling, but one which will rarely be encountered in exploratory operations, is drain-hole drilling. A conventional hole is drilled, and then the drilling of low-angle deviated holes actually penetrating the producing formation at various angles gives additional drainage into the main borehole. Drain-hole drilling is particularly effective in reservoirs of low pressure, low permeability, and containing low-gravity oil. The method also exposes more formation for hydraulic fracturing or treatment with acid. Drain holes over 150 feet in length have been drilled from the main borehole. The basic principles of directional drilling, using a whipstock and flexible drill collars, are used. The method has very limited application in exploratory operations.

Casing and Testing the Well

Once the hole has been cut to total depth and showings of oil and/or gas found, a decision must be made on the advisability of casing and testing. If the evidence is clear cut, with satisfactory drilling tests, electric log, and samples revealing a thick porous oil-bearing stratum, or if there is clear-cut evidence of a dry hole, there is no doubt about the procedure: either casing must be run and the well properly tested, or the well plugged and abandoned. Unfortunately, this ideal situation is not usually encountered, because tests are far too often contradictory, and someone must make a decision. In some organizations, the decision will be made by the geologist on the well. In others, he will make a recommendation which bounces from department to department for study and recommendations which often will conflict. The final decision will be made at top level on the basis of highly conflicting recommendations and by an individual with the least first-hand information in the entire chain of command. If the decision is made to run casing, and this often will depend on the perseverence of the well-site geologist, he must pick the casing point, that is, the depth where the casing is to be set. If the well is to be tested through perforations, he must be certain to pick a point at least one joint of casing below the oil-bearing stratum. Then a cement plug can be left in the casing below the perforations to ensure an adequate shutoff from fluids which might enter the pipe from below the casing point and to provide sufficient working room below the stratum to carry out the necessary procedures.

If it is decided to set on top of the potential pay horizon, other problems arise, particularly if the potential pay zone is thin and underlain by quantities of water. Then the casing must be carefully set in the very top of the formation so that uncased formations will not cave into the hole, but not so far in the formation that part of the pay is shut off. The well-site geologist should designate in writing the depth at which casing is to be set, and date and sign his recommendation. When setting on top, it is advisable to order the casing run to bottom and picked up the required number of feet in order to reduce the possibility of errors which may occur as a result of poor or conflicting measurements or stretch in the casing.

In some instances, when the geologist has selected the casing point, his work on the well site is completed. He may be asked to recommend the depths and intervals to be perforated. This is usually done in the office with all the data accumulated during the drilling of a well and while cement is hardening behind the casing. He may also be asked for recommendations for treating the formation with acid or by fracture. However, in a small organization, the geologist may be expected to supervise casing, testing, and even equipping and putting the well on production. This broad subject cannot be covered in detail in this chapter. Nevertheless, there are certain routine procedures of which all geologists should have a working knowledge.

Casing comes in a variety of sizes, ranging from 4½ to 20 inches in diameter; can weigh from 9½ to 94 pounds per foot; can be seamless, lap-weld, or electric-weld; and varies in length of joints, as does drillpipe. Each particular situation requires casing of certain specifications. The most important items to consider are the tensile and collapse strength. The diameter of the casing will depend on the character of the completion and the problems which are anticipated in completing or producing. Production casing generally selected usually has an outside diameter of 5½ or 7 inches. Tables are available showing the weight and quality of casing necessary to give adequate protection against collapsing or parting for various depths and conditions.

The actual running of a casing is usually handled by specialists, a service company which is called in to do the job. The problem of measurements becomes highly important. As the drillpipe is removed from the hole and laid down, each joint, collar and sub are carefully measured, preferably with a steel tape graduated in feet and tenths of feet, and recorded. Each joint of casing is measured and recorded in the same manner and numbered to be run in the hole by consecutive numbers chalked on the joints. If the measurements are made carefully, they will check within a few tenths of a foot. This is particularly important when casing is to be set in the top of producing formation to allow an open-hole completion. Before the casing is run in the hole, attachments are placed at certain points detected from the electric log and caliper survey. Centralizers, heavy steel spring guides, fit around the casing to keep it centered in the hole so that some joints will not lie against the side of the hole, thus preventing a good cement job. Scratchers are placed at certain points, and the casing spudded or rotated to remove thick mud filter cake which acts as a barrier and will not allow the cement to come in direct contact with the formation. Under certain circumstances of completion, this filter cake may give way and necessitate expensive recementing. The complete cementing process and centralizers and scratchers are shown in Figure 17-17. Figure 17-18 is a cutaway view of the cementing process.

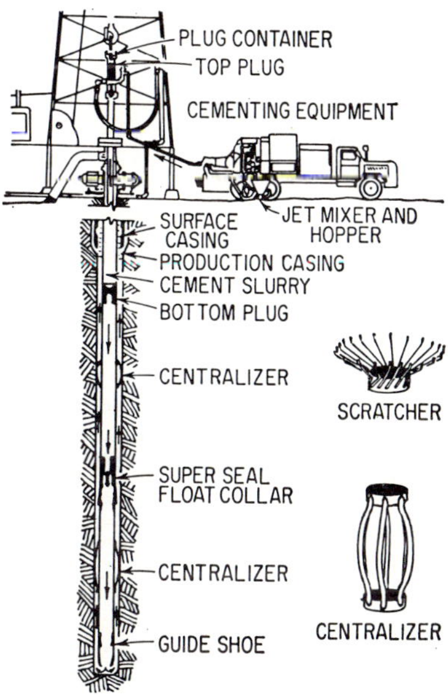

FIGURE 17-17. Diagram showing path of cement flow from surface through guide shoe and return; centralizer and scratcher also shown.

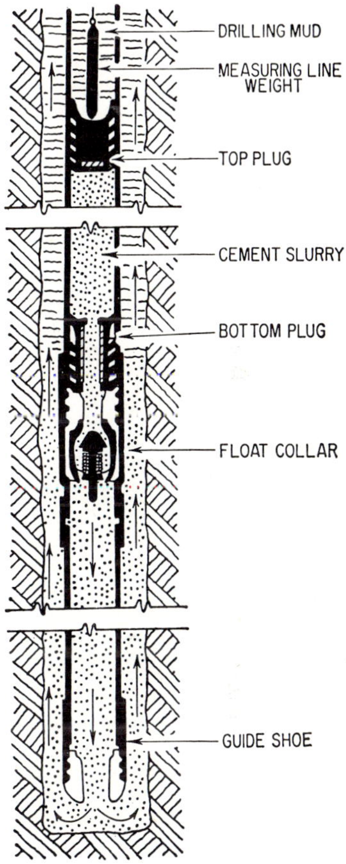

FIGURE 17-18. Cutaway diagram of cementing process.

Once the casing is placed at the desired setting depth, it is cemented to assure a firm bond between the formation and the pipe at the desired interval. Again tables are available to determine the amount of cement needed to fill the annulus between the casing and the hole to the required level. A service company performs the actual cementing. Additives may be blended into the cement to reduce the time for it to solidify or to allow it to remain in a fluid state for a longer period. The cement is pumped down the casing where it passes through the casing shoe, filling the annular space. As soon as the required amount of cement slurry has been pumped in, a drillable plug is inserted and drilling fluid pumped behind this plug to the bottom of the hole, thus forcing the cement out of the casing into the annular space. A few hours after the cementing operation is completed, a temperature survey, using a thermometer which is sensitive to the heat of the chemical action of the cement, is run to determine the exact level the cement has reached in the annulus.

When the cement has set the required time, tests are made to determine that it is properly set up, and the operation is ready for the next step.

If an open-hole completion is to be attempted, the procedure is rather simple. The plug must be drilled; the formation penetrated, cleaned up, and tested. There are three methods generally employed. The first is the use of a small portable cable-tool unit, generally termed a tailing-in unit. This is used when very little formation must be drilled and when high pressures and strong flows are not expected. The plug is drilled; the hole cleaned out with a bailer or sand pump and tested for oil and gas showings. Drilling proceeds cautiously a few feet at a time, and the hole is frequently bailed and tested. When sufficient oil to indicate a commercial well has been encountered, the formation may be treated with acid or fractured by a service company, the well cleaned up and tested. If sufficient oil or gas is present, the well is then put on production. This method of testing, once widely used, has generally been superseded by quicker, more effective techniques. Its principal advantage is that accurate measurements of the amount of fluid entering the hole can be made frequently. It is often slow and tedious but is still used extensively in many areas.

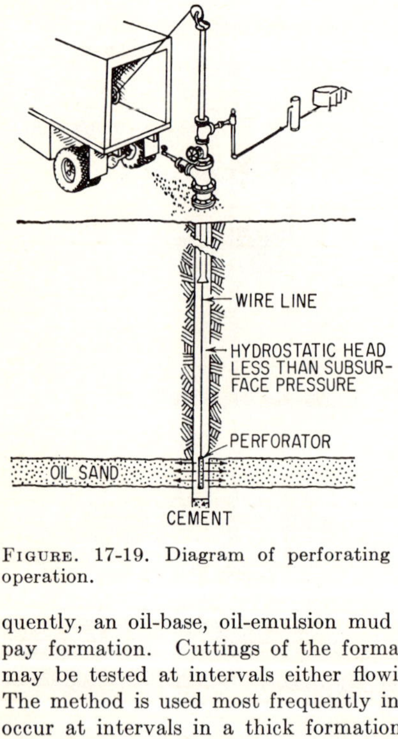

FIGURE. 17-19. Diagram of perforating operation.

In areas where considerable formation must be drilled after the plug is drilled, to effect an open-hole completion, rotary tools are used, either with a conventional rotary setup, using tubing or small-diameter drillpipe, or with a reverse circulator where the drilling fluid is pumped down the annulus and returns through the drillpipe or tubing. Conventional drilling fluid with mud base may be used and displaced. More frequently, an oil-base, oil-emulsion mud or oil is used to avoid contamination of the pay formation. Cuttings of the formation are caught and examined, and the well may be tested at intervals either flowing through the drill string or by swabbing. The method is used most frequently in areas where thin, porous, oil-bearing zones occur at intervals in a thick formation, such as in the Permian dolomite of west Texas, and when operators do not want to attempt to complete through perforations.

By far the most common technique for completing is by perforating casing opposite the potential producing formations. In effecting such a completion, the general practice is to run a gamma-ray neutron survey in the cased hole to locate accurately porous potential oil zones logged before casing was run. A typical perforating setup can be seen in Figure 17-19. A gun is then run on the same line, and bullets or shaped charges are fired electrically from the surface at the required depth to penetrate the casing and cement, allowing the formation to give up oil or gas through the perforations into the casing and well bore. Through the use of drillable plugs and/or squeezing off previous perforations, it is possible to test successively higher zones, starting with the lowest, for oil and gas, and a dual or triple completion may be effected.

Intervals to be perforated must be carefully selected, first, to avoid water entering through perforations and, second, to avoid excessive gas/oil ratios.

The ideal situation, of course, is that on drilling in or perforating, the well commences to flow in quantity. If no tubing is in the hole, tubing is run, surface hookup completed, permanent tankage erected, and the well is put on production. This is a rare occurrence in most areas. Usually some form of artificial stimulation must be used to obtain the necessary reaction. In the past, in open-hole completions, the common practice was to set off a charge of nitroglycerin opposite the pay formation. This was quite effective in stimulating production from sandstones but not nearly so effective in limestones and dolomites. The current technique employs artificial hydraulic fracturing in sandstones and acidization or a combination of acidization and

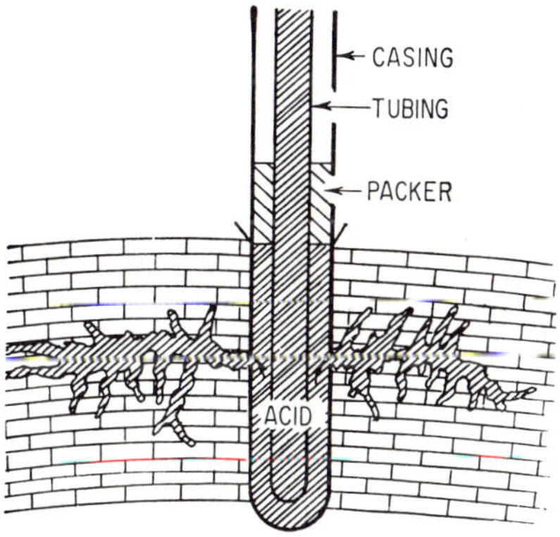

FIGURE 17-20. Cutaway diagram of acidizing limestone.

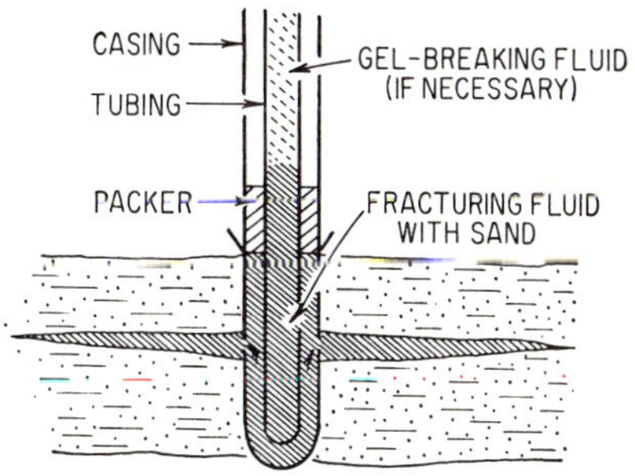

FIGURE 17-21. Cutaway diagram of hydraulic fracturing in sandstone.

fracturing in limestones and dolomites. Such treatments are routine and are performed by a number of service companies. Treatments may be massive, and formations which yield only a bare showing of oil or gas naturally may produce prolifically after such stimulation. Figure 17-20 shows the action of acid on the formation; Figure 17-21 shows the action of hydraulic fracture.

The first step is to determine through consultation with service company engineers the best type and method of treatment. Either treatment can be administered through casing or tubing. Highest injection rates and pressures can be secured through treatments down the casing. Each situation is unique and must be evaluated on the basis of information secured from the drilling and testing of the well and experience under similar circumstances.

Completing the Well

After treatment has been administered and the well cleaned up and tested, if commercial amounts of oil or gas are evident, the well is put on production. If natural flow is encountered, the well is tubed and surface equipment is installed. If two zones are to be produced, a packer may be set on the tubing and the lower zone produced through the tubing. The upper zone is produced through the annulus. In some cases, two strings of tubing may be necessary, and in the event of a triple completion, three strings of tubing may be run.

Surface equipment on a flowing well is called the "Christmas tree." This controls the flow of oil or gas from the well. The size and complexity of the installation depend on the pressures and volumes of fluid encountered. Again, the average geologist must call on experts to aid him in choosing the proper equipment. Many wells are pumped from the time of completion because there is insufficient gas or water drive to flow oil from the well. The choice of pumping equipment will depend on the depth and the volume of fluid anticipated. Figure 17-22 is an illustration of the surface and

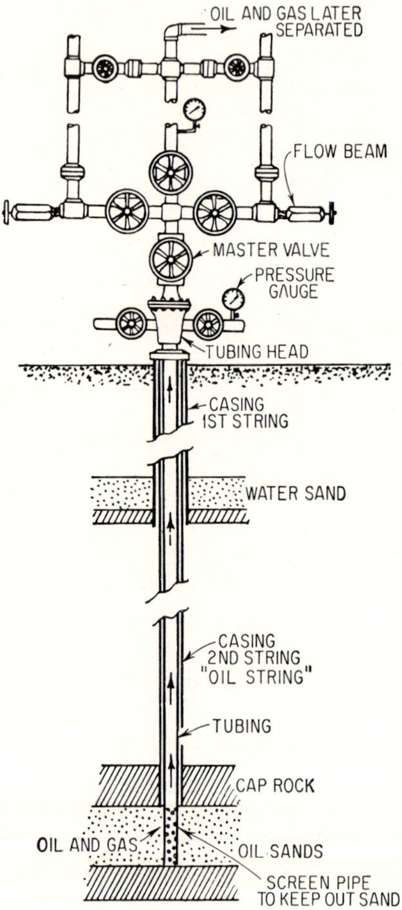

FIGURE 17-22. Typical surface and subsurface installation, flowing well.

subsurface equipment on a flowing well, and Figure 17-23 shows the typical surface and subsurface installation on a pumping well.

Finally, once the necessary surface installations have been made so that the well can be produced, a tank battery must be erected and flow lines laid from the well to the battery. The capacity and number of tanks will depend on the amount of oil to be produced over a given period and the frequency that the tanks are emptied into the purchaser's facilities. If the well flows, one or more separators to separate gas from the oil are installed between the well and the tank battery. The size and character of the separators again depend on the volume of gas and oil and the pressures. If the well pumps, a separator may also be necessary. If water is produced concurrently with the oil, a heater and treater installation to assure complete separation of water from the oil may be necessary. The oil may pass through a stock tank before going into the tank battery when special problems are encountered. Ordinarily, the average geologist will encounter few of these problems.

When the well is finally completed, either as a dry hole or as a producer, it is the duty of the well-site geologist to prepare a completion report. A typical completion report

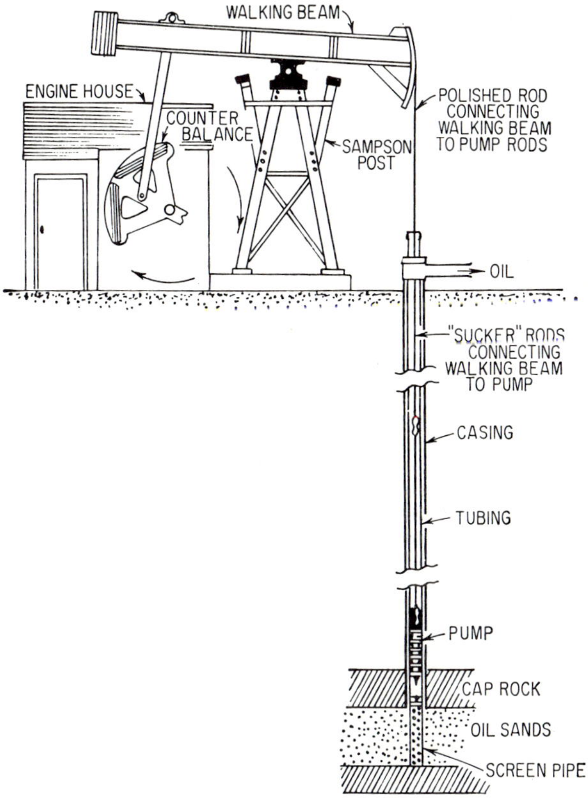

FIGURE 17-23. Typical surface and subsurface installation, pumping well.

is shown in Figure 17-24. The purpose of the completion report is to give a concise history of the drilling and testing of the well. This is not so important if the well is dry, but if it does produce, the completion report is highly important. It is often necessary years after a well has been completed to rework or recomplete in another zone. The conclusions and opinions of a man who actually supervised the drilling of a well are highly important years later when a stranger must make decisions on recompletion. It is therefore mandatory that the well-site geologist thoroughly describe and evaluate any oil or gas showings and make recommendations for testing potential producing zones before the well is abandoned. The completion report, copies of the graphic log made from drill cuttings, the written descriptions of cuttings, the time log, the electric log, radioactive log, and all other pertinent information of this type are placed in a permanent file. Even though many years elapse, a complete

Well Completion Report

Well ________ No. ________

Location: EL WL

NL SL

FILE ________

LEASE ________

POOL OR AREA ________

COUNTY STATE

Commenced: ________ Completed: ________

Net Interest: ________ Partner: ________

Operator: ________ Contractor: ________

Depth drilled with Rotary: ________ Cable Tools: ________

Method of drilling in ________

Sample Log: ________ Time Log: ________ Electric Log: ________

Core Analysis ________ Other: ________

Geologist: ________ Elevation: ________ Total Depth: ________

Casing: Surface: ________ Intermediate: ________ Oil String: ________

Stratigraphy:	ACTUAL DEPTH SAMPLES	ACTUAL DEPTH E-LOG	SUBSEA DEPTH SAMPLES	SUBSEA DEPTH E-LOG

IP: ________ Pump-Flow Gravity: ________

Pay (formation and depth): ________

Structural position: ________

FIGURE 17-24*a*. Completion report form—front.

stranger to the operation can then thoroughly evaluate the events which took place during the drilling and have the benefit of the recommendations made by the man who was actually on the ground.

It is quite obvious that the duties of the well-site geologist who supervises the drilling of an exploratory well may vary greatly. In many instances, his duties are to observe, carefully describe, and recommend. His principal responsibilities relate to observations and descriptions and the communication to his superiors. The other extreme is when the well-site geologist not only observes and describes but makes the decisions for every step in drilling and completing of the exploratory well. His duties are completed only when the well is placed on production. No matter what his responsibilities, he must constantly bear in mind certain important factors. First, there is no substitute for being on the ground and making first-hand observations. Second, every event must be carefully described and recorded at the time. Third, additional hole may always be cut, but once hole is cut, it can never be recovered. Finally, in case of doubt, shut the operation down and secure expert advice. If these procedures are followed, the well-site geologist will never have the ignominy of another operator entering a so-called "dry hole" and completing it as a producer, or the humiliation of setting casing and testing, at great expense, a failure.

Oil or gas showings (formation, depth, quality of showing) ____________________

Cores: ____________________

Drill stem tests (depths and results): ____________________

Testing record (including perforating, acids, shots, etc.): ____________________

Recommendation for further testing, deepening or plugging back prior to abandonment: ____________________

Reason for drilling: ____________________

Estimated gross cost: ____________ Estimated Net Cost ____________

Net Revenue Int.: ____________ Preliminary Est. Net Reserve: ____________

Remarks: ____________________

Date ____________ Prepared by ____________

All measurements rotary bushing unless otherwise noted.
Report must be prepared immediately following completion. Copies of written sample log, time log, electric log, core analyses, etc. must be filed with the report. If all information cannot be placed on form, use additional plain sheets and attach to report.

FIGURE 17-24*b*. Completion report form—back.

REFERENCES

"A Primer of Oil and Gas Production," 6th printing, American Petroleum Institute, Division of Production, New York, November, 1956.

'A Primer of Oil Well Drilling," 2d ed., American Association of Oil Well Drilling Contractors. and Petroleum Extension Division, University of Texas, Austin, June, 1957.

"A Primer of Oil Well Service and Work Over," American Association of Oil Well Drilling Contractors, and Petroleum Extension Division, University of Texas, Austin, July, 1954.

Ball, Max W.: "This Fascinating Oil Business," The Bobbs Merrill Company, Inc., Indianapolis, 1940.

Brantly, J. E.: "Rotary Drilling Hand Book," Russell Palmer, New York, 1936.

Buckley, Stuart E. (ed.): "Petroleum Conservation," chap. 3, pp. 47–83, Symposium of the American Institute of Mining and Metallurgical Engineers, 1951.

Clark, Stuart K., James I. Daniels, and J. T. Richards: Logging Rotary Wells from Drill Cuttings, *Bull. AAPG*, vol. 12, pp. 59–76, January, 1928.

DeGolyer, E. (ed.): "Elements of the Petroleum Industry," chap. 9, pp. 116–159; Chap. 10, pp. 160–250, Symposium of the American Institute of Mining and Metallurgical Engineers, 1940.

Engineering Fundamentals on Petroleum Reservoirs (a basic reference manual for production men), *Oil Gas J.*, 1953.

"Finding and Producing Oil" (a manual), sec. IV, Sampling, Coring and Bore Hole Surveying, pp. 125–139; sec. V, Production Methods, pp. 141–192, American Petroleum Institute, Division of Production, New York, 1939.

"Fundamentals of Rotary Drilling," 5th ed., by numerous authors. Reprinted from *Petrol. Engr.*, November, 1957.

Hager, Dorsey: "Practical Oil Geology," 6th ed., McGraw-Hill Book Company, Inc., New York, 1951.

Heistand, T. C., and P. B. Nichols: Drilling Time Data in Rotary Practice, *Bull. AAPG*, vol. 23, pp. 1820–1834, December, 1939.

Hills, John M.: Sampling and Examination of Well Cuttings, *Bull. AAPG*, vol. 33, pp. 73–92, January, 1949.

Jeffry, Walter H.: "Deep Well Drilling," 3d ed., The Gulf Publishing Company, Houston, 1931.

Kraus, Edgar: Logging Wells Drilled by the Rotary Method, *Bull. AAPG*, vol. 8, pp. 641–650, September–October, 1924.

Leroy, L. W., and Harry M. Crane (compilers and eds.): "Subsurface Geological Methods," chap. 1, pp. 1–25; chap. 4, pp. 297–432; chap. 5, pp. 433–594, Symposium of the Colorado School of Mines, Golden, Colo., 1949.

Leroy, L. W. (compiler and ed.): "Subsurface Geological Methods," 2d ed., chap. 1, pp. 1–11; chap. 5, pp. 344–503; chap. 6, pp. 504–774; chap. 9, pp. 810–855, Symposium of the Colorado School of Mines, Golden, Colo., 1950.

McCray, Arthur W., and Frank W. Cole: "Oil Well Drilling Technology," University of Oklahoma Press, Norman, Okla., 1958.

Martin, R. I.: Fundamentals of Electric Logging (a technical manual), *Oil Gas J.*, 1953.

Millikan, C. V.: Petroleum Engineering as an Aid in Exploration Geology, *Bull. AAPG*, vol. 24, pp. 1370–1376, August, 1940.

Moore, Carl A. (ed.): "A Symposium on Subsurface Logging Techniques," University Book Exchange, Norman, Okla., 1949.

"Petroleum, Discovery and Production" (pamphlet), American Petroleum Institute, New York, 1946.

Stearns, Glen: Engineering Fundamentals in Modern Drilling (a basic reference manual), *Oil Gas J.*, 1955.

Swain, James F.: Interpretation of Cable Tool Drilling Logs, *Bull. AAPG*, vol. 27, pp. 997–1000, July, 1943.

Whiteside, Robert M.: Geologic Interpretation from Rotary Well Cuttings, *Bull. AAPG*, vol. 16, pp. 653–674, July, 1932.

Chapter 18

CORING

By Clark Millison

Consulting Geologist
Englewood, Colo.

Prior to the use of rotary tools in drilling, the detection of oil and gas in a cable-tool borehole was not so difficult. The obvious advantages of rotary drilling increased its use, and since 1919, many devices have been introduced to determine what is present in the section at the bottom of the rotary bit. One of these devices and aids is the coring instrument.

The first coring apparatus appeared in Holland in 1908 (Figure 18-1*A*) when an ingenious driller improvised an inner steel tube to protect the core from the flushing action of the circulating fluid. Although this concept was not introduced into the United States until 1919, cores had been attempted by simply fashioning V-shaped sawlike teeth, bent slightly inward, on one end of a short length of drillpipe so that when rotated, it would cut a short piece of rock (Figure 18-1*B*). Weight was then applied which bent the teeth inward and held the small core sample as the drillpipe was removed from the hole. Cores obtained in this manner were often too burned to be of value. In 1921, J. E. Elliott combined the tooth-bit principle with the inner barrel for the first effective device. By 1925, the instrument had been improved with a removable core head with alloy steel teeth (Figure 18-1*C*), a core catcher, a stationary inner barrel, and a 20-foot core barrel. Only refinements have been added to this basic design. Figure 18-1*D*, *E*, and *F* shows modern cutter heads. Side-wall coring, which operates on a different principle and for a different reason, was introduced in 1940.

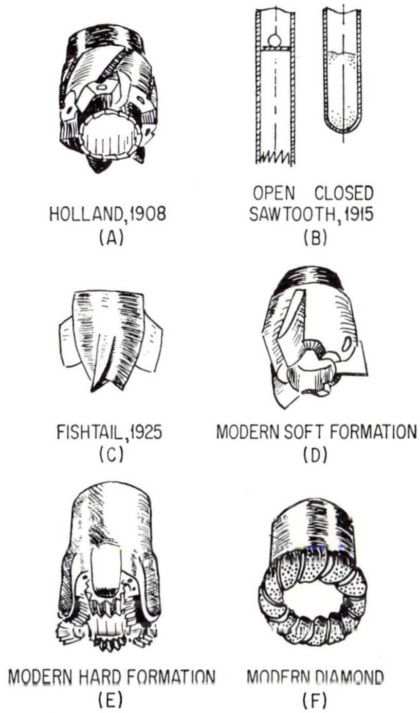

Figure 18-1. Evolution of rotary core bits. (*F*, *courtesy of World Oil, January, 1959*).

Coring is used in order to obtain a sample of rock from a borehole that can be seen, smelled, tasted, felt, and mechanically and chemically examined and analyzed. A core sample is superior to cuttings because of its natural state and known position. Since it is similar to examining a rock on the outcrop, it furnishes information that many times cannot be obtained from the many other instruments devised to determine what is actually going on in a hole full of rotary mud. However, it is always necessary to balance the value of the information that can be obtained against the cost of obtaining the core. Unfortunately, many times full recovery is not obtained, which in some cases defeats the purpose for which the coring operation was intended. There must be a constant weighing of the value received against the cost.

There are many tools now in use and yet to be devised and perfected for solving the problems that were raised by the use of rotary tools. Coring has its place in this family of instruments. Its use is determined by the particular circumstances of each well bore, and the relation to cost must always be considered. For example, core analyses to determine porosity and permeability of a reservoir rock can be many times more expensive than a quantitative analysis of an electric or gamma-ray log *provided* the analyses from these logs can be calculated with sufficient accuracy. Where unknown factors affect the characteristics of electric recordings, coring may be the only method to obtain this vital information. This is often the case.

Different Coring Uses

One of the earliest uses of coring was in core drill exploration. Where insufficient or irrelevant surface outcrops prevented mapping of structure, small mobile drilling units were used to drill shallow holes from which cores at predetermined depths or from the surface to total depth were taken. Accurate correlation of beds in these cores from either lithologic or paleontologic characteristics supplied mappable horizons to determine their structure. Usually a grid ranging from ½- to 2-mile intervals was used. With the advent of electric logging, portable logging devices have replaced coring because of the cheaper and faster technique and usually with excellent results.

Coring is one of the tools used in an exploratory well to determine the age of the rock, its content, character, and nature. Coring affords the only positive means in the initial exploratory well in an unexplored geologic province to determine the age and lithology of the rocks in a rotary hole. In a well-explored area, coring is one of the devices used to evaluate the potential oil and gas horizons. In these regions, the stratigraphy and lithologic nature of the rock column are usually well known. The principal concern, therefore, is to ascertain which reservoir bed contains oil and/or gas. Gas detectors, drill-stem tests, circulating of samples, electric logging, and many other instruments and devices are used. When there is any question that the other associated instruments will not evaluate a potential oil- or gas-producing zone, coring will usually bring to the surface an adequate sample of the rock for physical and chemical evaluation. Again, one of the considerations is the relative cost and the anticipated effectiveness of evaluation by each method.

Coring of the producing zones in confirmation wells in an oil field is undertaken for exploitation engineering information. Core analyses are extremely important for such subsequent decisions as methods and rates of producing the oil and/or gas, calculated ultimate reserves, secondary and tertiary recovery methods, and remedial work-overs.

Methods of Coring

Bailing in a cable-tool hole offered relatively accurate information of cuttings and reservoir content. There were, nevertheless, those who wanted to have a larger piece of rock in their hands for better examination and evaluation. To satisfy this need, cable-tool coring devices were developed. A cable-tool core barrel, shown in Figure 18-2, consists of an outer drilling barrel and an inner core-retaining tube which remains stationary on bottom during the coring operation. The outer drilling barrel slides up and down on the core-retaining tube and cuts away the rock around the core. As the rock is drilled away, the inner barrel moves down over the core. When coring is completed, core catchers prevent the core from dropping out of the inner barrel.

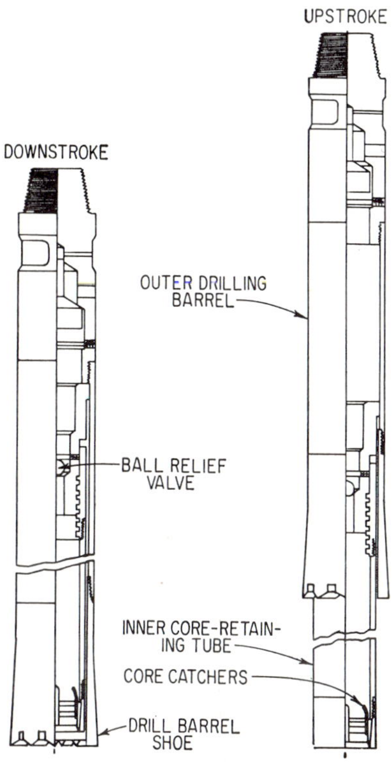

FIGURE 18-2. Cable-tool core barrel. (*Courtesy of Baker Oil Tools, Inc.*)

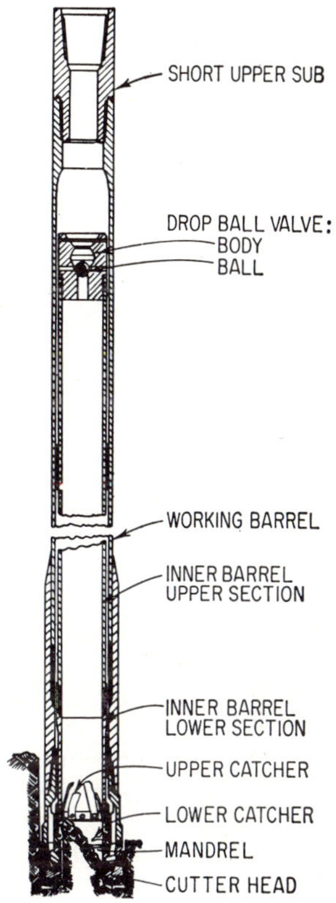

FIGURE 18-3. Modern conventional rotary core barrel. (*Courtesy of Hughes Tool Company.*)

Since the entire coring assembly is on the bull-wheel line, it can quickly be lowered and retrieved from the hole.

The earliest practical core barrel used with rotary tools is still known as the conventional barrel (see Figure 18-3). All makes are similar in principle. The entire assembly is attached to the end of the drillpipe with a sub of varying length. The assembly is made up on the walk before each core, and a new core head bit is generally used each time. Below the sub, the length of which is varied in order to cut 20 feet of core without making a connection, is the assembly which prevents the mud from passing from the drillpipe through the inner barrel but forces it to pass between the inner and outer or working barrel. In this assembly is also a ball-and-seat arrangement that allows the mud which is static within the inner barrel to drain upward and out as core enters the inner barrel. Below the inner and working barrels is the core head assembly. Inside are the core catchers. In soft formations, spring-type prongs are pushed flat against the wall of the inner barrel as the core enters but spring against the core or close shut if the core movement is downward. Below and within the core catcher is the mandrel, or "plug," which prevents loose cuttings and mud balls from entering the inner barrel as the drillpipe is lowered into the hole. Rivets, which shear when the mandrel is set on bottom, hold the mandrel in place. At the bottom of the core assembly is the cutter head. As shown in Figure 18-1*D*, *E*, and *F*, it may be of several types depending on the hardness, softness, or nature of the formation to be cored.

After assembly, the conventional barrel is gently lowered into the hole. As the barrel approaches bottom, circulation is established to flush out any cuttings or mud balls in the bottom of the hole. The drillpipe is then lowered until the core cutter head is firmly on bottom, which shears the pins in the mandrel and forces it into the inner barrel. Speed of rotation of the drillpipe, the weight, and pump pressure are controlled by experience with the physical characteristics of the zone to be cored in the region where the well is located. As much as 20 feet of core can be cut. The bit is allowed to drill off, and the speed of rotation is increased for a short time in order to "burn off" the core. The drillpipe is picked up off bottom gently with the pumps still running. Great care must be used in coming out of the hole with the drillpipe; there should be no rapid pickups but slow and consistent speed. When the core assembly is out of the hole, the cutter head is removed and the inner barrel gently removed. The core then can be removed from the bottom of the inner barrel, the bottom part of the interval cored appearing first.

Because a conventional core barrel, unless run after dulling a drilling bit, required two complete round trips with drillpipe to be made to obtain a core, the wire-line barrel was devised and introduced in 1928, illustrated in Figure 18-4. Instead of the core assembly being attached to the bottom of the drill string, the wire-line assembly is smaller in diameter and fits inside the drillpipe. The kelly is removed, and the assembly placed inside the drillpipe. Because of its weight, the barrel sinks rapidly through the column of mud inside the drillpipe until it seats itself at the bottom of the drill string. A dog latch which later can be released holds the assembly in place, or it is held in place by the pump pressure forcing mud down the inside of the drillpipe. The assembly is similar to the conventional core barrel. The wire-line barrel remains stationary, and the mud passes between the wire-line barrel and the drillpipe. The core cutter head cuts the rock surrounding the periphery of the core with the drill bit cutting the remainder of the hole. The core, usually up to 15 feet in length, passes into the wire-line barrel and is held in place by catchers similar to the conventional barrel. When the desired length of core has been cut, an overshot device on a wire line is lowered inside the drillpipe and grabs the pull bar. The tension on the pull bar releases the dog latch, and the assembly is pulled out of the pipe. The wire-line barrel has been used principally in soft formations such as the clastic Tertiary and Cretaceous rocks of South Central and Southeastern United States. Diamond bits, on both the bottom of the drillpipe and the wire-line barrel, allow the use of this type of coring assembly in hard rock formations. It is obvious that, if coring is required, the wire-line barrel is more economical with increase in depth. However, cores recovered are smaller in diameter for a given hole size.

Many statistics have been compiled on percentage recoveries from different types of coring. Comparison is difficult because there are so many variables such as geographical location, age and nature of the rocks, hole and mud condition, and even the driller's ability. However, it is probably correct to state that recoveries are better with a conventional barrel than a wire-line barrel and that diamond coring recovers a

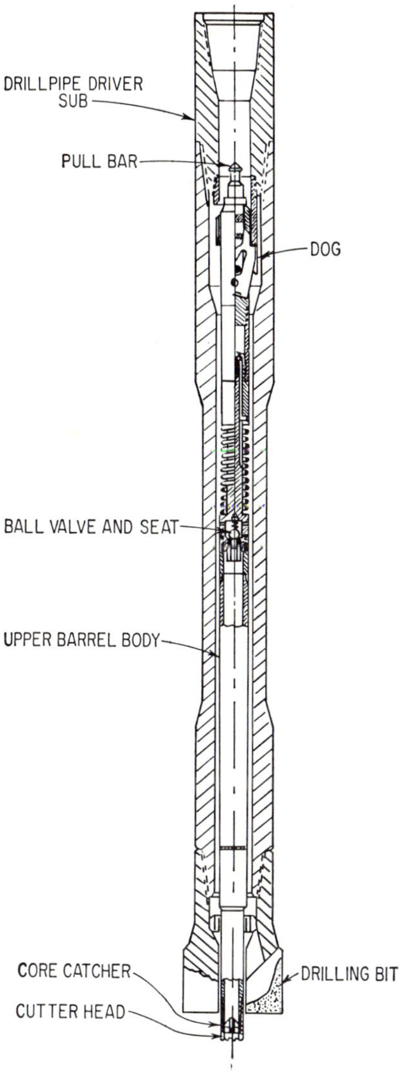

FIGURE 18-4. Wire-line rotary core barrel. (*Courtesy of Reed Roller Bit Company.*)

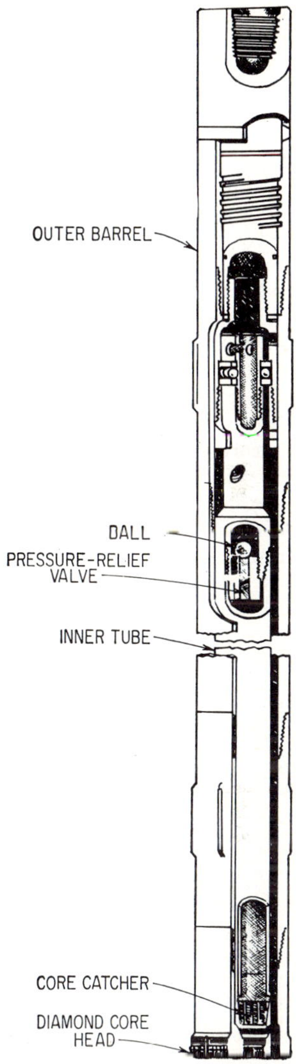

FIGURE 18-5. Rotary diamond core barrel. (*Courtesy of Christensen Diamond Products Company.*)

greater percentage than either of the others. Diamond core heads have been used for years, principally in mining operations and structural core drilling, but it has been only since about 1945 that the use of diamond core heads has become practical in the oil fields. Mechanical methods, instead of hand setting, for manufacturing diamond bits, which decreased the cost, are probably the principal reason for this increased use. A diamond-bit coring assembly is similar to the conventional barrel except that the inner barrel accommodates a 60-foot core and the cutter head is a tough metal alloy set with black commercial diamonds as shown in Figure 18-5. The diamond cutter head core barrel is attached to the end of the drillpipe, requiring two round trips to core, and the procedure is approximately the same. Diamond coring is used principally in hard rock formations.

The three types of coring instruments—conventional, wire line, and diamond—probably account for more than 90 per cent of the cores now being taken in the world. Generally speaking, diamond coring is more expensive than conventional and wire line, but in holes in which long cores are required and in hard formations, the cost per foot is about the same or in some instances less. Conventional coring requires no additional equipment except a core barrel, which can be rented but is usually purchased by the drilling contractor, who amortizes the cost over many coring operations. A new core bit for each core must be paid for by the operator and costs about the same as a drilling bit. Rig time is the most expensive item to consider. Wire-line coring requires a wire-line reel, a wire line, a prime mover, and a sheave on the derrick floor. In addition, a core assembly and a special drill collar and bit are usually owned by the drilling contractor and amortized over many coring operations. Wire-line coring, where the equipment is already available, in a soft-formation province is probably the cheapest coring operation. Diamond coring requires only the long 60-foot barrel plus either the rental or purchase of a diamond core head, which is very expensive when compared with the cost of other bits. As can be readily imagined, the loss of a diamond core head in a hole is a serious catastrophe.

There are four additional types of coring equipment which are used generally under particular conditions. In order of their use in the field, they are the side-wall core assembly, the rubber-sleeve core barrel, reverse circulation coring, and the pressure core barrel (see Figure 18-6*A*, *B*, and *C*).

Side-wall coring is used on two occasions. In some areas and by some oil operators, the hole is drilled to a predetermined total depth as rapidly as possible. Logs are run, and potential oil and gas zones chosen. Side-wall cores are then taken of these zones for examination of fluid content, formation composition, and paleontologic information. In some cases, after a well has reached total depth with careful supervision as it was drilled, it is learned that perhaps some zone or zones might contain oil or gas and were not properly evaluated as the well was drilled. In this case, side-wall cores are often taken. To side-wall-core, a cylindrical instrument is lowered into the hole on a wire line to the desired depth. Then either mechanically or with electrically detonated powder charges a small tube with an appropriate cutter head is driven into the wall of the hole and retrieved by a stout wire line or knuckled connection attached to the cylinder. The whole instrument is withdrawn from the hole after the desired number of cores has been taken. The most popular of these side-wall coring devices is one which is coordinated with the *S-P* curve on the electric log and activated by an electrically detonated powder charge at an exact depth in the potential zone. The core tubes can be fitted with either soft- or hard-formation cutter heads. Drawbacks of this type of core sampling, although the cost is relatively small, are the small core sample recovered, which is usually not sufficient for accurate core analysis; the flushing action of the mud filtrate; and the physical change in the core sample due to the method.

The rubber-sleeve barrel is similar in operation to a conventional barrel except that the core is captured and held securely in a rubber sleeve and the weight for coring is by the mud pressure rather than the weight of the drillpipe string. This special instrument is used to core unconsolidated formations, conglomerates, and badly fractured rocks.

Reverse circulation coring is generally used after the casing has been set on top of a producing formation. Mud circulates down the hole between the casing and the drillpipe as the core is cut with a conventional or diamond core barrel. The core, as it is cut, is circulated up to the surface by the mud stream inside the drillpipe. The cores are caught in a screen basket as the mud is returned to the pits. Excellent recoveries are usual, but the extra cost of equipment and the special conditions requiring its use limit its demand.

The pressure core barrel was devised to recover a core of a producing zone and retain the original pressure and content of the formation. As its use is limited to detailed

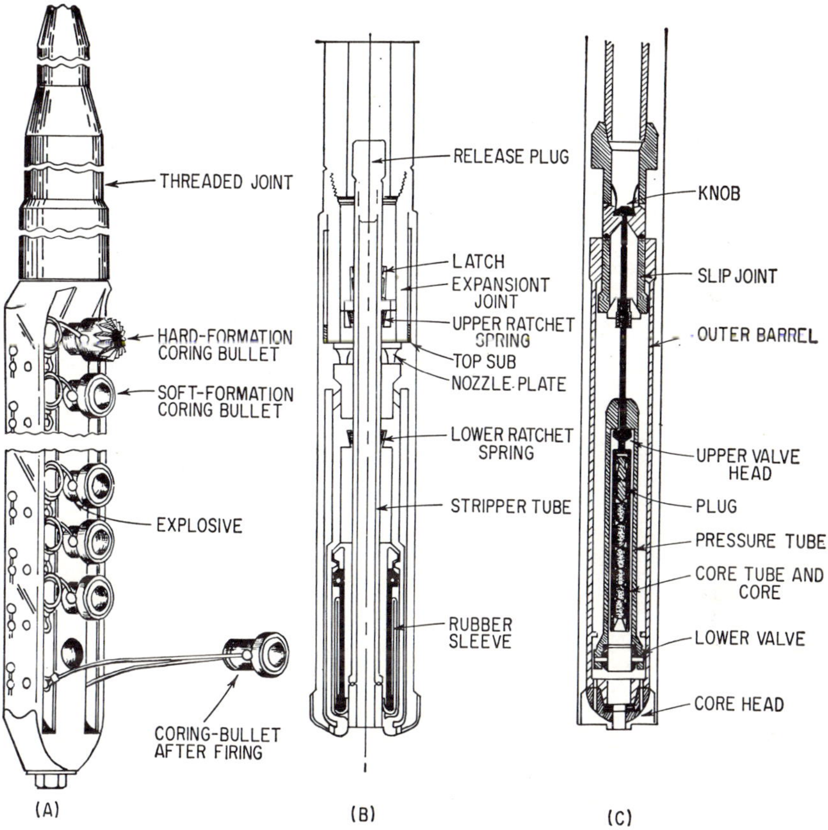

FIGURE 18-6. (*A*) Side-wall coring gun (*courtesy of Schlumberger Well Surveying Corporation*); (*B*) rotary rubber-sleeve core barrel (*courtesy of Christensen Diamond Products Company*); (*C*) rotary pressure core barrel (*courtesy of Jersey Production Research Company*).

exploitation engineering and the technique requires a special type of core barrel expertly handled, there has been little use for this type of core barrel since its introduction in 1939.

Examination and Handling of Cores

While coring is in progress, it is advisable to keep a record of the drilling time per foot and observe the action of the rotary table and pumps in order to determine, if possible, where any loss of core occurs. No instrument has been perfected to determine what part of the core is lost where there is not full recovery, although attempts to design such an instrument have been made.

The core is extracted from the barrel and laid out in the exact order in which it is recovered. It is then measured carefully with a steel line, usually marking the core in depths from the top down. Side-wall cores are usually placed in small bottles which show the name of the well, location, and depth. The core is then wiped thoroughly with dry cloths to clean away the mud. The core is now ready to be examined.

In the majority of cases, the core will be examined by the geologist or engineer on the well, and then representative samples will be saved. In many instances, however, the core will be subject to many different kinds of care. First, examination should be made as quickly as possible. A dry gas zone can be detected only if the core is smelled on a fresh break immediately after coming from the barrel. The release of pressure will quickly dispel the gas odor, and after a few minutes, no hydrocarbons can be detected by any means. Pieces of the core are then examined under the microscope, and a detailed lithologic description of the entire core should be made. As the examination continues, the core is tasted, smelled, and examined under ultraviolet light for fluorescence. If oil or gas is suspected and not definitely indicated by taste, smell, or fluorescence, ether or chloroform tests may be made. (See Chapter 17 for details of these tests.)

Many times, the core is considered sufficiently important to subject it to every test that can be made. In these instances, the core is broken in about 2- to 3-inch lengths and sealed in tin cans to prevent escape of volatile and liquid material. The core may be placed in plastic bags and frozen by the use of dry ice or refrigerating equipment. This helps to prevent any loss of gases or fluids. These cores are then taken to a laboratory for subsequent tests discussed later. Time may be such an important factor that a mobile laboratory in a panel truck is at the well and many of the tests are made within minutes after the core is extracted from the barrel. If the core is to be examined in the laboratory for only physical and chemical characteristics, the core is placed in wooden or aluminum boxes about 3 feet in length clearly marked as to depth, well name, and location. These are then taken to the laboratory for later examination and tests.

Detailed discussion is presented in Chapter 22 of methods and results of core analyses. In exploratory wells, the decision to set casing, which in many cases is as expensive as drilling the well, must be made immediately and cannot wait for laboratory core analysis except with an on-location truck. Faunal, floral, and lithologic determinations are also made from cores in order to locate the position of the core in the stratigraphic section. In exploitation wells, the nature of the producing horizon and new unexpected producing zones are determined from core analysis. For producing zones, porosity, permeability, oil saturation, and water saturation can be determined from cores. For stratigraphic and depositional studies, cores are used for studying the mineral content, stratification, chemical analysis, and included foreign materials. Correct interpretation of electric logs requires such information as water salinities, density, porosity, permeability, and formation factors. In new areas or new stratigraphic zones, these data can be determined only from cores.

Reasons for Coring

The principal reason that most cores are taken is to determine if a certain reservoir contains oil and gas in commercial quantities. Many other tools are available for the same purpose. Surprising as it may seem, even with these techniques, errors are made and casing is set on a failure, or casing is not set on a well that later is found to be productive. In many of these "mistake" wells, available tools are not used because the operator cannot justify the cost. If all the tools that are available to determine if oil or gas is present in a borehole are used, the cost would rarely exceed 20 per cent of the cost of drilling the hole. Mud logging, drill-stem tests, side-wall tests, and electric logs are complementary to cores, and each or all must be used as dictated by experience in the particular area.

In addition to evaluating the oil and gas content, the determination of the lithology and faunal, floral, and mineral constituents may lead to a solution of a geological problem. Cores from different parts of a bioherm will furnish evidence to outline the bioherm better. Facies characteristics in a clastic series can determine the direction to go for the productive facies. Microfossils and macrofossils from cores, particularly in unexplored areas, will define the location in the section and determine whether drilling should be continued. Cores are sometimes taken in exploratory holes for many minerals such as sulfur or radioactive compounds.

In complicated structural provinces, cores have been used to determine dip and strike of the rock at the coring interval. Since rocks usually have polarity which agrees essentially in direction with the magnetic field of the earth at that locality, pieces of core can, in the proper laboratory, be used to determine the direction of dip and strike of the sample. With great expense and tedious care, the drillpipe can be oriented so that the orientation of the core is known. Magnetic core orientation lacks the speed often required in petroleum exploration, and oriented cores are too expensive to compete with dip-meter surveys furnished by electrical logging services (see Chapter 21).

The use of data derived from cores in the laboratory has furnished much basic information for the quantitative analysis of electric and gamma-ray logs. Any new area, new stratigraphic horizon, or facies change from a known area requires core analyses for correct geophysical log interpretation. Shale and particularly bentonitic or montmorillonitic content of dirty sandstones, anhydrite or salt content of carbonate rocks, and permeability in fractured rocks are some of the factors which can be learned only from cores for correct electric or gamma-ray log interpretations.

Cores from boreholes aid immensely in the search for facies changes which has become so prominent in petroleum exploration in recent years. Many of the larger oil companies with complete research laboratories are relying more and more on cores to augment their studies of sedimentation in the search for oil and gas.

In exploitation wells principally, analyses of cores are a major source of data for petroleum engineering studies. Gas-oil and water-oil contacts are established. Lenticularity of porous and permeable streaks affects production methods. Programs for secondary recovery are based on residual oil and water saturation in partially drained areas. Artificial stimulation of producing zones and work-overs of older wells are more successful with core-analysis data.

Operators attempt to estimate primary ultimate oil and/or gas reserves and reserves which can be recovered using some type of secondary method. Core analyses are the chief basis for these calculations when production is prorated. Whether it is profitable to build a gas pipeline into an area is usually based on core analyses because sufficient productive history of the gas wells is lacking for other methods of calculating reserves. Since oil and/or gas in the ground are an asset, all financing and borrowing must be based on some estimate of reserves. Prudent estimates of oil and/or gas reserves from core analyses are accepted by most banks for oil loans.

Costs of Coring

So many factors enter into the costs of coring that it is impossible to generalize. A core from a 7,500-foot well in central Oklahoma costs much less than a core from a

7,500-foot test well in the Montaña of Peru. A wire-line core in east Texas costs much less than a diamond core to the same depth in west Texas.

When a well is coring, the costs are usually paid to the contractor as daywork separately from the drilling cost. Furthermore, except on some "turnkey jobs," any trouble such as a fishing job or lost barrel is the responsibility of the operator. A lost diamond core barrel costs from $2,500 to $3,000. Rig time, paid by the operator, ranges from $500 to $1,500 per day. Therefore, for a conventional core, two round trips and coring time might cost from $250 to $1,500 depending on the depth. In addition, conventional core bits cost from $200 to $300. Diamond bits rent for about $8 per foot and cost approximately $2,000. Wire-line core barrels rent for about $3.50 per foot plus bits, which cost $250 each. Side-wall cores cost about $20 per core plus rig time of about 2 hours for 30 cores at 10,000 feet. Specialized coring services depend on the location of the well and many other factors. In summary, coring costs range from about $500 to as much as $2,000 per core. Core analysis is in addition to these costs.

Summary

The decision to core or not to core should be made before the drilling starts. With side-wall and wire-line coring, the cost per core is reduced as more cores are taken. The characteristics of the potential producing zone will determine whether one of the other detecting methods is superior or satisfactory and costs less than coring to accomplish the objective desired. If coring is favored or also used, the objective depth is usually known in an explored area with reasonable accuracy. Drilling time in minutes per foot is recorded or preferably plotted, and usually a reservoir rock has a change in drilling rate that indicates the top of the zone. Coring should be used either when it is the only way the required information can be obtained or when essential as a required complement to one of the other methods. The cost of coring should be weighed against the information that is needed and anticipated. Coring will continue to be one of the best tools in rotary drilling procedure because it is the only way that one can see, feel, taste, smell, and analyze a sample of the subsurface formation.

REFERENCES

"Coring Catalogue," Christensen Diamond Products, Salt Lake City, Utah.

Engineering Fundamentals on Petroleum Reservoirs, *Oil Gas J.*, pp. 25–43, 1953.

"Finding and Producing Oil," pp. 125–131, American Petroleum Institute, 1939.

Kirkpatrick, C. F.: Rotary Coring, Fundamentals of Rotary Drilling, *Petrol. Engr.*, pp. 125–129, November, 1957.

LeRoy, L. W. (ed.): "Subsurface Geological Methods," Introduction, pp. 12–17, 20–21; Coring Analysis, pp. 295–319; Oriented Cores, pp. 591–595; Magnetic Core Orientation, pp. 596–608; Coring Techniques and Applications, pp. 609–624; Duties and Reports of Subsurface Geology, pp. 810–819, Colorado School of Mines, Golden, Colo., 1950.

Miller, C. R., and C. B. Evans: Porosity Determination from Microlog and Side-wall Cores, *Proc. 4th Symposium on Subsurface Geol.*, pp. 41–47, University of Oklahoma Press, Norman, Okla., 1955.

"A Primer of Oil Well Drilling," 2d ed., p. 72, American Association of Oil Well Drilling Contractors, and Petroleum Extension Division, University of Texas, Austin, June, 1957.

"Rotary Core Drilling," Hughes Tool Company, Houston, Tex.

"Schlumberger Side-wall Coring," Schlumberger Well Surveying Corporation, Houston, Tex.

Suman, John R.: Drilling, Testing, and Completion, "Elements of the Petroleum Industry," pp. 180–187, AIME, 1940.

Chapter 19

ELECTRICAL LOGGING*

By H. G. DOLL, *Chairman of the Board and Technical Advisor*

M. P. TIXIER, *Manager of Field Interpretation*

M. MARTIN, *Formerly Head of Publication and Interpretation Department*†

and

F. SEGESMAN, *Publication and Interpretation Department*

Schlumberger Well Surveying Corporation

FUNDAMENTALS

General. "Well logging" denotes an operation wherein a continuous recording is made, vs. depth, of some characteristic datum of the formations penetrated by a drill hole. The record is called a log.

Many types of well logs are recorded by means of appropriate down-hole instruments called *sondes,* lowered into the wellbore on the end of a cable. The winch of the logging cable is generally brought to the well on a special logging truck (Fig. 19-1) which also carries the recorders, power sources, and auxiliary equipment. The parameters being logged are measured *in situ* as the sonde is moved along the borehole. The resulting signals from the sonde are transmitted, via electrical conductors in the cable, to the surface, where the continuous recording, or log, is made.

Electrical logging is one of the important branches of well logging. Essentially, it is the recording, *in uncased sections of a borehole,* of the *resistivities* (or their reciprocals, the *conductivities*) of the subsurface formations, generally along with the *spontaneous potentials* (SP) generated in the borehole.

Electrical logging has been accepted as one of the most efficient tools in the search for and the production of oil and gas. When a hole has been drilled, or at intervals during the drilling, an electrical survey is run for the purpose of obtaining quickly and economically *a complete record of the formations penetrated.* This recording is of immediate value for the *geological correlation of the strata* and also for the *detection and evaluation of possibly productive horizons.* The information derived from the electrical logs may at the same time be supplemented by side-wall samples of the formations taken from the wall of the hole or by still other types of drill-hole investigations which can be performed by means of additional wire line equipment available for use with the logging truck [deviation surveys, caliper (hole-diameter) surveys, dipmeter surveys, temperature surveys, radioactivity (gamma-ray and neutron) surveys, wire-line formation tester, etc. . . .].

As will be explained later in this chapter (page 19-15) several types of resistivity measuring systems are used which have been designed in order to obtain the greatest

* Reprinted by permission from "Petroleum Production Handbook," edited by Thomas C. Frick, McGraw-Hill Book Company, Inc., New York, 1961.

† Now Executive Vice President, Géopétrole, Paris, France.

possible information under diverse conditions, e.g., conventional devices (normals and laterals), induction log, laterolog, microlog, and microlaterolog.

The typical appearance of a standard electrical log is illustrated in Fig. 19-2. The left-hand track of the log contains the SP curve, which may be taken as qualitatively indicative of the existence of permeability in the formations. The middle track contains a 16-in. short normal (shallow-investigation resistivity curve), recorded on both regular and amplified sensitivity scales as solid curves, and a 64-in. normal (medium-investigation resistivity curve, dashed curve). The right-hand track contains an 18 ft 8 in. lateral (deep-investigation curve).

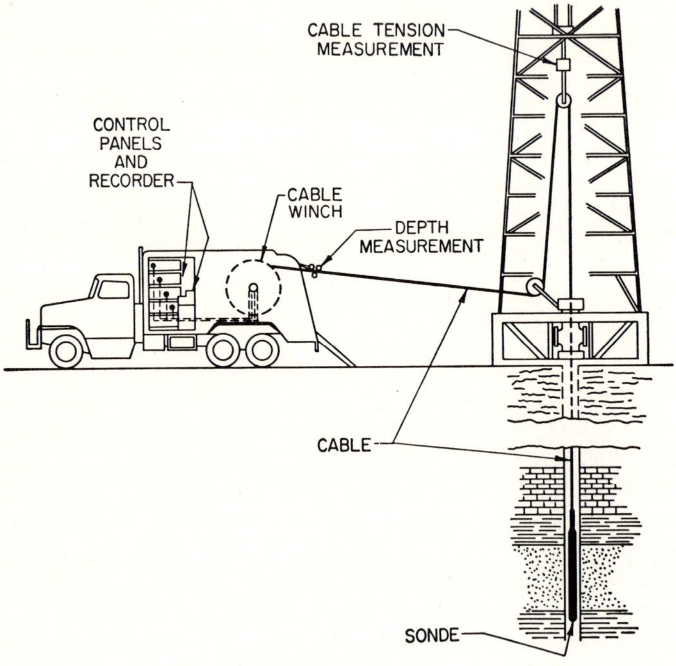

Fig. 19-1. Setup for wire-line logging operations in wells (schematic).

Logs recorded with other combinations of resistivity-measuring devices (page 19-38) have a similar general appearance, although the corresponding devices differ basically in principle and performance. Microresistivity logs (Figs. 19-27 and 19-29) generally include a microcaliper curve (hole-diameter recording), which is useful in the location of permeable zones and in the evaluation of formation porosity.

The curves are recorded on the most appropriate of several available sensitivity scales. The usual depths of scales are 2 in. = 100 ft (regular) and 5 in. = 100 ft (detail). Less frequently a 1 in. = 100 ft scale is used. For cases where great detail is involved, as in micrologging and dipmeter logging, special expanded scales are available.

Earth Resistivities. The resistivities of the formations are important clues to their probable lithology and fluid content. With a few exceptions rare in oil-field practice, such as metallic sulfides and graphite, dry rocks are very good insulators, but when their pores are impregnated with water they conduct electric current. Subsurface formations in general have finite measurable resistivities because of the water contained in their pores or adsorbed on their interstitial clay. Formation resistivity also depends on the shape and the interconnection of the pore spaces occupied by the water. These depend on the lithology of the formation and, in the case of reservoir rocks, on the presence of nonconductive oil or gas.

Units of Resistivity and Conductivity. In electrical logging the resistivity is usually measured. An exception is induction logging, in which the conductivity is recorded, along with its reciprocal, the resistivity.

The resistivity (specific resistance) of a substance, at any given temperature, is the resistance measured between opposite faces of a unit cube of that substance. In electrical-logging work, the meter was chosen as the unit of length; so the unit of resistivity is taken as the ohm-meter²/meter, or more simply, the *ohm-meter* (abbreviation, ohm-m).

Since conductivity is the reciprocal of resistivity ($\sigma = 1/R$), a possible unit of conductivity would be 1/ohm-meter, or mho/meter. However, since this unit would necessitate extensive use of decimal fractions, a unit one-thousandth as large, the *millimho/meter* (mmho/m), is employed. Thus formations having resistivities of 10, 100, or 1,000 ohm-m have conductivities of 100, 10, or 1 mmho/m, respectively.

Dependence of Water Resistivity on Salinity and Temperature. 1. *The resistivity of an electrolytic solution decreases as the amount of chemicals therein increases.* At any given temperature the electrical conductivity of a formation water or a drilling mud will depend on the concentration and nature of the dissolved chemicals.

In most cases the predominant solute is sodium chloride; therefore the NaCl conversion chart (Fig. 19-3) may generally be used to obtain resistivity from concentration. If other chemicals are present in relatively large amounts, it is possible to convert the concentrations of such chemicals into equivalent concentrations of sodium chloride to find the resistivity. To make the conversion, apply the appropriate multipliers following for the concentration of each separate ion [in parts per million (ppm) by weight, or in grains per gallon (gr/gal)], and add the products.[17]*

Cations		*Anions*	
Na	1.0	Cl	1.0
Ca	0.95	SO_4	0.5
Mg	2.0	CO_3	1.26
		HCO_3	0.27

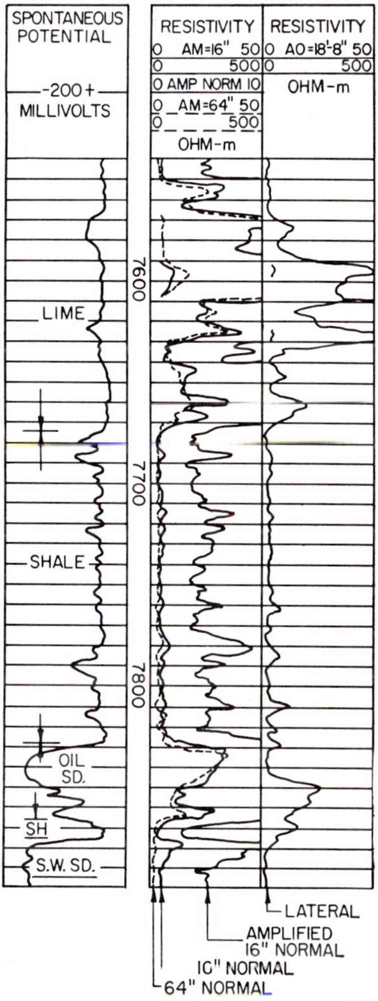

FIG. 19-2. A typical electrical log. This is the standard electrical-log combination described on page 19-15.

Note that concentrations expressed in milligrams per liter (mg/l) and in ppm may be appreciably different at high concentrations. Below about 50,000 ppm, however, measurements at room temperature in the two units may be used interchangeably without serious error.

2. *The resistivity of an electrolytic solution decreases as its temperature increases.* This is of great importance, since temperature in the earth increases with depth.

Before the resistivity of the drilling mud, measured at surface temperature, can be compared with that of a formation, measured at a much higher temperature in a deep well, the resistivities must be converted to values which would have been observed at a common temperature. The temperature conversion is accomplished by means

* Superscript numbers refer to references listed at the end of this chapter.

of Fig. 19-3, which shows for sodium chloride solutions the effects of both salinity and temperature on resistivity. Down-hole temperatures may be estimated from a so-called "bottom-hole temperature" (BHT) obtained by means of a maximum-reading thermometer inserted in the body of the sonde.

Resistivities of Formation Waters (see also Chap. 7*). Formation waters can vary remarkably with geographic location, depth, and geological age. Shallow ground waters are usually fresh (not saline), with resistivities sometimes exceeding 20 to 50 ohm-m at room temperature. They may also contain appreciable amounts of calcium and magnesium salts, which make them "hard." At great depths, formation waters

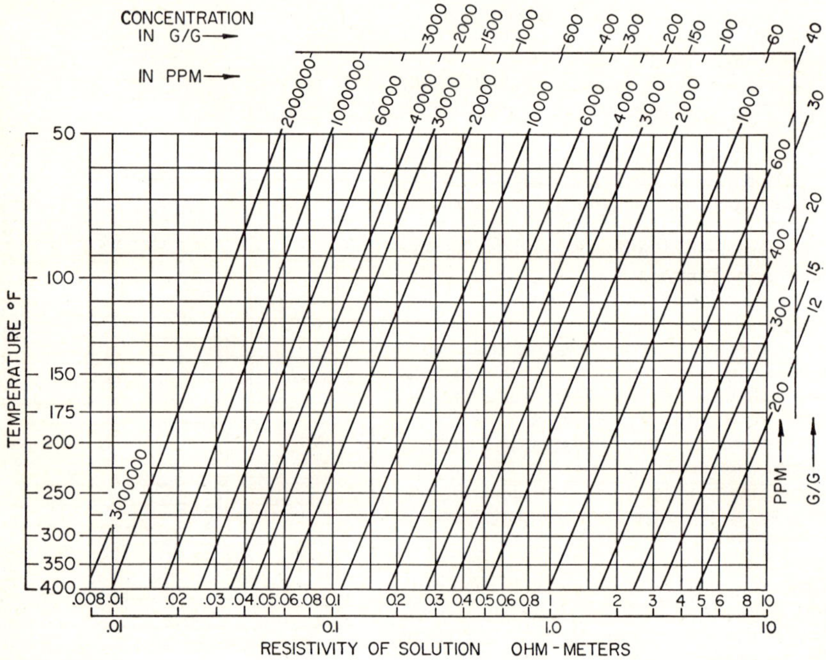

FIG. 19-3. Resistivity vs. concentration for sodium chloride solutions at various temperatures.

generally tend to be more saline. In deep wells, formation-water resistivities sometimes may correspond to complete saturation (0.014 ohm-m at 200°F).

A knowledge of R_w, the formation-water resistivity, is important in electrical-log interpretation. R_w may be obtained from the readings of the SP curve (page 19-14) or from resistivity measurements on samples of formation water recovered in drill-stem tests, or it may be estimated from measurements of the resistivity R_o of the permeable formations of interest when 100 per cent water-saturated if the porosity or formation factor is known [Eqs. (1) and (2), page 19-5]. R_w may also be computed, as has been explained, from analyses of formation waters.

Mud, Mud-cake, and Mud-filtrate Resistivities. The resistivities of the mud R_m, the mud cake R_{mc}, and the mud filtrate R_{mf} are all important in log interpretation. R_m is obtained by direct measurement on a mud sample. R_{mf} and R_{mc} are obtained by direct measurements on filtrate and mud cakes pressed from a sample of the mud, or they can be estimated from average statistical data based on the mud resistivity (Chart A-4 of Ref. 41 and Refs. 27 and 30). Correction for the variation of these resistivities with temperature is made by means of Fig. 19-3.

Formation Resistivity Factor. If R_o is the resistivity of a clean (nonshaly) formation completely saturated with water of resistivity R_w, the ratio R_o/R_w will be a con-

* Frick, *op. cit.*

stant which depends on the lithologic structure of the formation and not on the resistivity R_w of the saturating water. This constant is the *formation resistivity factor F*, commonly called "formation factor."

$$F = \frac{R_o}{R_w} \tag{1}$$

Dependence of Formation Factor on Porosity and Lithology. The formation factor F of a clean formation can be related to its porosity ϕ by an empirical formula of the form $F = a/\phi^m$, where a and m are constants. The exponent m, sometimes called the *cementation factor*, varies with the lithology.

In the construction of numerous graphs for log interpretation,[41] the "Humble formula" proposed by W. O. Winsauer et al.[51] has been quite generally adopted:

$$F = \frac{0.62}{\phi^{2.15}} \tag{2}$$

An early formula, proposed by G. E. Archie, which fits particularly well for consolidated formations such as hard sandstones and limestones is

$$F = \frac{1}{\phi^2} \tag{3}$$

Limestones often contain vugs, interconnected by fissures, which add their porosity to that of the matrix. Wherever the vugs and fissures are closely spaced compared with the spacings of the resistivity-measuring devices, Eq. (3) can often be used just as in the case of sandstones or limestones with only granular porosity. Nevertheless, it is sometimes advisable to use values of m greater than 2, as required to fit local observations.[2]

Shaly (Dirty) Formations. Shales and clays are themselves porous and are generally impregnated with mineralized water. They have, therefore, appreciable conductivity, which is enhanced by ion-exchange conduction through the shale matrix. (This shale conduction is sometimes, though not quite properly, referred to as being due to "conductive solids.") On the other hand, the size of the shale pores is so small that practically no movement of fluid is possible. Accordingly, shale, whether deposited in thin laminations or dispersed in the interstices of the sand, contributes to the conductivity of the formation without contributing to its effective porosity.

The relation between formation resistivity and porosity becomes more complex than for clean formations. Because of the additional shale conductance, the ratio of formation resistivity to water resistivity (i.e., the formation factor) is not constant when the resistivity of the impregnating water changes.[33] Nevertheless, if the shale content is not too great, experimental observations show that *for low enough values of water resistivity this ratio is almost constant,* as though the conductance of the shale were then negligible in comparison with that of the water; *and a limiting formation factor is found* which is related approximately to the effective porosity in the same way as the formation factor of a clean sand.

Relation between Formation Resistivity and Saturation. When a part of the pore space is occupied by an insulating material such as oil or gas, the resistivity R_t of the rock is greater than the resistivity R_o which it has when 100 per cent water bearing. The resistivity of such rock is a function of the fraction of the pore volume occupied by water.

For substantially clean formations, the water saturation S_w is related to R_t (resistivity of formation containing hydrocarbons and formation water, with a water saturation S_w) and R_o (resistivity of same formation when 100 per cent saturated with the same water) by an empirical relation known as the Archie equation[1]

$$S_w = \left(\frac{R_o}{R_t}\right)^{1/n} \tag{4}$$

Empirically determined values of n range between 1.7 and 2.2, depending upon the type of formation. Experience shows that taking $n = 2$ should give a sufficiently good approximation in general. Then, combining Eqs. (4) and (1)

$$S_w = \left(\frac{R_o}{R_t}\right)^{1/2} = \left(\frac{FR_w}{R_t}\right)^{1/2} \tag{5}$$

The ratio R_t/R_o is sometimes designated as the *resistivity index I;* accordingly, $S_w = (I)^{-1/n}$.

The relation between formation resistivity and water saturation is more complex when the formations contain some shale or clay, because of the additional conductance due to the interstitial shale.

Ranges of Resistivity—Classification of Formations. Clays and shales are porous and practically impervious formations and are often very uniform throughout their mass. Their resistivity is comparatively low and practically constant over wide intervals.

Compact and impervious rocks, such as gypsum, anhydrite, dense calcareous formations, or certain kinds of coal, are highly resistive because of their very small interstitial water content.

Resistivities of porous and permeable formations, such as sands, vary widely, depending on their lithology and fluid content. In electrical logging it is convenient to classify reservoir rocks as follows:

Soft Formations. Chiefly poorly consolidated sand-shale series. The porosity of the sands is intergranular and exceeds 20 per cent. Resistivities range from about 0.3 ohm-m for salt-water-bearing sands to a several ohm-meters for oil-saturated sands.

Intermediate Formations. Chiefly moderately consolidated sandstones, but frequently limestones and/or dolomites. Reservoir porosity is generally intergranular, ranging from about 15 to 20 per cent. The reservoir formations are interbedded with shales and very often with tight rocks. Resistivities range from 1 to about 50 to 100 ohm-m.

Hard Formations. Chiefly limestones and/or dolomites, and also consolidated sandstones. They consist mostly of tight rocks containing porous and permeable zones, and shale streaks. The porosity of reservoirs is below 15 per cent. Most often, the porous and permeable zones contain fissures and vugs. Resistivity range is from 2 or 3 ohm-m to several hundred. For the completely tight formations, such as salt and anhydrite, the resistivity may be practically infinite.

Anisotropy. In many sedimentary strata, the mineral grains have a flat or platelike shape with an orientation parallel to the sedimentation. Current travels with great facility along the water-filled interstices, which are mostly parallel to the stratification. These strata, therefore, do not possess the same resistivity in all directions. Such *microscopic anisotropy* is observed mostly in shales.

Moreover, in electrical logging, the distance between electrodes on the measuring devices is great enough that the volume of formation involved in the measurements very often includes sequences of interbedded resistive and conductive streaks. Since current flows more easily along the beds than perpendicular to them, the formation has *macroscopic anisotropy.*

Both kinds of anisotropy may add their respective effects to influence the apparent resistivity. The longitudinal, or horizontal, resistivity R_H, measured along the bedding planes, is always less than the transversal, or perpendicular, resistivity R_V.

Resistivity-measuring devices whose readings are not appreciably affected by the borehole (the induction log, the laterolog, and the long lateral when the ratio R_H/R_m is low or moderate) will read R_H. Because of the borehole effect, the short normal and, to a lesser extent, the long normal usually read values greater than R_H.[26]

Distribution of Fluids and Resistivities in Permeable Formations Invaded by Mud Filtrate. Inasmuch as the hydrostatic pressure of the mud is usually maintained greater than the natural pressure of the formations, mud filtrate, forced into the permeable beds, displaces the original formation fluids in the region close to the borehole. Solid materials from the mud deposited on the wall of the hole form a mud cake which tends to impede and reduce further infiltration.

The thickness and the nature of the mud cake depend on the nature of the mud and on the drilling conditions rather than on the formations. The thickness (t_{mc}) is usually between ⅛ and 1 in. For water-base muds the mud-cake resistivity R_{mc} is about equal to one or two times the mud resistivity R_m. In some oil-emulsion muds, R_{mc} may be somewhat greater.

Figure 19-4*a* represents a schematic cross section of an oil-bearing permeable bed penetrated by a borehole. Figure 19-4*c* shows the corresponding radial variations of resistivity.

As indicated in Fig. 19-4*a* the zones of different resistivity may be divided into the drilling mud within the borehole (of resistivity R_m); the mud cake R_{mc}, the flushed zone R_{xo}; a transition zone; in some cases an "annulus" (present only in certain oil- or gas-bearing formations); and the uncontaminated zone (of resistivity R_t). The invaded zone (of "average" resistivity R_i) includes the flushed zone and the transition zone.

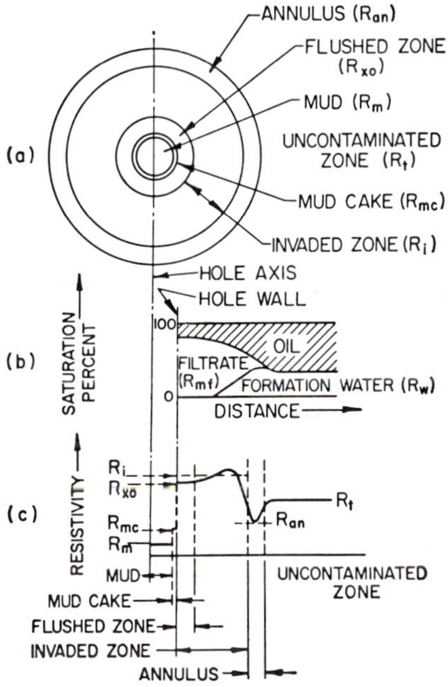

FIG. 19-4. (*a*) Horizontal section through a permeable oil-bearing bed ($S_w < 60$ per cent). (*b*) Radial distribution of fluids in formation (qualitative). (*c*) Radial distribution of resistivities.

Invaded Zone. Behind and close to the wall of the hole, it is believed that most of the original interstitial fluids have been flushed out by the mud filtrate. This flushed zone, of resistivity R_{xo}, is considered to extend, under usual conditions of invasion, at least 3 in. from the wall.

If the bed is water-bearing, the pores in the flushed zone are completely filled with the mud filtrate, and for clean formations R_{xo} is equal to FR_{mf}, F being the formation factor and R_{mf} the mud-filtrate resistivity.

If the bed is oil-bearing, the flushed zone contains some residual oil saturation (ROS). From Eq. (5), S_{xo}, the water saturation in the flushed zone, is

$$S_{xo} = \left(\frac{FR_{mf}}{R_{xo}}\right)^{1/2} \quad \text{or} \quad R_{xo} = \frac{FR_{mf}}{S_{xo}^2} \tag{6}$$

where $S_{xo} = 1 - \text{ROS}$

Beyond the region of maximum flushing R_{xo} there is a more or less extended transition region, the nature of which depends on the characteristics of the formation, the speed of invasion, and the hydrocarbon content. The invaded zone includes the flushed zone and the part of the transition zone invaded by filtrate. In the case of water-bearing sands and oil-bearing sands of high water saturation, the invaded zone extends up to the uncontaminated zone R_t.

There can be no exact definition of the depth of the invaded zone, but it is convenient to introduce a factor D_i, called the "electrically equivalent diameter of invasion," corresponding to an average invaded zone of resistivity R_i, which has the same effect as the actual invaded zone on measurements made in the borehole.

The depth of invasion is quite variable. It depends on the plastering properties of the mud, on the pressure differences between the mud column and the formation, on the time elapsed since the formation was drilled, on the porosity of the formation, on the proportion and nature of the fluids (water, oil, gas) present in the pores, on the reaction of any interstitial clays with the mud filtrate, etc.

All other conditions being the same, the greater the porosity, the smaller the depth of invasion. With usual muds, D_i seldom exceeds $2d$ (d = hole diameter) in high-porosity sands, but it may exceed $5d$ and even $10d$ in low-porosity formations such as consolidated sandstones or limestones. Invasion may be extremely shallow in case of low-water-loss or oil-base mud, and in gas-bearing formations.

In very permeable beds, when there is an appreciable difference between the specific gravities of the mud filtrate and the salt-laden connate water, gravity-segregation effects may occur, with the fresher filtrate tending to accumulate at the top boundary of the bed and a decrease in the depth of invasion in the lower part of the bed.[13]

In fissured formations, the permeability due to the fissures quite often is enormous, much greater than the permeability of the matrix material surrounding them.

Suppose that a formation is composed of a porous but relatively impermeable material, broken by networks of roughly parallel fissures. Mud filtrate penetrates the fissures easily and deeply, driving out much of the original fluids (oil and formation water). On the other hand, the matrix itself may be penetrated hardly at all by the filtrate. Since the fissures constitute a small part of the total pore volume, only a very small proportion of the total original fluids is displaced. As a result, R_{xo} is little different from R_t, and the ratio R_{xo}/R_{mf} is no longer representative of the formation factor.

The Annulus. When the formation *contains hydrocarbons*, the process of invasion is complex. The distribution of fluids is then affected by the two-phase permeabilities, relative gravities and viscosities of the fluids, capillary forces, etc.

When the initial water saturation is low (less than about 50 per cent), one important feature is the existence of an *annular region* just inside the uncontaminated zone, *containing mainly formation water and some residual oil.*[4] This "annulus" is explained as follows: the mud filtrate penetrates the formation radially, sweeping the removable oil and formation water ahead of it. For large oil saturation, the relative permeability to oil is appreciably greater than that to water. Therefore, the oil moves faster, leaving a zone (the annulus) enriched in formation water behind it.

It seems likely that, because of the effects of diffusion, capillary pressure, gravity, etc., the existence of a well-defined annulus is a transitory phenomenon. Field log experience nevertheless seems to show that the annulus does very often exist at the time the logs are run.

Computations have shown *that the presence of the annulus has a practically negligible effect on the response of the devices using electrodes* (*normals, laterals, and laterolog*). It may have an effect on the induction log, but this can be taken care of for practical purposes by means of appropriate interpretation charts (Chart D-12 of Ref. 41).

The Uncontaminated Zone. For clean formations, from Eq. (5),

$$R_t = \frac{FR_w}{S_w^2} \tag{7}$$

In the usual case, R_{mf} is 10 to 25 times as large as R_w. Thus comparing Eqs. (6) and (7) with usual values of S_w and S_{xo}, R_t, even in oil-bearing formations, is generally less than R_{xo}, as represented in Fig. 19-4*c*.

Apparent Resistivity. Since any resistivity measurement is affected in some degree by the resistivities of all the media in the immediate vicinity of the sonde (i.e., mud, different parts of the formation which vary in resistivity, adjacent formations if the bed measured is thin), any given device records an *apparent resistivity.* Each resistivity device is calibrated so that when the sonde is in a homogeneous medium (or in some other condition appropriate to practice, specified for the particular device) the apparent resistivity reading is equal to the actual resistivity.

Requirements for and Types of Resistivity Devices. Inspection of the basic relations [Eqs. (1), (2), (5), and (6)] shows that a determination of S_w and ϕ requires a knowledge of R_t and R_{xo} (or R_i, in certain cases where R_{xo} is not easily determined). Thus, for the reservoir-evaluation problem, it is necessary to have resistivity-measuring devices with different depths of investigation, in order to obtain values indicative of the resistivities of the invaded zone and the uncontaminated zone. The readings of the deep- and shallow-investigation curves may often be used to correct each other, through correction charts or departure curves, to obtain better values of R_t and R_i.

Another function of the resistivity recording is to provide an accurate definition of bed boundaries, particularly of permeable beds.

Finally, it is desirable that the readings not be influenced by the effect of the mud column or, in case of thin beds, by the adjacent formations.

With the "conventional" resistivity devices, the above requirements are only partly satisfied. The introduction of the microdevices and focused devices has brought about an appreciable improvement.

Resistivity devices in present use may be classified as follows:

1. Macrodevices, which derive their reading from about 10 to 100 cu ft of material around the sonde (useful for R_t and R_i evaluation). They include:
   a. Unfocused-electrode devices and focused-electrode devices
   b. Induction logging
2. Microdevices (also called wall resistivity devices), which derive their readings from a few cubic inches of material behind or close to the wall of the hole. Since the electrodes are mounted on an insulating rubber pad pressed against the wall of the hole, measurements are little affected by the mud column.

Microdevices are of unfocused and focused types.

Resistivity devices using electrodes may be used in holes filled with water or water-base drilling mud, which provides the electrical contact necessary between electrodes and formation. The induction log can also be used in empty holes or in holes filled with nonconductive oil-base mud. The various resistivity devices will be described on page 19-15.

THE SPONTANEOUS POTENTIAL (SP) LOG

The SP log is a record of the naturally occurring potentials in the mud at different depths in a borehole. The measurement is made *in uncased holes containing water-base or oil-emulsion muds* between an exploratory electrode on the sonde in the borehole and a stationary reference electrode at the surface.

Usually the SP curve (Fig. 19-2) consists of a more or less straight base line (corresponding to the shales) having excursions or peaks to the left (opposite the permeable strata). The shapes and the amplitudes of the excursions may be different, according to the formations; but there is no *definite correspondence between the magnitudes of the excursions and the values of permeability or porosity of the formation.*

The principal uses of the SP curve are to:

1. Detect the permeable beds
2. Locate their boundaries (except when the formations are too resistive)
3. Correlate such beds
4. Obtain good values for R_w, the formation-water resistivity

Origin of the SP. The character of the potentials measured in the mud results from ohmic drops produced by the flow of SP currents through the mud resistance. If the mud is extremely conductive these ohmic drops may be insignificant, and the variations in the SP curve may be too small to be useful.*

The SP currents flow as a result of electromotive forces (emfs) existing within the formations or at the boundaries between formations and mud.

One phenomenon which could cause an emf to appear across the mud cake opposite a permeable bed is *electrofiltration*. The mud filtrate, in being forced through the mud cake, would tend to produce an emf, positive in the direction of flow. According to recent experiments,[19] the emf across the mud cake may be quite sizable, but there is also an electrofiltration emf generated across the adjacent shales. Thus the

* In such a case the gamma-ray log, which distinguishes shales from nonshale beds, is sometimes recorded as a substitute for the SP.

net effect of electrofiltration in causing variations of SP is small and in most cases negligible for all practical purposes—a conclusion verified by field experience.*

Most important are the emfs of electrochemical origin which occur at the contacts between the drilling mud (or its filtrate) and the formation water, in the pores of the permeable beds, and across the adjacent shales.[53] In a clean sand lying between shale beds, all penetrated by a drill hole containing conductive (water-base) mud, the *total electrochemical emf* E_c is produced in the following chain (Fig. 19-5):

Mud/mud filtrate/formation water/shale/mud

The emf of the junction, mud/mud filtrate, is taken to be practically nil because, although the resistivities of the mud and its filtrate may differ, their electrochemical activities should be the same.

The part of the chain consisting of "formation water/shale/mud" gives rise to the *shale-membrane emf* E_M. The part "mud filtrate/formation water" gives rise to the *liquid-junction emf* E_J. For sodium chloride (monovalent-ion) solutions, at 75°F,

$$E_M = 59 \log_{10} \frac{a_w}{a_{mf}}$$

$$E_J = 12 \log_{10} \frac{a_w}{a_{mf}}$$

where a_w and a_{mf} are the chemical activities of the formation water and mud filtrate, respectively (at 75°F), and E_M and E_J are in millivolts. The total electrochemical emf E_c is the sum of E_M and E_J:

$$E_c = K \log_{10} \frac{a_w}{a_{mf}} \tag{8}$$

where $K = 71$ at 75°F

Equation (8) is general, provided both formation water and mud filtrate are essentially sodium chloride solutions of any concentration. The values of K are directly proportional to the absolute temperature. Thus at 150°F the coefficient K in Eq. (8) becomes 81 instead of 71, and at 300°F, it becomes 101 (see Fig. 19-8).

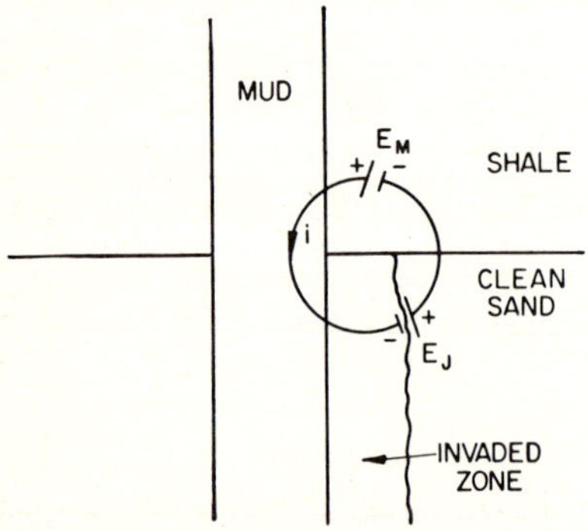

FIG. 19-5. Schematic representation of electrochemical chain and SP current path at boundary between permeable bed and adjacent shale.

From Eq. (8), in the usual case of a_w greater than a_{mf}, E_c is positive. However, if a_{mf} is greater than a_w, corresponding to *mud more saline than formation water*, then E_c is negative, and the *SP deflections corresponding to permeable beds are then reversed on the log.*

Effect of Interstitial Shales on the SP. Increasing amounts of shale or clay in a permeable bed effectively result in a reduction, then reversal, and eventually an increase in the opposing sense, of the liquid-junction emf E_J. Thus the total emf E_c $(= E_M + E_J)$ is reduced. In the limit, for 100 per cent shaliness, $E_J = -E_M$, and E_c becomes zero; that is, the "sand" is then all shale and indistinguishable from the surrounding shales.

The presence of oil in a shaly sand tends to enhance the effect of the shale on the liquid-junction emf. All other conditions being the same, the total E_c of a shaly sand will be smaller if oil-bearing than if water-bearing.

The effect of interstitial shale is also greater in low-porosity formations. In these cases, only a small amount of shale reduces the SP deflection appreciably. Conversely, the E_c of shaly water-bearing sands of high porosity remains practically equal to the E_c of a clean sand, as long as the shale content is reasonably low, i.e., does not exceed a few per cent.

* Further information on the electrofiltration emf, or streaming potential, may be found in Refs. 21, 54, and 57.

Circulation of the SP Current. The various emfs add their effects to generate the SP currents, which follow the paths represented schematically in Fig. 19-6*b* (right) by solid lines. Each current line encircles the junction of mud, invaded zone, and uncontaminated zone. In the usual case where the formation waters are saltier than the mud, E_c is positive, and the current circulates in the direction of the arrows. The potential of a point in the mud column opposite the sand is negative with respect to one opposite the shale.

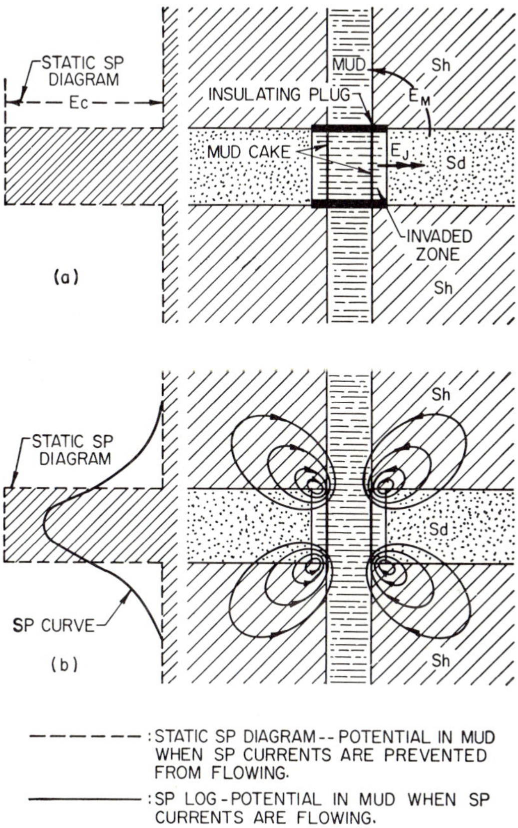

FIG. 19-6. (*a*) Static SP diagram (left) which would be observed in hole when current is prevented from flowing by means of insulating plugs (right). (*b*) Actual SP diagram (solid curve, left) and schematic representation of SP current distribution in and around permeable bed (right).

Along its path, the SP current forces its way through a series of resistances, both in the ground and in the mud. Along a closed line of current flow, the total of the ohmic-potential drops is necessarily equal to the algebraic sum of the electromotive forces encountered. Moreover, the total potential drop is divided between the different formations and the mud in proportion to the resistance of the path through each respective medium.

The Static SP (Clean Formations): Pseudostatic SP (Shaly Formations). It is convenient to use an idealized representation in which the SP current is prevented from flowing by means of insulating plugs, placed across hole and invaded zone, as shown in Fig. 19-6*a* (right). Under these conditions, a plot of the potential in the mud column would appear as shown as the dashed crosshatched curve on the left of Fig. 19-6*a*, with a *maximum negative deflection* opposite the permeable bed *equal to* *the algebraic sum of all the emfs of various origins.* This is the maximum SP that could be measured. It is therefore convenient to use this theoretical value as a reference. In the case of a clean sand, it is called the static SP (SSP). If the sand is shaly, it is called the pseudostatic SP (PSP).

For given values of the activities of mud and formation water, the pseudostatic SP of a shaly sand is smaller than the static SP of a clean sand. The ratio PSP/SSP is called the reduction factor and is designated by the symbol α.

The SP log records only that portion of the potential drop occurring in the mud. *When the bed is sufficiently thick the amplitude of the SP deflection approaches the static SP (or PSP in case of shaly formations),* because then the resistance offered to the current by the bed itself is negligible compared with the resistance of the path through the mud in the borehole.

Factors Influencing the Shape and Amplitude of the SP Deflections. As seen in Fig. 19-6*b*, the current circulates in the hole not only opposite the permeable formation but also a short distance beyond its boundaries. As a result, although on the static SP diagram the boundaries of a permeable bed are indicated by sharp breaks, those on the actual SP curve show a more gradual change in potential. An analysis of the circulation of the current[7] shows that, for uniform resistivity in the formations, the bed boundaries are located at the inflection points on the SP log. This fact provides a means of determining the thickness of a bed from the SP log.

Both the shape of the SP deflection and its relative amplitude (in fractional parts of the SSP or PSP) are influenced by the following factors which determine the conditions for the circulation of SP currents:

The thickness of the bed
The resistivities of the bed, of the adjacent formations, and of the mud
The diameter of the drill hole
The depth of invasion

All other factors remaining the same, a change of the total emfs affects the amplitude but does not modify the general shape of the SP log.

1. *Influence of the Mud Resistivity and Hole Diameter. The mud resistivity has a predominant influence on the SP curve.* If the mud is of about the same degree of salinity as the formation water, electrochemical emfs are small. If the mud is more saline than the formation water, the SP may be reversed (sand deflections toward the positive side of the log). Moreover, the lower the mud resistivity, in comparison with the formation resistivity, the broader the deflection above and below the permeable bed and, because the ohmic drops in the mud are decreased, the smaller the amplitude of the deflection.

An increase in hole diameter acts approximately like an increase in the ratio of formation resistivity to mud resistivity. It tends to round off the deflections on the SP log and reduce the amplitude of the deflections opposite thin beds. A decrease in hole diameter has the same effect as a decrease in the ratio of formation resistivity to mud resistivity.

The SP log would also be influenced by a lack of homogeneity of the mud—a change in salinity of the mud at a certain level would result in an SP base-line shift at that level. However, it has been found in practice that such changes in salinity are very rare.

2. *Effect of Invasion.* Permeable beds in general are invaded by mud filtrate. Because the boundary between mud filtrate and connate water is somewhere inside the formation, a fraction of the SP current flows directly from the shale into the invaded zone, without penetrating the mud column. As a result, *the presence of the invaded zone has an effect on the SP log similar to that of an increase in hole diameter.*

The SP in Soft Formations. 1. Theory and field experience have shown that the amplitude of the SP deflection is practically equal to the static SP (of a clean sand) or to the pseudostatic SP (of a shaly sand) when the permeable beds are thick and when the resistivities of the formations are not too great in comparison with that of the mud. Moreover, the SP curves define the boundaries of the bed with great accuracy.

The amplitude of the deflection is less than the static SP or pseudostatic SP for thin beds, and the thinner the bed, the smaller the deflection.

2. On the other hand, when the resistivity of the formation R_t is considerably greater than that of the mud R_m, the SP curves are rounded off, the boundaries are marked less accurately, and all other conditions being the same, the amplitude of the peak is less than when the ratio R_t/R_m is close to 1.

For the case of shaly sands, the SP curve may also be affected by the presence of oil. A change in the magnitude of the SP deflection occurs very often when passing an oil-water contact in a shaly sand. This change is not a positive criterion for the detection of oil, however, since the same effect would be obtained if the salinity of the interstitial water were reduced or if the percentage of shale were increased.

The SP in Hard Formations. Hard formations are highly resistive except for (1) permeable beds, whether oil- or water-bearing, and (2) shales, which are impervious.

The SP currents generated by the different emfs flow into the hole out of the shale sections and out of the hole into the permeable sections. In between, they flow through the mud rather than through resistive sections close to the borehole, because of the large resistances the latter paths offer. However, within the formation at a distance from the borehole, where the paths through the resistive beds have larger cross sections and hence lower resistances, the SP currents can complete their circuits from permeable beds to shale. They cannot return to the mud through adjacent permeable beds because there they encounter emfs opposing them.

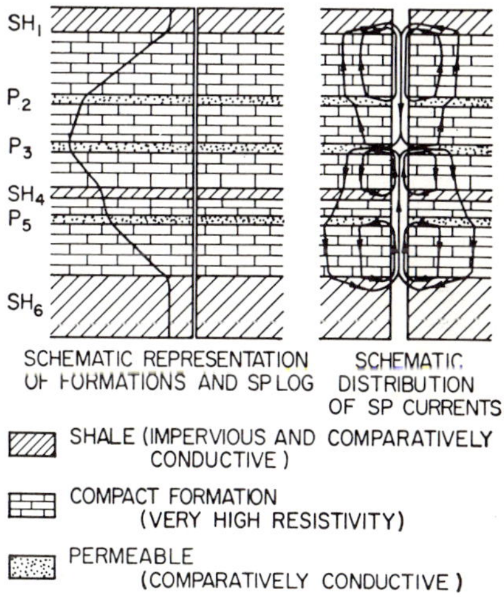

FIG. 19-7. SP phenomena in highly resistive formations (schematic).

Opposite a given resistive bed, the SP current in the mud column remains essentially constant along the borehole. This means that the potential drop per unit length of hole is also constant, thus giving a constant slope on the SP log as shown by the straight-line portions of the SP in Fig. 19-7 (left). At the level of each conductive bed, some SP current will enter or leave the mud column thus modifying the slope of the SP log. For instance the slope of the SP log changes at the level of the permeable bed P_2 because part of the current leaves the hole and flows into that bed.[7]

As a general rule in hard formations the *permeable beds are characterized on the SP log by slope changes or curvatures which are convex toward the negative* side of the log. *Shales are characterized by curvatures which are convex toward the positive* side of the log. *Highly resistive beds correspond to essentially straight parts* of the SP log.

Determination of the Static SP. The SP deflection is measured with respect to the *shale base line*, a reference line which can generally be traced along the extreme positive edges of the SP curve. Usually the shale line is a straight vertical line.[*]

In any given well since the mud salinity is constant and the connate waters may tend to be constant, there is often seen a definite tendency for the maximum SP deflections to be the same for the same types of permeable formations at comparable depths. Thus it is usually possible to draw, parallel to the shale line, a *sand line* on the log along the maximum negative deflections of the clean sands of sufficient thickness.

It is very likely that, for all the beds where the SP peaks reach the sand line, (1) the

[*] Field experience has shown that in certain regions there may be shifts of the shale line. Sometimes these shifts are found systematically at the same places in the geological column and can be used as markers.

formation-water resistivity is practically the same, (2) the beds are virtually free from shaly material, and (3) *the amplitude of the deflection is equal to the static SP.*

For thin beds in cases where the SSP cannot be determined as above (or for a thin shaly sand) the SP reading from the log must be corrected by means of appropriate charts in order to obtain the SSP (or PSP) (see Chart A-8 of Ref. 41 and Ref. 52).

Determination of R_w from the Static SP

Since the variations of electrofiltration potential from sand to shale can generally be neglected the *static SP is taken in practice as equal to the corresponding value of* $-E_c$.

It is convenient to replace Eq. (8) by

$$\text{SSP} = -K \log \frac{R_{mf}}{R_{we}} \tag{9}$$

where R_{we} is an equivalent formation-water resistivity. The computation of R_{we} is given in the chart of Fig. 19-8, and R_w is derived from R_{we} by means of the auxiliary chart at the lower right of Fig. 19-8. The solid curves on this auxiliary chart correspond to highly saline formation waters, where the presence of salts different from NaCl is negligible in practice. They are derived from the known activity-resistivity

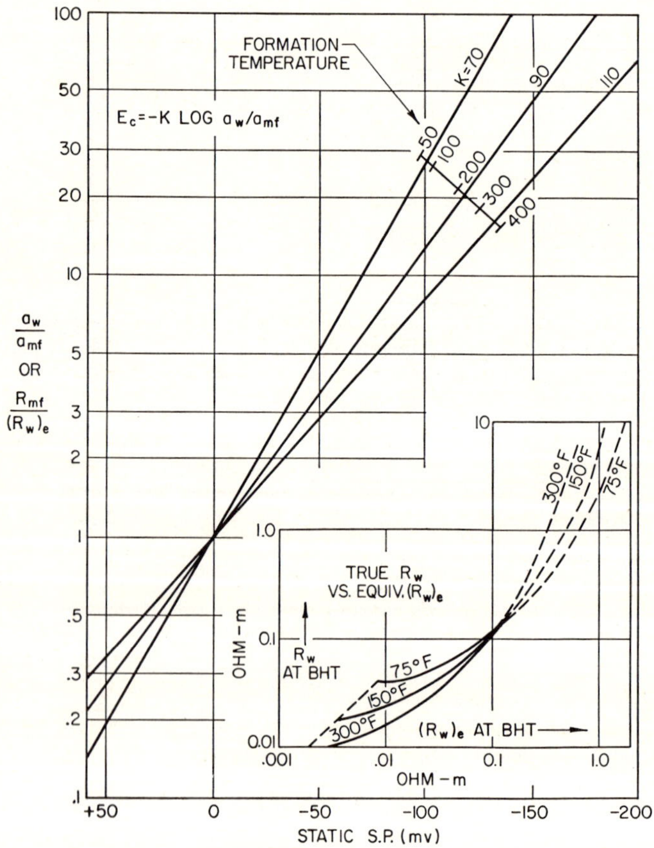

FIG. 19-8. R_w determination from the SP. The inset chart of true R_w vs. R_{we} applies to formation waters of average composition.

relationships for pure NaCl solutions. The dashed curves correspond to formation waters of low salinity, where the presence of other salts (calcium and magnesium chlorides, sulfates, and bicarbonates) have an important bearing on the activity values. These curves are derived from empirical observations and cover formation waters of average composition.[18]

Note that, for intermediate salinities ($0.08 < R_w < 0.3$ at 75°F), the value of R_{we} is practically equal to R_w.

The mud filtrate is taken here as a NaCl solution, and this is generally done in practice, except for muds containing gypsum, $CaCl_2$, or NaOH. In such cases, the determination of R_w from the SP curve requires the measurement of the activity of the mud. A field instrument is provided for this purpose.

RESISTIVITY LOGGING DEVICES

A general classification of the types of resistivity logging devices was given on page 19-9.

Electrical-log Combinations. In most locations, the *standard ES* (electrical survey) (Fig. 19-2) includes, *in addition to the SP*, usually three conventional (unfocused) resistivity curves, namely, a *short normal* curve ($AM = 16$ in.), a *long normal* ($AM = 64$ in.) or a short lateral ($AO = 6$ to 9 ft), and a *standard lateral* ($AO = 18$ ft 8 in. in general), all recorded simultaneously. In some regions, such as the Permian Basin (West Texas and New Mexico), the short-normal spacing is reduced to 10 in., and the limestone sonde is recorded instead of the long normal.

The *induction-electrical survey* (Figs. 19-27 and 19-29), a simultaneous recording of induction (conductivity and resistivity) curves, 16 in. normal, and SP curve, is the most modern combination for the logging of formations of low to moderate resistivities in fresh muds.

Fullest benefit of the above combinations is usually derived if a microresistivity survey is also run. *Microresistivity surveys* generally include a microcaliper curve (hole-diameter recording) (see Figs. 19-27 and 19-29).

In wells drilled with very saline mud, the *salt-mud combination* (e.g., Fig. 19-19) is used. It consists of a gamma-ray log (which distinguishes shales from nonshales), a laterolog or guard-electrode log, and a microlaterolog, with the recommended addition of a neutron or sonic log for the determination of porosity.

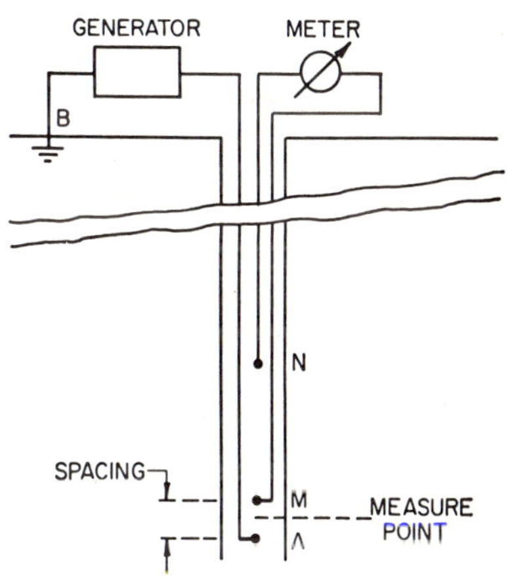

FIG. 19-9. Normal device (schematic).

Conventional Resistivity Devices

Principles: Normal and Lateral Devices. In conventional resistivity logging a current of known intensity is sent between two electrodes A and B (A on the sonde, B on the sonde or at the surface), and the resulting potential difference is measured between other electrodes M and N. The apparent resistivity is proportional to the measured potential difference.

For normal devices, the distance AM is small (1 to 6 ft) in comparison with MN, MB, and BN. In practice N or B may be placed in the hole at a large distance above A and M (Fig. 19-9). The voltage measured is practically the potential of M (due to current from A), referred to an infinitely distant point. The distance AM of a normal device is its *spacing*. The point of measurement is midway between A and M.

For lateral devices, measuring electrodes M and N are close to each other and located several feet below current electrode A. Current-return electrode B is at a

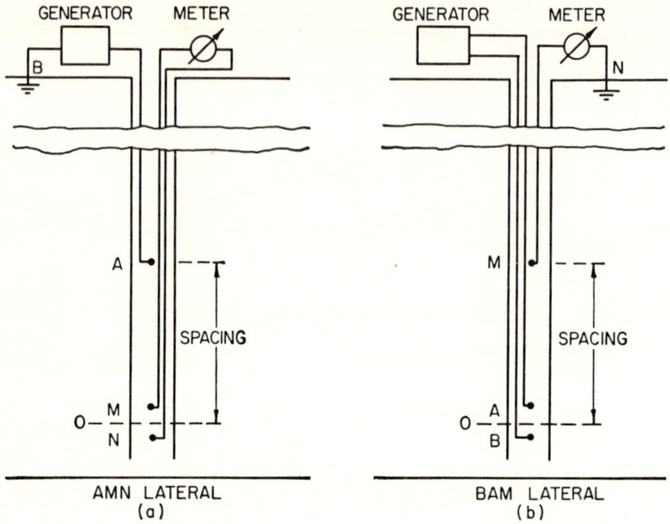

FIG. 19-10. Lateral device (schematic).

great distance above A or at the surface. The voltage measured is approximately equal to the potential gradient at the *point of measurement* O, midway between M and N. The distance AO is the *spacing of the lateral device.* The two arrangements shown in Fig. 19-10 (in which current and measuring electrodes are interchanged) are equivalent as regards measured potentials (and resistivities).

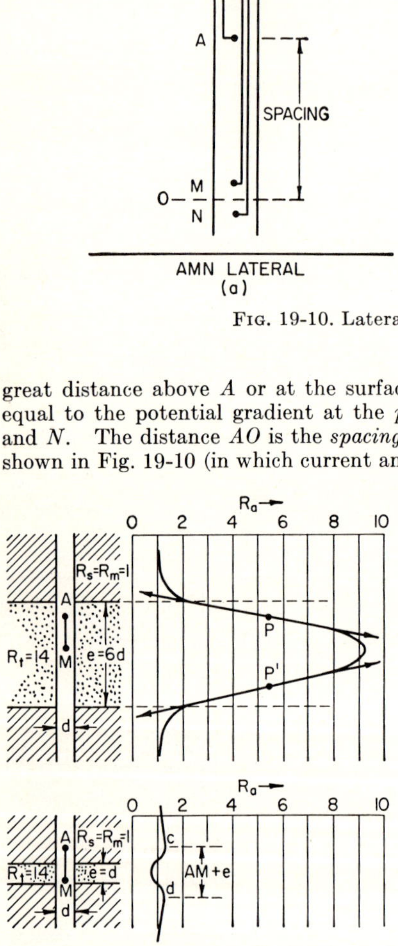

FIG. 19-11. Laboratory curves for normal sonde of spacing $AM = 2d$ through uninvaded beds more resistive than adjacent formations.

Curves Shapes—Laboratory Results. Figure 19-11 shows laboratory curves from a normal device for homogeneous resistive layers between adjacent beds of low resistivity. The curves are symmetrical with respect to the center planes of the layers. The same curves are recorded if M is above A instead of, as in the figure, A above M.

The upper part of Fig. 19-11 shows a resistive bed thicker than the spacing (bed thickness $e = 6d$; spacing $AM = 2d$; d = hole diameter). At the boundaries of the bed the curve tends to be rounded off owing to the influence of the drill hole. Moreover, the indicated bed thickness (distance between the inflection points P and P') is less than the actual thickness. Normal curves tend to show resistive beds thinner than they actually are (and conductive beds thicker than they actually are) by an amount equal to the spacing AM. The error in picking the boundaries of thick resistive beds is small for short-spacing normals—one reason for the recording of a short normal.

As shown in the lower part of Fig 19-11, for a resistive layer thinner than the spacing the curve shows a depression opposite the layer with two symmetrical small peaks c and d on either side. The main disadvantage of the normal device is that beds thinner than the spacing, no matter how resistive they may be, appear on the log as being conductive.

Figure 19-12 shows similar curves for a lateral. The lateral curves are markedly dissymmetrical, and their features are more complex. Again the transitions in the curves at the boundaries have been rounded off by the effect of the drill hole.

When the bed is thicker than the spacing, the upper boundary of the bed is not well defined on the lateral curve, and as a whole, the bed appears as being displaced downward by a distance equal to the spacing AO.

In the lower part of Fig. 19-12 the lateral indicates a resistive layer thinner than the spacing by a sharp peak of relatively low apparent resistivity. Below the layer is a low-resistivity "blind zone," followed by a "reflection peak" at a distance AO below the bottom boundary of the layer. The blind zone is recorded when the resistive streak is located between the current electrode and the measuring electrodes.

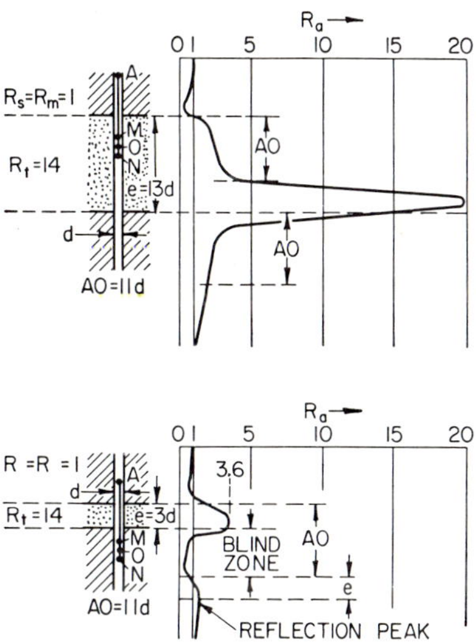

FIG. 19-12. Laboratory curves for lateral sonde of spacing $AO = 11d$ through uninvaded beds more resistive than adjacent formations.

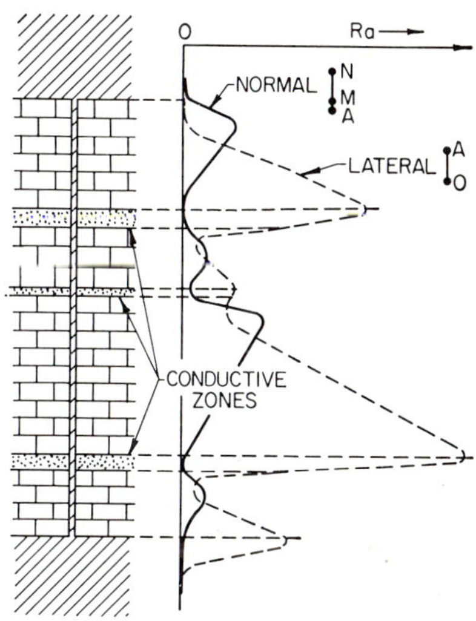

FIG. 19-13. Responses of normals and laterals in hard formations (qualitative).

The lateral is useful for the location of thin, highly resistive streaks, although interpretation may be difficult if several resistive streaks are close together. A lower streak located in the blind zone of an upper resistive streak may be missed, and the reflection peaks may be mistaken for actual resistive streaks in the formation. For a resistive layer of thickness approximately equal to the spacing (*critical thickness*), the lateral is almost completely flattened.

Similar generalizations are possible for lateral curves recorded for beds more conductive than the surrounding formations. Whether the layer is thick or thin, the shape of the curve is dissymmetrical; and the anomalies are spread downward, outside the bottom boundaries. The apparent increase of bed thickness is roughly equal to AO.

Normals and Laterals in Hard Formations. Figure 19-13 shows schematically the behavior of the normals and laterals in thick, highly resistive formations containing porous or shaly (that is, more conductive) zones.

In a highly resistive formation most of the current from electrode A flows up or down the borehole, dividing in inverse proportion to the resistances of the two paths, which are determined mostly by the resistance of the mud column in the hole between the current electrode and the nearest conductive beds. At the conductive beds, depending on their thickness and conductivity, the current has low-resistance paths from the hole. The lopsided appearance of the normal and lateral curves is explainable in terms of the unequal division of current flowing up and down the hole.

The normal, for example, has M and N above the current electrode. The voltage measured is the ohmic potential drop in the hole due to current flowing in the mud between M and N. When the device is near the bottom of a resistive bed, most of the current flows down to the conductive bed just below, and there is little potential drop between M and N because the current up is small. When the device has moved farther up in the bed the current down decreases because the resistance of that path has increased. Also, since the resistance of the upward path has decreased, the current up increases. Therefore, the potential drop between M and N increases as the device moves upward until electrode N reaches the next conductive bed, where the upward current is diverted from the hole. Above that level the normal reading decreases.

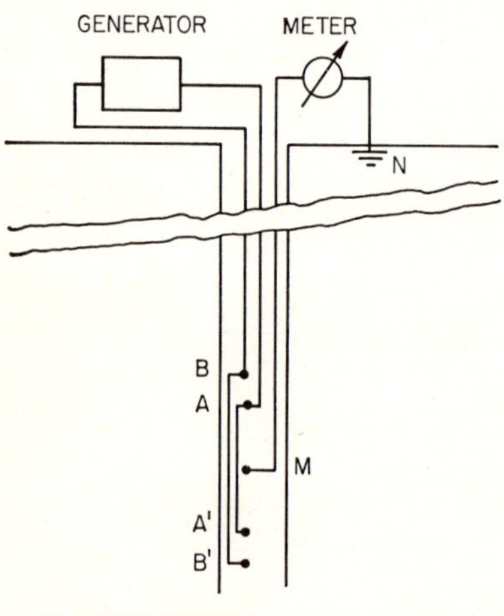

FIG. 19-14. Limestone sonde (schematic).

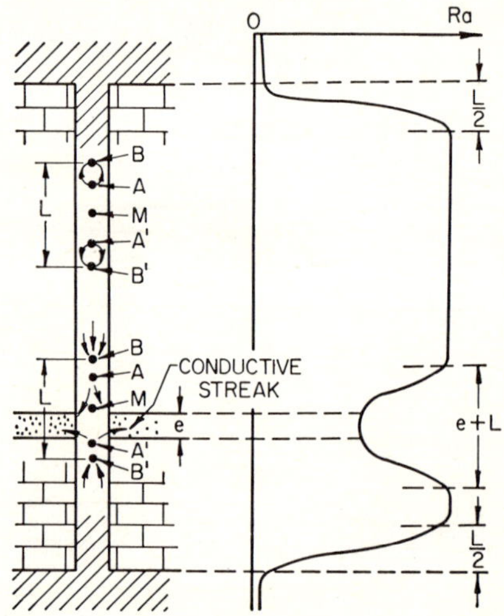

FIG. 19-15. Principle of limestone sonde (schematic).

The explanation of the shape of the lateral curve is similar. The direction of the lopsidedness for either device depends on whether the measuring electrodes are above or below the current electrode.

The depressions read on the curves opposite the conductive beds are smooth and, in the case of the lateral, much broadened and displaced downward. Accurate determinations of bed boundaries from the curves are practically impossible.

The Limestone Sonde. Four current electrodes A and A', B and B', connected as shown in Fig. 19-14 by means of insulated wires of negligible resistance, are symmetrically arranged so that $AB = A'B'$. A measuring electrode M is placed in the middle of the device. Depths are measured from electrode M. In practice $AM = A'M = 30$ or 35 in., and $AB = A'B' = 4$ or 5 in. The device is therefore a symmetrical double lateral.

Opposite a thick highly resistive layer (upper part of Fig. 19-15) practically all the flow of current is confined to the spaces between A and B and between A' and B'. No current flows from B or B', up or down the hole away from the device. Hence, from Ohm's law, B and B' are at zero potential. Similarly M is at the same potential as A and A'.

The potential of M is therefore equal to the potential drop in the mud, due to the flow of current, between A and B (or A' and B'). As long as all the electrodes of the devices are opposite the resistive formation this potential difference is dependent only on hole size and mud resistivity; if these are constant, a constant apparent resistivity is recorded.

If the device is located just above a conductive streak (as in the lower part of Fig. 19-15), the streak is effectively a low resistance connecting adjacent portions of the device to points at zero potential. Part of the current now flows in the paths indicated by the arrows, and the potential of electrode M is correspondingly decreased. The conductive streak is indicated on the log by a relatively sharp, symmetrical depression.

The limestone sonde gives clearer and simpler logs in hard formations, but measurements with the limestone sonde are strongly affected by the mud column. When the formations are much more resistive than the mud, the readings are appreciably lower than the formation resistivities.

Application of Conventional Resistivity Logs. The three devices which are generally recorded in the conventional electrical log (16 in. normal, 64 in. normal, and 18 ft 8 in. lateral) were designed to provide the most complete information with a system of nonfocused macrodevices.

The short normal is well adopted for bed definition, boundary determination, and correlation of formations of low or moderate resistivities (sand-shale series). The lateral generally shows sharp peaks at the level of thin resistive beds, but the definition of these beds is often obscured by the presence of blind zones and spurious peaks.

The precision of the normal and lateral curves for bed definition is limited in hard formations and is quite poor when salty muds are used. Somewhat better resolution is obtained with the limestone sonde.

In all cases, formation delineation is more detailed and accurate with the induction log and focused devices (laterolog), and with the microdevices.

In hard formations the 16-in. normal and the limestone sonde can provide an approximation to the value of R_i and hence an approach for formation-factor evaluation.

The capabilities of the conventional tools for the determination of R_t are discussed on page 19-33.

Induction Logging

Induction logging[8] is a method wherein the conductivity or its reciprocal, the resistivity, of the formations is measured by means of induced currents without the help of electrodes.

Principle of Induction Logging. The bottom-hole instrument comprises essentially a transmitter coil, fed with a constant alternating current from an oscillator, and a receiver coil connected through an amplifier to the measuring galvanometer at the surface (Fig. 19-16).

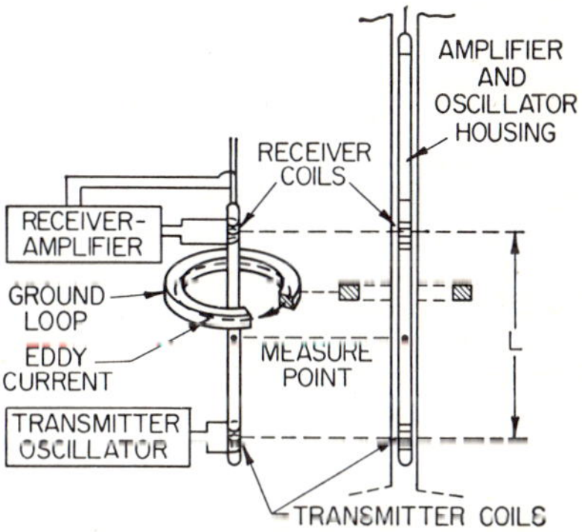

FIG. 19-16. Induction-logging equipment (schematic).

The alternating magnetic field set up by the transmitter coil induces eddy currents in the earth surrounding the coil. These eddy currents in turn have their own magnetic field, which induces an emf in the receiver coil. The magnitude of the eddy currents is proportional to the conductivity of the earth and so also is the emf induced in the receiver coil. Accordingly, the voltage recorded by the galvanometer at the surface is a determination of the conductivity of the surrounding earth.

In addition to the two "main coils" referred to above, sondes used in practice include several additional coils whose characteristics and positions are chosen to minimize as much as possible the influence on the measurements of the mud column and of the formations above and below the instrument. Also, the direct coupling between the transmitter and receiver coils is balanced out; i.e., if the coil system is surrounded by absolutely nonconductive media, as, for example, when it is suspended in the air high enough above the ground, the measuring apparatus will show a zero reading.

The recording equipment includes a reciprocal computer, making it possible to record a linear resistivity curve as well as the conductivity curve. On the induction-electrical survey, a 16-in. normal and an SP curve are also recorded (see Figs. 19-27 and 19-29).

The conductivity curve is recorded on a linear scale having its zero at the right side of the film; deflections to the left correspond to increasing conductivity. (In terms of resistivity units this curve would have a nonlinear scale which would tend to emphasize variations in the low resistivities but compress the variations in the high-resistivity range.) The unit of conductivity is the millimho per meter (mmho/m) (page 19-3).

The most usual type of instrument in the field at the present time is the 5FF40 sonde (five-coil sonde, main coils 40 in. apart).

Geometrical Factor—Skin Effect. The apparent conductivity σ_a, indicated by an induction-logging sonde, depends on the conductivities of all the media in the vicinity of the sonde. Assuming axial symmetry and negligible interaction between the currents of the different ground loops, σ_a is related to the conductivities of the media through *geometrical factors*, which depend on the geometry of the media and their location relative to the sonde. Thus, for a thin permeable bed,

$$\sigma_a = G_m\sigma_m + G_i\sigma_i + G_t\sigma_t + G_s\sigma_s \quad (10)$$

where σ_m, σ_i, σ_t, and σ_s are, respectively, the conductivities of the mud, invaded zone, uncontaminated zone, and adjacent formations, and where G_m, G_i, G_t, and G_s are the corresponding *geometrical factors*. The G factors are fractions, which add up to unity for all space.[8]

If formation conductivities are high, the interaction between the currents in the different ground loops may no longer be negligible, and the geometrical factor response of Eq. (10) is only a good first approximation; the induction-log conductivity reading is then in general reduced. The reduction in reading is known as *skin effect** (Appendix of Ref. 16).

Properties of Induction Log. The delineation of different beds and the determination of their boundaries are easier and more accurate with the induction log than with the conventional logs. With the induction log, bed definition is excellent down to about a 4-ft bed thickness and is still adequate for thinner beds.

1. *Effect of the Mud Column.* The conductivity signal due to the mud column increases with increasing hole size and with decreasing mud resistivity. Furthermore, the proportion of the total signal contributed by the mud column becomes greater as the formation resistivity increases.

With the present 5FF40 tool, for usual hole diameters (10 in. or less), and usual muds ($R_m > 0.5$ ohm-m), the borehole effect can be neglected when the sonde is approximately centered in the borehole by means of appropriate spacers.

For the use of the induction log in salty muds, it is preferable that the ratio of formation resistivity to mud resistivity be not greater than about 50.

In wells filled with oil-base mud, the correction for the mud column is usually negligible except in the rather rare cases of very deeply caved beds.

2. *Effect of Bed Thickness.* Induction-log measurements made at the level of a given bed are much less affected by the presence of the adjacent formations than are conventional resistivity logs.

For a given bed thickness, the influence of the adjacent formations differs according to whether they are more or less resistive than the bed. This comes from the fact that the eddy currents induced in the earth have a tendency to circulate through the most conductive media. If the resistivity R_s of the adjacent formations is greater

* This "skin effect" should not be confused with a different phenomenon, bearing the same name, which refers to alterations in permeability in the vicinity of the drill hole caused by drilling and completion operations.

than the resistivity R_t of the bed, the eddy currents concentrate within the boundaries of the bed. The signal contributed by the adjacent formations is then comparatively small. Computations and experience show that the measurements with the 5FF40 induction log are practically unaffected if the thickness of the bed is 6 ft or more. Approximate charts have been computed for thinner beds (Fig. 6 of Ref. 16).

Very thin resistive streaks within more conductive beds affect the response of the induction log very little. Such resistive streaks (such as lime or lignite) cause important distortions on the conventional logs.

3. *Effect of Invasion. The induction log is at its best for the investigation of permeable beds when R_i is greater than R_t.* In this case, most of the induced currents tend to flow in the uncontaminated zone, and the contribution of the latter to the total signal is large. When D_i does not exceed 25 in. (as is usual in case of high porosities), the 5FF40 induction-log reading is practically equal to R_t. For deeper invasion, the contribution of R_t to the measurement is still quite important. As a result, the induction log is well adapted to the detection of water-bearing beds in the usual case of R_{mf} greater than R_w.

For oil-bearing formations in which S_w is greater than about 60 per cent, R_t most often is appreciably lower than R_{xo}, and there is no annulus present (page 19-8). Accordingly, the conclusions are the same as in the preceding paragraph.

For lower water saturations, it becomes necessary to take the effect of the annulus into account. Computations based on usual conditions show that, when the diameter of invasion does not exceed about 16 in., the presence of an annulus has practically no influence on the present 40-in. induction log, and the invaded zone itself has negligible influence. The induction log accordingly gives the value of R_t without correction. For deeper invasion, the effect of the annulus increases and reaches a maximum when D_i is equal to about 30 to 40 in., then decreases very slowly with increasing values of D_i.

Furthermore, R_{xo} is usually greater than R_t, and R_{an} (resistivity of the annulus) is always less than R_t. As a result, the effects of invasion and of the annulus tend to compensate for each other. In fact, the compensation in certain regions is such that the induction-log reading can usually be taken as giving the value of R_t in oil- and gas-bearing beds with sufficient accuracy for all practical purposes.

4. *Applications.* The induction-electrical log (a combination of 5FF40 induction log, 16-in. normal, and SP) is the most reliable available at the present time, particularly for thin beds, in wells drilled with comparatively fresh muds, provided R_{mf}/R_w is not too small (greater than about 4) and the resistivities to be measured are not exceedingly high. (Excellent accuracy is assured up to 50 ohm-m, with a reasonable accuracy up to 200 ohm-m.)

The correlation between the new induction-electrical log and the older standard electrical survey (page 19-15) presents no problem inasmuch as the short normal and SP curves are common to both.

The induction log–short normal combination has some limitations in hard formations, essentially because the short normal is affected by the borehole and by the adjacent formations. An induction log–short laterolog combination now under development is expected to be superior in this respect.

The induction log is furthermore appropriate for resistivity measurements in empty holes and in holes drilled with oil-base mud.

The Laterolog

The laterolog[11] is a resistivity-measuring method wherein the current is forced to flow radially through the formations in a sheet of predetermined thickness by means of an appropriate electrode arrangement and an automatic control system. The measurement involves a portion of ground of limited vertical extent; the measured value is practically unaffected by the mud column.

Principle of Laterolog. Laterolog devices in use are laterolog 7 and laterolog 3.

Laterolog 7 uses small electrodes, as schematically represented in Fig. 19-17. The device comprises one central electrode A_0 and three pairs of electrodes M_1 and M_2,

M_1' and M_2', A_1 and A_2 placed symmetrically with respect to A_0, the electrodes of each pair being respectively short-circuited. A current of constant and calibrated intensity is sent through electrode A_0. Through electrodes A_1 and A_2 is fed additional current of the same polarity, the intensity of which is automatically and continuously adjusted in such a way that the difference of potential between the short-circuited pairs M_1M_2 and $M_1'M_2'$ is maintained practically equal to zero (i.e., there is no ohmic drop between M_1 and M_2 or between M_1' and M_2', hence no current flow between them). The potential of any one of these four M electrodes is measured for the resistivity determination.

Since the current from A_0 is prevented from flowing upward past electrodes M_1 and M_1' or downward past electrodes M_2 and M_2', it is therefore obliged to flow outward within a horizontal slice of formation whose thickness is about equal to the distance separating the mid-point O_1 of M_1M_1' from the mid-point O_2 of M_2M_2'. With most present equipments, O_1O_2, the width of the current sheet, is 32 in.

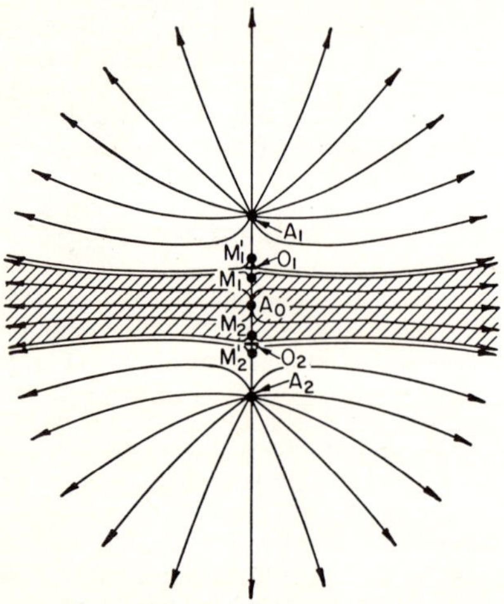

FIG. 19-17. Laterolog 7 electrode arrangement, showing current lines in a homogeneous medium. (Shaded area represents the sheet of measuring current.)

The distance the current travels across the mud is very small, and *the mud column has very little influence on the measurements except in the case of deep cavings.*

Laterolog 3 consists of a central small electrode, which emits a constant calibrated current, and long electrodes above and below, which are used to achieve focusing somewhat as in laterolog 7. A smaller beam width can be obtained; so that this device is useful for studying very thin beds.

Effect of the Invaded Zone. In permeable beds, the current has to cross the invaded zone before it reaches the uncontaminated zone. Thus the potential difference representing the apparent resistivity measured is the sum of the ohmic drops across each of these zones. When invasion is deep, as is usual in hard formations, the effect of the invaded zone on the apparent resistivity can become important.

Departure curves computed for thick beds,[38] giving the value of the true resistivity R_t corresponding to the apparent resistivity R_a from the laterolog 7, the invasion diameter D_i, and the invaded-zone resistivity, may be condensed into a few formulas for quick use. For all practical purposes these formulas also apply to the laterolog 3 in present use.

$$\begin{aligned} &\text{For } D_i = 20 \text{ in.:} && R_t = 1.25R_a - 0.25R_{xo} \\ &\text{For } D_i = 40 \text{ in.:} && R_t = 1.66R_a - 0.66R_{xo} \\ &\text{For } D_i = 80 \text{ in.:} && R_t = 2.5R_a - 1.5R_{xo} \end{aligned} \tag{11}$$

In the above formulas, R_{xo} can be obtained from the microlog or microlaterolog, or derived from the neutron log or the sonic log. The value of D_i has to be estimated from local experience.

Since R_{xo} and R_t are respectively proportional to R_{mf} and R_w, the effect of the invaded zone depends directly on the value of the ratio R_{mf}/R_w.

When R_{mf}/R_w is large (fresh muds and highly saline formation waters), R_{xo} is appreciably greater than R_t for water-bearing beds, and about equal to R_t in the case of oil-bearing beds. If invasion is deep, most of the resistance offered to the flow of current is located in the invaded zone; and the laterolog readings in each type of bed may not differ greatly. In this case it may be difficult to distinguish between oil and water zones.

When R_{mf}/R_w is small (fresh muds and fresh formation waters, or salty muds and saline waters), R_{xo} is close to R_t in the case of a water-bearing bed, and less than R_t in the case of an oil-bearing bed. In this case, the contribution of the invaded zone is comparatively less important, and the laterolog can readily distinguish between oil and water zones.

The laterolog is, therefore, widely used in those regions where salty muds are commonplace—a large part of the hard-formation territories. Such conditions are most unfavorable for the application of conventional logging devices.

This does not mean, however, that the use of salt-saturated muds should be recommended in order to be able to run the laterolog. Such muds have some undesirable features, among the most important of which is the reduction of the SP to almost a straight line, of no value for locating permeable beds. Taking all factors *into account,*

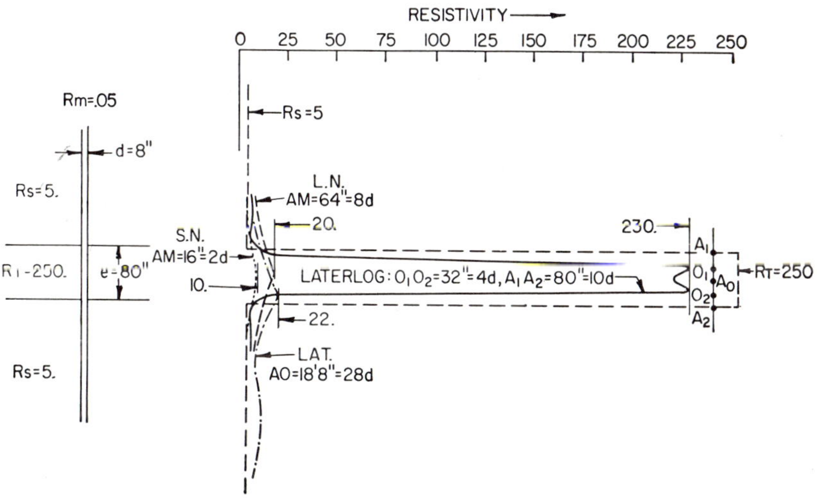

FIG. 19-18. Responses of laterolog and conventional devices in very salty mud opposite a resistive noninvaded bed (laboratory determination).

the optimum mud resistivity should be about two to four times that of the formation water. In fact, *this rule holds for all electrical-logging devices except the induction log.*

Effect of Bed Thickness. Because of its focusing system, the laterolog is better adapted to the investigation of thin beds than the conventional devices.

Figure 19-18 shows a typical example of curves recorded in the laboratory opposite a thin resistive noninvaded bed, penetrated by a borehole filled with conductive mud. The hole diameter d is 8 in. The bed thickness e is $10d$, or 80 in. The beam width of the laterolog device is $4d$ or 32 in., with A_1A_2 equal to 80 in. The bed resistivity R_t is 250 ohm-m, the mud resistivity R_m is 0.05 ohm-m, and the resistivity R_s of the adjacent formations is 5 ohm-m.

The curves recorded with a short normal, $AM = 2d$ (16 in.); a long normal, $AM = 8d$ (64 in.); and a lateral device, $AO = 28d$ (18 ft 8 in.), are also shown for comparison.

Figure 19-18 shows clearly that the laterolog indicates the boundaries of the bed much more sharply than the conventional devices. Moreover, the apparent resistivity read with the laterolog in the bed is 230 ohm-m compared with the R_t of 250 ohm-m, whereas the conventional devices show maxima of about 10, 20, and 22 ohm-m, respectively.

Applications. The laterolog has been accepted in hard-rock territories as an essential tool for the definition of beds, for correlation, and for reservoir evaluation. For the determination of R_t it is preferable that R_{mf}/R_w be small (less than 4).

High-salinity Connate Waters. 1. Salty muds—in this case R_{mf}/R_w is close to 1. The laterolog is the basic tool for the determination of R_t and saturation. The auxiliary tools are the microlaterolog and the gamma-ray log (Fig. 19-19), with the recommended addition of a neutron log or sonic log for the determination of porosity.

2. Fresh muds—the ratio R_{mf}/R_w is generally high. The laterolog is recommended as the auxiliary tool to be run in conjunction with the induction log and microlog. In this case the induction log is the basic tool for R_t determination.

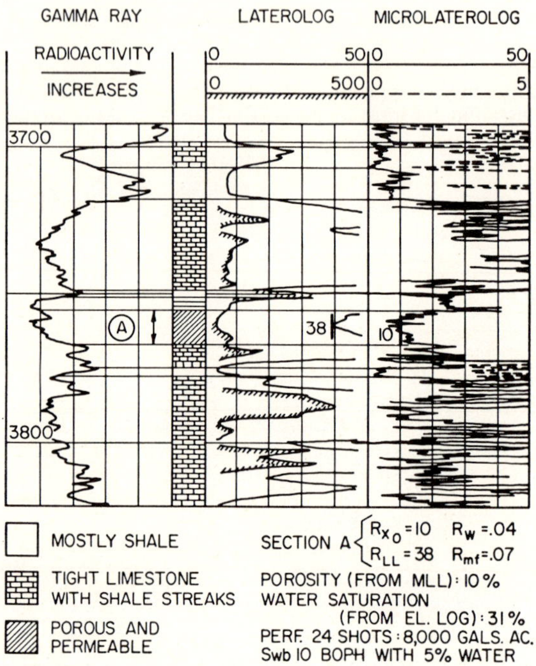

FIG. 19-19. Salt-mud logging combination, showing productive horizon at *A*.

Fresh Connate Waters. In this case the muds in general are also fresh and the ratio R_{mf}/R_w is low, a favorable condition for using the laterolog.

The Guard-electrode Device

In the guard-electrode system,[32] the surveying current flows into the adjacent formations from a measuring electrode disposed between relatively long upper and lower guard electrodes from which current also flows. The guard electrodes tend to confine the current from the measuring electrode to a generally horizontal path. The measuring and guard electrodes are connected through a very low impedance, as necessary to measure the surveying current supplied to the measuring electrode.

A resistivity value is obtained by recording the ratio of the voltage of an electrode in the assembly (referred to a distant point) to the current emitted from the measuring electrode.

The guard-electrode device is used mostly in hard-rock territories for detailed bed definition, correlation, and as a help in reservoir evaluation. For the determination of R_t it is preferable that R_{mf}/R_w be small (less than 4), as in the case of salty muds.

The Micrdisplay-Microcaliper Log

A micrologt[9] is a resistivity log recorded with electrodes mounted at short distances from each other in the face of a rubber pad. The insulating pad is pressed against the wall of the drill hole, thus reducing the short-circuiting action of the mud. The electrodes measure the average resistivity of a small volume of material in front of the pad.

The contact log[49] and the microsurvey are similar to the microlog.

Figure 19-20 shows a microlog pad, of a type in present field use, mounted on a powered microsonde.

A constant current is sent through electrode A, and the potential differences are measured (1) between electrodes M_1 and M_2 and (2) between electrode M_2 and a reference electrode N at the surface. These two measurements are referred to respectively as the "microinverse" and "micronormal." The volume of ground involved in the measurement is smaller for the microinverse than for the micronormal; in other words, the microinverse has a smaller "radius of investigation."

Most microresistivity sondes include a second "backup" pad on the opposite side of the sonde. Simultaneously with the microlog recording the distance between the faces of the two pads is also recorded. *The resulting microcaliper log is a detailed and accurate record of hole diameter*, which serves to indicate the presence of caved sections and to verify the presence of mud cake, indicative of permeable beds.

A micro SP may also be recorded using an electrode on the backup pad.

The microlog sonde may also be used to make an *in-hole measurement of mud resistivity* R_m. This is done while lowering the sonde into the hole collapsed to a minimum pad-to-pad extension. Since the electrodes are then generally at a distance from the hole wall, the readings are predominantly influenced by the mud resistivity, particularly opposite deeply caved sections. After reaching bottom, the pads are extended so that they press against the hole wall, and the microlog-microcaliper measurements are performed coming up.

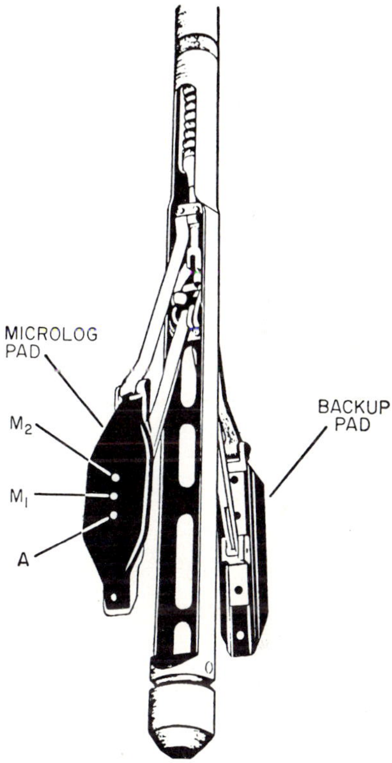

FIG. 19-20. Powered microsonde carrying microlog pad.

Porous and Permeable Beds. 1. When the pad is applied to a porous and permeable bed, the mud cake represents an important proportion of the small volume involved in the measurement. The other part of the volume is a portion of the flushed zone of resistivity R_{xo} [where R_{xo} is related to the porosity through Eqs. (6) and (2)]. For porosities of interest, the measured microresistivities do not usually exceed a maximum of fifteen or twenty times the resistivity of the mud.

2. Because of its smaller depth of investigation, the microinverse is more influenced by the mud cake than the micronormal, and the two readings are therefore usually different. The difference between them is called *separation*. Since in most cases the R_{xo} is greater than R_{mc} the micronormal gives a higher apparent resistivity than the microinverse, and the separation is said to be "positive."

Accordingly, *porous and permeable beds* are usually characterized by

Comparatively low microresistivity readings, not more than 20 times the mud resistivity

Positive separation between the microinverse and the micronormal

Confirmation of presence of mud cake (e.g., hole diameter less than bit size) on the microcaliper

3. The "positive-separation" rule should be used with caution when the mud cakes are very thin, as is generally the case with highly saline muds. In this case, the separation is comparatively small.

When the formation is salt-water-bearing and the depth of invasion is so small that the volume involved in the resistivity measurements includes a portion of the formation beyond the flushed zone, then, since the mud cake is more resistive than the formations behind, the micronormal reading will be lower than the microinverse—in other words, the separation is negative. Such shallow invasion prevails when low-water-loss mud is used and when the porosity and vertical permeability are very high, as in many places on the Gulf Coast.

4. The effect of the mud cake tends to level out the resistivity readings within a permeable bed; so that the curves do not show sharp variations, even when the formation factor changes appreciably.

5. The above criteria still apply to formations with well-distributed secondary porosity. However, in fissured formations where the matrix is not flushed, there is no mud cake and the microlog may not show positive separation.

6. The microlog can detect permeability even if very low—provided it is sufficient for the mud cake to build up (i.e., down to a fraction of 1 millidarcy). A bed can be indicated by the microlog as permeable when its permeability is too low for commercial production.

Tight Sections. In tight beds the electrode system is separated from the formation by only a thin mud film, perhaps $\frac{1}{16}$ in. thick or less on the average. The microlog readings accordingly are very high, equal to $20R_m$ or greater. Since the thickness of the mud film is not constant, because the wall of the hole is not perfectly smooth, the paths offered for the current to escape toward the mud column may be quite variable. As a result, the curves show numerous sharp peaks and depressions—and the separation may be positive or negative.

Shales. When the shales are not caved the pad again is separated from the formations by only a thin mud film. The reading is equal to or, more often, smaller than the resistivity of the shales. The separation may be negative, zero, or slightly positive. This behavior of the microlog opposite shales is not entirely explained, although experiments have shown that negative separation may be due partly to the anisotropy of the shales (page 19-6).

The additional information furnished by the SP curve and/or the gamma-ray curve may serve to resolve any ambiguity resulting from the response of the microlog opposite shales. The microcaliper is also extremely helpful here since a decrease in hole diameter is likely to correspond to a mud cake, hence to a permeable bed; furthermore increased hole diameter due to caving is usually seen in shale sections.

Hole Enlargements. Caved sections of hole are mostly encountered in shale beds, but they may also occur in other types of formations. In deeply caved holes, the pad generally does not apply to the wall, and both microlog curves are likely to indicate the same resistivity, equal to or close to the mud resistivity. However, if the cave is not too deep, the readings may still be affected by the formations, and since these are usually more resistive than the mud, a positive separation is observed; and again a confusion with permeable streaks is possible. Of course, the presence of caving is quickly seen by reference to the microcaliper. The SP and/or gamma-ray log may also assist the lithological interpretation in this case.

Determination of R_{xo} from the Micrology. Except in the case of very shallow invasion, the investigation of the microlog does not extend beyond the flushed zone. The measured resistivities are functions of the resistivity R_{xo} of the flushed zone, the hole diameter d, the resistivity R_{mc} of the mud cake, and the thickness t_{mc} of the mud cake. An interpretation chart (Fig. 19-21) prepared from laboratory measurements provides for the determination of R_{xo} and t_{mc} from two microlog readings when R_{mc} is known (page 19-4). The t_{mc} found can be checked against the microcaliper.

Depending on the character of the mud cake, reliable determinations of R_{xo} can be made from the microlog for porosities greater than about 12 to 15 per cent.

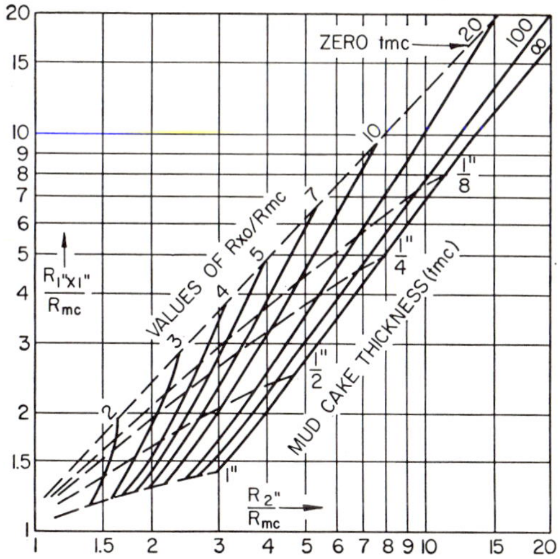

FIG. 19-21. Interpretation chart for 8-in.-hole series C microlog for which adjacent electrodes are 1 in. apart. $R_{2in.}$ designates reading of 2-in. micronormal (AM_2 = 2 in.) and $R_{1\times1in.}$ is 1- by 1-in. microinverse (AM_1 = 1 in., M_1M_2 = 1 in.). Type I hydraulic pad, noninsulated sonde.

Where it can give R_{xo} with sufficient accuracy, the microlog is also valuable when it is necessary, in the determination of water saturation, to correct the readings of the macrodevices.

The Microlaterolog

The microlaterolog device,[12] as presently used, comprises one center electrode A_0, of very small size, and three circular (ring) electrodes M_1, M_2, and A_1, concentric with A_0 and spaced with narrow gaps (about ½ to 1 in.) between successive rings [Fig. 19-22 (left)]. These electrodes are embedded in an insulating pad which is applied against the wall of the borehole.

A current of constant and known intensity is sent through the center electrode A_0. Another current of the same polarity, fed through the outer ring electrode A_1, is automatically adjusted in such a way that the difference of potential between rings M_1 and M_2 is kept practically zero. The potential then existing at either M_1 or M_2 is recorded as an indication of resistivity.

Current flowing from the center electrode A_0 is prevented by the current emitted by the ring electrode A_1 from diverging freely in all directions and is confined to a trumpet-shaped beam as indicated by the crosshatched area of Fig. 19-22 (right). That part of the formations located beyond about 3 in. from the wall has little influence on measurements with the present instrument.

In permeable formations there will be a mud cake in front of the pad. Since the current beam crosses the mud cake perpendicularly (i.e., via the shortest path) and also since the resistivity of the mud is usually less than the formation resistivity, the influence of the mud cake on the measurement is small compared with that of the formation and may be negligible for all practical purposes if the mud-cake thickness is not too great. It is recalled, by way of contrast, that for the microlog the mud cake provides a return path for a part of the current toward the mud column; consequently the mud cake affects the microlog measurements more than it does the microlaterolog —particularly for formations with low porosity (see Fig. 19-23). In case of thin mud cakes, the resistivity obtained with the microlaterolog is essentially R_{xo}, the flushed-zone resistivity.

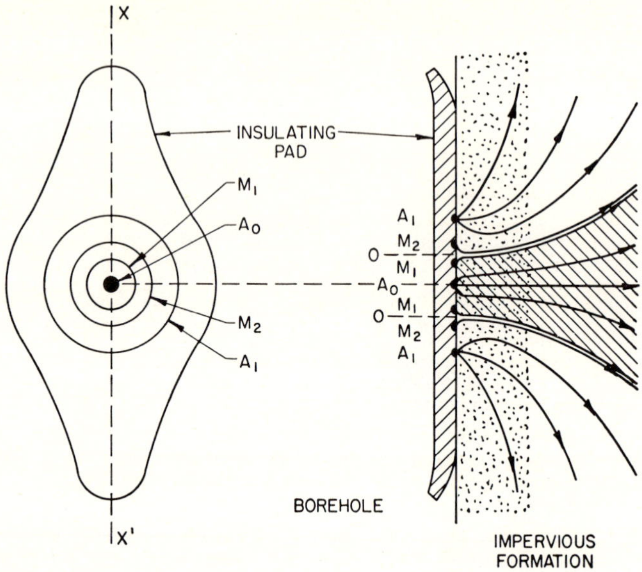

FIG. 19-22. Microlaterolog pad, showing electrode distribution (left), and section showing schematically the current lines (right). The shaded area represents the beam of measuring current.

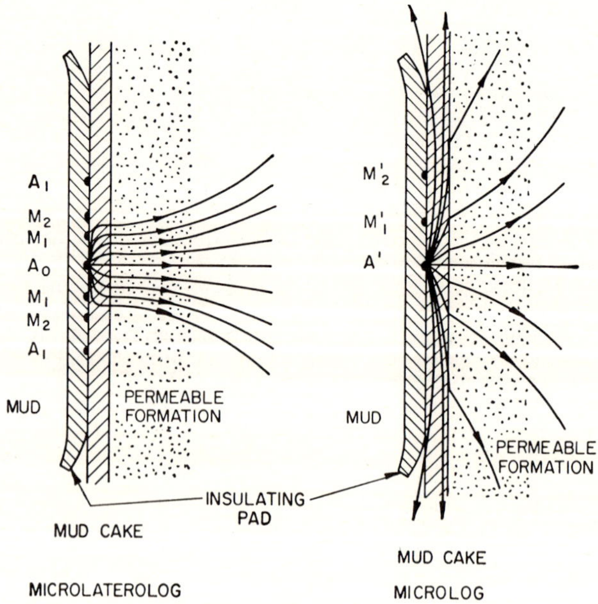

FIG. 19-23. Comparison of paths followed by surveying current for microlog and microlaterolog in case of a porous and permeable formation more resistive than the mud cake (qualitative).

Present equipment records a single microlaterolog curve with an electrode system referred to as type A (spacing between successive electrodes = $\frac{9}{16}$ in.). A microcaliper is also recorded.

Interpretation and Applications. Because the current beam has a very small diameter (about 2 in.), the resolution of the microlaterolog is very great. The recorded curves can give an accurate and detailed definition of the boundaries for very thin beds.

Tight Formations. Since tight formations usually do not cave much, the pad is seated against them, and very high readings are obtained.

Shales. When there is no caving, the pad is applied directly to the formation. Because of the focusing system, the presence of a thin mud film between the pad and the wall has no influence on the measurement of the shale resistivity.

As in the case of the microlog, a caved hole may result in curves similar to those for porous beds. In such cases, the SP, the gamma-ray, or the microcaliper curves can aid the interpretation.

Porous and Permeable Formations. 1. *Mud-cake thickness less than $\frac{3}{8}$ in.* Laboratory tests and field experience have proved that *the type A microlaterolog gives R_{xo} without correction when the mud-cake thickness is less than about $\frac{3}{8}$ in.*

Thin mud cakes are observed in wells drilled with highly saline muds. The microlaterolog accordingly is one of the components of the salt-mud combination (page 19-15) which is used extensively in hard-rock territories, where highly saline muds are common.

2. *Mud-cake thickness greater than $\frac{3}{8}$ in.* The effect of a thick mud cake is, in this case, no longer negligible, and resistivities recorded with the present type A device are generally lower than R_{xo}. Correction charts (Chart C-14 of Ref. 41), established in the laboratory, give R_{xo} from the apparent resistivity R_{MLL}, when the mud-cake resistivity R_{mc} and mud-cake thickness t_{mc} are known.

The value of t_{mc} can be obtained, in many instances, from the microcaliper log, but this must be done with caution. The difference between the nominal hole diameter, given by the drill-bit size, and the actual hole diameter shown by the microcaliper is equal to twice the mud-cake thickness *only* if the formation considered has not caved. Holes drilled in highly consolidated formations, such as limestones, are generally true to gauge, but softer formations frequently cave.

New Developments

Three paths of future development of the electrical log are obviously open and are being vigorously investigated at the present time.

The first of these is the *further development of interpretation techniques of the presently available logs.* This is a subject of constant study.

The second is *further combination of various tools to provide for simultaneous recording.* As additional services are combined, their effectiveness will be increased and their cost decreased.

The third and possibly the *most important is the trend toward the use of tools which minimize the effect of the borehole, invaded zone, and adjacent beds.* The laterolog 7, laterolog 3, guard-electrode device, and 5FF40 induction log have already proved their great advantages over the conventional resistivity devices. It is expected that further advances will be made with the introduction of improved systems now being field-tested.

At the time of writing, a deep-investigation induction log, also combined with the 16 in. normal and SP curve, is being introduced into field practice. This new tool has a radius of investigation about twice as great as that of the 5FF40 sonde, while its vertical resolution is about the same.

A new device, the proximity log, for the measurement of R_{xo} under usual conditions of mud-cake thickness is now becoming available. Also, a laterolog with a shorter radius of investigation, which can be recorded in combination with the induction log, is being field-tested.

USES AND INTERPRETATION OF ELECTRICAL LOGS

Bed Detection and Definition

Formations encountered in uncased boreholes may be detected and their boundaries defined by a number of different logs.

The SP and short-normal curves are most commonly used. In hard formations, better definition can be obtained by using the induction log and/or the laterolog. The long lateral is very useful for detecting thin beds more resistive than the surrounding formations.

For great detail, the microdevices are superior—microlog in fresh muds and microlaterolog in very salty muds.

A substitute for the SP in salt muds is the gamma-ray log, which distinguishes shales from nonshale beds. Also, the sonic log could be used in all types of formations, with any type of mud.

In holes filled with nonconductive muds, or in empty holes, the induction, radiation, temperature, sonic, and perhaps section-gauge logs are the only ones which can provide much useful information. Sometimes conventional devices using wall scratchers, for contact with the formation, may be run.

Porous and Permeable Beds—Sand Count. Porous and permeable beds are of primary interest since they are the potential oil and gas reservoirs. A *sand-count* determination of the total effective thickness of a permeable section, excluding shale streaks and other impermeable zones, can be derived from electrical logs.

Fresh Muds. The SP and microlog are the principal curves for locating and defining permeable beds. The SP has a good resolving power in formations of low and moderate resistivity. In very resistive formations it can still detect shales and permeable beds, but it cannot define their boundaries accurately.

The microlog is best in all types of formations for the detailed location of the boundaries of permeable beds.

The microcaliper helps in the location of permeable beds because it can usually detect the mud cakes, particularly if the adjacent formations are approximately at bit size. Shales, on the other hand, generally tend to cave or erode more than the permeable or hard impervious beds.

Very often, and mostly in sand-shale series, the difference between the readings of a short-investigation and a long-investigation device (e.g., a short normal vs. a long normal or vs. the induction log) is a clear indication of the presence of a permeable bed.

Salt Muds. The combination of very salty mud with moderate to high formation resistivity affects adversely the conventional (nonfocused) devices and the SP curve. Also, mud cakes are thin; as a consequence the microcaliper (or also the section gauge) is not of great help for the detection of permeable sections but detects clearly the caved shales.

In conjunction with the laterolog and microlaterolog the gamma-ray log distinguishes between shales and nonshales, and the neutron, sonic log, and gamma-gamma density log provide indications of porosity.

Correlation

The process of correlating two (or more) logs in different wells depends essentially upon the similarity in shapes or outlines of the curves (Fig. 19-24). In some regions such correlations may be easily made for wells miles apart; in other cases (where serious faulting, lenticular deposits, or unconformities are present) it may be difficult to correlate logs from wells only a few hundred feet apart. The knowledge of formation dip from dipmeter determinations assists in correlation.[43]

Correlations are facilitated when the curves show characteristic "markers" such as, for example, a well-known chalk or shale section. Thin resistive beds such as lignites frequently furnish valuable correlation points.

For long-range correlation, the devices which investigate large volumes of forma-

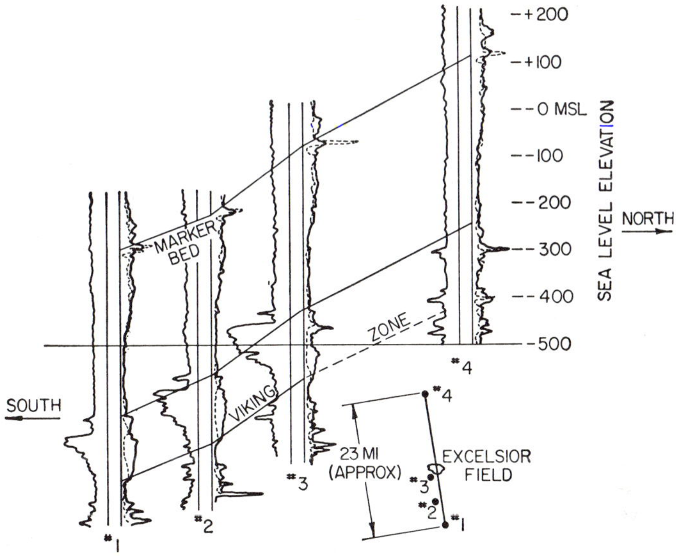

FIG. 19-24. Example of long-distance correlation.

tion are best. However, when the pay zones are known in a field, they can be correlated in great detail from well to well by comparing the micrologs or microlaterologs; thus the geometry of the reservoir can be determined.

Investigation of Porosity

Evaluation of formation porosity from electrical logs is based on determination of the formation factor F, which is related to porosity ϕ by Eq. (2).

1. F can be found from the value of R_{xo} as determined from the microlog (page 19-26) or microlaterolog (pages 19-27 to 19-29) using Eq. (6) rewritten as

$$F = (R_{xo}/R_{mf})S_{xo}^2 \tag{12}$$

where $S_{xo} = 1 - \text{ROS}$

In order to determine R_{xo} from microresistivity measurements, it is of course necessary that invasion be deep enough that the measurements are not affected by the formation resistivities beyond the flushed zone. An exceedingly great mud-cake thickness may also limit the accuracy of the R_{xo} determination (page 19-29). For reliable porosity interpretation from microlog data the porosity should be greater than about 12 to 15 per cent.

R_{mf} is determined as indicated on page 19-4.

In an oil-bearing sand, the residual oil saturation (ROS) must be surmised for use in Eq. (12). Taking a value of 20 per cent for the ROS would not usually entail too great an error in most cases, at least in formations with granular porosity and containing light oils. Greater residual oil saturations are frequent, however, chiefly in the cases of highly viscous oil. Gas-bearing formations also seem to display high residual (gas) saturations in the flushed zone if the permeability is high, because of segregation effects brought about by a combination of gravity and capillary forces. Because of these residual saturations, the microlog may show water contacts.

2. To check the results of the microlog, or to replace the microlog in case it is not available, F may be determined as equal to R_o/R_w, R_o being the true resistivity of the formation at a level where it is 100 per cent water-bearing. Extrapolation of a value of F determined at a water-bearing level to the oil-bearing section within the

same formation implies that the lithological characteristics are approximately constant, which is not always the case, however.

3. Still another approach often used in hard formations is based on the value of R_i, the average resistivity of the invaded zone estimated from the readings of the limestone sonde or the short normal.[45,47] This procedure has proved valuable in many cases.

Other Tools for Obtaining Porosity. Coming into increased importance as auxiliaries to the electrical logs for improved pososity determination are the sonic log,* the neutron log,* and the very new gamma-gamma density log.

Furthermore, with the present *large sample taker,* reliable determinations of formation factor and porosity are possible from side-wall cores.

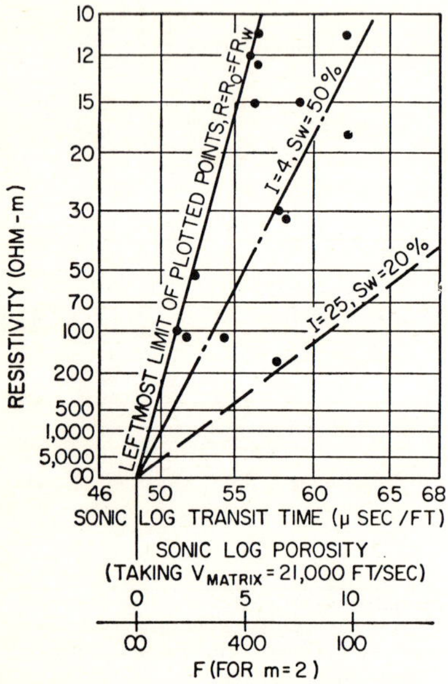

FIG. 19-25. Qualitative method for differentiating water-bearing and oil- or gas-bearing zones by plotting deep-investigation resistivity values against sonic-log transit times. The ordinate scale is proportional to $(1/R)^{1/2.15}$.

Investigation of Fluid Content

Quite often good qualitative judgments as to whether a formation is a potential producing zone can be had by direct visual inspection of the log. Basically such judgments usually depend on

1. Identification of a permeable formation by means of the SP, mud-cake indications on microcaliper, positive separation on microlog, indication of invasion by separation between shallow- and deep-investigation macroresistivity curves, etc.

2. Indication by the deep-investigation resistivity devices that R_t in the permeable formation is appreciably larger than the resistivity R_o that the formation would be expected to have if water-bearing.

When it can be assumed that there is no abrupt change in the salinity of the formation waters, the radius of invasion, etc., qualitative evaluations of saturation may be made by comparing a porosity log with a resistivity log over a large enough section of formations. Such a method, in which the neutron log was the source of the porosity data, was proposed several years ago.[55] Recently, the possibility of accurate measurements of porosity with the sonic log has prompted similar qualitative interpretation procedures.[3,48]

When invasion is not too deep, so that the deep-investigation resistivity devices read fairly close to R_t, the following procedures are applicable:

1. Sonic transit time may be plotted vs. induction-log resistivity (in case of fresh muds) or vs. the deep laterolog resistivity (in case of salty muds). By a proper choice of scales, lines of constant resistivity index I (page 19-6) become straight lines on the chart, as in Fig. 19-25. The line of 100 per cent water saturation ($I = 1$) is constructed as the line bounding the leftmost plotted points and passing through the point corresponding to infinite resistivity and zero sonic-log porosity. Lines for other values of I may then be constructed. It is possible to distinguish between oil- or gas-bearing and water-bearing zones by the relative positions of the plotted points. For a zone to be oil- or gas-productive the plotted point should fall appreciably to the right of and below the line of 100 per cent saturation. This method is useful even if the formation-water resistivity is not well known.

* The short-interval sonic velocity log (sonic log) is discussed in Chapter 21. The neutron log is discussed in Chapter 20.

2. Using porosity derived from the sonic log and the resistivity from the deep-investigation log, *apparent* values of the formation-water resistivity may be computed [by direct substitution into Eqs. (1) and (2)]. When tabulated in terms of depth, those apparent values which stand definitely above the average trend indicate the presence of saturation. This procedure is rapid. It is applicable in shaly sands. It is also applicable when formation-water salinity varies appreciably with depth, provided the variation is slow and continuous.

3. The readings of a short-investigation resistivity log should give indications of the variations in the formation factor (and porosity). Accordingly saturation can also be qualitatively determined by examining the ratios of the readings of a short-investigation device (such as a short normal, short laterolog, or proximity log) to those of the induction log. The two sets of readings may be used to locate points on a graphic plot, as in (1) above, or to construct a continuous curve of the ratio vs. depth. In either case the comparison is facilitated if the logs have been recorded on a logarithmic sensitivity scale.[14]

The following procedure is helpful when, because of deep invasion, the value of the true resistivity is in doubt.

The porosity from the sonic log is compared with the porosity computed from a shallow-investigation resistivity log (such as short normal, limestone sonde, short laterolog or proximity log). The latter porosity is affected by any residual oil saturation, whereas the sonic-log porosity is not. Zones where the two porosities differ should accordingly indicate potentially productive formations.

Also, in hard formations, where permeability is generally low, a resistivity gradient, corresponding to a saturation gradient, is observed on the logs between the zone of lowest water saturation and the level of 100 per cent water saturation. The existence of this gradient is the basis of a method for the delineation of intervals saturated with hydrocarbons.[44]

Quantitative Interpretation

Quantitative determinations of the hydrocarbon saturation from electrical logs are essentially determinations of the water saturation S_w in the uncontaminated zone. Recalling Eq. (5):

$$S_w = \left(\frac{R_0}{R_t}\right)^{1/2} = \left(\frac{FR_w}{R_t}\right)^{1/2} \tag{5}$$

The evaluation of R_t is a major step in the determination. Evaluation of F is equivalent to an evaluation of porosity, as already discussed. R_w is obtained as indicated on pages 19-3, 19-4, and 19-14, or R_o may be determined in a 100 per cent water-saturated section of the same sand or in a lithologically similar sand.

Some interpretation procedures do not require explicit evaluation of R_t and F. Along with a knowledge of the static SP, or equivalently the ratio R_{mf}/R_w, they are based on the ratios of readings taken with deep- and shallow-investigation devices (e.g., R_{xo}/R_t method, induction–electrical log interpretation, Rocky Mountain method). A special approach is applied in case of shaly formations.

Determination of R_t. The conventional resistivity logs can provide for the value of R_t only for thick beds, since the readings in general cannot be reasonably corrected for the effect of adjacent formations.

The long lateral, because of its large radius of investigation, is practically unaffected by the mud column and the invaded zone and gives a good approximation to R_t when the beds are reasonably uniform over a thickness of at least 30 ft. Such favorable conditions are very seldom met in practice.

The long-normal reading usually requires a correction for the borehole effect. This correction is sufficiently accurate in fresh mud. The reading is close to R_t when the bed thickness is at least 10 to 15 ft and invasion is shallow. The effect of deep invasion can be accounted for, to some extent, by means of departure curves.[36,39]

Correction for invasion requires the help of other devices with shorter radii of investigation (microlog and/or short normal), which provide values of the invaded-zone resistivity.

2. The advantage of the induction log and the laterolog is that under usual conditions of application their readings are practically unaffected by the mud column or by the adjacent formations for bed thicknesses greater than about 5 to 6 ft.

The induction log, with the present technique, is appropriate for logging in fresh mud. Its accuracy is excellent for formations reading up to 50 ohm-m and reasonable up to 200 ohm-m. Above that figure the accuracy is not so good. The induction log gives R_t when the invasion diameter does not exceed 15 to 25 in. and can be corrected for deep invasion. The auxiliary readings are the microlog and/or the short normal, or better, the laterolog, particularly in case of consolidated formations.

The laterolog is the essential tool for the logging of hard formations in salty mud. It gives R_t for shallow invasion (D_i less than 15 to 25 in.). Correction for invasion is made with the help of the microlaterolog.

With all the logging tools available *at present*, the correction for invasion implies some reasonable assumption, based on local experience, as to the depth of invasion. When invasion is very deep, correction for invasion becomes difficult. A deep-investigation induction log, which will minimize the effect of deep invasion in the evaluation of R_t, has recently been introduced into field practice.

Maximum Producible Oil Index

If the porosity ϕ of a formation is uniform and intergranular, the quantities $S_{xo}\phi$ and $S_w\phi$ represent the *amounts* of water per unit of bulk volume present in the pores of the flushed and uncontaminated zones, respectively. The difference $Y = (S_{xo} - S_w)\phi$ is the amount of oil per bulk volume displaced by the mud filtrate. Y is called the *maximum producible oil index* because it approximates the maximum amount of oil per bulk volume producible with water drive. Y is given approximately by the equation[15]

$$Y = \left(\frac{R_{mf}}{R_{xo}}\right)^{1/2} - \left(\frac{R_w}{R_t}\right)^{1/2} \tag{13}$$

Evaluation of Y by this equation does not require a direct knowledge of porosity ϕ, formation saturation S_w, or flushed-zone saturation S_{xo} (or residual oil saturation ROS).

Other Interpretation Techniques

It is impossible, in the space available, to present all log-interpretation techniques or to present all charts and tabulations made up therefor. For the practicing log interpreter most of the charts needed are included in Ref. 41. Additional information may also be found in Refs. 22, 34, 37, 40, 50, and 56.

The R_{xo}/R_t Method. For clean formations, saturation can be computed from the empirical relationship found by combination Eqs. (5) and (6),

$$S_w = S_{xo}\left(\frac{R_{xo}}{R_t}\,\frac{R_w}{R_{mf}}\right)^{1/2} \tag{14}$$

Any method which gives correct values for R_{xo}, R_t, R_{mf}, and R_w may be used to determine S_w, provided S_{xo} is known or can be reasonably surmised.

The chart of Fig. 19-26 provides a convenient way for solving Eq. (14). The ratio R_{mf}/R_w is entered in abscissas (upper scale) and the ratio R_{xo}/R_t in ordinates. From the point so plotted, an oblique line is extended up to its intersection with the edge of the chart; from this intersection, a horizontal line is drawn which gives the values of S_w for different values of S_{xo}.

The value of S_w found directly by interpolation between the two nearby oblique lines is based on the assumption that S_{xo}, on the average, is related to S_w through the empirical equation $S_{xo} = \sqrt[5]{S_w}$.

Within the limits where $R_w \approx R_{we}$ (i.e., R_w between 0.08 and 0.3 ohm-m at 75°F), the SSP (static SP) can be entered, instead of R_{mf}/R_w, using the lower grid.

Shaly-sand Method. The amplitudes of both the SP deflection and the resistivities are reduced by the presence of interstitial shale in a formation. In some extreme cases shaly oil sands are hardly differentiated from the adjacent shales on the logs. The microlog furthermore does not show much separation between the curves. Interpretations in shaly sands, therefore, are generally made under comparatively unfavorable conditions. The practical method of interpretation described hereafter is quite approximate and is applicable only when the amount of shaliness is not too great.

SATURATION DETERMINATION

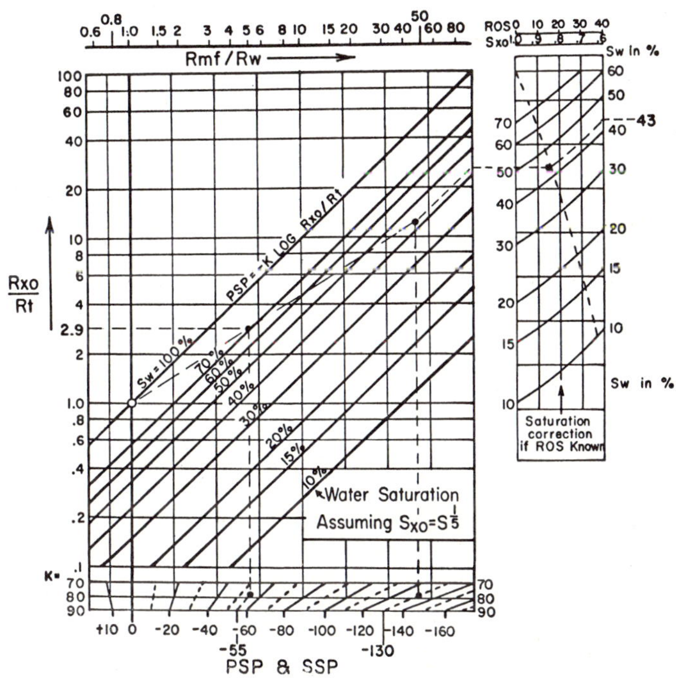

FIG. 19-26. Interpretation chart for R_{xo}/R_t and shaly-sand methods. (Working lines illustrate example of Fig. 19-27.)

The practical method of interpretation in shaly formations[35,*] is based on the observations made on field logs that in 100 per cent water-saturated shaly sands the PSP (pseudostatic SP) is given by

$$\text{PSP} = -K \log \frac{R_{xo}}{R_o} \tag{15}$$

For an interlaminated sand-shale formation containing oil and/or gas, the PSP can be expressed by the equation

$$\text{PSP} = -K \log \frac{R_{xo}}{R_t} - 2\alpha K \log \frac{S_{xo}}{S_w} \tag{16}$$

where α is the SP reduction factor and is defined as PSP/SSP (page 19-12).

* Some other shaly-sand references are Refs. 6, 10, 22, and 33.

For clean sands $\alpha = 1$ and Eq. (16) reduces to Eq. (15) as a special case.

According to Eq. (16) the determination of S_w in a shaly sand requires the knowledge of R_{xo}, S_{xo}, R_t, PSP, and SSP. The SSP can be determined from clean sands reasonably close to the shaly formations considered. The PSP is given by the deflection of

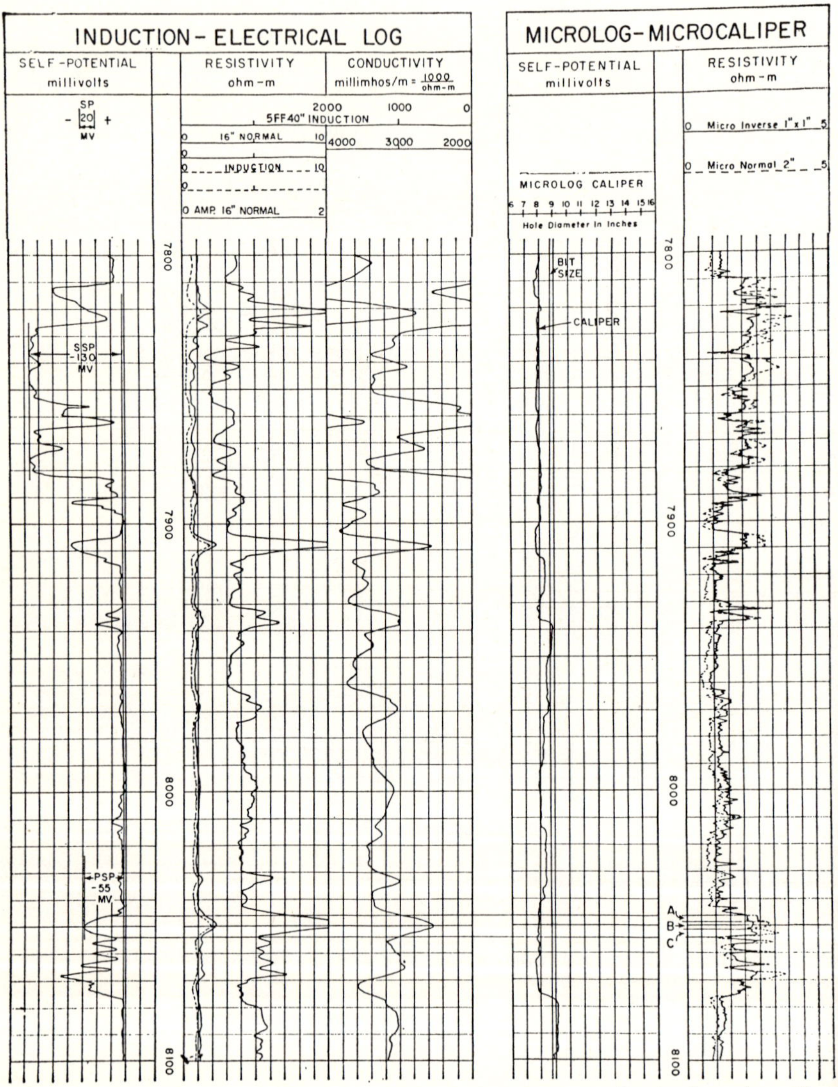

FIG. 19-27. Shaly-sand example: induction-electrical survey and microlog-microcaliper survey in well on Louisiana Gulf Coast. $R_m = 1.45$, $R_{mf} = 1.2$, and $R_{mc} = 1.3$ at BHT (140°F). $d = 8\frac{3}{4}$ in.

the SP curve at the level of the bed, after correction for bed thickness, if necessary. The value of S_{xo} (= 1 − ROS) again has to be surmised.

The solution of the equation for S_w is given by Fig. 19-26, which, it is seen, applies to both clean and shaly formations. For a shaly sand, the PSP and R_{xo}/R_t are first entered in abscissa and ordinate, respectively. This gives us the apparent saturation. To find the true saturation, a line is drawn through the origin (circled), and the point representing the apparent saturation; and this line is extended until it intercepts the SSP or the corresponding value of R_{mf}/R_w. The saturation at this point is treated in the same manner as in the R_{xo}/R_t method above.

If the SSP is not known and R_w is unobtainable, it is not possible to determine the water saturation S_w. Nevertheless, plotting the ratio R_{xo}/R_t vs. PSP on Fig. 19-26 will show whether or not the sand falls on the 100 per cent water-saturation line. If it falls some distance below this line, there is a chance for production.

It must also be remembered that in sand-shale laminations the value of S_w is the water saturation of the sand itself. In sands containing disseminated shale, it is the water bound by the quartz grains and does not include the water held by the colloids.

Example. Figure 19-27 shows the induction-electrical log and the microlog over a portion of a well in Jefferson Davis Parish, Louisiana. The shaly sand from 8,046 to 8,054 ft will be interpreted.

The mud resistivity at formation temperature of 140°F is 1.45 ohm-m, R_{mc} = 1.3 ohm-m, and R_{mf} = 1.2 ohm-m at 140°F. Bit size is 8¾ in.

From the induction-electrical log, the PSP is −55 mv. The SSP (sand at 7,830 to 7,850 ft) is −130 mv. Thus α (= PSP/SSP) is 0.42. The short-normal resistivity $R_{16in.}$ is 2.2 ohm-m, and the induction-log resistivity R_{IL} is 1.9 ohm-m.

Invasion is known to be quite shallow, as further substantiated by the readings of the microlog. Therefore, R_{IL} may be taken as equal to R_t.

As shown by the positive separation on the microlog, three distinct porous intervals A, B, and C appear between 8,046 and 8,054 ft. Using Fig. 19-21, the values of R_{xo} and t_{mc} for each interval are found. Thus, for interval A R_{xo} is 4.0, for interval B it is 5.6, and for interval C it is 6.8. The average R_{xo} is 5.5, and for all three intervals the mud-cake thickness is ½ in. Note that the microcaliper indicates a hole diameter of 7¾ in., exactly 1 in. smaller than bit size, corresponding to a ½-in. mud cake, thus verifying the t_{mc} values from Fig. 19-21.

All the necessary data are now at hand: R_{xo} = 5.5 ohm-m. R_t (= R_{IL}) = 1.9 ohm-m. PSP = −55 mv. SSP = −130 mv. α = 0.42. K (for the SP) at 140°F is close to 79 (page 19-10). R_{xo}/R_t = 2.9, and R_{xo}/R_{mf} = 4.6.

Figure 19-26 is entered with R_{xo}/R_t = 2.9 and PSP = −55 at K = 79, locating a point on the chart. Then a line is drawn through this point from the origin and extended until it intersects the vertical line erected at SSP = −130 mv for K = 79. This intersection gives a value of S_w equal to 43 per cent, assuming that ROS is 15 per cent, which is reasonable.

The sand was perforated from 8,046 to 8,050 ft and made 75 bbl of oil and 280,000 cu ft of gas per day through a 7⁄64-in. choke.

The Induction-electrical Log Interpretation Method. A chart for practical interpretation of fluid saturation from the readings of the induction–electrical log combination is shown in Fig. 19-28[16] and pages D-11 and D-12 of Ref. 41. The conditions limiting valid use of the chart are

1. The invasion diameter D_i must be between $2d$ and $10d$ (d is the hole diameter).
2. R_{xo} must be greater than R_t. This is generally fulfilled if R_{mf}/R_w is greater than 3 to 5, as for fresh muds and saline formation waters.
3. The beds must be fairly homogeneous, i.e., the short-normal reading must not be perturbed by the presence of resistive streaks.

The chart incorporates approximate compensation for the presence of an annulus (page 19-8), assuming equal viscosities for the oil and water and an ROS in the flushed zone of 15 to 30 per cent.

To use the chart the value of the ratio $R_{16in.}/R_{IL}$ (resistivity from 16 in. normal divided by the induction-log resistivity), corrected, if necessary, for effect of borehole and adjacent formations, is plotted either vs. the static SP (lower grid) entered opposite the appropriate formation temperature, or vs. R_{mf}/R_w (upper scale).

Points falling within the shaded area correspond to water-bearing sands. Points falling below the shaded area correspond to oil saturation. Approximate saturation scales are provided for D_i values of $3d$ and $5d$. The dashed lines represent lines of equal saturation on the chart. The critical saturation corresponds to S_w = 60 per cent in soft formations and 50 per cent in consolidated sandstones. These may not be the correct critical saturations for many limestones.

For shaly formations a construction like that used in the shaly-sand method can be used in Fig. 19-28. Although not rigorously correct, this procedure should give acceptable results if the shaliness is not too great.

Since it is not always known beforehand whether the interpretation chart (Fig. 19-28) is within its limits of applicability, it is useful to employ the value of porosity

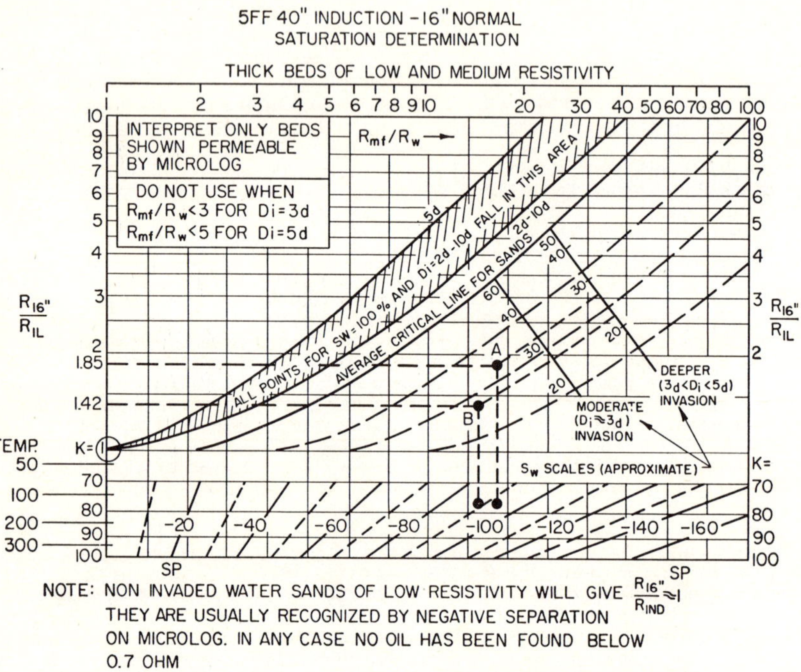

FIG. 19-28. 40-in. induction log, 16-in. normal interpretation chart. (Working lines illustrate example of Fig. 19-29.)

or formation factor, when known from independent sources, to check the results by means of Eq. (5). Details of this "porosity balance" check are given in Ref. 46.

Example. Figure 19-29 shows part of the induction-electrical log and microlog on a well in Young County, Texas. The sand at the bottom of the figure is divided into two parts, *A* and *B*, by a thin hard streak as shown on the microlog. *A* and *B* will be interpreted separately.

Necessary well data are R_m = 0.95 at BHT of 117°F; R_{mf} = 0.61 at BHT; hole diameter d is 7⅞ in.

Interval A. From the induction-electrical log, R_{IL} = 17.5 ohm-m, $R_{16in.}$ = 32.5 ohm-m, SP = −95 mv, and BHT is 117°F.

Applying Fig. 19-28, the ratio $R_{16in.}/R_{IL}$ = 32.5/17.5 = 1.85. The SP is −95 mv; so the point is fixed on the chart, giving an average S_w of about 30 per cent.

Interval B. Here the R_{IL} = 26.0 ohm-m and $R_{16in.}$ = 37.0 ohm-m. $R_{16in.}/R_{IL}$ = 37.0/26.0 = 1.42. The SP is −90 mv at 117°F. This fixes a point on Fig. 19-28 which gives a value of S_w of about 25 per cent.

Drill-stem tests on the two intervals showed that the sand contained gas and distillate, with a higher flowing pressure over interval *B*.

The Rocky Mountain Method. The Rocky Mountain method[45] (pages D-7 and D-8 of Ref. 41) was developed for the interpretation of conventional electrical logs to yield values of S_w and ϕ in invaded, clean hard formations when they are sufficiently

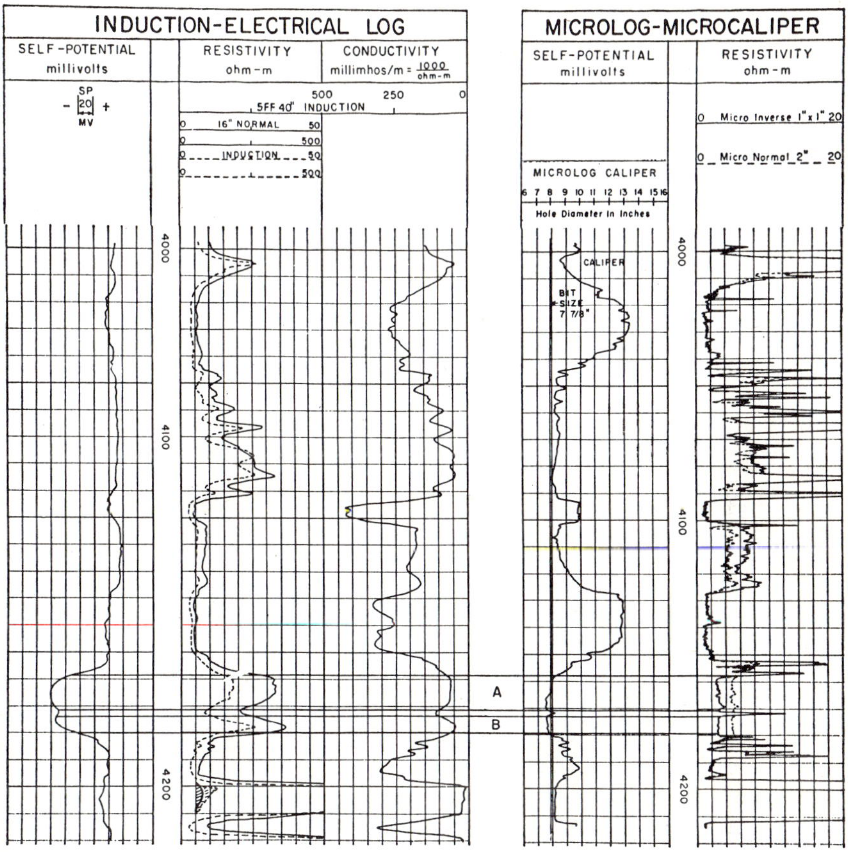

FIG. 19-29. Induction-electrical log and microlog-microcaliper log on a well in Young County, Texas. $R_m = 0.95$ and $R_{mf} = 0.61$ at BHT (117°F). $d = 7\frac{7}{8}$ in.

thick and homogeneous. It is necessary that the invasion be such that the average resistivities recorded by the short normal and the lateral (when corrected for borehole effect) approximate the average values of R_i and R_t, respectively. For this to be true, the short-normal readings should be at least ten times the mud resistivity. Only average values, over thick intervals, can be obtained with this method.

$$R_i = \frac{FR_z}{S_i^2} \tag{17}$$

where:

1. R_z is the resistivity of the water found in the invaded zone. It is usually made up of filtrate and connate water mixed in such proportion that

$$\frac{1}{R_z} = \frac{z}{R_w} + \frac{(1 - z)}{R_{mf}}$$

where z is the fraction of connate water in the total mixture (usually 5 to 10 per cent)

2. S_i is the water saturation in the invaded zone. It has been found by experience that $S_i^2 \approx S_w$; so that $R_i = FR_z/S_w$.

Inasmuch as $R_t = FR_w/S_w^2$, we obtain $S_w = (R_i/R_t)/(R_z/R_w)$.

Since, from above, $R_w/R_z = z + (1 - z)R_w/R_{mf}$, it follows that R_z/R_w is a function of the SP value. The upper part of Fig. 19-30 gives a graphical solution of the equation for S_w. The SP is entered in ordinate and the ratio R_i/R_t on oblique lines. The intersection gives the abscissa S_w.

The lower part of the chart is used to obtain the porosity by using the water saturation S_w, just found, with the value of R_t/R_w. The intersection falls on or between oblique lines that are graduated into porosity values, according to the Humble formula [Eq. (2), page 19-5].

This method should not be used in salt muds.

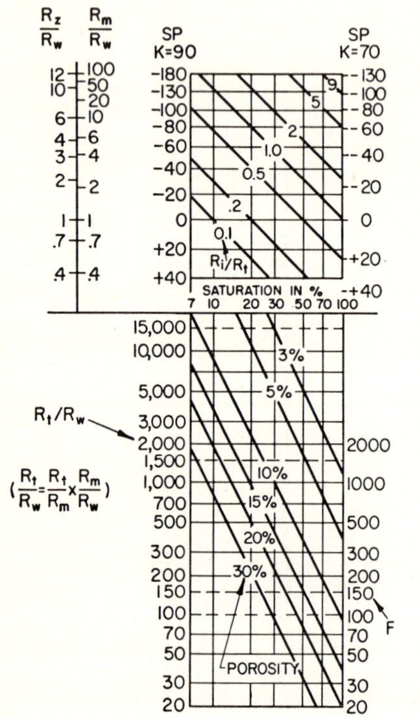

FIG. 19-30. Interpretation chart for Rocky Mountain method.

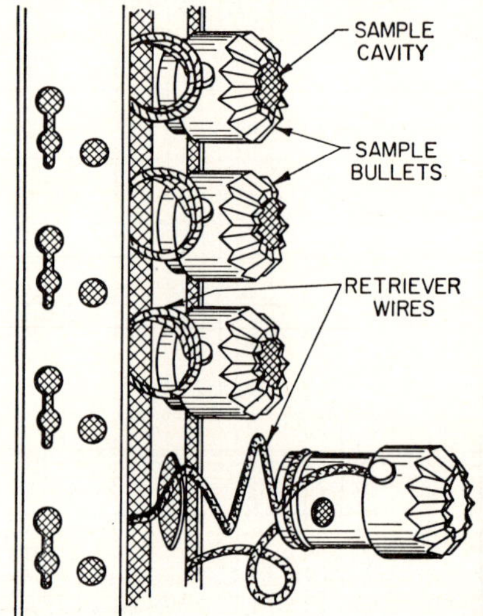

FIG. 19-31. Section of percussion-type side-wall sample-taker gun loaded with hard-formation sample bullets. Lower bullet is removed from gun to show means by which bullet is retrieved.

Side-wall Sampling and Testing—Important Auxiliaries

It is often desirable to supplement the electrical logs with data provided by samples and tests (information on permeability, rock texture, oil gravity, paleontological characteristics, formation pressure, etc.).

Wire-line side-wall samplers and testers, adapted for running on the cable of the logging truck, make such investigations possible immediately after the electrical logs have indicated the exact depths of the different beds of interest. Depths can be accurately pinpointed with reference to the electrical log by observation of an SP curve recorded during the sampling or testing run.

Side-wall Sample Taker (Fig. 19-31). Percussion-type side-wall sample takers[29] are in most general use. Not only can they be used to sample soft formations, but also, with the recent introduction of hard-formation bullets, they are now capable of taking samples from almost any kind of formation which can produce oil.

In the percussion side-wall sample gun, hollow cylinders, which serve as core barrels, are shot one at a time into the wall of the hole by means of electrically ignited powder charges. The cylinder remains attached to the gun by means of two lengths of wire.

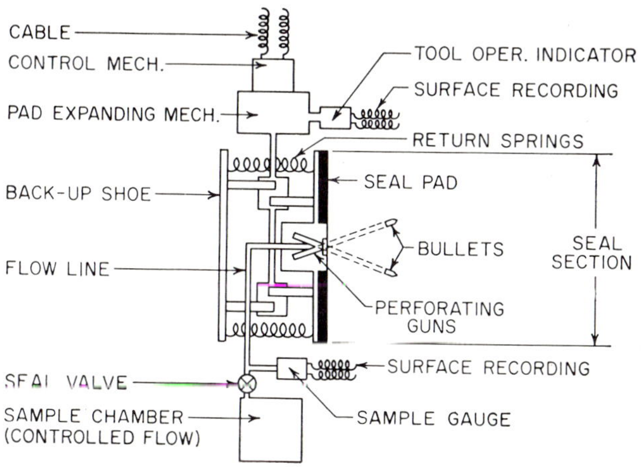

FIG. 19-32. Wire-line formation tester (schematic).

Thus, by applying a tension to the cable to which the gun is attached, the core-filled bullet may be pulled from the formation.

According to the particular gun used, percussion-type sample takers, on one trip in the hole, take, at selected depths, up to 30 cores ¾ to 1³⁄₁₆ in. in diameter. Core length is up to 2¼ in. maximum, depending on recovery.

Formation Tester (Fig. 19-32). The wire-line formation tester[28] can take samples of 1, 2.75, or 5.5 gal of formation fluid, depending on the sample chamber. It also records the formation shut-in pressure and the hydrostatic (mud) pressure.

The collapsed tester is lowered on the logging cable to the level of the zone to be tested. It is then expanded, forcing a pad tightly against the wall of the hole to form a seal separating the formation under the pad from the mud column. The formation is then perforated, and the formation fluids flow into a chamber in the tool. When the chamber is filled, a valve is closed, and the fluid sample is sealed in at maximum developed pressure. The pad and back-up shoe are then retracted and the tool is brought to the surface where the sample is removed for detailed study. Throughout the entire test, electric circuits permit a complete recording at the surface of the progress of the operation.

Over 60 per cent of tests attempted in soft formations are successful, and empirical rules have been worked out for predicting production on the basis of the fluids recovered. In hard formations some indication of formation permeability and static reservoir pressure can be derived from the test results by application of pressure build-up theory.

Table 19-1. Letter Symbols Used in Chapter 19*

a_{mf}	chemical activity of mud filtrate
a_w	chemical activity of formation water
d	hole diameter
D_i	electrically equivalent diameter of invaded zone
e	individual bed thickness
E_c	electrochemical component of the spontaneous electromotive force

Table 19-1. Letter Symbols Used in Chapter 19* (*Continued*)

E_k	electrokinetic (electrofiltration) component of the spontaneous electromotive force—usually neglected for practical purposes
E_J	liquid-junction emf—one component of E_c
E_M	shale emf—the other component of E_c
F	formation resistivity factor
I	resistivity index—equals R_t/R_o
K	coefficient in Eqs. (8) and (9), which give E_c
m	formation resistivity factor exponent (cementation factor) (see page 19-5)
n	saturation exponent [see Eq. (4)]
PSP	pseudostatic SP (of a shaly formation)
R	resistivity (in ohm-meters)
R_a	apparent resistivity—the reading of a given resistivity-logging device
$R_{16\text{in.}}$	apparent resistivity as given by the 16-in. normal
$R_{2\text{in.}}$	apparent resistivity as given by the 2-in. micronormal
$R_{1\times 1\text{in.}}$	apparent resistivity as given by the 1- by 1-in. microinverse
R_{IL}	apparent resistivity as given by the induction log
R_{MLL}	apparent resistivity as given by the microlaterolog
R_{an}	resistivity of the annulus
R_i	average resistivity of invaded zone
R_m	mud resistivity
R_{mc}	mud-cake resistivity
R_{mf}	mud-filtrate resistivity
R_o	resistivity of formation 100 per cent saturated with water of resistivity R_w
R_s	resistivity of adjacent formations (on either side of a bed under study)
R_t	true formation resistivity (in the uncontaminated zone)
R_w	resistivity of formation water
R_{we}	equivalent formation-water resistivity based on chemical activity [for insertion into Eq. (9), which gives SSP]
R_{xo}	resistivity of flushed zone, close to the wall of the hole
R_z	resistivity of the water in the invaded zone
ROS	residual oil saturation (in the flushed zone)
S	water saturation (usually expressed as a fraction of the pore volume, sometimes as per cent)
S_i	water saturation in the invaded zone
S_w	water saturation in the uncontaminated zone
S_{xo}	water saturation in the flushed zone
SSP	static SP (of a clean permeable formation)
t_{mc}	mud-cake thickness
z	the fraction of the formation water making up the mixture of filtrate and formation water in the invaded zone
Y	maximum producible oil index
α	SP reduction factor (of a shaly formation)—equal to PSP/SSP
σ	conductivity [the same subscripts as used with R (resistivity), to specify the meaning, may also be used with σ]
ϕ	porosity (expressed as a fraction of bulk volume)

* Although most of these symbols are from the ASME and AIME list of American Standard Symbols for Electrical Logging, some others, used in this chapter, have been added for the convenience of the reader.

REFERENCES†

1. Archie, G. E.: The Electrical Resistivity Log as an Aid in Determining Some Reservoir Characteristics. *Trans. AIME*, vol. 146, pp. 54–62, 1942. Also in *Pet. Tech.*, vol. 5, *Tech. Paper* 1422, 1942.
2. Archie, G. E.: Classification of Carbonate Reservoir Rocks and Petrophysical Considerations, *Bull. AAPG*, vol. 36, no. 2, pp. 278–298, February, 1952.
3. Berry, J. E.: Acoustic Velocity in Porous Media, *Trans. AIME*, vol. 216, 1959.
4. Campbell, W. M., and J. L. Martin: Displacement Logging—A New Exploratory Tool, *Trans. AIME*, vol. 204, pp. 233–239, 1955. Also in *J. Pet. Tech.*, vol. 7, no. 12, December, 1955.
5. de Witte, L., K. P. Fournier, and H. Tejada-Flores: Calculation of Guard Electrode Response Curves, *Geophysics*, vol. 22, no. 1, pp. 67–74, January, 1957.
6. de Witte, L.: A Study of Electric Log Interpretation Methods in Shaly Formations, *Trans. AIME*, vol. 204, pp. 103–110, 1955. Also in *J. Pet. Tech.*, vol. 7, no. 7, July, 1955.
7. Doll, H. G.: The S.P. Log: Theoretical Analysis and Principles of Interpretation, *Trans. AIME*, vol. 179, pp. 146–185, 1949. Also in *J. Pet. Tech.*, vol. 11, September, 1948.

† For a more general bibliography on wire-line well logging, see Ref. 40.

8. Doll, H. G.: Introduction to Induction Logging and Application to Logging of Wells Drilled with Oil-base Mud, *Trans. AIME*, vol. 186, pp. 148–162, 1949. Also in *J. Pet. Tech.*, vol. 1, no. 6, June, 1949.
9. Doll, H. G.: The Microlog—A New Electrical Logging Method for Detailed Determinations of Permeable Beds, *Trans. AIME*, vol. 189, pp. 155–164, 1950. Also in *J. Pet. Tech.*, vol. 2, no. 6, June, 1950.
10. Doll, H. G.: The S.P. Log in Shaly Sands, *Trans. AIME*, vol. 189, pp. 205–214, 1950. Also in *J. Pet. Tech.*, vol. 2, no. 7, July, 1950.
11. Doll, H. G.: The Laterolog—A New Resistivity Logging Method with Electrodes Using an Automatic Focusing System, *Trans. AIME*, vol. 192, pp. 305–316, 1951. Also in *J. Pet. Tech.*, vol. 3, no. 11, November, 1951.
12. Doll, H. G.: The MicroLaterolog, *Trans. AIME*, vol. 198, pp. 17–32, 1953. Also in *J. Pet. Tech.*, vol. 5, no. 1, January, 1953.
13. Doll, H. G.: Filtrate Invasion in Highly Permeable Sands, *Pet. Engr.*, vol. 27, no. 1, pp. B53–B66, January, 1955.
14. Doll, H. G., J. L. Dumanoir, and M. Martin: Present Trends in Electrical Logging, for presentation to SEG, Los Angeles, November, 1959.
15. Doll, H. G., and M. Martin: How to Use Electric Log Data to Determine Maximum Producible Oil Index in a Formation, *Oil and Gas J.*, vol. 53, no. 9, pp. 120–126, July 5, 1954.
16. Dumanoir, J. L., M. P. Tixier, and M. Martin: Interpretation of the Induction-electrical Log in Fresh Mud, *Trans. AIME*, vol. 210, pp. 202–217, 1957. Also in *J. Pet. Tech.*, vol. 9, no. 7, July, 1957.
17. Dunlap, H. F., and H. R. Hawthorne: The Calculation of Water Resistivities from Chemical Analysis, *Trans. AIME*, vol. 192, pp. 373–375, 1951. Also in *J. Pet. Tech.*, vol. 3, no. 3, p. 17, March, 1951.
18. Gondouin, M., M. P. Tixier, and G. L. Simard: An Experimental Study on the Influence of the Chemical Composition of Electrolytes on the S.P. Curve, *Trans. AIME*, vol. 210, pp. 58–72, 1957. Also in *J. Pet. Tech.*, vol. 9, no. 2, February, 1957.
19. Gondouin, M., and C. Scala: Streaming Potential and the S.P. Log, *Trans. AIME*, vol. 213, pp. 170–179, 1958. Also in *J. Pet. Tech.*, vol. 10, no. 8, August, 1958.
20. Hamilton, R. G., and P. Charrin: Quantitative Log Analysis, *Oil and Gas J.*, vol. 54, Nov. 12, Dec. 3 and 24, 1956; vol. 55, Feb. 18, Mar. 18, May 6 and 20, and June 3, 1957.
21. Hill, H. J., and A. E. Anderson: Streaming Potential Phenomena in S.P. Log Interpretation, *Trans. AIME*, vol. 216, 1959.
22. Jones, Park J.: Electric Log Resistivity, Porosity, and Water Saturation for Clean and Shaly Sandstones, *World Oil*, vol. 132, no. 2, pp. 140–148, February, 1951.
23. Keller, G. V.: An Improved Electrode System for Use in Electric Logging, *Prod. Monthly*, vol. 13, no. 10, pp. 12–15, August, 1949.
24. Keller, G. V.: Modified Mono-Electrodes for Improved Resistivity Logging, *Prod. Monthly*, vol. 14, no. 9, pp. 13–16, July, 1950.
25. Kerver, J. K., and C. L. Prokop: Effect of the Presence of Hydrocarbons on Well Logging Potential, SPE of AIME, Los Angeles meeting, October, 1957. Also in *Oil and Gas J.*, vol. 56, no. 50, pp. 102–106, Dec. 15, 1958.
26. Kunz, K., and J. Moran: Some Effects of Anisotropy on Resistivity Measurements in Boreholes, *Geophysics*, vol. 23, no. 4, pp. 770–794, October, 1958.
27. Lamont, N.: Relationship between the Mud Resistivity, Mud Filtrate Resistivity, and the Mud Cake Resistivity of Oil Emulsion Mud Systems, *Trans. AIME*, vol. 210, pp. 387–388, 1957. Also in *J. Pet. Tech.*, vol. 9, no. 8, pp. 51–52, August, 1957.
28. Lebourg, M., R. Q. Fields, and C. A. Doh: A Method of Formation Testing on Logging Cable, *Trans. AIME*, vol. 210, pp. 260–267, 1957. Also in *J. Pet. Tech.*, vol. 9, no. 9, September, 1957.
29. Leonardon, E. G., and D. C. McCann: Exploring Drill Holes by Sample-taking Bullets, *Trans. AIME*, vol. 132, pp. 141–155, 1939. Also in *J. Pet. Tech.*, vol. 2, no. 5, May, 1939.
30. Lipson, L. B., and H. L. Overton: The Effect of Treating Agents on the Electrochemical Activities of Drilling Mud Filtrates, SPE of AIME, Dallas meeting, October, 1957.
31. Martin, M., and K. Kunz: Why Those Low Lateral Readings? *Oil and Gas J.*, vol. 56, no. 6, pp. 88–92, Feb. 10, 1958.
32. Owen, J. E., and W. J. Greer: The Guard Electrode Logging System, *Trans. AIME*, vol. 192, pp. 347–356, 1951. Also in *J. Pet. Tech.*, vol. 3, no. 12, December, 1951.
33. Patnode, H. W., and M. R. J. Wyllie: The Presence of Conductive Solids in Reservoir Rocks as a Factor in Electric Log Interpretation, *Trans. AIME*, vol. 189, pp. 47–52, 1950. Also in *J. Pet. Tech.*, vol. 2, no. 2, February, 1950.
34. Pirson, S. J.: Formation Evaluation by Log Interpretation, *World Oil*, vol. 144, no. 5, pp. 159–168, April, 1957; no. 6, pp. 170–183, May, 1957.

35. Poupon, A., M. E. Loy, and M. P. Tixier: A Contribution to Electrical Log Interpretation in Shaly Sands, *Trans. AIME*, vol. 201, pp. 138–145, 1954. Also in *J. Pet. Tech.*, vol. 6, no. 6, June, 1954.

36. *a.* Schlumberger Well Surveying Corporation: Document 3, Resistivity Departure Curves, 1949 Issue.
b. Welex, Inc.: Interpretation Charts for Electric Logs and Contact Logs, *Bull.* A-101.

37. *a.* Schlumberger Well Surveying Corporation: Document 4, Interpretation Handbook for Resistivity Logs, 1951.
b. Welex, Inc.: Guyod's Electrical Well Logging, *Bull* A-132.

38. Schlumberger Well Surveying Corporation: Document 6, Departure Curves for Laterolog, August, 1952.

39. *a.* Schlumberger Well Surveying Corporation: Document 7, Resistivity Departure Curves (Beds of Infinite Thickness), 1955 Issue.
b. Welex, Inc.: Fundamentals of Quantitative Analysis of Electric Logs, *Bull.* A-120.

40. Schlumberger Well Surveying Corporation: Document 8, Introduction to Schlumberger Well Logging, 1958.

41. *a.* Schlumberger Well Surveying Corporation: Log Interpretation Charts, 1958 Issue.
b. Lane-Wells Co.: Electrolog Interpretation Manual.
c. Welex, Inc.: The Contact Log, *Bull.* A-124.
d. Welex, Inc.: Welex Radioactivity Logging with Series 2000 Tool, *Bull.* A-126.

42. Segesman, F., and M. P. Tixier: Some Effects of Invasion on the S.P. Curve, *Trans. AIME*, vol. 216, p. 138.

43. Stratton, E. F., and R. G. Hamilton: Application of Dipmeter Surveys, AIME meeting, Tulsa, Okla., October, 1947.

44. Tixier, M. P.: Evaluation of Permeability from Electric Log Resistivity Gradient, *Oil and Gas J.*, vol. 48, no. 6, pp 113–123, June 16, 1949.

45. Tixier, M. P.: Electric Log Analysis in the Rocky Mountains, *Oil and Gas J.*, vol. 48, no. 7, pp. 143–148, 217–219, June 23, 1949.

46. Tixier, M. P.: Porosity Balance Verifies Water Saturation Determined from Logs, *Trans. AIME*, vol. 213, pp. 161–169, 1958. Also in *J. Pet. Tech.*, vol. 10, no. 7, July, 1958.

47. Tixier, M. P.: Porosity Index in Limestone from Electrical Logs, *Oil and Gas J.*, vol. 50, pp. 140–142, 169–173, Nov. 15; pp. 63–68, 87, Nov. 23, 1951.

48. Tixier, M. P., R. P. Alger, and D. R. Tanguy: Recent Progress in Induction and Sonic Logging, for presentation to SPE of AIME, Dallas, October, 1959.

49. Winn, R. H.: The Contact Log, *Mines Mag.*, vol. 43, no. 12, pp. 16–19, December, 1953.

50. Winn, R. H.: Log Interpretation in Heterogeneous Carbonate Reservoirs, *Trans. AIME*, vol. 210, pp. 268–274, 1957. Also in *J. Pet. Tech.*, vol. 9, no. 9, September, 1957.

51. Winsauer, W. O., H. M. Shearin, Jr., P. H. Masson, and M. Williams: Resistivity of Brine-saturated Sands in Relation to Pore Geometry, *Bull. AAPG*, vol. 36, no. 2, pp. 253–277, February, 1952.

52. Worthington, A. E., and R. F. Meldau: Departure Curves for the Self-potential Log, *Trans. AIME*, vol. 213, pp. 11–16, 1958. Also in *J. Pet. Tech.*, vol. 10, no. 1, January, 1958.

53. Wyllie, M. R. J.: A Quantitative Analysis of the Electrochemical Component of the S.P. Curve, *Trans. AIME*, vol. 186, pp. 17–26, 1949. Also in *J. Pet. Tech.*, vol. 1, no. 1, January, 1949.

54. Wyllie, M. R. J.: An Investigation of the Electrokinetic Component of the Self-potential Curve, *Trans. AIME*, vol. 192, pp. 1–18, 1951. Also in *J. Pet. Tech.*, vol. 3, no. 1, January, 1951.

55. Wyllie, M. R. J.: Procedures for the Direct Employment of Neutron Log Data in Electric Log Interpretation, *Geophysics*, vol. 17, no. 4, pp. 790–805, October, 1952.

56. Wyllie, M. R. J.: *The Fundamentals of Electric Log Interpretation*, 2d ed., Academic Press, Inc., New York, 1954, rev. 1957.

57. Wyllie, M. R. J., A. J. de Witte, and J. E. Warren: On the Streaming Potential Problem in Well Logging, *Trans. AIME*, vol. 213, *Tech. Note* 2018, 1958.

Chapter 20

RADIOACTIVITY WELL LOGGING*

By JOHN L. P. CAMPBELL

Head Office Sales Engineer–Geologist
Lane-Wells Company

FOREWORD

Radioactivity well logging has progressed far since its original experimental stage. This chapter is published in the spirit of open-minded analysis of theories and techniques, without prejudice. Its purpose is to inform the user of the basis of radioactivity well logging, to describe the equipment used, present accepted principles of interpretation, illustrate a wide range of uses and applications, and point out known advantages and limitations.

BASIC PRINCIPLES OF RADIOACTIVITY

Structure of the Atom. The atom is the smallest unit into which we can divide any mass of a chemical unit without destroying the element. All atoms consist of a nucleus and varying numbers of electrons. The positive electric charge of the nucleus is precisely balanced by the required number of negatively charged electrons revolving around the nucleus. Every normal atom is therefore electrically neutral. The chemical characteristics of each atom depend upon the electron structure, which in turn is fixed by the number of positive charges in the nucleus.

We have to think of the atom as space—very little of which is occupied by anything! The diagrams of some of the simplest atoms are illustrated in Fig. 20-1.

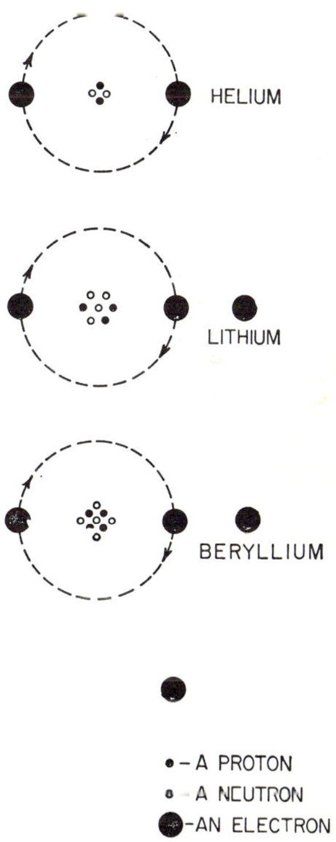

FIG. 20-1. Atomic structure of three atoms.

What Is Radioactivity? The well-known definition of radioactivity in most physics books explains the phenomenon in terms of the spontaneous disintegration of atoms, accompanied by the emission of radiation.

Alpha, Beta, Gamma, and Neutron Radiations. Since the invisible rays, alpha (α), beta (β), gamma (γ), are carriers of nearly all the energy emitted by radioactive substances and are the cause of most of

* Reprinted by permission from "Petroleum Production Handbook," edited by Thomas C. Frick, McGraw-Hill Book Company, Inc., New York, 1961.

the effects which can be observed from radioactivity, we shall pay particular attention to these.

Alpha, beta, and gamma rays differ enormously in the rate at which they are stopped by matter which lies in their path. Alpha rays are absorbed by a single sheet of paper. Beta rays may penetrate from 25 to 30 sheets of paper. High-energy gamma rays can penetrate through a deep stack of thick books or through 3 in. or more of lead.

In well logging, the radiations must penetrate from the rocks, through fluid and sometimes through cement and casing, into an instrument thick enough to withstand the hydrostatic pressure encountered at great depth. Here alpha and beta rays are not useful. Of the naturally emitted radiations only the gamma rays penetrate sufficiently. This is the problem which occurs in radioactivity well logging.

There is another kind of invisible radiation—the neutron—which is not emitted in any natural radioactive process but is produced by nuclear disintegration. This radiation can penetrate several feet of lead and is so difficult to observe directly that it remained undiscovered long after alpha, beta, and gamma rays were well known. Neutrons are applicable to well logging since they can be integrated conveniently, and their interaction with rock material produces effects which can be detected in the borehole.

Gamma-ray Well Logging. Differences in the radioactive content of the various rock layers surrounding a well result in corresponding differences in gamma radiation within the wellbore. By measuring gamma-ray intensity at any depth and plotting the data as a function of depth, a graph is obtained which depicts the radioactivity of the strata. This graph is known as the gamma-ray curve. Conventionally, it is recorded at the surface of the earth while continuously moving a detector upward along the wellbore. From this curve, the rock strata can be identified and their position and thickness can be determined. Broadly speaking, this permits the sandstones and limestones, which are of interest as possible oil-producing formations, to be distinguished from the clays and shales. However, no appreciable difference in radioactivity is found between oil-bearing and barren formations. Despite this limitation the gamma-ray curve finds wide application throughout the petroleum industry. Figure 20-2 illustrates a normal recording.

Neutron Well Logging. The importance of obtaining information through the casing relating to the fluid content of the rocks led to the development of the neutron curve. This curve is obtained by moving along the well a source which emits energetic neutrons together with a radiation detector at a fixed distance from the source. The source bombards the rocks with a constant flux of neutrons and the detector records the varying intensity of gamma radiation resulting from this bombardment. This method can therefore be properly described as a neutron-gamma (n-γ) process. The resulting curve is found to be influenced principally by the amount of hydrogen present in the vicinity, and this in turn depends upon the fluid content of the rocks. By substituting a detector which is sensitive to neutrons in place of the gamma-ray detector, a similar result is obtained. In this case the method comprises neutron bombardment and neutron detection; hence it is called a neutron-neutron (n-n) process.

The processes which have been described occur throughout the region which surrounds the neutron source. Neutrons are emitted into the well at high velocity, uniformly in all directions. As they travel outward they are scattered and slowed by collisions and finally are captured. The addition of some lighter atoms, particularly hydrogen, which is present to some degree throughout any well, shortens their average range; more hydrogen diminishes the range still further. The properties of the surrounding materials are such that in well logging the range of the majority of the neutrons is rarely, if ever, as great as the source-detector spacing. For these reasons, the response of the detector normally increases as the hydrogen content of the surroundings diminishes.

Rock materials vary in hydrogen content from essentially zero up to a maximum of about $\frac{1}{10}$ g/cu cm, depending upon their chemical composition, their porosity, and the nature of whatever substance occupies the pore space. Of the common rock types the shales and clays contain by far the most hydrogen. These materials almost invariably contain as much or more hydrogen than highly porous, fluid-filled sandstones or limestones. The additional hydrogen of shales and clays is chemically bound. Hydromites, for example, add appreciably.

Neutron sources commonly employed in well logging produce several million fast neutrons per second by the action of alpha rays on beryllium. Several hundred milligrams of radium, in equilibrium with its family of decay products and intimately mixed with a few grains of finely powdered beryllium in a sealed capsule, is considered to be the most convenient and effective neutron source for this purpose. Beryllium atoms react when struck by a sufficiently energetic alpha particle by absorbing the alpha particle and emitting an energetic (fast) neutron.

Radium, whose half life is approximately 1,600 years, will thus provide a neutron source which will diminish to half strength in that period of time. Other alpha-ray-emitting substances such as polonium, whose half life is 140 days, are sometimes

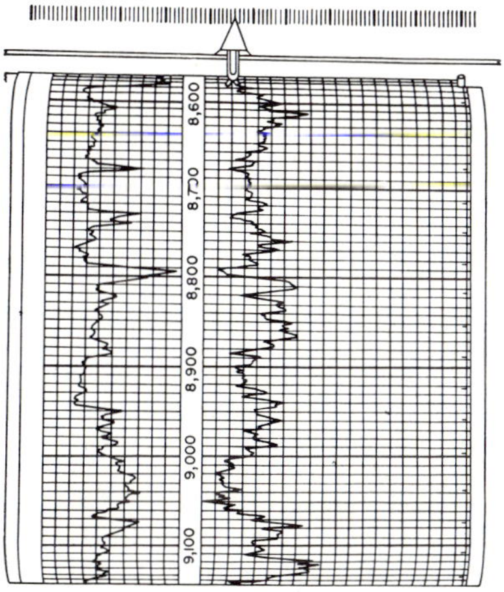

FIG. 20-2. Typical combination gamma-ray–neutron survey on recording chart.

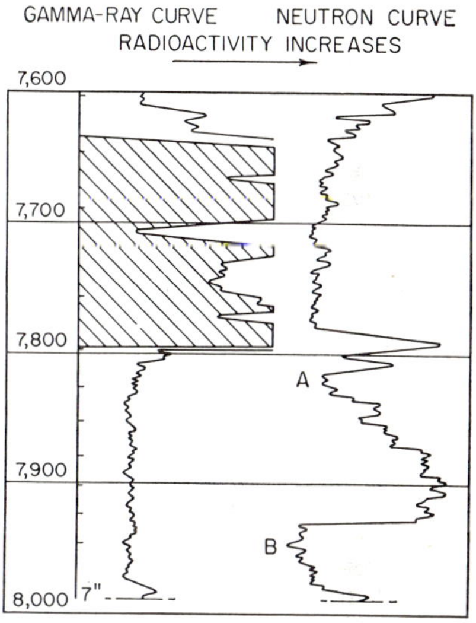

FIG. 20-3. Highly radioactive shale by crosshatched section vs. normal limestone logging at base.

employed in neutron sources, but the much longer half life of radium has the advantage of providing essentially constant emission.

In neutron-gamma well logging, it is desired to record only the gamma radiation produced by the emitted neutrons. To accomplish this satisfactorily, all other gamma radiation must be effectively prevented from reaching or registering in the detector. Natural gamma rays, those which are emitted by radioactive elements in the rocks, are obscured by using a neutron source of sufficient strength and a detector which is less sensitive and usually smaller than the type used for gamma-ray logging. The gamma rays which accompany neutron emission from the source are prevented from reaching the detector directly by interposing lead or other dense absorbing material within the instrument itself. Figure 20-3 compares very radioactive with normal formations.

Methods of Detecting Nuclear Radiations. All nuclear radiations, as previously stated, are detected through the indirect effects which they set in motion. The most commonly observed influence is the well-known phenomenon of ionization.

Among all the substances known, certain gases are the most nonconductive. Many of them (in the absence of nuclear radiation) are known to be nonconductors of electricity except at very high temperatures. Helium, argon, nitrogen, hydrogen, methane, the vapor of petroleum ether, carbon dioxide at low pressures, and many other common gases have this nonconducting property. *On the other hand, all these electrically insulating gases conduct electricity to a very measurable extent when in the presence of even very small traces of radioactive material.* This occurs because the radiations emitted by radioactive substances ionize the neutral atoms by releasing one or more electrons, each of which carries one unit of electrical charge. In gases, as opposed to solid insulating materials, these electrons are mobile and their flow represents an electric current.

All the radiation detectors which we shall discuss depend directly or indirectly upon the release of electric charge and the resulting conduction of electric current.

Ionization Chambers. A typical ionization chamber consists of a closed vessel containing an ionizable gas. An electric field is established throughout the enclosed column by applying an electrical potential. The positive ions move to the negatively polarized electrode and the negative ions to the positive electrode. A single incoming radiation liberates a large number of ions and thus results in an impulse of electric current which flows toward one or the other electrode.

Geiger counters similarly contain an ionizable gas, but a very much stronger electric field is employed. The ions are accelerated rapidly and soon gain sufficient energy to ionize other atoms. These in turn are accelerated and create more ions, thus producing an electric avalanche. The electric charge released in a given pulse in a geiger counter is always the same amount no matter what caused it.

A *proportional counter* resembles the geiger counter except that the opportunities for the initially liberated electric charge to free other charge are much less. In a proportional counter more than one electrically charged particle must be initially liberated to produce a big pulse. In this type of counter the total electric charge in each impulse is proportional to the initial amount of charge produced by each ionizing radiation.

The *scintillation counter* employs a complicated indirect method of generating the electrical signal. In the scintillation counter, the incident radiation sets electrons in motion in a crystal or other scintillation medium where there is no electric field. These moving particles act upon the crystal or other medium and cause light to be emitted. The light is transmitted into a photomultiplier tube where it impinges upon a thin layer of light-sensitive metal in a vacuum. The layer of metal is negatively charged and repels these electrons. The speed acquired by the repelled electrons enables them to liberate even more electrons from a metal target which they hit. The metal target also has a sufficient negative charge to be able to repel the additional electrons against a still further target which is similarly constructed. A succession of such targets tremendously multiplies the number of electrons received in the final target as compared with those released from the metal surface upon which the light fell.

INSTRUMENTATION AND LOGGING PROCEDURES

Recording Panels and Methods of Recording. Depending on the type or types of radiation detectors employed in the subsurface instruments and the method selected for transmitting the information over the cable to the surface apparatus, the signals in the instrument truck are recorded as deflections on a chart or film, customarily with increasing intensity toward the right. Commonly, in the United States, one vertical inch of chart or film corresponds to 20, 50, or 100 ft of depth in the well.

Subsurface Equipment. Two or more radioactivity-logging curves may be recorded simultaneously by means of separate curves (or separate galvanometers in the photographic type), and in addition, the indications of a casing collar are similarly registered on the same chart.

Dual recording of two radioactivity curves simultaneously is accomplished with this system by coupling together two subsurface instruments, plus casing-collar locator, and connecting them to separate conductors in the well cable. Duplicate surface apparatus is provided, except that a dual mechanism is employed rather than two separate recorders, in order to register all the data on a single chart.

Radioactivity-well-logging instruments must detect minute amounts of radiation with extreme accuracy and speed, while at the same time they must withstand the terrific extremes of temperature, pressure, and jarring which they encounter while traversing the well. The instruments are commonly constructed in several sizes, 3⅝ in. outside diameter being the most prevalent. Others of 2⅝ in. OD and 1¾ in. OD, while sacrificing some sensitivity and accuracy because of their smaller detectors, find application where casing or tubing size prevents the use of larger apparatus. Invariably the casing-collar locator is included or attached to the radioactivity instrument and forms part of the subsurface assembly. In this way the permanent depth markers provided by the casing or tubing are tied in to the radioactivity log.

This assembly can be exposed to elevated temperatures for many hours before any change of internal temperature occurs. Carefully selected components in the power supply and amplifier circuits provide the required stability without requiring thermal protection.

In extreme cases, elevated temperatures impair the performance and shorten the life of nearly all the materials of which logging instruments are constructed. The electronic components in the subsurface logging instruments are the most critical and must be built to the highest standards. They are very often especially designed and constructed for this service. Also, the circuits are designed to be highly independent of variations in the components. In this way commercial logging instruments are enabled to withstand temperatures up to 350°F for the length of time required to record the log. Long exposures or repeated exposures without adequate maintenance would eventually result in impaired performance.

Radioactivity Logging Systems. As many different systems can be created as there are ways of detecting radiations and ways of transmitting and recording information. In the future, as improved forms of radiation detectors are developed, it is certain that they will be applied to well logging. The performance of new systems can be expected to be limited *only by the detectors*, not by any lack of ability to transmit and record the information. One circuit system is diagramed in Fig. 20-4.

Statistical Fluctuations. The intensities of radiation encountered in gamma-ray well logging and radiation which it is feasible to obtain in neutron logging are ordinarily very low. The detected radiation may produce an energetic electrical output which will vary appreciably from moment to moment even though the radiation intensity is constant. These variations are not due to spurious effects within the detector or to deficiencies in the electric circuits. They are a natural consequence of the *number* of individual radiations. These variations are known as statistical fluctuations.

Effects of Logging Speed and Time Constant. All radioactivity measuring systems respond to a very sudden change of radiation intensity by a somewhat slower movement of the recording element. This is a consequence of measuring a rate, time being a necessary ingredient. The time response of most logging systems is exponential; they move toward a new position of rest, rapidly at first, then more and more slowly, traveling the same fraction of the remaining distance in each equal interval of time. The time for such a system to reach 63 per cent of final response is known as the time constant. The half time is defined as the interval required for 50 per cent response. This is equal to 0.69 × the time constant, or conversely, the time constant is 1.45 × the half time.

All slow speeds give similar results, within the limit of repeatability imposed by the statistical fluctuations, but at higher speeds the effect of the time constant distorts the shape of the radioactivity curve. At these speeds, the curve is smoothed. To avoid these effects, it is necessary to log at a velocity below what may be termed the "speed limit." A semiempirical relationship for determining the speed limit is

$$v \text{ max (ft/min)} = \frac{\text{detector length (ft)} + \frac{1}{2}\text{ (ft)}}{\text{half time (sec)}} \times 60 \left(\frac{\text{seconds}}{\text{minutes}}\right)$$

At all speeds other than zero the time constant introduces another effect, lag. As the radiation intensity increases and decreases the recording apparatus follows the variations but lags behind an appreciable length of time. In that time the detector pro-

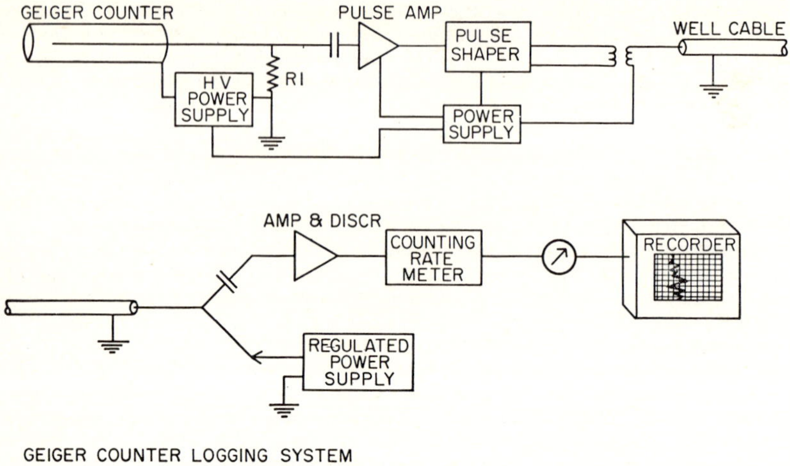

(*a*)

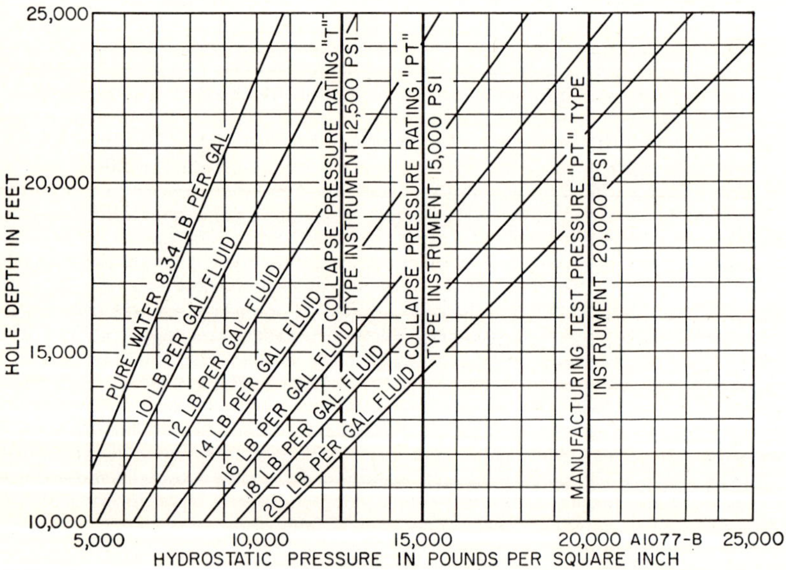

(*b*)

FIG. 20-4. (*a*) Typical circuit logging system; (*b*) collapse pressures of instruments vs. mud weights.

gresses upward a certain distance which is related to the hoisting speed and the time constant of the system. The distance traveled during an interval of time equal to the time constant is called the depth constant. The distance traversed during one half-time interval is very nearly seven-tenths as great. This distance may be taken as representative of the average amount of lag. The lag is thus given by

$$\text{Lag (ft)} = \frac{\text{actual velocity (ft-min)} \times \text{half time (sec)}}{60 \text{ (seconds/minute)}}$$

Under normal logging conditions, without exceeding the "speed limit," the lag may amount to as much as several feet, and proper correction must be made if the radioactivity curve is to be used for precise depth determinations. Ordinarily, logging speeds are held close to the permissible maximum, so that velocity and lag are both systematically related to time constant. At the maximum velocity, as given by the formula above, the lag is equal to the detector length plus ½ ft.

Statistical Error. A "statistical run" is accomplished by recording for a period of time with the detecting instrument stationary in the well. Such a recording, if all other factors remain constant, shows the magnitude of the random variations encountered at that particular radiation intensity.

Log Quality. The quality index of each curve depends upon conditions inherent in the particular well and upon the characteristics of the logging apparatus. It is generally agreed that the quality index must be 5 or greater in order to interpret the results reliably. In other words, the probable error is approximately 5 per cent as large as the curve.

Quality can be improved by a longer time constant to reduce the statistical fluctuations and correspondingly reduce the logging speed.

Sensitivity and Calibration. Sensitivity, in well logging, refers to the amount of response obtained for some standard increment of radiation intensity, generally the calibrating increment.

Calibration of the apparatus alone is relatively simple and straightforward. In practice it is quite convenient to perform this type of calibration which standardizes all factors within the detecting and recording systems.

Gamma-ray detectors and their associated recording systems are usually standardized by applying a minute quantity of radium at a fixed location with respect to the detector. Commonly a group of identical "standard" radium capsules is maintained in order that all instruments operated by one service organization can be calibrated alike.

An alternate method of gamma-ray calibration consists of surrounding the detector with a uniform field of radiation from a distributed source of radiation. This method is particularly desirable for use in radiation detectors whose sensitivity is not uniformly distributed throughout their useful volume.

Standardization of systems employing slow-neutron detectors requires some means for producing this type of radiation. A convenient way, which makes use of the well-logging neutron source and thereby includes it in the calibration, consists of a hollow cylinder containing hydrogenous material such as wax, plastic, or rubber. It is made sufficiently long to encompass the source and detector. Fast neutrons from the source are slowed down by the hydrogen and a fixed fraction of the slow neutron upon the detector. If external effects are negligible, it may be assumed that the response of the system will fall to zero upon removal of the cylinder.

By proper choice of dimensions and materials the same cylinder can be made to produce a suitable amount of gamma radiation in addition to the slow neutrons. The same device can then be employed to standardize neutron-gamma recording apparatus in addition to neutron-neutron equipment.

INTERPRETATION OF RADIOACTIVITY LOGS

Gamma-ray and Neutron Curves. These radioactivity curves indicate natural and artificially produced radiation intensities within the well. Were no other information sought, their interpretation would be simple, but radioactivity logs are much more than this, as indeed they must be to solve critical problems of the petroleum industry.

Experience, familiarity with local conditions, and an understanding of all the factors which affect the radioactivity log are requisites for the interpreter. So are drilling records, core data, and other types of logs of the same well and nearby wells, especially if quantitative conclusions are to be drawn from the indications of the radioactivity log.

This is the process of interpretation. When applied expertly it leads to knowledge otherwise unattainable and to an evaluation of the certainty of that knowledge.

To find the answers the interpreter must extract from the radioactivity log all the applicable information it contains. His familiarity with the typical gamma-ray and neutron-curve responses permits him to identify, tentatively, particular rocks or well conditions. Experience with local stratigraphy and any local exceptions from typical response enables him to confirm these identifications.

In general, interpretation provides answers about any factor that influences the response of the radioactivity log. Some answers, however, can be obtained from the log only if certain other questions can be resolved first. For any particular response of the radioactivity curves there are always several conceivable causes.

Reading the Log. The conventional form of heading for radioactivity logs conforms to standards adopted by the American Petroleum Institute. These require the inclusion of certain data concerning the pertinent characteristics of the logging instruments and the well. Additional data, beyond the minimum required by the API, are often supplied. A typical form of radioactivity-log heading is illustrated in Fig. 20-5.

The well is identified and its location defined. Some rather permanent features of the well such as the ground elevation or a specified wellhead flange provide the most satisfactory reference datum for the log.

The specific logging conditions and instrument characteristics are listed. *Type of log* may be indicated as G/R for gamma ray, N/L for neutron log, or Comb. for a combination of these two. This information may show the neutron as N/gamma or N/N, indicating that the neutron log is of the neutron-gamma type or the neutron-neutron type. *Run number* enables the user to determine whether an earlier radioactivity log was run. If so, its comparison with the later run may greatly aid the interpretation. *Total* and *Effective depth* (*driller*) refer to the driller's records concerning total depth reached. *Top* and *Bottom of logged interval* define the depths between which the log was actually recorded, not the maximum depth reached. *Type of fluid in hole* refers to the liquid which fills the wellbore below the fluid level. The fluid level is the position of the top of the liquid column at the time the log was recorded. *Maximum recorded temperature* is given in the heading if a thermometer was run with the radioactivity instruments. This is done when extreme temperatures are anticipated, in order to evaluate the instrument performance.

Neutron-source strength and type are generally stated in terms of the service company's standard designation for the particular neutron emitter used. *Source spacing* relates to the distance between the source and the detector and thus serves to distinguish between curves obtained under different circumstances. Within a single organization equal spacings are always designated by the same number. Any other spacing is indicated by adding or subtracting the appropriate difference in inches. Thus within one service company 8.25 in. represents a certain definite spacing and 3.75 in. represents a spacing 4.5 in. shorter.

Length of measuring device defines the effective length of the radiation detector in inches. This is a theoretical distance throughout which the logging instrument is sensitive to incident radiation. This length determines, in part, the thinnest beds which can properly be recorded by the apparatus. The *Time constant* is a measure of the speed of the response of the entire recording system while recording a particular curve.

Logging speed is stated in feet per minute, or sometimes in feet per hour. It frequently represents the logging speed employed throughout the entire logging operation.

Statistical variation is the peak-to-peak excursion caused by spontaneous fluctuations of the radiation intensity, as in Fig. 20-6.

Sensitivity reference describes the standard radiation with which the instrument was calibrated. The sensitivity expresses the response of the recorder to this standard.

Representatives of the service company and of the customer, who attended the logging operation, are named under *Recorded by* and *Witnessed by*, respectively.

Part of the heading provides space for the *Casing record.* This information is essential to a proper interpretation of the log. It includes the size, weight, and loca-

THIS FORM CONFORMS TO API RP 33

LANE WELLS COMPANY — *Radioactivity Log*

COMPANY: LUCKY OIL COMPANY
WELL: PROSPECT No. 1
FIELD: COW CREEK
LOCATION: SEC. 27 T7S R3E
COUNTY: CAMERON
STATE: TEXAS

	Well Location
COMPANY: LUCKY OIL COMPANY	
WELL: PROSPECT No. 1	
FIELD: COW CREEK	
LOCATION: 660' FSL & 990' FEL OF SEC. 27 T7S R3E	
COUNTY: CAMERON STATE: TEXAS	

LOG ZERO	KELLY DRIVE BUSHING	ELEV. 237'
DRLG. ZERO	KELLY DRIVE BUSHING	ELEV. 237'
PERM. DATUM	7" BRADENHEAD FLANGE	ELEV. 231'

TYPE OF LOG	GAMMA-RAY	N/Gamma		
RUN NO.	ONE - NW	ONE - NW		
DATE	9-6-56	9-6-56		
TOTAL DEPTH (DRILLER)	4585	4585		
EFFECTIVE DEPTH (DRILLER)	4585	4585		
TOP OF LOGGED INTERVAL	388	388		
BOTTOM OF LOGGED INTERVAL	4571	4582		
TYPE OF FLUID IN HOLE	11.2# MUD	11.2# MUD		
FLUID LEVEL	FULL	FULL		
MAXIMUM RECORDED TEMP.	185° F.	185° F.		
SOURCE STRENGTH & TYPE		S10A3		
SOURCE SPACING — IN.		18"		
LENGTH OF MEAS. DEVICE— IN.	4"	4"		
O.D. OF INSTRUMENT—IN.	3-5/8"	3-5/8"		
TIME CONSTANT — SECONDS	3	3		
LOGGING SPEED FT./MIN.	25	25		
STATISTICAL VARIATION — IN.	RECORDED	RECORDED		
SENSITIVITY REFERENCE	274 JIG	D368		
DETECTOR TYPE	D4G1	D4N1		
DETECTOR CLASS	SCINT.	SCINT		
RECORDED BY	SMITH	SMITH		
WITNESSED BY	JONES	JONES		

WELL RECORD

RUN	BIT SIZE	CASING	WT.—LB.	FROM WELL RECORD			FROM LOG		
1	10-3/4	9-5/8	54	SURF.	TO	407	SURF.	TO	406
1	8-3/4	7	26	SURF.	TO	4002		TO	
1	6-1/4	5	17	SURF.	TO	4585	SURF.	TO	4584
					TO			TO	
					TO			TO	

REMARKS OR OTHER DATA

Fig. 20-5. Typical API heading of log and well information.

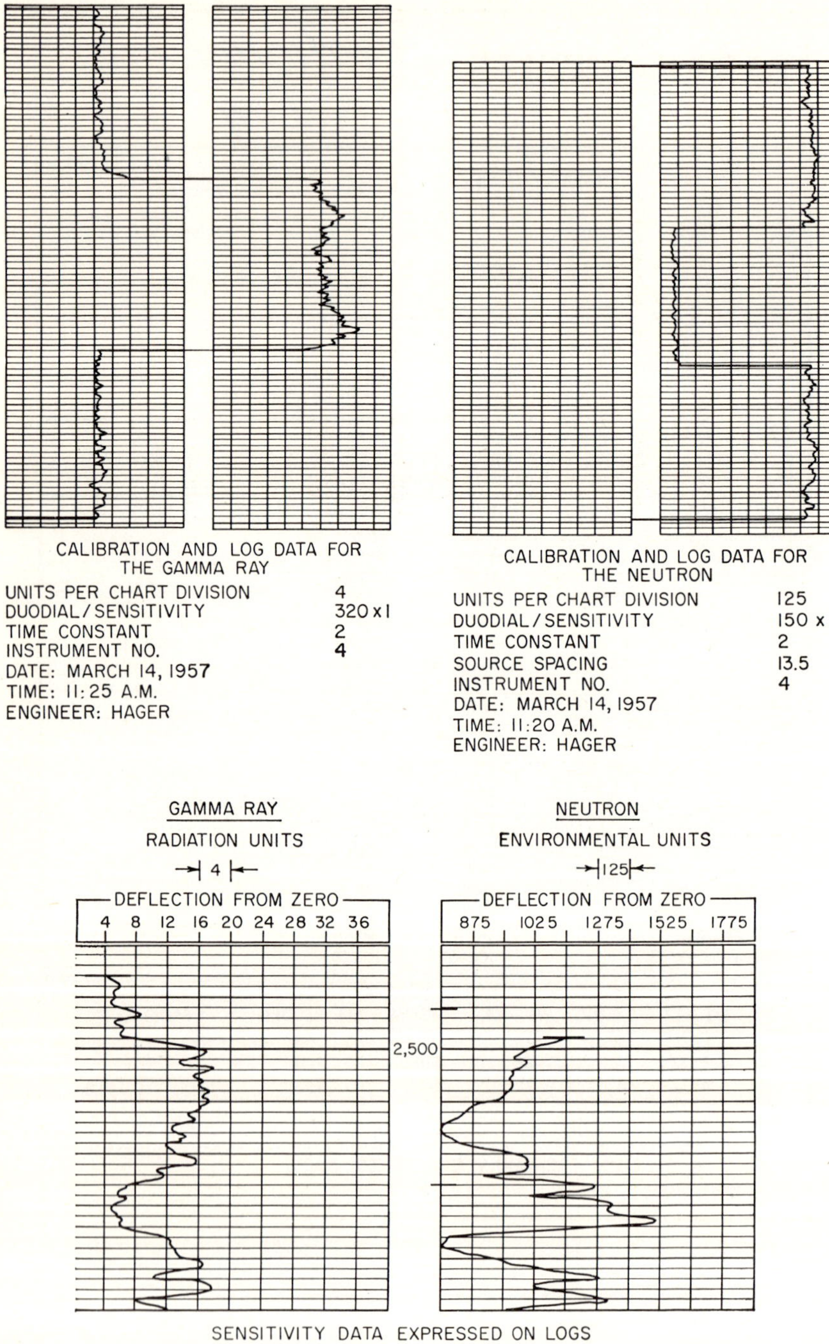

Fig. 20-6. Calibration system for sensitivity.

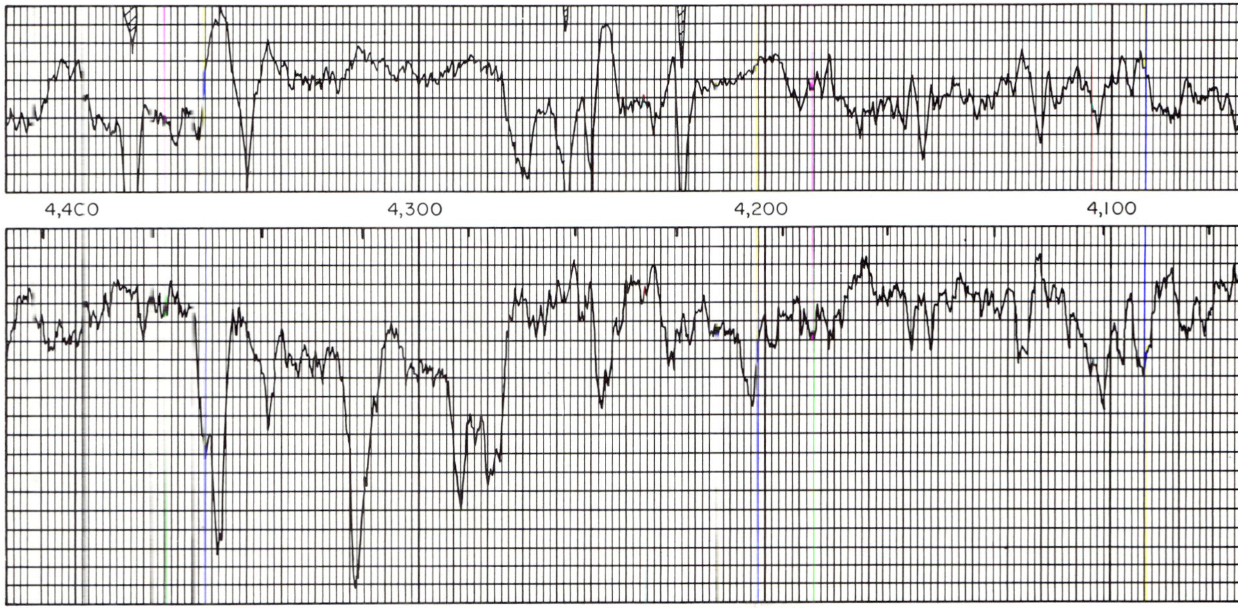

Fig. 20-6. (*Continued*)

tion of each casing or liner in the well at the time of logging. *The open-hole record* shows each size of bit used in drilling the well and the corresponding depth intervals. This information has an essential bearing on the interpretation, particularly in the comparison of logs from more than one well.

Under *Remarks* notations are made of unusual circumstances which may affect the interpretation of the log and any other pertinent data for which there is insufficient space elsewhere.

The log proper is displayed below the heading in one or more ruled columns, according to the number of radioactivity curves it contains. It is conventional to record the gamma-ray curve on the left and the neutron curve on the right, with intervening unruled columns for the depth numbers and the casing-collar log. Each of the columns of the log is appropriately captioned to indicate the type of curve and the direction in which increasing radiation intensity is recorded.

The positions of zero intensity and sensitivity are indicated by a symbol or as radiation units for gamma-ray calibration and environmental units for neutron evaluation. Other units used are counts per second or roentgens per hour.

The casing and collar log is drawn in diagrammatic form in the space between the radioactivity curves. In effect, it resembles a cross section of the well. Each casing or liner is represented by a vertical line, the innermost at the right, others successively to the left. The diameter of each, as obtained from the driller's records, is designated numerically. Their extent in depth is shown by terminating each line at the appropriate point along the log. Casing shoes are indicated by a triangular symbol.

It is common practice to provide a "detail section" at the bottom of the log. This is a log of a part of the well, generally 500 or 600 ft long, containing the zone of maximum interest. A vertical scale of 5 in. of chart for each 100 ft of well is usually employed.

At the bottom of each section of each curve, two depths are given in numerals. The greatest depth at which a radioactivity curve has been recorded is indicated as RD, meaning recording depth. The greatest depth reached by the bottom end of the logging instrument is designated TD, meaning total depth.

Determination of Depth and Thickness. Depths are determined from the log by noting the position of the mid-points of the deflections. These mid-points are chosen at the center of the deflection from left to right. The depths corresponding to each mid-point are then read from the depth scale of the log to locate the position of the top or bottom of each layer. Thickness may be obtained by these figures.

Assuming the transitions to be abrupt, the mid-points will be at the correct depth on the chart if the layer between them is sufficiently thick to be recorded at its proper intensity. This would correspond approximately to a thickness equal to the detector length or to the depth constant (time constant multiplied by logging speed) whichever is greater.

Marking off the mid-points of the major transitions and observing their depths are generally the first steps toward interpreting a radioactivity log. The next step is ordinarily that of discerning the identity of some or all of the intervals thus picked (Fig. 20-6).

Identification. Questions of identity fall into two groups. One seeks to locate on the radioactivity log some specific feature such as a particular formation or the casing shoe and the fluid level. The other group seeks information about the well conditions, the rock characteristics at a definite depth in the well or opposite a particular response on the log. Both types require the interpreter to draw upon his knowledge of the way the radioactivity curves respond to all the factors which influence them.

Figure 20-7 shows, in the form of a log, the typical types of rock commonly encountered in oil-producing regions. Alternate responses are shown for the neutron curve in some of these rocks, illustrating that it responds differently depending upon the fluid content or the shale content of these formations. Some alternate responses are shown for the gamma-ray curve, representing cases in which a sandstone, limestone, or other rock ordinarily low in radioactivity has acquired an abnormal content of radioactive material. Fortunately for the interpreter such anomalies are infrequent but they must be taken into consideration as being possible when examining logs obtained in any unfamiliar area.

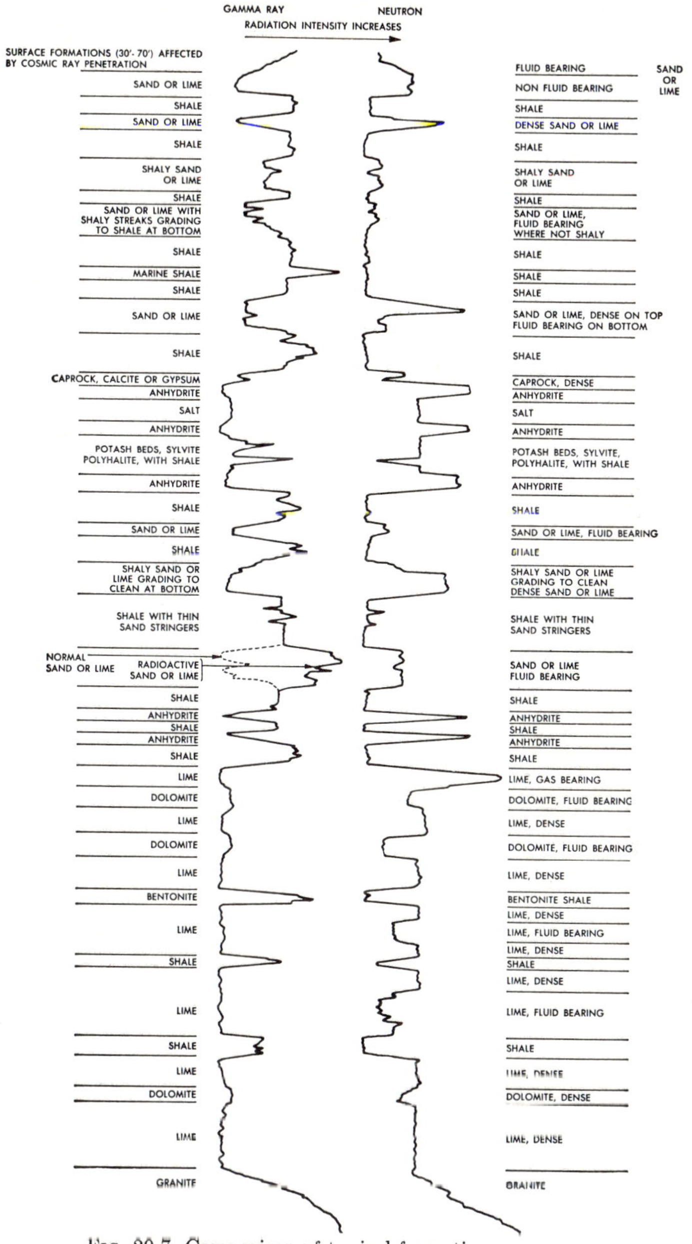

FIG. 20-7. Comparison of typical formation responses.

With relatively few exceptions similar rocks throughout the world provide similar responses on a radioactivity log. On this basis, a particular formation can usually be identified on the log with reasonable certainty. For example, if we seek a limestone, we expect to find it where the gamma-ray curve indicates a low intensity. This is the typical response of limestone. The neutron curve may have a very great intensity if the limestone lacks porosity, but even if it were extremely porous the neutron curve would be expected to register a somewhat higher intensity than in the shales. Unless the limestone is a very silty one, or one of the radioactive exceptions, the gamma-ray curve response will be low.

It can be seen from Fig. 20-7 that more than one type of rock can produce the same responses on the radioactivity log. It is also possible that the observed response on the log could be caused by some physical characteristics of this section of the well, such as a greatly enlarged diameter or an overlap of casing and liner. The radioactivity log itself may contain the clues which enable the interpreter to dismiss these as impossible, or other information about the well may contradict such a supposition. Proceeding in this fashion, the interpreter increases the certainty of the identification by taking into account every factor which could possibly affect his conclusion.

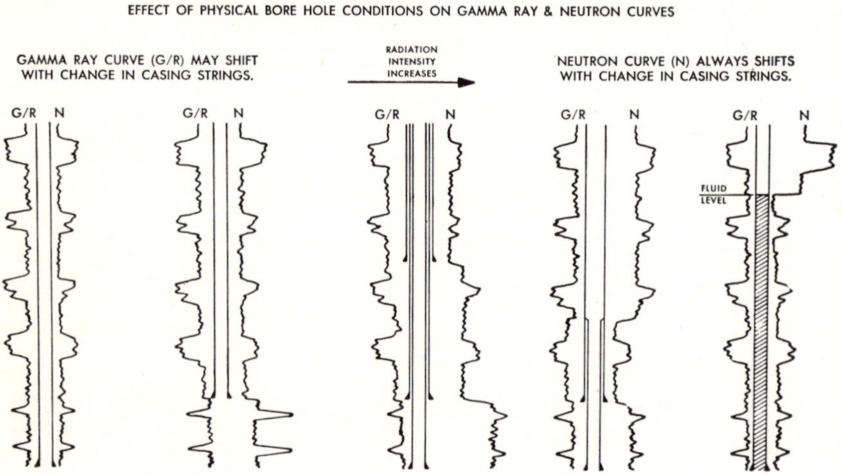

FIG. 20-8. Effect of physical borehole conditions on gamma-ray–neutron curves.

Identification of Borehole Effects. Many uses of the radioactivity log are based on its response to borehole conditions. Since these primarily affect the neutron curve the role of the gamma-ray curve is secondary in these cases. However, in seeking to learn about the borehole from the neutron curve, it is often essential to employ information from the gamma-ray curve in order to differentiate between formation effects and borehole effects. Some of the frequent and important uses of the radioactivity log in this group are

Location of casing seats
Location of fluid level
Location of liner tops
Location of aluminum or magnesium casing sections
Location of parted casing strings

Knowledge of these locations is important, from the standpoint both of deciding what and how to conduct further operations on the well and of interpreting the formation responses of the radioactivity log. Figure 20-8 illustrates several standard examples.

In many cases the physical borehole conditions are found from the radioactivity log to be quite different from those which were expected. Often borehole features such as casings and liners are found to be located at significantly different positions than believed or intended. Uncorrected, such errors might prevent the success of further recovery operations and result in the loss of a productive well. Figure 20-9 shows a portion of a log run to solve a problem in casing location.

Variations of well diameter alone, unaccompanied by variation of the thickness of

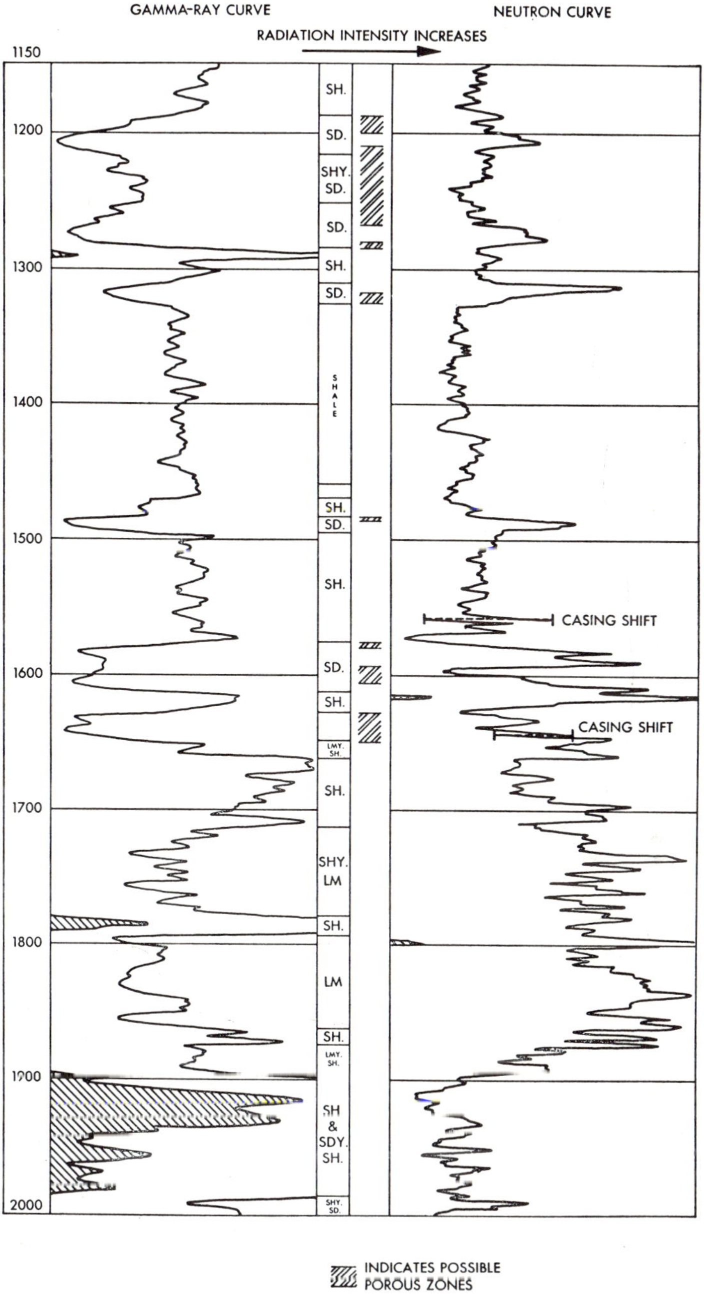

FIG. 20-9. Typical log in natural and radioactive formations showing physical casing shifts.

the absorbing materials within the hole, have no effect on the radiation intensity recorded by the gamma-ray curve.

The neutron curve is affected by borehole factors. Well fluid, steel casing, and cement all absorb neutrons.

The additional thickness of steel at the collars sometimes causes a visible decrease in the neutron-curve intensity, in air-filled holes, and particularly in drill pipe.

As well diameter is increased, a greater surface of rock is exposed to direct bombardment by the neutron source, the range of the neutron is extended, and a greater area of rock is within "view" of the detector. Thus the neutron-curve response tends to increase with enlargement of the well diameter.

Typical responses of the gamma-ray and neutron curves to variations of hole diameter, casing thickness, cement, borehole fluid, and other physical characteristics pertaining to the well are illustrated in Fig. 20-8. Fortunately, sufficient data are usually obtainable within the radioactivity log to enable the interpreter to separate the borehole effects from the formation effects.

Fluid level, the interface between air and liquid in the borehole, produces a larger abrupt deflection of the neutron curve than any other subsurface condition.

Identification of Formations. Radioactivity logs are especially effective for identifications. Electric logs are so greatly affected by salt water or oil in the fluid column within the well as to become inoperative in some wells. Information which might otherwise be unobtainable can be gained from the radioactivity log since it is equally effective through any ordinary well fluids.

Under certain circumstances the radioactivity log supplies more, or better, information than any other method and it then becomes the primary means of identification. Examples of this are found in many areas in which the production is from carbonate rocks. There the neutron curve serves better to identify the productive zones than any other logging method. Wells which must be completed with casing and cement immediately after drilling are another group in which the primary identification may be provided by the radioactivity curves. Still another is the vast number of completed wells for which the existing data are insufficient or inaccurate. In these wells the radioactivity log often provides the only practical way to obtain the necessary information.

The simplest formation identification by means of the radioactivity log is that of distinguishing between the shale, sandstone, and limestone. With all shales accurately defined, the remaining layers may be assumed to be nonshales and therefore of possible interest as potential producing zones. This, of course, depends upon reasonable knowledge of local stratigraphy. The identification of shale on the radioactivity log is ordinarily very simple. The gamma-ray curve with few exceptions shows medium to high intensity in shale. The neutron curve records at its lowest intensity. Ordinarily these sections of the log can be distinguished at a glance.

Sandstone, limestone, chert, salt, chalk, dolomite, and anhydrite all give similar (but not necessarily identical) responses. Contrasting with those listed above but similar to one another are shale, clay, siltstone, slate, and marl. The radioactivity log alone does not ordinarily permit differentiation within these groups.

Correlation. The process of comparing one log with another of the same type or of a different type enables the interpreter to obtain more information than can be derived from either log alone. The characteristic responses of the gamma-ray and neutron curves to various subsurface formations are found to be highly consistent over large areas. Certain formations are easily recognizable from the unique shape of the response they produce on the radioactivity curves and provide an accurate indication of the trend of the intervening strata. Several wells are correlated in Fig. 20-10.

Another useful type of correlation is "zoning," which consists of dividing the producing formation into sections or zones which extend from well to well but vary in thickness and in depth.

A comparison of the radioactivity logs of a number of wells throughout a given region, as shown in Fig. 20-10, makes it apparent that this formation is consistently differentiated into a number of separate zones. By means of the gamma-ray and neutron curves it is thus possible to correlate those data and bring about an orderly

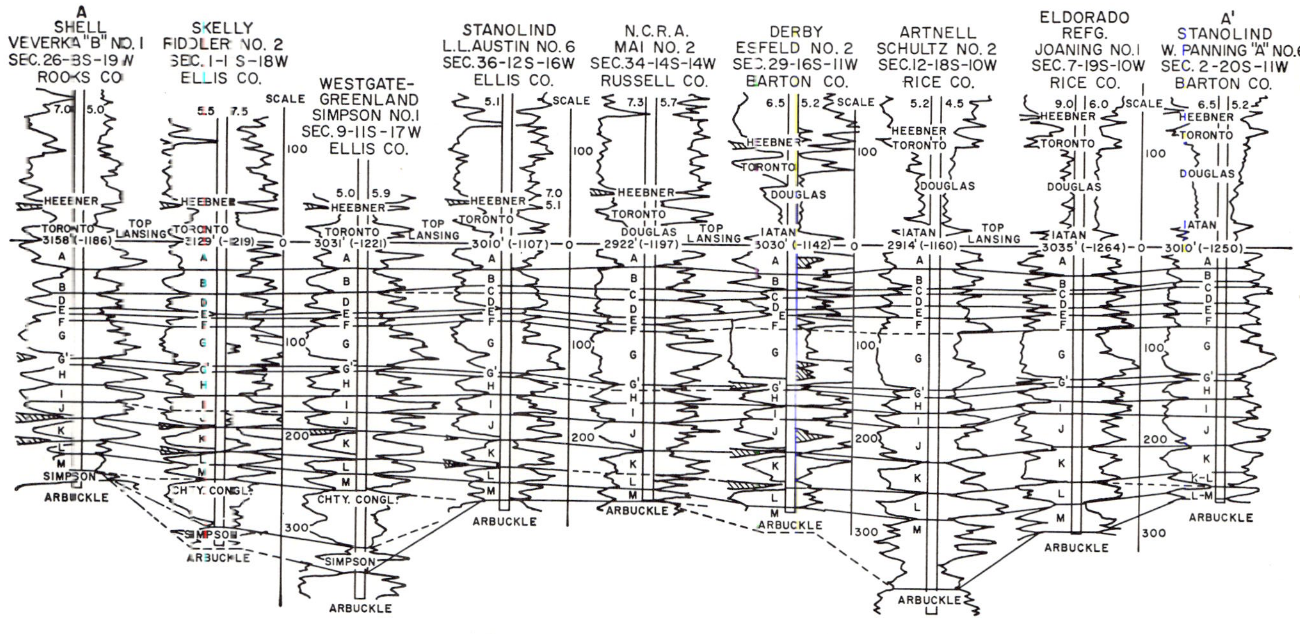

FIG. 20-10. Correlation by zoning. Gamma-ray–neutron cross section. Lower Pennsylvanian system, central Kansas.

analysis. Zoning is helpful in the exploitation of any thick producing sections which can be differentiated by means of radioactivity logs.

Correlations such as these need not be based upon radioactivity logs alone. Though the physical properties of the rock to which the electrical curves respond are completely different from the properties which produce the gamma-ray and neutron curves, a strong tendency exists for these factors to vary together. Thus a resemblance is usually found between a gamma-ray curve and the SP curve, the gamma radiation decreasing as the self-potential increases, while the resistivity curve generally shows high resistivity in those formations which produce a high neutron-curve response. Figure 20-11 shows this resemblance.

Studies of the geologic section may thus be extended or enhanced by including electrical logs from outlying or intervening wells.

Another type of correlation comprises a comparison of the depths indicated for the same formation on two or more different logs of the same well. If the radioactivity log has been recorded after casing and cementing, its depth indications are ordinarily more reliable than the depth indications of other logs which were made in open holes. This is especially true if depths are measured with reference to the casing joints as indicated by the casing-collar record on the radioactivity log. Then, upon correlating these logs, the identification of the strata may be obtained chiefly from the core log, cutting record, or electrical log, but with accurate depth and thickness information obtained from the radioactivity log. This eliminates errors resulting from estimating the depth from which certain cores or cuttings may have come.

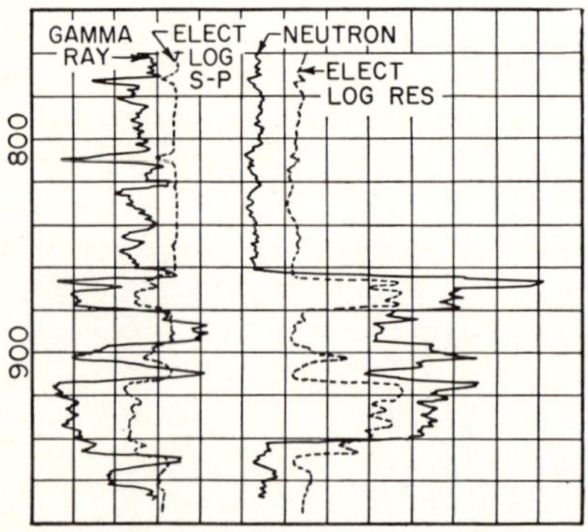

Fig. 20-11. Comparison of radioactivity and electrical log in the same hole.

These examples illustrate how correlation is used to enhance determinations of depth, thickness, and identity. The importance of comparing and correlating all available data with the indications of the gamma-ray and neutron curves during their interpretation cannot be overemphasized.

Determining Formation Characteristics. Questions regarding the condition or composition of rocks, their porosity and their content of oil, water, and gas, sometimes their shale or silt content, fall into this category. Such questions as these cannot ordinarily be answered directly from a radioactivity log alone, since the factors involved in these questions do not directly affect the gamma-ray or neutron curves. However, certain relationships between one rock property and another, which remain consistent over wide areas, enable the radioactivity curves to serve to great advantage in obtaining these answers. Porosity is a property which, in itself, has no direct effect on the radioactivity-logging processes other than to diminish slightly the amount of rock material within a given distance from the apparatus. However, within the earth the pore spaces are filled with either water, oil, or gas, and these materials affect the neutron curve in a way distinctly different from that of rock materials.

Petroleum Finding. While it was once thought that the natural radioactivity of petroleum should be high, the opposite is generally true. For this reason accumulations of petroleum cannot be observed on the gamma-ray curve. Also, no positive indication of petroleum occurs on the neutron curve, water and oil being nearly alike in their hydrogen content. The role of radioactivity logs in oil finding is thus an indirect one. Petroleum finding must then be done within the well by one or more of the accepted methods. Radioactivity and electrical logs are employed in this connection.

Liquid-filled Porosity. Pore spaces in rocks can be considered to be liquid-filled when they contain any of the following:

1. Fresh water
2. Salt water
3. Liquid petroleum
4. Butane or propane liquefied by pressure
5. Tar

The sense in which we are employing the term "liquid-filled" is to consider all those things which will exhibit the indications of liquid from the standpoint of a neutron log. In keeping with this definition there is a special case of unrecoverable liquid contained in pores which is chemically combined water. This corresponds with water of crystallization of hydrated salts and with water chemically bound in shales and fine sediments.

Porosity Determination. Since the neutron curve is principally affected by hydrogen in the surrounding space, it responds to variations of porosity. As long as the pores are liquid-filled, there will be more hydrogen as the porosity increases, and consequently the neutron curve will record a lower intensity. Less liquid, corresponding to less porosity, causes an increase in the neutron curve. This relationship is extremely useful in evaluating the content of producible fluid in reservoir rock, but before the neutron curve can be used for this purpose, it is necessary to distinguish between such rock and the shales. This is ordinarily done by means of the gamma-ray curve, which serves as a lithologic indicator to distinguish between the sandstones and limestones, and the shales. The gamma-ray curve does not indicate porosity with normal well conditions, but it does clearly differentiate between the reservoir rocks, which are more porous, and the shales, which are not. Then the high fluid indication of the neutron curve (low radiation intensity) in the shale sections can be disregarded, since it is merely an indication of their chemically bound hydrogen. Porous formations and shale are compared in Fig. 20-12.

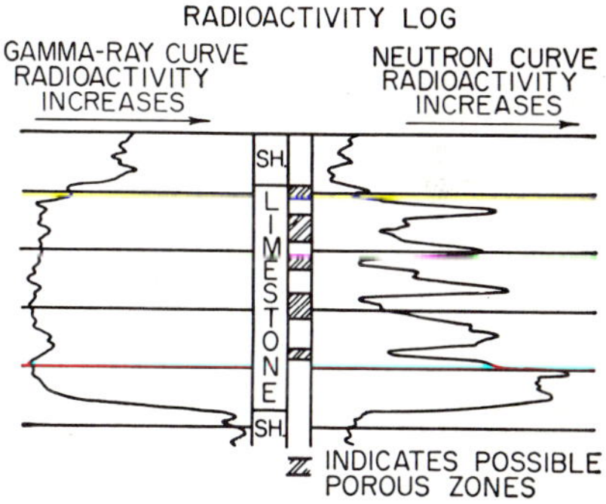

FIG. 20-12. Typical neutron response for fluid or porous zones.

Full attention is then devoted to the neutron-curve response in the reservoir rocks where the hydrogen content ordinarily corresponds to porosity filled with producible liquids. If the gamma-ray intensity is low the zone may be regarded as free of shale, and the entire amount of hydrogen indicated by the neutron curve will probably represent producible liquid-filled porosity. If, however, the gamma-ray curve shows an intermediate intensity, indicating that some shaly material is present, allowance should be made for its effects on the neutron curve. In partly shaly material the neutron curve will record a lower intensity value than for a shale-free formation having the same porosity.

In the simplest terms the neutron curve may be regarded as a shale-plus-liquid indicator. By identifying the shales and estimating the shaliness of the other zones by means of the gamma-ray curve, or on the basis of other information, the neutron curve can be used as an indication of liquid-filled porosity.

Quantitative Porosity Determination. Wherever the silt or shale content of the rock is negligible, and the pore space can be properly assumed to be filled with liquid, it is found that the neutron-curve response diminishes logarithmically as the amount of liquid, and consequently the amount of porosity, increases. This relationship is extremely useful, particularly in those areas where clean (nonsilty) limestones prevail. Its application to a particular log requires that the relationship between neutron-curve intensity and porosity first be established for this well by field or region. Once established, this relationship can be applied to other wells in which similar conditions exist. The procedure for obtaining the neutron-curve porosity relationship may be outlined as follows:

1. Recover representative cores throughout the zone of interest with accurate knowledge of the depths from which each core came.
2. Run a neutron curve in the same well.
3. Determine the porosity of the cores as accurately as possible. Form a weighted average of porosity over a depth interval comparable with the resolution of the neutron curve (generally 2 ft) and plot a porosity log.
4. Shift logs for best agreement in depth between the neutron log and the porosity log.
5. Plot on semilog graph paper, with porosity along the logarithmic axis, the neutron-curve intensity along the linear axis.

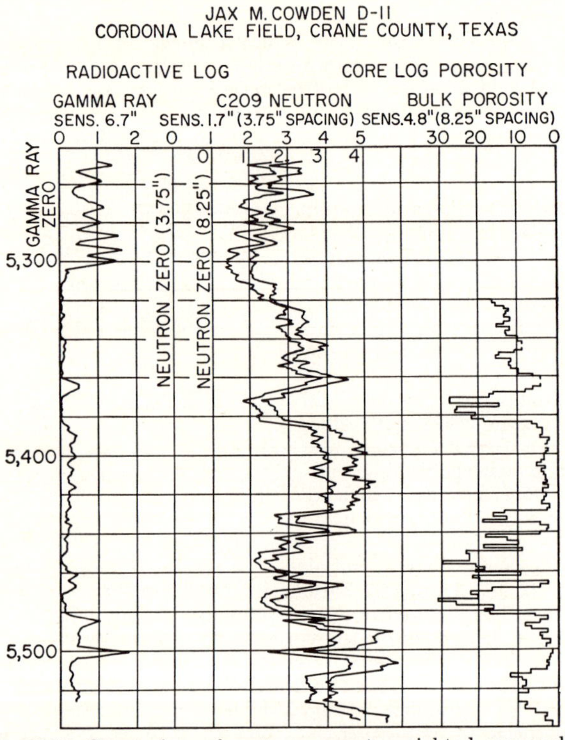

FIG. 20-13. Comparison of neutron curve to weighted core analysis.

6. Fit the best straight line through the scattering of data points. This line may be extended, if desired, through 1 per cent porosity and through 30 or 50 per cent porosity.

The calibration curve thus obtained is then applicable to other nearby wells, provided the same hole size, casing size, and well-fluid conditions prevail.

An example of this procedure is shown in Fig. 20-13. In this well two neutron curves were run with the same apparatus but with different spacings between the neutron source and detector. Consequently the relationship between the porosities as obtained from the core data and each of the neutron curves can be determined. These relationships can be used to determine porosities in other wells in the same field which are logged using either of the two spacings.

The bulk porosity is plotted at the right, exactly as obtained from the core analysis. This graph, while having a stepwise appearance, correlates reasonably well with the neutron curves. A much better correlation, however, is exhibited by the weighted-average core-porosity curve (Fig. 20-13). This curve has been plotted from the bulk-porosity data foot by foot, by weighted averages. The result is a smoother curve which, when plotted on a logarithmic scale, as has been done in this case, bears a remarkably close resemblance to the neutron curve.

From these same core data two charts (Figs. 20-14 and 20-15) have been constructed in accordance with the procedure outlined. These charts show the relationships existing between neutron-curve intensity and porosity for each of the differently spaced neutron curves. The recorded intensities have been transposed into standard

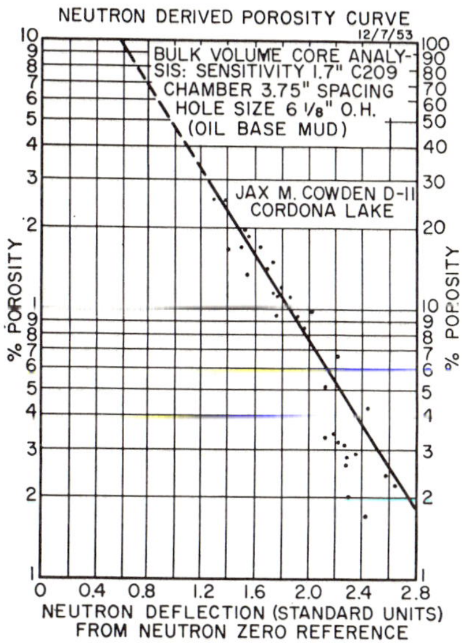

FIG. 20-14. Neutron-derived porosity curve —short-spaced neutron.

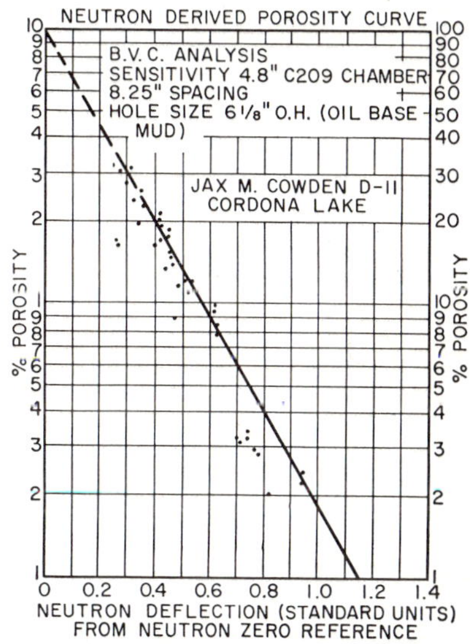

FIG. 20-15. Neutron-derived porosity curve —long-spaced neutron.

units by dividing them by the sensitivity. In this form the charts are applicable to neutron curves of other wells regardless of the choice of sensitivity used in recording them.

Estimating Porosity Values. Without the core data it is still often possible to estimate fluid-filled porosity. To do this without any additional information requires that an assumption be made regarding what porosity would be required to diminish the neutron-curve response to the intensity observed in the shale. In the absence of better information based on local experience, this may be taken as 40 per cent for

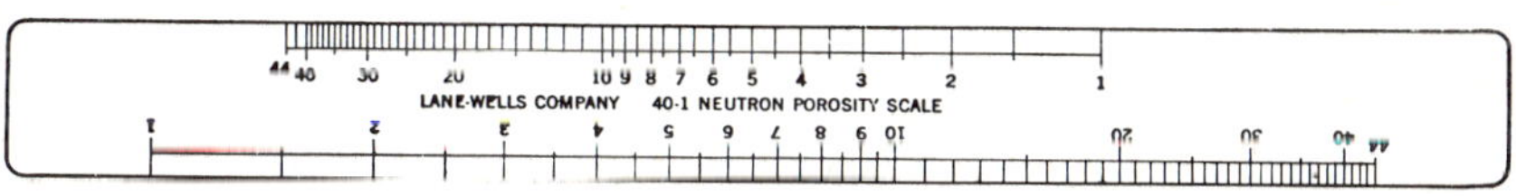

FIG. 20-16. Neutron porosity scale.

limestone or 32 per cent for sandstone. In this way two points are obtained, one corresponding to 1 or 2 per cent porosity with a relatively high neutron-curve intensity, the other corresponding to 40 or 32 per cent porosity with minimum neutron-curve intensity. These two points may be plotted and joined by a straight line to provide an approximate calibration curve.

As a convenient alternate to the calibration chart, one can employ a movable scale from which the porosity corresponding to any intensity on the neutron curve can be read directly. A transparent scale especially designed for this purpose is illustrated in Fig. 20-16, but any scale with logarithmic graduations, such as a pocket slide rule, will serve. To apply this scale, place it on the log with its numbers decreasing toward the right. Shift the scale until the number assigned for the porosity equivalent to shale (40 per cent for limestone, 32 per cent for sandstone unless local experience has shown a different value to be a better approximation) is located at the shale-intensity position on the log. With this point fixed, pivot the scale until the number representing the minimum porosity coincides with the corresponding high intensity value recorded by the neutron curve. Now the markings of the scale can be transferred to the log itself, thus providing it with a numerical scale of porosity. The porosities of the other shale-free zones can then be read by observing their response on the neutron curve relative to this scale. Borehole conditions such as fluid, casing, and cement must be the same or negligibly different at both points from which this scale was derived, and at all the points to which it is applied. It is necessary to observe the shale response and the minimum-porosity response under like well conditions; the shale applies where a shale response and a low-porosity response can both be observed in each section.

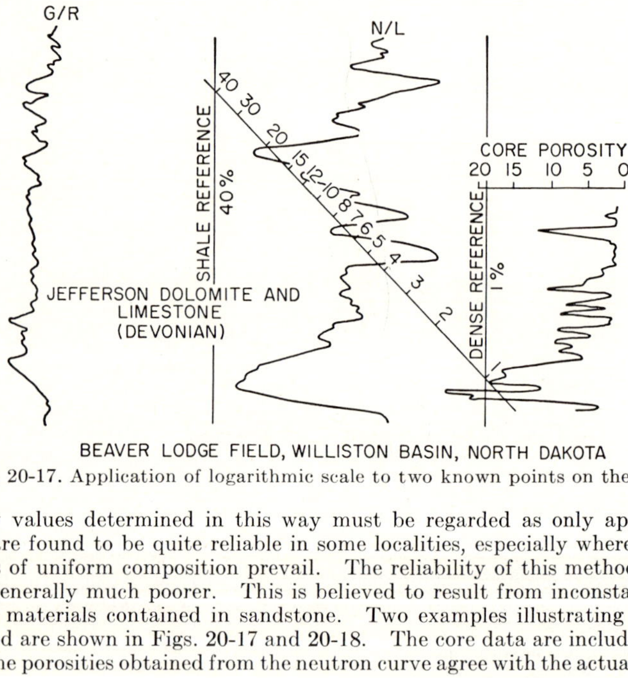

FIG. 20-17. Application of logarithmic scale to two known points on the log.

Porosity values determined in this way must be regarded as only approximate, but they are found to be quite reliable in some localities, especially where limestone formations of uniform composition prevail. The reliability of this method for sandstones is generally much poorer. This is believed to result from inconstancy of the cementing materials contained in sandstone. Two examples illustrating the use of this method are shown in Figs. 20-17 and 20-18. The core data are included to show how well the porosities obtained from the neutron curve agree with the actual measured values.

Gas-filled Porosity. When the pore space is occupied by fluid in the vapor state the amount of hydrogen present is ordinarily much less than if the same space were filled with liquid, since the density of most gaseous hydrocarbons is very much less than that of liquids. Consequently the neutron-curve response to vapor-filled porosity is high. In fact, highly porous rock containing only vapor provides a greater response than does dense (nonporous) rock. Confusion sometimes arises in connection with the interpretation of the neutron curve under these circumstances, but if a known "tight" zone exists within the well it can be safely assumed that any appreciably greater response than that of the "tight" zone represents vapor-filled porosity. Another factor which must be considered in connection with the interpretation of the neutron curve is that, at the pressure and temperature prevailing within the formation, the gas may be liquefied, while in the borehole at lower pressure, it may pass into the vapor state. Thus many wells which produce gas do so from a liquid-filled rather than a vapor-filled reservoir. The neutron curve in such wells shows the low-intensity response which is characteristic of liquid-filled porosity instead of the high intensity caused by vapor.

The only gases which do not become liquefied under conditions of a deep well are methane, ethane, and propane. Propane may be liquefied in certain cases—particularly when heavier, higher-molecular-weight compounds of carbon and hydrogen are present. The interpretation of known gas-filled horizons demands a knowledge of the composition of gas which is present and the temperature and pressure under which it exists. Formidable as these demands may seem they can often be fulfilled in a degree satisfactory for the interpretation of a neutron log. Pressure may frequently be determined from the depth at which the stratum is found and such determination may very well have enough accuracy for the purpose. Temperature can reasonably be inferred from the depth and from the geothermal gradient established in the region where the log is made. This latter inference must be entered into with

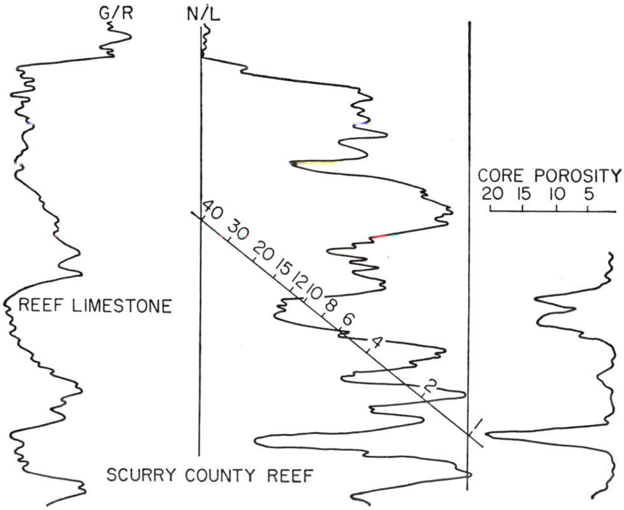

FIG. 20-18. Logarithmic scale applied to log and compared with cored porosity.

some degree of care because the geothermal gradient is often quite different from an oil-producing structure. Composition of the gas, of course, is most satisfactorily determined from a sample obtained from the given stratum and in the particular well in which the survey is being interpreted. Nevertheless composition can be taken in another well in a structurally similar position in the same oil field, provided that the sample from the other well is reliably known to correspond with the same porous medium. The difficulties of the simple interpretation of gas-filled porosity are illustrative of the need for the interpreter to secure every scrap of related information in order to derive the most from the neutron curve. Further, it is often the case that liquid and vapor phases exist together within the pores. In such cases the hydrogen content of the liquid phase will diminish the neutron-curve response making it appear that the rock is liquid-saturated but indicating less than the correct porosity value. Thus an accumulation of gas must be in the vapor state and essentially free of liquid in order to be identified on the neutron curve by way of its extra high response. In reservoirs where this condition is known to prevail the neutron curve provides an excellent indicator of gas and clearly denotes the gas liquid contact by an abrupt transition. The gas-liquid interface can thus be followed from well to well, or from time to time to provide important reservoir performance data.

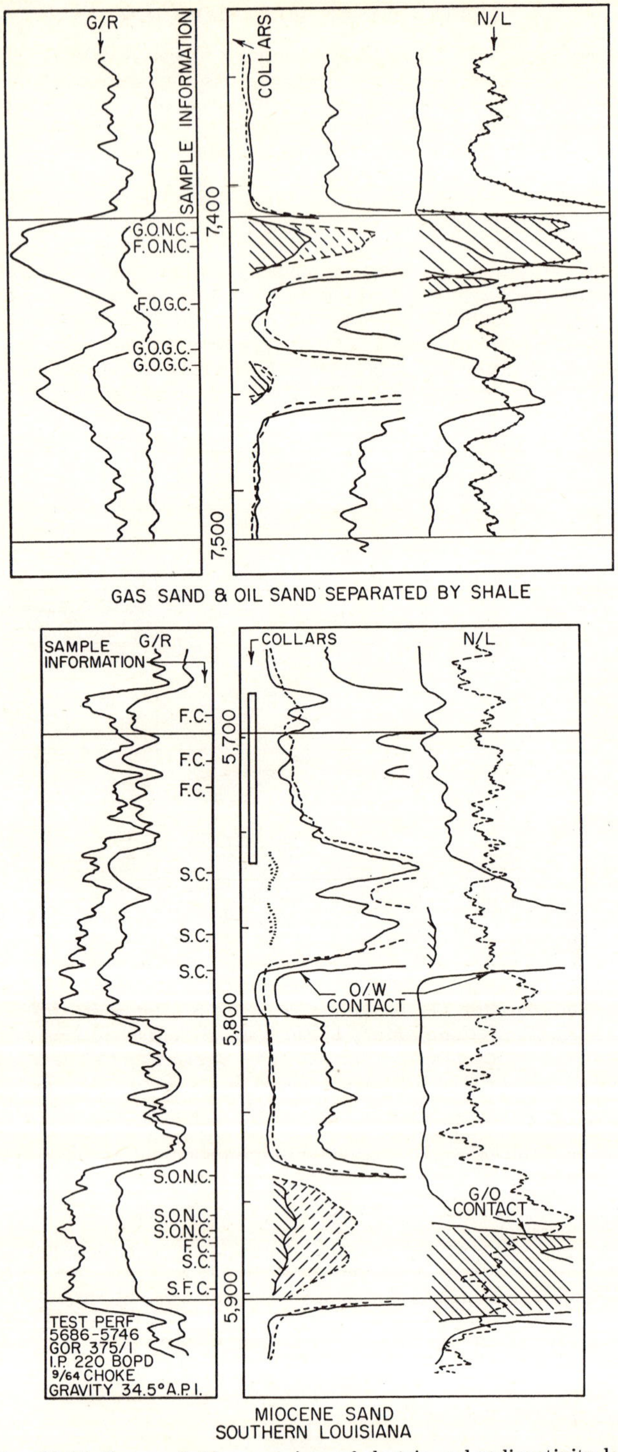

FIG. 20-19. Gas vs. fluid comparison of electric and radioactivity logs.

Oil-water Differentiation. Oil and water are not identical in their hydrogen content, but with the relatively small amounts contained in porous rock slight variations of porosity can alter the total amount of hydrogen to a greater extent than the difference in composition of these liquids. Where the amount of liquid present is relatively great the distinction becomes more apparent and more reliable.

Ordinarily the hydrogen content of oil-filled porous rock is so nearly identical to that of water-filled porosity that the neutron curve fails to differentiate between these two conditions. A small change in the porosity is sufficient to overshadow the slight difference occasioned by the change of fluid.

Determination of Saturation. Where oil and water together occupy the pore space, it is often important to know the proportion of each, in addition to knowing the porosity. A method which has been successfully applied to this problem makes use of the facts that the neutron-curve response depends primarily upon the total fluid content, irrespective of whether it is oil or water, but that the electrical conductivity of the formation depends principally upon its content of saline water. A resistivity

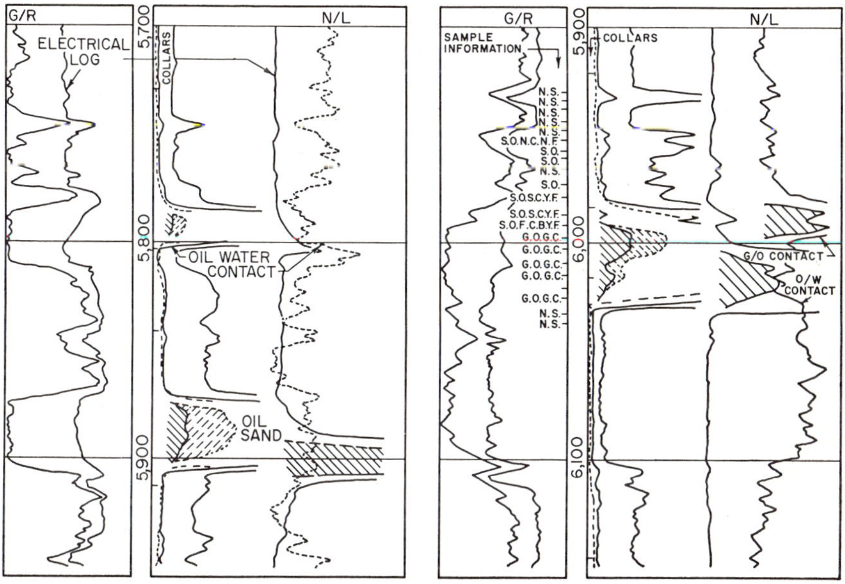

FIG. 20-20. Comparative interpretation of radioactivity and electrical log for oil-water and gas-oil contacts.

curve together with a neutron curve and a knowledge of the actual or expected salinity of the water provide the data required to determine the per cent of oil saturation. Almost always, the radioactivity log is correlated with the electrical log if, of course, one is available, as in Fig. 20-20.

Estimation of Shale Content. Since silt, shale, and clay contain hydrogen, they diminish the neutron-curve response in just the same way as fluid does. The gamma-ray curve, however, is not directly affected by hydrogen. In certain areas the gamma-ray-curve intensity is found to increase in direct proportion to the shale content of the formations as the consequence of a uniform distribution of radioactive material in the shale. Where experience has shown this relationship to be valid, it is possible to obtain an estimate of the shale content from the gamma-ray curve. This knowledge can then be applied in the interpretation of the neutron curve to correct the apparent porosity by an amount corresponding to the shale content. In the absence of data substantiating a different value, it may be assumed that each 2 per cent of shale causes an effect on the neutron curve equal to 1 per cent of liquid-filled porosity.

OTHER RADIOACTIVITY CURVES AND THEIR USES

There are several radioactivity-logging methods which differ from the more widely known gamma-ray and neutron curves. These methods serve a specialized set of purposes, but the information which they can provide is often of primary importance and utility.

The Photon Curve. In principle, this curve is the result of emitting photons (gamma radiations) into the well and detecting a portion of the scattered gamma rays which return to the apparatus. In practice, this is generally done by attaching a source which emits gamma radiations only, instead of a neutron source, to an ordinary neutron-logging instrument of the type which responds to gamma radiation.

Consequently this primary radiation will respond to changes in diameter and changes of fluid density along the well. The most prominent density changes are

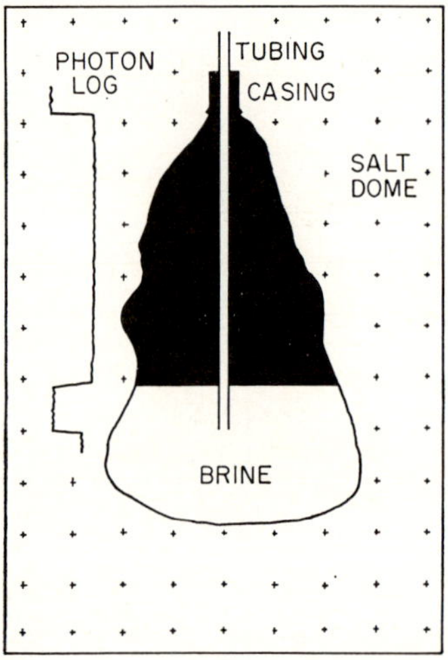

FIG. 20-21. Schematic diagram of typical LPG storage well. Gas-brine interface in LPG storage cavity with photon log.

those occasioned by additional strings of steel casing and by the fluid level in the borehole. Thus the photon curve responds primarily to important borehole characteristics such as casings, liners, fluid level, or diameter changes, and secondarily to formation characteristics. Figure 20-21 typifies one run in a storage well on a salt dome.

The Density Log. Another radioactivity-logging process known as the density log has been developed utilizing the principle of emitting and detecting gamma rays. It is practiced differently from the photon curve in order to emphasize the effect of density changes and minimize the effect of diameter variations. The density-logging apparatus is held against one side of the borehole by a spring arrangement to ensure close proximity to the formations. The gamma-ray source is mounted eccentrically and is heavily shielded in the direction of the wellbore. The gamma-ray detector is similarly shielded except for a "window" on the side adjacent to the formations. In this way the gamma radiation which reaches the detector is effectively limited to that which penetrates into the rock and is there scattered in the direction of the detector. Since the absorption and scattering of gamma radiation depends mainly upon the amount of matter in its path, the density-logging equipment responds primarily to variations in the specific gravity of the material with which it is in contact. Figure 20-22*a* illustrates density characteristics.

Calibration curves are available, based upon the results of experiments with rocks of known specific gravity, which enable the user to convert from the radiation inten-

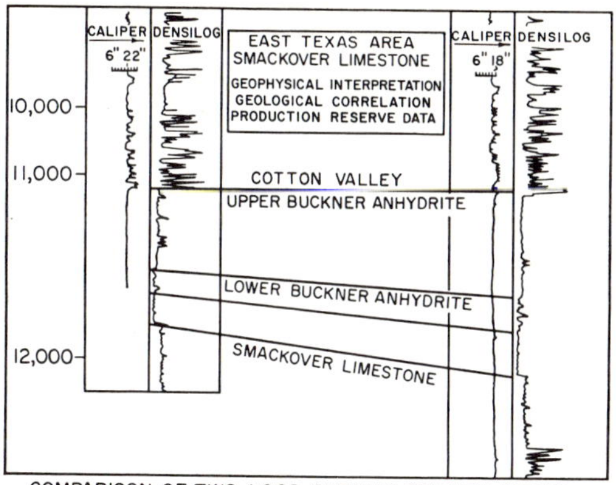

COMPARISON OF TWO LOGS SHOWING DENSITY DIFFERENCE LITHOLOGICALLY

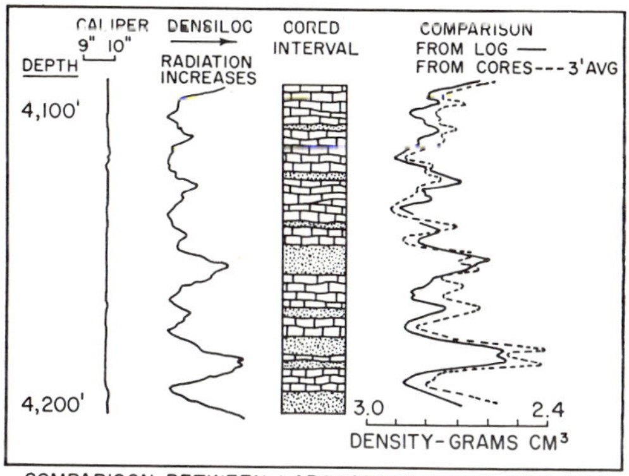

COMPARISON BETWEEN LABORATORY CORED DATA AND LOGGED DENSITIES

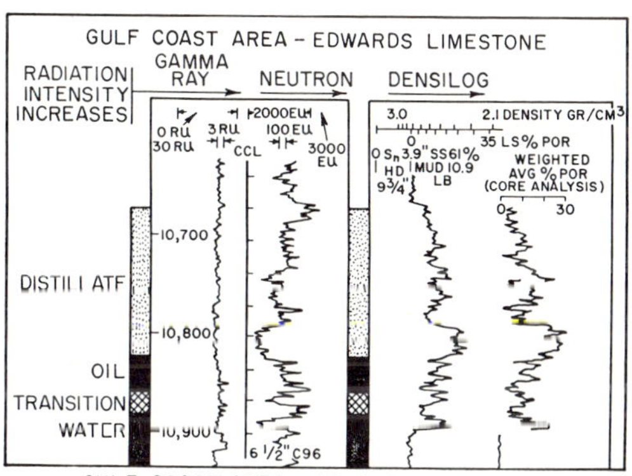

GULF COAST AREA - EDWARDS LIMESTONE

FIG. 20-22*a*. Density characteristics.

sities actually recorded by the density log to the densities of the formations which surround the well. Since it is impractical to eliminate all radiations traversing the borehole, a slightly different calibration curve holds for each diameter. This effect is quite small for any uniform diameter, but should the instrument lose contact with the wall, a large effect takes place. Figure 20-23 typifies a scale for calibration of logs in terms of density and porosity.

A decrease of density results in an increase of the returning gamma radiation and hence moves the curve toward the right. A larger borehole or a less dense fluid produces a similar, though small, change of radiation intensity.

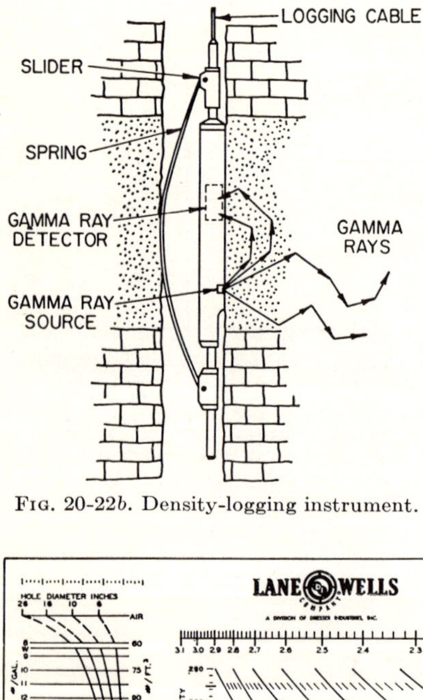

FIG. 20-22*b*. Density-logging instrument.

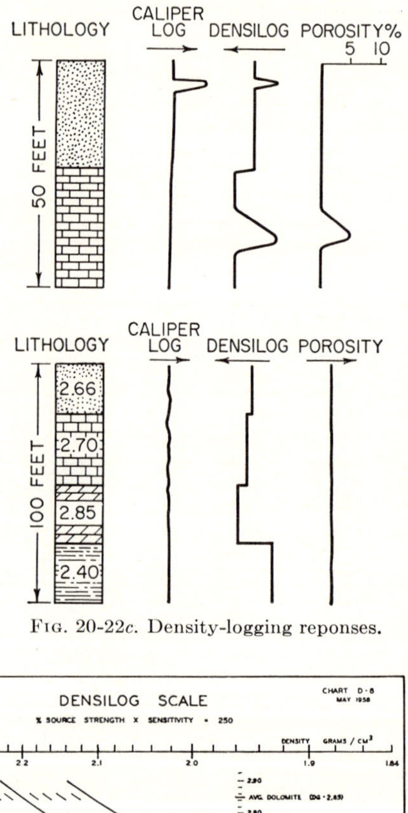

FIG. 20-22*c*. Density-logging reponses.

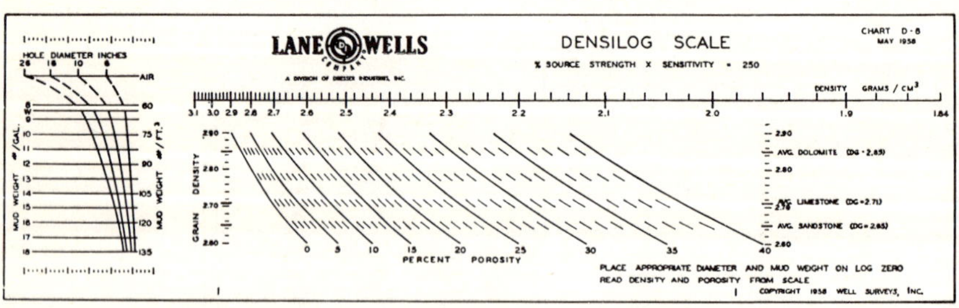

FIG. 20-23. Densilog scale for computing density and porosity.

The density log is applied only in open holes, since the presence of casing and cement would prevent contact with the rock itself. Despite this limitation it serves in many cases to provide valuable information. Density or specific gravity is an important rock property. By comparing the density of the various strata it is possible to identify them and also to infer some of their characteristics. Porosity, for example, is a property which correlates very well with density. It is obvious that the same rock matrix must decrease in density as its porosity increases. The density-log variations within a uniform sandstone or limestone body must therefore represent increases and decreases of its porosity, and this information is of primary concern to the reservoir engineer.

Shales can ordinarily be recognized immediately on the density-log record by their low density and by the fact that their irregular diameter gives the curve a rough appearance. A caliper survey in conjunction with the density log over the section of interest is considered good practice to enable better proper interpretation in terms of density, porosity, and lithology.

Radioactive Tracers. The subject of radioactive tracers is complicated indeed, but for basic simplicity, a few definitions and terms might serve to qualify the use to the petroleum industry.

A radioactive tracer is a material such as a gas, liquid, or solid which has high gamma-ray emission. Upon injection of the material to any portion of a wellbore, the placement or movement can be traced and recorded by a gamma-ray instrument. Originally amorphous tracers were used to determine channeling or travel of a squeezed cement through a section of gun-perforated casing. The trend then favored permeability studies for evaluation of relative permeability in a productive formation. The problem, once a well-to-well program, became a well-to-well program for production, waterflooding, acid and oil fracing, fluid levels, LPG to brine interface, thief zones, lost circulation, and gas tracers in gas wells.

Amorphous tracers could not fulfill the many requirements. Today tracers are available in capsules of gas, liquid, mixed solution, and particles.

Carnotite is a solid, mined as an earth of varying radioactive intensity, and serves as an excellent tracer for cement or plastic. Chemically, it is a hydrous potassium uranyl vanadate. The half life, in the order of 1,590 years, made its use in a well prohibitive for further normal gamma-ray logging.

Radioactive iodine is an isotope and is commonly used as a liquid, as the half life is 8 days and permits normal gamma-ray logging in a reasonable time.

Radioactive iodine is artificially produced, as is antimony with a 60-day half life, iridium with 75 days, and silver with 9 months half life. There are approximately twenty-five different radioisotopes suitable for oil or gas tracer logging by reason of their gamma radiation and ready availability. The type and half life may be considered dependent upon the problem involved.

Liquid tracers assist in evaluating relative permeability and enter the porous and permeable zones. Particle tracers, such as artificially radioactive charcoal, plate against permeable zones to assist in evaluating relative permeability and can be backwashed and cleaned from the wellbore.

Sand tracers are admixtures for injection, fracing, and gravel packing.

Gas tracers supply needed information for injection, production, channeling, and communication of gas-producing or storage wells.

Radioactive tracers are administered by varied methods:

1. Pumping runs vs. time logging to determine rate through casing.
2. Controlled pumping runs vs. time logging to determine rate through tubing.
3. Dump bailer with a frangible glass disk on bottom. The tracer is ejected by firing a blasting cap at the desired depth.

Most tracer problems confine the radioactive material to a short section of well-fluid column, such that the gamma-ray logging tool which is in the hole at the time of application may determine limits of height, travel, time, displacement, or loss by a quantitative or qualitative method.

Tracers for oil-field use range from 1 or 2 millicuries for a single-well study to 100 millicuries or more for interwell problems.

Safety is of prime importance. Safety equipment is carried to the well site in kit form by the logging company. This kit contains

1. Dosimeters for personal exposure
2. Film badges
3. Low-level survey meter for radioactive contamination

4. High-level survey for monitoring operations
5. Gloves, respirator, goggles
6. Also required, shovels, water supply, and waste containers

Tracers are commonplace in oil-field problems and are restricted only by the proper choice of isotope to satisfy an evaluation requirement completely.

Radioactive Tracers

Type of preparation	Isotope	Tracer material			
		Iodine 131	Antimony 124	Iridium 192	Silver 110
	Half life	8 days	60 days	75 days	9 months
	Relative radiation intensity	0.4	2.1	0.6	2.4
	Max per package	100 mc	50 mc	100 mc	25 mc
Solution tracer	....	✓	✓	✓	✓
Particle tracer	....	✓	✓	✓	✓
Sand tracer	....	...	✓	✓	✓
Gas tracer	....	✓			

Cement Log. The cement log results in a measurement of the cement behind the casing regardless of the age of the well. Temperature, time, chemical, and physical conditions of the wellbore have little effect upon the curve.

The cement log is a special form of photon log in which the basic measurement is the measurement of scattered gamma radiation from the source. The detecting of gamma radiation is primarily dependent upon the absorbing effect of the cement to photons.

Generally, well-cemented casing absorbs photons, and lack of cement increases photon effect. Therefore, the presence or lack of cement is qualitatively determined by the departure between a correlated or superimposed neutron and cement log.

The cement log, as a supplemental curve to the neutron log, measures the height of the cement column behind the casing and is further interpreted in qualitative terms of channeling.

Figure 20-24 shows the visible departure between the neutron and the cement log when directly compared.

Chlorine or Salinity Logging. For a number of years radioactivity-logging devices have been used to define lithology, bed thickness, permeable intervals, and porosity both behind casing and within open holes. The need has long existed of defining oil and water saturation either quantitatively or qualitatively within intervals already cased off, and it is now feasible to differentiate oil- or gas-bearing from salt-water-bearing zones with a combination neutron curve and a chlorine curve. The two curves fall along the same line opposite oil-filled or fresh-water-filled zones. Opposite gas-filled zones, the curves indicate very low hydrogen density with corresponding lower chlorine content. Opposite salt-water-filled intervals, the two curves depart widely from each other in recorded response and thus offer a graphic means of distinguishing hydrocarbon-bearing from nonproducible horizons. Figure 20-26 compares response for oil in the upper sand vs. water in the lower sand.

There are a number of methods that may be used to determine the presence of chlorine. Among these are the neutron-neutron method, the neutron-gamma method,

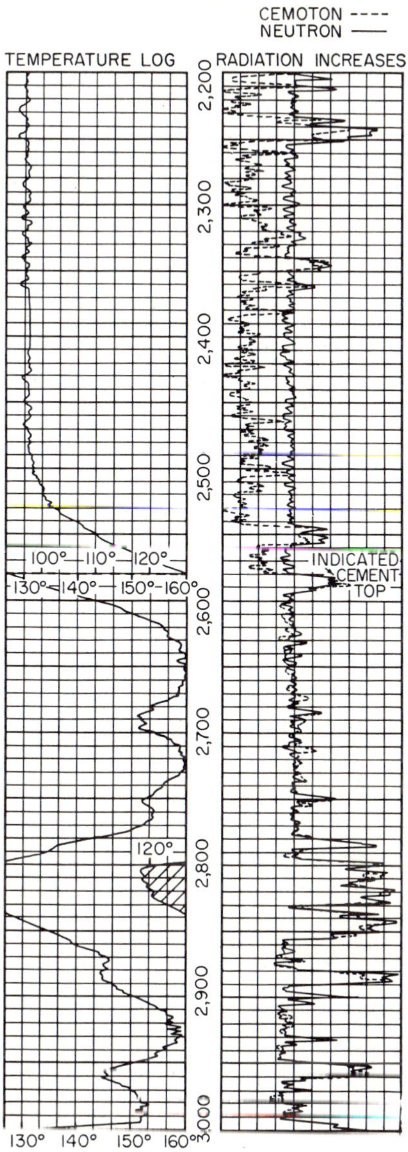

FIG. 20-24. Comparative departure between cement log and neutron log indicates presence or lack of cement behind the casing.

and the spectral-analysis method. The instrument used to record both the H curve and the Cl curve is of the scintillation type.

After a neutron becomes thermalized, hydrogen no longer has the dominating influence. Neutrons are more readily captured by the elements with the larger neutron-capture cross sections. The cross section of chlorine is approximately 100 times that of hydrogen; therefore it is apparent that a small amount of chlorine can have a relatively large effect.

In new wells, invading mud filtrate commonly fills the pore spaces of the formation near the borehole, and little distinction can be made between the standard neutron and chlorine curves. For this reason the log should be run in cased holes that have been cemented longer than 72 hours in order to allow equilibrium of filtrate and formation fluid to take place. Logging old wells is very common, and they are more easily analyzed qualitatively and quantitatively.

The chlorine log is presented as two curves. One, the neutron, responds primarily to the hydrogen content of the formation and can be used for porosity determination. The other, the chlorine, responds to the hydrogen content as well as to the chlorine

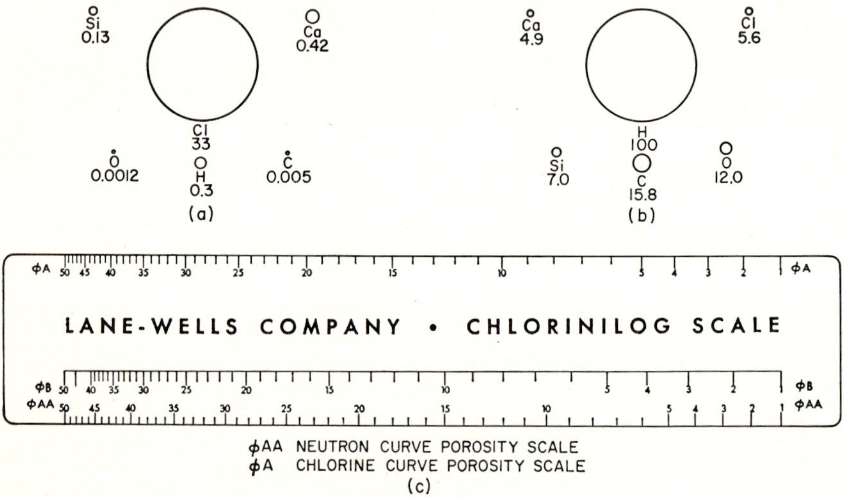

FIG. 20-25. (*a*) Relative cross section of neutron capture; (*b*) relative effectiveness in slowing down neutrons; (*c*) chlorinilog scale.

content of the formation. These curves have been designated as the H curve and the Cl curve, respectively. In zones where little or no chlorine exists, the two curves will essentially superimpose and similar porosities will be indicated by both curves. However, where chlorine exists the Cl curve will indicate a lower porosity than that indicated by the H curve. In salt-water sands, the two curves will not superimpose. This difference of apparent porosity is caused primarily by the amount of chlorine in the formation.

Proper recording permits a determination of approximate water saturation behind casing. Therefore, some particular zone or interval can be equated against another interval for the water saturation. Let us consider the following expression:

$$\left[\frac{\phi n\ (\text{ppm})\ Sw\%}{\Delta\phi}\right]_1 = \left[\frac{\phi n\ (\text{ppm})\ Sw\%}{\Delta\phi}\right]_2$$

where 1 = known sand
2 = unknown sand
ϕn = neutron-derived porosity
ppm = parts per million salinity of formation water
$Sw\%$ = water saturation, per cent
$\Delta\phi$ = difference in indicated porosity between neutron curve and chlorine curve

Porosities for the neutron and chlorine curves are read directly from a scale or chart. Therefore, it can be seen that a semiquantitative evaluation of fluid-bearing horizons can be achieved even though dissolved-salt content is not accurately known.

To evaluate apparent salt-water saturation in various sands or zones of interest, the neutron curve also can be plotted in units of amplitude vs. the chlorine curve in

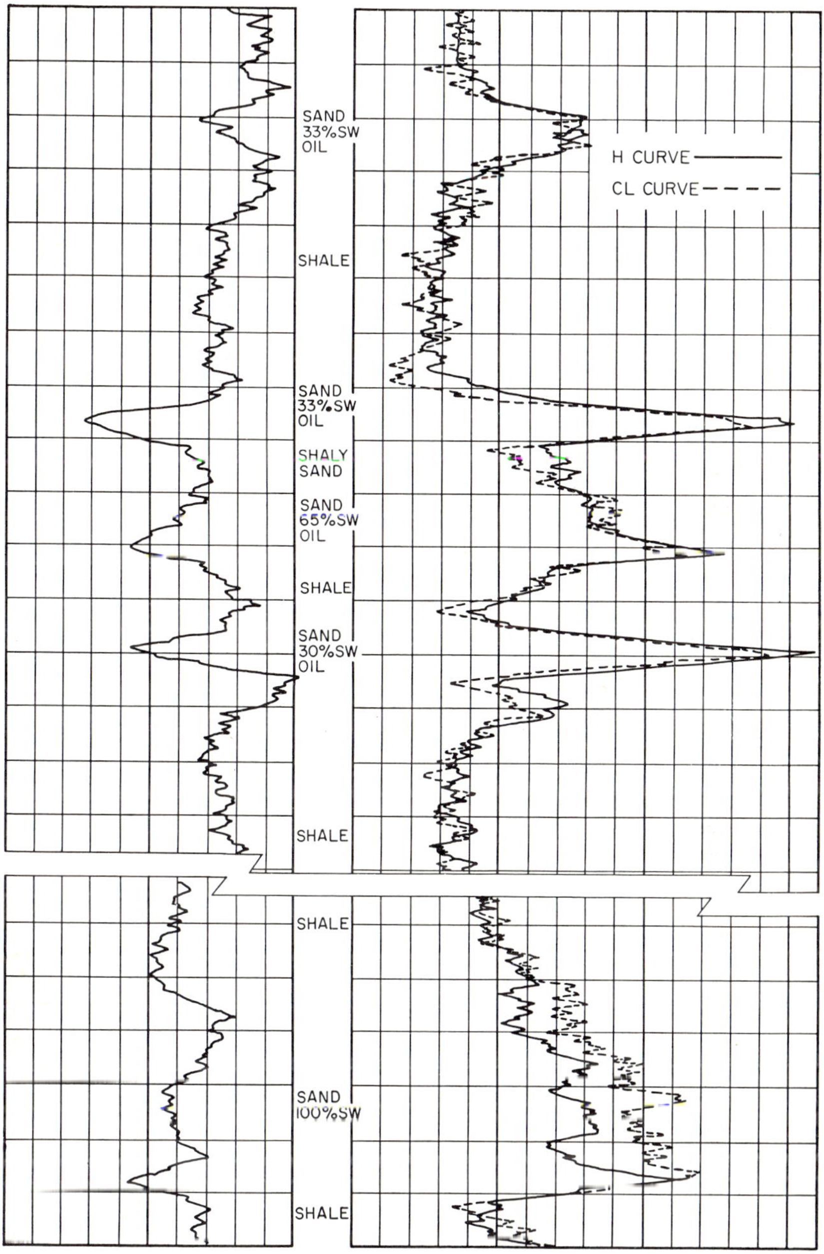

Fig. 20-26. Chlorinilog.

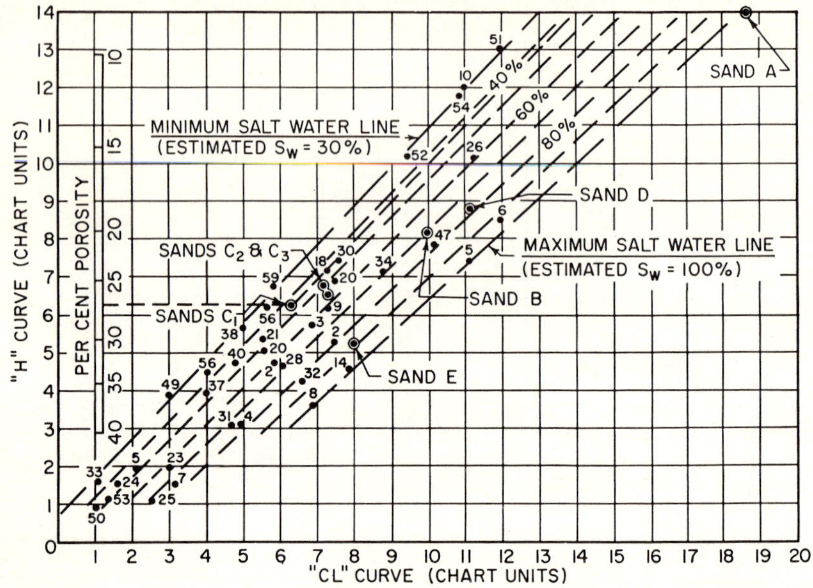

Fig. 20-27. Chlorinilog scatter diagram.

$\emptyset_N$ FROM $\emptyset_{AA}$ (FOR D_6N_1)
$\emptyset_{CL}$ FROM $\emptyset_A$ (FOR D_4N_5)
$\emptyset_{CL}$ FROM $\emptyset_B$ (FOR D_4N_4)
ABOVE $\emptyset_{AA}$, $\emptyset_A$, AND $\emptyset_B$ SCALES FROM LANE-WELLS RADIOACTIVITY SLIDE RULE OR POROSITY SCALE

KNOWN SAND DEPTH	E	UNKNOWN SANDS	D	C3	C2	C1	B	A
$\emptyset_{N_1}$	30.5	$\emptyset_{N_2}$	21.5	28	27	30	23.5	9
$\emptyset_{CL_1}$	24	$\emptyset_{CL_2}$	15.5	25	24	27	17.5	4.5
$\emptyset_{N_1}$-$\emptyset_{CL_1}$ OR $\Delta\emptyset_1$	6.5	$\Delta\emptyset_2$	6.0	3	3	3	60	4.5
Sw_1	100%	Sw_2	83%	45 TO 50%			90%	45 TO 50%

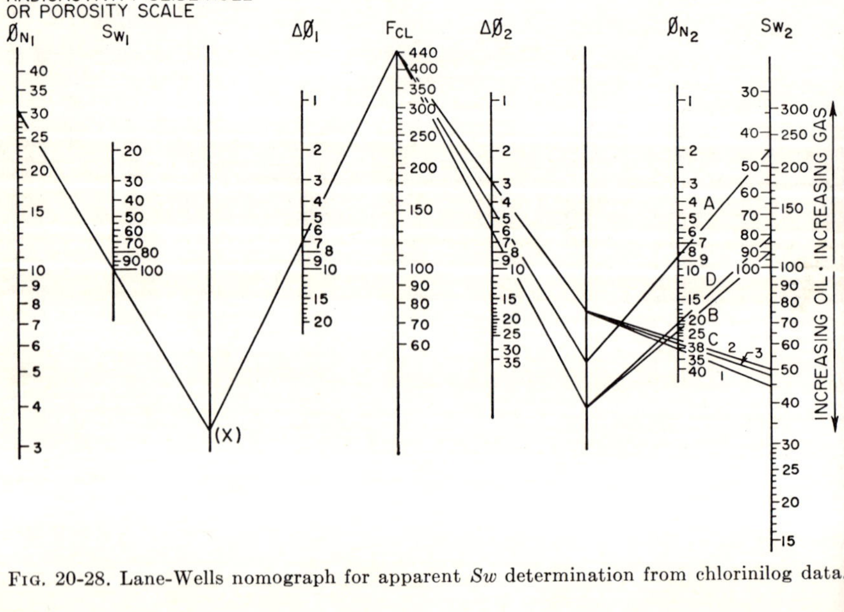

Fig. 20-28. Lane-Wells nomograph for apparent *Sw* determination from chlorinilog data.

like units. This provides a "scatter diagram" which properly divided in water saturation, depending upon the geologic sequence, evaluates all zones and permits zones for recommended tests for production. The "scatter diagram" (Fig. 20-27) evaluates gas, oil, water, and possible contacts. The neutron ordinate, in apparent hydrogen per cent porosity, assists in the solution of zones of interest. Particular sections are confirmed through the equation of evaluation or simplification by use of a nomograph (Fig. 20-28).

It is now possible, using a chlorine curve and a neutron curve, to perform some of the operations through the casing that formerly were possible only in open-hole wells, thus opening many old wells as well as newly drilled boreholes to evaluation.

RADIOACTIVITY-WELL-LOG APPLICATIONS

Engineering and Production Studies

1. Depth measurements
2. Cased hole, open hole, or through tubing
3. New wells or old wells
   a. Production
   b. Bottom-water shut-off
   c. Lower ratios
      (1) Gas-oil
      (2) Gas-water
      (3) Oil-water
   d. Casing seats and liners
   e. Casing and tubing collars
   f. Packers
   g. Salt-water disposal
   h. Fresh-water production
   i. Radioactive crusts or beds
   j. Flowing neutron for quantitative or qualitative ratios
   k. Subsequent gamma-ray log for producing zones
   l. Subsequent neutron log for apparent fluid or gas content

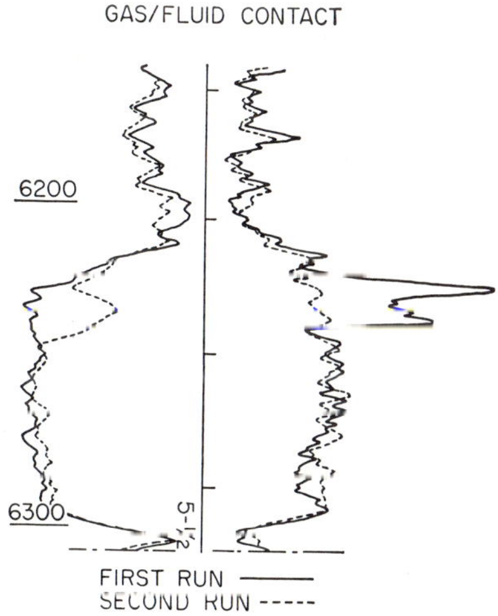

FIG. 20-29. First- and second-run gamma ray for radioactive crust, and neutron for comparative gas-fluid contact.

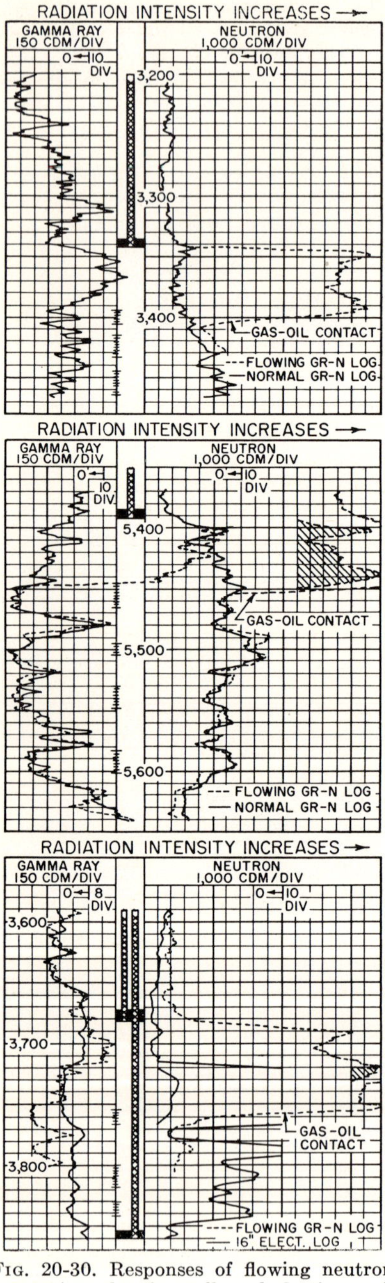

FIG. 20-30. Responses of flowing neutron comparing shut-in well and flowing well through perforations.

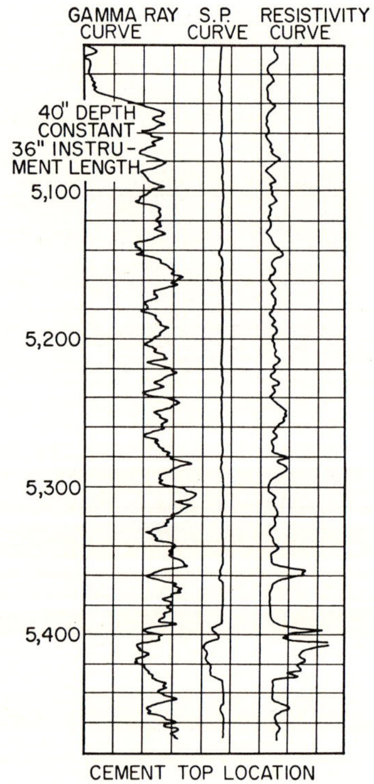

FIG. 20-31. Tracers.

4. Open hole or cased hole
   a. Contaminated mud, oil, fresh or salt water, empty
   b. Deepening, from correlation of logs
   c. Attempting blow-out; log through drill pipe
   d. Borehole changes

Tracers

1. Cement travel
   a. Channeling
2. Plastic travel
3. Injection
   a. Fluid (include isoflow)
   b. Gas
4. Fracing
5. Acidizing
6. Waterflooding
7. Permeability profile
8. Thief zones
9. Lost circulation
10. Casing or tubing leaks

Evaluation Studies

1. Porosity
   a. Quantitative
   b. Qualitative

Geological Studies

1. Structure
   a. New or old wells
   b. With or without other available logs
2. Correlation
   a. By wells
   b. Group section
   c. Zoning
   d. Sedimentary markers
3. Depth measurements
4. Mapping
   a. Local
   b. Regional
5. Sedimentation
   a. Shale content
   b. Radioactive bedding
   c. Marine or lignite markers
   d. Normal-bed definition
6. Thin-sand definition
7. Subsidence

Special Radioactivity Well Logs

1. Photon log
   a. Fluid level for LPG storage
   b. Brine storage
   c. Casing record
   d. Well diameters
   e. Cement top for old and new wells
2. Density log
   a. Gravity-meter studies
   b. Porosity, and rock-density studies
   c. Mining

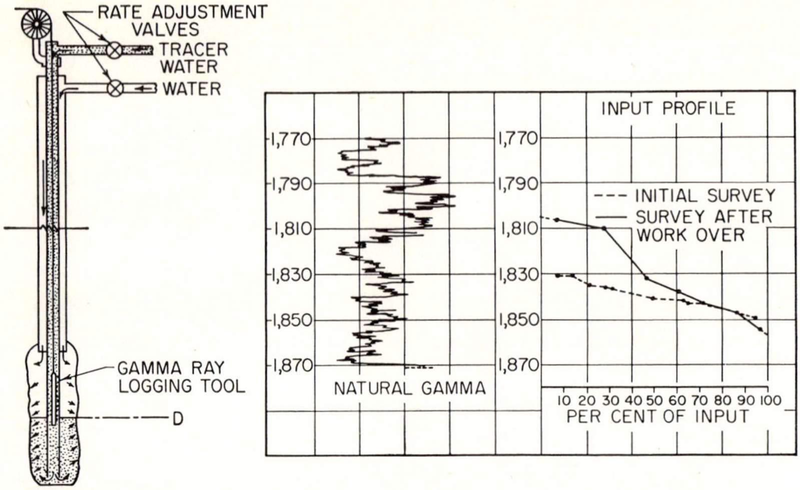

FIG. 20-32*a*. Tracers—isoflow logging. In this process, tubing is run to a level below the horizon to be surveyed, or as close to bottom as practical. The water through the tubing and casing is metered independently. A soluble radioactive material is then injected into either stream. This provides one stream free of radioactivity. An interface thus is established in the annulus above which the water routed through the tubing enters the formation. This interface is then located with a small gamma-ray counter which is operated in the tubing. By varying the per cent of total injection water entering the tubing and annulus the position of the interface may be changed at will. The amount of change, vertically in feet, is directly related to the change in input in barrels. By establishing a number of stations, a precise foot-by-foot profile of the borehole, telling where every barrel of fluid goes, is obtained. This profile is independent of hole size and injection rates.

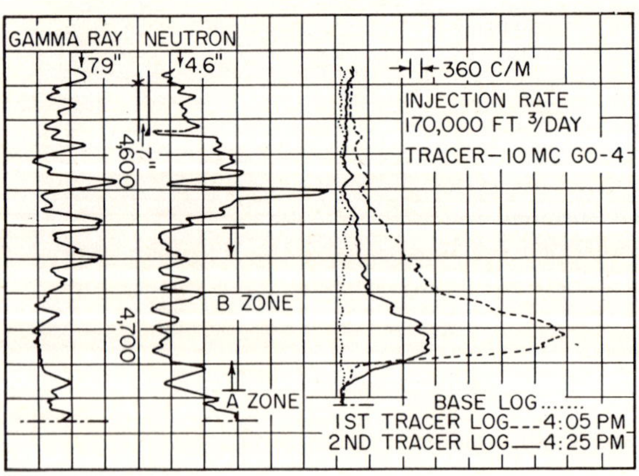

FIG. 20-32*b*. Tracers—gas-injection profile.

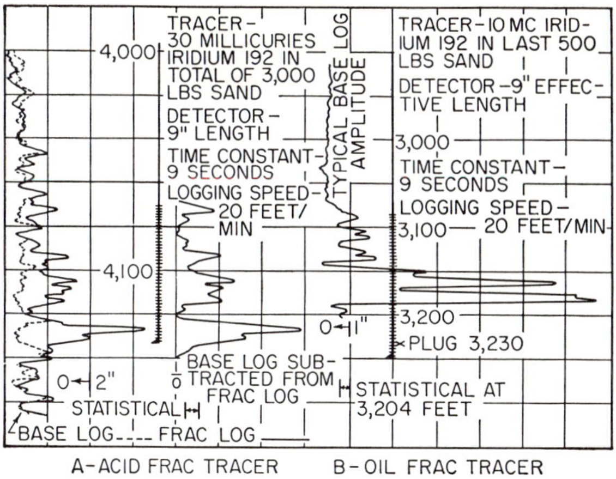

FIG. 20-33. Tracers.

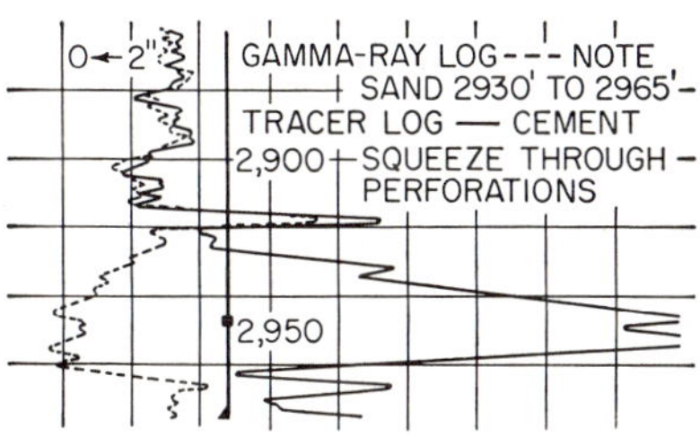

FIG. 20-34. Squeeze-cement tracer. Survey is obtained by mixing the radioactive tracer with the cement. The radioactive cement reveals to the detector its top, bottom, and extent of travel.

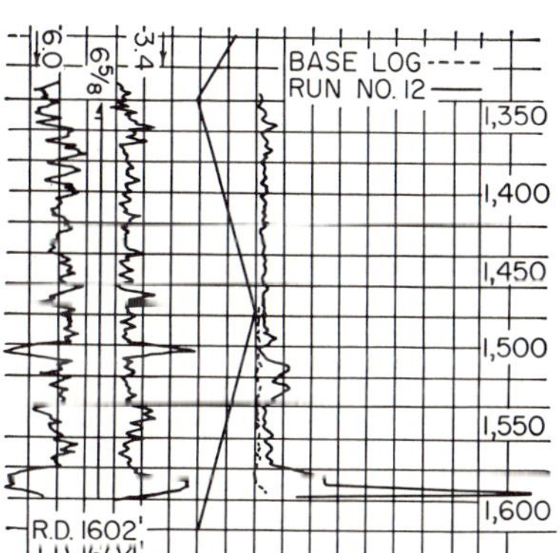

FIG. 20-35. Waterflood tracer. From this type of waterflood injectivity profile a definite measure of permeability is available to the operator.

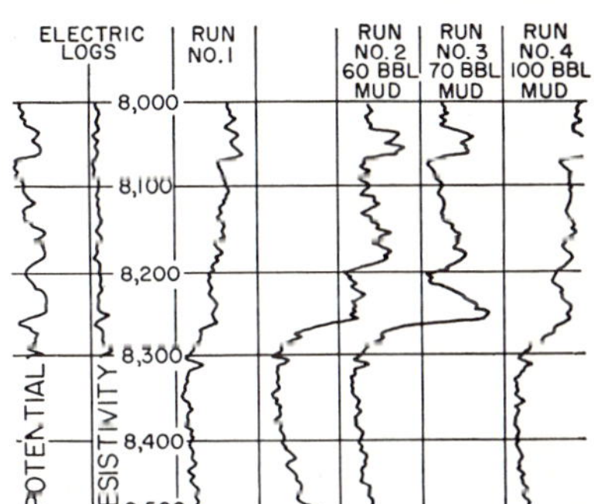

FIG. 20-36. Lost-circulation tracer. This survey gave the operator the exact information he needed to take remedial measures and continue drilling.

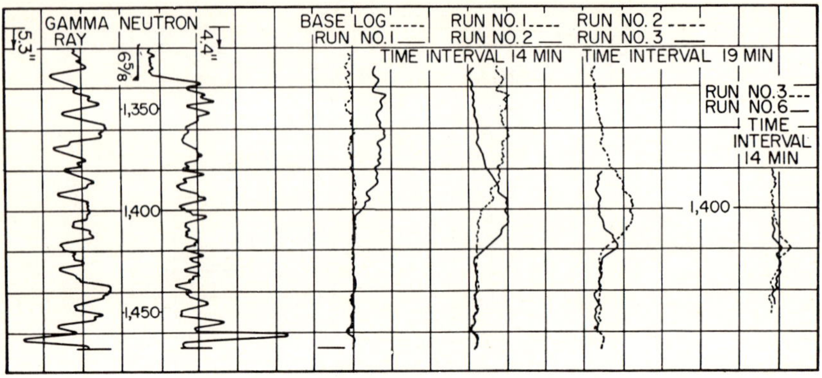

FIG. 20-37. Tracer particle loss.

REFERENCES

1. Atkins, E. R., Jr.: Radiation Logging, *Oil and Gas J.*, July 16, 1956, pp. 88–93.
2. Brown, A. A., and Bryan Bowers: The Relationship between Neutron Log Deflection and Porosity, *Can. Oil and Gas Inds.*, vol. II, no. 3, pp. 59–63, March, 1958.
3. Bush, R. E., and E. S. Mardock: The Quantitative Application of Radioactivity Logs, *J. Pet. Tech.*, vol. III, no. 7, pp. 191–198, July, 1951.
4. Fearon, R. E.: Gamma Ray Logging and Neutron Well Logging, *Nucleonics*, vol. 4, nos. 4, 6, April, June, 1949.
5. Flagg, A. H., J. P. Myers, J. L. P. Campbell, J. M. Terry, and E. S. Mardock: Radioactive Tracers in Oil Production Problems, *J. Pet. Tech.*, January, 1955.
6. Jackson, W. J., and J. L. P. Campbell: Some Practical Aspects of Radioactivity Well Logging, *J. Pet. Tech.*, September, 1945.
7. Halliburton Oil Well Cementing Company: Flow Logging, March, 1955.
8. Mercier, V. J.: Radioactivity Well Logs, Interpretation and Application, *Oil Weekly*, Oct. 14, 21, 1946.
9. Norelius, R. G.: Radioactivity Tracer Logging with Oil Well Cementing, *Pet. Engr.*, vol. 29, no. 12, November, 1957.
10. Russell, W. L.: Interpretation of Neutron Well Logs, *Bull. AAPG*, vol. 36, no. 2.
11. Swift, G., and R. G. Norelius: The Multispaced Neutron Log, *Oil and Gas J.*, Oct. 15, 1956, pp. 109–113.
12. Tittle, C. W., and M. R. J. Wyllie: How to Get the Most out of Electric and Radioactivity Logs, *Oil and Gas J.*, Nov. 21, 1955.
13. Wyllie, M. R. J.: Procedure for Direct Employment of Neutron Log Data in Electric Log Interpretation, *Geophysics*, vol. XVII, no. 4, October, 1952.
14. Stroud and Schaller: A New Nuclear Log for the Determination of Reservoir Salinity, SPE 1958 fall meeting, *J. Pet. Tech.*, February, 1960, revised.
15. Waddell and Wilson: Chlorinity and Porosity Determination in Cased and Open Holes by Nuclear Logs; SPE Midland, May, 1959.

Readers who look up these papers will find the numerous additional references which they contain.

Chapter 21

MISCELLANEOUS WELL LOGS*

By A. J. PEARSON

Petroleum Engineer
Continental Illinois National Bank
and Trust Company of Chicago

INTRODUCTION

Many well logs are not generally classified as electrical or nuclear logs. The most important of these logs are discussed in this chapter. Some of these logs actually are based on principles similar to electrical logs but, because of specialized use, are not classified as electrical logs.

Included as miscellaneous well logs are the drilling-time, geologic-sample, hydrocarbon-mud, directional, dipmeter, caliper, temperature, acoustic-velocity, and collar-locator logs. All these are used by the petroleum engineer on occasion. The logs are discussed in the order listed. This order was selected for convenience and has no relationship to the importance of the logs.

DRILLING-TIME LOG

Introduction

Drilling time is the time required to rotary-drill a linear unit of formation. Thus the drilling-time log is a foot-by-foot record of how fast a formation is drilled.

The drilling-time log is the first log recorded on a well. It is normally recorded 10 to 30 minutes before the sample log and hydrocarbon-mud log (because of the time required for the mud stream to carry cuttings from the bottom to the top of the hole). It may be recorded days before the electrical log. With present-day techniques most of the information supplied by the drilling-time log is provided more accurately by other logs. But the drilling-time log furnishes subsurface information *while* the well is being drilled.

Drilling-time logs have been recorded since the earliest days of the oil industry. Published references are available from 1880.

Methods of Recording

For many years drilling-time logs were made by marking the kelly in 1-ft intervals and recording the time when each mark reached the kelly bushing. This method is still used in some areas. Disadvantages of this method lie in inaccurate marking—some feet may be 10 in., others 14—and the human element in correctly recording the time.

* Reprinted by permission from "Petroleum Production Handbook," edited by Thomas C. Frick, McGraw-Hill Book Company, Inc., New York, 1961.

In 1943, mechanical methods of recording drilling time were introduced. The most common recorder is actuated by a cable which is fastened to the rotary hose-swivel coupling. The cable passes over a pulley at the crown and down to the recorder where it is held by a spring-activated drum. The cable operates a metering wheel in the recorder. At intervals of 1 ft (or 5 ft if desired), the recorder makes a mark on a time-actuated chart. A portion of a typical log is shown in Fig. 21-1. The spacing of the marks indicates how fast the hole was drilled. The close spacing of the marks over interval *A* indicates fast drilling and, in this case, a porous sand. Parallel to the drilling-time column, a second line records drilling operations. When the kelly is raised, this line moves to the right (as at 12:07). When drilling is resumed, the pen returns to its normal position.

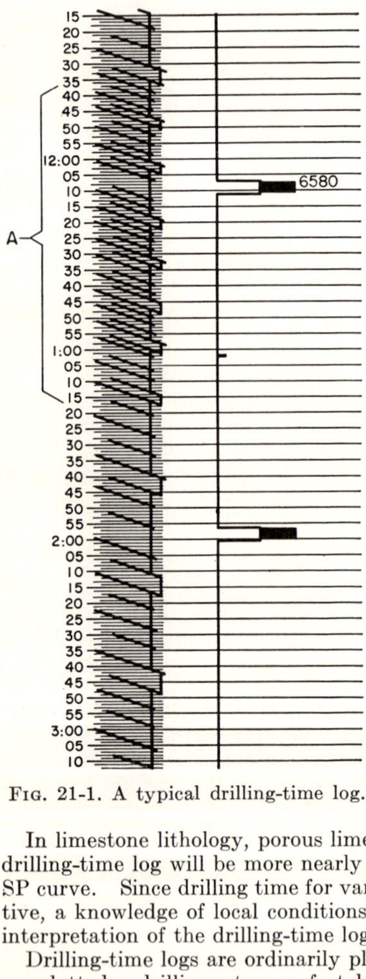

Fig. 21-1. A typical drilling-time log.

Another method uses hydraulic principles to record the drilling-time log. A small water reservoir is attached to the rotary swivel and connected to the recording mechanism. Lowering the kelly decreases fluid pressure on the recorder. The pressure on the recorder is correlated with depth.

Interpretation

The time required to rotary-drill a formation is a function of two things: the type of rock and the conditions under which the rock is drilled. Drilling time cannot be quantitatively correlated with formation properties because it varies with hole size, type and condition of the bit, drilling weight, speed, torque and friction, and mud type and condition. Hence we cannot state that a drilling time of 5 min/ft indicates a particular type of rock. But drilling conditions normally do not change radically while short intervals of a hole are drilled. So in a particular well a change in drilling time from 15 to 5 min/ft may show that a different type of rock has been penetrated.

Porous sands are usually drilled at a faster rate than shales. In areas of predominantly sand-shale lithology, the drilling-time log will closely correlate with the self-potential curve. An example of this comparison is shown in Fig. 21-2, which was prepared using the log shown in Fig. 21-1.

In limestone lithology, porous limes are drilled faster than nonporous limes and the drilling-time log will be more nearly comparable with the neutron log than with the SP curve. Since drilling time for various formations is relative rather than quantitative, a knowledge of local conditions or information from other logs is necessary for interpretation of the drilling-time log.

Drilling-time logs are ordinarily plotted as shown in Fig. 21-2. Occasionally they are plotted as drilling rate, e.g., feet drilled per hour Plotting as drilling rate averages the data and may be preferred for sections of rapid drilling. Plotting as drilling time (though tedious) gives a more detailed picture of lithology. Of course, visual inspection of the drilling-time record (see Fig. 21-1) may be sufficient for many uses.

Drilling-time logs are not always interpretable. For example, in drilling the shallow part of the hole in the Gulf Coast, the hole is jetted or washed down using a fishtail bit. In this case drilling time may not reflect changes in the formation. Likewise,

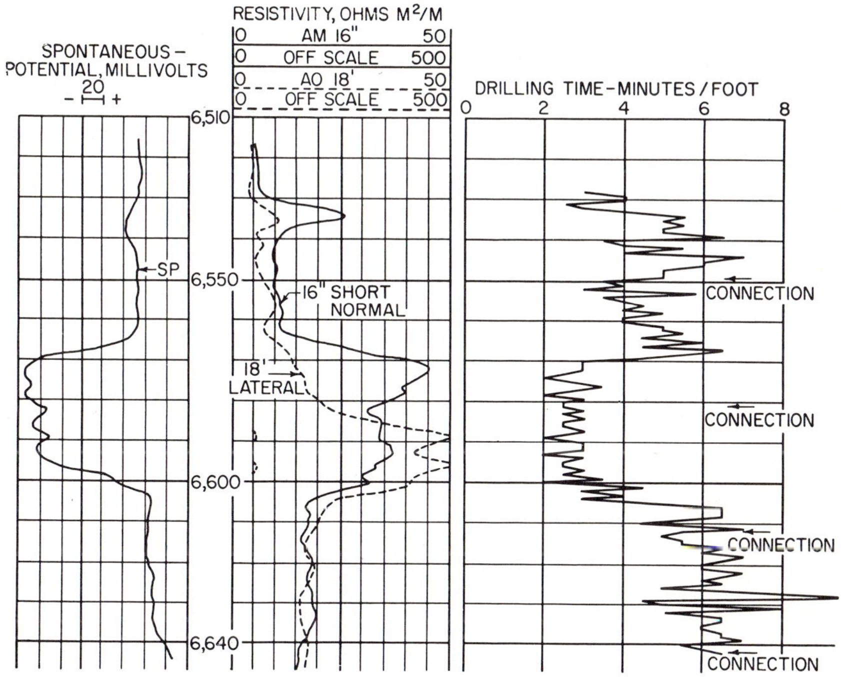

FIG. 21-2. Correlation of drilling time with electrical log.

when holding up on weight (as in trying to straighten a crooked hole) drilling time may be meaningless.

Use of the Drilling-time Log

In considering use of drilling-time logs, we must remember that it furnishes information while the well is being drilled—long before the electrical log is available. Many of its uses rely on this fact.

Perhaps the most important use is the detection of porous zones. If a drilling break (decrease in drilling time) is encountered on an exploratory well, drilling can be immediately halted and samples circulated out. If samples are favorable, a drill-stem test can be made of the porous interval already penetrated and the remainder of the zone cored.

Another important use is correlation—often in conjunction with the geologic-sample log. By plotting the drilling-time log and comparing with nearby wells, the well can be correlated as it is drilled. This immediate correlation has many uses: selecting coring and drill-stem test intervals, picking packer seats for open-hole drill-stem tests, determining whether well has been drilled to the desired total depth, determining casing point, and many others.

A few other uses of the drilling-time log are (1) establishing control points for the sample log, (2) determining sample lag, (3) checking to see that samples were caught properly at the rig, (4) interpreting lithology of unrecovered portions of a core, (5) providing correlative information if the hole is lost before it is electrically logged, (6) correlating driller's measurements with electrical log. (In Fig. 21-2, the electrical-log total depth was 6,645 ft; driller's total depth was 6,660. Since the drilling-time log correlates closely with the electrical log at the producing sand, we can conclude that the electrical log failed to reach total depth.)

In the preceding paragraphs we have covered the uses of the drilling-time log to the reservoir engineer and geologist. There is one other very important use for the drilling-time log. To the driller and drilling contractor, the drilling-time log is an account of the rig's performance. The time required to make trips, connections, and repairs is shown by the mechanical recorder. The data for analysis of both "down time" and "pay time" (for example, bit performance) are shown on this log. The importance of these data in the very competitive drilling industry is apparent.

GEOLOGIC-SAMPLE LOG

Introduction

The geologic-sample log is a record of the rocks penetrated in drilling a well. It is prepared by describing the rock type as determined from a microscopic examination of the rock cuttings or samples recovered at the surface.

In areas where "soft rocks" are found, electrical logs and side-wall cores have largely replaced the geologic-sample log. Samples in soft rocks are used primarily for paleontology studies. But in the "hard-rock" country, most oil and gas reservoirs are located by the geologist's inspection of rock cuttings (long before the electrical log is run).

The terms "soft" and "hard" rock are used widely but are not closely defined. "Soft rock" is usually used to refer to the post-Cretaceous sediments. These rocks are drilled relatively rapidly, and samples are difficult to classify as to exact depth. The samples are poor for pay formations because of the unconsolidated nature of the sands; often only sand grains will be recovered. Also the typical sand-shale lithology of these sediments provides the most easily interpreted electrical log. The "hard-rock," or pre-Cretaceous, sediments contain many limestones, dolomites, and anhydrites in addition to sands and shales. They are drilled relatively slowly, and rock cuttings are usually readily identified as to depth interval. Perhaps most important is the fact that rock cuttings from the "hard rocks" are usually consolidated—really small chips of the formation. Most of the following discussion of the geologic-sample log will be applicable primarily to the "hard-rock" country.

Collecting Samples

Samples from a rotary-drilled hole are caught from the mud stream returning from the bit. The samples are collected each time a specified interval of hole is drilled. In "hard-rock" country, samples are usually caught in 5- or 10-ft intervals but 1-ft intervals may be used in pay formation. In "soft-rock" areas samples are caught at intervals varying from 20 to 50 ft.

It takes the returning mud stream some time to reach the surface (1 min per 100 ft of depth as an order of magnitude). It is necessary to correct for this "sample lag" to determine the true depth of which the sample is representative. Sample lag can be determined by inserting some easily identified material in the mud column when adding a joint of drill pipe. Dyes are sometimes used but may be misleading since lag time of the cuttings may be greater than lag time of the mud. Hence a solid material of about the same size and density of the cuttings is most desirable. The time required for this material to return to the surface is recorded. By subtracting the time from the surface to the bit (easily calculated from pump pressure and volume), sample "lag" is determined. Samples may be more accurately positioned by comparison of the sample with the drilling-time log, as explained in discussion of that log.

Often when a drilling break is encountered in an exploratory well, drilling is immediately halted and samples circulated out. The time required is a good indication of sample lag. Of course, the sample from the zone of interest is accurately positioned when samples are circulated out.

Various methods are used actually to catch the sample from the mud stream. In general, all methods divert a small portion of the returning mud stream to a device for separating the rock cuttings from the mud. Such devices include rotary screens, vibrating screens, or small settling tanks. After the sample is caught, it is washed and dried.

Cable-tool samples are usually caught from the first bailer after each run of the bit. In zones of interest the bailer is usually run each time a driller detects a change in the formation being drilled. Hence cable-tool samples are not caught at regular intervals —one sample may represent 8 ft, the next 2 ft.

Sample Quality

The quality of samples from rotary-drilled holes varies. The most common cause of "bad" samples is contamination with rock cuttings other than those actually drilled in the interval sampled. If thick shale beds are present up the hole, there is almost always some contamination of samples from caving of these beds. (In some areas, such as the Permian Basin, much shale-free hole may be drilled, and there will be little contamination.) The amount of caving and resulting sample contamination can be reduced greatly by using a good mud to drill the hole. Another serious problem in "hard-rock" areas is lost circulation. Many times sections of the hole are drilled with intermittent loss of circulation. Samples from a well plagued with lost circulation are usually badly contaminated and may be completely unrepresentative of the section drilled. If a mud with insufficient gel strength to lift the cuttings is used for drilling, samples may be powdered—ground very fine by redrilling. Such samples are very difficult to interpret. If the mud pit is small and a shale shaker is not used, cuttings may even be recirculated and so add to sample contamination.

Another factor affecting sample quality is the human element. Samples are usually caught and washed by a roughneck. If he is not too conscientious, samples may not be caught properly. For example, if a roughneck forgets to catch a sample for a given interval, he is tempted to "spread out" the sample from the next interval. Of course, inadvertent errors in labeling may add to poor sample quality.

The fact that samples are contaminated does not mean that they are worthless. A trained geologist will not be troubled by a normal amount of contamination. But, as sample contamination increases, the reliability of the sample log decreases.

Cable-tool samples are usually relatively uncontaminated. This is true because caving in a cable-tool hole, if severe, prohibits drilling until a protective string of casing is run.

Examination of Samples

The geologist examines samples using a low-power microscope (usually 12- to 24-power). The power should be high enough that the structure and texture of the cuttings are readily visible. The samples may be examined either dry or covered with water. In the Mid-Continent area the samples are usually examined dry. In the Permian Basin where limestone, dolomites, and anhydrites prevail, the samples are usually examined wet. Under water some of the sample qualities such as oölitic structure and crystal formation are clarified by differential refraction.

Two methods of describing samples are used. In the Mid-Continent the interpretative method is generally used. Using this method, the geologist picks out those cuttings he believes to be representative of the formation and describes the entire sample as if it were composed of those cuttings. For example, a sample from 5,090 to 5,095 ft may be composed of 95 per cent shale and 5 per cent sand. If the sand did not appear in the preceding sample from 5,085 to 5,090 ft, the geologist would describe the sample from 5,090 to 5,095 ft as sand—ignoring the shale content. In areas where the geologist has a fair idea of the lithology and thickness of the subsurface formations, and where they do not change facies rapidly, the interpretative method works very well.

In the Permian Basin, where the combination of shale basins, limestone platforms, and reef deposits causes profound facies and thickness changes, the interpretative method does not work well. There most geologists use the percentage system. All cuttings in the sample (except obvious cavings) are described by percentage. For example, a sample might be composed 40 per cent of dolomite and 60 per cent of anhydrite. The geologist would describe both materials and show graphically the amounts of each on his sample log. Frequently the geologist will prepare an interpretative sample log by using the sample log with the electrical log. For example, the geologist may prepare a percentage sample log as the well is being drilled. When the well is completed, he will use the electrical log to correct the top and bottom of the formations shown on the sample log.

The Sample Log

A typical sample log is compared in Fig. 21-3 with the electrical log. As shown, the sample log consists of the plotted log and the accompanying written description. There is no industry-wide standard list of symbols for the plotted log—most companies have adopted their own system of nomenclature. Many geologists do not use symbols for the log plot; instead, they use a color code to identify various types of rocks. One of the shortcomings of the common geologic-sample log is that it is a hand-prepared log and, for duplication, must usually be copied by hand.

The descriptive portion of the log is fully as important as the log plot. It includes many factors, varying with the type of rock being described. The type of rock is usually repeated in the description although shown in the log plot. Rock types include limestone, dolomite, granite, anhydrite, coal or lignite, shale, sandstone, quartzite, chert, salt, schist, slate, marble, etc., plus many combinations (as shaly sand).

Shales are generally described as follows:

Color: red, green, etc.
Texture: waxy, velvety, papery
Fabric: laminated, splintery, flaky, fissile, blocky
Accessory minerals: micaceous, sandy, calcareous, bentonitic, fossiliferous, carbonaceous, glauconitic, bituminous

Sands are generally described as follows:

Color: brown, gray, etc.
Texture: very fine, fine, medium, coarse, very coarse-grained
Shape of grain: rounded, subrounded, angular, subangular
Sorting: well sorted, poorly sorted
Secondary material: clean, shaly, calcareous, dolomitic, silty, phosphatic, siliceous, tuffaceous, micaceous, glauconitic, pyritic, fossiliferous, carbonaceous, sideritic, ferruginous
Degree of cementation: friable, light, dense
Cementation material: calcareous, siliceous, shaly, ferrous
Porosity and permeability: poor, fair, good

Limestone and dolomites are generally described as follows:

Color: white, brown, etc.
Texture: dense, chalky, sucrosic, oölitic, oölicastic, coquinoid
Crystal size: mat, cryptocrystalline, microcrystalline, megacrystalline
Secondary material: shaly, sandy, dolomitic or limy, silty, cherty, siliceous, fossiliferous, ferruginous, anhydritic
Luster: dull, earthy, resinous
Type of porosity: intergranular, intercrystalline, pinpoint, oölicastic, vugular, fracture
Degree of porosity and permeability: poor, fair, good

One very important portion of the description concerns the shows of oil or gas noted in the samples. Oil stain is usually readily detected in limestones, dolomites, or sandstones. However, oil stain from high-gravity oil may be misinterpreted as gas. Most gas zones produce enough condensate to stain the rocks lightly, but a dry gas may be missed entirely. A fluoroscope or ultraviolet light may be used to detect shows that might otherwise be missed (because oil fluoresces when exposed to ultraviolet light). While it is certainly possible to miss gas or high-gravity oil shows in the cuttings,

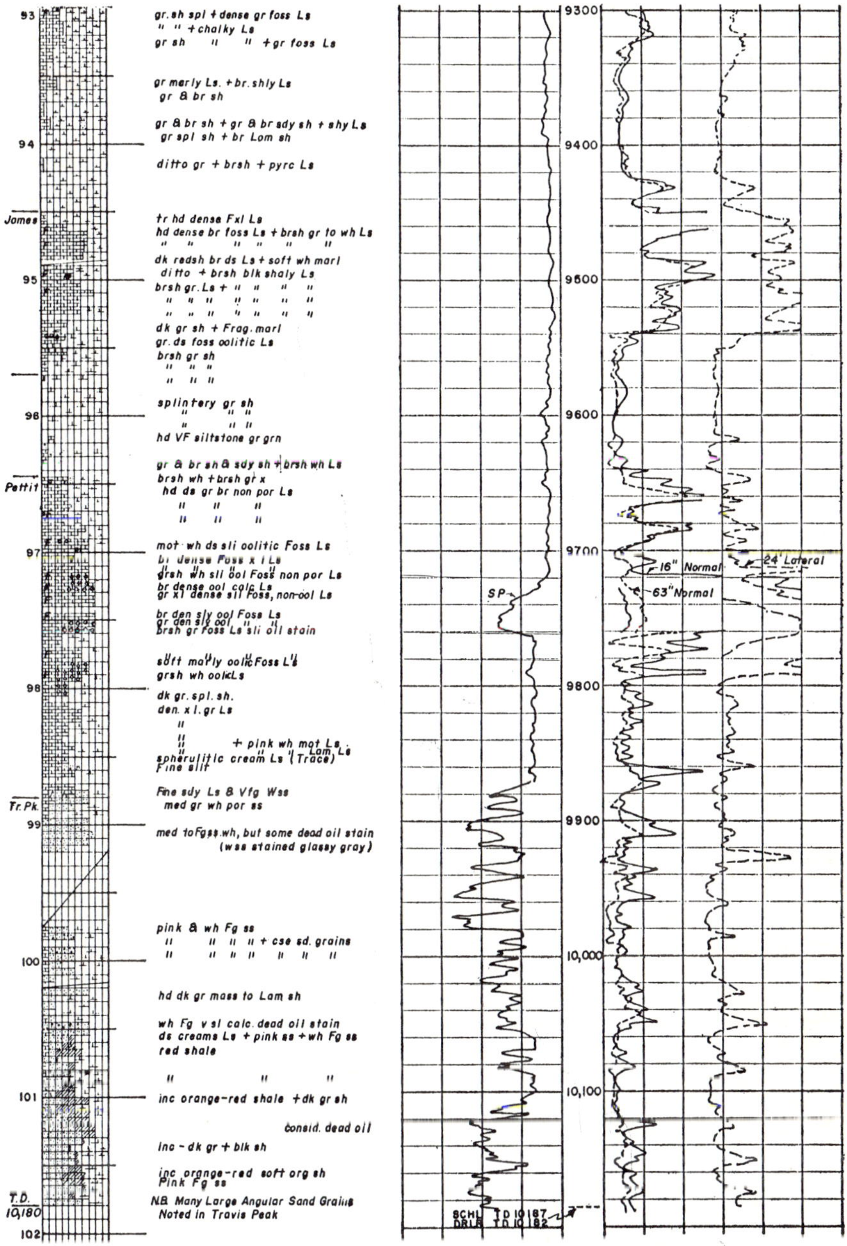

FIG. 21-3. Comparison of geologic-sample log with electrical log.

usually the trained geologist will have no difficulty. In most cases both gas-oil and water-oil contacts can be picked from the sample log by the change in oil stain.

Uses of the Geologic-sample Log

In the introduction we stated most oil and gas reservoirs in the "hard-rock" country are found by the geologist's inspection of rock cuttings—perhaps days before the electrical log is run. The location of oil and gas reservoirs is probably the most important use of the geologic-sample log. On exploratory wells the drilling-time log gives warning that a porous zone has been encountered. Samples are circulated out, and from inspection of these cuttings the geologist tells if a potential oil (or gas) zone has been encountered. Samples (if promising) may be confirmed by a drill-stem test before an additional hole is drilled. The well in Fig. 21-3 was completed from 9,750 to 9,760 ft. This zone was discovered from the samples before the electrical log was run.

Another use of sample logs is correlation—both regional and local. The electrical log provides more accurate correlation for most formations, but the geologic-sample log provides it while the well is being drilled. Then, too, some correlations are not evident from electrical logs. If there is lateral variation in formations, complicated faulting, etc., it is often impossible to correlate electrical logs without referring to the sample log. Other examples of uses of the sample log in correlation are picking coring points, drill-stem-test intervals, packer seats, casing points, etc. Much of this information could be obtained from frequent electrical-log runs—but at much greater cost.

In "soft-rock" areas the paleontologist uses samples as a source of fossils. Many of these fossils are found only in a relatively short time span. The paleontologist's log, like the sample log, is useful in correlation, particularly in structurally complicated areas or for regional correlation.

At first glance, it might appear that the sample log and electrical log in Fig. 21-3 do not correlate closely. But the formation tops picked from the sample log compare closely with the tops indicated by the electrical log.

Perhaps the most fundamental use of the geologic-sample log is in basic study of the rocks. From study of rock cuttings from several wells, the geologist finds the key to how the rocks were deposited and their subsequent history. Such study is basic in determining the geology of an area and finding oil. Today, stratigraphic traps are receiving more attention than ever before. The geologic-sample log is probably the basic tool in finding these reservoirs.

The quality of the samples available for the geologist varies. In one well, the samples and hence the sample log may be very reliable—in the next, they may be poor. Sometimes the geologist fails to note on his log that samples were poor over an interval and that the log may be unreliable. The quality of the sample log also varies with the training and experience (quality) of the geologist preparing the log. When the sample log is prepared from reasonably good samples by a competent geologist, it rates as one of the most important logs—perhaps still the most important in hard-rock country.

HYDROCARBON-MUD LOG

Introduction

When a rock is drilled, the bit breaks the rock into many small fragments or cuttings. If oil or gas is present in the rock, some of it is released into the mud stream, and some remains in the cuttings as residual oil. The hydrocarbon-mud log detects the presence of minute quantities of gas and oil both in the mud stream and in the cuttings. In addition, a drilling-time log and geologic-sample log are included in the hydrocarbon-mud log. A typical hydrocarbon-mud log is shown in Fig. 21-4.

The most important of the measurements recorded by the hydrocarbon-mud log is the amount of gas in the mud. The drilling-time log and geologic-sample log are ordinarily obtained even if a hydrocarbon-mud log is not recorded. The detection

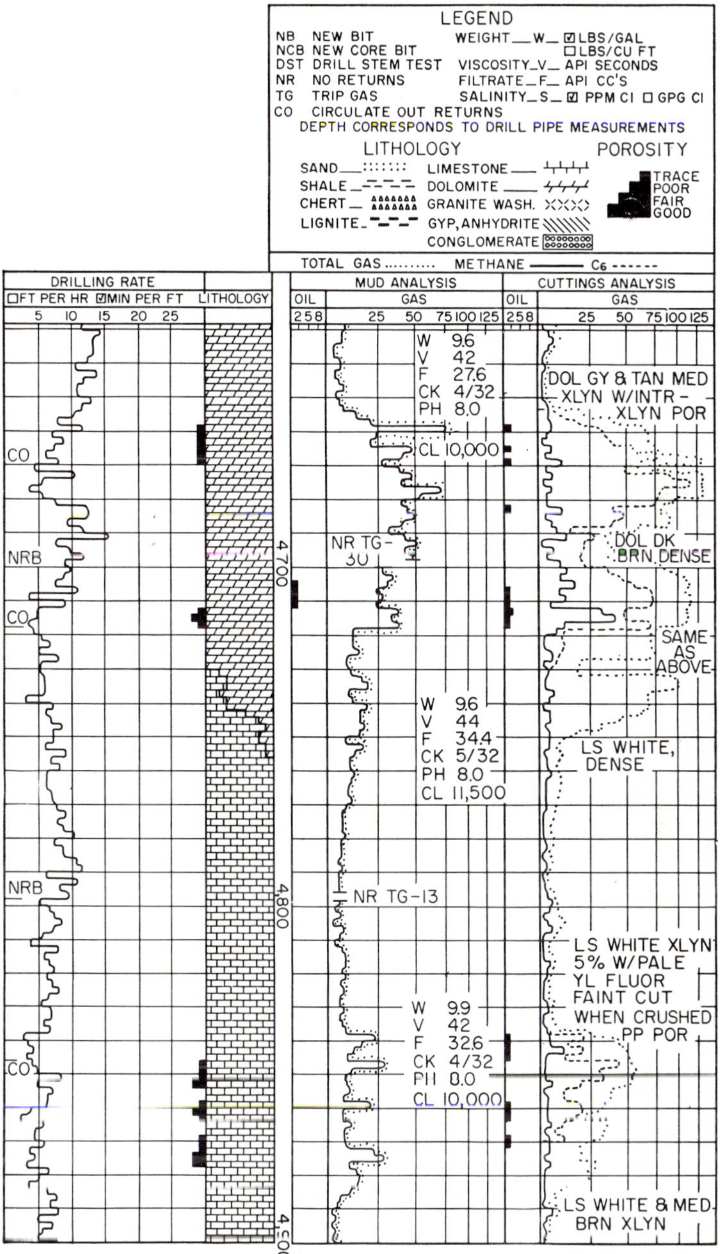

Fig. 21-4. A typical hydrocarbon mud log.

of oil and gas in the cuttings is normally a part of the geologic-sample log. (However, the hydrocarbon-mud log utilizes more mechanical aids than does the ordinary geologist. Thus, the hydrocarbon-mud log can record minute quantities of hydrocarbons that the geologic-sample log might miss.) But the detection of gas in the mud is unique with the hydrocarbon-mud log.

Hydrocarbon-mud logging was commercially introduced to the oil industry in 1939. It is now a very competitive field, and many companies provide this service to the oil operators.

Theory

Only a small volume of oil and gas is contained in the cylinder of rock drilled from an oil formation. For example, a 9-in. hole drilled through a 20-ft oil sand "drills up" about 9 cu ft of rock. Assuming 20 per cent porosity, 80 per cent initial oil saturation, a formation volume factor of 1.6, and a solution gas-oil ratio of 1,000 cu ft/bbl, this 9 cu ft of rock contains about 7 gal of oil and 167 standard cu ft of gas. Much of the oil and gas present in the rock never enters the mud stream. Perhaps 80 per cent is driven out of the drilled rock by invasion of mud filtrate ahead of the bit. If 50 per cent of the remaining oil stays in the cuttings, we have only 0.7 gal of oil and 33 standard cu ft of gas mixed in the mud stream. If it takes 60 min to drill the 20-ft oil sand, this oil and gas may be mixed with 24,000 gal of mud.

These small quantities may not be evident to the human senses but can be detected using sensitive instruments. Limits of detection for the hydrocarbon-mud log are usually estimated to be 0.01 cu ft of gas per hour or 1 part oil in 100,000 parts mud.

The preceding illustration is only an example of possible conditions. Actual conditions of oil in place, drilling rate, amount flushed, etc., vary greatly. This example does show that small but significant quantities of gas and oil are liberated in the mud stream (and remain in the cuttings).

Methods of Recording the Hydrocarbon-mud Log

The hydrocarbon-mud log is recorded as the well is drilled—on a 24-hr schedule. The essential parts of the logging equipment are (1) a system to "sample" the mud (and suspended cuttings) from the mud-return line, (2) a system to transport this sample and separate gas, mud, and cuttings, (3) methods of detecting the presence of gas and oil in the mud and cuttings, (4) a method of recording drilling time, and (5) equipment and personnel to prepare the geologic-sample log. This equipment is usually mounted in a trailer and operated by three men on 8-hr shifts.

Various methods are used to obtain samples of mud and cuttings from the mud-return line. Often a portion of the mud stream is diverted to a separator or gas trap. In the trap air is drawn countercurrent to the flow of mud with a slight vacuum. This vacuum combined with agitation of the mud stream separates the gas from the mud (and mixes it with air). After passing through a filter, the gas and air mixture goes to the detector. When this system is used, samples may be caught from the rig shale shaker or a special sample box. Many units pump a portion of the mud stream direct to the trailer. There the gas and cuttings are separated automatically, and the remaining mud is pumped to the mud pit. The method of sampling the mud stream is a very important part of hydrocarbon-mud logging. This portion of mud logging requires close attention regardless of the system used.

For many years the hot wire (Wheatstone-bridge-type) gas detector was standard for detecting hydrocarbon gas. (The log illustrated in Fig. 21-4 used this type of detector.) A schematic diagram of the hot-wire detector is shown in Fig. 21-5. An amount of current is supplied to the compensatory and detector filaments sufficient to heat them to a specified temperature. When a hydrocarbon gas is passed over the detector filament, it burns catalytically on the platinum filament. This increases the temperature and so the resistance of this arm. The increased resistance unbalances the circuit and causes a current to flow through the galvanometer. The current flowing through the galvanometer varies with the amount of hydrocarbon gas in the mixture. Usually two bridge circuits are used. In one circuit the detector filament is heated to a low temperature that is insufficient to ignite methane. In the second circuit the detector filament is heated to a temperature high enough to ignite all hydrocarbon gases including methane. Thus two measurements are made: (1) all hydrocarbons except methane, (2) all hydrocarbons including methane. The difference between these two readings is the methane content. In practice any two of the three values obtained may be presented.

The hot-wire detector is very sensitive. It will react to hydrogen, hydrogen sulfide, or any other combustible gas. Under certain conditions this may provide a high erratic background through which it is difficult to detect the presence of hydrocarbon gases. In addition, the detector filament may become "poisoned"—rendered insensitive by the deposition of sulfur or carbon on the filament. If the gas volume is very high, some of the gas may not be burned by the high-voltage detector filament and actually may tend to cool the filament (in some cases record zero gas concentration). Thus the magnitude of the reading may not be at all indicative of the amount of hydrocarbon gas present. Likewise, if a large amount of heavy hydrocarbons are present, they may heat the low-voltage detector filament to a temperature high enough for methane to ignite. In this case, both the high-voltage and low-voltage detectors will read the same. The principal advantage of the hot-wire technique is its simplicity. The equipment is inexpensive and does not require extensive personnel training for satisfactory operation.

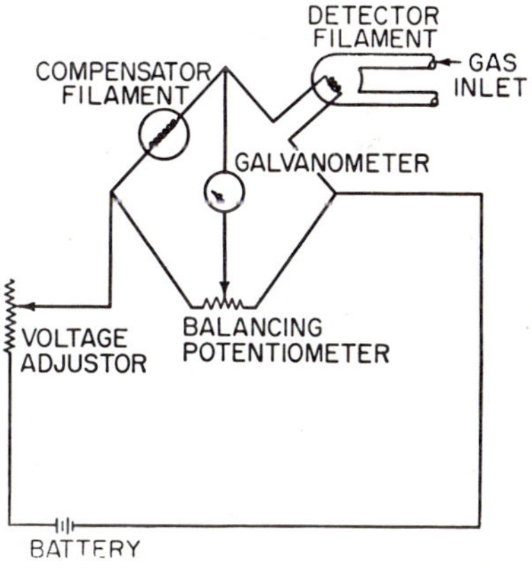

FIG. 21-5. Schematic diagram for hot-wire detector.

Of the newer methods of detecting hydrocarbon gas, gas chromatography is the most common. In this process a sample of gas from the drilling mud is mixed with an inert gas (helium, nitrogen, air) and flows through a "partition column." The "partition column" is packed with a material such as activated charcoal which absorbs hydrocarbon gases. These hydrocarbon gases are then released to the inert-gas stream in order of their boiling points. Changes in composition of the gas emerging from the column may be detected by measuring the thermal conductivity of the gas or by burning the gas on a platinum filament. Gas chromatography has been widely used in refinery operations but only recently has been applied to hydrocarbon-mud logging.

Another method uses an infrared detector to detect methane. Infrared radiation is passed through a sample cell and calibrated with air passing through the cell. When the gas from the mud stream is passed through the sample cell, any methane present will absorb some of the infrared radiation. When compared with a reference cell containing air, the unbalance of the detector can be recorded. When an infrared detector is used for methane, a chromatograph is usually used to detect the heavier hydrocarbons.

The mass spectrometer has also been used to a limited extent in hydrocarbon-mud logging.

To detect the presence of oil in the drilling mud, a sample of mud is first treated to reduce surface tension and gel strength. The sample is then examined under ultraviolet light. Almost every crude oil fluoresces under ultraviolet light. (Rig greases and oils in the mud may also fluoresce.)

To detect gas in the cuttings, a measured volume of cuttings and water is placed in a Waring blender or similar grinding device. The cuttings are ground and the gas drawn off. If hydrocarbons are present in the cuttings, the "cutting gas" is detected in the same way as is "mud gas."

Oil in the cuttings is found by examining a sample of the cuttings under fluorescent light. If oil-emulsion mud is used, cuttings are lightly washed with water first. The fluorescence, if any, is noted. Since some minerals fluoresce, it is difficult in some cases to determine whether oil in the cuttings or the cuttings themselves are causing fluorescence. So, if fluorescence is noted, the sample is treated with carbon tetrachloride (or another solvent). The solvent will leach oil from the cuttings. When the solvent dries, a residual oil ring is left on the sample glass. The residual oil ring is usually graded as: no, trace, fair, or good cut.

The geologic-sample log is described on page 21-4. For the hydrocarbon-mud log a percentage type of geologic-sample log is prepared even including known cavings. It is necessary to prepare a strict percentage log so that any shows of gas and oil in the cuttings can be correlated with the cuttings.

The drilling-time log is recorded as described on page 21-1. The mechanical drilling-time log is replotted as part of the hydrocarbon-mud log.

Factors Affecting Performance of the Hydrocarbon-mud Log

Before discussing interpretation of the hydrocarbon-mud log, it is well to consider some of the factors that affect its performance. Perhaps the most important of these is the mud used to drill the well. It is ironic that many times a company will pay thousands of dollars for the hydrocarbon-mud log—yet ignore the effect of their mud program on that log.

The value of drilling with oil-emulsion mud is well known. However, the use of oil-emulsion mud has increased the difficulty of hydrocarbon-mud logging. If crude oil is added to the mud (to free stuck drill pipe or prepare oil-emulsion mud), the hydrocarbon-mud log is adversely affected. If a diesel oil is added to make oil-emulsion mud, it is still possible to log effectively. However, the diesel oil will cause a background fluorescence and background gas readings (because formation gas will go into solution in the oil), and more careful handling of the cuttings samples will be required. Diesel oil reduces the sensitivity of the log. Nonfluorescent oils are available for use in drilling but have not been used widely because of the increased expense. In addition, nonfluorescent oils are very easily contaminated. When used, they should be handled very carefully to prevent contamination. When a mud-logging unit is used, nonfluorescent pipe lubricants should be used. The additional cost is minor for these lubricants.

In some areas water for drilling mud is obtained by gas-jetting water wells. The solution of gas in this water reduces the effectiveness of the mud log. Even the use of crude-oil-contaminated carriers for rig water will affect the mud log.

The depth of the hole may affect the magnitude of mud-log readings because of greater dilution and mixing on the way out of the hole. Drilling rates are usually quite slow in deep wells. Slow drilling rates are detrimental because the oil and gas are mixed with larger volumes of drilling fluid. In addition, very fast drilling rates may be detrimental since the logging personnel may have difficulty keeping up with their various measurements.

High-viscosity and high-gel-strength mud may cause recirculation of gas, which will cause a high background. The high viscosity prevents ready release of gas from the mud in the mud trap and in the pits and also tends to hold cuttings more tightly in suspension.

High-density mud, too much emulsifying agent in the mud (or too little if oil-emulsion mud is used), lost-circulation material in the mud, excessive sloughing, "swabbing" while pulling drill pipe, frequent drilling variations, variations in mud-pump pressure, and variations in weight on the bit all may adversely affect the mud log.

In using some types of detectors, hydrogen can be troublesome. Hydrogen evolution may be caused by action of high-pH muds on aluminum or magnesium alloys (contaminating mud by drilling plugs) or possibly by drill-pipe corrosion when a low-pH mud is used.

Another factor influencing the mud log is personnel. The personnel operating the log are vital—the difference between a good and bad log. Although many of the readings on modern mud logs are automatic, constant supervision of the equipment is essential for best results. Then, too, some of the measurements are directly dependent on how qualified and conscientious the operator is.

Interpretation of the Hydrocarbon-mud Log

Because of the many variables involved, it is impossible to correlate the magnitude of any show with the chances for commercial production. Any indication of hydrocarbons—whether from "mud gas," "cuttings gas," oil in mud, or oil in cuttings—may indicate commercial production. Any such zone must be investigated further with other formation-evaluation techniques such as electrical logs and side-wall cores to determine whether a productive zone has been encountered.

Further, the hydrocarbon-mud log may occasionally miss commercially productive zones—none of the detectors may indicate presence of hydrocarbons and yet a commercial oil zone may be present.

The hydrocarbon-mud log normally records shows for lignite, some shales, and some salt-water sands as well as for productive zones. The gas contained in lignite is usually entirely methane. While shale hydrocarbons are usually methane, in some cases shales contain both oil and heavier hydrocarbon gas. Salt water at depth often contains gas, usually, but not always, methane. In addition, frequently zones are found that contain only residual oil. These zones appear the same as oil-productive zones on the mud log.

Although many service companies have developed methods for rating shows from their log, these must be considered qualitative rather than quantitative. The safest way to interpret the hydrocarbon-mud log is to consider each increase of gas in the mud or indication of gas in the cuttings or oil in either mud or cuttings as a "show." Each show should be evaluated further using other well logs, cores, production tests, etc., to determine whether the zone is productive.

Use of the Hydrocarbon-mud Log

The purpose of the hydrocarbon-mud log is to increase the chances of finding oil or gas in an exploratory well. The log may be used in one of two ways. The first way is to drill ahead just as if the log were not installed. Then, when total depth is reached, evaluate all shows on the hydrocarbon-mud log using electrical logs, side-wall cores, formation tester, etc. (In addition, any zone of potential interest from other formation-evaluation techniques should be evaluated even if the mud log gives no show. In short, use the mud log as a signal to evaluate a zone; never use the mud log to condemn a zone.)

The second way to use the mud log is to stop drilling operations long enough to circulate each drilling break. If the mud log looks favorable for the circulated interval, a core could be cut or a drill-stem test made. If the drilling rate is not too fast, it may not be necessary to interrupt drilling operations and circulate—rather, it may be sufficient to drill ahead and then drill-stem test the zone if the mud log is favorable.

Because so many shows are encountered on the mud log, the second method may be quite expensive. Particularly in soft rocks it is probably not economic to stop drilling at each break or hydrocarbon-mud-log show. In hard rocks where it is difficult to determine productive possibilities without drill-stem tests and cores, the second method may be justified.

DIRECTIONAL LOG

Introduction

The directional log is a record of hole drift (the deviation of the wellbore from the vertical) and the direction of that drift. The necessity of determining hole drift to maintain a straight hole was recognized many years ago. The increasing use of intentionally deviated holes has greatly increased the use of the directional log in recent years.

Methods of Recording

A variety of methods are used to record directional logs. They may be classed as drift indicators, single-shot surveys, multiple-shot surveys, and continuous surveys. Drift indicators are not really directional logs but are included because of their similar application.

Drift indicators record only the amount of drift and not the direction of drift. They are usually run either on a wire line or as a go-devil. One of the first drift indicators was the acid bottle.

Although not generally used today, it is interesting from a historical standpoint. It consisted of a small amount of hydrofluoric acid in a small bottle. The bottle was lowered in the hole and allowed to remain motionless for a few minutes. Since the acid etched the glass bottle, the amount of drift could be determined by the level of the etched glass.

A common type of drift indicator uses a plumb bob to determine angle of drift. A clock actuates a stylus on the end of the plumb bob. In another type a clock actuates a light beam on a sensitized film. In another method the record is made by the fluid level of a chemical on a paper chart. Still another method uses a plumb bob which stays in contact with a disk. A small electric current circulating through the instrument produces a dot on the recording disk when the instrument is at rest.

Single-shot surveys are similar to drift indicators since they also make only one reading and may be run on a wire line or as a go-devil. In addition single-shot recorders are often run in drill pipe—positioning the instrument in a nonmagnetic drill collar. This allows periodic checks to be made to ensure that the hole is being properly deviated while directionally drilling. The single-shot instruments record both drift angle and direction of drift. One type of instrument uses a small compass bar which rotates on the head of the angle indicator. Needles in this bar are clock-actuated to punch a chart. The orientation of the punch marks shows the direction of magnetic north in reference to the hole drift. Another instrument uses a clock to operate a camera. The film is carried on a semifloating magnetic compass. Still other types use a camera with a compass and plumb-bob arrangement.

Multiple-shot surveys are frequently used to survey the entire uncased portion of a hole. They may be run as a go-devil and positioned in a nonmagnetic collar—then used to record periodic measurements as the drill pipe is pulled. Or they may be run on a wire line taking readings at 100- or 200-ft intervals.

In general they record similar to the single-shot instruments, but since a number of measurements are made all employ a camera. In one commonly used type a camera records a picture of both a compass and a steel ball rolling in a glass graduated cup.

Another type of particular interest employs a gyrocompass rather than a magnetic compass (with a plumb bob to determine drift angle). Unlike those instruments which use magnetic compasses (or magnetometers), it can be run inside casing.

Recently, *continuous-recording directional logs* have been developed. One type employs two pendulums that swing perpendicular to each other. Hole drift is thus resolved into two components—each proportional to the swing of its respective pendulum. The orientation of the plane of the pendulums and of the instrument is determined using magnetometers. The magnetometers consist of high-permeability alloy, inserted in two coaxial coils. When one coil is fed with an alternating current, a current is induced in the second coil. This current is a function of the orientation of the magnetometer in respect to the earth's magnetic field. Another continuous-recording directional log uses a rotary potentiometer with its wiper arm connected to a compass needle. The hole drift and direction are recorded from a magnetic plumb bob connected to a potentiometric arm.

Another type employs a magnetic compass to determine direction of deviation. A photoelectric scanning device sends impulses from the compass to the surface recorder.

Hole inclination is determined using a microammeter with a weighted field and scale portion. The microammeter acts similar to a plumb bob—seeking the low side of the hole and indicating both the direction and angle of hole drift (in respect to the instrument). A system of mirrors and a photoelectric cell are used to scan the driftmeter. The driftmeter scanner is synchronized with the compass scanner so that borehole direction and dip can be readily determined.

All the continuous-recording directional logs are surface-recording instruments.

Calculation of Directional Logs

If only a drift indicator is used, it is impossible to calculate the exact location of the bottom of the hole. Usually, the maximum possible deviation is computed by

Depth	Drift Angle	DEFLECTION True Bearing	Horizontal Footage	Vertical Footage
511	CSG.	SHOE		511.00
530	1-30	S71W	.52	529.99
600	2-00	S71W	2.43	599.98
800	1-30	S76W	5.24	799.91
1000	2-00	S51W	6.98	999.79
1200	3-30	S56W	12.20	1199.41
1400	4-30	S56W	15.70	1398.79
1600	5-00	S71W	17.44	1598.03
1800	5-00	S66W	17.44	1797.27
2000	5-45	S56W	20.04	1996.27
2200	6-15	S51W	21.78	2195.09
2400	6-45	S46W	23.50	2393.71
2600	7-00	S51W	24.37	2592.21
2800	7-30	S51W	26.11	2790.49
3000	7-30	S51W	26.11	2988.77
3200	7-45	S41W	26.97	3186.95
3400	7-15	S36W	25.24	3385.35
3600	6-00	S10W	20.91	3584.25
3800	5-45	S36W	20.04	3783.25
4000	5-00	S36W	17.43	3982.49
4200	4-30	S36W	15.69	4181.87
4400	4-15	S36W	14.82	4381.31
4600	4-15	S56W	14.82	4580.75
4800	6-45	S16W	23.51	4779.37
5000	6-30	S41W	22.64	4978.09
5200	6-15	S46W	21.77	5176.91
5400	5-15	S46W	18.30	5376.07
5600	5-00	S56W	17.43	5575.31
5800	3-45	S46W	13.08	5774.89
6000	3-45	S46W	13.08	5974.47
6200	3-30	S46W	12.21	6174.09
6400	3-30	S46W	12.21	6373.71
6600	3-15	S56W	11.34	6573.39
6800	3-00	S41W	10.17	6773.11
7000	2-00	S81W	6.98	6972.99
7148	2-15	S86W	5.80	7120.87

GRAPH OF DIRECTIONAL LOG
ALL DIRECTIONS ARE FROM TRUE NORTH

	DEPTH	
Position of bottom of hole at	7148	LIES 378'S & 397'W OR 548'S 46°24'W OF CSG. SHOE
Position of bottom of hole at		
Position of bottom of hole at		
Position of bottom of hole at		

GRAPH

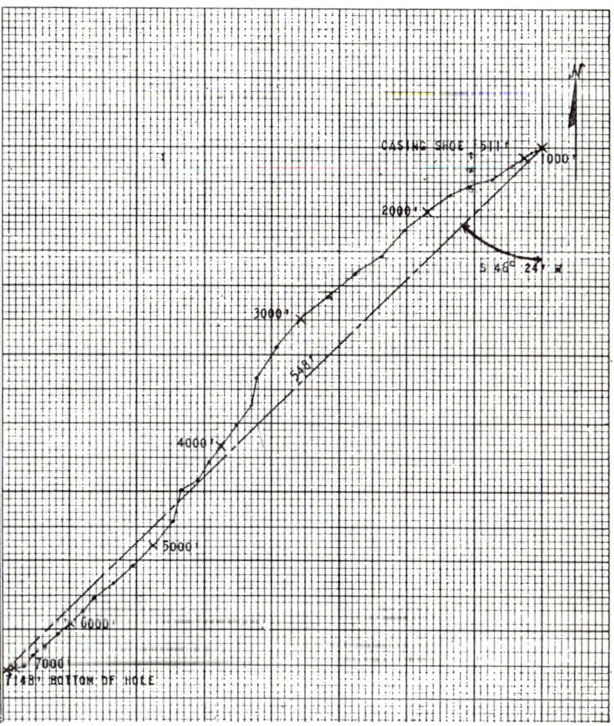

Fig. 21-6. An example of directional-log calculations.

assuming that all drift is in the same direction. If this deviation is excessive, a directional log should be run to determine true deviation.

Usually directional logs are computed by calculating the horizontal footage and vertical footage for each interval. The true vertical depth is determined by summation of the vertical increments using the proper drift angle. The bottom-hole location is determined by plotting the horizontal increments using the proper direction of drift. An example of the calculations is shown in Fig. 21-6. In this well the producing zone was found at 6,886 ft. The directional survey indicates a 27-ft correction at this depth, or a true vertical depth of 6,859 ft. The formation was penetrated at a point 387 ft west and 377 ft south of the surface location (assuming the hole was straight from 0 to 511 ft).

Uses of Directional Logs

Perhaps the first use of the directional log was to make sure that a straight hole was being drilled. The driller needs a cheap simple measurement to be sure he is not drilling a crooked useless hole. Most drilling contracts contain a straight-hole clause, specifying the maximum hole drift. The drift indicator supplies the information—before it is too late to correct the situation cheaply.

Many wells are directionally drilled. Usually the single-shot instruments are used in a magnetic collar as an aid to directional drilling. To deviate a hole to a desired location it is necessary to make frequent measurements of hole drift and direction. Modern survey instruments have been responsible at least in part for the widespread use of directional drilling in side tracking, straightening crooked wells, whipstocking the lower portion of the hole (to gain better structural position), and where surface-location cost is very high.

In deeper holes the driller needs to know both hole drift and direction of drift. A "dogleg" in the hole may cause keyseating and sticking of drill pipe unless it is reamed. The direction of drift may be more important than the angle. In areas where this drilling problem is encountered, frequent single-shot measurements or a multiple-shot survey may save an expensive fishing job.

Directional logs are useful to the production engineer. A dogleg in a hole may cause production problems (e.g., tubing failure due to rod wear) if preventive measures are not taken.

When wells are drilled near lease lines, a directional log is necessary to ensure that the producing formation is encountered on the proper lease instead of on a neighboring lease. Most state regulatory agencies require that directional logs be made on wells drilled near the lease line.

Directional logs are needed to interpret dipmeter surveys. To determine true dip of a formation it is necessary to correct the apparent dip for the amount and direction of hole drift. Continuous-recording directional logs are used with dipmeters since accuracy of the dipmeter is greatly improved if hole deviation is available for each point on the dipmeter log.

The reservoir engineer and geologist use the directional log to determine the true vertical depth of the pay zone and the true location where the pay is found. Usually it is not necessary to consider the correction for true vertical depth except in directionally drilled holes or where hole drift is large. However, the horizontal deviation from the surface location may be important even if drift is small. Ignoring the hole deviation will create erroneous structure maps.

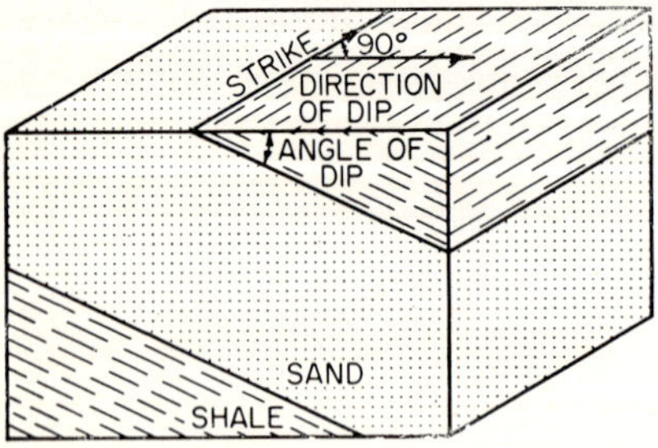

FIG. 21-7. Illustration of dip, strike, and direction of dip.

DIPMETER LOG

Introduction

The dipmeter is run to determine the direction and angle of formation dip from the survey of one borehole. This information is of obvious importance in the study of structural and stratigraphic problems.

As illustrated in Fig. 21-7, the angle of formation dip is the angle between a horizontal plane and the bedding plane of the formation. The strike of a formation is the direction of the horizontal line formed by the intersection of these two planes.

Although strike is a common geologic term (particularly in surface geology), it is more convenient to use "direction of dip" in discussing the dipmeter. The direction of dip is perpendicular to the strike. In the remainder of this section direction of dip will be used instead of strike. Although it will not be used here in this sense, azimuth is used synonymously with direction of dip in many references.

Instrumentation

The dipmeter records data so that both the amount and direction of formation dip can be computed. Since the inclination of the borehole from vertical must be considered, the dipmeter also measures amount and direction of hole drift. To compute the amount of dip it is necessary to know wellbore diameter, and so the dipmeter also includes a caliper log. In recent years all these measurements have been combined in one continuously recording instrument.

Formation dip is the angle between a horizontal plane and the formation bedding plane. Since any three points not in a straight line determine a plane, the bedding plane can be determined by locating three points on it. The instrument for measuring dip uses three electrodes (or sets of electrodes) spaced 120° apart in a plane perpendicular to the wellbore. (Centralizers hold the instrument parallel to the borehole.) Each of these three electrodes records a separate curve. From inspection of these three curves it is possible to correlate the curves—i.e., locate points common to all three curves. If the bedding plane is not the same as the plane of the three electrodes, the curves will be displaced; e.g., one curve may show the top of a formation several inches higher than the other two curves.

Various methods have been used to record the three dip curves for the dipmeter log. The first method in common use recorded three self-potential curves.* Later, a resistivity dipmeter which recorded three lateral curves (3-ft spacing) was used. Another instrument recorded three caliper curves instead of electrical curves. Today these instruments have been almost entirely supplanted by instruments which record either three contact-log curves (usually only the microinverse or micronormal curve) or three microlaterolog curves. In the present instruments the electrodes are embedded in pads which are pressed against the wall of the hole. One of these instruments is illustrated in Fig. 21-8.

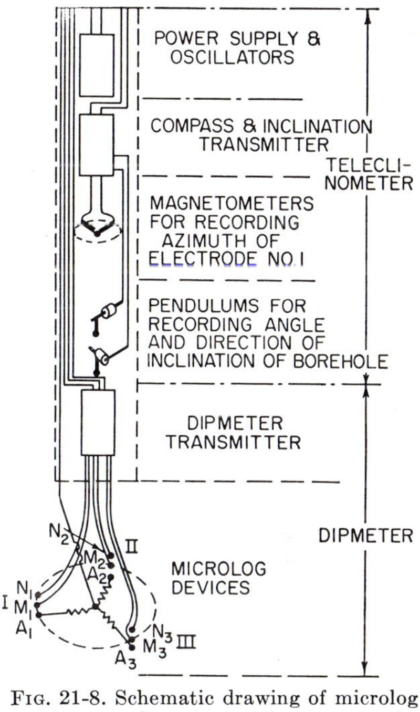

FIG. 21-8. Schematic drawing of microlog continuous dipmeter. (*From paper by Pierre de Chambrier presented to Society of Exploration Geophysicists*, 1953.)

Because the electrodes are continuously pressed against the borehole wall and because of the excellent resolution of the microdevices, sharp curves which are readily correlatable can be obtained. To determine the direction of dip the orientation of the electrode system must be determined. This is done using either a magnetic compass or magnetometers to determine the orientation of the instrument in respect to magnetic north.

The apparent dip and direction of dip are calculated from the dip curves and orientation of the electrode system. This apparent dip must be corrected for any deviation of the wellbore from vertical (since the dip electrodes are located in a plane

* For discussion of self-potential and other electrical log curves, see Chapter 19.

perpendicular to the wellbore). The borehole drift and direction of drift are measured using continuously recording directional logs. This portion of the instrument was discussed under Directional Logs.

Figure 21-8 illustrates one type of continuous dipmeter log. This instrument measures apparent formation dip, orientation of the electrodes and direction of dip, and borehole drift and direction of drift.

Computing the Dipmeter Log

The dipmeter log may be computed by the service company, by a consultant, or by the company for which it was run. In any case, close coordination is needed with the geological staff of the well operator if the data are to be used to the best advantage. This will be apparent in the following discussion.

The first step in computing the dipmeter log is to select the beds to be studied. When noncontinuous dipmeters were used these zones were selected before the dipmeter was run and only the best markers were recorded. Today the continuous-recording dipmeters have replaced the "station" or noncontinuous dipmeters, and so the zones can be selected after the dipmeter is run. If computed results are erratic, additional zones can always be calculated. In selecting the zones to be studied, the dipmeter should be compared with the electrical survey. Those zones which are continuous from well to well provide the best markers.

The zones which best reflect formation dip are listed in order of preference as follows:

1. Thin limes or sands (2 or 3 to 10 ft thick) in a shale section.
2. The base or top of a thick sand body. Usually the base of the sand will provide the sharpest contact.
3. Sandy or limy shales in a thick shale section.

Thin shales or limes in a thick sand body are quite likely to reflect cross bedding rather than formation dip. The same is true of thin shales in massive lime sections. While this is not necessarily the case, dips computed within thick sands or limes are usually not reliable and should not be used. Even in the best "marker" zones, dips should be computed for several levels. If the dips do not check, additional levels should be selected or, if necessary, that particular bed should be disregarded.

After the marker bed has been selected for study, the three curves of the dipmeter should be correlated. This correlation is aided by making two translucent prints of the original log. Using these prints superimposed on a light table, the curve for the No. 1 electrode may be correlated first with the No. 2 electrode and then with the No. 3 electrode. If the sonde is not rotating in the interval (as shown by the azimuth of the No. 1 electrode) a 10-ft interval can be studied conveniently. By careful study of this interval, it can usually be determined whether the correlation reflects formation dip or whether the correlation of individual "kicks" is erratic. The sharpest kicks are selected for further study.

A typical dipmeter record is shown in Fig. 21-9. The zone selected for study on this section of the log was found just above 9,202 ft. The displacement of the three curves from a convenient depth marker is then carefully measured—4.3, 3.2, and 2.4 in. in Fig. 21-9. It is very important that the relative displacement be measured precisely. A small measurement error can make a large error in dip and direction of dip.

The hole inclination on this particular log is recorded in degrees in the east-west and north-south planes, respectively. This inclination is recorded at the level of the No. 1 electrode. The azimuth of the No. 1 electrode is likewise read at this level as is hole diameter. As discussed under Directional Logs, the borehole drift and drift direction are recorded in different ways for different types of instruments. Hence all dipmeter logs are not identical with the one illustrated in Fig. 21-9.

After the basic readings are made the dip and direction of dip may be calculated in several ways. A mechanical three-dimensional model is available which duplicates the geometrical features of the hole. The mechanical model calculates dip and direction of dip directly. Nomograms have been developed for calculating the dip and strike and are distributed by most service companies running dipmeters. The dipmeter may also be solved graphically using descriptive geometry and the stereographic net. Since these solutions are somewhat tedious and nomograms are available for various instrument types from the service companies, the detailed calculation procedure will not be discussed here.

The accuracy of the computed results varys with the sharpness of the electrical markers on the correlation curves. The accuracy of the computed dips is about ±2 or 3° with the continuous-recording dipmeters if correlation markers are sharp

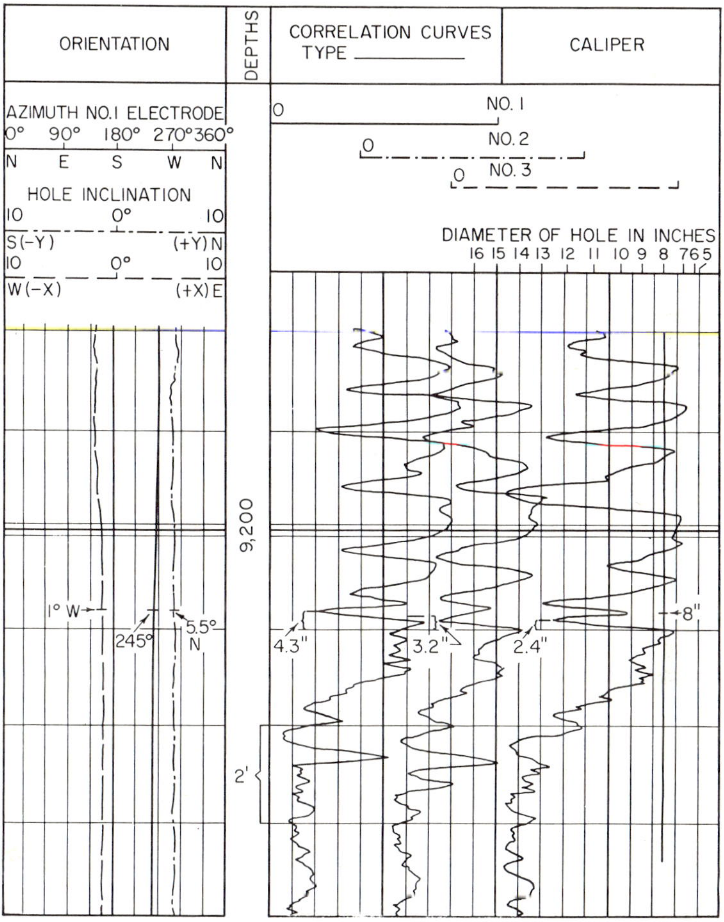

FIG. 21-9. Example of a dipmeter log.

and they are computed carefully. Accuracy of dip direction varies with the amount of dip. If the dip is less than 5°, only the general direction of dip can be determined. If it is greater than 5°, dip direction should be accurate within about ±30°.

Application of Dipmeter Results

The dipmeter is run to determine dip and direction of dip from the survey of one borehole. This information is vital to the study of structural and stratigraphic problems of all kinds. If the structure of a field is simple, dipmeters may be needed only on the first few wells. If it is complex, dipmeters may be useful in every well in the field.

One of the most important uses of the dipmeter log is to gain an idea of structure from an exploratory well. If the first well in an area is productive, the dip and dip direction can be used to select the best location for the next well. If the exploratory well is dry, the dip data may show that another well can be drilled at a more advantageous structural position. Or perhaps the dry hole can be whipstocked to a possible oil trap upstructure. For example, if a well finds only a few feet of oil over water in a thick sand, it might be whipstocked upstructure where the entire sand would contain oil. (Or a well might be whipstocked downstructure if only a few feet of oil sand was found in the base of a thick gas sand.)

As additional wells are drilled to develop a field, the dipmeter can be used to advantage in defining the structure. Provided the computed dip data are relatively accurate (this requires that formation dip be 3° or more), one well with a dipmeter provides as much structural information as two wells without dipmeters. Likewise,

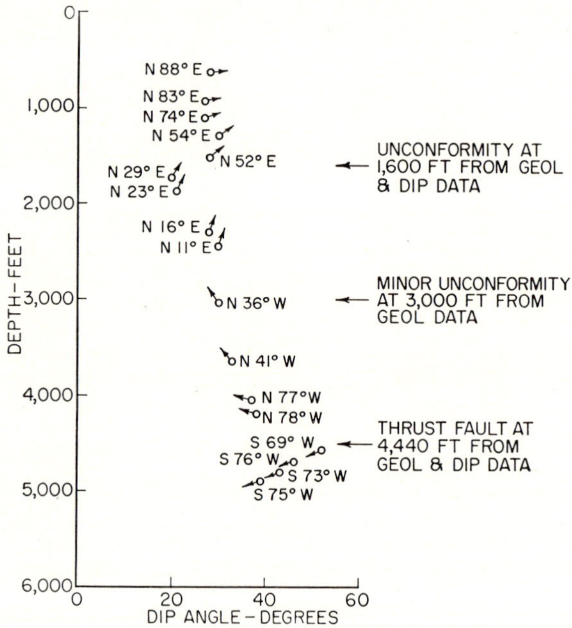

FIG. 21-10. Typical plot of dipmeter results.

two wells with dipmeters will provide as much or more structural information than three wells without dipmeters, and so on. Many a development dry hole can be avoided if dipmeter data are properly used.

Another use of dipmeter data is in correlation. Correlation is often difficult in areas where there is lateral change in formation, particularly if formation dip is high. The presence of angular unconformities may further complicate correlation. Dipmeter data can often be used to aid in solving a problem of this nature.

In many cases much information about structure and stratigraphy can be inferred from the dipmeter data on one well. The angle of dip and direction of dip should change little or at least change consistently with depth unless the well cuts an angular unconformity or cuts a fault. If an angular unconformity is penetrated the dip must change. Usually, although not necessarily so, different fault blocks penetrated in a well will have different angles of dip and direction of dip. Figure 21-10 shows a typical plot of dipmeter data—illustrating the effect of a fault and angular unconformity.

The general geology of the area should be known in order to obtain the greatest amount of information from the dipmeter. For example, from the dipmeter data alone it would be impossible to say whether the changes in dip in Fig. 21-10 are due to an unconformity or a fault. But when the dipmeter data are combined with regional geology, the interpretation shown in Fig. 21-10 is possible. As another example, dips generally increase as a piercement-type salt dome is approached. From study of dipmeter and geological data, the location of the dome and perhaps even the presence of salt overhangs may sometimes be inferred. Another use of dipmeter data is to locate stratigraphic traps. The thinning of beds indicated by gradual change of dip in a well way be the key to locating a possible oil trap.

In summary, a complete description of the application of dipmeter data would be a text in subsurface geology. The dipmeter is very valuable (in areas where dips are more than 3°) if properly used. But it must be computed carefully and used intelligently. It is essential that the person interpreting the data check the dipmeter data to make sure that enough markers have been computed, that they have been computed in the right place, and that they are computed properly. In many cases the geologist or petroleum engineer has been guilty of interpreting computed data (often insufficient data at that) with no thought as to the accuracy of the data. This often leads to errors in structural interpretation. But if full advantage is taken of the information available on the dipmeter log, the dipmeter may be used to great advantage in the structural and stratigraphic interpretation of an area.

CALIPER LOG

Introduction

The caliper log measures the diameter of the borehole. The first caliper logs were developed to determine borehole size in holes shot with nitroglycerin. These early logs showed large variations in hole size even in the portions of the hole that had not been shot. This illustrated the need for the caliper log over the entire hole.

Methods of Recording

Several types of caliper log are in current use. One type consists of three or four spring-driven arms which contact the wall of the borehole. The instrument is lowered to total depth, and the arms are released either mechanically or electrically. The spring tension against the arms centers the tool in the well. The arms move in and out with the change in wellbore diameter. The arm motion is transmitted to a rheostat so that change in the resistance of an electric circuit is proportional to the hole diameter. The borehole diameter is recorded at the surface by measuring the potential across this resistance.

Another instrument uses three flexible springs which contact the wall of the borehole. These springs are connected to a plunger which moves up or down as the springs expand or contract with changes in borehole diameter. The plunger passes through two coils. When an alternating current is passed through one coil, an electromotive force is induced in the other coil. The amount of this induced emf is a function of plunger position and is proportional to borehole diameter. Either of the preceding instruments may be adjusted so that they will record borehole area rather than hole diameter. If the caliper log is used to determine hole volume, it is desirable to record area on a linear scale. If the caliper log is used to determine hole configuration, the hole diameter is recorded on a linear scale.

A third type of caliper log is the microcaliper, which is discussed in connection with the electrical-log microdevices. This instrument uses two pads rather than arms or flexible springs. Hole diameter is determined by the movement of these pads, which are held against the borehole wall by springs.

Typical Configuration of the Borehole

A schematic drawing of a typical borehole is shown in Fig. 21-11. As illustrated in this figure, some formations cave considerably, causing enlarged holes. Other formations do not cave, and because of the presence of mud cake, the hole size may actually be reduced to less than bit size. Although not illustrated in Fig. 21-11, some formations may swell, causing reduction in hole size.

The primary cause of formation caving is action of the drilling fluid. Action of the bit and action of the drill pipe also have an effect. Most drilling muds are composed primarily of water. The chemical action of this water on shales (hydration of the shales) causes many shales to disintegrate and slough into the hole. The amount and rate of this sloughing depend on the nature of the mud and shale. Other shales (heaving shales) swell rather than disintegrate.

If a fresh-water mud is used to drill a salt section, it will dissolve salt until the mud becomes salt-saturated. The drilling fluid does not "react" with formation such as limestone, dolomite, and sandstone. However, if those formations are permeable, a mud cake will be formed as illustrated in Fig. 21-11. This mud cake forms rapidly. The character (density and thickness) of the mud cake varies with the mud used to drill the well. Of course, thickness of the mud cake is limited by erosion of the drilling fluid while circulating.

In some areas the shallow portion of the hole is drilled with water. If loosely cemented sands are encountered, they may cave under this condition.

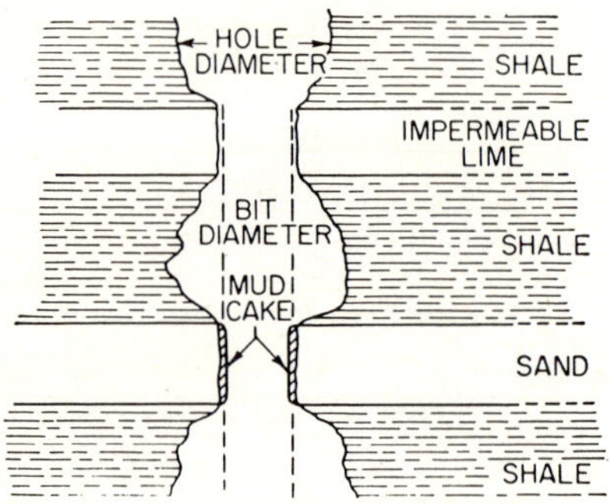

Fig. 21-11. A typical borehole.

The action of the bit is probably not too important. But if a thin sand is surrounded by shales that have caved, the bit probably knocks off part of the sand ledge with each round trip.

Action of the drill pipe against the side of the hole causes some enlargement even in sandstones and limestones. Usually this enlargement is not great enough to affect hole volume appreciably, but it may cause keyseating and a fishing job. Formation "wear" by the drill pipe will cause the hole to be noncylindrical and is not readily detected by present-day caliper logs.

Interpretation and Application of Caliper Logs

Caliper logs are usually recorded on vertical scales from

$$1 \text{ in.} = 100 \text{ ft to } 5 \text{ in.} = 100 \text{ ft}$$

The horizontal scale is selected to show a detail picture of hole diameter and is usually in the order of 1 in. = 4 in. Because of this difference in scales, it is easy to get the impression from caliper logs that tremendous cavities are created. When plotted on the same horizontal and vertical scales, it is evident that the normal borehole is quite "regular." This should be kept in mind when using the caliper log.

The primary uses of the caliper log are (1) to compute hole volume to determine the amount of cement needed to fill up to a certain depth, (2) to determine hole diameter accurately for use in interpreting other logs, and (3) to locate permeable zones as evidenced by the presence of a filter cake. Other applications of the caliper log include (1) as a guide in proper location of casing centralizers, (2) as an aid in determining top of cement from the temperature survey, (3) for accurate determination of the amount of plastic and cement to use in plugback jobs, and (4) to locate packer seats for open-hole drill-stem tests.

TEMPERATURE LOG

Introduction

The temperature log measures borehole temperature as a function of depth. It is likely that temperature was the first physical property measured in a borehole. Temperature measurements were made in boreholes as early as 1869. Temperature is an important factor in both production and reservoir engineering (e.g., equipment design, calculation of gas in place, etc.). However, temperature increases with depth in a borehole and with a rather consistent gradient over large areas. As a result, formation temperature for most production and reservoir-engineering uses can be determined adequately by using a maximum-recording thermometer run in conjunction with other surveys. Temperature logs are run primarily to determine temperature variations when a well is not at thermal equilibrium, e.g., to locate the top of cement by the temperature increase associated with the setting of cement.

Instrumentation

The equipment for making temperature surveys is quite simple in comparison with other logging instruments. Most instruments use a thermometer of the electrical-resistance type—i.e., the resistance of the element is sensitive to changes in temperature. If a constant current is passed through this resistance, the voltage drop will be proportional to the resistance and may be recorded directly as temperature.

Unlike most other logs the temperature survey is usually recorded going down the hole. This eliminates temperature changes that might be caused by disturbing the fluids in the borehole by "swabbing" action of the cable and sonde.

Uses of the Temperature Log

Probably the most common use of temperature logs is to locate the top of cement in the casing-wellbore annulus. Since cement generates considerable heat when it sets, a temperature increase above normal geothermal temperature is created in the interval where cement is located. Below the top of cement the temperature usually varies with hole size. In a caved zone there will be considerable cement present and hence a high temperature. In a zone where formations are not caved, there will be less cement and a comparatively lower temperature (although still higher than normal temperature).

Temperature logs to determine the top of cement are usually run about 12 hours after cementing. The optimum time will vary with the type of cement used; however, the time is ordinarily not too critical.

Another use of temperature logs is to locate a zone causing loss of circulation. If mud is pumped into a well with lost circulation, it will circulate only as deep as the "thief zone." Below that point the mud will be relatively stagnant and near thermal equilibrium. Thus the temperature survey will show an anomaly at the zone of lost circulation.

When gas is produced from a formation into the wellbore, the pressure drops rapidly, causing a decrease in temperature. Temperature surveys may be used to locate this cooling effect and so the entry of gas into the wellbore. The temperature survey may be used to locate the gas-oil contact, gas entering through a casing leak, etc. In some areas gas is used to drill low-permeability gas formations. A temperature survey recorded with gas flowing out of the formation will show which zones are producing gas. In ideal cases it may even be possible to compute the relative volumes of gas flowing from several sands from the temperature survey. Since formation crude oil contains gas which may be released from solution when the oil flows into the wellbore, it may also be possible to locate oil-producing zones from the temperature survey. If an oil well is producing salt water, the water will not contain much gas and will appear as a "hot" area relative to the crude oil. Thus, if several zones are perforated, the temperature survey may sometimes be used to locate the zone producing water.

The Differential-temperature Survey

Although the normal-temperature surveys are recorded with accuracy of 1° or less, in some cases it is desirable to detect and define very small temperature anomalies. The differential-temperature survey is used for this purpose. The differential-temperature survey employs two thermometers—usually spaced 6 ft apart. The difference in temperature between the two thermometers is recorded. In effect the differential-temperature survey measures the temperature gradient. As long as the gradient is consistent, the differential survey will show the same reading. But small changes in temperature gradient can be located precisely. Temperature anomalies of 0.1°F can be located using the differential-temperature survey.

Interpretation of Temperature Surveys

In most cases the interpretation of temperature surveys is qualitative rather than quantitative. Interpretation usually requires only a knowledge of well characteristics and the phenomena producing the temperature anomaly.

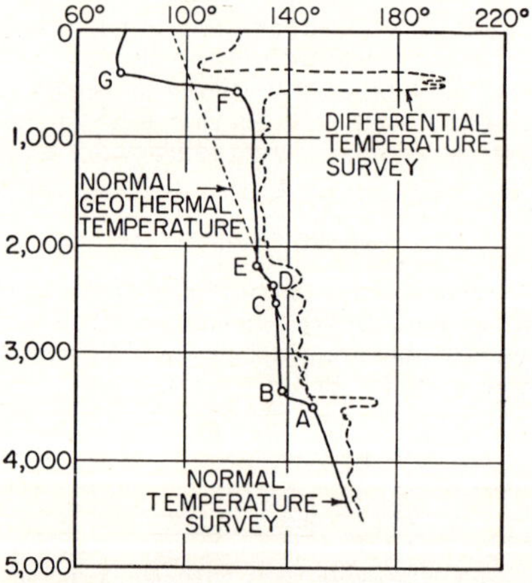

FIG. 21-12. Typical temperature surveys.

Figure 21-12 shows both the normal-temperature survey and a differential-temperature survey for a problem well. In this area a shallow fresh-water sand suddenly began producing gas in fresh-water wells. Temperature surveys were run in all oil and gas wells in the area in an attempt to locate the source of this troublesome and dangerous gas. The leak was located by the temperature surveys illustrated. This well was producing oil with a low gas-oil ratio, but gas was traveling between the casing and the borehole wall from a gas sand to the fresh-water sand.

Figure 21-12 is interpreted as follows:

1. Below point A the temperature gradient is normal and relatively consistent.

2. The normal-temperature survey shows a decrease in temperature from A to B, caused by gas entering the casing-wellbore annulus. The differential-temperature survey shows a change in gradient for this interval correlating with a thin sand on the electrical log.

3. From B to D the temperature gradient in the wellbore is fairly constant. From B to C the temperature is less than geothermal temperature. However, the gas continues to decrease in temperature showing continued decrease in pressure (possibly because it is flowing through a restricted channel or rising through mud or water). Many temperature surveys will indicate increase in temperature over the interval BC due to heat passing from the formation to the gas. At C the gas temperature and geothermal temperature are equal, and from C to D the gas temperature is greater than geothermal temperature. From C to D the temperature continues to decrease because of continued pressure drop and because heat is passing from the gas to the formations.

4. Additional cooling occurs from D to E, perhaps because of a restriction in the annulus, or perhaps gas is expanding into a low-pressure formation. Above E the gas is cooled slightly by geothermal temperature. But because it is cooled very slightly, a large volume of gas is flowing. It is important to remember that the amount of cooling anomaly on the routine temperature survey (as at AB and DE) is primarily a function of pressure drop. The slope of the curve above the anomaly reflects the quantity of gas. For example, the more nearly EF approaches constant temperature, the greater the volume of gas flowing.

5. From F to G the routine temperature survey shows a sharp decrease in temperature as the gas enters the low-pressure fresh-water sand.

The preceding discussion is illustrative of the type of interpretation that is usually used for temperature surveys.

ACOUSTIC-VELOCITY LOG

Introduction

The acoustic-velocity log measures the time required for a sound wave to pass through a definite length of formation. It was originally developed as a geophysical tool but has also proved to be valuable to the reservoir engineer and geologist as a porosity and correlation log. Its extensive use as a porosity and correlation tool began in 1958. Because of the relatively recent introduction of the velocity log as an exploitation log, many improvements in both the log and its interpretation are anticipated. The velocity log may be run in uncased holes containing oil, water, or any type of mud. A velocity log cannot be recorded in air- or gas-filled boreholes. It is not normally run in cased holes, although formation velocities have been experimentally recorded through casing where the casing was cemented.

Theory

The velocity of sound in a formation depends primarily on the velocity of sound through the rock matrix, the amount of porosity, and the velocity of sound through the fluid contained in the pore spaces.

The velocity of sound through the rock matrix depends on the elastic properties of the matrix. The normal range of matrix velocity for various types of sedimentary rocks is given in Table 21-1.

Table 21-1

Rock	*Matrix velocity, ft/sec*
Shale	6,000 to 16,000 depending on compaction
Salt	15,000
Anhydrite	20,500
Limestone	21,000 to 23,000
Dolomite	24,500
Sandstone	Up to 18,000 depending on compaction and shale content

The velocity of sound in various types of formation fluids depends on their density and compressibility. Normal ranges of velocity in fluids in sedimentary formations are given in Table 21-2.

Table 21-2

Fluid	*Velocity, ft/sec*
Water..........	5,000 to 5,500 depending on salinity
Oil.............	4,000 to 4,900 depending on amount of gas in solution
Gas............	1,100 to 2,000 depending on temperature and pressure

For rocks in which porosity is uniformly distributed, the time required for a sound wave to pass through a formation (transit time) appears to be affected primarily by the relative quantity of solid and fluid. Since velocity is inversely proportional to transit time,

$$\frac{1}{V_a} = \frac{\phi}{V_f} + \frac{1 - \phi}{V_m} \tag{1}$$

where V_a = apparent velocity through porous fluid-filled formation
V_f = velocity in the interstitial fluids
V_m = velocity in the matrix material of the rock
ϕ = porosity fraction

From this equation the application of the velocity log as a porosity tool is apparent.

For rocks in which porosity is not uniformly distributed, the transit time for a sound wave may also be affected by the distribution of the porosity, and Eq. (1) will not be strictly correct. However, distribution of porosity appears to be of comparatively minor importance.

Instrumentation

The velocity-log instrument consists of a source to create sound or acoustic energy and a receiver (or receivers) to detect this acoustic energy after it travels through the

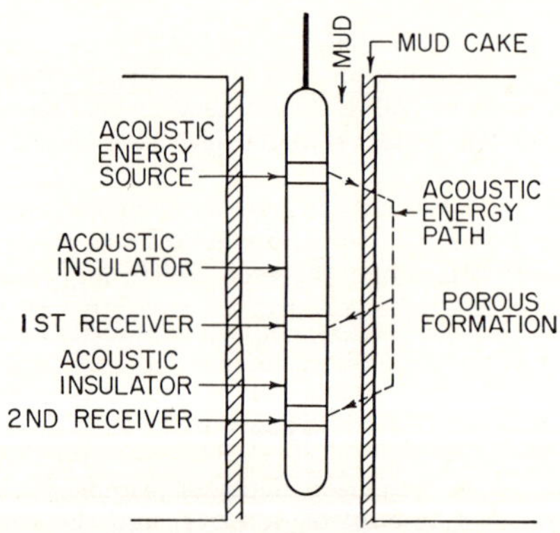

Fig. 21-13. Schematic illustration of velocity log.

formation. Two basic types of velocity instrument have been used, commonly called the one-receiver and the two-receiver types. Figure 21-13 shows a schematic illustration of the two-receiver type but may be used to explain both types. The source emits acoustic-energy pulses at a constant rate. The shortest travel path for a pulse is through the formation since the mud has low velocity and the sonde itself is constructed of low-velocity material. The first arrival of sound energy at each receiver actuates the receiver response system. The difference in time of arrival at the two receivers is recorded. This difference in arrival time (transit time) is the time required for the acoustic energy to pass through a length of formation equal to the spacing. As long as hole conditions are the same at the two receivers, the travel time through the mud and mud cake will cancel out. Spacings of 1 to 3 ft between receivers are used in most cases.

If a one-receiver system is used the transit time from the source to the receiver is recorded. This transit time includes time for the sound wave to pass through the mud and mud cake at the source and again at the receiver. Although these times are not large, they may be significant. Hence the two-receiver system, although more complicated, is a more accurate tool. Since the two-receiver system is displacing the single-receiver system, the single receiver will not be discussed further in this section.

Various types of acoustic-energy sources and detectors have been used. Sound sources include magnetostrictive alloys of metal, piezoelectric quartz crystals, electromechanical hammer and anvil, or an electrical spark gap. Detectors include geophones using piezoelectric quartz crystals, pressure-sensitive crystals, and magnetostrictive metal alloys.

Recording the Velocity Log

The velocity log records the time required for a sound wave to pass through a length of formation. With the two-receiver system this transit time is usually recorded as microseconds per foot. Transit time is readily converted to velocity

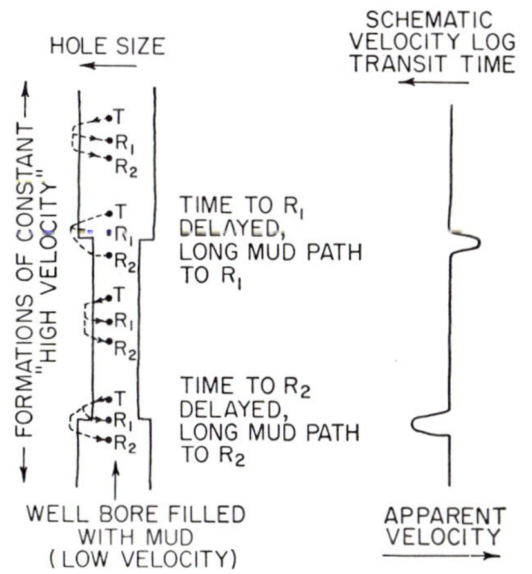

FIG. 21-14. Effect of caving on velocity log.

(velocity in ft/sec = 10^6 divided by transit time in microseconds). Most velocity logs also include the integrated time. This is shown on the log record as a pip or dot at convenient time intervals—usually 1,000 msec.

Previously, a statement was made that the first arrival of sound energy at each receiver actuates the receiver response system. While generally true, exceptions do occur—leading to the anomaly known as "skip." If the first pulse arrival does not trigger the receivers, they may both be triggered by a subsequent pulse. If so, the log will not be affected. But in some cases the receiver nearest the transmitter will be triggered by the first pulse, but the pulse will be too weak to trigger the second receiver. The second receiver may be triggered by a later pulse. As a result the transit time will be abnormally great and an abnormally low velocity will be indicated by the log. This anomaly is known as "cycle skip" and its significance will be discussed under use and interpretation of the velocity log.

The two-receiver velocity log is not affected by any formation except between the receivers. Likewise, it is not directly affected by mud and hole size except between the two receivers. (Indirectly, an enlarged wellbore between the transmitter and the upper receiver could increase the likelihood of "cycle skip.") Since the distance between receivers is short (1 to 3 ft), mud and hole conditions are generally about the same opposite the two receivers and cancel out as previously explained. However, if the borehole is enlarged opposite one receiver and not the other, the log will be affected. Figure 21-14 illustrates log response in this case. The abnormal readings at the caved zone average out when total travel time is determined since they are in opposite directions. The depth of investigation of the velocity log is believed to be quite shallow.

Use and Interpretation

Geophysical. The velocity log was developed as a geophysical tool, and this application is still one of the most important. The geophysicist uses the velocity log to determine the velocity of earth formations and to obtain vertical-travel-time data. These data are needed for seismic record-depth determination. Prior to the velocity log such information was obtained by recording a near-surface explosion with a geophone in the well. Such an operation was expensive and time-consuming.

Determination of Porosity. To the engineer the most important use of the velocity log is in determination of porosity. From Eq. (1) it is evident that porosity can be determined from the velocity log if the matrix velocity of the rock is known and the velocity of the fluids is known.

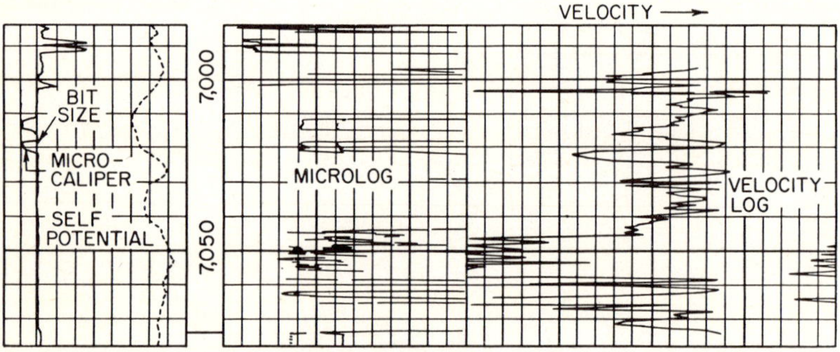

FIG. 21-15. Comparison of contact and velocity logs.

In limestones, dolomites, and hard sandstones (generally, formations of 5 to 15 per cent porosity), the matrix velocity can usually be estimated fairly accurately. As a first approximation, velocities as follows may be assumed:

Sandstone 18,000 ft/sec
Limestone 21,000 ft/sec
Dolomite 23,000 ft/sec

As more is learned regarding a particular formation in an area, more precise values of matrix velocity may be used to refine the answer. In formations of this nature, mud filtrate invasion is usually deep enough that the acoustic energy travels in the invaded zone. Thus the fluid velocity to be used in Eq. (1) approaches that of water whether the sand originally contained water, oil, or gas.

Figure 21-15 shows a typical velocity log in a limestone formation compared with a contact log. The porosity calculated from this velocity log (assuming 21,000 ft/sec matrix velocity and 5,300 ft/sec fluid velocity) compared almost exactly with porosity from core analysis. In higher-porosity sands (about 15 to 25 per cent) corrections may be necessary to obtain accurate porosity from the velocity log. Since associated shale will generally have a lower velocity than a clean sand of this porosity, it is difficult to determine whether a decrease in velocity in this type of sand is caused by an increase in porosity or an increase in shale content. If the sand is shaly, the porosity indicated by the velocity log will be too high unless a correction is made. In addition such sands are frequently not invaded deep enough for the acoustic energy to travel in the invaded zone. If this is the case, the velocity in oil (or gas for a gas sand) must be used in Eq. (1) to obtain porosity from the velocity log.

In unconsolidated sands (usually more than 25 per cent porosity), Eq. (1) is not accurate. Unconsolidated sands may often be recognized by the velocity of adjacent shales. If the velocity of adjacent shales is less than 10,000 ft/sec, an unconsolidated sand may be expected and the velocity log used cautiously to determine porosity. Possibly corrections will be developed in the future that will make the velocity log more useful in this range (some have already been proposed but have not been proved).

Some producing formations have less than 5 per cent porosity. Since an error in assumed matrix velocity of 1,000 ft/sec causes about 1 porosity per cent error, the velocity log will not have good accuracy in this range unless matrix velocity can be determined very accurately. However, in producing formations of less than 5 per cent porosity, fractures are usually present, and the velocity log may be of importance in locating fracture zones, as will be discussed later.

In summary, the velocity log appears to be an excellent porosity tool for hard formations with porosities in the 5 to 15 per cent range. It is fair to good in the 15 to 25 per cent range and below 5 per cent. In the high-porosity unconsolidated sands (over 25 per cent porosity), large corrections are required and its usefulness as a porosity tool is not fully established.

Correlation. The velocity log gives excellent bed definition and appears to be a very good correlation log, especially for field correlations where one velocity log may be correlated with another. For areal corrections, too few velocity logs have been run to determine its value at this time.

In evaporite beds (which have long been difficult to correlate with other types of logs) the velocity log is an excellent correlation log. Not only can velocity logs be correlated by curve appearance but also quantitative correlation appears possible. In Table 21-1 the velocity of salt is shown as 15,000 ft/sec and anhydrite as 20,500 ft/sec. Thus these two types of evaporite can be quantitatively distinguished from each other on the velocity log.

Quantitative correlation may also be possible in recent shale sections where velocity varies with depth; i.e., a shale may be identified as the same shale in a nearby well by a velocity comparison. Since "matrix velocity" is a formation characteristic, direct determination of formation type from the velocity log appears a definite possibility in many other cases.

Direct Indications—"Cycle Skip." "Cycle skips" have been discussed. Although these skips detract from the value of the log as a porosity tool, some field examples show they may have definite value. Cycle skips are caused when acoustic energy is sufficient to trip the first but not the second receiver. This occurs in zones of relatively low elasticity where the signal is attenuated (degraded) between the two receivers. "Cycle skips" in low-porosity limestones and sandstones have been found to correlate with fractured and vugular zones. Since cycle skips are readily apparent on the log, the velocity log may provide a means for locating such zones. In addition, unconsolidated sands often show cycle skip, especially if they contain gas. Sometimes skips are noted in oil sands but not usually in water sands. Evidently the water sand offers enough elasticity to conduct the acoustic energy, but the gas sand does not. The "skips" in these unconsolidated sands may provide a direct means of locating gas sands. It should be noted that, in some cases, gas sands do not have enough rigidity to conduct the acoustic energy and the energy is transmitted by the mud column. While not strictly a "skip," this occurrence shows the same log appearance and so is included under skips. Because of the variations in gas, oil, and water velocity, in some cases the velocity log may be used directly to pick gas-oil or oil-water contacts. The determination of gas-oil contacts is of particular importance since they are not apparent on electrical logs. Both gas-oil and water-oil contacts may be evident on the velocity logs in formations that are not deeply invaded.

COLLAR-LOCATOR LOG

The collar locator is used to locate casing collars—usually in conjunction with another cased-hole service such as a nuclear log or a perforating gun. Perhaps its most common use is in precisely locating perforating points. To do this, the collar locator is run with a nuclear log (either the gamma-ray or neutron log) after the casing is set. This survey accurately positions casing collars in reference to the nuclear log. By correlating the nuclear log with logs run in an open hole, casing collars can be accurately positioned in reference to the electrical log. The collar locator is then run with the perforating gun. The collars adjacent to the desired perforating interval are located and the desired interval perforated using the casing collars as reference points. Use of the collar locator makes it possible to locate perforations within a few inches of the desired interval—even 15,000 ft below the surface of the ground.

Various types of collar locators are now in use. Some of the collar locators are sensitive enough to locate old perforations in casing. The collar locator can also be used to locate the casing shoe in open-hole completions.

REFERENCES

1. Shepherd, G. Frederick, and Gordon I. Atwater: Geologic Use of Drilling-time Data, *Oil Weekly*, July 3, 10, 1944.
2. Shepherd, G. Frederick: Drilling-time Logging, *Subsurface Geologic Methods*, 2d ed., pp. 455–475, Colorado School of Mines, Golden, Colo.
3. Hills, John W.: Sampling and Examination of Well Cuttings, *Subsurface Geologic Methods*, 2d ed., pp. 344–364, Colorado School of Mines, Golden, Colo.
4. Gill, R. J.: Composite-cuttings-analysis Logging, *Subsurface Geologic Methods*, 2d ed., pp. 495–502, Colorado School of Mines, Golden, Colo.
5. Langton, Arthur: Well Logging by Drilling—Mud and Cuttings Analysis, *Subsurface Geologic Methods*, 2d ed., pp. 449–455, Colorado School of Mines, Golden, Colo.
6. de Chambrier, Pierre: The Microlog Continuous Dipmeter, Paper given at Society of Exploration Geophysicists in Houston, March, 1953.
7. Prescott, B. Osborne: Calculating Dip and Strike from Continuous Dipmeter, *Oil and Gas J.*, Mar. 7, 1955.
8. Doll, H. G.: The SP Dipmeter, *Pet. Tech.*, January, 1943.
9. Guyod, Hubert: Caliper Well Logging, *Oil Weekly*, Aug. 27, Sept. 3, 10, 17, 1945.
10. Guyod, Hubert: Temperature Well Logging, *Oil Weekly*, Oct. 21, 28, Nov. 4, 11, Dec. 2, 9, 16, 1946.
11. Wyllie, M. R. J.: *Electric Log Interpretation*, 2d ed., pp. 112–126, Academic Press, Inc., New York.
12. Peacock, Douglas R., and John A. Disch: Delta Log—Differential Type Temperature Survey, Paper at Third Annual Petroleum Conference on Well Logging Interpretations at McMurry College, Oct. 15–17, 1958.
13. Hardy, Hugh W.: The Continuous Velocity Logger—A Geophysical-geological Instrument, Paper at Third Annual Petroleum Conference on Well Logging Interpretations at McMurry College, Oct. 15–17, 1958.
14. Brand, H. C.: Application of the Sonic Log to Geologic Formation in North Central and East Texas, Paper at Third Annual Petroleum Conference on Well Logging Interpretations at McMurry College, Oct. 15–17, 1958.
15. Stripling, A. A.: Velocity Logging, Paper at Third Annual Petroleum Conference on Well Logging Interpretations at McMurry College, Oct. 15–17, 1958.
16. Tixier, M. P., R. P. Alger, and C. A. Doh: Sonic Logging, *Trans. AIME*, vol. 216, p. 106, 1959.

Chapter 22

LABORATORY PROCEDURES IN EXPLORATION

By H. J. Schaetti
Section Head
Charles V. Campbell
Research Geologist
C. G. Claypool
Group Head
William R. Evitt
Research Associate
and
William S. Hoffmeister
Group Leader (Paleontology)

Jersey Production Research Company
Tulsa, Okla.

Laboratory analyses are essential to the modern exploration geologist in his search for oil deposits. Much information contained in the rocks and fluids can be determined only by special procedures in a properly equipped laboratory. Often this special type of information provides the key to the solution of a particular geologic problem. However, no single analysis can be a panacea, and each piece of analytical data has to be considered in its proper relation to the whole problem and must ultimately be tied to the field evidence. When so used, laboratory techniques are an important and indispensable part of modern exploration geology.

A great variety of laboratory techniques are available for the qualitative or quantitative measurement of many different rock properties. The choice of a given analytical procedure is governed by the problem and the type of information sought for its solution. Some measurements are in common, everyday use. Porosity and permeability analyses, for example, and the microscopic examination of core samples are but two types of laboratory techniques familiar to all geologists. Other measurements, such as those of capillary pressure or the organic content of rocks, are used more rarely, for special purposes. It is certain that new and different analytical developments will become commonplace in the future. As our understanding of oil accumulations grows and the reasons for the pooling of oil under certain geologic conditions become more apparent, new information will be required by the exploration geologist and will be made available through new laboratory procedures.

In the following pages many of the laboratory techniques that are in use today are grouped according to the rock properties they are designed to measure. Since these techniques are often complicated and require specialized knowledge and training, they can be discussed here only in terms of the type of information they furnish. Where possible, the principles behind a given method and its limitations are pointed up. Some of the newer techniques that are still very much in the research phase have been omitted.

The extensive bibliography will serve as a guide to those who want more detailed information about the laboratory techniques described.

PHYSICAL PROPERTIES OF ROCKS

Core Analysis

Core analysis is the most accurate laboratory procedure for determining the porosity, permeability, and residual fluid content of rocks. Knowledge of these rock properties assists the well-site geologist and petroleum engineer in selecting potentially productive intervals and in locating the oil-water contacts. An approximation of the amount of oil in place and the recoverable reserves may also be made using porosity and fluid content values from core analyses. Specialized core-analysis techniques (*A*,* Pirson, 1958; Pollard and Reichertz, 1952) to determine wettability, water-flood susceptibility, capillary pressures, and relative permeabilities are of interest chiefly to the reservoir engineer and exploitation geologist after a reservoir has been discovered; these specialized techniques will not be discussed here.

Some oil companies have their own core-analysis laboratories. Most, however, depend upon service companies that specialize in core analysis and have laboratories conveniently located in nearly all regions of active exploration. These service companies usually provide the results of their work within hours after the core is recovered.

Core analyses are of two types. Routine procedures utilize a plug ¾ to 1 inch in diameter cut from the core with an appropriate diamond drilling device. This type of analysis is used for homogeneous materials. Fractured, vuggy, and otherwise nonhomogeneous core samples are analyzed by techniques requiring full-diameter samples of the core. Procedures used in full-diameter analyses are similar to those used for plugs except for modifications required to analyze larger samples (*A*, Stewart and Spurlock, 1952; Locke and Bliss, 1950; Kelton, 1950).

The way in which cores are sampled and preserved depends upon both the type of analysis and the particular information sought. If only porosity and permeability are required, no special precautions are needed to preserve the original fluid content of the samples. If fluid-content values are desired, the core samples should be sealed immediately after extraction from the core barrel. Samples may be sealed in cans or wrapped in aluminum foil and then cast in paraffin. Another common method of preserving cores practiced by some service companies at the well site is to freeze the samples with dry ice. The resulting expansion of the water in samples containing high concentrations of connate water and mud filtrate, however, may alter the original porosity and permeability (*A*, Lebeaux, 1952; Kelton, 1953).

Porosity, permeability, and fluid contents preferably are determined from the same core samples. This requirement, however, is not rigorously followed in most commercial core-analysis laboratories. Adjacent samples are used to provide more rapid service. Samples are prepared for porosity and permeability determinations by Soxhlet or other appropriate extraction techniques (*A*, Pirson, 1958).

* The letter *A* refers to part *A* of the reference list, Physical Properties of Rocks, at the end of the chapter.

Porosity

The porosity of a rock is the ratio of the volume of the pore spaces to the bulk volume and is expressed as the percentage of the bulk volume. Two types of porosity are recognized: (1) absolute porosity, or the total of all pore spaces relative to the bulk volume, and (2) effective porosity, or the total of all communicating pore spaces relative to bulk volume. Conventional core-analysis techniques measure the effective porosity, which is the porosity from which oil is produced. Porosity determinations are performed in two steps: (1) measurement of bulk volume and (2) measurement of either the grain volume or the pore volume.

Several procedures are used to measure bulk volume, which is the volume of the rock including both the grains and the pores between the grains (*A*, Pirson, 1958). The volume of regularly shaped samples can be measured with calipers. This method is recommended only for samples with large surface pores or vugs and is the only satisfactory method known for this type of sample. The simplest and most widely used procedure is to determine the amount of mercury displaced by the sample, using a pycnometer.

Pore volumes of rocks can be measured by two basically different techniques. In the first, the amount of liquid required to fill the pore space is measured (*A*, Pirson, 1958). Either kerosene or mercury may be used, with the mercury-injection method especially adaptable to low-porosity samples. The second procedure employs the Washburn-Bunting porosimeter (*A*, Washburn and Bunting, 1922) and is the most commonly used and fastest method. This porosimeter measures the amount of air extracted by evacuation from the pores in the sample.

Grain volumes of rocks can be determined by noting the reduction in void space in a pressure bomb of known volume compared with the volume of the bomb containing the sample (*A*, Taliaferro et al., 1937; Kaye and Freeman, 1948). This method is more accurate than the Washburn-Bunting method but is the slowest for determining porosity.

Permeability

Permeability of a rock is the measure of the capacity of the rock to transmit fluids and is calculated from Darcy's law: $K = \mu q L / \Delta p\, A$, where K is permeability in darcys, μ is the viscosity of fluid in centipoises, q is the volume rate of flow in cubic centimeters per second, L is the length of the sample in centimeters, Δp is the pressure difference across the core in atmospheres, and A is the cross-sectional area of the sample in square centimeters. The unit of measure is the darcy, which is defined as the permeability that will permit the flow of a homogeneous fluid of one centipoise viscosity at the rate of one milliliter per second per square centimeter cross-sectional area under a pressure gradient of one atmosphere per centimeter.

Permeabilities can be measured in plugs cut parallel to the bedding (horizontal permeability) or perpendicular to the bedding (vertical permeability). The former values are usually higher.

Permeabilities are measured in conventional core analyses by determining the rate of flow of dry air through an extracted plug. Ideally, permeability is an intrinsic property of the rock and is independent of the type of fluid flowing through the rock. Actually, however, permeabilities to gas are generally higher than to liquids. This difference is due to the Klinkenberg effect (*A*, Klinkenberg, 1941). When gas is flowing, the layer of molecules next to the surface slips along the surface, whereas with liquids in viscous flow, the layer of molecules next to the surface is stationary.

Permeability differences due to this effect are relatively greatest in low-permeability samples. Permeabilities to liquids may also vary if there is interaction between the liquid and the rock. This is most apparent in rocks which contain clays, especially of the montmorillonite variety. In such rocks permeabilities to water are lower than to air or oil and permeabilities to fresh water are much lower than to brine. Presence of clays in the formation can cause effective formation permeabilities to be much lower than indicated by routine core analysis using air.

Many different types of permeameters have been designed in accordance with *API Rept. Invest.* 27 (*A*, American Petroleum Institute, 1956). Essentially, a permeameter consists of a source of compressed air applied to the upstream face of the plug, a sample holder which prevents bypassing of air, a manometer to measure the pressure drop across the length of the core, and a flowmeter to measure the rate of flow through the sample.

Residual Fluid Saturations

Residual fluid saturations measured in routine core analyses are useful for comparative rather than absolute evaluations. However, fluid saturations can be measured more accurately when special precautions are taken in the coring operations to prevent contamination of the cores and when the cores are sealed so as to retain all fluids and gases. The residual oil plus residual water reported in routine analyses seldom equal 100 per cent; the difference between 100 per cent and the summation of the amounts of oil and water is presumed to be gas.

Various retort and extraction techniques are used to determine fluid saturations. Retort methods measure simultaneously the amounts of oil and water driven from the sample (*A*, Yuster and Levine, 1938). Extraction procedures involve refluxing of the sample with a boiling solvent such as toluene and measurement of the water driven from the sample (*A*, Yuster, 1944; Rall and Taliaferro, 1947). The extraction technique is more accurate and reliable; the retort method is more rapid. However, the latter may measure too great a volume of oil because of cracking of the crude when the sample is heated, or it may measure too great a volume of water because of dehydration of clay minerals.

Textural Analysis

Texture is the size, shape, and arrangement of the component elements of a sedimentary rock (*A*, Pettijohn, 1957). An analysis of texture includes determination of the size distribution, sorting, shape, sphericity, roundness, surface texture, fabric, and packing of the component grains. Other important textural properties of a porous rock are the size distribution and geometry of the pores as well as the network pattern formed by connected pores.

Evaluation of this combination of textural properties provides data for interpreting the history of formation of a sedimentary rock. Size distribution and sorting reflect the energy conditions at the site of deposition and, in some instances, the limited size range of the original grains supplied to the site of deposition. Shape commonly is related to the original geometric form of the grain modified by weathering or to the mode of genesis of the grains in carbonate rocks. The sphericity of a particle affects its hydraulic behavior during transportation and deposition. Roundness is a record of the abrasion and corrosion history of the particle. Surface textures result from abrasion and corrosion of the particle and may have genetic significance. Fabric of sandstones is related to the direction of movement of the depositing medium, but compaction may modify the original fabric. Packing suggests original depositional conditions but may be more a reflection of compaction. All these textural properties govern the distribution and geometry of the pores in the rock.

Procedures for analyzing texture are a combination of mechanical and microscopic techniques. The size distribution and sorting of readily disaggregated rock samples and unconsolidated materials are determined by mechanical methods; other textural properties have to be determined by microscopic examinations, as do the size distribution and sorting of indurated rocks.

Mechanical Methods

The mechanical method used to determine the size distribution and sorting of readily disaggregated rock samples and unconsolidated materials depends upon the grade sizes of the material. Pebbles and larger-size fragments can be measured by calipers (*A*, Krumbein and Pettijohn, 1938), sand-size components can be rapidly and conveniently sorted by sieving, and particles smaller than $\frac{1}{16}$ millimeter can be satisfactorily separated in sedimentation cylinders by the pipette or hydrometer methods. Many other mechanical procedures have also been used (*A*, Krumbein and Pettijohn, 1938; Twenhofel and Tyler, 1941; Poole, 1957).

Sieve Analyses. Sieve analyses are based on the known dry weight of the sample, which has been disaggregated and split by appropriate techniques (*A*, Krumbein and Pettijohn, 1938; Twenhofel and Tyler, 1941). The sample is placed in the top unit of a set of nested sieves selected to give separation into grade sizes representing the continuous nature of the frequency distribution. Shaking the nest of sieves by hand or in a suitable shaking device, which taps the sieves while rotating them, separates the different grade sizes. The weight collected on each screen is determined, and the data are plotted as histograms or frequency or cumulative curves which can be analyzed statistically (*A*, Krumbein and Pettijohn, 1938; Krumbein, 1938; Otto, 1939; Inman, 1952).

Pipette and Hydrometer Analyses. Particles less than $\frac{1}{16}$ millimeter in diameter are analyzed by the pipette (*A*, Krumbein and Pettijohn, 1938) and hydrometer (*A*, Casagrande, 1934; Day, 1950; Lambe, 1951; ASTM, 1958) methods. The sample is dispersed, placed in a liter graduate or a similar container, mixed thoroughly, and the amount of suspended material is measured by special procedures. These analyses are based on Stokes' law, which states that spheres settle at a rate proportional to the attraction of gravity, the square of the diameter of the particles, and the difference between the density of the particles and that of the fluid, and inversely proportional to the viscosity of the fluid. The chief objection to application of this law is that sedimentary particles are seldom spherical. Comparison of observed velocities of irregular grains and computed velocities of spheres from Stokes' law, however, agree closely for particles smaller than 0.05 millimeter (*A*, Krumbein and Pettijohn, 1938). Somewhat greater precision is obtained using Wadell's sedimentation formula (*A*, Wadell, 1934; Krumbein and Pettijohn, 1938), a modification of Stokes' law, to account for grain shape. Both of these formulas have proved satisfactory in practical application. Pipette analyses are somewhat more precise than hydrometer analyses, but the hydrometer method is more rapid (*A*, Jarrett and Heywood, 1954).

Microscope Methods

Both stereoscopic and petrographic types of microscopes are used in textural analyses. When precise determinations of optical properties of minerals are necessary to decipher textural features, a petrographic microscope is required. Many textural features of rocks, however, can be distinguished using stereoscopic microscopes.

Accessory equipment is usually necessary to adapt either type of microscope for a particular technique. Actual examination of grains or thin sections through the microscope can often be eliminated by use of enlarged photomicrographs or projections of thin sections or acetate peels (*A*, Sternberg and Belding, 1942; Bissell, 1957).

Grain Size and Sorting. Grain size and sorting of indurated rocks must be determined from thin sections or polished surfaces. In theory, comparative areas are measured that are directly proportional to corresponding volumes (*A*, Chayes, 1956; Glagolev, 1933). Operational problems arise, however, especially in the definitions of size used, the sampling method, and the conversion of the apparent sizes observed to the actual sizes that would be obtained in sieve analyses.

Several procedures have been used to measure grain size and sorting in thin sections. Intercept distances along lines of traverse have been measured and have been classed into grade sizes according to the short diameter of the grain (*A*, Packham, 1955) or according to intercept length (*A*, Roethlisberger, 1955). A third technique is to measure the apparent long diameters of the grains falling at points of intersection on a point-count grid (*A*, Friedman, 1958; Rosenfeld et al., 1953).

Different computational procedures are required for each method. The best procedure for correlation between grain size from thin section and from sieve analyses so far reported is by Friedman (*A*, 1958), but Rosenfeld et al. (*A*, 1953) caution against wide application of such corrections. Results from any of these methods should be comparable for a given suite of rocks.

Shape, Sphericity, and Roundness. The shape of a particle is its geometric form and can be classified in many ways. One of the simplest systems is to classify according to ratios of the intermediate to long and short to intermediate diameters of the particle. This method, developed by Zingg (*A*, 1935), classifies particles into geometric forms termed tabular or oblate, equant, bladed, and prolate. Such a shape classification is qualitative and is not amenable to statistical analysis. Krumbein, however, has related Zingg ratios to numerical values for sphericity (*A*, Krumbein, 1941; Krumbein and Sloss, 1951).

Sphericity and roundness are two geometrical parameters commonly used to characterize shape. True sphericity refers to the approximation of a particle to the shape of a sphere (Wadell, 1932). The definition of sphericity used in general practice is the ratio of the diameter of a sphere having the same volume as the particle ("nominal diameter") to the diameter of the smallest sphere circumscribing the particle (*A*, Wadell, 1934). Roundness refers to the sharpness of edges and corners of the grain. A two-dimensional image of the grain is used, and roundness is defined as the ratio of the average radius of the curvature of the corners to the radius of the maximum inscribed circle (*A*, Wadell, 1932, 1935).

Sphericity and roundness are most simply and rapidly determined by comparison of grain images with a set of reference images (sphericity, *A*, Rittenhouse, 1943; roundness, *A*, Powers, 1953; both combined, *A*, Krumbein and Sloss, 1951). Evaluation of this technique for loose grains (*A*, Rosenfeld and Griffiths, 1953) shows that valid differences for both roundness and sphericity can be estimated by this visual technique but that variation among operators may be high.

Sphericity can be computed from measurements of grain diameters in either two or three dimensions. In two-dimensional analyses, two axes of the grain image are measured. The square root of the ratio of the short to long diameter is a close approximation of true sphericity (*A*, Pye and Pye, 1943). Tridimensional analysis is applicable only to loose grains. These grains are mounted on a glass slide, and the third, or depth, dimension is measured by turning the slide on edge. Loose grains can also be photographed with lighting placed at such an angle that the grains cast shadows. The length of the shadow can be compared with the shadow from a particle of known size to give the depth dimension (*A*, Huitt, 1954; Wright, 1957). Stereophotomicrographic techniques (*A*, Aschenbrenner, 1955; Goodman, 1959) have also been used to measure the depth dimension of loose grains. A numerical value for sphericity can be obtained by using the chart (*A*, Krumbein, 1941; Krumbein and Sloss, 1951; Pettijohn, 1957) relating Zingg ratios to sphericity. A more precise evaluation of sphericity is suggested by Aschenbrenner (*A*, 1956); he relates the sphericity to the geometrical form of a tetrakaidekahedron having the axial dimensions measured for each grain.

Many laborious methods for measuring roundness have been used (*A*, Krumbein and Pettijohn, 1938). A recent technique requires a graticule upon which 50 circles of different sizes and known diameters have been inscribed (*A*, Robson, 1958). The average radius of corners and edges and the size of the largest inscribed circle are determined by comparison of the circles on the graticule with the grain.

Surface Textures. Pettijohn (*A*, 1957) defines surface textures as the minor features of the grain surface that are independent of size, shape, and roundness of the particle. Quantitative studies of surface features of grains are not reported in the literature. Two qualitative categories of surface textures are recognized. The first describes the degree of polish or lack of polish, and the second includes the types of markings on the grain surface. A binocular microscope is used to study surface textures.

Our knowledge of surface textures has been summarized by Pettijohn (*A*, 1957). An excellent bibliography of surface features of grains was presented by Williams (*A*, 1937). More recent studies treat frosting of quartz grains by carbonate replacement (*A*, Walker, 1957), and physical and chemical characteristics of desert varnish (*A*, Engel and Sharp, 1958).

Fabric. Fabric is the orientation or lack of orientation expressed by the nonspherical components of a rock.

Two basically different techniques are used to evaluate fabric. The first is a crystallographic method based on the assumption that the *c* axis in quartz parallels the long dimension of the detrital quartz grain. Evidence presented by some workers (*A*, Wayland, 1939; Krynine, 1940; Rowland, 1939, 1940) supports this assumption, but other studies (*A*, Ingerson and Ramisch, 1942; Rowland, 1946) suggest that this assumption is not always valid. The second method is dimensional and requires only measurement of the orientation of the long dimension of the grain. This latter technique is the simpler.

Crystallographic orientations are measured by either of two techniques. A universal stage is used in the first procedure to determine the orientation of the crystallographic direction assumed to parallel the long dimension of the grain (*A*, Fairbairn, 1949). A photometer is used in the second method (*A*, Martinez, 1958; Pierson, 1959). It consists of measuring the variation in the intensity of monochromatic light passed through a standard thin section of sandstone on the stage of a petrographic microscope. The minimum intensity of light will be transmitted when the trend of the optic axes of the quartz grains lies parallel with the slow direction of vibration of light in an accessory gypsum plate.

Dimensional measurements of the orientation of sedimentary particles are made directly or from projected images. The orientation of the long axis of pebbles can be determined by reorienting the pebbles taken from the outcrop relative to reference lines previously marked on the pebble (*A*, Krumbein and Pettijohn, 1938; Karlstrom, 1952, Schlee, 1957). Direct measurement of the grain orientation in a rock sample is made from rough surfaces or peels (*A*, Schwarzacher, 1951) or from thin sections (*A*, Curray, 1956*a*, 1956*b*; Griffiths, 1949; Griffiths and Rosenfeld, 1950, 1953). Grains are selected for measurement according to a random sampling pattern. The direction of the least-projection elongation (*A*, Dapples and Rominger, 1945) is measured rather than the orientation of the long dimension of the grain. This least-projection elongation is parallel to the two most closely spaced lines that can be drawn tangent to the outline of the grain and is theoretically the orientation in which the grain offers a minimum resistance to flow of the depositing medium.

Packing. Packing refers to the spacing of all the components in the rock. Techniques for evaluating packing are described by Kahn (*A*, 1956); he considers both unit and aggregate properties. One aspect of the unit properties, termed "packing proximity," is expressed as the percentage of grain-to-grain contacts. The other aspect of the unit properties, which is not evaluated by Kahn, is the measure of fit of the contacts between grains. The aggregate property is termed "packing density" and is expressed as the percentage of the length of the traverse of a thin section that is occupied by grains. Kahn's properties of packing can be treated by standard statistical methods.

Both packing proximity and packing density can be determined from simultaneous measurements made on the same traverse across a thin section. As each grain is crossed, the nature of the contact at only one boundary is noted. The length of the traverse is measured as well as the intercept distance for each grain along the line of traverse. From these data Kahn's two properties of packing can be computed. Kahn's method for determining packing proximity is adapted from Winsaur et al. (*A*, 1952), and packing proximity is the same as the packing index defined by Winsaur et al.

Another method for evaluating packing, which is not amenable to statistical analyses, was used by Taylor (*A*, 1950); she determined the average number of grain-to-grain contacts for randomly selected grains and noted the shape of the contact.

Porosity. The porosity of rocks has three aspects: (1) the amount and size distribution of pores, (2) the geometrical form of individual pores, and (3) the geometrical form of the network of interconnecting pores. Satisfactory microscope procedures for the quantitative study of the amount and size distribution of pores have been developed, but methods for describing the geometrical form of both individual pores and the pore network are qualitative.

The amount and size distribution of pores are determined from impregnated samples. Canada balsam (*A*, Waldo and Yuster, 1937) and plastic (*A*, Lockwood, 1950), each containing a colored dye, have been used satisfactorily. Any petroleum in the sample to be impregnated should be extracted. Standard thin sections of the impregnated sample are ground and studied with a petrographic microscope. Another impregnation technique is to inject a low-melting-point alloy such as Wood's metal into the sample contained in a suitable pressure device (*A*, Winsaur et al., 1952). The lowest possible injection pressure that will force the metal into the pore spaces should be used so that the particles composing the rock will not be fractured. A surface on the metal-injected plug is ground, polished, and studied with a metallographic attachment on a petrographic microscope. The amount of porosity and the size distribution of pores are measured by the methods previously noted for determining grain size and sorting from thin sections (*A*, Friedman, 1958; Roethlisberger, 1955; Packham, 1955). One should remember that these methods give the apparent pore-size distribution rather than the true pore-size distribution.

A method for measuring the porosity in samples containing large vugs and fractures is described by Waring and Layer (*A*, 1950). These workers placed clear plastic over sawed surfaces of cores and traced the outlines of vugs, fractures, and other components of the rock. The area occupied by these different components was measured with a planimeter. These areas were assumed to be proportional to the volumes of these different rock components. Chin and Rose (1951) suggest application of the Chalkley method (*A*, Chalkley et al., 1949), a statistical method for determination of volume from surface measurements, to vuggy and fractured rock samples.

The geometrical forms of individual pores and the pore network have been studied by first impregnating the sample and then dissolving out the rock material. Ryder (*A*, 1948, 1949) used the method first proposed by Nutting (*A*, 1929) and impregnated sandstone samples with wax. The quartz grains were dissolved out with hydrofluoric acid. Imbt and Ellison (*A*, 1946) impregnated carbonate rock samples with plastic and then used hydrochloric acid to remove the carbonate material and hydrofluoric acid to remove silica. The casts of the individual pores and the pore network were then studied with a binocular microscope. These studies provided only qualitative data; techniques for the quantitative measurement and statistical evaluation of parameters measured from casts of individual pores and the pore network have not been reported in the literature.

Another approach to tridimensional porosity studies is the use of models that have properties measured for rock samples. Fatt (*A*, 1958) constructed models using compressible spheres and cubes. His model composed of cubes simulates the porosity and electrical resistivity characteristics of the rock samples he tested. Thus, the geometrical form of individual pores and the pore network of some sandstones is believed to approximate the known patterns in the model. Model studies of pore configuration should be interpreted geologically with caution, however, for too many of nature's variables cannot be simulated in the models.

Other Physical Properties

Knowledge of other physical properties of rocks will sometimes assist the geologist in solving his problems. Engineering tests for soils (*A*, ASTM, 1958; Lambe, 1951) are adaptable to sediments, and those for natural building stones (ASTM, 1958) are adaptable to rocks. Of interest to geologists are tests for plasticity, elasticity, compactability, and compressive strength. Some other specific testing procedures are briefly summarized in the following paragraphs.

The density of a rock is the mass per unit of bulk volume. Rock densities are important in the interpretation of regional gravity surveys. Geologists also use density measurements, especially for shales, as a record of the degree of compaction of the rock (*A*, Athy, 1930; Hedberg, 1936; Johnson, 1950). Compaction reduces the volume of the rock, thus increasing its density. Either wet or dry densities may be sought. The wet density is the weight of the sample saturated with water divided by the volume of water that the saturated sample displaces. Dry densities are determined by first extracting by Soxhlet any petroleum in the sample, then drying the sample to constant weight at 105 to 110°C, next saturating the sample with a fluid such as kerosene, and finally measuring the volume of the saturating fluid displaced by the sample. Care must be exercised to ensure that the sample is saturated. A pycnometer is used to measure the displaced volume. Core-analysis reports commonly record bulk density values.

Rock samples are subjected to triaxial compression at high pressures in order to obtain ideas about their deformation under geologic conditions. Equipment and procedures for these tests are described by Handin (*A*, 1953) and Handin and Hager (*A*, 1957).

The thermal conductivity of a rock is a measure of the ease of heat flow through the rock. This property of rocks has been investigated for its effect on mechanical logs run in well bores. Apparatus and a procedure for measuring thermal conductivities of rocks are described by Zierfuss and van der Vliet (*A*, 1956).

The magnetic susceptibility of a rock is a measure of the ability of the rock to be magnetized. Knowledge of the magnetic properties of rocks assists the geologist in evaluating certain types of well logs and geophysical prospecting methods. Techniques for measuring magnetic susceptibility are discussed by Estermann (*A*, 1959), Spokes and Mitchell (*A*, 1958), Howell et al. (*A*, 1958), Bruckshaw and Robertson (*A*, 1948).

COMPOSITIONAL PROPERTIES OF ROCKS

Inorganic Components

Mineralogical Analyses

The mineralogical composition of rocks is a fundamental property that is determined as a matter of course in all geological work. It serves as the basis for describing and classifying rocks. It is used to interpret the environment in which rocks were formed and to recognize later diagenetic or metamorphic changes. In fact, there is no phase of geology that does not require a mineralogical analysis of the rocks.

Often it is sufficient to determine the major mineral component of a rock, as when differentiating between a limestone and a dolomite. In other cases a more complete analysis may be required, as when determining the effect of clay content in a sandstone on fluid flow. Tools for making mineralogical analyses accordingly range in cost and size from the hand lens to the electron microscope. Each has its application. The great majority of sedimentary rocks are not made up of a single mineral but are complex mixtures. For this reason it may be necessary to utilize several tools to make the necessary mineralogical analysis.

Microscope Methods. The microscopic examination of rocks is a basic and essential laboratory technique for making mineralogical analyses, despite many newer instrumental procedures. The petrographic microscope with the various lens combinations and accessory plates (gypsum plate, mica plate, quartz wedge), a set of standard immersion liquids, and a suitable light source are the equipment needed for routine optical mineral determinations. The universal stage can be extremely helpful in optical mineralogy but is usually not required in routine work.

Minerals may be studied in thin section or by immersion techniques. Although thin sections are extremely useful in observing the texture and grain-to-grain relationships in a rock, the immersion method, utilizing crushed mineral fragments, is most satisfactory for identification (*B*, Rogers and Kerr, 1942). A small amount of material, isolated from the rock mass and crushed to between 100 and 200 mesh, is placed on glass slides with drops of the various refractive index immersion media, and its physical and optical characteristics are determined.

The physical properties of the mineral fragments, such as cleavage, hardness, fracture, etc., can aid greatly in a mineral identification and are usually described early in the procedure.

The most important optical property possessed by all minerals which can be measured quantitatively is the index of refraction (*B*, Winchell, 1948). In isotropic minerals there will be a single refractive index regardless of the crystallographic direction. Anisotropic minerals may possess two or three refractive indices depending on whether they are uniaxial or biaxial. It is the determination of these refractive indices and other optical properties which requires the use of accessory equipment and calls for more refined techniques.

As the optical and physical properties are determined, they are compared with tables of these properties which are provided in most mineralogical references (*B*, Hurlbut, 1941; Palache et al., 1952; Rogers and Kerr, 1942; Winchell, 1951).

The optical and physical properties of pure, crystalline chemical compounds or minerals and of end members of a solid solution series are constant and diagnostic. However, many mixed crystal series, characterized by partial or complete isomorphism, will display a gradation of optical and physical properties, which makes identifications difficult. In spite of these difficulties, the optical method of identification still remains one of the better methods for nonopaque materials (*B*, Wahlstrom, 1956).

X-ray Diffraction. During the past few years the use of X rays has been developed as a powerful analytical tool for mineral and elemental analyses. For mineral analyses, X-ray diffraction is used, and for elemental analyses, the X-ray emission spectrograph. The combination of these two tools allows the analyst to determine the atomic, molecular, and crystal structure of a rock specimen.

X-ray dirffaction is a nondestructive method of analyzing the structure and composition of crystalline materials. A crystal is composed of atoms or groups of atoms arranged in a regular and repeated pattern. If such an arrangement of atoms is exposed to electromagnetic radiation of a wavelength of the same order of magnitude as the distance between the atoms, the rays will be diffracted at an angle to the primary beam. Since the distance between atoms in a crystal is of the order of 1×10^{-8} centimeter, X rays are necessary to fulfill the requirements for diffraction. The X rays derived from a copper target X-ray tube are those most used in diffraction work. These X rays have a wavelength of 1.54×10^{-8} centimeter. The diffracted X rays can be detected by either film or electronic counters. The details of X-ray diffraction theory and procedures are given by Bunn (*B*, 1946), Buerger (*B*, 1942), Klug and Alexander (*B*, 1954), Guinier (*B*, 1952), and Clark (*B*, 1955).

Mineral components of a rock can be identified by X-ray diffraction in a ½-gram sample down to concentrations of about 1 per cent. This sensitivity can be increased by utilizing mineral concentration techniques such as particle-size separation, mineral separation, insoluble-residue techniques, etc. The method can be made quantitative by the preparation of synthetic mixtures of minerals to construct calibration curves. Internal standard techniques can be used for refining the analysis.

In geological studies, X-ray diffraction is used as a supplementary petrographic tool. Mineral analyses accomplished by X-ray diffraction have been used for (1) sand-shale studies, (2) study of insoluble residues of carbonates, (3) calcite-dolomite ratio determinations, (4) correlation studies, (5) environmental studies, and (6) mineral identification. The study of clay minerals has probably been the largest field to which the technique has been applied. The extremely fine particle size of the clay minerals makes microscopic studies difficult but is not a problem to X-ray diffraction.

Heavy-mineral Analysis. Heavy minerals are defined as those with a specific gravity greater than about 2.90. Some of the more common heavy minerals include tourmaline, zircon, garnet, staurolite, and various pyroxenes and amphiboles.

The separation procedure for removing the light clastic materials is relatively simple and requires a minimum of equipment. The disaggregation of the rock particles in such a manner as not to alter mineral surfaces can be a more difficult problem. Heavy-mineral techniques are well described by Pettijohn (*B*, 1957) and by Twenhofel and Tyler (*B*, 1941).

Heavy minerals can be used in the correlation and identification of stratigraphic units. Suites of minerals can be established for various stratigraphic units. These techniques are also useful in outlining the source area of sediments.

Insoluble Residues. Insoluble residues are the insoluble material left after the digestion of a carbonate with acid. Hydrochloric acid ($4N$) is most often used, but others such as acetic acid can be used where a less severe digestion is required (*B*, Twenhofel and Tyler, 1941). By determining the gross amount and the mineralogy of the insoluble residue, it is often possible to correlate thick and monotonous limestone or dolomite sequences. The technique has been used widely and apparently successfully in the correlation of Paleozoic rocks in Texas, Oklahoma, Kansas, Missouri, and Illinois (*B*, Sloss and Hamblin, 1942; Ireland, 1936; McQueen, 1931; Grohskopf and McCracken, 1949; Stocker, 1954).

Recent work with insoluble residues has been in the direction of more gentle treatment of the sample to preserve some of the noncarbonate minerals dissolved or decanted in earlier procedures. Chief among these are the clay minerals. Some workers have employed ion-exchange resins to remove the carbonates and have been able to recover a greater proportion of the clays (*B*, Lloyd, 1954). Insoluble residues can be analyzed by any of the petrographic techniques. The cost of preparing the samples is small, but in the case of the more pure carbonates the sample-size requirements may become objectionable.

Staining Techniques. The main value of staining techniques is to determine mineral distribution and facilitate rapid identification of certain minerals (*B*, Twenhofel and Tyler, 1941). Staining techniques have been described for distinguishing (1) calcite from dolomite, (2) calcite from aragonite, (3) quartz from feldspar, (4) potash feldspar from plagioclase, (5) brucite from serpentine, and (6) clay minerals. Perhaps the most useful of the staining techniques are those applying to the carbonates and to feldspar and quartz. The stains are usually applied to the surfaces of etched or polished rock slabs but also can be used on thin-section surfaces or grain aggregates. The details of staining techniques have been described by various authors (*B*, Twenhofel and Tyler, 1941; Heinrich, 1956; Friedman, 1959).

Other Techniques. Among other techniques employed in mineral and rock analysis are differential thermal analysis, thermoluminescence, electron microscopy, radiography, and isotope methods. The method of differential thermal analysis determines the temperatures at which thermal reactions take place in a material when it is heated continuously to elevated temperatures (*B*, Grim, 1953). These temperature reactions may be endothermic, owing to loss of water and crystal structure, or exothermic, owing to the formation of new phases at elevated temperatures. Although the technique has been most used in the study of clays, it has also been successfully used in the study of hydrates, carbonates, sulfides, organics, etc. (*B*, Smothers and Yao, 1958).

Thermoluminescence is the emission of light produced by heating a solid to a temperature below that of incandescence. It is exhibited by crystals that have been exposed to X rays or radioactivity and then heated rapidly. The high-energy radiation dislodges electrons that become trapped in lattice imperfections. When sufficient kinetic energy is supplied in the form of heat, these trapped electrons are released and light is emitted (*B*, Daniels, 1953). The light intensity is plotted as a function of temperature to produce glow curves, which are then used to characterize the stratigraphic unit of interest. The method has been used with limited success in the stratigraphic correlation of carbonates (*B*, Parks, 1953; Saunders, 1953).

The electron microscope produces an image by the scattering of electrons as they pass through the specimen. Magnifications in the range of 3,000 to 20,000 diameters are obtainable, and with photographic enlargement over 100,000 diameters are practical. The application of the electron microscope can be likened to that of the light microscope, but in the range of large molecules, colloids, and other particles beyond the resolving power of the light microscope. The instrument has been used extensively in the study of clays to describe the particle shapes of these minerals (*B*, Grim, 1953; Fischer, 1954; Wyckhoff, 1958).

A radiograph is a photographic record produced when X rays or gamma rays are passed through a sample on to a film. The amount of radiation absorbed depends on the nature (absorption coefficient) and thickness of the material. Fine-grained films and photographic enlargement permit the study of elemental and mineral distribution in thin slices of rocks, which is not possible with light microscopy. Although the method appears to have merit for petrographic work, it has not been applied nearly so widely as in the metallurgical and medical fields (*B*, Cosslet et al., 1957; Sproull, 1946).

The isotopic composition of certain elements in rocks is increasingly being used in studies of depositional environment, correlation, and absolute-age dating. This new and rapidly growing area of work has come to be known as isotope geology (*B*, Rankama, 1954). Isotopes are different forms of the same element that have different atomic weights because of slight changes in the structure of the atomic nucleus. Many isotopes are stable, and recent work, made possible by the improvement of the complex analytical techniques required, indicates that the natural variations in the isotopic constitution of the elements may be of considerable consequence in understanding the history of lithologic units (*B*, National Research Council, 1958). A great amount of research is currently being applied to the extent and cause of isotopic fractionation and its interpretation in geological and geochemical terms. Most promising in this respect appear to be the stable isotopes of hydrogen, carbon, nitrogen, oxygen, and sulfur. Many elements, such as carbon, potassium, rubidium, or uranium, also have unstable isotopes which decay by natural radioactivity to form other atomic species. Because the decay rates of these radioactive isotopes are known, they can be used to determine the absolute age of the minerals in which they and their decay products occur. Although these methods of absolute-age dating have up to now been applied mostly to minerals of igneous origin, current work has extended their use to sediments.

Chemical Analyses

Chemical analysis of sedimentary rocks and their contained fluids can provide much information about the geochemical history of the sediment. The sediments are, in a sense, the results of a large-scale chemical and mechanical fractionation process (*B*, Pettijohn, 1957). To interpret these processes, the chemical composition of each product must be known.

Rock Analyses. Wet chemical, emission spectrographic, and X-ray fluorescence techniques have been applied to the elemental composition analysis of rocks. The extent of the analysis will be dictated in most cases by the nature of the problem. In some studies, a complete quantitative analysis of the elements present will be desired; in others, a qualitative analysis will suffice. Much recent work has been done on selected trace elements, where it is only necessary to analyze for the specific elements of interest. Accurate wet chemical techniques have been developed for most routine rock-analysis problems (*B*, Hillebrand et al., 1953). However, since some are quite time consuming, the analyst will probably wish to use instrumental methods if a large number of analyses are required.

The emission spectrograph has probably been the instrument most used in rock analysis. Emission spectrographic techniques vary widely but essentially consist of exciting the electrons of the atoms in the rock—by flame, electric arc, or spark—to a higher energy level. When the electrons return to their ground, or lowest, energy level, the characteristic line spectrum of the element is emitted. The spectrograph uses a slit; a dispersing medium, such as a prism or grating; and a recording device, such as film or photomultipliers. The resulting record is a series of lines, images of the slit, which can be attributed to wavelengths of the light emitted by the sample.

The wavelength of the lines identifies the material, and the intensity is a function of the amount present (*B*, Harrison et al., 1948). The electric arc, spark, or flame has only sufficient energy to remove the outer-shell electrons from an atom. However, if the atom is bombarded with high-energy electrons or X rays, the inner-shell electrons will be knocked out. Then when the atom returns to its ground level, X rays will be produced, which are unique for that particular atomic species. This phenomenon is utilized in X-ray fluorescence spectrography. The produced X rays are dispersed by diffracting them from a single crystal and are recorded electronically. The sample is destroyed in the emission spectrograph but not in the X-ray method. The former is best suited for the analysis of many elements in a sample and for low levels of concentration, whereas the latter is usable for few elements but over all concentration ranges (*B*, ASTM, 1953).

Elemental analyses of rocks have been used in correlation studies, studies of depositional environment, age determinations, compositional studies, and geochemical studies.

Water Analyses. The analyses of formation waters from drill-stem tests on exploration or field development wells can prove very helpful to geologists. The waters are usually analyzed by wet chemical methods, and the results are reported as concentrations of positive and negative ions. For geological interpretation and correlation work these values are then plotted graphically to yield patterns that are distinctive for a given formation water.

There are three prominent methods used for plotting water-analysis data. These are the Palmer, Stiff, and Tickell plots. The Palmer system (*B*, Palmer, 1911) plots the reacting values in per cent of the radicals. This plot separates waters into four classes based on primary and secondary salinity and primary and secondary alkalinity. The Stiff system (*B*, Stiff, 1951) gives the reaction values in milliequivalents per liter directly. Since actual concentrations are used, two waters of similar ion distribution but unlike concentrations can be differentiated. The Tickell (*B*, Tickell, 1921) method plots the per cent reaction values on a six-axis diagram, utilizing only five of the axes. Each axis represents 0 to 50 per cent. Ions are grouped as follows: Na and K, Ca and Mg, Cl, SO_4, CO_3 and HCO_3.

All the methods provide a means of plotting the chemical composition of waters so that they can be diagrammatically characterized and visually compared with other waters (*B*, Levorsen, 1954). Water-analysis data are used in interpreting well logs, correlating formations, detecting geologic environment changes, detecting water from formations other than producing formations, detecting salt domes, evaluating water floods, and combating corrosion.

Organic Components

The organic constituents of sediments have a great range of physical and chemical properties. More than a quarter of a million organic compounds are known to the organic chemists. The petroleum geologist's interests lie chiefly with the class of organics called hydrocarbons, those composed only of carbon and hydrogen. This somewhat reduces the number of organic species to be dealt with, but the known hydrocarbons number many thousands. These hydrocarbons occur in seeps, coals, solid hydrocarbons, and oil shales; as minor constituents in nonreservoir rocks; and reservoired as crude oils, gases, and natural asphalts. Analytical techniques have been developed to help the petroleum geologist identify and characterize hydrocarbons in all these situations. However, because of the complex mixture of molecules in naturally occurring organic materials, most of the analytical methods in general use delineate only gross characteristics (*B*, Levorsen, 1954).

Crude-oil Analyses

Crude oils are composed of a complex mixture of hydrocarbon molecules with varying but small amounts of inorganic and nonhydrocarbon impurities. Crude oils from different reservoirs vary widely in their compositions. A great amount of study and research has been carried out on these differences and their causes.

In the early days of petroleum production, crude oils were analyzed primarily to determine the refining characteristics and the products available. This is still of major importance, but the information is also used in petroleum exploration for correlating crude oils, for relating crude oils to stratigraphy (*B*, Hunt, 1953), and for studying the genesis of petroleum. The most commonly measured properties of crude oils are (1) gravity, (2) refractive index, (3) distillation characteristics, (4) residuum quantity and characteristics, (5) molecular composition, and (6) inorganic components.

Gravity. The gravity of crude oils is generally represented by either of two methods. In Europe it is expressed as specific gravity. In the United States a special gravity scale is used, the American Petroleum Institute, or API, gravity scale (*B*, ASTM, 1955). In this scale a specific gravity of 1.0 is equivalent to 10° API. The scale extends in practical use from 0 for a heavy asphalt to about 75 for a very light naphtha. In addition to the API gravity of a whole crude, specific gravities of different boiling ranges can be determined. The specific gravity of a crude-oil fraction indicates the type of hydrocarbon molecules present within a restricted boiling range. Paraffinic crudes have a low specific gravity, whereas naphthenic and aromatic crudes have a high specific gravity (*B*, Hunt, 1953).

Oil gravity in saturated cores or drill cuttings can be determined by extracting them with suitable solvents such as carbon tetrachloride or pentane, distilling off the solvent, measuring the refractive index of the residual oil, and converting this measurement to the desired gravity scale (*B*, Hedberg, 1937; *A*, Pirson, 1958). Because light ends are lost during this procedure, however, the results are only qualitative and must be corrected accordingly. In addition, the relationship between oil gravity and refractive index varies rather widely with the type of crude and must be determined experimentally for different areas.

Refractive Index. The refractive index of a liquid is the measure of the ratio of the velocity of light, of a specified wavelength, in a vacuum to its velocity in the liquid (*B*, ASTM, 1955). Refractive dispersion of a substance is the difference in its refractive indices for two different wavelengths of light. Specific dispersion is the refractive dispersion divided by the density of the substance, both measurements being made at the same temperature. These measurements are useful in determining the nature of simple organic mixtures such as oil fractions that boil within a restricted range. Aromatic compounds have high specific dispersions and refractive indices, saturated hydrocarbons have low ones, and naphthenic hydrocarbons have intermediate ones. These measurements are usually applied to distillation cuts but have also been used on whole crudes.

Distillation Analysis. The most frequently used distillation analysis for crude oils is the Hempel analysis as used by the U.S. Bureau of Mines (*B*, Smith et al., 1951). In this analysis 10 distillation cuts are taken at 25° intervals between 50 and 275°C at atmospheric pressure. The pressure is then reduced to 40 millimeters of mercury, and five cuts are taken between 200 and 300°C. The volumes and specific gravities are determined for each distillation cut. Type of material, such as gasoline, kerosene, gas oil, and lubricating distillates, can then be reported as percentages of the original crude oil.

Systematic procedures for the analysis of crude oils, such as the Bureau of Mines Hempel method, provide general descriptions of the analyzed oils in simple, comparable terms. Such analyses, however, emphasize similarities, whereas interesting and diagnostic differences may be enhanced by "index numbers" (*B*, Smith, 1940). The Bureau of Mines has developed a simple and useful function for characterizing crude oils known as the correlation index. The correlation index is calculated from the specific gravity and the average boiling point of a distillation fraction. Paraffinic hydrocarbons have index values from 0 to 15; values of 15 to 50 indicate naphthenes or mixtures of paraffins, naphthenes, and aromatics; and values above 50 indicate mixtures of naphthenes and aromatics. When the correlation indices are plotted against distillation fractions of each crude, a characteristic graph is obtained. These graphs have been widely used in crude-oil classification studies.

Molecular-type Analysis. Gas chromatography is used for the molecular analysis of both gaseous and liquid hydrocarbons. It is a separation technique in which a sample to be analyzed is introduced in vapor form into a moving gas stream and passed through a stationary column of adsorbent. The gas stream is usually helium, but other carrier gases such as carbon dioxide and nitrogen can be used. The column may consist of a solid adsorbent (such as alumina or silica gel) or a solid inert support (such as Celite or firebrick) coated with a high-boiling liquid adsorbent (*B*, Keulemans, 1957). As the sample passes through the column, the components partition themselves between the stationary phase and the carrier gas. Each component has a different partition coefficient, or affinity, for the column material and will therefore appear at the exit of the column at different times. A detector at the end of the column records the arrival of the different components and provides an index of the amount of each component present in a mixture (*B*, Hardy and Pollard, 1959).

Other Analyses. Other crude-oil components that are routinely determined are sulfur, carbon in the residuum, and trace metals, especially vanadium and nickel. The sulfur content of a crude oil is an index of oil quality. It is determined by standard ASTM techniques (*B*, ASTM Committee D-2, 1957). The carbon residue of the Hempel residuum indicates the abundance of fused-ring compounds. High-carbon residues indicate fused-ring hydrocarbon systems deficient in hydrogen; low residues indicate paraffinic hydrocarbons (*B*, Hunt, 1953). Trace metals can be determined by either wet chemical or spectrographic techniques on ashed crude samples. The trace metals have been investigated as a means of crude-oil identification (*B*, Witherspoon and Nagashima, 1957).

Gas Analyses

Natural gas consists of varying proportions of gaseous hydrocarbons like methane, ethane, propane, butane, and isobutane; it occasionally contains liquid hydrocarbons such as pentanes, hexanes, etc., and nonhydrocarbon gases such as carbon dioxide, hydrogen sulfide, nitrogen, and helium. So-called dry gas consists principally of methane. Wet gas contains a high proportion of the higher-molecular-weight, condensable hydrocarbons. Many methods of gas analysis have been developed (*B*, Altieri, 1958), but of primary interest in exploration are the gas analyses conducted at the well site rather than in the laboratory. Gas logging of the mud stream and drill cuttings with a hot-wire detector or gas chromatograph has become common practice as a means of detecting subsurface oil and gas shows. A mud gas log normally provides readings on both total gas and wet gas content plotted against depth. The proper interpretation of these data requires considerable care and experience. Also of interest may be analyses of the nonhydrocarbon gases. Helium has been found in significant quantities in some areas and is of commercial value. Nitrogen and carbon dioxide can render a natural gas of low value or even worthless as a fuel gas if they occur in sufficient concentrations. Hydrogen sulfide can be a serious hazard. It is rapidly fatal in concentrations of 0.1 per cent in air by volume and can cause painful and serious injury at 0.01 per cent or less. Fortunately it can be readily detected by its distinctive odor at concentrations of a few parts per million.

Seep Identification

A seep is a surface evidence of oil or gas. Nearly all oil-producing areas of the world contain some seeps, which often lead directly to the discovery of commercial oil deposits. This points up the importance of establishing the identity of suspected seeps. A seep may be live if it is continually replenished from an underground source, it may be dead if it has been cut off from its source, or it may be false, having the appearance of a true oil seep but actually resulting from other, even man-made causes. Field geologists will have little trouble recognizing a liquid oil seep or mineral wax deposit. More difficult to distinguish are the asphalts, asphaltites, and pyrobitumens. Boiling water will generally melt or soften an asphalt, whereas an asphaltite will remain unaltered. Pyrobitumens will only slightly color solvents such as carbon tetrachloride, whereas asphaltites quickly darken the solution. In the laboratory, the various materials can be identified by solubility characteristics, melting point, refractive index, boiling range, trace-metal content, infrared spectrophotometry, X-ray diffraction, and carbon and hydrogen analysis (*B*, Abraham, 1945).

FOSSIL CONTENT OF ROCKS

The fossil content of sediments can provide the geologist with essential information on the age of the rock and on the depositional environment under which it was formed. Fossils can also be used for the detailed zonation and correlation of rock units. Although every geologist should have a working knowledge of paleontology, the science that deals with ancient life as it is recorded in the fossil remains of plants and animals, the full and proper interpretation of the fossil content of rocks generally has to be provided by the specialist with adequate library, reference collections, and laboratory facilities at his disposal.

Fossils are found in many different types of rocks. They may be impressions, molds, or remnants of the whole or any part of an ancient plant or animal. They may vary in size from a few microns, as in the case of a discoasterid, up to several tens of feet, as in the case of a dinosaur skeleton. Macrofossils, or those larger than the arbitrary size limit of 10 millimeters, can normally be used only when outcrop samples or core material are available. Microfossils are much more widely used in the oil industry because they are not destroyed so readily by the drill bit and can be recovered from well cuttings. The smaller the fossil, the better its chance of escaping the destructive action of the bit.

Within the last decade a new branch of micropaleontology has assumed a prominent place in the petroleum industry. It has come to be known as palynology (from the Greek verb *palyno*, meaning to scatter or distribute) and is concerned with minute fossil organisms of both animal and plant origin which generally measure less than 150 microns and average about 50 microns. The study of these exceedingly small forms requires special sample preparation techniques and high-power biological microscopes.

The fossils in a rock sample are intimately related to the sedimentary materials surrounding them, as well as to the time and conditions of deposition. Their occurrence is controlled by three basic factors. One is a biologic or genetic factor that is passed on from each individual to its offspring. This is paramount in controlling the shape, size, composition, and appearance of organisms. Changes in this factor over long periods of time cause the changes in the organism called organic evolution. The second is the ecologic factor, i.e., the relation of the organism to its environment, both physicochemical and organic. This plays the chief role in determining the relative number and distribution of organisms. The third is the preservation factor, i.e., whether or not conditions were favorable for the preservation, as fossils, of the organisms that lived at any particular time.

Fossils are really a record in the rocks of the complex interaction of these three factors. Therefore, they can be used to interpret both geologic time and sedimentary environment. The principles that make this possible are simple: First, changes in the vertical sequence of fossils through a series of rock strata partly record evolutionary changes and are thereby a key to geologic age. Second, the combined horizontal and vertical distribution of fossils makes a three-dimensional record of shifting sedimentary environments with time.

The principles are simple, but the application is less so. Neither environment nor evolution works separately in determining the occurrence of a particular fossil in a particular sample of rock. Yet the accuracy of interpretations based on fossils often depends on distinguishing between the effects of these factors.

Age Dating

The age of a formation is usually needed in the relative terms of the geological time scale, not in absolute years. The ideal time indicator would be the fossil of an organism that combines a world-wide geographic distribution, a complete tolerance of varied environments, and a very short range in terms of geologic time. This is only a hypothetical ideal. All fossils are limited in varying degrees to particular environments (i.e., they are facies fossils) and are also geographically restricted. In addition, many fossil types have relatively long geologic ranges and are therefore not useful as time indicators.

The specific types of fossils most useful for dating a given sample depend upon many factors, including general geologic age, type of rock, size of sample, and precision of dating required. The quality of preservation is also important. This determines how readily the details critical for identification of the fossil can be studied.

Detailed information on the main groups of fossils can be found in basic texts such as those by Jones (*C*, 1956); Glaessner (*C*, 1945); Cushman (*C*, 1955); Swinnerton (*C*, 1947); Roemer (*C*, 1945); Arnold (*C*, 1947); Shrock and Twenhofel (*C*, 1953); Moore, Lalicker, and Fischer (*C*, 1952); and Howe (*C*, 1958). Principal index fossils of North America are described and illustrated by Schimer and Shrock (*C*, 1944). Neaverson (*C*, 1955) gives a comprehensive treatment of the fossil record, with emphasis on distribution in space and time, while the "Treatise on Invertebrate Paleontology" [*C*, R. C. Moore (ed.), 1954] gives a complete and up-to-date treatment of various groups of fossils.

A few examples will emphasize that each fossil group has varied aspects which must be understood before it can be used successfully.

Graptolites, floating inhabitants of Lower Paleozoic seas, were a rapidly evolving group whose distinctive species have very limited geologic ranges. Therefore, they are very precise indicators of geologic time. However, they occur abundantly only in dark-colored marine shales of Cambrian to Silurian range. Thus, the limited geologic distribution of the group in rocks of a restricted lithology severely limits their usefulness to the petroleum geologist.

The ammonites, a group of shelled Mesozoic marine mollusks related to the squids, are the basis for the standard time subdivision of Mesozoic marine strata. They are among the most valuable fossils for dating Mesozoic surface sections, but their large size (1 inch to several feet) precludes getting many identifiable specimens from cores or cuttings.

The bottom-dwelling foraminifera (the first microfossil group to be used extensively in subsurface geology) are small, easily recovered from cores and cuttings, and have a wide geographic distribution. They have played a large and important role in subsurface studies and, undoubtedly, will continue to do so. In many cases, however, they have led to erroneous conclusions about age because the extent to which bottom conditions control their distribution was not fully appreciated. In contrast, the pelagic foraminifera, living as floating organisms near the sea surface, are much more reliable age indicators because they were not controlled by bottom conditions.

Minute spores and pollen from land-dwelling plants have advantages shared by very few other fossils: They occur commonly in both continental and marine sediments, and they may be abundant in rocks devoid of other fossils. In Mesozoic rocks they are often precise age indicators on a world-wide basis. However, in Tertiary rocks they may have only local age significance because of a spotty and shifting distribution pattern controlled especially by complex patterns of temperature and rainfall that characterized the Cenozoic.

Environmental Interpretation

Every fossil tells something about the environment inhabited by the organism that made it. Abundant corals, crinoids, or brachiopods are a conclusive indication of relatively shallow marine conditions in waters of moderate to warm temperature. When supplemented by other samples in a regional study, they may be the basis for locating and tracing reef trends and ancient shore lines.

Paleoecology, the relation of fossil organisms to the environments in which they lived, is a new and rapidly growing phase of paleontology. For the petroleum geologist it has great potentials that are only beginning to be explored. A compendium of knowledge on the ecology and paleoecology of marine organisms is presented by Hedgpeth and Ladd (*C*, 1957), together with extensive references to earlier literature. A few examples will show how fossils used to interpret ancient environments can help the oil geologist.

Knowledge of the most basic requirements of an organism at once gives some indication of the environment represented by its fossil. The fundamental distinction between continental and marine deposits can be based on fossils. For example, an abundance of spores and pollen and complete absence of typical marine microfossils point to a continental deposition. A mixture of land-derived and marine forms suggests a marine realm close to shore, whereas an abundance of shells of floating organisms (pteropods, radiolaria, diatoms, pelagic foraminifera) and almost nothing else suggests a sea bottom remote from shore. Recent studies of oxygen isotope ratios (e.g., *C*, Lowenstam, 1954; Turekian, 1955) show that fossil shells can provide information on ancient water temperature and salinity, although the analytical techniques required are outside the field of the paleontologist.

Many stratigraphic features important in oil accumulation (e.g., sand wedge-outs, sand shale-outs, and offshore bars) are related to the position of ancient shore lines. A statistical study of the distribution of land-derived and sea-derived microfossils (*C*, Hoffmeister, 1954) can indicate the direction to ancient shore lines and suggest possible areas of oil and gas accumulation.

A fine example of a study involving paleoecology is that by Newell et al. (*C*, 1953) in the west Texas Permian Basin. This had significant results for the exploration geologist. It unraveled the depositional history of the area and explained the complex spatial relationships of many contrasting rock types in a deep reef-margined basin.

Together with extensive field work, detailed laboratory studies of the type, size, form, number, and preservation of fossils played a basic role in the study. This investigation also showed that a sound interpretation of past environments must be based on a thorough understanding of modern ones. Work on modern reefs, reef-dwelling organisms, and carbonate mud formation in both the Bahamas and South Pacific was an integral part of Newell's study. Although this was an expensive and long-range research project, it illustrates one kind of information that fossils can help provide. The same approach, scaled down in all respects, can contribute to a local exploration program.

Zonation and Correlation

The zonation, or subdivision, of a thick sedimentary sequence on the basis of fossils may be a virtual foot-by-foot guide to the driller. Correlation of rock units by means of fossils may provide the clue to regional or local structure, unconformities, and major stratigraphic changes. It may also make possible comparisons between surface and subsurface sections.

It is these applications of fossils that are the most commonplace and the most valuable in the everyday activities of petroleum exploration. They are considered last only because they combine the principles discussed in the two foregoing sections. In these applications fossils are useful because organisms changed with time and were restricted to special environments. It is not always possible or relevant to determine whether the time factor or the environmental factor is more important in controlling the occurrence or absence of particular fossils. Many fossils without geologic time significance on an interregional basis may be highly useful in relatively local studies. Even the extreme facies fossil, narrowly limited to one environment, may be useful if the distribution pattern of that environment helps in analyzing the stratigraphy and structure of the area.

To subdivide a body of rocks vertically on the basis of fossils requires that one discover discontinuities in the distribution of fossils through the rock unit. The ease and success with which this can be done depend on many variables, such as the thickness of the section, the spacing of the samples, the abundance of fossils, the distinctness of the types present, and the precision of zonation required. At the easy extreme would be the case where one has large and abundant samples from strata with ammonites (Mesozoic), over beds with fusulinid foraminifera (Upper Paleozoic), in turn underlain by graptolitic shale (Lower Paleozoic). Quite the other extreme would be a 12,000-foot subsurface section of Oligocene sands and clays whose only abundant fossils are microfossils of plants with a known Eocene to Miocene range. In the first case, a few readily recognized index fossils might provide an adequate zonation. In the second, zonation might necessitate detailed statistical study of ratios of numerous, difficult-to-distinguish fossil pollen and spores—and good luck.

In a similar fashion, it is sometimes possible to correlate units and identify strata after seeing only a few diagnostic index fossils. Usually, however, one must carefully consider many types of fossils along with what is known of their ecology and their geographic and stratigraphic ranges.

REFERENCES

A. Physical Properties of Rocks

American Petroleum Institute: Recommended Practice for Determining Permeability of Porous Media, *Am. Petrol. Inst. Rept. Invest.* 27, 3d ed., 1952, reissued 1956.

American Society for Testing Materials (ASTM): "1958 Book of ASTM Standards," part 4, Methods of Tests for Soils, pp. 1099–1230; part 5, Methods of Tests for Natural Building Stones, pp. 641–676, 1958.

Aschenbrenner, B. C.: A Photogrammetric Method for the Tri-dimensional Measurement of Sand Grains, *Photogrammetric Engr.*, vol. 21, no. 3, pp. 376–382, 1955.

Aschenbrenner, B. C.: A New Method of Expressing Particle Sphericity, *J. Sediment. Petrol.*, vol. 26, no. 1, pp. 15–31, 1956.

Athy, L. F.: Density, Porosity, and Compaction of Sedimentary Rocks, *Bull. AAPG*, vol. 14, pp. 1–35, 1930.

Bissell, H. J.: Combined Preferential Staining and Cellulose Peel Technique, *J. Sediment. Petrol.*, vol. 27, no. 4, pp. 417–420, 1957.

Bruckshaw, J. M., and E. I. Robertson: The Measurement of Magnetic Properties of Rocks, *J. Sci. Instr.*, vol. 25, no. 12, pp. 444–446, 1948.

Casagrande, A.: "Die Arämeter-Methode zur Bestimmung der Kornverteilung von Boden und andern Materialien," Springer-Verlag, Berlin, 1934.

Chalkley, H. W., Jerome Cornfield, and Helen Park: A Method for Estimating Volume-Surface Ratios, *Science*, vol. 110, no. 2856, pp. 295–297, 1949.

Chayes, Felix: "Petrographic Modal Analysis, an Elementary Statistical Appraisal," John Wiley & Sons, Inc., New York, 1956.

Chin, W. S., and Walter Rose: Examination of Components of Limestone Porosity by Chalkley Method (Geol. Note), *Bull. AAPG*, vol. 35, no. 3, pp. 615–618, 1951.

Curray, J. R.: Dimensional Grain Orientation Studies of Recent Coastal Sands, *Bull. AAPG*, vol. 40, no. 10, pp. 2440–2456, 1956.

Curray, J. R.: The Analysis of Two-dimensional Orientation Data, *J. Geol.*, vol. 64, no. 2, pp. 117–131, 1956.

Dapples, E. C., and J. F. Rominger: Orientation Analysis of Fine-grained Clastic Sediments, *J. Geol.*, vol. 53, no. 4, pp. 246–261, 1945.

Day, P. R.: Physical Basis of Particle Size Analysis by the Hydrometer Method, *Soil Sci.*, vol. 70, pp. 363–374, 1950.

Engel, C. G., and R. P. Sharp: Chemical Data on Desert Varnish, *Bull. Geol. Soc. Am.*, vol. 69, no. 5, pp. 487–518, 1958.

Estermann, Immanuel (ed.): "Methods of Experimental Physics," vol. 1, Classical Methods, Academic Press, Inc., New York, 1959.

Fairbairn, H. W.: "Structural Petrology of Deformed Rocks," Addison-Wesley Publishing Company, Reading Mass., 1949.

Fatt, I.: Pore Structure in Sandstones by Compressible Sphere-pack Models, *Bull AAPG*, vol. 42, no. 8, pp. 1924–1957, 1958.

Friedman, G. M.: Determination of Sieve-size Distribution from Thin-section Data for Sedimentary Petrological Studies, *J. Geol.*, vol. 66, no. 4, pp. 394–416, 1958.

Glagolev, A. A.: "On Geometrical Methods of Quantitative Mineralogical Analysis of Rocks," *Trudy Inst. Priklad. Mineral.*, No. 59, 1933; translated by J. B. Sykes, Atomic Energy Research Establishment Library Translation 567, 1955.

Goodman, R. E.: Determination of Sand Grain Sphericity by Stereo Photomicrography, *Photogrammetric Engr.*, vol. 25, no. 1, pp. 58–61, 1959.

Griffiths, J. C.: Directional Permeability and Dimensional Orientation in Bradford Sand, *Penn. State Coll. Mineral Ind. Expt. Sta. Bull.* 54, pp. 138–163, 1949. (Also in *Producers Monthly*, vol. 14, no. 8, pp. 26–32, 1950.)

Griffiths, J. C., and M. A. Rosenfeld: Progress in Measurement of Grain Orientation in Bradford Sand, *Penn. State Coll. Mineral Ind. Expt. Sta. Bull.* 56, pp. 202–236, 1950. (Also in *Producers Monthly*, vol. 15, no. 8, pp. 24–36, 1951.)

Griffiths, J. C., and M. A. Rosenfeld: A Further Test of Dimensional Orientation of Quartz Grains in Bradford Sand, *Am. J. Sci.*, vol. 251, no. 3, pp. 192–214, 1953.

Handin, J.: An Application of High Pressure in Geophysics: Experimental Rock Deformation, *Trans. ASME*, vol. 75, pp. 315–324, 1953.

Handin, J., and R. V. Hager, Jr.: Experimental Deformation of Sedimentary Rocks under Confining Pressure: Tests at Room Temperature on Dry Samples, *Bull. AAPG*, vol. 41, no. 1, pp. 1–50, 1957.

Hedberg, Hollis D.: Gravitational Compaction of Clays and Shales, *Am. J. Sci.*, ser. 5, vol. 31, no. 184, pp. 241–287, 1936.

Howell, L. G., J. D. Martinez, and E. H. Stratham: Some Observations on Rock Magnetism, *Geophysics*, vol. 23, no. 2, pp. 285–298, 1958.

Hultt, J. L.: Three-dimensional Measurement of Sand Grains, *Bull. AAPG*, vol. 38, no. 1, pp. 159–160, 1954.

Imbt, W. C., and S. P. Elison, Jr.: Porosity in Limestone and Dolomite Petroleum Reservoirs, "Drilling and Production Practice," pp. 364–371, American Petroleum Institute, 1946.

Ingerson, Earl, and J. L. Ramisch: Origin of Shapes of Quartz Sand Grains, *Am. Mineral.*, vol. 27, no. 9, pp. 595–606, 1942.

Inman, D. I.: Measures for Describing the Size Distribution of Sediments, *J. Sediment. Petrol.*, vol. 22, no. 3, pp. 125–145, 1952.

Jarrett, B. A., and H. Heywood: A Comparison for Particle Size Analysis, *Brit. J. Appl. Phys.*, Suppl. 3: Symposium—The Physics of Particle Size Analysis, pp. S21–S28, 1954.

Johnson, F. W.: Shale Density Analysis, in L. W. LeRoy (ed.), "Subsurface Geological Methods," Colorado School of Mines, Golden, Colo., 1950.

Kahn, J. S.: The Analysis and Distribution of the Properties of Packing in Sand-Size Sediments: 1. On the Measurement of Packing in Sandstones, *J. Geol.*, vol. 64, no. 4, pp. 385–395, 1956. 2. The Distribution of Packing Measurements and an Example of Packing Analysis, *Ibid.*, vol. 64, no. 6, pp. 578–606, 1956.

Karlstrom, T. N. V.: Improved Equipment and Techniques for Orientation Studies of Large Particles in Sediments, *J. Geol.*, vol. 60, no. 5, pp. 487–493, 1952.

Kaye, Emil, and M. L. Freeman: An Air Porosimeter, "Drilling and Production Practice," pp. 187–200, American Petroleum Institute, 1948.

Kelton, F. C.: Analysis of Fractured Limestone Cores, *Trans. Am. Inst. Mining Met. Engrs.*, vol. 189, pp. 225–234, 1950.

Kelton, F. C.: Effect of Quick-freezing vs. Saturation of Oil Well Cores, *J. Petrol. Technol.*, vol. 5, no. 1, sec. 1, p. 21, 1953.

Klinkenberg, L. J.: The Permeability of Porous Media to Liquids and Gases, "Drilling and Production Practice," pp. 200–213, American Petroleum Institute, 1941.

Krumbein, W. C.: Size Frequency Distributions of Sediments and the Normal Phi Curve, *J. Sediment. Petrol.*, vol. 8, no. 3, pp. 84–90, 1938.

Krumbein, W. C.: Measurement and Geologic Significance of Shape and Roundness of Sedimentary Particles, *J. Sediment. Petrol.*, vol. 11, no. 2, pp. 64–72, 1941.

Krumbein, W. C., and F. J. Pettijohn: "Manual of Sedimentary Petrography," Appleton-Century-Crofts, Inc., New York, 1938.

Krumbein, W. C., and L. L. Sloss: "Stratigraphy and Sedimentation," W. H. Freeman and Company, San Francisco, 1951.

Krynine, P. D.: Petrology and Genesis of the Third Bradford Sand, *Penn. State Coll. Bull.* 29, 1940.

Lambe, T. W.: "Soil Testing for Engineers," John Wiley & Sons, Inc., New York, 1951.

Lebeaux, J. M.: Some Effects of Quick-freezing upon the Permeability and Porosity of Oil Well Cores, *J. Petrol. Technol.*, vol. 4, no. 11, sec. 1, p. 19, 1952.

Locke, L. C., and J. E. Bliss: Core Analysis Technique for Limestone and Dolomite, *World Oil*, vol. 131, no. 4, pp. 204–207, September, 1950.

Lockwood, W. N.: Impregnating Sandstone Specimens with Thermosetting Plastics for Studies of Oil-bearing Formations, *Bull. AAPG*, vol. 34, no. 10, pp. 2061–2067, 1950.

Martinez, J. D.: Photometer Method for Studying Quartz Grain Orientation, *Bull. AAPG*, vol. 42, no. 3, pp. 588–608, 1958.

Nutting, P. G.: Some Physical Problems in Oil Recovery, *Oil Gas J.*, vol. 28, no. 27, p. 44, Nov. 21, 1929.

Otto, G. H.: A Modified Logarithmic Probability Graph for the Interpretation of Mechanical Analyses of Sediments, *J. Sediment. Petrol.*, vol. 9, no. 2, pp. 62–76, 1939.

Packham, G. H.: Volume-, Weight-, and Number-frequency Analysis of Sediments from Thin-section Data, *J. Geol.*, vol. 63, no. 1, pp. 50–58, 1955.

Pettijohn, F. J.: "Sedimentary Rocks," 2d ed., Harper & Brothers, New York, 1957.

Pierson, III, A. L.: A Photomultiplier Photometer for Studying Quartz Grain Orientation, *J. Sediment. Petrol.*, vol. 29, no. 1, pp. 98–103, 1959.

Pirson, S. J.: "Oil Reservoir Engineering," 2d ed., McGraw-Hill Book Company, Inc., New York, 1958.

Pollard, T. A., and P. P. Reichertz: Core Analysis Practice—Basic Methods and New Developments, *Bull. AAPG*, vol. 36, no. 2, pp. 230–252, 1952.

Poole, D. M.: Size Analysis of Sand by a Sedimentation Technique, *J. Sediment. Petrol.*, vol. 27, no. 4, pp. 460–468, 1957.

Powers, M. C.: A New Roundness Scale for Sedimentary Particles, *J. Sediment. Petrol.*, vol. 23, no. 2, pp. 117–119, 1953.

Pye, W. D., and M. H. Pye: Sphericity Determinations of Pebbles and Sand Grains, *J. Sediment. Petrol.*, vol. 13, no. 1, pp. 28–34, 1943.

Rall, C. G., and D. B. Taliaferro: A Method for Determining Simultaneously the Oil and Water Saturations of Oil Sands, *U.S. Bur. Mines, Rept. Invest.* 4004, 1947.

Rittenhouse, G.: A Visual Method of Estimating Two-dimensional Sphericity, *J. Sediment. Petrol.*, vol. 13, no. 2, pp. 79–81, 1943.

Robson, D. A.: New Technique for Measuring Roundness of Sand Grains, *J. Sediment. Petrol.*, vol. 28, no. 1, pp. 108–110, 1958.

Roethlisberger, H.: An Adequate Method of Grain-size Determinations in Sections, *J. Geol.*, vol. 63, no. 6, pp. 569–584, 1955.

Rosenfeld, M. A., and J. C. Griffiths: An Experimental Test of Visual Comparison Technique in Estimating Two Dimensional Sphericity and Roundness of Quartz Grains, *Am. J. Sci.*, vol. 251, no. 8, pp. 553–585, 1953.

Rosenfeld, M. A., Lynn Jacobsen, and J. C. Fern: A Comparison of Sieve and Thin-section Technique for Size Analysis, *J. Geol.*, vol. 61, no. 2, pp. 114–132, 1953.

Rowland, R. A.: Petrotectonic Analysis of Cleavage in Otherwise Unmetamorphosed Sediments, *J. Geol.*, vol. 47, no. 5, pp. 449–471, 1939.

Rowland, R. A.: Petrofabric Determination of Quartz Grain Orientation in Sediments (abstract), *Bull. Geol. Soc. Am.*, vol. 51, pp. 1941–1942, 1940.

Rowland, R. A.: Grain-shape Fabrics of Clastic Quartz, *Bull. Geol. Soc. Am.*, vol. 57, no. 6, pp. 547–564, 1946.

Ryder, H. M.: Character of Pores in Oil Sand, *World Oil*, vol. 127, no. 13, pp. 129–134, April, 1948.

Ryder, H. M.: Character of Pores in Oil Sand, in L. W. Leroy and H. M. Crain (eds.), Subsurface Geologic Methods, *Colo. School Mines Quart.*, vol. 44, no. 3, pp. 583–594, 1949.

Schlee, John: Fluvial Gravel Fabric, *J. Sediment. Petrol.*, vol. 27, no. 2, pp. 162–176, 1957.

Schwarzacher, W.: Grain Orientation in Sands and Sandstones, *J. Sediment. Petrol.*, vol. 21, no. 3, pp. 162–172, 1951.

Spokes, E. M., and D. R. Mitchell: Relation of Magnetic Susceptibility to Mineral Composition, *Mining Engr.*, vol. 10, no. 3, pp. 373–379, 1958.

Sternberg, R. M., and H. F. Belding: Dry Peel Technique, *J. Paleontol.*, vol. 16, no. 1, pp. 135–136, 1942.

Stewart, C. R., and J. W. Spurlock: How to Analyze Large Core Samples, *Oil Gas J.*, vol. 50, pp. 89–95, 120–124, Sept. 15, 1952.

Taliaferro, Jr., D. B., T. W. Johnson, and E. J. Dewees: A Method of Determining Porosity: A List of Porosities of Oil Sands, *U.S. Bur. Mines, Rept. Invest.* 3352, 1937.

Taylor, J. M.: Pore-space Reduction in Sandstones, *Bull. AAPG*, vol. 34, no. 4, pp. 701–716, 1950.

Twenhofel, W. H., and S. A. Tyler: "Methods of Study of Sediments," McGraw-Hill Book Company, Inc., New York, 1941.

Wadell, Hakon: Volume, Shape and Roundness of Rock Particles, *J. Geol.*, vol. 40, no. 5, pp. 443–451, 1932.

Wadell, Hakon: Shape Determinations of Large Sedimental Rock Fragments, *Pan Am. Geologist*, vol. 61, pp. 187–220, April, 1934.

Wadell, Hakon: Some New Sedimentation Formulas, *Physics*, vol. 5, pp. 459–490, 1934.

Wadell, Hakon: Volume, Shape, and Roundness of Quartz Particles, *J. Geol.*, vol. 43, no. 3, pp. 250–280, 1935.

Walker, T. R.: Frosting of Quartz Grains by Carbonate Replacement, *Bull. Geol. Soc. Am.*, vol. 68, no. 2, pp. 267–268, 1957.

Waldo, A. W., and S. T. C. Yuster: Method of Impregnating Porous Materials to Facilitate Pore Studies, *Bull. AAPG*, vol. 21, no. 2, pp. 259–267, 1937.

Waring, W. W., and D. B. Layer: Devonian Dolomitized Reef, D3 Reservoir, Leduc Field, Alberta, Canada, *Bull. AAPG*, vol. 34, no. 2, pp. 295–312, 1950.

Washburn, E. W., and E. N. Bunting: Determination of Porosity by the Method of Gas Expansion, *J. Am. Ceram. Soc.*, vol. 5, pp. 113–129, 1922.

Wayland, R. G.: Optical Orientation in Elongate Clastic Quartz, *Am. J. Sci.*, vol. 237, no. 2, pp. 99–109, 1939.

Williams, Lou: Classification and Selected Bibliography of Surface Textures of Sedimentary Fragments, *Natl. Research Council, Rept. Comm. Sedimentation* 1936–1937, pp. 114–128, 1937.

Winsaur, W. O., H. M. Shearin, Jr., P. H. Masson, and Milton Williams: Resistivity of Brine-saturated Sands in Relation to Pore Geometry, *Bull. AAPG*, vol. 36, no. 2, pp. 253–277, 1952.

Wright, A. E.: Three-dimensional Shape Analysis of Fine-grained Sediments, *J. Sediment. Petrol.*, vol. 27, no. 3, pp. 306–312, 1957.

Yuster, S. T.: Determining Saturations by the Extraction-Distillation Method, *Oil Weekly*, vol. 113, no. 3, pp. 20–30, Mar. 20, 1944.

Yuster, S. T., and J. B. Levine: Determination of Oil and Water Saturation by the Retort Method, *Oil Weekly*, vol. 101, pp. 22*ff*., May 23, 1938.

Zierfuss, H., and G. van der Vliet: Laboratory Measurements of Heat Conductivity of Sedimentary Rocks, *Bull. AAPG*, vol. 40, no. 10, pp. 2475–2488, 1956.

Zingg, T.: Beitrag zur Schotteranalyze, *Schweiz. mineral. petrog. Mitt.*, bd. 15, pp. 39–140, 1935.

B. Compositional Properties of Rocks

Abraham, Herbert: "Asphalts and Allied Substances," vols. 1 and 2, D. Van Nostrand Company, Inc., Princeton, N.J., 1945.

Altieri, V. J.: "Gas Analysis and Testing of Gaseous Materials," American Gas Association, Inc., New York, 1958.

American Society for Testing Materials: Symposium on Fluorescent X-ray Spectrograph Analysis, *ASTM Publ.* 157, Philadelphia, 1953.

American Society for Testing Materials: Significance of ASTM Tests for Petroleum Products, *ASTM Spec. Tech. Publ.* 7-B, 1955.

ASTM Committee D-2 on Petroleum Products and Lubricants: "ASTM Standards on Petroleum Products," 1957.

Buerger, M. J.: "X-ray Crystallography," John Wiley & Sons, Inc., New York, 1942.

Bunn, C. W.: "Chemical Crystallography," Oxford University Press, London, 1946.

Clark, George L.: "Applied X-Rays," 4th ed., McGraw-Hill Book Company, Inc., New York, 1955.

Cosslet, V. E., Arne Engstrom, and H. H. Pattee, Jr.: "X-ray Microscopy and Microradiography," Academic Press, Inc., New York, 1957.

Daniels, F., C. A. Boyd, and D. F. Saunders: Thermoluminescence as a Research Tool, *Science*, vol. 117, no. 3040, pp. 343–349, 1953.

Fischer, Robert B.: "Applied Electron Microscopy," Indiana University Press, Bloomington, Ind., 1954.

Friedman, Gerald M.: Identification of Carbonate Minerals by Staining Methods, *J. Sediment. Petrol.*, vol. 29, no. 1, pp. 87–97, 1959.

Grim, Ralph E.: "Clay Mineralogy," McGraw-Hill Book Company, Inc., New York, 1953.

Grohskopf, J. G., and E. McCracken: Insoluble Residues of Some Paleozoic Formations of Missouri, Their Preparation, Characteristics and Application, *Missouri Geol. Survey and Water Resources Rept. Invest.* 10, 1949.

Guinier, André: "X-ray Crystallographic Technology," Hilger and Watts, Ltd., London, 1952.

Hardy, C. J., and F. H. Pollard: *J. Chromatog.*, vol. 2, no. 1, pp. 1–43, 1959.

Harrison, George R., Richard C. Lord, and John R. Loofbourow: "Practical Spectroscopy," Prentice-Hall, Inc., Englewood Cliffs, N.J., 1948.

Hedberg, Hollis D.: Evaluation of Petroleum in Oil Sands by Its Index of Refraction, *Bull. AAPG*, vol. 21, no. 11, pp. 1464–1476, 1937

Heinrich, E. W.: "Microscopic Petrography," McGraw-Hill Book Company, Inc., New York, 1956.

Hillebrand, W. F., G. E. F. Lundell, H. A. Bright, and J. I. Hoffman: "Applied Inorganic Analysis," John Wiley & Sons, Inc., New York, 1953.

Hunt, John M.: Composition of Crude Oil and Its Relation to Stratigraphy in Wyoming, *Bull. AAPG*, vol. 37, no. 8, pp. 1837–1872, 1953.

Hurlbut, Cornelius S.: "Dana's Manual of Mineralogy," John Wiley & Sons, Inc., New York, 1941.

Ireland, H. A.: Use of Insoluble Residues for Correlation in Oklahoma, *Bull. AAPG*, vol. 20, no. 8, pp. 1086–1121, 1936.

Keulemans, A. I. M.: "Gas Chromatography," Reinhold Publishing Corporation, New York, 1957.

Klug, Harold P., and Leroy E. Alexander: "X-ray Diffraction Procedures," John Wiley & Sons, Inc., New York, 1954.

Levorsen, A. I.: "Geology of Petroleum," W. H. Freeman and Company, San Francisco, 1954.

Lloyd, R. Michael: A Technique for Separating Clay Minerals from Limestones, *J. Sediment. Petrol.*, vol. 24, no. 3, p. 218, 1954.

McQueen, H. S.: Insoluble Residues as a Guide in Stratigraphic Studies, *Missouri Bur. Geol. Mines, Bien. Rept.*, pp. 102–131, 1931.

National Research Council: Nuclear Science Series Report 23: Cosmological and Geological Implications of Isotope Ratio Variations, *Natl. Acad. Sci. Publ.* 572, 1958.

Palache, Charles, Harry Beeman, and Clifford Frondel: "Dana's System of Mineralogy," John Wiley & Sons, Inc., New York, 1952.

Palmer, Chase: The Geochemical Interpretation of Water Analyses, *U.S. Geol. Survey Bull.* 479, 1911.

Parks, Jr., J. M.: Use of Thermoluminescence of Limestones in Subsurface Stratigraphy, *Bull. AAPG*, vol. 37, no. 1, pp. 125–142, 1953.

Pettijohn, F. J.: "Sedimentary Rocks," Harper & Brothers, New York, 1957.

Rankama, Kalervo: "Isotope Geology," Pergamon Press, Ltd., London, 1954.

Rogers, Austin F., and Paul F. Kerr: "Optical Mineralogy," McGraw-Hill Book Company, Inc., New York, 1942.
Saunders, D. F.: Thermoluminescence and Surface Correlation of Limestones, *Bull. AAPG*, vol. 37, no. 1, pp. 114–124, 1953.
Sloss, L. L., and R. H. Hamblin: Stratigraphy and Insoluble Residue of Madison Group (Miss.) of Montana, *Bull. AAPG*, vol. 26, no. 3, pp. 305–335, 1942.
Smith, H. M.: Correlation Index to Aid in Interpreting Crude Oil Analyses, *U.S. Bur. Mines Tech. Paper* 610, 1940.
Smith, N. A. C., H. M. Smith, O. C. Blade, and E. L. Garton: The Bureau of Mines Routine Method for the Analysis of Crude Petroleum, *U.S. Bur. Mines Bull.* 490, 1951.
Smothers, W. J., and Chiang Yao: "Differential Thermal Analysis," Chemical Publishing Company, Inc., New York, 1958.
Sproull, Wayne T.: "X-Rays in Practice," McGraw-Hill Book Company, Inc., New York, 1946.
Stiff, Jr., Henry A.: The Interpretation of Chemical Water Analysis by Means of Patterns, *Tech. Note* 84, *Petrol. Technol.*, *Trans. AIMME*, vol. 192, pp. 376–379, 1951.
Stocker, G.: Correlations in Madison Formation by Insoluble Residues, *Bull. AAPG*, vol. 38, no. 6, pp. 1309–1310, 1954.
Tickell, F. G.: A Method for the Graphical Interpretation of Water Analyses, *Summary Operations, Calif. Oil Fields*, vol. 6, no. 9, p. 7, 1921.
Twenhofel, W. H., and S. A. Tyler: "Methods of Study of Sediments," McGraw-Hill Book Company, Inc., New York, 1941.
Wahlstrom, Ernest E.: "Optical Crystallography," John Wiley & Sons, Inc., New York, 1956.
Winchell, Alexander N.: "Elements of Optical Mineralogy," part I, John Wiley & Sons, Inc., New York, 1948.
Witherspoon, Paul A., and Kozo Nagashima: Use of Trace Metals to Identify Illinois Crude Oil, *Illinois State Geol. Survey Circ.* 239, 1957.
Wyckoff, Ralph W. G.: "The World of the Electron Microscope," Yale University Press, New Haven, Conn., 1958.

C. Fossil Content of Rocks

Arnold, C. A.: "An Introduction to Paleobotany," McGraw-Hill Book Company, Inc., New York, 1947.
Cushman, J. A.: "Foraminifera, Their Classification and Economic Use," Harvard University Press, Cambridge, Mass., 1955.
Glaessner, M. F.: "Principles of Micropaleontology," Oxford University Press, London, 1945.
Hedgpeth, J. W., and H. S. Ladd: Treatise on Marine Ecology and Paleoecology, *Geol. Soc. Am. Mem.* 67, vol. 1, Ecology, vol. 2, Paleoecology, 1957.
Hoffmeister, W. S.: Microfossil Prospecting for Petroleum, U.S. Patent 2,686,108, 1954.
Howe, H. V., and Laura Laurencich: "Introduction to the Study of Cretaceous Ostracoda," Louisiana State University Press, Baton Rouge, La., 1958.
Jones, D. J.: "Introduction to Microfossils," Harper & Brothers, New York, 1956.
Lowenstam, H. A.: "Status of Invertebrate Paleontology, 1953." part II, Systematic, Paleoecologic and Evolutionary Aspects of Skeleton-building Materials, *Bull. Museum Comp. Zool.*, vol. 112, no. 3, pp. 287–317, 1954.
Moore, R. C. (ed.): "Treatise on Invertebrate Paleontology," University of Kansas Press and Geological Society of America, 1954.
Moore, R. C., C. G. Lalicker, and A. G. Fischer: "Invertebrate Fossils," McGraw-Hill Book Company, Inc., New York, 1952.
Neaverson, Ernest: "Stratigraphical Paleontology," 2d ed., Oxford University Press, London, 1955.
Newell, N. D., et al.: "The Permian Reef Complex of the Guadalupe Mountain Region, Texas and New Mexico," W. H. Freeman and Company, San Francisco, 1953.
Roemer, A. S.: "Vertebrate Paleontology," 2d ed., University of Chicago Press, Chicago, 1945.
Schimer, H. W., and R. R. Shrock: "Index Fossils of North America," John Wiley & Sons, Inc., New York, 1944.
Shrock, R. R., and W. H. Twenhofel: "Principles of Invertebrate Paleontology," McGraw-Hill Book Company, Inc., New York, 1953.
Swinnerton, H. H.: "Outlines of Paleontology," Edward Arnold & Co. London, 1947.
Turekian, K.: Paleoecological Significance of the Strontium-Calcium Ratio in Fossils and Sediments, *Bull. Geol. Soc. Am.*, vol. 66, no. 1, pp. 155–158, 1955.

Chapter 23

EXPLORATION RESEARCH

By R. Dana Russell

Supervisor, Geology Department, Denver Research Center, The Ohio Oil Company Littleton, Colo.

INTRODUCTION

In this chapter the author will make no attempt to cite specifically all the work that has been done or even that currently under way in exploration research. To do so in even a semicomprehensive manner would mean a bibliography longer than the text of the chapter, and a bibliography, moreover, that would give a distorted picture of the amount of work in different fields. This distortion results from the fact that some areas of research have an immediate application and any results may be of competitive advantage, so that publications are rare even though a great amount of work is being done; others, more "fundamental" or long-range in nature, are represented by dozens or hundreds of published articles. The two reviews that the writer has found most helpful are Research Needs in Petroleum Geology, by the AAPG Research Committee (1957), and the talk Geological Research Developments, presented by Harold N. Fisk (1959) at the 1959 Annual Meeting of the AAPG in Dallas. For additional references the reader will have to consult the geological, geochemical, and geophysical journals. Rather than review the literature, the author considers that the function of this chapter, in a book on petroleum exploration, should be to attempt an answer to the following questions: What is "exploration research"? How does it differ from exploration in the usual sense? Why is it needed? Where and how is it done? What, in general, does it consist of?

To the man engaged in exploration research, these questions may seem naïve and their answers obvious, but this is not the case generally. Some geologists maintain that all geology is research and all petroleum geology is for purposes of exploration, so how can one separate exploration research? The National Science Foundation, on the other hand, at one time maintained that no aspect of geology is research, or at least "pure" research; as an applied science it should be classed with engineering, and all geological work considered "development." Moreover, the most cursory survey of petroleum research organizations shows that the work included under the title of "exploration research" varies so greatly that the different company managements obviously have quite different ideas of the meaning of the term. So this becomes more than an academic problem of semantics; it is tied up with budgets, and we had better start by defining what we are talking about.

"Fundamental" and "Applied" Research

To begin with, there is much confusion as to the meaning of the term "research," and particularly on the distinctions among "pure," "basic," "fundamental," "background," and "applied" research. One of the questions most frequently asked of a man in an industrial research laboratory is "Do you do fundamental or applied research?" Since the people concerned often have quite different pictures of the meaning of these terms, misunderstandings are common. The President's Scientific Research Board defined fundamental (also known as pure and basic) research as "research without specific practical ends. It results in general knowledge and understanding of nature and its laws. . . . The pure scientist may not be at all interested in the practical applications of his work . . . " (Steelman, 1947, p. 311). Applied research and "development," on the other hand, have a specific objective, which "can often be definitely mapped out beforehand. . . . If successful, the results of applied research are of a definitely practical or commercial value" (*ibid.*, p. 313). Background research is separated out and defined as the accumulation of background knowledge or data, such as the "preparation of accurate topographic and geologic maps, the collection of meteorological data, the determination of physical and chemical constants . . . " (*ibid.*, p. 312). Applied research is distinguished from development on the basis of "whether the work is done on a *novel* device, method or process as contrasted with refinement of an existing model, type, device, method or process" (*ibid.*, p. 314).

According to these definitions, then, the distinction between fundamental and applied research is primarily one of the establishment of practical objectives and of motivation. If a specific objective of potential commercial value is in view, the research is *applied*. From the viewpoint of management, therefore, all industrial research is applied, since management is not likely to invest funds unless there is at least some prospect of a return. The man doing the research, on the other hand, may not be concerned with practical applications of his work, and to him the research may be *fundamental*, whether done in an industrial laboratory or in a university. The distinction can perhaps be clarified by asking whether the research is reviewed and for what purpose. The man in the industrial research establishment may be doing exactly the same type of research and going about it in exactly the same way as his brother in a university. He may not even care whether practical applications are likely to result from his research. But his supervisors care, and if they review his work periodically, even if no oftener than every 5 years, and if they may decide, following such a review, that he should shift the direction or emphasis of his research, the research is applied according to our definition.

From this viewpoint all industrial research is applied. Many enlightened industrial research establishments encourage their research men to devote a part of their time to "free" research—research on problems of their own choice in contrast to "assigned" research. But few of these establishments would encourage indefinite continuance of work in a field completely unrelated to the purposes of the parent organization; not many petroleum research organizations, for example, would encourage a man to spend a considerable portion of company time over a period of years on a study of the taxonomy of butterflies, even if his conscience would permit him to do so. Even free research may thus be applied if the above definition is rigorously used, and the term loses essentially all its utility.

More important than definitions is the effect on the research man. Does he feel free to analyze and attack the problem in his own way, with adequate time to evaluate possible solutions, or does he feel under continual pressure to provide "some kind of an answer"? The exploration geologist is usually under this type of pressure, and it is in part responsible for the thrill and excitement of exploration. But it is not the ideal atmosphere for the solution of long-range research problems. Thus the man considering a career in industrial research would be well advised not to say to his prospective employer "Is your research fundamental or applied?" but to ask such questions as "How much freedom would I have in formulating my research objectives and methods?" and "What percentage of your research is long-range in nature compared with urgent problems for which an 'immediate' solution is sought?"

Exploration Research

If we accept the above definitions, all exploration research is applied, since it has the specific commercial objective of improving our ability to find oil. Distinctions between long-range and short-range research, development, and "technical service," on the other hand, are much more meaningful and very useful in administering a research and development program. For our present purpose, however, it will suffice to define exploration research as including all research and development whose objective is to develop new or improve existing methods, techniques, or devices designed to increase our ability to find oil or gas. We would thus include fundamental, long-range studies such as those on the origin of hydrocarbons and the effects of sedimentation processes upon sedimentary products, and such short-range, developmental-type projects as an improved filter system or recording method for the reflection seismograph. We shall exclude, however, the actual application of such information, methods, or techniques to the search for hydrocarbons—this is exploration.

Under this definition, the development of a new method of stratigraphic or lithofacies analysis would be classed as exploration research; the use of such a method, whether on a regional basis to delimit the most favorable areas in a sedimentary basin or on a more detailed, local level to pinpoint individual prospects, would be classed as exploration. Many companies have established special staff groups, sometimes called "exploration research" groups, to conduct regional stratigraphic or structural studies in advance of or concurrent with detailed exploration for prospects by the line exploration organization. Such studies are extremely valuable and are becoming more essential as good drillable prospects become even harder to find, but they are here considered as a phase of exploration—not exploration research.

Is there a need for exploration research as we have defined it? Certainly the consensus of thinking men in the petroleum industry is that it not only is needed but is essential to long-range success in our search for additional petroleum reserves. With the exception of surface geologic mapping, practically all our present exploration methods, techniques, and tools are the products of exploration research. The most obvious ones are the tools, such as logging devices and geophysical exploration instruments, but improvement in geologic interpretation resulting from studies of the physical and chemical characteristics of sediments, of certain aspects of their mineralogy, and of their faunal content and paleoecology may, though not so obvious, eventually prove more important. This is especially likely to be true in our search for the hard-to-find "stratigraphic" trap.

But, as the Red Queen told Alice, here it seems to take all the running you can do to keep in the same place. Increasing exploration costs and decreasing success ratios will soon make domestic exploration uneconomic unless we can improve our methods still more. The marked increase in exploration research effort by most of the major oil companies during the past 5 to 10 years indicates awareness of the need to run still faster if we are to stay ahead of rising exploration costs.

Where is exploration research done? A large part is carried on today by the major oil companies and by the larger petroleum service companies, with the latter concentrating primarily on logging methods and devices, geophysical exploration techniques and instrumentation, and drilling methods and equipment. But a considerable part, if not most, of the information and knowledge needed for the better geologic interpretation required for improved success ratios in oil finding is coming from research in geology, geophysics, and geochemistry in universities and in the United States and state geological surveys. The importance of such research has long been recognized by the petroleum industry, and support provided through the American Petroleum Institute's Research Committee and the Petroleum Research Fund of the American Chemical Society. In fiscal 1958–1959, these two agencies contributed nearly a half million dollars in support of fundamental exploration and production research, mostly in universities. The U.S. Geological Survey devotes a major part of its effort both to fundamental geological, geophysical, and geochemical research and to applied research on specific mining and petroleum exploration problems. Many of the state surveys are also spending increasing amounts for such purposes. Finally, many of the technical societies, through their research committees, are recognizing the urgency of the problem and are stimulating research efforts through symposia on major problems, surveys of research needs, and fellowships or grants-in-aid of research. It is therefore not practical to examine in detail where and how exploration research is done, but it may be worthwhile to examine briefly the present status of the exploration research effort within the petroleum industry.

PETROLEUM RESEARCH ORGANIZATIONS

The petroleum industry has long been a leader in industrial research, and most of the major companies have research organizations of long standing. Petroleum research (excluding marketing research) is commonly divided into refining or process research, product research, production research, and exploration research. These fields are listed in order of historical development; improved methods of refining crudes and the development of new products were essentially the exclusive targets of early research in the industry. Production research—the investigation of drilling and well-completion methods (including research on drilling muds and cements), of reservoir behavior, and of improved recovery methods—came more recently, and the rapid growth of exploration research has occurred chiefly since World War II, spurred by the increasing need for improved exploration methods.

The amount of effort on exploration research in petroleum research organizations varies greatly. Some still devote little or no attention to this field, concentrating instead on refining, product, and production research and perhaps counting on purchasing crude reserves if their exploration program is not successful. Other research organizations in the industry devote more than a third of their research effort to exploration research. The relative effort appears to be governed primarily by the character of the organization (whether chiefly production or refining and marketing) and secondarily by the amount of long-range planning or, contrariwise, the backgrounds and predilections of its management.

Personnel and Organization

One tremendous advantage that the centralized research organization has over the line operating organization is the ability to bring together specialists from many different fields to concentrate their special knowledge upon a particular problem. Similarly, the central research organization can provide its scientists with the most modern equipment at a fraction of the cost required to set up such equipment in all the widespread operating parts of the company. Thus physicists, chemists, engineers, and geologists can all contribute their special knowledge to the solution of an exploration problem in a research organization, whereas hiring all these special talents in a district exploration office and providing them with needed equipment would be prohibitively expensive and wasteful.

Administering such a varied group of people, however, has its problems. How should they be grouped—by their fields of training, by the objectives of their research, or by the functions they perform? All these methods have been tried and are in use. That none is preeminently best is shown by the fact that nearly every industrial

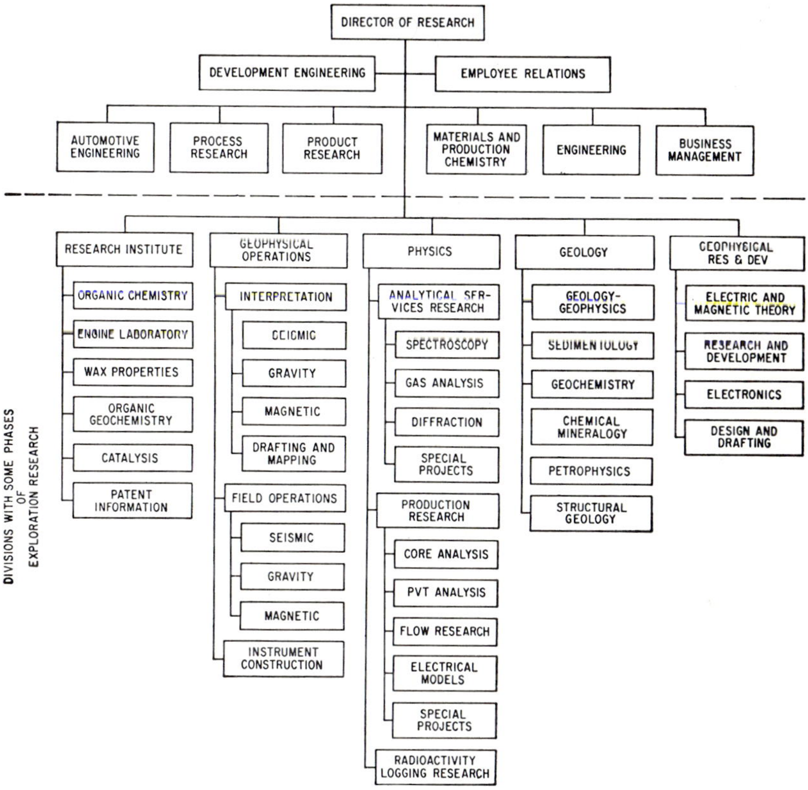

FIGURE 23-1. Organization of a large petroleum research establishment that grew "like Topsy." Major divisions are based partly on function, partly on objectives, and partly on fields of science or training. Note that there are 11 divisions and two staff groups reporting to the director of research.

research group of long standing has suffered several reorganizations (Figures 23-1 and 23-2). The most common type of research organization is one grouped by functions or objectives at the top and by subject fields at the next level, e.g., exploration research as a major group, subdivided into geology, geophysics, and geochemistry or into structural research, sedimentology, paleontology and stratigraphy, etc. (Figures 23-3 and 23-4). There are advantages in an organization where major grouping is by fields of science, with chemistry, engineering, physics, and geology departments, for example, so that each scientist reports to a man trained in his field and presumably much more familiar with his problems. In such an organization, coordination among the scientific disciplines can be achieved by organizing the research program on the basis of

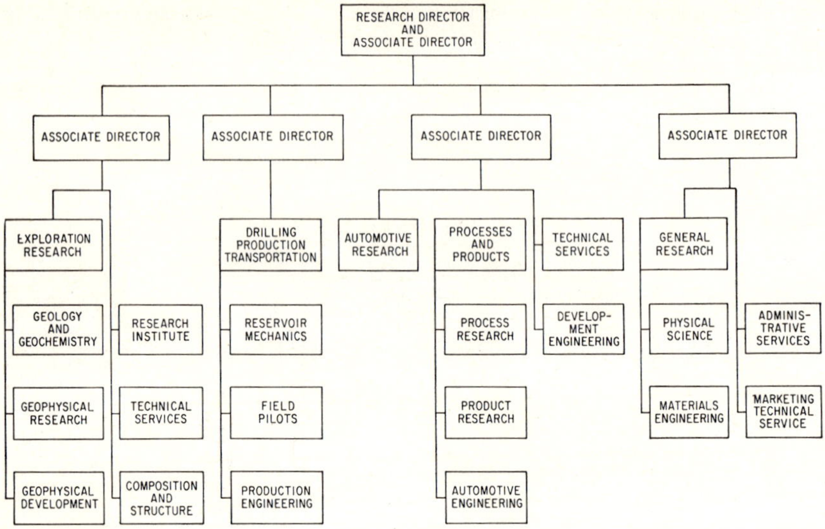

FIGURE 23-2. The same research establishment shown in Figure 23-1, 5 years later. There has been consolidation to provide more effective administration, but there is still no basic philosophy of organization, and it is probably too late to establish one.

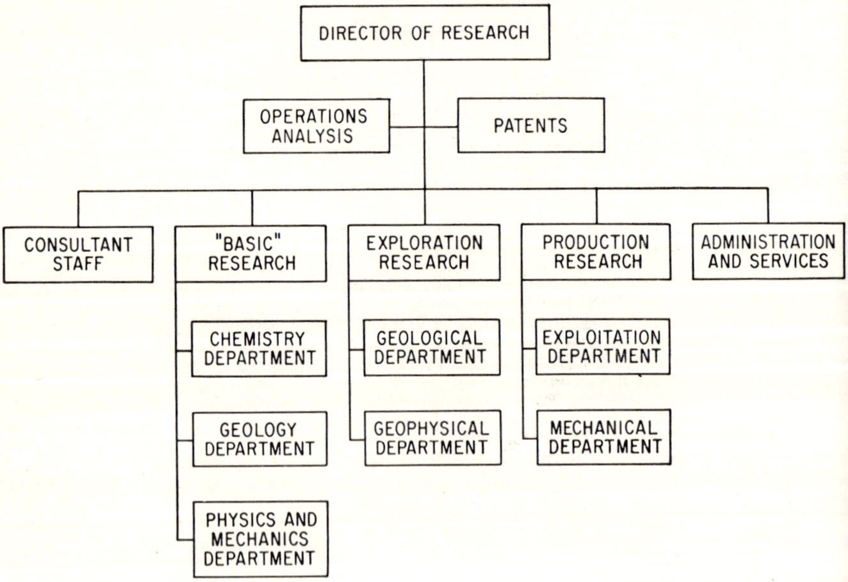

FIGURE 23-3. Organization of exploration and production research in a major oil company. Major divisions are by functions or objectives; the next level by fields of science in the "basic" research and exploration research divisions and by objectives or functions in the production research division.

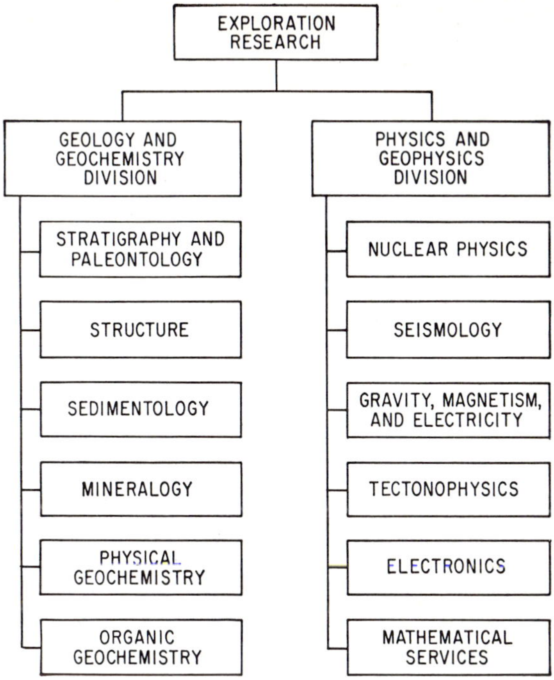

FIGURE 23-4. Breakdown of exploration research by fields of science in a large petroleum research organization (a composite of several existing organizations).

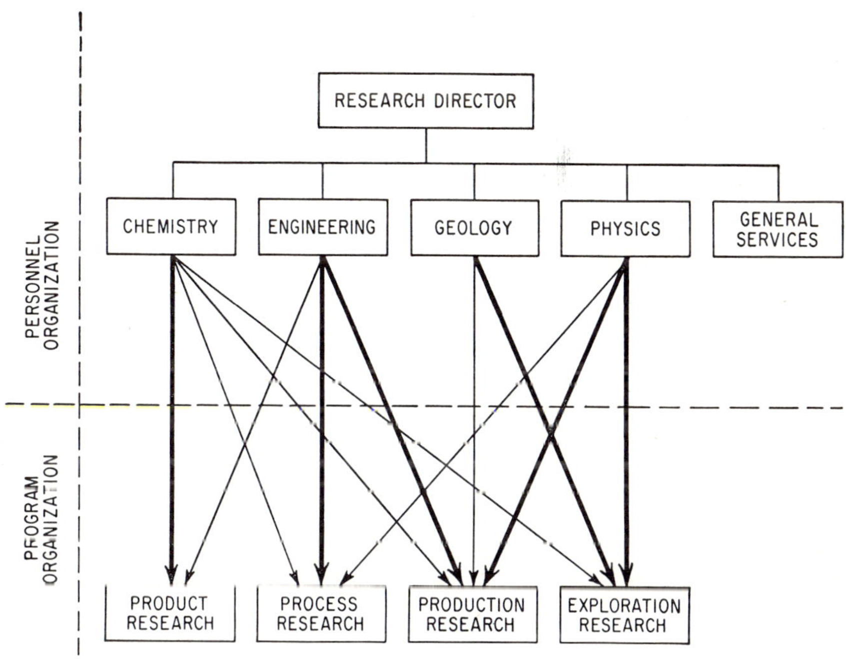

FIGURE 23-5. Personnel and program organization in a relatively small research establishment organized by fields of science, with program organization by objectives or functions. Major program responsibilities of each personnel division are shown by heavy arrows; minor responsibilities by light arrows.

functions or research objectives (Figure 23-5). The best method in any particular organization, however, will depend partly upon its size and functions and partly upon the characteristics of the top management personnel. Organizations are more often modeled to men than men to organizations.

Regardless of the type of organization, the special advantages of the central research group must be maintained if it is to function most effectively. These are (1) the free exchange of ideas among scientists of different backgrounds and training and the effective integration of their special contributions and (2) optimum use of expensive special facilities. This applies just as much to exploration research as to any other type.

Communications with Operations

If communication is important within a research organization, it is equally important, if not more so, between research and operations. The results of research are of no value if they are not used. Too often a potential competitive advantage resulting from a significant improvement in exploration methods developed by a research group has been lost through reluctance on the part of line exploration personnel to use this "new" and unfamiliar method. This can be very frustrating to the research personnel responsible for the development, but it is equally frustrating to the exploration man to see the people in his company's research organization working on problems that apparently (and perhaps actually) have no possible application to the urgent problems facing him. It is therefore essential that the research man or those guiding his research be familiar with the practical problems of the line organization and those in operations be familiar with the objectives of the research effort and how research results can be applied to their problems.

This is a continuing problem facing all companies with a central research organization, and many different techniques have been used in an attempt to overcome it. These include an intermediate organizational group to "interpret" research results to operations and operation problems to research, "liaison" personnel, advisory groups, exchange of personnel between research and operations, special meetings, seminars, etc. All may be useful but none is adequate by itself. Certainly the old standby, the technical report, even though very necessary, cannot do the job alone no matter how well it is written, nor can any form of written communication. The author is convinced that there is no adequate substitute for person-to-person communication. Frequent discussion of each other's problems by research and operating personnel can pay off many times over in the mutual understanding that results if each is truly interested in learning the other's point of view.

AREAS OF EXPLORATION RESEARCH

Essentially all of the surface and near-surface geologic structures in the United States have now been tested by the drill, so the major problems in domestic exploration today are (1) how to find stratigraphic traps, (2) how to locate deeply buried structural traps, and (3) how to interpret correctly the complexly faulted or folded structural trap. Of these, the first is generally considered by far the most important because the consensus is that there probably are far more undiscovered stratigraphic than structural traps. In foreign areas this is not yet true; many promising surface and near-surface structures have not yet been tested. But eventually we shall also be forced to search for these hard-to-find traps in foreign areas as well. The most active areas of exploration research naturally reflect concern with these problems, and essentially all the areas of research discussed below bear upon them in one way or another.

Interpretation of Sediment Characteristics and Sedimentary Facies

In 1947 the Research Committee of the AAPG, after reviewing research needs in the field of petroleum geology, concluded that research was most needed on various aspects of sedimentology. Ten years later, in 1957, the Research Committee repeated this review. The 1947 recommendation greatly stimulated research on sediments, and much had been accomplished in the intervening 10 years, yet the 1957 review reached nearly the same conclusions as the one 10 years earlier. This is not particularly surprising if one stops to think about it, for it is obvious that successful petroleum exploration, particularly for stratigraphic traps, is largely dependent upon successful interpretation of sedimentary rocks. In spite of all we have learned, however, we are still a long way from being able to read all we need to know from the properties of sediments, so research in this area is very active. It includes major efforts on both modern and ancient detrital and carbonate sediments and on stratigraphic analysis and lithofacies mapping methods.

Modern Detrital Sediments

It has long been evident that one of the most promising approaches to a better understanding of ancient sediments is to learn more about modern sediments and sedimentary processes. Research on modern sediments was quite active before World War II but has been greatly expanded since, with major efforts in several of the larger petroleum research organizations and in a number of universities. In the field of detrital sediments, a major stimulus to research throughout the world has been the work of scientists of the Scripps Institution of Oceanography of the University of California, sponsored by the American Petroleum Institute under its Project 51. Project 51 has been carried on by Scripps since 1950, with detailed investigation of sediments in modern depositional environments of the northern Gulf of Mexico, particularly in the active delta of the Mississippi River and along the Texas coast. The results of these Project 51 studies are scattered through a great many publications, but they now have been summarized in a special publication of the AAPG (Shepard et al., 1960). The work under Project 51 has now been transferred to the Gulf of California, where the structural framework, climatic conditions, and sedimentary processes are quite different from those in the northern Gulf of Mexico.

Other studies in this field worth special mention, outside those in petroleum research organizations, are the work at Louisiana State University on the delta of the Mississippi River, studies of the Orinoco shelf off eastern Venezuela by the staff and students of the University of Groningen sponsored by the Royal Dutch Shell group, and those off the California coast by scientists at the University of Southern California, supported by the Hancock Foundation and the Petroleum Research Fund of the American Chemical Society.

In all these studies, the objectives are to determine the characteristics of sediments deposited in different modern environments and the processes involved in their deposition and diagenesis. The work of biologists, oceanographers, chemists, and bacteriologists, as well as of geologists, is being coordinated to provide an over-all picture of the physical, chemical, and biological characteristics of modern sedimentary facies. The results of these studies, applied to the interpretation of ancient sediments, should provide powerful tools in our search for stratigraphic traps.

Ancient Detrital Sediments

In addition to the data being accumulated on modern detrital sediments, many direct studies of ancient sediment properties are proving extremely useful for interpretation. Especially noteworthy are new approaches to the study of grain-size distribution in detrital sediments, studies of their allogenic and authigenic mineralogy (including the clay minerals), of trace minerals and of isotope ratios, and of primary sedimentary structures, especially the directional ones.

Recent studies of grain-size distribution in sediments, in contrast to former studies that attempted to correlate size-distribution patterns with various environments, have concentrated on the significance of mean or median size and of maximum size. Vertical variations in these properties within a sedimentary formation may give important clues to its origin, as may areal variations in average values of the same properties. The latter, in addition to providing evidence on the transporting agent, may help in reconstructing the regional paleoslope and in determining the location and trend of ancient shore lines.

Studies of the detrital, or allogenic, minerals of sediments, for a long time the most active phase of sedimentology, are still important in determining provenance and dispersal of sediments and in studies of their "maturity." But their relative importance has decreased by reason of increased emphasis on other aspects of sediment mineralogy. The clay minerals, for example, have been, and are, receiving a great deal of attention. Studies of clay minerals in modern sediments show that some are altered in response to changes in the chemistry of their environment, so there is the possibility that they may be environmental indicators. Others appear to reflect primarily the character of the rocks and weathering conditions in their source areas. Since clay mineralogists are not agreed on the relative importance of source and environmental factors, more research is needed before the clays can be used in routine fashion in the interpretation of sediments.

More definitely authigenic minerals may offer more promise as environmental indicators, and consequently they are being actively investigated. Unfortunately, we still know relatively little about the physicochemical conditions controlling their formation. Recent studies of glauconite (the "old reliable" as an indicator of marine conditions), for example, have shown that there are a great many varieties of so-called glauconite and that they probably are formed from a number of original materials under a variety of environmental conditions.

A different, but related, approach to the problem of identification of environment that is now being actively investigated is the use of trace elements and isotope ratios. Studies at Pennsylvania State University, for example, suggest that boron is more abundant in marine sediments and gallium in fresh-water ones. Work at the Rice Institute suggests that thorium/uranium ratios may be indicative, with uranium more abundant in marine sediments than in nonmarine ones. Carbon 13/12 ratios also offer a promising approach, as shown by the work of Craig at Scripps, Silverman at the California Research Corporation, and Epstein at the California Institute of Technology. Marine organisms have a higher C-13/C-12 ratio than terrestrial plants, with coal about the same as the latter. Crude oils and hydrocarbons extracted from sediments are enriched in the lighter isotope, but the nonmarine ones are still markedly lighter than the marine. Another application of isotope ratios to the interpretation of sediments is the use of O-16/O-18 ratios to determine paleotemperatures.

One of the most interesting developments in the study of ancient sediments during the past decade has been that of the measurement and mapping of the directional properties of sediments, work pioneered by Pettijohn and his students. Directional properties include cross stratification, ripple marks, fabric (dimensional orientation of sand grains or pebbles), and a variety of minor structures and markings such as flute markings or flow casts, groove casts, slide marks and slump structures, and oriented plant fragments or other fossils. Such studies have proved useful in a variety of paleogeographic applications, such as reconstruction of the regional paleoslope, fixing the location and trend of ancient shorelines, locating source areas, and even in determining prevailing wind directions.

Flow casts, groove casts, flute markings, slide marks, pull-aparts, and slump structures have also been described in some detail in recent years in deposits ascribed to turbidity currents, and, along with graded bedding, are considered by many geologists to be typical of such deposits. Although turbidity currents have been known for 60 years, the concept that they are important factors in sedimentation is relatively new. If as common as their more ardent proponents would have us believe, they have been extremely important agents because they provide a mechanism for transporting coarse sediments to deep water, for the filling of deep sedimentary basins, and for providing oil and gas migration paths from deep-basin sediments to potential reservoir sands deposited in shallower water. More research is needed, however, in the author's opinion, before we shall have sure criteria for the recognition of ancient turbidity-current deposits and can evaluate their importance throughout past geologic time.

Modern Carbonate Sediments

Although the total effort on modern carbonate sediments has been considerably less than on modern detritals, much is being learned through studies in the Florida Keys, Bermudas, and Bahamas by Illing and Ginsburg, by Newell and associates at Columbia University, and by groups at the U.S. Geological Survey. The Survey has also been very active in undertaking or sponsoring work on carbonate sediments of the Pacific islands. In addition, several oil company research laboratories are active in modern carbonate research. Considerable progress is being made on the correlation of biotic components, textures, and structures with environments of deposition, such as the protected areas of back reef, bay, or lagoon, compared with the more exposed environments of reefs and open shelves. Variations in the chemistry of these sediments, including trace elements and isotope ratios, also offer considerable promise of help in identifying the environments of deposition of comparable ancient carbonate rocks.

Ancient Carbonate Sediments

For decades carbonate rocks have been dismissed with a cursory description as marl, limestone, or dolomite, with perhaps a notation as to color, crystallinity, the relative amount of detrital components, and the presence of the more obvious fossils. Within the past 10 years, this picture has almost completely changed. The work of Cuvillier in France, of Wanless and Carozzi at the University of Illinois, of Newell and his students at Columbia University, of Harlan Johnson at the Colorado School of Mines, of Folk at the University of Texas, and of Lowenstam at the California Institute of Technology has shown that the carbonate rocks will abundantly repay detailed examination. Thin sections, acetate peels, and even polished slabs, when examined under the microscope, reveal a wealth of detail extremely useful in interpreting the origin and history of these rocks. The biotic constituents can be identified —not, of course, as to genus and species in most cases, but as to major group, which is usually sufficient for paleoecologic interpretations. The general size of the original carbonate particles, e.g., mud, sand, or larger, can usually be determined, and hence the degree of turbulence in the environment of deposition can be inferred. Some limestones may be completely recrystallized or dolomitized, and much of this evidence destroyed, but it is surprising how many carbonate rocks show a wealth of detail under the microscope. Consequently, research programs on the interpretation of carbonate rocks are active in most petroleum research organizations, and the results of this research are already finding applications in exploration.

Minor Sedimentary Types

We still know relatively little about the precise conditions of formation of evaporites, cherts, sedimentary iron ores, phosphates, and coals. Of these, evaporites are the most important to petroleum exploration because of the world-wide association of oil occurrence with evaporite deposits. Some effort is being devoted to this problem in a few petroleum research organizations and universities, but the total effort relative to work on detrital and carbonate sediments is rather small.

Diagenesis

One of the largest gaps in our knowledge of sediments, as pointed out by the AAPG Research Committee (1957, pp. 1859, 1863), concerns the changes taking place during and after deposition, including compaction and cementation. Not only is such information needed for optimum use of data on modern sediments in the interpretation of ancient ones, but it is critical to our understanding of hydrocarbon source, migration, and entrapment problems. Although some work is under way in a few universities and petroleum research laboratories on diagenesis, the effort is not so great as would seem to be justified by the importance of the problems involved.

Stratigraphic Analysis and Facies Mapping

One of the most powerful tools developed in recent years to synthesize information on all types of sedimentary rocks and to reveal patterns of sedimentary facies and of ancient environments is facies or stratigraphic analysis and its accompaniment, the various types of facies maps. Krumbein and his associates at Northwestern University have pioneered in the development of these methods for analyzing and portraying facies information. Sand-shale and other types of lithofacies ratio maps are the most familiar applications of these techniques, but many other possible applications are being studied. The most recent development is the use of high-speed electronic data-processing and computing equipment to expedite handling of the data and computation of percentages, ratios, and the significance of these and other parameters.

Correlation and Age Dating

Since correlation of sedimentary horizons and their dating are essential to many phases of the interpretation of sediments, this area of research has always been active and is especially so now that isotope techniques give promise of precise correlation and dating, independent of the relative results from conventional paleontologic and lithologic methods. Were it possible to obtain accurate figures on the distribution of effort in petroleum research organizations, this area might well prove to be second only to studies of sediment characteristics and facies in the amount of work currently devoted to it.

Paleontologic Methods

Conventional use of Foraminifera and other common microfossils for correlation is being supplemented by studies aimed at evaluating planktonic Foraminifera for interregional correlation and by investigations of ultramicroscopic forms such as the coccoliths, tintinnids, discoasterids, microforaminifera, hystrichospherids, and pollen and spores. Many of these occur in sediments where the better known fossils are rare or absent, so their value is correspondingly enhanced. The acid-insoluble forms, such as pollen and spores, can be concentrated by treatment of sediments with hydrofluoric or other strong acids to remove practically all of the rock, leaving a residue consisting essentially of these fossils. Palynology (the study of pollen and spores) has shown an extremely rapid growth during the past decade and is already being applied to semiroutine correlation by some oil companies.

Lithologic and Mineralogic Methods

"Heavy" minerals are no longer used to any great extent for correlation, but more refined methods for characterizing the mineralogy or chemistry of sediments are being actively investigated. These include X-ray and spectroscopic analyses to identify characteristic clay minerals or trace elements and mass-spectrographic analysis of stable isotopes. "Microfacies analysis," or characterization of sediments by typical features seen under the microscope, is also proving a useful tool for correlation under some conditions, especially with carbonate rocks.

Radioactive Isotopes

The greatest effort and the greatest promise, in the area of age dating and correlation, lie in the field of radioactive isotopes. Research on methods of isotope age dating is active in a number of universities (partly supported by the American Petroleum Institute, the Petroleum Research Fund of the American Chemical Society, and several government agencies, particularly the Atomic Energy Commission), the U.S. Geological Survey, and most of the larger petroleum research organizations. Successful application of isotope methods could provide the petroleum geologist with such information as the absolute age of unfossiliferous sediments, rates of sedimentation and compaction, time of structural growth, and many other important geologic events involved in the origin, migration, and accumulation of petroleum.

Unfortunately, none of the radioactive isotope methods are yet ready for routine application nor is any a panacea. The most widely publicized, the carbon-14 method, can be used only for the last 30,000 or possibly 50,000 years of geologic time. The ionium method is limited to deep-sea sediments younger than 400,000 years. The potassium-argon method, utilizing glauconite and authigenic mica and micaceous clay minerals, may make possible the dating of some types of sediments older than 1.5 million years. The rubidium-strontium and the various lead and uranium-lead methods are applicable only to certain sediments more than 20 million years old. Even if all these methods could be successfully applied, there would thus be a gap consisting of most of the Pleistocene and part of the late Tertiary where none of these methods is currently applicable. Recently, investigations have started on beryllium-10 as a possible bridge for this gap.

Most of the above methods are still in the research stage; dates obtained by one method frequently cannot be confirmed by other methods. Part of this difficulty results from problems of technique, from the very small amount of material available for analysis, or from insufficient knowledge of decay rates or other geochemical factors not yet well understood. But much of the difficulty results from the fact that, to date a sediment, the minerals used in the dating must be the same age as the sediment—only authigenic, not allogenic, minerals can be used. Mineralogists and petrographers must develop methods for identifying, and perhaps isolating, the useful authigenic minerals before the geochemists can successfully perfect their techniques. In spite of these difficulties this whole field of research appears most promising.

Other Methods

Paleomagnetism, or studies of remanent magnetism in rocks and changes in the direction of magnetization, resulting from changes in the direction of the earth's magnetic field and reversals of the field, offers some hope of providing a correlation tool as well as furnishing new information on earth history. Thermoluminescence in carbonate rocks is also being evaluated as a correlation tool. New and improved logging and geophysical techniques are of potentially great importance; these are discussed in a later section.

Source, Migration, and Accumulation of Petroleum

This area of petroleum exploration research has predominated, along with studies of modern sediments, in most reviews of research needs (AAPG Research Committee, p. 1870). The origin of oil is one of the oldest problems of petroleum geology and one on which, until recently, comparatively little progress had been made. The first research sponsored by the Research Committee of the AAPG, and later by the American Petroleum Institute, was in this field; also David White's research, resulting in the carbon-ratio theory; Parker Trask's comprehensive studies of organic matter in sediments; and later, work by ZoBell and others on the role of bacteria in petroleum genesis. These studies resulted in the accumulation of a vast amount of data without providing definitive answers; recent studies on trace elements and stable isotopes in petroleum and analyses of potential source rocks show more promise. One of the most interesting developments is the study of hydrocarbons in modern sediments. This work indicates that hydrocarbons begin forming from organic matter in sediments almost immediately after burial.

Some of the current work on the source of hydrocarbons is being done in universities and in government organizations such as the U.S. Bureau of Mines and the Research Council of Alberta, Canada, but the greatest effort is in the major oil company research organizations and involves the closely coordinated teamwork of chemists, biologists, and geologists. Much of the work is considered confidential, so the number of publications is not indicative of the relative effort.

Considerable research is also under way, chiefly in industry research laboratories, on the ways that hydrocarbons migrate, on the effects of compaction of sediments on migration, and on the various factors involved in the entrapment of hydrocarbons. This research includes the role of hydrodynamics in migration and accumulation.

Structural Analysis

Current research in structural geology applicable to petroleum exploration problems includes a wide variety of geological and geophysical investigations. These range from general studies of the architecture of the continents and oceans and of sedimentary basins to fundamental investigations in petrophysics needed to provide essential information on the rock properties involved in fracture and flow. Such work is active in universities, in Federal and state surveys, and in petroleum research organizations. Recent studies include new information on the role and mechanism of gravity in folding and faulting; local and regional studies of thrust faults, high-angle reverse faults, and normal faults developing in basins during sedimentation; model studies, including investigation of scaling factors; and microstructure analysis of rocks in oriented thin sections to deduce the presence of faults not detected by other means.

Photogeology

The use of aerial photographs in geologic mapping dates back to the 1920s, but a very rapid increase in the use of this tool followed World War II, when great advances were made in the quality of film and equipment. Photogeologic mapping has now become an essential phase of petroleum exploration, but research on interpretive photogeology (a phase of geomorphology) is still active in the expectation of improving our ability to read subsurface geology from surface expressions.

Logging Methods

The introduction of electrical logging in 1929 provided a tremendous shot in the arm to subsurface geologic interpretation. Perhaps the new tool was too successful, because the ease and rapidity with which subsurface formations could be correlated by electric logs caused a rapid decline in the study of cores and cuttings, previously used for both correlation and interpretation of sedimentary horizons. Exploration geologists used the logs almost exclusively as an empirical correlation device—comparing one set of squiggles with another—and left to the engineers the refinements of interpretation. It is the reservoir and production engineers, interested in more accurate determination of porosity, permeability, and fluid content, who have pushed the development of new and improved logging methods; exploration geologists have failed to press for the development of logs that would give them more specific information on the lithologic characteristics of formations penetrated. It has even reached the point where cores are requested more frequently by engineers, interested in the properties of the reservoir, than by exploration geologists. This trend must be reversed if we are to provide the better geologic interpretation needed to find stratigraphic traps. As noted by the AAPG Research Committee (1957, p. 1867), "we are not getting as much geological information as we should from logging devices. Geologists must condition themselves to think in terms of how much geological information they can get from a borehole, and logging devices as their geologic probes."

Meanwhile research and development on new logging methods and devices continue at a high level, almost entirely in petroleum research organizations and in service logging companies. Any new physical principle or process is quickly exploited, in spite of the difficulty in adapting it to borehole use. New logging methods introduced since the electric log arec overed in Chapters 19–21 of this book; some of those currently under development include the following: a device that measures the energy spectrum of neutron-induced gamma rays and which thereby identifies some of the chemical elements around the borehole (notably chlorine, for identification of salt water). With an experimental particle accelerator, capable of being lowered into a borehole, it is possible to turn off the source of neutrons and make further identifications of chemical elements from the decay rates of induced gamma rays. Two approaches to the long-standing problem of measuring rock densities in place are being pursued. A density logger using scattered gamma rays is nearing commercial acceptance, and work is continuing on the development of a down-hole gravimeter which would be little influenced by borehole effects. A recently announced development is that of an electric logging tool that can be pumped down the drill stem, without removal of the drill string, and will log a freshly drilled zone before mud invasion is fully developed.

To make optimum use of these and other new logging tools, as well as the old standby, the electric log, cooperative research on log interpretation by geologists, physicists, and engineers is badly needed. Fortunately, such research is under way in several petroleum research organizations.

Seismic Exploration

The chief method for locating buried structures for the past 25 or 30 years has been the reflection seismograph. The most elaborate and most expensive of all geophysical methods, it has also proved to be the most effective, with more discoveries to its credit than any other geophysical technique. Refraction shooting is still used for special applications, but these are comparatively minor in quantity.

The search for deeper structures and for buried structures in areas where reflections are poor because of "weathering," near-surface scattering layers, or other complicating factors has kept research on seismic interpretation and the development of instrumental techniques at a high level. "Fundamental" research on the behavior of seismic waves under varied conditions and in different media has been carried on primarily in universities (sponsored in part by the American Petroleum Institute) and in petroleum research organizations. The development of new instrumentation and techniques has been undertaken chiefly by petroleum company research laboratories and the seismograph service companies. Recent developments include the use of dropped weights instead of explosives; magnetic tape recording of seismic data, particularly broad-band recording; automatic playback, filtering, and plotting of data; the use of high frequencies with close geophone spacing for the mapping of steep dips; the development of techniques for continuous measurement of shallow seismic profiles in water-covered areas; and the construction of synthetic seismograms from continuous-velocity logs to aid in interpretation. New methods of filtering and correlating data to increase signal-to-noise ratios are also being investigated, and many of these techniques are now being directed specifically at the problem of locating facies changes indicative of potential stratigraphic traps.

Other Geophysical Techniques

These constitute a minor part of the total geophysical exploration effort (anon., 1959, p. 156) but are still important, particularly for reconnaissance surveys in relatively unexplored areas. The air-borne magnetometer, developed during World War II, has been proved a valuable reconnaissance tool that can cover vast areas quickly and inexpensively, regardless of terrain conditions. Promising new developments in the field of magnetics include the nuclear-precession magnetometer (based upon the principle of nuclear-magnetic resonance) and the Zeeman-spectrum magnetometer.

Gravity measurements were the first geophysical technique applied to exploration for petroleum, and gravity methods still constitute nearly 15 per cent of the total geophysical exploration effort. Research on better methods for interpreting gravity data and the development of improved instruments are still active. Ultrasensitive instruments have been developed for the measurement of very small gravity changes, such as are caused by earth tides, and more rugged instruments, with improved mountings and movements, permit shipboard use. An air-borne gravimeter is now under development.

Various electrical and radioactivity methods for geophysical prospecting have been proposed, and some are being actively investigated, but it seems unlikely that any of them will replace the seismograph or the gravimeter as the chief geophysical exploration tools.

Exploratory Drilling

No geophysical or other indirect subsurface method can substitute entirely for actual sampling of subsurface formations and their fluid contents. Core drilling of "stratigraphic tests," most active in the 1920s, has dropped off markedly in recent years because of increased drilling costs and because the depth to which "strat-tests" must be drilled to provide needed information has also increased. As pointed out by the AAPG Research Committee (1957, p. 1875), this is one of the most critical problems in petroleum exploration; unless exploratory drilling costs can be reduced by cheaper and better methods, the cost of drilling may become a barrier to further exploration. Research and development in this area are therefore very active, primarily by the larger oil companies, but no major break-through is currently in sight.

DIRECT OIL-FINDING METHODS

No discussion of exploration research would be complete without some mention of direct oil-finding methods. Men have always been so intrigued by the possibility of locating valuable deposits by a direct method that they have been suckers for a wide variety of "doodle-bug" devices from the willow wand to complicated stacks of electronic equipment purporting to measure mysterious emanations from oil accumulations, gold deposits, or buried treasure.

Some direct methods are based on valid scientific principles but purport to measure physical or chemical effects that are known to be too weak to be determined. The most promising appear to be those based upon analysis of soil samples for hydrocarbon residues or for bacteria living upon them. These methods have been known for a number of years and have been extensively tested with inconclusive results. Although there is always a possibility that a successful direct method will be discovered, at present it looks as if the discovery of additional reserves must continue to depend upon geologic interpretation, aided by all that modern science can bring to bear in the way of geological, geophysical, and geochemical knowledge and techniques.

CONCLUSIONS

With the discovery and development of the more easily located petroleum accumulations, exploration geologists are confronted with an increasingly more difficult problem. To locate the ever-harder-to-find trap, they must use more scientific and more precise techniques, and above all, they must develop increasingly better geologic interpretation. This means better-trained exploration geologists, equipped to use modern techniques, and increased knowledge and improved methods for them to use—developed through exploration research.

REFERENCES

AAPG Research Committee: Research Needs in Petroleum Geology, *Bull. AAPG*, vol. 41, no. 8, pp. 1854–1876, 1957.

Anon.: A Century of Oil, chap. 2, Oil Discovery Techniques, *World Oil*, vol. 148, no. 1, pp. 148–158, January, 1959.

Fisk, H. N.: Geological Research Developments, *Program AAPG—SEPM Ann. Meetings 1959*, p. 18, Dallas, Tex., March, 1959.

Shepard, F. P., F. B. Phleger, and Tj. H. van Andel (eds.): "Recent Sediments, Northwest Gulf of Mexico," American Association of Petroleum Geologists, Tulsa, Okla., 1960.

Steelman, John R.: "Science and Public Policy: A Report to the President by John R. Steelman, Chairman, The President's Scientific Research Board," vol. 2, The Federal Research Program, Government Printing Office, Washington, D.C., September, 1947.

Chapter 24

EXPLORATION SCOUTING

By HARVEY W. LEE

Administrative Assistant, Foreign Operations, Union Oil Company of California
Los Angeles, Calif.

Scouting dates back practically to the birth of the oil industry shortly after the drilling of the Drake well in 1865. In the period that followed operators made every effort to conceal facts and withhold information because a dry hole or a new producer would materially affect the price of crude and the price of oil company stock. Conditions have changed considerably since that time, but the increased need for reliable information has continued with the growth of the industry. Today, all major oil companies, many of the minor oil companies, and large independent operators employ scouts and maintain a scouting department within their organizational framework. If such is not the case, per se, then undoubtedly this function is the responsibility of some other department of a company or large operator engaged in oil exploration and production. The necessity and value of obtaining accurate, current, technical information on wildcatting operations or field developments are well known to industry. It enables an operator to make important decisions regarding his exploratory operations and assists the oil-field engineer in the solution of many drilling and mechanical problems.

Duties

It is evident, therefore, that a very important niche in the oil industry and in any exploration or production company is filled by the oil scout. He is the eyes and ears of the exploration team and endeavors to see all, hear all, and know all about exploratory activities in this highly competitive business of finding and producing oil. The duties of an oil-field scout are so many and varied that even the industry often has but a vague idea of his numerous activities. He is concerned with all geological or geophysical exploration, any leasing activity, prospective drilling, current well developments, or anything that pertains in any way to the oil-finding business. He is the first to enter a new potential oil province or territory and usually the last to leave, for he remains to watch and to report as long as there is anything of interest.

Responsibility to Company

The scout's chief function is to alert the headquarters of his company concerning any event that might indicate the discovery of a new oil field, any geological data or information that would change the trend of exploration thinking in an area, and any development such as a lease play, geophysical activity, or other information that would prove of value. In addition, the scout prepares periodic reports to his company concerning land acquisitions by competitors, prospective "plays," geological and engineering data on drilling wells, and production statistics and often includes refinery, pipeline, and transportation data, all of which information the industry depends upon for its guidance.

In some areas, particularly the Pacific Coast, he may be assigned to the geological department of his company, where a knowledge of regional oil geology, prospective producing horizons, and anticipated pays is a necessary requirement. In the Gulf Coast, Rocky Mountain, and mid-continental areas, he is more often an integral part of the land department and, in order to take immediate advantage of important information, is furnished with a checkbook so that he can lease land for his company on the spot when authorized.

Scout Meeting

In order that his company may keep abreast with the current drilling and geological information, reports are usually made weekly to the division headquarters or to the home office. The scout is probably in daily communication with his supervisors and more frequently if the need arises. In sparsely settled areas some automobiles driven by scouts are equipped with mobile telephone for the immediate transmittal of important messages or urgent instructions.

The weekly route of the oil scout may vary from a comparatively small area, covering a limited oil basin or several counties in a densely drilled region, up to an area embracing many thousands of square miles in areal extent, involving perhaps 1,500 miles a week of hard, tortuous driving as is the case in many of the mid-continental and Rocky Mountain areas. The country may be highly populated near metropolitan centers like Los Angeles or Oklahoma City, or the terrain can be wild and barren, rugged and mountainous, as in the plains and hilly country of the Middle West or the dry desert areas of west Texas and New Mexico.

Much of the information furnished by him in report form is obtained from scout checks or scout meetings. The enormous area usually embraced within a single scouting district and the large number of wells to be visited along with the endless miles traveled make it necessary for the scout to obtain some of his well data by other means. At these gatherings, scouts from the various companies in the same district assemble weekly to exchange certain information with one another. This is not to the detriment of his own company as one might at first surmise. For information to be of real value, it has to be "spot news," current, right up-to-the-minute. This same information a day or so later might prove to be interesting news but would be received too late for a company to act upon and in all probability would be available through the press or other channels.

The district scout meeting is headed by a chairman, or "bull scout," elected from one of the members. This position rotates periodically so that the responsibility load is evenly distributed and does not become burdensome on any one individual or company. One of the duties of the chairman is to assign specific routes to the various members so that there is complete coverage of the district. These routes are also rotated periodically so that all members become thoroughly familiar with the entire district. This assignment is the individual scout's responsibility to the meeting and may or may not happen to coincide with the specific areas of interest of his company. He, therefore, has something of a dual allegiance to his company and to his scout check.

Personality

Needless to say, in wildcat territory many of these data are difficult to obtain. Sometimes the operator has a good and legitimate reason for drilling his well "tight," the vernacular for "confidential" in the oil industry. It is then up to the resourcefulness and ingenuity of the scout to obtain this information through personal contacts or perhaps by watching an exploratory well through field glasses from a great distance. He is proud of his ability to ferret out in an ethical manner information considered confidential by competitors and of the accuracy of the reports to his company.

If the scout is exceptionally successful, he may have obtained a copy of the drilling log, a core description, some formation tops or correlation points, perchance a copy of one or more of the various types of electric logs, and possibly core samples for paleontological determination and correlation. If production is obtained in a new area, a sample of oil is often secured and sent to his company's refinery for analysis to determine its economic value based on resultant refinery products. The engineer will also want to know the porosity and permeability of the producing horizon and any other data that may assist in the technical evaluation of this discovery. These facts and many more the scout jots down in his notebook so that the information will ultimately repose in the scouting office for handy reference.

For obvious reasons, the scout should be a very friendly, versatile individual with a persuasive personality. On his weekly rounds he becomes well acquainted with landowners and farmers, service-station operators and local government officials, oil-field service employees and equipment personnel, drillers and drilling contractors, bartenders and café proprietors, in fact, almost anyone who will pass the time of day. Unless on an urgent mission he usually finds time for a "coke" or a cup of coffee at many of his regular stops. He is a unique person, being equally at home on a fishing or hunting trip, on a golf course, or in the office of an oil executive, and his career is strange indeed. A casual, apparently insignificant remark from any of these people may give a hint or clue to some new exploratory operation. Possibly some landowner has recently leased his acreage, and the scout would like to know the exact location, the price, and lease terms. Maybe a right of way has been granted for a road or a pipeline or a geophysical option has been signed. The scout's natural inquisitiveness leads him to find out the facts about all these operations and many more besides. With experience as his guide and an imaginative mind, unrelated bits of information are gradually blended into a complete picture as one would solve a jigsaw puzzle.

Art of Scouting

There are certain qualities in a scout that contribute to the finesse so necessary in the art of scouting. He should possess a pleasant smile, a cordial greeting, a hearty handshake and be fully conversant on daily topics, matters of public interest, and possibly politics, all of which are an asset to a good conversationalist. He should endeavor to learn something of a personal nature about the people from whom he is trying to obtain information, their hobbies, sports, or other interests. He must have deep interest in and liking for people. This will permit a common subject of discussion whenever they meet.

A person is usually more affable when there is a common interest, and before long some of the information the scout desired is offered voluntarily without too much questioning. He should also be a judge of human nature, cognizant of a person's whims and fancies, and should realize that certain times are more propitious for securing the information than others. He may not agree with the ideas and opinions expressed by the person giving information, but it is politic to submerge his own opinion in order to prevent argument. The scout may also possess other favorable qualities, but above all else, he must be an extrovert to be successful in his profession.

Information Obtained

The importance of all exploratory drilling in the never-ending search for oil is indicative of the scout's interest in wells, where they will be drilled, when they will be drilled, the proposed objective, current drilling depths, the presence of any oil or gas showings, in fact, all there is to know about the activity. He must be able to gain the confidence of a driller or tool pusher and visually check the information told him. The number of stands of drillpipe in a rig, the presence of certain tools on the rig floor, his familiarity with drilling techniques, a casual glance at the mud pit, all serve him in assessing the importance of any development. At scout meetings the data recorded include the name of the operator, location of the well, elevation of the ground, rig floor or kelly bushing, spud or abandonment date, casing sizes and their cementing points, together with the number of sacks of cement used, water-shutoff data, drilling depths, whether the well has been drilled or cored, the types of bit used, formation tests, surveys run, in fact, all possible engineering and geologic data up to the actual completion or abandonment of the well.

If the well is successful, additional information required will be size and depth of tubing; whether the well is flowing, pumping, or on gas lift; completion date; initial gauge and subsequent gauges of the oil; gravity and cut, together with the size and number of chokes or beans to restrict the flow, gas production, as well as tubing and casing pressures. It may be of interest to know who is purchasing or transporting the oil and gas. Other pertinent development information would include the probable well spacing, offset requirements, and the time between well commitments. All this adds to the store of knowledge collected by the scout and is systematically filed or catalogued in the office.

The advance of electronic data-processing equipment has reached a point where some companies are already applying this type of data storage as a ready reference to augment their regular files. However, no such substitute has yet been found for the filing of the detailed log and history or any of the various types of electric logs currently in use.

Public Relations

Perhaps the scout more than any other person in the company is in a better position to serve as a public relations man, with the possible exception of the neighborhood service-station operator. His endless travels and his many contacts with people in all walks of life make him truly a direct representative of his company.

To small independent operators the company scout is often a welcome friend. Through his regular visits to a well or office he is frequently able to offer advice as to where a well should start coring, whether a showing encountered is worthy of testing, or perhaps assist in the evaluation and interpretation of an electric log. It also may be possible for him to obtain cores or ditch samples for an examination by his company's paleontological laboratory and thus materially aid in well correlation. In this manner both the operator and the company profit from the additional knowledge obtained.

Commercial Services

In later years commercial scouting services with coverage in all areas have been established with daily, weekly, or monthly reports. This is particularly true in foreign countries. As previously indicated this may not be "spot" news, as some company scouts have no doubt already secured this same information before the report went to press. It does, however, serve a good purpose in maintaining a continual record of drilling progress and current well data.

International Oil Scouts Association

The scouting fraternity of the United States and Canada has been organized and is united at the present time under the banner of the International Oil Scouts Association, which lends professional status to this unique group. Their slogan "How Deep Are You?" is legendary with all oil scouts. This outstanding organization was founded in 1923 under the name of the National Oil Scouts Association of America, Inc., but was subsequently changed to the National Oil Scouts and Landmen's Association about 1940. Effective January 1, 1960, the name was again changed, to International Oil Scouts Association.

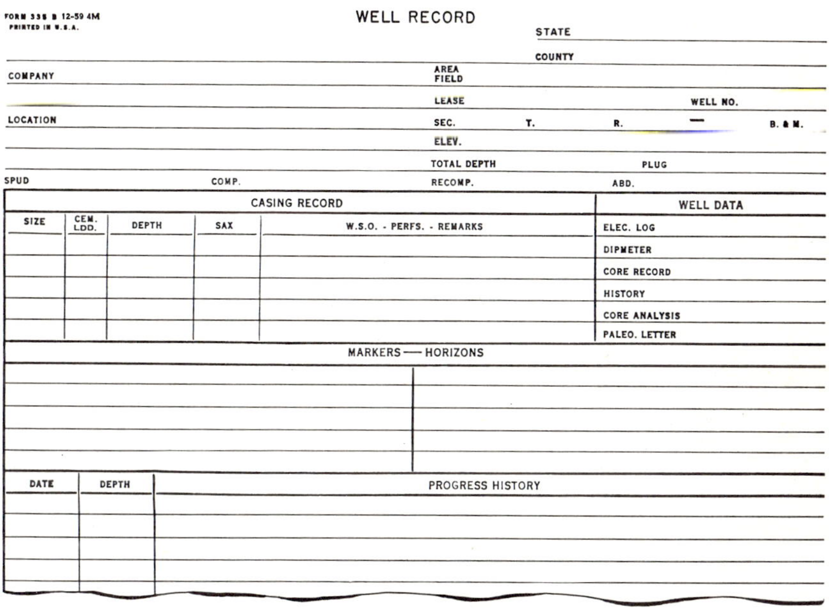

FORM 335 B 12-59 4M
PRINTED IN U.S.A.

WELL RECORD

STATE
COUNTY
COMPANY
AREA FIELD
LEASE
WELL NO.
LOCATION
SEC. T. R. — B. & M.
ELEV.
TOTAL DEPTH
PLUG
SPUD
COMP.
RECOMP.
ABD.

CASING RECORD					WELL DATA
SIZE	CEM. LDD.	DEPTH	SAX	W.S.O. - PERFS. - REMARKS	ELEC. LOG
					DIPMETER
					CORE RECORD
					HISTORY
					CORE ANALYSIS
					PALEO. LETTER

MARKERS — HORIZONS

DATE	DEPTH	PROGRESS HISTORY

FIGURE 24-1. Scout report form much used in California. This lacks headings for "Fracturing" and "Acidizing" which would be on forms for other parts of the country.

Commencing in 1930 the Association published a summary of all drilling and producing operations, including other pertinent statistical data on the southern part of the mid-continent area. Every year since then, the scope of this book has been enlarged until now it probably represents the most comprehensive and complete compilation of oil-field data and statistical review of the known oil- and gas-producing areas in the world. The yearbook now prepared by the International Oil Scouts Association, in cooperation with the Society of Petroleum Engineers of the American Institute of Mining, Metallurgical, and Petroleum Engineers, reports both domestic and foreign petroleum activity in a two-part annual volume, which includes a worldwide review of geological and geophysical prospects, land and leasing activity, wildcat exploration, proved field developments, oil and gas production, pipeline and refinery statistics.*

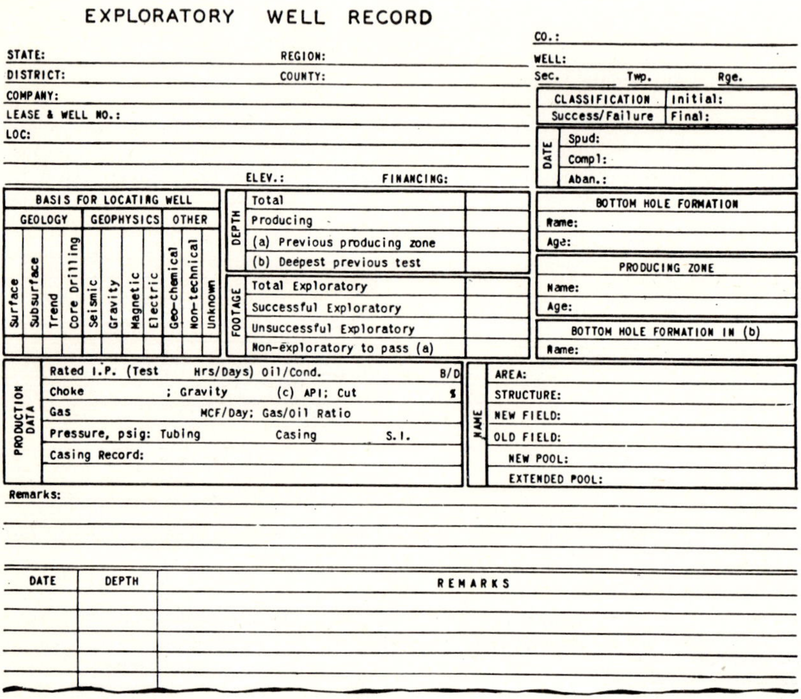

EXPLORATORY WELL RECORD

CO.:
STATE: REGION: WELL:
DISTRICT: COUNTY: Sec. Twp. Rge.
COMPANY:
LEASE & WELL NO.:
LOC:
CLASSIFICATION | Initial:
Success/Failure | Final:
DATE: Spud: / Compl: / Aban.:
ELEV.: FINANCING:

BASIS FOR LOCATING WELL
GEOLOGY: Surface, Subsurface, Trend, Core Drilling
GEOPHYSICS: Seismic, Gravity, Magnetic, Electric
OTHER: Geo-chemical, Non-technical, Unknown

DEPTH: Total / Producing / (a) Previous producing zone / (b) Deepest previous test
FOOTAGE: Total Exploratory / Successful Exploratory / Unsuccessful Exploratory / Non-exploratory to pass (a)

BOTTOM HOLE FORMATION — Name: / Age:
PRODUCING ZONE — Name: / Age:
BOTTOM HOLE FORMATION IN (b) — Name:

PRODUCTION DATA
Rated I.P. (Test Hrs/Days) Oil/Cond. B/D
Choke ; Gravity (c) API; Cut %
Gas MCF/Day; Gas/Oil Ratio
Pressure, psig: Tubing Casing S.I.
Casing Record:

NAME: AREA: / STRUCTURE: / NEW FIELD: / OLD FIELD: / NEW POOL: / EXTENDED POOL:

Remarks:

DATE	DEPTH	REMARKS

FIGURE 24-2. Scout form used by Pacific Coast Subcommittee of the Committee on Statistics of Exploratory Drilling of the AAPG. Note "Classification" box in upper right corner for indication of kind of exploratory well being drilled (see Chapter 4).

Scout Report Forms

There are many types of forms used by scouts in recording information on exploratory wells. Two examples are given in Figures 24-1 and 24-2. Figure 24-1 is typical of those prevalent in California. An analogous one for other parts of the country would carry headings for fracturing and acidizing information. Figure 24-2 is a form devised particularly for the Pacific Section of the Committee on Statistics of Exploratory Drilling of the AAPG. It provides space under "Classification" to indicate the kind of exploratory hole being drilled (see Chapter 4).

* This reference book is on standing order to many banks, university and public libraries, governmental agencies, and publishing institutions. It can be obtained from the International Oil Scouts Association, P.O. Box 1095, Austin 66, Tex.

Chapter 25

ESTIMATION OF PETROLEUM RESERVES DISCOVERED

By Charles E. Webber

Director, Natural Gas Administration, Sun Oil Company. President and Director of Natural Gasoline Association of America

Philadelphia, Pa.

GENERAL

The determination of reserves crowns the work of the oil and gas explorer; that a discovery has been made is a credit to his effort; the reserves are his legacy to the producer. Early wildcatters measured their success by the daily production of their wells, whether crude oil or gas. They were not so much concerned with the ultimate size of their discovery—indeed, they were ill-equipped to measure it. Their interest was centered in the question: How much oil (or gas) will I produce *today?*

Attempts to estimate the magnitude of oil and gas reserves were prompted first by natural curiosity. Most of the early appraisals were made by comparing declining productivity with that of fields which had already neared depletion. In gas fields, the rate of decline in production provided a good indication of the extent of the reserves.

Later, estimates of reserves became matters of business necessity. The Revenue Act of 1918 imposed, retroactive to 1917, the requirement that a producer determine his recoverable reserves of oil and gas on every lease. In calculating the tax on producers' income, the Act allowed a deduction for depletion. This depletion allowance (not to be confused with statutory depletion, which came later) was determined for each unit of oil and gas removed from the reservoir and sold, based on cost and on value, as of some given date. The cost of a property was relatively easy to compute. But its value could be determined only by an appraisal of the recoverable reserves. Thus the determination of reserves took on new significance and entered a new phase of development in the petroleum industry.

Because of the income tax law, one of the early comprehensive publications on the subject of reserves was published in 1919 by the U.S. Treasury Department (U.S. Treasury, 1919) with the help of engineers and geologists of the Bureau of Mines and the Geological Survey. It was entitled "Manual for the Oil and Gas Industry

Under the Revenue Act of 1918." In this were outlined many of the concepts which are used today, particularly with respect to natural-gas-production decline, comparison with similar fields which have approached exhaustion, the size of the reservoir, and so on.

As the industry matured, close estimates of reserves were demanded by other considerations—some political, some operational. What should be the sharing arrangement in a unitized field, for example, in the absence of detailed knowledge of reserves? On the political front, great alarm arose in the twenties that the nation was "running out of oil." Drastic governmental action to solve this apparent problem through the imposition of controls on the industry was a conceivable course of events unless it could be demonstrated that the alarm was without foundation. To allay the fears, as well as to provide the industry a definite annual appraisal of proved reserves, the American Petroleum Institute in 1936 created a Committee on Petroleum Reserves. Subsequently, each year this committee has computed the crude-oil reserves of the United States recoverable under existing economic and operating conditions. This annual report is considered the bible of the nation's proved reserves of oil. (See Appendix for tabulation of annual reserves.)

Early estimates of the nation's natural-gas reserves were made by various individuals associated with the natural-gas business. In 1945 and annually thereafter, the Committee on Natural Gas Reserves of the American Gas Association has published its authoritative study of proved natural-gas reserves. American Petroleum Institute and American Gas Association Committees jointly prepare estimates of natural-gas liquids reserves. In recent years the importance of petroleum, including natural gas, in Canada has created a demand for reserves studies in that country, and the Central Reserves Committee of the Canadian Petroleum Association annually publishes computed recoverable reserves of both oil and gas.

An accurate determination of the recoverable reserves is required on numerous occasions during the life of an oil or natural-gas field. Upon its discovery, management wants to know immediately what has been found. As development gets under way, different segments of the business may want reserves estimates for a variety of purposes. The exploration geologist or engineer may be called upon to make reserves estimates for reasons such as the following:

To estimate the extent and importance of a discovery

To appraise an exploration program for an area in which a particular well is an important part

To justify additional exploration work and field development

To appraise a property for purchase or sale or to obtain loans

To furnish reserves estimates for accounting and tax purposes in computing Federal income tax

To settle estates

To serve as a basis for participation in some type of unit operation with other producers

This chapter will deal with reserves estimating during this early field development. Therefore, it will relate more to procedures and techniques employed during such period by the exploration man rather than to those more complicated and highly specialized methods employed by the reservoir engineer during the exploitation period of production from the field. Because development drilling in a reservoir and production experience always yield additional data and tools for evaluating reserves and production performance, the later reserves estimates by the reservoir engineer are unquestionably more accurate but at the same time are highly technical in nature and require both production history and many basic data. Electronic computers and analyzers are frequently utilized in reducing the inherent complexities of such calculations. This discussion will be limited to the less complicated procedures which, because of limitations in basic data, must be employed by the exploration geologist or engineer.

The early studies should be approached from a geologic viewpoint. A few questions are sometimes helpful. For example: Why is this oil or gas where it is? What kind of a structure has this well penetrated? Is the structure highly faulted? Could there be a water drive? Is it in an area known to contain blanket-type reservoirs or small random ones? A sound appraisal of these and other pertinent questions can minimize many of the challenging uncertainties in deciding how much of the oil or gas trapped in the reservoir can be produced.

CRUDE-OIL RESERVES ESTIMATES

Volumetric Procedure

At best, the geologist or engineer launching into a reserves study will have his own log made from on-the-spot examination of cuttings and cores, the driller's log prepared from cuttings and the "feel of the bit," plus a variety of electric logs readily available to industry today and possibly some cores which have been examined and analyzed in the field or laboratory. From a study of these he arrives at the thickness of the producing formation. The areal extent of the reservoir must be an assumption in the early stages, relying upon whatever geophysical data are available, the subsurface geology of the general area, and a knowledge of similar structures, formations, or reservoirs near by. Realizing the potential errors involved in estimating the areal extent of a reservoir at this initial period, many companies arbitrarily consider the productive acreage to be not more than the anticipated well-spacing pattern of the area until additional wells are drilled; i.e., if 40-acre spacing is planned, the proved acreage estimated for the discovery is only 40 acres.

By far the most widely used procedure for estimating reserves is some form of volumetric procedure. In simplest terms, the bulk reservoir volume is computed and a factor applied to convert it to stock-tank oil recoverable at the surface.

$$\text{Recoverable stock-tank oil (bbl)} = AtF \tag{25-1}$$

where A = area of the producing reservoir, acres
t = effective thickness, ft
F = recoverable oil, bbl per acre-ft

Often industry must rely, at least temporarily, on reserves calculations made in this simple manner. They are at best only estimates, but they do afford an acceptable measure of the magnitude of success of the exploratory well. Many times the purchase of a lease or block of leases or the bids on governmental lands have no more verification than a reserves estimate made in this manner.

As development wells are drilled, a thorough analysis of additional data permits a more accurate determination of the thickness of the reservoir or separate reservoirs of the field. The laboratory may have determined experimentally the porosity, water saturation (interstitial water), and permeability relations and obtained the physical characteristics of the oil.

The volumetric formulas applicable in the light of these additional data are still simple from a mathematical standpoint.

$$\text{Recoverable stock-tank oil} = \frac{At \times 7{,}758 \times \phi(1 - S_w)R}{FVF} \tag{25-2}$$

$$= \frac{V_0 \times 7{,}758 \times \phi(1 - S_w)R}{FVF} \tag{25-3}$$

where A = area of the producing reservoir, acres
t = effective thickness, ft
V_0 = bulk volume of the reservoir, acre-ft
7,758 = number of barrels in 1 acre-ft
ϕ = effective porosity expressed as a decimal fraction
S_w = interstitial water expressed as a decimal fraction
FVF = formation volume factor (ratio of oil in place to oil recovered in stock tank)

Each factor entering into these two formulas is important and will be discussed in some detail with information as to how each one can be obtained and how it applies in the calculation of recoverable reserves.

Productive Area *A*

There are times when the reserves of a whole reservoir are desired, but seldom does one operator own the whole reservoir. More frequently a company is interested only in the reserves of its own leases or in some particular lease ownership of a part of the reservoir which the company is considering for purchase or in which it is acquiring an interest. In these cases the determination of an area A is not usually difficult. When the property in question falls within the productive limits of the field, the area A is obviously the area in acres of the tract or tracts involved. When the properties are at the edge or extend beyond the productive limits, or when the area of the entire reservoir is desired, a subsurface geological map must be made as described in Chapter 13. This may be contoured on the top of the producing formation if enough data are available, but it may be necessary to contour on some reliable marker or other formation above but close to the producing zone. It is very likely that during early development the oil-water interface will have been determined, and this is usually assumed to have a horizontal position. These data coupled with the subsurface map will give an outline of the productive limits of the reservoir. A planimeter will determine the area A.

Net Effective Thickness *t*

Reservoir thickness normally is determined from various types of logs. The self-potential curve of the ordinary electrical survey in many cases indicates top and bottom of the producing section. Doll (1949) discussed this technique in detail. Continuous and extensive research has improved upon both the instruments for making electrical surveys and the art of interpreting the curves. In drilling through limestones or reefs or when salt-base muds are present in the well bore, the self-potential curve is considered unreliable. The neutron log, which generally indicates porosity, is beneficial in limestones, reefs, and lime-cemented sands in estimating thickness. The gamma-ray log is helpful in determining tops and bottoms of sands and limestones which are bounded by shales. The resistivity curve of an ordinary electric log will indicate the interface between the gas and oil in sand formation but leaves considerable doubt when applied to limestones. A comprehensive discussion of the various techniques and applications of such surveys is given in Chapters 19, 20, and 21.

Where there is complete recovery throughout the pay section, cores, of course, offer excellent evidence of the thickness of the producing formation and will show the oil-water interface.

Drill-stem-test records contribute to the general knowledge of the location of the producing formation and can be beneficial in estimating producing formation thickness. In many cases they will tie down the oil-water contact.

Bulk Volume of the Reservoir V_0

Estimating area and thickness in order to determine the bulk volume of the reservoir is frequently the only method available during the initial drilling in a field. Where the area in question is only a small part of the over-all field, the thickness of the productive reservoir under that small area can be estimated with reasonable accuracy. As soon as it becomes necessary to appraise reserves for the whole reservoir or a large part thereof, a more reliable procedure must be employed.

Such a procedure is to calculate the reservoir volume from an isopachous map (see Chapter 13). In reservoir work, the isopachous map is constructed to show the areal extent and thickness of the productive reservoir. The area bounded by each successive contour indicates the thickness of the reservoir within the area.

Paper maps shrink or expand as humidity changes. Before a planimeter is used to determine the areas bounded by the various contours, it must be calibrated. This is

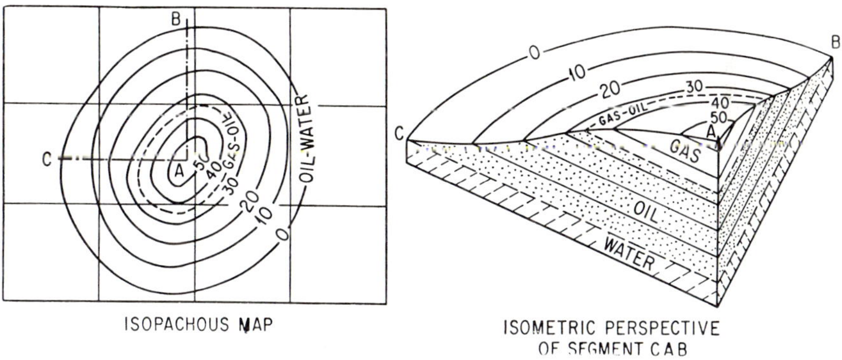

FIGURE 25-1. Isopachous map and enlarged perspective segment of a simplified reservoir

done by circumscribing a plot on the map whose area is designated, i.e., a 640-acre section, and applying the appropriate correction to the readings of the contour areas. The areas inside the zero contour and each successive contour of the isopachous map are then determined and recorded as $a_0, a_1, a_2, \ldots, a_n$. The bulk volume of the reservoir is calculated by the use of the trapezoidal rule, which states that

$$V_0 = Ch\left(\frac{a_0}{2} + a_1 + a_2 + \cdots + a_{n-1} + \frac{a_n}{2}\right) + t_n a_n \tag{25-4}$$

where V_0 = bulk volume of the reservoir, acre-ft
C = correction factor from calibration of planimeter, 1 plus or minus error in decimal fraction
h = contour interval, ft
a_0 = area within the zero or base contour, acres
a_1, a_2, etc. = area within successive contours, acres
a_n = area within the top contour, acres
t_n = average thickness of the formation above the n contour, ft

Figure 25-1 shows an isopachous map of a simple reservoir occurring above a water table and having a gas cap. Also shown is a perspective view of a segment CAB of the reservoir. In the determination of the bulk volume of the oil reservoir, the oil water contact is the zero contour and the 30-foot contour is the n contour. The contour intervals h are 10 feet. The average thickness of the formation above the n contour is the thickness of that layer of oil reservoir between the 30-foot or n contour and the gas-oil contact at approximately 34 feet. It is obvious that the smaller the contour interval h on the isopachous map, the greater is the accuracy of the bulk volume determination.

As the field is further developed, more accurate means are employed in calculating the volume of the reservoir. For example, Simpson's rule for irregular areas (Perry, 1950) is considered more accurate than the trapezoidal rule. Then sometimes an improvement is made over the isopachous map wherein an isovol map is constructed. On this, the contour lines represent equal values of the product of porosity and effective pay thickness. Using a planimeter on this type of map and applying the trapezoidal rule give an answer which is the over-all reservoir effective pore volume rather than just bulk volume. These refinements usually must wait for greater development of the field and are not widely used by the exploration geologist or engineer.

Effective Porosity ϕ

This important factor converts the over-all bulk volume of the reservoir to the void or pore space which contains the fluids water, oil, or gas. Here relatively minute samples in proportion to the total reservoir are tested and the results are assumed to be representative of the entire reservoir. This is risky. Extreme care must be exercised in selecting the core samples for testing so that to the greatest degree possible they are reasonably representative of the entire reservoir.

There are two kinds of porosity—absolute and effective. Absolute porosity is total void space without regard to the interconnection of the pore voids. Effective porosity is that pore volume which is interconnected and not sealed off and thus can yield oil or gas. *Effective porosity* is used in reserves calculations. Numerous methods are available for measuring effective porosity. Some are more adaptable to the laboratory, and others to rapid field tests. Some apply to sands, and others are best suited to large core samples, particularly limestones. In the calculation of reserves, porosity ϕ is effective pore volume divided by bulk volume. Acceptable methods for determining effective porosity are described in Chapter 22 and by Pirson (1958).

Geologists trained in the art of interpreting electric logs do a creditable job of estimating porosity from logs. The neutron log measures hydrogen saturation or, in turn, saturation of the rock. By calibration of neutron readings of rocks (Bush and Murdoch, 1950) with their known physical characteristics, as determined in the laboratory, the porosity can be estimated. By similar appropriate calibration the microlog (Guerrero and Stewart, 1958) and microlaterolog (Doll, 1953) can yield estimates of porosity. But it is highly advisable to obtain a few cores from a reservoir in question for actual porosity determination in order to verify the results obtained from the various types of logs. Chapters 19, 20, and 21 discuss the application of various logging procedures.

Interstitial Water S_w

The water which occupies some of the pore space at the time the reservoir is discovered is termed interstitial water S_w. This water reduces the pore space available for the accumulation of oil or gas. The most reliable data on the interstitial water content of the pore volume come from determinations on actual cores. Several methods are used in making these determinations as described in Chapter 22. A very widely accepted procedure is to measure the bulk volume of a core sample and then subject it to heat in a retort or to vacuum distillation. All fluids are driven out of the core.

The oil and water vapors are condensed and measured separately, thus yielding data on the interstitial water which was present in the core.

Obtaining appropriate cores for these water-saturation tests is difficult. Much of the outer surface of the core is contaminated with drilling fluids and water from other formations during the process of coring the well. As a remedial measure operators sometimes resort to oil-base drilling fluid during coring so that foreign waters cannot contact the cores. Cores taken with such special care are usually sealed, coated, and even quick-frozen to preserve as nearly as possible their original content.

Where the special coring techniques are not feasible, many laboratories use a "restored-state" procedure for determining interstitial water. The many variations of this procedure are complicated but generally involve complete extraction of the core, resaturation with salt water, and then permitting an appropriate capillary pressure to be exerted, thereby removing the salt water to a point corresponding to initial reservoir conditions (Pirson, 1958).

The exploration man frequently resorts to interpretations from electrical surveys. Several combinations of curves under different conditions permit reasonable estimates of interstitial water content using empirical correlations of log readings vs. experimental water-saturation data. Interstitial water content of cores usually averages between 0.2 and 0.4, but there are exceptions. In very thick and permeable reservoirs, the capillary forces which help to hold the water within the pore space are small. In such reservoirs, except for that zone just above the oil-water interface, interstitial water frequently drops as low as 0.05.

Formation Volume Factor *FVF*

Crude oil in the reservoir, depending to a considerable degree upon the subsurface temperature and pressure and the gravity of the oil, contains dissolved gas. Producing this oil at the surface permits much of this gas to escape, thus reducing the volume of the remaining oil. Oil which yields income to a producer is the oil in the stock tank, and it has become customary in speaking of oil reserves to mean stock-tank oil. The ratio of the volume of original reservoir oil to the shrunken volume of stock-tank oil is termed formation volume factor (*FVF*). If the reservoir oil has a very low gravity and is at shallow depths which do not permit high subsurface pressures, the formation volume factor exceeds unity by only a small amount. Very volatile oils under great pressure in the reservoir may shrink upon reaching the surface and the stock tank to less than half their original volume and, in rare cases, even more. Therefore, formation volume factors can be as high as 2 or even higher.

Formation volume factor can be determined accurately from a laboratory investigation of an actual subsurface sample of oil. In the absence of such an actual subsurface sample, a synthetic subsurface sample can be prepared by obtaining samples of separator oil and gas at the well and combining proper proportions of each in a bomb or cell in the laboratory. Such synthetic samples are termed "recombined samples." This type of laboratory investigation is done in a high-pressure container known as a pressure-volume-temperature cell (*PVT* cell) in which the pressure is reduced at proper temperature and the volume of each phase, gas and liquid, is measured. Too frequently, however, this laboratory work is not completed and the prediction of *FVF* must be made from other data which have been taken during initial production tests and flowing of the well.

A reasonable prediction of *FVF* can be made from gas/oil ratio of the well effluent. Where there is reasonable assurance that the ratios being determined are solution gas ratios (not including a large amount of free gas from the gas cap), the following tabulation gives a sufficiently accurate estimate for general use.

Table 25-1

Gas/oil ratio, cu ft per bbl	*FVF*
100	1.10
200	1.15
300	1.20
400	1.26
500	1.31
600	1.36
700	1.42
800	1.47
900	1.52
1,000	1.59
1,500	1.85
2,000	2.00

Frequently used for closer estimates is a correlation of formation volume factors with gas/oil ratio, specific gravity of the gas, API gravity of the oil, and the reservoir temperature (Standing, 1947).

Recovery Factor *R*

Few subjects have had more attention in the technical literature than recovery factors—the ratio of the oil in place to that which can be recovered in the stock tanks. The influence of the many properties of both the formation and the fluids therein and the degree of such influences on oil recovery have brought about a new, highly specialized profession—that of reservoir engineering. The geologist or petroleum engineer in exploration work seldom has either sufficient basic data or the time to apply these elaborate engineering procedures. He must rely upon his wealth of experience and general knowledge and, with prudence and wisdom, just predict unhesitatingly the probable recovery factor.

Oil pools have been classified according to the following categories: solution gas drive, gas-cap drive, and water drive. These categories are descriptive and practical. The type of drive establishes the operating characteristics of the reservoir and to a large degree determines the ultimate recovery.

1. In solution gas drive (also called depletion drive) reservoirs, the force driving the oil to the well bore is developed by the volumetric expansion of the gas as pressure is reduced and it escapes from the oil. Gravity drainage enters the picture as the energy of the expanding solution gas declines. Solution gas drive yields low recoveries. Predominant characteristics of this type of production are rapid and continuous decline in pressure, a sharp rise in gas/oil ratios, and the production of little water.

2. In gas-cap drive reservoirs, the predominant force is derived from the expansion of the gas overlying the oil. Obviously more energy is available and recovery of oil is somewhat greater than from solution gas drive reservoirs. Predominant characteristics of this type of production are moderate but continuous decline in pressure, increasing gas/oil ratios as the expanding gas front reaches the wells, and little water production.

3. In water drive reservoirs the predominant producing force is the water front encroaching into the oil portion of the reservoir. This type of drive affords much higher recoveries of oil than solution gas or gas-cap drive. However, to take full advantage of the natural water drive it is important to determine the proper reservoir pressure balance and adjust rate of oil withdrawal to be commensurate with the rate of water encroachment. Thus, the rate of production becomes an influencing factor on ultimate recovery.

As a practical matter many reservoirs are produced by combinations of these driving mechanisms. For example, a deep reservoir in the Gulf Coast area might have a gas cap as well as an effective water drive. Obviously all three producing forces take part in the over-all process of producing the oil.

If no gas cap is encountered in drilling into the productive formation, solution gas drive mechanism is assumed. If the reservoir is known to have a gas cap, the somewhat higher recovery of such reservoirs is assumed. If efficient water drive is anticipated, still higher recovery can be expected. Establishment of the existence of a water drive on the other hand is not simple. In general, blanket producing formations have water drives more times than not. This statement applies to sands and includes the Ellenberger and Devonian limestone formations of the Permian Basin. Small, lenticular, isolated, or closed reservoirs, including reefs, seldom have natural water drives. One characteristic of a reservoir which usually rules out hope of a water drive is an abnormal pressure, one which does not conform to the normal depth gradient. A thorough knowledge of the performance of similar reservoirs cannot be overemphasized in deciding whether water drive will be effective. If doubt exists, consideration should be given only to the lower recoveries obtained by solution gas or gas-cap drive mechanism. Little criticism will come to one who later revises a reserves estimate upward, but the reverse is seldom true. The goal, however, is to approach as closely as possible the correct answer on the first estimate.

Guidance is needed by the less experienced person in predicting probable oil recovery. There is no single property or group of simple characteristics of a reservoir or fluids therein which concisely defines the recovery of the oil from a reservoir. But, as guidance, there is value in a knowledge of known recoveries from older existing fields where similarities exist in their properties and those of a new field in question. From this standpoint, unfortunately, few depleted fields exist on which complete original reservoir data and accurate reserves computations are available. Most of the country's major oil fields still produce. Therefore, reliance must be placed on calculated recovery data on fields which have been thoroughly investigated and for which accurate recovery estimates have been made. Such an investigation is summarized in part in Table 25-2, where pertinent data and estimated oil recovery are shown for a number of fields having solution gas drive, gas-cap drive, and water drive. This is a partial list selected from an investigation of a well-spacing study sponsored by the American Petroleum Institute and conducted by Craze and Buckley (1945). In the absence of any other criteria and without a vast background of experience, considerable help can be had in predicting recovery of oil by comparing characteristics of reservoirs being appraised with tabulations such as Table 25-2.

Table 25-3 also may be useful in helping those who are less experienced in making reserves estimates. It shows some of the average basic characteristics of several well-known formations and average approximate recoveries which may be anticipated from each. Recovery in barrels per acre-foot F or as a decimal fraction R are shown in both Tables 25-2 and 25-3. If only the bulk volume of the reservoir, area A times thickness t, is known, the recovery F expressed as barrels per acre-foot can be used in formula (25-1). However, if the porosity and interstitial water content are known, the bulk volume of the reservoir can be converted to that part which contains oil. Here the recovery factor R as a decimal fraction can be applied in formulas (25-2) and (25-3) to provide greater accuracy.

The calculations of reserves described here relate to original stock-tank oil in place in the reservoir, and obviously, if a reservoir has produced prior to any reserve estimate, the cumulative oil withdrawn is subtracted from the estimates of total ultimate recovery to attain current reserves.

Table 25-2. Characteristics of a Random Selection of Reservoirs and Their Estimated Oil Recoveries

Sample No.*	Type of formation	Gravity of stock-tank oil, °API	Porosity ϕ	Interstitial water S_w	Permeability, millidarcys	Formation volume factor *FVF*	Estimated recovery, bbl/acre-ft *F*	Recovery factor *R*
				Solution Gas Drive				
1	Sand	38	0.20	0.15	90	1.25	368	0.35
12	Sand	41	0.153	0.40	6.8	1.28†	138	0.25
15	Sand	40	0.17	0.35	125	1.14	187	0.25
18	Sand	32–34	0.202	0.35	395	1.45	237	0.34
36	Sand	39	0.162	0.265	25–228	1.14	354	0.33
57	Sand	46.4	0.28	0.25	600	1.18	460	0.33
67	Sand	44.2	0.255	0.30	300	1.47	343	0.36
69	Sand	40	0.218	0.25	400–500	1.10	285	0.25
70	Sand	22.5	0.22	0.30	500	1.10	453‡	0.42
86	Sand	33.1	0.23	0.35	500	1.03	225	0.20
94	Sand	36.5	0.235	0.292	130	1.50	126	0.15
95	Sand	36.0	0.222	0.31	85	1.54	180	0.23
				Gas-cap Drive				
5	Sand and lime	41	0.125	0.20	150	1.24	232	0.37
19	Sand	42	0.18	0.32	80	1.10	297	0.34
21	Sand	31	0.20	0.20	400	1.06	530	0.45
22	Sand	37.2	0.182	0.25	356	1.10	213	0.22
24	Sand	41	0.13	0.20	206	1.67	233	0.48
25	Sand	38.5	0.15	0.20	....	1.25	247	0.33
28	Sand	32	0.179	0.35	100	1.08	180	0.22
42	Sand	44	0.27	0.35	....	1.25	547‡	0.50
46	Sand	38.6	0.18	0.38	100	1.15	256	0.34
47	Sand and lime	38	0.18	0.40	90	1.28	165	0.25
48	Sand and lime	37	0.172	0.355	75	1.20	248	0.35
49	Sand	22.5	0.23	0.25	....	1.05†	500‡	0.39
50	Sand	24.5	0.29	0.20	....	1.08†	443‡	0.27
				Water Drive				
2	Sand	31	0.30	0.35	1,000	1.20	600	0.47
3	Sand	31	0.18	0.20	75	1.07	410	0.55
6	Sand	38–39.8	0.252	0.17	2,000–3,000	1.20	1,058	0.78
10	Sand	23	0.32	0.30	50–5,000	1.20	629	0.43
16	Sand	35	0.34	0.33	3,000	1.29	958	0.70
20	Sand	39	0.14	0.25	118	1.07	306	0.40
23	Lime	39	0.176	0.30	1,010	1.49	407	0.73
26	Sand	33.2	0.15	0.15	328	1.1	437	0.49
38	Sand	20	0.28	0.10	5,000	1.09†	1,121	0.61
43	Sand	39	0.269	0.259	1,335	1.51	617	0.60
44	Sand	36.6	0.179	0.40	104	1.04	429	0.53
89	Sand	37	0.327	0.20	1,865	1.43	1,030	0.73

SOURCE: From a tabulation of 103 samples of an American Petroleum Institute study by Craze and Buckley, A Factual Analysis of the Effect of Well Spacing on Oil Recovery, in "Drilling and Production Practices," American Petroleum Institute, 1945.

* Sample numbers refer to the original publication.

† Estimated.

‡ Well spacing of 10 acres or less.

Table 25-3. Generalized Average Oil Recoveries

Producing formation	Average reservoir conditions				Stock-tank oil in place, bbl/acre-ft F	Recovery of oil from reservoir					
	Formation condition*	Porosity ϕ	Interstitial water S_w	Formation volume factor FVF		Solution gas drive: Bbl/acre-ft F	Solution gas drive: Recovery factor R	Gas-cap drive: Bbl/acre-ft F	Gas-cap drive: Recovery factor R	Water drive: Bbl/acre-ft F	Water drive: Recovery factor R
Miocene	Good	0.32	0.22	1.35	1,435	460	0.32	574	0.40	900	0.63
	Fair	0.27	0.26	1.35	1,150	320	0.27	356	0.31	600	0.57
	Poor	0.23	0.35	1.35	860	180	0.21	215	0.25	350	0.41
Frio	Good	0.31	0.24	1.40	1,306	418	0.32	470	0.36	785	0.60
	Fair	0.26	0.30	1.40	1,010	252	0.25	308	0.28	590	0.52
	Poor	0.22	0.40	1.40	732	132	0.18	154	0.21	340	0.46
Hackberry and Lower Frio (Nodosaria)	Good	0.28	0.28	1.70	920	293	0.32	331	0.36	535	0.58
	Fair	0.25	0.35	1.70	742	192	0.26	222	0.30	356	0.48
	Poor	0.20	0.45	1.70	502	112	0.22	125	0.25	200	0.40
Cockfield and Yegua	Good	0.26	0.30	1.45	975	309	0.32	351	0.36	556	0.57
	Fair	0.22	0.35	1.45	765	230	0.30	260	0.34	382	0.50
	Poor	0.18	0.50	1.45	482	133	0.28	150	0.31	203	0.42
Wilcox (upper Gulf Coast)	Good	0.22	0.35	1.80	617	153	0.25	185	0.30	418	0.66
	Fair	0.17	0.45	1.80	403	80	0.20	97	0.24	169	0.42
	Poor	0.13	0.55	1.80	253	43	0.17	51	0.20	86	0.34
Limestones, dolomites, and reefs	Good	0.20	0.35	1.45	694	118	0.17	187	0.27	382	0.55
	Fair	0.16	0.40	1.45	515	76	0.15	106	0.21	204	0.40
	Poor	0.07	0.45	1.45	207	29	0.14	37	0.18	62	0.30

* *Good formation condition:* Uniform blanket formation, thick, not lenticular or broken by shale or impermeable streaks or having sections of poor porosity. If a sand, not highly cemented. If sand, permeabilities of several hundred millidarcys and greater. If limestone, permeabilities of 100 and greater.

Fair formation condition: Intermediate between good and poor, considering all characteristics.

Poor formation condition: Nonuniform, lenticular, thin, or shaly. Poor effective porosity, vugular, or highly cemented. If sand, permeabilities of 100 millidarcys or less. If limestone, permeabilities of 15 millidarcys or less.

In the space permitted, an attempt has been made to cover the important aspects of oil reserves estimates. Additional information and concepts might be obtained from some of the published articles available in recent technical literature. For example, Arps (1956) presented an excellent discussion on estimating oil reserves during the early life of the field. McLendon (1957) described in some detail the estimating of oil reserves and presented a nomograph for simplifying the detailed work. This publication emphasized the use of various types of electrical survey in estimating formation thickness, porosity, and interstitial water saturation.

Example of Oil Reserves Computation. *Assumptions:*

$$
\begin{aligned}
\text{Reservoir area } A &= 500 \text{ acres}\\
\text{Thickness } t &= 40 \text{ ft}\\
\text{Bulk volume } V_0 &= 20{,}000 \text{ acre-ft}\\
\text{Porosity } \phi &= 30 \text{ per cent or } 0.3\\
\text{Interstitial water } S_w &= 40 \text{ per cent or } 0.4\\
\text{Formation volume factor } FVF &= 1.4\\
\text{Recovery } R &= 32 \text{ per cent or } 0.32
\end{aligned}
$$

Using formula (25-2),

$$
\begin{aligned}
\text{Recoverable oil} &= \frac{At \times 7{,}758 \times \phi(1 - S_w)R}{FVF}\\
&= \frac{500 \times 40 \times 7{,}758 \times 0.3 \times (1 - 0.4) \times 0.32}{1.4}\\
&= 6{,}383{,}726 \text{ bbl of stock-tank oil}
\end{aligned}
$$

NATURAL-GAS RESERVES ESTIMATES

Nonassociated and Associated Gas

A procedure similar to that already described for the estimation of oil reserves is also used in predicting recoverable natural-gas reserves. Briefly, the bulk volume of the reservoir is determined and the necessary factors for converting this to gas in place are applied.

Before discussing the details of natural-gas reserves estimating procedures, it is necessary to define the three recognized classifications for natural gas. The report entitled "Proved Reserves of Crude Oil, Natural Gas Liquids and Natural Gas," prepared jointly by the American Gas Association, the American Petroleum Institute, and the Canadian Petroleum Association, defines these three classifications as follows:

Nonassociated gas is free gas not in contact with crude oil in the reservoir and free gas in contact with oil where the production of such gas is not significantly affected by the production of crude oil.

Associated gas is free gas in contact with crude oil in the reservoir where the production of such gas is significantly affected by the production of crude oil.

Dissolved gas, or solution gas, is gas in solution with crude oil in the reservoir. Nonassociated and associated gas occupy their own separate portions within a reservoir and can be determined much as if they existed alone, ignoring the presence of any possible oil underneath. Solution gas, on the other hand, must be calculated from the oil reserves when enough is known about the characteristics of the reservoir oil to determine how much gas can be liberated from each barrel of oil by the time all the recoverable oil has been produced.

The following discussion, embodying volumetric formulas, applies to *nonassociated* and *associated gas.* Solution gas will be covered later. Natural gas reserves are tabulated as cubic feet at some standard condition of temperature and pressure.

$$\text{Recoverable gas (cu ft)} = \left[43{,}560 \times At\phi(1 - S_w)\frac{460 + 60}{460 + T_r}\frac{P_r}{P_b}\frac{1}{Z_r}\right] - \left[43{,}560 \times At\phi(1 - S_w)\frac{460 + 60}{460 + T_r}\frac{P_a}{P_b}\frac{1}{Z_a}\right] \quad (25\text{-}5)$$

$$\text{Recoverable gas (cu ft)} = \left[43{,}560 \times V_0\phi(1 - S_w)\frac{460 + 60}{460 + T_r}\frac{P_r}{P_b}\frac{1}{Z_r}\right] - \left[43{,}560 \times V_0\phi(1 - S_w)\frac{460 + 60}{460 + T_r}\frac{P_a}{P_b}\frac{1}{Z_a}\right] \quad (25\text{-}6)$$

$$\text{Recoverable gas (cu ft)} = 43{,}560(At \text{ or } V_0)\phi(1 - S_w)\frac{460 + 60}{460 + T_r}\frac{P_r}{P_b}\frac{1}{Z_r}R \quad (25\text{-}7)$$

where 43,560 = number of cubic feet in 1 acre-ft
460 = number added to °F to get °R
A = area of the producing reservoir, acres
t = effective thickness, ft
V_0 = bulk volume of the reservoir, acre-ft
ϕ = effective porosity expressed as a decimal fraction
S_w = interstitial water expressed as a decimal fraction
T_r = average reservoir temperature, °F
P_b = standard gas measurement pressure base, psia
P_r = average initial reservoir pressure, psia
Z_r = ideal-gas law correction (compressibility factor at original reservoir conditions)
P_a = average reservoir pressure at abandoned conditions, psia
Z_a = ideal-gas law correction (compressibility factor at abandoned reservoir conditions)
R = recovery factor expressed as a decimal fraction

Formulas (25-5) and (25-6), although appearing complicated at first glance, are actually simple mathematical computations of gas in place at *original reservoir pressure* less gas remaining in place at some *assumed future abandoned pressure.* The only difference between them is that in formula (25-5) the bulk volume of the reservoir is determined by multiplying area A and thickness t whereas in formula (25-6) bulk volume of the reservoir is determined directly from an isopachous map as previously described.

Formulas (25-5) and (25-6) are applicable only where the original pore space of the reservoir containing gas will not change appreciably during its productive life—indicating usually little or no water encroachment. Reservoir pore space can change where a very active water drive moves in as gas is produced. In such reservoirs with active water drives, pressure is maintained to a large degree. Thus the recoverable reserve of gas is not measurable by relating original pressure to an assumed pressure at abandonment. Here, formula (25-7) applies. It is a mathematical computation of original gas in place times a recovery factor R. Recovery from reservoirs having very active water drives is high, probably exceeding 80 per cent.

The factors used in formulas (25-5), (25-6), and (25-7) now will be discussed in detail to explain their function and how they can be utilized. Area A, thickness t, bulk reservoir volume V_0, porosity ϕ, and interstitial water S_w have already been discussed in relation to oil reserves. The factor V_0 is expressed in acre-feet and if multiplied by 7,758 will give barrels. In formulas (25-6) and (25-7) it is multiplied by 43,560 to obtain cubic feet.

Reservoir Temperature T_r

Natural-gas reserves are expressed as cubic feet at some acceptable temperature and pressure measurement base. In this country the standard temperature is 60°F. Because gas-producing formations are invariably warmer than 60°F, the gas in the reservoir when produced will shrink upon cooling to 60°F in accordance with Charles's law: At constant pressure the volume of a gas will always be directly proportional to its absolute temperature. Therefore, in order to correct the reservoir volume of gas to that volume which would exist at the surface, we must know the reservoir temperature.

Along the Louisiana and Texas Gulf coasts, with a few exceptions, there exists a fairly uniform subsurface temperature gradient with depth, averaging about 1°F below average atmospheric temperature for every 70 feet of depth. This does not prevail in other areas. It is therefore dangerous to rely on a depth-temperature correlation in calculating gas reserves. Usually in running an electric log, a maximum reading thermometer is or can be attached to the instrument. A reasonably accurate reservoir temperature thus is available. Where no such reading is made, the only recourse is to estimate a temperature from some nearby field of comparable depth and formation.

Pressure Base P_b

There is no universally accepted gas-measurement pressure base. Normally the standard cubic foot is thought of as being a cubic foot of space at atmospheric pressure at sea level (14.73 psia). But several states have legally defined different pressure bases: Texas uses 14.65 psia and Louisiana 15.025 psia. Thus the governing regulatory body or the company's own rules may determine this.

Boyle's law is as follows: At constant temperature, the absolute pressure of a gas is inversely proportional to its volume, or restated, at constant temperature, the volume of a gas is inversely proportional to the absolute pressure. Gas is under pressure in the formation, and when produced it will expand. To calculate this expansion the reservoir pressure must be known.

Fortunately, and with the exception of mountainous areas, subsurface pressures vary rather uniformly with depth. If no subsurface pressure measurement is available, no great error will result if reservoir pressure is estimated from depth: 0.465 psi for each foot of hole. An 8,000-foot reservoir will probably have 3,720 psia pressure. In mountains the source of hydrostatic pressure does not always come from just beneath the oil or gas reservoir and a normal gradient does not exist. Occasionally a supercharged reservoir is encountered where the gas is held under greater pressure than would be expected from the normal pressure gradient with depth. If this higher pressure is not known and a normal subsurface pressure is used in converting reservoir gas volume to standard atmospheric pressure base, the error is fortunately on the low side. An upward revision in reserves can be made later when the actual reservoir pressure is measured.

Compressibility Factors—Original Reservoir Z_r and Abandoned Conditions Z_a

The perfect-gas law is a combination of Charles's and Boyle's laws and is expressed as

$$PV = ZNRT$$

where P = pressure, psia
V = volume, cu ft
Z = compressibility factor deviation from the ideal gas
N = pounds of gas divided by its molecular weight and thus a constant for any given gas
R = 10.73 for all gases (R used here is not to be confused with the recovery factor R used in the formulas for estimating reserves)
T = temperature of the gas, °R (460 + °F)

N and R are constants and can be dropped from the equation when dealing with the same gas. Thus more simply the ideal-gas law in this instance becomes $PV = ZT$. Using the subscripts r and b to signify *reservoir pressure* and *base pressure* for measurement, respectively, the equation $P_rV_r = Z_rT_r$ defines the gas at reservoir conditions and $P_bV_b = Z_bT_b$ defines the same gas at measurement base conditions. Because it is the same over-all amount of gas

$$\frac{P_rV_r}{Z_rT_r} = \frac{P_bV_b}{Z_bT_b}$$

Solving for the volume at the surface, $V_b = V_rT_bP_rZ_b/T_rP_bZ_r$. But the compressibility factor correction to the ideal-gas law Z_b at and near atmospheric pressure base is unity. The equation is written as

$$V_b = V_r \frac{T_b}{T_r} \frac{P_r}{P_b} \frac{1}{Z_r}$$

This is the way the temperatures, pressures, and compressibility factors appear in the volumetric formulas (25-5), (25-6), and (25-7) for converting the gas volume under reservoir conditions to the volume at the base pressure at the surface.

It is recognized that the compressibility factors for light hydrocarbon natural gases can be correlated accurately with two hypothetical properties of the gases, *pseudoreduced pressure* and *pseudoreduced temperature.*

$$\text{Pseudoreduced pressure} = \frac{\text{actual pressure of the gas (psia)}}{\text{pseudocritical pressure}}$$

$$\text{Pseudoreduced temperature} = \frac{\text{actual temperature of the gas (460 + °F)}}{\text{pseudocritical temperature}}$$

Figure 25-2 is such a correlation, showing Z plotted against pseudoreduced pressures and temperatures (Standing and Katz, 1942). Utilization of this correlation requires a method of determining the pseudocritical pressure and temperature of the gas in question.

It has been found that natural gases containing less than 2 or 3 per cent of nonhydrocarbons from fields all over the country closely fit a single curve when specific gravity of the gas is plotted against *pseudocritical pressure* and *pseudocritical temperature.* Gas condensates having heavier components fit a slightly different curve (Brown et al., 1948). The California Natural Gasoline Association also adapted a similar correlation (CNGA, 1948). All three curves are shown in Figure 25-3 and permit the selection of appropriate pseudocritical pressure and temperature from the specific gravity of the gas.

Example Computation. *Assumptions:*

Reservoir pressure P_r = 4,000 psia
Reservoir temperature T_r = 160°F (620°R)
Specific gravity of gas = 0.72
Reservoir material is gas condensate

Find the compressibility factor Z_r.

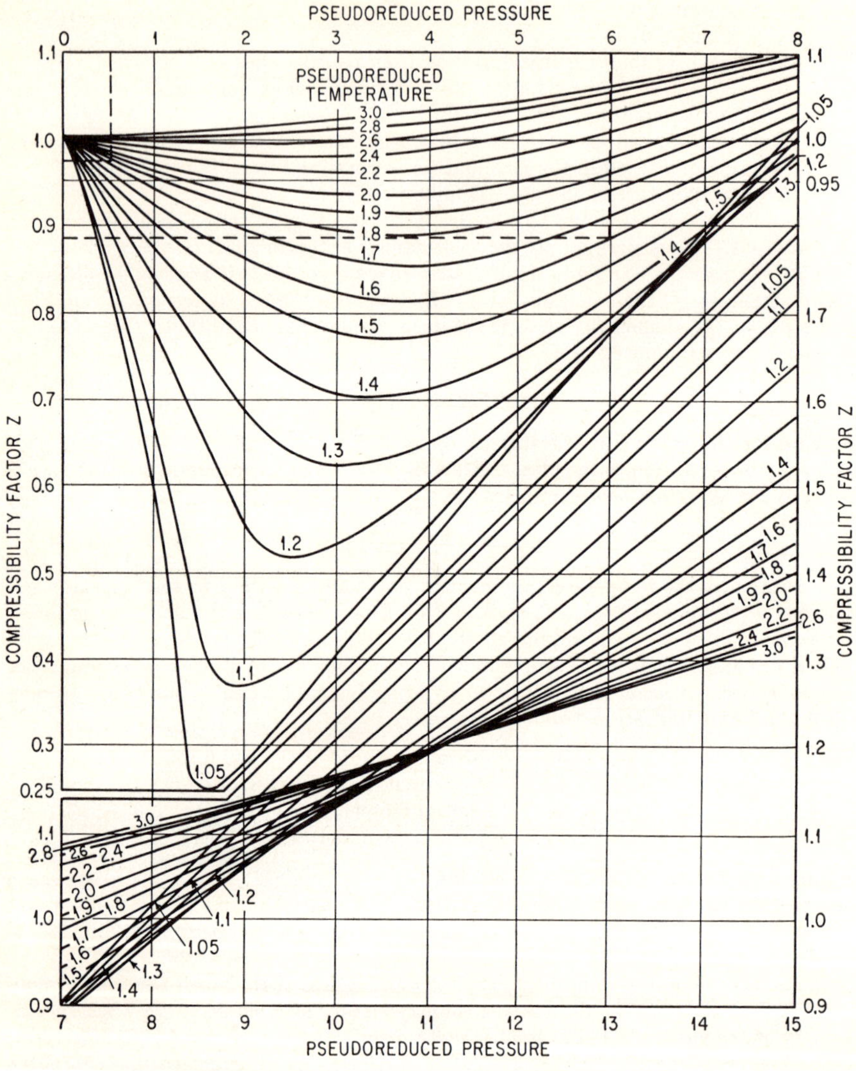

FIGURE 25-2. Compressibility factors for natural gases (*M. B. Standing and D. L. Katz, Trans. AIME, 1942*).

In Figure 25-3 opposite specific gravity of 0.72 and from lines labeled Condensate Well Fluids, read the pseudocritical pressure = 665 psia and the pseudocritical temperature = 386°R.

$$\text{Pseudoreduced pressure} = \frac{4{,}000}{665} = 6.0$$

$$\text{Pseudoreduced temperature} = \frac{620}{386} = 1.6$$

The value Z_r corresponding to the above pseudoreduced pressure and temperature in Figure 25-2 is 0.89.

In determining the compressibility factor at abandoned conditions Z_a, the procedure is the same except the assumed abandoned pressure P_a is used instead of original reservoir pressure P_r.

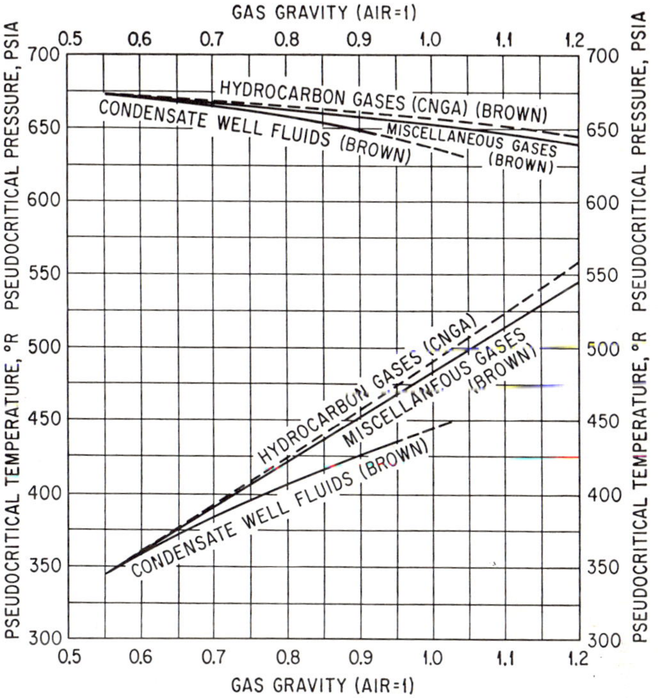

FIGURE 25-3. Specific gravity of gas vs. pseudocritical temperature and pressure. (*Calif. Nat. Gas. Assoc. Bull. TS-461; C. G. Brown, D. L. Katz, C. G. Oberfell, and R. C. Alden, NGAA Special Publication, 1948.*)

Abandoned Pressure P_a

As gas is withdrawn from a reservoir, pressure declines. The decline rate depends upon the size of the reservoir, the production rate, and whether or not an active water drive (encroachment) is helping to retard pressure decline. At some future date this pressure will reach the point where gas will not enter the customer's pipeline from its own energy. Augmenting this energy by compressing the gas finally becomes too costly. To a large degree, large-diameter gas lines operate in the vicinity of 700 to 1,000 pounds pressure. With no water drive or compressor facilities, a pressure in this vicinity is the "end of the line" for a gas reservoir. Obviously, though, an operator is not going to abandon a field if it still contains a sizable gas volume and if the gas price is such that compressors can be installed and operated with some profit left. Thus the prediction of an abandoned pressure is based upon economics.

No chart, table, or rule of thumb can answer the question. Many of the consulting firms which specialize in computing gas reserves for purposes of negotiating gas purchase contracts use 700 to 1,000 psi. Many oil and gas producers work on the theory that compressors will eventually be installed and therefore use an abandonment pressure of 300 to 500 psi in formulas (25-5) and (25-6).

Recovery Factor R

To those who estimate recoverable gas reserves as being the difference between original gas in place and the remainder in place at abandonment, formula (25-5) or (25-6), the recovery factor obviously need not be predicted. But if formula (25-7) is used whereby in-place gas is multiplied by a recovery factor, a prediction of R is necessary. Fortunately, recovery factors for gas are not so erratic as for oil. In general they are much higher, frequently running from 0.70 to 0.85. Estimates in this vicinity can be used with safety except in extremely poor reservoirs or where economic conditions will not justify continuing production beyond the point where the reservoir pressure will force the produced gas into the pipeline.

Example of Gas Reserves Computation for Nonassociated or Associated Gas. *Assumptions:*

$$\begin{aligned}
\text{Reservoir area } A &= 500 \text{ acres}\\
\text{Thickness } t &= 40 \text{ ft}\\
\text{Bulk volume } V_0 &= 20{,}000 \text{ acre-ft}\\
\text{Porosity } \phi &= 30 \text{ per cent or } 0.3\\
\text{Interstitial Water } S_w &= 40 \text{ per cent or } 0.4\\
\text{Reservoir temperature } T_r &= 160°\text{F or } 620°\text{R}\\
\text{Measurement pressure base } P_b &= 14.65 \text{ psia}\\
\text{Initial reservoir pressure } P_r &= 4000 \text{ psia}\\
\text{Abandoned pressure } P_a &= 600 \text{ psia}\\
\text{Specific gravity of gas} &= 0.72
\end{aligned}$$

Determine the recoverable gas. Using formula (25-6),

$$\begin{aligned}
\text{Recoverable gas} &= \left[43{,}560 \times V_0\phi(1 - S_w)\frac{460 + 60}{460 + T_r}\frac{P_r}{P_b}\frac{1}{Z_r}\right]\\
&\qquad - \left[43{,}560 \times V_0\phi(1 - S_w)\frac{460 + 60}{460 + T_r}\frac{P_a}{P_b}\frac{1}{Z_a}\right]\\
&= \left[43{,}560 \times 20{,}000 \times 0.3 \times (1 - 0.4) \times \frac{520}{620} \times \frac{4000}{14.65} \times \frac{1}{0.89}\right]\\
&\qquad - \left[43{,}560 \times 20{,}000 \times 0.3 \times (1 - 0.4) \times \frac{520}{620} \times \frac{600}{14.65} \times \frac{1}{0.93}\right]\\
&= [40{,}343{,}265{,}000] - [5{,}791{,}192{,}000]\\
&= 34{,}552{,}073{,}000 \text{ cu ft}
\end{aligned}$$

In this case at an abandoned pressure of 600 psia, the recovery factor R calculates to be 0.86 (34,552 ÷ 40,343).

NOTE: Z_r and Z_a are determined in the manner described in the section, Compressibility Factors—Original Reservoir Z_r and Abandoned Conditions Z_a, and shown in Figures 25-2 and 25-3.

Solution (Dissolved) Gas

Solution gas has been defined previously. Solution gas reserves depend upon the volume of initial oil in the reservoir and the extent to which that oil is saturated with gas. Consequently, estimates of solution gas reserves must follow or be made in conjunction with the computation of the oil reserves.

$$\text{Recoverable solution gas} = \text{recoverable stock-tank oil} \times G_s R_g \qquad (25\text{-}8)$$

where G_s = original solution gas/oil ratio (gas volume which will be liberated from each barrel of oil in place as it is produced as a barrel of stock-tank oil)

R_g = recovery factor expressed as a decimal fraction

The recoverable stock-tank oil is computed by formulas (25-1), (25-2), and (25-3).

Original Solution Gas/Oil Ratio G_s

This factor should be determined in the laboratory in a *PVT* cell from a subsurface sample of oil or a synthetic recombined sample. The sample of oil, actual or recombined, is placed in the cell. A valve is opened and the oil escapes, liberating the gas as it leaves the cell. The volumes of oil and gas separated by this flashing are measured at standard conditions, and their ratio in terms of cubic feet per barrel is the factor G_s.

In the absence of samples for a *PVT* cell investigation, actual field separator gas/oil ratios may be used as the factor G_s.

Recovery Factor R_g

The recovery factor R_g is estimated using as a guide the recovery factor previously selected from the recovery of oil in the oil reserves computation. Sometimes this factor R_g will be identical with R for oil, sometimes greater, and occasionally less, depending principally on the nature of the producing mechanism of the reservoir.

Solution (or *depletion*) *gas drive* reservoirs immediately begin to liberate gas as oil is withdrawn. This continues until the economical limit of oil production is reached. The depleted reservoir still contains oil, some gas remaining in solution at final reservoir pressure and free gas filling the pore space originally occupied by oil. If the original reservoir pressure and the bubble-point pressure of the oil were high, a lot of gas was originally in solution. A large part of it was produced with the oil, and the remainder, which is dissolved in the oil or occupies the voids, is proportionately less. In such reservoirs the gas-recovery factor is usually slightly higher than that for oil. A minor upward adjustment is in order.

On the other hand, if the bubble-point pressure is low, little gas is present in solution with the oil. At abandonment much of it may still be in the void spaces vacated by oil. Recoveries of gas will probably be correspondingly lower than for oil.

As a practical matter these adjustments in gas recovery from the figure chosen for the oil are difficult without prior experience. By many they are ignored.

Gas-cap expansion drive reservoirs behave differently, depending largely on the relative magnitude of the volume of free gas above the oil to the volume of oil in place. The internal mechanics and redistribution of solution gas liberated as free gas are complicated. The exploration geologist or engineer seldom has sufficient data (which can come only from production history) to predict them. Thus he usually chooses the same recovery factor for gas (R_g) as he does for oil (R).

Water drive reservoirs, if not produced too rapidly, will have little pressure drop. If the pressure does not drop below the bubble-point pressure of the oil, practically no gas is ever liberated from the oil within the reservoir.

When the oil is depleted, water will continue to encroach, bypass the residual oil, and force the gas toward producing wells. Under these conditions ultimate recovery of gas is higher than that for oil. An upward adjustment is in order.

NATURAL-GAS LIQUID RESERVES ESTIMATES

Natural-gas liquids are defined in the joint American Gas Association, American Petroleum Institute, Canadian Petroleum Association annual reports on reserves as "those hydrocarbon liquids which are gaseous or in solution with crude oil in the reservoir and which are recoverable as liquids by the process of condensation or absorption which takes place in field separators, scrubbers, gasoline plants or cycling plants. Natural gasoline, condensate, and liquefied petroleum gases fall in this category."

Natural gas upon production is capable of yielding two generalized but closely related types of gas liquids. The first type of liquids is that which does not separate from the gas in passing through the ordinary field separator but which must be recovered by refrigerating, compressing, or stripping in a natural-gas processing plant. The gas liquids thus obtained are normally thought of as natural gasoline and liquefied petroleum gas. The other type of liquid separates from the gas of its own volition (retrograde condensation) in passing through the ordinary separator where its pressure and temperature are naturally reduced during the course of production. This type of liquid, termed gas-condensate, is heavier than natural gasoline and resembles a very volatile crude oil. In color it may vary from water white to amber. Gas which yields a high volume of liquid, whether it be natural gasoline and liquefied petroleum gas or condensate, is called "rich" or "wet" gas. Gas which yields little or no liquids of any type is referred to as "lean" or "dry."

During the early stages of field development, reserves estimates of gas liquids of the first type, being those which include natural gasoline and liquefied petroleum gas, are not normally made. However, sometimes a gas will be so rich that it obviously would be further processed for recovery of natural gasoline and liquefied petroleum gas. In these cases the gas liquid content is determined by calculating the volume of propane, butane, pentane, and heavier hydrocarbons from an analysis of the natural gas. This calculation is very simple and merely involves the summation of the quotients obtained by dividing the mole per cent or volume per cent for each of the components of the gas (propane, butane, etc.) by the factor "cubic feet of vapor per gallon of liquid" (NGAA, 1957) for that particular component, which is done frequently by the laboratory.

Example:

	Mole or vol. %	÷ Cu ft vapor/gal liquid	= Liquid content/100 cu ft
Methane	80.0		
Ethane	10.0		
Propane	5.0	36.35	0.138
i-Butane	0.6	30.59	0.020
Butane	1.4	31.75	0.044
i-Pentane	1.0	27.40	0.036
Pentane	1.0	27.68	0.036
Heavier	1.0	20.00 (est.)	0.050
Total	100.0		0.324

The gasoline content is 0.324 gallon per 100 cubic feet or 3.24 gallons per 1,000 cubic feet or

$$\frac{3.24 \times 1,000}{42} = 77 \text{ bbl per million cu ft}$$

Multiplying this value by the recoverable gas reserves determined from formula (25-5), (25-6), or (25-7) gives the estimated recoverable gas liquid. It is obvious that for this type of gas, where its liquid content is assumed to remain constant, the recovery factor for gas liquid (natural gasoline and liquefied petroleum gas) must be the same as the recovery factor for the gas.

Condensate-yield gases can present an added complication. If a very efficient water drive exists, gas will be produced with little or no pressure drop in the reservoir. No retrograde condensation of condensate will occur within the reservoir, and the liquid content of the gas will remain substantially constant throughout the producing life of the field.

More often than not, the volume of production exceeds the volume of existing water encroachment. Therefore reservoir pressure will drop during the life of the field and liquid will condense in the reservoir pore space, some of which will remain forever. In these reservoirs the recovery factor for the condensate will be less than the recovery factor for the gas. As a broad general statement, the magnitude of this loss of

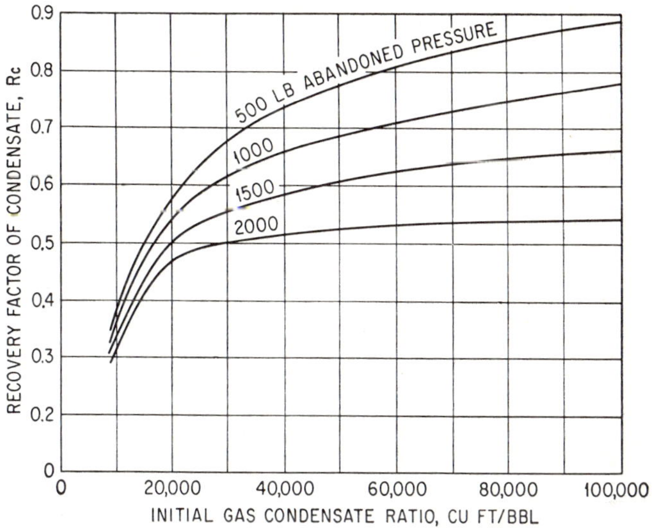

FIGURE 25-4. Recovery factor R_c stock-tank condensate vs. initial producing gas/condensate ratio.

liquid due to retrograde condensation in the reservoir is dependent upon the initial gas/liquid ratio and the anticipated reduction in pressure down to abandonment.

Figure 25-4 shows recovery factors of condensate R_c as a function of gas/condensate ratio and abandoned pressure. These data are calculated by a procedure outlined by Brinkley and Curtis (1951).

Gas-condensate content is normally determined experimentally by measuring the liquid condensate accumulation which is produced with some measured volume of the gas. This figure is the initial gas/condensate ratio. In conducting such a full-scale field gas/condensate ratio test, it is important that the well flows sufficiently long prior to testing to ensure that drilling fluid has been cleaned from the well and that the flow conditions have stabilized. This usually takes a minimum of 24 hours at a reasonably rapid flow rate. The initial gas/condensate ratio, multiplied by the volume of natural gas calculated to be in place in the reservoir, gives the condensate in place. Formula (25-7) would be applicable for this type of computation except that the recovery factor R as shown for gas would be replaced by the recovery factor R_c for condensate from Figure 25-4.

OTHER TECHNIQUES USED IN RESERVES ESTIMATING

Two other techniques are used in estimating oil and gas reserves: decline curves and material balance equations. Both require a considerable amount of production history and therefore are not so widely used by exploration people as by reservoir engineers. They have wide application, and a familiarity with them is important to the exploration geologist or engineer.

Decline Curves

Studies of decline curves are one of the oldest of the various methods of estimating oil and gas reserves. They present a graphic history of the productive record of a well, group of wells, or a reservoir. Prior to the advent of proration, the free and uninterrupted production of both gas and oil wells during their early life permitted plotting a curve which could be extrapolated into the future and serve as a basis for reserves and productivity forecasts. Today, decline curves have less utility. Production generally is prorated to market demand. Water encroachment in many fields helps to maintain reservoir pressure, and production is frequently curtailed to aid this process. Decline curves mean little until enough time has elapsed to establish a definite trend or rate of decline. If this trend can be established early in the life of the field, the exploration geologist or engineer can avail himself of this method of estimating reserves. If many years must elapse before such a trend develops, the exploration man probably will have parted company with the field in question and turned it over to the reservoir engineers.

In estimating reserves, the most convenient type of graph is a plot of producing rate, i.e., average barrels per day vs. cumulative production. Extrapolation of the line to a point of "no economic return" gives a direct reading of total ultimate cumulative production. This, of course, is recoverable reserves. In all cases the aim is to select graph scales which will result in a line as nearly straight as possible. Obviously this results in the most accurate extrapolation.

There are three general types of decline curves, each of which can be represented by a mathematical formula (Arps, 1956).

Constant Percentage Decline. Productivity decline is a constant percentage as a function of time. For example, at the beginning of the first full year of life a reservoir produces an average of 1,000 barrels daily. At the end of this first year it produces 900 barrels daily and at the end of the second year 810 barrels daily. This reservoir declines at 10 per cent annually. The decline curve of average rate vs. cumulative production straightens out on regular coordinate paper.

Hyperbolic Decline. Production decline is proportional to a fractional power of the production rate. This type of decline with average producing rate plotted against cumulative production approaches a straight line on log-log paper. Some shifting of the scale will further straighten the graph.

Harmonic Decline. Production decline is proportional to the producing rate. This type of decline where average producing rate is plotted against cumulative production straightens out on semilog paper. Production rate is plotted on the log scale and cumulative production on the regular coordinate.

Figure 25-5 shows how decline data appear when plotted in the usual manner. Actual average daily producing rate for a small oil field is plotted on three types of graph scales. The decline is of the harmonic type, and a straight line extrapolation on semilog paper indicates a reserve of approximately 7 million barrels. It is obvious that the other two curves could not be extrapolated with any accuracy.

So far this discussion of decline curves has been devoted to plots of rate of production vs. cumulative production. Although this type of curve extrapolates directly to recoverable reserves, it reveals nothing with respect to productive life of the field and not always does it point out a need for some means of augmenting production, such as pressure maintenance or water flooding. Neither does it reveal any data on the economics of the project. There are many different types of curves for these purposes which also yield information on reserves. A few examples will be mentioned briefly. Production vs. time extrapolates to show ultimate productive life and anticipated producing rate during future years. Plots of pressure decline vs. time indicate whether or not pressure maintenance is required. Plots of rate of production vs. income or profit can result in reserves estimates based on economic factors. In all cases it is desirable to plot the data, whatever they may be, so that the trend is a straight line

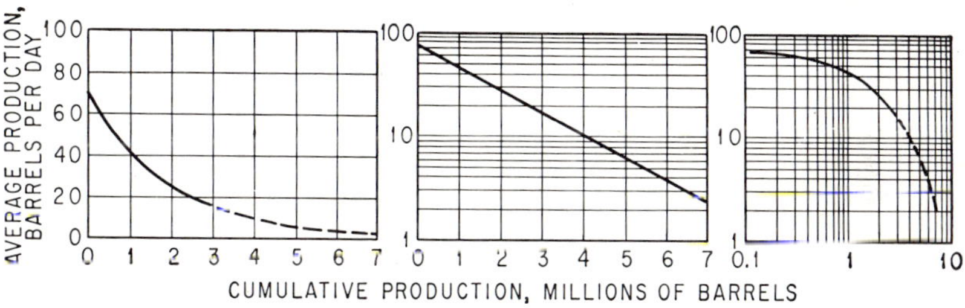

FIGURE 25-5. Identical production decline data plotted on three types of graphs.

for extrapolation. Excellent detailed discussions of these additional types of curves and their interpretation have been published (Clark, 1951; Menzie, 1953).

Material Balance Equation

To an increasing degree, those dealing with reserves estimating are being confronted with material balance concepts as a solution to many of the problems involved. Material balance methods of estimating oil in place are normally applied to a reservoir as a whole. Seldom is it certain that there is no migration of reservoir fluids across lease or property lines. The material balance equations in mathematical terms state that the *original oil in place* at original reservoir conditions is *equal to oil remaining* at certain subsequent reservoir conditions *plus the net withdrawals of reservoir* fluids during the elapsed time interval. This general statement is based upon the assumption that (1) the bulk volume of the reservoir pore space is constant and only the volume and nature of the reservoir fluids change and (2) substantial equilibrium exists at all times within the reservoir.

All forms of the material balance equations are merely a different way of expressing the following fundamental relationship:

$$\text{Original stock-tank oil in place} = \frac{\text{net volumetric fluid withdrawals}}{\text{volumetric gas expansion of reservoir fluid per unit of stock-tank oil}}$$

Between some initial period in the productive life of a reservoir (frequently termed original) and some subsequent period, production from the reservoir results in pressure decline unless water encroachment maintains it. If the pressure declines, certain changes occur in the physical character of the oil and gas remaining in the reservoir at the subsequent time period. A study and evaluation of these various changes in the character of the reservoir fluids in relation to the volumes withdrawn as production permit computation of the initial or original oil in place. To this, the estimator of reserves must apply a recovery factor.

Obviously, the changes in reservoir fluid properties must be both recognizable and measurable. This requires either a high production rate, an appreciable time interval of several years, or perhaps both. The engineer using only a slide rule or mechanical calculator frequently faces a prodigious task in detecting and determining the minor changes of reservoir fluid properties and mathematically applying them in the material balance equation. The problem is complicated further where reservoirs produce by combinations of the driving mechanism: solution gas drive, gas-cap expansion, or water drive. Consequently the engineer has resorted to electronic analyzers. These ingenuous devices inherently detect and apply many of the minor changes in reservoir fluid properties which might be overlooked and neglected in ordinary computations. Thus the resulting solution of the material balance equation can be done with great speed and at a much earlier period of elapsed productive life of the reservoir than otherwise possible (Bruce, 1943; Patterson et al., 1951).

Space does not permit deriving the basic material balance equation or describing all the factors that are involved. This discussion does no more than attempt to give the exploration geologist or engineer a general idea of the basic concept. Several excellent texts cover the subject in detail (Pirson, 1958; Muskat, 1949; Calhoun, 1953).

REPORTING RESERVES

Although the industry-wide oil and gas reserves are reported in a consistent manner by the American Petroleum Institute and the American Gas Association, there is little uniformity in the methods used by different companies in reporting reserves. Some companies estimate reserves on a basis of "proved" and also "probable" or "possible." A few companies, in addition to their own reserves, make estimates of reserves of all producers within their operating territory.

Company-wide reserves estimates are most useful when they can be incorporated with studies or reports showing financial status of the property as well as the recoverable oil and gas. Such a system not only presents to management just a reserves study by lease, field, area, and state but also points out pertinent financial data. Management can readily appraise success or failure of exploration in any given area. It can also relate newly discovered oil to the costs incurred. These types of presentation are becoming more and more important.

In addition to the over-all reserves of a producer, determined and reported usually on an annual basis, managements frequently desire to view each well drilled in the light of what it added to the company's books. Well completion reports filed on every well as it is completed are required. A single-sheet report can show the cost of the well; proved reserves; the basic factors such as formation thickness, porosity, recovery factor, etc.; and also data on other possible producing formations penetrated up the well bore for future completion.

CONCLUSIONS

The exploration man, as he gains experience in reserves estimating, will discover within his company additional empirical techniques, correlations, and short cuts which will help him in his computations. Space does not permit inclusion of many of them here. An attempt has been made to furnish sufficient basic knowledge and data to perform a reasonable job of reserves estimating. Beyond this, experience is invaluable.

Numerous individual factors, area, effective formation thickness, porosity, etc., enter a reserves computation. An inherent human tendency in predicting these factors is to "play safe" and be conservative. This can be overdone very easily, resulting in an abnormally low estimate. Recklessness, on the other hand, to enhance an exploration effort is just as dangerous. The exploration geologist or engineer is a technical man doing a technical job. His results should be as he interprets the data, using his best judgment, and should be as realistic as possible.

REFERENCES

Amyx, J. W., D. M. Bass, Jr., and R. L. Whiting: "Petroleum Reservoir Engineering," McGraw-Hill Book Company, Inc., New York, 1960.

Arps, J. J.: Estimation of Primary Oil Reserves, *Trans. AIME*, vol. 207, pp. 182–191, 1956.

Brinkley, T. W., and R. C. Curtis: Calculation of Natural Condensate Recovery, *World Oil*, vol. 132, no. 7, pp. 178–182, 1951.

Brown, C. G., D. L. Katz, C. G. Oberfell, and R. C. Alden, "Natural Gasoline and the Volatile Hydrocarbons," NGAA Spec. Publ., sec. 1, p. 44, 1948.

Bruce, W. A.: An Electrical Device for Analyzing Oil-reservoir Behavior, *Petrol. Develop. Technol.*, vol. 151, p. 112, 1943.

Bush, R. E., and E. S. Murdoch: Some Preliminary Investigations of Quantitative Interpretations of R/A Logs, *J. Petrol. Technol.*, vol. 189, p. 25, 1950.

Calhoun, Jr., J. C.: "Fundamentals of Reservoir Engineering," University of Oklahoma Press, Norman, Okla., 1953.

CNGA: California Natural Gasoline Association Bulletin TS-461, 1948.

Clark, Jr., R. W.: How to Estimate Crude Oil Reserves by Decline Trend Method, *Petrol. Engr.*, p. B-7, September, 1951.

Craze, R. C., and S. E. Buckley: A Factual Analysis of the Effect of Well Spacing on Oil Recovery, in "Drilling and Production Practices," pp. 144–155, American Petroleum Institute, 1945.

Doll, H. G.: The Self Potential Log: Theoretical Analyses and Principles of Interpretation, *Petrol. Develop. Technol.*, vol. 179, pp. 146–185, 1949.

Doll, H. G.: The Microlaterolog, *Trans. AIME*, vol. 198, p. 17, 1953.

Guerrero, E. T., and F. M. Stewart: Effective Porosity from Microlog and Microlaterolog Data, *Oil Gas J.*, vol. 56, no. 51, pp. 81–82, 1958.

McLendon, Dan H.: Oil Reserve Nomograph, "Tomorrow's Tools—Today," Lane Wells Company, First Quarter, 1957.

Menzie, D. E.: How to Select the Best Decline Curve, *World Oil*, p. 180, December, 1953.

Muskat, M.: "Physical Principles of Oil Production," McGraw-Hill Book Company, Inc., New York, 1949.

NGAA: Physical Constants of Paraffin Hydrocarbons and Other Components of Natural Gas, *Publ. 2145*, revised 1957.

Patterson, O. L., K. E. Montague, and B. Wiess, Jr.: High-speed Electronic Reservoir Analyzer, in "Drilling and Production Practices," p. 47, American Petroleum Institute, 1951.

Perry, John H. (ed.): "Chemical Engineers' Handbook," 3d ed., p. 58, McGraw-Hill Book Company, Inc., New York, 1950.

Pirson, S. J.: "Oil Reservoir Engineering," 2d ed., chap. 2, McGraw-Hill Book Company, Inc., New York, 1958.

Standing, M. B.: A Generalized Pressure-Volume-Temperature Correlation for Mixtures of California Oils and Gases, *Oil Gas J.*, vol. 46, no. 2, pp. 95–101, 1947.

Standing, M. B., and D. L. Katz: Density of Natural Gases, *Trans. AIME*, vol. 146, p. 140, 1942.

U.S. Treasury: "Manual for the Oil and Gas Industry Under the Revenue Act of 1918," 1919.

BIBLIOGRAPHY

Abelson, P. H. (ed.): "Researches in Chemistry," John Wiley & Sons, Inc., New York, 1959.

Abrams, T.: "The Essentials of Aerial Surveying and Photo Interpretation," McGraw-Hill Book Company, Inc., New York, 1944.

Alger, P. A.: "Mathematics for Science and Engineering," McGraw-Hill Book Company, Inc., New York, 1957.

Allen, E. S.: "Six-Place Tables," 7th ed., McGraw-Hill Book Company, Inc., New York, 1947.

Allendoerfer, C. B., and C. O. Oakley: "Fundamentals of Freshman Mathematics," McGraw-Hill Book Company, Inc., New York, 1959.

American Association of Petroleum Geologists Committee on Preparation of Projection Slides (W. R. Moran, Chairman): "Slide Manual," American Association of Petroleum Geologists, Tulsa, Okla., 1960.

AAPG Committee on Preservation of Samples and Cores (R. E. Taylor, Chairman): "Sample and Core Repositories of the United States, Alaska, and Canada," American Association of Petroleum Geologists, Tulsa, Okla., 1957. (Reproduced and distributed through the courtesy of Core Laboratories, Inc.)

American Geological Institute Glossary Committee (J. V. Howell, Coordinating Chairman): "Glossary of Geology and Related Sciences," American Geological Institute, Washington, D.C., 1957.

AGI Glossary Review Committee (J. M. Weller, Chairman): "Supplement to the Glossary of Geology and Related Sciences," American Geological Institute, Washington, D.C., 1960.

American Petroleum Institute: "Petroleum Facts and Figures," American Petroleum Institute, New York, 1959.

API Division of Refining: "Glossary of Terms Used in Petroleum Refining," American Petroleum Institute, New York, 1953.

American Society of Photogrammetry: "Manual of Photogrammetry," 2d ed., American Society of Photogrammetry, Washington, D.C., 1952.

Amyx, J. W., D. M. Bass, Jr., and R. L. Whiting: "Petroleum Reservoir Engineering," McGraw-Hill Book Company, Inc., New York, 1960.

Anderson, E. M.: "Dynamics of Faulting and Dyke Formation," 2d ed., Oliver & Boyd, Ltd., London, 1951.

Badgley, P. C.: "Structural Methods for the Exploration Geologist," Harper & Brothers, New York, 1959.

Ball, M.: "This Fascinating Oil Business," The Bobbs-Merrill Company, Inc., Indianapolis, 1940.

"Bibliography of Geochronology," Geochron Laboratories, Inc., Cambridge, Mass., 1961.

Bishop, M. S.: "Subsurface Mapping," John Wiley & Sons, Inc., New York, 1960.

Blickle, M. D.: "Reports for Science and Industry," Henry Holt and Company, Inc., New York, 1958.

Bowie, W., and O. A. Adams: "Grid System for Progressive Maps in the United States," Dept. of Commerce Spec. Publ. No. 59, Washington, D.C., 1919.

Brinkmann, R.: "Geologic Evolution of Europe," Hafner Publishing Company, New York, 1960. Translated by J. E. Sanders from "Abriss der Geologie," 2d vol., Ferd. Enke Verlag, Stuttgart, 1959.

Burcik, E. J.: "Properties of Petroleum Reservoir Fluids," John Wiley & Sons, Inc., New York, 1957.

Cairnes, C. E., and H. M. A. Rice: "Guide for the Preparation of Geological Maps and Reports," Geological Survey of Canada, Ottawa, 1957.

Calhoun, Jr., J. C.: "Fundamentals of Reservoir Engineering," University of Oklahoma Press, Norman, Okla., 1958.

Campbell, J. M.: "Oil Property Evaluation," Prentice-Hall, Inc., Englewood Cliffs, N.J., 1959.

Carozzi, A. V.: "Microscopic Sedimentary Petrography," John Wiley & Sons, Inc., New York, 1960.

Clark, S. K.: Classification of Faults, *Bull. AAPG*, vol. 27, pp. 1245–1265, 1943.

Continental Oil Company: "Landman's Legal Handbook: A Practical Guide in Leasing for Oil and Gas," F. H. Gower, Denver, 1957.

Coqueron, F. G., H. D. Hammar, and J. G. Winger: "Future Growth of the World Petroleum Industry," Chase Manhattan Bank, New York, 1958.

Daly, R. A.: "Our Mobile Earth," Charles Scribner's Sons, New York, 1926.

Deetz, C. H., and O. S. Adams: "Elements of Map Projection," Coast and Geodetic Survey, Washington, D.C., 1945.

De Golyer, E. L.: "Elements of the Petroleum Industry," American Institute of Mining and Metallurgical Engineers, New York, 1940.

De Sitter, L. U.: "Structural Geology," McGraw-Hill Book Company, Inc., New York, 1956.

Dix, C. H.: "Seismic Prospecting for Oil," Harper & Brothers, New York, 1952.

Dobrin, M. B.: "Introduction to Geophysical Prospecting," 2d ed., McGraw-Hill Book Company, Inc., New York, 1960.

Dott, R. H., and G. E. Murray (eds.): "Geological Cross Section of Paleozoic Rocks: Central Mississippi to Northern Michigan," American Association of Petroleum Geologists, Tulsa, Okla., 1954.

Dunbar, C. O., and J. Rodgers: "Principles of Stratigraphy," John Wiley & Sons, Inc., New York, 1957.

Du Toit, A. L.: "The Geology of South Africa," Hafner Publishing Company, New York, 1954.

Eardley, A. J.: "Structural Geology of North America," Harper & Brothers, New York, 1951.

Easton, W. H.: "Invertebrate Paleontology," Harper & Brothers, New York, 1960.

Emery, K. O.: "The Sea off Southern California," John Wiley & Sons, Inc., New York, 1960.

Emmons, W. H., G. A. Thiel, C. R. Stauffer, and I. S. Allison: "Geology: Principles and Processes," 5th ed., McGraw-Hill Book Company, Inc., New York, 1960.

Ewing, W. M., W. S. Jardetzky, and F. Press: "Elastic Waves in Layered Media," McGraw-Hill Book Company, Inc., New York, 1957.

Faul, H. (ed.): "Nuclear Geology," John Wiley & Sons, Inc., New York, 1954.

Freeman, O. W., and J. W. Morris: "World Geography," McGraw-Hill Book Company, Inc., New York, 1958.

Gignoux, M.: "Stratigraphic Geology," W. H. Freeman and Company, San Francisco, 1955. Translation of 4th ed. by G. G. Woodford, "Geologie Stratigraphique," Masson et Cie, Paris, 1950.

Gilluly, J., A. C. Waters, and A. O. Woodford: "Principles of Geology," W. H. Freeman and Co., San Francisco, 1951.

Graves, E. C.: "Ulrich's Periodical Directory," 9th ed., R. R. Bowker Company, New York, 1959.

Gutenberg, B.: "Physics of the Earth's Interior," Academic Press, Inc., New York, 1959.

Guthrie, V. B. (ed.): "Petroleum Products Handbook," McGraw-Hill Book Company, Inc., New York, 1960.

Haun, J. D., and L. W. LeRoy (eds.): "Subsurface Geology," Colorado School of Mines, Golden, Colo., 1958.

Heath, D. W., and J. McFarland: "Comprehensive Index" (of AAPG publications, 1946–1957), American Association of Petroleum Geologists, Tulsa, Okla., 1957.

Heezen, B. C., M. Tharp, and M. Ewing: "The Floors of the Oceans: I. The North Atlantic," Geological Society of America, New York, 1959.

Heiskanen, W. A., and F. A. Vening Meinesz: "The Earth and Its Gravity Field," McGraw-Hill Book Company, Inc., New York, 1958.

Hill, M. L.: Classification of Faults, *Bull. AAPG*, vol. 31, p. 1669, 1947.

Hoffman, L. E.: "Oil and Gas Leasing on the Public Domain," F. H. Gower, Denver, 1957.

Howell, Jr., B. F.: "Introduction to Geophysics," McGraw-Hill Book Company, Inc., New York, 1959.

Howell, J. V., and A. I. Levorsen, with R. H. Dott and J. W. Wilds of AAPG: "Directory of Geological Material in North America," American Geological Institute, Washington, D.C., 1957.

Hull, J. P. D., and J. V. Howell (eds.): "Structure of Typical American Oil Fields," vol. I, 1957; vol. II, 1957; vol. III, 1948, American Association of Petroleum Geologists, Tulsa, Okla.

Irving, R. H.: "Accounting Practices in the Petroleum Industry," The Ronald Press Company, New York, 1958.

Jacobs, J. A., R. D. Russell, and J. T. Wilson: "Physics and Geology," McGraw-Hill Book Company, Inc., New York, 1959.

Jeffreys, H.: "The Earth, Its Origin, History, and Physical Constitution," 4th ed., Cambridge University Press, New York, 1959.

Kartsev, A. A., Z. A. Tabasaranski, M. I. Subbota, and G. A. Mogilevski (P. A. Witherspoon and W. D. Romey, tr., from 1954 Russian ed.): "Geochemical Methods of Prospecting for Petroleum and Natural Gas," University of California Press, Berkeley, 1959.

King, R. R., V. M. Jussen, E. S. Loud, and G. D. Conant: "Bibliography of North American Geology," *U.S. Geol. Survey Bull.* 1095, 1960.

Krumbein, W. C., and L. L. Sloss: "Stratigraphy and Sedimentation," W. H. Freeman and Company, San Francisco, 1951.

Kulp, J. L.: Absolute Age Determination of Sedimentary Rocks, *Proc. of Fifth World Petrol. Congr.*, sec. 1, paper 37, 1959.

Kulp, J. L.: Geologic Time Scale, *Science*, vol. 133, no. 3459, pp. 1105–1114, 1961.

Lahee, F. H.: "Field Geology," 6th ed., McGraw Hill Book Company, Inc., 1961.

Landes, K. K.: "Petroleum Geology," 2d ed., John Wiley & Sons, New York, 1959.

Lange, N. A.: "Handbook of Chemistry," 10th ed., McGraw-Hill Book Company, Inc., New York, 1961.

Leith, C. K.: "Structural Geology," Henry Holt and Company, Inc., New York, 1923.

Lenczowski, G.: "Oil and the State in the Middle East," Cornell University Press, Ithaca, N.Y., 1960.

LeRoy, L. W., and J. W. Low: "Graphic Problems in Petroleum Geology," Harper & Brothers, New York, 1954.

Levorsen, A. I.: "Geology of Petroleum," W. H. Freeman and Company, San Francisco, 1954.

Levorsen, A. I.: "Paleogeologic Maps," W. H. Freeman and Company, San Francisco, 1960.

Levorsen, A. I., et al. (eds.): "Stratigraphic Type Oil Fields," American Association of Petroleum Geologists, Tulsa, Okla., 1941 and 1955.

Link, T. A.: Some Applications of the Strain Ellipsoid, *Bull. AAPG*, vol. 13, pp. 1449–1466, 1929.

Lobeck, A. K.: "Geomorphology," McGraw-Hill Book Company, Inc., New York, 1939.

Lobeck, A. K.: "Block Diagrams and Other Graphic Methods Used in Geology and Geography," 2d ed., Emerson-Trussel Book Company, Amherst, Mass., 1958.

Low, J. W.: "Geologic Field Methods," Harper & Brothers, New York, 1957.

Lueder, D. R.: "Aerial Photographic Interpretation," McGraw-Hill Book Company, Inc., New York, 1959.

Mason, B.: "Principles of Geochemistry," John Wiley & Sons, Inc., New York, 1958.

"McGraw-Hill Author's Book," McGraw-Hill Book Company, Inc., New York, 1955.

Middlemiss, R. R.: "Algebra for College Students," McGraw-Hill Book Company, Inc., New York, 1953.

Miller, V. C., and C. F. Miller: "Photogeology," McGraw-Hill Book Company, Inc., 1961.

Mitchell, H. C., and L. G. Simmons: "The State Coordinate Systems," Coast and Geodetic Survey, Washington, D.C., 1945.

Moffitt, F. H.: "Photogrammetry," International Textbook Company, Scranton, Pa., 1959.

Moody, J. D., and M. J. Hill: Wrench-fault Tectonics, *Bull. Geol. Soc. Am.*, vol. 67, pp. 1207–1246, 1956.

Moore, R. C.: "Introduction to Historical Geology," 2d ed., McGraw-Hill Book Company, Inc., New York, 1958.

Muskat, M.: "Physical Principles of Oil Production," McGraw-Hill Book Company, Inc., New York, 1949.

Neaverson, E.: "Stratigraphic Paleontology," 2d ed., Oxford University Press, New York, 1955.

Novitzky, A.: "Diccionario Minero-Metalurgico-Geologico-Mineralogico-Petrografico y de Petroleo: Ingles-Espanol-Frances-Alemen-Ruso," published by the author, Buenos Aires, 1951–1958.

Nuclear Science Series Rept. No. 23: "Cosmological and Geological Implications of Isotope Ratio Variations," National Academy of Sciences–National Research Council Publ. 572, 1958.

Paine, P.: "Oil Property Evaluation," John Wiley & Sons, Inc., New York, 1942.

Pearl, R. M.: "Guide to Geologic Literature," McGraw-Hill Book Company, Inc., New York, 1951.

Persch, F.: "Oil Dictionary, German, English, French, and Spanish," Industrieverlag von Hernhaussen, 1944.

Persch, F.: "English-German Oil Dictionary," Rudolph Schmidt, Berlin, 1955.

Petroleum Educational Institute: "Illustrated Petroleum Dictionary and Products Manual," Petroleum Educational Institute, Los Angeles, 1952.

Pettijohn, F. J.: "Sedimentary Rocks," 2d ed., Harper & Brothers, New York, 1957.

Phillips, F. C.: "The Use of Stereographic Projection in Structural Geology," Edward Arnold & Co., London, 1954.

Pirson, S. J.: "Oil Reservoir Engineering," 2d ed., McGraw-Hill Book Company, Inc., New York, 1958.

Poldervaart, A. (ed.): "Crust of the Earth," Geol. Soc. Am. Spec. Paper 62, Geological Society of America, New York, 1955.

Purdy, G. A.: "Petroleum," McGraw-Hill Book Company, Inc., New York, 1958.

Raisz, E.: "General Cartography," 2d ed., McGraw-Hill Book Company, Inc., New York, 1948.

Richter, C. F.: "Elementary Seismology," W. H. Freeman and Company, San Francisco, 1958.

Rock-color Chart Committee (E. M. Goddard, Chairman): "Rock-color Chart," National Academy of Science–National Research Council, Washington, D.C., 1948.

Russell, W. L.: "Principles of Petroleum Geology," 2d ed., McGraw-Hill Book Company, Inc., New York, 1960.

Russell, W. L.: "Structural Geology for Petroleum Geologists," McGraw-Hill Book Company, Inc., New York, 1955.

Shepard, F. P.: "Submarine Geology," Harper & Brothers, New York, 1948.

Shepard, F. P.: "The Earth Beneath the Sea," Johns Hopkins Press, Baltimore, 1959.

Shrock, R. R., and W. H. Twenhofel: "Principles of Invertebrate Paleontology," McGraw-Hill Book Company, Inc., New York, 1953.

Shwadran, B.: "The Middle East Oil and the Great Powers," 2d ed., Council for Middle East Affairs Press, 1959.

Siegrist, M., M. C. Grier, et al.: "Bibliography and Index of Geology Exclusive of North America," vol. 23 (for 1958), Geological Society of America, New York, 1960. (See vols. 1–22 for material published prior to 1958.)

Sloss, L. L.: "Lithofacies Maps: An Atlas of the United States and Southern Canada," John Wiley & Sons, Inc., New York, 1960.

Sloss, L. L., E. C. Dapples, and W. C. Krumbein: "Lithofacies Maps," John Wiley & Sons, Inc., New York, 1960.

Smith, C. C.: "Accounting for Oil and Gas Producers," Prentice-Hall, Inc., Englewood Cliffs, N.J., 1959.

Society of Exploration Geophysicists: "Geophysical Case Histories," vol. 1 (L. L. Nettleton, ed.), 1948; vol. 2 (P. Lyons, ed.), 1956, Society of Exploration Geophysicists, Tulsa, Okla.

Souther, J. W.: "Technical Report Writing," John Wiley & Sons, Inc., New York, 1957.

Struth, H. J.: What It Costs to Find Oil, *Petrol. Engr.*, vol. 32, no. 12, pp. B-19–B-25, 1960.

Sullivan, R. E.: "Handbook of Oil and Gas Law," Prentice-Hall, Inc., Englewood Cliffs, N.J., 1959.

Swanson, E. B.: "A Century of Oil and Gas in Books," Appleton-Century-Crofts, Inc., New York, 1960.

Thornbury, W. D.: "Principles of Geomorphology," John Wiley & Sons, Inc., New York (Chapman & Hall, Ltd., London), 1957.

Trask, P. D., and H. W. Patnode: "Source Beds of Petroleum," American Association of Petroleum Geologists, Tulsa, Okla., 1942.

Turner, F. J., and J. Verhoogen: "Igneous and Metamorphic Petrology," 2d ed., McGraw-Hill Book Company, Inc., New York, 1960.

Twenhofel, W. H.: "Principles of Sedimentation," 2d ed., McGraw-Hill Book Company, Inc., New York, 1950.

U.S. Bureau of Land Management: "Ephemeris of the Sun, Polaris, and Other Selected Stars," Department of the Interior, Washington, D.C., 1958.

U.S. Bureau of Land Management: "Manual of Instructions for the Surveying of Public Lands of the United States," Department of the Interior, Washington, D.C., 1947.

U.S. Bureau of Land Management: "Standard Field Tables and Trigonometric Formulas," 7th ed., Department of the Interior, Washington, D.C., 1955.

U.S. Geological Survey: "Mapping Procedures," chap. 3G1: Planimetric Map Compilation with Trimetrogon Photographs, Department of the Interior, Washington, D.C., 1960.

U.S. Geological Survey: "Publications of the Geological Survey" (to 1958), Department of the Interior, Washington, D.C., 1958.

U.S. Geological Survey: "Publications of the Geological Survey" (June, 1958, to May, 1960), Department of the Interior, Washington, D.C., 1960.

U.S. Geological Survey: "State Coordinates and Polyconic Maps," Department of the Interior, Washington, D.C., 1949.

U.S. Geological Survey: "Suggestions to Authors of the Reports of the United States Geological Survey," 5th ed., Department of the Interior, Washington, D.C., 1958.

U.S. Government Printing Office: "Style Manual," rev. ed., Washington, D.C., 1953.

Vance, H.: "Appraisal of Oil and Gas Properties from the Viewpoint of a Banker and an Oil Man," Bank of the Southwest, Houston, 1959.

Ver Wiebe, W. A.: "North American Petroleum," published by the author, Ann Arbor, Mich., 1952. (Includes chapter on the oil fields of the Middle East.)

Visser, W. A.: Nomographic Solutions of Dip Computations, *Bull. AAPG*, vol. 31, no. 10, pp. 1878–1886, 1947.

Weeks, L. G. (ed.): "Habitat of Oil and Some Factors That Control It," American Association of Petroleum Geologists, Tulsa, Okla., 1958.

Weller, J. M.: "Stratigraphic Principles and Practice," Harper & Brothers, New York, 1960.

Wilmarth, M. G.: Lexicon of Geologic Names of the United States (including Alaska) (up to Jan. 1, 1936), *U.S. Geol. Survey Bull.* 896, 1938.

Wilson, D., W. J. Sando, and R. W. Kopf: Geologic Names of North America (geologic names of North America introduced in 1936–1955; continuation of *U.S Geol. Survey Bull.* 896), *U.S. Geol. Survey Bull.* 1056-1, 1957.

Wolfe, J. H., W. F. Mueller, and S. D. Mullikin: "Industrial Algebra and Trigonometry with Geometrical Applications," McGraw-Hill Book Company, Inc., New York, 1945.

Wrather, W. E., and F. H. Lahee (eds.): "Problems of Petroleum Geology," American Association of Petroleum Geologists, Tulsa, Okla., 1949.

APPENDIXES

Table A-1. Tectono-Environmental

	Fluvial-lacustrine-eolian environment	Transitional environment (fluvial-lagoonal-littoral)	Epineritic environment	Infraneritic environment
General conditions; types of specific environments	Alluvial plains, stream channels, lakes, swamps, local or extensive wind action.	Alluvial plains, lagoons, marshes or swamps, barrier beaches, deltaic conditions.	Shallow wave and current-agitated marine waters, open circulation.	Water depths exceed 120 feet; relatively quiet offshore zones.
General colors and properties of sediments; main faunal elements	White, gray, yellow, red, brown, maroon, mottled. Lenticular or sheet sandstones. Blocky to poorly bedded shales. Limestones very subordinate; coal and lignite beds. Plant impressions and remains; land vertebrates; fresh-water gastropods and pelecypods.	Gray, brown, red, green, blue, dark gray, and black. Lenticular sandstones; some sheet sands. Shales blocky to well bedded; coal and lignite beds. Limestones subordinate; here and there a tongue of marine limestone. Plants; land vertebrates, gastropods, pelecypods, phosphatic brachiopods, and ostracodes common.	Gray, light brown, greenish gray, bluish gray, dark gray to black. Sheet sandstones; local linear bodies. Shales commonly bedded; limestones range from argillaceous to fossiliferous-fragmental types. Great variety of stout-shelled benthonic invertebrates (mollusks, brachiopods, echinoderms, corals, etc.).	Sediment colors gray, greenish gray, bluish gray, brownish gray, dark gray, black. Reds and browns subordinate. Sheet sands mainly fine-grained; shales well bedded. Limestones show wide variety; normal marine types dominant. Great variety of benthonic and nektobenthonic types, including more delicate forms. Significant per cent planktonic types.
Stable shelf occurrence	Quartzose sandstones, cross-bedded. Massive clay shales, commonly mottled. Carbonaceous; seldom calcareous. Fresh-water limestones, nodular, dense, local.	Quartzose sandstones, cross-bedded, commonly lenticular. Clay shales dominant, brackish varieties bedded; carbonaceous, locally calcareous, marly. Fresh-water limestone subordinate.	Quartzose sandstone cross-bedded. Siltstones ripple-marked. Shales commonly fine clayey, and greenish. Lms. with clastic textures, evenly bedded or locally cross-bedded. Ss. may grade directly to lms.	Fine-grained quartzose sandstones. Clay shales common, well bedded; calcareous, carbonaceous. Limestones normal marine, fusulinid, chalk; numerous planktonic components.
Unstable shelf occurrence	Subgraywacke sandstones, some linear quartzose channel sands. Shales mainly siltstones, massive to banded, mic., carb., seldom calc. Fresh-water lms. subordinate, nodular dense to uneven sugary texture.	Lenticular quartzose sandstones; sheet sands are subgraywacke. Shales mainly siltstone; claystones commonly bedded; micaceous, carbonaceous, calcareous. Fresh-water to marly limestones.	Quartzose to subgraywacke sandstones, cross-bedded, ripplemarked. Shales commonly siltstones, carbonaceous, calcareous, light colors. Lms. thicker stable neritic types, argillaceous, locally denser.	Fine-grained to subgraywacke sandstones, evenly bedded. Silty claystones, carbonaceous, calcareous. Limestones as above, locally denser, less widespread.
Intracratonic basin occurrence	Subgraywacke sandstones, local arkose associations. Shales mainly uneven-textured siltstones; micaceous, carbonaceous, semiwaxy, seldom calcareous. Lms. nodular to uneven thin-bedded.	Quartzose, subgraywacke, arkosic sandstones; thick shales commonly change characteristics from top to bottom. Micaceous, carbonaceous, calcareous. Fresh-water lms. may locally be thick.	Subgraywacke sandstones, arkosic sheets or wedges. Shales silty, carbonaceous, micaceous, calcareous. Limestones, nodular, uneven, may be argillaceous or dense.	Subgraywacke sandstones, thin-bedded. Shales fine siltstones to clay shales; dark colors common. Limestones thin, commonly dark; may be nodular.
Geosynclinal occurrence	Graywacke sandstones with subordinate subgraywacke channels. Shales massive to banded, chloritic or feldspathic. Fresh-water lms. may form thick lenses.	Graywacke sandstones with subordinate subgraywackes. Massive to banded shales, mainly silty to sandy. Local thick fresh-water limestones.	Graywacke sandstones with thinner subgraywackes. Shales commonly uneven-bedded; may show dark colors. Limestones subordinate, nodular.	Graywacke sandstones subordinate. Shales mainly siltstones with uneven texture. Limestones rare; siliceous, dark.

* Simplified for ease of reference. Adapted from W. C. Krumbein, L. L. Sloss, and E. C. Dapples, ations: mic., micaceous; calc., calcareous; carb., carbonaceous; lms., limestone; ss., sandstone.

LOGICAL MATERIAL

Classification of Sedimentary Rocks*

Epineritic biostromal environment	Bathyal-abyssal environment	Restricted lagoonal humid	Restricted lagoonal arid
Shallow clear waters, open circulation, little or no land-derived sediment. Controlling factors: temp., salinity, oxygen content, depth.	Water depths exceed 600 feet; these conditions may locally be fulfilled in rapidly subsiding geosynclines. Rare or absent in craton.	Mainly neritic depths; volume may be constant, but circulation restricted by barriers, sills, or biohermal control.	Mainly epineritic depths; evaporation exceeds inflow. Circulation restricted by barriers, sills, or biohermal control.
Light colors, with tan, bluish, cream dominant. Sandstones and shales subordinate; main bulk of sediments are carbonates, with abundant evidence of life forms and associated debris. Corals, bryozoans, algae, oysters, specialized brachiopods and larger foraminifera, crinoids, etc.	Sediment colors commonly dark; blue, green, red, dark gray to black. Land-derived sediments relatively rare; shales siliceous, diatomaceous. Limestones dark siliceous. Chiefly planktonic types, smaller foraminifera, diatoms, pteropods, etc.	Sediment colors commonly dark gray to black. Sandstones rare; shales dominate; bituminous, pyritic; calcareous, mainly claystones. Limestones dark, bituminous, thin-bedded. Phosphatic brachiopods, conodonts, certain mollusks, spores, algae.	Sediment colors commonly light; white, cream, brownish, greenish, bluish, pink, red. Sandstones rare. Clay shales dominate, gypsiferous, calcareous. Limestones dense; primary dolomites, nodular to thin-bedded. Evaporites may range from subordinate to dominant. Fauna aberrant, depauperate or lacking.
This is a typical environment of stable to mildly unstable shelves, in widespread shallow seas. The biostromal area may occur locally or over large areas. Sporadic bioherms common. Limestones include fossiliferous-fragmental, reefoid, oölitic, and chalks. Subordinate clay shales, marls, and thin quartzose sands in the association.	These environmental conditions doubtfully present on shelf areas; may occur locally in intracratonic basins. Typical but rare occurrence in eugeosynclinal association at times of rapid subsidence and slow deposition. No sandstones known; shales very siliceous, splintery; primary chert beds or nodules; limestone dark, dense, very siliceous.	In shelf occurrences, this environment may form widespread black shales behind barrier beaches or fringing reefs. The sequence may grade into continental sands and shales on one or more sides. Within the environment sandstones and limestones are rare.	In shelf occurrences this environment may form widespread evaporitic sequences dominated by gypsiferous shales and thin evaporite beds, grading, landward to typical redbed continental and transitional sequences. Limestones thin and subordinate within sequence.
Under basin or geosynclinal conditions, biostromes may rarely occur in local optimum areas. Significant is the growth of biohermal zones along tectonic hinge lines at edges of intracratonic basins. Here reefoid, fragmental, oölitic limestones are the rule, with other types subordinate.		Thick sequences of dark shales, bituminous, waxy, mainly clay types. Sandstones rare, fine-grained. Limestones subordinate, dark, bituminous varieties.	Thick sequences of typical evaporite associations, gypsum, anhydrite, salt, thin limestones, and dolomites; bright-colored shales, commonly clayey and gypsiferous. Cyclical evaporites common.
		Dark gray, brown, greenish silty shales, uneven textures. Some fine graywackes or subgraywacke sandstones. Limestones rare or absent.	It is doubtful whether true evaporite basins occur in geosynclinal conditions, although local sheets of evaporitic beds and gypsiferous shales may be associated with deltaic parts of geosynclinal sedimentation.

Sedimentary Tectonics and Environments, *Bull. AAPG*, vol. 33, no. 11, pp. 1876–1877, 1949. Abbrevi-

Table A-2. Megascopic Classification of Normal Sedimentary Rocks*

	TEXTURES: MAIN TEXTURE	TEXTURES: SUB TEXTURE	BONDING: CHEMICAL CEMENT OR MATRIX		COMPOSITION OF RECOGNIZABLE CLASTIC SILICATE FRACTION: QUARTZ ± CHERT (QUARTZITE SERIES)			QUARTZ + ROCK FRAGMENTS + CHERT + MICACEOUS OR CHLORITIC CLAY (GRAYWACKE SERIES): −feldspar (<10%)	+feldspar (>10%)		QUARTZ + FELDSPAR (>20%) ∓ KAOLINITIC CLAY ∓ impurities (<20%) (ARKOSE SERIES)	
DETRITAL ROCKS (>50% clastic silicates)	COARSE CLASTIC (average >2.0 mm) Conglomerate and Breccia class	Boulder Conglomerate (>64 mm)	— cement (<10%)	Loose		Quartz or Chert Gravel		Graywacke Gravel (Till — Boulder Clay)			Feldspathic Gravel (Feldspathic Till)	
		CONGLOMERATE or BRECCIA (4–64 mm)		Consolidated A, M–C	∓	QUARTZ OR CHERT CONGLOMERATE	+	Common	Feldspathic	+	ARKOSIC CONGLOMERATE OR BRECCIA	+
		Fine Conglomerate (<4 mm)	+ cement (>10%)	SiO_2	+		−	Graywacke		∓		∓
		Sandy, Silty, or Clayey Conglomerate		CO_3	+		±	Conglomerate or Breccia		−		∓
		(>20% sand, silt, or clay)		Fe, G, P	∓		−	Fanglomerate or Tillite		−	Fanglomerate or Tillite	−
	MEDIUM CLASTIC (0.0625–2.0 mm) Sandstone class	Conglomeratic Sandstone (>20% pebbles)	— cement (<10%)	Loose		Quartz Sand		Quartz–Chert and Mica Sand ("Pepper and salt sand")			Feldspathic or Arkosic Sand	
		Pebbly Sandstone (>10% pebbles)		Consolidated A, M–C	∓	"Sandstone" rare, usually passes into→	+	gray or red	black or dark gray	+	gray or red	+
		SANDSTONE	+ cement (>10%)	SiO_2		QUARTZITE (Gannister)	−	Common (low rank)	Feldspathic (high rank)	±		−
		Silty Sandstone (>20% silt)		CO_3	+	IMPURE (CO_3, Fe) QUARTZITE (50–75% SiO_2 in cements) QUARTZITIC SANDSTONE (25–50% SiO_2 in cements)	±	GRAYWACKE		−	ARKOSE (<20% impurities)	±
		Clayey Sandstone (>20% clay)		Fe, G, P	∓	QUARTZOSE SANDSTONE (<25% SiO_2 in cements)	∓			−	Sub Arkose (20–40% impurities)	∓

Rock group	Class	Subdivisions	Cement	Type	Rocks
DETRITAL ROCKS (>50% clastic silicates)	FINE CLASTIC (average <0.0625 mm) Siltstone — Shale class	Sandy Siltstone (>20% sand); SILTSTONE (gritty): a. Siltstone proper, b. Micro-conglomerate or micro-breccia; Silty Shale (semi-gritty); SHALE (smooth)	— cement (<10%)	Loose	(Laterite) ← Silt [+] and Clay → (Loess) (Quartzose, micaceous-chloritic, or feldspathic-kaolinitic)
				Consolidated	— SILTSTONE OR SHALE (Quartzose, micaceous-chloritic, or feldspathic-kaolinitic) [++] [+] ±
			+ cement (>10%)	SiO_2	± Siliceous, Opaline, or Chalcedonic Shale or Siltstone [−] [+] ±
				CO_3	+ ← MARL → [+] [−] ∓
				Fe, G, P	∓ Ferruginous, Glauconitic or Phosphatic Siltstone or Shale [+] [−] ∓
CHEMICAL ROCKS (<50% clastic silicates)	SANDY (5–50% clastic silicates)	CLASTIC: Coarse: >2.0 mm; Medium: 0.0625–2.0 mm; Fine: <0.0625 mm		SiO_2	Sandy and Oolitic Cherts / (rare) / Sandy Bedded Volcanic Cherts / Sandy Diatomites
				CO_3	++ SANDY LIMESTONE AND DOLOMITE [−] (± ← oolitic → ∓) [−] ∓
				Miscellaneous	SANDY: Gypsum, Anhydrite, Glauconite, Phosphates, Chemical Iron Ores, Salt (rare)
	PURE (not sandy) (<5% clastic silicates)	I. Clastic or arenitic if more than 10% of fragmental particles of any kind (sizes as above); II. CRYSTALLINE: Coarse: >4.0 mm; Medium: 1.0–4.0 mm; Fine: <1.0 mm		SiO_2	CHERT (nodular + bedded) / (rare) / CHERT (bedded) / Diatomite
				CO_3	++ LIMESTONE AND DOLOMITE (clastic or crystalline) [−+] [−] ∓
				Miscellaneous	SALT, GYPSUM, ANHYDRITE, PHOSPHATES, CHEMICAL IRON ORES, COAL

1. CHEMICAL CEMENTS

Fe Ferruginous; G Glauconitic; P Phosphatic; SiO_2 Siliceous; CO_3 Calcareous or Dolomitic

SYMBOLS

+ and − signs indicate frequency of occurrence and are quantitatively defined in the original text

2. MATRIX

A Argillaceous; M-C Micaceous or chloritic

* A simplified, integrated tabulation compiled by P. D. Krynine (*J. Geol.*, vol. 56, no. 2, 1948), showing commonly used boundaries among the gradational sedimentary rocks. Most of the concentrates and placers are omitted or left undifferentiated in this table.

Table A-3. Field Terms for Glassy Rocks
(Prepared by Wolfgang E. Elston, University of New Mexico)

Fracture	Felsic	Mafic
Conchoidal	Obsidian or pitchstone	Trachylite (rare)
Perlitic	Perlite	Rare

Table A-4. Field Terms for Dike Rocks
(Prepared by Wolfgang E. Elston, University of New Mexico)

Texture	Felsic	Mafic
Aphanitic	Felsite	Diabase or dolerite or trap
Aphanitic-Porphyritic		
Phenocrysts may include feldspars, quartz, ferromagnesian minerals only	Felsite porphyry	Diabase porphyry or dolerite porphyry
Phenocrysts of ferromagnesian minerals only		Lamprophyre
Phaneritic		
Aplitic	Aplite	Rare
Micrographic	Granophyre	
Pegmatitic	Pegmatite	Mafic pegmatite (rare)

Table A-5. Identification of Metamorphic Rocks*

(Compiled by J. Paul Fitzsimmons, University of New Mexico)

THERMALLY METAMORPHOSED ROCKS
Recrystallized rocks, products of thermal (static) metamorphism. Banding may be present but is a relict feature; sedimentary bedding or, possibly, flow structure in igneous rocks

Rock name	Characteristic minerals	Distinctive features		Original rocks
Marble Forsterite marble, tremolite marble, wollastonite marble, etc.	Calcite Calcite and forsterite, tremolite, wollastonite, etc.	Distinguished from limestone by recrystallization fabric; banded marble exhibits relict bedding		Limestone Impure limestone
Lime-silicate hornfels		Fine-grained		Very impure limestone, or calcareous-argillaceous-siliceous sediments
Lime-silicate granulite	Various lime silicates such as diopside, tremolite, wollastonite, idocrase, monticellite, etc.	Medium to coarse-grained	The mineral assemblage is distinctive of this group of rocks	
Tactite or skarn	Iron silicates may be abundant	A contact rock; coarse to very coarse-grained		Tactite may come from above or from more or less pure limestone modified by additions from igneous rocks
Quartzite	Quartz may contain small quantities of mica, actinolite, etc.	Recrystallization fabric banded quartzite exhibits relict bedding		Sandstone
Hornfels	Quartz, mica, may be feldspar; possibly andalusite, cordierite	Usually aphanitic or very fine-grained; is harder than shale or mudstone; banded hornfels commonly shows relict bedding May be characterized by spotty porphyroblasts, commonly amebic (but may be idioblastic), which give rise to spotted hornfels		Shale or mudstone
Greenstone	Chlorite, epidote, calcite	Generally fine-grained; greenish to greenish-black color is distinctive		Mafic igneous rocks or dolomitic shales

CATACLASTIC ROCKS
Due to mechanical shearing; temperature low, insufficient to produce recrystallization, but may cause local fusion

Rock name	Character
Cataclasite Metamorphic breccia usually called friction breccia or crush breccia Fault breccia in fault zone	Angular rock fragments, banding faint
Kakirite or flaser rock. Original rock can usually be recognized and is used with the term flaser (flaser gabbro, flaser rhyolite, etc.)	"Eyes" or lenticles of rock in wavy fine-grained matrix, the trend of the eyes and of the wavy matrix indicating direction of shearing
Mylonite. Usually a compact, hard rock, commonly found along faults, but may occur in larger masses; if found in soft masses along a fault, it is called fault gouge	Finely grained and laminated, a finely ground rock
Ultra mylonite or pseudotachylite	Very fine-grained, locally fused

Migmatite
An intimate mixture of igneous or igneouslike rock with other rock such as schist or gneiss. The mixture is such that the units cannot be mapped separately. The mixture may be intricate and haphazard, or may occur in bands to produce composite gneisses or lit-par-lit gneisses

* Metamorphic rocks are generally distinguished from other kinds of rock by distinctive minerals, by distinctive structures, or by both. Fundamental processes effecting metamorphism are recrystallization and shearing; the first is frequently accompanied by addition of material—metasomatism. Some minerals are distinctively metamorphic, and metamorphic origin should be considered and investigated if the following are present: garnet, wollastonite, kyanite, andalusite, sillimanite, cordierite, staurolite, monticellite, chlorite, chloritoid, chondrodite, and scapolite.

Table A-5. Identification of Metamorphic Rocks* (*Continued*)

FOLIATED ROCKS

Due to recrystallization and shearing actfon, with resulting orientation of tabular and prismatic minerals; foliated structure is diagnostic of these rocks. Grade of metamorphism generally increases downward from slate to gneiss

Rock name	Characteristic minerals	Distinctive features	Original rocks
Slate	Sericite and quartz	Aphanitic, excellent cleavage; colors generally black, red, green	Shale
Phyllite	Sericite and quartz	Aphanitic with lustrous and crenulated surface; cleavage good	Shale
Schist	Mica and quartz, possibly chlorite, garnet, staurolite, tourmaline (named for distinctive minerals: mica schist, garnet schist, etc.)	Generally fine-grained; micaceous and prismatic minerals show planar or linear arrangement; cleavage fair but not smooth	Shale, possibly rhyolite or other silicic igneous rocks
Calc-schist	Calcite and lime silicates		Calcareous sediments
Green-schist	Actinolite, epidote, chlorite		Mafic igneous rocks or dolomitic shales
Gneiss	Feldspar characteristic, quartz usually present; other high grade minerals may be present (rock named for distinctive mineral, or, if composition is similar to a grained igneous rock such as hornblende gneiss, sillimanite gneiss, granite gneiss, or diorite gneiss, it is named for this rock	Lenticular or parallel bands of schistose and granulose aggregates, or chiefly granulose with occasional or scattered oriented tabular or prismatic minerals; cleavage poor	Shale or silicic igneous rocks
Lime-silicate gneiss	Lime silicate		Calcareous sediments
Amphibolite	Plagioclase and hornblende		Mafic igneous rocks, or dolomitic shales
Marble gneiss	Calcite and usually some tabular or prismatic mineral		Impure limestones
Quartzite gneiss	Quartz and some tabular or prismatic mineral		Impure sandstones
Conglomerate gneiss	Quartz, shows relict of pebbles or cobbles		Conglomerate

Table A-6. Classification of Pyroclastic Rocks*

(Prepared by Wolfgang E. Elston, University of New Mexico)

A. Descriptive factors

1. Size and shape of pyroclastic particles:

Size of particles, mm	Shape of particles	Name of particles	Name of rock composed of particles
>32	Rounded to subangular, but not water-worn; material partly or entirely molten at time of eruption	Bombs	Agglomerate
	Angular; material solid at time of eruption	Blocks	Volcanic breccia
4–32	Any shape, massive	Lapilli	Lapilli-tuff
	Any shape, vesicular	Cinders	Cindery lapilli-tuff. Varieties are scoriaceous lapilli-tuff (common in mafic rocks) and pumiceous lapilli-tuff (common in felsic rocks)
$\frac{1}{4}$–4	Any shape	Coarse ash	Coarse tuff
$<\frac{1}{4}$	Any shape; rock is either unconsolidated, or held together by a cement or matrix	Fine ash *or* volcanic dust	Fine tuff
	Glassy constituents are welded or fused together	....	Welded tuff

2. Size, shape, composition, and abundance of admixed clastic sedimentary particles: Use nomenclature of sedimentary rocks to describe size, shape, and composition of sedimentary particles
   a. *Volcanic sedimentary rock* contains more than 50 per cent sedimentary material (Example: tuffaceous sandstone)
   b. *Sedimentary pyroclastic rock* contains less than 50 per cent sedimentary material (Example: sandy tuff)
3. Color: Self-explanatory
4. Source of constituent particles: *Essential* or *juvenile* constituents are formed directly by crystallization of magma. *Accessory* constituents are particles of previously consolidated volcanic rocks, or related hypabyssal rocks, torn loose from the walls of a volcanic vent in a subsequent eruption. *Accidental* constituents are inclusions of rocks (igneous, sedimentary, or metamorphic) derived from the prevolcanic basement
5. Classification of pyroclastic particles by igneous rock type: Use nomenclature for volcanic igneous rocks (Example: rhyolite tuff). Name prominent minerals (Example: biotite latite lapilli-tuff) and igneous textures in coarse particles (Example: porphyritic hornblende andesite agglomerate)
6. Devitrification and alteration: Name textures resulting from devitrification (Example: spherulitic rhyolite welded tuff). Name minerals resulting from alteration and/or devitrification (Example: bentonitic rhyolite tuff)
7. Primary (depositional) structures: Use terminology of sedimentary rocks to describe bedding, cross-bedding, ripple marks, mud cracks, and other primary structures
8. See Table A-7, Relative Proportions of Glass Shards, Crystal Fragments, and Rock Fragments in Tuffs

Table A-6. Classification of Pyroclastic Rocks* (*Continued*)

B. Genetic factors: mode of transportation and deposition

1. Pyroclastic rocks deposited subaerially
   a. *Explosion pyroclastic rocks* are transported directly by the force of a volcanic explosion. They form poorly sorted deposits, or deposits with graded bedding, around volcanic vents. *Ignimbrites* are rocks deposited by fast-moving clouds of hot gases and incandescent pyroclastic fragments (nuées ardentes); they are usually welded tuffs. *Wind-blown tuffs* are fine-grained pyroclastic rocks transported by wind currents. They resemble aeolian sedimentary rocks in textures and primary structures
2. Pyroclastic rocks deposited subaqueously
   a. *Explosion pyroclastic rocks deposited in water* are usually altered by reaction between hot volcanic fragments and water. *Palagonite* is a variety of hydrated basaltic glass formed in this manner
   b. *Water-laid pyroclastic rocks*, originally deposited on land but later eroded by running water and redeposited subaqueously, resemble fluviatile, lacustrine, or marine sedimentary rocks in textures and primary structures

C. Examples of pyroclastic nomenclature

1. *Tuff*, light brown, sandy; 60% essential andesite vitric-crystal tuff, slightly altered; 15% accessory andesite porphyry lapilli; 5% accidental granite and limestone blocks; 20% rounded quartz sand grains, beds 2–8 inches thick, cross-bedded, water-laid. *A short field description of this rock might be:* bedded, light brown andesite tuff
2. *Welded tuff*, cream; 85% essential rhyolite, vitric-crystal welded tuff with spherulitic devitrification structures; 15% accessory pumice and andesite porphyry lapilli. Single massive 50-foot bed with columnar jointing probably an ignimbrite deposited from a nuée ardente during a Pelean eruption. *A short field description of this rock might be:* massive cream-colored rhyolite welded tuff

* As far as possible the terms used here follow the recommendations of C. K. Wentworth and Howel Williams, The Classification and Terminology of the Pyroclastic Rocks, *Natl. Research Council Bull*. 89, pp. 19–52, 1932.

Table A-7. Relative Proportions of Glass Shards, Crystal Fragments, and Rock Fragments in Tuffs

(Prepared by Wolfgang E. Elston, University of New Mexico)

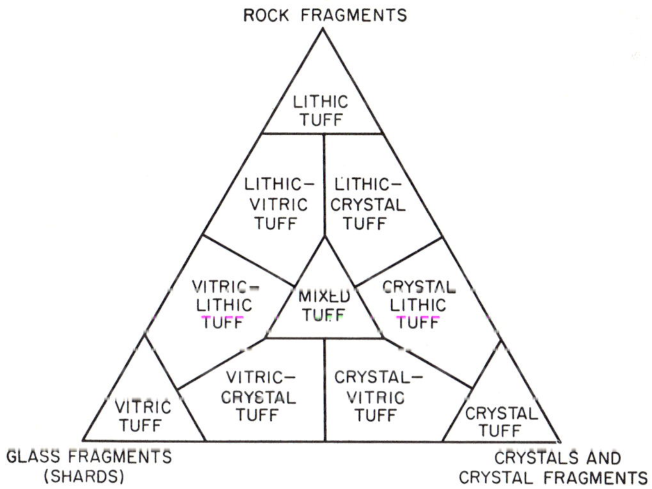

Table A-8. Classification of Common Igneous Rocks
(Prepared by Wolfgang E. Elston, University of New Mexico)

	Degree of saturation with SiO_2:	Oversaturated (with quartz, without olivine or feldspathoid)				Saturated (without quartz, olivine, or feldspathoid)			
	Type of plagioclase:	Albite-oligoclase	Oligoclase andesine		Labradorite	Albite-oligoclase	Oligoclase	Andesine	Labradorite-bytownite
	Plagioclase–potash feldspar ratio:	Potash feldspar > plagioclase	Potash feldspar ≅ plagioclase	Plagioclase > potash feldspar	Little or no potash feldspar	Potash feldspar > plagioclase	Potash feldspar ≅ plagioclase	Plagioclase > potash feldspar	Little or no potash feldspar
% 33	Felsic ("Acidic") Dark minerals usually biotite and/or amphibole	GRANITE[1] Rhyolite				SYENITE Trachyte[5]			ANORTHOSITE
66	Intermediate Dark minerals usually amphibole and/or pyroxene		QUARTZ[2] MONZONITE Quartz latite	GRANO-DIORITE[3] Dacite			MONZONITE Latite[6]	DIORITE Andesite	
90	Mafic ("Basic") Dark minerals usually pyroxene and/or olivine				QUARTZ[4] GABBRO Quartz[8] basalt				GABBRO[7] Basalt[8]
100	Ultramafic Dark minerals olivine and/or orthopyroxene								PYROXENITE

NOTE: For *PLUTONIC ROCKS* use names in all capital letters.
For *Volcanic rocks* use names in initial capital and lower case letters.
For *HYPABYSSAL ROCKS* use name in capital letters if texture is phaneritic or phaneritic-porphyritic; use name in capital and lower case letters if texture is aphanitic or aphanitic-porphyritic.

[1] *Alaskite* is a variety of granite containing almost no dark minerals. *Charnockite* is a granite containing hypersthene. *Alkali granite* contains little or no plagioclase, *soda granite* contains abundant plagioclase. Granite aplite has sugary texture, *granophyre* has micrographic texture.
[2] *Adamellite* is a synonym for quartz monzonite.
[3] *Quartz diorite* or *tonalite* is a variety of granodiorite with little or no potash feldspar, transitional to quartz gabbro.
[4] Quartz gabbro and quartz basalt usually contain only a small amount of quartz.
[5] *Keratophyre* is a variety of trachyte in which potash feldspar has been replaced by albite. It is usually associated with spilite and geosynclinal sedimentary rocks.
[6] *Trachyandesite* is a synonym for latite.
[7] *Norite* is a variety of gabbro containing hypersthene. *Ferrogabbro* is a variety of gabbro in which the dark minerals are high in iron, low in magnesium.
[8] *Tholeiite basalt* is a saturated or slightly oversaturated basalt low in alkalis, particularly common in continental flood basalt plateaus. *Spilite* is a variety of basalt in which labradorite has been hydrothermally (?) replaced by albite-oligoclase. It is usually associated with keratophyre and geosynclinal sedimentary rocks.

Table A-8. Classification of Common Igneous Rocks (*Continued*)

Degree of saturation with SiO_2:		Undersaturated (without quartz, with olivine and/or feldspathoid)				
		With feldspathoid, without olivine		With feldspathoid ± olivine	With olivine, without feldspathoid	
Type of plagioclase:		Albite-oligoclase	Oligoclase-andesine	Labradorite-bytownite		
Plagioclase–potash feldspar ratio:		Potash feldspar > plagioclase	Potash feldspar ≅ plagioclase	Plagioclase > potash feldspar		Little or no potash feldspar
% 33	Felsic ("Acidic") Dark minerals usually biotite and/or amphibole	FELDSPATHOIDAL[9] SYENITE Phonolite				
66	Intermediate Dark minerals usually amphibole and/or pyroxene		FELDSPATHOIDAL[10] MONZONITE Feldspathoidal latite			
90	Mafic ("Basic") Dark minerals usually pyroxene and/or olivine			ALKALI-GABBRO[11] Alkali-basalt[12]		OLIVINE GABBRO Olivine basalt
100	Ultramafic Dark minerals olivine and/or orthopyroxene				ALKALI-PERIDOTITE	PERIDOTITE[13]

[9] *Nepheline syenite* is the most abundant rock in this category, but there are many other varieties. In some varieties, such as *ijolite*, abundant feldspathoid (especially nepheline) takes the place of feldspar. A few varieties are intermediate rather than felsic.

[10] Rocks in this category are rare.

[11] This category includes many varieties, many of them rare. Unlike other mafic rocks, they are relatively rich in alkalis and usually contain one or more of the following minerals in addition to the ones indicated on the chart: biotite, soda amphibole, and soda pyroxene. The feldspathoid usually is nepheline or analcite. *Essexite*, *shonkinite teschenite*, and *theralite* are all varieties of alkali-gabbro. *Kentellanite* is a rock transitional to monzonite. *Picrite* is a rock transitional to alkali peridotite.

[12] *Tephrite* (= basalt containing nepheline or leucite), *basanite* (= olivine tephrite) and analcite or melilite-bearing basalt are the most common varieties. The terms *nepheline basalt*, *leucite basalt*, *analcite basalt*, and *melilite basalt* refer to varieties without feldspar. *Oceanite* and *picrite basalt* are varieties rich in olivine and low in plagioclase. *Mugearite* contains oligoclase rather than labradorite-bytownite, and is a rock transitional between alkali-basalt and trachyte. *Trachybasalt* is a general term for olivine-bearing alkali-basalt. *Lamprophyre* is a term used for hypabyssal alkali-basalt, usually in dikes, with phenocrysts of ferromagnesian minerals.

[13] *Dunite* is a variety of peridotite consisting almost entirely of olivine. The olivine and enstatite in peridotite are commonly hydrated to form serpentine. The rock is then called *serpentinite*. The term *ophiolite* is used, mainly by European geologists, to describe submarine serpentinite, spilite, keratophyre, and other altered volcanic rocks associated with eugeosynclinal sedimentary rocks.

Table A-9. Field Classification of Carbonate Rocks*

(Prepared by Robert L. Folk, University of Texas)

Lithology:	Limestones					Dolomites
Percentages:	More than 10 % allochems		Less than 10 % allochems		Allochems unidentifiable	
Composition of allochems:	I. Sparry allochemical rocks (usually turbulent-water environment)	II. Microcrystalline allochemical rocks (transitional or calm-water environment)	III. Microcrystalline rocks (calm-water environment)		IV. Biolithite (specify organism—coral or algal reef rock)	Specify crystal size as shown in Table A-10*B*. If allochems are evident, these may be added to the name; for example, oolitic finely crystalline dolomite, fossiliferous coarsely crystalline dolomite
Cement	Sparry calcite dominant as cement	Microcrystalline calcite dominant as matrix	1–10 % allochems	0–10 % allochems	Organic structures in place	
Over 25 % intraclasts	*Intrasparite* [Limestone conglomerate]	*Intramicrite* (rare)	*Intraclast-bearing micrite*	*Micrite* [Lithographic or aphanitic limestone]		
Under 25 % intraclasts Over 25 % oolites	*Oosparite* [Oolitic limestone]	*Oomicrite* (rare)	*Oolite-bearing micrite*			
Under 25 % oolites Fossil > pellets	*Biosparite* (specify fossils) [Coquina, shelly limestone]	*Biomicrite* (specify fossils) [Shelly limestone]	*Fossiliferous micrite*	If with irregular patches of spar, *dismicrite* [Birdseye limestone]		
Pellets > fossils	*Pelsparite*	*Pelmicrite*	*Pellet-bearing micrite*			

* Field names in common use are indicated by brackets.

Limestones are very largely products of mechanical deposition just as are the sandstones and shales. Terrigenous rocks are divisible into three classes: first, silt, sand, or gravel grains may be deposited in an environment of turbulent waves or currents, where clay matrix is winnowed out and pore spaces are later filled with chemically precipitated cement such as quartz or calcite. Second are those rocks wherein sand grains are embedded in a clay matrix, reflecting a lower-energy environment (deeper neritic, lagoonal, floodplain). Third, there are rocks (shales) which consist largely of fine clay with little or no admixed sand, usually formed in calm-water environments. Limestones form an exactly analogous series, consisting of organized aggregate grains, allochems, that form the structural framework of the rock (like sand grains in a sandstone), together with microcrystalline calcite ooze analogous to clay, and sparry calcite cement. In type I limestones, the allochemical carbonate grains are cemented with clear, crystalline, sparry, calcite; the transported carbonate grains are often well-sorted and are obviously calcarenites in the field. These usually form in high-energy zones such as carbonate beaches, bars, shoals, and swift tidal channels. Type II limestones contain carbonate grains in a lime mud, micrite, matrix, which is dull-lustered and opaque, and may vary from white to nearly black or brownish. These form in transitional, relatively low energy environments where the lime mud is not winnowed out. Type III limestones contain less than 10 per cent carbonate grains and over 90 per cent lime mud, forming the typical "lithographic" or "aphanitic" limestones. These form in shallow lagoonal or in deeper marine waters where there is little current or wave action. Quite commonly type I limestones grade through type II to type III as the environment changes from a beach or bar through transitional to calm-water environments.

Determination of the three major textural groups as outlined above is usually inadequate in defining the specimen. It is very important in limestones to specify the nature of the transported carbonate grains, or allochems, just as in sandstones one wishes to describe mineralogy of the sand grains. Only four types are of any volumetric importance: oolites, fossils, pellets, and intraclasts. Oolites and fossils are well-understood, but pellets are ovoid particles of lime mud about 0.05–0.15 mm in diameter, usually very uniform in size and shape. They probably represent fecal pellets of invertebrates but unfortunately cannot be recognized easily in the field or in hand specimens; an acetate peel or thin section is required to see them. Intraclast is a term used to cover any penecontemporaneous, semilithified sediment that is torn up and redeposited. For example, a storm on a lime mud bottom may tear up the sediment into variously sized lumps, round them, and redeposit them. Intraclasts range from 0.1 mm up to several inches long as in the edgewise conglomerates of the Paleozoic.

These four types of allochem grains may each occur either with sparry calcite cement or in a lime mud, micrite, matrix, giving rise to eight limestone types. Types are named by combining the allochem term with the cement-matrix term, e.g., intrasparite for intraclasts in sparry calcite cement, or biomicrite for fossils in a lime mud matrix. Type III limestones, the essentially pure lime muds, are called micrite or, if disturbances by burrowers or other means have produced "eyes" of sparry calcite, dismicrite. One type of limestone remains: that constructed in place by organisms such as corals, algae, and others. These are termed biolithite.

Modification of the main rock terms may be made by addition of a grain-size term such as calcarenite or calcirudite, or by adding such adjectives as dolomitized, glauconitic, phosphatic, pyritic, sandy, silty, clayey, etc., if pertinent. For type I (sparry) limestones, grain size of the allochems should also be specified (see Table A-10*B*). Biosparite and biomicrite should always be further qualified by naming the dominant fossil. Examples of usage: dolomitized intrasparite, crinoid biomicrite, pyritic pelmicrite, sandy glauconitic oosparite, etc.

Table A-10. Bases for Simplified Carbonate Classification

(Compiled by Robert L. Folk, University of Texas)

A. Comparison of Limestone and Terrigenous Rock Classifications

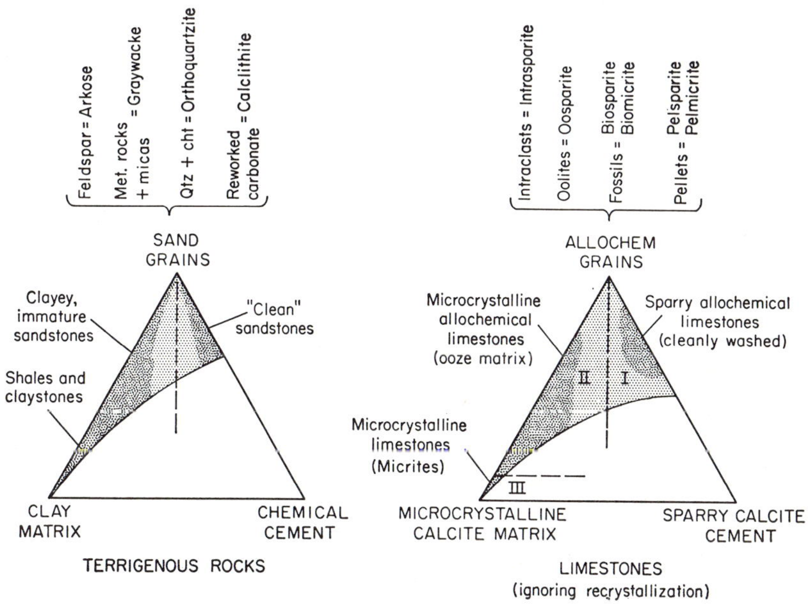

Diagram comparing limestone classification with analogous classification of terrigenous rocks. Shaded areas are those parts of composition triangle which occur most commonly.

Terrigenous rocks could be classified by proportions of sand grains (structural framework fraction), clay matrix, and chemical cement, the proportions of the last two being an index to degree of sorting. Nonrecrystallized limestone can be classified by the proportion of allochems (structural framework fraction), microcrystalline calcite matrix, and sparry calcite cement, the proportions of the last two also being an index of sorting.

Three basic limestone families are proposed: sparry allochemical limestone (type I), representing good sorting; microcrystalline allochemical limestone (type II), representing poorly winnowed sediments; and microcrystalline limestone (type III), analogous with claystone in terrigenous triangle. Just as one uses composition of sand grains for further classification of terrigenous rocks into arkose, graywacke, orthoquartzite, and calclithite, each ranging from clayey to nonclayey, so one uses composition of allochems for division of limestones into subvarieties such as intrasparite or biomicrite.

Table A-10. Bases for Simplified Carbonate Classification (*Continued*)
(Compiled by Robert L. Folk, University of Texas)
B. Grain-size Scale for Carbonate Rocks

Size	Transported Constituents	Authigenic Constituents	Size
	Very coarse calcirudite	Extremely coarsely crystalline	
64 mm	Coarse calcirudite		
16 mm	Medium calcirudite		
4 mm	Fine calcirudite	Very coarsely crystalline	4 mm
1 mm	Coarse calcarenite	Coarsely crystalline	1 mm
0.5 mm	Medium calcarenite		
0.25 mm	Fine calcarenite	Medium crystalline	0.25 mm
0.125 mm	Very fine calcarenite		
0.062 mm	Coarse calcilutite	Finely crystalline	0.062 mm
0.031 mm	Medium calcilutite		
0.016 mm	Fine calcilutite	Very finely crystalline	0.016 mm
0.008 mm	Very fine calcilutite		
0.004 mm		Aphanocrystalline	0.004 mm

Carbonate rocks contain both physically transported particles (oolites, intraclasts, fossils, and pellets) and chemically precipitated minerals (either as pore-filling cement, primary ooze, or as products of recrystallization and replacement). Therefore, the size scale must be a double one, so that one can distinguish which constituent is being considered (e.g., coarse calcirudites may be cemented with very finely crystalline dolomite, and fine calcarenites may be cemented with coarsely crystalline calcite). The size scale for transported constituents uses the terms of Grabau but retains the finer divisions of Wentworth except in the calcirudite range; for dolomites of obviously allochemical origin, the terms "dolorudite," "dolarenite," and "dololutite" are substituted for those shown. The most common crystal size for dolomite appears to be between 0.062 and 0.25 mm and for this reason that interval was chosen as the "medium crystalline" class.

Table A-10. Bases for Simplified Carbonate Classification (*Continued*)
(Compiled by Robert L. Folk, University of Texas)
C. Graphic Representation of Major Limestone Types I–IV
(Folk, 1959)

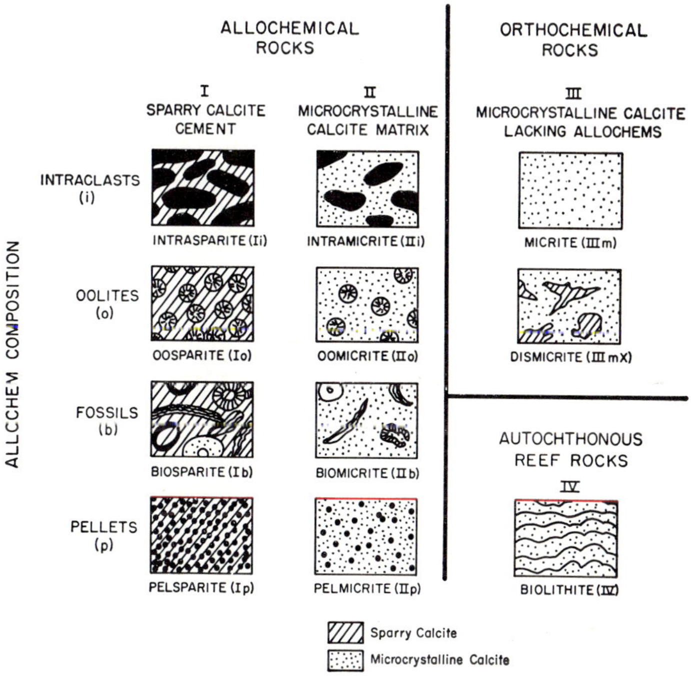

Table A-11. Suggested Classification of Detrital Particles*

(Based on work by the Committee on Sedimentation, National Research Council. This Radix-2 tabulation is an expansion of Wentworth's classification.)

Measurements			Suggested name
Millimeters	Microns	Inches	
2048–4096		80–160	Very large boulders
1024–2048		40–80	Large boulders
512–1024		20–40	Medium-sized boulders
256–512		10–20	Small boulders
128–256		5–10	Large cobbles
64–128		2.5–5	Small cobbles
32–64		1.3–2.5	Large pebbles
16–32		0.6–1.3	Medium-sized pebbles
8–16		0.3–0.6	Small pebbles
4–8		0.16–0.3	Very small pebbles
2–4		0.08–0.16	Extremely small pebbles (or perhaps granules or grit)
1–2	1000–2000		Very coarse sand
½–1	500–1000		Coarse sand
¼–½	250–500		Medium grained sand
⅛–¼	125–250		Fine sand
1/16–⅛	64–125		Very fine sand
1/32–1/16	32–64		Coarse silt
1/64–1/32	16–32		Medium grained silt
1/128–1/64	8–16		Fine silt
1/256–1/128	4–8		Very fine silt
1/512–1/256	2–4		Coarse clay
1/1024–1/512	1–2		Medium grained clay
1/2048–1/1024	0.5–1		Fine clay
1/4096–1/2048	0.25–0.5		Very fine clay

* Excerpts from letter of April 3, 1942, from Prof. E. W. Lane to P. D. Trask.

Table A-12. Tentative Grade Scale of Carbonate and Clastic Rocks
(Modified from DeFord, 1946; compiled from many sources by R. V. Hollingsworth, Paleontological Laboratory, Midland, Texas.)

Radix 10				No. units using X12 eyepiece and X1 obj. 50 unit micrometer	Old terms	Alling's classification			Wentworth's classification
						1000	Coarse	Boulderstone	Boulders
						560	Medium	Boulderstone	Boulders
						320	Fine	Boulderstone	Cobbles
						180	Very fine	Boulderstone	Cobbles
						100	Coarse	Cobblestone	Pebbles
						56	Medium	Cobblestone	Pebbles
						32	Fine	Cobblestone	Pebbles
						18	Very fine	Cobblestone	Pebbles
PHANERIC (Grains distinct to normal eye)	Megograined	Coarse	10mm	100	Very coarse crystalline	10	Coarse	Gravelstone	Granule
PHANERIC (Grains distinct to normal eye)	Megograined	Medium	5.6	56	Very coarse crystalline	5.6	Medium	Gravelstone	Granule
PHANERIC (Grains distinct to normal eye)	Megograined	Fine	3.2	32	Very coarse crystalline	3.2	Fine	Gravelstone	Granule
PHANERIC (Grains distinct to normal eye)	Megograined	Very fine	1.8	18	Very coarse crystalline	1.8	Very fine	Gravelstone	Very coarse (Sand)
PHANERIC (Grains distinct to normal eye)	Mesograined	Coarse	1.0	10	Coarse crystalline	1.0	Coarse	Sandstone	Coarse (Sand)
PHANERIC (Grains distinct to normal eye)	Mesograined	Medium	.56	5.6	Coarse crystalline	.56	Medium	Sandstone	Medium (Sand)
PHANERIC (Grains distinct to normal eye)	Mesograined	Fine	.32	3.2	Medium crystalline	.32	Fine	Sandstone	Fine (Sand)
PHANERIC (Grains distinct to normal eye)	Mesograined	Very fine	.18	1.8	Medium crystalline	.18	Very fine	Sandstone	Very fine (Sand)
APHANIC (Grains not distinct to normal eye)	Paurograined	Coarse	0.1	1.0	Fine crystalline	0.1	Coarse	Siltstone	Silt
APHANIC (Grains not distinct to normal eye)	Paurograined	Medium	.056	.56	Fine crystalline	.056	Medium	Siltstone	Silt
APHANIC (Grains not distinct to normal eye)	Paurograined	Fine	.032	.32	Very fine crystalline	.032	Fine	Siltstone	Silt
APHANIC (Grains not distinct to normal eye)	Paurograined	Very fine	.018	.18	Very fine crystalline	.018	Very fine	Siltstone	Silt
APHANIC (Grains not distinct to normal eye)	Micrograined	Coarse	.01	0.1	Microcrystalline (Lithographic)	.01	Coarse	Claystone	Silt
APHANIC (Grains not distinct to normal eye)	Micrograined	Medium	.0056	.056	Microcrystalline (Lithographic)	.0056	Medium	Claystone	Silt
APHANIC (Grains not distinct to normal eye)	Micrograined	Fine	.0032	.032	Microcrystalline (Lithographic)	.0032	Fine	Claystone	Clay
APHANIC (Grains not distinct to normal eye)	Micrograined	Very fine	.0018	.018	Microcrystalline (Lithographic)	.0018	Very fine	Claystone	Clay
APHANIC (Grains not distinct to normal eye)	Cryptograined		.001	.01	Cryptocrystalline	.001	Coarse	Colloid stone	Clay
APHANIC (Grains not distinct to normal eye)	Cryptograined				Cryptocrystalline	.00056	Medium	Colloid stone	Clay
APHANIC (Grains not distinct to normal eye)	Cryptograined				Cryptocrystalline	.00032	Fine	Colloid stone	Clay
APHANIC (Grains not distinct to normal eye)	Cryptograined				Cryptocrystalline	.00018	Very fine	Colloid stone	Clay

Note: With X12 eyepiece and X3 objective 1 unit=33.3 microns

Table A-13. Symbols for Sedimentary Rocks Based On Lithology Only and Derived from Symbols for Constituent Particles
(After J. C. Maher and L. H. Lukert, Suggestions for Preparation of Regional Stratigraphic Cross Sections, *Bull. AAPG*, vol. 39, no. 8, pp. 1664–1665, 1955.)

		MODIFYING CONSTITUENT									
	PARTICLES	CLAY	SILT	SAND	GRANULE	PEBBLE	COBBLE AND BOULDER	CALCITE AND/OR ARAGONITE	DOLOMITE	ANHYDRITE OR GYPSUM	CHERT
	BASIC SYMBOL →			·	•	∘		⊥		////	△
NONBEDDED SEDIMENTS											
AGGREGATES	CLAY	CLAY	SILTY CLAY	SANDY CLAY	GRAVELLY CLAY GRANULE CLAY	PEBBLY CLAY	COBBLY OR BOULDERY CLAY	LIMY CLAY	DOLOMITIC CLAY	GYPSIFEROUS CLAY	CHERTY CLAY
	SILT	CLAYEY SILT	SILT	SANDY SILT	GRAVELLY SILT GRANULE SILT	PEBBLY SILT	COBBLY OR BOULDERY SILT	LIMY SILT	DOLOMITIC SILT	GYPSIFEROUS SILT	CHERTY SILT
	SAND	CLAYEY SAND	SILTY SAND	SAND	GRAVELLY SAND GRANULE SAND	PEBBLY SAND	COBBLY OR BOULDERY SAND	LIMY SAND	DOLOMITIC SAND	GYPSIFEROUS SAND	CHERTY SAND
	GRAVEL	CLAYEY GRAVEL	SILTY GRAVEL	SANDY GRAVEL	GRAVEL GRANULE GRAVEL	PEBBLY GRAVEL	COBBLY OR BOULDERY GRAVEL	LIMY GRAVEL	DOLOMITIC GRAVEL	GYPSIFEROUS GRAVEL	CHERTY GRAVEL
BEDDED SEDIMENTS											
ROCKS	SHALE	SHALE		SANDY SHALE	CONGLOMERATIC SHALE GRANULE SHALE	PEBBLY SHALE	COBBLY OR BOULDERY SHALE	LIMY SHALE	DOLOMITIC SHALE	ANHYDRITIC SHALE	CHERTY SHALE
	CLAYSTONE (OR CLAY SHALE)	CLAYSTONE	SILTY CLAYSTONE	SANDY CLAYSTONE	CONGLOMERATIC CLAYSTONE GRANULE CLAYSTONE	PEBBLY CLAYSTONE	COBBLY OR BOULDERY CLAYSTONE	LIMY CLAYSTONE	DOLOMITIC CLAYSTONE	ANHYDRITIC CLAYSTONE	CHERTY CLAYSTONE

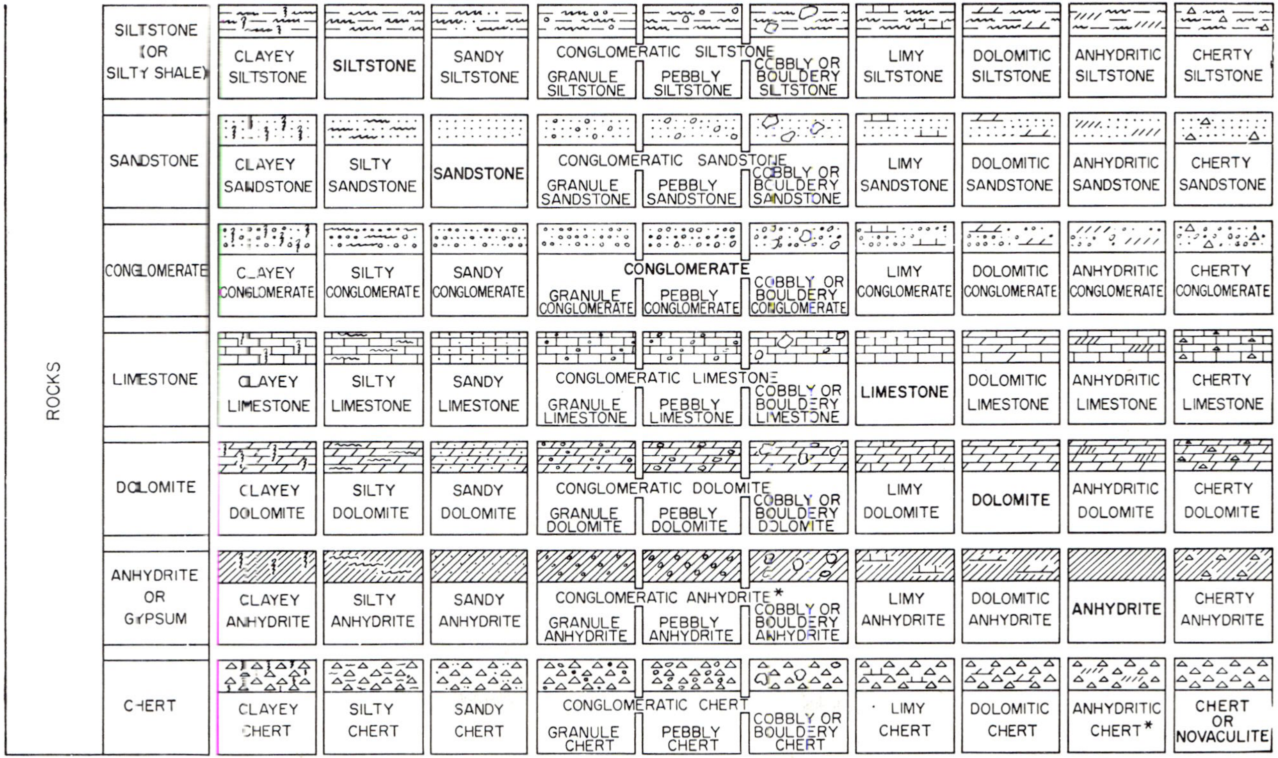

NOTE: "Limy" and "anhydritic" may be used interchangeably with "calcareous" and "gypsiferous."
* Unlikely lithologic combination.

Table A-14. Symbols for Rock Particles, Bedding, Fossils, and Miscellaneous Data
(After J. C. Maher and L. H. Lukert, Suggestions for Preparation of Regional Stratigraphic Cross Sections, *Bull. AAPG*, vol. 39, no. 8, p. 1662, 1955.)

PARTICLE SYMBOL ALPHABET

BASIC PARTICLES OF ROCKS		SPECIAL PARTICLES OF ROCKS	
1. Clay		1. Glauconite	
2. Silt		2. Pyrite (or sulfides in general)	
3. Sand		3. Hematite (or oxides in general)	
4. Gravel		4. Phosphate	
5. Pebbles		5. Feldspar	
6. Cobbles		6. Mica	
7. Boulders		7. Oolites	
8. Calcite or aragonite		8. Concretions*	
9. Dolomite		9. Fossils (macro)*	
10. Anhydrite or gypsum	////	10. Fossils (micro)*	
11. Salt	#	11. Fossils (plant)*	Z
12. Chert			
13. Coal			
14. Volcanic ash			

BEDDING SYMBOLS

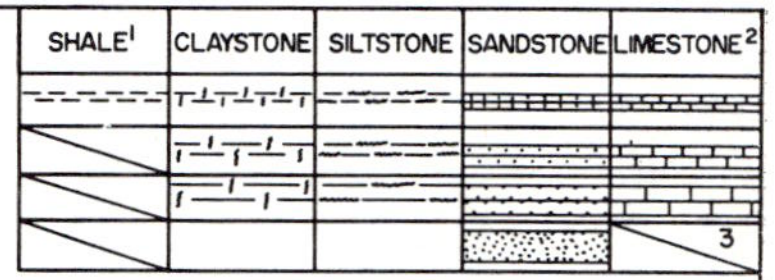

1. Fissile and platy
2. Very thin, thin-, and medium-bedded
3. Thick-bedded and massive
4. Cross-bedded

[1] Shale limited here to fissile and platy bedding
[2] Dolomite, anhydrite, gypsum, and chert can follow limestone example
[3] Cross-bedding in limestone probably will require sand symbol similar to sandstone example

Bedding terms and definitions

1. Fissile, less than 1/16 in. thick. Platy, 1/16 in. to 1/2 in. thick
2. Very thin bedded, 1/2 in. to 2 in. thick. Thin-bedded, 2 in. to 4 in. thick. Medium-bedded, 4 in. to 12 in. thick
3. Thick-bedded, 12 in. to 36 in. thick. Massive, more than 36 in. thick

PALEONTOLOGIC SYMBOLS

(All symbols to be placed outside lithologic column)			
1. Fossils (macro)		11. Corals	
2. Fossils (micro)		12. Sponges	
3. Fossils (plant)	Z	13. Spines	
4. Brachiopods		14. Foraminifera	
5. Pelecypods		15. Fusulinids	
6. Bryozoa		16. Fish	
7. Crinoids		17. Fish plates	
8. Cephalopods		18. Conodonts	
9. Gastropods		19. Ostracods	
10. Trilobites			

*Symbol to be placed outside of lithologic column

MISCELLANEOUS SYMBOLS

STRUCTURE		FLUIDS	
1. Stylolites*		1. Oil stain, or production*	
2. Slickensides*	//	2. Slight oil stain, or show*	
3. Cone-in-cone*		3. Gas*	
IGNEOUS AND METAMORPHIC ROCKS		4. Gas show*	S
1. Acidic igneous rocks		5. Fluorescence*	F
2. Basic igneous rocks	+ + +	6. Slight fluorescence*	SF
3. Metamorphic rocks		7. Water*	W
SAMPLES		8. Oil and water*	OW
1. No samples, or interval covered	NS	9. Salt water*	SW
2. Core*	CR-[	10. Fresh water*	FW

Table A-15. Geologic Column and Scale of Time

(Courtesy of American Geological Institute. The tables of this data sheet were prepared by Chester R. Longwell in cooperation with the Data Sheet Committee of the American Geological Institute. The radiogenic dates were selected and checked by A. Knopf. March, 1958.)

Era	System and Period	Series and Epoch	Stage and Age		Absolute age
			North America	Europe	
Cenozoic	Quaternary	Recent			(Duration in years)[a] Approximately the last 10,000 years
		Pleistocene	In glaciated regions (Glacial stages in *italics*)		
			Wisconsin[d] Sangamon *Illinoisan* Yarmouth *Kansan* Aftonian *Nebraskan*	*Würm*[e] Würm-Riss *Riss* Riss-Mindel *Mindel* Mindel-Günz *Günz*	10,000 ± to > 35,000 years ago
	Tertiary	Pliocene	(Atlantic and Gulf Coast)[f] Upper	(Europe) Astian	(Millions of years ago)
			Lower	Plaisancian	
		Miocene	Upper	Sahelian Pontian Sarmatian	21[b]
			Middle	Tortonian Helvetian	
			Lower	Burdigalian Aquitanian	
		Oligocene	Upper Middle Lower	Chattian Rupelian Tongrian	
		Eocene	Jackson	Ludian Bartonian	39[b]
			Claiborne	Auversian Lutetian	
			Wilcox	Cuisian Ypresian	
		Paleocene	Midway	Thanetian Montian	60[c]

Table A-15. Geologic Column and Scale of Time (*Continued*)

Era	System and Period	Series and Epoch	Stage and Age: North America	Stage and Age: Europe	Absolute age
Mesozoic	Cretaceous	Upper (Late)	No accepted classification for North America generally. These European names commonly used in North America also	Maestrichtian	70[c]
				Campanian	70[b]
				Santonian	
				Coniacian	
				Turonian	
				Cenomanian	90[b]
		Lower (Early)		Albian	
				Aptian	
				Barremian	
				Hauterivian	
				Valanginian	
				Berriasian	
	Jurassic	Upper (Late)		Purbeckian	
				Portlandian	
				Kimeridgian	
				Oxfordian	140[c]
		Middle (Middle)		Callovian	
				Bathonian	
				Bajocian	
		Lower (Early)		Toarcian	
				Pliensbachian	
				Sinemurian	
				Hettangian	
	Triassic	Upper (Late)		Rhaetian	
				Norian	
				Carnian	
		Middle (Middle)		Ladinian	
				Anisian	
		Lower (Early)		Scythian	

[a] Based on hundreds of radiocarbon dates. [b] Determined by argon method. [c] Determined by strontium method. [d] Terminology for central North America. [e] Terminology for Alps. Pleistocene marine stages also are recognized. [f] Other classifications are established for continental and Pacific Coast units. [g] Determined by lead-uranium method. [h] Represents date checked by two or more methods. [i] These provincial divisions are being extended widely in North America. [j] Divisions in Western Europe differ somewhat from those in Russia.

Table A-15. Geologic Column and Scale of Time (*Continued*)

Era	System and Period	Varied subdivisions			Absolute age (Millions of years ago)
Paleozoic	Permian	Series and Epoch	Stage and Age		
		(West Texas)[i] Ochoa Guadalupe Leonard Wolfcamp	Not established in North America	(Russia)[j] Kazanian Kungurian Artinskian Sakmarian	220[g]
	Carboniferous Systems Pennsylvanian	(Central North America) Virgil Missouri Des Moines Atoka Morrow	Not established	(Europe) Stephanian Westphalian Upper Namurian	
	Mississippian	Chester Meramec Osage Kinderhook	Not established	Lower Namurian Viséan Tournaisian	
	Devonian	Series and Epoch	Stage and Age		
		(Eastern United States)		(Europe)	
		Bradfordian	Conewango	Famennian	
		Chautauquan	Cassadaga Chemung		
		Senecan	Finger Lakes Taghanic	Frasnian	270[b]
		Erian	Tioughnioga Cazenovia Onesquethaw	Givetian Eifelian	
		Ulsterian	Deerpark Helderberg	Coblenzian Gedinnian	
	Silurian	Series and Epoch		Stage and Age	
		(North America) Cayugan Niagaran Albion	(Britain) Downtonian Ludlovian Wenlockian Valentian	Not established	
	Ordovician	Series and Epoch	Stage and Age	Series and Epoch	
		(North America)		(Britain)	
		Cincinnatian	Gamachian Richmondian Maysvillian Edenian	Ashgillian Caradocian	375[h]
		Champlainian	Mohawkian Chazyan	Llandeilian Skiddavian	
		Canadian	Not established	Tremadocian	
	Cambrian	Upper (Late) Middle (Middle) Lower (Early)	Not established	Not established	440[h] 470[c]

Table A-15. Geologic Column and Scale of Time (*Continued*)

Era	System and Period	Varied subdivisions	Absolute age (Millions of years ago)
Precambrian	(No bases for worldwide divisions)	Latest Precambrian (?) (Wichita Mountains, Oklahoma)	550[h]
		(Minneosta Scale)	
		Keweenawan Group	1100[b]
		Pre-Keweenawan orogeny	1700[b]
		Animikie Group	
		Algoman orogeny	2300[b]
		Knife Lake Group	
		"Laurentian" orogeny	2700[b]
		Keewatin Group	
		Oldest radiogenic date reported (Southern Rhodesia)	3310[b]

[a] Based on hundreds of radiocarbon dates. [b] Determined by argon method. [c] Determined by strontium method. [d] Terminology for central North America. [e] Terminology for Alps. Pleistocene marine stages also are recognized. [f] Other classifications are established for continental and Pacific Coast units. [g] Determined by lead-uranium method. [h] Represents date checked by two or more methods. [i] These provincial divisions are being extended widely in North America. [j] Divisions in Western Europe differ somewhat from those in Russia.

Chief sources of information on terminology in these tables are: Correlation charts and articles contributed by sections of Committee on Stratigraphy, National Research Council, and published in *Geol. Soc. Am. Bulletin:* Cenozoic, vol. 54, p. 1713f.; Cretaceous, vol. 53, p. 435f., vol. 63, p. 1011f., vol. 65, p. 223f., vol. 55, p. 1005f.; Jurassic, vol. 63, p. 953f.; Triassic, vol. 68, p. 1451f.; Pennsylvanian, vol. 55, p. 657f.; Mississippian, vol. 59, p. 91f.; Devonian, vol. 53, p. 1729f.; Silurian, vol. 53, p. 533f.; Ordovician, vol. 65, p. 247f. Other sources: Mesozoic and Permian, M. Gignoux's "Stratigraphic Geology," 4th ed., 1950 (English trans. by G. G. Woodford), chaps. 6–8; also "Treatise on Invertebrate Paleontology, Part L, Mollusca," R. C. Moore, ed., Geol. Soc. Am., 1957, p. 124–129; Outline of Precambrian, Minnesota Scale, from F. F. Grout et al., Precambrian Stratigraphy of Minnesota, *Geol. Soc. Am. Bull.*, vol. 62, p. 1017f., 1951.

NOTE: For later work see Kulp, 1961, in Bibliography.

APPENDIX B. MAPPING AND SURVEYING

Table B-1. Scale Conversions*

Fraction or ratio	Inches to miles	Inches to feet
1:120	1 in. = 0.0019 mi.	(1 in. = 10 ft.)
1:600	1 in. = 0.009 mi.	(1 in. = 50 ft.)
(1:1,200)	1 in. = 0.0189 mi.	(1 in. = 100 ft.)
(1:2,400)	1 in. = 0.038 mi.	(1 in. = 200 ft.)
1:6,000	1 in. = 0.094 mi.	(1 in. = 500 ft.)
(1:12,000)	1 in. = 0.189 mi.	(1 in. = 1,000 ft.)
1:12,672	(1 in. = 0.20 mi.)	1 in. = 1,056 ft.
1:15,840	(1 in. = 0.25 mi.)	1 in. = 1,320 ft.
1:21,120	(1 in. = 0.333 mi.)	1 in. = 1,760 ft.
(1:24,000)	1 in. = 0.378 mi.	1 in. = 2,000 ft.
1:31,680	(1 in. = 0.500 mi.)	1 in. = 2,640 ft.
(1:48,000)	1 in. = 0.757 mi.	1 in. = 4,000 ft.
(1:62,500)	1 in. = 0.986 mi.	1 in. = 5,208 ft.
1:63,360	(1 in. = 1.000 mi.)	1 in. = 5,280 ft.
(1:96,000)	1 in. = 1.515 mi.	1 in. = 8,000 ft.
(1:125,000)	1 in. = 1.972 mi.	1 in. = 10,416 ft.
(1:250,000)	1 in. = 3.945 mi.	1 in. = 20,833 ft.
(1:500,000)	1 in. = 7.891 mi.	1 in. = 41,666 ft.
(1:1,000,000)	1 in. = 15.782 mi.	1 in. = 83,333 ft.
(1:2,000,000)	1 in. = 31.565 mi.	1 in. = 166,666 ft.
(1:2,500,000)	1 in. = 39.456 mi.	1 in. = 208,333 ft.
(1:3,000,000)	1 in. = 47.348 mi.	1 in. = 250,000 ft.

( ) Indicates the form most commonly used.

* From J. W. Low, "Plane Table Mapping," Harper & Brothers, New York, 1952.

Table B-2. Dip, Depth, and Thickness of Inclined Strata*

(The thickness of inclined strata or the depth of a point in an inclined stratum may be determined if the dip and the breadth of outcrop on a horizontal surface are known. Divide the breadth of outcrop by 100 and multiply the result by the constant for thickness, or depth, for the given dip.)

Dip	Thickness	Depth	Dip	Thickness	Depth
1°	1.75	1.75	46°	71.93	103.55
2°	3.49	3.49	47°	73.14	107.24
3°	5.23	5.24	48°	74.31	111.06
4°	6.98	6.99	49°	75.47	115.04
5°	8.72	8.75	50°	76.60	119.18
6°	10.45	10.51	51°	77.71	123.49
7°	12.19	12.28	52°	78.80	127.99
8°	13.92	14.05	53°	79.86	132.70
9°	15.64	15.84	54°	80.90	137.64
10°	17.36	17.63	55°	81.92	142.81
11°	19.08	19.44	56°	82.90	148.26
12°	20.79	21.26	57°	83.87	153.99
13°	22.50	23.09	58°	84.80	160.03
14°	24.19	24.93	59°	85.72	166.43
15°	25.88	26.79	60°	86.60	173.21
16°	27.56	28.67	61°	87.46	180.40
17°	29.24	30.57	62°	88.29	188.07
18°	30.90	32.49	63°	89.10	196.26
19°	32.56	34.43	64°	89.88	205.03
20°	34.20	36.40	65°	90.63	214.45
21°	35.84	38.39	66°	91.35	224.60
22°	37.46	40.40	67°	92.05	235.59
23°	39.07	42.45	68°	92.72	247.51
24°	40.67	44.52	69°	93.36	260.51
25°	42.26	46.63	70°	93.97	274.75
26°	43.84	48.77	71°	94.55	290.42
27°	45.40	50.95	72°	95.11	307.77
28°	46.95	53.17	73°	95.63	327.09
29°	48.48	55.43	74°	96.13	348.74
30°	50.00	57.74	75°	96.59	373.21
31°	51.50	60.09	76°	97.03	401.08
32°	52.99	62.49	77°	97.44	433.15
33°	54.46	64.94	78°	97.81	470.46
34°	55.92	67.45	79°	98.16	514.46
35°	57.36	70.02	80°	98.48	567.13
36°	58.78	72.65	81°	98.77	631.38
37°	60.18	75.36	82°	99.03	711.54
38°	61.57	78.13	83°	99.25	814.43
39°	62.93	80.98	84°	99.45	951.44
40°	64.28	83.91	85°	99.62	1143.01
41°	65.61	86.93	86°	99.70	1430.07
42°	66.91	90.04	87°	99.86	1908.11
43°	68.20	93.25	88°	99.94	2863.63
44°	69.47	96.57	89°	99.98	5729.00
45°	70.71	100.00			

* From F. H. Lahee, "Field Geology," 6th ed., McGraw-Hill Book Company, Inc. New York, 1961.

Table B-3. Correction for Dip in Directions not Perpendicular to Strike*

Angle of full dip	Angle between strike and direction of section							
	80°	75°	70°	65°	60°	55°	50°	45°
10°	9°51′	9°40′	9°24′	9° 5′	8°41′	8°13′	7°41′	7° 6′
15°	14°47′	14°31′	14° 8′	13°39′	13°34′	12°28′	11°36′	10° 4′
20°	19°43′	19°23′	18°53′	18°15′	17°30′	16°36′	15°35′	14°25′
25°	24°48′	24°15′	23°39′	22°55′	22° 0′	20°54′	19°39′	18°15′
30°	29°37′	29° 9′	28°29′	27°37′	26°34′	25°18′	23°51′	22°12′
35°	34°36′	34° 4′	33°21′	32°24′	31°13′	29°50′	28°12′	26°20′
40°	39°34′	39° 2′	38°15′	37°15′	36° 0′	34°30′	32°44′	30°41′
45°	44°34′	44° 1′	43°13′	42°11′	40°54′	39°19′	37°27′	35°16′
50°	49°34′	49° 1′	48°14′	47°12′	45°54′	44°17′	42°23′	40° 7′
55°	54°35′	54° 4′	53°19′	52°18′	51° 3′	49°29′	47°35′	45°17′
60°	59°37′	59° 8′	58°26′	57°30′	56°19′	54°49′	53° 0′	50°46′
65°	64°40′	64°14′	63°36′	62°46′	61°42′	60°21′	58°40′	56°36′
70°	69°43′	69°21′	68°49′	68° 7′	67°12′	66° 8′	64°35′	62°46′
75°	74°47′	74°30′	74° 5′	73°32′	72°48′	71°53′	70°43′	69°14′
80°	79°51′	79°39′	79°22′	78°59′	78°29′	77°51′	77° 2′	76° 0′
85°	84°56′	84°50′	84°41′	84°29′	84°14′	83°54′	83°29′	82°57′
89°	88°59′	88°58′	88°56′	88°54′	88°51′	88°47′	88°42′	88°35′

Angle of full dip	Angle between strike and direction of section								
	40°	35°	30°	25°	20°	15°	10°	5°	1°
10°	6°28′	5°46′	5° 2′	4°15′	3°27′	2°37′	1°45′	0°53′	0°10′
15°	9°46′	8°44′	7°38′	6°28′	5°14′	3°33′	2°40′	1°20′	0°16′
20°	13°10′	11°48′	10°19′	8°45′	7° 6′	5°23′	3°37′	1°49′	0°22′
25°	16°41′	14°58′	13° 7′	11° 9′	9° 3′	6°53′	4°37′	2°20′	0°28′
30°	20°21′	18°19′	16° 6′	13°43′	11°10′	8°30′	5°44′	2°53′	0°35′
35°	24°14′	21°53′	19°18′	16°29′	13°28′	10°16′	6°56′	3°30′	0°42′
40°	28°20′	25°42′	22°45′	19°31′	16° 0′	12°15′	8°17′	4°11′	0°50′
45°	32°44′	29°50′	26°33′	22°55′	18°53′	14°30′	9°51′	4°59′	1° 0′
50°	37°27′	34°21′	30°47′	26°44′	22°11′	17° 9′	11°41′	5°56′	1°11′
55°	42°33′	39°20′	35°32′	31° 7′	26° 2′	29°17′	13°55′	7° 6′	1°26′
60°	48° 4′	44°47′	40°54′	36°14′	30°29′	24° 8′	16°44′	8°35′	1°44′
65°	54° 2′	50°53′	46°59′	42°11′	36°15′	29° 2′	20°25′	10°35′	2° 9′
70°	60°29′	57°36′	53°57′	49°16′	43°13′	35°25′	25°30′	13°28′	2°45′
75°	67°22′	64°58′	61°49′	57°37′	51°55′	44° 1′	32°57′	18° 1′	3°44′
80°	74°40′	72°75′	70°34′	67°21′	62°43′	55°44′	44°33′	26°18′	5°31′
85°	82°15′	81°20′	80° 5′	78°19′	75°39′	71°20′	63°15′	44°54′	11°17′
89°	88°27′	88°15′	88° 0′	87°38′	87° 5′	86° 9′	84°15′	78°41′	44°15′

* This table has been adapted from Appendix 1, on p. 128 in Dr. A. R. Dwerryhouse's "Geological and Topographical Maps," published by Messrs. Edward Arnold, London. Adaptation is reprinted from F. H. Lahee, "Field Geology," 6th ed., McGraw-Hill Book Company, Inc., New York, 1961.

Table B-4. Equivalent Dips Expressed in Angles and in Feet per Mile*

Degrees	Minutes	Feet per mile	Degrees	Minutes	Feet per mile	Degrees	Minutes	Feet per mile
0	01	1.54	0	42	64.51	1	46	162.8
..	02	3.07	..	43	66.04	..	48	165.9
..	03	4.61	..	44	67.57	..	50	169.0
..	04	6.14	..	45	69.11	..	52	172.0
..	05	7.68	..	46	70.64	..	54	175.1
..	06	9.22	..	47	72.18	..	56	178.2
..	07	10.75	..	48	73.72	..	58	181.3
..	08	12.29	..	49	75.26	2	00	184.4
..	09	13.82	..	50	76.80	..	10	199.8
..	10	15.36	..	51	78.33	..	10+	200.0
..	11	16.90	..	52	79.87	..	20	215.1
..	12	18.43	..	53	81.40	..	30	230.5
..	13	19.96	..	54	82.94	..	40	245.9
..	14	21.50	..	55	84.47	..	43	250.0
..	15	23.04		56	86.01	..	50	261.3
..	16	24.58	..	57	87.54	3	00	276.7
..	17	26.11	..	58	89.08	..	10	292.1
..	18	27.64	..	59	90.62	..	15	300.0
..	19	29.17	1	00	92.16	..	20	307.5
..	20	30.72	..	02	95.23	..	30	322.9
..	21	32.26	..	04	98.30	..	40	338.4
..	22	33.80	..	05	100.00	..	47+	350.0
..	23	35.33	..	06	101.4	..	50	353.8
..	24	36.86	..	08	104.5	4	00	369.2
..	25	38.40	..	10	107.5	..	10	384.6
..	26	39.94	..	12	110.6	..	20	400.1
..	27	41.47	..	14	113.6	.	30	415.5
..	28	43.01	..	16	116.7	..	40	431.0
..	29	44.54	..	18	119.8	..	50	446.5
..	30	46.08	..	20	122.9	..	52	450.0
..	31	47.62	..	22	126.0	5	00	461.9
..	32	49.16	..	24	129.1	..	10	477.4
..	33	50.69	..	26	132.1	..	20	492.9
..	34	52.23	..	28	135.2	..	25	500.0
..	35	53.76	..	30	138.3	..	30	508.4
..	36	55.30	..	32	141.3	..	40	523.9
..	37	56.83	..	34	144.4	..	50	539.4
..	38	58.37	..	36	147.4	..	57	550.0
..	39	59.90	..	38	150.5	6	00	555.0
..	40	61.44	..	40	153.6	..	..	.....
..	41	62.97	..	42	156.6	..	..	.....
..	..	.....	..	44	159.7	..	..	.....

* From F. H. Lahee, "Field Geology," 6th ed., McGraw-Hill Book Company, Inc., 1961.

Table B-5. Field Conversion Tables
(Courtesy of American Geological Institute)

A. Contour spacing from slope angles

Contour interval, ft =	1	2	5	10	20	25	50	100
Slope or dip angle, deg	Contour spacing in feet on the ground							
1	57.3	104.6	286.4	573	1046	1432	2864	5729
2	28.6	56.3	143.2	286	563	716	1432	2864
3	19.1	38.2	95.4	191	382	477	954	1908
4	14.3	28.6	71.5	143	286	357	715	1430
5	11.4	22.9	57.1	114	229	285	571	1143
6	9.5	19.0	47.5	95	190	237	475	951
7	8.1	16.3	40.7	81	163	203	407	814
8	7.1	14.2	35.5	71	142	177	355	711
9	6.3	12.6	31.5	63	126	157	315	631
10	5.6	11.3	28.3	57	113	141	283	567
11	5.1	10.3	25.7	51	103	128	257	514
12	4.7	9.4	23.5	47	94	117	235	470
13	4.3	8.7	21.6	43	87	108	216	433
14	4.0	8.0	20.0	40	80	100	200	401
15	3.7	7.5	18.6	37	75	93	186	373
16	3.5	7.0	17.4	35	70	87	174	349
17	3.3	6.5	16.3	33	66	81	163	327
18	3.1	6.2	15.4	31	62	77	154	308
19	2.9	5.8	14.5	29	58	72	145	290
20	2.7	5.5	13.7	27	55	68	137	270
21	2.6	5.2	13.0	26	52	65	130	260
22	2.5	4.9	12.3	25	49	62	123	247
23	2.3	4.7	11.7	23	47	59	117	235
24	2.2	4.5	11.2	22	45	56	112	225
25	2.1	4.3	10.7	21	43	53	107	214
26	2.0	4.1	10.2	20	41	51	102	204
27	2.0	3.9	9.8	20	39	49	98	196
28	1.9	3.8	9.4	19	37	47	94	188
29	1.8	3.6	9.0	18	36	45	90	180
30	1.7	3.5	8.6	17	35	43	86	173
31	1.7	3.3	8.3	16	33	41	83	166
32	1.6	3.2	8.0	16	32	40	80	160
33	1.5	3.1	7.7	15	31	38	77	154
34	1.5	3.0	7.4	15	30	37	74	148
35	1.4	2.9	7.1	14	29	35	71	143
36	1.4	2.8	6.9	14	28	34	69	138
37	1.3	2.7	6.6	13	27	33	66	133
38	1.3	2.6	6.4	13	26	32	64	128
39	1.2	2.5	6.1	12	25	31	61	123
40	1.2	2.4	5.9	12	24	30	59	119
41	1.1	2.3	5.7	11	23	29	57	115
42	1.1	2.2	5.5	11	22	28	55	111
43	1.1	2.1	5.3	11	21	27	53	107
44	1.0	2.0	5.1	10	20	26	51	103
45	1.0	2.0	5.0	10	20	25	50	100

EXAMPLE: To find the contour spacing for a slope or dip angle of 24° with a contour interval of 20 ft, locate line 24 in the left column and number 20 on the top line. At the intersection of these two lines, read the proper value for the contour spacing, which is 45 ft. One may interpolate for values between those shown on the data sheet. For instance, a contour spacing of 96 ft falls midway between 27 and 28° in the left column; hence the slope angle is 27½°.

Table B-5. Field Conversion Tables (*Continued*)
(Courtesy of American Geological Institute)

B. Conversion from per cent of grade to vertical angles in degrees and minutes of arc

Per cent grade	Vertical angle	Per cent grade	Vertical angle	Per cent grade	Vertical angle
1	00°35′	35	19°17′	68	34°13′
2	01 09	36	19 48	69	34 36
3	01 44	37	20 18	70	34 59
4	02 18	38	20 48	71	35 22
5	02 52	39	21 18	72	35 45
6	03 26	40	21 48	73	36 08
7	04 00	41	22 18	74	36 30
8	04 35	42	22 47	75	36 52
9	05 09	43	23 16	76	37 14
10	05 43	44	23 45	77	37 36
11	06 17	45	24 13	78	37 57
12	06 51	46	24 42	79	38 18
13	07 24	47	25 11	80	38 39
14	07 58	48	25 39	81	39 00
15	08 32	49	26 06	82	39 21
16	09 05	50	26 34	83	39 41
17	09 39	51	27 01	84	40 01
18	10 12	52	27 28	85	40 22
19	10 45	53	27 56	86	40 42
20	11 19	54	28 22	87	41 02
21	11 51	55	28 49	88	41 21
22	12 24	56	29 15	89	41 40
23	12 57	57	29 41	90	41 59
24	13 30	58	30 07	91	42 19
25	14 02	59	30 33	92	42 37
26	14 34	60	30 58	93	42 55
27	15 06	61	31 23	94	43 14
28	15 39	62	31 48	95	43 32
29	16 10	63	32 13	96	43 50
30	16 42	64	32 37	97	44 07
31	17 13	65	33 02	98	44 25
32	17 44	66	32 36	99	44 43
33	18 16	67	33 49	100	45 00
34	18 46				

C. Conversion from degrees to per cent of grade

Vertical angle, deg	Per cent grade	Vertical angle, deg	Per cent grade	Vertical angle, deg	Per cent grade
1	1.7	18	32.5	35	70.0
2	3.5	19	34.4	36	72.6
3	5.2	20	36.4	37	75.3
4	7.0	21	38.4	38	78.1
5	8.7	22	40.4	39	81.0
6	10.5	23	42.4	40	83.9
7	12.3	24	44.5	41	86.9
8	14.0	25	46.6	42	90.0
9	15.8	26	48.8	43	93.2
10	17.6	27	50.9	44	96.6
11	19.4	28	53.2	45	100.0
12	21.2	29	55.4	46	103.5
13	23.1	30	57.7	47	107.2
14	24.9	31	60.1	48	111.1
15	26.8	32	62.5	49	115.5
16	28.7	33	64.9	50	110.2
17	30.6	34	67.4		

Table B-6. Horizontal Distances and Elevations from Stadia Readings*

Minutes	0°		1°		2°		3°		4°		5°	
	Horizontal distances	Difference of elevation	Horizontal distances	Difference of elevation	Horizontal distances	Difference of elevation	Horizontal distances	Difference of elevation	Horizontal distances	Difference of elevation	Horizontal distances	Difference of elevation
0	100.00	0.00	99.97	1.74	99.88	3.49	99.73	5.23	99.51	6.96	99.24	8.68
2	100.00	0.06	99.97	1.80	99.87	3.55	99.72	5.28	99.51	7.02	99.23	8.74
4	100.00	0.12	99.97	1.86	99.87	3.60	99.71	5.34	99.50	7.07	99.22	8.80
6	100.00	0.17	99.96	1.92	99.87	3.66	99.71	5.40	99.49	7.13	99.21	8.85
8	100.00	0.23	99.96	1.98	99.86	3.72	99.70	5.46	99.48	7.19	99.20	8.91
10	100.00	0.29	99.96	2.04	99.86	3.78	99.69	5.52	99.47	7.25	99.19	8.97
12	100.00	0.35	99.96	2.09	99.85	3.84	99.69	5.57	99.46	7.30	99.18	9.03
14	100.00	0.41	99.95	2.15	99.85	3.90	99.68	5.63	99.46	7.36	99.17	9.08
16	100.00	0.47	99.95	2.21	99.84	3.95	99.68	5.69	99.45	7.42	99.16	9.14
18	100.00	0.52	99.95	2.27	99.84	4.01	99.67	5.75	99.44	7.48	99.15	9.20
20	100.00	0.58	99.95	2.33	99.83	4.07	99.66	5.80	99.43	7.53	99.14	9.25
22	100.00	0.64	99.94	2.38	99.83	4.13	99.66	5.86	99.42	7.59	99.13	9.31
24	100.00	0.70	99.94	2.44	99.82	4.18	99.65	5.92	99.41	7.65	99.11	9.37
26	99.99	0.76	99.94	2.50	99.82	4.24	99.64	5.98	99.40	7.71	99.10	9.43
28	99.99	0.81	99.93	2.56	99.81	4.30	99.63	6.04	99.39	7.76	99.09	9.48
30	99.99	0.87	99.93	2.62	99.81	4.36	99.63	6.09	99.38	7.82	99.08	9.54
32	99.99	0.93	99.93	2.67	99.80	4.42	99.62	6.15	99.38	7.88	99.07	9.60
34	99.99	0.99	99.93	2.73	99.80	4.48	99.62	6.21	99.37	7.94	99.06	9.65
36	99.99	1.05	99.92	2.79	99.79	4.53	99.61	6.27	99.36	7.99	99.05	9.71
38	99.99	1.11	99.92	2.85	99.79	4.59	99.60	6.33	99.35	8.05	99.04	9.77
40	99.99	1.16	99.92	2.91	99.78	4.65	99.59	6.38	99.34	8.11	99.03	9.83
42	99.99	1.22	99.91	2.97	99.78	4.71	99.59	6.44	99.33	8.17	99.01	9.88
44	99.98	1.28	99.91	3.02	99.77	4.76	99.58	6.50	99.32	8.22	99.00	9.94
46	99.98	1.34	99.90	3.08	99.77	4.82	99.57	6.56	99.31	8.28	98.99	10.00
48	99.98	1.40	99.90	3.14	99.76	4.88	99.56	6.61	99.30	8.34	98.98	10.05
50	99.98	1.45	99.90	3.20	99.76	4.94	99.56	6.67	99.29	8.40	98.97	10.11
52	99.98	1.51	99.89	3.26	99.75	4.99	99.55	6.73	99.28	8.45	98.96	10.17
54	99.98	1.57	99.89	3.31	99.74	5.05	99.54	6.78	99.27	8.51	98.94	10.22
56	99.97	1.63	99.89	3.37	99.74	5.11	99.53	6.84	99.26	8.57	98.93	10.28
58	99.97	1.69	99.88	3.43	99.73	5.17	99.52	6.90	99.25	8.63	98.92	10.34
60	99.97	1.74	99.88	3.49	99.73	5.23	99.51	6.96	99.24	8.68	98.91	10.40
c = 0.75	0.75	0.01	0.75	0.02	0.75	0.03	0.75	0.05	0.75	0.06	0.75	0.07
c = 1.00	1.00	0.01	1.00	0.03	1.00	0.04	1.00	0.06	1.00	0.08	0.99	0.09
c = 1.25	1.25	0.02	1.25	0.03	1.25	0.05	1.25	0.08	1.25	0.10	1.24	0.11

* From W. Gurley and L. E. Gurley, "Manual of Surveying Instruments," W. and L. E. Gurley, New York, 1941.

Table B-6. Horizontal Distances and Elevations from Stadia Readings (*Continued*)

Minutes	6° Horizontal distances	6° Difference of elevation	7° Horizontal distances	7° Difference of elevation	8° Horizontal distances	8° Difference of elevation	9° Horizontal distances	9° Difference of elevation	10° Horizontal distances	10° Difference of elevation	11° Horizontal distances	11° Difference of elevation
0	98.91	10.40	98.51	12.10	98.06	13.78	97.55	15.45	96.98	17.10	96.36	18.73
2	98.90	10.45	98.50	12.15	98.05	13.84	97.53	15.51	96.96	17.16	96.34	18.78
4	98.88	10.51	98.48	12.21	98.03	13.89	97.52	15.56	96.94	17.21	96.32	18.84
6	98.87	10.57	98.47	12.26	98.01	13.95	97.50	15.62	96.92	17.26	96.29	18.89
8	98.86	10.62	98.46	12.32	98.00	14.01	97.48	15.67	96.90	17.32	96.27	18.95
10	98.85	10.68	98.44	12.38	97.98	14.06	97.46	15.73	96.88	17.37	96.25	19.00
12	98.83	10.74	98.43	12.43	97.97	14.12	97.44	15.78	96.86	17.43	96.23	19.05
14	98.82	10.79	98.41	12.49	97.95	14.17	97.43	15.84	96.84	17.48	96.21	19.11
16	98.81	10.85	98.40	12.55	97.93	14.23	97.41	15.89	96.82	17.54	96.18	19.16
18	98.80	10.91	98.39	12.60	97.92	14.28	97.39	15.95	96.80	17.59	96.16	19.21
20	98.78	10.96	98.37	12.66	97.90	14.34	97.37	16.00	96.78	17.65	96.14	19.27
22	98.77	11.02	98.36	12.72	97.88	14.40	97.35	16.06	96.76	17.70	96.12	19.32
24	98.76	11.08	98.34	12.77	97.87	14.45	97.33	16.11	96.74	17.76	96.09	19.38
26	98.74	11.13	98.33	12.83	97.85	14.51	97.31	16.17	96.72	17.81	96.07	19.43
28	98.73	11.19	98.31	12.88	97.83	14.56	97.29	16.22	96.70	17.86	96.05	19.48
30	98.72	11.25	98.29	12.94	97.82	14.62	97.28	16.28	96.68	17.92	96.03	19.54
32	98.71	11.30	98.28	13.00	97.80	14.67	97.26	16.33	96.66	17.97	96.00	19.59
34	98.69	11.36	98.27	13.05	97.78	14.73	97.24	16.39	96.64	18.03	95.98	19.64
36	98.68	11.42	98.25	13.11	97.76	14.79	97.22	16.44	96.62	18.08	95.96	19.70
38	98.67	11.47	98.24	13.17	97.75	14.84	97.20	16.50	96.60	18.14	95.93	19.75
40	98.65	11.53	98.22	13.22	97.73	14.90	97.18	16.55	96.57	18.19	95.91	19.80
42	98.64	11.59	98.20	13.28	97.71	14.95	97.16	16.61	96.55	18.24	95.89	19.86
44	98.63	11.64	98.19	13.33	97.69	15.01	97.14	16.66	96.53	18.30	95.86	19.91
46	98.61	11.70	98.17	13.39	97.68	15.06	97.12	16.72	96.51	18.35	95.84	19.96
48	98.60	11.76	98.16	13.45	97.66	15.12	97.10	16.77	96.49	18.41	95.82	20.02
50	98.58	11.81	98.14	13.50	97.64	15.17	97.08	16.83	96.47	18.46	95.79	20.07
52	98.57	11.87	98.13	13.56	97.62	15.23	97.06	16.88	96.45	18.51	95.77	20.12
54	98.56	11.93	98.11	13.61	97.61	15.28	97.04	16.94	96.42	18.57	95.75	20.18
56	98.54	11.98	98.10	13.67	97.59	15.34	97.02	16.99	96.40	18.62	95.72	20.23
58	98.53	12.04	98.08	13.73	97.57	15.40	97.00	17.05	96.38	18.68	95.70	20.28
60	98.51	12.10	98.06	13.78	97.55	15.45	96.98	17.10	96.36	18.73	95.68	20.34
$c = 0.75$	0.75	0.08	0.74	0.10	0.74	0.11	0.74	0.12	0.74	0.14	0.73	0.15
$c = 1.00$	0.99	0.11	0.99	0.13	0.99	0.15	0.99	0.16	0.98	0.18	0.98	0.20
$c = 1.25$	1.24	0.14	1.24	0.16	1.23	0.18	1.23	0.21	1.23	0.23	1.22	0.25

Table B-6. Horizontal Distances and Elevations from Stadia Readings (*Continued*)

Minutes	12°		13°		14°		15°		16°		17°	
	Horizontal distances	Difference of elevation	Horizontal distances	Difference of elevation	Horizontal distances	Difference of elevation	Horizontal distances	Difference of elevation	Horizontal distances	Difference of elevation	Horizontal distances	Difference of elevation
0	95.68	20.34	94.94	21.92	94.15	23.47	93.30	25.00	92.40	26.50	91.45	27.96
2	95.65	20.39	94.91	21.97	94.12	23.52	93.27	25.05	92.37	26.55	91.42	28.01
4	95.63	20.44	94.89	22.02	94.09	23.58	93.24	25.10	92.34	26.59	91.39	28.06
6	95.61	20.50	94.86	22.08	94.07	23.63	93.21	25.15	92.31	26.64	91.35	28.10
8	95.58	20.55	94.84	22.13	94.04	23.68	93.18	25.20	92.28	26.69	91.32	28.15
10	95.56	20.60	94.81	22.18	94.01	23.73	93.16	25.25	92.25	26.74	91.29	28.20
12	95.53	20.66	94.79	22.23	93.98	23.78	93.13	25.30	92.22	26.79	91.26	28.25
14	95.51	20.71	94.76	22.28	93.95	23.83	93.10	25.35	92.19	26.84	91.22	28.30
16	95.49	20.76	94.73	22.34	93.93	23.88	93.07	25.40	92.15	26.89	91.19	28.34
18	95.46	20.81	94.71	22.39	93.90	23.93	93.04	25.45	92.12	26.94	91.16	28.39
20	95.44	20.87	94.68	22.44	93.87	23.99	93.01	25.50	92.09	26.99	91.12	28.44
22	95.41	20.92	94.66	22.49	93.84	24.04	92.98	25.55	92.06	27.04	91.09	28.49
24	95.39	20.97	94.63	22.54	93.81	24.09	92.95	25.60	92.03	27.09	91.06	28.54
26	95.36	21.03	94.60	22.60	93.79	24.14	92.92	25.65	92.00	27.13	91.02	28.58
28	95.34	21.08	94.58	22.65	93.76	24.19	92.89	25.70	91.97	27.18	90.99	28.63
30	95.32	21.13	94.55	22.70	93.73	24.24	92.86	25.75	91.93	27.23	90.96	28.68
32	95.29	21.18	94.52	22.75	93.70	24.29	92.83	25.80	91.90	27.28	90.92	28.73
34	95.27	21.24	94.50	22.80	93.67	24.34	92.80	25.85	91.87	27.33	90.89	28.77
36	95.24	21.29	94.47	22.85	93.65	24.39	92.77	25.90	91.84	27.38	90.86	28.82
38	95.22	21.34	94.44	22.91	93.62	24.44	92.74	25.95	91.81	27.43	90.82	28.87
40	95.19	21.39	94.42	22.96	93.59	24.49	92.71	26.00	91.77	27.48	90.79	28.92
42	95.17	21.45	94.39	23.01	93.56	24.55	92.68	26.05	91.74	27.52	90.76	28.96
44	95.14	21.50	94.36	23.06	93.53	24.60	92.65	26.10	91.71	27.57	90.72	29.01
46	95.12	21.55	94.34	23.11	93.50	24.65	92.62	26.15	91.68	27.62	90.69	29.06
48	95.09	21.60	94.31	23.16	93.47	24.70	92.59	26.20	91.65	27.67	90.66	29.11
50	95.07	21.66	94.28	23.22	93.45	24.75	92.56	26.25	91.61	27.72	90.62	29.15
52	95.04	21.71	94.26	23.27	93.42	24.80	92.53	26.30	91.58	27.77	90.59	29.20
54	95.02	21.76	94.23	23.32	93.39	24.85	92.49	26.35	91.55	27.81	90.55	29.25
56	94.99	21.81	94.20	23.37	93.36	24.90	92.46	26.40	91.52	27.86	90.52	29.30
58	94.97	21.87	94.17	23.42	93.33	24.95	92.43	26.45	91.48	27.91	90.48	29.34
60	94.94	21.92	94.15	23.47	93.30	25.00	92.40	26.50	91.45	27.96	90.45	29.39
$c = 0.75$	0.73	0.16	0.73	0.17	0.73	0.19	0.72	0.20	0.72	0.21	0.72	0.23
$c = 1.00$	0.98	0.22	0.97	0.23	0.97	0.25	0.96	0.27	0.86	0.28	0.95	0.30
$c = 1.25$	1.22	0.27	1.21	0.29	1.21	0.31	1.20	0.34	1.20	0.35	1.19	0.38

Table B-6. Horizontal Distances and Elevations from Stadia Readings (*Continued*)

Minutes	18°		19°		20°		21°		22°		23°	
	Horizontal distances	Difference of elevation	Horizontal distances	Difference of elevation	Horizontal distances	Difference of elevation	Horizontal distances	Difference of elevation	Horizontal distances	Difference of elevation	Horizontal distances	Difference of elevation
0	90.45	29.39	89.40	30.78	88.30	32.14	87.16	33.46	85.97	34.73	84.73	35.97
2	90.42	29.44	89.36	30.83	88.26	32.18	87.12	33.50	85.93	34.77	84.69	36.01
4	90.38	29.48	89.33	30.87	88.23	32.23	87.08	33.54	85.89	34.82	84.65	36.05
6	90.35	29.53	89.29	30.92	88.19	32.27	87.04	33.59	85.85	34.86	84.61	36.09
8	90.31	29.58	89.26	30.97	88.15	32.32	87.00	33.63	85.80	34.90	84.57	36.13
10	90.28	29.62	89.22	31.01	88.11	32.36	86.96	33.67	85.76	34.94	84.52	36.17
12	90.24	29.67	89.18	31.06	88.08	32.41	86.92	33.72	85.72	34.98	84.48	36.21
14	90.21	29.72	89.15	31.10	88.04	32.45	86.88	33.76	85.68	35.02	84.44	36.25
16	90.18	29.76	89.11	31.15	88.00	32.49	86.84	33.80	85.64	35.07	84.40	36.29
18	90.14	29.81	89.08	31.19	87.96	32.54	86.80	33.84	85.60	35.11	84.35	36.33
20	90.11	29.86	89.04	31.24	87.93	32.58	86.77	33.89	85.56	35.15	84.31	36.37
22	90.07	29.90	89.00	31.28	87.89	32.63	86.73	33.93	85.52	35.19	84.27	36.41
24	90.04	29.95	88.96	31.33	87.85	32.67	86.69	33.97	85.48	35.23	84.23	36.45
26	90.00	30.00	88.93	31.38	87.81	32.72	86.65	34.01	85.44	35.27	84.18	36.49
28	89.97	30.04	88.89	31.42	87.77	32.76	86.61	34.06	85.40	35.31	84.14	36.53
30	89.93	30.09	88.86	31.47	87.74	32.80	86.57	34.10	85.36	35.36	84.10	36.57
32	89.90	30.14	88.82	31.51	87.70	32.85	86.53	34.14	85.31	35.40	84.06	36.61
34	89.86	30.19	88.78	31.56	87.66	32.89	86.49	34.18	85.27	35.44	84.01	36.65
36	89.83	30.23	88.75	31.60	87.62	32.93	86.45	34.23	85.23	35.48	83.97	36.69
38	89.79	30.28	88.71	31.65	87.58	32.98	86.41	34.27	85.19	35.52	83.93	36.73
40	89.76	30.32	88.67	31.69	87.54	33.02	86.37	34.31	85.15	35.56	83.89	36.77
42	89.72	30.37	88.64	31.74	87.51	33.07	86.33	34.35	85.11	35.60	83.84	36.80
44	89.69	30.41	88.60	31.78	87.47	33.11	86.29	34.40	85.07	35.64	83.80	36.84
46	89.65	30.46	88.56	31.83	87.43	33.15	86.25	34.44	85.02	35.68	83.76	36.88
48	89.61	30.51	88.53	31.87	87.39	33.20	86.21	34.48	84.98	35.72	83.72	36.92
50	89.58	30.55	88.49	31.92	87.35	33.24	86.17	34.52	84.94	35.76	83.67	36.96
52	89.54	30.60	88.45	31.96	87.31	33.28	86.13	34.57	84.90	35.80	83.63	37.00
54	89.51	30.65	88.41	32.01	87.27	33.33	86.09	34.61	84.86	35.85	83.59	37.04
56	89.47	30.69	88.38	32.05	87.24	33.37	86.05	34.65	84.82	35.89	83.54	37.08
58	89.44	30.74	88.34	32.09	87.20	33.41	86.01	34.69	84.77	35.93	83.50	37.12
60	89.40	30.78	88.30	32.14	87.16	33.46	85.97	34.73	84.73	35.97	83.46	37.16
c = 0.75	0.71	0.24	0.71	0.25	0.70	0.26	0.70	0.27	0.69	0.29	0.69	0.30
c = 1.00	0.95	0.32	0.94	0.33	0.94	0.35	0.93	0.37	0.92	0.38	0.92	0.40
c = 1.25	1.19	0.40	1.18	0.42	1.17	0.44	1.16	0.46	1.15	0.48	1.15	0.50

Table B-6. Horizontal Distances and Elevations from Stadia Readings (*Continued*)

Minutes	24°		25°		26°		27°		28°		29°		30°	
	Horizontal distances	Difference of elevation	Horizontal distances	Difference of elevation	Horizontal distances	Difference of elevation	Horizontal distances	Difference of elevation	Horizontal distances	Difference of elevation	Horizontal distances	Difference of elevation	Horizontal distances	Difference of elevation
0	83.46	37.16	82.14	38.30	80.78	39.40	79.39	40.45	77.96	41.45	76.50	42.40	75.00	43.30
2	83.41	37.20	82.09	38.34	80.74	39.44	79.34	40.49	77.91	41.48	76.45	42.43	74.95	43.33
4	83.37	37.23	82.05	38.38	80.69	39.47	79.30	40.52	77.86	41.52	76.40	42.46	74.90	43.36
6	83.33	37.27	82.01	38.41	80.65	39.51	79.25	40.55	77.81	41.55	76.35	42.49	74.85	43.39
8	83.28	37.31	81.96	38.45	80.60	39.54	79.20	40.59	77.77	41.58	76.30	42.53	74.80	43.42
10	83.24	37.35	81.92	38.49	80.55	39.58	79.15	40.62	77.72	41.61	76.25	42.56	74.75	43.45
12	83.20	37.39	81.87	38.53	80.51	39.61	79.11	40.66	77.67	41.65	76.20	42.59	74.70	43.47
14	83.15	37.43	81.83	38.56	80.46	39.65	79.06	40.69	77.62	41.68	76.15	42.62	74.65	43.50
16	83.11	37.47	81.78	38.60	80.41	39.69	79.01	40.72	77.57	41.71	76.10	42.65	74.60	43.53
18	83.07	37.51	81.74	38.64	80.37	39.72	78.96	40.76	77.52	41.74	76.05	42.68	74.55	43.56
20	83.02	37.54	81.69	38.67	80.32	39.76	78.92	40.79	77.48	41.77	76.00	42.71	74.49	43.59
22	82.98	37.58	81.65	38.71	80.28	39.79	78.87	40.82	77.42	41.81	75.95	42.74	74.44	43.62
24	82.93	37.62	81.60	38.75	80.23	39.83	78.82	40.86	77.38	41.84	75.90	42.77	74.39	43.65
26	82.89	37.66	81.56	38.78	80.18	39.86	78.77	40.89	77.33	41.87	75.85	42.80	74.34	43.67
28	82.85	37.70	81.51	38.82	80.14	39.90	78.73	40.92	77.28	41.90	75.80	42.83	74.29	43.70
30	82.80	37.74	81.47	38.86	80.09	39.93	78.68	40.96	77.23	41.93	75.75	42.86	74.24	43.73
32	82.76	37.77	81.42	38.89	80.04	39.97	78.63	40.99	77.18	41.97	75.70	42.89	74.19	43.76
34	82.72	37.81	81.38	38.93	80.00	40.00	78.58	41.02	77.13	42.00	75.65	42.92	74.14	43.79
36	82.67	37.85	81.33	38.97	79.95	40.04	78.54	41.06	77.09	42.03	75.60	42.95	74.09	43.82
38	82.63	37.89	81.28	39.00	79.90	40.07	78.49	41.09	77.04	42.06	75.55	42.98	74.04	43.84
40	82.58	37.93	81.24	39.04	79.86	40.11	78.44	41.12	76.99	42.09	75.50	43.01	73.99	43.87
42	82.54	37.96	81.19	39.08	79.81	40.14	78.39	41.16	76.94	42.12	75.45	43.04	73.93	43.90
44	82.49	38.00	81.15	39.11	79.76	40.18	78.34	41.19	76.89	42.15	75.40	43.07	73.88	43.93
46	82.45	38.04	81.10	39.15	79.72	40.21	78.30	41.22	76.84	42.19	75.35	43.10	73.83	43.95
48	82.41	38.08	81.06	39.18	79.67	40.24	78.25	41.26	76.79	42.22	75.30	43.13	73.78	43.98
50	82.36	38.11	81.01	39.22	79.62	40.28	78.20	41.29	76.74	42.25	75.25	43.16	73.73	44.01
52	82.32	38.15	80.97	39.26	79.58	40.31	78.15	41.32	76.69	42.28	75.20	43.18	73.68	44.04
54	82.27	38.19	80.92	39.29	79.53	40.35	78.10	41.35	76.64	42.31	75.15	43.21	73.63	44.07
56	82.23	38.23	80.87	39.33	79.48	40.38	78.06	41.39	76.59	42.34	75.10	43.24	73.58	44.09
58	82.18	38.26	80.83	39.36	79.44	40.42	78.01	41.42	76.55	42.37	75.05	43.27	73.52	44.12
60	82.14	38.30	80.78	39.40	79.39	40.45	77.96	41.45	76.50	42.40	75.00	43.30	73.47	44.15
$c = 0.75$	0.68	0.31	0.68	0.32	0.67	0.33	0.66	0.35	0.66	0.36	0.65	0.37	0.65	0.38
$c = 1.00$	0.91	0.41	0.90	0.43	0.89	0.45	0.89	0.46	0.88	0.48	0.87	0.49	0.86	0.51
$c = 1.25$	1.14	0.52	1.13	0.54	1.12	0.56	1.11	0.58	1.10	0.60	1.09	0.62	1.08	0.64

Table B-7. Geologic Map Symbols

(American Geological Institute data sheet listing map symbols commonly used in the preparation of geologic maps, compiled principally from published lists of the U.S. Geological Survey, Map Symbol Committee, and recently published geologic maps.)

Part 1

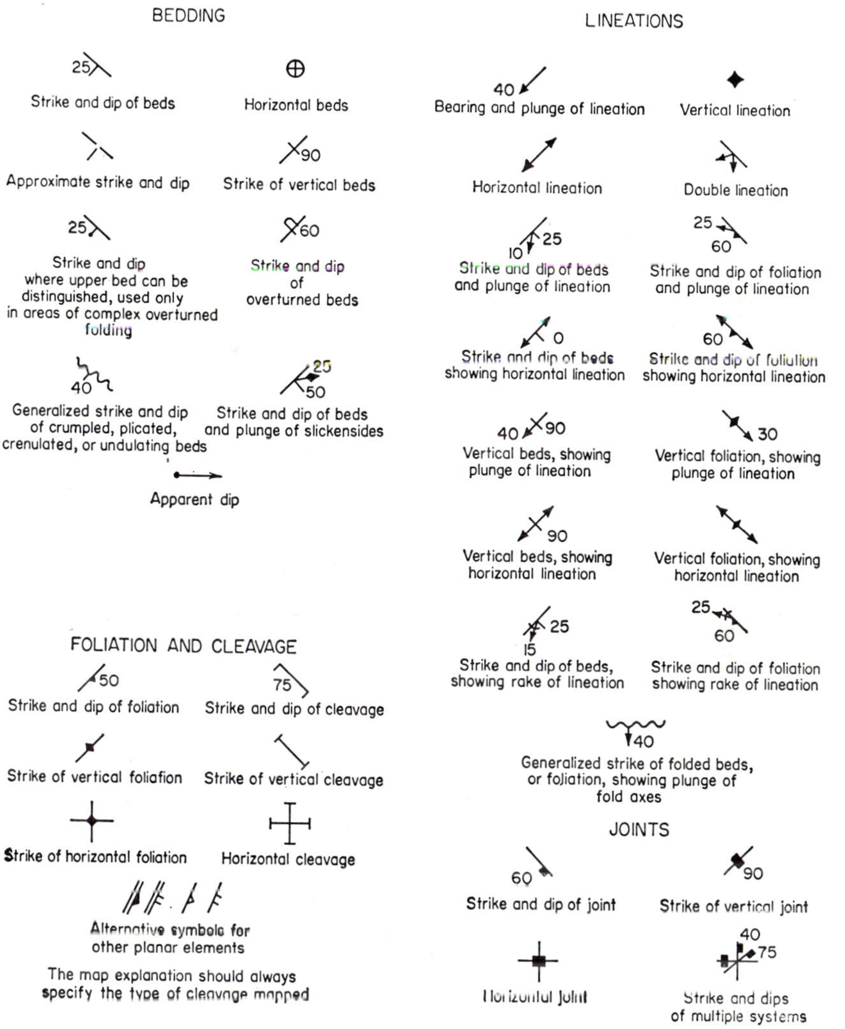

Table B-7. Geologic Map Symbols (*Continued*)

(American Geological Institute data sheet listing map symbols commonly used in the preparation of geologic maps, compiled principally from published lists of the U.S. Geological Survey, Map Symbol Committee, and recently published geologic maps.)

Part 2

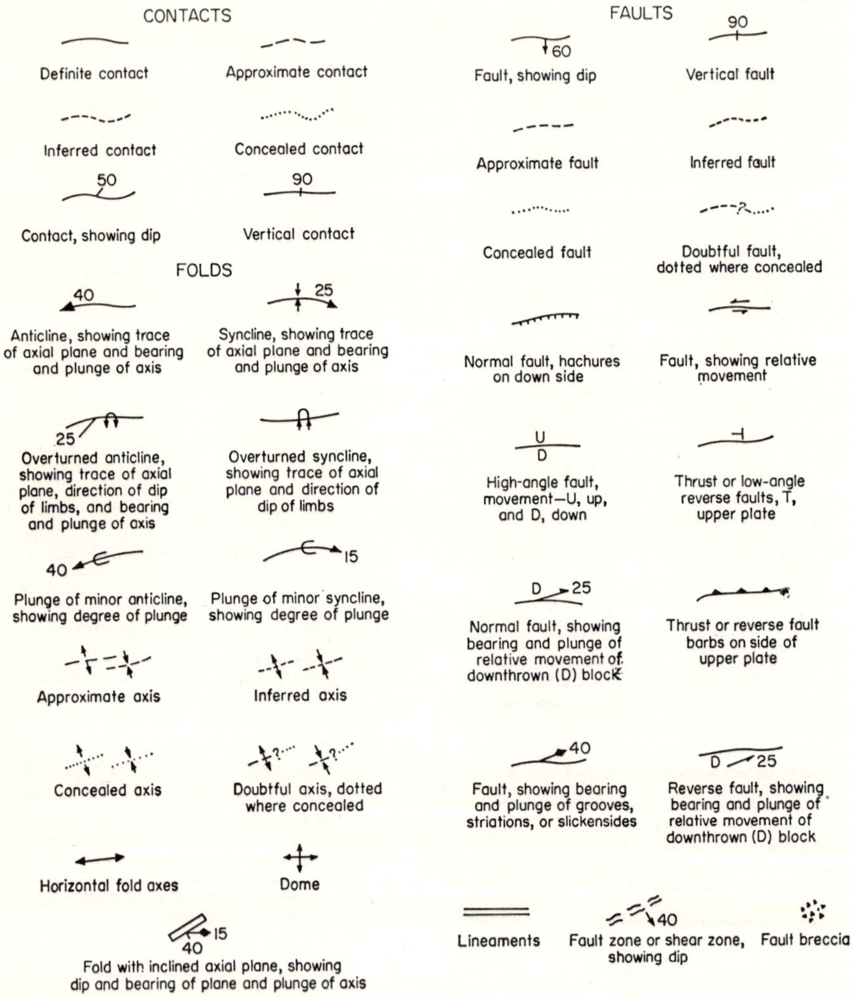

Table B-7. Geologic Map Symbols (*Continued*)

(American Geological Institute data sheet listing map symbols commonly used in the preparation of geologic maps, compiled principally from published lists of the U.S. Geological Survey, Map Symbol Committee, and recently published geologic maps.)

Part 3

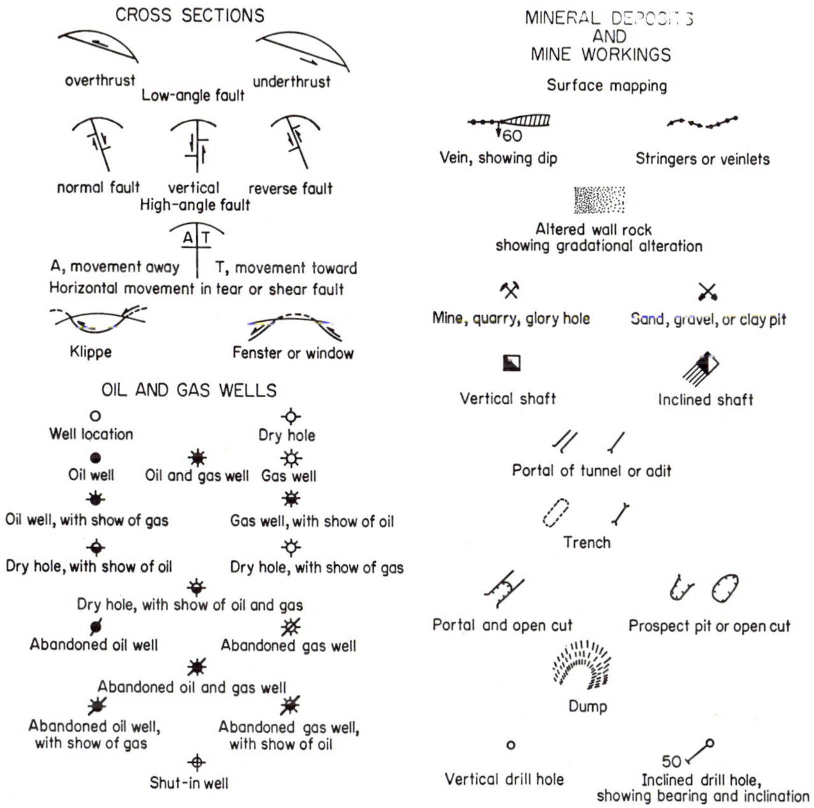

Table B-8. Fault Symbols

(American Geological Institute data sheet by Mason L. Hill)

Introduction. The following fault symbols are designed to remove the ambiguity resulting from failure of traditional symbols to distinguish between fault *slip* and fault *separation.* Where a linear geologic element is displaced the *actual* relative movement (slip) can be determined (e.g., displaced intersection of dike and bed). Generally, however, where a tabular geologic element is displaced only *apparent* relative movement (separation) can be determined. Thus, for example, these symbols provide for the important distinction between normal fault (only separation known) and normal slip fault (slip known). Refer to "Dual Classification of Faults," Mason L. Hill, *AAPG Bull.*, vol. 43, pp. 217–221, 1959.

GENERAL SYMBOLS

————	Fault trace, for maps and sections.
— — — —	Approximately located trace, for maps and sections.
—?— — —	Conjectural trace, for maps and sections.
....	Concealed map trace; conjectural (. . ? . .).
	Dip direction; amount (60), approximate amount (25±), conjectural direction (?).
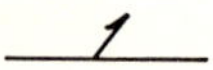	Relative slip direction; conjectural (?); slip plunge ( ), approximate plunge (25±).

NOTE: Fault trace may be distinguished from other geologic contacts by weight or color of line, or by labeling with name or symbol, as desired. Slip plunge is vertical angle measured downward from horizontal to net slip.

SLIP SYMBOLS FOR MAPS

(Add direction and amount of dip, direction of relative slip, and slip plunge, if and where known.)

	Thrust slip fault. Triangles on relatively overthrust block; fault dips <45°.
	Reverse slip fault. Rectangles on relatively elevated hanging wall block; fault dips >45°. Dip direction is shown here.
	Normal slip fault. Barbs on relatively depressed hanging wall block. Fault dip and direction of relative slip are shown here.
	Right-lateral slip fault. Arrows shown sidewise relative movement of block opposite the observer.
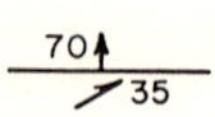	Left-lateral slip fault. Fault dip and slip plunge are shown here. If dip-slip and strike-slip components were nearly equal, the name reverse left-lateral slip fault would be appropriate.

NOTE: Triangles, rectangles, and barbs may be shown as appropriate and convenient along the map trace of the fault. However, none of these symbols should be used on maps unless some evidence of at least the approximate orientation of slip is obtained.

Table B-8. Fault Symbols (*Continued*)

SLIP SYMBOLS FOR SECTIONS

Thrust slip fault. Arrow shows principal relative movement component. Fault dips <45°

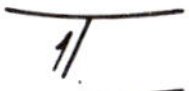
Reverse slip fault. Fault dips >45°.

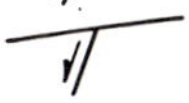
Normal slip fault.

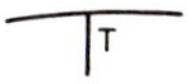

Right-lateral slip fault. Principal relative movement component of block toward observer is shown by the letter T.

Left-lateral slip fault. Letter A (away) and arrow (downward) show relative movement components. If these components are nearly equal, the name normal left-lateral slip fault is used.

NOTE: Single barb arrows and letters (T and A) may be shown on either side of the section trace of the fault, as appropriate and convenient. However, none of these symbols should be used on sections if only separation is determined.

SEPARATION SYMBOLS FOR MAPS

(Add direction and amount of dip, if and where known.)

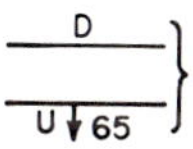

Dip separation-apparent relative movement in fault dip. D, downthrown; U, upthrown. Normal fault has dip toward downthrown block; reverse fault has >45° dip toward upthrown block (illustrated); thrust fault has <45° dip toward overthrown block.

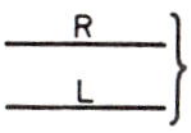

Strike separation-apparent relative movement in fault strike of block opposite the observer. R, right-lateral fault; L, left-lateral fault.

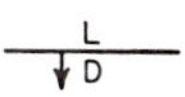

Dip and strike separations nearly equal. A normal left-lateral fault is illustrated.

NOTE: Letters indicating separation may be shown as appropriate and convenient on either side of the fault trace. The symbols (+) and (−) may be substituted for U and D but none represents any component of slip. Separation symbols are not needed for sections, and are only occasionally necessary for maps because the displacement of tabular geologic elements is usually obvious.

Remarks. The essential function of these proposed fault symbols is to let geologists clearly indicate where information on fault slip has been determined, and not allow them to indicate slip where only separation is known. As customary, only those symbols which are used on a particular geologic illustration need be shown in the legend.

APPENDIX C. MATHEMATICAL TABLES

Table C-1. Miscellaneous Data

A. Greek Alphabet*

Alpha	= A, α	= A, a	Nu	= N, ν	= N, n
Beta	= B, β	= B, b	Xi	= Ξ, ξ	= X, x
Gamma	= Γ, γ	= G, g	Omicron	= O, o	= O, o
Delta	= Δ, δ	= D, d	Pi	= Π, π	= P, p
Epsilon	= E, ϵ	= E, e	Rho	= P, ρ	= R, r
Zeta	= Z, ζ	= Z, z	Sigma	= Σ, σ	= S, s
Eta	= H, η	= E, e	Tau	= T, τ	= T, t
Theta	= Θ, θ	= Th, th	Upsilon	= Υ, υ	= U, u
Iota	= I, ι	= I, i	Phi	= Φ, ϕ	= Ph, ph
Kappa	= K, κ	= K, k	Chi	= X, χ	= Ch, ch
Lambda	= Λ, λ	= L, l	Psi	= Ψ, ψ	= Ps, ps
Mu	= M, μ	= M, m	Omega	= Ω, ω	= O, o

B. Numerical Constants*

Constant	Value	Constant	Value	Constant	Value
π	3.141593	$1/\pi^3$	0.032252	$1/e^2$	1.135335
$1/\pi$	0.318310	$\sqrt[3]{\pi}$	1.464592	$\sqrt{e}$	1.648721
π^2	9.869604	$1/\sqrt[3]{\pi}$	0.682784	1 radian	57.295780 deg.
$1/\pi^2$	0.101321	$\log_e \pi$	1.144730	1 radian	3,437.7468 min.
$\sqrt{\pi}$	1.772454	$\log_{10} \pi$	0.497150	1 radian	206,264.81 sec.
$1/\sqrt{\pi}$	0.564190	e	2.718282	1 deg.	0.017453 radian
π^3	31.00628	$1/e$	0.367879	1 min.	0.0002909 radian
		e^2	7.389056	1 sec.	0.00000485 radian

* From John H. Perry (ed.), "Chemical Engineers' Handbook," 3d ed., McGraw-Hill Book Company, Inc., New York, 1950.

Table C-1. Miscellaneous Data *(Continued)*
C. Solution of Triangles*

Right Triangle (Fig. 635):

$$\sin A = \frac{a}{b}. \quad \cos A = \frac{c}{b}. \quad \tan A = \frac{a}{c}.$$

$$\sin C = \frac{c}{b}. \quad \cos C = \frac{a}{b}. \quad \tan C = \frac{c}{a}.$$

$$C = 90° - A. \quad b = \sqrt{a^2 + c^2}. \quad c = \sqrt{(b + a)(b - a)}.$$

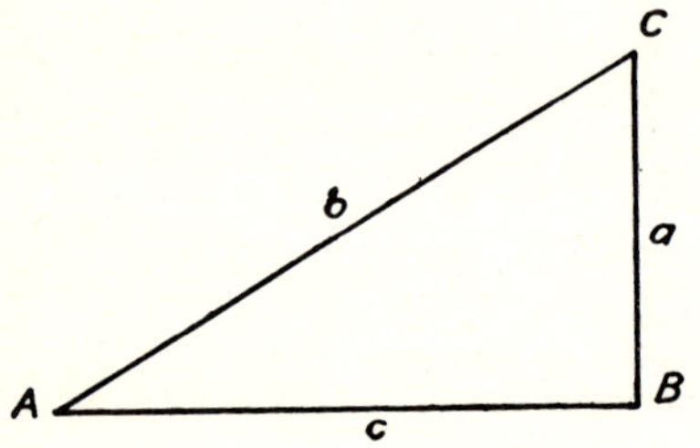

FIG. 635

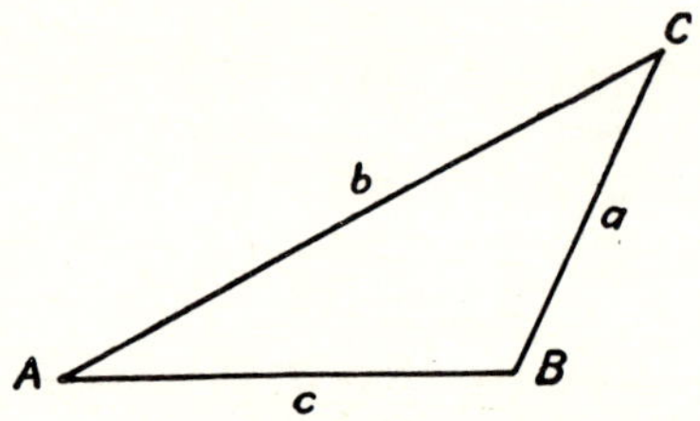

FIG. 636

Oblique Triangle (Fig. 636):

Given	Required	Formula
A, B, a	b	$b = \frac{a \cdot \sin B}{\sin A}.$
A, a, b	B	$\sin B = \frac{b \cdot \sin A}{a}.$
C, a, b	B	$\tan B = \frac{b \cdot \sin C}{a - b \cdot \cos C}.$
a, b, c	A	If $s = \frac{1}{2}(a + b + c)$, $\sin \frac{1}{2}A = \frac{\sqrt{(s - b)(s - c)}}{bc}$; or $\cos \frac{1}{2}A = \frac{\sqrt{s(s - a)}}{bc}.$
$a, b, c,$	Area	Area $= \sqrt{s(s - a)(s - b)(s - c)}$
A, B, c	Area	Area $= \frac{1}{2}\, bc \cdot \sin A.$

* From F. H. Lahee, "Field Geology," 6th ed., McGraw-Hill Book Company, Inc., 1961.

Table C-2. Trigonometric Functions (at intervals of 10′)†

Annex—10 in columns marked *

Degrees	Radians	Sines		Cosines		Tangents		Cotangents			
		Nat.	Log *	Nat.	Log *	Nat.	Log *	Nat.	Log		
0° 00′	0.0000	.0000	∞	1.0000	0.0000	.0000	∞	∞	∞	1.5708	90° 00′
10	0.0029	.0029	7.4637	1.0000	.0000	.0029	7.4637	343.77	2.5363	1.5679	50
20	0.0058	.0058	.7648	1.0000	.0000	.0058	.7648	171.89	.2352	1.5650	40
30	0.0087	.0087	.9408	1.0000	.0000	.0087	.9409	114.59	.0591	1.5621	30
40	0.0116	.0116	8.0658	.9999	.0000	.0116	8.0658	85.940	1.9342	1.5592	20
50	0.0145	.0145	.1627	.9999	.0000	.0145	.1627	68.750	.8373	1.5563	10
1° 00′	0.0175	.0175	8.2419	.9998	9.9999	.0175	8.2419	57.290	1.7581	1.5533	89° 00′
10	0.0204	.0204	.3088	.9998	.9999	.0204	.3089	49.104	.6911	1.5504	50
20	0.0233	.0233	.3668	.9997	.9999	.0233	.3669	42.964	.6331	1.5475	40
30	0.0262	.0262	.4179	.9997	.9999	.0262	.4181	38.188	.5819	1.5446	30
40	0.0291	.0291	.4637	.9996	.9998	.0291	.4638	34.368	.5362	1.5417	20
50	0.0320	.0320	.5050	.9995	.9998	.0320	.5053	31.242	.4947	1.5388	10
2° 00′	0.0349	.0349	8.5428	.9994	9.9997	.0349	8.5431	28.636	1.4569	1.5359	88° 00′
10	0.0378	.0378	.5776	.9993	.9997	.0378	.5779	26.432	.4221	1.5330	50
20	0.0407	.0407	.6097	.9992	.9996	.0407	.6101	24.542	.3899	1.5301	40
30	0.0436	.0436	.6397	.9990	.9996	.0437	.6401	22.904	.3599	1.5272	30
40	0.0465	.0465	.6677	.9989	.9995	.0466	.6682	21.470	.3318	1.5243	20
50	0.0495	.0494	.6940	.9988	.9995	.0495	.6945	20.206	.3055	1.5213	10
3° 00′	0.0524	.0523	8.7188	.9986	9.9994	.0524	8.7194	19.081	1.2806	1.5184	87° 00′
10	0.0553	.0552	.7423	.9985	.9993	.0553	.7429	18.075	.2571	1.5155	50
20	0.0582	.0581	.7645	.9983	.9993	.0582	.7652	17.169	.2348	1.5126	40
30	0.0611	.0610	.7857	.9981	.9992	.0612	.7865	16.350	.2135	1.5097	30
40	0.0640	.0640	.8059	.9980	.9991	.0641	.8067	15.605	.1933	1.5068	20
50	0.0669	.0669	.8251	.9978	.9990	.0670	.8261	14.924	.1739	1.5039	10
4° 00′	0.0698	.0698	8.8436	.9976	9.9989	.0699	8.8446	14.301	1.1554	1.5010	86° 00′
10	0.0727	.0727	.8613	.9974	.9989	.0729	.8624	13.727	.1376	1.4981	50
20	0.0756	.0756	.8783	.9971	.9988	.0758	.8795	13.197	.1205	1.4952	40
30	0.0785	.0785	.8946	.9969	.9987	.0787	.8960	12.706	.1040	1.4923	30
40	0.0814	.0814	.9104	.9967	.9986	.0816	.9118	12.251	.0882	1.4893	20
50	0.0844	.0843	.9256	.9964	.9985	.0846	.9272	11.826	.0728	1.4864	10
5° 00′	0.0873	.0872	8.9403	.9962	9.9983	.0875	8.9420	11.430	1.0580	1.4835	85° 00′
10	0.0902	.0901	.9545	.9959	.9982	.0904	.9563	11.059	.0437	1.4806	50
20	0.0931	.0929	.9682	.9957	.9981	.0934	.9701	10.712	.0299	1.4777	40
30	0.0960	.0958	.9816	.9954	.9980	.0963	.9836	10.385	.0164	1.4748	30
40	0.0989	.0987	.9945	.9951	.9979	.0992	.9966	10.078	.0034	1.4719	20
50	0.1018	.1016	9.0070	.9948	.9977	.1022	9.0093	9.7882	0.9907	1.4690	10
6° 00′	0.1047	.1045	9.0192	.9945	9.9976	.1051	9.0216	9.5144	0.9784	1.4661	84° 00′
10	0.1076	.1074	.0311	.9942	.9975	.1080	.0336	9.2553	.9664	1.4632	50
20	0.1105	.1103	.0426	.9939	.9973	.1110	.0453	9.0098	.9547	1.4603	40
30	0.1134	.1132	.0539	.9936	.9972	.1139	.0567	8.7769	.9433	1.4574	30
40	0.1164	.1161	.0648	.9932	.9971	.1169	.0678	8.5555	.9322	1.4544	20
50	0.1193	.1190	.0755	.9929	.9969	.1198	.0786	8.3450	.9214	1.4515	10
7° 00′	0.1222	.1219	9.0859	.9925	9.9968	.1228	9.0891	8.1443	0.9109	1.4486	83° 00′
10	0.1251	.1248	.0961	.9922	.9966	.1257	.0995	7.9530	.9005	1.4457	50
20	0.1280	.1276	.1060	.9918	.9964	.1287	.1096	7.7704	.8904	1.4428	40
30	0.1309	.1305	.1157	.9914	.9963	.1317	.1194	7.5958	.8806	1.4399	30
40	0.1338	.1334	.1252	.9911	.9961	.1346	.1291	7.4207	.8709	1.4370	20
50	0.1367	.1363	.1345	.9907	.9959	.1376	.1385	7.2687	.8615	1.4341	10
8° 00′	0.1396	.1392	9.1436	.9903	9.9958	.1405	9.1478	7.1154	0.8522	1.4312	82° 00′
10	0.1425	.1421	.1525	.9899	.9956	.1435	.1569	6.9682	.8431	1.4283	50
20	0.1454	.1449	.1612	.9894	.9954	.1465	.1658	6.8269	.8342	1.4254	40
30	0.1484	.1478	.1697	.9890	.9952	.1495	.1745	6.6912	.8255	1.4224	30
40	0.1513	.1507	.1781	.9886	.9950	.1524	.1831	6.5606	.8169	1.4195	20
50	0.1542	.1536	.1863	.9881	.9948	.1554	.1915	6.4348	.8085	1.4166	10
9° 00′	0.1571	.1564	9.1943	.9877	9.9946	.1584	9.1997	6.3138	0.8003	1.4137	81° 00′
		Nat.	Log *	Nat.	Log *	Nat.	Log *	Nat.	Log		
		Cosines		Sines		Cotangents		Tangents		Radians	Degrees

† From Marks' "Mechanical Engineers' Handbook," Theodore Baumeister (ed.), 6th ed., McGraw-Hill Book Company, Inc., New York, 1958.

Table C-2. Trigonometric Functions (*Continued*)
Annex—10 in columns marked *

Degrees	Radians	Sines Nat.	Sines Log *	Cosines Nat.	Cosines Log *	Tangents Nat.	Tangents Log *	Cotangents Nat.	Cotangents Log		
9° 00′	0.1571	.1564	9.1943	.9877	9.9946	.1584	9.1997	6.3138	0.8003	1.4137	81° 00′
10	0.1600	.1593	.2022	.9872	.9944	.1614	.2078	6.1970	.7922	1.4108	50
20	0.1629	.1622	.2100	.9868	.9942	.1644	.2158	6.0844	.7842	1.4079	40
30	0.1658	.1650	.2176	.9863	.9940	.1673	.2236	5.9758	.7764	1.4050	30
40	0.1687	.1679	.2251	.9858	.9938	.1703	.2313	5.8708	.7687	1.4021	20
50	0.1716	.1708	.2324	.9853	.9936	.1733	.2389	5.7694	.7611	1.3992	10
10° 00′	0.1745	.1736	9.2397	.9848	9.9934	.1763	9.2463	5.6713	0.7537	1.3963	80° 00′
10	0.1774	.1765	.2468	.9843	.9931	.1793	.2536	5.5764	.7464	1.3934	50
20	0.1804	.1794	.2538	.9838	.9929	.1823	.2609	5.4845	.7391	1.3904	40
30	0.1833	.1822	.2606	.9833	.9927	.1853	.2680	5.3955	.7320	1.3875	30
40	0.1862	.1851	.2674	.9827	.9924	.1883	.2750	5.3093	.7250	1.3846	20
50	0.1891	.1880	.2740	.9822	.9922	.1914	.2819	5.2257	.7181	1.3817	10
11° 00′	0.1920	.1908	9.2806	.9816	9.9919	.1944	9.2887	5.1446	0.7113	1.3788	79° 00′
10	0.1949	.1937	.2870	.9811	.9917	.1974	.2953	5.0658	.7047	1.3759	50
20	0.1978	.1965	.2934	.9805	.9914	.2004	.3020	4.9894	.6980	1.3730	40
30	0.2007	.1994	.2997	.9799	.9912	.2035	.3085	4.9152	.6915	1.3701	30
40	0.2036	.2022	.3058	.9793	.9909	.2065	.3149	4.8430	.6851	1.3672	20
50	0.2065	.2051	.3119	.9787	.9907	.2095	.3212	4.7729	.6788	1.3643	10
12° 00′	0.2094	.2079	9.3179	.9781	9.9904	.2126	9.3275	4.7046	0.6725	1.3614	78° 00′
10	0.2123	.2108	.3238	.9775	.9901	.2156	.3336	4.6382	.6664	1.3584	50
20	0.2153	.2136	.3296	.9769	.9899	.2186	.3397	4.5736	.6603	1.3555	40
30	0.2182	.2164	.3353	.9763	.9896	.2217	.3458	4.5107	.6542	1.3526	30
40	0.2211	.2193	.3410	.9757	.9893	.2247	.3517	4.4494	.6483	1.3497	20
50	0.2240	.2221	.3466	.9750	.9890	.2278	.3576	4.3897	.6424	1.3468	10
13° 00′	0.2269	.2250	9.3521	.9744	9.9887	.2309	9.3634	4.3315	0.6366	1.3439	77° 00′
10	0.2298	.2278	.3575	.9737	.9884	.2339	.3691	4.2747	.6309	1.3410	50
20	0.2327	.2306	.3629	.9730	.9881	.2370	.3748	4.2193	.6252	1.3381	40
30	0.2356	.2334	.3682	.9724	.9878	.2401	.3804	4.1653	.6196	1.3352	30
40	0.2385	.2363	.3734	.9717	.9875	.2432	.3859	4.1126	.6141	1.3323	20
50	0.2414	.2391	.3786	.9710	.9872	.2462	.3914	4.0611	.6086	1.3294	10
14° 00′	0.2443	.2419	9.3837	.9703	9.9869	.2493	9.3968	4.0108	0.6032	1.3265	76° 00′
10	0.2473	.2447	.3887	.9696	.9866	.2524	.4021	3.9617	.5979	1.3235	50
20	0.2502	.2476	.3937	.9689	.9863	.2555	.4074	3.9136	.5926	1.3206	40
30	0.2531	.2504	.3986	.9681	.9859	.2586	.4127	3.8667	.5873	1.3177	30
40	0.2560	.2532	.4035	.9674	.9856	.2617	.4178	3.8208	.5822	1.3148	20
50	0.2589	.2560	.4083	.9667	.9853	.2648	.4230	3.7760	.5770	1.3119	10
15° 00′	0.2618	.2588	9.4130	.9659	9.9849	.2679	9.4281	3.7321	0.5719	1.3090	75° 00′
10	0.2647	.2616	.4177	.9652	.9846	.2711	.4331	3.6891	.5669	1.3061	50
20	0.2676	.2644	.4223	.9644	.9843	.2742	.4381	3.6470	.5619	1.3032	40
30	0.2705	.2672	.4269	.9636	.9839	.2773	.4430	3.6059	.5570	1.3003	30
40	0.2734	.2700	.4314	.9628	.9836	.2805	.4479	3.5656	.5521	1.2974	20
50	0.2763	.2728	.4359	.9621	.9832	.2836	.4527	3.5261	.5473	1.2945	10
16° 00′	0.2793	.2756	9.4403	.9613	9.9828	.2867	9.4575	3.4874	0.5425	1.2915	74° 00′
10	0.2822	.2784	.4447	.9605	.9825	.2899	.4622	3.4495	.5378	1.2886	50
20	0.2851	.2812	.4491	.9596	.9821	.2931	.4669	3.4124	.5331	1.2857	40
30	0.2880	.2840	.4533	.9588	.9817	.2962	.4716	3.3759	.5284	1.2828	30
40	0.2909	.2868	.4576	.9580	.9814	.2994	.4762	3.3402	.5238	1.2799	20
50	0.2938	.2896	.4618	.9572	.9810	.3026	.4808	3.3052	.5192	1.2770	10
17° 00′	0.2967	.2924	9.4659	.9563	9.9806	.3057	9.4853	3.2709	0.5147	1.2741	73° 00′
10	0.2996	.2952	.4700	.9555	.9802	.3089	.4898	3.2371	.5102	1.2712	50
20	0.3025	.2979	.4741	.9546	.9798	.3121	.4943	3.2041	.5057	1.2683	40
30	0.3054	.3007	.4781	.9537	.9794	.3153	.4987	3.1716	.5013	1.2654	30
40	0.3083	.3035	.4821	.9528	.9790	.3185	.5031	3.1397	.4969	1.2625	20
50	0.3113	.3062	.4861	.9520	.9786	.3217	.5075	3.1084	.4925	1.2595	10
18° 00′	0.3142	.3090	9.4900	.9511	9.9782	.3249	9.5118	3.0777	0.4882	1.2566	72° 00′
		Nat.	Log *	Nat.	Log *	Nat.	Log *	Nat.	Log		
		Cosines		Sines		Cotangents		Tangents		Radians	Degrees

Table C-2. Trigonometric Functions (*Continued*)
Annex—10 in columns marked *

Degrees	Radians	Sines		Cosines		Tangents		Cotangents			
		Nat.	Log *	Nat.	Log *	Nat.	Log *	Nat.	Log		
18° 00′	0.3142	.3090	9.4900	.9511	9.9782	.3249	9.5118	3.0777	0.4882	1.2566	72° 00′
10	0.3171	.3118	.4939	.9502	.9778	.3281	.5161	3.0475	.4839	1.2537	50
20	0.3200	.3145	.4977	.9492	.9774	.3314	.5203	3.0178	.4797	1.2508	40
30	0.3229	.3173	.5015	.9483	.9770	.3346	.5245	2.9887	.4755	1.2479	30
40	0.3258	.3201	.5052	.9474	.9765	.3378	.5287	2.9600	.4713	1.2450	20
50	0.3287	.3228	.5090	.9465	.9761	.3411	.5329	2.9319	.4671	1.2421	10
19° 00′	0.3316	.3256	9.5126	.9455	9.9757	.3443	9.5370	2.9042	0.4630	1.2392	71° 00′
10	0.3345	.3283	.5163	.9446	.9752	.3476	.5411	2.8770	.4589	1.2363	50
20	0.3374	.3311	.5199	.9436	.9748	.3508	.5451	2.8502	.4549	1.2334	40
30	0.3403	.3338	.5235	.9426	.9743	.3541	.5491	2.8239	.4509	1.2305	30
40	0.3432	.3365	.5270	.9417	.9739	.3574	.5531	2.7980	.4469	1.2275	20
50	0.3462	.3393	.5306	.9407	.9734	.3607	.5571	2.7725	.4429	1.2246	10
20° 00′	0.3491	.3420	9.5341	.9397	9.9730	.3640	9.5611	2.7475	0.4389	1.2217	70° 00′
10	0.3520	.3448	.5375	.9387	.9725	.3673	.5650	2.7228	.4350	1.2188	50
20	0.3549	.3475	.5409	.9377	.9721	.3706	.5689	2.6985	.4311	1.2159	40
30	0.3578	.3502	.5443	.9367	.9716	.3739	.5727	2.6746	.4273	1.2130	30
40	0.3607	.3529	.5477	.9356	.9711	.3772	.5766	2.6511	.4234	1.2101	20
50	0.3636	.3557	.5510	.9346	.9706	.3805	.5804	2.6279	.4196	1.2072	10
21° 00′	0.3665	.3584	9.5543	.9336	9.9702	.3839	9.5842	2.6051	0.4158	1.2043	69° 00′
10	0.3694	.3611	.5576	.9325	.9697	.3872	.5879	2.5826	.4121	1.2014	50
20	0.3723	.3638	.5609	.9315	.9692	.3906	.5917	2.5605	.4083	1.1985	40
30	0.3752	.3665	.5641	.9304	.9687	.3939	.5954	2.5386	.4046	1.1956	30
40	0.3782	.3692	.5673	.9293	.9682	.3973	.5991	2.5172	.4009	1.1926	20
50	0.3811	.3719	.5704	.9283	.9677	.4006	.6028	2.4960	.3972	1.1897	10
22° 00′	0.3840	.3746	9.5736	.9272	9.9672	.4040	9.6064	2.4751	0.3936	1.1868	68° 00′
10	0.3869	.3773	.5767	.9261	.9667	.4074	.6100	2.4545	.3900	1.1839	50
20	0.3898	.3800	.5798	.9250	.9661	.4108	.6136	2.4342	.3864	1.1810	40
30	0.3927	.3827	.5828	.9239	.9656	.4142	.6172	2.4142	.3828	1.1781	30
40	0.3956	.3854	.5859	.9228	.9651	.4176	.6208	2.3945	.3792	1.1752	20
50	0.3985	.3881	.5889	.9216	.9646	.4210	.6243	2.3750	.3757	1.1723	10
23° 00′	0.4014	.3907	9.5919	.9205	9.9640	.4245	9.6279	2.3559	0.3721	1.1694	67° 00′
10	0.4043	.3934	.5948	.9194	.9635	.4279	.6314	2.3369	.3686	1.1665	50
20	0.4072	.3961	.5978	.9182	.9629	.4314	.6348	2.3183	.3652	1.1636	40
30	0.4102	.3987	.6007	.9171	.9624	.4348	.6383	2.2998	.3617	1.1606	30
40	0.4131	.4014	.6036	.9159	.9618	.4383	.6417	2.2817	.3583	1.1577	20
50	0.4160	.4041	.6065	.9147	.9613	.4417	.6452	2.2637	.3548	1.1548	10
24° 00′	0.4189	.4067	9.6093	.9135	9.9607	.4452	9.6486	2.2460	0.3514	1.1519	66° 00′
10	0.4218	.4094	.6121	.9124	.9602	.4487	.6520	2.2286	.3480	1.1490	50
20	0.4247	.4120	.6149	.9112	.9596	.4522	.6553	2.2113	.3447	1.1461	40
30	0.4276	.4147	.6177	.9100	.9590	.4557	.6587	2.1943	.3413	1.1432	30
40	0.4305	.4173	.6205	.9088	.9584	.4592	.6620	2.1775	.3380	1.1403	20
50	0.4334	.4200	.6232	.9075	.9579	.4628	.6654	2.1609	.3346	1.1374	10
25° 00′	0.4363	.4226	9.6259	.9063	9.9573	.4663	9.6687	2.1445	0.3313	1.1345	65° 00′
10	0.4392	.4253	.6286	.9051	.9567	.4699	.6720	2.1283	.3280	1.1316	50
20	0.4422	.4279	.6313	.9038	.9561	.4734	.6752	2.1123	.3248	1.1286	40
30	0.4451	.4305	.6340	.9026	.9555	.4770	.6785	2.0965	.3215	1.1257	30
40	0.4480	.4331	.6366	.9013	.9549	.4806	.6817	2.0809	.3183	1.1228	20
50	0.4509	.4358	.6392	.9001	.9543	.4841	.6850	2.0655	.3150	1.1199	10
26° 00′	0.4538	.4384	9.6418	.8988	9.9537	.4877	9.6882	2.0503	0.3118	1.1170	64° 00′
10	0.4567	.4410	.6444	.8975	.9530	.4913	.6914	2.0353	.3086	1.1141	50
20	0.4596	.4436	.6470	.8962	.9524	.4950	.6946	2.0204	.3054	1.1112	40
30	0.4625	.4462	.6495	.8949	.9518	.4986	.6977	2.0057	.3023	1.1083	30
40	0.4654	.4488	.6521	.8936	.9512	.5022	.7009	1.9912	.2991	1.1054	20
50	0.4683	.4514	.6546	.8923	.9505	.5059	.7040	1.9768	.2960	1.1025	10
27° 00′	0.4712	.4540	9.6570	.8910	9.9499	.5095	9.7072	1.9626	0.2928	1.0996	63° 00′
		Nat.	Log *	Nat.	Log *	Nat.	Log *	Nat.	Log		
		Cosines		Sines		Cotangents		Tangents		Radians	Degrees

Table C-2. Trigonometric Functions (*Continued*)

Annex—10 in columns marked *

Degrees	Radians	Sines		Cosines		Tangents		Cotangents			
		Nat.	Log *	Nat.	Log *	Nat.	Log *	Nat.	Log		
27° 00′	0.4712	.4540	9.6570	.8910	9.9499	.5095	9.7072	1.9626	0.2928	1.0996	63° 00′
10	0.4741	.4566	.6595	.8897	.9492	.5132	.7103	1.9486	.2897	1.0966	50
20	0.4771	.4592	.6620	.8884	.9486	.5169	.7134	1.9347	.2866	1.0937	40
30	0.4800	.4617	.6644	.8870	.9479	.5206	.7165	1.9210	.2835	1.0908	30
40	0.4829	.4643	.6668	.8857	.9473	.5243	.7196	1.9074	.2804	1.0879	20
50	0.4858	.4669	.6692	.8843	.9466	.5280	.7226	1.8940	.2774	1.0850	10
28° 00′	0.4887	.4695	9.6716	.8829	9.9459	.5317	9.7257	1.8807	0.2743	1.0821	62° 00′
10	0.4916	.4720	.6740	.8816	.9453	.5354	.7287	1.8676	.2713	1.0792	50
20	0.4945	.4746	.6763	.8802	.9446	.5392	.7317	1.8546	.2683	1.0763	40
30	0.4974	.4772	.6787	.8788	.9439	.5430	.7348	1.8418	.2652	1.0734	30
40	0.5003	.4797	.6810	.8774	.9432	.5467	.7378	1.8291	.2622	1.0705	20
50	0.5032	.4823	.6833	.8760	.9425	.5505	.7408	1.8165	.2592	1.0676	10
29° 00′	0.5061	.4848	9.6856	.8746	9.9418	.5543	9.7438	1.8040	0.2562	1.0647	61° 00′
10	0.5091	.4874	.6878	.8732	.9411	.5581	.7467	1.7917	.2533	1.0617	50
20	0.5120	.4899	.6901	.8718	.9404	.5619	.7497	1.7796	.2503	1.0588	40
30	0.5149	.4924	.6923	.8704	.9397	.5658	.7526	1.7675	.2474	1.0559	30
40	0.5178	.4950	.6946	.8689	.9390	.5696	.7556	1.7556	.2444	1.0530	20
50	0.5207	.4975	.6968	.8675	.9383	.5735	.7585	1.7437	.2415	1.0501	10
30° 00′	0.5236	.5000	9.6990	.8660	9.9375	.5774	9.7614	1.7321	0.2386	1.0472	60° 00′
10	0.5265	.5025	.7012	.8646	.9368	.5812	.7644	1.7205	.2356	1.0443	50
20	0.5294	.5050	.7033	.8631	.9361	.5851	.7673	1.7090	.2327	1.0414	40
30	0.5323	.5075	.7055	.8616	.9353	.5890	.7701	1.6977	.2299	1.0385	30
40	0.5352	.5100	.7076	.8601	.9346	.5930	.7730	1.6864	.2270	1.0356	20
50	0.5381	.5125	.7097	.8587	.9338	.5969	.7759	1.6753	.2241	1.0327	10
31° 00′	0.5411	.5150	9.7118	.8572	9.9331	.6009	9.7788	1.6643	0.2212	1.0297	59° 00′
10	0.5440	.5175	.7139	.8557	.9323	.6048	.7816	1.6534	.2184	1.0268	50
20	0.5469	.5200	.7160	.8542	.9315	.6088	.7845	1.6426	.2155	1.0239	40
30	0.5498	.5225	.7181	.8526	.9308	.6128	.7873	1.6319	.2127	1.0210	30
40	0.5527	.5250	.7201	.8511	.9300	.6168	.7902	1.6212	.2098	1.0181	20
50	0.5556	.5275	.7222	.8496	.9292	.6208	.7930	1.6107	.2070	1.0152	10
32° 00′	0.5585	.5299	9.7242	.8480	9.9284	.6249	9.7958	1.6003	0.2042	1.0123	58° 00′
10	0.5614	.5324	.7262	.8465	.9276	.6289	.7986	1.5900	.2014	1.0094	50
20	0.5643	.5348	.7282	.8450	.9268	.6330	.8014	1.5798	.1986	1.0065	40
30	0.5672	.5373	.7302	.8434	.9260	.6371	.8042	1.5697	.1958	1.0036	30
40	0.5701	.5398	.7322	.8418	.9252	.6412	.8070	1.5597	.1930	1.0007	20
50	0.5730	.5422	.7342	.8403	.9244	.6453	.8097	1.5497	.1903	0.9977	10
33° 00′	0.5760	.5446	9.7361	.8387	9.9236	.6494	9.8125	1.5399	0.1875	0.9948	57° 00′
10	0.5789	.5471	.7380	.8371	.9228	.6536	.8153	1.5301	.1847	0.9919	50
20	0.5818	.5495	.7400	.8355	.9219	.6577	.8180	1.5204	.1820	0.9890	40
30	0.5847	.5519	.7419	.8339	.9211	.6619	.8208	1.5108	.1792	0.9861	30
40	0.5876	.5544	.7438	.8323	.9203	.6661	.8235	1.5013	.1765	0.9832	20
50	0.5905	.5568	.7457	.8307	.9194	.6703	.8263	1.4919	.1737	0.9803	10
34° 00′	0.5934	.5592	9.7476	.8290	9.9186	.6745	9.8290	1.4826	0.1710	0.9774	56° 00′
10	0.5963	.5616	.7494	.8274	.9177	.6787	.8317	1.4733	.1683	0.9745	50
20	0.5992	.5640	.7513	.8258	.9169	.6830	.8344	1.4641	.1656	0.9716	40
30	0.6021	.5664	.7531	.8241	.9160	.6873	.8371	1.4550	.1629	0.9687	30
40	0.6050	.5688	.7550	.8225	.9151	.6916	.8398	1.4460	.1602	0.9657	20
50	0.6080	.5712	.7568	.8208	.9142	.6959	.8425	1.4370	.1575	0.9628	10
35° 00′	0.6109	.5736	9.7586	.8192	9.9134	.7002	9.8452	1.4281	0.1548	0.9599	55° 00′
10	0.6138	.5760	.7604	.8175	.9125	.7046	.8479	1.4193	.1521	0.9570	50
20	0.6167	.5783	.7622	.8158	.9116	.7089	.8506	1.4106	.1494	0.9541	40
30	0.6196	.5807	.7640	.8141	.9107	.7133	.8533	1.4019	.1467	0.9512	30
40	0.6225	.5831	.7657	.8124	.9098	.7177	.8559	1.3934	.1441	0.9483	20
50	0.6254	.5854	.7675	.8107	.9089	.7221	.8586	1.3848	.1414	0.9454	10
36° 00′	0.6283	.5878	9.7692	.8090	9.9080	.7265	9.8613	1.3764	0.1387	0.9425	54° 00′
		Nat.	Log *	Nat.	Log *	Nat.	Log *	Nat.	Log		
		Cosines		Sines		Cotangents		Tangents		Radians	Degrees

Table C-2. Trigonometric Functions (*Continued*)

Annex—10 in columns marked *

Degrees	Radians	Sines		Cosines		Tangents		Cotangents			
		Nat.	Log *	Nat.	Log *	Nat.	Log *	Nat.	Log		
36° 00′	0.6283	.5878	9.7692	.8090	9.9080	.7265	9.8613	1.3764	0.1387	0.9425	54° 00′
10	0.6312	.5901	.7710	.8073	.9070	.7310	.8639	1.3680	.1361	0.9396	50
20	0.6341	.5925	.7727	.8056	.9061	.7355	.8666	1.3597	.1334	0.9367	40
30	0.6370	.5948	.7744	.8039	.9052	.7400	.8692	1.3514	.1308	0.9338	30
40	0.6400	.5972	.7761	.8021	.9042	.7445	.8718	1.3432	.1282	0.9308	20
50	0.6429	.5995	.7778	.8004	.9033	.7490	.8745	1.3351	.1255	0.9279	10
37° 00′	0.6458	.6018	9.7795	.7986	9.9023	.7536	9.8771	1.3270	0.1229	0.9250	53° 00′
10	0.6487	.6041	.7811	.7969	.9014	.7581	.8797	1.3190	.1203	0.9221	50
20	0.6516	.6065	.7828	.7951	.9004	.7627	.8824	1.3111	.1176	0.9192	40
30	0.6545	.6088	.7844	.7934	.8995	.7673	.8850	1.3032	.1150	0.9163	30
40	0.6574	.6111	.7861	.7916	.8985	.7720	.8876	1.2954	.1124	0.9134	20
50	0.6603	.6134	.7877	.7898	.8975	.7766	.8902	1.2876	.1098	0.9105	10
38° 00′	0.6632	.6157	9.7893	.7880	9.8965	.7813	9.8928	1.2799	0.1072	0.9076	52° 00′
10	0.6661	.6180	.7910	.7862	.8955	.7860	.8954	1.2723	.1046	0.9047	50
20	0.6690	.6202	.7926	.7844	.8945	.7907	.8980	1.2647	.1020	0.9018	40
30	0.6720	.6225	.7941	.7826	.8935	.7954	.9006	1.2572	.0994	0.8988	30
40	0.6749	.6248	.7957	.7808	.8925	.8002	.9032	1.2497	.0968	0.8959	20
50	0.6778	.6271	.7973	.7790	.8915	.8050	.9058	1.2423	.0942	0.8930	10
39° 00′	0.6807	.6293	9.7989	.7771	9.8905	.8098	9.9084	1.2349	0.0916	0.8901	51° 00′
10	0.6836	.6316	.8004	.7753	.8895	.8146	.9110	1.2276	.0890	0.8872	50
20	0.6865	.6338	.8020	.7735	.8884	.8195	.9135	1.2203	.0865	0.8843	40
30	0.6894	.6361	.8035	.7716	.8874	.8243	.9161	1.2131	.0839	0.8814	30
40	0.6923	.6383	.8050	.7698	.8864	.8292	.9187	1.2059	.0813	0.8785	20
50	0.6952	.6406	.8066	.7679	.8853	.8342	.9212	1.1988	.0788	0.8756	10
40° 00′	0.6981	.6428	9.8081	.7660	9.8843	.8391	9.9238	1.1918	0.0762	0.8727	50° 00′
10	0.7010	.6450	.8096	.7642	.8832	.8441	.9264	1.1847	.0736	0.8698	50
20	0.7039	.6472	.8111	.7623	.8821	.8491	.9289	1.1778	.0711	0.8668	40
30	0.7069	.6494	.8125	.7604	.8810	.8541	.9315	1.1708	.0685	0.8639	30
40	0.7098	.6517	.8140	.7585	.8800	.8591	.9341	1.1640	.0659	0.8610	20
50	0.7127	.6539	.8155	.7566	.8789	.8642	.9366	1.1571	.0634	0.8581	10
41° 00′	0.7156	.6561	9.8169	.7547	9.8778	.8693	9.9392	1.1504	0.0608	0.8552	49° 00′
10	0.7185	.6583	.8184	.7528	.8767	.8744	.9417	1.1436	.0583	0.8523	50
20	0.7214	.6604	.8198	.7509	.8756	.8796	.9443	1.1369	.0557	0.8494	40
30	0.7243	.6626	.8213	.7490	.8745	.8847	.9468	1.1303	.0532	0.8465	30
40	0.7272	.6648	.8227	.7470	.8733	.8899	.9494	1.1237	.0506	0.8436	20
50	0.7301	.6670	.8241	.7451	.8722	.8952	.9519	1.1171	.0481	0.8407	10
42° 00′	0.7330	.6691	9.8255	.7431	9.8711	.9004	9.9544	1.1106	0.0456	0.8378	48° 00′
10	0.7359	.6713	.8269	.7412	.8699	.9057	.9570	1.1041	.0430	0.8348	50
20	0.7389	.6734	.8283	.7392	.8688	.9110	.9595	1.0977	.0405	0.8319	40
30	0.7418	.6756	.8297	.7373	.8676	.9163	.9621	1.0913	.0379	0.8290	30
40	0.7447	.6777	.8311	.7353	.8665	.9217	.9646	1.0850	.0354	0.8261	20
50	0.7476	.6799	.8324	.7333	.8653	.9271	.9671	1.0786	.0329	0.8232	10
43° 00′	0.7505	.6820	9.8338	.7314	9.8641	.9325	9.9697	1.0724	0.0303	0.8203	47° 00′
10	0.7534	.6841	.8351	.7294	.8629	.9380	.9722	1.0661	.0278	0.8174	50
20	0.7563	.6862	.8365	.7274	.8618	.9435	.9747	1.0599	.0253	0.8145	40
30	0.7592	.6884	.8378	.7254	.8606	.9490	.9772	1.0538	.0228	0.8116	30
40	0.7621	.6905	.8391	.7234	.8594	.9545	.9798	1.0477	.0202	0.8087	20
50	0.7650	.6926	.8405	.7214	.8582	.9601	.9823	1.0416	.0177	0.8058	10
44° 00′	0.7679	.6947	9.8418	.7193	9.8569	.9657	9.9848	1.0355	0.0152	0.8029	46° 00′
10	0.7709	.6967	.8431	.7173	.8557	.9713	.9874	1.0295	.0126	0.7999	50
20	0.7738	.6988	.8444	.7153	.8545	.9770	.9899	1.0235	.0101	0.7970	40
30	0.7767	.7009	.8457	.7133	.8532	.9827	.9924	1.0176	.0076	0.7941	30
40	0.7796	.7030	.8469	.7112	.8520	.9884	.9949	1.0117	.0051	0.7912	20
50	0.7825	.7050	.8482	.7092	.8507	.9942	.9975	1.0058	.0025	0.7883	10
45° 00′	0.7854	.7071	9.8495	.7071	9.8495	1.0000	0.0000	1.0000	0.0000	0.7854	45° 00′
		Nat.	Log *	Nat.	Log *	Nat.	Log *	Nat.	Log		
		Cosines		Sines		Cotangents		Tangents		Radians	Degrees

Table C-3. Common Logarithms of Numbers*

N	0	1	2	3	4	5	6	7	8	9
10	0000	0043	0086	0128	0170	0212	0253	0294	0334	0374
11	0414	0453	0492	0531	0569	0607	0645	0682	0719	0755
12	0792	0828	0864	0899	0934	0969	1004	1038	1072	1106
13	1139	1173	1206	1239	1271	1303	1335	1367	1399	1430
14	1461	1492	1523	1553	1584	1614	1644	1673	1703	1732
15	1761	1790	1818	1847	1875	1903	1931	1959	1987	2014
16	2041	2068	2095	2122	2148	2175	2201	2227	2253	2279
17	2304	2330	2355	2380	2405	2430	2455	2480	2504	2529
18	2553	2577	2601	2625	2648	2672	2695	2718	2742	2765
19	2788	2810	2833	2856	2878	2900	2923	2945	2967	2989
20	3010	3032	3054	3075	3096	3118	3139	3160	3181	3201
21	3222	3243	3263	3284	3304	3324	3345	3365	3385	3404
22	3424	3444	3464	3483	3502	3522	3541	3560	3579	3598
23	3617	3636	3655	3674	3692	3711	3729	3747	3766	3784
24	3802	3820	3838	3856	3874	3892	3909	3927	3945	3962
25	3979	3997	4014	4031	4048	4065	4082	4099	4116	4133
26	4150	4166	4183	4200	4216	4232	4249	4265	4281	4298
27	4314	4330	4346	4362	4378	4393	4409	4425	4440	4456
28	4472	4487	4502	4518	4533	4548	4564	4579	4594	4609
29	4624	4639	4654	4669	4683	4698	4713	4728	4742	4757
30	4771	4786	4800	4814	4829	4843	4857	4871	4886	4900
31	4914	4928	4942	4955	4969	4983	4997	5011	5024	5038
32	5051	5065	5079	5092	5105	5119	5132	5145	5159	5172
33	5185	5198	5211	5224	5237	5250	5263	5276	5289	5302
34	5315	5328	5340	5353	5366	5378	5391	5403	5416	5428
35	5441	5453	5465	5478	5490	5502	5514	5527	5539	5551
36	5563	5575	5587	5599	5611	5623	5635	5647	5658	5670
37	5682	5694	5705	5717	5729	5740	5752	5763	5775	5786
38	5798	5809	5821	5832	5843	5855	5866	5877	5888	5899
39	5911	5922	5933	5944	5955	5966	5977	5988	5999	6010
40	6021	6031	6042	6053	6064	6075	6085	6096	6107	6117
41	6128	6138	6149	6160	6170	6180	6191	6201	6212	6222
42	6232	6243	6253	6263	6274	6284	6294	6304	6314	6325
43	6335	6345	6355	6365	6375	6385	6395	6405	6415	6425
44	6435	6444	6454	6464	6474	6484	6493	6503	6513	6522
45	6532	6542	6551	6561	6571	6580	6590	6599	6609	6618
46	6628	6637	6646	6656	6665	6675	6684	6693	6702	6712
47	6721	6730	6739	6749	6758	6767	6776	6785	6794	6803
48	6812	6821	6830	6839	6848	6857	6866	6875	6884	6893
49	6902	6911	6920	6928	6937	6946	6955	6964	6972	6981
50	6990	6998	7007	7016	7024	7033	7042	7050	7059	7067
51	7076	7084	7093	7101	7110	7118	7126	7135	7143	7152
52	7160	7168	7177	7185	7193	7202	7210	7218	7226	7235
53	7243	7251	7259	7267	7275	7284	7292	7300	7308	7316
54	7324	7332	7340	7348	7356	7364	7372	7380	7388	7396

* From A. E. Knowlton (ed.), "Standard Handbook for Electrical Engineers," 9th ed., McGraw-Hill Book Company, Inc., New York, 1957.

Table C-3. Common Logarithms of Numbers (*Continued*)

N	0	1	2	3	4	5	6	7	8	9
55	7404	7412	7419	7427	7435	7443	7451	7459	7466	7474
56	7482	7490	7497	7505	7513	7520	7528	7536	7543	7551
57	7559	7566	7574	7582	7589	7597	7604	7612	7619	7627
58	7634	7642	7649	7657	7664	7672	7679	7686	7694	7701
59	7709	7716	7723	7731	7738	7745	7752	7760	7767	7774
60	7782	7789	7796	7803	7810	7818	7825	7832	7839	7846
61	7853	7860	7868	7875	7882	7889	7896	7903	7910	7917
62	7924	7931	7938	7945	7952	7959	7966	7973	7980	7987
63	7993	8000	8007	8014	8021	8028	8035	8041	8048	8055
64	8062	8069	8075	8082	8089	8096	8102	8109	8116	8122
65	8129	8136	8142	8149	8156	8162	8169	8176	8182	8189
66	8195	8202	8209	8215	8222	8228	8235	8241	8248	8254
67	8261	8267	8274	8280	8287	8293	8299	8306	8312	8319
68	8325	8331	8338	8344	8351	8357	8363	8370	8376	8382
69	8388	8395	8401	8407	8414	8420	8426	8432	8439	8445
70	8451	8457	8463	8470	8476	8482	8488	8494	8500	8506
71	8513	8519	8525	8531	8537	8543	8549	8555	8561	8567
72	8573	8579	8585	8591	8597	8603	8609	8615	8621	8627
73	8633	8639	8645	8651	8657	8663	8669	8675	8681	8686
74	8692	8698	8704	8710	8716	8722	8727	8733	8739	8745
75	8751	8756	8762	8768	8774	8779	8785	8791	8797	8802
76	8808	8814	8820	8825	8831	8837	8842	8848	8854	8859
77	8865	8871	8876	8882	8887	8893	8899	8904	8910	8915
78	8921	8927	8932	8938	8943	8040	8954	8960	8965	8971
79	8970	8982	8987	8993	8998	9004	9009	9015	9020	9025
80	9031	9036	9042	9047	9053	9058	9063	9069	9074	9079
81	9085	9090	9096	9101	9106	9112	9117	9122	9128	9133
82	9138	9143	9149	9154	9159	9165	9170	9175	9180	9186
83	9191	9196	9201	9206	9212	9217	9222	9227	9232	9238
84	9243	9248	9253	9258	9263	9269	9274	9279	9284	9289
85	9294	9299	9304	9309	9315	9320	9325	9330	9335	9340
86	9345	9350	9355	9360	9365	9370	9375	9380	9385	9390
87	9395	9400	9405	9410	9415	9420	9425	9430	9435	9440
88	9445	9450	9455	9460	9465	9469	9474	9479	9484	9489
89	9494	9499	9504	9509	9513	9518	9523	9528	9533	9538
90	9542	9547	9552	9557	9562	9566	9571	9576	9581	9586
91	9590	9595	9600	9605	9609	9614	9619	9624	9628	9633
92	9638	9643	9647	9652	9657	9661	9666	9671	9675	9680
93	9685	9689	9694	9699	9703	9708	9713	9717	9722	9727
94	9731	9736	9741	9745	9750	9754	9759	9763	9768	9773
95	9777	9782	9786	9791	9795	9800	9805	9809	9814	9818
96	9823	9827	9832	9836	9841	9845	9850	9854	9859	9863
97	9868	9872	9877	9881	9886	9890	9894	9899	9903	9908
98	9912	9917	9921	9926	9930	9934	9939	9943	9948	9952
99	9956	9961	9965	9969	9974	9978	9983	9987	9991	9996

Table C-4. Natural Logarithms*

These two pages give the natural or Napierian logarithms (ln) of numbers between 1 and 10, correct to four places. Moving the decimal point n places to the right [or left] in the number is equivalent to adding n times 2.3026 [or n times $\bar{3}.6974$] to the logarithm. Base $e = 2.71828+$

n	n (2.3026)	n (0.6974–3)
1	2.3026	0.6974–3
2	4.6052	0.3948–5
3	6.9078	0.0922–7
4	9.2103	0.7897–10
5	11.5129	0.4871–12
6	13.8155	0.1845–14
7	16.1181	0.8819–17
8	18.4207	0.5793–19
9	20.7233	0.2767–21

Number	0	1	2	3	4	5	6	7	8	9	Avg diff
1.0	0.0000	0100	0198	0296	0392	0488	0583	0677	0770	0862	95
1.1	0953	1044	1133	1222	1310	1398	1484	1570	1655	1740	87
1.2	1823	1906	1989	2070	2151	2231	2311	2390	2469	2546	80
1.3	2624	2700	2776	2852	2927	3001	3075	3148	3221	3293	74
1.4	3365	3436	3507	3577	3646	3716	3784	3853	3920	3988	69
1.5	0.4055	4121	4187	4253	4318	4383	4447	4511	4574	4637	65
1.6	4700	4762	4824	4886	4947	5008	5068	5128	5188	5247	61
1.7	5306	5365	5423	5481	5539	5596	5653	5710	5766	5822	57
1.8	5878	5933	5988	6043	6098	6152	6206	6259	6313	6366	54
1.9	6419	6471	6523	6575	6627	6678	6729	6780	6831	6881	51
2.0	0.6931	6981	7031	7080	7129	7178	7227	7275	7324	7372	49
2.1	7419	7467	7514	7561	7608	7655	7701	7747	7793	7839	47
2.2	7885	7930	7975	8020	8065	8109	8154	8198	8242	8286	44
2.3	8329	8372	8416	8459	8502	8544	8587	8629	8671	8713	43
2.4	8755	8796	8838	8879	8920	8961	9002	9042	9083	9123	41
2.5	0.9163	9203	9243	9282	9322	9361	9400	9439	9478	9517	39
2.6	9555	9594	9632	9670	9708	9746	9783	9821	9858	9895	38
2.7	0.9933	9969	*0006	*0043	*0080	*0116	*0152	*0188	*0225	*0260	36
2.8	1.0296	0332	0367	0403	0438	0473	0508	0543	0578	0613	35
2.9	0647	0682	0716	0750	0784	0818	0852	0886	0919	0953	34
3.0	1.0986	1019	1053	1086	1119	1151	1184	1217	1249	1282	33
3.1	1314	1346	1378	1410	1442	1474	1506	1537	1569	1600	32
3.2	1632	1663	1694	1725	1756	1787	1817	1848	1878	1909	31
3.3	1939	1969	2000	2030	2060	2090	2119	2149	2179	2208	30
3.4	2238	2267	2296	2326	2355	2384	2413	2442	2470	2499	29
3.5	1.2528	2556	2585	2613	2641	2669	2698	2726	2754	2782	28
3.6	2809	2837	2865	2892	2920	2947	2975	3002	3029	3056	27
3.7	3083	3110	3137	3164	3191	3218	3244	3271	3297	3324	27
3.8	3350	3376	3403	3429	3455	3481	3507	3533	3558	3584	26
3.9	3610	3635	3661	3686	3712	3737	3762	3788	3813	3838	25
4.0	1.3863	3888	3913	3938	3962	3987	4012	4036	4061	4085	25
4.1	4110	4134	4159	4183	4207	4231	4255	4279	4303	4327	24
4.2	4351	4375	4398	4422	4446	4469	4493	4516	4540	4563	23
4.3	4586	4609	4633	4656	4679	4702	4725	4748	4770	4793	23
4.4	4816	4839	4861	4884	4907	4929	4951	4974	4996	5019	22
4.5	1.5041	5063	5085	5107	5129	5151	5173	5195	5217	5239	22
4.6	5261	5282	5304	5326	5347	5369	5390	5412	5433	5454	21
4.7	5476	5497	5518	5539	5560	5581	5602	5623	5644	5665	21
4.8	5686	5707	5728	5748	5769	5790	5810	5831	5851	5872	20
4.9	5892	5913	5933	5953	5974	5994	6014	6034	6054	6074	20

$\ln x = (2.3026) \log_{10} x \qquad \log_{10} x = (0.4343) \ln x$

where $2.3026 = \ln 10$ and $0.4343 = \log_{10} e$

* From Marks' "Mechanical Engineers' Handbook," Theodore Baumeister (ed.), 6th ed., McGraw-Hill Book Company, Inc., New York, 1958.

Table C-4. Natural Logarithms (*Continued*)

Number	0	1	2	3	4	5	6	7	8	9	Avg diff
5.0	1.6094	6114	6134	6154	6174	6194	6214	6233	6253	6273	20
5.1	6292	6312	6332	6351	6371	6390	6409	6429	6448	6467	19
5.2	6487	6506	6525	6544	6563	6582	6601	6620	6639	6658	19
5.3	6677	6696	6715	6734	6752	6771	6790	6808	6827	6845	18
5.4	6864	6882	6901	6919	6938	6956	6974	6993	7011	7029	18
5.5	1.7047	7066	7084	7102	7120	7138	7156	7174	7192	7210	18
5.6	7228	7246	7263	7281	7299	7317	7334	7352	7370	7387	18
5.7	7405	7422	7440	7457	7475	7492	7509	7527	7544	7561	17
5.8	7579	7596	7613	7630	7647	7664	7681	7699	7716	7733	17
5.9	7750	7766	7783	7800	7817	7834	7851	7867	7884	7901	17
6.0	1.7918	7934	7951	7967	7984	8001	8017	8034	8050	8066	16
6.1	8083	8099	8116	8132	8148	8165	8181	8197	8213	8229	16
6.2	8245	8262	8278	8294	8310	8326	8342	8358	8374	8390	16
6.3	8405	8421	8437	8453	8469	8485	8500	8516	8532	8547	16
6.4	8563	8579	8594	8610	8625	8641	8656	8672	8687	8703	15
6.5	1.8718	8733	8749	8764	8779	8795	8810	8825	8840	8856	15
6.6	8871	8886	8901	8916	8931	8946	8961	8976	8991	9006	15
6.7	9021	9036	9051	9066	9081	9095	9110	9125	9140	9155	15
6.8	9169	9184	9199	9213	9228	9242	9257	9272	9286	9301	15
6.9	9315	9330	9344	9359	9373	9387	9402	9416	9430	9445	14
7.0	1.9459	9473	9488	9502	9516	9530	9544	9559	9573	9587	14
7.1	9601	9615	9629	9643	9657	9671	9685	9699	9713	9727	14
7.2	9741	9755	9769	9782	9796	9810	9824	9838	9851	9865	14
7.3	1.9879	9892	9906	9920	9933	9947	9961	9974	9988	*0001	13
7.4	2.0015	0028	0042	0055	0069	0082	0096	0109	0122	0136	13
7.5	2.0149	0162	0176	0189	0202	0215	0229	0242	0255	0268	13
7.6	0281	0295	0308	0321	0334	0347	0360	0373	0386	0399	13
7.7	0412	0425	0438	0451	0464	0477	0490	0503	0516	0528	13
7.8	0541	0554	0567	0580	0592	0605	0618	0631	0643	0656	13
7.9	0669	0681	0694	0707	0719	0732	0744	0757	0769	0782	12
8.0	2.0794	0807	0819	0832	0844	0857	0869	0882	0894	0906	12
8.1	0919	0931	0943	0956	0968	0980	0992	1005	1017	1029	12
8.2	1041	1054	1066	1078	1090	1102	1114	1126	1138	1150	12
8.3	1163	1175	1187	1199	1211	1223	1235	1247	1258	1270	12
8.4	1282	1294	1306	1318	1330	1342	1353	1365	1377	1389	12
8.5	2.1401	1412	1424	1436	1448	1459	1471	1483	1494	1506	12
8.6	1518	1529	1541	1552	1564	1576	1587	1599	1610	1622	12
8.7	1633	1645	1656	1668	1679	1691	1702	1713	1725	1736	11
8.8	1748	1759	1770	1782	1793	1804	1815	1827	1838	1849	11
8.9	1861	1872	1883	1894	1905	1917	1928	1939	1950	1961	11
9.0	2.1972	1983	1994	2006	2017	2028	2039	2050	2061	2072	11
9.1	2083	2094	2105	2116	2127	2138	2148	2159	2170	2181	11
9.2	2192	2203	2214	2225	2235	2246	2257	2268	2279	2289	11
9.3	2300	2311	2322	2332	2343	2354	2364	2375	2386	2396	11
9.4	2407	2418	2428	2439	2450	2460	2471	2481	2492	2502	11
9.5	2.2513	2523	2534	2544	2555	2565	2576	2586	2597	2607	10
9.6	2618	2628	2638	2649	2659	2670	2680	2690	2701	2711	10
9.7	2721	2732	2742	2752	2762	2773	2783	2793	2803	2814	10
9.8	2824	2834	2844	2854	2865	2875	2885	2895	2905	2915	10
9.9	2925	2935	2946	2956	2966	2976	2986	2996	3006	3016	10
10.0	2.3026										

Moving the decimal point n places to the right [or left] in the number requires adding n times 2.3026 [or n times (0.6974–3)] in the body of the table. See auxiliary table of multiples on top of the preceding page.

Table C-5. Reciprocals, Squares, and Cubes*

n	$1/n$	n^2	n^3	$\sqrt{n}$	$\sqrt[3]{n}$
1	1.	1	1	1	1.
2	0.50000	4	8	1.414	1.260
3	.33333	9	27	1.732	1.442
4	.25000	16	64	2.000	1.587
5	.20000	25	125	2.236	1.710
6	.16667	36	216	2.449	1.817
7	.14286	49	343	2.646	1.913
8	.12500	64	512	2.828	2.000
9	.11111	81	729	3.000	2.080
10	.10000	100	1,000	3.162	2.154
11	.09091	121	1,331	3.317	2.224
12	.08333	144	1,728	3.464	2.289
13	.07692	169	2,197	3.606	2.351
14	.07143	196	2,744	3.742	2.410
15	.06667	225	3,375	3.873	2.466
16	.06250	256	4,096	4.000	2.520
17	.05882	289	4,913	4.123	2.571
18	.05556	324	5,832	4.243	2.621
19	.05263	361	6,859	4.359	2.668
20	.05000	400	8,000	4.472	2.714
21	.04762	441	9,261	4.583	2.759
22	.04545	484	10,648	4.690	2.802
23	.04348	529	12,167	4.796	2.844
24	.04167	576	13,824	4.899	2.884
25	.04000	625	15,625	5.000	2.924
26	.03846	676	17,576	5.099	2.962
27	.03704	729	19,683	5.196	3.000
28	.03571	784	21,952	5.291	3.037
29	.03448	841	24,389	5.385	3.072
30	.03333	900	27,000	5.477	3.107
31	.03226	961	29,791	5.568	3.141
32	.03125	1,024	32,768	5.657	3.175
33	.03030	1,089	35,937	5.745	3.208
34	.02941	1,156	39,304	5.831	3.240
35	.02857	1,225	42,875	5.916	3.271
36	.02778	1,296	46,656	6.000	3.302
37	.02703	1,369	50,653	6.083	3.332
38	.02632	1,444	54,872	6.164	3.362
39	.02564	1,521	59,319	6.245	3.391
40	.02500	1,600	64,000	6.325	3.420
41	.02439	1,681	68,921	6.403	3.448
42	.02381	1,764	74,088	6.481	3.476
43	.02326	1,849	79,507	6.557	3.503
44	.02273	1,936	85,184	6.633	3.530
45	.02222	2,025	91,125	6.708	3.557
46	.02174	2,116	97,336	6.782	3.583
47	.02128	2,209	103,823	6.856	3.609
48	.02083	2,304	110,592	6.928	3.634
49	.02041	2,401	117,649	7.000	3.659
50	.02000	2,500	125,000	7.071	3.684

* F. Daniels, "Mathematical Preparation for Physical Chemistry," McGraw-Hill Book Company, Inc., New York, 1928.

Table C-5. Reciprocals, Squares, and Cubes (*Continued*)

n	$1/n$	n^2	n^3	$\sqrt{n}$	$\sqrt[3]{n}$
51	.01961	2,601	132,651	7.141	3.708
52	.01923	2,704	140,608	7.211	3.733
53	.01887	2,809	148,877	7.280	3.756
54	.01852	2,916	157,464	7.348	3.780
55	.01818	3,025	166,375	7.416	3.803
56	.01786	3,136	175,616	7.483	3.826
57	.01754	3,249	185,193	7.550	3.849
58	.01724	3,364	195,112	7.616	3.871
59	.01695	3,481	205,379	7.681	3.893
60	.01667	3,600	216,000	7.746	3.915
61	.01639	3,721	226,981	7.810	3.936
62	.01613	3,844	238,328	7,874	3.958
63	.01587	3,969	250,047	7.937	3.979
64	.01563	4,096	262,144	8.000	4.000
65	.01538	4,225	274,625	8.062	4.021
66	.01515	4,356	287,496	8.124	4.041
67	.01493	4,489	300,763	8.185	4.062
68	.01471	4,624	314,432	8.246	4.082
69	.01449	4,761	328,509	8.307	4.102
70	.01429	4,900	343,000	8.367	4.121
71	.01408	5,041	357,911	8.426	4.141
72	.01389	5,184	373,248	8.485	4.160
73	.01370	5,329	389,017	8.544	4.179
74	.01351	5,476	405,224	8.602	4.198
75	.01333	5,625	421,875	8.660	4.217
76	.01316	5,776	438,976	8.718	4.236
77	.01299	5,929	456,533	8.775	4.254
78	.01282	6,084	474,552	8.832	4.273
79	.01266	6,241	493,039	8.888	4.291
80	.01250	6,400	512,000	8.944	4.309
81	.01235	6,561	531,441	9.000	4.327
82	.01220	6,724	551,368	9.055	4.344
83	.01205	6,889	571,787	9.110	4.362
84	.01191	7,056	592,704	9.165	4.380
85	.01177	7,225	614,125	9.220	4,397
86	.01163	7,396	636,056	9.274	4.414
87	.01149	7,569	658,503	9.327	4.431
88	.01136	7,744	681,472	9.381	4.448
89	.01124	7,921	704,969	9.434	4.465
90	.01111	8,100	729,000	9.487	4.481
91	.01099	8,281	753,571	9.539	4.498
92	.01087	8,464	778,688	9.592	4.514
93	.01075	8,649	804,357	9.644	4.531
94	.01064	8,836	830,584	9.695	4.547
95	.01053	9,025	857,375	9.747	4.563
96	.01042	9,216	884,736	9.798	4.579
97	.01031	9,409	912,673	9.849	4.595
98	.01020	9,604	941,192	9.899	4.610
99	.01010	9,801	970,299	9.950	4.626
100	.01000	10,000	1,000,000	10.000	4.642

Table C-6. Circumferences of Circles by Hundredths*

D	0	1	2	3	4	5	6	7	8	9	Avg diff
1.0	3.142	3.173	3.204	3.236	3.267	3.299	3.330	3.362	3.393	3.424	31
.1	3.456	3.487	3.519	3.550	3.581	3.613	3.644	3.676	3.707	3.738	
.2	3.770	3.801	3.833	3.864	3.896	3.927	3.958	3.990	4.021	4.053	
.3	4.084	4.115	4.147	4.178	4.210	4.241	4.273	4.304	4.335	4.367	
.4	4.398	4.430	4.461	4.492	4.524	4.555	4.587	4.618	4.650	4.681	
1.5	4.712	4.744	4.775	4.807	4.838	4.869	4.901	4.932	4.964	4.995	
.6	5.027	5.058	5.089	5.121	5.152	5.184	5.215	5.246	5.278	5.309	
.7	5.341	5.372	5.404	5.435	5.466	5.498	5.529	5.561	5.592	5.623	
.8	5.655	5.686	5.718	5.749	5.781	5.812	5.843	5.875	5.906	5.938	
.9	5.969	6.000	6.032	6.063	6.095	6.126	6.158	6.189	6.220	6.252	
2.0	6.283	6.315	6.346	6.377	6.409	6.440	6.472	6.503	6.535	6.566	
.1	6.597	6.629	6.660	6.692	6.723	6.754	6.786	6.817	6.849	6.880	
.2	6.912	6.943	6.974	7.006	7.037	7.069	7.100	7.131	7.163	7.194	
.3	7.226	7.257	7.288	7.320	7.351	7.383	7.414	7.446	7.477	7.508	
.4	7.540	7.571	7.603	7.634	7.665	7.697	7.728	7.760	7.791	7.823	
2.5	7.854	7.885	7.917	7.948	7.980	8.011	8.042	8.074	8.105	8.137	
.6	8.168	8.200	8.231	8.262	8.294	8.325	8.357	8.388	8.419	8.451	
.7	8.482	8.514	8.545	8.577	8.608	8.639	8.671	8.702	8.734	8.765	
.8	8.796	8.828	8.859	8.891	8.922	8.954	8.985	9.016	9.048	9.079	
.9	9.111	9.142	9.173	9.205	9.236	9.268	9.299	9.331	9.362	9.393	
3.0	9.425	9.456	9.488	9.519	9.550	9.582	9.613	9.645	9.676	9.708	
.1	9.739	9.770	9.802	9.833	9.865	9.896	9.927	9.959	9.990	10.022	31
.1										10.02	3
.2	10.05	10.08	10.12	10.15	10.18	10.21	10.24	10.27	10.30	10.34	
.3	10.37	10.40	10.43	10.46	10.49	10.52	10.56	10.59	10.62	10.65	
.4	10.68	10.71	10.74	10.78	10.81	10.84	10.87	10.90	10.93	10.96	
3.5	11.00	11.03	11.06	11.09	11.12	11.15	11.18	11.22	11.25	11.28	
.6	11.31	11.34	11.37	11.40	11.44	11.47	11.50	11.53	11.56	11.59	
.7	11.62	11.66	11.69	11.72	11.75	11.78	11.81	11.84	11.88	11.91	
.8	11.94	11.97	12.00	12.03	12.06	12.10	12.13	12.16	12.19	12.22	
.9	12.25	12.28	12.32	12.35	12.38	12.41	12.44	12.47	12.50	12.53	
4.0	12.57	12.60	12.63	12.66	12.69	12.72	12.75	12.79	12.82	12.85	
.1	12.88	12.91	12.94	12.97	13.01	13.04	13.07	13.10	13.13	13.16	
.2	13.19	13.23	13.26	13.29	13.32	13.35	13.38	13.41	13.45	13.48	
.3	13.51	13.54	13.57	13.60	13.63	13.67	13.70	13.73	13.76	13.79	
.4	13.82	13.85	13.89	13.92	13.95	13.98	14.01	14.04	14.07	14.11	
4.5	14.14	14.17	14.20	14.23	14.26	14.29	14.33	14.36	14.39	14.42	
.6	14.45	14.48	14.51	14.55	14.58	14.61	14.64	14.67	14.70	14.73	
.7	14.77	14.80	14.83	14.86	14.89	14.92	14.95	14.99	15.02	15.05	
.8	15.08	15.11	15.14	15.17	15.21	15.24	15.27	15.30	15.33	15.36	
.9	15.39	15.43	15.46	15.49	15.52	15.55	15.58	15.61	15.65	15.68	

Explanation of Table of Circumferences

This table gives the product of π times any number *D* from 1 to 10; that is, it is a table of multiples of π. (*D* = diameter.)

Moving the decimal point ONE place in column *D* is equivalent to moving it ONE place in the body of the table.

$$\text{Circumference} = \pi \times \text{diam} = 3.141593 \times \text{diam}$$

Conversely,

$$\text{Diameter} = \frac{1}{\pi} \times \text{circum} = 0.31831 \times \text{circum}$$

* From Marks' "Mechanical Engineers' Handbook," Theodore Baumeister (ed.), 6th ed., McGraw-Hill Book Company, Inc., New York, 1958.

Table C-6. Circumferences of Circles by Hundredths (*Continued*)

D	0	1	2	3	4	5	6	7	8	9	Avg diff
5.0	15.71	15.74	15.77	15.80	15.83	15.87	15.90	15.93	15.96	15.99	3
.1	16.02	16.05	16.08	16.12	16.15	16.18	16.21	16.24	16.27	16.30	
.2	16.34	16.37	16.40	16.43	16.46	16.49	16.52	16.56	16.59	16.62	
.3	16.65	16.68	16.71	16.74	16.78	16.81	16.84	16.87	16.90	16.93	
.4	16.96	17.00	17.03	17.06	17.09	17.12	17.15	17.18	17.22	17.25	
5.5	17.28	17.31	17.34	17.37	17.40	17.44	17.47	17.50	17.53	17.56	
.6	17.59	17.62	17.66	17.69	17.72	17.75	17.78	17.81	17.84	17.88	
.7	17.91	17.94	17.97	18.00	18.03	18.06	18.10	18.13	18.16	18.19	
.8	18.22	18.25	18.28	18.32	18.35	18.38	18.41	18.44	18.47	18.50	
.9	18.54	18.57	18.60	18.63	18.66	18.69	18.72	18.76	18.79	18.82	
6.0	18.85	18.88	18.91	18.94	18.98	19.01	19.04	19.07	19.10	19.13	
.1	19.16	19.20	19.23	19.26	19.29	19.32	19.35	19.38	19.42	19.45	
.2	19.48	19.51	19.54	19.57	19.60	19.63	19.67	19.70	19.73	19.76	
.3	19.79	19.82	19.85	19.89	19.92	19.95	19.98	20.01	20.04	20.07	
.4	20.11	20.14	20.17	20.20	20.23	20.26	20.29	20.33	20.36	20.39	
6.5	20.42	20.45	20.48	20.51	20.55	20.58	20.61	20.64	20.67	20.70	
.6	20.73	20.77	20.80	20.83	20.86	20.89	20.92	20.95	20.99	21.02	
.7	21.05	21.08	21.11	21.14	21.17	21.21	21.24	21.27	21.30	21.33	
.8	21.36	21.39	21.43	21.46	21.49	21.52	21.55	21.58	21.61	21.65	
.9	21.68	21.71	21.74	21.77	21.80	21.83	21.87	21.90	21.93	21.96	
7.0	21.99	22.02	22.05	22.09	22.12	22.15	22.18	22.21	22.24	22.27	
.1	22.31	22.34	22.37	22.40	22.43	22.46	22.49	22.53	22.56	22.59	
.2	22.62	22.65	22.68	22.71	22.75	22.78	22.81	22.84	22.87	22.90	
.3	22.93	22.97	23.00	23.03	23.06	23.09	23.12	23.15	23.18	23.22	
.4	23.25	23.28	23.31	23.34	23.37	23.40	23.44	23.47	23.50	23.53	
7.5	23.56	23.59	23.62	23.66	23.69	23.72	23.75	23.78	23.81	23.84	
.6	23.88	23.91	23.94	23.97	24.00	24.03	24.06	24.10	24.13	24.16	
.7	24.19	24.22	24.25	24.28	24.32	24.35	24.38	24.41	24.44	24.47	
.8	24.50	24.54	24.57	24.60	24.63	24.66	24.69	24.72	24.76	24.79	
.9	24.82	24.85	24.88	24.91	24.94	24.98	25.01	25.04	25.07	25.10	
8.0	25.13	25.16	25.20	25.23	25.26	25.29	25.32	25.35	25.38	25.42	
.1	25.45	25.48	25.51	25.54	25.57	25.60	25.64	25.67	25.70	25.73	
.2	25.76	25.79	25.82	25.86	25.89	25.92	25.95	25.98	26.01	26.04	
.3	26.08	26.11	26.14	26.17	26.20	26.23	26.26	26.30	26.33	26.36	
.4	26.39	26.42	26.45	26.48	26.52	26.55	26.58	26.61	26.64	26.67	
8.5	26.70	26.73	26.77	26.80	26.83	26.86	26.89	26.92	26.95	26.99	
.6	27.02	27.05	27.08	27.11	27.14	27.17	27.21	27.24	27.27	27.30	
.7	27.33	27.36	27.39	27.43	27.46	27.49	27.52	27.55	27.58	27.61	
.8	27.65	27.68	27.71	27.74	27.77	27.80	27.83	27.87	27.90	27.93	
.9	27.96	27.99	28.02	28.05	28.09	28.12	28.15	28.18	28.21	28.24	
9.0	28.27	28.31	28.34	28.37	28.40	28.43	28.46	28.49	28.53	28.56	
.1	28.59	28.62	28.65	28.68	28.71	28.75	28.78	28.81	28.84	28.87	
.2	28.90	28.93	28.97	29.00	29.03	29.06	29.09	29.12	29.15	29.19	
.3	29.22	29.25	29.28	29.31	29.34	29.37	29.41	29.44	29.47	29.50	
.4	29.53	29.56	29.59	29.63	29.66	29.69	29.72	29.75	29.78	29.81	
9.5	29.85	29.88	29.91	29.94	29.97	30.00	30.03	30.07	30.10	30.13	
.6	30.16	30.19	30.22	30.25	30.28	30.32	30.35	30.38	30.41	30.44	
.7	30.47	30.50	30.54	30.57	30.60	30.63	30.66	30.69	30.72	30.76	
.8	30.79	30.82	30.85	30.88	30.91	30.94	30.98	31.01	31.04	31.07	
.9	31.10	31.13	31.16	31.20	31.23	31.26	31.29	31.32	31.35	31.38	
10.0	31.42										

Moving the decimal point ONE place in *D* requires moving it ONE place in body of table

Table C-7. Areas of Circles by Hundredths*

D	0	1	2	3	4	5	6	7	8	9	Avg diff
1.0	0.785	0.801	0.817	0.833	0.849	0.866	0.882	0.899	0.916	0.933	16
.1	0.950	0.968	0.985	1.003	1.021	1.039	1.057	1.075	1.094	1.112	18
.2	1.131	1.150	1.169	1.188	1.208	1.227	1.247	1.267	1.287	1.307	20
.3	1.327	1.348	1.368	1.389	1.410	1.431	1.453	1.474	1.496	1.517	21
.4	1.539	1.561	1.584	1.606	1.629	1.651	1.674	1.697	1.720	1.744	23
1.5	1.767	1.791	1.815	1.839	1.863	1.887	1.911	1.936	1.961	1.986	24
.6	2.011	2.036	2.061	2.087	2.112	2.138	2.164	2.190	2.217	2.243	26
.7	2.270	2.297	2.324	2.351	2.378	2.405	2.433	2.461	2.488	2.516	27
.8	2.545	2.573	2.602	2.630	2.659	2.688	2.717	2.746	2.776	2.806	29
.9	2.835	2.865	2.895	2.926	2.956	2.986	3.017	3.048	3.079	3.110	31
2.0	3.142	3.173	3.205	3.237	3.269	3.301	3.333	3.365	3.398	3.431	32
.1	3.464	3.497	3.530	3.563	3.597	3.631	3.664	3.698	3.733	3.767	34
.2	3.801	3.836	3.871	3.906	3.941	3.976	4.011	4.047	4.083	4.119	35
.3	4.155	4.191	4.227	4.264	4.301	4.337	4.374	4.412	4.449	4.486	37
.4	4.524	4.562	4.600	4.638	4.676	4.714	4.753	4.792	4.831	4.870	38
2.5	4.909	4.948	4.988	5.027	5.067	5.107	5.147	5.187	5.228	5.269	40
.6	5.309	5.350	5.391	5.433	5.474	5.515	5.557	5.599	5.641	5.683	42
.7	5.726	5.768	5.811	5.853	5.896	5.940	5.983	6.026	6.070	6.114	43
.8	6.158	6.202	6.246	6.290	6.335	6.379	6.424	6.469	6.514	6.560	45
.9	6.605	6.651	6.697	6.743	6.789	6.835	6.881	6.928	6.975	7.022	46
3.0	7.069	7.116	7.163	7.211	7.258	7.306	7.354	7.402	7.451	7.499	48
.1	7.548	7.596	7.645	7.694	7.744	7.793	7.843	7.892	7.942	7.992	49
.2	8.042	8.093	8.143	8.194	8.245	8.296	8.347	8.398	8.450	8.501	51
.3	8.553	8.605	8.657	8.709	8.762	8.814	8.867	8.920	8.973	9.026	53
.4	9.079	9.133	9.186	9.240	9.294	9.348	9.402	9.457	9.511	9.566	54
3.5	9.621	9.676	9.731	9.787	9.842	9.898	9.954	10.010			56
								10.01	10.07	10.12	6
.6	10.18	10.24	10.29	10.35	10.41	10.46	10.52	10.58	10.64	10.69	6
.7	10.75	10.81	10.87	10.93	10.99	11.04	11.10	11.16	11.22	11.28	
.8	11.34	11.40	11.46	11.52	11.58	11.64	11.70	11.76	11.82	11.88	
.9	11.95	12.01	12.07	12.13	12.19	12.25	12.32	12.38	12.44	12.50	
4.0	12.57	12.63	12.69	12.76	12.82	12.88	12.95	13.01	13.07	13.14	7
.1	13.20	13.27	13.33	13.40	13.46	13.53	13.59	13.66	13.72	13.79	
.2	13.85	13.92	13.99	14.05	14.12	14.19	14.25	14.32	14.39	14.45	
.3	14.52	14.59	14.66	14.73	14.79	14.86	14.93	15.00	15.07	15.14	
.4	15.21	15.27	15.34	15.41	15.48	15.55	15.62	15.69	15.76	15.83	
4.5	15.90	15.98	16.05	16.12	16.19	16.26	16.33	16.40	16.47	16.55	
.6	16.62	16.69	16.76	16.84	16.91	16.98	17.06	17.13	17.20	17.28	
.7	17.35	17.42	17.50	17.57	17.65	17.72	17.80	17.87	17.95	18.02	
.8	18.10	18.17	18.25	18.32	18.40	18.47	18.55	18.63	18.70	18.78	8
.9	18.86	18.93	19.01	19.09	19.17	19.24	19.32	19.40	19.48	19.56	

Explanation of Table of Areas of Circles

Moving the decimal point ONE place in column *D* is equivalent to moving it TWO places in the body of the table. (*D* = diameter.)

$$\text{Area of circle} = \frac{\pi}{4} \times (\text{diam}^2) = 0.785398 \times (\text{diam}^2)$$

Conversely,

$$\text{Diam} = \sqrt{\frac{4}{\pi}} \times \sqrt{\text{area}} = 1.128379 \times \sqrt{\text{area}}$$

* From Marks' "Mechanical Engineers' Handbook," Theodore Baumeister (ed.), 6th ed., McGraw-Hill Book Company, Inc., New York, 1958.

Table C-7. Areas of Circles by Hundredths (*Continued*)

D	0	1	2	3	4	5	6	7	8	9	Avg diff
5.0	19.63	19.71	19.79	19.87	19.95	20.03	20.11	20.19	20.27	20.35	8
.1	20.43	20.51	20.59	20.67	20.75	20.83	20.91	20.99	21.07	21.16	
.2	21.24	21.32	21.40	21.48	21.57	21.65	21.73	21.81	21.90	21.98	
.3	22.06	22.15	22.23	22.31	22.40	22.48	22.56	22.65	22.73	22.82	
.4	22.90	22.99	23.07	23.16	23.24	23.33	23.41	23.50	23.59	23.67	9
5.5	23.76	23.84	23.93	24.02	24.11	24.19	24.28	24.37	24.45	24.54	
.6	24.63	24.72	24.81	24.89	24.98	25.07	25.16	25.25	25.34	25.43	
.7	25.52	25.61	25.70	25.79	25.88	25.97	26.06	26.15	26.24	26.33	
.8	26.42	26.51	26.60	26.69	26.79	26.88	26.97	27.06	27.15	27.25	
.9	27.34	27.43	27.53	27.62	27.71	27.81	27.90	27.99	28.09	28.18	
6.0	28.27	28.37	28.46	28.56	28.65	28.75	28.84	28.94	29.03	29.13	10
.1	29.22	29.32	29.42	29.51	29.61	29.71	29.80	29.90	30.00	30.09	
.2	30.19	30.29	30.39	30.48	30.58	30.68	30.78	30.88	30.97	31.07	
.3	31.17	31.27	31.37	31.47	31.57	31.67	31.77	31.87	31.97	32.07	
.4	32.17	32.27	32.37	32.47	32.57	32.67	32.78	32.88	32.98	33.08	
6.5	33.18	33.29	33.39	33.49	33.59	33.70	33.80	33.90	34.00	34.11	
.6	34.21	34.32	34.42	34.52	34.63	34.73	34.84	34.94	35.05	35.15	
.7	35.26	35.36	35.47	35.57	35.68	35.78	35.89	36.00	36.10	36.21	11
.8	36.32	36.42	36.53	36.64	36.75	36.85	36.96	37.07	37.18	37.28	
.9	37.39	37.50	37.61	37.72	37.83	37.94	38.05	38.16	38.26	38.37	
7.0	38.48	38.59	38.70	38.82	38.93	39.04	39.15	39.26	39.37	39.48	
.1	39.59	39.70	39.82	39.93	40.04	40.15	40.26	40.38	40.49	40.60	
.2	40.72	40.83	40.94	41.06	41.17	41.28	41.40	41.51	41.62	41.74	
.3	41.85	41.97	42.08	42.20	42.31	42.43	42.54	42.66	42.78	42.89	12
.4	43.01	43.12	43.24	43.36	43.47	43.59	43.71	43.83	43.94	44.06	
7.5	44.18	44.30	44.41	44.53	44.65	44.77	44.89	45.01	45.13	45.25	
.6	45.36	45.48	45.60	45.72	45.84	45.96	46.08	46.20	46.32	46.45	
.7	46.57	46.69	46.81	46.93	47.05	47.17	47.29	47.42	47.54	47.66	
.8	47.78	47.91	48.03	48.15	48.27	48.40	48.52	48.65	48.77	48.89	
.9	49.02	49.14	49.27	49.39	49.51	49.64	49.76	49.89	50.01	50.14	
8.0	50.27	50.39	50.52	50.64	50.77	50.90	51.02	51.15	51.28	51.40	13
.1	51.53	51.66	51.78	51.91	52.04	52.17	52.30	52.42	52.55	52.68	
.2	52.81	52.94	53.07	53.20	53.33	53.46	53.59	53.72	53.85	53.98	
.3	54.11	54.24	54.37	54.50	54.63	54.76	54.89	55.02	55.15	55.29	
.4	55.42	55.55	55.68	55.81	55.95	56.08	56.21	56.35	56.48	56.61	
8.5	56.75	56.88	57.01	57.15	57.28	57.41	57.55	57.68	57.82	57.95	
.6	58.09	58.22	58.36	58.49	58.63	58.77	58.90	59.04	59.17	59.31	14
.7	59.45	59.58	59.72	59.86	59.99	60.13	60.27	60.41	60.55	60.68	
.8	60.82	60.96	61.10	61.24	61.38	61.51	61.65	61.79	61.93	62.07	
.9	62.21	62.35	62.49	62.63	62.77	62.91	63.05	63.19	63.33	63.48	
9.0	63.62	63.76	63.90	64.04	64.18	64.33	64.47	64.61	64.75	64.90	
.1	65.04	65.18	65.33	65.47	65.61	65.76	65.90	66.04	66.19	66.33	15
.2	66.48	66.62	66.77	66.91	67.06	67.20	67.35	67.49	67.64	67.78	
.3	67.93	68.08	68.22	68.37	68.51	68.66	68.81	68.96	69.10	69.25	
.4	69.40	69.55	69.69	69.84	69.99	70.14	70.29	70.44	70.58	70.73	
9.5	70.88	71.03	71.18	71.33	71.48	71.63	71.78	71.93	72.08	72.23	
.6	72.38	72.53	72.68	72.84	72.99	73.14	73.29	73.44	73.59	73.75	
.7	73.90	74.05	74.20	74.36	74.51	74.66	74.82	74.97	75.12	75.28	
.8	75.43	75.58	75.74	75.89	76.05	76.20	76.36	76.51	76.67	76.82	
.9	76.98	77.13	77.29	77.44	77.60	77.76	77.91	78.07	78.23	78.38	16

Moving the decimal point ONE place in *D* requires moving it TWO places in body of table

Table C-8. Weights and Measures of Various Systems*

United States Customary System

Linear Measure

12 inches (in.) or (″)	= 1 foot (ft.) or (′)
3 feet	= 1 yard (yd.)
16.5 feet, 5.5 yards	= 1 rod (rd.)
5280 feet, 320 rods	= 1 mile (mi.)
1 mil	= 0.001 inch (in.)

Nautical:

6080.2 feet	= 1 nautical mile
6 feet	= 1 fathom
120 fathoms	= 1 cable length
1 knot	= 1 nautical mile per hour
60 knots	= 1° (measured at equator)

Square Measure

144 sq. inches (sq. in.) or (in.²) or (□″)	= 1 sq. foot (ft.²) or (□′)
9 sq. feet (ft.²) (□′)	= 1 sq. yard (yd.²)
30.25 sq. yards	= 1 sq. rod, pole, or perch
160 sq. rods = {10 sq. chains, 43,560 sq. ft.}	= 1 acre
640 acres = 1 sq. mile	= 1 section
1 circular inch (area of circle of 1 inch diameter)	= 0.7854 sq. inch
1 sq. inch	= 1.2732 circular inch
1 circular mil	= area of circle of 0.001 inch diameter
1,000,000 circular mils	= 1 circular inch

Circular Measure

60 seconds (″) (sec.)	= 1 minute (min.) or (′)
60 minutes (′)	= 1 degree (°)
90 degrees (°)	= 1 quadrant
360 degrees (°)	= 1 circumference
57.29578 degrees	= 1 radian (rad.) = 57° 17′ 44.81″

Volume Measure

Solid:

1728 cubic in. (cu. in.) (in.³)	= 1 cubic foot (cu. ft.) (ft.³)
27 cu. ft.	= 1 cubic yard (cu. yd.)

Dry Measure:

2 pints	= 1 quart
8 quarts	= 1 peck
4 pecks	= 1 bushel
1 United States Winchester bushel	= 2150.42 cubic inches

Liquid:

4 gills	= 1 pint (pt.)
2 pints	= 1 quart (qt.)
4 quarts	= 1 gallon (gal.)
7.4805 gallons	= 1 cubic foot

Apothecaries' Liquid:

60 minims (min. or ♏)	= 1 fluid dram or drachm
8 drams (ʒ)	= 1 fluid ounce
16 ounces (oz. or ℥)	= 1 pint

Water Measure:

1 miner's inch = amount of water flowing through an orifice of 1 sq. in. cross section under a head varying from 4 to 6.5 in. (as fixed by state law). Units now most generally used are: 1 cu. ft. per sec. (1 ft.³/sec.) or 1 gal. per sec. (1 gal./sec.).

* Prepared by Dr. Lewis V. Judson, National Bureau of Standards, for "Chemical Engineers' Handbook," John H. Perry (ed.), 3d ed., McGraw-Hill Book Company, Inc., New York, 1950.

Table C-8. Weights and Measures of Various Systems (*Continued*)

Avoirdupois Weight

16 drams	=	437.5 grains	= 1 ounce (oz.)
16 ounces	=	7000 grains	= 1 pound (lb.)
100 pounds	=	1 hundredweight (cwt.)	
2000 pounds	=	1 short ton; 2240 pounds	= 1 long ton

Troy Weight

24 grains	=	1 pennyweight (dwt.)
20 pennyweights	=	1 ounce (oz.)
12 ounces	=	1 pound (lb.)

Apothecaries' Weight

20 grains (gr.)	=	1 scruple (℈)
3 scruples	=	1 dram (ʒ)
8 drams	=	1 ounce (℥)
12 ounces	=	1 pound (lb.)

Board Measure (B.M.)

1 board foot = product of 1 foot length, 1 foot breadth, and 1 inch thickness.

Metric System

Linear Measure

1 micromicron ($\mu\mu$)	=	0.000001 micron
1 Ångström unit (Å.)	=	0.0001 micron
1 millimicron (mμ)	=	0.001 micron (μ)
1 micron (μ)	=	0.001 millimeter (mm.)
10 millimeters (mm.)	=	1 centimeter (cm.)
10 centimeters (cm.)	=	1 decimeter (dm.)
10 decimeters (dm.) } 100 centimeters (cm.) }	=	1 meter (m.) = 39.37 inches
1 dekameter (dkm.)	=	10 meters (m.)
1 hectometer (hm.)	=	100 meters (m.)
1 kilometer (km.)	=	1,000 meters (m.)
1 myriameter	=	10,000 meters (m.)
1 megameter	=	1,000,000 meters (m.)

Square Measure

1 sq. millimeter (mm.^{2})	=	0.01 sq. centimeter (sq. cm.)
1 sq. centimeter (cm.^{2})	=	0.01 sq. decimeter (sq. dm.)
1 sq. decimeter (dm.^{2})	=	0.01 sq. meter (sq. m.)
1 sq. meter (centiare) or (m.^{2})	=	0.01 sq. dekameter (sq. dkm.)
1 sq. dekameter (are)	=	0.01 sq. hectometer (sq. hm.)
1 hectare (ha.)	=	10,000 sq. meters
1 sq. kilometer (km.^{2})	=	100 sq. hectometers

Volume Measure

1 cubic millimeter	=	10^{-6} liter = 0.001 cu. cm.
1 cubic centimeter	=	10^{-3} liter = 0.001 cu. dm.
1 cubic decimeter	=	1 liter† = 0.001 cu. m.
1 decistere	=	0.1 cubic meter
1 stere	=	1 cubic meter
1 dekastere	=	10 cubic meters
1 microliter (μl. or λ)	=	10^{-6} liter = 0.000001 liter
1 milliliter (ml.)	=	10^{-3} liter = 0.001 liter
1 centiliter (cl.)	=	10^{-2} liter = 0.01 liter
1 deciliter (dl.)	=	10^{-1} liter = 0.1 liter
1 dekaliter (dkl.)	=	10 liters
1 hectoliter (hl.)	=	10^2 liters = 100 liters

† Accurately: 1 liter = 1.000028 cubic decimeters.

Table C-8. Weights and Measures of Various Systems (*Continued*)

Mass Measure

1 microgram (μg. or γ)	=	10^{-6} gram
1 milligram (mg.)	=	10^{-3} gram
1 centigram (cg.)	=	10^{-2} gram
1 decigram (dg.)	=	10^{-1} gram = 0.1 gram
15.432 grains (gr.)	=	1 gram
1 dekagram (dkg.)	=	10 grams
1 hectogram (hg.)	=	10^2 grams = 100 grams
1 kilogram (kg.)	=	10^3 grams = 1,000 grams
1 metric ton	=	10^6 grams = 1,000,000 grams
1 metric carat	=	200 milligrams

Table C-9. Conversion Tables*

Length Equivalents

Centimeters	Inches	Feet	Yards	Meters	Chains	Kilometers	Miles
1	**0.3937** $\bar{1}.59517$	**0.03281** $\bar{2}.51598$	**0.01094** $\bar{2}.03886$	**0.01** $\bar{2}.00000$	**0.0_34971** $\bar{4}.69644$	**10^{-5}** $\bar{5}.00000$	**0.0_56214** $\bar{6}.79335$
2.540 0.40483	**1**	**0.08333** $\bar{2}.92082$	**0.02778** $\bar{2}.44370$	**0.0254** $\bar{2}.40483$	**0.001263** $\bar{3}.10127$	**0.0_4254** $\bar{5}.40483$	**0.0_41578** $\bar{5}.19818$
30.48 1.48402	**12** 1.07918	**1**	**0.3333** $\bar{1}.52288$	**0.3048** $\bar{1}.48402$	**0.01515** $\bar{2}.18046$	**0.0_33048** $\bar{4}.48402$	**0.0_31894** $\bar{4}.27736$
91.44 1.96114	**36** 1.55630	**3** 0.47712	**1**	**0.9144** $\bar{1}.96114$	**0.04545** $\bar{2}.65758$	**0.0_39144** $\bar{4}.96114$	**0.0_35682** $\bar{4}.75449$
100 2.00000	**39.37** 1.59517	**3.281** 0.51598	**1.0936** 0.03886	**1**	**0.04971** $\bar{2}.69644$	**0.001** $\bar{3}.00000$	**0.0_36214** $\bar{4}.79335$
2012 3.30356	**792** 2.89873	**66** 1.81954	**22** 1.34242	**20.12** 1.30356	**1**	**0.02012** $\bar{2}.30356$	**0.0125** $\bar{2}.09691$
100000 5.00000	**39370** 4.59517	**3281** 3.51598	**1093.6** 3.03886	**1000** 3.00000	**49.71** 1.69644	**1**	**0.6214** $\bar{1}.79335$
160935 5.20665	**63360** 4.80182	**5280** 3.72263	**1760** 3.24551	**1609** 3.20665	**80** 1.90309	**1.609** 0.20665	**1**

The equivalents are given in the heavier type. Logarithms of the equivalents are given immediately below. In some cases the equivalents have been rounded off, while the logarithm corresponds to the equivalent carried to a greater number of decimal places.

Subscripts after any figure, 0_3, 9_4, etc., mean that that figure is to be repeated the indicated number of times.

* From Marks' "Mechanical Engineers' Handbook," Theodore Baumeister (ed.), 6th ed., McGraw-Hill Book Company, Inc., New York, 1958.

Table C-9. Conversion Tables (*Continued*)
Conversion of Lengths *

	Inches to millimeters	Millimeters to inches	Feet to meters	Meters to feet	Yards to meters	Meters to yards	Miles to kilometers	Kilometers to miles
1	25.40	0.03937	0.3048	3.281	0.9144	1.094	1.609	0.6214
2	50.80	0.07874	0.6096	6.562	1.829	2.187	3.219	1.243
3	76.20	0.1181	0.9144	9.842	2.743	3.281	4.828	1.864
4	101.60	0.1575	1.219	13.12	3.658	4.374	6.437	2.485
5	127.00	0.1968	1.524	16.40	4.572	5.468	8.047	3.107
6	152.40	0.2362	1.829	19.68	5.486	6.562	9.656	3.728
7	177.80	0.2756	2.134	22.97	6.401	7.655	11.27	4.350
8	203.20	0.3150	2.438	26.25	7.315	8.749	12.87	4.971
9	228.60	0.3543	2.743	29.53	8.230	9.842	14.48	5.592

* EXAMPLE: 1 in. = 25.40 mm.

COMMON FRACTIONS OF AN INCH TO MILLIMETERS (FROM 1/64 TO 1 IN.)

64ths	Millimeters	64ths	Millimeters	64ths	Millimeters	64ths	Millimeters	64ths	Millimeters	64ths	Millimeters
1	0.397	13	5.159	25	9.922	37	14.684	49	19.447	57	22.622
2	0.794	14	5.556	26	10.319	38	15.081	50	19.844	58	23.019
3	1.191	15	5.953	27	10.716	39	15.478	51	20.241	59	23.416
4	1.588	16	6.350	28	11.113	40	15.875	52	20.638	60	23.813
5	1.984	17	6.747	29	11.509	41	16.272	53	21.034	61	24.209
6	2.381	18	7.144	30	11.906	42	16.669	54	21.431	62	24.606
7	2.778	19	7.541	31	12.303	43	17.066	55	21.828	63	25.003
8	3.175	20	7.938	32	12.700	44	17.463	56	22.225	64	25.400
9	3.572	21	8.334	33	13.097	45	17.859				
10	3.969	22	8.731	34	13.494	46	18.256				
11	4.366	23	9.128	35	13.891	47	18.653				
12	4.763	24	9.525	36	14.288	48	19.050				

DECIMALS OF AN INCH TO MILLIMETERS (FROM 0.01 IN. TO 0.99 IN.)

	0	1	2	3	4	5	6	7	8	9
.0		0.254	0.508	0.762	1.016	1.270	1.524	1.778	2.032	2.286
.1	2.540	2.794	3.048	3.302	3.556	3.810	4.064	4.318	4.572	4.826
.2	5.080	5.334	5.588	5.842	6.096	6.350	6.604	6.858	7.112	7.366
.3	7.620	7.874	8.128	8.382	8.636	8.890	9.144	9.398	9.652	9.906
.4	10.160	10.414	10.668	10.922	11.176	11.430	11.684	11.938	12.192	12.446
.5	12.700	12.954	13.208	13.462	13.716	13.970	14.224	14.478	14.732	14.986
.6	15.240	15.494	15.748	16.002	16.256	16.510	16.764	17.018	17.272	17.526
.7	17.780	18.034	18.288	18.542	18.796	19.050	19.304	19.558	19.812	20.066
.8	20.320	20.574	20.828	21.082	21.336	21.590	21.844	22.098	22.352	22.606
.9	22.860	23.114	23.368	23.622	23.876	24.130	24.384	24.638	24.892	25.146

MILLIMETERS TO DECIMALS OF AN INCH (FROM 1 TO 99 MM)

	0	1.	2.	3.	4.	5.	6.	7.	8.	9.
0		0.0394	0.0787	0.1181	0.1575	0.1968	0.2362	0.2756	0.3150	0.3543
1	0.3937	0.4331	0.4724	0.5118	0.5512	0.5906	0.6299	0.6693	0.7087	0.7480
2	0.7874	0.8268	0.8661	0.9055	0.9449	0.9842	1.0236	1.0630	1.1024	1.1417
3	1.1811	1.2205	1.2598	1.2992	1.3386	1.3780	1.4173	1.4567	1.4961	1.5354
4	1.5748	1.6142	1.6535	1.6929	1.7323	1.7716	1.8110	1.8504	1.8898	1.9291
5	1.9685	2.0079	2.0472	2.0866	2.1260	2.1654	2.2047	2.2441	2.2835	2.3228
6	2.3622	2.4016	2.4409	2.4803	2.5197	2.5590	2.5984	2.6378	2.6772	2.7165
7	2.7559	2.7953	2.8346	2.8740	2.9134	2.9528	2.9921	3.0315	3.0709	3.1102
8	3.1496	3.1890	3.2283	3.2677	3.3071	3.3464	3.3858	3.4252	3.4646	3.5039
9	3.5433	3.5827	3.6220	3.6614	3.7008	3.7402	3.7795	3.8189	3.8583	3.8976

Table C-9. Conversion Tables (*Continued*)

Area Equivalents

(1 hectare = 100 ares = 10,000 centiares or square meters)

Square meters	Square inches	Square feet	Square yards	Square rods	Square chains	Roods	Acres	Square miles or sections
1	1550	10.76	1.196	0.0395	0.002471	0.0_39884	0.0_32471	0.0_63861
	3.19033	1.03197	0.07773	$\bar{2}.59699$	$\bar{3}.39288$	$\bar{3}.99494$	$\bar{4}.39288$	$\bar{7}.58670$
0.0_36452	**1**	0.006944	0.0_37716	0.0_42551	0.0_51594	0.0_66377	0.0_61594	0.0_92491
$\bar{4}.80967$		$\bar{3}.84164$	$\bar{4}.88740$	$\bar{5}.40667$	$\bar{6}.20255$	$\bar{7}.80461$	$\bar{7}.20255$	$\overline{10}.39637$
0.09290	144	**1**	0.1111	0.003673	0.0_32296	0.0_49184	0.0_42296	0.0_73587
$\bar{2}.96803$	2.15836		$\bar{1}.04576$	$\bar{3}.56503$	$\bar{4}.36091$	$\bar{5}.96297$	$\bar{4}.36091$	$\bar{8}.55473$
0.8361	1296	9	**1**	0.03306	0.002066	0.0_38264	0.0002066	0.0_63228
$\bar{1}.92227$	3.11260	0.95424		$\bar{2}.51927$	$\bar{3}.31515$	$\bar{4}.91721$	$\bar{4}.31515$	$\bar{7}.50898$
25.29	39204	272.25	30.25	**1**	0.0625	0.02500	0.00625	0.0_59766
1.40300	4.59333	2.43497	1.48072		$\bar{2}.79588$	$\bar{2}.39794$	$\bar{3}.79588$	$\bar{6}.98970$
404.7	627264	4356	484	16	**1**	0.4	0.1	0.0001562
2.60712	5.79745	3.63909	2.68484	1.20412		$\bar{1}.60206$	$\bar{1}.00000$	$\bar{4}.19382$
1012	1568160	10890	1210	40	2.5	**1**	0.25	0.0_33906
3.00506	6.19539	4.03703	3.08278	1.60206	0.39794		$\bar{1}.39794$	$\bar{4}.59176$
4047	6272640	43560	4840	160	10	4	**1**	0.001562
3.60712	6.79745	4.63909	3.68484	2.20412	1.00000	0.60206		$\bar{3}.19382$
2589998		27878400	3097600	102400	6400	2560	640	**1**
6.41330		7.44527	6.49102	5.01030	3.80618	3.40824	2.80618	

Volume and Capacity Equivalents

Cubic inches	Cubic feet	Cubic yards	U.S. Apothecary fluid ounces	U.S. quarts		U.S. gallons	Bushels U.S.	Liters
				Liquid	Dry			
1	0.0_35787	0.0_42143	0.5541	0.01732	0.01488	0.0_24329	0.0_34650	0.01639
	$\bar{4}.76246$	$\bar{5}.33109$	$\bar{1}.74360$	$\bar{2}.23845$	$\bar{2}.17263$	$\bar{3}.63639$	$\bar{4}.66748$	$\bar{2}.21450$
1728	**1**	0.03704	957.5	29.92	25.71	7.481	0.8036	28.32
3.23754		$\bar{2}.56864$	2.98114	1.47599	1.41017	0.87393	$\bar{1}.90502$	1.45205
46656	27	**1**	25853	807.9	694.3	202.2	21.70	764.5
4.66891	1.43136		4.41251	2.90736	2.84153	2.30530	1.33638	2.88340
1.805	0.001044	0.0_43868	**1**	0.03125	0.02686	0.007812	0.0_38392	0.02957
0.25640	$\bar{3}.01886$	$\bar{5}.58749$		$\bar{2}.49485$	$\bar{2}.42903$	$\bar{3}.89279$	$\bar{4}.92388$	$\bar{2}.47091$
57.75	0.03342	0.001238	32	**1**	0.8594	0.25	0.02686	0.9463
1.76155	$\bar{2}.52401$	$\bar{3}.09264$	1.50515		$\bar{1}.93418$	$\bar{1}.39794$	$\bar{2}.42903$	$\bar{1}.97604$
67.20	0.03889	0.001440	37.24	1.164	**1**	0.2909	0.03125	1.101
1.82737	$\bar{2}.58983$	$\bar{3}.15847$	1.57097	0.06582		$\bar{1}.46376$	$\bar{2}.49485$	0.04188
231	0.1337	0.004951	128	4	3.437	**1**	0.1074	3.785
2.36361	$\bar{1}.12607$	$\bar{3}.69470$	2.10721	0.60206	0.53624		$\bar{1}.03109$	0.57812
2150	1.244	0.04609	1192	37.24	32	9.309	**1**	35.24
3.33252	0.09498	$\bar{2}.66362$	3.07612	1.57097	1.50515	0.96891		1.54703
61.03	0.03532	0.001308	33.81	1.057	0.9081	0.2642	0.02838	**1**
1.78551	$\bar{2}.54796$	$\bar{3}.11659$	1.52909	0.02394	$\bar{1}.95812$	$\bar{1}.42188$	$\bar{2}.45297$	

Conversion of Areas *

	Sq in. to sq cm	Sq cm to sq in.	Sq ft to sq m	Sq m to sq ft	Sq yd to sq m	Sq m to sq yd	Acres to hectares	Hectares to acres	Sq mi to sq km	Sq km to sq mi
1	6.452	0.1550	0.0929	10.76	0.8361	1.196	0.4047	2.471	2.590	0.3861
2	12.90	0.3100	0.1858	21.53	1.672	2.392	0.8094	4.942	5.180	0.7722
3	19.35	0.4650	0.2787	32.29	2.508	3.588	1.214	7.413	7.770	1.158
4	25.81	0.6200	0.3716	43.06	3.345	4.784	1.619	9.884	10.360	1.544
5	32.26	0.7750	0.4645	53.82	4.181	5.980	2.023	12.355	12.950	1.931
6	38.71	0.9300	0.5574	64.58	5.017	7.176	2.428	14.826	15.540	2.317
7	45.16	1.085	0.6503	75.35	5.853	8.372	2.833	17.297	18.130	2.703
8	51.61	1.240	0.7432	86.11	6.689	9.568	3.237	19.768	20.720	3.089
9	58.06	1.395	0.8361	96.87	7.525	10.764	3.642	22.239	23.310	3.475

* EXAMPLE: 1 sq in. = 6.452 sq cm.

Table C-9. Conversion Tables (*Continued*)

Conversion of Volumes or Cubic Measure

	Cu in. to cu cm	Cu cm to cu in.	Cu ft to cu m	Cu m to cu ft	Cu yd to cu m	Cu m to cu yd	Gallons to cu ft	Cu ft to gallons
1	16.39	0.06102	0.02832	35.31	0.7646	1.308	0.1337	7.481
2	32.77	0.1220	0.05663	70.63	1.529	2.616	0.2674	14.96
3	49.16	0.1831	0.08495	105.9	2.294	3.924	0.4010	22.44
4	65.55	0.2441	0.1133	141.3	3.058	5.232	0.5347	29.92
5	81.94	0.3051	0.1416	176.6	3.823	6.540	0.6684	37.40
6	98.32	0.3661	0.1699	211.9	4.587	7.848	0.8021	44.88
7	114.7	0.4272	0.1982	247.2	5.352	9.156	0.9358	52.36
8	131.1	0.4882	0.2265	282.5	6.116	10.46	1.069	59.84
9	147.5	0.5492	0.2549	317.8	6.881	11.77	1.203	67.32

Conversion of Volumes or Capacities

	Fluid ounces to cu cm	Cu cm to fluid ounces	Liquid pints to liters	Liters to liquid pints	Liquid quarts to liters	Liters to liquid quarts	Gallons to liters	Liters to gallons	Bushels to hecto-liters	Hecto-liters to bushels
1	29.57	0.03381	0.4732	2.113	0.9463	1.057	3.785	0.2642	0.3524	2.838
2	59.15	0.06763	0.9463	4.227	1.893	2.113	7.571	0.5284	0.7048	5.676
3	88.72	0.1014	1.420	6.340	2.839	3.170	11.36	0.7925	1.057	8.513
4	118.3	0.1353	1.893	8.454	3.785	4.227	15.14	1.057	1.410	11.35
5	147.9	0.1691	2.366	10.57	4.732	5.284	18.93	1.321	1.762	14.19
6	177.4	0.2092	2.839	12.68	5.678	6.340	22.71	1.585	2.114	17.03
7	207.0	0.2367	3.312	14.79	6.624	7.397	26.50	1.849	2.467	19.86
8	236.6	0.2705	3.785	16.91	7.571	8.454	30.28	2.113	2.819	22.70
9	266.2	2.3043	4.259	19.02	8.517	9.510	34.07	2.378	3.171	25.54

Mass Equivalents

Kilograms	Grains	Ounces		Pounds		Tons		
		Troy and apoth	Avoir-dupois	Troy and apoth	Avoir-dupois	Short	Long	Metric
1 $\bar{3}.00000$	**15432** 4.18843	**32.15** 1.50719	**35.27** 1.54745	**2.6792** 0.42801	**2.205** 0.34333	**$0.0_{2}1102$** $\bar{3}.04230$	**$0.0_{3}9842$** $\bar{4}.99309$	**0.001** $\bar{3}.00000$
$0.0_{4}6480$ $\bar{5}.81157$	**1**	**$0.0_{2}2083$** $\bar{3}.31876$	**$0.0_{2}2286$** $\bar{3}.35902$	**$0.0_{3}1736$** $\bar{4}.23958$	**$0.0_{3}1429$** $\bar{4}.15490$	**$0.0_{7}7143$** $\bar{8}.85387$	**$0.0_{7}6378$** $\bar{8}.80465$	**$0.0_{7}6480$** $\bar{8}.81157$
0.03110 $\bar{2}.49281$	**480** 2.68124	**1**	**1.09714** 0.04026	**0.08333** $\bar{2}.92082$	**0.06857** $\bar{2}.83614$	**$0.0_{4}3429$** $\bar{5}.53511$	**$0.0_{4}3061$** $\bar{5}.48590$	**$0.0_{4}3110$** $\bar{5}.49281$
0.02835 $\bar{2}.45255$	**437.5** 2.64098	**0.9115** $\bar{1}.95974$	**1**	**0.07595** $\bar{2}.88056$	**0.0625** $\bar{2}.79588$	**$0.0_{4}3125$** $\bar{5}.49485$	**$0.0_{4}2790$** $\bar{5}.44563$	**$0.0_{4}2835$** $\bar{5}.45255$
0.3732 $\bar{1}.57199$	**5760** 3.76042	**12** 1.07918	**13.17** 1.11944	**1**	**0.8229** $\bar{1}.91532$	**$0.0_{3}4114$** $\bar{4}.61429$	**$0.0_{3}3673$** $\bar{4}.56508$	**$0.0_{3}3732$** $\bar{4}.57199$
0.4536 $\bar{1}.65667$	**7000** 3.84510	**14.58** 1.16386	**16** 1.20412	**1.215** 0.08468	**1**	**0.0005** $\bar{4}.69897$	**$0.0_{3}4464$** $\bar{4}.64975$	**$0.0_{3}4536$** $\bar{4}.65667$
907.2 2.95770	**140_{6}** 7.14613	**29167** 4.46489	**320_{3}** 4.50515	**2431** 3.38571	**2000** 3.30103	**1**	**0.8929** $\bar{1}.95078$	**0.9072** $\bar{1}.95770$
1016 3.00691	**15680_{4}** 7.19535	**32667** 4.51411	**35840** 4.55437	**2722** 3.43492	**2240** 3.35025	**1.12** 0.04922	**1**	**1.016** 0.00691
1000 3.00000	**15432356** 7.18843	**32151** 4.50719	**35274** 4.54745	**2679** 3.42801	**2205** 3.34333	**1.102** 0.04230	**0.9842** $\bar{1}.99309$	**1**

The equivalents are given in the heavier type. Logarithms of the equivalents are given immediately below. In some cases the equivalents have been rounded off, while the logarithm corresponds to the equivalent carried to a greater number of decimal places.

Subscripts after any figure, 0_3, 9_4, etc., mean that that figure is to be repeated the indicated number of times.

Table C-9. Conversion Tables (*Continued*)

Conversion of Masses

	Grains to grams	Grams to grains	Ounces (avdp) to grams	Grams to ounces (avdp)	Pounds (avdp) to kilograms	Kilograms to pounds (avdp)	Short tons (2000 lb) to metric tons	Metric tons (1000 kg) to short tons	Long tons (2240 lb) to metric tons	Metric tons to long tons
1	0.06480	15.43	28.35	0.03527	0.4536	2.205	0.907	1.102	1.016	0.984
2	0.1296	30.86	56.70	0.07055	0.9072	4.409	1.814	2.205	2.032	1.968
3	0.1944	46.30	85.05	0.1058	1.361	6.614	2.722	3.307	3.048	2.953
4	0.2592	61.73	113.40	0.1411	1.814	8.818	3.629	4.409	4.064	3.937
5	0.3240	77.16	141.75	0.1764	2.268	11.02	4.536	5.512	5.080	4.921
6	0.3888	92.59	170.10	0.2116	2.722	13.23	5.443	6.614	6.096	5.905
7	0.4536	108.03	198.45	0.2469	3.175	15.43	6.350	7.716	7.112	6.889
8	0.5184	123.46	226.80	0.2822	3.629	17.64	7.257	8.818	8.128	7.874
9	0.5832	138.89	255.15	0.3175	4.082	19.84	8.165	9.921	9.144	8.858

Velocity Equivalents

Centimeters per sec	Meters per sec	Meters per min	Kilometers per hour	Feet per sec	Feet per min	Miles per hour	Knots
1	0.01	0.6 $\bar{1}.77815$	0.036 $\bar{2}.55630$	0.03281 $\bar{2}.51598$	1.9685 0.29414	0.02237 $\bar{2}.34965$	0.01943 $\bar{2}.28836$
100 2.00000	**1**	60 1.77815	3.6 0.55630	3.281 0.51598	196.85 2.29414	2.237 0.34965	1.943 0.28836
1.667 0.22184	0.01667 $\bar{2}.22184$	**1**	0.06 $\bar{2}.77815$	0.05468 $\bar{2}.73783$	3.281 0.51598	0.03728 $\bar{2}.57150$	0.03238 $\bar{2}.51022$
27.78 1.44370	0.2778 $\bar{1}.44370$	16.67 1.22184	**1**	0.9113 $\bar{1}.95968$	54.68 1.73783	0.6214 $\bar{1}.79335$	0.53959 $\bar{1}.73205$
30.48 1.48402	0.3048 $\bar{1}.48402$	18.29 1.26217	1.097 0.04032	**1**	60 1.77815	0.6818 $\bar{1}.83367$	0.59209 $\bar{1}.77238$
0.5080 1.70586	0.005080 $\bar{3}.70586$	0.3048 $\bar{1}.48402$	0.01829 $\bar{2}.26217$	0.01667 $\bar{2}.22185$	**1**	0.01136 $\bar{2}.05553$	0.00987 $\bar{3}.99423$
44.70 1.65035	0.4470 $\bar{1}.65035$	26.82 1.42850	1.609 0.20670	1.467 0.16633	88 1.94448	**1**	0.86839 $\bar{1}.93871$
51.479 1.71163	0.51479 $\bar{1}.71163$	30.887 1.48978	1.8532 0.26793	1.68894 0.22761	101.337 2.00577	1.15155 0.06128	**1**

Conversion of Linear and Angular Velocities

	Cm per sec to feet per min	Feet per min to cm per sec	Cm per sec to miles per hour	Miles per hour to cm per sec	Feet per sec to miles per hour	Miles per hour to feet per sec	Radians per sec to rev per min	Rev per min to radians per sec
1	1.97	0.508	0.0224	44.70	0.682	1.47	9.55	0.1047
2	3.94	1.016	0.0447	89.41	1.364	2.93	19.10	0.2094
3	5.91	1.524	0.0671	134.1	2.045	4.40	28.65	0.3142
4	7.87	2.032	0.0895	178.8	2.727	5.87	38.20	0.4189
5	9.84	2.540	0.1118	223.5	3.409	7.33	47.75	0.5236
6	11.81	3.048	0.1342	268.2	4.091	8.80	57.30	0.6283
7	13.78	3.556	0.1566	312.9	4.773	10.27	66.84	0.7330
8	15.75	4.064	0.1790	357.6	5.455	11.73	76.39	0.8378
9	17.72	4.572	0.2013	402.3	6.136	13.20	85.94	0.9425

Table C-9. Conversion Tables (*Continued*)
Conversion of Pressures *

	Pounds per sq in. to kilograms per sq cm	Kilograms per sq cm to pounds per sq in.	Atmospheres to pounds per sq in.	Pounds per sq in. to atmospheres	Atmospheres to kilograms per sq cm	Kilograms per sq cm to atmospheres
1	0.0703	14.22	14.70	0.0680	1.033	0.9678
2	0.1406	28.45	29.39	0.1361	2.066	1.936
3	0.2109	42.67	44.09	0.2041	3.100	2.904
4	0.2812	56.89	58.78	0.2722	4.133	3.871
5	0.3515	71.12	73.48	0.3402	5.166	4.839
6	0.4218	85.34	88.18	0.4083	6.199	5.807
7	0.4921	99.56	102.9	0.4763	7.233	6.775
8	0.5625	113.8	117.6	0.5444	8.266	7.743
9	0.6328	128.0	132.3	0.6124	9.299	8.711

* EXAMPLE: 1 lb per sq in. = 0.0703 kg per sq cm.

Pressure Equivalents
(For conversion table see immediately above)

Megabars or megadynes per sq cm	Kilograms per sq cm (metric atmospheres)	Pounds per sq in.	Short tons per sq ft	Atmospheres	Columns of mercury at temperature 0 C and g = 980.665 cm per sec^2		Columns of water at temperature 15 C and g = 980.665 cm per sec^2		
					Meters	Inches	Meters	Inches	Feet
1	1.0197 0.00848	14.50 1.16148	1.044 0.01882	0.9869 $\bar{1}$.99427	0.7501 $\bar{1}$.87510	29.53 1.47025	10.21 1.00886	401.8 2.60402	33.49 1.52485
0.9807 $\bar{1}$.99152	1	14.22 1.15300	1.024 0.01034	0.9678 $\bar{1}$.98579	0.7356 $\bar{1}$.86662	28.96 1.46177	10.01 1.00038	394.1 2.59556	32.84 1.51636
0.06895 $\bar{2}$.83852	0.07031 $\bar{2}$.84700	1	0.072 $\bar{2}$.85733	0.06805 $\bar{2}$.83280	0.05171 $\bar{2}$.71360	2.036 0.30876	0.7037 $\bar{1}$.84738	27.70 1.44254	2.309 0.36336
0.9576 $\bar{1}$.98119	0.9765 $\bar{1}$.98966	13.89 1.14267	1	0.9451 $\bar{1}$.97547	0.7183 $\bar{1}$.85628	28.28 1.45143	9.774 0.99006	384.8 2.58521	32.07 1.50604
1.0133 0.00573	1.0332 0.01420	14.70 1.16722	1.058 0.02453	1	0.76 $\bar{1}$.88081	29.92 1.47598	10.34 1.01459	407.1 2.60975	33.93 1.53058
1.3332 0.12490	1.3595 0.13338	19.34 1.28640	1.392 0.14373	1.316 0.11919	1	39.37 1.59517	13.61 1.13378	535.7 2.72894	44.64 1.64976
0.03386 $\bar{2}$.52975	0.03453 $\bar{2}$.53823	0.4912 $\bar{1}$.69124	0.03536 $\bar{2}$.54857	0.03342 $\bar{2}$.52402	0.02540 $\bar{2}$.40484	1	0.3456 $\bar{1}$.53861	13.61 1.13378	1.134 0.05460
0.09798 $\bar{2}$.99114	0.09991 $\bar{2}$.99962	1.421 0.15262	0.1023 $\bar{1}$.00996	0.09670 $\bar{2}$.98541	0.07349 $\bar{2}$.86622	2.893 0.46139	1	39.37 1.59517	3.281 0.51598
0.002489 $\bar{3}$.39598	0.002538 $\bar{3}$.40446	0.03609 $\bar{2}$.55745	0.002599 $\bar{3}$.41479	0.002456 $\bar{3}$.39024	0.001867 $\bar{3}$.27106	0.07349 $\bar{2}$.86622	0.02540 $\bar{2}$.40484	1	0.08333 $\bar{2}$.92082
0.02986 $\bar{2}$.47516	0.03045 $\bar{2}$.48364	0.4331 $\bar{1}$.63663	0.03119 $\bar{2}$.49397	0.02947 $\bar{2}$.46942	0.02240 $\bar{2}$.35024	0.8819 $\bar{1}$.94540	0.3048 $\bar{1}$.48402	12 1.07918	1

Table C-10. Temperature Conversion*

RELATION OF TEMPERATURE SCALES

Temperature scale	Melting point of ice, deg	Boiling point of water, deg	Transformation equation, deg
Fahrenheit (F)	32	212	F = 1.8C + 32
Centigrade (C)	0	100	C = 5/9(F − 32)
Rankine (R)	491.7	671.7	R = F + 459.69
Kelvin (K)	273.2	373.2	K = C + 273.16

Conversion of Thermometer Readings

DEGREES CENTIGRADE TO DEGREES FAHRENHEIT

C	*F*	*C*	*F*	*C*	*F*	*C*	*F*	*C*	*F*	*C*	*F*
−40	−40.0	+5	+41.0	+40	+104.0	+175	+347	+350	+662	+750	+1382
−38	−36.4	6	42.8	41	105.8	180	356	355	671	800	1472
−36	−32.8	7	44.6	42	107.6	185	365	360	680	850	1562
−34	−29.2	8	46.4	43	109.4	190	374	365	689	900	1652
−32	−25.6	9	48.2	44	111.2	195	383	370	698	950	1742
−30	−22.0	10	50.0	45	113.0	200	392	375	707	1000	1832
−28	−18.4	11	51.8	46	114.8	205	401	380	716	1050	1922
−26	−14.8	12	53.6	47	116.6	210	410	385	725	1100	2012
−24	−11.2	13	55.4	48	118.4	215	419	390	734	1150	2102
−22	− 7.6	14	57.2	49	120.2	220	428	395	743	1200	2192
−20	− 4.0	15	59.0	50	122.0	225	437	400	752	1250	2282
−19	− 2.2	16	60.8	55	131.0	230	446	405	761	1300	2372
−18	− 0.4	17	62.6	60	140.0	235	455	410	770	1350	2462
−17	+ 1.4	18	64.4	65	149.0	240	464	415	779	1400	2552
−16	3.2	19	66.2	70	158.0	245	473	420	788	1450	2642
−15	5.0	20	68.0	75	167.0	250	482	425	797	1500	2732
−14	6.8	21	69.8	80	176.0	255	491	430	806	1550	2822
−13	8.6	22	71.6	85	185.0	260	500	435	815	1600	2912
−12	10.4	23	73.4	90	194.0	265	509	440	824	1650	3002
−11	12.2	24	75.2	95	203.0	270	518	445	833	1700	3092
−10	14.0	25	77.0	100	212.0	275	527	450	842	1750	3182
− 9	15.8	26	78.8	105	221.0	280	536	455	851	1800	3272
− 8	17.6	27	80.6	110	230.0	285	545	460	860	1850	3362
− 7	19.4	28	82.4	115	239.0	290	554	465	869	1900	3452
− 6	21.2	29	84.2	120	248.0	295	563	470	878	1950	3542
− 5	23.0	30	86.0	125	257.0	300	572	475	887	2000	3632
− 4	24.8	31	87.8	130	266.0	305	581	480	896	2050	3722
− 3	26.6	32	89.6	135	275.0	310	590	485	905	2100	3812
− 2	28.4	33	91.4	140	284.0	315	599	490	914	2150	3902
− 1	30.2	34	93.2	145	293.0	320	608	495	923	2200	3992
0	32.0	35	95.0	150	302.0	325	617	500	932	2250	4082
+ 1	33.8	36	96.8	155	311.0	330	626	550	1022	2300	4172
2	35.6	37	98.6	160	320.0	335	635	600	1112	2350	4262
3	37.4	38	100.4	165	329.0	340	644	650	1202	2400	4352
4	39.2	39	102.2	170	338.0	345	653	700	1292	2450	4442

TABLE OF VALUES FOR INTERPOLATION IN THE ABOVE TABLE

Degrees centigrade	1	2	3	4	5	6	7	8	9
Degrees Fahrenheit	1.8	3.6	5.4	7.2	9.0	10.8	12.6	14.4	16.2

* From Marks' "Mechanical Engineers' Handbook," Theodore Baumeister (ed.), 6th ed., McGraw-Hill Book Company, Inc., New York, 1958.

Table C-10. Temperature Conversion (*Continued*)

Conversion of Thermometer Readings (*Continued*)

DEGREES FAHRENHEIT TO DEGREES CENTIGRADE*

F	*C*	*F*	*C*	*F*	*C*	*F*	*C*	*F*	*C*	*F*	*C*
−40	−40.00	+30	−1.11	+80	+26.67	+250	+121.11	+500	+260.00	+900	+482.22
−38	−38.89	31	−0.56	81	27.22	255	123.89	505	262.78	910	487.78
−36	−37.78	32	0.00	82	27.78	260	126.67	510	265.56	920	493.33
−34	−36.67	33	+0.56	83	28.33	265	129.44	515	268.33	930	498.89
−32	−35.56	34	1.11	84	28.89	270	132.22	520	271.11	940	504.44
−30	−34.44	35	1.67	85	29.44	275	135.00	525	273.89	950	510.00
−28	−33.33	36	2.22	86	30.00	280	137.78	530	276.67	960	515.56
−26	−32.22	37	2.78	87	30.56	285	140.55	535	279.44	970	521.11
−24	−31.11	38	3.33	88	31.11	290	143.33	540	282.22	980	526.67
−22	−30.00	39	3.89	89	31.67	295	146.11	545	285.00	990	532.22
−20	−28.89	40	4.44	90	32.22	300	148.89	550	287.78	1000	537.78
−18	−27.78	41	5.00	91	32.78	305	151.67	555	290.55	1050	565.56
−16	−26.67	42	5.56	92	33.33	310	154.44	560	293.33	1100	593.33
−14	−25.56	43	6.11	93	33.89	315	157.22	565	296.11	1150	612.11
−12	−24.44	44	6.67	94	34.44	320	160.00	570	298.89	1200	648.89
−10	−23.33	45	7.22	95	35.00	325	162.78	575	301.67	1250	676.67
− 8	−22.22	46	7.78	96	35.56	330	165.56	580	304.44	1300	704.44
− 6	−21.11	47	8.33	97	36.11	335	168.33	585	307.22	1350	732.22
− 4	−20.00	48	8.89	98	36.67	340	171.11	590	310.00	1400	760.00
− 2	−18.89	49	9.44	99	37.22	345	173.89	595	312.78	1450	787.78
0	−17.78	50	10.00	100	37.78	350	176.67	600	315.56	1500	815.56
+ 1	−17.22	51	10.56	105	40.55	355	179.44	610	321.11	1550	843.33
2	−16.67	52	11.11	110	43.33	360	182.22	620	326.67	1600	871.11
3	−16.11	53	11.67	115	46.11	365	185.00	630	332.22	1650	898.89
4	−15.56	54	12.22	120	48.89	370	187.78	640	337.78	1700	926.67
5	−15.00	55	12.78	125	51.67	375	190.55	650	343.33	1750	954.44
6	−14.44	56	13.33	130	54.44	380	193.33	660	348.89	1800	982.22
7	−13.89	57	13.89	135	57.22	385	196.11	670	354.44	1850	1010.00
8	−13.33	58	14.44	140	60.00	390	198.89	680	360.00	1900	1037.78
9	−12.78	59	15.00	145	62.78	395	201.67	690	365.56	1950	1065.56
10	−12.22	60	15.56	150	65.56	400	204.44	700	371.11	2000	1093.33
11	−11.67	61	16.11	155	68.33	405	207.22	710	376.67	2050	1121.11
12	−11.11	62	16.67	160	71.11	410	210.00	720	382.22	2100	1148.89
13	−10.56	63	17.22	165	73.89	415	212.78	730	387.78	2150	1176.67
14	−10.00	64	17.78	170	76.67	420	215.56	740	393.33	2200	1204.44
15	− 9.44	65	18.33	175	79.44	425	218.33	750	398.89	2250	1232.22
16	− 8.89	66	18.89	180	82.22	430	221.11	760	404.44	2300	1260.00
17	− 8.33	67	19.44	185	85.00	435	223.89	770	410.00	2350	1287.78
18	− 7.78	68	20.00	190	87.78	440	226.67	780	415.56	2400	1315.56
19	− 7.22	69	20.56	195	90.55	445	229.44	790	421.11	2450	1343.33
20	− 6.67	70	21.11	200	93.33	450	232.22	800	426.67	2500	1371.11
21	− 6.11	71	21.67	205	96.11	455	235.00	810	432.22	2550	1398.89
22	− 5.56	72	22.22	210	98.89	460	237.78	820	437.78	2600	1426.67
23	− 5.00	73	22.78	215	101.67	465	240.55	830	443.33	2650	1454.44
24	− 4.44	74	23.33	220	104.44	470	243.33	840	448.89	2700	1482.22
25	− 3.89	75	23.89	225	107.22	475	246.11	850	454.44	2750	1510.00
26	− 3.33	76	24.44	230	110.00	480	248.89	860	460.00	2800	1537.78
27	− 2.78	77	25.00	235	112.78	485	251.67	870	465.56	2850	1565.59
28	− 2.22	78	25.56	240	115.56	490	254.44	880	471.11	2900	1593.33
29	− 1.67	79	26.11	245	118.33	495	257.22	890	476.67	2950	1621.11

TABLE OF VALUES FOR INTERPOLATION IN THE ABOVE TABLE

Degrees Fahrenheit......	1	2	3	4	5	6	7	8	9
Degrees centigrade*......	0.56	1.11	1.67	2.22	2.78	3.33	3.89	4.44	5.00

* All decimals in the table are repeating decimals; 37.78 is really 37.777 . . . .

Table C-11. Gravity, Weight, and Volume Conversions for Petroleum Products*
(All measurements at 60°F)

Gravity, degrees API	Specific gravity	Gallons per pound	Pounds per gallon	Pounds per barrel	Barrels per short ton	Barrels per metric ton	Barrels per long ton
0	1.0760	0.1116	8.962	376.40	5.31	5.86	5.95
10	1.0000	0.1201	8.328	349.78	5.72	6.30	6.40
15	0.9659	0.1243	8.044	337.85	5.92	6.52	6.63
18	0.9465	0.1269	7.882	331.04	6.04	6.66	6.77
20	0.9340	0.1286	7.778	326.68	6.12	6.75	6.86
22	0.9218	0.1303	7.676	322.39	6.20	6.84	6.95
24	0.9100	0.1320	7.578	318.28	6.28	6.93	7.04
26	0.8984	0.1337	7.481	314.20	6.36	7.02	7.13
28	0.8871	0.1354	7.387	310.25	6.45	7.10	7.22
30	0.8762	0.1371	7.296	306.43	6.53	7.19	7.31
32	0.8654	0.1388	7.206	302.65	6.61	7.28	7.40
34	0.8550	0.1405	7.119	299.00	6.69	7.37	7.49
36	0.8448	0.1422	7.034	295.43	6.77	7.46	7.58
38	0.8348	0.1439	6.951	291.94	6.85	7.55	7.67
40	0.8251	0.1456	6.870	288.54	6.93	7.64	7.76
42	0.8156	0.1473	6.790	285.18	7.01	7.73	7.85
44	0.8063	0.1490	6.713	281.95	7.09	7.82	7.94
46	0.7972	0.1507	6.637	278.75	7.17	7.91	8.04
48	0.7883	0.1524	6.563	275.65	7.26	8.00	8.13
50	0.7796	0.1541	6.491	272.62	7.34	8.09	8.22
55	0.7587	0.1583	6.316	265.27	7.54	8.31	8.44
60	0.7389	0.1626	6.151	258.34	7.74	8.53	8.67
65	0.7201	0.1668	5.994	251.75	7.94	8.76	8.90
70	0.7022	0.1711	5.845	245.49	8.15	8.98	9.12
75	0.6852	0.1753	5.704	239.57	8.35	9.20	9.35
80	0.6690	0.1796	5.569	233.90	8.55	9.43	9.58
85	0.6536	0.1838	5.440	228.48	8.75	9.65	9.80
90	0.6388	0.1881	5.317	223.31	8.96	9.87	10.03
95	0.6247	0.1923	5.199	218.36	9.16	10.10	10.26
100	0.6112	0.1966	5.087	213.65	9.36	10.32	10.48

Authority: American Society for Testing Materials; American Petroleum Institute.
* From "Petroleum Facts and Figures," American Petroleum Institute, 1959.

APPENDIX D. TABLES OF ESTIMATED AND PROVED RESERVES IN THE UNITED STATES AND CANADA

Table D-1. Estimated Proved Reserves of Crude Oil in the United States[a]

(Barrels of 42 U.S. gallons)

	Proved reserves as of December 31, 1959 (1)	Changes in proved reserves due to extensions and revisions during 1960 (2)	Proved reserves discovered in new fields and in new pools in old fields in 1960[b] (3)	Production during 1960[c] (4)	Proved reserves as of December 31, 1960 (cols. 1 + 2 + 3 less col. 4) (5)	Changes in reserves during 1960 (col. 5 less col. 1) (6)
Alabama	34,248,000	6,387,000	—	7,330,000	33,305,000	(—) 943,000
Arkansas	312,911,000	14,416,000	820,000	26,150,000	301,997,000	(—) 10,914,000
California[d]	3,762,507,000	189,168,000	10,916,000	304,049,000	3,658,542,000	(—)103,965,000
Colorado	380,952,000	28,128,000	1,510,000	46,731,000	363,859,000	(—) 17,093,000
Illinois	594,407,000	37,694,000	1,473,000	77,648,000	555,926,000	(—) 38,481,000
Indiana	73,751,000	3,338,000	947,000	11,985,000	66,251,000	(—) 7,700,000
Kansas	917,503,000	68,198,000	11,165,000	113,017,000	883,849,000	(—) 33,654,000
Kentucky	136,324,000	14,146,000	122,000	21,685,000	128,907,000	(—) 7,417,000
Louisiana[d]	4,660,088,000	417,301,000	54,275,000	346,284,000	4,785,380,000	125,292,000
Michigan	55,335,000	37,752,000	625,000	15,890,000	77,822,000	22,487,000
Mississippi	389,337,000	64,643,000	3,238,000	50,120,000	407,098,000	17,761,000
Montana	309,268,000	(—)16,890,000	4,590,000	30,281,000	266,687,000	(—) 42,581,000
Nebraska	80,558,000	20,828,000	8,335,000	23,547,000	86,174,000	5,616,000
New Mexico	1,025,789,000	142,951,000	19,187,000	104,269,000	1,083,658,000	57,869,000
New York	34,197,000	—	—	1,814,000	32,383,000	(—) 1,814,000
North Dakota	381,909,000	63,962,000	6,580,000	21,930,000	430,521,000	48,612,000
Ohio	74,137,000	5,590,000	—	5,050,000	74,677,000	540,000
Oklahoma	1,864,749,000	106,157,000	9,248,000	189,654,000	1,790,500,000	(—) 74,249,000
Pennsylvania	113,574,000	—	—	5,942,000	107,632,000	(—) 5,942,000
Texas[d]	14,859,674,000	703,918,000	90,959,000	896,059,000	14,758,492,000	(—)101,182,000
Utah	194,539,000	50,068,000	1,332,000	37,475,000	208,464,000	13,925,000
West Virginia	51,259,000	2,449,000	—	2,275,000	51,433,000	174,000
Wyoming	1,402,938,000	146,690,000	8,519,000	130,772,000	1,427,375,000	24,437,000
Miscellaneous[e]	9,193,000	4,578,000	20,015,000	1,507,000	32,279,000	23,086,000
Total United States	31,719,347,000	2,111,472,000	253,856,000	2,471,464,000	31,613,211,000	(—)106,136,000

[a] Prepared by API Committee. The API Committee includes in its crude oil figures all condensate which comes out of the separator with the crude oil and is run with the crude as part of the crude oil stream. All other condensate is included by the Natural Gas Reserves Committee in its figures on Natural Gas Liquids.

[b] Only a limited area is assigned to each new discovery, even though the committee may believe that eventually a much larger area will produce; for, in this report, the concern is only with actually proved reserves.

[c] The 1960 production figures were compiled by the committee from records of actual production for whatever period such were available, with estimates for the remainder of the year. Any variance between the actual production, as later reported, and the figures used herein will be compensated for through revision when the following year's reserve report is compiled. These revisions have in the past been very small. Because, in each annual report by this committee, the production figure for the year is an estimate, as just explained, correct total cumulative production cannot be obtained by adding together the figures listed in successive reports of this committee, or in the columns marked Production of Table D-3.

[d] Includes off-shore reserves.

[e] Under Miscellaneous are included Alaska, Arizona, Florida, Missouri, Nevada, South Dakota, Tennessee, Virginia and Washington.

Table D-2. Estimated Proved Reserves of Natural Gas Liquids in the United States[a]
(Includes condensate,[a] natural gasoline and liquefied petroleum gases)[g]
(Barrels of 42 U.S. gallons)

	Proved reserves as of December 31, 1959 (1)	Changes in proved reserves due to extensions (new natural gas liquids) and revisions during 1960 (2)	Proved reserves discovered in new fields and in new pools in old fields in 1960[b] (3)	Production during 1960[c] (4)	Proved reserves as of December 31, 1960 (cols. 1 + 2 + 3 less col. 4) (5)	Changes in reserves during 1960 (col. 5 less col. 1) (6)
Alabama[d]	—	—	—	—	—	—
Arkansas	32,017,000	(—) 2,640,000	15,000	1,895,000	27,497,000	(—) 4,520,000
California[e]	325,318,000	16,607,000	365,000	28,429,000	313,861,000	(—) 11,457,000
Colorado	24,914,000	2,699,000	360,000	3,949,000	24,024,000	(—) 890,000
Illinois	10,754,000	850,000	27,000	1,578,000	10,053,000	(—) 701,000
Indiana	124,000	5,000	1,000	20,000	110,000	(—) 14,000
Kansas	196,912,000	7,680,000	930,000	7,119,000	198,403,000	1,491,000
Kentucky	49,290,000	2,000,000	1,000,000	3,238,000	49,052,000	(—) 238,000
Louisiana[e]	1,357,798,000	113,002,000	44,466,000	82,291,000	1,432,975,000	75,177,000
Michigan	1,562,000	118,000	156,000	217,000	1,619,000	57,000
Mississippi	40,944,000	(—) 4,598,000	2,340,000	2,505,000	36,181,000	(—) 4,763,000
Montana	13,511,000	(—) 2,297,000	1,101,000	627,000	11,688,000	(—) 1,823,000
Nebraska	5,432,000	1,450,000	25,000	752,000	6,155,000	723,000
New Mexico	423,040,000	86,134,000	3,972,000	28,347,000	484,799,000	61,759,000
North Dakota	97,928,000	(—) 5,066,000	—	1,644,000	91,218,000	(—) 6,710,000
Ohio	1,304,000	(—) 1,304,000	—	—	—	(—) 1,304,000
Oklahoma	367,569,000	(—) 5,374,000	4,601,000	28,483,000	338,313,000	(—) 29,256,000
Pennsylvania	3,677,000	(—) 1,500,000	—	67,000	2,110,000	(—) 1,567,000
Texas[e]	3,430,375,000	335,403,000	58,537,000	228,141,000	3,596,174,000	165,799,000
Utah	32,376,000	20,370,000	—	2,044,000	50,702,000	18,326,000
West Virginia	34,166,000	10,828,000	3,610,000	3,870,000	44,734,000	10,568,000
Wyoming	73,279,000	29,252,000	3,000	6,159,000	96,375,000	23,096,000
Miscellaneous[f]	18,000	2,000	—	4,000	16,000	(—) 2,000
Total United States	6,522,308,000	603,621,000	121,509,000	431,379,000	6,816,059,000	293,751,000

[a] Prepared by API and AGA Committees. The API Committee includes in its crude oil figures all condensate which comes out of the separator with the crude oil and is run with the crude as part of the crude oil stream. All other condensate is included by the Natural Gas Reserves Committee in its figures on Natural Gas Liquids.

[b] Only a limited area is assigned to each new discovery, even though the committee may believe that eventually a much larger area will produce; for, in this report, the concern is only with actually proved reserves.

[c] The 1960 production figures were compiled by the committee from records of actual production for whatever period such were available, with estimates for the remainder of the year. Any variance between the actual production, as later reported, and the figures used herein will be compensated for through revision when the following year's reserve report is compiled. These revisions have in the past been very small. Because, in each annual report by this committee, the production figure for the year is an estimate, as just explained, correct total cumulative production cannot be obtained by adding together the figures listed in successive reports of this committee, or in the columns marked Production of Table D-3.

[d] Included in Miscellaneous.

[e] Includes off-shore reserves.

[f] Includes Alabama and Florida.

[g] Proved recoverable reserves of natural gas liquids are those contained in the recoverable gas reserves subject to being produced as natural gas liquids by separators or extraction plants, now in operation, under construction or planned for the immediate future. For purposes of developing reserve estimates, natural gas liquids are defined as those hydrocarbon liquids which, in the reservoir, are either gaseous or in solution with crude oil and which are recoverable as liquids by the processes of condensation or absorption which take place in field separators, scrubbers, gasoline plants, or cycling plants. Natural gasoline, condensate, and liquefied petroleum gases fall in this category. While the liquids so collected and the products derived from them in some of the modern plants are known by a variety of names, they have been grouped together here under the general heading Natural Gas Liquids.

Table D-3. Summary of Proved Reserves as Reported for 1946 and Thereafter
(Barrels of 42 U.S. gallons)

	New oil added during year					
	Through revisions of previous estimates and extensions to known fields	Through discoveries of new fields and of new pools in old fields	Total through new discoveries, extensions, and revisions (cols. 1 + 2)	Production during year*	Estimated proved reserves as of end of year	Increase over previous year
	(1)	(2)	(3)	(4)	(5)	(6)
Crude oil only (API Committee)						
1946	2,413,628,000	244,434,000	2,658,062,000	1,726,348,000	20,873,560,000	931,714,000
1947	2,019,140,000	445,430,000	2,464,570,000	1,850,445,000	21,487,685,000	614,125,000
1948	3,398,726,000	396,481,000	3,795,207,000	2,002,448,000	23,280,444,000	1,792,759,000
1949	2,297,428,000	890,417,000	3,187,845,000	1,818,800,000	24,649,489,000	1,369,045,000
1950	1,997,769,000	564,916,000	2,562,685,000	1,943,776,000	25,268,398,000	618,909,000
1951	4,024,698,000	389,256,000	4,413,954,000	2,214,321,000	27,468,031,000	2,199,633,000
1952	2,252,860,000	496,428,000	2,749,288,000	2,256,765,000	27,960,554,000	492,523,000
1953	2,704,450,000	591,680,000	3,296,130,000	2,311,856,000	28,944,828,000	984,274,000
1954	2,287,231,000	585,806,000	2,873,037,000	2,257,119,000	29,560,746,000	615,918,000
1955	2,393,767,000	476,957,000	2,870,724,000	2,419,300,000	30,012,170,000	451,424,000
1956	2,507,114,000	467,222,000	2,974,336,000	2,551,857,000	30,434,649,000	422,479,000
1957	2,008,603,000	416,197,000	2,424,800,000	2,559,044,000	30,300,405,000	(—)134,244,000
1958	2,293,513,000	314,729,000	2,608,242,000	2,372,730,000	30,535,917,000	235,512,000
1959	3,297,383,000	369,362,000	3,666,745,000	2,483,315,000	31,719,347,000	1,183,430,000
1960	2,111,472,000	253,856,000	2,365,328,000	2,471,464,000	31,613,211,000	(—)106,136,000

Table D-3. Summary of Proved Reserves as Reported for 1946 and Thereafter *(Continued)*

	New oil added during year					
	Through revisions of previous estimates and extensions to known fields	Through discoveries of new fields and of new pools in old fields	Total through new discoveries, extensions, and revisions (cols. 1 + 2)	Production during year*	Estimated proved reserves as of end of year	Increase over previous year
	(1)	(2)	(3)	(4)	(5)	(6)
Natural gas liquids only (API and AGA Committees)						
1946	†	†	†	†	3,163,219,000	†
1947	192,237,000	59,301,000	251,538,000	160,782,000	3,253,975,000	90,756,000
1948	405,874,000	64,683,000	470,557,000	183,749,000	3,540,783,000	286,808,000
1949	294,211,000	92,565,000	386,776,000	198,547,000	3,729,012,000	188,229,000
1950	707,879,000	58,183,000	766,062,000	227,411,000	4,267,663,000	538,651,000
1951	648,497,000	75,494,000	723,991,000	267,052,000	4,724,602,000	465,939,000
1952	475,170,000	81,668,000	556,838,000	284,789,000	4,996,651,000	272,049,000
1953	648,047,000	95,922,000	743,969,000	302,698,000	5,437,922,000	441,271,000
1954	20,830,000	86,520,000	107,350,000	300,815,000	5,244,457,000	(—)193,465,000
1955	447,160,000	67,348,000	514,508,000	320,400,000	5,438,565,000	194,108,000
1956	715,764,000	94,056,000	809,820,000	346,053,000	5,902,332,000	463,767,000
1957	8,884,000	128,508,000	137,392,000	352,364,000	5,687,360,000	(—)214,972,000
1958	749,956,000	108,250,000	858,206,000	341,548,000	6,204,018,000	516,658,000
1959	593,905,000	109,539,000	703,444,000	385,154,000	6,522,308,000	318,290,000
1960	603,621,000	121,509,000	725,130,000	431,379,000	6,816,059,000	293,751,000

* The 1960 production figures were compiled by the committee from records of actual production for whatever period such were available, with estimates for the remainder of the year. Any variance between the actual production, as later reported, and the figures used herein will be compensated for through revision when the following year's reserve report is compiled. These revisions have in the past been very small. Because, in each annual report by this committee, the production figure for the year is an estimate, as just explained, correct total cumulative production cannot be obtained by adding together the figures listed in successive reports of these committees.

† This detail not available for 1946.

Table D-4. Estimated Proved Recoverable Reserves of Natural Gas in the United States[a]

(Millions of cubic feet—14.65 psia, at 60°F)

	Reserves as of December 31, 1959[b]	Changes in reserves during 1960				Reserves[b] as of December 31, 1960				
		Extensions and revisions[b]	Discoveries of new fields and new pools in old fields[b]	Net change in underground storage[c]	Net production[d]	Total (cols. 7 + 8 + 9 + 10 also cols. 1 + 2 + 3 + 4 less col. 5)	Non-associated[e]	Associated[f]	Dissolved[g]	Underground storage[h]
	(1)	(2)	(3)	(4)	(5)	(6)	(7)	(8)	(9)	(10)
Alaska	57,975	36,993	13,000	0	305	107,663	95,811	0	11,852	0
Arkansas	1,422,817	83,115	36,360	211	82,793	1,459,710	957,514	287,932	209,222	5,042
California[i]	8,593,447	395,351	324,694	30,875	500,472	8,843,895	2,509,684	1,962,792	4,249,526	121,893
Colorado	2,496,159	(—) 474,828	121,520	565	100,662	2,042,754	1,294,053	258,434	489,538	729
Illinois	175,110	8,900	297	7,461	18,633	173,135	9,349	0	111,185	52,601
Indiana	35,088	1,001	284	1,692	3,945	34,120	1,100	1,200	19,875	11,945
Kansas	19,981,403	128,848	156,452	19,866	666,345	19,620,224	18,735,589	527,181	274,886	82,568
Kentucky	1,159,381	46,048	9,020	1,918	72,518	1,143,849	1,041,373	0	76,365	26,111
Louisiana[i]	59,853,920	4,333,983	2,254,600	0	3,056,362	63,386,141	50,527,589	8,956,486	3,902,066	0
Michigan	515,403	7,549	31,200	56,057	24,451	585,758	116,015	71,793	56,634	341,316
Mississippi	2,486,524	31,804	210,390	(—) 73	186,307	2,542,338	1,906,135	340,020	290,718	5,465
Montana	665,491	(—) 17,245	0	14,126	36,194	626,178	465,058	29,138	76,799	55,183
Nebraska	132,719	(—) 3,057	861	0	12,695	117,828	93,277	8,673	15,878	0
New Mexico	17,912,798	(—) 1,654,454	127,457	(—) 3,317	778,760	15,603,724	11,214,132	2,753,119	1,592,049	44,424
New York	106,519	(—) 7,588	420	1,662	4,812	96,201	36,841	0	138	59,222
North Dakota	1,206,542	(—) 35,415	0	0	20,106	1,151,021	9,453	340,400	801,168	0
Ohio	748,766	34,297	9,250	12,549	39,309	765,553	326,987	0	89,251	349,315
Oklahoma	16,651,292	975,342	665,460	13,283	993,975	17,311,402	12,275,573	1,943,428	2,973,501	118,900
Pennsylvania	1,051,972	191,836	18,000	50,343	119,671	1,192,480	714,941	0	22,940	454,599
Texas[i]	120,475,783	2,471,209	2,428,403	16,787	5,902,789	119,489,393	78,214,180	27,137,031	14,077,131	61,051
Utah	1,264,250	260,445	63,734	0	62,289	1,526,140	1,007,349	31,476	487,315	0
Virginia	38,632	(—) 3,300	300	0	2,342	33,290	33,290	0	0	0
West Virginia	1,593,551	318,967	71,488	37,001	189,882	1,831,125	1,472,197	0	63,633	295,295
Wyoming	3,847,064	199,836	93,719	1,148	207,018	3,934,749	3,197,676	157,435	558,066	21,572
Miscellaneous[j]	123,987	3,203	100	20,654	7,815	140,129	48,172	0	15,184	76,773
Total	262,596,593	7,332,840	6,637,009	282,808	13,090,450	263,758,800	186,303,338	44,806,538	30,464,920	2,184,004

[a] Prepared by AGA Committee.
[b] Excludes gas loss due to natural gas liquids recovery.
[c] The net difference between gas stored in and gas withdrawn from underground storage reservoirs, inclusive of adjustments and native gas transferred from other reserve categories.
[d] Net production equals gross withdrawals less gas injected into producing reservoirs. Changes in underground storage and gas loss due to natural gas liquids recovery are excluded. Fourth quarter production estimated in some instances.
[e] Nonassociated gas is free gas not in contact with crude oil in the reservoir; and free gas in contact with oil where the production of such gas is not significantly affected by the production of crude oil.
[f] Associated gas is free gas in contact with crude oil in the reservoir where the production of such gas is significantly affected by the production of crude oil.
[g] Dissolved gas is gas in solution with crude oil in the reservoirs.
[h] Gas held in underground reservoirs (including native and net injected gas) for storage purposes.
[i] Includes off-shore reserves.
[j] Includes Alabama, Arizona, Florida, Iowa, Maryland and Missouri.

Table D-5. Summary of Annual Estimates of Natural Gas Reserves for Period December 31, 1945 to December 31, 1960[a]
(Millions of cubic feet—14.65 psia, at 60°F)

Year	Natural gas added during year			Net change in underground storage	Net production during year	Estimated proved reserves as of end of year	Increase over previous year
	Extensions and revisions	Discoveries of new fields and new pools in old fields	Total of discoveries, revisions, and extensions				
1945						147,789,367	
1946	[b]	[b]	17,729,152	[b]	4,942,617	160,575,901	12,786,534
1947	7,570,654	3,410,170	10,980,824	[b]	5,629,811	165,926,914	5,351,013
1948	9,769,483	4,129,089	13,898,572	51,482	6,007,628	173,869,340	7,942,426
1949	8,061,429	4,612,870	12,674,299	82,746	6,245,041	180,381,344	6,512,004
1950	9,172,381	2,877,351	12,049,732	54,301	6,892,678	185,592,699	5,211,355
1951	13,013,606	3,039,385	16,052,991	132,751	7,966,941	193,811,500	8,218,801
1952	8,934,470	5,411,043	14,345,513	198,850	8,639,638	199,716,225	5,904,725
1953	13,371,355	7,081,661	20,453,016	516,431[c]	9,238,540	211,447,132	11,730,907
1954	4,632,309	4,966,894	9,599,203	90,906	9,426,509	211,710,732	263,600
1955	16,298,125	5,719,069	22,017,194	87,637	10,118,118	223,697,445	11,986,713
1956	19,214,604	5,636,476	24,851,080	133,970	10,907,926	237,774,569	14,077,124
1957	11,118,319	8,998,993	20,117,312	179,733	11,502,359	246,569,255	8,794,686
1958	13,388,808	5,611,098	18,999,906	57,902	11,485,026	254,142,037	7,572,782
1959	14,933,109	5,800,749	20,733,858	161,326	12,440,628	262,596,593	8,454,556
1960	7,332,840	6,637,009	13,969,849	282,808	13,090,450	263,758,800	1,162,207

[a] Prepared by AGA Committee.
[b] Not estimated.
[c] All native gas in storage reservoirs formerly classified as a natural gas reserve is included in this figure.

Table D-6. The 1960 Report on Estimated Proved Reserves of Crude Oil and Natural Gas Liquids of the Canadian Petroleum Association

(In 35-Imperial-gallon barrels which are equivalent to 42-U.S.-gallon barrels)

	Proved reserves as of December 31, 1959 (1)	Changes in proved reserves due to extensions (new oil) and revisions duirng 1960 (2)	Proved reserves discovered in new fields and in new pools in old fields in 1960* (3)	Production during 1960 (4)	Proved reserves as of December 31, 1960 (cols. 1 + 2 + 3 less col. 4) (5)	Changes in reserves during 1960 (col. 5 less col. 1) (6)
Crude oil						
Northwest Territories	51,970,000	—	—	472,000	51,498,000	(—) 472,000
British Columbia	19,401,000	24,529,000	1,895,000	869,000	44,956,000	25,555,000
Alberta	2,898,878,000	268,807,000	15,394,000	131,887,000	3,051,192,000	152,314,000
Saskatchewan	495,787,000	53,643,000	4,755,000	52,107,000	502,078,000	6,291,000
Manitoba	23,054,000	2,446,000	—	4,750,000	20,750,000	(—) 2,304,000
Eastern Canada	8,034,000	1,053,000	—	1,019,000	8,068,000	34,000
Total crude oil	3,497,124,000	350,478,000	22,044,000	191,104,000	3,678,542,000	181,418,000
Natural gas liquids						
British Columbia	23,297,000	9,160,000	1,502,000	977,000	32,982,000	9,685,000
Saskatchewan	14,925,000	6,410,000	—	862,000	20,473,000	5,548,000
Alberta	463,923,000	18,479,000	9,379,000	6,715,000	485,066,000	21,143,000
Total natural gas liquids	502,145,000	34,049,000	10,881,000	8,554,000	538,521,000	36,376,000
Total liquid hydrocarbons in Canada	3,999,269,000	384,527,000	32,925,000	199,658,000	4,217,063,000	217,794,000

* Only a limited area is assigned to each new discovery even though the committee may believe that eventually a much larger area will produce, because in this report the committee is concerned only with actually proved reserves. Hence, figures in this column may be considered minimum figures.

Table D-7. Estimated Proved Recoverable Reserves of Natural Gas in Canada[a]

(Millions of cubic feet—14.65 psia, at 60°F)

	Reserves as of Dec. 31, 1959	Changes in reserves during 1960				Reserves as of December 31, 1960				
		Extensions and revisions	Discoveries of new fields and new pools in old fields	Net change in underground storage	Net production	Total cols. 7, 8, 9 and 10; also cols. 1, 2, 3, and 4 less col. 5	Non-associated	Associated	Dissolved	Under-ground storage
	(1)	(2)	(3)	(4)	(5)	(6)	(7)	(8)	(9)	(10)
Northwest Territories	32,063	—	5,680	—	377	37,366	11,080	—	26,286	—
British Columbia	1,825,238	1,215,682	139,088	—	82 078	3,097,930	2,997,012	54,960	45,958	—
Alberta	23,300,669	2,970,987	142,516	2,260	402 062	26,014,370	17,557,690	5,740,866	2,679,873	35,941
Saskatchewan	1 235,592	117,021	2,251	175	49 280	1,305,759	323,535	53,176	923,777	5,271
Manitoba	1,959	—	—	—	400	1,559	—	—	1,559	—
Total Western Canada	26 395,521	4,303,690	289,535	2,435	534,197	30,456,984	20,889,317	5,849,002	3,677,453	41,212
Total Eastern Canada[b]	209,815	9,125	523	3,000	5,395	217,068	184,165[c]	—	769	32,134
Total Canada	26 605,336	4,312,815	290,058	5,435	539,592	30,674,052	21,073,482	5,849,002	3,678,222	73,346

[a] Prepared by Canadian Petroleum Association.
[b] Includes New Brunswick and Ontario.
[c] Some Associated figures included in these totals.

INDEX